29th International Conference on Superconductivity (ISS 2016)

Journal of Physics: Conference Series Volume 871

Tokyo, Japan
13 - 15 December 2016

Editors:

Hiraku Ogino
Yasuhiro Iijima
Nobuyuki Yoshikawa
Akihiko Kandori
Mitsuho Furuse

ISBN: 978-1-5108-4558-9
ISSN: 1742-6588

Printed from e-media with permission by:

Curran Associates, Inc.
57 Morehouse Lane
Red Hook, NY 12571

CURRAN ASSOCIATES INC. proceedings .com

Some format issues inherent in the e-media version may also appear in this print version.

Printed by Curran Associates, Inc. (2017)

Institute of Physics
Dirac House, Temple Back
Bristol BS1 6BE UK

Phone: 44 1 17 929 7481
Fax: 44 1 17 920 0979

techtracking@iop.org

Additional copies of this publication are available from:

Curran Associates, Inc.
57 Morehouse Lane
Red Hook, NY 12571 USA
Phone: 845-758-0400
Fax: 845-758-2633
Email: curran@proceedings.com
Web: www.proceedings.com

Table of contents

Volume 871

29th International Symposium on Superconductivity
13–15 December 2016, Tokyo, Japan

Accepted papers received: 30 June 2017
Published online: 26 July 2017

Preface

Physics and Chemistry

Novel materials and new superconductors

Cuprate superconductors

Iron-based superconductors

Vortex physics – experimental studies

Vortex physics – theoretical studies

Application

Wires and Bulk

Coated conductors

REBCO bulk materials

MgB_2 bulk materials

Iron-based superconducting materials and wires

Electronic Devices

Digital systems

Detectors

SQUID

Device physics

Large Scale System Applications

Superconducting magnets

Magnetic bearing and levitation

Cables and tansformers

Magnetic separation and drug delivery

IOP Conf. Series: Journal of Physics: Conf. Series **871** (2017) 011001 doi:10.1088/1742-6596/871/1/011001

29th International Symposium on Superconductivity (ISS2016)

Preface

It is our pleasure to report that the Proceedings of the 29th International Symposium on Superconductivity (ISS2016) held at Tokyo International Forum, Japan, December 13–15, 2016, are now published in *Journal of Physics: Conference Series* (JPCS). ISS2016, organized by the National Institute of Advanced Industrial Science and Technology (AIST), was aimed to gather many scientists, engineers, academic students, corporate executives and other participants from all over the world, and to facilitate fruitful discussions to promote superconductivity technologies. 427 scientists and engineers from 15 countries participated in ISS2016, and the symposium was successfully held. Total of 321 papers were presented, which includes 6 plenary lectures, 38 invited talks, 67 contributed oral presentations and 210 poster presentations. The papers published in JPCS were categorized into the following four fields; (a) Physics and Chemistry, (b) Wires and Bulk, (c) Electronic Devices, (d) Large Scale System Applications, and arranged in such a way that oral presentations come first, followed by poster presentations in relevant topics.

To ensure the high publication standard mandated by JPCS, every paper was peer reviewed by a reviewer with expertise, and in some cases by two reviewers before it was accepted for publication. As editors of the Proceedings, we would like to express our sincere appreciation to all the reviewers involved in the evaluation of the papers for their invaluable contribution. Our special thanks go also to session chairs who recommended appropriate reviewers.

The Editors
Physics and Chemistry: Atsutaka Maeda (The University of Tokyo), Hiraku Ogino (AIST)
Wires and Bulk: Yoshiyuki Yoshida (AIST), Yasuhiro Iijima (Fujikura Ltd.)
Electronic Devices: Nobuyuki Yoshikawa (Yokohama National University), Akihiko Kandori (Hitachi Ltd.)
Large Scale System Applications: Mitsuho Furuse (AIST), Tsuyoshi Wakuda (Hitachi Ltd.)

IOP Conf. Series: Journal of Physics: Conf. Series **871** (2017) 011001 doi:10.1088/1742-6596/871/1/011001

29th International Symposium on Superconductivity

I 超 S 電 S 導 2016

TOKYO INTERNATIONAL FORUM, JAPAN
December 13-15, 2016

IOP Conf. Series: Journal of Physics: Conf. Series **871** (2017) 011001 doi:10.1088/1742-6596/871/1/011001

Organization of ISS2016

General Chair:

Dr. Toshihiko Kanayama

Senior-Vice-President, National Institute of Advanced Industrial Science and Technology (AIST)

Local Organizing Committee:

Secretary-General: Michiya Okada, TIA Central Office, AIST

Members:

Haruhiko Obara, Fumio Takemura, Teruhisa Horita, Mitsuho Furuse (Department of Energy and Environment, AIST)

Satoshi Haraichi, Masataka Ohkubo, Mutsuo Hidaka, Yoshiyuki Yoshida, Hiraku Ogino (Department of Electronics and Manufacturing, AIST)

Kenshiro Oki, Hirofumi Yamasaki, Kiyoshi Takagi, Susumu Ogawa (TIA Central Office, AIST)

Supported by:

Applied Superconductivity Constellations of Tsukuba (ASCOT)

TIA—a platform for open innovation

Ministry of Economy, Trade and Industry (METI)

New Energy and Industrial Technology Development Organization (NEDO)

Cryogenics and Superconductivity Society of Japan (CSSJ)

The Institute of Electrical Engineers of Japan (IEEJ)

Committees of ISS2016

Program committee

Physics and Chemistry

Co-Chair: Atsutaka Maeda (The Univ. of Tokyo), Sub-Chair: Hiraku Ogino (AIST)

Members: Ryotaro Arita (RIKEN), Xiao Hu (National Inst. for Materials Science), Takekazu Ishida (Osaka Prefecture Univ.), Minoru Nohara (Okayama Univ.), Tsutomu Nojima (Tohoku Univ.), Takao Sasagawa (Tokyo Inst. of Technology), Takasada Shibauchi (The Univ. of Tokyo)

Wires and Bulk

Co-Chair: Yoshiyuki Yoshida (AIST), Sub-Chair: Yasuhiro Iijima (Fujikura Ltd.)

Members: Satoshi Awaji (Tohoku Univ.), Hiroshi Ikuta (Nagoya Univ.), Takeshi Kato (Sumitomo Electric Industries, Ltd.), Takanobu Kiss (Kyushu Univ.), Akiyoshi Matsumoto (National Inst. for Materials Science), Kaname Matsumoto (Kyushu Inst. of Technology), Shin-ichi Mukoyama (Furukawa Electric Co., Ltd.), Jun-ichi Shimoyama (Aoyama Gakuin Univ.), Hideki Tanaka (Hitachi Ltd.)

Electronic Devices

Co-Chair: Nobuyuki Yoshikawa (Yokohama National Univ.), Sub-Chair: Akihiko Kandori (Hitachi Ltd.)

Members: Yoshimi Hatsukade (Kinki Univ.), Masashi Ono (The Univ. of Tokyo), Shigeo Sato (Tohoku Univ.), Naoto Sekiya (Yamanashi Univ.), Masamitsu Tanaka (Nagoya Univ.), Hirotake Yamamori (AIST)

Large Scale System Application

Co-Chair: Mitsuho Furuse (AIST), Sub-Chair: Tsuyoshi Wakuda (Hitachi Ltd.)

Members: Kazuhiro Kajikawa (Kyushu Univ.), Shinji Matsumoto (National Inst. for Materials Science), Naoko Nakamura (MAYEKAWA MFG. Co., Ltd.), Taketsune Nakamura (Kyoto Univ.), Shoichi Yokoyama (Mitsubishi Electric Corporation), Tomonori Watanabe (Chubu Electric Power Co., Inc.)

International Advisory Committee

Naoyuki Amemiya (Kyoto Univ.), Hideo Aoki (The Univ. of Tokyo), Bob Buckley (Victoria Univ. of Wellington), David Cardwell (Univ. of Cambridge), Leonardo Civale (Los Alamos National Laboratory), Michael Eisterer (Vienna Univ. of Technology), Pascal Febvre (Univ. of Savoie Mont Blanc), Toru Fukushima (SuperPower Inc.), Timothy J. Haugan (Air Force Research Laboratory), Mutsuo Hidaka (AIST), Bernhard Holzapfel (Karlsruhe Inst. of

IOP Conf. Series: Journal of Physics: Conf. Series **871** (2017) 011001 doi:10.1088/1742-6596/871/1/011001

Technology), Teruo Izumi (AIST), Hiroaki Kumakura (National Inst. for Materials Science), Wai-Kwong Kwok (Argonne National Laboratory), David Larbalestier (Florida State Univ.), Qiang Li (Brookhaven National Laboratory), Mathias Noe (Karlsruhe Inst. of Technology), Xavier Obradors (Institut Ciencia de Materials de Barcelona), Masataka Ohkubo (AIST), Hiroyuki Ohsaki (The Univ. of Tokyo), Satoshi Okuma (Tokyo Inst. of Technology), Minwon Park (Changwon National Univ.), Christopher M. Rey (Energy-to-Power Solutions), Horst Rogalla (Univ. of Colorado at Boulder), Venkat Selvamanickam (Univ. of Houston), Michael Sumption (Ohio State Univ.), Setsuko Tajima (Osaka Univ.), Keiichi Tanabe (SUSTERA), Pascal Tixador (Grenoble-INP), Masaru Tomita (Railway Technical Research Inst.), Hai-Hu Wen (Nanjing Univ.), Liye Xiao (Chinese Academy of Sciences), Yutaka Yoshida (Nagoya Univ.)

Peer review statement

All papers published in this volume of *Journal of Physics: Conference Series* have been peer reviewed through processes administered by the proceedings Editors. Reviews were conducted by expert referees to the professional and scientific standards expected of a proceedings journal published by IOP Publishing.

IOP Conf. Series: Journal of Physics: Conf. Series **871** (2017) 012001 doi:10.1088/1742-6596/871/1/012001

Paramagnetic and diamagnetic pair-breaking effect in electric-field-induced surface superconductivity under parallel magnetic fields

Masanori Ichioka[1,2], Masahiro Nabeta[1], Kenta K Tanaka[1] and Seiichiro Onari[1,2]

[1] Department of Physics, Okayama University, Okayama 700-8530, Japan
[2] Research Institute for Interdisciplinary Science (RIIS), Okayama University, Okayama 700-8530, Japan

E-mail: ichioka@okayama-u.ac.jp

Abstract. Electric-field-induced surface superconductivity is studied by Bogoliubov-de Gennes equation under magnetic fields parallel to the surface. We estimate the pair-breaking effects by the paramagnetic Zeeman shift and by diamagnetic screening current. We find that the depth dependences of pair potential, screening current, spin current, and paramagnetic moment under the magnetic fields reflect the multi-gap superconductivity in the sub-band structure.

1. Introduction

By strong electric field of the field-effect-transistor or the electric-double-layer-transistor, carriers are induced and trapped in the confinement potential of the electric field near the surface of insulators or semiconductors. In the surface metallic states, superconductivity appears at low temperature [1], as performed in $SrTiO_3$ [2], ZrNCl [3], $KTaO_3$ [4], and MoS_2 [5, 6]. A unique nature of the surface metallic state is that sub-bands are formed by the quantum confinement of carriers near the surface due to the strong electric field [2, 7, 8]. By a previous theoretical study, the surface superconductivity is expected to be multi-gap superconductivity depending on the sub-bands [9]. The multi-gap structure of sub-band system is related to the depth-dependence of the superconducting state.

In order to examine properties reflecting the multi-gap superconductivity in the electric-field-induced surface metallic state, we theoretically study electronic states when magnetic fields are applied parallel to the surface. In this configuration, experimental observation of the upper critical field H_{c2} suggests that vortices do not penetrate into the superconducting region near the surface [10]. The effects of parallel magnetic fields on the superconductivity occur by two contributions of paramagnetic and diamagnetic pair-breakings. Under the magnetic field, Fermi surface is split between up- and down-spin electrons. This split induces paramagnetic moment and the paramagnetic pair-breaking of the superconductivity. In reference [11], the paramagnetic pair-breaking effect was reported in the surface superconductivity in the confinement potential including screening effect of electric fields by carriers. On the other hand, by the contribution of vector potential, screening supercurrent flows near the surface so that magnetic field is

IOP Conf. Series: Journal of Physics: Conf. Series **871** (2017) 012001 doi:10.1088/1742-6596/871/1/012001

decreased inside the superconducting region. The diamagnetic pair-breaking occurs by the flow of supercurrent.

In this work, wave functions in the superconducting states are calculated from Bogoliubov-de Gennes (BdG) equation, assuming s-wave pairing interaction [9, 12]. Here, we concentrate on the case when confinement potential by electric field is given by triangular potential. We study spatial structure of the superconducting state under parallel magnetic fields, and estimate effects of paramagnetic and diamagnetic pair-breakings comparatively in the pair potential, screening current, spin current and paramagnetic moment. By the sub-band decomposition of these quantities, we can see properties of multi-gap superconductivity, because superconductivity in higher-level sub-band is more suppressed with increasing magnetic fields.

This paper is organized as follows. After the introduction, we explain our theoretical formulation by the BdG equation under parallel magnetic fields in section 2. We study the depth z dependence of pair potential $\Delta(z)$, supercurrent $J(z)$, and spin current $J_s(z)$ in section 3. The z-dependence of paramagnetic moment $m(z)$ is studied in section 4. Last section is devoted to summary.

2. Formulation by BdG equation

We use coordinate $\mathbf{r} = (x, y, z)$, where z (≥ 0) is depth from the surface at $z = 0$. The confinement potential near the surface is assumed to be triangular potential $V(z) = |e|F_0 z$. We typically consider the case of sheet carrier density $n_{2D} = 6.5 \times 10^{13}[\mathrm{cm}^{-2}]$, electric field $F_0 = 1.4 \times 10^{-3}[\mathrm{V/nm}]$, and single band with effective mass $m^* = 4.8m_0$, where m_0 is free electron's mass. This corresponds to one of the cases in reference [9].

Solving the BdG equation [12], we determine the pair potential $\Delta(\mathbf{r})$, and the wave functions $u_\epsilon(\mathbf{r})$, $v_\epsilon(\mathbf{r})$ for the eigen-energy E_ϵ. When we consider diamagnetic current, we set the vector potential as $\mathbf{A} = (0, A_y, 0)$ with $A_y = -Hz$, so that the magnetic field parallel to the surface is applied along the x direction, and the screening current flows along the y direction. In this situation, we can set $\Delta(\mathbf{r}) = \Delta(z)\mathrm{e}^{\mathrm{i}2qy}$ and

$$\begin{pmatrix} u_\epsilon(\mathbf{r}) \\ v_\epsilon(\mathbf{r}) \end{pmatrix} = \frac{1}{\sqrt{S}} \mathrm{e}^{\mathrm{i}(k_x x + k_y y)} \begin{pmatrix} u_\epsilon(z)\mathrm{e}^{\mathrm{i}qy} \\ v_\epsilon(z)\mathrm{e}^{-\mathrm{i}qy} \end{pmatrix}, \tag{1}$$

where S is unit area of surface. Thus the BdG equation is reduced to

$$\begin{pmatrix} K_+ & \Delta(z) \\ \Delta(z) & -K_- \end{pmatrix} \begin{pmatrix} u_\epsilon(z) \\ v_\epsilon(z) \end{pmatrix} = E_\epsilon \begin{pmatrix} u_\epsilon(z) \\ v_\epsilon(z) \end{pmatrix}, \tag{2}$$

with the kinetic term

$$K_\pm = \frac{-\hbar^2}{2m^*} \left(k_x^2 + (\pm k_y + q + \frac{\pi}{\phi_0} A_y)^2 - \partial_z^2 \right) + V(z) \mp \mu_\mathrm{B} H - \mu, \tag{3}$$

where ϕ_0 is a flux quanta, and $\pm\mu_\mathrm{B} H$ is Zeeman energy with the Bohr magneton $\mu_\mathrm{B} = 5.7883 \times 10^{-5}$ [eV/T]. The eigen-states of equation (2) are labeled by $\epsilon \equiv (k_x, k_y, i_z)$. i_z ($= 1, 2, \cdots$) indicates label for sub-bands coming from quantization by confinement in the z-direction. As the boundary condition at the surface $z = 0$, we set $u_\epsilon(z) = v_\epsilon(z) = 0$. Throughout this paper, energy, length, magnetic field, and local carrier densities are, respectively, presented in unit of eV, nm, T, and nm^{-3}. The chemical potential μ is determined to fix the carrier density as

$$n_{2D} = \int_0^\infty \{n_\uparrow(z) + n_\downarrow(z)\} \,\mathrm{d}z, \tag{4}$$

where local carrier density for up- and down-spins are, respectively, given by

$$n_\uparrow(z) = \sum_\epsilon |u_\epsilon(z)|^2 f(E_\epsilon), \quad n_\downarrow(z) = \sum_\epsilon |v_\epsilon(z)|^2 f(-E_\epsilon) \tag{5}$$

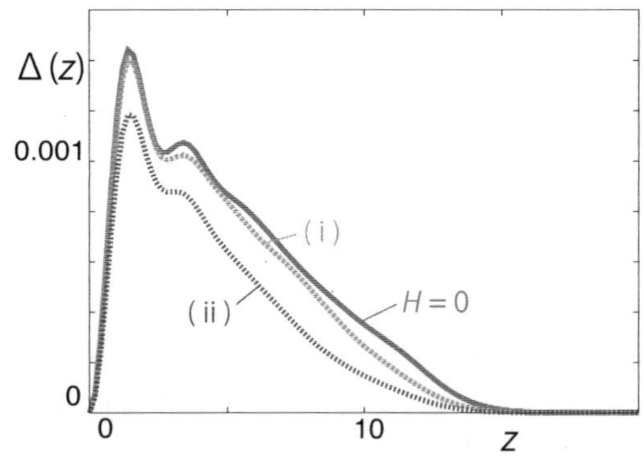

Figure 1. Profile of pair potential $\Delta(z)$ as a function of depth z. Solid line is for a zero magnetic field. Dashed lines are for $H = 10$ in the cases (i) and (ii) explained in text.

from the eigen states, with the Fermi distribution function $f(E)$.

The pair potential is calculated by the gap equation

$$\Delta(z) = V_{\rm pair}{\sum_{\epsilon}}' u_\epsilon(z)v_\epsilon(z)f(-E_\epsilon). \tag{6}$$

In equation (6), the energy cutoff $E_{\rm cut}$ of the pairing interaction is considered in the summation $\sum'$. Here, we consider a conventional case of spin-independent isotropic s-wave pairing. We typically use $V_{\rm pair} = 0.08$, and $E_{\rm cut} = 0.01$. Iterating calculations of equations (2) and (6), we obtain selfconsistent results of $\Delta(z)$ and wave functions. We study the superconducting state at low temperature $T = 1.16 \times 10^{-2}$K $\ll T_{\rm c}$ and $H = 10$T.

To study effects of paramagnetic and diamagnetic pair-breakings under parallel magnetic fields, we perform calculations in the following two cases.

Case (i): We neglect the diamagnetic effect, setting $A_y = q = 0$, and consider only the paramagnetic effect. This is to clarify contributions of the paramagnetic pair-breaking effect.

Case (ii): We consider both diamagnetic and paramagnetic pair-breaking effects, setting $A_y = -Hz$. q is tuned to satisfy the current conservation $\int_0^\infty J(z){\rm d}z = 0$ with local screening current density along the y direction, $J(z) = J_\uparrow(z)+J_\downarrow(z)$. The up- and down-spin contributions are, respectively, calculated as

$$J_\uparrow(z) = \frac{e\hbar}{2m}\sum_\epsilon \left(k_y + q + \frac{\pi}{\phi_0}A_y\right)|u_\epsilon(z)|^2 f(E_\epsilon), \tag{7}$$

$$J_\downarrow(z) = \frac{e\hbar}{2m}\sum_\epsilon \left(-k_y + q + \frac{\pi}{\phi_0}A_y\right)|v_\epsilon(z)|^2 f(-E_\epsilon) \tag{8}$$

from the eigen states. From comparison of the cases (i) and (ii), we find contributions of diamagnetic pair-breaking effect in the surface superconductivity.

3. Depth dependence of pair potential, screening current, and spin current

To study the pair-breaking effects by parallel magnetic fields, we compare the spatial structure of superconductivity at a magnetic field $H = 10$ and a zero field. The critical field to the normal state is the first order transition at $H \sim 12$. In figure 1, we present the profile of pair potential $\Delta(z)$ obtained from selfconsistent calculation of BdG equation in the cases (i) and (ii). Compared to the zero field case, we see suppression of $\Delta(z)$ under magnetic field $H = 10$. In the

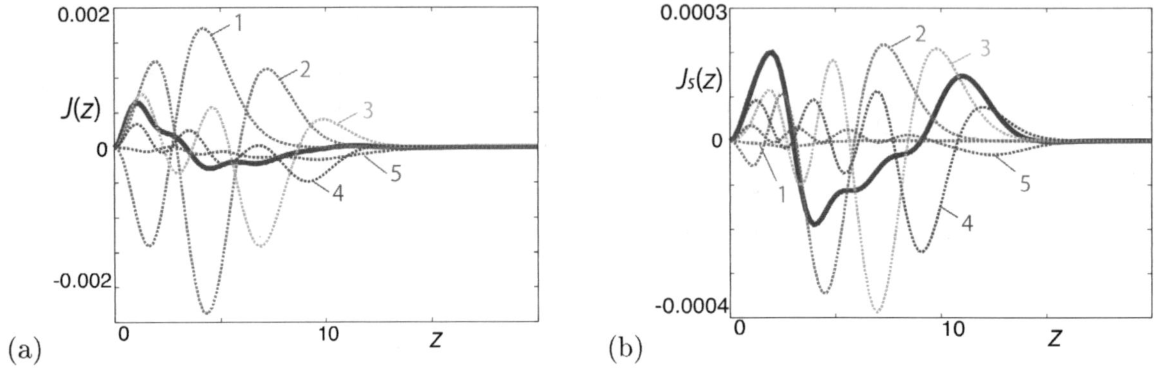

Figure 2. (a) Local current density $J(z) = J_\uparrow(z) + J_\downarrow(z)$ and (b) local spin current density $J_s(z) = J_\uparrow(z) - J_\downarrow(z)$ are presented by solid lines as a function of depth z at $H = 10$ in the case (ii). Dashed lines indicate contributions from i_z-th sub-band to $J(z)$ and $J_s(z)$. i_z =1, 2, $\cdots$, 5.

case (i) considering only paramagnetic pair-breaking effect, the suppression of $\Delta(z)$ is eminent at deeper z region. In the case (ii) considering both paramagnetic and diamagnetic pair-breaking effects, $\Delta(z)$ is further suppressed than that in case (i) due to the additional pair-breaking effect by the diamagnetic current. In this case, we see the suppression until smaller z region.

Profile of the diamagnetic screening current $J(z)$ along the y-direction parallel to the surface is presented by a solid line in figure 2(a) for the case (ii). Near the surface $z \sim 0$, $J(z) > 0$ to suppress the penetration of the magnetic field from the surface side. On the other hand, $J(z) < 0$ at $z \sim 5$ to screen the magnetic field from the deeper side of the superconducting region. The sub-band contributions to $J(z)$ are calculated by restricting the ϵ-summation to one sub-band i_z. In figure 2(a), sub-band contributions from lower three sub-bands $i_z = 1$, 2, 3 are large. By the summation of the oscillating wave function of the sub-band contributions, profile of $J(z)$ is constructed. We see small positive $J(z)$ at $z \sim 11$.

When both paramagnetic and diamagnetic pair-breaking effects work, there appears spin current $J_s(z) = J_\uparrow(z) - J_\downarrow(z)$, as shown in figure 2(b) for the case (ii). In the sub-band decomposition of $J_s(z)$, contributions from $i_z = 2$, 3, 4 are dominant. We note that contribution from lowest sub-band $i_z = 1$ is very small, because superconductivity at the sub-band $i_z = 1$ is not broken as seen later in figure 3(b).

4. Depth dependence of paramagnetic moment

In the presence of Zeeman shift, local paramagnetic moment $m(z) = n_\uparrow(z) - n_\downarrow(z)$ appears, as shown in figure 3. The moment $m(z)$ comes from contribution of sub-bands where superconductivity is largely suppressed, and electronic state becomes normal-state-like with small superconductivity. When diamagnetic pair-breaking effect is neglected, $m(z)$ consists of contributions from higher-level sub-bands $i_z = 4$ and 5, as shown in figure 3(a) for the case (i). Thus $m(z)$ has a main peak at a deep position $z \sim 11$. There are no contributions from lower-level sub-bands $i_z = 1$, 2, 3, where superconductivity is not yet broken. If the diamagnetic pair-breaking effect is included as in figure 3(b) for the case (ii), contributions from sub-bands $i_z = 2$ and 3 also appear in $m(z)$. Therefore region of large $m(z)$ extends towards smaller z, and $m(z)$ becomes larger compared to the case of figure 3(a). Even in this case, lowest sub-band $i_z = 1$ gives no contributions to $m(z)$ since it keeps large superconductivity.

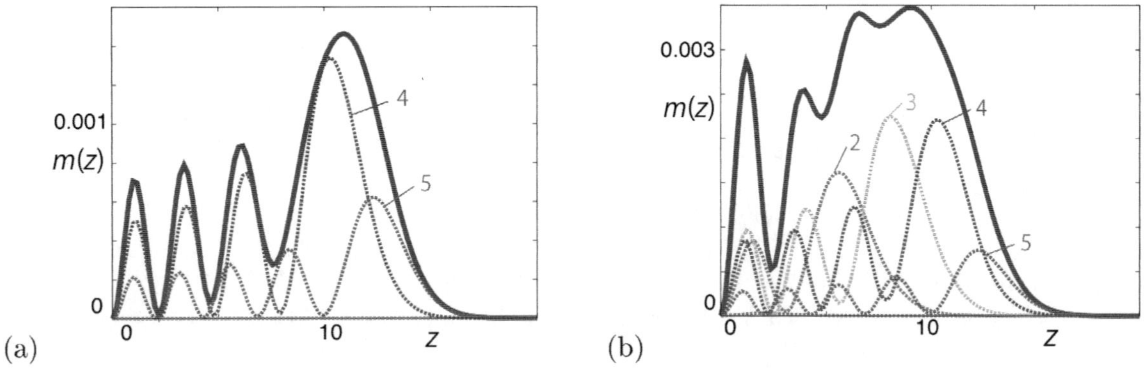

Figure 3. Local paramagnetic moment $m(z) = n_\uparrow(z) - n_\downarrow(z)$ is presented by solid lines as a function of depth z at $H = 10$. (a) The case (i) considering only paramagnetic pair-breaking effect. (b) The case (ii) including both paramagnetic and diamagnetic pair-breaking effects. Dashed lines indicate contributions from i_z-th sub-band to $m(z)$.

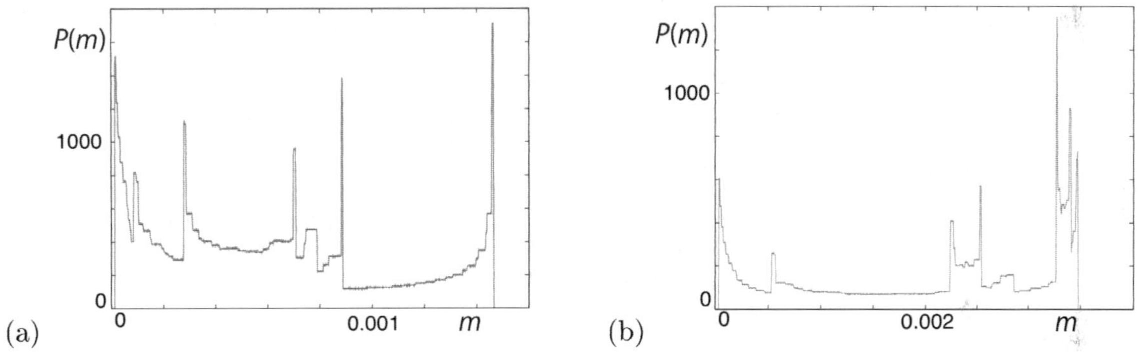

Figure 4. Distribution $P(m)$ estimated from $m(z)$ at $H = 10$. (a) The case (i) with paramagnetic effect only, obtained from figure 3(a). (b) The case (ii) with both paramagnetic and diamagnetic effects, obtained from figure 3(b).

We also calculate distribution of $m(z)$ as

$$P(m) = \int_0^\infty \delta(m - m(z))\mathrm{d}z, \tag{9}$$

which gives volume satisfying $m = m(z)$. $P(m)$ in figures 4(a) and 4(b) are, respectively, calculated from figures 3(a) and 3(b). Peaks in $P(m)$ come from local maximum or local minimum of $m(z)$ in figure 3. In the case (ii), because $m(z)$ has large value at many local minimum positions, weight of $P(m)$ is located at higher m in figure 4(b), compared to that in figure 4(a). The distribution $P(m)$ corresponds to the NMR or μSR spectra in bulk superconductors. In the surface superconductivity, NMR spectrum may selectively detect local $m(z)$ depending position z of nuclei used in NMR experiments.

5. Summary

Solving BdG equation selfconsistently, we studied paramagnetic and diamagnetic pair-breaking effects in electric-field-induced surface superconductivity under parallel magnetic fields. With increasing magnetic fields, superconductivity in higher-level sub-bands is suppressed and becomes normal-state-like with small superconductivity. We found that sub-band contributions to the pair potential and screening current dominantly come from lower-level sub-bands keeping large superconductivity. On the other hand, main contributions to paramagnetic moment and spin current are from occupied higher-level sub-bands with partially suppressed superconductivity. These properties come from multi-gap structure of superconductivity in the sub-band system of the surface metallic states.

Acknowledgments

This work was supported by JSPS KAKENHI Grant Number 25400373.

References

[1] Ueno K, Shimotani H, Yuan H, Ye J T, Kawasaki M and Iwasa Y 2014 *J. Phys. Soc. Jpn.* **83** 032001
[2] Ueno K, Nakamura S, Shimotani H, Ohtomo A, Kimura N, Nojima T, Aoki H, Iwasa Y and Kawasaki M 2008 *Nature Mater.* **7** 855
[3] Ye J T, Inoue S, Kobayashi K, Kasahara Y, Yuan H T, Shimotani H and Iwasa Y 2011 *Nature Mater.* **9** 125
[4] Ueno K, Nakamura S, Shimotani H, Yuan H T, Kimura N, Nojima T, Aoki H, Iwasa Y and Kawasaki M 2011 *Nature Nanotechnol.* **6** 408
[5] Taniguchi K, Matsumoto A, Shimotani H and Takagi H 2012 *Appl. Phys. Lett.* **101** 042603
[6] Ye J T, Zhang Y J, Akashi R, Bahramy M S, Arita R and Iwasa Y 2012 *Science* **338** 1193
[7] Santander-Syro A F, Copie O, Kondo T, Fortuna F, Pailhès S, Weht R, Qiu X G, Bertran F, Nicolaou A, Taleb-Ibrahimi A, Le Fèvre P, Herranz G, Bibes M, Reyren N, Apertet Y, Lecoeur P, Barthélémy A and Rozenberg M J 2011 *Nature* **469** 189
[8] King P D C, Walker S M, Tamai A, De la Torre A, Eknapakul T, Buaphet P, Mo S-K, Meevasana W, Bahramy M S and Baumberger F 2014 *Nat. Commun.* **5** 3414
[9] Mizohata Y, Ichioka M and Machida K 2013 *Phys. Rev.* B **87** 014505
[10] Ueno K, Nojima T, Yonezawa S, Kawasaki M, Iwasa Y and Maeno Y 2014 *Phys. Rev.* B **89** 020508
[11] Nabeta M, Tanaka K K, Onari S and Ichioka M 2016 *Physica* C **530** 8
[12] De Gennes P G 1989 *Superconductivity of Metals and Alloys* (Reading, Massachusetts: Addison-Wesley)

IOP Conf. Series: Journal of Physics: Conf. Series **871** (2017) 012002 doi:10.1088/1742-6596/871/1/012002

The electrical resistivity of epitaxially deposited chromium films

Masaki Sawabu[1], Masashi Ohashi[1,2], Kohei Ohashi[1], Masahiro Miyagawa[1], Takahide Kubota[3] and Koki Takanashi[3]

[1]Graduate School of Natural Science and Technology, Kanazawa University, Kakuma-machi, Kanazawa, 920-1192, Japan

[2]Institute of Science and Engineering, Kanazawa University, Kakuma-machi, Kanazawa, 920-1192, Japan

[3]Institute for Materials Research, Tohoku University, 2-1-1 Katahira, Aoba-ku, Sendai, 980-8577, Japan

E-mail: msawabu@stu.kanazawa-u.ac.jp

Abstract. We studied the electrical resistance and crystal structure of epitaxial chromium (Cr) films. The lattice constant of the Cr films is larger than that of the bulk Cr because the sample and the substrate MgO have different lattice constants. An chromium oxide layer having a thickness of 1 nm was found on all films from the result of X-ray reflectivity measurements. Although there is sample dependence in electric resistance of Cr films, it is remarkable that a zero resistance was observed for one film of 50 nm thick. T_C is obtained to be 3.2 K, which is about twice the previous reports. The large upper critical field is obtained to be H_{C2} = 13 kOe, and it suggests that Cr may belongs to a type II superconductor.

1. Introduction

Since magnetic ordering and superconductivity apparently compete in conventional superconductors, some magnetic materials do not exhibit superconductivity. For example, iron (Fe) is a typical magnetic metal element that shows ferromagnetism at room temperature and ambient pressure. However, superconductivity is observed in Fe under high pressure between 15 and 30 GPa at 2 K[1, 2]. Such behavior is related to the structural phase transition under pressure from the ferromagnetic bcc (α-Fe) phase to the paramagnetic hcp (ϵ-Fe) phase[3].

Chromium (Cr), an antiferromagnet below the Néel temperature T_N = 311 K[4] at ambient pressure, doesn't exhibit superconductivity even under pressure[1]. This may be attributed to the fact that T_N decreases with increasing pressure but tends to saturate. Such behavior can be explained by taking into account of a two-band model of itinerant antiferromagnetism[5, 6].

On the other hand, Schmidt *et al.* reported that thin films of Cr metal suppress the antiferromagnetic ordering and become superconductive at $T_C \sim 1.5$ K[7, 8]. From the point of view not only of pure research but also of applied and practical one, there are many examples of superconducting thin films. But almost all of them were those in which superconductivity disappears when the film thickness is made thin[9, 10]. It is likely to make a superconducting electron pair when it becomes thinner than the mean free path of electrons. Taking account that the bulk Cr does not show superconductivity, the relationship between film thickness and

IOP Conf. Series: Journal of Physics: Conf. Series **871** (2017) 012002 doi:10.1088/1742-6596/871/1/012002

existence of superconductivity of Cr is opposite to the other examples of superconducting thin films.

Assuming that the magnetic correlation interaction is suppressed by controlling the film thickness, Cr may show a magnetic order-disorder transition tuned by the film thickness, and unconventional superconductivity may occur at the quantum critical point of $T_N \to 0$, where an electron pair appears due to quantum fluctuation. Such quantum phase transitions have been reported in strongly correlated electron systems, where various parameters such as pressure and magnetic field have been used by a driving force[11, 12, 13, 14, 15, 16].

However, quantum phase transition using the film thickness as a driving force has not been discovered in thin films at all. Generally, the three-dimensional electron system such as bulk compounds and two-dimensional one such as thin films are described by completely different Hamiltonian. That is, film thickness is an effective parameter that can directly control dimensionality, and it is interested in whether Cr is the first case of the new quantum phase transition tuned by film thickness.

It seems that there is still insufficient data for Cr films because neither the resistivity drop nor the Meissner effect has been reported yet[7, 8]. Recently we have studied the electrical resistance of polycrystalline Cr films, but it is difficult to conclude whether a superconducting transition occurs because of the large residual resistivity at low temperature[17]. In the present study, we prepare single crystal films of Cr, and perform precise electrical resistance measurements of Cr thin films in a wide temperature range and a magnetic field.

2. Experimental method

Single-crystal Cr films of 50 nm thick were fabricated epitaxially using a conventional magnetron sputtering device in ultrahigh vacuum below 2×10^{-6} Pa. The Ar pressure during deposition was 0.1 Pa. The substrate for growing Cr epitaxially (001) is MgO. X-ray diffraction measurements were carried out using Rigaku MiniFlex II. The lattice constant a is obtained to be between 2.974 and 2.976 Å which is larger than that of the bulk Cr, 2.880 Å[18]. It may come from the fact that the lattice constant of MgO, a = 4.213 Å is larger than $\sqrt{2}a$ = 4.073 Å in bulk Cr. It suggests that the surface of the Cr layer can be subjected to the (110) orientation on the (001) MgO substrate. X-ray reflectance measurements were also carried out using the Rigaku SmartLab to estimate the film thickness, density, surface roughness, interface roughness, etc.

The electrical resistance was measured by a four-point collinear four-probe dc method with the current direction on the film plane. Since the chromium oxide layer is uncongenial to the gold wires, aluminum wires were bonded on the film plane by wire bonding. The temperature dependence of the electrical resistivity was measured using the Quantum-Design PPMS between 0.5 and 350 K in the low-temperature laboratory, Kanazawa University. The direction of the applied field was perpendicular to the film plane and the electrical current.

3. Result and discussion

3.1. X-ray reflectivity

Figure 1 shows the X-ray reflectivity profile of the epitaxial Cr film of 50 nm thick. Vibration appears in the reflected intensity due to the interference of the reflected light for each layer constituting the film. Figure 2 shows the fast Fourier transform (FFT) on this profile. If only the Cr film is present on the substrate, only one peak should appear at the position of 50 nm due to the interference between the Cr thin film and the substrate. However, in this result, two peaks were observed not only at around 50 nm but also at around 1 nm, which corresponds to the film thickness of the chromium oxide film such as CrO_2 generated by the oxidation of Cr on the surface. It means that an chromium oxide layer of 1 nm thick was formed on a Cr film of 50 nm thick.

IOP Conf. Series: Journal of Physics: Conf. Series **871** (2017) 012002 doi:10.1088/1742-6596/871/1/012002

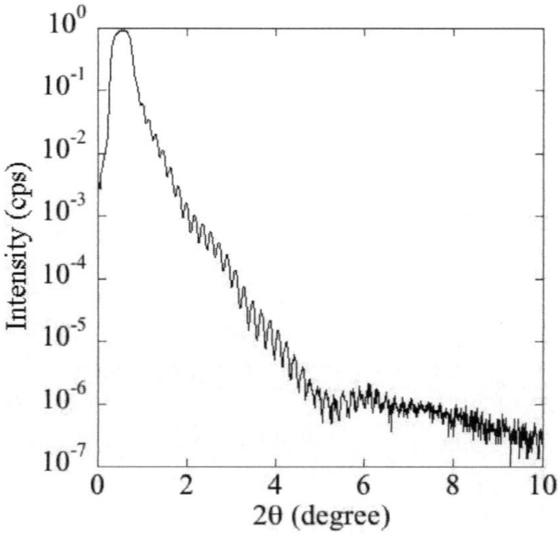

Figure 1. X-ray reflectivity profile of Cr film of 50 nm thick.

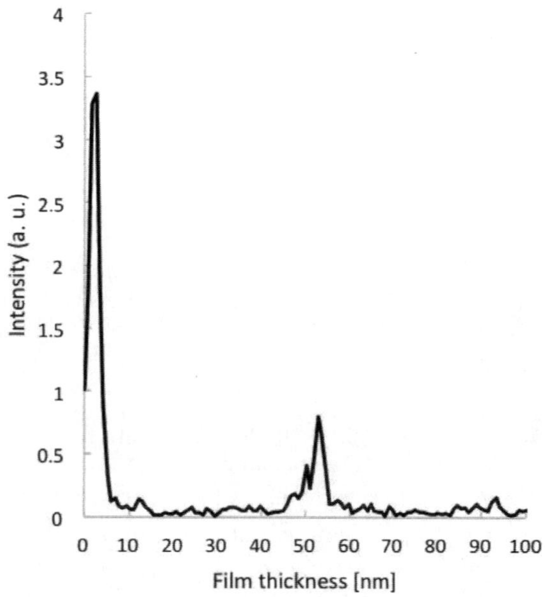

Figure 2. Relative Fourier transform magnitude calculated from the result of Figure 1.

3.2. Electrical resistivity

Figure 3 shows the electrical resistivity $\rho(T)$ of several epitaxial Cr films of 50 nm thick as a function of temperature between 0.5 and 350 K. For Cr#1 and Cr#2 sample, ρ is 14.0 and 16.6 $\mu\Omega$cm, respectively. Both $\rho(T)$ curves decrease with decreasing temperature and show a hump around 300 K. Such behavior compares with previous studies for epxtaxial Cr films of 200 and 400 nm thick[17] and for the bulk one[19]. It is consistent with the fact that the bulk Cr is an itinerant antiferromagnet with T_N below which the incommensurate spin density wave is stabilized[4, 5, 6]. Both for Cr#1 and Cr#2 sample, T_N is obtained to be 288 K, where a minimum is observed in the temperature derivative of the resistivity $d\rho/dT$ as shown in the inset of Fig. 4. Note that T_N is lower than that of the bulk Cr[5, 6].

Next, we focus the sample dependence observed in $\rho(T)$ curve. $\rho(T)$ curve for Cr#3 sample also decreases with decreasing temperature and shows a hump at around 300 K. But the magnitude of ρ is much larger than that of Cr#1 and Cr#2 samples and of previous reports. T_N is obtained to be 285 K for Cr#3 sample. The data of Cr#4 shows the result of Cr#3 sample after four wires are removed and re-bonded again. For Cr#4 sample, on the other hand, the magnitude of ρ is the largest in that of all samples, and no anomaly due to any magnetic phase transition is observed in the $\rho(T)$ curve.

Figure 4 shows the electrical resistivity of epitaxial Cr films of 50 nm thick at low temperature. For Cr#1 and Cr#2 samples, $\rho(T)$ becomes almost constant below 5 K within the experimental error, indicating that superconducting transition does not occur. For Cr#3 sample, on the other hand, the resistance drops is observed as shown in the inset of Fig. 4. T_C is obtained to be 1.36 K which is close to that of the previous reports[7, 8]. But it is difficult to conclude whether superconducting transition occurs in Cr#3 sample since the electrical resistivity is not zero. Moreover, the magnitude of the resistivity drop ratio 0.34 % is very small. Such small resistance drop is also observed in the previous study for polycrystalline films[17]. The following possibilities can be considered as factors causing such behavior. First, some chromium oxide

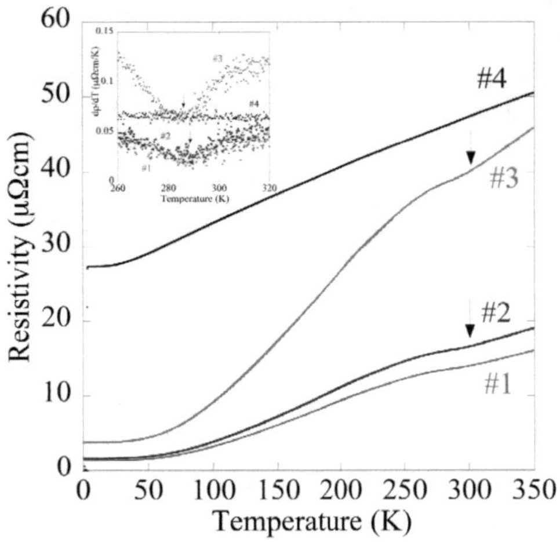

Figure 3. Temperature dependence of electrical resistivity ρ of Cr films of 50 nm thick. Inset shows the temperature derivative of ρ. Arrow shows an anomaly caused by an antiferromagnetic phase transition.

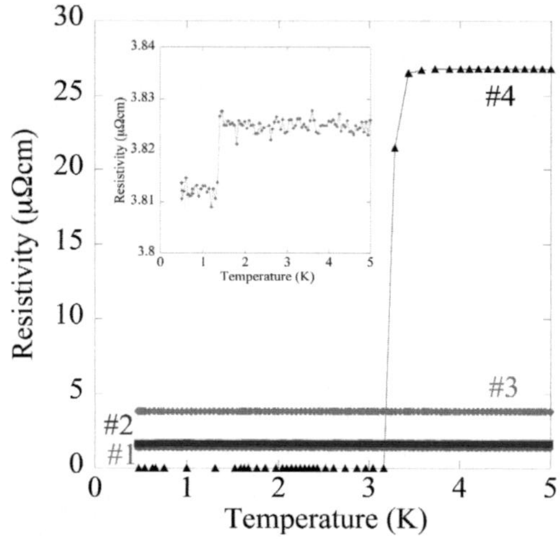

Figure 4. Temperature dependence of electrical resistivity ρ of Cr films of 50 nm thick at low temperature. Inset shows the detail of the data for Cr#3 sample.

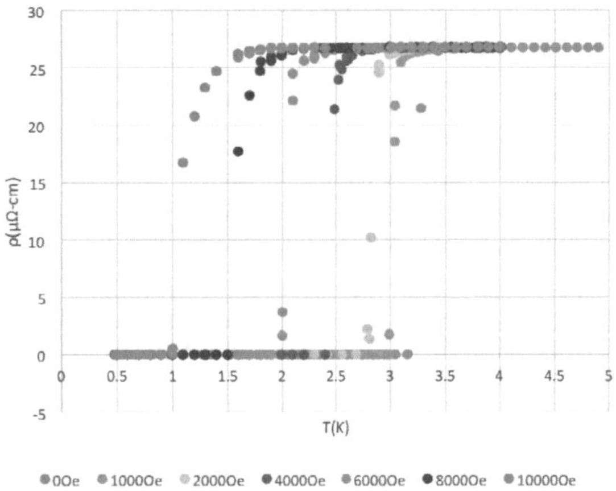

Figure 5. Temperature dependence of electrical resistivity ρ of Cr#4 sample at several magnetic fields.

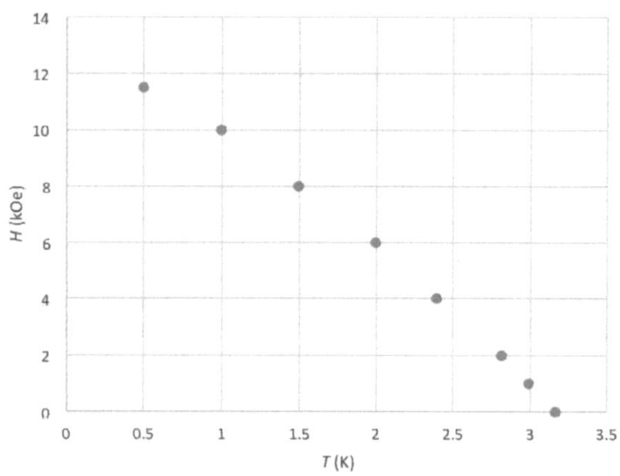

Figure 6. Magnetic phase diagram of Cr#4 sample.

layer produced on a film surface is partially superconducting. Second, crystalline impurities in Cr may be superconducting or show some magnetic transition. Third, taking account that T_C is enhanced in granular aluminum films[20], aluminum wires as electrodes may create grains on the surface of Cr film.

It is remarkable that zero resistivity is observed for Cr#4 sample, for which an anomaly in the $\rho(T)$ curve due to the antiferromagnetic transition is suppressed as shown in Fig. 3. Such behavior can not be explained by superconductivity of aluminum wires, and it is highly likely that Cr or its oxide shows superconductivity below T_C. The superconducting transition temperature is obtained to be T_C = 3.2 K, which is nearly the twice that of the previous report[7, 8]. Taking account that Cr#3 and Cr#4 are same sample but the different results are shown in the electrical resistivity, it suggests that the sample is partially superconductivity in Cr#3 sample. Here, we attempted to measure the electrical resistivity at magnetic field as shown in Figure 5. T_C shifted to the lower temperature side as the magnetic field is increased. Figure 6 shows the temperature and magnetic field phase diagram of superconductivity obtained from Fig. 5.

Extrapolated from the plotted data, the upper critical field at absolute zero was determined as H_{C2} = 13 kOe. Generally, most of the superconducting metal elements such as Al and Sn belongs to the type I superconductor, and their H_{C2} is about several hundreds. However, taking account that H_{C2} of Cr is one order of magnitude larger than that of the type II superconductor such as Nb (H_{C2} = 1.980 kOe) and V (H_{C2} = 1.420 kOe)[18], Cr#4 sample may also belongs to a type II superconductor.

4. Summary

We have fabricated epitaxial chromium (Cr) films of 50 nm thick using magnetron sputtering. The lattice constant of the Cr films is larger than that of the bulk Cr. Since it is well known that lattice constant is strongly correlated to physical parameters such as magnetic exchange interaction and electric conduction, it is necessary to study whether the similar behavior occurs when the lattice constant varies depending on the film thickness in the magnetic film. From the results of the X-ray reflectivity measurements, it was found that the Cr film used in this study also has an chromium oxide layer of about 1 nm thick. Therefore, the influence of the oxide layer of the surface of Cr films can not be ignored even in the previous reports in which it was not discussed until now.

Although there is sample dependence in electric resistance of Cr films, it is remarkable that a zero resistance was observed for one sample. The upper critical field H_{C2} of this sample is one to two orders of magnitude larger than the other metal element superconductor. The measurement of Meissner effect is planed in future although it is difficult to detect a signal because the thin film has a small sample volume the magnetism of the substrate becomes a large background. Moreover, it is needed to clear the condition such as the film thickness, density, surface roughness, interface roughness and the lattice constant under which zero resistance appears.

This research is a joint use study (Proposal No. 15 G0039) at the New Materials Collaborative Research and Development Center attached to the Metal Materials Research Institute, Tohoku University.

References

[1] Shimizu K, Amaya K and Suzuki N 2005 *J. Phys. Soc. Jpn.* **74** 1345
[2] Shimizu K, Kimura T, Fujimoto S, Takeda K, Kontani K, Ōnuki Y and Amaya K 2001 *Nature* **412** 316
[3] Takeda K and Mito M 2002 *J. Phys. Soc. Jpn.* **71** 729
[4] Koehler W C, Moon R M, Trego A L and Machintosh A R 1966 *Phys. Rev.* **151** 405
[5] McWhan D B and Rice T M 1967 *Phys. Rev. Lett.* **19** 846
[6] Ohashi M and Oomi G 2009 *Jpn. J. Appl. Phys.* **48** 070221
[7] Schmidt P H, Castellano R N, Barz H and Matthias B T 1972 *Physics Letters* **41A** 367
[8] Schmidt P H, Castellano R N, Barz H, Cooper A S and Spencer E G 1973 *J. Appl. Phys.* **44** 1833
[9] Jaeger H M, Haviland D B, Orr B G and Goldman A M 1989 *Phys. Rev. B* **40** 182
[10] Haviland D B, Liu Y, Nease B and Goldman A M 1990 *Phisica B* **165-166** 1457
[11] Uwatoko Y, Umehara I, Ohashi M, Nakano T and Oomi G 2012 *Handbook on the Physics and Chemistry of Rare Earths* vol 2 (Amsterdam: North-Holland) chap 252, p 1
[12] Nakano T, Ohashi M, Oomi G, Matsubayashi K and Uwatoko Y 2009 *Phys. Rev. B* **79** 172507
[13] Miyagawa H, Oomi G, Ohashi M, Satoh I, Komatsubara T, Hedo M and Uwatoko Y 2008 *Phys. Rev. B* **78** 064403
[14] Ohashi M, Oomi G, Koiwai S, Hedo M and Uwatoko Y 2003 *Phys. Rev. B* **68** 144428
[15] Ohashi M, Oomi G and Satoh I 2007 *J. Phys. Soc. Jpn.* **76** 114712

IOP Conf. Series: Journal of Physics: Conf. Series **871** (2017) 012002 doi:10.1088/1742-6596/871/1/012002

[16] Ohashi M, Miyagawa H, Nakano T, Oomi G, Sechovsky V, Satoh I and Komatsubara T 2014 *J. Phys. Soc. Jpn.* **83** 024701
[17] Ohashi M, Ohashi K, Sawabu M, Miyagawa M, Kubota T and Takanashi K 2016 *Physics Letters A* **380** 3133
[18] Kittel C 1996 *Introduction to Solid State Physics* 7th ed (John Wiley & Sons, Inc.)
[19] Yeh A, Soh Y A, Booke J, Aeppll G, Rosenbaum T F and Hayden S M 2002 *Nature* **419** 459
[20] Kobayashi S, Nakamura A and Komori F 1990 *J. Phys. Soc. Jpn.* **59** 4219

Effect of Fe-doping on CDW state in 1*T*-TaS₂ investigated by STM/STS

Y Fujisawa, T Shimabukuro, H Kojima, K Kobayashi, S Demura and H Sakata

Department of Physics, Tokyo University of Science, Shinjuku-ku, Tokyo 162-8601, Japan

E-mail : 1215702@ed.tus.ac.jp

Abstract. 1*T*-TaS_2 has been known to undergo Mott transition from nearly commensurate charge density wave (CDW) state to commensurate CDW state at about 200 K. Recently, Fe doping was found to suppress the Mott transition and induce superconductivity. In this study, we report on the scanning tunneling microscopy (STM) and spectroscopy (STS) study on Fe doped 1*T*-TaS_2 with different Fe concentration. STM observations reveal that the CDW superlattice becomes irregular as increasing Fe concentration. STS measurements uncover the spatial change in the local density of state (LDOS). We find two types of David-star: one has symmetric LDOS with respect to Fermi energy, and the other has asymmetric one. We identify that the asymmetric spectrum is due to an Fe dopant.

1. Introduction

1*T*-TaS_2 is one of the transition metal dichalcogenides and is known to form charge density wave (CDW) states that originates from low dimensionality of the crystal structure. Below 200 K, it becomes commensurate CDW (CCDW) state. In the CCDW state, 13 Ta atoms form a "David-star" cluster and it forms triangular lattice with $\sqrt{13}a_0 \times \sqrt{13}a_0$ periodicity (where a_0 is in-plane lattice constant). This phase is the Mott state due to the strong correlation effect between Ta 5*d* electrons at the center of the David-star [1]. Above 200 K, the CCDW state melts into nearly commensurate CDW (NCCDW) state. In the NCCDW state, the hexagonal array of the CDW domains are separated by discommensurations. Accompanied with this structural change, the electronic structure changes from the Mott state to metallic one.

Not only the CDW order, but superconductivity has been realized by several methods, such as chemical doping or physical pressure [2-4]. Especially, Fe substitution influences on electronic properties of 1*T*-TaS_2 effectively. In $Ta_{1-x}Fe_xS_2$ system, only 1 % Fe doping suppresses the Mott transition, and 2 % doping induces superconductivity at about 3K [3]. Further doping reduces the electrical conductivity and $Ta_{0.95}Fe_{0.05}S_2$ shows insulating behavior at low temperature due to Anderson localization [3, 5]. Thus, this system is suitable for investigate the competition between superconductivity and CDW state. In order to disentangle the relation between CDW state and superconductivity, the role of Fe dopant should be important.

In this paper, we report on scanning tunneling microscopy and spectroscopy (STM/STS) measurements on 1*T*-$Ta_{1-x}Fe_xS_2$ with different Fe concentration. STM observation reveals that CDW superlattice becomes irregular as increasing Fe concentration. STS measurements on $Ta_{0.98}Fe_{0.02}S_2$ uncover the spatial change in the local density of state (LDOS). We find two types of David-star: one has symmetric LDOS with respect to Fermi energy (E_F), and the other has asymmetric one. Because the number of the David-star that has asymmetric LDOS increases as increasing Fe concentration, we conclude that asymmetric LDOS is due to the dopant Fe.

2. Experimental

Single crystals of $Ta_{1-x}Fe_xS_2$ (nominal x = 0, 0.02, and 0.05) were grown by chemical vapor transport method. For the pristine TaS_2, Mott transition was observed at about 220 K in heating process in the electrical resistivity measurements. For the sample with x = 0.02, Mott transition was not observed, but onset of superconducting transition was observed at 4 K. For x = 0.05, in contrast to the previous report [7, 9], the insulating behaviour was not observed. This seems to indicate that our samples have slightly less Fe concentration than the nominal one. The STM/STS measurements were performed by using a laboratory-build scanning tunneling microscope. All of the measurements were done at 4.2 K. Electrochemically polished Au wire was used as STM tip. Clean surface was obtained by cleavage at 4.2 K in a few torr He gas environment. STM images were obtained by constant-current mode. Tunneling spectra were obtained by the numerical differentiation of the I-V curves.

3. Results and Discussion

Figures 1(a) to (c) show typical STM images of $Ta_{1-x}Fe_xS_2$ (nominal x = 0, 0.02, and 0.05), respectively. In all images, bright points, which correspond to David-stars can be seen. Figure 1(a) shows almost perfect triangular lattice composed of David-stars. On the other hand, in the sample with x = 0.02, the CDW superlattice distorts. Furthermore, in the sample with x = 0.05, the distortion becomes significant. Here, it should be noted that the defects of David-star lattice that can be seen as dark points in figures 1(b) and (c) do not correspond to Fe site, because such defect exists even in the pristine sample (not shown) and its concentration is not proportional to the nominal Fe concentration.

Figure. 1. (a)-(c) STM images of $Ta_{1-x}Fe_xS_2$ (nominal x = 0, 0.02, and 0.05) obtained at 4.2 K on a 14 × 14 nm^2 field of view, respectively. The sample bias voltage is 500 mV and the tunneling current is 200 pA.

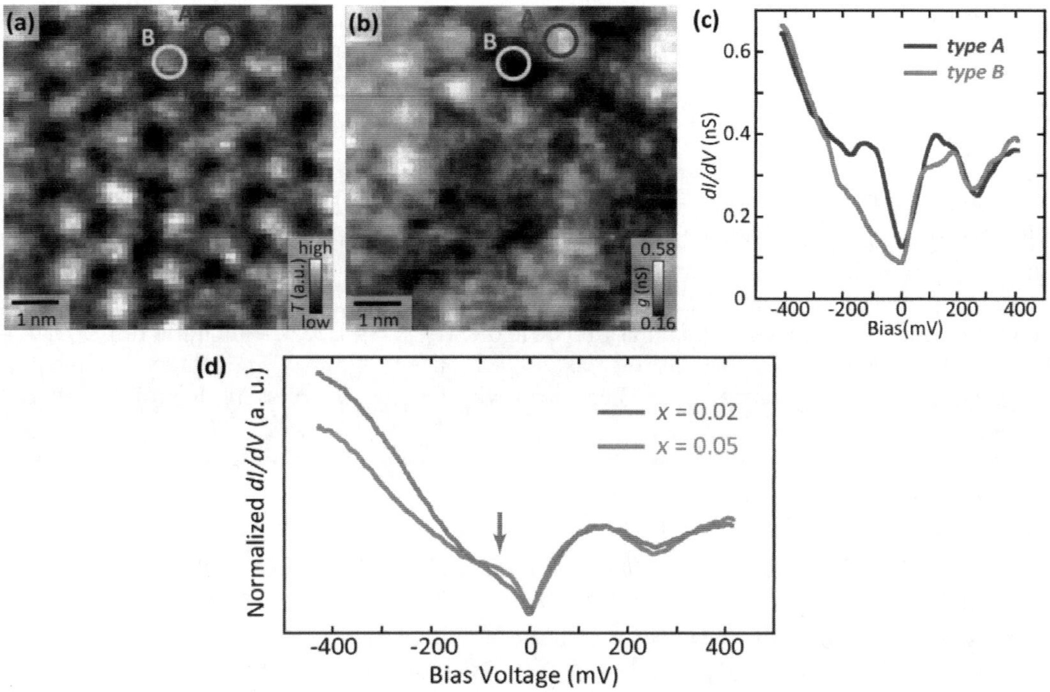

Figure. 2. (a); STM image of $Ta_{0.98}Fe_{0.02}S_2$ that is simultaneously obtained by STS measurements on a 7 × 7 nm² field of view. The sample bias voltage is 500 mV and the tunneling current is 200 pA. (b); The conductance (g) map at -120 mV on the same field of view as (a). (c); Tunneling spectra obtained in the circles shown in (a) and (b). (d); Spatially averaged tunneling spectra on a 14×14 nm² field of view. Gray and red lines indicate the spectra taken on the sample with x = 0.02 and 0.05, respectively. Each spectrum is normalized at +300 mV.

Figures 2(a) and (b) show a STM image and the simultaneously obtained conductance (g) map at -120 mV in x = 0.02 sample, respectively. The g map (figure 2(b)) indicates the existence of the spatial inhomogeneity of the LDOS while the STM image shows the periodic array of David-stars. This is in contrast to pristine 1T-TaS_2 [1, 6] where each David star shows the same tunneling spectrum. This implies that Fe dopant affects the LDOS. To investigate the effect of Fe doping, we examined the spatial change in LDOS at each David-star. Because the STM image shows the array of David-stars, we averaged the tunneling spectra at each David-star. We found the existence of two types of tunneling spectrum. Typical spectra are shown in figure 2 (c). These tunneling spectra were obtained in the regions indicated by circles shown in figure 2(c). These spectra have finite conductance at 0 mV. This is consistent with the metallic behavior in the electric conductivity. The spectrum averaged in purple circle shown in figure (type A) has two peak structures around -100 mV and 100 mV, and thus shows symmetry with respect to E_F. We think that this structure represents the Mott gap reduced by local structural strain or reduction of the coherency of the CDW [7]. On the other hand, the spectrum averaged in blue circle (type B) shows suppression of the peaks, especially at negative bias, and thus shows asymmetry with respect to E_F. It is plausible that type B spectrum is caused by the existence of Fe dopant because the peaks caused by Mott gap are reduced.

To investigate the Fe concentration dependence of the tunneling spectra, we spatially averaged on a 14×14 nm² field of view in x = 0.02 and 0.05 samples. Gray and red spectra in figure 2(d) represent the results taken on the sample with x = 0.02 and 0.05, respectively. Both spectra show peak structure at about 100 mV. This can be explained by the existence of the peak structure at the similar bias voltage in both type A and B spectrum. On the other hand, the peak structure at negative bias voltage becomes less conspicuous in the tunneling spectrum obtained in the sample with x = 0.05. This can be

IOP Conf. Series: Journal of Physics: Conf. Series **871** (2017) 012003 doi:10.1088/1742-6596/871/1/012003

recognized as the ratio of type B spectrum to that of type A spectrum increased in the sample with x = 0.05. This result confirms that type B spectrum is caused by the existence of Fe dopant. Precise site where Fe is doped and the reason why Fe site shows type B spectrum are not clear at present. Further STM/STS and other experimental and theoretical investigations are required.

4. Summary

We have performed STM/STS measurements on $1T$-$Ta_{1-x}Fe_xS_2$ with different Fe concentration in order to clarify the effect of dopant Fe on $1T$-TaS_2. STM images show that the CDW superlattice becomes irregular as increasing Fe concentration. From the STS measurements, we found the spatial inhomogeneity of electronic state that is represented by two types of tunneling spectrum: symmetric spectrum with respect to E_F (type A), and asymmetric one (type B). Because the ratio of the David-star that has type B spectrum increases as increasing Fe concentration, we conclude that the type B tunneling spectrum is due to dopant Fe. The reason why Fe site shows such electronic feature is an open question.

References

[1] Kim J-J, Yamaguchi W, Hasegawa T and Kitazawa K, 1994 *Phys. Rev. Lett.* **73** 2103
[2] Sipos B, Kusmartseva A F, Akrap A, Berger H, Forró L and Tutiš E 2008 *Nat. Mater.* **7** 960
[3] Li L J, Lu W J, Zhu X D, Ling L S, Qu Z and Sun Y P 2012 *Euro. Phys. Lett.* **97** 67005
[4] Liu Y, Ang R, Lu W J, Song W H, Li L J and Sun Y P 2013 *App. Phys. Lett.* **102** 192602
[5] DiSalvo F J, Wilson J A and Waszczak 1976 *Phys. Rev. Lett.* **36** 885
[6] Cho D, Cho Y M, Cheong S W, Kim K S and Yeom H W 2015 *Phys. Rev. B* **92** 085132
[7] Chatterjee U *et al* 2015 *Nat. Commun.* **6** 6313

IOP Conf. Series: Journal of Physics: Conf. Series **871** (2017) 012004 doi:10.1088/1742-6596/871/1/012004

Superconducting gap of the single crystal β-$PdBi_2$

H Matsuzaki[1], K Nagai[2], N Kase[1*], T Nakano[1], and N Takeda[2]

[1]Graduate School of Science and Technology, Niigata University, Niigata 950-2102, Japan
[2]Department of Materials Science and Technology, Niigata University, Niigata 950-2102, Japan

E-mail: `n-kase@eng.niigata-u.ac.jp`

Abstract. We investigate superconducting and normal properties of the single crystal of β-$PdBi_2$. The electrical resistivity $\rho(T)$ shows superconductivity at $T_c = 5.0$ K. Residual resistivity ratio (RRR) is estimated to be 2.9 obtained from $\rho(300\ \mathrm{K})/\rho(5.0\ \mathrm{K})$. The H_{c2} curve obtained from $\rho(T)$ in magnetic fields shows cleat enhancement from the Wertharmer-Helfand-Hohenberg theory in dirty limit. Specific heat $C(T)$ measurement shows that clear jump is observed at T_c = 4.8 K. T-dependence of the electronic specific heat $C_e(T)$ suggests full-gap symmetry with a single gap and strong coupling with $\Delta C_e/\gamma T_c = 1.8$.

1. Introduction

Several superconductors developed by utilizing topological insulators are reported, such as Cu-intercalated Bi_2Se_3[1, 2], In-doped $SnTe_{13}$[3] and M_2Te_3 (M = Bi, Sb) under pressure[4, 5]. The recent discovery of the topologically protected surface states in β-$PdBi_2$ has reignited the research interest in this class of superconductors[6]. β-$PdBi_2$ is reported to be a superconductor with superconducting transition temperature (T_c) = 5.2 K. Interestingly, specific heat measurement shows clear hump below T_c, suggesting that β-$PdBi_2$ is a multiple-band/multiple-gap superconductor[7]. On the other hand, μSR measurement suggests that superconductivity is full-gap symmetry with a single gap[8]. Thus, superconducting gap symmetry of β-$PdBi_2$ remains to be clarified at this stage.

In this article, we discuss the superconducting gap symmetry of the single-crystalline β-$PdBi_2$. To reveal superconducting gap symmetry of β-$PdBi_2$, we performed specific heat measurement down to 0.28 K with a quasi-adiabatic method.

2. Experimental Details

Single crystals of β-$PdBi_2$ were grown by a melt growth method[7]. The stoichiometric mixture of Pd (powder, 99.9%) and Bi (grain, 99.999%) was sealed in an evacuated quartz tube (total mass: 2 g). The ampule was heated at 800 ℃ maintained for 24 h, and then cooling down to 450 ℃ at a down rate of 4 ℃/h before being quenched in ice water. The obtained single crystals had good cleavage, producing flat surfaces as large as 1×1 cm^2.

The crystal structure of the synthesized single-crystalline sample was examined by the powder X-ray diffraction (XRD) technique using a conventional X-ray spectrometer equipped with Cu-Kα radiation and a monochromator (RAD-2X, Rigaku). The intensity data were collected over a 2θ range of 10°–80° with a step width of 0.01° and a counting rate of 4°/minutes.

Electrical resistivity $\rho(T)$ was measured by a dc-four-probe method with a current source (Model 6221, Keithley) and a nano voltmeter (Model 2182A, Keithley), down to 0.3 K in zero and various applied magnetic fields. Specific heat $C(T)$ was measured by an adiabatic heat-pulse and relaxation method down to 0.3 K using a commercial ^{3}He refrigerator (Heliox-VL, Oxford Instruments).

3. Experimental Results

3.1. X-ray diffraction pattern

Figure 1 shows X-ray diffraction patterns of the single-crystalline β-$PdBi_2$ along c-axis at room temperature. These peaks can be indexed with a tetragonal unit cell with the space group of $I4/mmm$. The lattice constants estimated from all indexes are $a = 0.3620$ nm, $c = 1.2984$ nm for β-$PdBi_2$. These values are almost consistent with those of the previous report[7].

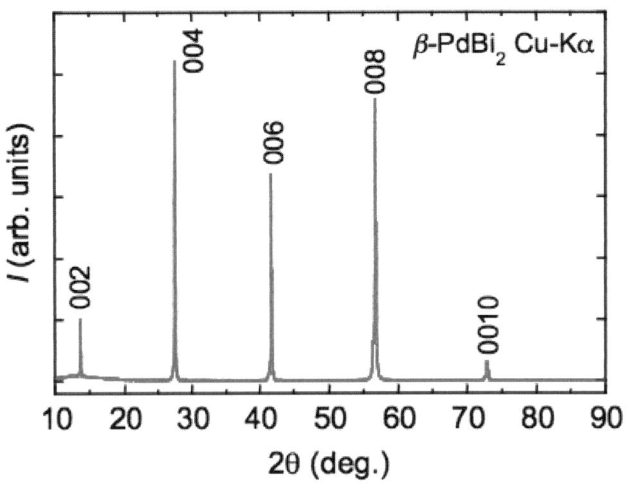

Figure 1. X-ray diffraction patterns of the single crystal β-$PdBi_2$ along c-axis at room temperature.

3.2. electrical resistivity

Figure 2 shows the T-dependence of the electrical resistivity $\rho(T)$ of β-$PdBi_2$. The $\rho(T)$ shows typical metallic behavior with a residual resistivity ρ_0 of 22 $\mu\Omega$cm. The residual resistivity ratio (RRR) is estimated to be 2.9 obtained from $\rho(300\text{ K})/\rho(5.0\text{ K})$, as shown in the inset of Fig. 2. This value is smaller than that of the previous report[7]. The $\rho(T)$ of β-$PdBi_2$ shows superconductivity at $T_c = 5.1$ K. Zero resistivity was observed at around 5.0 K in $\rho(T)$ with the superconducting transition width ΔT of about 0.02 K. The sharp transition suggests good quality of sample.

3.3. specific heat

Figure 3 shows the T-dependence of the specific heat $C(T)$ divided by temperature T of β-$PdBi_2$ in 0 and 1 T. The normal state specific heat can be fitted by the equation $C/T = \gamma + \beta T^2 + \gamma T^4$, where γ is the electronic specific heat coefficient (Sommerfeld constant), and $C_{\text{ph}} = \beta T^2 + \gamma T^4$ is the phonon contribution. The parameters were obtained as $\gamma = 7.56$ mJ/(mol·K^2), $\beta = 2.121$ mJ/(mol·K^4) for β-$PdBi_2$. In a simple Debye model of the phonon contribution, the β coefficient is related to the Debye temperature Θ_D according to the equation $\Theta_D = (\frac{12\pi^4}{5\beta} nR)^{1/3}$, where $R = 8.314$ J/(mol·K) and $n = 3$ for β-$PdBi_2$. From this relationship, we calculated $\Theta_D =$

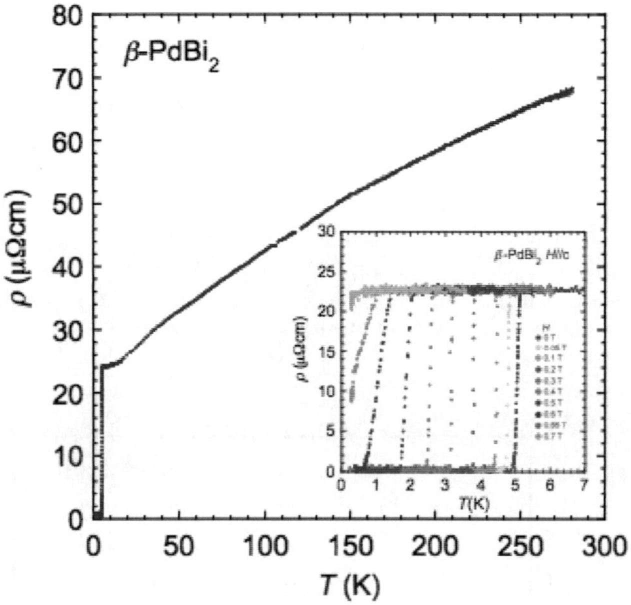

Figure 2. T-dependence of the electrical resistivity $\rho(T)$ of β-$PdBi_2$. The inset shows $\rho(T)$ of β-$PdBi_2$ in several magnetic fields.

149 K, for β-$PdBi_2$. From specific heat measurement, clear jump was observed at $T_c = 4.8$ K. T-dependence of the electronic specific heat $C(T)$ suggests that full-gap symmetry with a single gap and strong coupling with $\Delta C_e/k_B T_c = 1.8$.

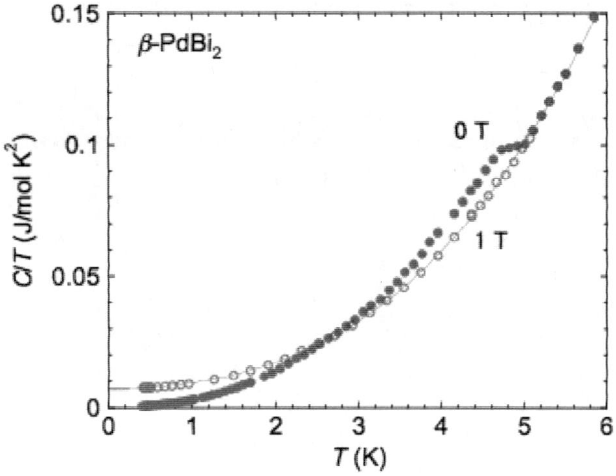

Figure 3. T-dependence of the specific heat $C(T)$ divided by T of β-$PdBi_2$ in magnetic fields of 0 and 1 T. Solid line shows the fitted result of $C(T)/T = \gamma + \beta T^2 + \gamma T^4$.

IOP Conf. Series: Journal of Physics: Conf. Series **871** (2017) 012004 doi:10.1088/1742-6596/871/1/012004

3.4. upper critical field

From $\rho(T)$ in several magnetic fields, the upper critical field $\mu_0 H_{c2}(T)$ is determined. T_c of $\rho(T)$ is defined as the temperature at which the resistivity drops to 50% of ρ_0. The H-T phase diagram is presented in Fig. 4. The H_{c2} curve shows cleat enhancement from the Wertharmer-Helfand-Hohenberg (WHH) theory in dirty limit ($h^* = 0.69$)[9, 10]. Such a behavior is seen in a multi gap superconductor. However, our specific heat measurements show a single gap superconductor. The enhancement seems to be generated from strong electron-phonon coupling, and it is often observed in strong electron-phonon coupling superconductors[11].

To estimate H_{c2}, we fitted $\mu_0 H_{c2}(T)$ using polynomial function presented by solid line. From the analysis, $\mu_0 H_{c2}(0)$ is determined to be 1.07 (ab-plane) and 0.72 T (c-axis) for β-$PdBi_2$. From the relationship $\mu_0 H_{c2}(0) = \Phi_0/2\pi\xi(0)^2$, $\xi(0)$ was calculated to be 24.5 (ab-plane) and 30.7 nm (c-axis).

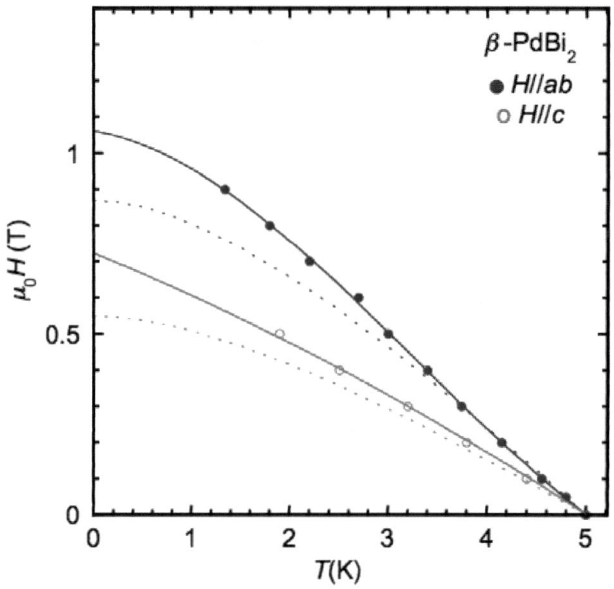

Figure 4. T-dependence of the upper critical field $\mu_0 H_{c2}(T)$ of β-$PdBi_2$. The dotted lines are the WHH prediction with dirty limit. The solid lines show the polynomial function.

4. Summary

We studied superconducting properties of the single crystal of β-$PdBi_2$ at $T_c = 4.9$ K. Electrical resistivity $\rho(T)$ shows metallic behavior with a residual resistivity ρ_0 of 22 $\mu\Omega$cm and the residual resistivity ratio (RRR) is estimated to be 2.9. The upper critical field $\mu_0 H_{c2}(0)$ of β-$PdBi_2$ is estimated to be 1.07 (ab-plane) and 0.72 T (c-axis), which are good agreement with the previous report. Specific heat measurements show a single gap superconductor with strong electron-phonon coupling.

Acknowledgments

This work was partially supported by the Grant for Promotion of Niigata University Research Projects (No. 524501), the Grant-in-Aid for Young Scientists (B) (No. 26800183) from the Ministry of Education, Culture, Sports, Science, and Technology, Japan, the Sasakawa Scientific

IOP Conf. Series: Journal of Physics: Conf. Series **871** (2017) 012004 doi:10.1088/1742-6596/871/1/012004

Research Grant (26-207) from the Japan Science Society, and the Uchida Energy Science Promotion Foundation (26-1-22).

References

[1] Hor Y S, Williams A J, Checkelsky J G, Roushan P, Seo J, Xu Q, Zandbergen H W, Yazdani A, Ong N P, and Cava R 2010 *J, Phys. Rev. Lett.* **104** 057001

[2] Kriener M, Segawa K, Ren Z, Sasaki S, and Ando Y 2011 *Phys. Rev. Lett.* **106** 127004

[3] Sasaki S, Ren Z, Taskin A A, Segawa K, Fu L, and Ando Y 2012 *Phys. Rev. Lett.* **109** 217004

[4] Zhanga J L, Zhanga S J, Wenga H M, Zhanga W, Yanga L X, Liua Q Q, Fenga S M, Wanga X C, Yua R C, Caob L Z, Wangc L, Yangc W G, Liud H Z, Zhaoe W Y, Zhang S C, Daia X, Z. Fanga Z, and Jin C Q 2011 *Proc. Natl Acad. Sci. USA* **108** 24-28

[5] Zhu J, Zhang J L, Kong P P, Zhang S J, Yu X H, Zhu J L, Liu Q Q, Li X, Yu R C, Ahuja R, Yang W G, Shen G Y, Mao H K, Weng H M, Dai X, Fang Z, Zhao Y S, and Jin C Q 2013 *Sci. Rep.* **3** 2016

[6] Sakano M, Okawa K, Kanou M, Sanjo H, Okuda T, Sasagawa T, and Ishizaka K 2015 *Nat. Commun.* **6** 8595

[7] Imai Y, Nabeshima F, Yoshinaka T, Miyatani K, Kondo R, Komiya S, Tsukada I, and Maeda A 2012 *J. Phys. Soc. Jpn* **81** 113708

[8] Biswas P K, Mazzone D G, Sibille R, Pomjakushina E, Conder K, Luetkens H, Baines C, Gavilano J L, Kenzelmann M, Amato A, and Morenzoni E 2016 *Phys. Rev.* B **93** 220504(R)

[9] Helfand E and Werthamer N R 1966 *Phys. Rev.* **147** 288

[10] Maki K 1966 *Phys. Rev.* **148** 362

[11] Kase N, Hayamizu H, and Akimitsu 2011 *Phys. Rev.* B **83** 184509

Upper critical field of the single crystal La(O,F)$BiSe_2$

Y Terui[1], N Kase[1*], T Nakano[1], and N Takeda[2]

[1]Graduate School of Science and Technology, Niigata University, Niigata 950-2102, Japan
[2]Department of Materials Science and Technology, Niigata University, Niigata 950-2102, Japan

E-mail: n-kase@eng.niigata-u.ac.jp

Abstract. Single crystal growth and superconducting properties of $LaO_{0.5}F_{0.5}BiSe_2$ are studied. Single crystals of $LaO_{0.5}F_{0.5}BiSe_2$ were grown by a CsCl-flux method. Electrical resistivity $\rho(T)$ measurements were performed to elucidate its superconducting properties. The ρ(T) shows metallic behavior and superconducting transition at 4.0 K. From $\rho(T)$ measurements in several magnetic fields, upper critical fields $H_{c2}^{||ab}(0)$ and $H_{c2}^{||c}(0)$ are estimated to be 7.05 T and 0.52 T, respectively. The superconducting anisotropic parameter Γ is determined to be 13.6.

1. Introduction

Since the discovery of superconductors in the layered bismuth oxysulfide $Bi_4O_4S_3$ and $LaO_{0.5}F_{0.5}BiS_2$, BiS_2-based superconductors have attracted much attention[1, 2]. The crystal structure of $LaOBiS_2$ consists of alternating superconducting BiS_2 layers and blocking LaO layers with a space group $P4/nmm$ (No. 129). Interestingly, the crystal structure is very similar to the Fe-based superconductors. In the BiS_2-based superconductors, it is known that substituting F for O makes insulating material induce superconductivity at $T_c = 2.5$ K. Same effect is also seen in replacing La with tetravalent elements (Ti, Zr, Hf and Th)[3]. In addition, ascension of T_c is found by replacing S atoms to the congeneric atom of Se ($T_c = 4.2$ K for $LaO_{0.5}F_{0.5}BiSS$[4], 3.7 K for $LaO_{0.5}F_{0.5}BiSe_2$[5]). The recent research suggests that the enhancement of T_c in Se doped compounds is attributable to an in-plane chemical pressure[6].

In this article, we report the single crystal growth of $LaO_{0.5}F_{0.5}BiSe_2$ and its superconducting properties. Although single crystal growth of $LaO_{0.5}F_{0.5}BiSe_2$ is reported by several groups, however, low-temperature properties are not reported at this stage. In order to elucidate superconducting state, we measured the electrical resistivity of $LaO_{0.5}F_{0.5}BiSe_2$ down to 0.4 K in magnetic fields up to 11 T.

2. Experimental Details

Single crystals of $LaO_{0.5}F_{0.5}BiSe_2$ were grown by a CsCl-flux method similar to the previous reports[5, 6, 7, 8]. The mixture of the starting materials of 0.8 g and CsCl as flux of 5 g are sealed in an evacuated quartz tube. The tube was heated up to 880 ℃ in 10 h and to 900 ℃ in 2 h, then kept for 10 h and cooled down to 650 ℃ for 125 h. The flux materials were easily removed from the crystals by water, ethanol, and acetone. The single crystals were plate-like shape with the size of approximately 1.0×1.0 mm^2. The crystal structure of the single crystal was determined by the X-ray diffraction (XRD) method using a conventional X-ray spectrometer equipped with

IOP Conf. Series: Journal of Physics: Conf. Series **871** (2017) 012005 doi:10.1088/1742-6596/871/1/012005

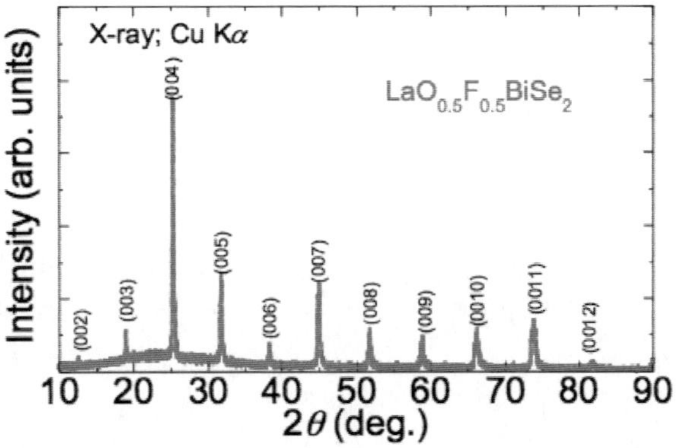

Figure 1. X-ray diffraction patterns of $LaO_{0.5}F_{0.5}BiSe_2$ along *ab*-plane at room temperature.

Cu-Kα radiation and a monochromator (RAD-2X, Rigaku). The intensity data were collected over a 2θ range of 10°–80° with a step width of 0.01° and a counting rate of 4°/minutes.

Electrical resistivity $\rho(T)$ was measured along the *ab*-plane and *c*-plane by the dc-four-probe method with a current source (Model 6221, Keithley) and a nano voltmeter (Model 2182A, Keithley), down to 0.5 K in zero and magnetic fields up to 11 T. Au wires (ϕ = 20 μm) were attached to sample surface as *ab*-plane by Ag paste to form electrical contact.

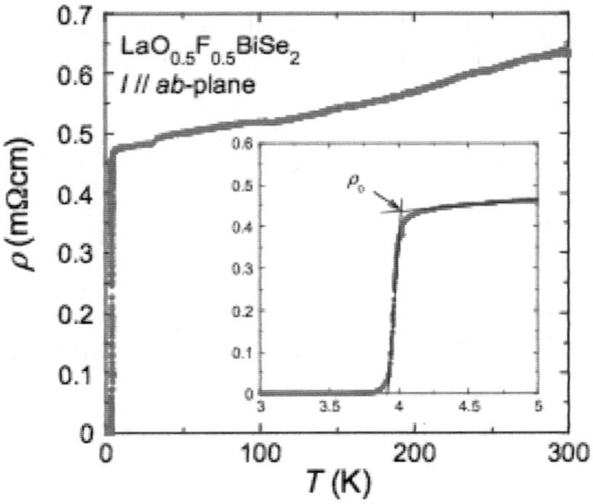

Figure 2. Temperature dependence of the electrical resistivity $\rho(T)$ of $LaO_{0.5}F_{0.5}BiSe_2$ with current along *ab*-plane. The inset shows $\rho(T)$ of $LaO_{0.5}F_{0.5}BiSe_2$ at low temperatures.

3. Experimental Results

3.1. X-ray diffraction pattern

Figure 1 shows the XRD diffraction pattern of the single crystals $LaO_{0.5}F_{0.5}BiSe_2$ at room temperature. Only $(00l)$ diffraction peaks were detected from the XRD pattern, indicating the crystallographic *c*-axis is perpendicular to the crystal plane. All the reflections can be indexed as the space group of $P4/nmm$. To estimate lattice constants, powder X-ray diffraction was performed by using several single crystals which are crashed into fine powder. The lattice constants of $LaO_{0.5}F_{0.5}BiSe_2$ are obtained to be $a = 0.4161$ nm and $c = 1.4380$ nm. The cell volume of $LaO_{0.5}F_{0.5}BiSe_2$ is calculated to be 0.2490 nm^3. The cell value is slightly larger than that of the previous reports[7, 8].

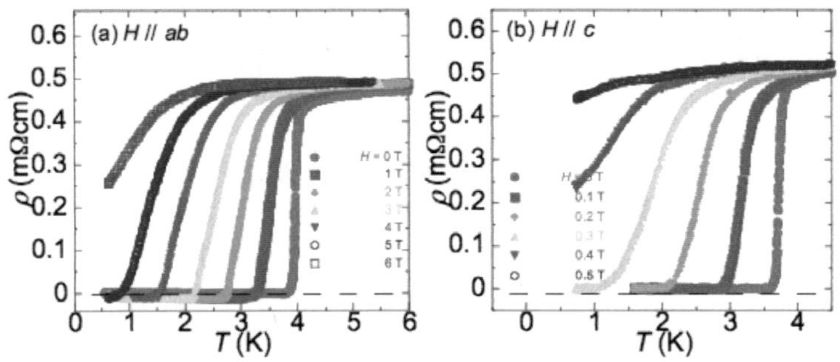

Figure 3. Temperature dependence of the electrical resistivity $\rho(T)$ of $LaO_{0.5}F_{0.5}BiSe_2$ in several magnetic fields along *ab*-plane (a) and *c*-plane (b) at low temperatures.

3.2. electrical resistivity

Figure 2 shows the temperature dependence of the electrical resistivity $\rho(T)$ of the single crystal $LaO_{0.5}F_{0.5}BiSe_2$ down to 0.5 K with electrical current along the *ab*-plane. $\rho(T)$ showed a metallic behavior but residual resistivity ρ_0 was large as single crystal. ρ_0 is defined as just above T_c and $\rho_0 = 0.47$ mΩcm was obtained. Sudden decrease originated from superconducting transition was observed around 4 K. T_c is defined as 50% drop from ρ_0 and determined to be 3.4 K. Although the behavior of $\rho(T)$ and the ρ_0 is almost the same value compared with polycrystalline sample of $LaO_{0.5}F_{0.5}BiSe_2$, T_c is slightly higher than that of polycrystalline sample ($T_c \sim 2.5$ K)[9].

Figure 3 shows $\rho(T)$ of the single crystal $LaO_{0.5}F_{0.5}BiSe_2$ in several magnetic fields along *ab*-plane (a) and *c*-plane (b). T_c shifts toward lower temperature in proportion to applied magnetic field. Superconductivity is insensitive to magnetic field applying along *ab*-plane. On the other hand, when applying along *c*-plane superconductivity is easily destroyed by the magnetic field. Such a behavior suggests that anisotropy exists in the superconductor.

3.3. upper critical field

Figures 4(a) and (b) show the temperature dependence of the upper critical fields $H_{c2}^{||ab}(0)$ and $H_{c2}^{||c}(0)$ of $LaO_{0.5}F_{0.5}BiSe_2$. Error bars are determined by the temperature of 10% and 90% drop from ρ_0. For *ab*-plane, the $\mu_0 H_{c2}^{||ab}(T)$ curve shows non-T-liner behavior. This behavior could not be explained with the Wertharmer-Helfand-Hohenberg (WHH) theory at the dirty limit ($h^* = 0.69$) as presented by dashed line[10, 11]. To obtain $H_{c2}(0)$ value, we employed the type-II superconductor approximation $\mu_0 H_{c2}(T) = \mu_0 H_{c2}(0)[1 - (T/T_c)^{3/2}]^{3/2}$[12], which is in a good agreement with the experimental result, as shown in the solid line of Fig. 4. $\mu_0 H_{c2}^{||ab}(0)$ is estimated to be approximately 7.05 T. For *c*-plane, $\mu_0 H_{c2}^{||c}(0)$ is estimated to be 0.52 T.

The anisotropic parameter Γ was determined using the anisotropic Ginzburg-Landau (GL) formula; $\Gamma = m_c/m_{ab} = H_{c2}^{||ab}/H_{c2}^{||c} = \xi_{ab}/\xi_c$, where m_i ($i = ab$, c) are the effective mass, and ξ_i ($i = ab$, c) are GL-coherence length. We calculated the ξ using $H_{c2}^{||ab} = \Phi_0/(2\pi\xi_{ab}\xi_c)$ and $H_{c2}^{||c} = \Phi_0/(2\pi\xi_{ab}^2)$, where Φ_0 is the quantum flux. ξ_{ab} and ξ_c are estimated to be 9.66 and

131 nm. The superconducting anisotropic parameter $\Gamma = H_{c2}^{||ab}/H_{c2}^{||c}$ is determined to be 13.6, which suggests that the superconductor has remarkably high anisotropy. The value is relatively smaller than that of other $RO_{1-x}F_xBiS_2$ compounds ($\Gamma \sim$ 30–40)[13], and is comparable to that of electric-field-induced superconductivity in MoS_2 and $NbSe_2$ atomic layers ($\Gamma \sim$ 10–20)[14, 15]

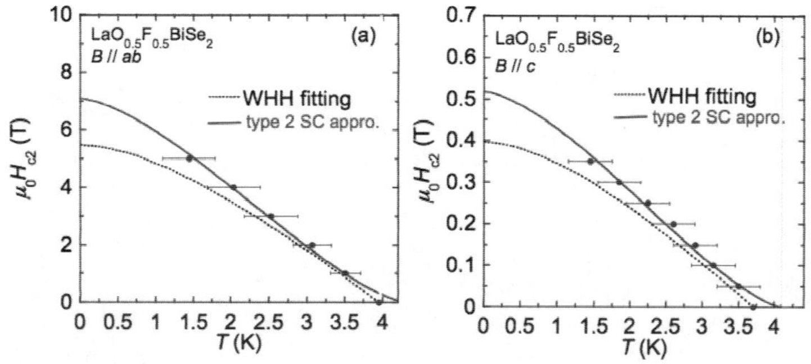

Figure 4. Temperature dependence of the upper critical fields $H_{c2}^{||ab}(T)$ (a) and $H_{c2}^{||c}(T)$ (b) of $LaO_{0.5}F_{0.5}BiSe_2$. The dotted lines show the WHH prediction. The Solid lines represent type-II superconductor approximation.

4. Summary

We investigated superconducting properties of the single crystal of $LaO_{0.5}F_{0.5}BiSe_2$. The growth single crystal shows a metallic behavior with a large residual resistivity $\rho_0 = 0.47$ mΩcm The upper critical fields $H_{c2}^{||ab}(0)$ and $H_{c2}^{||c}(0)$ obtained from electrical resistivity $\rho(T)$ in several magnetic fields are obtained to be 7.05 and 0.52 T. The $H_{c2}^{||ab}(0)$ is relatively smaller than that of the BiS_2-compounds. The superconducting anisotropic parameter $\Gamma = H_{c2}^{||ab}/H_{c2}^{||c}$ is determined to be 13.6.

Acknowledgments

This work was partially supported by the Grant for Promotion of Niigata University Research Projects (No. 524501), the Grant-in-Aid for Young Scientists (B) (No. 26800183) from the Ministry of Education, Culture, Sports, Science, and Technology, Japan, the Sasakawa Scientific Research Grant (26-207) from the Japan Science Society, and the Uchida Energy Science Promotion Foundation (26-1-22).

References

[1] Mizuguchi Y, Fujihisa H, Gotoh Y, Suzuki K, Usui H, Kuroki K, Demura S, Takano Y, Izawa H, and Miura O, 2012 *Phys. Rev.* B **86** 220510
[2] Mizuguchi Y, Demura S, Deguchi K, Takano Y, Fujihisa H, Gotoh Y, Izawa H, and Miura O, 2012 *J. Phys. Soc. Jpn.* **81** 114725
[3] Yazici D, Huang K, White B D, Jeon I, Burnett V W, Friedman A J, Lum I K, Nallaiyan M, Spagna S, and Maple M B, 2013 *Phys. Rev.* B **87** 174512
[4] Tanaka M, T. Yamaki T, Y. Matsushita Y, M. Fujioka M, S. J. Denholme S J, T. Yamaguchi T, H. Takeya H, and Y. Takano Y, 2015 *Appl. Phys. Lett.* **106** 112601
[5] Tanaka M, Nagao M, Matsushita Y, Fujioka M, Denholme S J, Yamaguchi T, Takeya H, and Takano Y, 2014 *J. Solid State Chem.* **219** 168
[6] Mizuguchi Y, Miura A, Kajitani J, Hiroi T, Miura O, Tadanaga K, Kumada N, Magome E, Moriyoshi C, and Kuroiwa Y, 2015 *Sci. Rep.* **5** 14968

[7] Wu S L, Sun Z A, Chiang F K, Ma C, Tian H F, Zhang R X, Zhang B, Li J Q, and Yang H X 2015 *Solid State Comm.* **205** 14
[8] Nagao M, Tanaka M, Watauchi S, Tanaka I, and Takano Y 2014 *J. Phys. Soc. Jpn.* **83** 114709
[9] Krzton-Maziopa A, Guguchia Z, Pomjakushina E, Pomjakushin V, Khasanov R, Luetkens H, Biswas P K, Amato A, Keller H, and Conder K, 2014 *J. Phys.:Condens. Matter* **26** 215702
[10] Helfand E and Werthamer N R 1966 *Phys. Rev.* **147** 288
[11] Maki K 1966 *Phys. Rev.* **148** 362
[12] Micnas R, Ranninger J and Robaszkiewicz S 1990 *Rev. Mod. Phys.* **62** 113
[13] Nagao M 2015 *Nov. Supercond. Mater.* **1** 64
[14] Saito Y, Nakamura Y, Bahramy M S, Kohama Y, Ye J, Kasahara Y, Nakagawa Y, Onga M, Tokunaga M, Nojima T, Yanase Y, and Iwasa Y, 2016 *Nat. Phys.* **12** 144
[15] Xi X, Wang Z, Zhao W, Park J H, Law K T, Berger H, Forro L, Shan J, and Mak K F, 2016 *Nat. Phys.* **12** 139

Evaluation of Bi defect concentration in $\boldsymbol{Ln}\mathbf{O_{1\text{-}x}F_xBi}\boldsymbol{Ch}_\mathbf{2}$ by scanning tunneling microscopy

S Demura, N Ishida, Y Fujisawa and H Sakata

Department of Physics, Tokyo University of Science, Shinjyuku-ku, Tokyo 162-8601, Japan

E-mail : demura@rs.tus.ac.jp

Abstract. We examined a concentration of Bi defects in layered BiCh_2 based superconductors LnO$_{1-x}$F$_x$BiCh_2 (Ln = La, Ce, Nd Ch = S, Se). These materials show superconductivity by electron carrier doping into BiCh_2 layer. Since Bi defects affect the carrier concentration directly, an examination of the concentration is important to evaluate an actual carrier concentration. In this paper, the concentration of Bi defects on the BiCh_2 layers in BiCh_2 based superconductors was evaluated by scanning tunneling microscopy measurements in real space. We found the samples with $BiSe_2$ layers have less Bi defects than those with BiS_2 layers. Furthermore, the concentration of Bi defects was found to be almost constant regardless of F concentration and Ln ion.

1. Introduction

Recently discovered BiS_2-based superconductors are attractive materials because they have a similar layered structure to cuprates and iron-based superconductors [1-12]. Among them, LnOBiS$_2$ (Ln = La, Pr, Ce, Nd, Yb, Bi) shows superconductivity by electron carrier doping by a partial substitution of F^- ions for O^{2-} ions. The superconducting transition temperature (T_c) varies with the F concentration. The F concentration is often analysed by an energy dispersive X-ray spectrometry (EDS) or electron probe micro analyser (EPMA). On the other hand, the carrier concentration induced by the substitution of F^- ions is estimated from a size of Fermi surface observed by an angle resolved photoemission spectroscopy (ARPES) measurement. In $LaO_{1-x}F_xBiS_2$ and $LaO_{1-x}F_xBiSe_2$, the carrier concentration estimated from a size of Fermi surface is consistent with those obtained from EDS or EPMA measurements [13, 14]. However, in $NdO_{1-x}F_xBiS_2$ these values are inconsistent with each other [15, 16]. A suggested reason about this inconsistency is an existence of Bi defects. Indeed, the existence of Bi defects was confirmed by an analysis of the crystal structure and by the scanning tunneling microscopy/spectroscopy (STM/STS) observations [17-19]. However, the density of Bi defects obtained by both techniques was approximately 3 %, while the expected concentration from previous EDS report was 15% [15].

Thus, it is important for understanding electronic properties in LnO$_{1-x}$F$_x$BiCh_2 to evaluate the concentration of Bi defects. Although, the Bi defect concentration has been evaluated in real space in some samples, a systematic evaluation of the concentration of Bi defects has not been done [18, 19]. The samples that have $BiSe_2$ plane should also be investigated. In this paper, F dependence of Bi defects concentration in $NdO_{1-x}F_xBiS_2$ was examined. In addition, we examined the concentration of Bi defects in $LaO_{1-x}F_xBiSe_2$ in real space. We found the samples with $BiSe_2$ layers have less Bi defects than those with BiS_2 layers. Furthermore, the concentration of Bi defects was found to be almost constant regardless of F concentration.

2. Experimental

Single crystals of $NdO_{0.1}F_{0.9}BiS_2$ were synthesized by a CsCl and KCl flux method in vacuumed quartz tubes using Nd_2S_3, Bi_2S_3, Bi_2O_3, BiF_3, and Bi as starting materials [21]. A detail of this synthesis procedure is described in reference 21.

Single crystals of $LaO_{0.5}F_{0.5}BiSe_2$ were synthesized by a CsCl flux method in vacuumed quartz tubes [22-24]. Mixtures of Bi (Mituswa Chemicals Co. Ltd., 99.9%), Se (Kojyundo Chemical Laboratory Co. Ltd., 99.99%), Bi_2O_3 (Kojyundo Chemical Laboratory Co. Ltd., 99.99%), and BiF_3 (Stella Chemifa Co. Ltd., 99%) were ground with nominal compositions of $LaO_{0.5}F_{0.5}BiSe_2$ except La. After grind, La grains (Rare Metallic Co. Ltd., 99.9 %) were put into this mixture. The mixture of 0.8 g was mixed with CsCl powder (Kojyundo Chemical Laboratory Co. Ltd., 99.9%) of 5 g, and sealed in an evacuated quartz tube. The tube was heated at 800 °C for 10 hours and cooled down to 630 °C at a rate of 0.5 °C/h. After this thermal process, the sintered materials were washed by distilled water to remove the flux.

The surface structure of single crystals was observed by a laboratory-build scanning tunneling microscope (STM) in the He gas at 4 K. The clean surface was prepared by cleaving the single crystal at 4 K *in situ*. A bias voltage was applied to the sample in all measurements.

3. Result

Figure 1 shows a typical STM image on a surface of $LaO_{0.5}F_{0.5}BiSe_2$ single crystal. The observed surface is a $BiSe_2$ plane because of the existence of the cleavage plane between two $BiSe_2$ planes [18, 19]. The observed atoms on the surface are Bi atoms, which form a square lattice with a lattice constant of approximately 0.5 nm. This value almost corresponds to a previous report [22-24].

Bi defects are imaged as black spots as shown in Fig. 1. As in the case of $NdO_{1-x}F_xBiS_2$ or $CeO_{1-x}F_xBiS_2$, streaks run along the diagonal directions of Bi square lattice from the defect sites [18, 19]. The observed concentration of Bi defects was approximately 1.0 %. This value is smaller than that observed in BiS_2 based superconductors [18, 19]. This result is consistent with the report that $BiSe_2$ based samples have better crystalline than BiS_2 based ones [14].

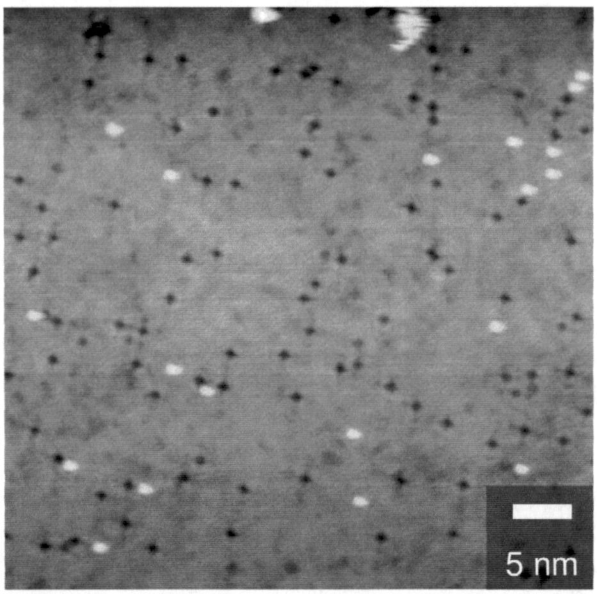

Figure 1. (Color online) The surface structure of $LaO_{0.5}F_{0.5}BiSe_2$ single crystal.

In order to investigate F dependence of the Bi defect concentration in BiS_2 compounds, we examined $NdO_{0.1}F_{0.9}BiS_2$, and obtained the concentration of about 2.5%. Compared with the result in $NdO_{0.5}F_{0.5}BiS_2$ [18], the concentration of Bi defects was almost same. This indicates that the concentration of Bi defect is almost independent of F concentration.

In Table 1, we summarized the observed and reported concentration of Bi defects observed by STM measurements. Regardless of *Ln*, *Ch* or F concentration, the concentration of Bi defects in $LnO_{1-x}F_xBiCh_2$ is in a range between 1 – 3%. The large deviation from this range such as 15% was not observed [15]. Since the defect concentration is almost independent of F concentration, this result certifies that F concentration is proportional to the carrier concentration.

STM measurements examine the Bi defects on the outer most layer that prepared by the cleavage. Because the cleavage may induce the Bi defects on the top most layer, the observed concentration of Bi defects is possibly the upper most one. However, because the cleavage was performed at low temperature of 4.2K, we believe the observed concentration is bulk one, which correspond to the value observed by EDS or EPMA.

Table 1. The summary of the concentration of Bi defects for $LnO_{1-x}F_xBiCh_2$.

	$F_{nominal}$	S	Se
La	0.5	-	1.0 %
Ce	0.5	2.0 % [19]	-
	0.7	3.0 % [19]	
Nd	0.3	3.0 % [18]	-
	0.9	2.5 %	

4. summary

We examined concentration of Bi defects in layered $BiCh_2$ based superconductors $LnO_{1-x}F_xBiCh_2$ (*Ln* = La, Ce, Nd *Ch* = S, Se) by STM measurements in real space. We found the sample with $BiSe_2$ layer has less Bi defects than that with BiS_2 layers. Furthermore, the concentration of Bi defects was found to be almost constant regardless of F concentration and *Ln* ion.

Acknowledgements

We are grateful to Dr. T. Machida and Dr. M. Nagao. This work was partly supported by a Grant-in-Aid for Scientific Research from the Ministry of Education, Culture, Sports, Science and Technology (KAKENHI).

References

[1] Mizuguchi Y et *al* 2012 *Phys. Rev. B* **86** 220510
[2] Mizuguchi Y, Demura S, Deguchi K, Takano Y, Fujihisa H, Gotoh Y, Izawa H and Miura O 2012 *J. Phys. Soc. Jpn.* **81** 114725
[3] Demura S et *al* 2013 *J. Phys. Soc. Jpn.* **82** 033708
[4] Lei H, Wang K, Abeykoon M, Bozin E S and Petrovic C 2013 *Inorg. Chem.* **52** 10685
[5] Xing J, Li S, Ding X, Yang H and Wen H H 2012 *Phys. Rev. B* **86** 214518
[6] Jha R, Kumar A, Singh S K and Awana V P S 2013 *J. Sup. and Novel Mag.* **26** 499
[7] Yazici D, Huang K, White B D, Chang A H, Friedman A J, Maple M B 2012 *Philosophical Magazine* **93** 673
[8] Yazici D et *al Phys. Rev. B* **87** 174512
[9] Lin X et *al* 2013 *Phys. Rev. B* **87** 020504
[10] Zhai H F et *al Phys. Rev. B* **90** 064518
[11] Krzton-Maziopa A et *al* 2014 *J. Phys.: Condens. Matter* **26** 215702
[12] Thakur G S, Selvan G K, Haque Z, Gupta L C, Samal S L, Arumugam S and Ganguli A K 2015 *Inorg. Chem.* **54** 1076
[13] Terashima K *et al* 2014 *Phys. Rev. B* **90** 220512
[14] Saini N L *et al* 2014 *Phys. Rev. B* **90** 214517
[15] Ye Z R *et al* 2014 *Phys. Rev. B* **90** 045116
[16] Zeng L K *et al* 2014 *Phys. Rev. B* **90** 054512
[17] Miura A, Nagao M, Takei T, Watauchi S, Tanaka I and Kumada N 2014 *J. Solid State Chem.* **212** 213
[18] Machida T, Fujisawa Y, Nagao M, Demura S, Deguchi K, Mizuguchi Y, Takano Y and Sakata H 2014 *J. Phys. Soc. Jpn* **83** 113701
[19] Demura S, Y.Fujisawa Y, Machida T, Nagao M, Takano Y and Sakata H 2016 *Phys. Proc.* **81** 49
[20] Nagao M, Demura S, Deguchi K, Miura A, Watauchi S, Takei T, Takano Y, Kumada N and Tanaka I 2012 *J. Phys. Soc. Jpn.* **81** 103702
[21] Nagao M, Miura A, Demura S, Deguchi K, Watauchi S, Takei T, Takano Y, Kumada N and Tanaka I 2014 *Solid State. Commun.* **178** 33
[22] Nagao M, Tanaka M, Watauchi S, Tanaka I and Takano Y 2014 *J. Phys. Soc. Jpn.* **83** 114709
[23] Tanaka M, Nagao M, Matsushita Y, Fujioka M, Denholme S J, Yamaguchi T, Takeya H, and Takano Y 2014 *J. Solid State Chem.* **219** 168
[24] Shao J, Liu Z, Yao X, Zhang L, Pi L, Tan S, Zhang C and Zhang Y 2014 *Europhysics Lett.* **107** 37006

IOP Conf. Series: Journal of Physics: Conf. Series **871** (2017) 012007 doi:10.1088/1742-6596/871/1/012007

Synchrotron powder X-ray diffraction and structural analysis of $Eu_{0.5}La_{0.5}FBiS_{2-x}Se_x$

K. Nagasaka[1], G. Jinno[1], O. Miura[1], A. Miura[2], C. Moriyoshi[3], Y. Kuroiwa[3], Y. Mizuguchi[1]*

1. Department of Electrical and Electronic Engineering, Tokyo Metropolitan University, 1-1, Minami-osawa, Hachioji 192-0397, Japan.

2. Faculty of Engineering, Hokkaido University, Kita-13, Nishi-8, Kita-ku, Sapporo 060-8628 Japan.

3. Department of Physical Science, Hiroshima University, 1-3-1 Kagamiyama, Higashihiroshima, Hiroshima 739-8526 Japan.

mizugu@tmu.ac.jp

Abstract. $Eu_{0.5}La_{0.5}FBiS_{2-x}Se_x$ is a new BiS_2-based superconductor system. In $Eu_{0.5}La_{0.5}FBiS_{2-x}Se_x$, electron carriers are doped to the BiS_2 layer by the substitution of Eu by La. Bulk superconductivity in this system is induced by increasing the in-plane chemical pressure, which is controlled by the Se concentration (x). In this study, we have analysed the crystal structure of $Eu_{0.5}La_{0.5}FBiS_{2-x}Se_x$ using synchrotron powder diffraction and the Rietveld refinement. The precise determination of the structural parameters and thermal factors suggest that the emergence of bulk superconductivity in $Eu_{0.5}La_{0.5}FBiS_{2-x}Se_x$ is achieved by the enhanced in-plane chemical pressure and the decrease in in-plane disorder.

1. Introduction

The BiS_2-based superconductors have been drawing much attention as a new layered superconductor family [1-3]. The typical crystal structure is tetragonal (*P4/nmm*) and is composed of alternate stacks of the electrically conducting BiS_2 layers and the insulating (blocking) layers, which resembles those of the Cu-based and the FeAs-based high transition temperature (high-T_c) superconductors [4,5]. The parent phase of the BiS_2-based compounds is a semiconductor with a band gap [1,6,7]. Upon the generation of the electron carriers in the BiS_2 layers, metallic conductivity is induced, and superconductivity is observed in the electron-doped compounds. With this scenario that the superconductivity can be induced by the electron carrier doping, many BiS_2-based superconductors have been discovered [3]. Recent studies, however, have suggested that the emergence of superconductivity in BiS_2-based compounds is related with not only the carrier doping but also the crystal structure optimization. For example, although the superconducting transitions are observed in $LaO_{1-x}F_xBiS_2$ when electron carriers were doped by substitution of F for O, bulk superconductivity never appears while filamentary superconductivity with a small shielding volume fraction is observed. In addition, T_c in the filamentary superconducting phase is less than 3 K for optimally doped $x = 0.5$ [2]. The absence of bulk superconductivity in $LaO_{0.5}F_{0.5}BiS_2$ would be caused by the long bond length of in-plane Bi-S, in which orbital overlaps between Bi and S are small. According to our recent report, the enhancement of in-plane chemical pressure, which is related to the enhancement of orbital overlaps, are required to induce bulk superconductivity in the $LaO_{0.5}F_{0.5}BiS_2$ system [8].

Similar situation is observed in $Eu_{0.5}La_{0.5}FBiS_2$, which has a lattice structure similar to the $LaO_{1-x}F_xBiS_2$ system [9]. Although the electron carriers are doped by the substitution of Eu by La, bulk superconductivity is not observed. Instead, superconductivity is induced by the external pressure effect. T_c is also enhanced to ~10 K by the application of external pressure [9]. The bulk superconductivity states of BiS_2-based compounds can be induced by not only the external pressure effect but also the chemical pressure effect. We demonstrated the Se substitution for the $Eu_{0.5}La_{0.5}FBiS_2$ system. Polycrystalline samples of $Eu_{0.5}La_{0.5}FBiS_{2-x}Se_x$ were synthesized up to $x = 1$, and the emergence of bulk superconductivity with a T_c of ~ 4 K for $x \geq 0.6$ was confirmed by the electrical resistivity and magnetic susceptibility measurements [10]. From powder X-ray diffraction with the conventional laboratory X-ray source (a CuKα radiation) and the Rietveld refinements, we have suggested that the Se substitution resulted in the enhancement of the in-plane chemical pressure, and hence, the bulk superconductivity is induced. This result is consistent with previous studies on $LaO_{0.5}F_{0.5}BiS_{2-x}Se_x$ and $CeO_{0.5}F_{0.5}BiS_{2-x}Se_x$, in which the enhancement of in-plane chemical pressure induced bulk superconductivity [8,11,12]. To confirm that the in-plane chemical pressure scenario is universal to the BiS_2-based superconductors, precise analyses of the sample purity and crystal structure parameters of the newly discovered superconductor $Eu_{0.5}La_{0.5}FBiS_{2-x}Se_x$ are needed. In this study, we have performed synchrotron powder X-ray diffraction for $Eu_{0.5}La_{0.5}FBiS_{2-x}Se_x$. The evolution of crystal structure is analyzed by the Rietveld method.

2. Experimental details

The polycrystalline samples of $Eu_{0.5}La_{0.5}FBiS_{2-x}Se_x$ used in this study were prepared using a conventional solid-state-reaction method as described in Ref. 10. Synchrotron X-ray diffraction experiment was performed at the beam line BL02B2 of Spring-8 under a proposal No. 2016B1078. The synchrotron X-ray with an energy of 25 keV was used. Using the one-dimensional semiconductor detector (MYTHEN), diffraction data with a step of 0.006° were collected. All the experiments were performed at room temperature. The Rietveld refinements were performed using the RIETAN software [13]. For the in-plane Bi and chalcogen (Ch) sites, the Rietveld analyses were performed with the anisotropic thermal factors. The crystal structure image was drawn using the VESTA software [14].

3. Results and discussion

The X-ray diffraction patterns for $x = 0$–1 were refined using a tetragonal *P4/nmm* space group. For $x = 0$, 0.2, 0.4, and 0.6, fluoride impurities (BiF_3 and LaF_3) with populations of 4%, 4%, 5%, and 7% were found, respectively. For $x = 0.8$ and 1, small impurity peaks of the fluorides and unidentified broad peaks at $2\theta = 6.2°$ and 9.3° were observed. Although the broad impurity peaks would be selenides because of the appearance at higher nominal concentration of Se, we could not refine the impurity phase. Figure 1 displays the typical synchrotron X-ray diffraction pattern and Rietveld refinement fitting result for $x = 0.6$. Assuming the major phase ($x = 0.6$) and two fluoride impurities, the diffraction pattern is well fitted, and the resulting reliability factor (R_{wp}) is 8%.

With the obtained structural parameters, we discuss the evolution of crystal structure of Se-substituted $Eu_{0.5}La_{0.5}FBiS_{2-x}Se_x$. Figures 2(a) and 2(b) show the Se concentration dependences of lattice constant of a and c. The lattice constant of a monotonically increases with increasing x. In contrast, the lattice constant of c does not change for $x = 0$–0.6 and slightly increases at $x \geq 0.8$. These results are consistent with the previous study performed with a laboratory X-ray diffractometer [10] and suggests that the Se is selectively occupies the in-plane Ch1 site (see the inset of Fig. 1 for the definition of the Ch1 and Ch2 site). As experimental facts, we have not succeeded the refinement of Se occupancy at the Ch1 and the Ch2 sites because of the resulting negative small value of Se occupancy at the Ch2 site, which may be due to the fact that almost 100% Se goes to the in-plane Ch1 site. Therefore, we did the Rietveld refinements with the fixed Se occupancy as the nominal values x.

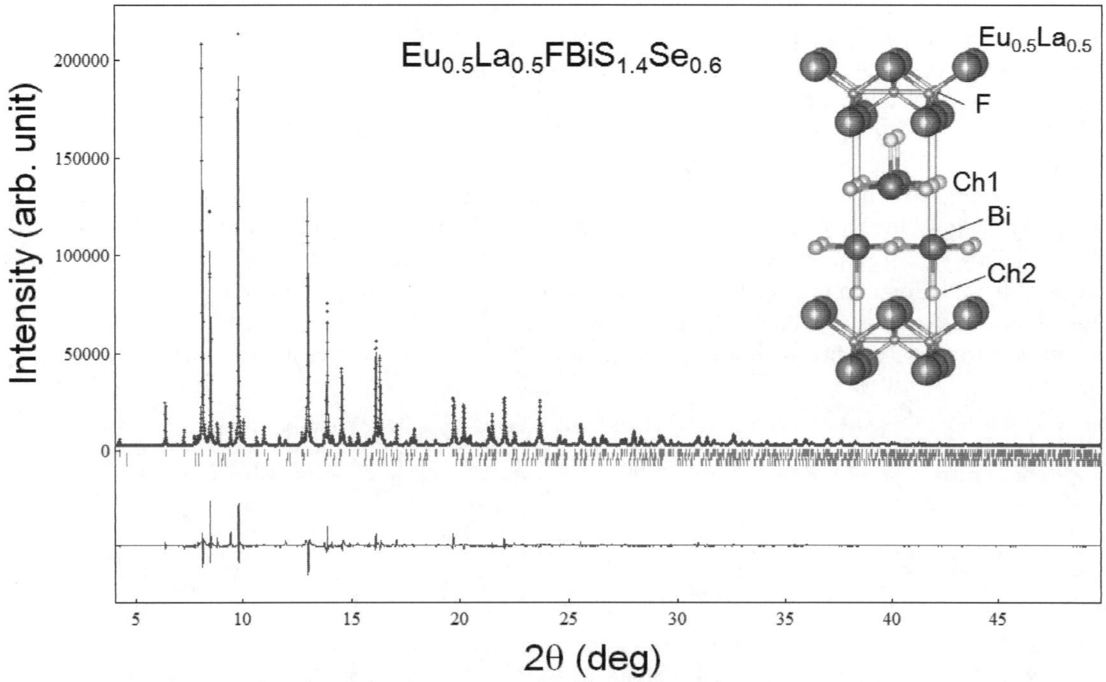

Figure 1. X-ray diffraction pattern and the Rietveld fitting with a three-phase ($Eu_{0.5}La_{0.5}FBiS_{1.4}Se_{0.6}$, BiF_3, and LaF_3) analysis method for $Eu_{0.5}La_{0.5}FBiS_{1.4}Se_{0.6}$ ($x = 0.6$).

Figure 2(c) shows the Se concentration dependence of the Ch1-Bi-Ch1 bond angle. With increasing x, the bond angle decreases. Therefore, the flatness of the Bi-Ch1 plane becomes worse by Se substitution. This tendency is consistent with the structural evolution in the sister system $LaO_{0.5}F_{0.5}Bi(S,Se)_2$ [11]. Figures 2(d)–2(f) show the in-plane Bi-Ch1 bond distance, Bi-Ch2 bond distance, and inter-plane Bi-Ch1 bond distance. The evolution of the in-plane Bi-Ch1 distance seems to be correlating with the lattice constant of a. In contrast, the changes in the Bi-Ch2 distance and the inter-plane Bi-Ch1 distance seem to be correlating with the lattice constant of c. These direct correlation between the local structure parameters and the lattice constant also suggest the selective occupancy of Se at the in-plane Ch1 site.

As demonstrated in Refs. 8 and 10, the in-plane chemical pressure parameters (CP) were calculated using the obtained in-plane Bi-Ch1 distance and the ionic radii using the equation, $CP = (R_{Bi}$ and $R_{Ch}/$(in-plane Bi-Ch1 distance). R_{Bi} is the ionic radius of $Bi^{2.5+}$, estimated in a previous structural analysis of a single crystal of $La(O,F)BiS_2$ [15]. R_{Ch} (= 104.19 pm) is the average value of the ionic radius at the Ch1 site, which is calculated using nominal x and ionic radii of S (184 pm) and Se (198 pm). With increasing Se concentration, the in-plane chemical pressure increases. Although the in-plane Bi-Ch1 distance increases by Se substitution (Fig. 2(d)), the in-plane packing density increases by Se substitution, which results in the enhancement of orbital overlap between Bi and Se. As proposed in Refs. 8 and 10, we have confirmed that the enhancement of the in-plane chemical pressure is essential for the emergence of bulk superconductivity in this system as well.

Finally, we discuss the evolution of the in-plane disorder by Se substitution. One of the advantages of synchrotron X-ray diffraction is the precise determination of the thermal factor, which can represent information about structural disorder. Since the superconductivity is induced in the Bi-Ch1 plane, we performed the Rietveld refinements with the anisotropic thermal factors for Bi and Ch1. Figure 2(h) shows the Se concentration dependences of the anisotropic thermal factor U_{11} for in-plane Bi and Ch1 sites. U_{11} for Bi does not show a remarkable changes with increasing x. In contrast, U_{11} for Ch1 shows a notable dependence on x. With increasing Se concentration, U_{11} for Ch1 decreases, which indicates that the in-plane disorder is suppressed by Se substitution. Generally in the BiS_2-based compounds, the existence of huge in-plane disorders has been detected using several probes (neutron diffraction, X-ray diffraction, and X-ray absorption spectroscopy) [16-20]. We consider that the effect of the in-plane chemical pressure on the emergence of superconductivity in Se-substituted $Eu_{0.5}La_{0.5}FBiS_{2-x}Se_x$ would be the suppression of the in-plane disorder. This scenario is consistent with the results of extended X-ray absorption fine structure studies on BiS_2 compounds, in which the in-plane Bi-S1 bond intensity is enhanced by the in-plane chemical pressure effect generated by the small RE (rare earth ion) substitution in $REO_{0.5}F_{0.5}BiS_2$ [8,20]. Therefore, with the present structural studies on the $Eu_{0.5}La_{0.5}FBiS_{2-x}Se_x$ system, we propose the relationship between the local in-plane disorder, the in-plane chemical pressure effect, and the emergence of superconductivity in the $BiCh_2$-based superconductor family. To further confirm this scenario, structural studies with local scale and/or single crystal studies are needed.

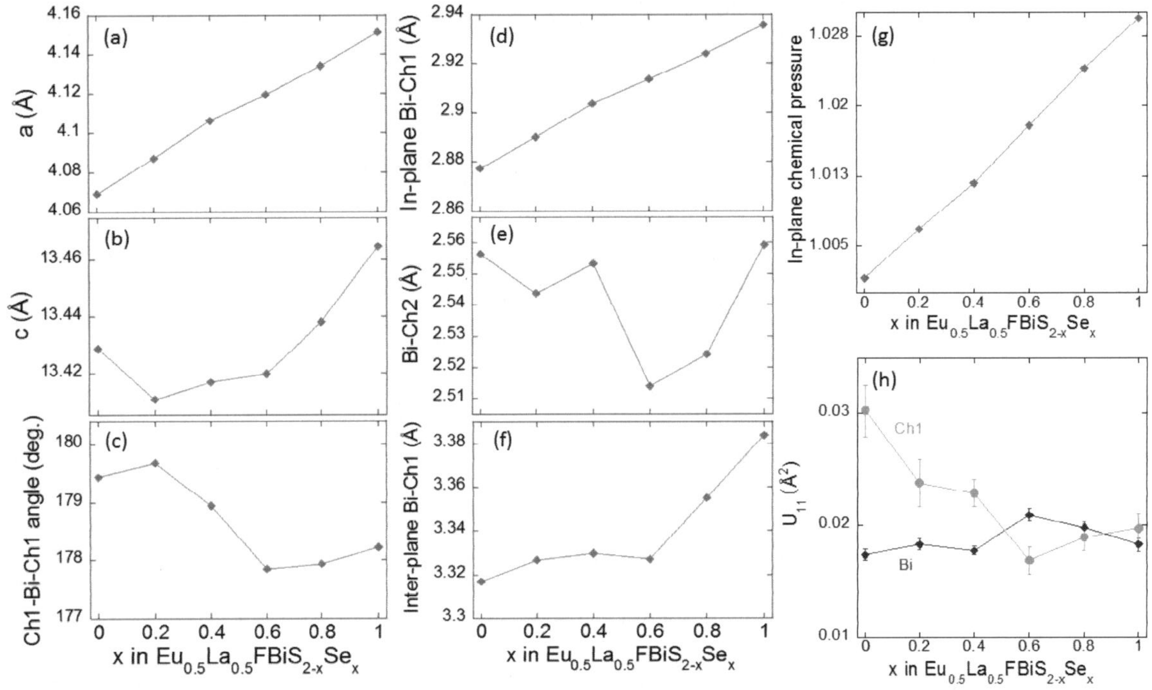

Figure 2. Se concentration dependences of structural parameters for $Eu_{0.5}La_{0.5}FBiS_{2-x}Se_x$: (a) lattice constant of a, (b) lattice constant of c, (c) Ch1-Bi-Ch1 bond angle, (d) in-plane Bi-Ch1 bond distance, (e) Bi-Ch2 bond distance, (f) inter-plane Bi-Ch1 bond distance, (g) in-plane chemical pressure, and (h) anisotropic thermal factor of U_{11} for in-plane Bi and Ch1 sites.

4. Summary

In this study, we have performed synchrotron X-ray powder diffraction for the new BiS2-based superconductor system $Eu_{0.5}La_{0.5}FBiS_{2-x}Se_x$, in which the bulk superconductivity was induced by Se substitution. The obtained structural parameters suggest four structural tendencies: (1) the selective occupancy of Se at the in-plane Ch1 site, (2) the direct correlation between Bi-Ch bond distances and the lattice constants, (3) the enhancement of the in-plane chemical pressure, and (4) the suppression of the in-plane disorder by Se substitution. These results will be useful for understanding the relationship between superconductivity and crystal structure in the $BiCh_2$-based superconductor family.

Acknowledgements
The authors would like to thank the staffs of BL02B2 of SPring-8 for the experimental support. This work was partly supported by Grants-in-Aid for Scientific Research (Nos. 16H04493 and 15H05886) and SPring-8 (proposal No. 2016B1078).

References
[1] Mizuguchi Y, Fujihisa H, Gotoh Y, Suzuki K, Usui H, Kuroki K, Demura S, Takano Y, Izawa H, and Miura O 2012 *Phys. Rev. B* vol 86 pp 220510
[2] Mizuguchi Y, Demura S, Deguchi K, Takano Y, Fujihisa H, Gotoh Y, Izawa H, and Miura O 2012 *J. Phys. Soc. Jpn.* vol 81 pp 114725
[3] Mizuguchi Y 2015 *J. Phys. Chem. Solids* vol 84 pp 34
[4] Bednorz J B and Müller K 1986 *Z. Phys. B* vol 64 pp 189
[5] Kamihara Y, Watanabe T, Hirano M, and Hosono H 2008 *J. Am. Chem. Soc.* Vol 130 pp 3296
[6] Usui H, Suzuki K, and Kuroki K 2012 *Phys. Rev. B* vol 86 pp 220501
[7] Usui H and Kuroki K 2015 *Nov. Supercond. Mater.* vol 1 pp 50
[8] Mizuguchi Y, Miura A, Kajitani J, Hiroi T, Miura O, Tadanaga K, Kumada N, Magome E, Moriyoshi C, and Kuroiwa Y 2015 *Sci. Rep.* vol 5 pp 14968
[9] Thakur G S, Jha R, Haque Z, Awana V P S, Gupta L C, and Ganguli A K 2015 *Supercond. Sci. Technol.* vol 28 pp 115010
[10] Jinno G, Jha R, Yamada A, Higashinaka R, Matsuda T D, Aoki Y, Nagao M, Miura O, and Mizugchi Y 2016 *J. Phys. Soc. Jpn.* vol 85 pp 124708
[11] Hiroi T, Kajitani J, Omachi A, Miura O, and Mizuguchi Y 2015 *J. Phys. Soc. Jpn.* vol 84 pp 024723
[12] Mizuguchi Y, Hiroi T, and Miura O 2016 *J. Phys.: Conf. Ser.* vol 683 pp 012001
[13] Izumi F and Momma K 2007 *Solid State Phenom.* vol 130 pp 15
[14] Momma K and Izumi F 2008 *J. Appl. Crystallogr.* vol 41 pp 653
[15] Miura A, Nagao M, Takei T, Watauchi S, Tanaka I, and Kumada N 2014 *J. Solid State Chem.* vol 212 pp 213
[16] Athauda A, Yang J, Lee S, Mizuguchi Y, Deguchi K, Takano Y, Miura O, and Louca D 2014 *Phys. Rev. B* vol 91 pp 144112
[17] Sagayama R, Sagayama H, Kumai R, Murakami Y, Asano T, Kajitani J, Higashinaka R, Matsuda T D, and Aoki Y 2015 *J. Phys. Soc. Jpn.* vol 84 pp 123703
[18] Sugimoto T, Joseph B, Paris E, Iadecola A, Mizokawa T, Demura S, Mizuguchi Y, Takano Y, and Saini N L 2014 *Phys. Rev. B* vol 89 pp 201117
[19] Paris E, Joseph B, Iadecola A, Sugimoto T, Olivi L, Demura S, Mizuguchi Y, Takano Y, Mizokawa T, and Saini N L 2014 *J. Phys.: Condens. Matter* vol 26 pp 435701
[20] Mizuguchi Y, Paris E, Sugimoto T, Iadecola A, Kajitani J, Miura O, Mizokawa T, and Saini N L 2015 *Phys. Chem. Chem. Phys.* vol 17 pp 22090

Study of microwave-induced phase switches from the finite voltage state in $Bi_2Sr_2CaCu_2O_y$ intrinsic Josephson junctions

Haruhisa Kitano*, Ayami Yamaguchi, Yusaku Takahashi, Daiki Kakehi and Shin-ya Ayukawa

Department of Physics and Mathematics, Aoyama Gakuin University, Fuchinobe 5-10-1, Chuo-ku, Sagamihara, Kanagawa 252-5258, Japan

hkitano@phys.aoyama.ac.jp

Abstract. We study the microwave-induced phase switches from the finite voltage state for the underdamped intrinsic Josephson junctions (IJJs) made of $Bi_2Sr_2CaCu_2O_y$ (Bi2212). We observe the resonant double-peak structure in the switching current distribution at low temperatures. This feature is successfully explained by a quantum mechanical model where the strong microwave field effectively suppresses the potential barrier for the phase escape from a potential well and the macroscopic quantum tunneling (MQT) is resonantly enhanced. The detailed analyses considering the effects of multiple phase retrapping processes after the phase escape strongly suggest that the intense microwave field suppresses the energy-level spacing in the potential well, by effectively decreasing the fluctuation-free critical current and the Josephson plasma frequency. This effect also reduces the number of photons required for the multiphoton transition between the ground and the first excited states, making it possible to observe the energy level quantization in the MQT state. The temperature dependence of the resonance peak emerging in the switching rate clearly demonstrates that the quantized energy state can be survived up to ~10 K, which is much higher than a crossover temperature predicted by the conventional Caldeira-Leggett theory.

1. Introduction

The phase switching dynamics in the current-biased intrinsic Josephson junctions (IJJs) of high-T_c cuprate superconductors have much attention, since the macroscopic quantum tunneling (MQT) in the phase switches from the zero voltage state was discovered in the underdamped IJJs made of $Bi_2Sr_2CaCu_2O_y$ (Bi2212) [1]. A gigantic critical current density and a perfectly smooth tunnel junction in IJJs benefit an increase of crossover temperature T_{cr} to the MQT region [2], expected to be much higher than a typical value of T_{cr} for the conventional Al- or Nb-based Josephson junctions (JJs). As well known, the current-voltage (*I-V*) curves of IJJs show a multiple-branched feature, suggesting that the neighboring superconducting layers involved into IJJs are strongly coupled with each other. Thus, it is quite important to investigate the phase switches from the finite voltage state as well as those from the zero voltage state in IJJs, toward the complete understanding of the complicated phase dynamics in IJJs.

In the phase switching events from the finite voltage state, such as the switch from the first to the second voltage state (2nd SW) and that from the second to the third voltage state (3rd SW) in the multiple branched *I-V* curves, the influence of the phase running state before the occurrence of the phase switch should be carefully considered, in contrast to the phase switch from the zero to the first voltage state (1st SW). Recently, we discovered that the multiple phase retrapping (PR) processes, which were observed as a negative curvature in the switching rate $\Gamma(I)$ as a function of bias current, were systematically enhanced as the order of the phase switch was increased, even for the underdamped Bi2212-IJJs showing large hysteresis in the *I-V* curve [3]. Our results clearly show that the switching current distribution $P(I)$ and $\Gamma(I)$ for the higher-ordered phase switches should be analyzed by considering the multiple PR processes, which have been discussed for the moderately damped JJs [4, 5], in addition to the conventional thermally-activated (TA) escape processes from a potential well. Furthermore, we demonstrated that the resonant double-peak structure in $P(I)$ under the strong microwave irradiation was observed for the 2nd SW of Bi2212-IJJs below T^*(~10 K) [6] and that it was successfully reproduced by a *quantum-mechanical* model proposed by Fistul *et al* [7]. These results strongly suggest that the MQT-like behavior in the 2nd SW of Bi2212-IJJs, observed as a clear saturation of the effective phase escape temperature T_{esc} below T^* [8, 9], essentially has the same origin as that for the MQT phenomena established in the 1st SW of Bi2212-IJJs [1, 10].

In this paper, we present the detailed analyses of $P(I)$ and $\Gamma(I)$, which were measured for the 2nd SW of Bi2212-IJJs under the microwave irradiation [6], by considering both the TA escape and the multiple PR processes for the principal peak in the double-peaked $P(I)$. It is shown that the strong microwave field suppresses the energy-level spacing between the quantized energy states in the potential well by effectively decreasing the fluctuation-free critical current I_{c0} and the Josephson plasma frequency ω_p. This effect reduces the number of photons (n=5 to 3) required for the multiphoton transition between the ground state and the first excited state, making it possible to observe the energy level quantization (ELQ) through the multiphoton transition. We also discuss the temperature dependence of the resonance peak emerging in $\Gamma(I)$, providing clear evidence for the formation of ELQ surviving up to T^*.

2. Experiments and analyses

Bi2212 single crystals were grown by the floating zone method and annealed to be optimal doped. The bridge-type Bi2212-IJJs shown in the inset of Fig. 1 were fabricated by using the focused ion beam (FIB) milling techniques [11]. The lateral sizes and the junction number of the IJJs were 1.2 × 1.1 μm^2 and 6, respectively. The superconducting transition temperature T_c of the fabricated IJJs was estimated as ~85 K by the zero resistance. $P(I)$ for the 2nd SW was measured from 4.3 K to 30 K by applying a bias current with a constant rate dI/dt=3 mA/s under 5,000 or 10,000 switching events. Other details are described in our previous paper [6].

The switching rate $\Gamma(I)$ was obtained from the measured $P(I)$ [12] and analyzed by considering both the TA escape and the subsequent PR processes. In the limit of $\Delta U/k_BT \gg 1$, $\Gamma(I)$ is expressed by [13]

$$\Gamma(I) = \Gamma_T(1 - W_R)\frac{\ln(1-W_R)^{-1}}{W_R}. \qquad (1)$$

Here, ΔU is the potential barrier for an escape process, $\Gamma_T = (\omega_p/2\pi)\exp[-\Delta U/k_BT]$ is the TA escape rate, and W_R is the PR probability that the phase particle escapes from a potential minimum and is retrapped in the next minimum.

The influence of PR processes to $\Gamma(I)$ is evaluated by using a characteristic ratio [3],

$$\frac{I_{PR}-I_{SW}^{min}}{I_{SW}^{max}-I_{SW}^{min}}. \qquad (2)$$

Here I_{PR} is a maximum current where the multiple PR processes occur after the phase escape from the potential well. I_{SW}^{max} and I_{SW}^{min} are maximum and minimum values of the measured switching current, respectively. Note that the PR processes are dominant in a whole range of the observed switching current if this ratio larger than unity, since I_{PR} larger than I_{SW}^{max} suggests that the measured phase switches are always influenced by the PR processes.

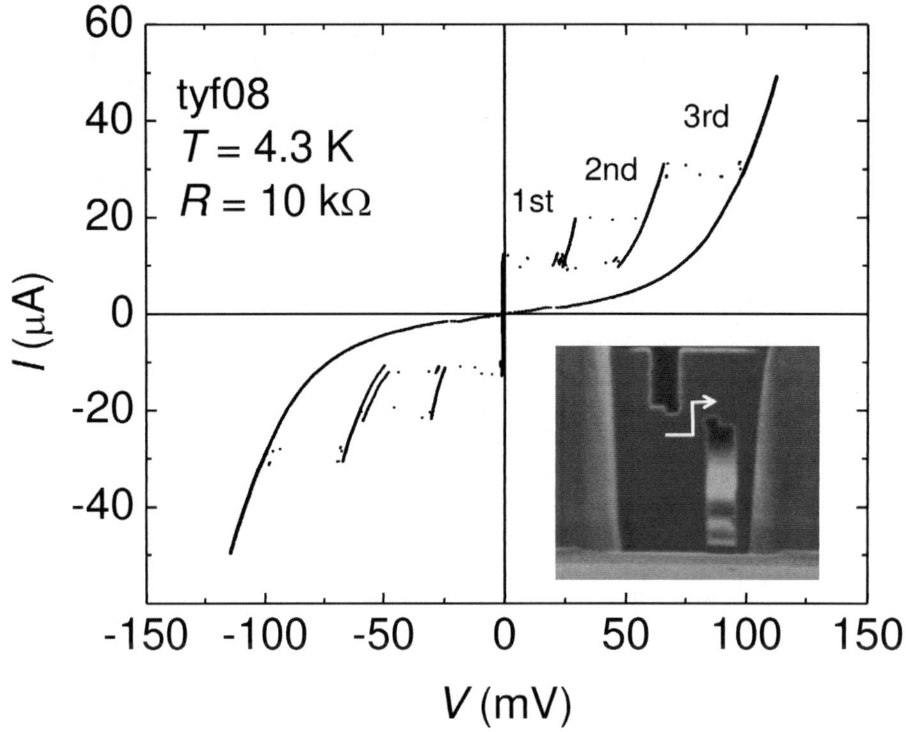

Figure 1. *I-V* characteristic of the bridge-type Bi2212-IJJs measured at 4.3 K. The inset shows an image of scanning ion microscope of the IJJ device. A white arrow shows a current flow.

The quantized energy levels E_n in the potential well for a quantum particle are numerically obtained by using the previous approach [14]. The numbers of photons required for the multiphoton transition between the ground state and the first excited state was estimated from such numerical calculations of $E_{01} \equiv E_1 - E_0$.

3. Results and discussion

Figure 1 shows the *I-V* characteristics, which were measured at 4.3 K by using triangular ac voltage waveform and a load resistor of 10 kΩ. We observed that each resistive branch from the 1st to the 3rd almost equally spaced. Note that the switching current from the 1st SW to the 3rd SW is monotonically increased with increasing the order of switch, showing a contrast to the previous measurements reporting the anomalous switching order [15]. The McCumber parameters $\beta_C = 2eI_{c0}CR_N{}^2/\hbar$ was ~60 for the 1st SW, ~110 for the 2nd SW, and ~160 for the 3rd SW, respectively. C (~60 fF) and R_N (~190 Ω) are a capacitance and the normal-state resistance for a single junction in IJJs.

Figure 2 shows the plots of $P(I)$ for the 2nd SW under the microwave field at several temperatures. We observe that the principal peak in $P(I)$ located at a higher current region, is shifted to the resonant peak located at a lower current region, with increasing the microwave irradiation power. The double-peak structure in $P(I)$ is clearly observed at the intermediate irradiation power (~−2.5 dBm) up to 10 K, while it is broadened at 20 K, as shown in Fig. 2.

The data set of $\Gamma(I)$, which was obtained from the measured $P(I)$ under the microwave field, was

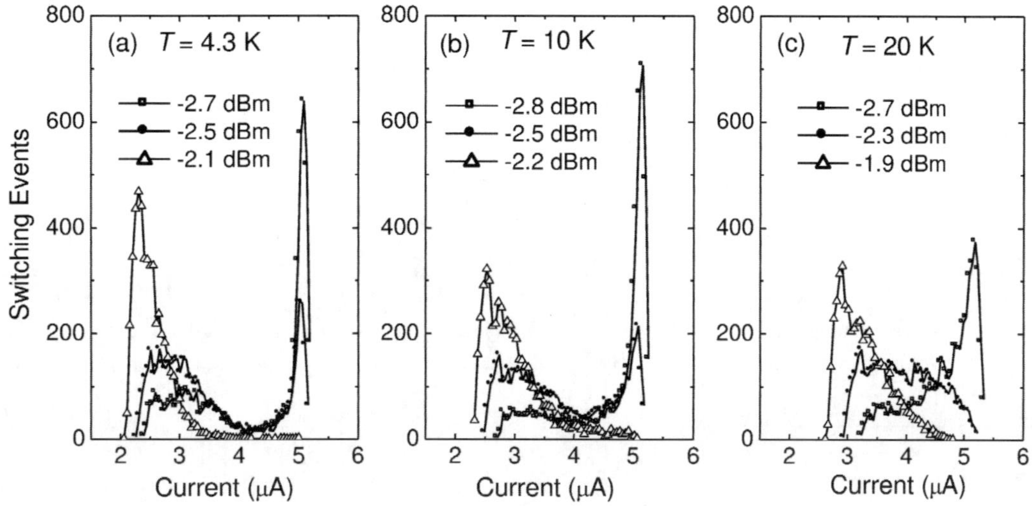

Figure 2. The switching current distribution for the 2nd SW under several levels of microwave irradiation, measured at (a) 4.3 K, (b) 10 K, and (c) 20 K, respectively.

analyzed by calculating Eq. (1) with the fitting parameters such as T_{esc} and I_{c0}. Figure 3 shows the microwave power dependence of several parameters for the measurements at 4.3 K under the microwave irradiation at 40.5 GHz up to -2.3 dBm, at which the double-peak structure in $P(I)$ is not prominent yet. We found that I_{c0} is strongly decreased by the microwave irradiation while T_{esc} is slightly decreased, as shown in Figs. 3(a) and 3(c). In addition, as shown in Fig. 3(b), the ratio estimated by Eq. (2) becomes larger than unity with increasing the microwave power, suggesting that the PR processes are enhanced by the microwave irradiation. These results clearly show that the strong microwave irradiation decreases the effective depth of the potential well (in other words, effectively assists the phase escape events) and that a rapid oscillation of the tilted washboard potential induced by the microwave field intensely increases the PR process after the phase escape.

Although the decrease of ΔU with the microwave power shows a good agreement with the concept proposed by Fistul *et al* [7] where ΔU is effectively suppressed by the strong microwave field, the observed decrease of ΔU is very weak, as shown in Fig. 3(e), compared to the strong decrease of I_{c0}. Such a moderate decrease of ΔU is qualitatively explained by the fact that the normalized switching current I_{sw}/I_{c0} is also decreased with the microwave power, as shown in Fig. 3(d), since ΔU is increased with decreasing I_{sw}/I_{c0} while it is decreased with decreasing I_{c0} [12]. However, the decrease of both I_{c0} and I_{sw}/I_{c0} leads to a depression of ω_p and E_{01}, as shown in Fig. 3(f). The numbers of photons for the multiphoton transition is consequently reduced from 5 down to 3 with the microwave power. This effect seems to increase the probability to detect the quantum phase escape from the first excited state through the multiphoton transition. Our results obtained by the detailed analyses clearly demonstrate that the microwave field strongly affects the phase switching dynamics occurring in IJJs and that it can effectively change the energy level spacing in the quantized potential well.

Finally, we investigate the temperature-dependent properties of the resonant peak observed in $\Gamma(I)$ under the microwave irradiation. Figure 4 shows the enhancement of the switching rate $\Delta\Gamma(I) \equiv [\Gamma(P_{mw}, I) - \Gamma(0, I)]/\Gamma(0, I)$ at several temperatures, where $\Gamma(0, I)$ and $\Gamma(P_{mw}, I)$ are the switching rate without and with the microwave irradiation, respectively. In these plots, we use the microwave

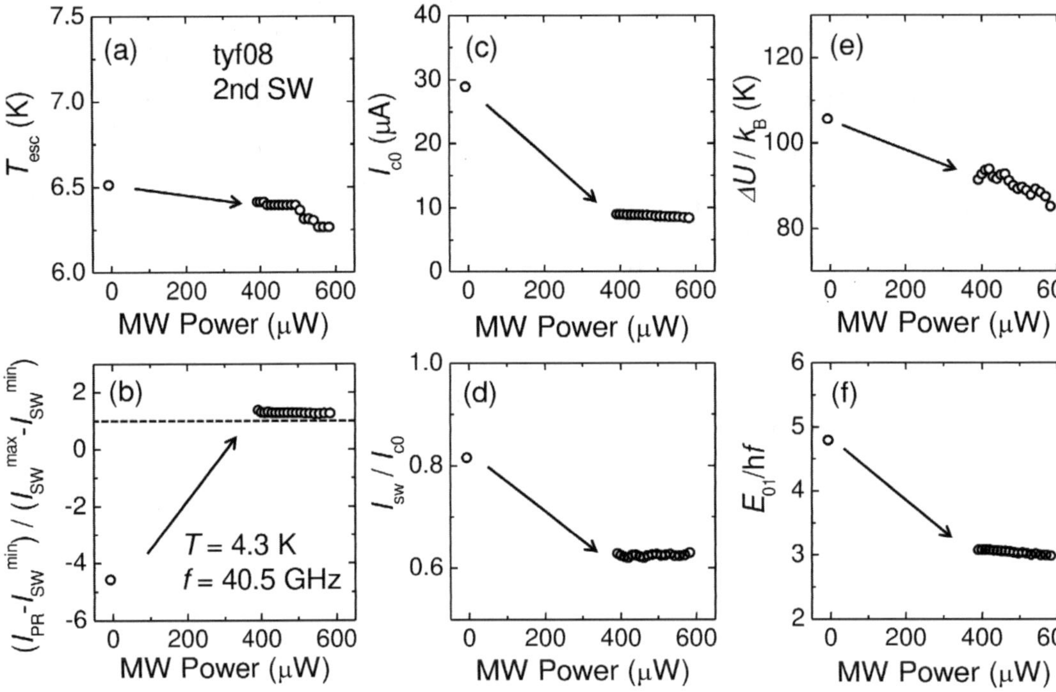

Figure 3. The microwave power dependence of (a) T_{esc}, (b) $(I_{PR} - I_{SW}^{min}) / (I_{SW}^{max} - I_{SW}^{min})$, (c) I_{c0}, (d) the normalized switching current I_{sw}/I_{c0}, (e) ΔU, and (f) E_{01}, respectively. They were obtained by analyzing $\Gamma(I)$ for the 2nd SW, measured at 4.3 K and under the microwave irradiation at 40.5 GHz.

irradiation power at which a positional shift from the principal peak to the excited peak in $P(I)$ is clearly observed, and the magnitude of $\Delta\Gamma(I)$ is appropriately normalized. Below 10 K where the MQT-like behavior appears in the temperature dependence of T_{esc} [6], we observe a clear resonance peak in $\Delta\Gamma(I)$ which can be fitted to the Lorentzian curve with the quality factor Q (=2.5 at 4.3 K and 1.7 at 10 K). On the other hand, above 16 K, the peak is not observed, as shown in Figs. 4(d)-4(f). Such a temperature-dependent feature in $\Delta\Gamma(I)$ is characteristic for the crossover from the MQT regime to the classical TA regime in a current-biased JJ [16]. Thus, our results provide clear evidence that we successfully detect the three-photon transition between the energy level spacing E_{01} in the 2nd SW of IJJs, which survives up to T^*(~10 K).

4. Conclusion

We present the detailed analyses of the switching current distribution $P(I)$ and the switching rate $\Gamma(I)$ for the 2nd SW of Bi2212-IJJs under the microwave irradiation, by considering both the TA escape and the multiple PR process in the phase-running state before the detection of the phase switching event. The strong microwave irradiation is found to assist the phase escape by effectively decreasing the fluctuation-free critical current I_{c0} and to enhance the PR process after the phase escape by rapidly oscillating the tilted washboard potential. We confirmed that the potential barrier ΔU for the phase escape is decreased with increasing the microwave irradiation power, supporting the previous proposal by Fistul *et al* [7].

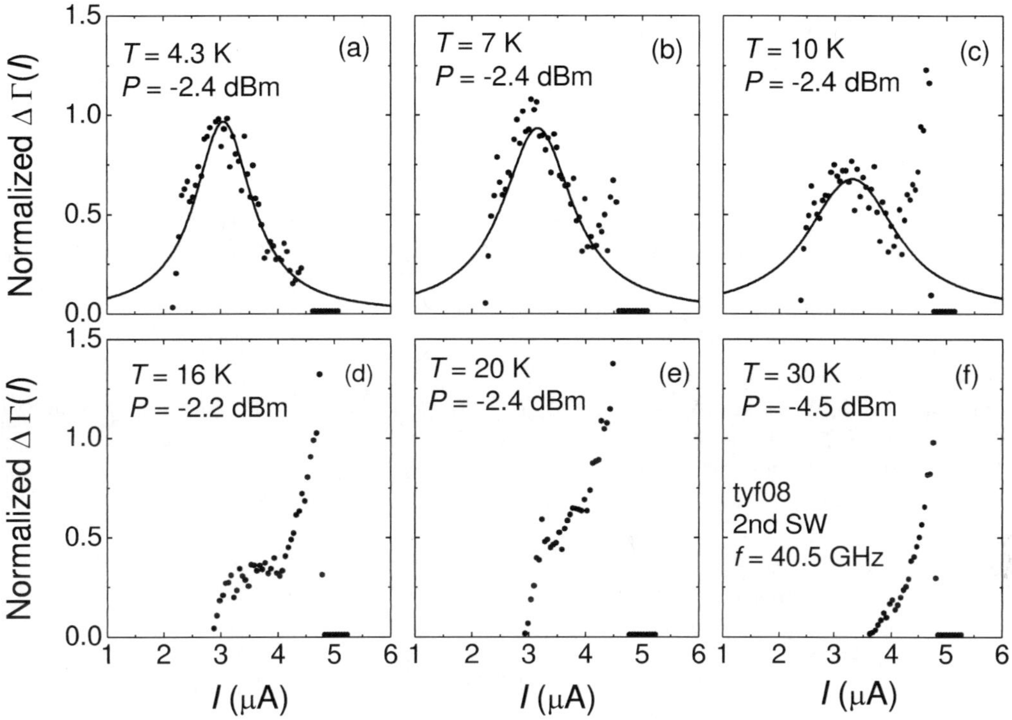

Figure 4. Enhancement of the switching rate $\Delta\Gamma(I)$ for the 2nd SW at (a) 4.3 K, (b) 7 K, (c) 10 K, (d) 16 K, (e) 20 K, and (f) 30 K, respectively. Solid lines are the Lorentzian fits.

The intense microwave field also suppresses the Josephson plasma frequency ω_p and the energy-level spacing E_{01} due to the strong decrease of both I_{c0} and I_{sw}/I_{c0}, giving a large reduction in the number of photons required for the multiphoton transition between the quantized energy levels. We observe the clear resonance peak in the enhancement of the switching rate $\Delta\Gamma(I)$ for the 2nd SW, which can be fitted to the Lorentzian curves with Q~1.7-2.5 by assuming the three-photon transition. The temperature dependence of $\Delta\Gamma(I)$ strongly suggests that the microwave-induced resonance in $\Delta\Gamma(I)$ has the quantum-mechanical origin and survives up to T^*(~10 K). These results strongly support the occurrence of the MQT state below T^* in the 2nd SW of Bi2212-IJJs, together with the recent report on the microwave power dependence of I_{sw} giving maxima of $P(I)$ [6]. Thus, the dissipation of quasiparticle tunneling in the finite voltage state seems to be too small to prevent the occurrence of the MQT state in the higher-ordered phase switches, as was previously predicted for the 1st SW in IJJs [17,18]. The influence of the phase-running state before the occurrence of the phase switch from the finite voltage state is rather the generation of ac Josephson current, as discussed in Ref. [3]. This feature is suggested to be unique to the phase switching phenomenon from the voltage state in the IJJ systems with the superconducting single and double layers such as $Bi_2Sr_2CuO_{6+\delta}$ (Bi2201) [19] and Bi2212, showing a sharp contrast to IJJs made of $Bi_2Sr_2Ca_2Cu_3O_{10+\delta}$ (Bi2223) with the superconducting triple layers [20].

A remained problem in the MQT phenomena in the phase switch from the finite voltage state is that T^* is much higher than T_{cr} calculated by the conventional theory. At the present, there are two possible scenarios to resolve this issue. One is the contribution of ac Josephson current generated in the phase-running state, which can depress ΔU as like as the microwave irradiation. The other is the fact that $\Gamma(I)$ for the 2nd SW is more enhanced than that for the 1st SW. Such an enhancement of $\Gamma(I)$ is very similar to the feature of $\Gamma(I)$ for the uniform phase switch, where several junctions involved into IJJs simultaneously switch from the zero voltage state [10]. Since the issue on the anomalously-enhanced MQT rate has extensively been discussed by several theoretical studies [21, 22], one of them is possible to resolve the problem remained in the phase switches from the voltage state.

Acknowledgments
We thank I. Kakeya, H. Asai, S. Kawabata, Y. Ota, and T. Koyama for fruitful discussion and valuable comments. This work was partly supported by Grants-in-Aid for Scientific Research from JSPS and MEXT of Japan (Grant Nos. 21560020, 24560017) and MEXT-Supported Program for the Strategic Research Foundation at Private Universities (2013-2017).

References
[1] Inomata K, Sato S, Nakajima K, Tanaka A, Takano Y, Wang H B, Nagao M, Hatano H and Kawabata S 2005 *Phys. Rev. Lett.* **95** 107005
[2] Caldeira A O and Leggett A J 1981 *Phys. Rev. Lett.* **46** 211
[3] Kitano H, Takahashi Y, Kakehi D, Yamaguchi H, Koizumi S and Ayukawa S 2016 *J. Phys. Soc. Jpn.* **85** 054703
[4] Bae M H, Sahu M, Lee H J and Bezryadin A 2009 *Phys. Rev.* B **79** 104509
[5] Kautz R L and Martinis J M, 1990 *Phys. Rev.* B **42** 9903
[6] Takahashi Y, Kakehi D, Takekoshi S, Ishikawa K, Ayukawa S and Kitano H 2016 *J. Phys. Soc. Jpn.* **85** 073702
[7] Fistul M V, Wallraff A and Ustinov A V 2003 *Phys. Rev.* B **68** 060504
[8] Kashiwaya H, Matsumoto T, Shibata H, Kashiwaya S, Eisaki H, Yoshida Y, Kawabata S and Tanaka Y, 2008 *J. Phys. Soc. Jpn.* **77** 104708
[9] Ota K, Hamada K, Takemura R, Ohmaki M, Machi T, Tanabe K, Suzuki M, Maeda A and Kitano H 2009 *Phys. Rev.* B **79** 134505
[10] Jin X Y, Koval Y, Lukashenko A, Ustinov A V, and Müller P 2006 *Phys. Rev. Lett.* **96** 177003
[11] Kakehi D, Takahashi Y, Yamaguchi H, Koizumi S, Ayukawa S and Kitano H 2016 *IEEE Trans. Appl. Supercond.* **26** (3) 1800204
[12] Fulton T A and Dunkleberger L N 1974 *Phys. Rev.* B **9** 4760
[13] Männik J, Li S, Qiu W, Chen W, Patel V, Han S and Lukens J E2005 *Phys. Rev.* B **71** 220509
[14] Kopietz P and Chakravarty S 1988 *Phys. Rev.* B **38** 97
[15] For example, Warburton P A, Saleem S, Fenton J C, Korsah M and Gronvenor C R M 2009 *Phys. Rev. Lett.* **103**, 217002
[16] Martinis J, Devoret M and Clarke J 1985 *Phys. Rev. Lett.* **55** 1543
[17] Kawabata S, Kashiwaya S, Asano Y and Tanaka Y 2004 *Phys. Rev.* B **70** 132505
[18] Yokoyama T, Kawabata S, Kato T and Tanaka Y 2007 *Phys. Rev.* B **76** 134501
[19] Nomura Y, Mizuno T, Kambara H, Nakagawa Y and Kakeya I 2015 *J. Phys. Soc. Jpn.* **84** 013704
[20] Nomura Y, Mizuno T, Kambara H, Nakagawa Y, Watanabe T, Kakeya I and Suzuki M 2014 *J. Phys: Conference Series* **507** 012038
[21] Sevel'ev S, Rakhmanov A L and Nori F 2007 *Phys. Rev. Lett.* **98** 077002; 2007 *Erratum*: **98** 269901
[22] Koyama T and Machida M 2008 *Physica C* **468** 695

IOP Conf. Series: Journal of Physics: Conf. Series **871** (2017) 012009 doi:10.1088/1742-6596/871/1/012009

An effective Hamiltonian and its phase diagram for T'-structure cuprates

Kunito Yamazaki[1] , **Takuya Yoshioka**[1] , **Hiroki Tsuchiura**[1,2] , **Masao Ogata**[3]

[1]Department of Applied Physics, Tohoku University, Sendai 980-8579, Japan
[2]Center for Spintronics Research Network, Tohoku University, Sendai 980-8577, Japan
[3]Department of Physics, University of Tokyo, Bunkyo, Tokyo 113-0033, Japan

E-mail: yamazaki@olive.apph.tohoku.ac.jp

Abstract. We study the superconducting transition near the half-filling of electron-doped cuprates with the so-called T'-structure based on an effective two-band model using fluctuation-exchange approximation and solving the linearized Eliashberg equation. We find d-wave superconductivity even at the half-filling of this model when the energy difference between the upper Hubbard level of Cu $d_{x^2-y^2}$ orbitals and O 2p levels is less than about 1.5eV, which is plausible for the T'-type cuprates.

1. Introduction

It has long been considered that the phase diagrams of hole- and electron-doped cuprates were well-established; there exists antiferromagnetism (AF) around the half-filling of their phase diagram, and then d-wave superconductivity (dSC) emerges with carrier doping. In particular, on the phase diagram of the electron-doped cuprates such as $Nd_{2-x}Ce_xCuO_4$ (NCCO), which is of the so-called T'-structure cuprates, AF phase overwhelmingly appears in the underdoped region of $x < 0.14$ [1]. Tsukada *et al.* claimed that, however, thin films of the T'-structure cuprates exhibit superconductivity without chemical doping of electrons, only when excess oxygen in as-grown samples are properly removed [2]. This observation has been followed by several experimental studies based on T'–NCCO thin films [3], polycrystalline samples [4, 5], and even on single crystals [6].

Theoretically, however, it is rather inconceivable that dSC appears around the half-filling of the phase diagrams of the t-J or the strong coupling regime of the Hubbard model on a square lattice. Recently, Weber *et al.* have shown that the parent compounds of T'-cuprates do not have the Mott gap based on DFT+DMFT calculations [7], which means the charge-transfer gap commonly found in the T-cuprates are vanished. If this is the case, we should consider the superconducting transition of T'-cuprate based on the d-p model to treat the O 2p degrees of freedom explicitly.

Thus in this paper, we propose an effective two-band model for T'-structure electron-doped cuprates. This two-band model is derived from the d-p model essentially along with the idea of Zhang and Rice [8], but explicitly consists of Cu d-orbitals and symmetrically hybridized O 2p orbitals. We also study the phase diagram of this two-band model using fluctuation exchange approximation (FLEX). We find that the d-wave superconductivity appears even at the half-filling when the charge transfer gap is small enough.

2. model and formalism

We start with the d-p model or three band Hubbard model that consists of the O 2p-orbitals in the CuO_2 planes as well as the Cu 3d-orbitals, written as

$$\begin{aligned}\mathcal{H}_{dp} &= -t_{dp}\sum_{\langle i,j\rangle\sigma}\left(d^\dagger_{i\sigma}p_{j\sigma}+h.c.\right) - t_p\sum_{\langle j,j'\rangle\sigma}\left(p^\dagger_{j\sigma}p_{j'\sigma}+h.c.\right)\\ &\quad +\Delta\sum_{j\sigma}n_{p,j\sigma}+U_d\sum_i n_{d,i\uparrow}n_{d,i\downarrow} \end{aligned} \qquad (1)$$

where $d_{i\sigma}$ is an annihilation operator of a hole with spin σ on the Cu $d_{x^2-y^2}$-orbital at site i, $p_{j\sigma}$ is that on the O 2pσ-orbital at site j, and $\langle i,j\rangle$ means the summation runs over nearest-neighbor Cu-O bonds, while $\langle j,j'\rangle$ the nearest-neighbor O-O bonds in the CuO_2 plane. The number operators for d- and p-holes are defined as $n_{d,i\sigma}=d^\dagger_{i\sigma}d_{i\sigma}$ and $n_{p,j\sigma}=p^\dagger_{j\sigma}p_{j\sigma}$. The Coulomb repulsion on Cu site is represented as U_d, and $\Delta=U_d-\Delta_p$ is the lowest excitation energy in the localized limit, where Δ_p is the energy difference between the O 2p and Cu 3d levels. We note here that the Coulomb repulsion on O site is omitted for simplicity.

Following the idea of Zhang and Rice [8, 9], we derive an effective Hamiltonian for the T'-type cuprates by a perturbative procedure, but this time, we explicitly keep active a part of the O 2p degrees of freedom because the energy difference between the upper Hubbard band and the O 2p level can be smaller than that of the T-structure cuprates. The detail of the derivation will be published elsewhere [10], here we roughly describe the steps which are important for the derivation. We consider the symmetric linear combination of the pσ-orbitals surrounding a Cu, and then construct Wannier orbitals of oxygen holes localized around each Cu based on these combinations. Omitting the anti-symmetric linear combinations of the pσ-orbitals, and also leaving the interaction terms as it is, now we obtain a two-band effective Hamiltonian given as

$$\begin{aligned}\mathcal{H}^{\text{eff}} &= \sum_{\boldsymbol{k}\sigma}\varepsilon_p(\boldsymbol{k})p^{(s)\dagger}_{\boldsymbol{k}\sigma}p^{(s)}_{\boldsymbol{k}'\sigma} + \sum_{\boldsymbol{k}\sigma}t_{dp}t_{\boldsymbol{k}}(d^\dagger_{\boldsymbol{k}\sigma}p^{(s)}_{\boldsymbol{k}\sigma}+h.c.)\\ &\quad +\frac{U_d}{N_L}\sum_{\boldsymbol{k},\boldsymbol{k}',\boldsymbol{q}}d^\dagger_{\boldsymbol{k}+\boldsymbol{q},\uparrow}d^\dagger_{\boldsymbol{k}'-\boldsymbol{q},\downarrow}d_{\boldsymbol{k}',\downarrow}d_{\boldsymbol{k},\uparrow}\end{aligned} \qquad (2)$$

where

$$\begin{aligned}\varepsilon_p(\boldsymbol{k}) &= \left\{\Delta-32t_p\frac{1}{t_{\boldsymbol{k}^2}}\sin^2(\frac{k_x}{2})\sin^2(\frac{k_y}{2})\right\},\\ t_{\boldsymbol{k}} &= 2\left\{\sin^2(\frac{k_x}{2})+\sin^2(\frac{k_y}{2})\right\}^{1/2},\end{aligned} \qquad (3)$$

N_L is the number of sites, and $p^{(s)}_{\boldsymbol{k}\sigma}$ is the annihilation operator of the Bloch state of symmetrized Wannier orbitals [9].

Here we carry out the FLEX calculation to the present two-band model to obtain the Green's function renormalized by the many-body self-energy correction. The effective interaction for singlet pairing is obtained by taking account of the contributions from bubble and ladder diagrams in the self-energy, as $V^{(2)}(k-k')=U_d+\frac{3}{2}U_d^2\chi_d^{(s)}(k-k')-\frac{1}{2}U_d^2\chi_d^{(c)}(k-k')$, where the

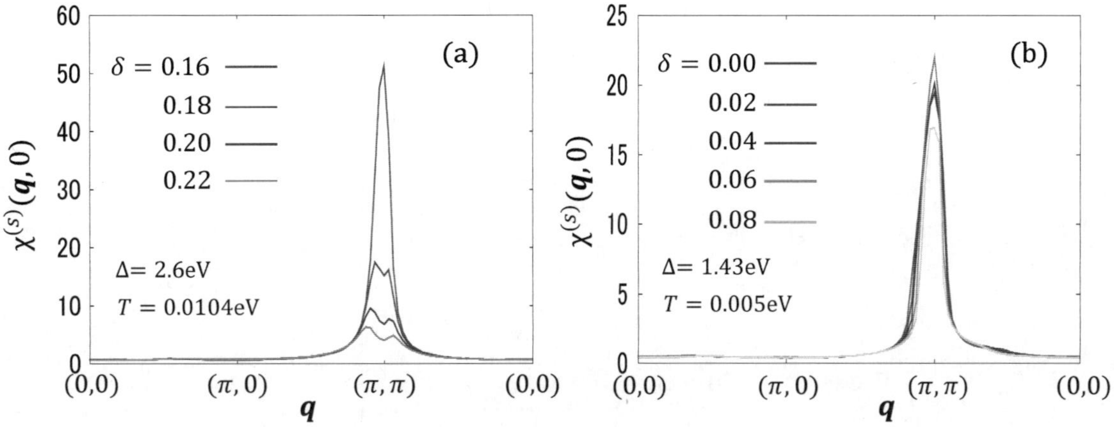

Figure 1. The static spin susceptibility $\chi^s(\boldsymbol{q},0)$ for (a) $\Delta = 2.6$eV and (b) $\Delta = 1.43$eV.

spin and charge susceptibilities for the d-orbital $\chi_d^{(s)}$ and $\chi_d^{(c)}$. Then we substitute the Green's function to linearized Eliashberg equation

$$\phi(\boldsymbol{k}, i\epsilon_n) = -\frac{T}{N_L}\sum_{\boldsymbol{k}',m} V^{(2)}(\boldsymbol{k}-\boldsymbol{k}', i\epsilon_n - i\epsilon_m) G_d(\boldsymbol{k}', i\epsilon_m) G_d(-\boldsymbol{k}', -i\epsilon_m)\phi(\boldsymbol{k}' i\epsilon_m) \tag{4}$$

to evaluate the transition temperature.

3. Results

As we mentioned in the Introduction, the charge-transfer gap can be vanished in the T'-cuprates. Thus we should assume that the excitation energy parameter Δ in the Hamiltonian (1) and (2) is rather small as estimated for T-cuprates (Δ = 2-4eV). Here we take $\Delta = 1.43$eV, and also 2.6eV for comparison. We also take $U_d = 6.5$eV for the present calculations. For FLEX calculation, we take up to 64×64 k-point meshes, and up to 8192 Matsubara frequencies for $\Delta = 1.43$eV, while 4096 for 2.6eV.

First, we look at the $\boldsymbol{q}$-dependences of the static susceptibility $\chi(\boldsymbol{q},0)$ just above T_c for several electron doping concentrations in the two-band model. Figure 1 shows $\chi(\boldsymbol{q},0)$ for (a) $\Delta = 2.6$eV (T-cuprates) and (b) $\Delta = 1.43$eV (T'-cuprates). We can clearly see in Fig. 1 (a) that an antiferromagnetic peak is found at $\delta = 0.16$, and then the peak split into two incommensurate peaks with increasing δ. This is consistent with the experimental observations in ref. [1], and also previous theoretical studies for T-cuprates [11, 12]. In contrast, when $\Delta = 1.43$eV (for T'-cuprates), the peak structure found at (π,π) in Fig. 1 (b) holds for $0 < \delta < 0.08$.

Next, we show the phase diagram for the two-band model in Fig. 2. We find that d-wave superconducting state appears in the overdoped region ($0.16 < \delta < 0.22$) for $\Delta = 2.6$eV, while in the underdoped region ($0 < \delta < 0.11$) for $\Delta = 1.43$eV. T_c's for $\Delta = 1.43$eV are lower than those for $\Delta = 2.6$eV, however, it is worth noting here that the T_c at the half-filling is not very low, which is consistent with the recent experimental results for T'-cuprates [2, 3, 4, 6].

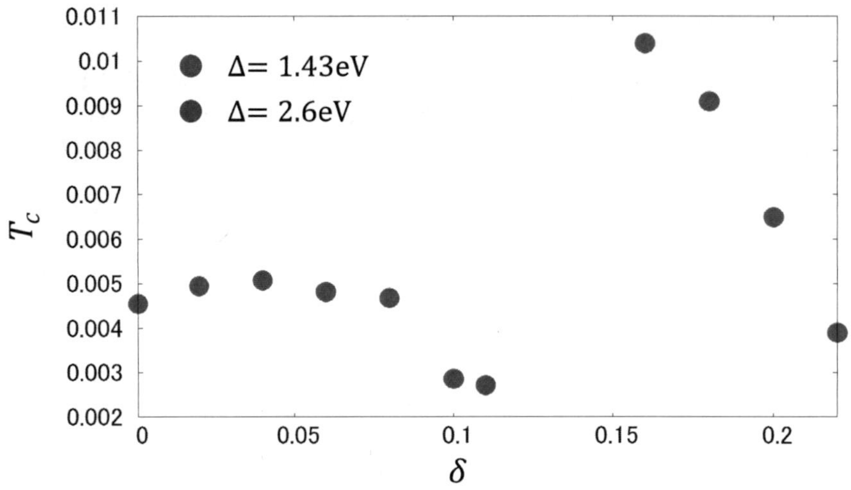

Figure 2. T_c of the effective two-band model for $\Delta = 1.43$eV and 2.6eV as functions of electron doping concentration δ.

4. Summary and discussions

In this paper, we have found that the *d*-wave superconducting state appears in the low electron-doping region, and even at the half-filling, of the effective two-band model derived from the d-p model when the energy level difference between the upper Hubbard level of Cu and O 2p level is smaller than 1.5eV. This result is consistent with the recent experimental observations for T'-cuprates. The detailed information about the two-band model and the nature of the superconducting state will be reported in the forthcoming paper [10].

5. Acknowledgments

The numerical computations were carried out at the Cyber-science Center, Tohoku University.

References

[1] Luke G M, Le L P, Sternlieb B J, Uemura Y J, Brewer J H, Kadono R, Kiefl R F, Kreitzman S R, Riseman T M, Stronach C E, Davis M R, Uchida S, Takagi H, Tokura Y, Hidaka Y, Murakami T, Gopalakrishnan J, Sleight A W, Subramanian M A, Early E A, Markert J T, Maple M B, and Seaman C L 1990 Phys. Rev. B **42** 7981
[2] Tsukada A, Krockenberger Y, Noda M, Yamamoto H, Manske D, Alff L, and Naito M 2005 Solid State Commun. **133** 427
[3] Matsumoto O, Utsuki A, Tsukada A, Yamamoto H, Manabe T, and Naito M 2009 Physica C **469** 924
[4] Asai S, Ueda S, and Naito M 2011 Physica C **471** 682
[5] Takamatsu T, Kato M, Noji T, and Koike Y 2012 Appl. Phys. Express **5** 073101
[6] Adachi T, Mori Y, Takahashi A, Kato M, Nishizaki T, Sasaki T, Kobayashi N, and Koike Y 2013 J. Phys. Soc. Jpn. **82** 063713
[7] Weber C, Haule K, and Kotliar G 2010 Nat. Phys. **6** 574
[8] Zhang F C and Rice T M 1988 Phys. Rev. B **37** 3759
[9] Ogata M and Fukuyama H 2008 Rep. Prog. Phys. **71** 036501
[10] Yamazaki K, Yoshioka T, Tsuchiura H, and Ogata M, unpublished.
[11] Koikegami S, Fujimoto S, and Yamada K 1997 J. Phys. Soc. Jpn. **66** 1438
[12] Takimoto T and Moriya T 1997 J. Phys. Soc. Jpn. **66** 2459

Oxidation and reduction effects of successive superconducting transitions in ultra-fine $YBa_2Cu_4O_8$ ceramics

Akihiko Hisada[1, 2], Kie Muranaka[2], Kuniyuki Koyama[1, 2], Ko-ichi Magishi[1, 2], Takahito Saito[2, 3], and Makoto Hagiwara[4]

1. Faculty of Science and Technology, Tokushima University, Tokushima 770-8506, Japan
2. Graduate School of Integrated Arts and Sciences, Tokushima University, Tokushima 770-8502, Japan
3. Institute of Liberal Arts and Sciences, Tokushima University, Tokushima 770-8502, Japan
4. Faculty of Electrical Engineering and Electronics, Kyoto Institute of Technology, Kyoto 606-8585, Japan

E-mail: a-hisada@tokushima-u.ac.jp

Abstract. We investigated the effects of oxidation and reduction on the successive superconducting transition in a ceramic material composed of $YBa_2Cu_4O_8$ (Y124) submicron grains by linear and non-linear resistivity measurements. This material shows two-step superconducting transition and the lower one is identified as an inter-grain ordering. A sintering process fuses and joins the adjacent grains, so the lower transition was suppressed with increasing sintered time. Although Y124 is a stoichiometric ceramic, a reduction treatment suppressed the temperature of zero resistivity and broadened the non-linear resistivity peak. This suggests an oxygen vacancy was induced on the grain surface.

1. Introduction

The ceramics system composed of submicron superconductive grains has been studied focusing on the transition process from intra-grain to inter-grain orderings. It forms a Josephson-coupled inter-grain network and a novel ordering phenomenon such as a chiral glass transition is expected. [1]

A ceramic material of stoichiometric superconductor $YBa_2Cu_4O_8$ (Y124) shows successive two-step superconducting transition. [2] Y124 is composed of the superconducting CuO_2 planes and the edge-sharing CuO double chains which are stable and considered to be free from microscopic randomness due to the oxygen vacancy. The temperature dependence of the electrical resistivity $\rho(T)$ shows a rapid drop at about 80 K but it decreases gently with decreasing temperature below 76 K, and then reaches zero at around 27 K. [3] The upper transition was identified as the intra-grain superconductive ordering, and the lower one was identified as the inter-grain ordering. These features are in agreement with the magnetic measurements. [2]

The inter-grain transition is characterized by a random Josephson-coupled network containing the so-called π junctions. The zero resistivity temperature depends on the sintering condition, which suggests the circumstance of the inter-grain boundary can be changed by the sintering treatment. [2]

Recently, the authors found that the lower transition is influenced by the oxidation and reduction treatment. This implies the inter-grain ordering is affected by the oxygen vacancy despite the stoichiometric composition. In the present work, we performed linear and non-linear resistivity measurements for the ultra-fine Y124 ceramics with different sintered time and investigated the oxidation and reduction effects of the successive superconducting transition.

2. Experimental

Ultra-fine Y124 ceramics were prepared by the citrate pyrolysis precursor method which can make homogeneous grains of submicron size. [4] The precursor was calcined at 778 °C for 120 h, and sintered at 780 °C for 12.5 to 200 h.

In the present work, for the annealing treatment, the samples were oxidized at 500 °C for 24 h flowing 0.2 l/min oxygen gas. After electrical resistivity measurements, each sample was vacuum-heated at 450 °C for 24 h as a reduction treatment. Although the attained oxygen degree was estimated by measuring weight loss of the sample, obvious mass change was not observed.

The electrical resistivity measurements were performed by 4-probe method applying AC current of 23.0 Hz. Linear resistivity and non-linear resistivity are defined as the coefficients of the first and the third power term of the voltage response, respectively.

3. Results and Discussion

We performed scanning electron microscope (SEM) observations and X-ray diffraction (XRD) analysis for the samples with different sintered time to estimate the grain size. Figure 1 shows SEM images and XRD patterns for the samples sintered for 25 h and 200 h. SEM images show a weak sintered structure with submicron scale grains. The sintering process fuses and joins the adjacent grains. The estimated grain size was less than 0.8 μm for 25 h sintered sample and 1.2 μm for 200 h sintered one. The XRD peaks tend to sharpen with increasing sintered time, which also reflects the microscopic crystal growth.

Figure 2 shows the temperature dependence of the linear resistivity $\rho(T)$ for oxidized and reduced Y124 ceramics. The onset of the superconducting transition temperature T_C^{onset} is indicated in each figure. T_C^{onset} was about 81 K for all samples, regardless of whether the samples were oxidized or reduced. This suggests the intra-grain ordering is independent of the reduction treatment.

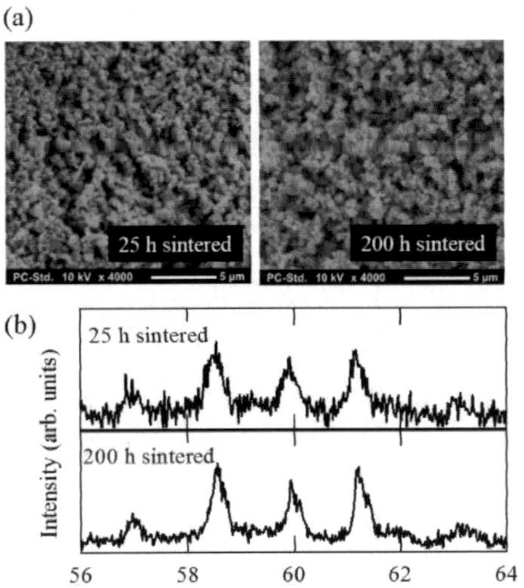

Figure 1. (a) SEM images and (b) XRD patterns for the Y124 ultra-fine ceramics sintered for 25 h and 200 h. To compare the peak width, XRD patterns are expanded.

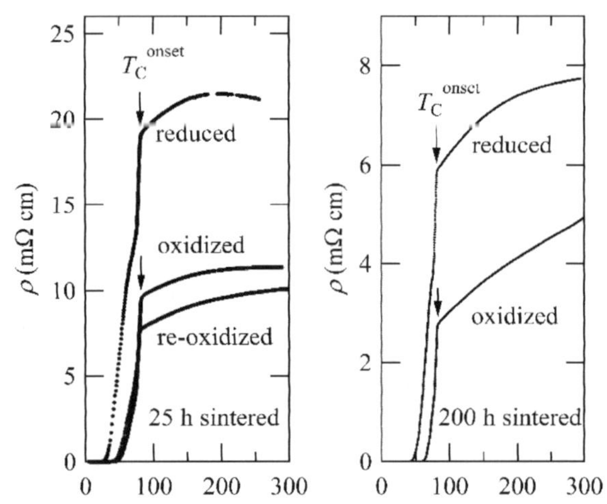

Figure 2. Temperature dependence of the linear resistivity up to the room temperature. The sintered time is shown in each view.

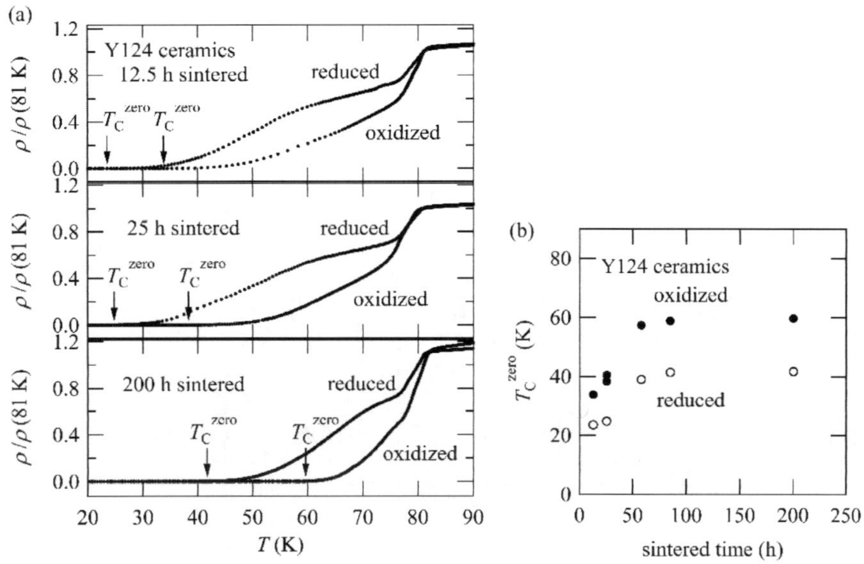

Figure 3. (a) Expanded views of the linear resistivity near the superconducting transition temperature. The sintered time is shown in each view. T_C^{zero} indicates the temperature of zero resistivity. (b) Sintered time dependence of T_C^{zero}.

$\rho(T)$ for the oxidized sample shows T-linear behaviour above T_C^{onset}. The reduced sample also shows metallic behaviour but the $\rho(T)$ curve is convex, which indicates a small part of the sample changes to semiconductive. This reduction effect is reversible; the $\rho(T)$ curve returns to the original shape by re-oxidizing treatment.

Figure 3 (a) shows the expanded views of the temperature dependence of the linear resistivity for Y124 ceramics with different sintered time near the superconducting transition temperature. In these figures, the linear resistivity is normalized to the value at 81 K and the temperature of zero resistivity T_C^{zero} is indicated. Focusing on the sintered time, the 12.5 h sintered sample shows the most gradual transition. This tendency is remarkable in the lower superconducting transition and T_C^{zero} decreases by more than 25 K. As mentioned above, the sintering process makes the grain size bigger. Putting these results together, the lower transition seems to be affected by the total surface area of the grains, which is in agreement with the previous report that T_C^{zero} is attributed to the inter-grain ordering among the superconducting grains by Josephson-coupling through the boundary. [1, 3]

Interestingly, T_C^{zero} shows a further decrease by the reduction treatment for every sintering condition specimens. This nature suggests the oxygen vacancy is induced on the grain surface area which governs the inter-grain couplings. T_C^{zero} recovered to the initial temperature by the re-oxidizing treatment, as is the case with the metallic behaviour above T_C^{onset}. T_C^{zero} is shown against the sintered time in Figure 3 (b). T_C^{zero} increases with increasing the sintered time at first, but it shows little change above about 85 h. This suggests each grain reaches mature size.

We performed non-linear resistivity measurements to investigate the transition process from intra-grain to inter-grain orderings. The third harmonic voltage response $V_{3\omega}$ shows a remarkable negative peak at around T_{peak}. Figure 4 (a) shows the temperature dependence of $V_{3\omega}$ which is normalized to the value at T_{peak}. We estimated a full width at half maximum (FWHM) as shown in Figure 4 (b).

The peak width reflects the inter-grain ordering process. FWHM decreases with increasing the sintered time, but it shows little change above 85 h, as is the case with T_C^{zero}. Because the inter-grain ordering is suggestive of the chiral glass ordering which is characteristic of the ceramic system composed of submicron grains, the crystal growth would suppress FWHM.

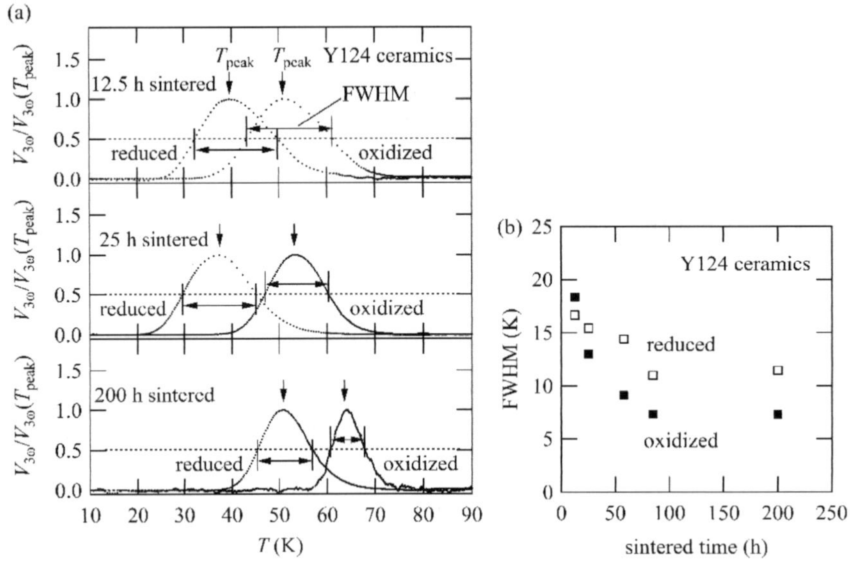

Figure 4. (a) Temperature dependence of the non-linear resistivity. The sintered time is shown in each view. T_{peak} indicates the peak temperature of $V_{3\omega}$. (b) Sitered time dependence of FWHM.

By the reduction treatment, FWHM tends to increase but that for the 12.5 h sintered sample shows no change within the experimental accuracy. This indicates the grain size has a stronger influence on the inter-grain ordering than the reduction effect.

Our experimental results may suggest the ultra-fine Y124 ceramics have peculiar surface on which oxygen vacancy can be induced. Because the inter-grain ordering among the superconducting grains is caused by Josephson-coupling on the grain surface, the oxidation and reduction effects could be observed.

Recently, the superconducting behaviour was found in reduction-treated $PrBa_2Cu_4O_8$ / $PrBa_2Cu_3O_{7-\delta}$ ceramics mixtures, which is considered to be caused by the inter-grain charge transfer effect from CuO single chain to CuO double chain. [5] The similar charge transfer might be induced by the reduction effect on the surface of the submicron grains.

4. Conclusions

We performed linear and non-linear electrical resistivity measurements for the ceramic material composed of Y124 submicron grains and investigated the oxidation and reduction effects of the successive superconducting transition. As the submicron grains increase in size by the sintering process, the inter-grain ordering was suppressed. The reduction effect was observed for the inter-grain ordering, although Y124 is stoichiometric. This suggests that the oxygen vacancy was induced on the grain surface.

References

[1] Yamao T, Hagiwara M, Koyama K and Matsuura M 1999 *J. Phys. Soc. Jpn.* **68** 871-75
[2] Kawachi M, Hagiwara M, Koyama K and Matsuura M 1994 *J. Phys. Soc. Jpn.* **63** 3405-11
[3] Yamao T, Kawachi M, Hagiwara M and Matsuura M 1996 *Physica* C **263** 464-67
[4] Hagiwara M, Yamao T and Matsuura M 2003 *Physica* C **392-396** 66-70
[5] Koyama K, Nakashima R, Magishi K, Saito T, Shima T and Hagiwara M 2011 *J. Phys. Conf. Series* **320** 012077

Field–Temperature Phase Diagram of Intergrain Ordering in Superconducting Ceramic YBCO

H Deguchi[1], R Warabino[1], S Ka[1], M Mito[1], M Hagiwara[2] and K Koyama[3]

[1]Faculty of Engineering, Kyushu Institute of Technology, Kitakyushu 804-8550, Japan

[2]Faculty of Engineering and Design, Kyoto Institute of Technology, Kyoto 606-8585, Japan

[3]Faculty of Science and Technology, Tokushima University, Tokushima 770-8502, Japan

E-mail: deguchi@mns.kyutech.ac.jp

Abstract. The ceramic $YBa_2Cu_4O_8$ superconductor composed of submicron grains is considered a random Josephson-coupled network containing the so-called π junctions and shows successive phase transitions. With decreasing temperature, first the intragrain superconductive transition occurs inside each grain at T_{c1} and then the chiral-glass transition occurs among the grains at T_{c2} (< T_{c1}). The third transition at T_{c3} (< T_{c2}) is the intergrain superconducting transition. We measured the nonlinear susceptibility and resistivity of the ceramic $YBa_2Cu_4O_8$ superconductor to determine the field dependences of the transition temperatures T_{c2} and T_{c3}. The phase diagram of the intergrain ordering is discussed in light of the result predicted by Kawamura.

1. Introduction

Ceramic cuprate superconductors composed of submicron grains are considered as random Josephson-coupled networks of 0 and π junctions. The circulation of a local loop-supercurrent and an orbital moment are generated spontaneously when there is an odd number of π junctions in a closed loop (π loop). The frustration effect due to the random distribution of π loops should lead to the chiral-glass state predicted by Kawamura and Li.[1] Kawamura investigated the ordering of ceramic cuprate superconductors, which is well described by the isotropic 3D *XY* spin-glass model.[2-3] Namely, the spin in a magnet corresponds to the phase of the superconducting order parameter at each grain, while the chirality corresponds to the circulation of superconducting current loops flowing in a Josephson-coupled network. At the chiral-glass transition, the chirality is frozen, i.e., the circulation of superconducting current loops is frozen, which can be observed experimentally as a negative divergence of the nonlinear susceptibility. At the spin-glass transition, the phase of the superconducting order parameter is frozen and the system becomes a true superconductor. This can be observed experimentally as the onset of zero resistivity. Kawamura predicted that successive chiral- and spin-glass transitions occur at $T = T_{CG}$ and $T = T_{SG}$ with $T_{CG} > T_{SG}$ in an isotropic system. [2] It is interesting to check experimentally whether the spin-chirality decoupling phenomenon is realized in the ordering of ceramic cuprate superconductors. Through experimental studies for ceramic $YBa_2Cu_4O_8$, the critical temperature

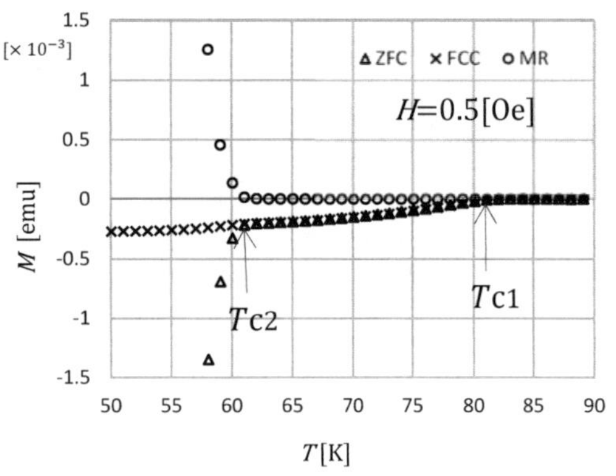

Figure 1. Temperature dependence of the zero-field magnetization (ZFC), field-cooled magnetization (FCC), and thermoremanent magnetization (MR) at $H = 0.5$ Oe

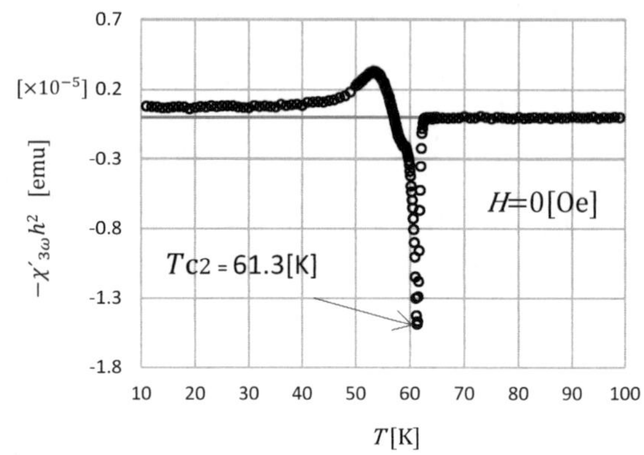

Figure 2. Temperature dependence of the nonlinear susceptibility at 1.0 Hz with the ac field amplitude of 0.1 Oe at $H = 0$ Oe

T_{c2} of the chiral-glass transition was detected using the nonlinear susceptibility [4-5] and the critical temperature T_{c3} of the intergrain superconductive transition was estimated using the onset of the zero linear resistivity at zero field.[6-7] In the chiral-glass phase below T_{c2}, the magnetic glass behavior and memory effects, similar to those of spin-glasses, were observed in ceramic $YBa_2Cu_4O_8$.[8-9]

In this work, we investigated the field dependence of T_{c2} and T_{c3} of ceramic $YBa_2Cu_4O_8$ and discussed the field–temperature phase diagram of the intergrain ordering.

2. Sample and experiments

The sample was prepared using the citrate pyrolysis method.[10] The precursor was calcined for 100 h at 777 °C to yield a pure $YBa_2Cu_4O_8$ phase, which was sieved and pressed, and then sintered for 85 h at 780 °C, to became a pure $YBa_2Cu_4O_8$ ceramic with submicron grains. The dc magnetization and the ac susceptibility were measured with a SQUID magnetometer (Quantum Design MPMS-5). The linear resistivity was measured by low-level pulsed electrical characterization with a current source (KEITHLEY 6221) and a nanovoltmeter (KEITHLEY 2182A) combination. The delta method for low voltage measurements was used to eliminate the constant thermoelectric voltage and to minimize the amount of power dissipated in the sample. An external field was applied perpendicular to the dc current.

3. Results

We measured the temperature dependences of the dc magnetization and ac susceptibility to determine the transition temperatures T_{c1} and T_{c2}. Figure 1 shows the temperature dependences of the zero-field-cooled, field-cooled, and thermoremanent magnetizations at $H = 0.5$ Oe. The transition at $T_{c1} = 81$ K was identified as the intragrain superconducting ordering, in which small diamagnetism due to the Meissner effect appears in zero-field-cooled and field-cooled magnetizations. The zero-field-cooled magnetization steeply branches away from the field-cooled one at the glass transition temperature $T_{c2} = 61$ K. This branching temperature coincides with thermoremanent magnetization vanishing point, which must be caused by the release of magnetic flux trapped in the ordered network.

The nonlinear susceptibility estimated from the first term of the series of in-phase odd-harmonic responses at 1.0 Hz with an ac field amplitude of 0.1 Oe is shown in Figure 2. A negative peak of nonlinear susceptibility was observed at $T_{c2} = 61.3$ K; at this temperature, the ceramic $YBa_2Cu_4O_8$ underwent a chiral-glass transition at zero applied field. We measured the temperature dependence of nonlinear susceptibility at some applied fields of magnitude up to 300 Oe and estimated T_{c2} from the peak temperature of nonlinear susceptibility in applied fields. The negative peak of nonlinear

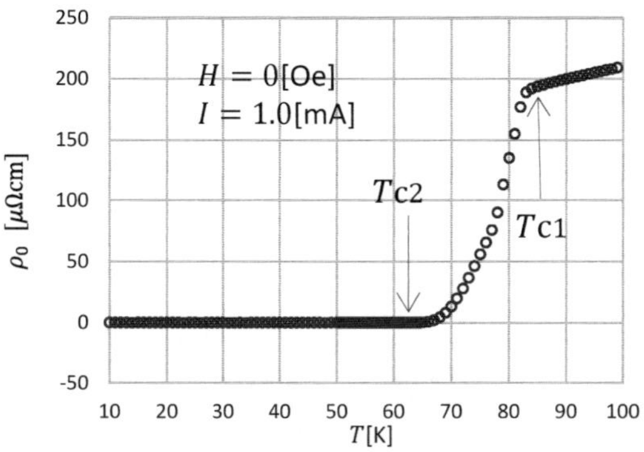

Figure 3. Temperature dependence of the linear resistivity for the applied dc current of 1.0 mA at $H = 0$ Oe

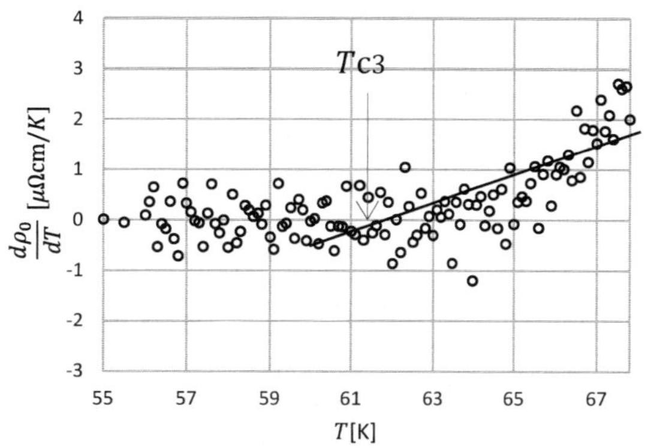

Figure 4. Temperature dependence of $d\rho_0/dT$ for the applied dc current of 1.0 mA at $H = 0$ Oe

susceptibility shifts slightly toward higher temperatures with increasing the field magnitude up to 30 Oe, and the magnitude of the peak gradually increases. However, the peak on the high-field side shifts remarkably toward lower temperatures with increasing the field magnitude up to 300 Oe and the magnitude of the peak remains almost constant. The field dependence of T_{c2} at low fields is not normal in chiral-glass ordering. In the fully isotropic Heisenberg case, the chiral-glass transition temperature should be independent of the field magnitude and should be constant at low fields. [2]

Next, we measured the temperature dependence of the linear resistivity ρ_0 for the applied dc current of 1.0 mA at some applied fields up to 100 Oe. Figure 3 shows the temperature dependence of ρ_0 at zero field. As the temperature was reduced, ρ_0 decreased rapidly at T_{c1}, and then decreased monotonously and almost disappeared around T_{c2}. It is difficult to estimate the transition temperature T_{c3} for $\rho_0 = 0$ because ρ_0 diminishes continuously with decreasing temperature. Therefore, T_{c3} was estimated as the temperature at which $d\rho_0/dT = 0$, as shown in Figure 4. At zero field, the intergrain superconducting transition temperature was identified as $T_{c3} = 61.8$ K; this temperature is close to the chiral-glass temperature $T_{c2} = 61.3$ K at zero field. The critical temperature T_{c3} shifts remarkably to the lower temperature side with increasing field magnitude and decreases 59.8 K at $H = 100$ Oe. The change and field dependence of T_{c3} in applied fields are very different from those of T_{c2}.

4. Discussion and conclusions

The transition temperatures T_{c2} and T_{c3} were obtained at zero field and some applied fields. Figure 5 shows the field–temperature phase diagram for the ceramic $YBa_2Cu_4O_8$. The diagram is divided into three phases, namely, the intragrain superconducting phase, the chiral-glass phase, and the intergrain superconducting phase, by T_{c2} and T_{c3}.

At zero field, chiral-glass ordering takes place at the same temperature as intergrain superconducting ordering ($T_{c2} = T_{c3}$). This means that the chirality is not decoupled from the phase of the superconducting order parameter (corresponding to the spin in a spin-glass system) in the ceramic $YBa_2Cu_4O_8$ at zero field. However, when a field is applied, both the temperature value and the field dependence of T_{c2} are different from those of T_{c3}. Thus, "spin-chirality decoupling" occurs in an applied field in the ceramic $YBa_2Cu_4O_8$.

Kawamura predicted that the "spin-chirality decoupling" phenomenon is realized in the ordering of granular cuprate superconductors in zero and applied fields.[2] The effects of magnetic fields on the spin-glass ordering at T_{SG} (corresponding to T_{c3} in the ceramic $YBa_2Cu_4O_8$) and on the chiral-glass ordering at T_{CG} (corresponding to T_{c2} in the ceramic $YBa_2Cu_4O_8$) were studied by Kawamura.[2] According to the Kawamura, the spin-glass transition line should exhibit a singular form, $|T_{SG}(0) - T_{SG}(H)|^2 \propto H$. We estimated the exponent from the field dependence of $|T_{c3}(0) - T_{c3}(H)|$ at high fields.

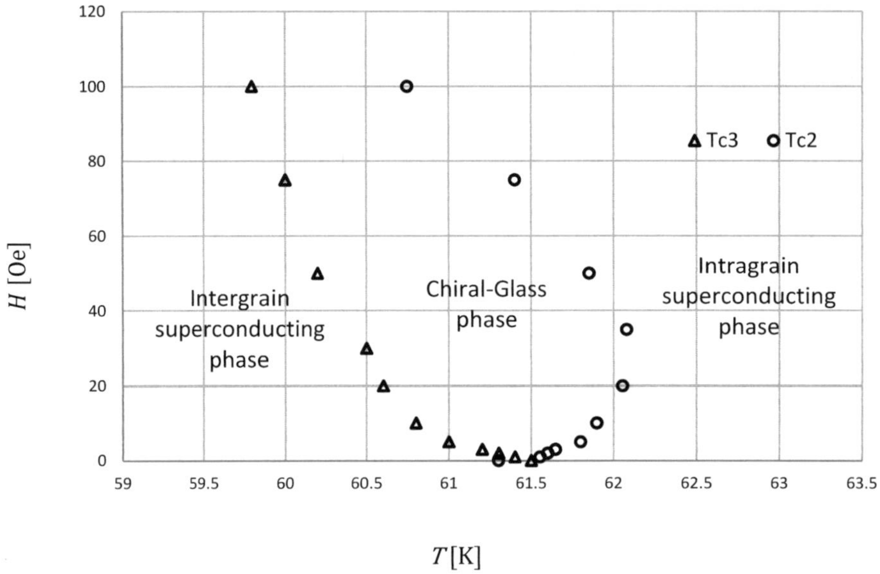

Figure 5. Phase diagram showing intergrain superconducting, chiral-glass, and intragrain superconducting phases as a function of temperature and field in the ceramic $YBa_2Cu_4O_8$. Triangle and circles correspond to T_{c3} and T_{c2}, respectively.

The estimated value of 2.3 is within the range expected from the theoretical value of 2. The chiral-glass transition temperature $T_{CG}(H)$ under applied fields should behave as the so-called the Gabay-Toulouse (GT) line of the mean-field model, $|T_{CG}(0) - T_{CG}(H)| \propto H^{1/2}$.[11] At a field magnitude lower than H= 30 Oe, T_{c2} increases gradually with increasing field magnitude, which is different from the field dependence of the GT line. However, the obtained exponent from the field dependence of $|T_{c2}(0) - T_{c2}(H)|$ at high fields is 0.53, which is consistent with the theoretical value of 1/2. Therefore, the transition lines of $T_{c2}(H)$ and $T_{c3}(H)$ at high fields seem to correspond with the chiral-glass and the intergrain superconducting transition lines, respectively. In an applied field, the chirality is decoupled from the phase of superconducting order and chiral-glass ordering takes place at a temperature higher than intergrain superconducting ordering, $T_{c2}(H) > T_{c3}(H)$.

In conclusion, we investigated the temperature dependences of nonlinear susceptibility and linear resistivity in zero and applied fields, and estimated the transition temperatures T_{c2} and T_{c3} at zero and some applied fields in the ceramic $YBa_2Cu_4O_8$. The field–temperature phase diagram of the intergrain ordering was obtained and compared with the expected phase diagram from the chirality scenario described by Kawamura. The results suggest that spin-chirality decoupling takes place at finite fields in granular cuprate superconductors.

Acknowledgements

The SQUID measurement was carried out at the Center for Instrumental Analysis, Kyushu Institute of Technology. This work was supported by Grant-in-Aids for Scientific Research on Priority Areas (No.19052006) and for Scientific Research (C) (No.26390010) from the Ministry of Education, Culture, Sports, Science and Technology (MEXT) of Japan.

References

[1] Kawamura H and Li M S 1997 *Phys. Rev. Lett.* **78** 1556

[2] Kawamura H 2010 *J. Phys. Soc. Jpn.* **79** 011007
[3] Obuchi T and Kawamura H 2013 *Phys. Rev.* B **87** 174438
[4] Matsuura M, Kawachi M, Miyoshi K, Hagiwara M and Koyama K 1995 *J. Phys. Soc. Jpn.* **65** 4540
[5] Deguchi H, Koyama K, Yasunaga M, Takagi S, Nagano N and Mizuno K 2002 *J. Phys. Chem. Solids* **63** 1081
[6] Yamao T, Hagiwara M, Koyama K and Matsuura M 1999 *J. Phys. Soc. Jpn.* **68** 871
[7] Deguchi H, Nawa D, Hashimoto Y, Mito M, Takagi S, Hagiwara M and Koyama K 2009 *J. Phys. Conf. Series* **150** 052043
[8] Deguchi H, Ashida T, Shoho T, Kato Y, Mito M, Takagi S, Hagiwara M and Koyama K 2011 *J. Phys. Conf. Series* **320** 012076
[9] Deguchi H, Syudo M, Ashida T, Sasaki Y, Mito M, Takagi S, Hagiwara M and Koyama K 2013 *Phys. Procedia* **45** 129
[10] Yamao T, Hagiwara M, Miyoshi K and Matsuura M 2004 *Physica* C **412-414** 98
[11] Gabay M and Toulouse G 1981 *Phys. Rev. Lett.* **47** 201

IOP Conf. Series: Journal of Physics: Conf. Series **871** (2017) 012012 doi:10.1088/1742-6596/871/1/012012

STM measurements of Ni impurity effects in Bi2212

A Nakagawa, T Sakaidani, Y Kiguchi, T Uto and A Matsuda

Department of Physics, Schools of Advanced Science and Engineering, Waseda University, Okubo 3-4-1, Shinjuku, Tokyo 160-8555, Japan

naktme@gmail.com

Abstract. We report on the STS measurements for the temperature dependences of density-of-states for optimally-doped and overdoped samples. In the tunnelling spectra above T_C in an overdoped sample, we found two types of a pseudogap; the large one, which had a typical pseudogap size and decreased with increasing a temperature, and the small one, which seemed almost temperature independent in our measured temperature range. The large gap seemed to switch to the small one with some overlap region with increasing a temperature. Similar behaviour was found in the optimally doped samples with an enhanced gap size and the temperature range. We identified the large gap as a conventional pseudogap and the small one as a new pseudogap, which is associated with Ni impurity. This shows a peculiarity of Ni impurity, and suggests an unusual interaction between an impurity and the Cu-O electronic system.

1. Introduction

In the high-temperature superconductor, so called a pseudogap state develops in the underdoped region above a superconducting transition temperature T_c. Although its origin has not been identified yet, it is intimately related to the superconducting state, since their gaps are smoothly connected each other at T_c. Recently, they are considered to be mutually competitive. Nevertheless, the origin of the pseudogap thought to be strongly related to the mechanism of high-temperature superconductivity.

We have been studied impurity effects on the pseudogap states, because they discriminate the difference and/or the similarity of the superconducting and the pseudogap states. One of the features of superconducting gap/pseudogap in Bi2212 is that its value continues to increase with increasing a temperature up to the pseudogap opening temperature T^* in the wide doping levels from the underdoped to the overdoped region. But, in the heavily overdoped region, it decreases with increasing a temperature and disappears at T_c (Figure 1) [1]. Previously, we reported transport properties and temperature and carrier doping dependences of the density-of-states (DOS) in the pseudogap states of Co-doped Bi2212s. We found that the Co^{3+} substitution into the Cu^{2+} sites essentially preserved the T_c vs carrier density (p) relation, so-called Tallon's empirical law, with some doping level shift due to the difference in the valence value. On the other hand, for overdoped samples, the Co impurity significantly enhances pseudogap value and alters the temperature dependence of the pseudogap from that of the pure-Bi2212 samples [2], while its T^* is unchanged. There is rather strong impurity effect on the pseudgap, in contrast to the common sense that the pseudogap shows a little impurity effect.

It is important to know that these results are universal or not. Here, we report the result of the Ni impurity effect on the pseudogap states. In the case of the Ni substitution, we found somewhat different behavior in the T_c vs p relation; (1) in the overdoped region, the T_c vs p relation followed

Talon's law, (2) in the underdoped region, superconductivity was more strongly destroyed by the Ni impurity [3]. These results indicate that there is large impurity species dependence in the impurity effects. However, a little has been known about the electronic state of Ni substituted samples.

In this paper, we concentrate on the impurity effect in the temperature dependence of a pseudogap in the Bi2212 and compare the results of Co and Ni. The DOSs were measured by a scanning tunneling microscope (STM). We will show a new type of a pseudogap specifically associated to Ni impurity.

a

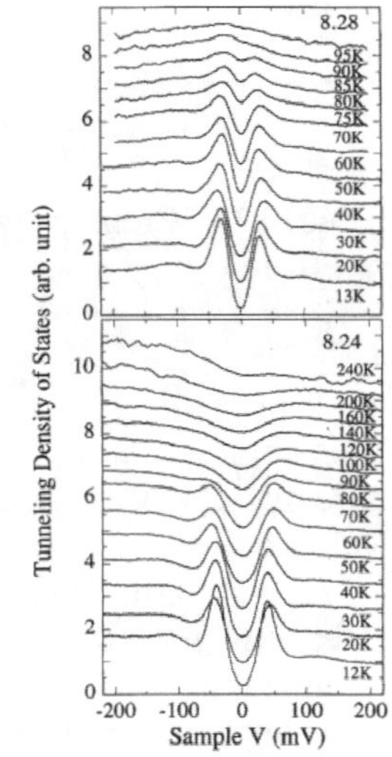

b

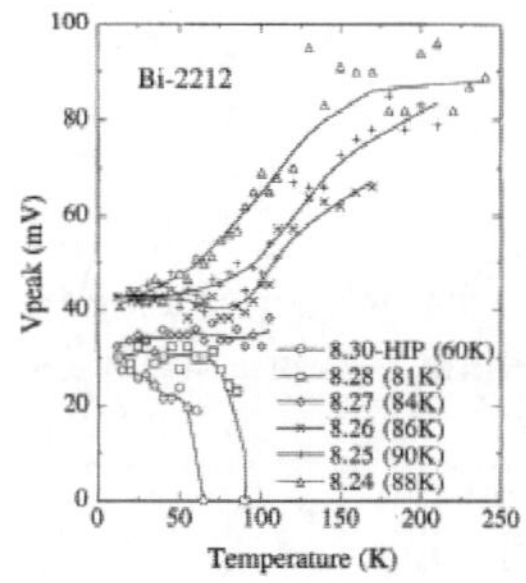

Figure 1. typical temperature dependence of DOS in pure-Bi2212; (a) The excess oxygen δ = 0.28 (upper panel), δ = 0.24 (lower panel), (b)temperature dependence of gap values estimated as positive bias peak position or shoulder position [1]

2. Experiments

The Ni substituted Bi2212 ($Bi_{2.1}Sr_{1.9}Ca(Cu_{1-x}Ni_x)_2O_{8+\delta}$; Ni-Bi2212, x=0.03) single crystals were grown by the traveling solvent floating zone method. They were annealed under the atmosphere of Ar and O_2 gas mixture, with a proper oxygen partial pressure. The excess oxygen δ was adjusted to 0.25 for optimally doped (OP) samples and 0.27 for overdoped one. The T_cs of both samples determined by the DC magnetic susceptibility measurements were 72 K and 71 K, respectively [6].

STM measurements were performed in the temperature range from 79 K to 230 K with an electro-chemically etched W tip under a pressure of $\sim 10^{-10}$ Torr. The Ni-Bi2212 crystals were cleaved under the ultra-high vacuum atmosphere at around 79 K. We acquired I-V characteristics in each 8x8 points of 10x10 nm areas. Then density-of-states were obtained by numerically differenciating the averaged I-V characteristics. The topography of Ni-Bi2212 surface is shown in Figure 2.

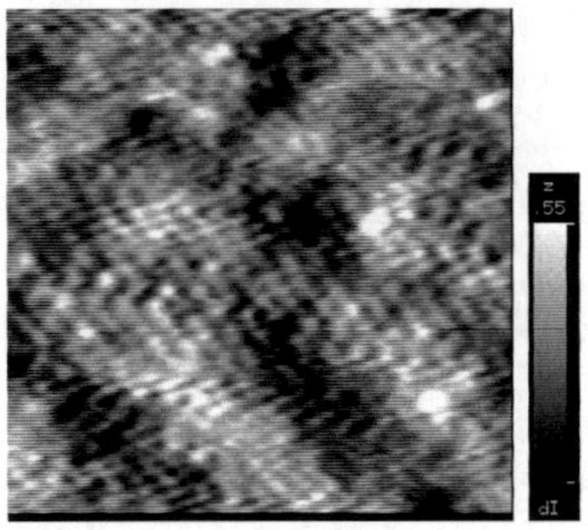

Figure 2. Topography of Ni-Bi2212 in 10x10 nm^2 areas.

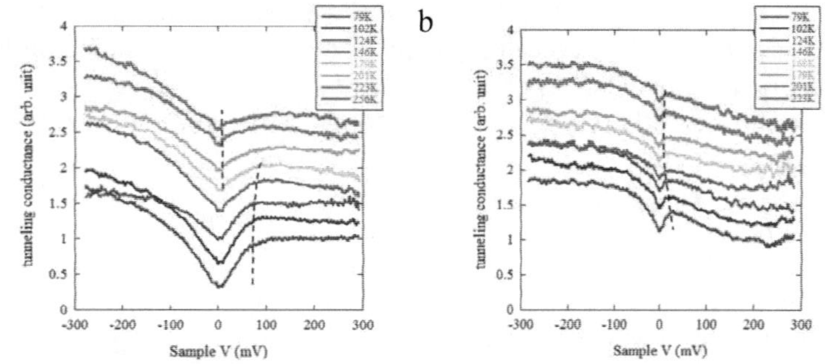

Figure 3. Temperature dependence of DOS; (a) OP sample and (b) OD sample. The dushed line shows the estimated gap value.

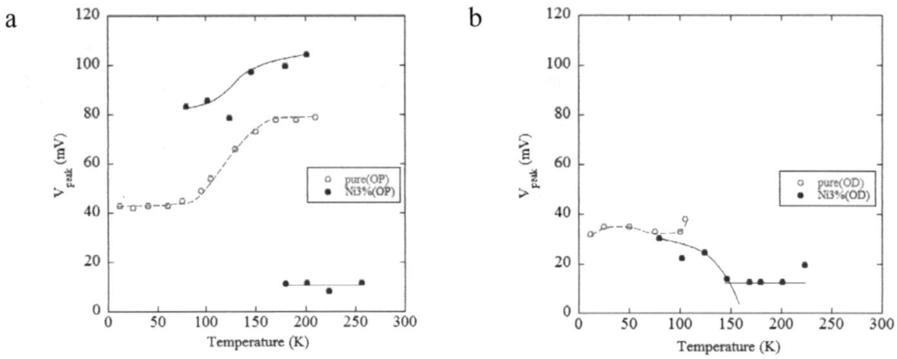

Figure 4. Temperature dependence of gap value in Bi2212 of Fig.2 responsibility, compared between pure and Ni substitution.

3. Results and Discussions

Figure 3 shows the tunnelling spectra of Ni-Bi2212s. In the optimally doped sample (Fig.3a), we found that the pseudogap value increases with increasing a temperature. However, we see additional structures above 179 K, which was never observed in the pure and Co substituted samples above T_c. We identify this structure as a small pseudgap with a gap value of ~10 mV. Smallness of the signal should be due to a small gap value and a high observing temperature. In the overdoped sample (Fig.3b), the same structure was confirmed at somewhat lower temperatures. Furthermore, the small pseudgap remains open above the pseudogap opening temperature T^* (~ 160 K), which was determined by the c-axis resistivity up-turn (see Fig. 5) [4]. The appearance of the small pseudogap is a new result and discriminate Ni impurity from Co impurity

We show temperature dependences of these pseudogap values in Fig. 4. The results of the pure-Bi2212 were also shown for the comparison. We adopted the gap value as a shoulder voltage in the DOS. In the optimally doped samples, we see that the pseudogap is enhanced and increasing behaviour with a temperature, similar to the results of the pure sample. In the overdoped samples, the pseudogap value decreases with increasing a temperature. Both tendencies are essentially same as the Co substituted one. We naturally identify the large sized gap as a conventional pseudogap and these characteristic behaviors are universal independent of impurity species. However the existence of small pseudogap is Ni impurity specific. They are temperature independent in our measured temperature range and coexist with a conventional pseudogap above certain temperature but disappear at low temperatures

In these samples, it is hard to consider that Ni substitution modify the band structure to produce the observed anomaly because the amount of Ni is as small as 3%. However, it is known that the Zn impurity strongly modify surrounding superconducting as well as normal electrons and form the Swiss-cheese-like non-uniform electronic state in the superconducting state [5]. Such a effect has not been examined and reported for Co and Ni impurity. However, our experiments suggest that the Ni impulity locally suppress surrounding density-of-states in the normal state, as Zn did in the superconducting state. Since small pseudogap feature was observed almost unformily in the measured area, the DOS suppressed area should be larger than 6x6 Cu atoms per Ni impurity.

Finally, Fig. 5 shows a comparison between the c-axis resistivity and the inverse of the density-of-states at a Fermi level. If the c-axis transport is goverened by a tunneling process between 2D superconducting layers, it should sensitively reflects opening a gap, and give an upturn at a gap opening temperature. As can be seen in Fig. 5, the upturns in the resistivity and the DOS plots coincide well and they assure the cleaved surface is doped to the same level as a bulk sample. They also provide good T^* estimations.

IOP Conf. Series: Journal of Physics: Conf. Series **871** (2017) 012012 doi:10.1088/1742-6596/871/1/012012

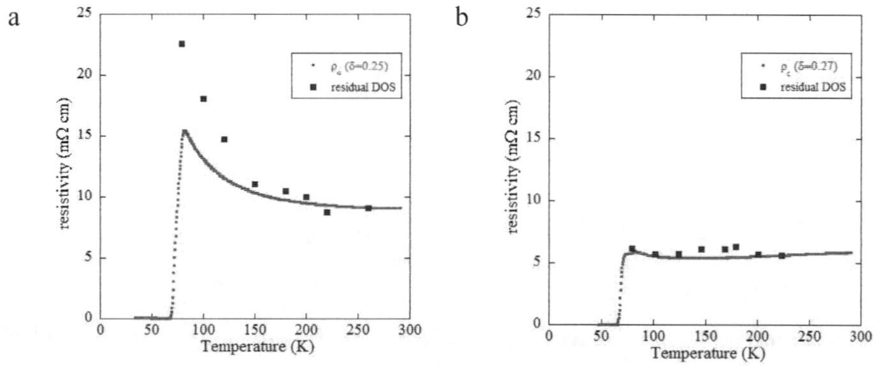

Figure 5. Comparison between inter-planar resistivity and residual DOS, (a) OP sample and (b) OD sample.

In summary, we reported the results of scanning tunneling spectroscopy measurements on the pseudogap states of Ni3%-Bi2212 samples. We found that Ni impurity induce the new small gap-like feature with weak temperature dependence, as well as the conventional pseudogap. The conventional pseudogap underwent similar impurity effects to the Co doping case, indicating their universal nature. The small gap-like feature is specific to Ni impurity and suggests to its anomalous interaction to the surrounding Cu. They can be related to the steep decrease in T_c in the Ni doping in the underdoped region.

References

[1] A.Matsuda et al., Journal of physics and Chemistry of solids 62, (2001) 65
[2] Yoshinori Numata et al., Physica C 524, (2016) 28
[3] Yuta Kiguchi et al., Physics Proceedia 58, (2014) 54
[4] Takao Watanabe et al., Phys. Rev. Lett. 84, (2000) 5848
[5] B. Nachumi et al., Phys. Rev. Lett. 77, (1996) 5421
[6] C. Izutani et al., Journal of Chemical Education 93, (2016) 1667

Spin injection effect in thin Bi2212 single crystal

Kenichiro Murata[1], Kazuto Otaka[2], Kazuhiro Yamaki[1,2], and Akinobu Irie[1,2]

[1]Department of Innovation Systems Engineering, Utsunomiya University, 7-1-2 Yoto, Utsunomiya 321-8585, Japan.
[2]Department of Electrical and Electronic Systems Engineering, Utsunomiya university, 7-1-2 Yoto, Utsunomiya 321-8585, Japan

E-mail address: dt147122@cc.utsunomiya-u.ac.jp

Abstract. The influence of spin-injection on the in-plane transport properties of thin $Bi_2Sr_2CaCu_2O_y$ (BSCCO) single crystal has been investigated. The in-plane transport measurements without and with spin injection were carried out at 77 K by four terminal method. The in-plane critical current was strongly reduced by injecting the current from Co/Au electrodes formed on the BSCCO bridge with 50 mm wide and 450 nm thick. Furthermore, it was observed that magnetic field dependence of the magnetoresistance shows a hysteresis loop. These results indicate that the in-plane superconductive transport property is affected by the spin-injection related to the magnetization of Co.

1. Introduction

The control of superconductivity by spin-polarized quasiparticle injection (spin-injection) has been a subject of great interest from the point of view of development of novel superconducting spintronic devices as well as understanding of fundamental physics in a ferromagnet-superconductor hybrid system [1]-[4]. Although there have now been many works related to the spin-injection effect in high temperature superconductors (HTS) [2],[5],[6], in the case of HTS, it is important to classify the direction of spin-injection because superconducting properties of HTS are anisotropic. However, this may not be so simple for HTS film samples which have been used in most of spin-injection experiments because of interface roughness. Therefore, in order to understand the influence of the spin-injection effect on transport in detail, it is required to use the high-quality samples with flat surface in experiments.

It is well known that high T_c superconductor $Bi_2Sr_2CaCu_2O_y$ (BSCCO) single crystal can be regarded as a stack of two dimensional superconducting layer of 0.3 nm and insulating layer of 1.2 nm. Due to the layered crystal structure, the properties of BSCCO are strongly anisotropic. As a result, a single crystal of BSCCO acts as a densely packed naturally formed stack of intrinsic Josephson junctions (IJJs) [7, 8]. Thus, large spin injection effect can be expected in BSCCO single crystals because of extremely thin superconducting layer. From this point of view, in our previous work, we studied the transport properties of IJJs in Co/BSCCO hybrids under applied magnetic fields parallel to the layer and observed the hysteretic field dependence of critical current I_c of IJJs due to spin injection effect [9]. This suggests that the in-plane transport property of BSCCO is also affected by spin injection. In this paper, we report on the influence of the spin injection effect for the in-plane transport properties of thin flake BSCCO.

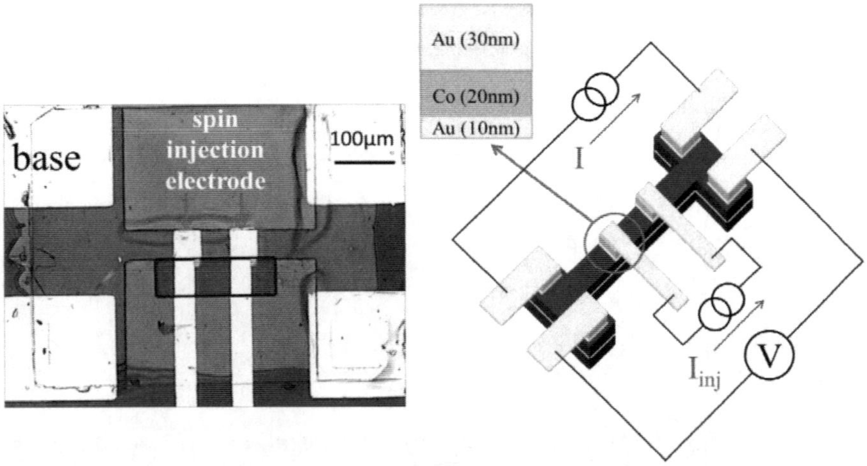

Figure 1 Optical image and a schematic view of the fabricated sample.

2. Experimental

An optical image and a schematic view of the fabricated sample are shown in figure 1. The sample is composed of BSCCO bridge and two Co/Au electrodes as spin injector. The width and thickness of BSCCO bride are 50 mm and 450 nm, respectively, and Co (20 nm)/Au (10 nm) electrodes with lateral dimensions of 50 × 50 μm^2 are formed on the bridge separated by a distance of 50 μm.

BSCCO single crystals with critical temperatures of 85-90K were grown by conventional melting method [10]. A piece of BSCCO single crystal of 400 × 600 μm^2 was adhered to glass substrate and was cleaved until its thickness was around 450 nm. Then, thin Au (10 nm) and Co (20 nm) layers were immediately deposited by thermal evaporation and DC sputtering. Here, Au layer which is thinner than the spin diffusion length of ~ 60 nm [11] was used to reduce the contact resistance between Co and BSCCO surface without spin scattering. The device structure shown in figure.1 was fabricated by photolithography and Ar ion milling.

The in-plane transport measurements without and with spin injection were carried out at 77 K by four terminal methods. The magnetic field was applied to parallel of the layer by solenoid coil and swept in both direction.

3. Results and Discussion

Figures 2(a) and 2(b) show the current-voltage (*I*-*V*) characteristics at 77 K for I_{inj}=0, 0.4, 0.5, and

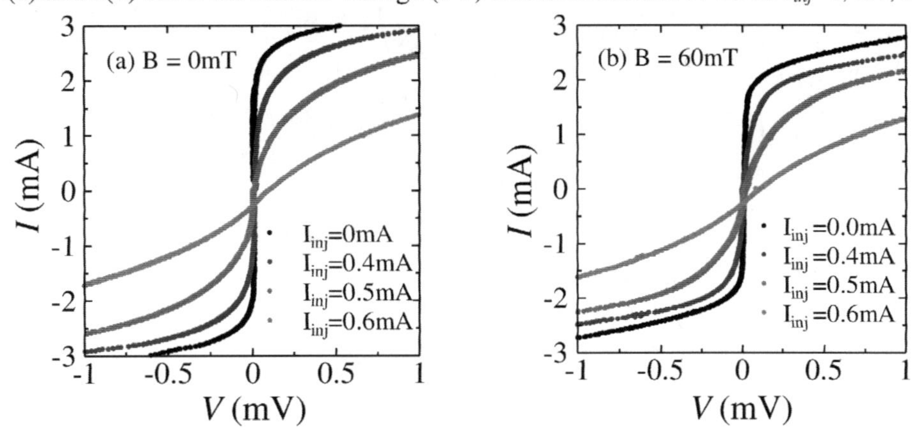

Figure 2 *I*-*V* characteristics for different injection currents in the magnetic fields of (a) 0 mT and (b) 60 mT.

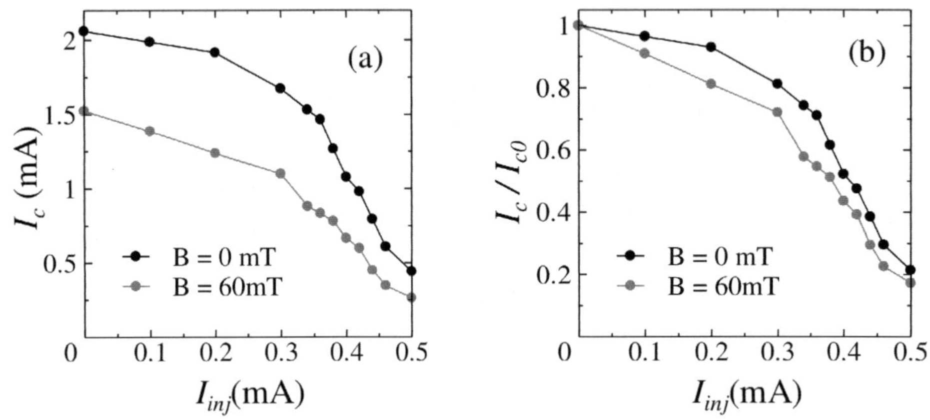

Figure 3 (a) Critical current and (b) normalized critical current as a function of the injection current for B=0 and 60 mT.

0.6 mA in the magnetic field B of zero and 60 mT, respectively. The critical current I_c without injection is 2.1mA for B=0 and 1.5 mA for B=60 mT. Here, we estimated the critical current I_c from $(I_c^+ - I_c^-)/2$ using I_c^+ and I_c^- defined as the current values at $V = \pm$ 15 μV. One can see that the I-V curves show considerably change with increasing I_{inj} , and nearly full suppression of I_c could be achieved at I_{inj} = ~0.6 mA. In figures 3(a) and 3(b), I_c and I_c / I_{c0} for B=0 and 60 mT are plotted as a function of I_{inj}, where I_{c0} is the critical current for I_{inj}=0 mA. As expected, I_c decrease with increasing I_{inj} and this indicates that the in-plane superconductive transport property is affected by the spin-injection related to the magnetization of Co. In this work, we did not demagnetize the sample before measurement, so that Co has some residual magnetization even without applied magnetic field. Therefore, the difference between the dependences without and with magnetic field comes from the magnetic property of Co, or polarization of Co layer.

As mentioned above, we observed the suppression of critical current by injecting the current from Co and attributed it to non-equilibrium superconductivity due to spin-injection. However, such a non-equilibrium state can be caused by the injection of non-polarized quasiparticle. Therefore, in order to check whether the observed suppression of I_c is related with spin-injection, we measured field dependence of transport properties at constant injection current because the spin-polarlized nature of

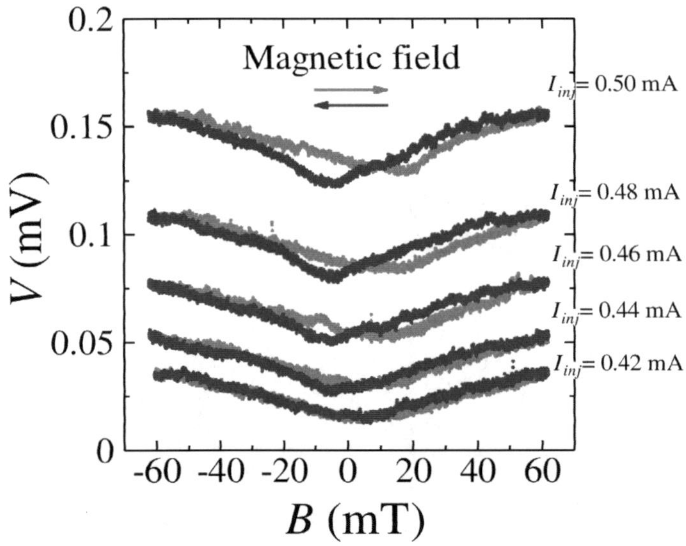

Figure 4 Magnetic field dependence of the sample voltage for I = 1mA.

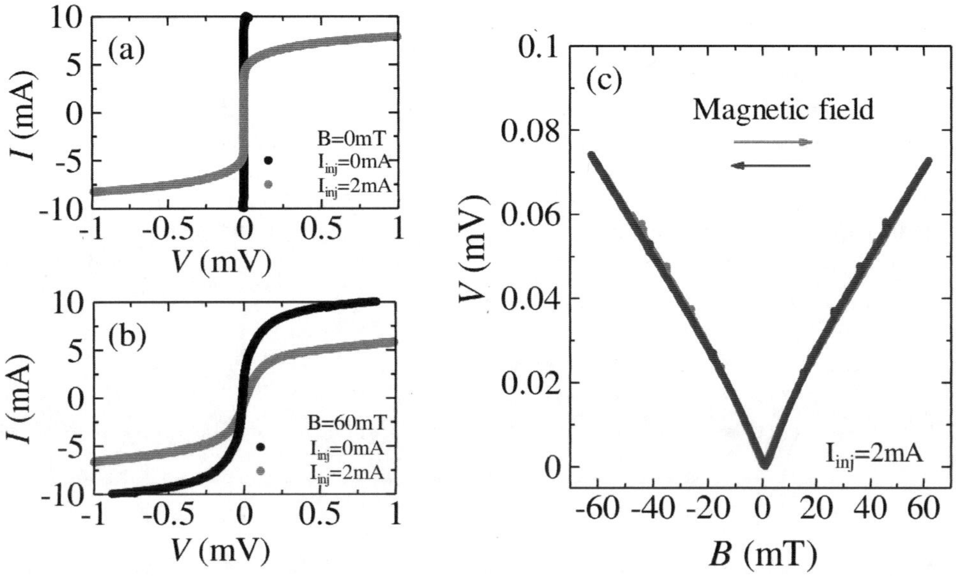

Figure 5 *I-V* characteristics of the device without Co layer measured at 77 K in the magnetic fields of (a) 0 mT and (b) 60 mT. (c)*V-B* characteristic for I = 2mA and I_{inj}=2mA.

Co has a hysteresis. Figure 4 shows the field dependence of the sample voltage V at different injection currents when the sample was biased at I =1 mA. The field was swept up and down between B = -60 mT and 60 mT. As seen from figure 4, the V (B) behaviour is hysteretic, and the field at which V becomes minimum is -4 and + 17 mT independent of I_{inj} in contrast to the voltage change. Therefore, the most likely explanation is that the observed V (B) behaviour reflects the magnetic hysteresis loop of Co layer and the voltages with minima are observed at the coercive field of Co because a higher I_c could be expected at the coercive field where the average spin polarization of Co becomes zero. On the other hand, in the case of non-polarized quasiparticle injection, we expect the hysteresis-free behaviour of V (B). Indeed, we observed the non-hysteretic $V(B)$ behaviour for the device without Co layer. Figure 5 shows the *I-V* and *V-B* characteristics of the device without Co layer measured at 77 K. It is found from figure 5 that the critical current is suppressed by the non-polarised quasiparticle injection similar to that of the device with Co shown in figure 2 but the *V-B* characteristic does not show hysteresis. Therefore, we believe that the observed suppression of superconductivity is caused by the spin-injection although we cannot exclude the possibility of heating-effect due to non-polarized quasiparticle injection completely.

4. Conclusion

We fabricated the in-plane spin-injection devices using thin flake BSCCO single crystals. The critical current of the sample decreases with increasing the spin-injection current. Magnetic field dependence of voltage showed hysteresis and the dip structure. It is found that spin injection into thin flake BSCCO affects superconductivity in surface of BSCCO transport.

References

[1] Vas'ko V A, Larkin V A, Kraus P A, Nikolaev K R, Grupp D E, Nordman C A and Goldman A M, 1997, *Phys. Rev. Lett.* **78** 1134
[2] Fratila L, maurin I, Dubourdieu C and villĕgir J C 2005 *Appl. Phys. Lett.* **86** 122505
[3] Dong Z W, Ramesh R, Venkatesan T, Johnson M, Chen Z Y, Pai S P, Talyansky V, Sharma R P, Shreekala R, Lobb C J and Greene R L 1997 *Appl. Phys. Lett.* **71** 1718
[4] Soltan S, Albrecht J and Habermeier H-U 2004 *Phys. Rev. B* **70**, 144517
[5] Yeh N-C, Vasquez R P, Fu C C, Samoilov, Li Y, Vakili K 1999 *Phys. Rev. B* **60**, 10522

IOP Conf. Series: Journal of Physics: Conf. Series **871** (2017) 012013 doi:10.1088/1742-6596/871/1/012013

[6] Plausinaitiene V, Abrutis A, Vengalis B, Butkute R, Senateur J P, Saltyte Z and Kubilius V 2001 *Physica C* **351**, 13
[7] Kleiner R, Steinmeyer F, Kunkel G and Müller P 1992 *Phys. Rev. Lett.* **68** 2394
[8] Oya G, Aoyama N, Irie A, Kishida S and Tokutaka H 1992 *Jpn. J. Appl. Phys.* **31** L829
[9] Irie A, Otsuka M, Murata K, Yamaki K and Kitamura M 2015 *IEEE Trans. Appl. Supercond.* **25** 3
[10] Irie A, Sakakibara M and Oya G 1994 *IEICE Trans. Electron.* **E77-C** 1191
[11] Ji Y, Hoffmann A, Jiang J S and Bader S D 2004 *Appl. Phys. Lett.* **85** 6218

Band-renormalization effects on antiferromagnetism and d-wave superconductivity in two-dimensional t-J model

Ryo Sato and Hisatoshi Yokoyama
Department of Physics, Tohoku University, Sendai 980-8578, Japan

E-mail: satoryo@cmpt.phys.tohoku.ac.jp

Abstract. Recent studies reported that antiferromagnetic (AF) states become stable and *d*-wave superconducting (*d*-SC) states are almost excluded in underdoped strongly correlated Hubbard models in two dimensions. In a viewpoint of many-body variational theory, this result stems from a band-renormalization effect (BRE) in AF states. Here, using a variational Monte Carlo method, BRE on AF and *d*-SC states are studied in the *t*-*t*'-*J* model, in which *d*-SC (AF) states have been considered to be more stable (fragile) than in the Hubbard model. Many features are qualitatively similar to the Hubbard model including effectiveness of BRE and the existence of a Lifshitz transition in AF states. However, a *d*-SC state becomes more stable than an AF state widely in the underdoped regime for J/t=0.3, in contrast to the Hubbard model.

1. Introduction

Until a few years ago, it had been considered that the strongly correlated Hubbard model yields results consistent with the features of cuprate superconductors (SCs). However, recent studies using various refined methods [1], including ours [2], has argued that antiferromagnetic (AF) states (and phase separation) widely prevail in the underdoped regime of two-dimensional Hubbard (*t*-*t*'-*U*) model for $U \gtrsim W$ (W: band width), and homogeneous $d_{x^2-y^2}$-wave superconductivity (*d*-SC) hardly appears. This outcome apparently disagrees with the features of cuprates. Thus, it is urgent to clarify the origin of this discrepancy.

In the preceding study [2], we showed using a variational Monte Carlo (VMC) method that the above results are reached by taking account of band-renormalization effects (BRE) in the AF state. Meanwhile, it has been reported in the *t*-*J* model, another reduced model of cuprate SCs, that AF states are rapidly destabilized by doping and *d*-SC states becomes stable in a wide range of doping rate ($\delta = 1 - N_e/N_s$; N_e: number of electrons, N_s: number of sites) [3, 4, 5]. The *t*-*J* model seems more favorable for *d*-SC than the Hubbard model. However, in these VMC studies, BRE was not introduced. Thus, the purpose of this study is to check whether the *t*-*J* model preserves such features favorable to *d*-SC even by introducing BRE, and to grasp to what extent the properties of the *t*-*J* model are inherited from those of the Hubbard model.

2. Model and Method

We consider the *t*-*J* model on a square lattice with diagonal hopping,

$$\begin{aligned}\mathcal{H} &= \mathcal{H}_t + \mathcal{H}_{t'} + \mathcal{H} \\ &= -t\sum_{\langle i,j\rangle\sigma}\left(\tilde{c}^\dagger_{i\sigma}\tilde{c}_{j\sigma} + \mathrm{H.c.}\right) - t'\sum_{(i,j)\sigma}\left(\tilde{c}^\dagger_{i\sigma}\tilde{c}_{j\sigma} + \mathrm{H.c.}\right) + J\sum_{\langle i,j\rangle}\left(\mathbf{S}_i\cdot\mathbf{S}_j - \frac{1}{4}\tilde{n}_i\tilde{n}_j\right), \qquad (1)\end{aligned}$$

where $c_{i\sigma}$ is an operator which annihilates an electron of spin σ at site i, $\tilde{c}_{i\sigma} = c_{i\sigma}(1-n_{i-\sigma})$ with $n_{i\sigma} = c^\dagger_{i\sigma}c_{i\sigma}$, $\mathbf{S}_i = \frac{1}{2}\sum_{\alpha,\beta} c^\dagger_{i\alpha}\sigma_{\alpha\beta}c_{i\beta}$ (σ: Pauli matrix), $\tilde{n}_i = \sum_\sigma \tilde{c}^\dagger_{i\sigma}\tilde{c}_{i\sigma}$, and $\langle i,j\rangle$ and (i,j) indicate a nearest-neighbor and next-nearest-neighbor pair, respectively. We use t and the lattice spacing as the units of energy and length, respectively. As a value plausible to cuprates, we use J/t =0.3 in this paper.

To this model, we apply a VMC method, in which variational expectation values with respect to many-body trial wave functions are computed in a virtually exact manner, using a Monte Carlo algorithm; local correlation factors are treated exactly. As a trial wave function, we adopt Jastrow-type ones [6]: $|\Psi\rangle \equiv \mathcal{P}_\mathrm{J}\mathcal{P}_\mathrm{G}|\Phi\rangle$, where$\Phi$ is a Hartree-Fock-type one-body state and $\mathcal{P}_\mathrm{J}$ and $\mathcal{P}_\mathrm{G} = \prod_j(1-n_{j\uparrow}n_{j\downarrow})$ are intersite and onsite correlation factors, respectively. For $\mathcal{P}_\mathrm{J}$, we consider both charge density and spin correlations between nearest-neighbor sites: $\mathcal{P}_\mathrm{J} = \mathcal{P}_\mathrm{h}\mathcal{P}_\mathrm{s}$, where

$$\mathcal{P}_\mathrm{h} = \prod_{\langle i,j\rangle}[1-(1-\alpha)h_ih_j] \tag{2}$$

which $h_j = (1-n_{j\uparrow})(1-n_{j\downarrow})$, and

$$\mathcal{P}_\mathrm{s} = \prod_{\langle i,j\rangle}[1-(1-\beta_1)(n_{i\uparrow}n_{j\uparrow}+n_{i\downarrow}n_{j\downarrow})][1-(1-\beta_2)(n_{i\uparrow}n_{j\downarrow}+n_{i\downarrow}n_{j\uparrow})] \tag{3}$$

In Eqs. (2) and (3), α, β_1 and β_2 are variational parameters. Using $\mathcal{P}_\mathrm{h}$ and $\mathcal{P}_\mathrm{s}$, we can assign distinct weights to all different nearest-neighbor bond configurations. $\mathcal{P}_\mathrm{s}$ plays a corrective role for the (often overestimated) AF order in Φ_AF discussed below.

For the one-body part $|\Phi\rangle$, we consider states with *d*-wave SC and AF orders independently similar to those used in Ref. [2]. A mixed state of the two orders will be discussed elsewhere. As a *d*-SC wave function, we use a canonical BCS wave function with fixed electron numberN_e:

$$|\Phi_{d\mathrm{-SC}}\rangle = \left(\sum_\mathbf{k} A(\mathbf{k})\, c^\dagger_{\mathbf{k}\uparrow}c_{\mathbf{k}\downarrow}\right)^{\frac{N_\mathrm{e}}{2}} |0\rangle, \tag{4}$$

$$A(\mathbf{k}) = \frac{\Delta_d}{\varepsilon^\mathrm{SC}_\mathbf{k} - \varsigma + \sqrt{(\varepsilon^\mathrm{SC}_\mathbf{k}-\varsigma)^2 + (\Delta_d)^2}}, \tag{5}$$

where we assume a $d_{x^2-y^2}$ symmetry SC gap $\Delta_d = \Delta_\mathrm{SC}(\cos k_x - \cos k_y)$. Δ_SC and ς are variational parameters that correspond to the singlet gap and chemical potential, respectively, for $U/t \to 0$. We will explain BRE owing to $\varepsilon^\mathrm{SC}_\mathbf{k}$ shortly. As an AF state, we use

$$|\Phi_\mathrm{AF}\rangle = \prod_{\mathbf{k}\in\{\mathbf{k}_\mathrm{occ}\}} a^\dagger_{\mathbf{k}\uparrow}a^\dagger_{\mathbf{k}\downarrow}|0\rangle \tag{6}$$

with AF Hartree-Fock quasi –particle operators

$$a^\dagger_{\mathbf{k},\sigma} = \alpha_\mathbf{k}c^\dagger_{\mathbf{k},\sigma} + \mathrm{sgn}(\sigma)\beta_\mathbf{k}c^\dagger_{\mathbf{k+Q},\sigma} \quad (\mathbf{k} \in \text{magnetic BZ}) \tag{7}$$

$$a^\dagger_{\mathbf{k+Q},\sigma} = -\mathrm{sgn}(\sigma)\beta_\mathbf{k}c^\dagger_{\mathbf{k},\sigma} + \alpha_\mathbf{k}c^\dagger_{\mathbf{k+Q},\sigma} \quad (\mathbf{k+Q} \notin \text{magnetic BZ}), \tag{8}$$

where $\mathbf{Q} = (\pi,\pi)$, $\mathrm{sgn}(\sigma) = 1(-1)$ for $\sigma =\uparrow(\downarrow)$, and

$$\alpha_{\mathbf{k}}(\beta_{\mathbf{k}}) = \frac{1}{\sqrt{2}}\sqrt{1-(+)\frac{\varepsilon_{\mathbf{k}}^{\mathrm{AF}}}{\sqrt{(\varepsilon_{\mathbf{k}}^{\mathrm{AF}})^2+(\Delta_{\mathrm{AF}})^2}}}. \quad (9)$$

As emphasized in Ref. [2], it is vital, especially for Φ_{AF}, to take account of BRE owing to electron correlation, namely, energy dispersions $\varepsilon_{\mathbf{k}}^{\Lambda}$ (Λ =AF or *d*-SC) in $|\Phi_{\mathrm{AF}}\rangle$ and $|\Phi_{d\text{-SC}}\rangle$ should be also optimized. To this end, we fit $\varepsilon_{\mathbf{k}}^{\Lambda}$ to a form of tight-binding bands up to the three-step distance as

$$\varepsilon_{\mathbf{k}}^{\Lambda} = -2t(\cos k_x+\cos k_y) - 4t_1^{\Lambda}\cos k_x \cos k_y - 2t_2^{\Lambda}(\cos 2k_x + \cos 2k_y) \\ - 4t_3^{\Lambda}(\cos 2k_x \cos k_y + \cos k_x \cos 2k_y) - 2t_4^{\Lambda}(\cos 3k_x + \cos 3k_y). \quad (10)$$

Here, four hopping integrals t_j^{Λ} ($j = 1-4$) work as band-adjusting parameters that are optimized with the other variational parameters at the same time. The reason why almost all previous studies have not treated the optimization of $\varepsilon_{\mathbf{k}}^{\mathrm{AF}}$ lies in the difficulty in the minimization of E/t; E/t is not only extremely slowly changing but also discontinuous as a function of t_j^{AF} for finite-size systems. This discontinuous behavior originates in discontinuous changes of occupied
k points (or Fermi surface) as t_j^{AF} change. Therefore, in this paper, we optimize the parameters with a fixed occupied **k**-point set $\{\mathbf{k}_{\mathrm{occ}}\}$ and find the global energy minimum among various $\{\mathbf{k}_{\mathrm{occ}}\}$'s. Here, for simplicity, we create $\{\mathbf{k}_{\mathrm{occ}}\}$'s according to

$$\varepsilon_{\mathbf{k}} = -2t(\cos k_x + \cos k_y) - 4t_{\mathrm{p}}\cos k_x \cos k_y, \quad (11)$$

where t_{p} is varied in a moderate range [see Fig. 1 for instance]; note that t_{p} is used only for creating $\{\mathbf{k}_{\mathrm{occ}}\}$ and is independent of t_1 in Eq. (10) as well as t' in Eq. (1). We confirmed that the optimized result $\{\mathbf{k}_{\mathrm{occ}}\}$ using $\varepsilon_{\mathbf{k}}$ in Eq. (11) becomes identical to that using a type of $\varepsilon_{\mathbf{k}}^{\mathrm{AF}}$ in Eq. (10) within the range of the system size and model parameters used in this paper.

Thus, because the variational energy E/t becomes a smooth function of variational parameters, optimization methods using differentiation become applicable such as the quasi-Newton method [7] and the stochastic reconfiguration method [8]. We mainly used the latter. The VMC calculations were performed on $N_{\mathrm{s}}(= L\times L)$-site lattices with $L = 10-16$ and the periodic-antiperiodic boundary conditions. Typically, 2.5×10^5 samples are used in estimating various physical quantities.

3. Results and discussions

To begin with, we discuss the optimization of AF state. Figure 1 shows the total energy per site E/t as a function of $\{\mathbf{k}_{\mathrm{occ}}\}$-determining parameter t_{p} for two values of doping rate δ. E/t discontinuously changes when $\{\mathbf{k}_{\mathrm{occ}}\}$ changes. Because t_{p} determines only $\{\mathbf{k}_{\mathrm{occ}}\}$, E/t is constant in a range of the same $\{\mathbf{k}_{\mathrm{occ}}\}$. The other parameters are optimized in each at range of t_{p}. For each δ, data for seven values of $t'/t(-0.3\le t'/t\le 0.3)$ are shown. The solid (open) arrows indicate the optimal range of t_{p}/t for positive (negative) values of t'/t. The range of t_{p} or $\{\mathbf{k}_{\mathrm{occ}}\}$ that gives the lowest energy is primarily determined by the sign of t'/t and not by the absolute value of t'/t. For a negative t'/t, the minimal E/t is given in narrow ranges of $-0.2\le t_{\mathrm{p}}/t\le -0.1$) for both doping rates. On the other hand, for a positive t'/t, the minimal E/t is given in relatively wide ranges including $t'/t = 0$. Thus, the renormalized $\{\mathbf{k}_{\mathrm{occ}}\}$ is distinct between the two cases. E/t is greatly improved compared to the energy without BR at $t_{\mathrm{p}} = t'$, especially for a large $|t'/t|$ ($\Delta E\sim 0.01$).

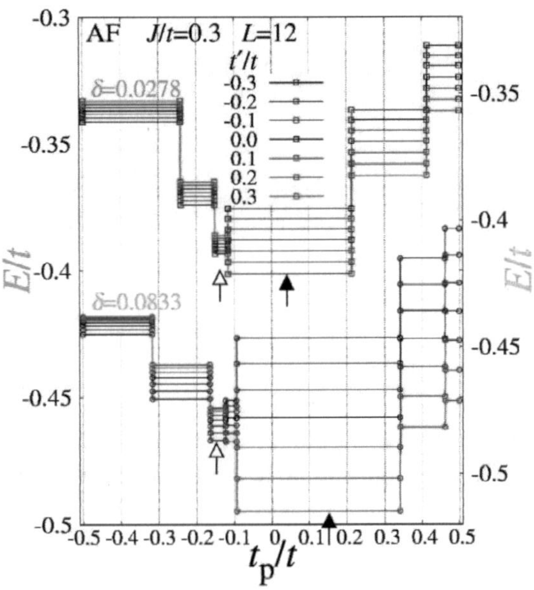

Figure 1. Variational energy per site of the AF state is plotted as a function of the $\{\mathbf{k}_{\rm occ}\}$-determining parameter $t_{\rm p}$for two doping rates. The scale for $\delta = 0.0278$ (0.00833) is given in the left (right) axis. In each δ, data for seven values of a model parameter t'/t are compared. Empty (solid) arrows show the lowest energy region of $t_{\rm p}/t$ for negative (positive) values of t'.

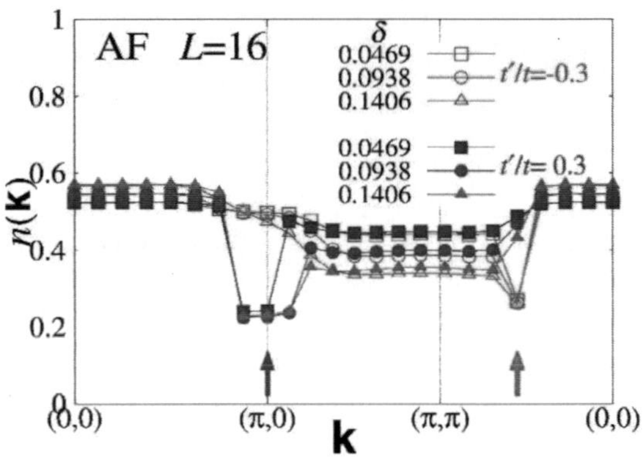

Figure 2. The momentum distribution functions along the path $(0,0)-(\pi,0)-(\pi,\pi)-(0,0)$ are compared between $t'/t = -0.3$ (red, type II) and 0.3 (blue, type I). A Lifshitz transition occurs at $t'/t{\sim}0$. For each value of t'/t, data of three values of doping rate are plotted. The arrows indicate the centers of pocket Fermi surfaces for the two t'/t values J/t=0.3.

This different behavior according to the sign of t'/t is probably related to the Lifshitz transition found in strongly correlated AF states for the Hubbard model[2]. Next, we discuss this point. As mentioned in Ref. [2], in the Hubbard model, a pocket-like Fermi surface centered at $(\pi/2,\pi/2)$ for $t' < t'_L$ (we called it type II AF) in doped AF states is switched to a one centered at $(\pi,0)$ (similarly type I AF) at $t'_L/t{\sim}-0.05$ as t'/t increases. Actually, a corresponding difference in the position of pocket Fermi surface is observed between hole-doped and electron-doped cuprates by angle resolved photoemission spectroscopy experiments [9]. We find a corresponding Lifshitz transition in thd *t-J* model at $t'_L/t{\sim}0$, where various quantities exhibit anomalous behavior (not shown). In Fig. 2, we show a momentum distribution function $n(\mathbf{k})$ along the path $(0,0)-(\pi,0)-(\pi,\pi)-(0,0)$ for $t'/t = 0.3(-0.3)$ as a typical case of the type I (II) AF state. It shows that a pocket Fermi surface centered at $(\pi/2,\pi/2)$ $[(\pi,0)]$ appears for a negative [positive] value of t'/t, irrespective of the doping rate. This aspect already arises in theories for weak correlations [2], but the sharpness of the transition becomes conspicuous for strongly correlated cases. The present result in the *t-J* model is consistent with that of the Hubbard model. Strictly speaking, the Lifshitz transition point in the *t-J* model is situated at $t'/t{\sim}0$ rather than ~-0.05; for $t'/t = 0$, the position of Fermi surface depends on the system size and doping rate. Anyway, a type I (II) AF state is realized for $t'/t > 0$ $(t'/t < 0)$ in the *t-J* model.

We turn to the behavior of order parameters in the AF and *d*-SC states: the staggered magnetization

$$m = \frac{2}{N_s}\left|\sum_j e^{i\mathbf{Q}\cdot\mathbf{r}_j}\langle S_j^z\rangle\right| \tag{12}$$

and P_d, the nearest-neighbor pair correlation function at the farthermost distance $\mathbf{r} = (L/2, L/2)$ (for the definition of P_d, see Ref. [10]). In Fig. 3, we show δ dependence of m and P_d calculated using Ψ_{AF} and Ψ_{d-SC}, respectively, for some values of t'/t. The SC order parameter exhibits a domelike shape [3] and its area extends as t'/t decreases [5]. This t'/t dependence is understood as loss in kinetic energy (not shown). Note that the behavior of P_d changes only a little even if BRE is introduced, as seen for $t'/t = -0.3$ in Fig. 3. On the other hand, m is greatly affected by BRE. In previous studies (without BRE) [3], AF state is rapidly vanished as δ increases, as shown by solid symbols in Fig. 3. When BRE is introduced, however, the range of m greatly extends to the overdoped regime. The tendencies of d-SC and AF orders are basically consistent with those in the Hubbard model [2]. In both models, BRE is vital to stabilize, especially, AF states.

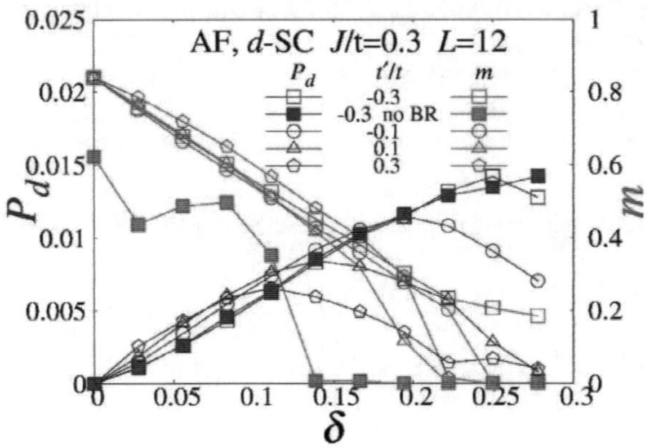

Figure 3. AF and d-SC order parameters, namely, the SC correlation function of nearest-neighbor singlet pairs at the farthermost distant point P_d (blue) and staggered magnetization m (red) are shown as functions of doping rate for several values of t'/t. Solid symbols indicate the data for cases without BR for $t'/t = -0.3$, where we set $t_p/t = t_1/t = -0.3$ and $t_2 = t_3 = t_4 = 0$. The system size is $L = 12$.

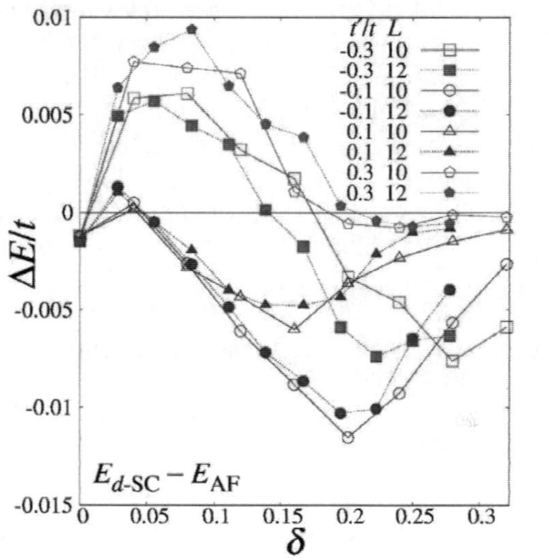

Figure 4. Energy difference between SC and AF states is shown as a function of doping rate. For $\Delta E > 0$, the AF state is more stable than the d-SC state Data of four values of t'/t are plotted; for each t'/t, date of two system sizes ($L = 10{,}12$) are compared.

Finally, we consider the energy difference between the AF and d-SC states:

$$\Delta E = E_{d-\mathrm{SC}} - E_{\mathrm{AF}}, \tag{13}$$

If $\Delta E/t < 0$ ($\Delta E/t > 0$), the d-SC (AF) state is more stable. Figure 4 shows the δ dependence of $\Delta E/t$. As mentioned, in the Hubbard model, the area in the model parameter space where the d-SC state is more stable than the AF state is limited to a narrow range in the overdoped density [2]. However, in the t-J model with $J/t = 0.3$, $\Psi_{d-\mathrm{SC}}$ is more stable than Ψ_{AF} in a much wider range of δ for small values of $|t'/t|$. This tendency is consistent with many previous studies. Thus, a phase diagram in the *t-J* model for $J/t = 0.3$ will be at least quantitatively different from that obtained for the Hubbard model with $U/t = 12$.

4. Summary

In this paper, we studied properties of an AF and *d*-SC states independently in the *t-J* model, for comparing with those in the Hubbard model [2]. We use a VMC method which takes account of BRE. The AF (*d*-SC) state is markedly (a little) stabilized by BRE. In the AF state, a Lifshitz transition was found at $t'/t{\sim}0$, corresponding to that in the Hubbard model. In contrast to the Hubbard model with $U/t = 12$, the *d*-SC state becomes more stable in a wide area in the underdoped regime for $J/t = 0.3$. It is important to pursue the origin of this stability of *d*-SC state. In future publications, we would like to consider J/t dependence, treat a mixed state ofAF and SC orders, and check the stability against inhomogeneous phases in the *t-J* model.

References

[1] Otsuki J, Hafermann H and Lichtenstein A I 2014 *Phys. Rev. B* **90** 235132; Misawa T and Imada M 2014 *Phys. Rev. B* **90** 115137; Zheng B-X, Chan G K-L 2016 *Phys. Rev. B* **93** 035126.
[2] Sato R and Yokoyama H 2016 *J. Phys. Soc. Jpn.* **85** 074701.
[3] Yokoyama H and Shiba H 1988 *J. Phys. Soc. Jpn.* **57** 2482; Yokoyama H and Ogata M 1996 *J. Phys. Soc. Jpn.* **65** 3615.
[4] Gros C 1988 *Phys. Rev. B* **38** 931.
[5] Shih C T, Lee T K, Eder R, Mou C -Y and Chen Y C 2004 *Phys. Rev. Lett.* **92** 227002.
[6] Gutzwiller M C 1963 *Phys. Rev. Lett.* **10** 159; Jastrow R 1955 *Phys. Rev.* **98** 1479; Yokoyama H and Shiba H, 1990 *J. Phys. Soc. Jpn.* **59** 3669.
[7] Umrigar C J, Wilson K G, Wilkins J W 1998 *Phys. Rev. Lett.* **60** 1719.
[8] Sorella S 2001 Phys. Rev. B 64 024512.
[9] For instance, Ronning F et al. 1998 *Science* **282** 2067; Armitage N P et al. 2002 *Phys. Rev. Lett.* **88** 257001.
[10] Yokoyama H, Ogata M, Tanaka Y, Kobayashi K and Tsuchiura H 2013 *J. Phys. Soc. Jpn.* **82** 014707.

Magnetsim, Fluctuations and Mechanism of High-Temperature Superconductivity

[1]Takashi Yanagisawa, [1]Izumi Hase, [2]Mitake Miyazaki, [1]Kunihiko Yamaji

[1]Electronics and Photonics Research Institute, National Institute of Advanced Industrial Science and Technology (AIST), Tsukuba Central 2, 1-1-1 Umezono, Tsukuba 305-8568, Japan
[2]Hakodate National College of Technology, 14-1 Tokura, Hakodate 042-8501, Japan

E-mail: t-yanagisawa@aist.go.jp

Abstract. We investigate the ground state of the two-dimensional d-p model (three-band Hubbard model) by using a variational Monte Carlo method. The superconducting condensation energy is evaluated for the Gutzwiller-BCS wave function. We show that there is a crossover between strongly and weakly correlated regions as the level difference between d and p orbitals increases. The gap function and the condensation energy can be large in the crossover region. This result indicates a possibility of high-temperature superconductivity in the two-dimensional d-p model.

1. Introduction

The research of mechanism of high-temperature superconductivity has attracted much attention since the discovery of cuprate high-temperature superconductors[1, 2]. Because it has been established that the Cooper pairs of cuprate high-temperature superconductors have the d-wave symmetry, the electron correlation plays an important role for the appearance of superconductivity. It is primarily important to clarify the phase diagram of electronic states in the CuO_2 plane contained commonly in cuprate high-temperature superconductors. The CuO_2 plane consists of oxygen atoms and copper atoms. The electronic model for this plane is the d-p model (or three-band Hubbard model)[3, 4, 5, 6, 7, 8, 9, 10, 11, 12, 13, 14, 15, 16, 17]. An interaction with a large energy scale is necessary for the realization of high-temperature superconductivity. It has been argued whether the on-site Coulomb repulsive interaction induces superconductivity for the two-dimensional single-band Hubbard model[18, 19, 20, 21, 22, 23, 24, 25, 26, 27, 28, 29, 30] and ladder model[31, 32, 33].

It is also expected that there is a superconducting phase in the two-dimensional d-p model. It is, however, difficult to obtain a clear evidence of superconductivity in the two-dimensional d-p model because the energy gain by introducing the superconducting gap is very small for the Gutzwiller-BCS wave function[8, 13]. Does this mean that there is no superconductivity in the d-p model? The purpose of this paper is to reexamine the stability of the superconducting state in the d-p model by using the variational Monte Carlo method. Numerous works performed so far have focused on the region where the energy difference between d and p levels are located closely each other. This is because there has been a suggestion that the superconducting critical temperature T_c increases as the d-p level difference decreases[34].

2. Hamiltonian

The three-band model that explicitly includes oxygen p and copper d orbitals contains the parameters U_d, U_p, t_{dp}, t_{pp}, t'_d, ϵ_d and ϵ_p. Our study is within the hole picture and the Hamiltonian is written as

$$\begin{aligned} H_{dp} &= \epsilon_d \sum_{i\sigma} d^\dagger_{i\sigma} d_{i\sigma} + \epsilon_p \sum_{i\sigma} (p^\dagger_{i+\hat{x}/2\sigma} p_{i+\hat{x}/2\sigma} + p^\dagger_{i+\hat{y}/2\sigma} p_{i+\hat{y}/2\sigma}) \\ &+ t_{dp} \sum_{i\sigma} [d^\dagger_{i\sigma}(p_{i+\hat{x}/2\sigma} + p_{i+\hat{y}/2\sigma} - p_{i-\hat{x}/2\sigma} - p_{i-\hat{y}/2\sigma}) + \text{h.c.}] \\ &+ t_{pp} \sum_{i\sigma} [p^\dagger_{i+\hat{y}/2\sigma} p_{i+\hat{x}/2\sigma} - p^\dagger_{i+\hat{y}/2\sigma} p_{i-\hat{x}/2\sigma} \\ &- p^\dagger_{i-\hat{y}/2\sigma} p_{i+\hat{x}/2\sigma} + p^\dagger_{i-\hat{y}/2\sigma} p_{i-\hat{x}/2\sigma} + \text{h.c.}] \\ &+ t'_d \sum_{\langle\langle ij\rangle\rangle\sigma} (d^\dagger_{i\sigma} d_{j\sigma} + \text{h.c.}) + U_d \sum_i d^\dagger_{i\uparrow} d_{i\uparrow} d^\dagger_{i\downarrow} d_{i\downarrow}. \end{aligned} \quad (1)$$

$d_{i\sigma}$ and $d^\dagger_{i\sigma}$ represent the operators for the d hole. $p_{i\pm\hat{x}/2\sigma}$ and $p^\dagger_{i\pm\hat{x}/2\sigma}$ denote the operators for the p holes at the site $R_{i\pm\hat{x}/2}$, and in a similar way $p_{i\pm\hat{y}/2\sigma}$ and $p^\dagger_{i\pm\hat{y}/2\sigma}$ are defined. t_{dp} is the transfer integral between adjacent Cu and O orbitals and t_{pp} is that between nearest p orbitals. $\langle\langle ij\rangle\rangle$ denotes a next nearest-neighbor pair of copper sites. t'_d was introduced to reproduce the Fermi surface[35] in several cuprate superconductors such as $Bi_2Sr_2CaCu_2O_{8+\delta}$[36] and $Tl_2ba_2CuO_{6+\delta}$[37]. U_d is the strength of the on-site Coulomb repulsion between d holes. In this paper we neglect U_p among p holes because U_p is small compared to U_d[38, 39, 40, 41]. In the low-doping region, U_p will be of minor importance because the p-hole concentration is small[42]. The parameter values were estimated as, for example, $U_d = 10.5$, $U_p = 4.0$ and $U_{dp} = 1.2$ in eV[39] where U_{dp} is the nearest-neighbor Coulomb interaction between holes on adjacent Cu and O orbitals. We neglect U_{dp} because U_{dp} is small compared to U_d. We use the notation $\Delta_{dp} = \epsilon_p - \epsilon_d$. The number of sites is denoted as N, and the total number of atoms is $N_a = 3N$. The energy unit is given by t_{dp}.

3. Superconducting wave function

We examine the superconducting ground state of the two-dimensional d-p model by using the variational Monte Carlo method. The wave function is the Gutzwiller-projected wave function given as

$$\psi_G = P_G \psi_0, \quad (2)$$

where P_G is the Gutzwiller operator to control the double occupancy of d holes:

$$P_G = \prod_i [1 - (1-g) n_{di\uparrow} n_{di\downarrow}]. \quad (3)$$

g is a variational parameter in the range from 0 to unity. $n_{di\sigma} = d^\dagger_{i\sigma} d_{i\sigma}$ is the number operator for d holes. ψ_0 is the Fermi sea where the lowest band is occupied up to the Fermi energy μ. To represent a superconducting state, we take ψ_0 as the BCS wave function

$$\psi_{BCS} = \prod_{\mathbf{k}} (u_{\mathbf{k}} + v_{\mathbf{k}} \alpha^\dagger_{\mathbf{k}\uparrow} \alpha^\dagger_{-\mathbf{k}\downarrow}) |0\rangle, \quad (4)$$

where $\alpha^\dagger_{\mathbf{k}\sigma}$ indicates the creation operator of the state in the lowest band with the momentum $\mathbf{k}$ which is represented by a linear combination of d and p electron operators $d^\dagger_{\mathbf{k}\sigma}$ and $p^\dagger_{\mathbf{k}\sigma}$. The

BCS parameters $u_{\mathbf{k}}$ and $v_{\mathbf{k}}$ are given by the ratio $v_{bfk}/u_{\mathbf{k}} = \Delta_{\mathbf{k}}/(\xi_{\mathbf{k}}+\sqrt{\xi_{\mathbf{k}}^2+\Delta_{\mathbf{k}}^2})$ where $\xi_{\mathbf{k}}$ is the dispersion relation of the lowest band. We assume the d-wave symmetry $\Delta_{\mathbf{k}} = \Delta(\cos k_x - \cos k_y)$ and we regard Δ as the superconducting gap. Δ is a variational parameter that is optimized to give the lowest ground energy. The Projected-BCS wave function is written as

$$\psi_S = P_{N_e} P_G \psi_{BCS}, \tag{5}$$

where P_{N_e} is a projection operator which extracts only the states with a fixed total hole number. In actual evaluations, we can use the method in Ref.[13]. We can also use the wave function obtained by performing the particle-hole transformation for down-spin holes[30, 43]:

$$P_G \psi_{BCS} = P_G \prod_{\mathbf{k}} (u_{\mathbf{k}} \beta_{\mathbf{k}}^\dagger + v_{\mathbf{k}} \alpha_{\mathbf{k}}^\dagger)|\tilde{0}\rangle, \tag{6}$$

where $\beta_{\mathbf{k}}^\dagger = \alpha_{-\mathbf{k}\downarrow}$ and $\alpha_{\mathbf{k}}^\dagger = \alpha_{\mathbf{k}\uparrow}^\dagger$. $|\tilde{0}\rangle$ denotes the vacuum for newly defined α and β particles satisfying $\alpha_{\mathbf{k}}|\tilde{0}\rangle = \beta_{\mathbf{k}}|\tilde{0}\rangle = 0$.

We use the variational Monte Carlo method in the evaluation of the ground-state energy[8]. We calculate the superconducting condensation energy $\Delta E = E(\Delta \to 0) - E(\Delta_{opt})$ where $E(\Delta)$ is the ground-state energy with the gap Δ and Δ_{opt} is the value of optimized superconducting gap Δ.

We show the condensation energy as a function of Δ in Fig.1 where calculations were carried out on 8×8 lattice with 192 atoms. The parameters that we used are $t_{pp} = 0.4$, $u_d = 10$, $U_p = 0$ and $t'_d = 0$ in units of t_{dp}. We put 76 holes on the lattice and we set $\Delta_{dp} = \epsilon_p - \epsilon_d = 2$, 4 and 8. As shown in Fig.1, the condensation energy ΔE becomes extremely large as the level difference Δ_{dp} increases. When $\Delta_{dp} = 2$, ΔE is very small and thus it is not easy to determine the condensation energy by numerical calculations. In contrast, surprisingly, ΔE turns out to be very large when Δ_{dp} is large. This indicates that the superconducting state is more stabilized in the strongly correlated region in accordance with the phenomenon in the single-band Hubbard model[30]. The existence of high-temperature superconductivity is suggested when the level difference Δ_{dp} is large.

Let us investigate the behavior of ΔE when the level difference Δ_{dp} increases further. We show the optimized gap amplitude Δ as a function of Δ_{dp} in Fig.2 where the band parameters are the same as in Fig.1. The figure indicates that there is a maximum in the optimized gap function as a function of Δ_{dp} in the large-Δ_{dp} region.

4. Discussion

The condensation energy ΔE exhibits a maximum with a large value, indicating that there occurs a crossover between strongly and weakly correlated regions. This crossover is quite similar to that in the two-dimensional Hubbard model[30]. A large fluctuation presumably exists in the crossover region. This indicates a possibility of high-temperature superconductivity in the d-p model. A crossover from weakly to strongly coupled systems is universal phenomenon that exists ubiquitously in the world. For example, the Kondo effect exhibits a universal logarithmic anomaly that appears as a crossover when the system approaches the strong coupling region (low temperature region) from the weak coupling region (high temperature region)[44, 45, 46, 47, 48, 49]. A two-impurity Kondo problem also shows a crossover[50, 51, 52]. There may a class of phenomena that shows a crossover between weakly and strongly interacting regions. This class may include, for example, QCD[53], BCS-BEC crossover[54], Hubbard model, sine-Gordon model[55, 56, 57] and the Kosterlitz-Thouless transition.

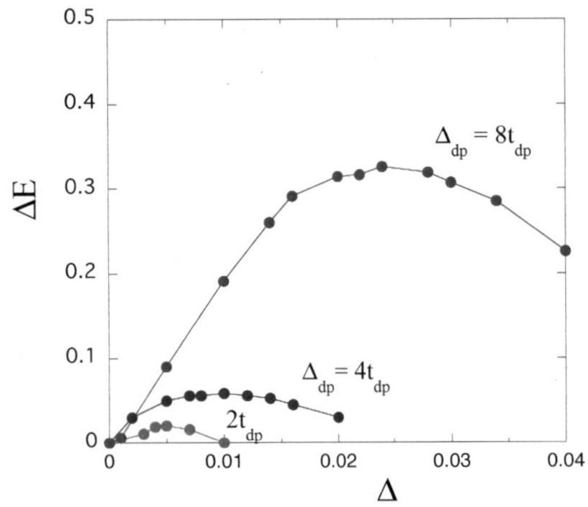

Figure 1. Superconducting condensation energy as a function of the gap function for $\Delta_{dp} = 2$, 4 and 8 in units of t_{dp}. Numerical calculations were carried on 8×8 lattice with 76 holes. The band parameters are $t_{pp} = 0.4$, $t'_d = 0$, $U_d = 10$ and $U_p = 0$. We used the Gutzwiller-BCS wave function in eq.(5) where the hole number is fixed to be 76.

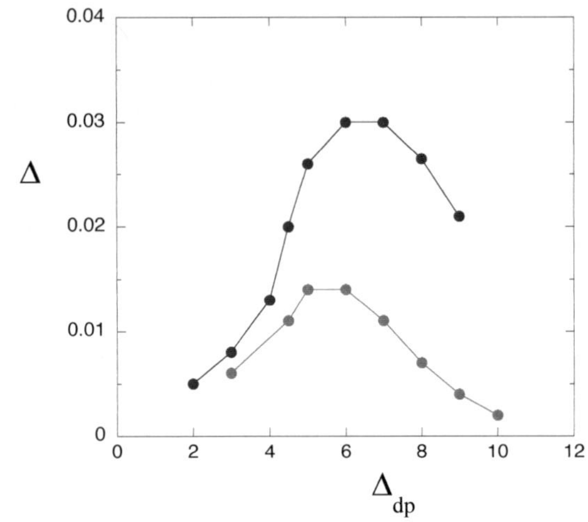

Figure 2. Superconducting gap as a function of the level difference $\Delta_{dp} \equiv \epsilon_p - \epsilon_d$. Numerical calculations were performed on 8×8 lattice with 76 holes. The band parameters are $t_{pp} = 0.4$, $t'_d = 0$, $U_d = 10$ and $U_p = 0$. The upper curve was calculated by using the Gutzwiller-BCS function in eq.(5) and the lower one is by the wave function in eq.(6).

5. Summary

We investigated the ground state of the two-dimensional Hubbard model on the basis of the variational Monte Carlo method. The superconducting condensation energy was calculated by using the Gutzwiller-projected BCS wave function. Although the condensation energy ΔE is very small when the level difference between d and p orbitals is small, ΔE increases as the level difference increases, suggesting a possibility of high-temperature superconductivity. It would be better to examine the d-p model by using improved wave functions[30, 35, 58, 59, 60, 61]. The Mott transition based on the d-p model was investigated with improved wave function[35]. A crossover for the antiferromagnetic correlation can be explored on the basis of improved wave functions.

References

[1] J. G. Bednorz and K. A. Müller 1986 Z. Phys. B64 189.
[2] *The Physics of Superconductor* (Vol.I and Vol.II) edited by K. H. Bennemann and J. B. Ketterson (Springer-Verlag, Berlin, 2003).
[3] V. J. Emery 1987 Phys. Rev. Lett. 58 2794.
[4] J. E. Hirsch, E. Y. Loh, D. J. Scalapino and S. Tang 1989 Phys. Rev. B39 243.
[5] R. T. Scalettar, D. J. Scalapino, R. L. Sugar, and S. R. White 1991 Phys. Rev. B44 770.
[6] A. Oguri, T. Asahata and S. Maekawa 1994 Phys. Rev. B49 6880.
[7] S. Koikegami and K. Yamada 2000 J. Phys. Soc. Jpn. 69 768.
[8] T. Yanagisawa, S. Koike and K. Yamaji 2001 Phys. Rev. B64 184509.
[9] S. Koikegami and T. Yanagisawa 2001 J. Phys. Soc. Jpn. 70 3499.
[10] T. Yanagisawa, S. Koike and K. Yamaji 2003 Phys. Rev. B67 132408.
[11] S. Koikegami and T. Yanagisawa 2003 Phys. Rev. B67 134517.
[12] S. Koikegami and T. Yanagisawa 2006 J. Phys. Soc. Jpn. 75 034715.
[13] T. Yanagisawa, M. Miyazaki and K. Yamaji 2009 J. Phys. Soc. 78 013706.

[14] C. Weber, A. Lauchi, F. Mila and T. Giamarchi 2009 Phys. Rev. Lett. 102 017005.
[15] B. Lau, M. Berciu and G. A. Sawatzky 2011 Phys. Rev. Lett. 106 036401.
[16] C. Weber, T. Giamarchi and C. M. Varma 2014 Phys. Rev. Lett. 112 117001.
[17] S. Tamura and H. Yokoyama 2016 Phys. Procedia 81 5.
[18] S. Zhang, J. Carlson and J. E. Gubernatis 1997 Phys. Rev. B55 7464.
[19] S. Zhang, J. Carlson and J. E. Gubernatis 1997 Phys. Rev. Lett. 78 4486.
[20] T. Nakanishi, K. Yamaji and T. Yanagisawa 1997 J. Phys. Soc. Jpn. 66 294.
[21] K. Yamaji, T. Yanagisawa, T. Nakanishi and S. Koike 1998 Physica C304 225.
[22] K. Yamaji, T. Yanagisawa and S. Koike 2000 Physica B284-288 415.
[23] N. Bulut 2002 Advances in Phys. 51 1587.
[24] H. Yokoyama, Y. Tanaka, M. Ogata and H. Tsuchiura 2004 J. Phys. Soc. Jpn. 73 1119.
[25] H. Yokoyama, M. Ogata and Y. Tanaka 2006 J. Phys. Soc. Jpn. 75 114706.
[26] T. Aimi and M. Imada 2007 J. Phys. Soc. Jpn. 76 113708.
[27] T. Yanagisawa 2008 New J. Phys. 10 023014.
[28] T. Yanagisawa 2013 New J. Phys. 15 033012.
[29] T. Yanagisawa, M. Miyazaki and K. Yamaji 2014 J. Mod. Phys. 4 33.
[30] T. Yanagisawa 2016 J. Phys. Soc. Jpn. 85 114707.
[31] R. M. Noack, S. R. White and D. J. Scalapino 1995 EPL 30 163.
[32] T. Yanagisawa, Y. Shimoi and K. Yamaji 1995 Phys. Rev. B52 R3860.
[33] T. Nakano, K. Kuroki and S. Onari 2007 Phys. Rev. B76 014515.
[34] G.-Q. Zheng, Y. Kitaoka, K. Ishida amd K. Asayama 1995 J. Phys. Soc. Jpn. 64 2524.
[35] T. Yanagisawa and M. Miyazaki 2014 EPL 107 27004.
[36] K. McElroy et al. 2003 Nature 422 592.
[37] N. E. Hussey et al. 2003 Nature 425 814.
[38] C. Weber, K. Haule and G. Kotliar: Phys. Rev. B78 (2008) 134519.
[39] M. S. Hybertsen, M. Schluter and N. E. Christensen: Phys. Rev. B39 (1989) 9028.
[40] H. Eskes, G. A. Sawatzky and L. F. Feiner: Physica C160 (1989) 424.
[41] A. K. McMahan, J. F. Annett and R. M. Martin: Phys. Rev. B42 (1990) 6268.
[42] H. Eskes and G. Sawatzky: Phys. Rev. B43 (1991) 119.
[43] T. Yanagisawa, S. Koike and K. Yamaji 1999 J. Phys. Soc. Jpn. 68 3608.
[44] J. Kondo 2012 *The Physics of Dilute Magnetic Alloys* (Cambridge University Press, Cambridge).
[45] K. G. Wilson 1975 Rev. Mod. Phys. 47 773.
[46] Y. Nagaoka 1965 Phys. Rev. 138 A1112.
[47] J. Zittartz and E. Müller-Hartmann 1968 Z. Phys. 212 380.
[48] T. Yanagisawa 2012 J. Phys. Soc. Jpn. 81 094713.
[49] T. Yanagisawa 2015 J. Phys. Soc. Jpn. 84 074705.
[50] C. Jayaprakash, H. R. Krishna-murthy and J. W. Wilkins 1981 Phys. Rev. Lett. 47 737.
[51] B. A. Jons and C. M. Varma 1989 Phys. Rev. B40 324.
[52] T. Yanagisawa 1989 J. Phys. Soc. Jpn. 60 29.
[53] R. K. Ellis, W. J. Stirling and B. R. Webber 1996 *QCD and Collider Physics* (Cambridge University Press, Cambridge).
[54] P. Nozieres and S. Schmitt-Rink 1985 J. Low Temp. Phys. 59 195.
[55] R. Rajaraman 1989 *Solitons and Instantons* (North-Holland, Amsterdam).
[56] J. Solyom 1979 Adv. Phys. 28 201.
[57] T. Yanagisawa 2016 EPL 113 41001.
[58] T. Yanagisawa, S. Koike and K. Yamaji 1998 J. Phys. Soc. Jpn. 67 3867.
[59] T. Yanagisawa 2007 Phys. Rev. B75 224503.
[60] D. Baeriswyl, D. Eichenberger and M. Menteshashvii 2009 New J. Phys. 11 075010.
[61] R. Sato and H. Yokoyama 2016 J. Phys. Soc. Jpn 85 074701.

IOP Conf. Series: Journal of Physics: Conf. Series **871** (2017) 012016 doi:10.1088/1742-6596/871/1/012016

Two-gap features revealed by specific heat measurements in FeSe

J T Chen*, Y Sun, T Yamada, S Pyon, and T Tamegai

Department of Applied Physics, The University of Tokyo, 7-3-1 Hongo, Bunkyo-ku, Tokyo 113-8656, Japan

chenjt910314@yahoo.co.jp

Abstract. The symmetry of superconducting gap is a key to understand the mechanism of superconductivity. FeSe, which has the simplest structure among iron-based superconductors, is most suitable for the understanding of the superconducting mechanism. Specific heat measurements have been performed on a high-quality FeSe single crystal with $T_c \sim 8.8$ K and $\Delta C/\gamma_n T_c = 1.62$. A sharp jump of specific heat at T_c and the second drop around 1.2 K are observed, suggesting the presence of two superconducting gaps in FeSe. Furthermore, the electronic specific heat under zero field can be well fitted by the two-gap model. Besides, the kink structure observed in the magnetic-field dependence of the electronic specific heat coefficient $\gamma(H)$ at low temperatures also support the two-gap nature of FeSe.

1. Introduction

A central issue concerning the investigation of unconventional superconductors is their superconducting gap symmetry and structure. Among iron-based superconductors (IBSs), FeSe has attracted most attention recently. This is because the crystal structure of FeSe is regarded as representative of the entire family of IBSs [1]. Scanning tunneling microscopy (STM) measurements on FeSe single crystals reveal the presence of distinct peak and shoulder structures in the spectra, which indicates the presence of two superconducting gaps [2]. Furthermore, low-temperature thermal conductivity under magnetic fields has also proved that it has multi-gap nature [3]. On the other hand, the residual linear component of the thermal conductivity κ_0/T reveals contradicting results on the presence or absence of nodes in the gap structure [3,4]. Therefore, it is necessary to perform another bulk-sensitive sensitivity experiments on FeSe to solve these controversies. Specific heat measurements on FeSe have claimed that it belongs to two-gap superconductors [1,5]. In order to confirm it as well as figure out whether nodes exist in the gap or not, we performed low temperature specific heat measurements on a high-quality FeSe single crystal, which shows sharp transition width $\Delta T_c < 0.5$ K from susceptibility measurement and large residual resistivity ratio ~33 [6,7].

2. Experimental

High-quality FeSe single crystals were grown by the vapor transport method [8]. Fe and Se powders were thoroughly mixed by grinding them in a high-purity nitrogen-filled glove box for more than 30 min, and sealed in an evacuated quartz tube together with mixture of $AlCl_3$ and KCl powders. The quartz tube with chemicals was loaded into a horizontal tube furnace with one end heated up to ~400 ℃, while the other end was kept at ~250℃. After more than 35 days, single crystals with dimensions over 1 x 1 x 0.1 mm^3 can be obtained in the cold end. The weight of FeSe single crystal that was used in the specific heat measurement is over 2 mg with dimensions of $4\times2\times0.1$ mm^3. Specific heat was measured using relaxation method with physical property measurement system (PPMS, Quantum Design), equipped with ^{3}He option, which allowed us to measure specific heat down to 0.5 K.

3. Result and discussion

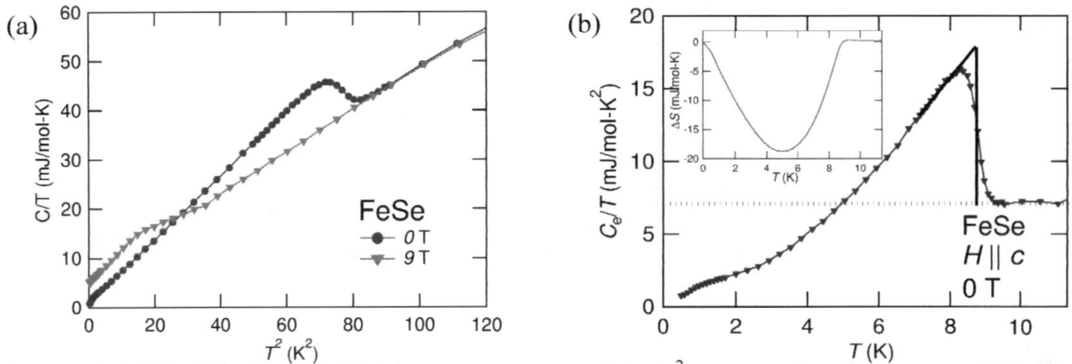

Figure 1. (a) Specific heat divided by temperature, C/T vs T^2, measured under zero and 9 T. (b) Zero field electronic specific heat (with the normal-state specific heat being subtracted) divided by temperature. The inset shows the entropy conservation required for a second-order phase transition.

Figure 1(a) shows specific heat of FeSe single crystal measured under 0 T and 9 T in the plot of C/T versus T^2. A clear jump associated with the superconducting transition is observed around 9 K under zero field, which is consistent with the susceptibility measurement. We fit the normal state specific heat C_n below 10 K by $C_n = \gamma_n T + C_{lat}(T)$, where $\gamma_n T$ is the normal electronic contribution and $C_{lat}(T) = \beta_3 T^3 + \beta_5 T^5 + \beta_7 T^7$ represents the phonon contribution. According to our fitting with the normal-state specific heat under 9 T, it yields that $\gamma_n = 6.85$ mJ/molK^2, $\beta_3 = 0.37$ mJ/molK^4, $\beta_5 = 0.001$ mJ/molK^6 and $\beta_7 = -5.7 \times 10^{-6}$ mJ/molK^8. The normalized specific heat jump at T_c, $\Delta C/\gamma_n T_c$ is estimated to be 1.62, which is larger than the weak-coupling value 1.43 of Bardeen-Cooper-Schrieffer (BCS) theory [9], implying that the superconductivity in FeSe is in the strong-coupling. Zero-field electronic specific heat C_e/T, is obtained from the subtraction of phonon terms, is shown in figure 1(b). Other than the specific heat jump at $T = 8.8$ K, a second drop at around 1.2 K is also observed. This is a typical behavior of a two-gap superconductor such as MgB_2 [10], $Lu_2Fe_3Si_5$ [11], suggesting the presence of the second smaller superconductor gaps in FeSe.

In order to get deeper understanding of the gap structure in FeSe, the data of zero-field electronic specific heat is first fitted by the following formula [12] based on the BCS theory,

$$C_{es} = \frac{4N(0)}{k_B T^2} \int_0^{\hbar\omega_D} d\varepsilon \left[\varepsilon^2 + \Delta^2 - \frac{T}{2} \frac{d\Delta^2}{dT} \right] (1 + e^{E/k_B T})^{-2} e^{E/k_B T},$$

where $N(0)$ is the density of state at the Fermi level, $E = (\varepsilon^2 + \Delta^2)^{1/2}$, $\Delta = \Delta_0$ is the superconducting energy gap for isotropy s-wave (see figure 2(a)). Obviously the single-gap fitting has significant deviations from the experimental data especially at low temperatures. Thus the two-gap model is further applied to fit the experimental data. The case of s + s-wave, s + d-wave, and s + (extended s-wave) are shown in figures 2(b), 2(c), and 2(d), respectively. The order parameters used to fit the data are $\Delta = \Delta_0 \cos 2\theta$ for d-wave and $\Delta = \Delta_0(1+\alpha\cos 2\theta)$ (where α donates the gap anisotropy) for extended s-wave. Apparently, the s + (extended s-wave) model leads to more satisfactory fits of the data than others. The gap values are $2\Delta_s/kT_c = 4.38$ and $2\Delta_{es}/kT_c = 0.86$ ($\alpha = 0.73$) with the relative weigh of 71% and 29%. Since $\alpha = 0.73$, the result illustrates a very anisotropy order parameter existing in FeSe.

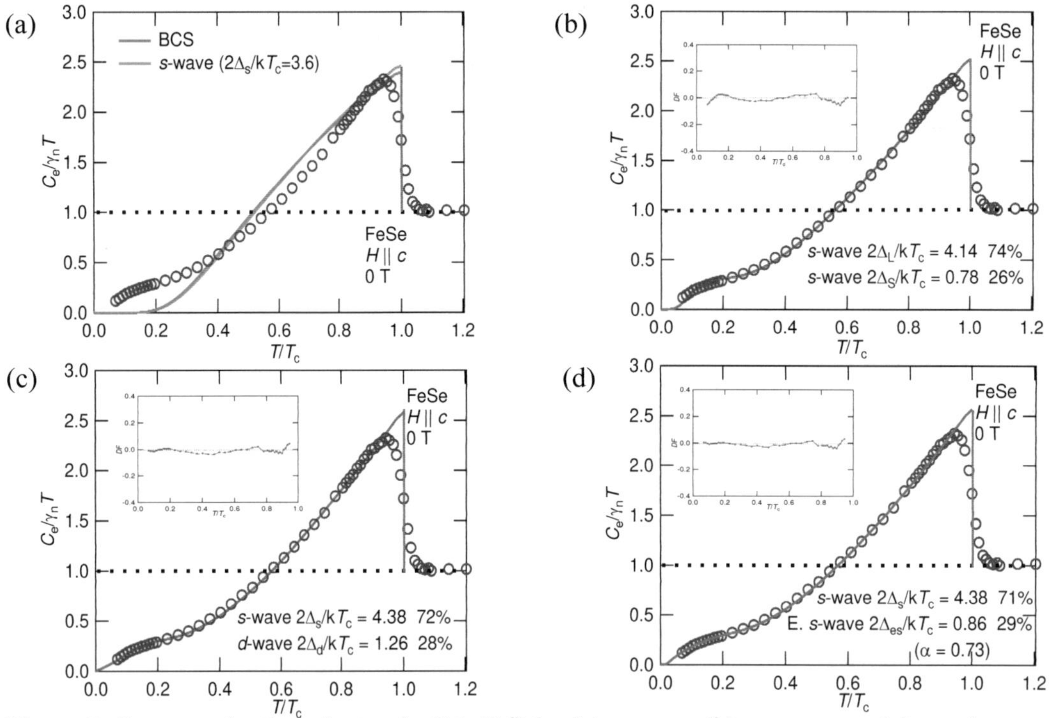

Figure 2. Superconducting electronic $C_e/\gamma_n T$ fit by (a) *s*-wave, (b) *s* + *s*-wave, (c) *s* + *d*-wave, and (d) *s* + extended *s*-wave models. The insets denote the deviations of the fits from the experimental data.

The presence of smaller gap can be confirmed by Arrhenius plot of the specific heat at low temperatures as shown in figure 3. In the weak-coupling BCS theory, specific heat should follow the following formula $C_{es}(T) \propto \exp(-\alpha T_c/T)$ in the (T_c/T) range between1.44 and 1.62 [13]. In contrast to BCS weak-coupling limit, $C_e/\gamma_n T$ for FeSe, shown in figure 3, decreases slowly with decreasing temperature from T_c to $T_c/5$. Below $T_c/5$, $C_e/\gamma_n T$ follows the above formula with a much smaller α of ~0.26. Very similar behavior has been also observed in typical two-gap superconductors MgB_2 [14] and $Lu_2Fe_3Si_5$ [15], where the values of α are 0.38 and 0.29, respectively.

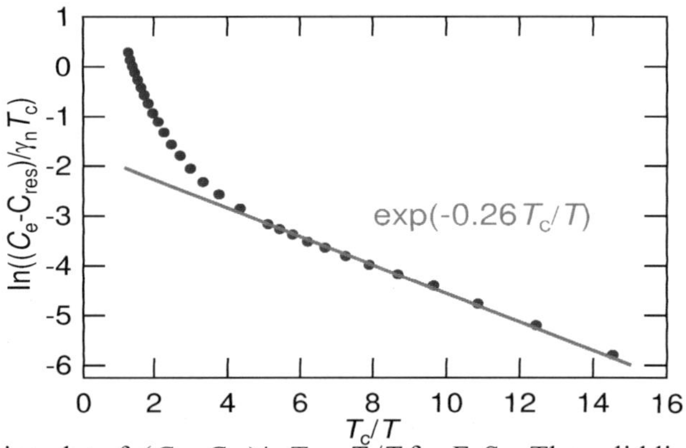

Figure 3. Arrhenius plot of $(C_e - C_{res})/\gamma_n T$ vs T_c/T for FeSe. The solid line is a linear fit above T_c /5. C_{res} is the residual specific heat in $T \rightarrow 0$ limit.

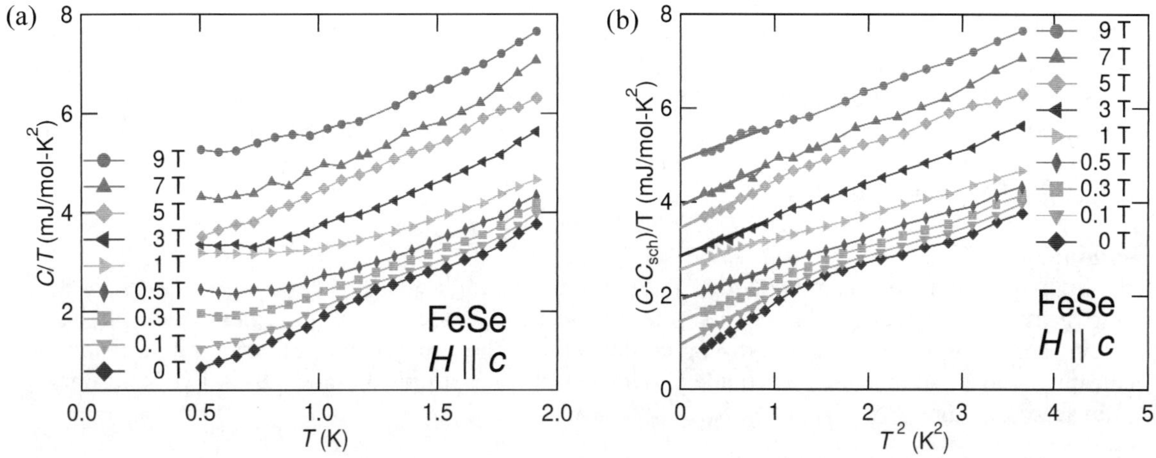

Figure 4. (a) Specific heat C/T at several fields up to 9 T in FeSe, (b) Specific heat corrected for the Schottky contribution $(C\text{-}C_{sch})/T$ as a function of T^2. The solid lines represent the linear fits to the low-temperature data.

Figure 4(a) shows the specific heat C/T for FeSe as a function of T at several fields up to 9 T. It should be noted that C/T shows an upturn under magnetic field at low temperatures, which probably comes from the Schottky term, originating from impurities. Thus, the specific heat C/T can be described by

$$C/T = C_{sch}(H,T) + \gamma_n + \beta T^2,$$

where the first term is the Schottky term, the second term is the electronic contribution and the third term is the implies phonon contribution, respectively. The Schottky anomaly term can be described as $C_{sch}(H,T)/T = A(H)/T^3$ [11]. The magnetic field dependence of $C/T = \gamma(H)$ at low temperatures is instructive for the quasiparticle excitation. We plot the low temperature $(C - C_{sch})/T$ as a function of T^2 under several magnetic fields in figure 4(b). By extrapolating to $T = 0$ K using a linear fit, we can obtain the field dependence of $\gamma(H)$, shown in figure 5.

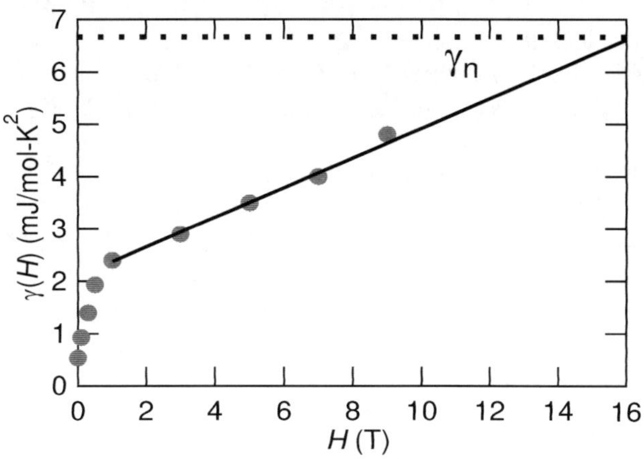

Figure 5. Magnetic field dependence of electronic specific heat coefficient in FeSe. The dotted line represents the normal-state value γ_n and the solid line shows the trend of $\gamma(H)$ when it reaches to its upper critical field.

IOP Conf. Series: Journal of Physics: Conf. Series **871** (2017) 012016 doi:10.1088/1742-6596/871/1/012016

In fully-gapped conventional superconductors, $\gamma(H)$ is proportional to H, which is due to the localized quasiparticle states in vortex core. By contrast, in nodal superconductors, such as *d*-wave superconductors, $\gamma(H)$ is proportional to √H, which originates from extended quasiparticle states near nodal directions of the order parameter. However, the kink structure in $\gamma(H)$ at low fields in FeSe is totally different from both the conventional *s*-wave and nodal superconductors but very similar to the typical two-gap superconductors [11,16,17]. The steep increase at low fields indicates the suppression of smaller gap by magnetic fields.

4. Summary

We have investigated the superconducting gap structure of high-quality FeSe single crystal through specific heat measurements. The results provide decisive evidence for two-gap nature in FeSe. Moreover, according to our several two-gap model fittings, the gap structure of FeSe is a combination of isotropy *s*-wave and anisotropy extended *s*-wave, and the isotropic *s*-wave gap is five times larger than the anisotropy one.

References

[1] Jiao L, Huang C-L, Rößler S, Koz C, RößlerR U K, Schwarz U and Wirth S 2016 arXiv:1605.01908

[2] Kasahara S, Watashige T, Hanaguri T, Kohsaka Y, Yamashita T, Shimoyama Y, Mizukami Y, Endo R, Ikeda H, Aoyama K, Terashima T, Uji S, Wolf T, von Löhneysen H, Shibauchi T and Matsuda Y 1992 *Proc. Natl. Acad. Sci. U.S.A.* **111,** 16309

[3] Bourgeois-Hope P, Chi S, Bonn D A, Liang R , Hardy W N, Wolf T, Meingast C, Doiron-Leyraud N, and Louis Taillefer 2016 *Phys. Rev. Lett.* **117,** 097003

[4] Dong J K, Guan T Y, Zhou S Y, Qiu X, Ding L, Zhang C, Patel U, Xiao Z L, and Li S Y 2009 *Phys. Rev. B* **80,** 024518

[5] Lin J Y, Hsieh Y S, Chareev D A, Vasiliev A N, Parsons Y, and Yang H D 2011 *Phys. Rev. B* **84,** 220507 (R)

[6] Sun Y, Pyon S, Tamegai T, Kobayashi R, Watashige T, Kasahara S, Matsuda Y, and Shibauchi T 2015 *Phys. Rev. B* **92,** 144509

[7] Sun Y, Pyon S, and Tamegai T 2016 *Phys. Rev. B* **93,** 104502

[8] Böhmer A E, Hardy F, Eilers F, Ernst D, Adelmann P, Schweiss P, Wolf T, and Meingast C 2013 *Phys. Rev. B* **87,** 180505

[9] Bardeen J, Cooper L N, and Schrieffer R J 1957 *Phys. Rev.* **108,** 1175

[10] Bouquet F, Fisher R A, Phillips N E, Hinks D G, and Jorgensen J D 2001 *Phys. Rev. Lett.* **87,** 047001

[11] Nakajima Y, Hidaka H, Nakagawa T, Tamegai T, Nishizaki T, Sasaki T, and Kobayashi N 2012 *Phys. Rev. B* **85,** 174524

[12] Huang C L, Lin L Y, Sun C P, Lee T K, Kim J D, Choi E M, Lee S I, and Yang H D 2006 *Phys. Rev. B* **73,** 012502

[13] Gladstone G, Jensen M A, and Schrieffer J R, in Superconductivity, edited by R. D. Parks (Marcel Dekker, New York, 1969), Vol. 2.

[14] Yang H D, Lin L Y, Li H H, Hsu F H, Liu C J, Li S C, Yu R C, and Jin C Q 2001 *Phys. Rev. Lett.* **87,** 167003

[15] Nakajima Y, Nakagawa T, Tamegai T, and Harima H 2008 *Phys. Rev. Lett.* **100,** 157001

[16] Bouquet F, Wang Y, Sheikin I, Plackowski T, Junod A, Lee S, and Tajima S 2002 *Phys. Rev. Lett.* **89,** 257001

[17] Nakajima Y, Hidaka H, Nakagawa T, Tamegai T, Nishizaki T, Sasaki T, and Kobayashi N 2012 *Phys. Rev. B* **85,** 174524

IOP Conf. Series: Journal of Physics: Conf. Series **871** (2017) 012017 doi:10.1088/1742-6596/871/1/012017

ARPES study of cesium-coated FeSe thin films on $SrTiO_3$

G. N. Phan[1], K. Nakayama[1], S. Kanayama[1], M. Kuno[1], K. Sugawara[2], T. Sato[1], and T. Takahashi[1,2]

[1]Department of Physics, Tohoku University, Sendai 980-8578, Japan
[2]WPI Research Center, Advanced Institute for Materials Research, Tohoku University, Sendai 980-8577, Japan

g.phan@arpes.phys.tohoku.ac.jp

Abstract. We have performed *in-situ* angle-resolved photoemission spectroscopy measurements of cesium(Cs)-deposited FeSe thin films on $SrTiO_3$. We found that Cs deposition enables heavily electron doping into the FeSe layer. In properly doped films, we also revealed the occurrence of superconductivity accompanied by the suppression of electronic nematicity. The present result demonstrates that Cs deposition is an effective way to realize and enhance superconductivity in FeSe.

1. Introduction

Recent discovery of possible high-temperature (T_c) superconductivity in one-monolayer (1-ML) FeSe film on $SrTiO_3$ [1] has brought great excitements. The reported T_c value by *ex-situ* transport measurements reaches around 40 K [1-4], which is much higher than T_c (8 K) of bulk FeSe [5]. Angle-resolved photoemission spectroscopy (ARPES) [6, 7] and two-coil mutual inductance measurements [8] reported a higher T_c value of 65 K, which exceeds the highest T_c (56 K) [9] for bulk iron-based superconductors discovered so far. Possible further enhancement of T_c up to 109 K has been suggested by *in-situ* transport study [10]. These observations generated fierce debates on the origin of the T_c enhancement in 1-ML FeSe film.

Besides the origin of high-T_c superconductivity in 1-ML film, another mystery is that multilayer (2-ML or thicker) FeSe films have been reported to be non-superconducting [1]. Previous ARPES measurements revealed a striking difference in the electron carrier density between 1-ML and multilayer films [7]; namely, 1-ML film is heavily electron-doped due to the electron transfer from the substrate, whereas multilayer films are almost non-doped because the electron transfer between FeSe layers is very small. This finding provides a question as to whether high-T_c superconductivity emerges in multilayer films if one can dope sufficient electrons. However, the difficulty in controlling the carrier density has hindered the elucidation of this essential question.

A novel approach to control the electron doping of FeSe thin films has been recently proposed by some of the present authors. They demonstrated that *in-situ* deposition of potassium (K) atoms onto the film markedly expands the accessible doping range [11]. Using this technique, non-superconducting multilayer films have been converted into superconductors with T_c values as high as 48 K [11, 12], establishing the importance of electron doping to realize high-T_c superconductivity. Subsequent study utilized Na atoms to control the carrier density but obtained low T_c of 20 K with the reduced superconducting pairing [13]. Significantly different T_c values obtained by K and Na deposition call for the further investigation on the suitable electron donor to achieve high-T_c superconductivity.

In this paper, we report our high-resolution ARPES study on Cs-deposited multilayer FeSe films. We have succeeded in achieving heavily electron doping and realizing superconductivity. We present the obtained electronic phase diagram and compare it with the previous results for the K- and Na-deposited samples.

2. Experiment

20-ML FeSe films were fabricated on TiO_2-terminated Nb(0.05wt%)-doped $SrTiO_3$ substrates (SHINKOSHA) by molecular-beam-epitaxy (MBE) technique, as described in detail in our previous work [11]. Cs atoms were deposited at room temperature using a Cs dispenser (SAES Getters). *In-situ* ARPES measurements were performed in an ultrahigh vacuum better than 8×10^{-11} Torr using a Scienta-Omicron SES2002 electron analyzer with a He discharge lamp ($h\nu$ = 21.218 eV) at Tohoku University. The energy and angular resolutions were set to 4-30 meV and 0.2°, respectively. The Fermi level (E_F) of the film was referenced to that of a gold film evaporated onto the sample holder.

3. Results and discussion

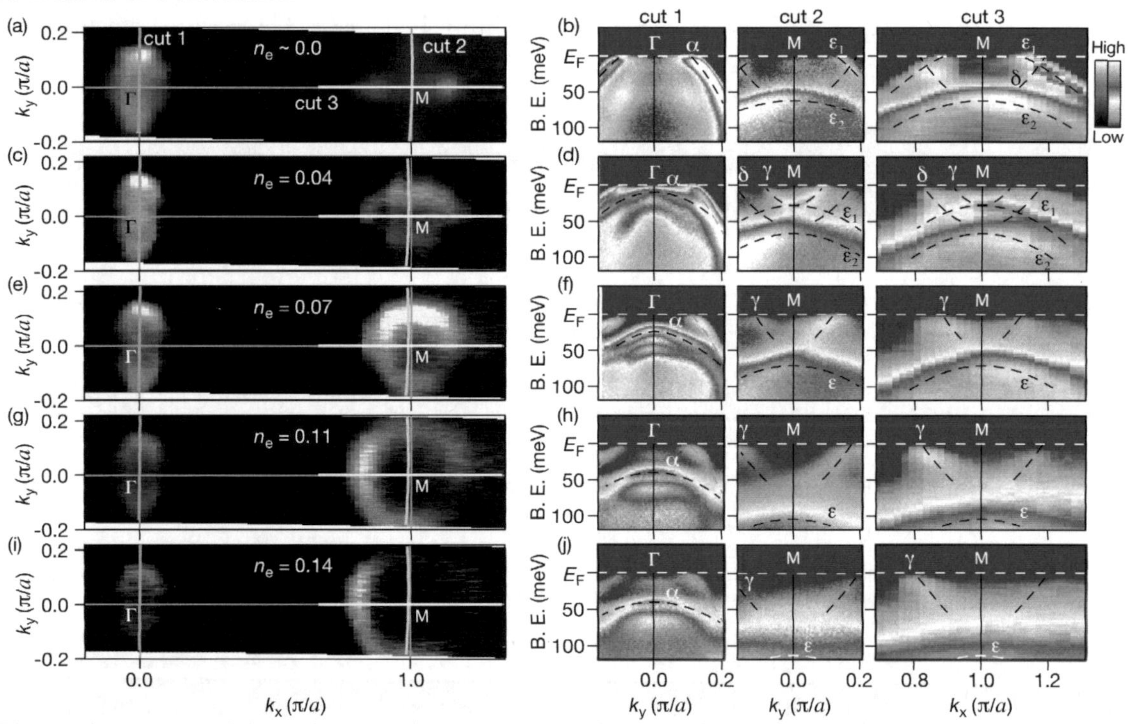

Figure 1. (a) ARPES intensity mapping at T = 30 K for the pristine 20-ML film plotted as a function of a two-dimensional wave vector. The intensity is obtained by integrating the spectral intensity within ±5 meV of E_F. (b) ARPES intensity plot near E_F as a function of binding energy and wave vector measured along the cuts 1-3 shown in (a). (c) and (d) Same as (a) and (b), but for Cs-deposited 20-ML film with n_e = 0.04 electrons/Fe. (e)-(j) Same as (c) and (d) but for n_e = 0.07 ((e) and (f)), 0.11 ((g) and (h)), and 0.14 ((i) and (j)).

Figures 1(a) and 1(b) show the Fermi surface and the band dispersion, respectively, for pristine 20-ML FeSe film measured at T = 30 K. We found signatures of the nematicity around the M point such as the two hole-like band dispersions originating from the lifting of d_{xz}/d_{yz} orbital degeneracy [see ε_1 and ε_2

IOP Conf. Series: Journal of Physics: Conf. Series **871** (2017) 012017 doi:10.1088/1742-6596/871/1/012017

bands in the right panel of Fig. 1(b)] [14, 15] and Dirac-cone-derived small electron-like Fermi surfaces [Fig. 1(a)] [16]. We also observed a hole-like pocket (α) around the Γ point, consistent with a semimetallic character of the pristine film. After deposition of Cs atoms onto the FeSe surface, we found sinking of the α band below E_F and the appearance of a larger electron-like Fermi surface (γ) centered at the M point [Fig. 1(c)], accompanied by a decrease of the energy separation between the ε_1 and ε_2 bands (ΔE_{nem}) [compare the right panels of Figs. 1(b) and 1(d)]. These observations indicate a successful electron doping by Cs deposition and the resultant suppression of the nematicity. It is noted that we still see the band structure of non-doped FeSe, e.g. a hole-like band crossing E_F around the Γ point indicated by gray dashed curves in the left panel of Fig. 1(d). This band structure would originate from interior FeSe layers which are kept non-doped even after Cs deposition, as discussed in the Na-deposited case [13]; namely, doped electrons would be confined in the topmost FeSe layer. Further Cs deposition results in a systematic downward shift of the α band as well as an expansion of the γ Fermi surface [Figs. 1(e)-1(j)]. The electron carrier concentrations (n_e) estimated from the Fermi-surface volume in Figs. 1(c), 1(e), 1(g), and 1(i) are 0.04, 0.07, 0.11, and 0.14 electrons/Fe, respectively. The largest n_e of 0.14 achieved by Cs deposition exceeds the accessible limit in 1-ML FeSe (~ 0.12) [6], demonstrating that Cs deposition is useful to control the electron doping over a wide range.

Next we performed high-resolution measurements at the Fermi wave vector (k_F) of the pristine and Cs-deposited films at low temperature (T = 13 K) to investigate the doping-induced evolution of superconductivity. As seen in Fig. 2(a), the leading-edge midpoint, which is located around E_F in the pristine (n_e ~ 0) and lightly doped (0.04) films, is clearly shifted toward higher binding energy for the doping range between 0.06 and 0.11, indicative of a superconducting-gap opening in properly doped films. The gap opening is confirmed by the observation of two-peaked structure in the symmetrized spectra in Fig. 2(b), where the influence of the Fermi-Dirac distribution function is eliminated. It is also noted that the superconducting gap is suppressed by heavily electron doping (n_e = 0.14).

Figure 3 shows the magnitude of the superconducting gap (Δ_{SC}) and the strength of the nematicity (ΔE_{nem}) plotted as a function of n_e. It is clearly seen that superconductivity is realized when the nematicity is suppressed by an electron doping through Cs deposition, pointing to the anti-correlation between superconductivity and nematicity. With increasing electron doping, the superconducting

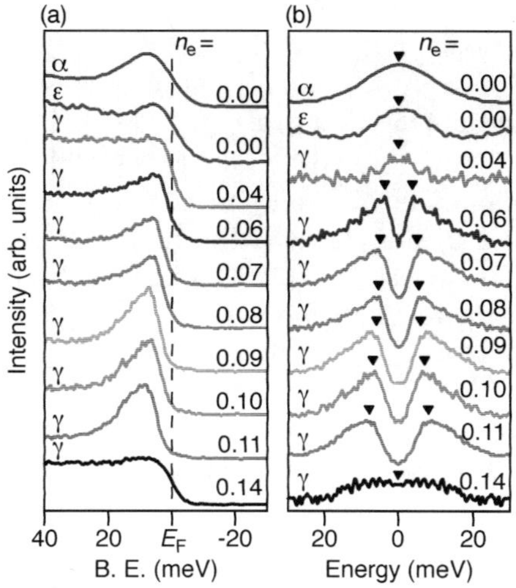

Figure 2. (a) Representative ARPES spectra measured at the k_F point of the α and ε_1 bands at T = 30 K for pristine 20-ML film (upper two spectra) and those of the γ band at T = 13 K for Cs-deposited films

with different doping levels (lower eight spectra). (b) Symmetrized spectra of (a). Triangles are a guide for the eyes to trace the peak position.

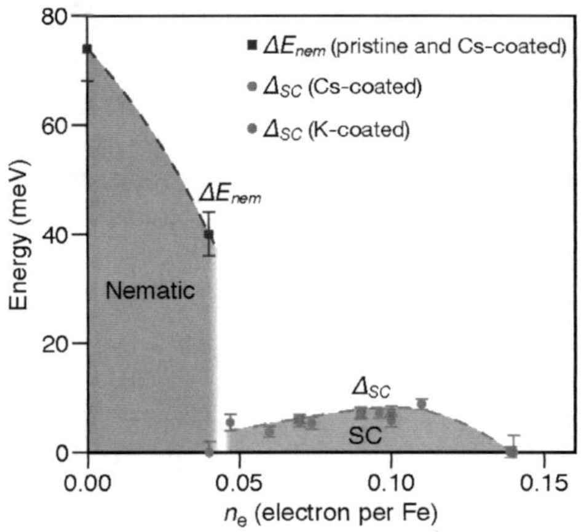

Figure 3. Electronic phase diagram for Cs- and K-deposited 20-ML FeSe films. Red and blue circles represent experimentally determined superconducting-gap size (Δ_{SC}) for Cs- and K-deposited 20-ML films, respectively. Each data has been extracted from the ARPES spectra shown in Figs. 2(b) and S1(b). Energy difference between the ε_1 and ε_2 bands at the M point (ΔE_{nem}) at T = 30 K, extracted from the data in Fig. 1, is plotted in blue square.

pairing is gradually enhanced until reaching the optimal doping ($n_e \sim 0.11$) and then suppressed in the overdoped regime, forming a dome-shaped superconducting phase which is a characteristic of high-T_c superconductors.

Now we discuss the implication of the present results in relation with the alkali-metal dependence. As described above, electron doping to a thick film or bulk crystal has been achieved by deposition of K or Na atoms [11-13]. However, the reported Δ_{SC} value is significantly different between the K- and Na-deposited cases (the maximum Δ_{SC} is ~10 meV [12] and ~4 meV [13], respectively), suggesting that the search for a suitable electron donor is of crucial importance to realize high-T_c superconductivity. In the present study, we have succeeded in achieving heavy electron doping by Cs deposition and realizing the superconductivity with the maximum Δ_{SC} up to 10 meV. The observed Δ_{SC} is much larger than that for the Na-deposited sample and is quantitatively similar to that for the K-deposited sample (compare red and green circles in Fig. 3) [17]. This result demonstrates that Cs is one of the suitable electron donors to realize and enhance superconductivity in FeSe. We think that the alkali-metal dependence of the maximum Δ_{SC} value may be related to a difference in the chemical stability of the surface. While the Cs- and K-deposited surfaces were found to be stable in an ultrahigh vacuum for at least 24 h, the lifetime of Na-deposited surface has been reported to be very short [13]. This suggests the occurrence of a chemical reaction at the Na-deposited surface, which may result in the smaller Δ_{SC} value in Na-deposited FeSe.

4. Summary

We reported high-resolution ARPES results on Cs-deposited 20-ML FeSe films grown on $SrTiO_3$ substrate. We found that Cs deposition enables the systematic investigation on the evolution of superconductivity and nematicity as a function of electron carrier density. By using this technique, we determined the electronic phase diagram. The observed characteristics in the phase diagram, such as the anti-correlation between nematicity and superconductivity and the dome-shaped nature of the superconducting region, would provide key insights into the origin of high-T_c superconductivity.

Acknowledgements

We thank N. Yamamura and M. Sato for their assistance in the ARPES measurements. This work was supported by grants from JSPS KAKENHI Grant Numbers JP15H02105, JP25287079, JP15H05853, JP25107003, the Program for Key Interdisciplinary Research, and the Murata Science Foundation.

References

[1] Wang Q-Y *et al.* 2012 *Chin. Phys. Lett.* **29** 037402
[2] Zhang W-H *et al.* 2014 *Chin. Phys. Lett.* **31** 017401
[3] Zhang A W *et al.* 2014 *Phys. Rev. B* **89** 060506(R)
[4] Sun A Y *et al.* 2014 *Sci. Rep.* **4** 6040
[5] Hsu F-C *et al.* 2008 *Proc. Natl. Acad. Sci.* **105** 14262-14164
[6] He S *et al.* 2013 *Nat. Mater.* **12** 605-610
[7] Tan S *et al.* 2013 *Nat. Mater.* **12** 634-640
[8] Zhang Z *et al.* 2015 *Sci. Bull.* **60** 1301-1304
[9] Wang C *et al.* 2008 *Europhys. Lett.* **83** 67006
[10] Ge J-F *et al.* 2015 *Nat. Mater.* **14** 285-289
[11] Miyata Y *et al.* 2015 *Nat. Mater.* **14** 775-779
[12] Wen C H *et al.* 2016 *Nat. Commun.* **7** 10840
[13] Seo J J *et al.* 2016 *Nat. Commun.* **7** 11116
[14] Nakayama K *et al.* 2014 *Phys. Rev. Lett.* **113** 237001
[15] Shimojima T *et al.* 2014 *Phys. Rev. B* **90** 121111(R)
[16] Tan S Y *et al.* 2016 *Phys. Rev. B* **93** 104513
[17] While a superconducting gap was not clearly observed in our previous study on 20-ML film [11], recent data with higher resolution show a clear gap opening as shown in Supplementary Information (Figure S1).
[18] Fan Q *el at.* 2015 *Nat. Phys.* **11** 946-952

Field-driven transition in $Ba_{1-x}K_xFe_2As_2$ with splayed columnar defects

N Ito[1], A Park[1], S Pyon[1], T Kambara[2] and T Tamegai[1]

[1] Department of Applied Physics, The University of Tokyo, 7-3-1 Hongo, Bunkyo-ku, Tokyo 113-8656, Japan

[2] Nishina Center, RIKEN, 2-1 Hirosawa, Wako, Saitama 351-0198, Japan

E-mail: ito-nozomu815@g.ecc.u-tokyo.ac.jp

Abstract. Columnar defects in type-II superconductors serve as artificial pinning centers, which lead to enhancement of critical current density. However, little is known about the details of vortex motion in iron-based superconductors with columnar defects. In this study, we performed magnetization measurements in $Ba_{0.6}K_{0.4}Fe_2As_2$ single crystals with splayed columnar defects for several directions of magnetic field in order to investigate an anomalous peak effect in the system. We report results of detailed magnetization measurements, and discuss possible origins of the field-driven transition.

1. Introduction

Columnar defects in type-II superconductors serve as artificial pinning centers, which lead to enhancement of critical current density J_c [1, 2, 3]. It has been proposed that a further enhancement of J_c is possible by dispersing the direction of the columnar defects [4]. In such a system with splayed columnar defects, enhancement of J_c has been confirmed in $YBa_2Cu_3O_{7-\delta}$ [5] and $Ba_{1-x}K_xFe_2As_2$ [6] single crystals. Due to the competing effects of suppression of vortex motion by unequal lengths of vortex segments between splayed columnar defects and detrimental effect of misalignment of vortices, the optimal splay angles were found to be $\approx \pm 5°$ in both of the above systems. However, little is known about the detail of vortex motion in superconductors with columnar defects. Actually, we found an anomalous peak effect when the splay angle of columnar defects is increased further [6]. On the other hand, it has been reported that a field-driven transition occurs in $Bi_2Sr_2CaCu_2O_{8+y}$ [7, 8, 9, 10] and $YBa_2Cu_3O_{7-\delta}$ [11] when columnar defects are introduced. Theoretically, a Monte Carlo simulation for $Bi_2Sr_2CaCu_2O_{8+y}$ with columnar defects suggests a field-driven transition with enhancement of vortex trapping rate by columnar defects at 1/3 of matching field B_Φ [12]. The relationship between the anomalous peak effect in $Ba_{1-x}K_xFe_2As_2$ with splayed columnar defects, and the field-induced transition in cuprates is not clear.

Here, we report results of detailed magnetization measurements in $Ba_{0.6}K_{0.4}Fe_2As_2$ single crystals with splayed columnar defects, and discuss possible origins of the field-driven transition.

2. Experimental Methods

$Ba_{0.6}K_{0.4}Fe_2As_2$ single crystals ($T_c \sim 38$ K) were synthesized by FeAs flux method [13, 14, 15], and were shaped into rectangular parallelepiped of length and width of approximately 500 μm and of thickness of 30 μm or less. Columnar defects were installed into the crystals by 2.6 GeV ^{238}U irradiation at the RIKEN Ring Cyclotron with a total dose of $B_\Phi = 8$ T. The incident directions of U ions were splayed by tilting the crystals about their *ab*-plane using a well-controlled servo-motor, and the splay angle ($\theta_{\rm CD}$) is denoted by the angle between their *c*-axis and the ion beam.

Magnetization measurements were performed with a superconducting quantum interference device (SQUID) magnetometer. For several directions of magnetic field, the component of magnetization parallel to the field was measured. For these measurements, an acrylic sample holder with a rotating sample slot was employed. Here we define ϕ_H as the angle between the splay-plane and the field-plane and θ_H as the angle between *c*-axis and the field. Magnetization measurements were performed for fields parallel to the *c*-axis ($\theta_H = 0°$), fields tilted in the splay-plane ($\phi_H = 0°, \theta_H \neq 0°$), and fields tilted in a plane perpendicular to the splay-plane ($\phi_H = 90°, \theta_H \neq 0°$). The critical current density, J_c, was calculated from the results of the

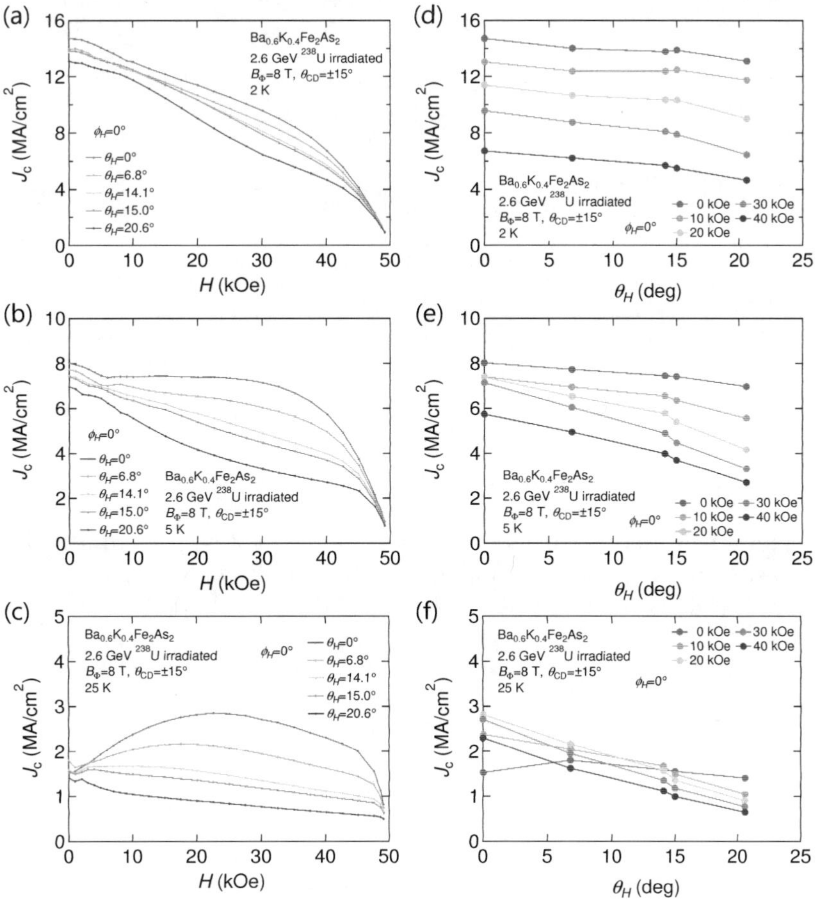

Figure 1. θ_H and magnetic field dependence of J_c for fields in the splay-plane ($\phi_H = 0°$) in $Ba_{0.6}K_{0.4}Fe_2As_2$ single crystals with splayed columnar defects of $\theta_{\rm CD} = \pm 15°$. Magnetic field dependence of J_c at several θ_H at 2 K, 5 K, and 25 K are shown in (a), (b), and (c), respectively. J_c as functions of θ_H at several fields at 2 K, 5 K, and 25 K are shown in (d), (e), and (f), respectively.

IOP Conf. Series: Journal of Physics: Conf. Series **871** (2017) 012018 doi:10.1088/1742-6596/871/1/012018

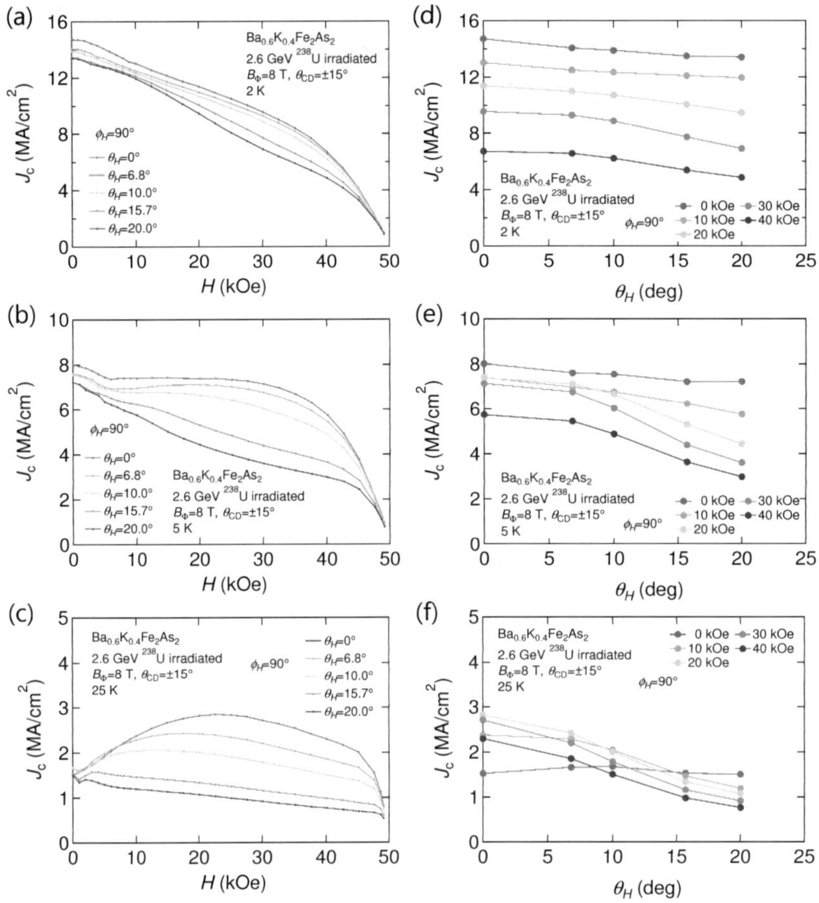

Figure 2. θ_H and magnetic field dependence of J_c for fields in a plane perpendicular to the splay-plane ($\phi_H = 90°$) in $Ba_{0.6}K_{0.4}Fe_2As_2$ single crystals with splayed columnar defects of $\theta_{CD} = \pm 15°$. Magnetic field dependence of J_c at several θ_H at 2 K, 5 K, and 25 K are shown in (a), (b), and (c), respectively. J_c as functions of θ_H at several fields at 2 K, 5 K, and 25 K are shown in (d), (e), and (f), respectively.

magnetization measurements using extended Bean model.

3. Results and Discussion

Figures 1(a)-(c) show magnetic field dependence of J_c at several θ_H for fields in the splay-plane ($\phi_H = 0°$) in $Ba_{0.6}K_{0.4}Fe_2As_2$ single crystals with splayed columnar defects of $\theta_{CD} = \pm 15°$. J_c as functions of θ_H at several fields are extracted from figures 1(a)-(c) and shown in figures 1(d)-(f). At 2 K, $J_c - H$ is weakly dependent on θ_H (figure 1(a)), and J_c is almost independent of θ_H as shown in figure 1(d). When the temperature is increased to 5 K, an anomalous peak effect starts to show up at around 30 kOe (figure 1(b)). This anomalous peak effect is suppressed as θ_H is increased. At higher temperature of 25 K, the anomalous peak effect shows up more clearly at around 25 kOe for $\theta_H = 0$, and it is suppressed quickly as θ_H is increased.

Figures 2(a)-(c) show magnetic field dependence of J_c at several θ_H for fields in a plane perpendicular to the splay-plane ($\phi_H = 90°$) in $Ba_{0.6}K_{0.4}Fe_2As_2$ single crystals with splayed

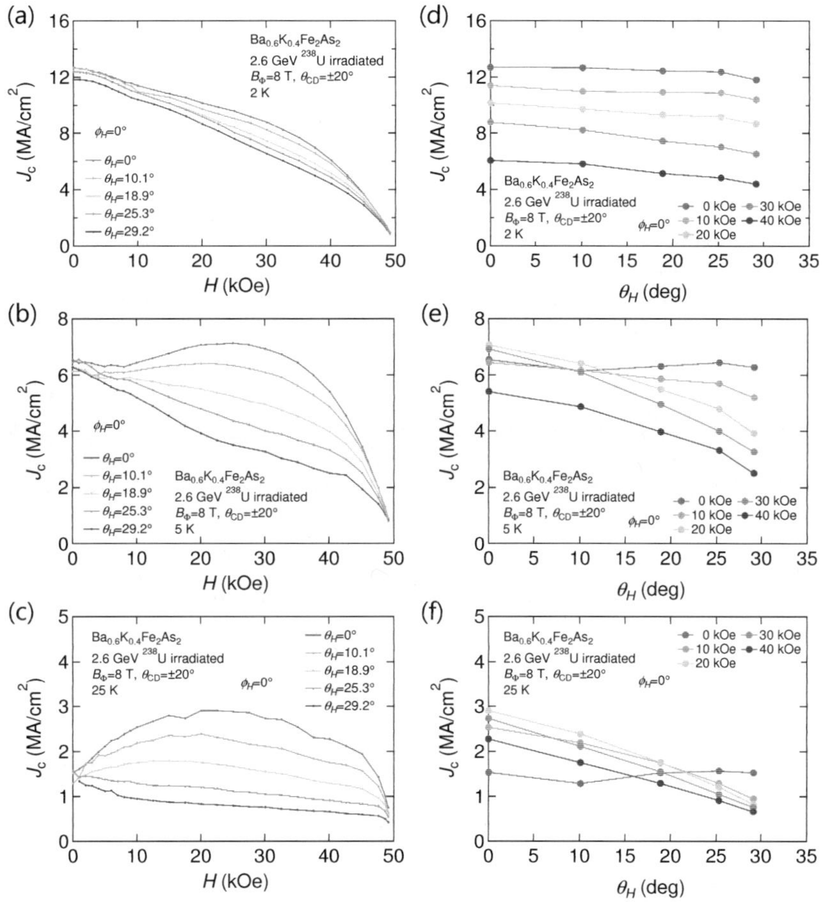

Figure 3. θ_H and magnetic field dependence of J_c for fields in the splay-plane ($\phi_H = 0°$) in $Ba_{0.6}K_{0.4}Fe_2As_2$ single crystals with splayed columnar defects of $\theta_{CD} = \pm 20°$. Magnetic field dependence of J_c at several θ_H at 2 K, 5 K, and 25 K are shown in (a), (b), and (c), respectively. J_c as functions of θ_H at several fields at 2 K, 5 K, and 25 K are shown in (d), (e), and (f), respectively.

columnar defects of $\theta_{CD} = \pm 15°$. J_c as functions of θ_H at several fields are extracted from figures 2(a)-(c) and shown in figures 2(d)-(f). At a glance, $J_c - H$ for $\phi_H = 90°$ is very similar to that for $\phi_H = 0°$ at the same θ_H. However, a close inspection makes it clear that the suppression of the anomalous peak effect with an increase in θ_H is weaker for $\phi_H = 90°$ as evidenced by the shallower initial slope of $J_c - \theta_H$ in figures 2(e) and (f).

Figures 3(a)-(c) show magnetic field dependence of J_c at several θ_H for fields in the splay-plane ($\phi_H = 0°$) in $Ba_{0.6}K_{0.4}Fe_2As_2$ single crystals with splayed columnar defects of $\theta_{CD} = \pm 20°$. J_c as functions of θ_H at several fields are extracted from figures 3(a)-(c) and shown in figures 3(d)-(f). Anomalous peak effects are also observed in these samples similar to those with $\theta_{CD} = \pm 15°$. However, there is a difference in that at 25 K $J_c - H$ in samples with $\theta_{CD} = \pm 20°$ still has an anomalous peak when the direction of field is closer to one families of columnar defects ($\theta_H \approx |\theta_{CD}|$) (figure 3(c)), while for $J_c - H$ in samples with $\theta_{CD} = \pm 15°$, the anomalous peak is already suppressed at $\theta_H \approx |\theta_{CD}|$ (figure 1(c)).

All the obtained results can be summarized as follows. In systems with splayed columnar defects, an additional component of J_c appears with a peak at $1/5-1/3$ of B_Φ, when the external field is applied along the average direction of the columnar defects. This additional component of J_c becomes larger as θ_{CD} is increased at least up to $\theta_{CD} = \pm 20°$. A clear peak effect of J_c has been observed at around $1/3$ of B_Φ also in $YBa_2Cu_3O_{7-\delta}$ with columnar defects created by irradiating 580 MeV Sn ions when the field is applied along the direction of columnar defects [1]. However, in the case of 580 MeV Sn, the direction of columnar defects is naturally splayed due to the deflection of ion motion owing to its relatively low energy. Therefore, the actual situation is very similar to the present case of intentional splay of the direction of columnar defects. These results are interpreted as follows. In a system with regular array of columnar defects, vortex pinning becomes most effective at the matching field B_Φ. On the other hand, random distribution of columnar defects in heavy-ion irradiated system, the optimal vortex pinning is realized at some fraction of B_Φ because of the strong repulsive interactions between vortices trapped in columnar defects. Splay in the arrangement of columnar defects makes the pinning stronger due to the suppression of kink motion and entanglement of vortices when the magnetic field is applied along the average direction of columnar defects. When the magnetic field is tilted away from the average direction in a system with splayed columnar defects, one set of columnar defects are more preferentially occupied and the advantageous effect of splayed columnar defects mentioned above is weakened, and thus the anomalous peak effect is suppressed.

4. Summary

We performed detailed magnetization measurements in $Ba_{0.6}K_{0.4}Fe_2As_2$ single crystals with splayed columnar defects, and clarified external-field-direction dependence of J_c. The results suggest that in systems with splayed columnar defects there exists an additional component of J_c with a peak at $1/5-1/3$ of B_Φ, when the external field is applied along the average direction of the columnar defects. In contrast to a system with regular array of columnar defects, in a system with random distribution of columnar defects produced by heavy-ion irradiation, the optimal vortex pinning is realized at some fraction of B_Φ because of the strong repulsive interactions between vortices trapped in columnar defects. In a system with splayed columnar defects, pinning of vortices becomes stronger due to the suppression of kink motion and entanglement of vortices when the magnetic field is applied along the average direction of columnar defects. When the magnetic field is tilted away from the average direction in such a system, one set of columnar defects are more preferentially occupied and the advantageous effect of splayed columnar defects mentioned above is weakened, and thus the anomalous peak effect is suppressed.

References

[1] Civale L, Marwick A D, Worthington T K, Kirk M A, Thompson J R, Krusin-Elbaum L, Sun Y, Clem J R and Holtzberg F 1991 *Phys. Rev. Lett.* **67** 648
[2] Nakajima Y, Tsuchiya Y, Taen T, Tamegai T, Okayasu S and Sasase M 2009 *Phys. Rev.* B **80** 012510
[3] Tamegai T, Taen T, Yagyuda H, Tsuchiya Y, Mohan S, Taniguchi T, Nakajima Y, Okayasu S, Sasase M, Kitamura H, Murakami T, Kambara T and Kanai Y 2012 *Supercond. Sci. Technol.* **25** 084008
[4] Hwa T, Doussal P L, Nelson D R and Vinokur V M 1993 *Phys. Rev. Lett.* **71** 3545
[5] Krusin-Elbaum L, Marwick A D, Wheeler R, Feild C, Vinokur V M, Leaf G K and Palumbo M 1996 *Phys. Rev. Lett.* **76** 2563
[6] Park A, Pyon S, Tamegai T and Kambara T 2016 *Physica* C **530** 58
[7] Sato M, Shibauchi T, Ooi S, Tamegai T and Konczykowski M 1997 *Phys. Rev. Lett.* **79** 3759
[8] Chikumoto N, Kosugi M, Matsuda Y, Konczykowski M and Kishio K 1998 *Phys. Rev.* B **57** 14507
[9] Kameda N, Tokunaga M, Tamegai T, Konczykowski M and Okayasu S 2004 *Phys. Rev.* B **69** 180502
[10] Kameda N, Shibauchi T, Tokunaga M, Ooi S, Tamegai T and Konczykowski M 2005 *Phys. Rev.* B **72** 064501
[11] Itaka K, Shibauchi T, Yasugaki M, Tamegai T and Okayasu S 2001 *Phys. Rev. Lett.* **86** 5144
[12] Sugano R, Onogi T, Hirata K and Tachiki M 1998 *Phys. Rev. Lett.* **80** 2925

[13] Kihou K, Saito T, Ishida S, Nakajima M, Tomioka Y, Fukazawa H, Kohori Y, Ito T, Uchida S I, Iyo A, Lee C H and Eisaki H 2010 *J. Phys. Soc. Jpn.* **79** 124713
[14] Taen T, Ohtake F, Akiyama H, Inoue H, Sun Y, Pyon S and Tamegai T 2013 *Phys. Rev.* B **88** 224514
[15] Taen T, Ohtake F, Pyon S, Tamegai T and Kitamura H 2015 *Supercond. Sci. Technol.* **28** 085003

Vortex states in micron-sized $Bi_2Sr_2CaCu_2O_{8+y}$ crystals

Shuuichi Ooi, Takashi Mochiku, Minoru Tachiki, Kazuto Hirata
National Institute for Materials Science, 1-2-1 Sengen, Tsukuba, Ibaraki, 305-0047, Japan
E-mail: OOI.Shuuichi@nims.go.jp

Abstract.
Large thermal fluctuation, owing to high superconducting transition temperature, short coherence length and quasi-two-dimensionality, brings about a rich variety of vortex phases in $Bi_2Sr_2CaCu_2O_{8+y}$ (Bi2212). To study how the vortex states and the transitions would be modified when vortices are confined in a small-sized crystal, we have measured the c-axis resistance using a stack of the intrinsic Josephson junctions (IJJs) of Bi2212. In tiny Bi2212 crystals of several micron, it is possible to observe the series of vorticity transitions using the c-axis resistance measurements. Combining the observation of vorticity changes and detection of a melting transition, we found an oscillating behavior of melting transition temperatures $T_{\rm m}$ as a function of magnetic field (or number of vortices) in small squared Bi2212 with a lateral dimension of 5-10 μm. In the case of the square-shaped crystals, it seems that $T_{\rm m}$ is enhanced around the vortex numbers $N_{\rm v}$ of i^2 (i: integer), indicating a matching of square vortex lattices in the square boundary. However, the frustration between the square boundary shape and vortex lattice that prefers a triangular lattice complicates the situation. A deformed square lattice without topological defects is probably realized at large i^2 as a geometrical matching state.

1. Introduction

Upon reducing the size of a superconductor, an Abrikosov vortex lattice realized in a clean bulk superconductor is strongly influenced by the screening current circulating around the sample edge. The modification of the vortex state confined in a small area has been frequently studied in various mesoscopic conventional superconductors in past decades. In a superconductor whose lateral dimension is comparable to the coherence length, a giant vortex sustaining a multiquanta fluxoid of $2\Phi_0$ or more has been predicted theoretically [1] and explored by various experimental approaches [2–4]. In mesoscopic superconducting disks with a larger lateral dimension, concentric shell structures of vortices (vortex shells) have been studied theoretically [5] and experimentally by the Bitter decoration technique [6] and a scanning SQUID microscope [7]. Direct visualization of vortex configurations has also been attempted in samples with other shapes, *e.g.*, triangles, squares, and pentagons [8–11].

Also, in mesoscopic high-T_c superconductors in a field perpendicular to the superconducting planes, the vortex state of a small number of vortices confined in a restricted area should be different from that in bulk samples. This has been investigated in $Bi_2Sr_2CaCu_2O_{8+y}$ (Bi2212) using magnetical techniques [12–15]. The first-order melting transition of the Bragg glass and the disorder-induced phase transition (the second magnetization peak effect) in mesoscopic Bi2212

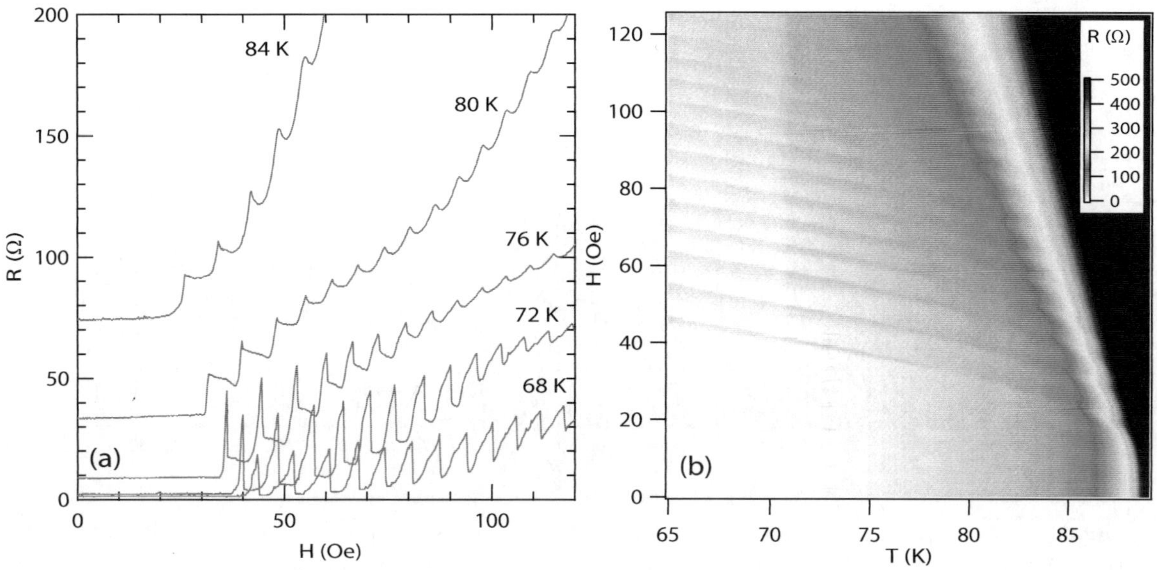

Figure 1. (a) Resistive steps, peaks and spikes in *R*-*H* curves due to the penetration of individual vortices observed in a 2.3×2.2 μm^2 IJJ stack. (b) Mapping of the *c*-axis resistance on *H*-*T* plane for the same sample.

down to 30 μm were originally explored by Wang *et al.* [12]. They found the disappearance of the second magnetization peak in samples smaller than a critical length, while the melting transition was insensitive to the sample size. The persistence of the melting transition line in the vortex phase diagram was also confirmed in smaller samples down to 20 μm by magnetic permeability measurements using a Hall probe with an excitation microcoil [16].

As an alternative method, transport measurements in a stack of IJJs facilitate the detection of the melting transition in a very small crystal [17]. The transitions have been successfully observed as a resistive jump of the *c*-axis resistance in a 47 × 79 μm^2 stack [17] and as an anomaly in the *c*-axis critical current in a 7.3 × 13 μm^2 stack [18]. Furthermore, it is possible to detect the penetration of a single vortex into a mesoscopic stack of IJJs of Bi2212 by *c*-axis transport measurement, which was first found by Kakeya *et al.* [19]. Therefore, simultaneous observation of the melting transition and the penetration of individual vortices in the same IJJ stack can provide another route to unveiling the vortex states in mesoscopic Bi2212.

2. Experimental

High-quality single crystals of Bi2212 were grown by the traveling-solvent floating-zone method [20]. Our fabrication process to form a z-shaped bridge containing a stack of IJJs of Bi2212 is based on the double-side etching technique [12], and it is also similar to the method used to fabricate whiskers by Latyshev and co-workers [21]. Details of the fabrication are described in ref. [22]. Transport measurements were performed using the four-probe technique with a current source (Keithley 6430) and a nanovoltmeter (Keithley 2182) in a chamber with a cryocooler. Data of the *c*-axis resistance shown here were acquired in repeats of field-scans at various fixed temperatures. Magnetic fields were applied parallel to the *c* axis.

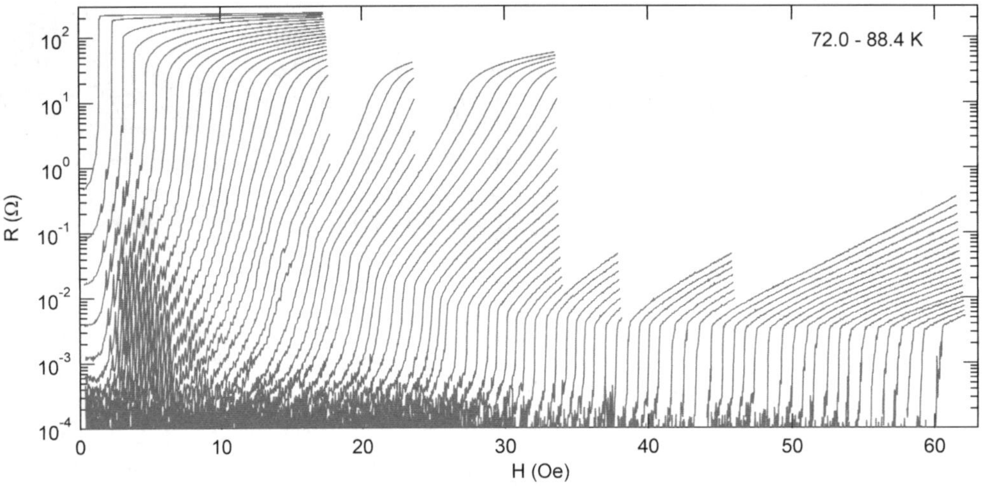

Figure 2. c-axis resistance of a 10×10 μm^2 IJJ stack as a function of magnetic field from 72.0 to 88.4 K at intervals of 0.2 K. With increasing temperature, the curves shift from right to left. Resistive jumps of the melting transition are observed more clearly at lower temperatures, and fine spikes caused by the penetration of individual vortices appear at low fields and high temperatures.

3. Results and discussion

3.1. Penetration of individual vortices

Primarily, the penetration of individual pancake vortices into IJJ stacks has been found in a series of experiments to determine the out-of-plane angular dependence of a Josephson vortex flow [19]. As the magnetic field is tilted from parallel to the superconducting planes, oscillating behavior of the c-axis resistance appears away from the lock-in angle in mesoscopic IJJ stacks, which is induced by the penetrations of pancake vortices. The suppression of the Josephson vortex-flow resistance by individual fluctuating pancake vortices near the lock-in angle has been studied theoretically and experimentally [23]. The experimentally-observed exponential dependence of the c-axis conductivity was well explained by the theoretical model.

Figure 1(a) shows c-axis resistance as a function of magnetic field for a $2.3\times2.2\mu$m IJJ stack, where the penetration of each vortex can be observed as a step, spike, or sawtooth shape, depending on the temperature and magnetic field, on a gradually increasing background. At higher temperatures, the resistance suddenly increases once a vortex penetrates and the R-H curves show step-like changes. At lower temperatures, vortex penetration brings about a sudden decrease in the resistance, causing the curve to become sawtooth shaped. At intermediate temperatures, for example, 76 K in Fig. 1(a), both increases and decreases in the resistance sequentially occur, leading spike-like shapes. Kakeya *et al.* found step-like changes in tiny stacks with in-plane areas of less than 2 μm^2 [19]. Thereafter, the penetrations of individual vortices were confirmed in samples with much larger sizes of up to 100 μm^2, and the appearance of R-H and I_c-H curves during the penetration of vortices was studied in samples with different sizes [24]. A vortex phase diagram constructed from the R-H curves of Fig. 1(a) is presented in Fig. 1(b). Transitions of the vortex numbers are clearly visualized, enabling us to determine the number of vortices in the stack at a certain point on the diagram.

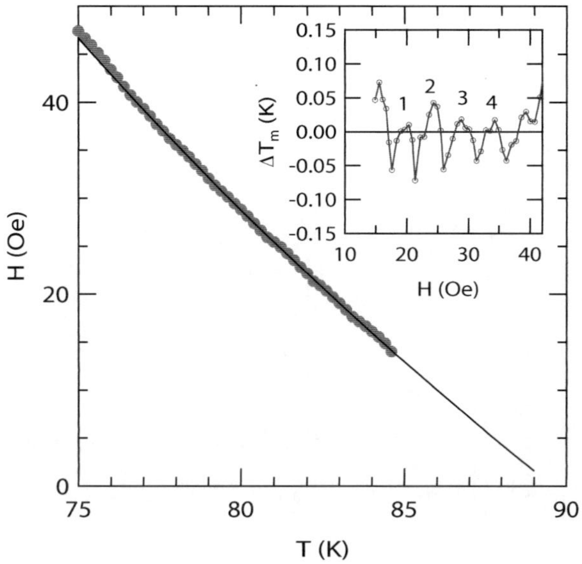

Figure 3. Temperature dependence of the melting transition field in a 10×10 μm^2 IJJ stack extracted from the onset of the resistive jumps in Fig. 2. The solid line is a fitting curve with the formula $H_m = H_0(T_c/T - 1)$ from decoupling theory. The inset shows the difference between the experimental melting temperature and the fitting curve as a function of magnetic field. The oscillation periods, which are numbered in the inset, are obtained from the intervals of the periodic dips.

3.2. Oscillations of the melting transition line in confined vortex matter

In very small stacks such as the sample used in Fig. 1, it has been difficult to find evidence of a melting transition. In the first place, its existence is not self-evident in a system containing only a small number of vortices less than 10. In relatively large stacks, however, a resistive jump accompanied by a melting transition is clearly observed as shown in Fig. 2. The sharp jumps at lower temperatures gradually broaden at higher temperatures and finally disappear near T_c. One can also observe very fine spikes below 10 Oe corresponding to the penetration of individual vortices. Extracted melting transition fields H_m at the onset of jumps are plotted in Fig. 3 as a function of temperature. Figure 3 appears to reproduce the behavior of the melting transition line in bulk crystals, and the temperature dependence can be fitted by $H_m = H_0(T_c/T - 1)$, theoretically predicted for the decoupling transition of vortex lines [25, 26]. However, by subtracting the fitting curve as a background, oscillatory behavior of the melting transition line is unveiled as shown in the inset of Fig. 4. This oscillating phenomenon has been confirmed in other square samples of a side length l from 5 to 10 μm [22]. In 5μm square stack, the penetrations of over 100 individual vortices were observed, making it possible to count the number of vortices in the vortex phase diagram. Interestingly, enhancements of the melting temperature occurred around square numbers of vortices, especially in higher fields [22].

In the case of a system of Josephson vortices in underdoped Y123 single crystals, the oscillation of the melting transition line is a manifestation of commensurability and incommensurability between the regularity of a vortex lattice and the underlying intrinsic pinning potential [27]. In a confined vortex system like present study, the two-dimensional geometrical matching of vortex lattice with the boundary shape is an important factor for the stability of a vortex crystal (or cluster) for each vortex number.

IOP Conf. Series: Journal of Physics: Conf. Series **871** (2017) 012019 doi:10.1088/1742-6596/871/1/012019

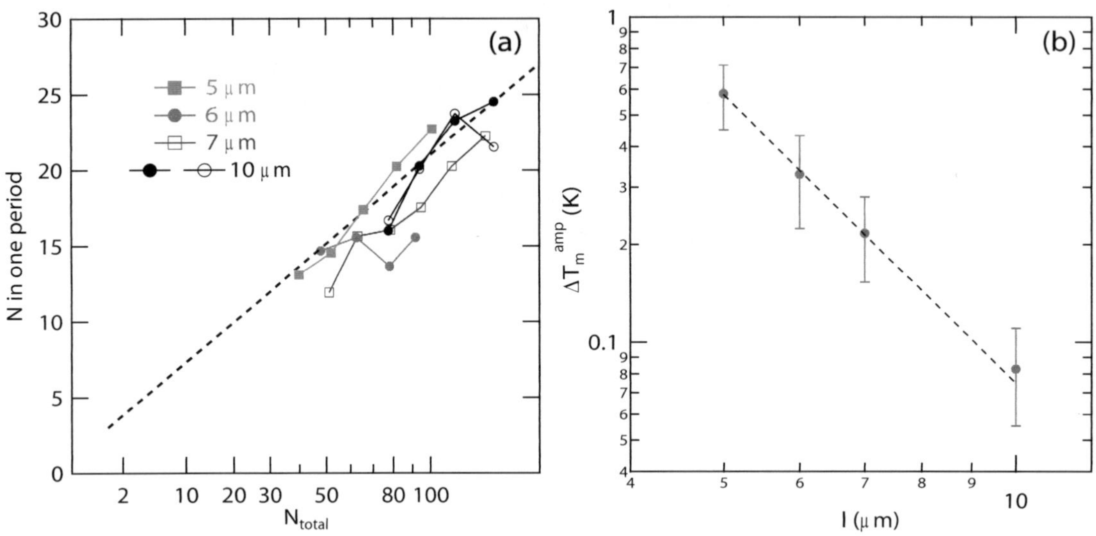

Figure 4. (a) Vortex number in one period of oscillations as a function of total vortex number in a square-root scale for four samples of different sizes. The vortex numbers were estimated from the magnetic-field dependence of the oscillation periods by dividing both axes by the period for ΔH_p [22]. For the 10 μm stack, measurements were performed twice (solid and open circles) to check for reproducibility. (b) Size dependence of the amplitude of ΔT_m oscillations. The amplitudes are evaluated by taking an average of the amplitude of each oscillation, and the error bar shows the standard deviation. The dashed line shows a power function of $\Delta T_m^{amp} = 66l^{-2.95}$ obtained by a fitting.

The sample-size dependence of the oscillations of the melting transition line supports the scenario of the matching effect by the square lattice structure [22]. There is a tendency for the oscillation period to increase with increasing magnetic field and with decreasing sample size. By converting from the magnetic field H to the roughly estimated vortex number N using $N = H/\Delta H_p$ with a field interval ΔH_p of the penetration of individual vortices, all the results for different samples appear to collapse onto a single curve as shown in Fig. 4(a). If the oscillation period is determined by the formation of square lattices by 4, 9, 16, ... , i^2 vortices (where i is an integer), the periodicity is expected to be $2\sqrt{N_{total}} + 1$ as a function of the total number of vortices $N_{total} = i^2$, which is indicated as a dashed line in Fig. 4(a). This line is in reasonable agreement with the experimental results. The amplitude of the ΔT_m oscillations (ΔT_m^{amp}) increases with reducing sample size as shown in Fig. 4(b). In the *log-log* plot, the data aligns on a linear line, indicating a power law relation. The dashed line in Fig. 4(b) is a fitting result of $\Delta T_m^{amp}[K] = 66.6 \cdot (l[\mu m])^{-2.95}$. This indicates that the influence of the confinement by a screening current on the melting becomes more noticeable in smaller Bi2212, especially less than 10 μm.

The vortex configurations and the filling rule in mesoscopic superconducting squares have been studied for patterned Nb films by the Bitter decoration technique with numerical simulations [9]. The formation of shell structures and the periodic filling of the outermost and internal shells were found. With increasing N_{total}, the outermost shell is filled until the number of vortices in it becomes $4i$, i.e., commensurate with a square shape. The vortices then fill the internal shells until their number becomes sufficiently large to create the next outermost shell. Whereas at small N_{total}, i.e., 4, 9, and 16, perfect square lattices are observed, for sufficiently large $N_{\rm total}$ the vortices do not form perfect square lattices even with square numbers in the simulation, and the lattices are deformed because the vortices tend to recover the triangular structure far from the boundary. Thus, it is difficult to expect the matching effect for perfect square lattices at large vortex numbers. Probably, the configuration of a deformed square lattice at large i^2 be related to the oscillations. Recently, the structural properties of a vortex solid confined in 30 μm Bi2212 squares and disks have been investigated by the Bitter decoration technique with single-vortex resolution [15]. An increase in the number of topological defects in the vortex solid was observed, particularly near the edge. A study on the relation between the defect density and the melting transition will be of interest to elucidate the mechanism of the melting-temperature oscillations.

An alternative explanation of the oscillations is the periodic filling of the outermost and internal shells. Dolz *et. al.* found oscillatory behavior in the magnetic transmittivity as a function of magnetic field in rather large Bi2212 disks of 50 μm diameter, which was interpreted as the discretized entrance of single shells [14]. In their simple model, $\sqrt{B}$ dependence of the oscillation period is expected, which resembles the oscillations of the melting transition. Although the numbers of vortices involved in both oscillations are considerably different, an intriguing issue is whether there is a correlation between these phenomena.

4. Summary

We have studied the vortex states confined in micron-sized Bi2212 single crystals by c-axis resistance measurements of a stack of IJJs. In Bi2212 crystals of about 2 μm, the penetrations of individual vortices were observed in a wide range of magnetic field at high temperatures. For square-shaped samples of 5-10 μm in side length, oscillating behavior in the melting transition line emerges. Results of the simultaneous observations of the melting and the penetrations of individual vortices suggest that a geometrical matching of deformed square vortex lattices with the square boundary shape cause the enhancement of the melting temperature at the square numbers of vortex.

Acknowledgements

Present studies were partially supported by KAKENHI Grants-in-Aid from Japan Society for the Promotion of Science (JSPS).

References

[1] Fink H J and Presson A G 1966 *Phys. Rev.* **151** 219
[2] Geim A K, Grigorieva I V, Dubonos S V, Lok J G S, Maan J C, Filippov A E and Peeters F M 1997 *Nature* **390** 259
[3] Kanda A, Baelus B J, Peeters F M, Kadowaki K and Ootuka Y 2004 *Phys. Rev. Lett.* **93** 257002
[4] Cren T, Serrier-Garcia L, Debontridder F and Roditchev D 2011 *Phys. Rev. Lett.* **107** 097202
[5] Baelus B, Cabral L and Peeters F 2004 *Phys. Rev. B* **69** 064506
[6] Grigorieva I, Escoffier W, Richardson J, Vinnikov L, Dubonos S and Oboznov V 2006 *Phys. Rev. Lett.* **96** 077005
[7] Kokubo N, Okayasu S, Kanda A and Shinozaki B 2010 *Phys. Rev. B* **82** 014501
[8] Zhao H J, Misko V R, Peeters F M, Dubonos S, Oboznov V and Grigorieva I V 2008 *Europhys. Lett.* **83** 17008
[9] Zhao H, Misko V, Peeters F, Oboznov V, Dubonos S and Grigorieva I 2008 *Phys. Rev. B* **78** 104517
[10] Huy H T, Kato M and Ishida T 2013 *Supercond. Sci. Technol.* **26** 065001
[11] Kokubo N, Okayasu S, Nojima T, Tamochi H and Shinozaki B 2014 *J. Phys. Soc. Jpn.* **83** 083704
[12] Wang Y, Fuhrer M, Zettl A, Ooi S and Tamegai T 2001 *Phys. Rev. Lett.* **86** 3626
[13] Connolly M R, Milošević M V, Bending S J, Clem J R and Tamegai T 2009 *Europhys. Lett.* **85** 17008
[14] Dolz M I, Fasano Y, Bolecek N R C, Pastoriza H, Konczykowski M and Beek C J v d 2014 *J. Phys: Conf. Ser.* **568** 022010
[15] Bolecek N R C, Dolz M I, Kolton A, Pastoriza H, van der Beek C J, Konczykowski M, Menghini M, Nieva G and Fasano Y 2015 *J. Low Temp. Phys.* **179** 35

IOP Conf. Series: Journal of Physics: Conf. Series **871** (2017) 012019 doi:10.1088/1742-6596/871/1/012019

[16] Dolz M, Fasano Y, Bolecek N C, Pastoriza H, Mosser V, Li M and Konczykowski M 2015 *Phys. Rev. Lett.* **115** 137003
[17] Ooi S, Mochiku T and Hirata K 2007 *Phys. Rev. B* **76** 224527
[18] Ooi S, Mochiku T and Hirata K 2002 *Physica C* **378** 523
[19] Kakeya I, Fukui K, Kawamata K, Yamamoto T and Kadowaki K 2008 *Physica C* **468** 669
[20] Mochiku T, Hirata K and Kadowaki K 1997 *Physica C* **282-287** 475
[21] Latyshev Y I and et al 1999 *Pis'ma Zh. Eksp. Teor. Fiz.* **69** 75
[22] Ooi S, Mochiku T, Tachiki M and Hirata K 2015 *Phys. Rev. Lett.* **114** 087001
[23] Koshelev A E, Buzdin A I, Kakeya I, Yamamoto T and Kadowaki K 2011 *Phys. Rev. B* **83** 224515
[24] Ooi S, Mochiku T, Tachiki M and Hirata K 2015 *Physics Procedia* **65** 109
[25] Zeldov E, Majer D, Konczykowski M, Geshkenbein V B, Vinokur V M and Shtrikman H 1995 *Nature* **375** 373
[26] Glazman L and Koshelev A 1991 *Phys. Rev. B* **43** 2835
[27] Gordeev S N, Zhukov A A, de Groot P A J, Jansen A G M, Gagnon R and Taillefer L 2000 *Phys. Rev. Lett.* **85** 4594

IOP Conf. Series: Journal of Physics: Conf. Series **871** (2017) 012020 doi:10.1088/1742-6596/871/1/012020

Random organization of vortices under an anisotropic condition

K Ienaga, M Dobroka, Y Shirahata, Y Kawamura, S Kaneko and S Okuma

Department of Physics, Tokyo Institute of Technology, 2-12-1 Ohokayama, Meguro-ku, Tokyo 152-8551 Japan

E-mail: `ienaga.k.aa@m.titech.ac.jp`

Abstract. Many colliding particles that are periodically sheared by ac drive self-organize to avoid future collisions, which is known as random organization. Recently, we have observed the random organization in the vortex system of a strip-shaped amorphous Mo_xGe_{1-x} film, where the vortices experience periodic local shear from ac drive and the random pinning potential. In this work, we study how random organization changes in the vortex system under the tilted field, where an anisotropic vortex-vortex interaction is introduced. We find that characteristic times of random organization for the vortices driven in the tilted direction are significantly smaller than those in the untilted field.

1. Introduction

A many particle system under an external driving force exhibits rich intriguing phenomena and has attracted much attention in recent years [1-7]. When particles are subjected to a periodic shear force, they collide with each other and gradually self-organize to avoid future collisions. This phenomenon is known as random organization [2-7]. In the final steady state, all the particles return to their initial position at the end of each cycle of a periodic shear, as long as the displacement amplitude of the ac shear is smaller than a threshold. This is called a reversible flow state. On the other hand, when the amplitude exceeds the threshold value, some particles do not return to their initial position, indicating that the collisions between the particles persist even in the steady state. This is called an irreversible flow state. A relaxation time of random organization, τ, that is, the time required for the system to settle into the steady state shows a power-law divergence at the threshold, indicating a dynamic transition from reversible to irreversible flow. The reversible-irreversible transition (RIT) is identified as the absorbing transition from non-fluctuating (reversible) to fluctuating (irreversible) states [8-14].

The RIT has been studied originally in periodically driven colloidal suspensions [1, 2] and then superconducting vortices in a Corbino disk where a global shear is applied [4, 5]. Recently, we have shown that random organization and RIT also occur in a strip-shaped superconducting film of amorphous Mo_xGe_{1-x} where a local shear due to random pinning is present [6, 7]. In the strip-shaped film, the spatial configuration of vortices and the vortex-vortex interaction are of course isotropic. However, when the magnetic field is tilted from normal to the sample plane, the vortex configuration and the resultant inter-vortex interaction become anisotropic. Recent studies have revealed that when the magnetic field is tilted at θ from normal to the sample plane, hexagonal vortex lattice is expanded by a factor of $1/\cos\theta$ along the tilt direction [15-18]. More recently, we have found that the relation between the dynamic melting field and the velocity of the driven anisotropic lattice changes depending on the direction of the dc drive [18], indicating that not only the static vortices but also the moving vortices feel the anisotropic interaction. Therefore, it is reasonable to expect that random organization of the vortices mentioned above may also be influenced by the introduction of the anisotropy. This issue is of general interest, because the self-organization in an anisotropic condition is found in different physical systems, which include growth of atomic-scale patterns on an anisotropic substrate [19-21]. However, a detailed mechanism describing the transient process has not been clarified.

In this work, we study how random organization observed in the vortex system under the perpendicular (i.e., untilted) field changes in the tilted field, where an anisotropic vortex-vortex interaction as well as an anisotropic vortex configuration is introduced. We find that relaxation times of random organization for the vortices driven in the tilt direction are much smaller than those driven under the untilted field. The results indicate that the anisotropy introduced by the tilt field significantly limits possible configurations of vortices in the transient state and expedites the random organization process.

2. Experimental

A strip-shaped film of amorphous Mo_xGe_{1-x} with thickness of 350 nm was prepared by rf-sputtering on a Si substrate installed on a rotating stage kept at room temperature with water cooling. The transition temperature where the resistivity falls to zero is $\sim$ 7 K at zero field. Since pinning potentials are shallow and randomly distributed in this amorphous sample, we can precisely detect detailed vortex motion using transport measurement. We measured the time evolutions of the ac voltage $V(t)$ responding to the square wave of ac current I_{ac} using a fast-Fourier transform spectrum analyzer. In each measurement, before applying I_{ac}, we prepared an initial vortex state with a random configuration by applying dc current yielding 100 μV, where the vortex-flow state is plastic flow [5]. Then, we applied I_{ac} yielding $V(t)$ whose amplitude in the steady state is $|V(t \to \infty)|(\equiv V^{\infty}) = 100$ μV. The displacement of a vortex per cycle is calculated from $d = \frac{v}{2f} = \frac{V^{\infty}}{Bl\cos\theta}\frac{1}{2f}$, where v is an average velocity of vortices, f a frequency of I_{ac}, l a distance between voltage terminals of the sample, and θ an angle of the field direction from normal to the sample plane. We changed d by varying f while keeping V^{∞} constant. All measurements were performed at 4.1K, where the sample was immersed into liquid ^{4}He. The applied magnetic field was $B = 2.0$ T and $\theta = 0°$ and $36°$.

3. Results and discussion

First, we show the results for the field applied perpendicular to the sample plane ($\theta = 0°$). The representative data of $V(t)$ for $d = 6.0$ and 2.2 μm, corresponding to $f = 3500$ and 9500 Hz, are presented in figures 1(a) and (b), respectively. $V(t)$ in the long time duration reaching to the steady state is shown in the insets, where the high V region is enlarged and shown. We observe a transient behavior of $V(t)$ for all d studied. With increasing the cycle number, the amplitude of the voltage, $|V(t)|$, gradually increases and saturates to the steady-state value $V^{\infty} = 100$ μV. These results indicate random organization of the vortex system, where vortices become more mobile by repeating the periodic drive [4-6].

The amplitude of the voltage, $|V(t)|$, is fitted to the following relaxation function [2, 3],

$$|V(t)| = V^{\infty} - (V^{\infty} - V_0)\exp(-t/\tau)/t^{\alpha}, \quad (1)$$

where V_0 is $|V(t)|$ of the first pulse and α is a numerical parameter. In each figure, the result of the fit is indicated with a green line, from which we obtain the characteristic time of random organization to be $\tau = 60$ and 300 cycles for $d = 6.0$ and 2.2 μm, respectively. Here, the values

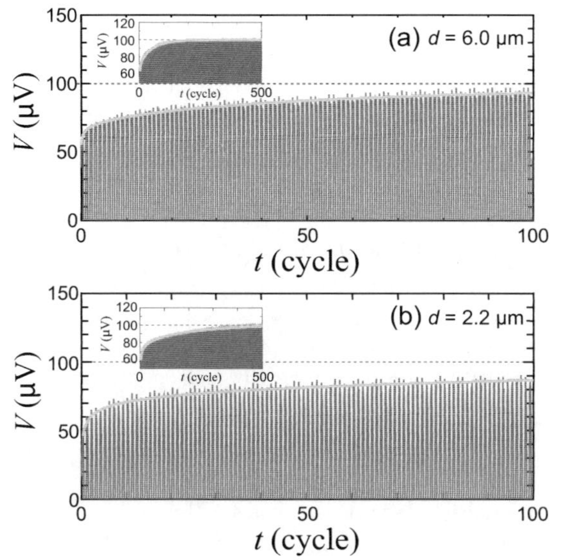

Figure 1. The time evolution of the voltage $V(t)$ responding to the square wave I_{ac} yielding (a) $d = 6.0$ and (b) 2.2 μm measured at 4.1 K under the untilted ($\theta = 0$) field of 2 T. Green lines represent the fit of $|V(t)|$ to eq. (1). Horizontal dashed lines indicate the location of $V^{\infty} = 100$ μV. Insets: The high V region of $V(t)$ is enlarged and $V(t)$ in the long time duration is shown.

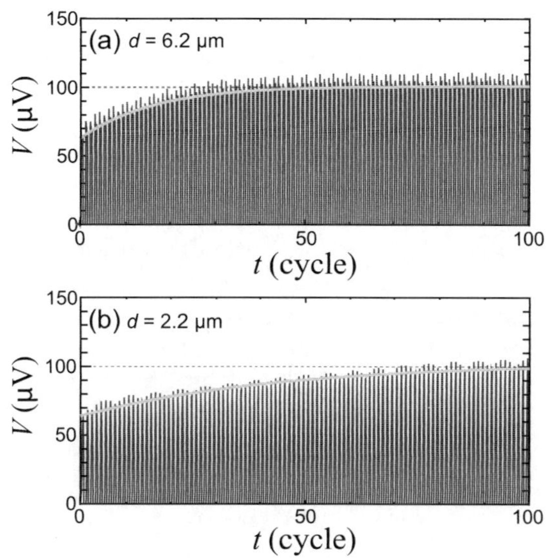

Figure 2. $V(t)$ responding to the square wave I_{ac} yielding (a) $d = 6.2$ and (b) 2.2 μm measured at 4.1 K under the tilted field ($\theta = 36°$) of 2 T. The direction of I_{ac} (/vortex motion) is perpendicular (/parallel) to the tilt direction. Green lines are the fit of $|V(t)|$ to eq. (1). Horizontal dashed lines indicate the location of $V^{\infty} = 100$ μV.

of α used in the fit are close to zero: $\alpha = 0.1$ and 0.18 for $d = 6.0$ and 2.2 μm, respectively. This fact is consistent with the theoretical prediction that the value of α only becomes relevant very close to the RIT. With decreasing d, τ seems to increase remarkably. The similar feature of $\tau(d)$ has been reported in our previous experiment carried out at 4.1 K in 4.0 T, where the critical divergence of τ has been observed with decreasing d in the irreversible phase [6].

Next, we show the results in the tilted ($\theta = 36°$) field. Here, the direction of the applied ac drive is parallel to the tilt direction, while the values of d as well as $V^{\infty}(= 100$ μV) are kept nearly same as the ones used in the untilted field. We again observe a transient behavior of $V(t)$, as shown in figures 2(a) and (b). However, the relaxation seems to be much faster. The values of τ derived from the fitting of $|V(t)|$ to eq. (1) are 15 and 38 cycles for $d = 6.2$ and 2.2 μm, respectively, which are significantly smaller than those obtained in the untilted field. We also use $\alpha \approx 0$ in the fitting for both cases. The fits are again satisfactory, as shown with green lines in figures 2(a) and (b). We have confirmed that under the perpendicular ($\theta = 0°$) field, $\tau(d)$ shows a trend to increase with decreasing the magnetic field. This result reflects the weakened random organization effect by the decreased vortex density. Thus, the marked suppression of τ by tilting the field observed here cannot be explained in terms of the decreased average vortex density (by 19 %).

We discuss the origin of the different values of τ under the untilted ($\theta = 0°$) field and tilted field. In the untilted field, the periodically sheared vortices show random organization from the randomly distributed initial configuration toward the less disordered steady-state configuration. In the transient process, the vortex-vortex (v-v) interaction is considered to be isotropic. On the other hand, under the tilted ($\theta = 36°$) field, an ordered vortex configuration both for the undriven case and fast-driven steady-state case shows an expanded hexagonal lattice by a factor of $1/\cos\theta(\approx 1.24)$ along the tilt direction [15-18]. Thus, the v-v interaction along the perpendicular direction (p-direction) to the tilt direction is stronger than that along the tilt direction (t-direction). In the case of the transient situation from a disordered initial state to a less disordered final state, such an anisotropy is likely to dominate random organization. This is because for the disordered initial state in the tilted field, the average distance between the vortices along the p-direction is expected to be smaller than that along the t-direction and hence the v-v interaction along the p-direction is stronger than that along the t-direction. This means that during the transient process, possible vortex configurations along the p-direction are limited compared to those along the t-direction. This limited degree-of-freedom of the vortex motion, which is introduced by tilting the field, may be a possible origin for the decreased relaxation time τ in the tilted field.

Acknowledgements
This work was supported by a Grant-in-Aid for Scientific Research (Grant Nos. 25287084 and 15K13516) from the Japan Society for the Promotion of Science. Y. K. acknowledges the financial support from the Global Center of Excellence Program by MEXT, Japan through the "Nanoscience and Quantum Physics" Project of the Tokyo Institute of Technology.

References

[1] Pine D J, Gollub J P, Brady J F and Leshansky A M 2005 *Nature* **438** 997
[2] Corté L, Chaikin P M, Gollub J P and Pine D J 2008 *Nature Phys.* **4** 420
[3] Reichhardt C, Reichhardt C J O 2009 *Phys. Rev. Lett.* **103** 168301
[4] Okuma S, Tsugawa Y and Motohashi A 2011 *Phys. Rev.* B **83** 012503
[5] Okuma S, Kawamura Y and Tsugawa Y 2012 *J. Phys. Soc. Jpn.* **81** 114718
[6] Nitta R, Kawamura Y, Kaneko S and Okuma S 2015 *Phys. Procedia* **65** 105
[7] Mangan N, Reichhardt C and Reichhardt C J O 2008 *Phys. Rev. Lett.* **100** 187002
[8] Hinrichsen H 2000 *Adv. Phys.* **49** 815
[9] Takeuchi K A, Kuroda M, Chate H and Sano M 2007 *Phys. Rev. Lett.* **99** 234503
[10] Zhang W, Zhou W and Luo M 2010 *Phys. Lett.* A **374** 3666
[11] Barghathi H and Vojta T 2012 *Phys. Rev. Lett.* **109** 170603
[12] Zhou C, Reichhardt C J O, Reichhardt C and Beyerlein I J 2015 *Sci. Rep.* **5** 8000
[13] Takeuchi K A, Kuroda M, Chaté H and Sano M 2009 *Phys. Rev.* E **80** 051116
[14] Jeanneret R and Bartolo D 2014 *Nat. Comms.* **5** 3474
[15] Cambell L J, Doria M M and Kogan V G 1988 *Phys. Rev.* B **38** 2439
[16] Kokubo N, Yoshimura T and Shinozaki B 2013 *J. Phys. Soc. Jpn.* **82** 094702
[17] Kamimura K, Nagatani K and Okuma S 2014 *JPS Conf. Proc.* **2** 010203
[18] Ochi A, Kawamura Y, Inoue T, Kaji T, Dobroka M, Kaneko S, Kokubo N and Okuma S 2016 *J. Phys. Soc. Jpn.* **85** 034712
[19] Barth J V, Costantini G and Kern K 2005 *Nature* **437** 671
[20] Classen T, Fratesi G, Costantini G, Fabris S, Stadler F L, Kim C, Gironcoli S, Baroni S and Kern K 2005 *Angew. Chem.* **117** 6298
[21] Liu C, Uchihashi T and Nakayama T 2008 *Phys. Rev. Lett.* **101** 146104

Numerical restoration of surface vortices in Nb films measured by a scanning SQUID microscope

Atsuki Ito[1], Ho Thanh Huy[1,2], Vu The Dang[1,2], Hiroki Miyoshi[1], Masahiko Hayashi[3], and Takekazu Ishida[1,4,*]

[1]Department of Physics and Electronics, Osaka Prefecture University, Sakai 599-8531, Japan

[2]University of Sciences, Vietnam National University HCMC, Ho Chi Minh, Viet Nam

[3]Faculty of Education and Human Studies, Akita University, Akita 010-8502, Japan

[4]Institute for Nanofabrication Research, Osaka Prefecture University, Sakai 599-8531, Japan

*E-mail: ishida@center.osakafu-u.ac.jp

Abstract. In the present work, we investigated a vortex profile appeared on a pure Nb film (500 nm in thickness, 10 mm x 10 mm) by using a scanning SQUID microscope. We found that the local magnetic distribution thus observed is broadened compared to a true vortex profile in the superconducting film. We therefore applied the numerical method to improve a spatial resolution of the scanning SQUID microscope. The method is based on the inverse Biot-Savart law and the Fourier transformation to recover a real-space image. We found that the numerical analyses give a smaller vortex than the raw vortex profile observed by the scanning microscope.

1. Introduction

While the penetration depth λ is one of the fundamental material parameters in a bulk type-II superconductor, a penetration depth λ is prolonged to an effective penetration depth Λ_{eff} in a film of small thickness d [1]. In our preceding studies, we investigated a vortex distribution in a concave decagon (star) and a Pacman superconducting plate by scanning SQUID microscopy for amorphous $Mo_{80}Ge_{20}$ films [2] [3] [4], the penetration depth in $Mo_{80}Ge_{20}$ film is rather extended compared to a sample size. This gives us an opportunity of investigating a *de facto* mesoscopic superconducting property even by using a micrometer-sized superconducting system.

In the present work, we measured vortex profiles on a pure Nb film (10 mm x 10 mm) by using a scanning SQUID microscope. The thickness of the Nb film is chosen to be 500 nm so as to minimize the enlarging effect of the penetration depth in a very thin film. We apply the numerical method to improve the spatial resolution of the scanning SQUID microscope in terms of the inverse Biot-Savart law [5]. We found the penetration depth of Nb film by correcting the effect of broadening of the vortex profile when the observation point is deviated from the surface. The penetration depth can be determined by the measurement of lower critical magnetic field H_{c1} while it would be difficult to determine it definitely due to the demagnetization effect. We consider that our method alternatively offers a direct method to determine the penetration depth in the superconducting film.

2. Vortex in a thin superconductor

In the mixed state of type-II superconductors, the magnetic flux penetrates into the superconductor in the form of flux quanta. By combining Maxwell's equation and the second London equation, one obtains an expression for spatial dependence of the local magnetic field $\vec{h}$ as

$$\nabla^2\vec{h} - \frac{\vec{h}}{\lambda^2} = -\frac{\Phi_0}{\lambda^2}\delta(\vec{r} - \vec{r_0})\boldsymbol{e}_z \quad (1)$$

where a center of the vortex is located at r_0. The solution for this equation is given by the zero-th order Bessel function:

$$h(r) = \frac{\Phi_0}{2\pi\lambda^2}K_0\left(\frac{r}{\lambda}\right) , \quad (2)$$

where K_0 is a Hankel function. At a large distance from the vortex origin $(r \rightarrow \infty)$, $h(r)$ decays exponentially as

$$h(r) \approx \frac{\Phi_0}{2\pi\lambda^2}\left(\frac{\pi}{2}\frac{\lambda}{r}\right)e^{-\left(\frac{r}{\lambda}\right)} \quad (3)$$

where Φ_0 (=2.07 x 10^{-15} Tm^2) is the flux quantum. On the other hand, the distribution of the magnetic flux line tends to spread over quickly as an observation point departs from the sample surface.

In Figure 1, we show the flux line distribution both inside and outside of the thin film as originally elucidated by Carneiro and Brandt [6]. The vortex profile inside and outside of a superconducting film is very dependent on the film thickness. Vortex profile inside of the thinner film is remarkably different from that of the bulk superconductor. This is the origin of the prolonged penetration depth in a thinner film. This is the reason why we choose an enough thickness of 500 nm for the $Mo_{80}Ge_{20}$ films used in the present studies. We find that the magnetic field penetrates into the superconducting film so as to give a radial distribution from the vortex center. It is difficult to measure a true local magnetic field by using a SQUID microscope because an observation point is somewhat deviated from the sample surface. Therefore, it is extremely difficult to

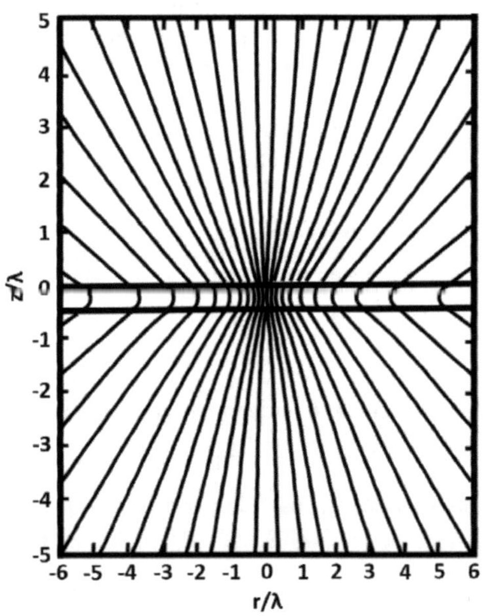

Figure 1. The calculations of the flux line distribution both inside and outside of the thin film in the *r-z* plane. The boundary surface of the superconducting film is shown by two horizontal lines [6].

observe a true vortex profile because the magnetic flux lines become broad very quickly when the observation point departs from the sample surface. In this work, we attempt to recover a true magnetic field distribution by using the inverse Biot-Savart law, which has been demonstrated to work properly to refine the vortex distribution in our relevant publication [5].

3. The inverse Biot-Savart Law

We use the algorithm of using the inverse Biot-Savart law algorithm developed by our group [5]. The magnetic flux measured by a SQUID coil is given by

$$\Phi(\vec{R}) = \frac{\mu_0}{2\cos\theta}\int \frac{d^3\vec{k}}{(2\pi)^2}(ik_x \sin\theta + k_\perp \cos\theta) \times \tilde{m}_z(\tilde{k}_\perp)e^{-k_\perp R_z + i\vec{k}_\perp \cdot \vec{R}_\perp}\mathcal{F}(\vec{k}_\perp) \tag{4}$$

where $\vec{R} = (R_x, R_y, R_z)$, θ is a declined angle of the SQUID sensor from the sample surface, and $\vec{k}$ is the wave number vector. We do not denote some of notations due to the limited pages allotted to this paper, and the reader should refer to Ref. [5] to identify the definition of some terminologies. The integral kernel is given by the following formula as

$$\mathcal{F}(\vec{k}_\perp) = \int_{-\infty}^{\infty} dx \int_{-\infty}^{\infty} dy\, f\left(\frac{x}{\cos\theta}, y\right) e^{k_\perp x \tan\theta + i\vec{k}_\perp \cdot (\frac{x}{\cos\theta}, y)} \tag{5}$$

The Fourier transform $m_z(k)$ of the magnetic moment $m_z(r)$ is expressed by

$$\tilde{m}_z(\vec{k}_\perp) = \frac{2\cos\theta}{\mu_0}\frac{\tilde{\Phi}(\vec{k}_\perp)e^{k_\perp R_z}\{\mathcal{F}(\vec{k}_\perp)\}^{-1}}{ik_x \sin\theta + k_\perp \cos\theta} \tag{6}$$

where $\tilde{\Phi}(\vec{k}_\perp)$ is the Fourier transform of the measured flux image. The real space image can be obtained by the inverse Fourier transformation of $\tilde{m}_z(\vec{k}_\perp)$. We obtain the current distribution on the surface with the aid of the Maxwell equation.

We consider that this algorithm [5] outlined here is quite useful to recover a true local field around a vortex. This is also useful to estimate the penetration depth λ by processing the broadening effect of the local magnetic flux measured by the scanning SQUID microscope at the observation point.

4. Measurements of Nb film by SQUID microscope

We measured a 500-nm thick Nb film (10 mm ×10 mm in size) with a scanning SQUID microscope (SQM-2000, Seiko Instruments, Inc.) with a pickup coil (10 μm in diameter). The reason we used the thick Nb film is to avoid a prolonged penetration depth, which appears in a very thin film as a Pearl vortex [1].

In Figure 2(a), we show an original image of a single vortex observed by the scanning SQUID microscope at 4K. If we observe the several vortices in a single image, it would be very difficult to discuss the quantization condition and the profile determined by the penetration depth. Therefore, we

IOP Conf. Series: Journal of Physics: Conf. Series **871** (2017) 012021 doi:10.1088/1742-6596/871/1/012021

searched for an image where a single vortex can be seen in the scanning area. After the several attempts of the in field-cooled measurements, we were successful in obtaining such an image of an isolated vortex as shown in Figure 2(a).

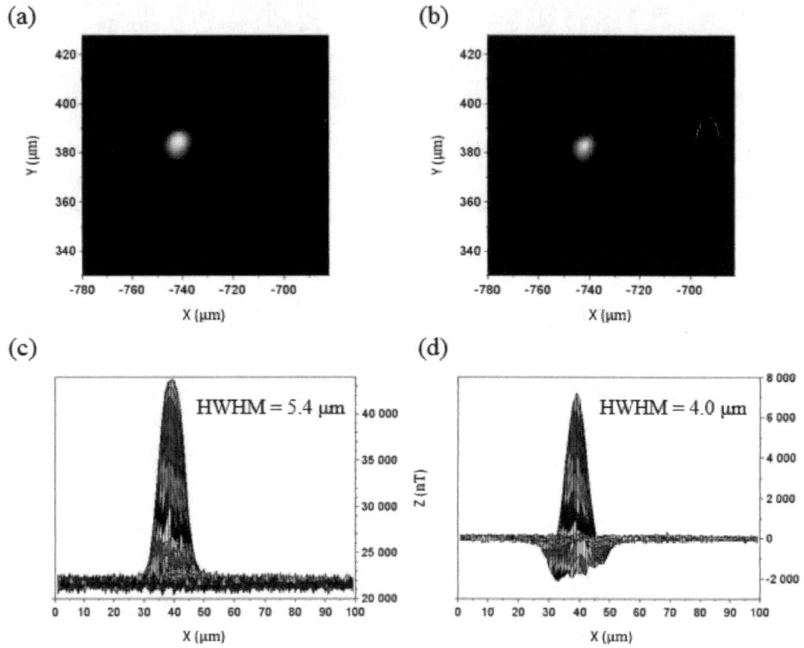

Figure 2. The vortex image in Nb film. (a) Original image of vortex by SQUID microscope. (b) Restored image of vortex by SQUID microscope. (c) Original image of vortex on the *x-z* plane (d) Restored image of vortex on the *x-z* plane.

We converted Figure 2(a) into a contour presentation to see a three-dimensional *XYZ* distribution. Then, we rotated the contour so as to be able to see it in the *X-Z* plane from the front side. In Figure 2(c), we show a front-side view of the single vortex obtained in such a manner. The vortex has a half width at half maximum (HWHM) of 5.4 µm. We may regard this value of 5.4 µm as an effective penetration depth if we consult a form of Eq. (3) having an exponential decay factor. Note that the local magnetic field distribution at observation points of Figure 2(a) is conducted not on the sample-surface plane but on a scanning plane somewhat lifted from the surface. This means that the observed vortex image is influenced by the broadening effect compared to a virtual true vortex image in the Nb film.

Hayashi *et al.* [5] successfully developed an algorithm to recover virtual true local magnetic fields on the sample surface. We applied the inverse Biot-Savart law to the original image of Figure 2(a), and obtained a restored image as shown in Figure 2(b), where a lift-off distance is taken as 1.6 µm. When we compare the restored image of Figure 2(b) with the original image of Figure 2(a), the profile of vortex becomes clearer. We take into account the deviation of the magnetic field distribution from the surface of Nb film in recovering an image on the sample-surface plane.

On the other hand, in Figure 2(c), we show an original image of vortex projected on the x-z plane while a restored image is projected on the x-z plane. In Figure 2(d), we assume that a distance from pickup coil to the surface of sample is 1.4 µm. We estimate a half width at half maximum (HWHM) as a measure of the penetration depth. The HWHM of the original image 5.4 µm, while the HWHM of the restored image is 4.0 μm. We note that the width of the peak decreased by 25%. We conclude that our numerical method is useful for extracting a true image coming from a difference of the sample surface and the observation point.

In Figure 3, we show the FWHM as a function of distance between pickup coil and sample surface. We found that HWHM increases rather monotonically as a function of a lift-off distance. This is consistent with the idea of local field broadening as indicated in Figure 1.

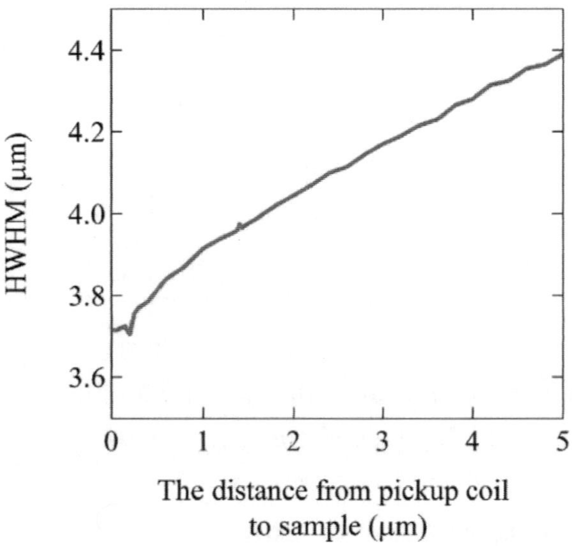

Figure 3. Relationship between The distance from pickup coil to sample and HWHM

5. Flux quantization of vortex

In investigating the penetration depth in Nb film, we first confirmed the validity of our original data by the SQUID microscope by evaluating the quantization condition of the magnetic flux in the mixed state for a single quantum of flux as [7]

$$\Phi_0 = \frac{hc}{2e} = 2.0678 \times 10^{-15}\ \mathrm{Tm}^2\ . \tag{7}$$

We investigated the Φ_0 value of the image observed by SSM satisfies the condition of the flux quantization with the following method.

In Figure 4, the red-colored square, which appeared at the surrounding area of the vortex, gives the magnetic flux density area of the flux quantization while the four areas indicated in green are used to evaluate an averaged background field. We can get a value of the flux quantization for the obtained image with the follow equation

$$\Phi = \int (B - \langle B \rangle) \mathrm{d}S \tag{8}$$

where B means the magnetic flux density of the measured flux quantization and $\langle B \rangle$ means the average of the background magnetic flux density. The calculated value of B was found to be 3.135×10^{-5} T and an averaged magnetic flux density $\langle B \rangle$ was found to be 2.166×10^{-5} T. We obtained the square area of $2.100 \times 10^{-10}\ \mathrm{m}^2$, and found the Φ_0 value of the image observed by SSM as $2.035 \times 10^{-15}\ \mathrm{T} \cdot \mathrm{m}^2$. This satisfied the condition of the flux quantization well. We found that the quantization condition is in good agreement with the actual flux evaluated by our SQUID microscope for a single isolated vortex appeared in the Nb film (see Figure 4).

6. Conclusion

We restore the penetration depth from the vortex profile when the observation point is deviated from the surface in the thick Nb film. We applied the Hayashi method [5] of the inverse Biot-Savart law analysis to recover a vortex image on the surface of the film. According to this numerical processing, we recovered a compressed vortex image in size, and confirmed that this process is useful to recover the deviation from a true local magnetic field. We investigated the Φ_0 value for the image obtained by SSM, and confirmed that it satisfies the flux quantization well. Further systematic studies are needed to elucidate remaining subjects on the SQUID microscopy.

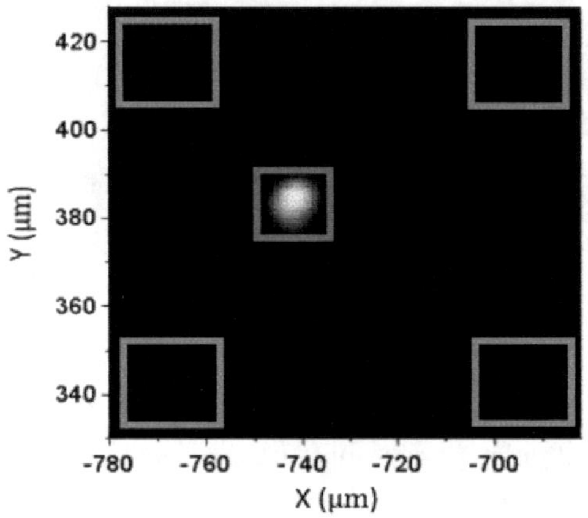

Figure 4. Original data used to confirm the flux quantization.

7. Acknowledgement

This work is partially supported by a Grant-in-Aid for Young Scientists (B) (No.26800192) and a Grant-in-Aid for Scientific Research (A) (No.16H02450) from JSPS.

References

[1] Pearl J *Appl Phys Lett* 1964 **5** 65
[2] Ho T H, Kato M, and Ishida T, *Supercond Sci Technol* 2013 **26** 065001
[3] Ho H T, Shishido H, Hayashi M, Yotsuya T, Kato M, and Ishida T, *Physica C* 2013 **484** 86
[4] Matsumoto H, Ho H T, Miyosi M, Okamoto T, Vu T D, Kato M, and Ishida T, *Physica C* 2016 **530** 46. (doi: j.physc.2016.06.015).
[5] Hayashi M, Ebisawa H, Ho H T, and Ishida T, *Appl Phys Lett* 2012 **100** 182601
[6] Carneiro G and Brandt E H, *Phys Rev B* 200 **61** 6370
[7] Abrikosov A A *Zh Eksperim iTeo Fiz* 1957 **32** 1442

IOP Conf. Series: Journal of Physics: Conf. Series **871** (2017) 012022 doi:10.1088/1742-6596/871/1/012022

Vortex states near absolute zero in a weak-pinning amorphous Mo_xGe_{1-x} film probed by pulsed mode-locking resonance

N Sohara[1], A Ochi[1], E Murakami[1], K Ienaga[1], S Kaneko[1], N Kokubo[2] and S Okuma[1]

[1]Department of Physics, Tokyo Institute of Technology, 2-12-1, Ohokayama, Meguro-ku, Tokyo 152-8551, Japan

[2]Department of Engineering Science, The University of Electro-Communications, 1-5-1 Chofugaoka, Chofu, Tokyo 182-8585, Japan

E-mail: sokuma@o.cc.titech.ac.jp

Abstract. We have developed measurements of the mode-locking (ML) resonance with pulsed currents, which generates much less heat than the conventional one with continuous currents. Here, we present the experimental details of the pulsed ML measurement. Using this technique, we have succeeded in determining the dynamic melting field of a driven vortex lattice for a weak-pinning thick amorphous Mo_xGe_{1-x} film down to 0.05 K. We construct an *ideal* vortex phase diagram in the *absence* of pinning near zero temperature as a function of magnetic field.

1. Introduction

Since the discovery of high-T_c cuprates, equilibrium vortex states in the field-temperature ($B-T$) plane have been thoroughly studied. While melting of the vortex lattice with increasing T and B has been well established, the true vortex-lattice melting in the limit of $T = 0$, which is induced by strong quantum fluctuations [1-17], remains unclarified despite its fundamental significance. Part of the reason is that the pinning effect unavoidably contained in actual samples gives rise to the order-disorder transition from the ordered (or weakly disordered) vortex-lattice phase (OP) to the disordered vortex-glass phase (DP) [18-20]. This is associated with the peak effect at B_p, where B_p is determined from a peak in the depinning current [21-27]. The emergence of DP masks the observation of the true melting transition of the vortex lattice. To overcome this difficulty, in our earlier work we have performed the measurement of the mode-locking (ML) resonance [28-38], which enables to probe melting of the moving lattice decoupled from the underlying pinning potential [28-30,39-41].

However, the ML measurement could not be carried out at low enough T, e.g., $T < 1$ K, because the fast vortex motion generates more heat than the conventional transport measurement. Quite recently, we have developed the *pulsed* ML measurement that generates much less average heat, by three orders of magnitude, than the conventional one [42]. In this paper, we present the experimental details of the pulsed ML measurement. Hopefully, this method will be widely used to detect the dynamics of the fast-driven elastic object, as well as vortex matter, interacting with the substrates that may generate large heat.

By using this method, we determine the dynamic melting field $B^{\infty}_{c,dyn}(0)$ of the driven vortex lattice in a weak-pinning amorphous (a-)Mo_xGe_{1-x} film near $T = 0$. We construct an *ideal* vortex phase diagram in the *absence* of pinning, at $T = 0$ as a function of magnetic field B, which will be compared with the conventional one with weak pinning. Since the a-Mo_xGe_{1-x} film is a typical, conventional type-II superconductor with weak pinning, the results obtained in this work are helpful to understand the vortex states at low T for a variety of superconductors. More detailed data, analysis, and discussion have been given in Ref. [42].

2. Experimental details

An a-Mo_xGe_{1-x} film with thickness of 330 nm was fabricated by rf sputtering on a Si substrate held at room temperature [29-32]. The linear resistivity ρ and dc current-voltage ($I - V$) characteristics were measured using a standard four-terminal method. The mean-field transition temperature T_{c0} defined by a 95% criterion of the normal-state resistivity [11, 12], $\rho(T_{c0}) = 0.95\rho_n$, and the zero-resistivity temperature T_c are 6.1 and 6.0 K, respectively. The equilibrium melting field $B_c(T)$ and the upper critical field $B_{c2}(T)$ are determined from $\rho(B_c) \to 0$ and $\rho(B_{c2}) = 0.95\rho_n$, respectively.

Figures 1(a) and (b) illustrate schematically the time evolution of the applied current I in the continuous-current mode used previously and the pulsed-current mode developed in this work, respectively. In the continuous mode, the dc current with fixed magnitude superimposed with the ac current I_{rf} is applied to the vortex system. The dc voltage V induced by the vortex motion is averaged over ten seconds and recorded. Then, the magnitude of the dc current I is increased in a step-by-step manner, as illustrated in figure 1(a). In the pulsed mode, on the other hand, the square pulse with duration of 1 ms is applied and its repetition time is set to 1 s, as shown in figure 1(b), which makes the duty ratio as small as 0.1 %. The voltage signal due to the vortex motion is again averaged over 10 s and recorded. In our ML measurement, the ac current I_{rf} with a frequency f_{ext} of up to 50 MHz is superimposed on the dc I pulse with 1 ms width. Hence, the number of cycles for superimposed I_{rf} should be adjusted depending on f_{ext}. For f_{ext} =50 MHz used in this work, for example, we apply I_{rf} over a duration of 50,000 cycles.

Figure 2 shows schematically the electrical circuit for the pulsed ML measurement, where the independent ac- and dc-pulse sources are used. The voltage enhanced with a preamplifier is recorded with a fast-Fourier transform (FFT) spectrum analyzer, which enables to measure $V(t)$

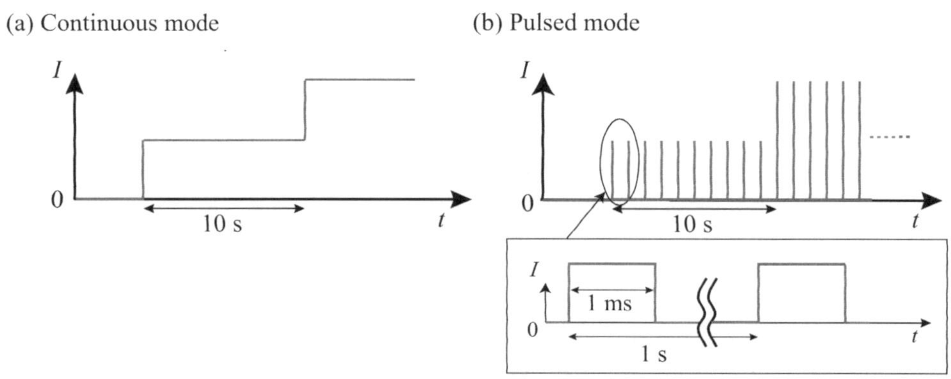

Figure 1. Schematic diagrams of the time evolution of the applied current I in (a) the continuous-current mode used conventionally and (b) the pulsed-current mode developed in this work, respectively. The ac current I_{rf} superimposed on I is not shown in (a) and (b). The lower diagram in (b) schematically illustrates the enlarged view of individual I pulses.

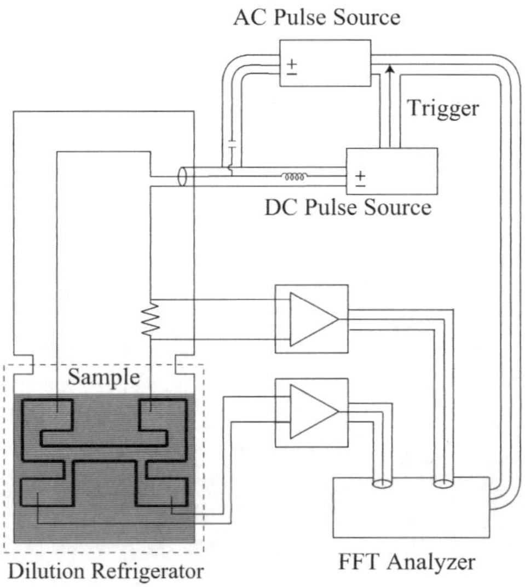

Figure 2. Schematic illustration of the electrical circuit used for the pulsed ML measurement.

with a voltage resolution as small as 0.01 mV. Since the time resolution of our FFT analyzer is 25 μs, the duration of the dc pulse should be longer than about 1 ms to ensure the accuracy of the ML measurement. The dc pulse is also used as a trigger signal to synchronize all the devices. In addition, we connect a capacitor and a coil with the ac and dc pulse sources, respectively, to prevent each pulse signal from penetrating the other pulse source.

Figures 3(a) and (b), respectively, show the set of waveforms for the voltage pulse $V(t)$ and current pulse $I(t)$ with different amplitudes superimposed with fixed 50 MHz I_{rf} measured in the pulsed current mode at 0.05 K in 8.0 T. Here, $I(t)$ of a given amplitude and corresponding $V(t)$ are indicated with the same color. A representative waveform of $V(t)$ selected from the data of figure 3(a) is shown in figure 3(c). The waveform of $V(t)$ as well as that of $I(t)$ exhibits a plateau-like behavior and hence $I-V$ characteristics are readily obtained from the plateau values. In the pulsed mode, the heat dissipated in the sample is estimated to be a few tens nW at around the first ML resonance step (see below). This leads to an increase in the local temperature of a few tens mK at 0.05 K [42], which will not affect the discussion that follows.

Now, we examine the experimental resolution of the ML measurement in the pulsed current mode in comparison with that in the continuous current mode. The solid and open circles in figure 4(a) show the dc I-V characteristics superimposed with 40 MHz I_{rf} at 4.1 K in 4.0 T, corresponding to OP near B_p, measured in the pulsed and continuous modes, respectively. Figure 4(b) displays $\mathrm{d}I/\mathrm{d}V$ versus V, where symbols correspond to those in figure 4(a). In this temperature and current-voltage range, heating effects caused by the vortex motion are negligibly small even if the continuous current mode is used. Figures 4(a) and (b) clearly show that the pulsed ML measurement yields the same result as the continuous one as long as the heating effects are not important and that the experimental resolution of ML in the pulsed mode is of the same level as that in the continuous mode [42].

3. Results and discussion

When a vortex lattice is driven at a given dc velocity v through a random pinning potential, it experiences a velocity modulation with a frequency f_{int} determined by the ratio of v to a lattice constant in the flow direction a, $f_{int} = v/a$ [41]. When the lattice is driven under the

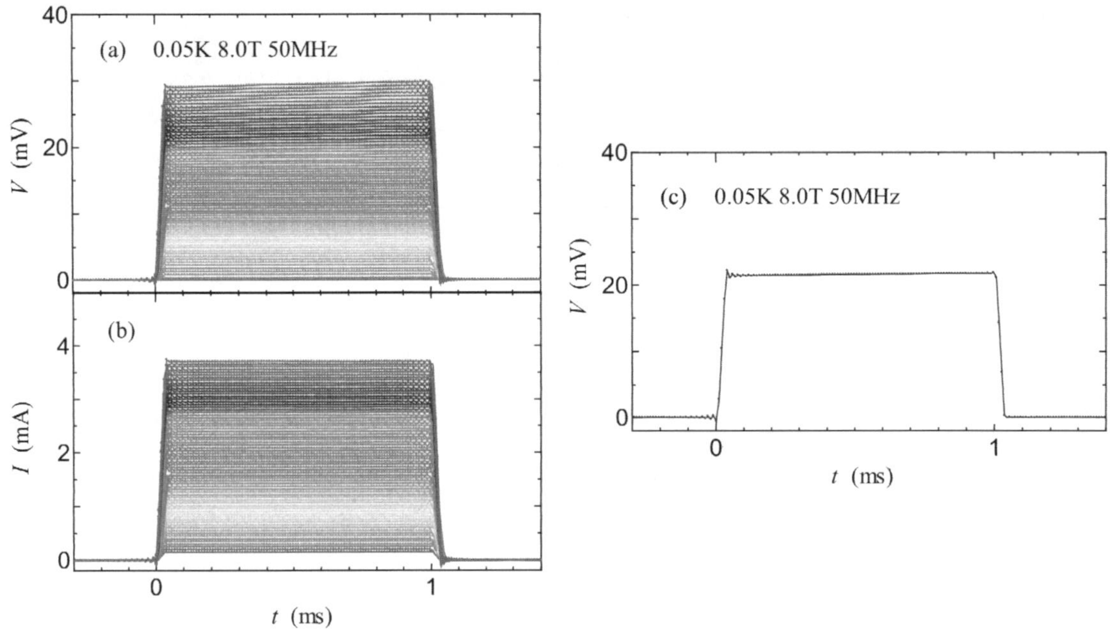

Figure 3. The set of waveforms for (a) the voltage pulse $V(t)$ generated by the vortex motion at 0.05 K in 8.0 T and (b) the current pulse $I(t)$ with different amplitudes on which 50 MHz I_{rf} with fixed amplitude is superimposed. Here, $I(t)$ of a given amplitude and corresponding $V(t)$ are indicated with the same color. (c) A typical waveform of $V(t)$ selected from the data in (a). All the waveforms show a plateau-like structure.

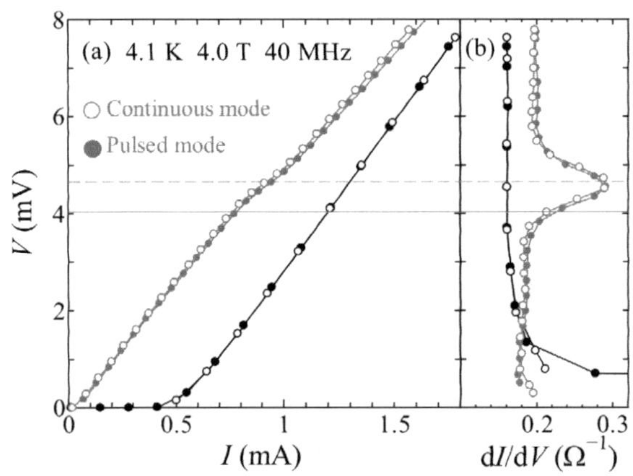

Figure 4. (a) Dc I-V characteristics at 4.1 K in 4.0 T measured in the pulsed (black solid circles) and continuous (black open circles) current modes, and those taken with superimposed 40 MHz I_{rf} in pulsed (red solid circles) and continuous (red open circles) current modes. (b) $\mathrm{d}I/\mathrm{d}V$ versus V, where symbols correspond to those in (a). Horizontal full and dashed lines mark the location of the ML resonance expected for the perpendicular and parallel orientations, respectively. Other lines are guides for the eye [42].

influence of externally applied dc and ac forces, a steplike structure analogous to Shapiro steps found in Josephson junctions appears in the $I-V$ characteristics [28-30,33-38], where I and V correspond to the dc driving force and the average velocity of vortices, respectively. The steps occur when the internal frequency $f_{int} = v/a$ of the system locks to the external frequency f_{ext} of the ac drive I_{rf} or when the relation $qf_{int} = pf_{ext}$ is satisfied, where p and q are integers.

If we assume a triangular vortex array, i.e., Abrikosov lattice, moving in the direction perpendicular to one side of the triangle(s), which we call the perpendicular orientation, we can calculate the value of the fundamental voltage step ($V_{p/q}^{perp}$) for a given B, satisfying the subharmonic resonant condition of $p/q = 1/2$, to be $V_{1/2}^{perp} = lf_{ext}(\sqrt{3}\Phi_0 B/2)^{1/2}$, where l is the distance between the voltage contacts and Φ_0 is the flux quantum [29, 31, 32, 36, 37]. The location of $V_{1/2}^{perp}$ is indicated by a full horizontal line in figures 4 and 5. When the vortex lattice is moving in the direction parallel to one side of the triangle(s), which we call the parallel orientation, the fundamental voltage step is calculated to be $V_{1/1}^{para} = (2/\sqrt{3})V_{1/2}^{perp}$. This is illustrated with a dashed horizontal line in figures 4 and 5. The observed ML resonance voltages in figures 4(a) and 4(b) nearly agree with the calculated values for the parallel orientation. This is consistent with previous work in which the lattice orientation favors the parallel one when the field is increased to around B_p [32, 43].

Let us present the data of the ML resonance acquired using pulsed currents at T =0.05 K, corresponding to $T/T_{c0} = 0.01$, which is the lowest temperature attained in this work. Figure 5(a) shows the I-V curves measured with superimposed 50 MHz I_{rf} of different amplitudes listed in the figure ($I_{rf} = 0.31$ - 2.77 mA) at 0.05 K in 8.0 T, which corresponds to OP prior to

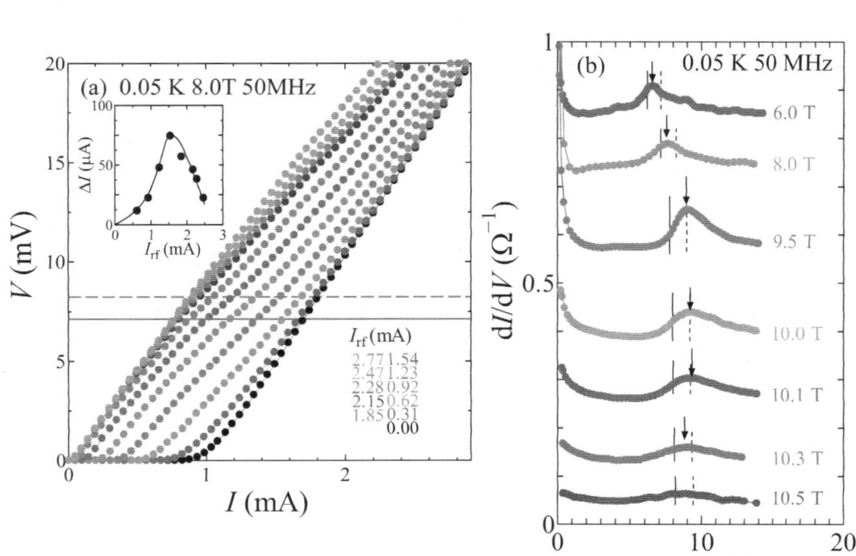

Figure 5. (a) I-V curves at 0.05 K in 8.0 T taken with superimposed 50 MHz I_{rf} of different amplitudes listed in the figure. Black circles represent the I-V data measured in the absence of superimposed I_{rf}. The location of $V_{1/2}^{perp}$ and $V_{1/1}^{para}$ is indicated with horizontal full and dashed lines, respectively. Inset: ΔI plotted as a function of I_{rf}. (b) dI/dV versus V curves measured with superimposed 50 MHz I_{rf} for different B [42]. A short arrow marks the location of the observed ML peak at each B. Short full and dashed lines indicate the location of $V_{1/2}^{perp}$ and $V_{1/1}^{para}$, respectively.

$B_p(\approx 10$ T). Black circles depict the I-V data taken in the absence of superimposed I_{rf}, showing the smooth curvature. In the presence of superimposed I_{rf}, by contrast, we observe the step-like structure at around a voltage marked with a horizontal full line. Here, the location of the calculated $V_{1/2}^{perp}$ and $V_{1/1}^{para}$ is indicated with the horizontal full and dashed lines, respectively. The ML resonance is more clearly seen by plotting dI/dV against V, as shown in figure 5(b). To quantify the strength of the ML resonance, we extract the width of the current step ΔI at the ML resonance in the I-V characteristics by integrating the peak of dI/dV versus V curves with respect to the flux-flow baseline [31, 32, 36, 37]. As shown in the inset of figure 5(a), thus obtained ΔI exhibits a maximum as a function of I_{rf}, consistent with earlier work that $\Delta I(I_{rf})$ is given by a squared Bessel function of the first kind [28, 36]. In the following discussion, we will deal with the data corresponding to the maximum resonance ΔI_{max}.

Now, we focus on the field dependence of the ML resonance at 0.05 K. Shown in figure 5(b) are the dI/dV versus V curves measured with superimposed 50 MHz I_{rf} for different B. Short vertical full and dashed lines represent the location of $V_{1/2}^{perp}$ and $V_{1/1}^{para}$ expected for the perpendicular and parallel orientations, respectively. A short arrow marks the position of the observed ML peak in individual fields indicated in the figure. The peak structure is visible over the broad B up to about 10.3 T but vanishes at $B \gtrsim 10.5$ T. By plotting ΔI_{max} against B, we find that at both 0.05 and 0.4 K, ΔI_{max} takes a sharp peak around B_p (≈ 10 T) and rapidly falls to zero at a certain field just above B_p [42]. The sharp peak structure in ΔI_{max} found around near B_p implies the temporal recovery of the ML resonance, which is most likely attributed to the growth in the portion of the ordered domains with parallel orientation [43]. The subsequent sudden drop in ΔI_{max} shows the rapid collapse of the lattice order, indicating a sharp melting from a moving lattice into a moving liquid. To identify the dynamic melting field $B_{c,dyn}$ at each temperature, we determine $B_{c,dyn}$ from a simple linear extrapolation of ΔI_{max} to zero. Then, we evaluate the dynamic melting field $B_{c,dyn}^{\infty}$ in the limit of the infinite velocity ($v \to \infty$) as a field at which f_{ext}^{-1} versus $B_{c,dyn}(f_{ext})$ extrapolates linearly to $f_{ext}^{-1} \to 0$. Thus obtained values of $B_{c,dyn}^{\infty}$ at 0.05 and 0.4 K are 10.5 and 10.4 T, respectively.

By plotting $B_{c,dyn}^{\infty}(T)$ and $B_{c2}(T)$ in the $B-T$ plane, we now complete the dynamic as well as the static vortex phase diagram over the entire B and T range [42]. From a simple extrapolation of the $B_{c,dyn}^{\infty}(T)$ curve to $T = 0$, we are able to obtain $B_{c,dyn}^{\infty}(0)$, as well as $B_p(0)$, $B_c(0)$ and $B_{c2}(0)$, thus constructing the phase diagram of an *ideal* vortex system *without* pinning at $T = 0$ as a function of B. This is illustrated in figure 6(a). For comparison, the similar $T = 0$ phase

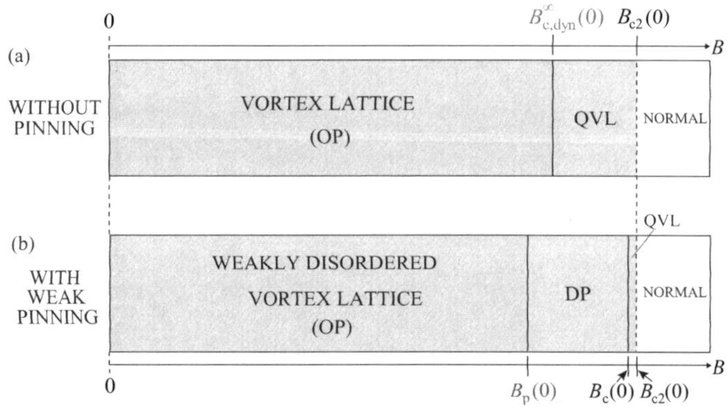

Figure 6. (a) The $T = 0$ equilibrium phase diagram with respect to B for an ideal vortex system without pinning and (b) that in an actual system with weak pinning [42].

diagram for a real system with weak pinning, which is constructed based on the data of $B_p(T)$, $B_c(T)$, and $B_{c2}(T)$ at $T \to 0$, is shown in figure 6(b).

From figures 6(a) and (b), one can trace how the equilibrium phase diagram for the ideal pinning free system alters as a small amount of pinning centers is introduced into the system. In the absence of pinning, only the vortex-lattice phase (OP) and the vortex-liquid phase exist, where the melting field of the vortex lattice is identified as $B^{\infty}_{c,dyn}(0)$ and the vortex-liquid phase at $T \to 0$ is considered to be a quantum vortex liquid (QVL) [11, 12]. When the weak pinning centers are introduced into the system, DP (or the vortex-glass phase) emerges just above OP. The upper bound of OP is slightly suppressed from $B^{\infty}_{c,dyn}(0)$ =10.5 T to $B_p(0)$ =10.0 T, indicative of pinning-induced disordering of the vortex lattice. On the other hand, the "melting" field between the vortex solid phase (DP) and the QVL phase is remarkably pushed up from $B^{\infty}_{c,dyn}(0)$ =10.5 T to $B_c(0)$ =12.4 T, leading to a significant suppression of the QVL phase. This result is explained in terms of the enhanced pinning effect at low T. We have indeed found that at high temperatures ($T \gtrsim 3$ K), $B^{\infty}_{c,dyn}(T)$ is very close to $B_c(T)$. This is because at high T, thermal-fluctuation effects on vortex pinning are so strong that the melting field $B^{\infty}_{c,dyn}(T)$ in the absence of pinning stays almost unaffected by the introduction of a small amount of pinning.

4. Conclusion

We have developed and presented the pulsed ML measurement that generates much less heat than the conventional one and successfully determined the dynamic melting field $B^{\infty}_{c,dyn}(T)$ of the driven vortex lattice for the thick a-Mo_xGe_{1-x} film in the limit $T \to 0$ and infinite velocity $v \to \infty$. From the dynamic as well as equilibrium $B - T$ phase diagram at low T, we have mapped out an ideal vortex phase diagram in the absence of pinning, at $T \to 0$ as a function of B. We hope that the pulsed ML technique developed and detailed in this paper may be widely used to explore the dynamics of the fast driven vortex matter and elastic objects interacting with random substrates that may generate large heat.

Acknowledgments

This work was supported by a Grant-in-Aid for Scientific Research from the Ministry of Education, Culture, Sports, Science, and Technology of Japan. A. O. acknowledges the financial support from the Global Center of Excellence Program by MEXT, Japan through the "Nanoscience and Quantum Physics" Project of the Tokyo Institute of Technology.

References

[1] Blatter G, Ivlev B, Kagan Y, Theunissen M, Volokitin Y and Kes P 1994 *Phys. Rev.* B **50** 13013
[2] Blatter G and Ivlev B 1993 *Phys. Rev. Lett.* **70** 2621
[3] Fisher M P A 1990 *Phys. Rev. Lett.* **65** 923
[4] Ikeda R 1996 *Int. J. Mod. Phys.* B **10** 601
[5] Myojin K, Ikeda R and Koikegami S 2008 *Phys. Rev.* B **78** 014508
[6] Chudnovsky E M 1995 *Phys. Rev.* B **51** 15351
[7] Rozhkov A and Stroud D 1996 *Phys. Rev.* B **54** R12697
[8] Chervenak J A and Valles Jr J M 1996 *Phys. Rev.* B **54** R15649
[9] Sasaki T, Biberacher W, Neumaier K, Hehn W, Andres K and Fukase T 1998 *Phys. Rev.* B **57** 10889
[10] Shibauchi T, Krusin-Elbaum L, Blatter G and Mielke C H 2003 *Phys. Rev.* B **67** 064514
[11] Okuma S, Imamoto Y and Morita M 2001 *Phys. Rev. Lett.* **86** 3136
[12] Okuma S, Togo S and Morita M 2003 *Phys. Rev. Lett.* **91** 067001
[13] Bustarret E, Kačmarčik J, Marcenat C, Gheeraert E, Cytermann C, Marcus J and Klein T 2004 *Phys. Rev. Lett.* **93** 237005
[14] Taylor B J, Scanderbeg D J, Maple M B, Kwon C and Jia Q X 2007 *Phys. Rev.* B **76** 014518
[15] Taylor B J and Maple M B 2007 *Phys. Rev.* B **76** 014517
[16] Kuwasawa Y, Kato K, Matsuo M and Nojima T 1998 *Physica* C **305** 95
[17] Saito Y, Nojima T and Iwasa Y 2016 *Supercond. Sci. Technol.* **29** 093001
[18] Okuma S, Kashiro K, Suzuki Y and Kokubo N 2008 *Phys. Rev.* B **77** 212505

IOP Conf. Series: Journal of Physics: Conf. Series **871** (2017) 012022 doi:10.1088/1742-6596/871/1/012022

[19] Paltiel Y *et al.* 2000 *Phys. Rev. Lett.* **85** 3712
[20] Paltiel Y *et al.* 2000 *Nature* **403** 398
[21] Kes P H and Tsuei C C 1983 *Phys. Rev.* B **28** 5126
[22] Bhattacharya S and Higgins M J 1993 *Phys. Rev. Lett.* **70** 2617
[23] Marley A C, Higgins M J and Bhattacharya S 1995 *Phys. Rev. Lett.* **74** 3029
[24] Ghosh K *et al.* 1996 *Phys. Rev. Lett.* **76** 4600
[25] Rosenstein B and Zhuravlev V 2007 *Phys. Rev.* B **76** 014507
[26] Bermúdez M M, Eskildsen M R, Bartkowiak M, Nagy G, Bekeris V and Pasquini G 2015 *Phys. Rev. Lett.* **115** 067001
[27] Guillamón I, Córdoba R, Sesé J, De Teresa J M, Ibarra M R, Vieira S and Suderow H 2014 *Nat. Phys.* **10** 851
[28] Schmid A and Hauger W 1973 *J. Low Temp. Phys.* **11** 667
[29] Kokubo N, Asada T, Kadowaki K, Takita K, Sorop T G and Kes P H 2007 *Phys. Rev.* B **75** 184512
[30] Okuma S, Inoue J and Kokubo N 2007 *Phys. Rev.* B **76** 172503
[31] Okuma S, Imaizumi H and Kokubo N 2009 *Phys. Rev.* B **80** 132503
[32] Okuma S, Imaizumi H, Shimamoto D and Kokubo N 2011 *Phys. Rev.* B **83** 064520
[33] Fiory A T 1971 *Phys. Rev. Lett.* **27** 501
[34] Higgins M J, Middleton A A and Bhattacharya S 1993 *Phys. Rev. Lett.* **70** 3784
[35] Reichhardt C, Scalettar R T, Zimányi G T and Grϕnbech-Jensen N 2000 *Phys. Rev.* B **61** R11914
[36] Kokubo N, Kadowaki K and Takita K 2005 *Phys. Rev. Lett.* **95** 177005
[37] Okuma S, Shimamoto D and Kokubo N 2012 *Phys. Rev.* B **85** 064508
[38] Reichhardt C and Olson Reichhardt C J 2015 *Phys. Rev.* B **92** 224432
[39] Koshelev A E and Vinokur V M 1994 *Phys Rev. Lett.* **73** 3580
[40] Olson C J, Reichhardt C and Nori F 1998 *Phys. Rev. Lett.* **81** 3757
[41] Togawa Y, Abiru R, Iwaya K, Kitano H and Maeda A 2000 *Phys. Rev. Lett.* **85** 3716
[42] Ochi A, Sohara N, Kaneko S, Kokubo N and Okuma S 2016 *J. Phys. Soc. Jpn.* **85** 044701
[43] Zhuravlev V and Maniv T 2006 *Europhys. Lett.* **73** 955

Simulation of Field Dependence of Critical Current Densities of Bulk High T_c Superconducting Materials regarding Thermally Activated Flux Motion

M Santosh[1*], S Pavan Kumar Naik[2] and M R Koblischka[3]

[1] E300 Huron Street, M5S 3J6, Faculty of Arts and Science, University of Toronto, Canada
[2] Superconducting Materials Laboratory, Department of Materials Science and Engineering, Shibaura Institute of Technology, 3-7-5 Toyosu, Koto-ku, Tokyo 135-8548, Japan
[3]Experimental Physics, Saarland University, Campus C 6 3, 66123 Saarbrücken, Germany

[*]santoshjuly@msn.com

Abstract. In the upcoming generation, bulk high temperature superconductors (HTS) will play a crucial and a promising role in numerous industrial applications ranging from Maglev trains to magnetic resonance imaging, etc. Especially, the bulk HTS as permanent magnets are suitable due to the fact that they can trap magnetic fields being several orders of magnitude higher than those of the best hard ferromagnets. The bulk HTS $LREBa_2Cu_3O_{7-\delta}$ (LREBCO or LRE-123, LRE: Y, Gd, etc.,) materials could obtain very powerful compact superconducting super-magnets, which can be operated at the cheaper liquid nitrogen temperature or below due to higher critical temperatures (i.e., ~90 K). As a result, the new advanced technology can be utilized in a more attractive manner for a variety of technological and medical applications which have the capacity to revolutionize the field. An understanding of the magnetic field dependence of the critical current density ($J_c(H)$) is important to develop better adapted materials. To achieve this goal, a variety of $J_c(H)$ behaviours of bulk LREBCO samples were modelled regarding thermally activated flux motion. In essence, the $J_c(H)$ curves follows a certain criterion where an exponential model is applied. However, to fit the complete $J_c(H)$ curve of the LRE-123 samples an unique model is necessary to explain the behavior at low and high fields. The modelling of the various superconducting materials could be understood in terms of the pinning mechanisms.

1. Introduction

The physical properties of high critical temperature (T_c) cuprate superconductors are highly susceptible to impurities and crystalline defects present in the final composites [1]. In cuprate superconductors, $LREBa_2Cu_3O_{7-\delta}$ (LREBCO or LRE-123, with light rare earth elements such as Y, Nd, Sm, Gd, etc.,) materials were found to be well suited for technological applications due to low flux creep effects as compared to Bi-Sr-Ca-Cu-O compounds. The important applications of the superconducting materials can be realized in many fields such as SQUIDS which is a highly sensitive magnetometer, superconducting electromagnets which are more powerful than ferromagnetic materials. Furthermore, magnetic resonance imaging and nuclear magnetic resonance imaging in the medical field and fusion reactors based on magnetic confinement, particle accelerators, fault current limiters, electric motors and generators, high sensitivity particle detectors, etc., are possible applications. All these applications clearly demonstrate a necessity to develop high quality superconductors. A high critical current density (J_c) is the prime requirement; designing a suitable microstructure of the final composition is elementary to enhance the magnetic properties at the boiling point of liquid nitrogen, i.e., 77 K. However, several reports from the literature show that it is essential to process the LREBCO compounds, except YBCO in reduced oxygen atmospheres or addition of extra Ba to prevent the formation of unwanted $LRE_{1-x}Ba_{2+x}Cu_3O_y$ solid solutions (LRE/Ba–ss) type with lower T_c [1-4]. The LRE^{3+} ions can occupy the Ba^{2+} sites because of similar ionic radii. Increasing replacement of Ba^{2+} sites can cause a reduction of T_c and lead to a deterioration of the superconducting properties [5-8]. The oxygen controlled atmospheres and addition of Ba not only suppresses the formation of solid solutions of the LRE/Ba–ss, but permits local compositional fluctuations of the $(LRE_{1-x}, LRE^1{}_x)Ba_2Cu_3O_{7-\delta}$ type (i.e. LRE/LRE^1 –ss, where LRE and LRE^1 are different light rare earth elements) in the processed samples which leads to enhanced flux pinning [8]. This process has suppressed the rapid fall of the J_c values at higher fields, often reported in MG processed samples.

In the case of melt-grown superconductors, further increase of the J_c values were reported by introducing nano-sized inclusions and mixed rare earth elements into the superconducting matrix phase [9]. Studies in the literature on mixed LREBCO superconductors [10-14] report strong flux pinning due to lattice mismatch defects arising from the presence of different LRE elements in the matrix. Hence, the high J_c is maintained up to high fields due to enhanced pinning which could be related to the following factors: firstly, LRE sites are substituted by LRE^1 elements, due to which the lattice mismatch arises among the neighbouring unit cells of LRE-123 and LRE^1-123.

The source of flux pinning in different field regions in the LREBCO materials have been investigated widely in recent years. The presence of low-T_c clusters due to the formation of different chemical fluctuations both the LRE/Ba-ss and LRE/LRE^1-ss has been confirmed through various methods. Muralidhar *et al.* [9] have proposed that in LREBCO composites nanometric sized LRE-rich low-T_c clusters act as field-induced pinning centres (ΔT_c pinning centres) and thus result in the peak effect. It is known that to enhance the J_c-H performance of superconductors, both the T_c and the number of effective pinning centres should be increased. The coherence length of the LREBCO superconductors is on the nanoscale; thus the increment of defects or size of compositional fluctuation in the nano-scale is beneficial for flux pinning. Including the interfacial defects generated due to 211/123 interface, twins, dislocations, chemical compositional fluctuations created due to the presence of different 123 unit cells, and stacking faults have been proven to be effective flux pinning centers [13-19].

The microstructural features in the LREBCO superconductors prevent the movements of the fluxiods. Depending on the size and type of the defect, the pinning will be effective at different fields. It is necessary to understand the basic mechanism playing different roles in LREBCO superconductors which are pivotal for enhanced magnetic properties. For this purpose, we have undertaken modelling of the J_c curves of different bulk LREBCO superconductors, which may help to understand the different pinning mechanisms and by which one can produce high quality LREBCO superconductors.

2. Experimental details

The synthesis of all precursor powders and fabrication of all the YBCO bulks are discussed elsewhere [20]. Sample A was fabricated with ball milled Y-211 precursor powders. Sample B was fabricated with as synthesized Y-211 phase particles. Both samples A and B were processed employing the infiltration growth process. Sample C is also YBCO superconductor, but processed via the melt-growth technique. The field dependence of critical current density ($J_c(H)$) of the processed LREBCO superconductors was calculated using extended Beans critical state model by utilizing the *M-H* loops recorded up to 6 T field at 77 K. To evaluate the different pinning mechanisms, present in the LREBCO bulk high-T_c superconductors and the $J_c(H)$ curves are fitted with an exponential expression.

3. Results and discussion

From the literature [21], it is well known that the size of the defects in the superconductors play a role influencing J_c at different applied magnetic fields. The presence of such defects can give rise to strong flux pinning or a peak in J_c at a particular field. This peak field is related to the vortex lattice spacing as shown in equation (1).

$$H_p = \frac{\emptyset_o}{\sqrt{3}(a_f^2)} \quad (1)$$

where ϕ_o is the flux quantum and a_f is the vortex lattice spacing. Equation (1) suggests that the flux pinning at fields less than 1 T can be caused by defects of a size more than 50 nm. The interfacial defects due to LRE-211/LRE-123 boundaries would give rise to flux pinning at such low fields. In the literature, correlations were made on the flux pinning at low fields due to LRE-211 content and the defects originating at the LRE-211/LRE-123 interfaces [22-25]. Figure 1 represents the possible dependence of the peak field on the defect spacing at different fields as given by equation (1). From this figure, it is clear that the smaller the defect size, the larger the peak field will be. Therefore, it is crucial to create fine sized defects or to control the size of LRE-211 particles within the LRE-123 matrix to make the interface defect density large in the LRE-123 matrix.

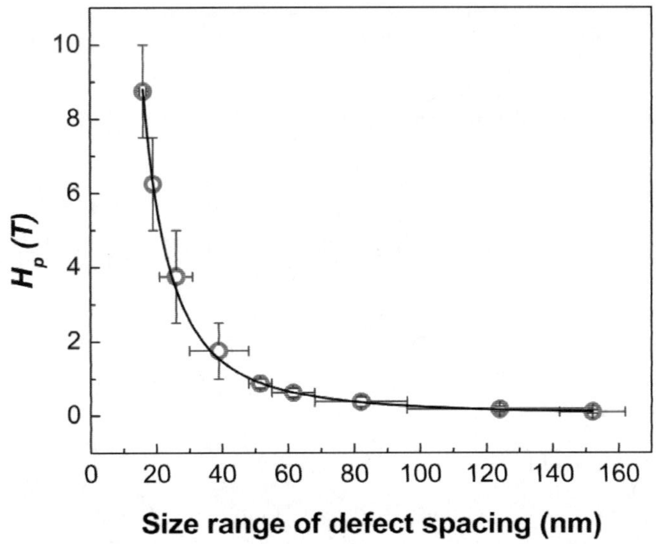

Figure 1. The relation of the peak field (H_p) with the size of the defect spacing in cuprate superconductors is illustrated as schematic diagram.

To account for high J_c observed up to large (1-10 T) fields, the defect spacing should vary from 15 – 50 nm. Researchers believe that the pinning at low field depends on the density of the interfacial defects due to LRE-211/LRE[1]-123 interfaces present in the matrix. In the present work, three different YBCO superconductors are being discussed which exhibited the enhancement of J_c up to high fields as compared to the literature [26-29]. From the above discussion, the presence of different $J_c(H)$ behaviour in LREBCO superconductors could be attributed to different types of pinning centres which are induced by various additives/defects. Therefore, $J_c(0)$ of sample A is superior to that of sample B and C. The rapid fall in the $J_c(H)$ curves of the melt-textured YBCO with large J_c values is reported to exhibit an exponential variation with magnetic field [30-33]. Since the present $J_c(H)$ curves show faster reduction at low fields, but slow decrease at high fields, we assume that there are two different mechanisms operative in describing the $J_c(H)$ behaviour across the field range up to 5 T. As a result, we consider two terms to fit the measured data to equation (2) given below.

$$\frac{J_c(H)}{J_c(0)} = A_1 \exp\left(-\frac{\mu_o H}{t_1}\right) + A_2 \exp(-\frac{\mu_o H}{t_2}) + A_3 \exp(-\frac{\mu_o H}{t_3}) + y_0 \quad (2)$$

where A_1, t_1, A_2, t_2, A_3, t_3, and y_0 are the fit parameters.

In order to investigate the nature of flux pinning that occurs up to high fields at 77 K, we have normalized the $J_c(H)$ with respect to zero field J_c, i.e., (J_c(H)/J_c(0)), which is plotted as a function of the applied field in figures 2 (a)-(c) for sample A, B and C, respectively. In each panel, both the experimental J_c/J_c(0) curve and the detailed fitting of the measured field dependence of the normalized J_c to equation (2) is shown.

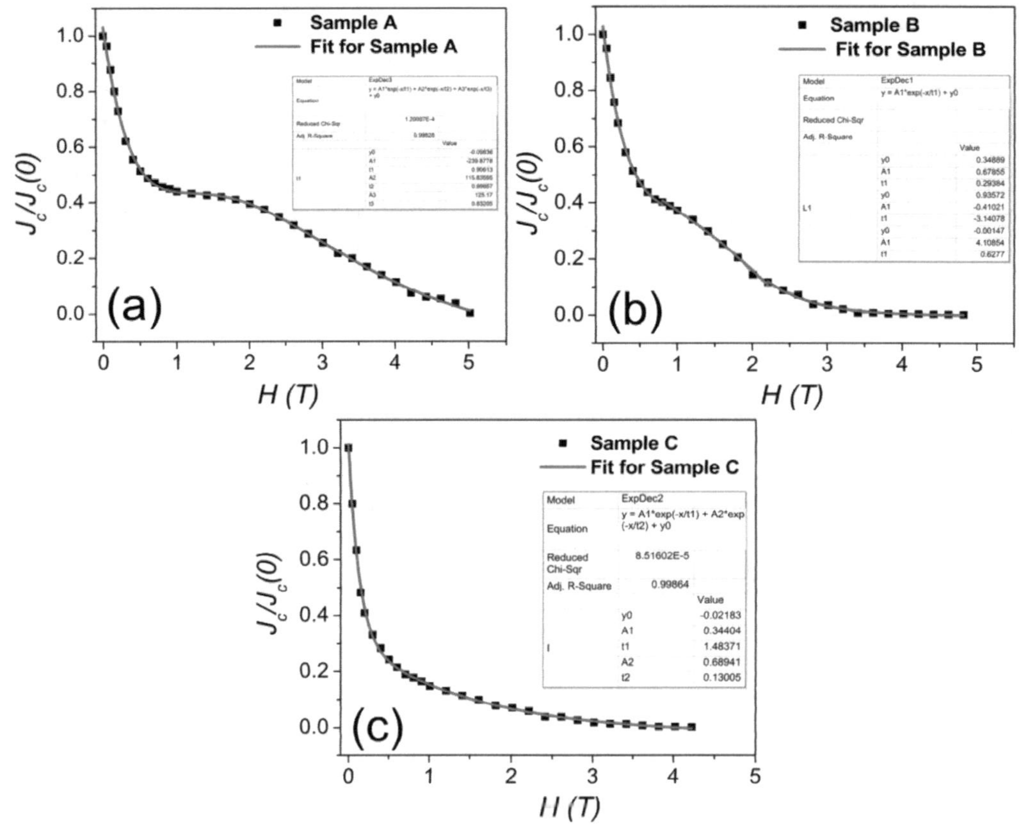

Figure 2. Panel (a) represents the normalized current density J_c(H)/J_c(0) and fitted curves of sample A, similar curves are given in panel (b) for sample B and in panel (c) for sample C.

Due to the two competing mechanisms of flux pinning due to precipitates and interfaces (grain boundaries and cell boundaries), it is difficult to separate the contribution of each mechanism. However, it is now accepted that the pinning is nearly proportional to the volume fraction of precipitates, and is inversely proportional to the cell size or grain size. The proportion of the pinning centers depends on the processing conditions. It is found that at the optimum size, the density of the precipitates is about ten times the density of grain boundaries.

The details of the fit parameters obtained from the $J_c(H)$ curves at 77 K for samples A, B and C superconductors are listed in Table 1. $J_c(H)$ at low fields fits well to the first term. The parameters t_1, t_2 and t_3 represent the decay rate of the J_c curve with A_i (i=1,2 and 3) as the amplitude. The parameter t_1 is applicable at low fields and its value is smaller for sample C as compared to the other two samples. This explains why the J_c curve of sample C decays more rapidly as compared to the other samples. The peaks in the J_c curves are observed for only sample A and B at intermediate fields, whereas no peak is observed in sample C. Due to this, samples A and B are fitted with 3 exponential terms and sample C is fitted with only two terms. The large t_2 values compared to t_1 represents the slow decay of the J_c curves.

Table 1. The fitting parameters for samples A, B and C.

	A_1	t_1	A_2	t_2	A_3	t_3	y_0
Sample A	-239.8778	0.9061	125.17	0.83205	115.83	0.9965	-0.09836
Sample B	0.67855	0.2938	4.1085	0.6277	-0.41021	-3.1407	0.34889
Sample C	0.68941	0.1300	0.34404	1.4837	--	--	-0.02183

Conclusions

In this paper, the flux pinning nature of different LREBCO superconductors at different applied magnetic fields (zero to high fields) were analyzed with an aid of a model. An exponential equation is utilized for analyzing the J_c curves of different bulk LREBCO superconductors. The exponential term (first term) in the expression fits the lower field region of the $J_c(H)$ curves. This is ascribed to flux pinning at interfacial defects. The t_1 parameter is smaller for sample C, due to fast decay of the $J_c/J_c(0)$ curve. The samples B and C show the peak effect at intermediate fields due to strong ΔT_c pinning provided by nano-sized defects within the superconducting matrix and this behavior is in agreement with the first term of the expression. Three exponential term equation is employed to explain the lower field, peak field and higher field behavior of J_c curves.

Acknowledgments

MS would like to thank Professor M. Murakami, President of Shibaura Institute of Technology (SIT), for giving me an opportunity to work at Superconductivity Research Laboratory, SIT. Special thanks to Dr. Milos Jirsa (Institute of Physics, ASCR) for his valuable suggestions and advising this project. Lastly, I would like to thank my loving family members for their endless encouragement and love.

References

[1]. Skakle J M S 1998 *Mat. Sci. Eng.* **R23** 1
[2]. Murakami M, Sakai N, Higuchi T and Yoo S I 1996 *Supercond. Sci. Technol.* **9** 1015
[3]. Oda M, Yao X, Yoshida X and Ikuta H 2009 *Supercond. Sci. Technol.* **22** 075012
[4]. Muralidhar M, Koblischka M R and Murakami M 2000 *Studies of high temperature superconductors* (ed. A. V. Narlikar), vol. 31, p. 89
[5]. Osabe G, Takizawa T, Yoo S I, Sakai N, Higuchi T and Murakami M 1999 *Mater. Sci. Eng. B-Solid State* **65** 11
[6]. Muralidhar M, Koblischka M R, Diko P and Murakami M 2000 *Appl. Phys. Lett.* **76** 91
[7]. Maňka J, Cigáň A, Polovková J, Koňakovský A and Prnová A 2011 Meas. Sci. Rev. **11** 1
[8]. Schätzle P, Bieger W, Wiesner U, Verges P and Krabbes G 1996 *Supercond. Sci. Technol.* **9** 869
[9]. Muralidhar M, Nariki S, Jirsa M, Wu Y and Murakami M 2002 *Appl. Phys. Lett.* **80** 1016
[10]. Sano M, Hayakawa Y and Kumagawa M 1996 *Supercond. Sci. Technol.* **9** 478
[11]. Cai C, Holzapfel B and Hänisch J 2004 *Phys. Rev. B* **69** 104531
[12]. Lee D F, Selvamanickam V and Salama K 1992 *Physica C* **202** 83
[13]. Muralidhar M, Chauhan H S, Saitoh T, Kamada K, Segawa K and Murakami M 1997 *Supercond. Sci. Technol.* **10** 663
[14]. Muralidhar M, Koblischka M R, Saitoh T and Murakami M 1998 *Supercond. Sci. Technol.* **11** 1349
[15]. Pi U H, Khim Z G and Kim D H 2004 *Phys. Rev. B* **69** 094518
[16]. Ramesh R, Jin S, Nakahara S and Tiefel T H 1990 *Appl. Phys. Lett.* **57** 1458
[17]. Chen S-Y, Chen I-G, Liao Y-C and Wu M-K 2005 *J. Mater. Res.* **20** 2
[18]. Pavan Kumar Naik S, Missak Swarup Raju P, Rajasekharan T and Seshu Bai V 2016 *IEEE Trans. Appl. Supercond.* **26** 8002206

[19]. Devendra Kumar N, Rajasekharan T, Ravi C Gundakaram and Seshu Bai V 2011 *IEEE Trans. Appl. Supercond.* **21** 3612
[20]. Muralidhar M, Ide N, Koblischka M R, Diko P, Inoue K and Murakami M 2016 *Supercond. Sci. Technol.* **29** 054003
[21]. Nakamura M, Hirayama T, Yamada Y, Ikuhara Y and Shiohara Y 1996 *Jpn. J. Appl. Phys.* **35** 3882
[22]. Muralidhar M (2012). *Superconductivity: Recent developments and new production technologies*. Nova Science Publishers, Inc
[23]. Salama K and Lee D F 1994 *Supercond. Sci. Technol.* **7** 177
[24]. Lee D F, Selvamanickam V and Salama K 1992 *Physica C* **202** 83
[25]. Kim C J and Hong G W 1999 *Supercond. Sci. Technol.* **12** R27
[26]. Pavan Kumar Naik S, Missak Swarup Raju P and Seshu Bai V 2016 *Mater. Chem. Phys.* **182** 503
[27]. Pavan Kumar Naik S, Missak Swarup Raju P, Seshu Bai V and Rajasekharan T 2016 *Supercond. Sci. Technol.* **29** 115001
[28]. Pavan Kumar Naik S and Seshu Bai V 2017 *J. Phys. Chem. Solids* **101** 65
[29]. Pavan Kumar Naik S and Seshu Bai V 2017 *Cryogenics* **81** 47
[30]. Jirsa M, Pust L, Dlouhy D and Koblischka M R 1997 *Phys. Rev. B* **55** 3276
[31]. Ganesan V, Srinivasan R, Gopalan R and Rajasekharan T 1993 *Pramana J. Phys.* **41** 61
[32]. Yang T, Wang Z H, Zhang H, Nie Y and Qiu L 2003 *Physica C* **386** 69
[33]. Santosh M 2014 *Acta Physica Polonica A* **126** 3

IOP Conf. Series: Journal of Physics: Conf. Series **871** (2017) 012024 doi:10.1088/1742-6596/871/1/012024

Spin-polarized local density of states around vortex in helical p-wave superconductors

Kenta K Tanaka[1], **Masanori Ichioka**[1,2] **and Seiichiro Onari**[1,2]

[1] Department of Physics, Okayama University, Okayama 700-8530, Japan

[2] Research Institute for Interdisciplinary Science (RIIS), Okayama University, Okayama 700-8530, Japan

E-mail: ktanaka@mp.okayama-u.ac.jp

Abstract. Based on the quasi-classical Eilenberger theory, we investigate the magnetic field dependence of order-parameters and spin-polarized local density of states (LDOS) in the vortex lattice state of helical p-wave superconductors. The spin-polarized LDOS is induced by the vorticity coupling to the chirality of up-spin pair or down-spin pair, even when Knight shift does not change. We clarify the instability of the helical p-wave state at high field, and that the spin-polarized LDOS shows the unique behaviors of the helical p-wave state.

1. Introduction

The ruthenate superconductor Sr_2RuO_4 being a candidate for the spin-triplet chiral p-wave superconductor [1, 2] has attracted great interest in condensed matter physics, since topological quantum states such as a Majorana state are expected. However, the detail structure of d-vector in Sr_2RuO_4 remains unclear in spite of various studies both experimentally and theoretically. Thus, the helical p-wave state also has been studied as another scenario [3, 4, 5, 6]. In addition, according to the theoretical study employing the multi-orbital model of Sr_2RuO_4 [7], the condensation energy of chiral and helical p-wave states are very close. Therefore, it is important that we investigate an behavior of physical quantity depending on the d-vector symmetry to distinguish between the chiral and helical states in Sr_2RuO_4 or other candidate superconductors.

In the bulk state of chiral superconductor, the superconducting state with broken time-reversal symmetry is realized because of finite angular momentum of Cooper pair ($L_z \neq 0$). The chirality of Cooper pair in the chiral p-wave superconductor, i.e., $L_z = \pm 1$ can be distinguished via coherence effect in the vortex state. From the theoretical studies for the impurity effects on the local density of states (LDOS) and the local NMR relaxation rate [8, 9, 10, 11, 12], the chirality dependence of physical quantities is appeared by the vorticity coupling to the chirality, depending on whether the chirality is parallel ($L_z = +1$) or anti-parallel ($L_z = -1$) to the vorticity ($W = +1$) [13, 14]. On the other hand, the bulk state of helical p-wave superconductor is time-reversal-invariant. This is because the chirality $L_z = \pm 1$ are quenched with the degeneracy between up-spin and down-spin pairs. The order-parameter of up-spin (down-spin) pair $\Delta_{\uparrow\uparrow}(\Delta_{\downarrow\downarrow})$ has chirality $L_z = -1(+1)$ so that $L_z + S_z = 0$ [3]. $S_z = +1(-1)$ is quantum number of spin state in up-spin (down-spin) pair. Therefore, in the vortex state of helical p-wave superconductor, spin components of LDOS are expected to show an unique behaviors, reflecting the vorticity coupling to the chirality of $\Delta_{\uparrow\uparrow}(L_z = -1)$ or $\Delta_{\downarrow\downarrow}(L_z = +1)$.

The site- and energy-dependence of LDOS in the vortex state can be detected by the scanning tunneling microscopy and spectroscopy (STM/STS) measurements in the various superconductors [15, 16]. Recently, the STM/STS mesurement in the vortex state of topological insulator-superconductor $Bi_2Te_3/NbSe_2$ heterostructure has also performed [17], and theoretical analyses for the measurement have supported the existence of Majonara state at the inside of the vortex core [18, 19]. The spin-polarized Majonara state in the vortex state of heterostructure has been also examined by using the spin-polarized STM/STS method [20], which can selectively observe the spin-dependent conductance. In addition, the spin-polarized Majonara state in the vortex state of $Cu_xBi_2Si_3$ has been studied by solving the Bogoliubov-de Gennes equation [21].

In this work, we study the magnetic field H dependence of the helical p-wave superconductor, and focus on the order-parameters and spin-polarized LDOS in the vortex lattice state. We especially calculate the site r- and H-dependences of the spin-polarized LDOS, and also the energy E spectra in order to clarify the unique behavior of the helical state. We expect that these calculation results help to investigate the vortex state of helical p-wave superconductor and the Majorana state by using the spin-polarized STM/STS method.

This paper is organized as follows. After the introduction, we explain formulation of the Eilenberger equation in the vortex lattice state and the spin-resolved LDOS in section 2. In section 3, we study the H-dependence of order-parameter. In section 4, we show the H-dependence of spin-resolved and spin-polarized LDOS. The E-dependence of LDOS are presented in section 5. The last section is devoted to the summary.

2. Formulation of Eilenberger equation

We calculate the spatial structure of the vortex lattice state by the quasi-classical Eilenberger theory, which is valid for $k_F\xi \gg 1$ (k_F is the Fermi wave number and ξ is the coherence length). For various superconductors including Sr_2RuO_4, the quasi-classical condition is well satisfied [1, 2].

For simplicity, we assume the helical p-wave pairing on the cylindrical Fermi surface, $\boldsymbol{k} = (k_x, k_y) = k_{\rm F}(\cos\theta_k, \sin\theta_k)$, and the Fermi velocity $\boldsymbol{v}_{\rm F} = v_{\rm F0}\boldsymbol{k}/k_{\rm F}$. In the following, the hat $\hat{}$ indicates the 2×2 matrix in spin space and the check $\check{}$ indicates the 4×4 matrix in particle-hole and spin spaces.

We obtain quasi-classical Green's functions $\check{g}(\mathrm{i}\omega_n, \boldsymbol{r}, \boldsymbol{k})$ in the vortex lattice state by solving the Riccati equation derived from Eilenberger equation [22, 23]

$$-\mathrm{i}\boldsymbol{v}\cdot\nabla\check{g}(\mathrm{i}\omega_n,\boldsymbol{r},\boldsymbol{k}) = \frac{1}{2}[(\mathrm{i}\omega_n - \boldsymbol{v}\cdot\boldsymbol{A})\check{\sigma}_z - \check{\Delta}(\boldsymbol{r},\boldsymbol{k}), \check{g}(\mathrm{i}\omega_n,\boldsymbol{r},\boldsymbol{k})] \tag{1}$$

in the clean limit, where $\boldsymbol{v} = \boldsymbol{v}_{\rm F}/v_{\rm F0}$, $\boldsymbol{r}$ is the center-of-mass coordinate of the Cooper pair, $\check{\sigma}_z$ is the Pauli matrix, and ω_n is Matsubara frequency. The quasi-classical Green's function and order parameter are defined by

$$\check{g}(\mathrm{i}\omega_n,\boldsymbol{r},\boldsymbol{k}) = -\mathrm{i}\pi\begin{bmatrix} \hat{g}(\mathrm{i}\omega_n,\boldsymbol{r},\boldsymbol{k}) & \mathrm{i}\hat{f}(\mathrm{i}\omega_n,\boldsymbol{r},\boldsymbol{k}) \\ -\mathrm{i}\underline{\hat{f}}(\mathrm{i}\omega_n,\boldsymbol{r},\boldsymbol{k}) & -\underline{\hat{g}}(\mathrm{i}\omega_n,\boldsymbol{r},\boldsymbol{k}) \end{bmatrix}, \tag{2}$$

$$\check{\Delta}(\boldsymbol{r},\boldsymbol{k}) = \begin{bmatrix} 0 & \hat{\Delta}(\boldsymbol{r},\boldsymbol{k}) \\ -\hat{\Delta}^\dagger(\boldsymbol{r},\boldsymbol{k}) & 0 \end{bmatrix} \tag{3}$$

where $\check{g}^2 = -\pi^2\check{1}$. The matrix elements of $\hat{g}$ and $\hat{\Delta}$ are defined by $g_{\sigma\sigma'}(\mathrm{i}\omega_n,\boldsymbol{r},\boldsymbol{k}) = [g_0(\mathrm{i}\omega_n,\boldsymbol{r},\boldsymbol{k})\hat{1} + \sum_{\mu=x,y,z} g_\mu(\mathrm{i}\omega_n,\boldsymbol{r},\boldsymbol{k})\hat{\sigma}_\mu]_{\sigma\sigma'}$ and $\Delta_{\sigma\sigma'}(\boldsymbol{r},\boldsymbol{k}) = [\mathrm{i}\sum_{\mu=x,y,z}(d_\mu(\boldsymbol{r},\boldsymbol{k})\cdot\hat{\sigma}_\mu)\hat{\sigma}_y]_{\sigma\sigma'}$ where $\sigma, \sigma' = \uparrow$(up-spin) or $\downarrow$(down-spin), and d_μ is μ-component of d-vector. On the other hand, when we consider the order-parameter $\Delta_{\pm,\sigma\sigma'}(\boldsymbol{r})$ and pairing function $\phi_{p\pm}(\boldsymbol{k}) = k_x \pm \mathrm{i}k_y$ for $p_\pm$-state, the matrix elements of order-parameter are also defined by

$$\Delta_{\sigma\sigma'}(\boldsymbol{r},\boldsymbol{k}) = \Delta_{+,\sigma\sigma'}(\boldsymbol{r})\phi_{p+}(\boldsymbol{k}) + \Delta_{-,\sigma\sigma'}(\boldsymbol{r})\phi_{p-}(\boldsymbol{k}). \tag{4}$$

We assume the condition $\Delta_{\downarrow\uparrow} = \Delta_{\uparrow\downarrow} = 0$. Length, temperature, and magnetic field are measured in unit of ξ_0, T_c, and B_0, respectively. Here, $\xi_0 = \hbar v_{F0}/2\pi k_B T_c$, $B_0 = \phi_0/2\pi\xi_0^2$ with the flux quantum ϕ_0. T_c is superconducting transition temperature at zero field. The energy E, pair potential Δ and ω_n are normalized by $\pi k_B T_c$. In the following, we set $\hbar = k_B = 1$. In this work, all calculations are set at $T = 0.5T_c$.

As magnetic fields are applied to the $\hat{\boldsymbol{z}}$ direction, the vector potential is set by $\boldsymbol{A}(\boldsymbol{r}) = \frac{1}{2}\boldsymbol{H}\times\boldsymbol{r} + \boldsymbol{a}(\boldsymbol{r})$ in the symmetric gauge, where $\boldsymbol{H} = (0,0,H)$ is a average flux density, and $\boldsymbol{a}(\boldsymbol{r})$ is related to the internal magnetic field $\boldsymbol{B}(\boldsymbol{r}) = (0,0,B(\boldsymbol{r})) = \boldsymbol{H} + \nabla\times\boldsymbol{a}(\boldsymbol{r})$. We set the square vortex lattice as the unit cell [1].

In order to obtain the spatial structure of the quasi-classical Green's functions and the order-parameter $\hat{\Delta}(\boldsymbol{r})$ selfconsistently, we calculate $\hat{\Delta}_\pm(\boldsymbol{r})$ by the gap equation

$$\hat{\Delta}_\pm(\boldsymbol{r}) = gN_0T \sum_{|\omega_n|\le\omega_{\rm cut}} \left\langle \phi^*_{p\pm}(\boldsymbol{k})\hat{f}(i\omega_n,\boldsymbol{r},\boldsymbol{k})\right\rangle_{\boldsymbol{k}}, \tag{5}$$

where $(gN_0)^{-1} = \ln T + 2T\sum_{0<\omega_n\le\omega_{\rm cut}}\omega_n^{-1}$, $\langle\ldots\rangle_{\boldsymbol{k}}$ indicates Fermi surface average, and we set $\omega_{\rm cut} = 20k_BT_c$. In equation (5), p-wave pairing interaction is isotropic in spin space. We use the current equation $\nabla\times(\nabla\times\boldsymbol{A}) = -\frac{2T}{\kappa^2}\sum_{0<\omega_n}\langle\boldsymbol{v}{\rm Im}\{g_0\}\rangle_{\boldsymbol{k}}$ with the Ginzburg-Landau parameter $\kappa = B_0/\pi k_BT_c\sqrt{8\pi N_0}$ for the selfconsistent calculation of the vector potential. In this work, we set $\kappa = 2.7$ appropriate to Sr_2RuO_4 as a candidate superconductor for the chiral or helical state. Equations (1)-(5) for ω_n are calculated selfconsistently until the order-parameter converges. From this calculation, we obtain the results of $\boldsymbol{A}(\boldsymbol{r})$, $\hat{\Delta}(\boldsymbol{r})$ and the quasi-classical Green's functions in the vortex lattice state.

d-vector in the helical p-wave superconductors is defined by $\boldsymbol{d}(\boldsymbol{k})\propto k_x\hat{x} + k_y\hat{y} = \phi_{p_+}(\boldsymbol{k})\boldsymbol{d}_- + \phi_{p_-}(\boldsymbol{k})\boldsymbol{d}_+$ at zero field $H = 0$, with $\boldsymbol{d}_\pm(\boldsymbol{k}) = \frac{1}{2}(1,\pm{\rm i},0)$. Thus, when we iterate calculations of equations (1)-(5), as an initial value we set to be $\boldsymbol{d}(\boldsymbol{r},\boldsymbol{k}) = d(\boldsymbol{r})(k_x\hat{x}+k_y\hat{y})$ where $d(\boldsymbol{r})$ is Abrikosov vortex lattice solution.

Next, we calculate $\check{g}(E\pm{\rm i}\eta,\boldsymbol{r},\boldsymbol{k})$ for real energy E by solving Eilenberger equation (1) with ${\rm i}\omega_n \to E\pm{\rm i}\eta$, and using the selfconsistently obtained spatial structure of $\boldsymbol{A}(\boldsymbol{r})$ and $\Delta(\boldsymbol{r})$. η is a small parameter as a smearing factor. The spin-resolved LDOS $N_\sigma(E,\boldsymbol{r})$ is given by

$$N_\sigma(E,\boldsymbol{r}) = \langle{\rm Re}\{[\hat{g}(E+{\rm i}\eta,\boldsymbol{r},\boldsymbol{k})]_{\sigma\sigma}\}\rangle_{\boldsymbol{k}}. \tag{6}$$

We define the LDOS $N(E,\boldsymbol{r}) = N_\downarrow(E,\boldsymbol{r}) + N_\uparrow(E,\boldsymbol{r})$, and spin polarized LDOS $M(E,\boldsymbol{r}) = N_\downarrow(E,\boldsymbol{r}) - N_\uparrow(E,\boldsymbol{r})$.

3. H-dependence of order-parameter

As shown in figure 1(a), we present the H-dependence of spatial average of the order-parameter amplitude, $\langle|\Delta_{\pm,\sigma\sigma'}(\boldsymbol{r})|\rangle_{\boldsymbol{r}}$ given by equation (4). In the vortex state of helical p-wave superconductor at $H < 0.35H_{c2}$, up-spin pair is described by $\Delta_{\uparrow\uparrow}(\boldsymbol{r},\boldsymbol{k}) = \Delta_{-,\uparrow\uparrow}(\boldsymbol{r})\phi_{p_-}(\boldsymbol{k}) + \Delta_{+,\uparrow\uparrow}(\boldsymbol{r})\phi_{p_+}(\boldsymbol{k})$ with small induced component $\Delta_{+,\uparrow\uparrow}(\boldsymbol{r})$. The small component $\Delta_{+,\uparrow\uparrow}(\boldsymbol{r})$ is induced around the vortex core, as shown in figure 1(b). The main component $\Delta_{-,\uparrow\uparrow}(\boldsymbol{r})$ has chirality $L_z = -1$, anti-parallel to vorticity $W = +1$ as $L_z + W = 0$. From the previous thoretical studies for the chiral p-wave superconductor [13, 14], the anti-parallel vortex state ($L_z + W = 0$) is stable compared to the parallel vortex state ($L_z + W = +2$). Therefore, the H-dependence of $\langle|\Delta_{-,\uparrow\uparrow}|\rangle_{\boldsymbol{r}}$ and $\langle|\Delta_{+,\uparrow\uparrow}|\rangle_{\boldsymbol{r}}$ survive until H_{c2}. On the other hand, down-spin pair is described by $\Delta_{\downarrow\downarrow}(\boldsymbol{r},\boldsymbol{k}) = \Delta_{+,\downarrow\downarrow}(\boldsymbol{r})\phi_{p_+}(\boldsymbol{k}) + \Delta_{-,\downarrow\downarrow}(\boldsymbol{r})\phi_{p_-}(\boldsymbol{k})$ with small induced $\Delta_{-,\downarrow\downarrow}(\boldsymbol{r})$. In figure 1(a), $\Delta_{\downarrow\downarrow}(\boldsymbol{r},\boldsymbol{k})$ is rapidly suppressed with increasing H, since the chirality $L_z = +1$ of main $\Delta_{+,\downarrow\downarrow}(\boldsymbol{r})$ is parallel to vorticity as $L_z + W = +2$. Besides, the r-dependence of order-parameter at low H, main components $|\Delta_{+,\downarrow\downarrow}|$ and $|\Delta_{-,\uparrow\uparrow}|$ show almost same behavior, and the induced component $\Delta_{-,\downarrow\downarrow}(\boldsymbol{r})$ is very small amplitude at every position, as shown in figure. 1(b).

In addition, we find that the helical p-wave state is unstable at higher H. At $H \sim 0.35 H_{c2}$, our calculation result shows the change of chirality $L_z = +1 \to -1$ in $\Delta_{\downarrow\downarrow}(\boldsymbol{r}, \boldsymbol{k})$, and that $\Delta_{-,\downarrow\downarrow}(\boldsymbol{r}, \boldsymbol{k})$ becomes main part of $\Delta_{\downarrow\downarrow}(\boldsymbol{r}, \boldsymbol{k})$. At $H > 0.35 H_{c2}$, $\langle|\Delta_{-,\downarrow\downarrow}|\rangle_{\boldsymbol{r}}$ ($\langle|\Delta_{+,\downarrow\downarrow}|\rangle_{\boldsymbol{r}}$) becomes equal to $\langle|\Delta_{-,\uparrow\uparrow}|\rangle_{\boldsymbol{r}}$ ($\langle|\Delta_{+,\uparrow\uparrow}|\rangle_{\boldsymbol{r}}$) for main (small) components, so that the order-parameter is chiral form. Here, it remains $\boldsymbol{d} \perp \boldsymbol{H}$. Therefore, the transition at $H \sim 0.35 H_{c2}$ is helical-chiral transition. Note that this critical magnetic field value can shift if there are other contributions to stabilization of helical or chiral state such as spin-orbit coupling [4].

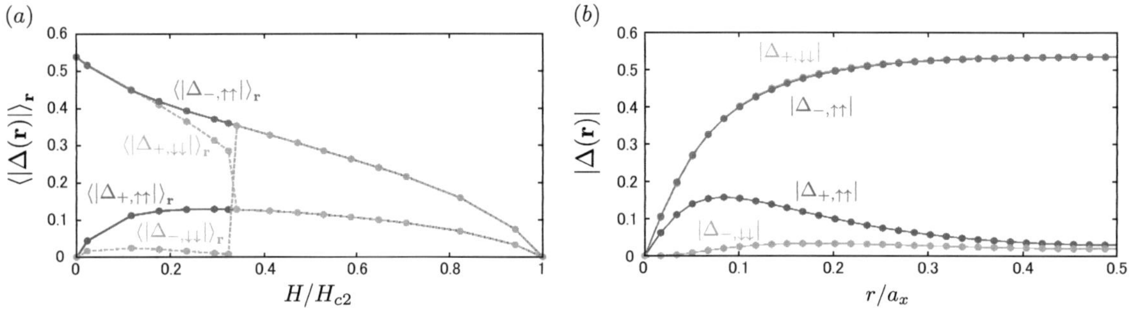

Figure 1. (a) H-dependence of the spatial average of the order-parameter amplitudes $\langle|\Delta_{-,\downarrow\downarrow}|\rangle_{\boldsymbol{r}}$, $\langle|\Delta_{+,\downarrow\downarrow}|\rangle_{\boldsymbol{r}}$, $\langle|\Delta_{-,\uparrow\uparrow}|\rangle_{\boldsymbol{r}}$ and $\langle|\Delta_{+,\downarrow\downarrow}|\rangle_{\boldsymbol{r}}$ defined by equation (4) in the helical p-wave state ($H < 0.35 H_{c2}$) and chiral p-wave state ($H > 0.35 H_{c2}$). (b) r-dependence of the order-parameter amplitudes $|\Delta_{-,\downarrow\downarrow}|$, $|\Delta_{+,\downarrow\downarrow}|$, $|\Delta_{-,\uparrow\uparrow}|$ and $|\Delta_{+,\downarrow\downarrow}|$ from the vortex center along the next-nearest-neighbor (NNN) direction at $H/H_{c2} \simeq 0.02$. a_x is NNN intervortex distance.

4. H-dependence of spin-resolved and spin-polarized LDOS

In figure 2, we present the H-dependence of zero-energy $M(E = 0, \boldsymbol{r})$ and $N_\sigma(E = 0, \boldsymbol{r})$ at some positions on a line between NNN vortices at $H < 0.7 H_{c2}$. At the far-site of the vortex center such as $r/a_x = 0.5$ which is midpoint of between NNN vortices and $r/a_x = 0.2$, $N_\downarrow(E = 0, H) > N_\uparrow(E = 0, H)$ and monotonically increase as a function of H. Thus, $M(E = 0, \boldsymbol{r})$ has a finite value and monotonically increase as a function of H, and jumps to zero at the helical-chiral transition. On the other hand, at the vortex core region in figures 2(c) and 2(d), $M(E = 0, \boldsymbol{r})$ at $H < 0.35 H_{c2}$ show a convex curve with large amplitude, where the first order transition is not obvious. In particular, at the vortex center in figure 2(e), H-dependence of $M(E = 0, H)$ shows the monotonic decreasing behavior with sign change. However, the sign change of $M(E = 0, H)$ vanishes at the $r/a_x = 0.025$, as shown in figure 2(d). These unique behaviors of the helical state are expected to be observed by spin-polarized STM measurement.

Moreover, we study the H-dependence of the zero-energy DOS $\langle N(E = 0, \boldsymbol{r})\rangle_{\boldsymbol{r}}$ and spin-polarized DOS $\langle M(E = 0, \boldsymbol{r})\rangle_{\boldsymbol{r}}$ with spin-resolved DOS $\langle N_\sigma(E = 0, \boldsymbol{r})\rangle_{\boldsymbol{r}}$. As shown in figure 2(f), the H-dependence of $\langle N_\downarrow(E = 0, \boldsymbol{r})\rangle_{\boldsymbol{r}}$ and $\langle M(E = 0, \boldsymbol{r})\rangle_{\boldsymbol{r}}$ show a similar behavior to the figures 2(a) and 2(b). In particular, the $\langle M(E = 0, \boldsymbol{r})\rangle_{\boldsymbol{r}}$ jumps to zero at the helical-chiral transition. At $H > 0.35 H_{c2}$ as the vortex state of chiral p-wave superconductor, where $\Delta_{\downarrow\downarrow} = \Delta_{\uparrow\uparrow}$, M vanishes. We expect that the jump behavior is observed by the low temperature specific heat measurement. The jump of specific heat is large (small) when the helical-chiral transition occurs at high (small) H.

In addition, since the zero energy states localized inside the vortex core in the chiral and helical superconductors correspond to Majorana zero energy mode, figure 2(f) shows that the spin-polarized Majorana state is induced in the helical p-wave superconductors. This is another type of spin-polarized Majorana state than that supposed in the vortex state of $Bi_2Te_3/NbSe_2$ [19] or $Cu_xBi_2Si_3$ [21].

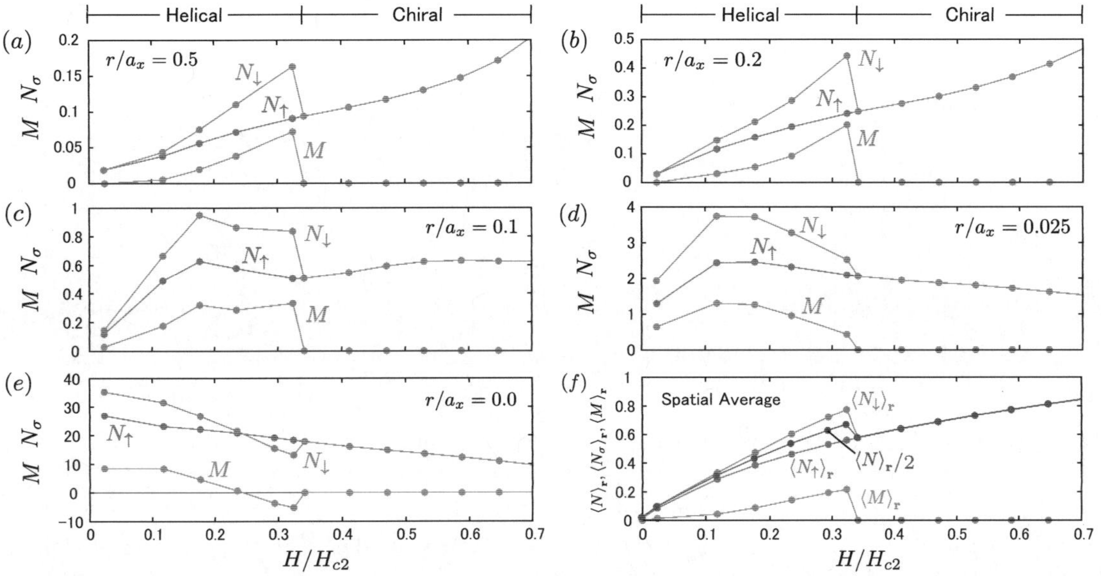

Figure 2. (a), (b), (c), (d), and (e) H-dependence of spin-resolved LDOS $N_\sigma(E = 0, \boldsymbol{r})$ and spin-polarized LDOS $M(\boldsymbol{r})$ at radius $r/a_x = 0.5$, 0.2, 0.1, 0.025, and 0.0 from the vortex center along the NNN direction, respectively. (f) H-dependence of DOS $\langle N(E = 0, \boldsymbol{r})\rangle_{\boldsymbol{r}}/2$, spin-resolved DOS $\langle N_\sigma(E = 0, \boldsymbol{r})\rangle_{\boldsymbol{r}}$ and spin-polarized DOS $\langle M(E = 0, \boldsymbol{r})\rangle_{\boldsymbol{r}}$. we use $\eta = 0.01$.

5. E-dependences of spin-polarized LDOS

We study the E-dependence of $N_\sigma(E, \boldsymbol{r})$ and $M(E, \boldsymbol{r})$ at the some positions on a line between NNN vortices, in order to study the LDOS spectrum of spin-polarized STS measurement. As shown in upper panel of figures 3(a)-(c), $N_\uparrow(E{\sim}\pm0.5, \boldsymbol{r})$ are larger than $N_\downarrow(E{\sim}\pm0.5, \boldsymbol{r})$, and the $N_\uparrow$ peaks of the gap edge ($E{\sim}\pm0.5$) outside vortices can survive until the vortex center. Thus, $M(E, \boldsymbol{r})$ at all positions are negative at $E{\sim}\pm0.5$, as shown in lower panel of figures 3(a)-(c). Moreover, we focus on the low- or zero-energy peaks of N_σ at the vortex core region, coming from the vortex bound state. As shown in figures 3(b) and 3(c), the height of low- or zero-energy peaks in $N_\uparrow$ are smaller than $N_\downarrow$. Thus, $M(E, \boldsymbol{r} = \boldsymbol{0})$ is positive around $E = 0$, and $M(E, \boldsymbol{r})$ at all positions show the sign change as a function of E. Note that, these weights cancel each other because of the condition of total spin polarization $\int_{-\infty}^{0} M(E, \boldsymbol{r})dE = 0$. This condition indicates that the Knight shift does not work in the helical p-wave state, where $\boldsymbol{d}\perp\boldsymbol{H}$. In order to detect the spin-polarized LDOS in the helical state, we have to perform E-resolved measurement such as a spin-polarized STM/STS method.

As shown in figures 4(a)-(c), we also investigate the higher field $H{\simeq}0.29H_{c2}$ cases. In figure 4(a) at outside of the vortex core $r/a_x = 0.5$, amplitude of low-energy $M(E, \boldsymbol{r})$ has larger than low field $H{\simeq}0.02H_{c2}$ case, since low-energy $N_\sigma(E, \boldsymbol{r})$ are induced and spin-polarization is reinforced by the magnetic field. Besides, in upper panel of figure 3(a), E-dependence of $N_\sigma(E, \boldsymbol{r})$ almost correspond to the zero magnetic state. In figure 4(b) around the vortex core, we can see some sign change behavior of $M(E, \boldsymbol{r})$ since the E-dependence of $N_\sigma(E, \boldsymbol{r})$ continuously change at all E range. The E-dependence of $M(E, \boldsymbol{r})$ at the vortex center is notably different from the low field $H \simeq 0.02 H_{c2}$ case in lower panel of figure 3(c). In figure 4(c), the hight of zero-energy peak of $N_\uparrow$ becomes larger than $N_\downarrow$, resulted in negative $M(E = 0, \boldsymbol{r} = \boldsymbol{0})$. To compensate negative value at $E = 0$ and at the gap edge, $M(E, r = 0)$ becomes positive for in-gap states for $0 < |E| < 0.5$.

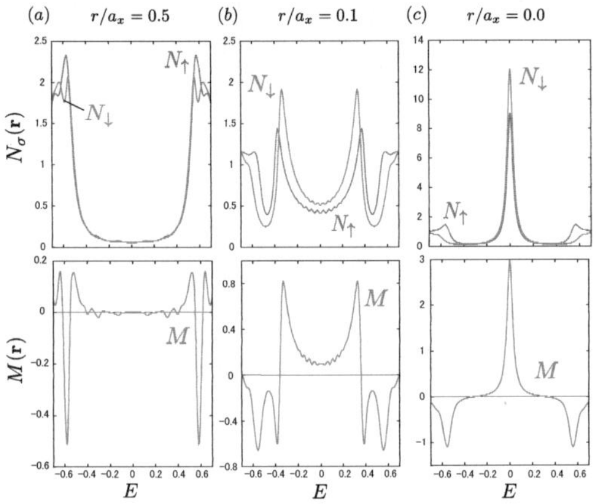

Figure 3. (a), (b), and (c) E-dependence of spin-resolved LDOS $N_\downarrow$, $N_\uparrow$ (upper panels), and spin-polarized LDOS M (lower panels) at $H/H_{c2} \simeq 0.02$ at $r/a_x = 0.5, 0.1$, and 0.0 from the vortex center along the NNN direction, respectively. we use $\eta = 0.03$.

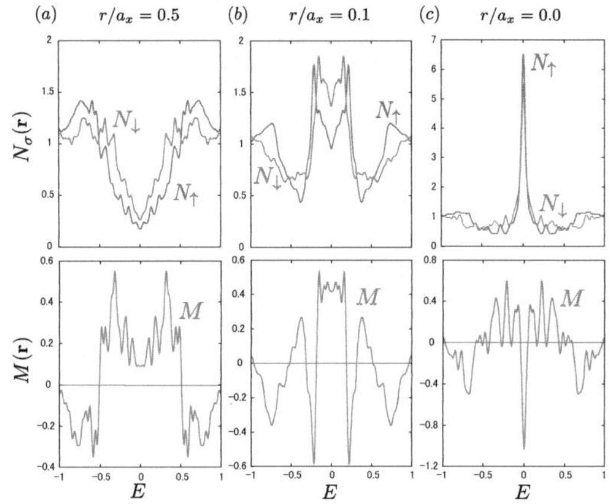

Figure 4. (a), (b), and (c) E-dependence of spin-resolved LDOS $N_\downarrow$, $N_\uparrow$ (upper panels), and spin-polarized LDOS M (lower panels) at $H/H_{c2} \simeq 0.29$ at $r/a_x = 0.5, 0.1$, and 0.0 from the vortex center along the NNN direction, respectively. we use $\eta = 0.03$.

6. Summary

We studied the H-dependence of helical p-wave superconductors, and calculated the order-parameters $\Delta_{\pm,\sigma\sigma'}(\boldsymbol{r})$ and spin-polarized LDOS $M(E, r)$ by solving the quasi-classical Eilenberger equation in the vortex lattice state. From the calculation results of H-dependence of order-parameter, the helical p-wave state is unstable at high fields and changes to the chiral p-wave state. In addition, we revealed that the spin-polarized LDOS $M(E, r)$ is induced by the vorticity coupling to the chirality, while the Knight shift does not change. In order to identify the helical p-wave state at low fields, we especially investigated the H-dependence of zero-energy $M(E = 0, \boldsymbol{r})$ related to Majorana state, and found that $M(E = 0, \boldsymbol{r})$ shows various behavior as a function H according to its position. In fact, H-dependence of $M(E = 0, \boldsymbol{r})$ shows monotonically increasing behavior with small value at the far-site of the vortex center, convex curve around the vortex core, and monotonically decreasing behavior with sign change at the vortex center. Moreover, we presented the E-dependence of $N_\downarrow(E, \boldsymbol{r})$, $N_\uparrow(E, \boldsymbol{r})$ and $M(E, \boldsymbol{r})$ at the some positions at low and high fields. The E-dependences of $M(E, \boldsymbol{r})$ at some positions show the unique sign change behavior. These calculation results of spin-polarized LDOS are expected to be examined and be used for detecting the spin-polarized Majorana state by the spin-polarized STM/STS measurement.

Acknowledgments

This work was supported by JSPS KAKENHI Grant Number JP16J05824.

IOP Conf. Series: Journal of Physics: Conf. Series **871** (2017) 012024 doi:10.1088/1742-6596/871/1/012024

References

[1] Mackenzie A P and Maeno Y 2003 *Rev. Mod. Phys.* **75** 657
[2] Maeno Y, Kittaka S, Nomura T, Yonezawa S and Ishida K 2012 *J. Phys. Soc. Jpn.* **81** 011009
[3] Rice T M and Sigrist M 1995 *J. Phys.: Condens. Matter* **7** L643
[4] Takamatsu S and Yanase Y 2013 *J. Phys. Soc. Jpn.* **82** 063706
[5] Scaffidi T, Romers J C and Simon S H 2014 *Phys. Rev.* B **89** 220510(R)
[6] Zhang J, Lörscher C, Gu Q and Klemn R A 2014 *J. Phys.: Condens. Matter* **26** 252201
[7] Tsuchiizu M, Yamakawa Y, Onari S, Ohno Y and Kontani H 2015 *Phys. Rev.* B **91** 155103
[8] Kato Y and Hayashi N 2003 *Physica* C **388** 519
[9] Tanuma Y, Hayashi N, Tanaka Y and Golubov A A 2009 *Phys. Rev. Lett.* **102** 117003
[10] Kurosawa N, Hayashi N and Kato Y 2015 *J. Phys. Soc. Jpn.* **84** 114710
[11] Hayashi N and Kato Y 2003 *Physica* C **388** 513
[12] Tanaka K K, Ichioka M and Onari S 2016 *Phys. Rev.* B **93** 094507
[13] Ichioka M and Machida K 2002 *Phys. Rev.* B **65** 224517
[14] Heeb R and Agterberg D F 1999 *Phys. Rev.* B **59** 7076
[15] Hess H F, Robinson R B, Dynes R C, Valles J M, Jr. and Waszczak J V 1989 *Phys. Rev. Lett.* **62** 214
[16] Fischer Ø, Kugler M, Maggio-Aprile I, Berthod C and Renner C 2007 *Rev. Mod. Phys.* **79** 353
[17] Xu J -P, Wang M -X, Liu Z L, Ge J -F, Yang X, Liu C, Xu Z A, Guan D, Gao C L, Qian D, Liu Y, Wang Q -H, Zhang F -C, Xue Q -K and Jia J -F 2015 *Phys. Rev. Lett.* **114** 017001
[18] Li Z -Z, Zhang F -C and Wang Q -H 2014 *Sci. Rep.* **4**, 6363
[19] Kawakami T and Hu X 2015 *Phys. Rev. Lett.* **115** 177001
[20] Sun H -H, Zhang K -W, Hu L -H, Li C, Wang G -Y, Ma H -Y, Xu Z -A, Gao C -L, Guan D -D, Li Y -Y, Liu C, Qian D, Zhou Y, Fu L, Li S -C, Zhang F -C and Jia J -F 2016 *Phys. Rev. Lett.* **116** 257003
[21] Nagai Y, Nakamura H and Machida M 2014 *J. Phys. Soc. Jpn.* **83** 064703
[22] Eilenberger G 1968 *Z. Phys.* **214** 195
[23] Tsutsumi Y, Kawakami T, Shiozaki K, Sato M and Machida K 2015 *Phys. Rev.* B **91** 144504

Spin-current induced around half-quantum vortices in chiral p-wave superconducting states

R Asaoka[1,2], **H Tsuchiura**[1,2], **and M Sigrist**[3]

[1]Department of Applied Physics, Tohoku University, Sendai 980-8579, Japan
[2]Center for Spintronics Research Network, Tohoku University, Sendai 980-8577, Japan
[3]Theoretiche Physik, Eidgenössisch Technische Hochschule, CH-8093 Zurich, Switzerland

E-mail: asaoka@olive.apph.tohoku.ac.jp

Abstract. We study the electronic state around a half-quantum vortex (HQV) in a chiral p-wave superconductor based on a square lattice three band tight-binding model by means of the Bogoliubov-de Gennes theory. In particular, the spatial distribution of charge and spin currents are mainly discussed. This analysis shows that the spin current is strengthened between the neighboring HQVs, resulting in the energy cost for HQV formation.

1. Introduction

Non-integer quantum vortices are allowed in multicomponent superfluids or superconductors[1]. In particular, a half-quantum vortex (HQV) expected in the chiral p-wave spin-triplet pairing superconducting state is an exciting topic due to Majorana fermion zero modes predicted at the vortex cores[2]. A HQV is formed by a π-rotation of the superconducting phase and also a π-rotation of the d-vector around the vortex core. Thus the magnetic flux through HQV is one-half of the flux quantum. HQVs are topologically allowed in spin-triplet superconducting states, however, whether the energy of a HQV or a pair of HQVs is lower than that of an integer quantum vortex (IQV) is a question to be carefully considered, largely because the spin current accompanied by a HQV increases the energy of a system[3].

Stability conditions of HQVs have been analyzed mainly based on the phenomenological Ginzburg-Landau (GL) theory. These studies have predicted that HQVs are unstable unless the ratio of the spin fluid density to superfluid density ρ_{sp}/ρ_s is sufficiently small[3], and unless the spin-orbit coupling is sufficiently small[4, 5]. To argue these conditions in the systems of realistic materials, features of materials should be considered, for example, the anisotropy of the pairing interaction and the Fermi surface structure. Appropriate consideration of these features needs microscopic analyses in parallel with the GL theory.

In this paper, as a first step, we investigate the distribution of spin current around a HQV in a chiral p-wave equal spin pairing superconducting state in a square lattice three band tight-binding model with the pairing interaction between nearest-neighbor sites based on the Bogoliubov-de Gennes (BdG) theory. We also compare the electronic energy of a pair of HQVs with that of an IQV in our system. In this study, we suppose the system of a two-dimensional RuO_2 plane in Sr_2RuO_4, which is strongly believed that a chiral p-wave superconducting state is realized[6]. The Fermi surfaces of Sr_2RuO_4 are formed by three band, α, β band, largely derived from the d_{yz} and d_{zx} orbitals of Ru, and γ band, from the d_{xy} orbital of Ru[6]. We consider all three bands and introduce the spin-orbit coupling to reproduce the Fermi surface of Sr_2RuO_4.

2. Model and Method

The Hamiltonian of a three band Hubbard model used in this study is given as[7]

$$\mathcal{H} = \mathcal{H}_{\alpha\beta} + \mathcal{H}_{\gamma} + \mathcal{H}_{W} + \mathcal{H}_{\rm SO} + \mathcal{H}_{\mu} \quad (1)$$

$$\begin{aligned}\mathcal{H}_{\alpha\beta} &= -\sum_{i\sigma}\left(t_{\alpha\beta i+\hat{x}i}c^{\dagger}_{1i+\hat{x}\sigma}c_{1i\sigma} + t_{\alpha\beta i+\hat{y}i}c^{\dagger}_{2i+\hat{y}\sigma}c_{2i\sigma} + \text{H.c.}\right) \\ &\quad - \sum_{m=1,2}\sum_{i\sigma}\left(t'_{\alpha\beta i+\hat{x}+\hat{y}i}c^{\dagger}_{mi+\hat{x}+\hat{y}\sigma}c_{\bar{m}i\sigma} + t'_{\alpha\beta i+\hat{x}-\hat{y}i}c^{\dagger}_{mi+\hat{x}-\hat{y}\sigma}c_{\bar{m}i\sigma} + \text{H.c.}\right) \quad (2)\end{aligned}$$

$$\begin{aligned}\mathcal{H}_{\gamma} &= -\sum_{i\sigma}\left(t_{\gamma i+\hat{x}i}c^{\dagger}_{3i+\hat{x}\sigma}c_{3i\sigma} + t_{\gamma i+\hat{y}i}c^{\dagger}_{3i+\hat{y}\sigma}c_{3i\sigma} + \text{H.c.}\right) \\ &\quad - \sum_{i\sigma}\left(t'_{\gamma i+\hat{x}+\hat{y}i}c^{\dagger}_{3i+\hat{x}+\hat{y}\sigma}c_{3i\sigma} + t'_{\gamma i+\hat{x}-\hat{y}i}c^{\dagger}_{3i+\hat{x}-\hat{y}\sigma}c_{3i\sigma} + \text{H.c.}\right) \quad (3)\end{aligned}$$

$$\mathcal{H}_{W} = -W\sum_{i\sigma\sigma'}\left(\sum_{m=1,3} n_{mi+\hat{x}\sigma}n_{mi\sigma'} + \sum_{m=2,3} n_{mi+\hat{y}\sigma}n_{mi\sigma'}\right) \quad (4)$$

$$\mathcal{H}_{\rm SO} = -i\lambda\sum_{mm'm''i}\epsilon_{mm'm''}\sum_{\sigma\sigma'}c^{\dagger}_{mi\sigma}\sigma^{m''}_{\sigma\sigma'}c_{m'i\sigma} \quad (5)$$

$$\mathcal{H}_{\mu} = -\mu\sum_{mi}\hat{n}_i - \delta\sum_{i} n_{3i}, \quad (6)$$

where m denotes the orbital number (m=1, 2, 3), and $\hat{x}$ ($\hat{y}$) denotes the difference of nearest-neighbor sites in the x (y) direction. Numbers 1, 2, 3 correspond to d_{yz}, d_{zx}, d_{xy} orbitals, respectively. Site symbol i involves a two-dimensional position, i.e. $i = (i_x, i_y)$. Nearest-neighbor (next-nearest-neighbor) hopping amplitude in d_{yz}, d_{zx} orbitals and d_{xy} orbitals are denoted as $t_{\alpha\beta ij}$ ($t'_{\alpha\beta ij}$) and $t_{\gamma ij}$ ($t'_{\gamma ij}$), respectively. The uniform magnetic field is introduced in terms of the Peierls phase of the hopping term as

$$t_{\alpha\beta(\gamma)ij} = t_{\alpha\beta(\gamma)}\exp\left[i\frac{\pi}{\Phi_0}\int_{\mathbf{r}_i}^{\mathbf{r}_j}\mathbf{A}(\mathbf{r})\cdot d\mathbf{r}\right] \quad \text{and} \quad t'_{\alpha\beta(\gamma)ij} = t'_{\alpha\beta(\gamma)}\exp\left[i\frac{\pi}{\Phi_0}\int_{\mathbf{r}_i}^{\mathbf{r}_j}\mathbf{A}(\mathbf{r})\cdot d\mathbf{r}\right]. \quad (7)$$

λ and δ stand for the the amplitude of the spin-orbit coupling and the energy difference between the d_{yz} or d_{zx} orbital and the d_{xy} orbital. $\epsilon_{mm'm''}$ and σ in $\mathcal{H}_{\rm SO}$ are the Levi-Civita symbol and the Pauli matrix, respectively. μ is the chemical potential. We set $t'_{\alpha\beta} = 0.1t_{\alpha\beta}$, $t_{\gamma} = 0.7t_{\alpha\beta}$, $t'_{\gamma} = 0.3t_{\alpha\beta}$, $\lambda = 0.1t_{\alpha\beta}$, $\delta = 0.065t_{\alpha\beta}$, and $\mu = t_{\alpha\beta}$ then a filling $n \simeq 4$ to reproduce the Fermi surface topology of Sr_2RuO_4[7]. The third term in the Hamiltonian (1) represents nearest-neighbor attractive interactions, then $W > 0$. We introduce the BCS-type mean field using superconducting order parameters $\Delta_{mij\sigma} = W\langle c_{mi\sigma}c_{mj\sigma}\rangle$, which are corresponding to the chiral p-wave gap function written as $\Delta(\mathbf{k}) = \sin k_x + i\sin k_y$ in wavenumber space. Total order parameter is defined as $\Delta_{ij\sigma} = \Delta_{1ij\sigma} + \Delta_{2ij\sigma} + \Delta_{3ij\sigma}$. We assume here for simplicity the d-vector lies within the ab-plane despite that it is parallel to the c-axis in Sr_2RuO_4[6].

We numerically solve the Bogoliubov-de Gennes (BdG) and the gap equation derived from the Hamiltonian (1) and the superconducting order parameters self-consistently, with $W = 0.75t_{\alpha\beta}$. To make the Peierls phase compatible with vortex lattice symmetry, we need a magnetic unit cell with $2N \times N$ sites including two vortices[8]. In this case, by the appropriate choice of gauge, the order parameter $\Delta_{ij\sigma}$ has a translational symmetry with respect to the magnetic unit cell. A position of a vortex center is located just on a site if N is odd, and in the middle of square plaquette surrounded by four sites if N is even. Here we take $N = 15$ with periodic boundary condition.

3. Results and Discussion

Figure 1 shows the local density of states (LDOS) without magnetic field, i.e. in a uniform chiral p-wave system. The LDOS exhibits a nodeless (full gap) structure of the superconducting gap expected in the chiral p-wave superconductor. In Fig. 2, we present the spatial distributions

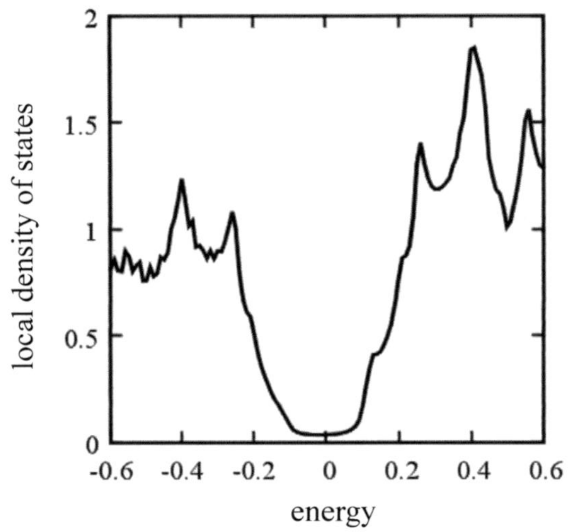

Figure 1. Local density of states without magnetic field, i.e. in a uniform chiral p-wave system.

of the amplitude of the superconducting order parameters around the cores of HQVs with the distributions of the up- and down-spin current. We define the average of the superconducting order parameters around a site $\Delta_{j\sigma} = (\Delta_{jj+\hat{x}\sigma} + \Delta_{j-\hat{x}j\sigma} - i\Delta_{jj+\hat{y}\sigma} - i\Delta_{j-\hat{y}j\sigma})/4$. We see that up-spin component and down-spin component form HQVs independently in Fig. 2 accompanied by the each spin current around the core of HQVs. We note here that the position of HQVs depend on the initial conditions for BdG calculations. The furrows of the superconducting order parameter parallel to the axial directions result from the (quasi-) one-dimensional property of α and β bands. Superconducting order parameters $\Delta_{1i\sigma\sigma}$ and $\Delta_{2i\sigma\sigma}$ collapse in the x direction and y direction respectively, since the superconducting components derived from the d_{yz} and d_{zx} orbitals have the long coherent lengths in the each direction. The electronic energy of this system is calculated, $E_{\rm HQV} = -10.020t_{\alpha\beta}$, while the energy of the system where an IQV is formed in our calculation $E_{\rm IQV} = -10.004t_{\alpha\beta}$. The difference between the pair of HQV and the IQV, $\Delta E = |E_{\rm HQV} - E_{\rm IQV}|/|E_{\rm HQV} \sim 0.002$, is fairly small, therefore, it is not suitable to discuss the energetic stability of HQVs within the present calculation framework, although the energy of HQVs is slightly lower than that of an IQV. Figure 3 shows the spatial distribution of the total charge and spin currents around the cores of HQVs. The charge current is weakened between the HQVs and is strengthened outside, while the spin current is strengthened between the HQVs and is weakened outside as prefigured from Fig 2. The free energy of spin current is represented as $T_{\rm sp} = J_{\rm sp}^2/2\rho_{\rm sp}$ ($J_{\rm sp}$ and $\rho_{\rm sp}$ are amplitude of spin current and spin current density), therefore, the enhance of spin current between the neighboring HQVs results in the energy cost for HQV formation.

4. Summary

We have investigated the distribution of spin current, which makes a HQV unstable, around a HQV in a chiral p-wave equal spin pairing superconducting state in a square lattice three band

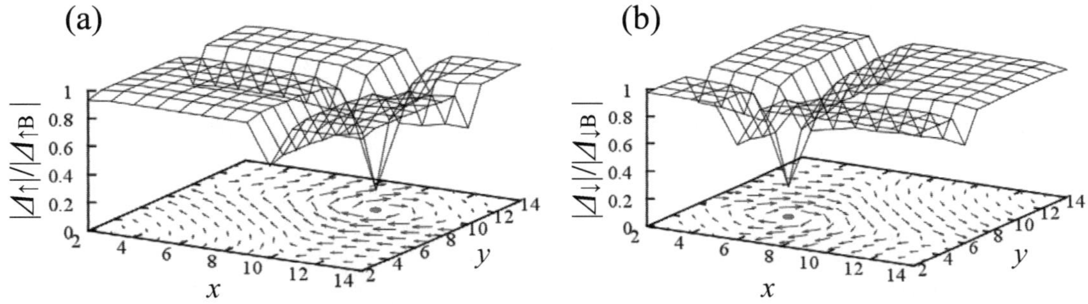

Figure 2. Spatial distributions of the amplitude of the superconducting order parameters, $|\Delta_{j\uparrow}|$ (a) and $|\Delta_{j\downarrow}|$ (b), around the cores of HQVs. The amplitude is normalized by that in a system without magnetic field, $|\Delta_{\sigma B}|$. Blue vectors in the $z = 0$ plane represent the up-spin (a) and down-spin (b) currents. Red dots indicate the cores of HQVs.

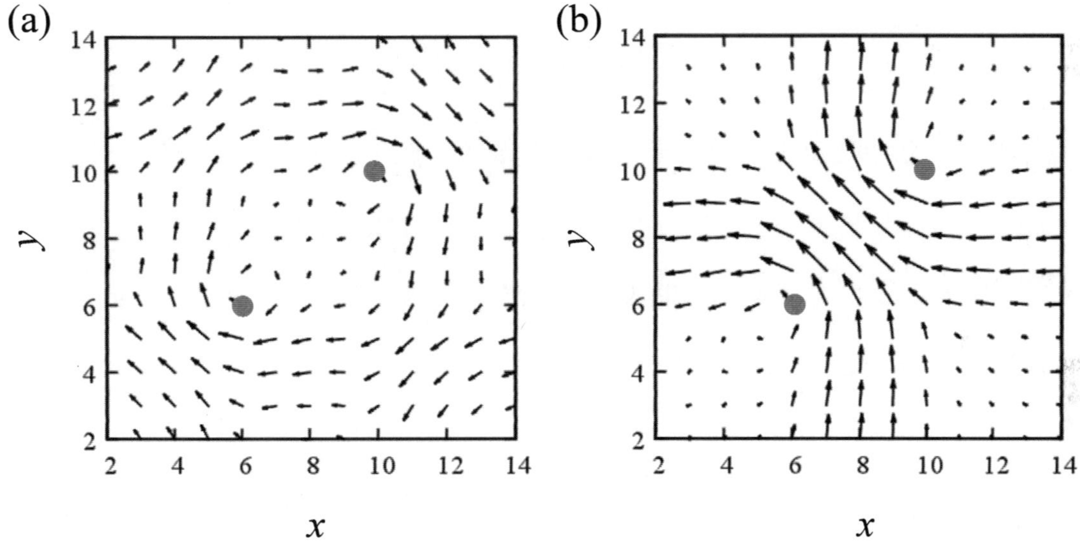

Figure 3. Distribution of the total charge (a) and spin (b) current around half quantum vortecies. Red dots indicate the cores of HQVs.

tight-binding model based on the Bogoliubov-de Gennes (BdG) theory. We confirmed that the spin current is strengthened between the HQVs and is weakened outside, therefore, the energy of a system tends to become high between HQVs.

Acknowledgments
The numerical computations were carried out at the Cyber-science Center, Tohoku University.

References

[1] M. M. Salomaa and G. E. Volovik, *Phys. Rev. Lett.* **55**, 1184 (1985).
[2] D. A. Ivanov, *Phys. Rev. Lett.* **86**, 268-271 (2001).
[3] S. B. Chung, H. Bluhm, and E. -A. Kim, *Phys. Rev. Lett.* **99**, 197002 (2007).
[4] H. -T. Kee and M. Sigrist, arXive:1307.5859.
[5] V. Vakaryuk and A. J. Leggett, *Phys. Rev. Lett.* **103**, 057003 (2009).

[6] A. P. Mackenzie and Y. Maeno, *Rev. Mod. Phys.* **75**, 657-712 (2003).
[7] Y. Imai, K. Wakabayashi, and M. Sigrist, *Phys. Rev. B*, **88** 144503 (2013).
[8] H. Tsuchiura, M. Ogata, Y. Tanaka, and S. Kashiwaya, *Phys. Rev. B* **68** 012509 (2003).

Magnetic field dependence of most stable vortex states in the chiral helimagnet / superconductor bilayer system

S Fukui[1], M Kato[1], and Y Togawa[2]

[1]Department of Mathematical Sciences, Osaka Prefecture University, 1-1, Gakuencho, Nakaku, Sakai, Osaka 599-8531, Japan

[2]Department of Physics and Electronics, Osaka Prefecture University, 1-1, Gakuencho, Nakaku, Sakai, Osaka 599-8531,Japan

E-mail: `st110035@edu.osakafu-u.ac.jp`

Abstract. We have studied vortex states in chiral helimagnet / superconductor bilayer systems. An effect of a chiral helimagnet on a superconductor is taken as an external magnetic field. Solving the Ginzburg-Landau equations, we obtain various vortex configurations. In order to find a stable vortex state, we compare the Ginzburg-Landau free energies of these states. Under the large magnetic field from the chiral helimagnet, the state in which vortices appear in a row becomes more stable.

1. Introduction

A vortex state in a superconductor is affected by a magnet, in particular, a ferromagnet [1]. In a ferromagnet / superconductor bilayer structure, a magnetic structure of a ferromagnet causes a magnetic field. In a superconducting layer, vortices appear due to the magnetic field from the ferromagnet and vortex configuration is affected by the magnetic domain of the ferromagnet. So, the ferromagnet / superconductor bilayer structure enhances superconductivity, in particular, its critical current due to a pininng of vortices.

We focus on a chiral helimagnet / superconductor bilayer system. A chiral helimagnet has spins that form a helical rotation along a helical axis [2],[3], which is shown in figure 1. This helical configuration of spins comes from a competition between the ferromagnetic exchange interaction and the Dzyaloshinsky-Moriya interaction. The ferromagnetic exchange interaction causes nearest neighbor spins to be parallel. On the other hand, the Dzyaloshinsky-Moriya interaction causes nearest neighbor spins to be perpendicular to each other [4],[5]. Due to the competition between these two interactions, the helical configuration of spins is slightly deviated from the ferromagnetic configuration of spins. Then, all spins rotates along the helical axis. We expect that this magnetic structure of the chiral helimagnet affects the superconductor strongly. These influences may be different from those of the ferromagnet.

In this paper, we investigate effects of the chiral helimagnet on the superconductor in the chiral helimagnet / superconductor bilayer system. In particular, we focus on vortex states of a superconducting layer. To achieve our purpose, we solve the Ginzburg-Landau equations with the finite element method [6]-[8].

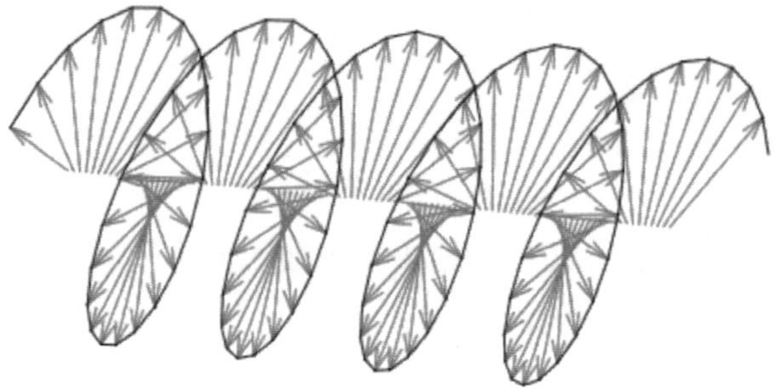

Figure 1. Spin configurations in the chiral helimagnet.

2. Methods

To obtain vortex states, we solve the Ginzburg-Landau equations,

$$\alpha|\psi|^2 + \beta|\psi|^2\psi + \frac{1}{2m^*}\left(\frac{\hbar}{i}\nabla - \frac{e^*}{c}\boldsymbol{A}\right)^2\psi = 0, \tag{1}$$

$$\mathrm{curl}(\mathrm{curl}\boldsymbol{A} - \boldsymbol{H}_{\mathrm{ext}}) = \frac{4\pi}{c}\boldsymbol{J} = \frac{4\pi}{c}\left\{\frac{e^*\hbar}{2m^*i}(\psi^*\nabla\psi - \psi\nabla\psi^*) - \frac{e^{*2}}{m^*c}\psi^*\psi\boldsymbol{A}\right\}, \tag{2}$$

where $\alpha = \alpha_0(T - T_c)$, T is a temperature, T_c is a critical temperature. α_0, β are constants. ψ is an superconducting order parameter, m^* is an effective mass, e^* is an effective charge, $\boldsymbol{A}$ is a magnetic vector potential, and $\boldsymbol{J}$ is a supercurrent density. As a system, we consider a chiral helimagnet / superconductor bilayer. The effect of the chiral helimagnet on the superconductor is taken as an external magnetic field $\boldsymbol{H}_{\mathrm{ext}}$. Effects of the superconductor on the chiral helimagnet are neglected. In this study, we take into account only the perpendicular component of $\boldsymbol{H}_{\mathrm{ext}}$, $(\boldsymbol{H}_{\mathrm{ext}})_z$. Then, our model is equivalent to a superconductor under an oscillating magnetic field $(\boldsymbol{H}_{\mathrm{ext}})_z$ in figure 2.

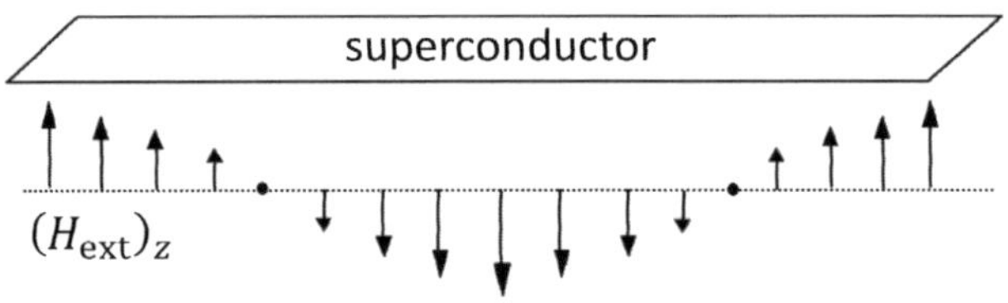

Figure 2. The superconductor under the magnetic field from the chiral helimagnet $(\boldsymbol{H}_{\mathrm{ext}})_z$

We solve the Ginzburg-Landau equations (1) and (2) with the finite element method [9]. First, we give the initial values of the order parameter ψ randomly. Then, we obtain several stable solutions by iterations, which leads to various vortex states. In order to find the most stable state, we calculate the Ginzburg-Landau free energy,

$$\mathcal{F}(\psi, \boldsymbol{A}) = \int_\Omega \left[\frac{1}{\xi(T)^2}\frac{1}{2}\left(|\psi|^2 - 1\right)^2 + |(i\nabla - \tilde{\boldsymbol{A}})\psi|^2\right] d\Omega + \kappa^2\xi^2(T)\int_\Omega \left|\mathrm{curl}\boldsymbol{A} - \frac{2\pi}{\Phi_0}\boldsymbol{H}_{\mathrm{ext}}\right|^2 d\Omega, \tag{3}$$

where $\xi(T)$ is a coherence length and $\tilde{\boldsymbol{A}} = (2\pi/\Phi_0)\boldsymbol{A}$.

$\boldsymbol{H}_{\rm ext}$ is expressed with the analytical solution of the chiral helimagnet, which is obtained from the hamiltonian[3],

$$\mathcal{H} = -J\sum_n \boldsymbol{S}_n \cdot \boldsymbol{S}_{n+1} + \boldsymbol{D}\cdot\sum_n \boldsymbol{S}\times\boldsymbol{S}_{n+1} + 2\mu_B \boldsymbol{H}_{\rm appl}\cdot\sum_n \boldsymbol{S}_n^z, \tag{4}$$

where J is a strength of the exchange interaction, $\boldsymbol{S}_n$ is a $n-$th spin, $\boldsymbol{D} = (D, 0, 0)$ is a DM vector, μ_B is the Borh magneton, and $\boldsymbol{H}_{\rm appl} = (0, 0, H_{\rm appl})$ is a uniform applied magnetic field. The three terms represent the ferromagnetic exchange interaction, the Dzyaloshinsky-Moriya interaction, and the Zeeman energy, respectively. From this Hamiltonian, we obtain an angle between nearest spins $\theta(x)$,

$$\theta(x) = 2\sin^{-1}\left[\mathrm{sn}\left(\frac{(H^*)^{1/2}}{k}x \mid k\right)\right] + \pi, \tag{5}$$

where $\mathrm{sn}(u|k)$ is the Jacobi's elliptic function and k is a modulus of the elliptic function. k is determined from the following relation,

$$\frac{\pi\alpha}{4(H^*)^{1/2}} = \frac{E(k)}{k}, \tag{6}$$

where $\alpha = \tan^{-1}(D/J)$ and $E(k)$ is the complete elliptic integral of the second kind. H^* is normalized applid magnetic field,

$$H^* = \frac{2\mu_B H_{\rm appl}}{a^2S^2(J^2+D^2)^{1/2}}, \tag{7}$$

where a is a lattice constant. We assume that the order of the lattice constant a is equal to the cohelence length at $T = 0$, ξ_0. Using equation (5), we obtain $(\boldsymbol{H}_{\rm ext})_z$,

$$(\boldsymbol{H}_{\rm ext})_z(x) = H_0\cos\theta(x) + H_{\rm appl}. \tag{8}$$

In this equation, the first term is a magnetic field from the chiral helimagnet and the second term is the uniform applied magnetic field. The period of the helical rotation L' is given by,

$$L' = \frac{2kK(k)}{(H^*)^{1/2}}, \tag{9}$$

where $K(k)$ is the complete elliptic integral of the first kind.

3. Results and Discussions

We solve the Ginzburg-Landau equations and obtain various vortex states. We take the Ginzburg-Landau parameter $\kappa = \lambda_0/\xi_0 = 10$ where λ_0 is a penetration depth at $T = 0$. The ratio between two interaction strengths is taken from the $Cr_{1/3}NbS_2$, $D/J = 0.16$ [10]. The temperature T is $0.3T_c$. In this study, we use only the magnetic field from the chiral helimagnet. System size is $5.0L'\xi_0 \times 20\xi_0$, where L' is given by equation (9). For $D/J = 0.16$ and $H_{\rm appl}/(\Phi_0/\xi_0^2) \sim 0.00$, L'/ξ_0 becomes 39.2699.

We show vortex states without the applied magnetic field in figure 3. These figures represent distributions of order parameter, phase, and magnetic field. In equation (8), $H_0/(\Phi_0/\xi_0^2)$ and $H_{\rm appl}/(\Phi_0/\xi_0^2)$ are 0.04, 0.00, respectively. $\Phi_0 = \frac{hc}{2e}$ is an quantum flux. Figures 3 (a)-(d) are obtained from different initial distributions for the order parameter. In these states in figure 3, the numbers of vortices are (a) 0, (b) 2, (c) 4, and (d) 9. Each vortices appear around the peaks

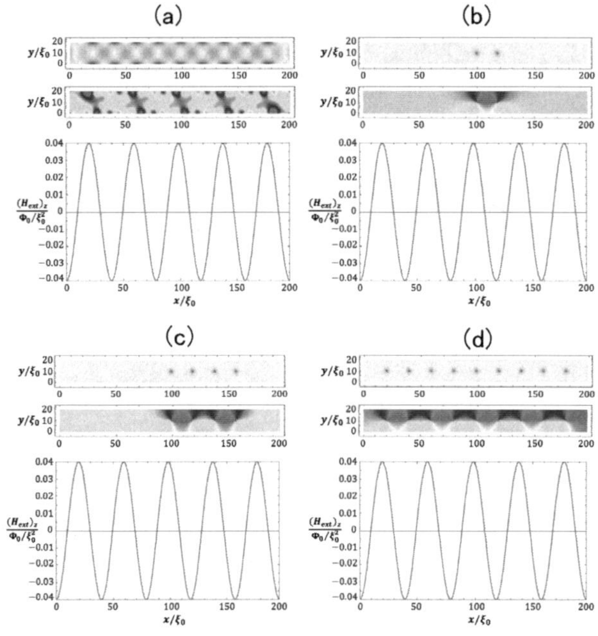

Figure 3. Distributions of order parameter, phase, and magnetic field. Magnetic fields in equation (8) are $(H_0/(\Phi_0/\xi_0^2))_z = 0.04$, $H_{\rm appl} = 0.00$. (a)-(d) are different from initial values of the order parameter.

of the magnetic field $|(\boldsymbol{H}_{\rm ext})_z/(\Phi_0/\xi_0^2)| = 0.04$ and these vortices are parallel to directions of the magnetic field.

Next, we calculate the Ginzburg-Landau free energy (equation (3)) in all states of figure 3. Dependences of free energies on magnetic fields are shown in figure 4. In figure 4, (a) red, (b) green, (c) blue, and (d) purple lines correspond to the states in figures 3 (a)-(d), respectively. For the small magnetic field $(\boldsymbol{H}_0/(\Phi_0/\xi_0^2))_z$, the state without vortices (= the Meissner state) has minimum free energy. However, for $(\boldsymbol{H}_0/(\Phi_0/\xi_0^2))_z \geq 0.64$, the state with nine vortices has minimum free energy.

From this result, we find that the most stable state of figures 3 (a)-(d) is the state without vortices (a) for the small magnetic field. Then, the most stable state changes into the state with vortices in a row (d) for the large magnetic field. Moreover, the states with two (b) and four (c) vortices are not the most stable states for all magnetic field. This transition of the most stable vortex state is different from those under the uniform magnetic field. Increasing the uniform magnetic field, the number of vortices increases as one, two, three, and so on. This difference can be explained considering the distributions of the magnetic field in two cases. In our model, the magnetic field $(\boldsymbol{H}_{\rm ext})_z$ sinusoidally oscillates. Therefore, in the superconductor, there are two types of domain where the magnetic field is upward or downward. In other words, we regard this superconductor layer as an ensemble of small superconductors, whose size is $\frac{1}{2}L'\xi_0 \times 20\xi_0$. Each small superconductor is equivalent. When vortices appear in the superconductor, the state with a vortex in each domain has the same free energy, which is lower than that for Meissner state. Therefore, the state with vortices in a row (figure 3 (d)) is more stable than states with two and four vortices (figure 3 (b), (c)).

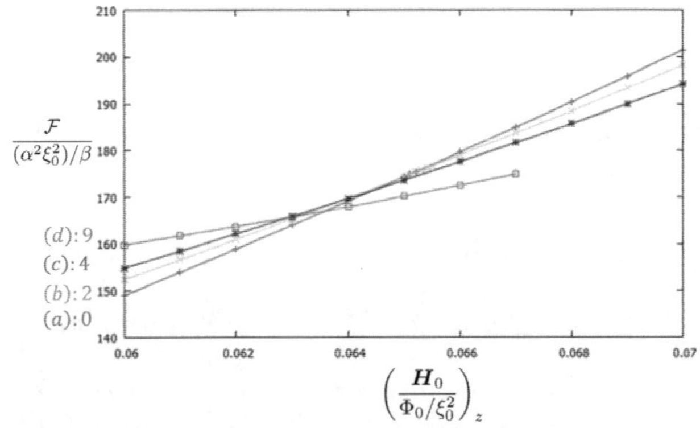

Figure 4. Magnetic field dependences of free energies. (a) Red, (b) green, (c) blue, and (d) purple lines correspond to the states in figure 3 (a)-(d), respectively.

4. Summary and Future

We have studied vortex states in the chiral helimagnet / superconductor bilayer system numerically. To investigate vortex states, we have solved the Ginzburg-Landau equations with the finite element method. We have obtained various vortex configurations. In order to find stable state, we have calculated the Ginzburg-Landau free energy. For the small magnetic field, the state without vortices is the stable state (Meissner state). On the other hand, for the large magnetic field, the state with vortices in all domains becomes stable.

Acknowledgments

This work was supported by JSPS KAKENHI Grant Number JP26400367.

References

[1] Lyuksyutov I F and Pokrovsky V L 2005 *Adv. Phys.* **54** 67
[2] Togawa Y, Koyama T, Takayanagi K, Mori S, Kousaka Y, Akimitsu J, Nishihara S, Inoue K, Ovschinnikov A S, and Kishine J 2012 *Phys. Rev. Lett.* **108** 107202
[3] Kishine J, Inoue K, and Yoshida Y 2005 *Prog. Theor. Phys. Sppl.* **159** 82
[4] Dzyaloshinskii I E 1958 *J. Phys. Chem. Solids* **4** 241
[5] Moriya T 1960 *Phys. Rev.* **120** 91
[6] Fukui S, Kato M, and Togawa Y 2015 *Physics Procedia* **65** 85
[7] Fukui S, Kato M, and Togawa Y 2016 *Supercond. Sci. Technol.* **29** 125008
[8] Fukui S, Kato M, and Togawa Y 2016 *Physica C*, **530**, 51
[9] Kato M, Ishida T, Koyama T, and Machida M 2013 *Superconductors-Materials, Properties and Applications, INTECH* 319-342
[10] Benjamin C, Alexander B, Nirmal G, David M, Minbhyea L 2014 *Appl. Phys. Lett.* **105** 072405

IOP Conf. Series: Journal of Physics: Conf. Series **871** (2017) 012027 doi:10.1088/1742-6596/871/1/012027

Simulations of vortices in a star-shaped plate with an artificial pin

Hiroki Miyoshi[1], Atsuki Ito[1], Vu The Dang[1,2], Ho Thanh Huy[1,2], Masahiko Hayashi[3], Masaru Kato[4,5], and Takekazu Ishida[1,5,*]

[1]Department of Physics and Electronics, Osaka Prefecture University, Sakai 599-8531, Japan
[2]University of Sciences, Vietnam National University HCMC, Ho Chi Minh, Viet Nam
[3]Faculty of Education and Human Studies, Akita University, Akita 010-8502, Japan
[4]Department of Mathematical Science, Osaka Prefecture University, Sakai 599-8531, Japan
[5]Institute for Nanofabrication Research, Osaka Prefecture University, Sakai 599-8531, Japan

*E-mail: ishida@center.osakafu-u.ac.jp

Abstract. Although a triangular vortex lattice is stable in a bulk type-II superconductor, exotic vortex configurations are expected to appear in a small superconducting plate. Theoretical calculations on vortex structures in a star-shaped superconducting plate have been given in our preceding work. In this work, we extended our theoretical studies to the case of having an artificial pin. We performed the Ginzburg-Landau (GL) calculations systematically to compare with the pin-free case by using the finite element method. We found that a vortex tends to accommodate preferentially in an aritificial pin in the star-shaped plate. We found a systematic evolution of vortex structure with increaseing magnetic field. We compare our theoretical calculations with vortices in a star-shaped $Mo_{80}Ge_{20}$ plate with an artificial pin and without an artificial pin obtained by a scanning SQUID microscope. We reconstructed the vortex image on the sample surface by using the inverse Biot-Savart law and the Fourier transformation.

1. Introduction

Interplay relationship between the magnetic field penetration depth λ and the sample size a of a bulk sample ($\lambda \ll a$) yields to form a triangular lattice in type II superconductors as revealed originally by Abrikosov [1]. However, a situation is subjected to change drastically when one considers a small-sized superconductor ($\lambda \simeq a$). Fertile features of having exotic vortex patterns would appear due to the vortex-vortex interaction and the vortex-boundary interaction, and the vortex-current interaction. Choice of sample geometry is also crucial to consider the possible fascinating configurations. We consider that some of exotic vortex patterns may serve as the basis of utilizing various superconducting

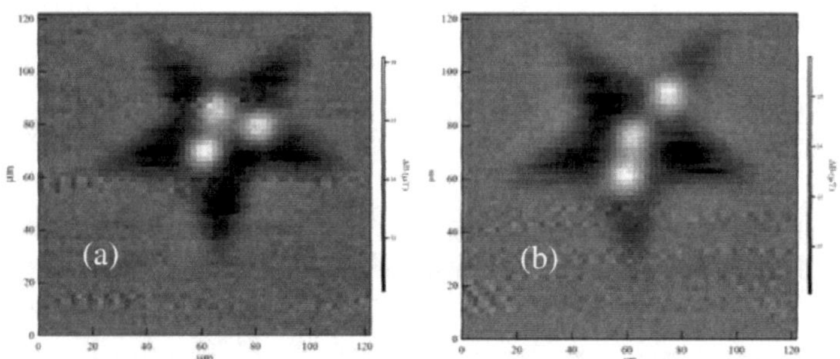

Figure 1. Vortex images taken by scanning SQUID microscopy by using a star-shaped $Mo_{80}Ge_{20}$ plate. (a) the star-shaped $Mo_{80}Ge_{20}$ film with no artificial pin (b) the star-shaped $Mo_{80}Ge_{20}$ plate with an artificial pin at the

devices. The symmetric changes in vortex structures have been investigated extensively by using the regular polygons with mirror symmetric lines. For instance, with the same vorticity *L*, the vortex configuration is different from each other between square and triangle plates [2]. Both theoretical and experimental approaches have used to explore the distribution of vortices. The nonlinear Ginzburg–Landau theory and the London approximation have been often used for such purposes [3] [4] [5]. In Figure 1, we show the vortex images of a star-shaped $Mo_{80}Ge_{20}$ amorphous thin plate with an artificial pinning center obtained by a scanning superconducting quantum interference device (SQUID) microscope in our group [6] [7]. In this work, we extended our theoretical studies to the case of having an artificial pin. We performed the Ginzburg-Landau (GL) calculations systematically to compare with the pin-free case by using the finite element method. We performed experiments on controlling vortices in a star-shaped superconductor containing a pinning center. It is also meaningful to investigate the nature of vortex profile under the presence of the pinning center in view of theoretical approach. In this study, we aim at answering a question of whether or not vortex control can be realized by introducing a pin. In addition, we reconstruct a clear vortex image on the sample surface from the observed image by applying the inverse Fourier transformation method.

2. Theoretical

It is interesting to consider the results obtained by scanning SQUID microscopy on star-shaped $Mo_{80}Ge_{20}$ plate films with artificial pinning center [7]. We performed the Ginzburg-Landau calculations of star-shaped superconducting small plates systematically to compare with experimental findings. The non-linear Ginzburg-Landau free energy used for the calculations is given by

$$F(\Delta, A) = \int d\Omega \left[\frac{1}{2}(\sqrt{\beta}|\Delta|^2 + \frac{\alpha}{\sqrt{\beta}})^2 + \frac{1}{4m}\left|\left(\frac{\hbar}{i}\nabla + \frac{2e}{c}A\right)\Delta\right|^2 + \frac{|\nabla \times A - H|^2}{8\pi} + \frac{1}{8\pi}(\mathrm{div}A)^2\right] \quad (1)$$

where $\Delta(r)$ is an order parameter of a superconductor, *A* is a vector potential and *H* is the external magnetic field. The parameter β is a positive constant, and α depends on temperature as $\alpha = \alpha(0)(1- T/T_c)$. We consider that a temperature of the system is well below the transition temperature T_c to reflect an experimental condition while the Ginzburg-Landau equation is applicable to temperatures near T_c. We note that our preceding studies showed that the extensive application of the Ginzburg-Landau equation to temperatures somewhat lower than T_c. We used the finite element method [8], where the model consists of a number of triangular elements to specify an order parameter Δ, a vector potential *A*, and a temperature *T* locally as

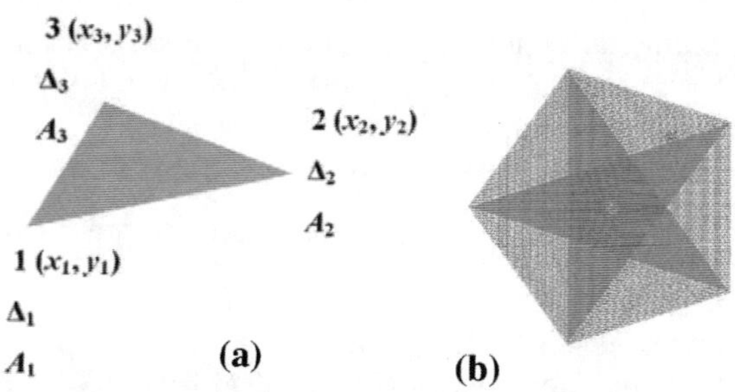

Figure 2. (a) A triangular element in FEM with an order parameter Δ, a vector potential **A**; (b) FEM mesh of a star-shaped plate. Superconducting part is shown in blue while normal state part is shown in white. Central part is normal to play as a pin.

$$\Delta(x,y) = N_1(x,y)\,\Delta_1 + N_2(x,y)\,\Delta_2 + N_3(x,y)\,\Delta_3 \quad (2)$$

$$A(x,y) = N_1(x,y)\,A_1 + N_2(x,y)\,A_2 + N_3(x,y)\,A_3 \quad (3)$$

$$N_i(x,y) = (\,a_i + b_i x + c_i y\,)\,/\,2S_e \qquad (\,i = 1,2,3\,) \quad (4)$$

where Δ_i and A_i are the values of the order parameter and the magnetic vector potential at the *i*-th node, respectively. The relations $a_i = x_j y_k - x_k y_j$, $b_i = y_j - y_k$, $c_i = x_j - x_k$, are cyclically defined using coordinates of nodes (x_i, y_i) of the triangular element and S_e is an area of the element.

IOP Conf. Series: Journal of Physics: Conf. Series **871** (2017) 012027 doi:10.1088/1742-6596/871/1/012027

3. Results and discussion

We calculated the distribution of vortex in a star-shaped with inscribed circle radius $r = \lambda_0$ and circumscribed circle radius $R = 2.61\ \lambda_0$. The size (radius) of pinning center is chosen as $0.16\ \lambda_0$. The value of the Ginzburg-Landau parameter is $\kappa = 10$ (type-II superconductors). The magnetic field is applied in perpendicular to the star-shaped plate at a temperature of $T/T_c = 0.5$. This is because T_c of $Mo_{80}Ge_{20}$ is 7.3 K and our experimental measurements are performed at $T = 4$ K ($T/T_c = 0.55$). Applied magnetic field H was represented by a dimensionless magnetic field of $h = H/H_{c2}$ reduced by the upper critical field H_{c2}. Figures 3 and 4 show the comparison of theoretical distribution with experimental vortex images. Figure 3 is comparative studies on the case of no artificial pin. Figure 4 is comparative studies in the presence of central pin. The systematic evolution of vortex configuration can be seen as a function of the magnetic field. We also compare the numerical calculations with the experimental images in Figure 3 and 4. One finds that there are in reasonable agreements between theoretical calculations and experimental data. Therefore, we consider that it is very useful to conduct such researches by comparing both theoretical and experimental approaches with each other in the field of the mesoscopic superconductivity. Further increase of the magnetic field shows an interesting feature in the theoretical vortex images. In Figure 3, at L=1, vortex appears at the center. At L=2, vortex distribution has the two-fold symmetry.

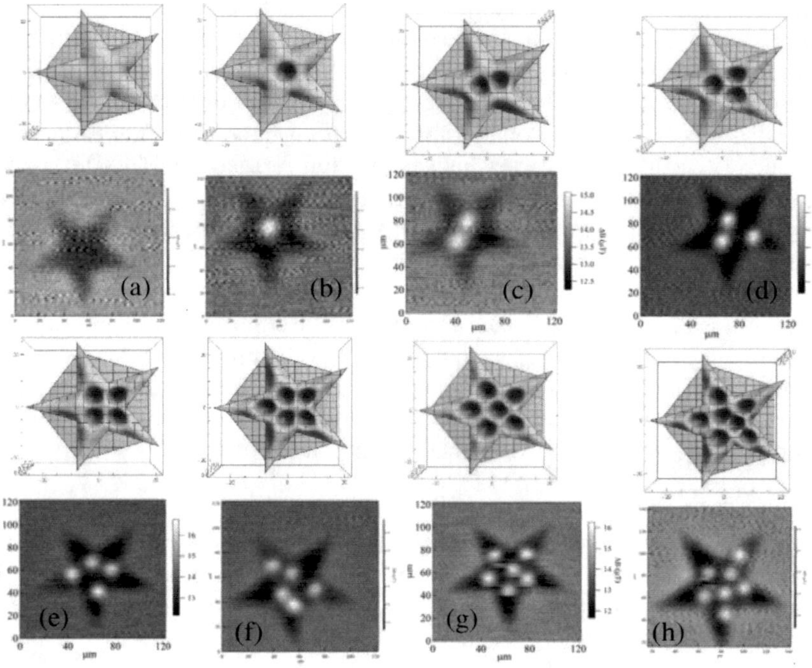

Figure 3. Comparative studies on the case of no artificial pin. (a) L= 0; (b) L=1; (c) L=2; (d) L=3; (e) L=4; (f) L= 5; (g) L=6; (h) L=7.

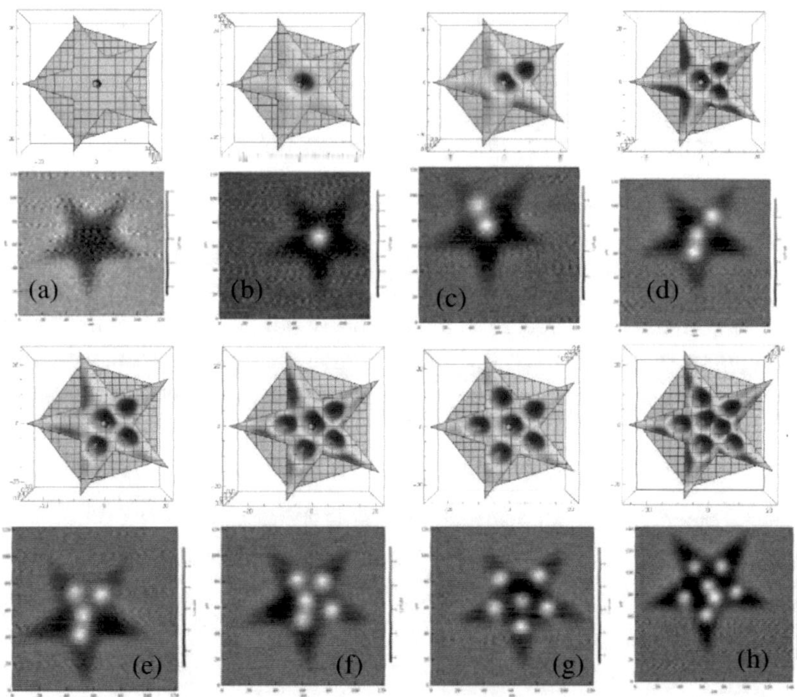

Figure 4. Comparative studies on the case of central pin. (a) L= 0; (b) L=1; (c) L=2; (d) L=3; (e) L=4; (f) L= 5; (g) L=6; (h) L=7.

At *L*=3, vortices form a triangular. At *L*=4, vortices form a cross. This distribution has the five-fold symmetry axis. At *L*=5, the vortex distribution has the five-fold symmetry. At *L*=6, vortices form a shell structure of *L*=(1,5). At *L*=7, vortices form a shell structure of *L* = (2,5). In Figure 4, at *L*=1, vortex appears at the center. This result is similar to the case obtained for the mode without pinning center. At *L*=2, one vortex appears at the center. At *L*=3, one vortex appears at the center, but it is not in good agreement with experience result. At *L*=4, one vortex appears at the center, but it is different from that observed in experimental results. At *L*=5, one vortex appears at the center. This is different from the case without pinning center. This result is different from the case without pinning center results. At *L*=6, one vortex appears at the center. This result is similar to the case of the experimental results. At *L*=7, one vortex appears at the center. From these results, we found a reasonable agreement between experiment and theory.

4. Processing images

We try to improve vortex images by considering the following items. Magnetic flux from tiny-sized magnetic moment spreads quickly with increasing distance from the surface. Demagnetization effect of the superconducting pick-up coil also affects the local field distribution around the coil. The current distribution on the surface can be refined by applying the inverse Fourier transformation to raw data of local magnetic field distribution. Figure 5 comparatively shows that original images and processing images. According to our studies, we not three remarks. First, noise was fairly removed in the processed image. Second, the spatial resolution was remarkably improved after conducting data procession. Third, the position of vortices becomes clear due to enhanced shapeless of the image.

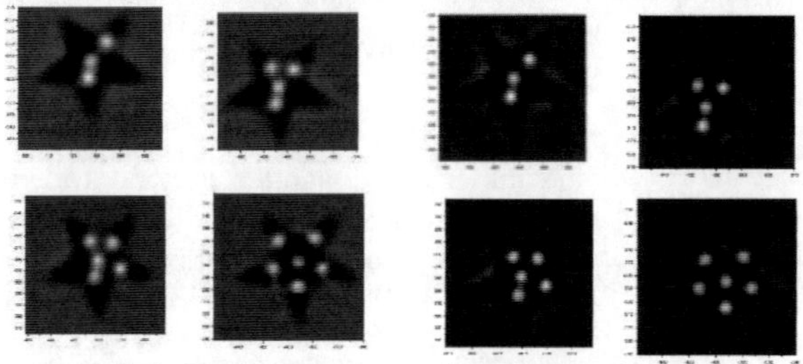

(a) Original data (b) Restored data

Figure 5. Original data versus processed data.

5. Conclusion

We developed theoretical calculations of the star-shaped plate with pinning systematically. We conclude based on theoretical calculation that a pin can control the vortex distribution. Although the agreement between theory and experiment is fair while it seems to be difficult to obtain the stable (reproducible) distribution for vortices *L* = 3 and 4. We demonstrated to improve the clearness of vortex images by using the inverse Biot-Savart-law algorithm.

Acknowledgment

This work is supported by Grant-in-Aid (No.26800192, No. 23226019, No.16H02450) from JSPS.

References

[1] Abrikosov A A 1957 *Sov. Phys. JETP* **5** 1174

[2] Matsumoto H, Miyoshi H, Okamoto T, Vu T D, Ho H T, Miyajima S, Shishido H, Kato M, Fujita N, Hayashi M, and Ishida T Vortices in small concave decagon $Mo_{80}Ge_{20}$ plate, *COST MP1201 Workshop Advances in Studies of Superconducting Hybrids: Theory and Modeling vs Experiment* (Arcachon, France, 2015)

[3] Kato M, Sato O, Ho H T, and Ishida T 2013 *Physica C* **494** 124

[4] Misko V R, Xu B and Peeters F M 2007 *Phys Rev B* **76** 024516

[5] Cabral L R E and Aguiar A J, 2009 *Phys Rev B* **80** 214533

[6] Kokubo N, Okayasu S, Nojima T, Tamochi H, and Shinozaki B 2014 *J Phys Soc Jpn* **83** 083704

[7] Matsumoto H, Ho H T, Miyoshi H, Okamoto T, Vu D T, Kato M, and Ishida T 2016 *Physica C* **530** 46

[8] Zienkiewicz O C and Taylor R L The Finite Element Method (Butterworth-Heinimann, Oxford, U.K., (2000) 5th ed

Molecular Dynamics Simulation on Vortex Lattice Melting in Meso-scopic Superconductors

Masaru Kato and Harutaka Kitago

Department of Mathematical Sciences, Osaka Prefecture University, 1-1, Gakuencho, Nakaku, Sakai, Osaka 5995-8531, Japan

kato@ms.osakafu-u.ac.jp

Abstract. Vortex lattice melting transitions in meso-scopic superconductors are studied using the molecular dynamics method for vortices. Temperature dependence of vortex dynamics comes from the fluctuation force and penetration dependence in the vortex-vortex interaction. Using the standard deviation of the positions of vortices, we can determine the melting temperature. Changing the vortex numbers, the melting temperature oscillates with increasing the vortex number.

1. Introduction

After the discovery of cuprate High-Tc superconductors [1], vortex states in the H-T phase diagram was explored in detail. The cuprate superconductors have many unconventional properties other than its higher transition temperature. Their Cooper pair symmetry is the d-wave and their conduction electron states are almost two dimensional and they have intrinsic Josephson junctions between superconducting layers. All of these properties are interesting and may useful for future applications. But also vortex states in the High-Tc cuprate superconductors are unique. In the pure High-Tc cuprate superconductor at low temperature and for higher field than lower critical field H_{C1}, vortices form a lattice. But for higher temperature this vortex lattice changes into a vortex liquid state [2]. There is a melting line of the vortex lattice. This melting behaviour comes from weak pinning, large thermal and quantum fluctuations in the cuprate High-Tc superconductors.

Recently, Ooi et al. showed the melting temperature in meso-scopic cuprate superconducting square plate oscillates with increasing magnetic field [3]. They explained this oscillation comes from the stability of configuration of n^2 vortices (n=1,2,3,4,5,...).

In order to confirm this phenomenon, we study the vortex lattice melting in meso-scopic superconductors using the molecular dynamics method for vortices [4,5]. Temperature dependence of vortex dynamics comes from the fluctuation force and penetration dependence in the vortex-vortex interaction. Using the standard deviation of the positions of vortices, we can determine the melting temperature. Changing the vortex numbers, we find the melting temperature depends on the vortex number.

In section 2, our method is explained and in section 3 we show numerical results. Section 4 is devoted to the summary.

2. Method

In the molecular dynamics method, a vortex are treated as a point-particle in a two dimensional space [3,4]. Previously we extended the molecular dynamics method including heat transport [5] and relaxation of superconducting order parameter [6]. In order to study the melting transition of a vortex lattice, we use a conventional molecular dynamics method because vortex motion is slow and heat generation and relaxation of order parameter are not important.

So, we solve following equation of motion for i-th vortex of which position is $\boldsymbol{r}_i$.

$$\eta \frac{\mathrm{d}\boldsymbol{r}_i}{\mathrm{d}t} = \boldsymbol{f}_{\mathrm{p}i}^{\mathrm{imp}} + \boldsymbol{f}_{\mathrm{v}i} + \boldsymbol{f}_{\mathrm{f}i} \tag{1}$$

Here η is the viscosity. The inertia term can be ignored comparing the viscosity term. The forces from pinning site is given as,

$$\boldsymbol{f}_{\mathrm{p}i}^{\mathrm{imp}} = \sum_{j}^{\mathrm{pinningsite}} \boldsymbol{f}_{ij}^{\mathrm{p}} \tag{2}$$

$$\boldsymbol{f}_{ij}^{\mathrm{p}} = \frac{f_{\mathrm{p}}}{r_{\mathrm{p}}} \left| \boldsymbol{r}_i - \boldsymbol{r}^{\mathrm{p}}{}_j \right| \Theta\left(\frac{r_{\mathrm{p}} - \left| \boldsymbol{r}_i - \boldsymbol{r}^{\mathrm{p}}{}_j \right|}{\lambda} \right) \hat{\boldsymbol{r}}_{ij} \tag{3}$$

where $\boldsymbol{r}_j^{\mathrm{p}}$ is the position of j-th pinning site, λ is the penetration depth and f^{p} is the strength of the pinning force. The vortex-vortex interaction is given as,

$$\boldsymbol{f}_{\mathrm{v}i} = \sum_j \boldsymbol{f}_{ij}^{\mathrm{vv}} \tag{4}$$

$$\boldsymbol{f}_{ij}^{\mathrm{vv}} = f_0 \sum_j K_1\left(\frac{r_{ij}}{\lambda} \right) \hat{\boldsymbol{r}}_{ij} \tag{5}$$

where $f_0 = \Phi_0^2 / 8\pi^2\lambda^3$ is the strength of the vortex-vortex and $K_1(k)$ is the complete elliptic integral of the 1st kind. The thermal fluctuation force $\boldsymbol{f}_i^{\mathrm{f}}$ should satisfy following fluctuation-dissipation theorem,

$$\left\langle \boldsymbol{f}_i^{\mathrm{f}}(t_1) \cdot \boldsymbol{f}_j^{\mathrm{f}}(t_2) \right\rangle = 2\eta k_B T \delta_{ij} \delta(t_1 - t_2) \tag{6}$$

In this study, we consider meso-scopic superconducting square plate. Although the vortices are confined small superconducting plate, we use a periodic boundary condition. Temperature dependence of the vortex motion is taken into the penetration depth λ and the fluctuation force $\boldsymbol{f}_i^{\mathrm{f}}$.

3. Numerical Results

First we consider a small ($3\lambda \times 3\lambda$) square superconducting plate, and put the fixed number of vortices randomly and then under the constant temperature we simulate the vortex motion using equation (1). In figure 1, we show the trajectories of vortices at the temperature $T/T_C = 0.02$. In this figure, each color of trajectories corresponds to each vortex. Because of the periodic boundary condition, trajectories of vortices are extended over the superconducting region ($3\lambda \times 3\lambda$). From this figure, we can see all of the vortices remain in the small regions, which do not overlap with each other. Therefore this state can be considered as a solid of vortices.

Increasing the temperature, the motion of vortices becomes wild at $T/T_C = 0.5$ (figure 2). The vortex solid gradually melts. Further increasing the temperature, at $T/T_C = 0.9$, vortices move randomly over the superconducting region ($3\lambda \times 3\lambda$), as shown in figure 3. In this state, now vortices become liquid.

In order to find the melting transition temperature, we calculate the variance of time averaged vortex positions.

IOP Conf. Series: Journal of Physics: Conf. Series **871** (2017) 012028 doi:10.1088/1742-6596/871/1/012028

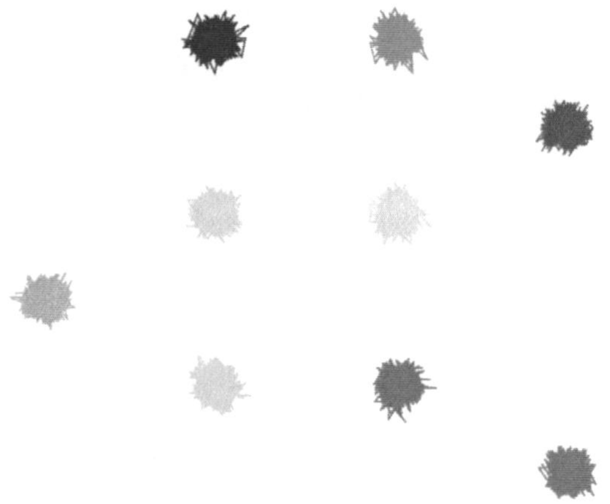

Figure 1. Trajectories of 9 vortices in a $3\lambda \times 3\lambda$ square superconducting plate. Temperature is $T/T_c = 0.01$.

$$\sigma_i^2 = \left\langle \left(\boldsymbol{r}_i - \langle \boldsymbol{r}_i \rangle \right)^2 \right\rangle \tag{7}$$

where average is taken over time interval from 0 to t_{TOTAL}. In this study, it happens that all of the vortices move to the same direction because of the periodic boundary condition, as seen in figure 2. In such case the variance of the vortex position becomes large, because the average of the position is the center of the trajectories in figure 2. But if we take the fixed boundary condition, the variance becomes rather small. In order to avoid this discrepancy, we divide the time interval into a series of short time intervals, calculate the variance for each short time interval and take average of the variances for all short time intervals. In figure 4, we show temperature dependence of the standard deviation σ, which is an average of the standard deviation σ_i for 9 vortices. In this figure we see σ abruptly becomes large around $T/T_C = 0.8$. This behavior corresponds to the vortex lattice melting.

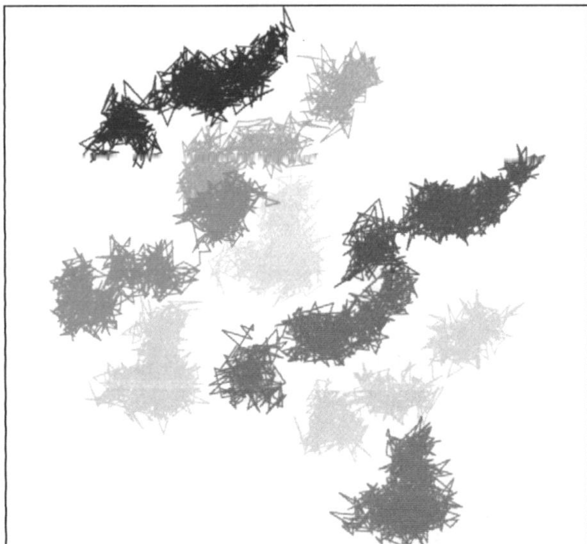

Figure 2. Trajectories of 9 vortices in a $3\lambda \times 3\lambda$ square superconducting plate. Temperature is $T/T_c = 0.5$.

Figure 3. Trajectories of 9 vortices in a $3\lambda \times 3\lambda$ square superconducting plate. Temperature is $T/T_c = 0.9$.

In figure 5, we show temperature dependences of the standard deviations for vortex number from 7 to 18. We can see that this temperature dependence does not monotonically depend on the vortex number.

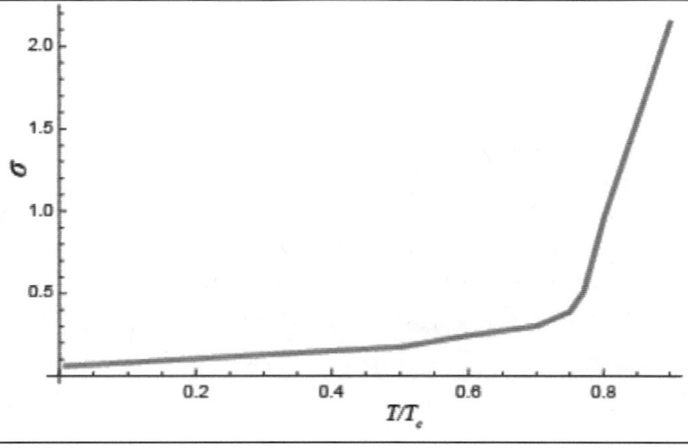

Figure 4. Average of standard deviations of 9 vortices in the $3\lambda \times 3\lambda$ square superconducting plate as a function of temperature.

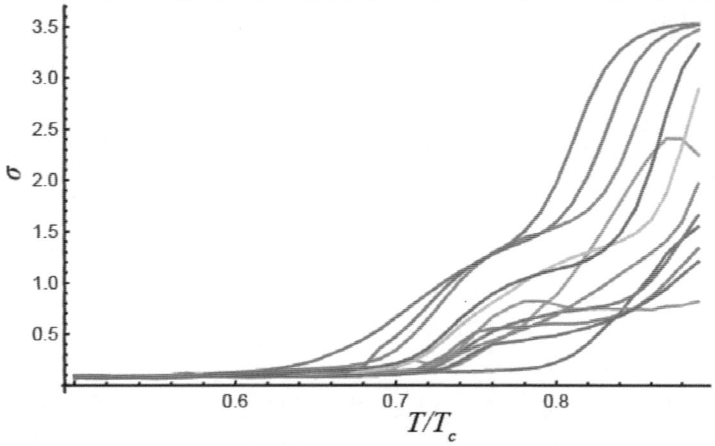

Figure 5. Average of standard deviations of vortices in the $3\lambda \times 3\lambda$ square superconducting plate as a function of temperature. Numbers of vortices are 7, 8, 9, 10, 11, 12, 13, 14, 15, 16, 17 and 18.

4. Conclusion

In order to clarify the oscillation of the melting temperature as a function of the magnetic field in the meso-scopic superconducting square plate, numerical simulations of the molecular dynamics have been done. Temperature dependence is included in the penetration length and the fluctuation force in the simulation. Numerical results show that clear increase of the standard deviation of vortex positions with increasing the temperature. This increase means vortex lattice melting. We also have found that the standard deviation curves as a function of temperature show non-monotonic behavior, which may correspond to the oscillation behavior of the melting temperature as a function of the magnetic field in Reference [3].

In this study, we consider a model with the periodic boundary condition. But in the real system, vortices are confined in the finite region. So we will do similar simulations on a model with a fixed boundary condition including the effect of the Meissner current.

Acknowledgments
This work was supported by JSPS KAKENHI Grant Number JP26400367.

References

[1] Bednorz J G and Muller K A 1986 *Z. Phys.* **64** 189
[2] Blatter G, Feigel'man M V, Geshkenbein V B, Larkin A I and Vinokur V M 1994 *Rev. Mod. Phys.* **66** 1225
[3] Ooi S, Mochiku T, Tachiki M and Hirata H 2015 *Phys. Rev. Lett.* **114** 087001
[4] Reichhardt C, Olson C J, Groth J, Field S, and Nori F 1996 *Phys. Rev.* B **53** R8898
[5] Misko V R and Peeters F M 2006 *Phys. Rev.* B **74** 174507
[6] Fujibayashi D E and Kato M 2013 *Physica* C **484** 94
[7] Kato M and Fujibayashi D E 2013 *Physics Procedia* **45** 133

A variety of vortex state solutions of Ginzburg-Landau equation on superconducting mesoscopic plates

O Sato[1] and M Kato[2]

[1]Department of Liberal Arts, Osaka Prefecture University College of Technology, Neyagawa, Osaka 572-8572, Japan
[2]Department of Mathematical Sciences, Osaka Prefecture University, Sakai, Osaka 599-8531, Japan

sato@osaka-pct.ac.jp

Abstract We report numerical solutions of the Ginzburg-Landau equation for superconducting mesoscopic square plate under a gradient magnetic field that changes linearly in strength in one direction. All the obtained vortex configurations have mirror symmetry. We found two different solutions containing the same number of the total flux quantum under the same field. Our calculation results imply that the obtained metastable states can be actually observed at almost the same frequency as the most stable state.

1. Introduction

In mesoscopic superconductors, effects on vortex states caused by shapes of the samples cannot be negligible. It is observed experimentally that flux quantum configurations have the same symmetry as the symmetry of the shape of the sample[1][2]. Especially, in the case of square mesoscopic sample in a homogeneous magnetic field, it has been studied experimentally in detail [1]. Near and below the transition temperature, theoretical calculations show that anti-vortex and giant-vortex can appear in order to maintain the symmetric vortex configuration[3][4]. This is because, near and below the transition temperature, the Ginzburg-Landau (GL) equation can be linearlized, since the order parameter $\psi(\boldsymbol{r})$ becomes very small and higher order terms of $\psi(\boldsymbol{r})$ can be neglected. In a general quantum mechanics, symmetry of the Hamiltonian determines symmetry of eingen functions of the Schrödinger equation. The linearlized GL (LGL) equation has a same form as the Schrödinger equation. Therefore, the LGL equation has vortex states solutions of having symmetry reflected by the symmetry of the sample shape. At low temperatures, the vortex configuration does not have to match the symmetry of the system because higher order terms of $\psi(\boldsymbol{r})$ in GL equation cannot be negligible. Here we discuss the vortex state of a mesoscopic square superconducting plate under a gradient magnetic field that changes linearly in strength in the direction of a side of the square. The gradient magnetic field changes the symmetry of the system from the four-fold symmetry into the mirror symmetry. If the square sample is in a homogeneous field, breaking of the four-fold symmetry of the vortex configuration occurs with lowering the temperature, but the sample is in the gradient field, we consider that mirror symmetry of the vortex configuration maintains in a wide temperature range. And we expect to find peculiar vortex state that cannot be found in the homogeneous field case.

2. Formalism

Here we consider the superconducting 2D square plate is in an external field $\boldsymbol{H}(\boldsymbol{r})$. The whole area that we will simulate is denoted by Ω. The modified Ginzburg-Landau free energy functional[5] $F(\psi, \boldsymbol{A})$ of the superconducting order parameter ψ and vector potential $\boldsymbol{A}$ with constraints $\nabla \cdot \boldsymbol{A} = 0$ is,

$$F(\psi, \boldsymbol{A}) = \int_\Omega \left[\frac{1}{2}\left(\sqrt{\beta}\left|\psi(\boldsymbol{r})\right|^2 + \frac{\alpha(T)}{\sqrt{\beta}} \right)^2 + \frac{1}{2m_s}\left|\left(-i\hbar\nabla - \frac{e_s \boldsymbol{A}(\boldsymbol{r})}{c} \right)\psi(\boldsymbol{r})\right|^2 + \frac{\left|\boldsymbol{h}(\boldsymbol{r}) - \boldsymbol{H}(\boldsymbol{r})\right|^2 + \left(\nabla \cdot \boldsymbol{A}(\boldsymbol{r})\right)^2}{8\pi} \right] d\Omega, \quad (1)$$

where $\alpha(T)$ is the temperature dependent coefficient as

$$\alpha(T) = a(T - T_c) \qquad (a > 0), \quad (2)$$

β is a positive constant, e_s and m_s are charge and mass of the Cooper pair respectively, and the microscopic magnetic field is denoted by $\boldsymbol{h}(\boldsymbol{r})$. Rewriting (1) by

$$\tilde{F}\left(\tilde{\psi}, \tilde{\boldsymbol{A}}\right) = \frac{F\left(\psi, \boldsymbol{A}\right)}{\alpha(T)\xi(T)^2 / \beta}, \quad (3)$$

$$\tilde{\psi}(\boldsymbol{r}) = \frac{\psi(\boldsymbol{r})}{\sqrt{\left|\alpha(T)\right| / \beta}} \text{ and } \tilde{\boldsymbol{A}}(\boldsymbol{r}) = \frac{2\pi \boldsymbol{A}(\boldsymbol{r})}{\Phi_0}, \quad (4)$$

we obtain,

$$\tilde{F}(\tilde{\psi}, \tilde{\boldsymbol{A}}) = \int_\Omega \left[\frac{1}{\xi(T)^2}\frac{1}{2}\left(\left|\psi(\boldsymbol{r})\right|^2 - 1 \right)^2 + \left|\left(-i\nabla - \tilde{\boldsymbol{A}}(\boldsymbol{r}) \right)\psi\right|^2 d\Omega + \kappa^2\xi(T)^2 \int_\Omega \left|\nabla \times \tilde{\boldsymbol{A}}(\boldsymbol{r}) - \frac{2\pi \boldsymbol{H}(\boldsymbol{r})}{\Phi_0}\right|^2 + \left(\nabla \cdot \tilde{\boldsymbol{A}}(\boldsymbol{r})\right)^2 \right] d\Omega, \quad (5)$$

where κ is the GL parameter and $\xi(T)$ is the temperature dependence GL coherence length which obeys $\xi(T) = \xi_0 / \sqrt{1 - T / T_{C0}}$. Here the transition temperature at zero field is denoted by T_{C0}.

Taking a variation of $\delta \tilde{F} = 0$, we obtain a couple of equations,

$$\int_\Omega \left[\left(i\nabla\psi(\boldsymbol{r}) - \tilde{\boldsymbol{A}}(\boldsymbol{r})\tilde{\psi}(\boldsymbol{r}) \right)\left(-i\nabla\delta\psi^*(\boldsymbol{r}) - \tilde{\boldsymbol{A}}(\boldsymbol{r})\delta\tilde{\psi}^*(\boldsymbol{r}) \right) + \left(i\nabla\delta\psi(\boldsymbol{r}) - \tilde{\boldsymbol{A}}(\boldsymbol{r})\delta\tilde{\psi}(\boldsymbol{r}) \right)\left(-i\nabla\psi^*(\boldsymbol{r}) - \tilde{\boldsymbol{A}}(\boldsymbol{r})\tilde{\psi}^*(\boldsymbol{r}) \right) \right] d\Omega$$

$$+ \int_\Omega \frac{1}{\xi(T)^2}\left(\left|\psi(\boldsymbol{r})^2 - 1\right| \right)\left(\psi(\boldsymbol{r})\delta\psi^*(\boldsymbol{r}) + \psi^*(\boldsymbol{r})\delta\psi(\boldsymbol{r}) \right) d\Omega = 0 \quad , (6)$$

$$\int_\Omega \kappa^2\xi(T)^2 \left[\left(\nabla \cdot \tilde{\boldsymbol{A}}(\boldsymbol{r}) \right)\left(\nabla \cdot \delta\tilde{\boldsymbol{A}}(\boldsymbol{r}) \right) + \left(\nabla \times \tilde{\boldsymbol{A}}(\boldsymbol{r}) \right)\left(\nabla \times \delta\tilde{\boldsymbol{A}}(\boldsymbol{r}) \right) \right.$$

$$\left. + \left|\tilde{\psi}(\boldsymbol{r})\right|^2 \tilde{\boldsymbol{A}} \cdot \delta\tilde{\boldsymbol{A}} - \frac{i}{2}\left\{ \tilde{\psi}^*(\boldsymbol{r})\nabla\tilde{\psi}(\boldsymbol{r}) - \tilde{\psi}(\boldsymbol{r})\nabla\tilde{\psi}^*(\boldsymbol{r}) \right\}\delta\tilde{\boldsymbol{A}}(\boldsymbol{r}) \right] d\Omega \quad . (7)$$

$$= \kappa^2\xi(T)^2 \int_\Omega \frac{2\pi}{\Phi_0}\boldsymbol{H}(\boldsymbol{r}) \cdot \left(\nabla \times \delta\tilde{\boldsymbol{A}}(\boldsymbol{r}) \right) d\Omega$$

To solve these equations numerically, we divide the area Ω into small triangular finite elements as Fig. 1. Using the Galerkin's method of the first order, a type of the finite element method (FEM), the functions $\tilde{\psi}(\boldsymbol{r})$ and $\tilde{\boldsymbol{A}}(\boldsymbol{r})$ in the j-th element are expressed by a linear combination of their values at the vertices i of the j-th element $\tilde{\psi}_i^{(j)}$, $\tilde{\boldsymbol{A}}_i^{(j)}$ as follows,

$$\tilde{\psi}(x, y) = \sum_{i=1}^{3} N_i^{(j)}(x, y)\tilde{\psi}_i^{(j)} \quad , \quad \tilde{\boldsymbol{A}}(x, y) = \sum_{i=1}^{3} N_i^{(j)}(x, y)\tilde{\boldsymbol{A}}_i^{(j)}. \quad (8)$$

Here we supposed Ω is in x-y plane. The area coordinates $N_i(x,y)^{(j)}$ $(j = 1,2,3)$ are equal to the area of three small triangles, which are made by drawing three line segments from the point (x, y) to the three vertices of the j-th finite element, and the definition of the area coordinate is shown in figure 2. By iterative solving the set of equations (6) and (7) self-consistently, we have to set the initial values of $\tilde{\boldsymbol{A}}(\boldsymbol{r})$ and $\tilde{\psi}(\boldsymbol{r})$. The initial value of $\tilde{\boldsymbol{A}}(\boldsymbol{r})$ was determined by the external magnetic field $\boldsymbol{H}(\boldsymbol{r})$, and the initial value of $\tilde{\psi}(\boldsymbol{r})$ was given by using a random number. Since the iterative calculation starts from various initial values, the same converged solution cannot always be obtained each time, however they are solutions of the most stable state or at least metastable states. Therefore, we calculate over 20 times in each set of physical parameters.

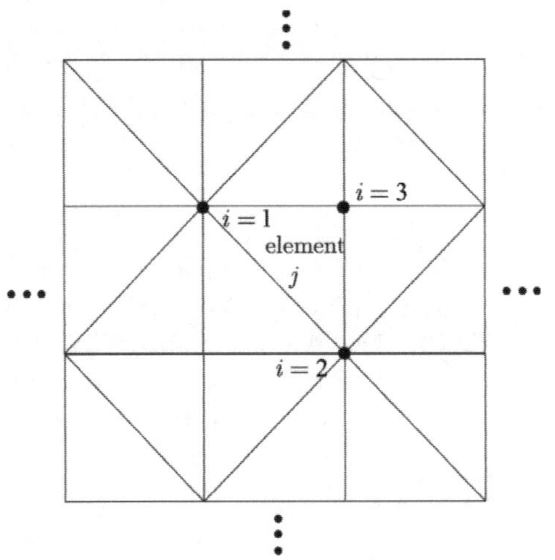

Figure 1. Schematic illustration of deviding the area into finite elements. The index j distinguishes elements, i denotes a vertex of the element.

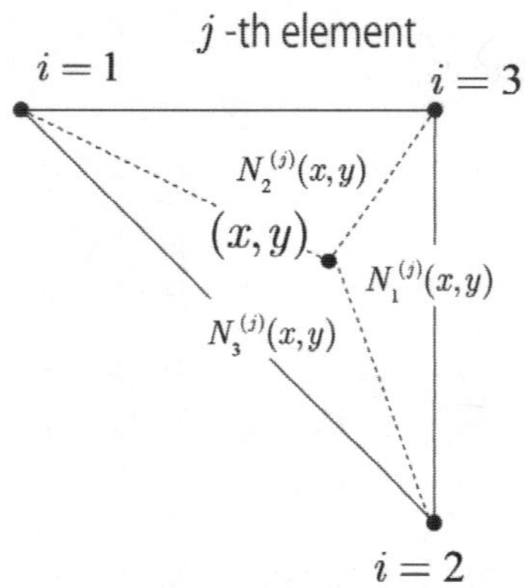

Figure 2. Description of area coordinate.

3. Results and discussions

We consider a mesoscopic square plate with a side length of $L = 29\xi_0$ placed on the x-y plane. The region of the whole area Ω is $30\xi_0 \le x \le 30\xi_0$, $30\xi_0 \le y \le 30\xi_0$. The superconducting plate occupies the reagion of $0.5\xi_0 \le x \le 29.5\xi_0$, $0.5\xi_0 \le y \le 29.5\xi_0$, and the rest is the vacuum area. A magnetic field is applied along the z-axis. In following discussion, field strength is denoted by renormalized field strength that is defined by $f = HL^2 / \Phi_0$. The field strength has a linear gradient toward the x-axis direction as

$$f = f_{\min} + (f_{\max} - f_{\min})x / 30\xi_0 . \qquad (9)$$

Figure 3 shows vortex distributions of the plate in various field strengths. Here, the GL parameter of the plate is $\kappa = 10$ and the temperature is $T = 0.80T_{C0}$. We set the gradient field parameter as $f_{\min} = 0$ and $f_{\max}$ =8.03, 10.07, 13.38, 16.6. There is more than one solution for each $f_{\max}$. All the obtained vortex distributions have mirror symmetry respect to the straight line of $y = 15\xi_0$. In the most stable state, as the f_{max} increases, the number of flux quanta passing through the plate also monotonically increases. It is noteworthy here that there are more than one state that have equal numbers of total flux that passing through the plate in the same field ($f_{max} = 16.06$ case shown in figure 3), which cannot be seen in the case of uniform magnetic field.

In solving the equations (5) and (6), by changing the initial value of order parameter, resulting solution also may change, and it is not necessary to converge to the most stable solution. In some cases, a specific metastable solution may be obtained more frequently than the most stable solution. Therefore, it is incorrect to abandon the metastable solutions obtained here, and we consider that the metastable states can be realized as well as the most stable state. However, in order to obtain strong evidence of the "stability" of metastable state, numerical studies using the TDGL theory would also be necessary.

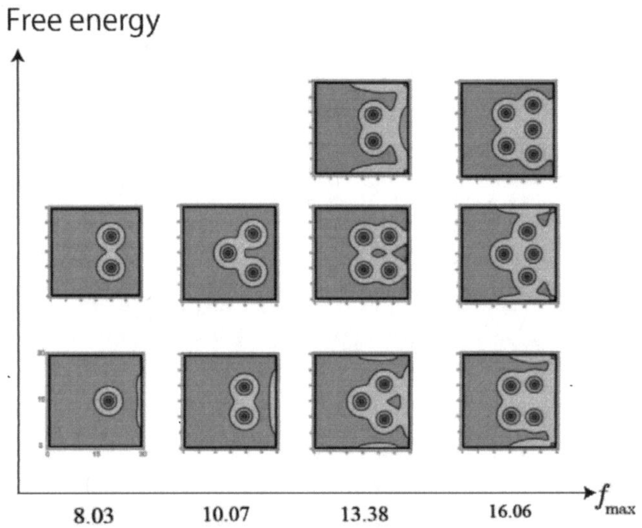

Figure 3. Vortex distribution in gradient field. Dark circles denote vortices. All of the obtained vortex distributions have mirror symmetry.

4. Conclusions

We obtained vortex states of the mesoscopic square superconducting plate under the gradient magnetic field by solving the GL equations make use of the FEM. When a gradient magnetic field is applied, a combined system of the square plate and the gradient magnetic field has the mirror symmetry. Then the obtained vortex configurations including metastable states of them have the mirror symmetry. We also found two different solutions containing the same number of the total flux quantum under the same gradient field, which cannot be seen under the homogeneous field. When solving the GL equation self-consistently, we obtained the metastable state solutions at almost the same frequency as the stable solution, which implies that the metastable states obtained here can be observed as the most stable state.

References

[1] Kokubo N, Okayasu S, Nojima T, Tamochi Sand Shiozaki B 2014 *J. Phys. Soc. Jpn.* **83** 083704
[2] Huy T H, Kato M and Sato O 2013 *Physica C Superconductivity* **494** 99
[3] Chibotaru L F, Ceulemans A, Bruyndoncx and Moshchalkov V V 2000 *Nature* **408** 833
[4] Sato O and Kato M 2011 *Physica C Superconductivity* **471(21)** 817
[5] Du Q, Gunzburger M D and Peterson J S 1992 *SIAM REVIEW* **34-1** 5

IOP Conf. Series: Journal of Physics: Conf. Series **871** (2017) 012030 doi:10.1088/1742-6596/871/1/012030

Evolution of the CDW gap in Valence Skipper $RbTlX_3$ (X=F,Cl,Br): A first-principle study

I Hase[1], T Yanagisawa[1] and K Kawashima[2]

[1]National Institute of Advanced Industrial Science and Technology (AIST), Tsukuba, 305-8568, Japan
[2]IMRA Material R&D Co. Ltd., Kariya, Aichi 448-0032, Japan

e-mail: i.hase@aist.go.jp

Abstract. In the present study we investigated the electronic structure of the valence skipping compound $RbTlX_3$ (X=F,Cl,Br) from first principles. Combined with structural optimization, we confirmed that the charge-density-wave (CDW) is formed in all of these three compounds, and the Tl atoms in two different sites take the valence Tl^{1+} and Tl^{3+}. We investigated the evolution of the CDW gap as a function of the volume and the atomic position, and found that this CDW gap can be collapsed by applying pressure with only few GPa for $RbTlBr_3$. In this metallic phase a large charge-fluctuation and a large electron-phonon interaction is expected.

1. Introduction

$BaBiO_3$, which is the parent compound of the high-T_c superconductor $(Ba,K)BiO_3$, is a so-called valence skipper[1]. Since the formal valence of Bi in $BaBiO_3$ is the unstable 4+ state, Bi atoms spontaneously disproportionate to Bi^{3+} and Bi^{5+}, accompanied with the displacement of anions[2]. These Bi^{3+} and Bi^{5+} ions are ordered like NaCl, and this charge order is also known as a charge density wave (CDW). Crystal symmetry is also lowered in this CDW phase. The hole doping in $BaBiO_3$ restores the high crystal symmetry, collapses the CDW gap, and induces high T_c superconductivity[3]. Since the valence skip ion can be an attractive center of conduction electrons, the induced superconductivity can be an isotropic one. This property is advantageous not only for high T_c, but also for various applications. Therefore it is highly desirable to find other superconductors induced by valence skip fluctuation.

Recently such a kind of candidate compound is proposed by Schoop *et al.*[4] They investigated the electronic structure of $ATlX_3$ (A=Rb,Cs; X=F,Cl,Br) and found that the electronic structure of $ATlX_3$ and $BaBiO_3$ quite resemble each other. They also predicted that carrier doping can melt the CDW order in $RbTlCl_3$, and the resulting metallic state is possibly superconducting.

In general, insulator can be metallic by carrier doping or pressure. In the present study we focus on the pressure effect. We found that $RbTlCl_3$ and $RbTlBr_3$ are very compressible, and a tractable (~several GPa) pressure may induce metal-insulator transition.

2. Details of Calculation

We have calculated the electronic structure of $RbTlX_3$ (X=F,Cl,Br) from first-principles. As for the crystal structure, we assume ordered perovskite structure which is seen in $CsTlCl_3$[5]: Space group #225 Fm-3m; Rb(1/4,1/4,1/4), Tl1(0,0,0), Tl2(0,0,1/2), X(0,0,z). This structure is essentially the same with that of $BaBiO_3$. Since the energy gap and the total energy are mainly determined by the breathing motion[4], we ignore tilting mode and concentrate on breathing mode. In this structure, the only one free structural parameters is the internal parameter z(X). Moreover, the lattice constant a, or the same thing, the unit cell volume V $(=a^3/4)$ is also a changeable parameter. These two parameters z(X) and V determine the whole electronic structure, including the total energy and the CDW gap Δ. We first set V by hand, and optimized z(X) for each V.

We have used a full-potential augmented plane wave (FLAPW) method within the general gradient approximation (GGA)[6]. The computer program package WIEN2k[7] was used in this work. The parameter RK_{max} is chosen as 7.0. We took ~1000 k-point mesh in the first Brillouin zone for the self-consistent calculations.

3. Results and Discussions

Figure 1 shows the energy dispersion curves of $RbTlX_3$ (X=F,Cl,Br) for the optimized V (i.e. ambient pressure) and z(X). The optimized value of z(X) is 0.2246, 0.2362, 0.2371 for X=F,Cl,Br, respectively. We can see that a band gap opens between the LUMO mainly consisting of Tl1–s state and the HOMO mainly consisting of Tl2-s state. Since LUMO is empty and HOMO is fully occupied, Tl1 atom becomes Tl^{3+} and Tl2 atom becomes Tl^{1+}. This is consistent with the simple ionic picture that Tl1-X distance should be shorter than Tl2-X distance. The band gap Δ is an indirect one for all these three compounds, and the magnitude of Δ is 2.25eV, 0.74eV, 0.39eV for X=F, Cl, Br, respectively. As the size of anion increases, the hybridization between Tl-s and X-p orbitals increases, and z(X) approaches 0.25 (center of the two Tl atoms) and eventually Δ decreases. This kind of gap structure is also seen in some valence skippers, for example, SnF_3 and $BaBiO_3$ [8].

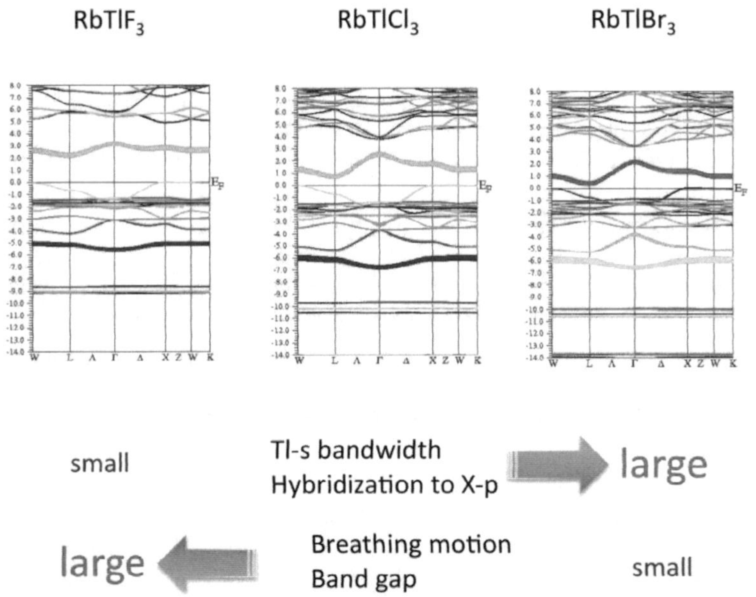

Figure 1 Energy dispersion curves of $RbTlX_3$ (X=F,Cl,Br). The units of vertical axis are eV. The thickness of curves denotes the contribution of Tl1-s orbital.

Since the magnitude of the CDW gap is quite small in $RbTlCl_3$ and $RbTlBr_3$, it is expected that external pressure can collapse this gap. We changed the lattice volume, and calculated the magnitude of the CDW gap. We also calculated the pressure at each volume by fitting the energy-volume plot to the Murnaghan equation of state.

Figure 2 shows the CDW gap of $RbTlBr_3$ with changing lattice volume. The dotted line shows Δ when z is fixed to z_0=0.2371, which is the z value at ambient pressure. In this case, Δ hardly changes with decreasing volume (or increasing pressure). However in fact, z is a function of V, i.e. z decreases with decreasing volume. And Δ is rather sensitive to the change of z, eventually Δ rapidly decreases and reaches zero when V decreases ~8 percent from the volume at ambient pressure. This is shown in the solid line of figure 2. Similarly, the CDW gap of $RbTlCl_3$ collapses at ~13% volume reduction. The point that the CDW gap becomes zero is called a quantum critical point (QCP), which is a central interest in condensed matter physics. We can see that this pressure induced metal-insulator phase transition in $RbTlCl_3$ and $RbTlBr_3$ is a second-order one.

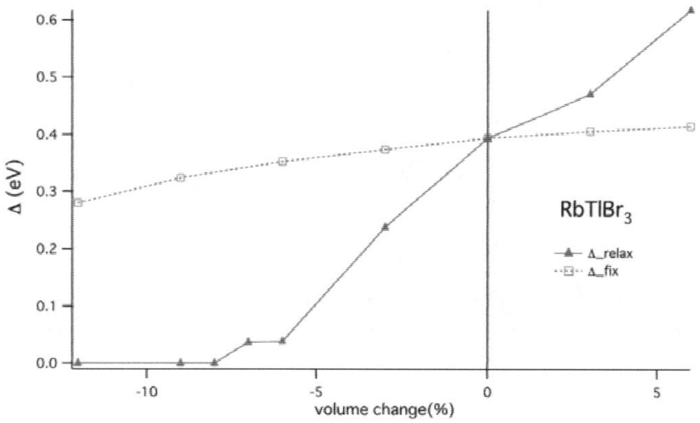

Figure 2 The CDW gap of $RbTlBr_3$ as a function of volume reduction (ΔV). ΔV=0 denotes the volume at the ambient pressure. The dotted line is for fixed z=z_0, and the solid line is for optimized z for each V. For example, at the QCP (ΔV=8%) z becomes 0.2412.

Figure 3 is the energy-volume plot of $RbTlBr_3$ fitted with the Murnaghan equation of state. Similarly to figure 2, the dotted line is fixing z=z_0 and the solid line is optimizing z for each V. The solid line is always below the dotted line as expected. The obtained bulk modulus of $RbTlCl_3$ and $RbTlBr_3$ is 28GPa and 18GPa, respectively. These values are much smaller than that of $BaBiO_3$, i.e. 157GPa which is estimated by the high-pressure X-ray diffraction experiment[9].

The lower panel of figure 3 shows the pressure P=dE/dV. Combined with the result of figure 2, we can obtain the pressure at which the CDW gap closes. Since the bulk modulus is very small, the pressure at the QCP is ~3.0GPa and ~1.4GPa for $RbTlCl_3$ and $RbTlBr_3$, respectively. Considering the general tendency that GGA overestimates the itineracy, this pressure may become larger. Nevertheless, we predict that applying pressure with several GPa can collapse the CDW gap of $RbTlBr_3$.

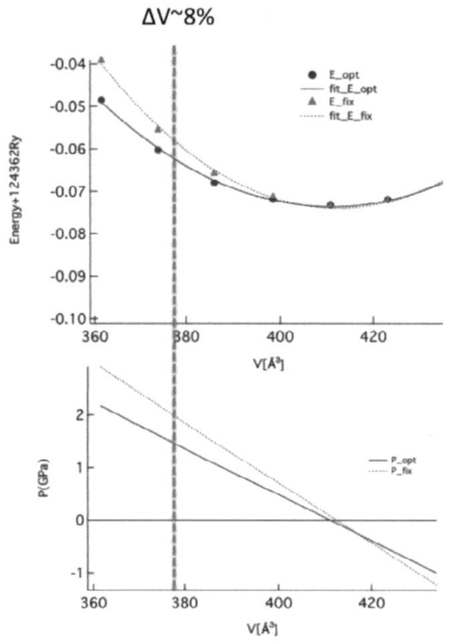

Figure 3 Upper panel: energy-volume plot of $RbTlBr_3$ fitted with Murnaghan equation of state. The dotted line is for fixed $z=z_0$, and the solid line is for optimized z for each V. Lower panel: Pressure at each volume calculated by $P=dE/dV$ for the fitted $E(V)$. The vertical broken line denotes the volume at QCP.

Although the band structure of $RbTlBr_3$ is similar to $BaBiO_3$, $BaBiO_3$ does not show such a pressure-induced metal-insulator transition because of its large bulk modulus. Another example is InTe, which shows CDW at ambient pressure and superconductivity at very high pressure (37GPa)[10]. $RbTlBr_3$ may be the first case to show metal-insulator transition induced by a tractable pressure in the valence-skip compound.

We thank to H. Eisaki and K. Odagiri for useful discussions. This work was partially supported by KAKENHI (Grant No. JP26400379) from Japan Society for the Promotion of Science (JSPS).

References

[1] C M Varma, Phys. Rev. Lett. **61** (1988) 2713.
[2] A W Sleight, J L Gibson and P E Bierstedt, Solid State Commun. **17** (1975) 27.
[3] L F Mattheiss, E M Gyorgy and D W Johnson Jr., Phys. Rev. B**37**(1988) 3745.
[4] L M Schoop, L Müchler, C Felser and R J Cava, Inorg. Chem. **52** (2013) 5479.
[5] R Ackermann, Thesis, Freiburg im Breisgau, Germany (2001).
[6] J P Perdew, K Burke and M Ernzerhof, Phys. Rev. Lett. **77** (1996) 3865.
[7] P Blaha, K Schwaltz, G K H Madsen, D Kvashnicka and J Luitz, WIEN2k, An Augmented Plane Wave + Local Orbitals Program for Calculating Crystal Properties, Vienna University of Technology, Vienna (2001).

[8] I Hase, T Yanagisawa and K Kawashima, Physica C**530** (2016) 11.
[9] H Sugiura and R Yamadaya, Solid State Commun. **49** (1984) 499.
[10] M D Banus, R E Hanneman, M Strongin and K Gooen, Science **142** (1963) 662.

Interplay between staggered flux and d-wave superconducting states in Hubbard model

Kenji Kobayashi[1] **and Hisatoshi Yokoyama**[2]

[1]Department of Natural Science, Chiba Institute of Technology, Shibazono, Narashino 275-0023, Japan

[2]Department of Physics, Tohoku University, Aoba-ku, Sendai, 980-8578, Japan

E-mail: kenji.kobayashi@it-chiba.ac.jp

Abstract. With cuprate superconductors in mind, we check whether the d-wave superconducting (dSC) state coexists with or excludes the staggered flux (SF) state, which was revealed to be the strong candidate for the pseudogap state. To this end, we use a variational Monte Carlo method for the square Hubbard model with diagonal transfer t'. In the trial wave function, dSC and SF orders coexist, which allows a continuous description of their interplay; essential factors for Mott physics and band renormalization are also included. It is found that the SF state does not coexist with dSC and is unstable for $U/t = 12$ regardless of the value of t'/t within the present work.

1. Introduction

In understanding superconductivity (SC) in high-T_c cuprates, what is the origin of pseudogap phenomena, which behave like a shadow of SC, is a long-standing problem [1, 2]. Among various candidates for the pseudogap state proposed so far, a staggered flux (SF) state was found to be moderately stable and have properties consistent with the behaviors in the pseudogap phase of cuprates, using a variational Monte Carlo (VMC) method for the square Hubbard model with diagonal transfer t' [3]. In this study, it is shown that the SF state becomes more stable than the normal state (projected Fermi sea) for a strongly correlated and underdoped area by introducing a configuration-dependent phase factor, which is vital to treat a current-carrying state for a large U/t [4], although the $d_{x^2-y^2}$-wave SC (dSC) state always has lower energy. The area where the SF state is more stable than the normal state expands in a wide doping range and the energy of SF state approaches to that of dSC, especially, for $t'/t = -0.3$ [3]. Meanwhile, some experiments reported coexistence of the pseudogap and dSC gaps [5].

Thus, it is important to check directly whether the SF and dSC orders coexist or are mutually exclusive, because the SF and dSC states have been independently treated so far. In the present study, we will directly check this problem using a mixed state of the SF and dSC orders. We apply a VMC method to the Hubbard (t-t'-U) model on a square lattice with the fixed value of $U/t = 12$, and for $t'/t = 0$ and -0.3 for comparison, to cope with strong correlations. In the trial wave function, we introduce the following features:

(i) We allow for the coexistence of SF and dSC orders, by which we can treat a continuous description of their interplay.

(ii) The configuration-dependent phase-adjusting factor, vital in SF state for a large U/t [3, 4], is introduced in addition to the ordinary Gutzwiller correlation [6] and refined doublon-holon (D-H) binding factors, which are crucial to describe Mott physics [7, 8, 9, 10].

(iii) The band renormalization effect is considered by adjusting the parameters of hopping integrals up to the fifth-neighbor sites; this effect is crucial near half filling [11, 12, 13].

On the basis of the data for optimized states, we find that the SF order competes with the dSC order and is not stabilized.

2. Method

We consider the Hubbard model on a square lattice with the next-nearest-neighbor transfer t', shown in Fig. 1 (a),

$$\begin{aligned}\mathcal{H} &= \mathcal{H}_t + \mathcal{H}_{t'} + \mathcal{H}_U \\ &= -t\sum_{\langle \mathbf{r}\mathbf{r}'\rangle,\sigma}\left(c^\dagger_{\mathbf{r},\sigma}c_{\mathbf{r}',\sigma} + \text{h.c.}\right) - t'\sum_{(\mathbf{r}\mathbf{r}'),\sigma}\left(c^\dagger_{\mathbf{r},\sigma}c_{\mathbf{r}',\sigma} + \text{h.c.}\right) + U\sum_{\mathbf{r}} n_{\mathbf{r}\uparrow}n_{\mathbf{r}\downarrow},\end{aligned} \tag{1}$$

where $c^\dagger_{\mathbf{r},\sigma}$ creates an electron of spin σ at site $\mathbf{r}$, $n_{\mathbf{r}\sigma} = c^\dagger_{\mathbf{r},\sigma}c_{\mathbf{r},\sigma}$, and $\langle \mathbf{r}\mathbf{r}'\rangle$ and $(\mathbf{r}\mathbf{r}')$ denote indices of nearest-neighbor and diagonal-neighbor pairs, respectively. In the following, we use the square lattice of N_s ($= L \times L$) sites with periodic-antiperiodic boundary conditions, and the unit of distance is the lattice constant.

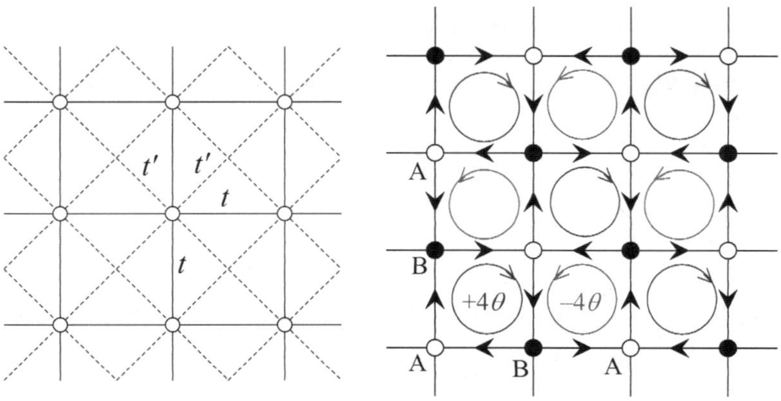

(a) Hubbard (t-t'-U) model (b) Staggered Flux

Figure 1. (a) Hubbard square lattice. (b) Schematic figure of the staggered flux state. Arrows denote the local circular current, which flows in alternate directions in each plaquette.

As a variational wave function, we use a Jastrow type, $\Psi = \mathcal{P}\Phi$, where $\mathcal{P}$ is a many-body correlation factor as introduced below. For the one-body part Φ, we adopt a coexisting state of SF and dSC for a continuous description of their interplay. For N_e electrons, Φ with a nearest-neighbor-pairing gap Δ_d is written as [14, 15, 16],

$$\Phi = \left(\sum_{\mathbf{k}}\sum_{\lambda=\pm}\lambda\varphi_{\mathbf{k}}\alpha^\dagger_{\lambda,\mathbf{k},\uparrow}\alpha^\dagger_{\lambda,-\mathbf{k},\downarrow}\right)^{\frac{N_e}{2}}|0\rangle\,, \tag{2}$$

where

$$\varphi_{\mathbf{k}} = \frac{\Delta(\mathbf{k})}{\varepsilon^{\rm SC}(\mathbf{k}) - \zeta + \sqrt{(\varepsilon^{\rm SC}(\mathbf{k}) - \zeta)^2 + |\Delta(\mathbf{k})|^2}}, \tag{3}$$

with $\Delta(\mathbf{k}) = \Delta_d(\cos k_x - \cos k_y)$. $\alpha^\dagger_{\lambda,\mathbf{k},\sigma}$ in Eq. (2) is a creation operator of a quasi-particle in SF state that is constructed by diagonalizing a SF Hamiltonian $\mathcal{H}^{\rm SF}(\theta)$ as follows. The noninteracting SF Hamiltonian in the sublattice (A,B) representation, shown in Fig. 1 (b), is written as [3],

$$\mathcal{H}^{\rm SF} = -t \sum_{\mathbf{r}\in {\rm A},\sigma} \left[e^{i\theta} a^\dagger_{\mathbf{r},\sigma} (b_{\mathbf{r}+\mathbf{x},\sigma} + b_{\mathbf{r}-\mathbf{x},\sigma}) + e^{-i\theta} a^\dagger_{\mathbf{r},\sigma} (b_{\mathbf{r}+\mathbf{y},\sigma} + b_{\mathbf{r}-\mathbf{y},\sigma}) + {\rm h.c.} \right]. \tag{4}$$

Here, $a^\dagger_{\mathbf{r},\sigma}$ ($b^\dagger_{\mathbf{r},\sigma}$) creates an electron with spin σ on a sublattice A (B), and $\mathbf{x}$ ($\mathbf{y}$) is the unit vector in the x (y) direction [$\mathbf{x} = (1,0)$, $\mathbf{y} = (0,1)$].

By applying a Bogoliubov transformation,

$$a_{\mathbf{k},\sigma} = \frac{\Gamma(\mathbf{k},\theta)}{\sqrt{2}} (\alpha_{-,\mathbf{k},\sigma} - \alpha_{+,\mathbf{k},\sigma}), \qquad b_{\mathbf{k},\sigma} = \frac{1}{\sqrt{2}} (\alpha_{-,\mathbf{k},\sigma} + \alpha_{+,\mathbf{k},\sigma}), \tag{5}$$

with

$$\Gamma(\mathbf{k},\theta) = \frac{u(\mathbf{k},\theta)}{S(\mathbf{k},\theta)}, \tag{6}$$

$$u(\mathbf{k},\theta) = e^{i\theta}\cos k_x + e^{-i\theta}\cos k_y, \tag{7}$$

$$S(\mathbf{k},\theta) = \sqrt{u(\mathbf{k},\theta)\ u^*(\mathbf{k},\theta)} = \sqrt{\cos^2 k_x + 2\cos(2\theta)\cos k_x \cos k_y + \cos^2 k_y}, \tag{8}$$

and $a_{\mathbf{k},\sigma}$ ($b_{\mathbf{k},\sigma}$) being a Fourier transformation of $a_{\mathbf{r},\sigma}$ ($b_{\mathbf{r},\sigma}$), $\mathcal{H}^{\rm SF}$ in Eq. (4) is diagonalized as

$$\mathcal{H}^{\rm SF} = \sum_{\mathbf{k},\sigma} \sum_{\lambda=\pm} \varepsilon^{\rm SF}_\lambda(\mathbf{k},\theta)\, \alpha^\dagger_{\lambda,\mathbf{k},\sigma}\alpha_{\lambda,\mathbf{k},\sigma}, \tag{9}$$

where the upper and lower band dispersions are given by

$$\varepsilon^{\rm SF}_\pm(\mathbf{k},\theta) = \pm 2tS(\mathbf{k},\theta). \tag{10}$$

Corresponding creation operators of the quasi-particle for the upper and lower band, which constitute Cooper pairs in Eq. (2), are obtained by the inverse transformation of Eq. (5) as

$$\alpha^\dagger_{\pm,\mathbf{k},\sigma} = \frac{1}{\sqrt{2}} \left(\mp\Gamma(\mathbf{k},\theta) a^\dagger_{\mathbf{k},\sigma} + b^\dagger_{\mathbf{k},\sigma} \right), \tag{11}$$

where the phase θ, which is introduced into Eq. (2) through $\alpha^\dagger_{\lambda,\mathbf{k},\sigma}$, is a variational parameter to be optimized together with the other parameters.

In addition, band renormalization effect owing to electron correlation is introduced. To this end, we assume that $\varepsilon^{\rm SC}(\mathbf{k})$ in Eq. (3) is fitted by a tight-binding band composed of transfers up to fifth-neighbor sites as shown in Fig. 2,

$$\begin{aligned} \varepsilon^{\rm SC}(\mathbf{k}) = & -2(t_{1x}\cos k_x + t_{1y}\cos k_y) - 4t_2 \cos k_x \cos k_y \\ & -2t_3[\cos(2k_x) + \cos(2k_y)] - 4t_4[\cos(2k_x)\cos(k_y) + \cos(k_x)\cos(2k_y)] \\ & -2t_5[\cos(3k_x) + \cos(3k_y)], \end{aligned} \tag{12}$$

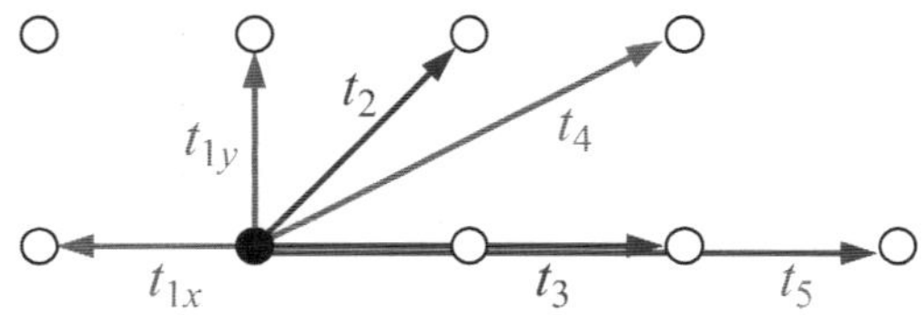

Figure 2. Hopping parameters in $\varepsilon^{\rm SC}(\mathbf{k})$, which are optimized to adjust the band renormalization effect owing to electron correlation.

where t_{1y} and t_2-t_5 are variational parameters to be optimized and t_{1x}/t is fixed to 1. In Eq. (12), asymmetry in t_1 is also introduced between x and y directions (t_{1x} and t_{1y}) to deal with the Pomeranchuk instability [11, 12].

For the Jastrow correlation factor $\mathcal{P}$ in Ψ, we assume the form $\mathcal{P} = \mathcal{P}_\phi\ \mathcal{P}_{\rm Q}\ \mathcal{P}_{\rm G}$, where $\mathcal{P}_{\rm G}$ is a usual on-site (Gutzwiller) projection [6], $\mathcal{P}_{\rm G} = \prod_{\mathbf{r}}[1-(1-g)n_{\mathbf{r}\uparrow}n_{\mathbf{r}\downarrow}]$, and $\mathcal{P}_Q$ is an extended projection between a doubly occupied site (doublon, D) and an empty site (holon, H), which controls the probability weight of D-H binding classified by the surrounded configuration, namely, the number of H neighboring D and *vice versa* [9, 10].

$\mathcal{P}_\phi$ is a configuration-dependent phase-adjusting factor that partially cancels out the phase θ in the processes of creating or annihilating D-H pairs [3, 4]:

$$\mathcal{P}_\phi = \exp\left[i\phi\sum_{\mathbf{r}}\lambda_{\rm AB}(\mathbf{r})\ d_{\mathbf{r}}\left(-h_{\mathbf{r+x}}-h_{\mathbf{r-x}}+h_{\mathbf{r+y}}+h_{\mathbf{r-y}}\right)\right], \tag{13}$$

where $d_{\mathbf{r}} = n_{\mathbf{r}\uparrow}n_{\mathbf{r}\downarrow}$, $h_{\mathbf{r}} = (1-n_{\mathbf{r}\uparrow})(1-n_{\mathbf{r}\downarrow})$, and $\lambda_{\rm AB}(\mathbf{r}) = +1(-1)$ if $\mathbf{r} \in$ A(B) lattice. The value of ϕ is determined variationally. $\mathcal{P}_\phi$ plays a vital role in the strongly correlated regime because a hopping process of annihilating a D-H pair such as $|\uparrow\downarrow, 0\rangle \rightarrow |\uparrow,\downarrow\rangle$ does not necessarily occur immediately after a D-H pair creating process, in which extra phase θ is added, and then θ should be canceled in each hopping process. Unless $\mathcal{P}_\phi$ is introduced, the energy of the SF state does not decrease at all from that of the normal state [3].

In Φ, variational parameters to be optimized are Δ_d, ζ, θ, t_{1y} and t_2-t_5. Besides these eight in Φ, variational parameters in the correlation factor $\mathcal{P}$ are g in $\mathcal{P}_{\rm G}$, ϕ in $\mathcal{P}_\phi$, and 20 parameters in $\mathcal{P}_{\rm Q}$. According to a standard procedure of VMC, we obtain the optimal set of the 30 parameters by minimizing the energy. Using the optimized wave functions, we calculate various quantities. In particular, we take notice of the order parameters of dSC and SF orders in this work. The real-space dSC correlation function for the nearest-neighbor pairing is defined by,

$$P_d(\mathbf{r}) = \frac{1}{N_{\rm s}}\sum_{\mathbf{r}'}\sum_{\boldsymbol{\tau},\boldsymbol{\tau}'}\lambda_{xy}(\boldsymbol{\tau})\lambda_{xy}(\boldsymbol{\tau}')\langle\Delta^\dagger(\mathbf{r}'+\mathbf{r},\boldsymbol{\tau})\Delta(\mathbf{r}',\boldsymbol{\tau}')\rangle, \tag{14}$$

with

$$\Delta(\mathbf{r},\boldsymbol{\tau}) = \frac{1}{\sqrt{2}}\left(c_{\mathbf{r}\uparrow}c_{\mathbf{r}+\boldsymbol{\tau}\downarrow} - c_{\mathbf{r}\downarrow}c_{\mathbf{r}+\boldsymbol{\tau}\uparrow}\right),$$

$\lambda_{xy}(\boldsymbol{\tau}) = +1$ for $\boldsymbol{\tau} = (\pm1, 0)$, and $\lambda_{xy}(\boldsymbol{\tau}) = -1$ for $\boldsymbol{\tau} = (0, \pm1)$. We use $P_d(\mathbf{r})$ at the farthermost distance $\mathbf{r} = (L/2, L/2)$ as a measure of dSC order. As the order parameter of the SF phase, we calculate a circular current defined as

$$J_{\rm c}/t = \frac{1}{N_{\rm s}}\sum_{\mathbf{r}\in{\rm A},\sigma}\sum_{\boldsymbol{\tau}}\lambda_{xy}(\boldsymbol{\tau})\ {\rm Im}\langle c^\dagger_{\mathbf{r}+\boldsymbol{\tau},\sigma}c_{\mathbf{r},\sigma} - c^\dagger_{\mathbf{r},\sigma}c_{\mathbf{r}+\boldsymbol{\tau},\sigma}\rangle. \tag{15}$$

As pilot calculations, we have fixed the value of U/t at 12 as a typical case of strongly correlated regime, and compare the results of $t'/t = 0$ and $t'/t = -0.30$ for systems of $L = 10$ and $L = 12$.

3. Results and discussion

To begin with, let us look at the optimized values of variational parameters in Ψ. In Fig. 3, the optimized values of the relevant variational parameters, the pairing gap Δ_d for dSC and two phases θ and ϕ for SF, are plotted as a function of doping rate δ for (a) $t'/t = 0$ and (b) $t'/t = -0.30$. For all range within the present calculations, Δ_d has finite values but both phases of SF state vanish, meaning that a SF order is absent and a pure dSC is realized.

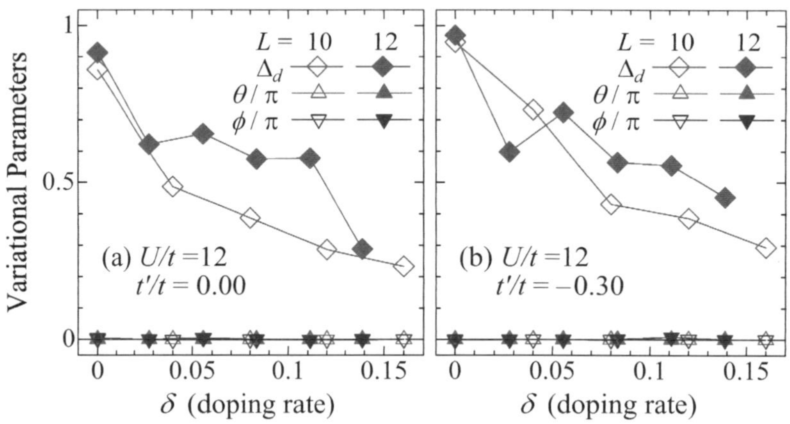

Figure 3. Optimized values of relevant variational parameters in Ψ, the pairing gap Δ_d and SF phases θ and ϕ, are plotted as a function of doping rate δ for (a) $t'/t = 0$ and (b) $t'/t = -0.30$.

Corresponding to these optimized parameters, the expectation values of order parameters estimated by the optimized Ψ show the same tendency as seen in Fig. 4: the circular current of SF $|J_c|/t$ scarcely flows, whereas the pair correlation function $P_d(L/2, L/2)$ has finite values for all range within the present calculations. This indicates that the pure dSC is realized, whose properties are described in detail in Refs.[13, 17].

From these results, SF and dSC states are mutually exclusive and SF is unstable in a wide range of δ for strongly correlated Hubbard model. In Ref.[3], we argued that the reason why dSC and SF orders do not coexist is that the SF state has no Fermi surface around the antinodes $(\pm\pi, 0), (0, \pm\pi)$, which is necessary to induce dSC. However, now we deduce from the behavior of different mixed states [18] that the above explanation might be incorrect. The problem of compatibility and exclusivity of the two orders in a mixed state can be considered as follows.

First, we define the predominant (subordinate) state in a mixed state as its pure state has a lower (higher) energy than the pure state of the other. (i) The order of the predominant state necessarily arises, if its pure state exhibits an order at a given model parameter. (ii) If the resultant electronic state is (in)compatible with the subordinate state, the two orders are coexistent (the pure state with the predominant order appears without subordinate order).

In the present case, the dSC (SF) state is predominant (subordinate). Therefore, a more appropriate explanation for the exclusive feature is that when dSC is realized, the Fermi surface along $(\pi, 0)$–$(0, \pi)$ is smeared out by the dSC gap except for the point in the nodal direction. This is incompatible with the claim that the SF state has a pocket Fermi surface along $(\pi, 0)$–$(0, \pi)$.

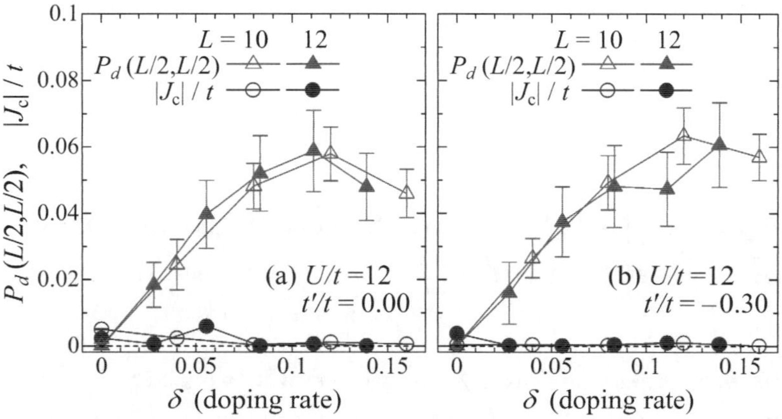

Figure 4. The order parameters of the two states are compared, which are obtained using the optimized Ψ. Namely, d-wave nearest-neighbor pair correlation function $P_d(\mathbf{r})$ for farthermost distance $\mathbf{r} = (L/2, L/2)$ (triangles) and circular current of SF $|J_c|/t$ (circles) as a function of δ for (a) $t'/t = 0$ and (b) $t'/t = -0.30$.

4. Summary

We have studied the interplay between SF and dSC in strongly correlated Hubbard model, using a coexisting wave function of SF and dSC orders in VMC calculations. These VMC results claim that the SF state is unstable and does not coexist with dSC for all range within the present calculations. It will be intriguing to search for interplay between SF and other orders because the properties of the SF state are mostly consistent with the behaviors in the pseudogap phase of cuprates as discussed in Ref.[3]. These are left for future studies.

References

[1] Yoshida T, Hashimoto M, Vishik I M, Shen Z -X and Fujimori A 2012 *J. Phys. Soc. Jpn.* **81** 011006
[2] Fujita K, Schmidt A R, Kim E -A, Lawler M J, Lee D H, Davis J C, Eisaki H and Uchida S 2012 *J. Phys. Soc. Jpn.* **81** 011005
[3] Yokoyama H, Tamura S and Ogata M 2016 *J. Phys. Soc. Jpn.* **85** 124707
[4] Tamura S and Yokoyama H 2015 *J. Phys. Soc. Jpn.* **84** 064707
[5] For instance, Liu Y H, Toda Y, Shimatake K, Momono N, Oda M and Ido M 2008 *Phys. Rev. Lett.* **101** 137003; Uykur E, Tanaka K, Masui T, Miyasaka S and Tajima S 2013 *J. Phys. Soc. Jpn.* **82** 033701; Yoshida T *et al.* 2016 *Phys. Rev. B* **93** 014513
[6] Gutzwiller M C 1963 *Phys. Rev. Lett.* **10** 159
[7] Kaplan T A, Horsch P and Fulde P 1982 *Phys. Rev. Lett.* **49** 889
[8] Yokoyama H and Shiba H 1990 *J. Phys. Soc. Jpn.* **59** 3669
[9] Kobayashi K and Yokoyama H 2007 *Physica C* **463-465** 141
[10] Kobayashi K and Yokoyama H 2008 *J. Phys. Chem. Solids* **69** 3274
[11] Kobayashi K and Yokoyama H 2015 *Phys. Procedia* **65** 9
[12] Kobayashi K, Sato R and Yokoyama H 2016 *Phys. Procedia* **81** 1
[13] Sato R and Yokoyama H 2016 *J. Phys. Soc. Jpn.* **85** 074701
[14] Giamarchi T and Lhuillier C 1991 *Phys. Rev. B* **42** 12943
[15] Himeda A and Ogata M 1999 *Phys. Rev. B* **60** R9935
[16] Pathak S, Shenoy V B, Randeria M and Trivedi N 2009 *Phys. Rev. Lett.* **102** 027002
[17] Yokoyama H, Ogata M, Tanaka Y, Kobayashi K and Tsuchiura H 2013 *J. Phys. Soc. Jpn.* **82** 014707
[18] Toga Y and Yokoyama H 2016 *Phys. Procedia* **81** 13

IOP Conf. Series: Journal of Physics: Conf. Series **871** (2017) 012032 doi:10.1088/1742-6596/871/1/012032

Effect of impurity potentials on antiferromagnetism in two-dimensional Hubbard model

Hisatoshi Yokoyama[1], Ryo Sato[1] and Kenji Kobayashi[2]

[1] Department of Physics, Tohoku University, Aoba-ku, Sendai, 980-8578, Japan
[2] Department of Natural Science, Chiba Institute of Technology, Narashino, 275-0023, Japan

E-mail: yoko@cmpt.phys.tohoku.ac.jp

Abstract. Recent numerical studies have shown that antiferromagnetism (AF) prevails widely near half filling in the two-dimensional Hubbard (t-t'-U) model, contrary to the experiments on cuprate superconductors. In cuprates, disorders owing to the dopants, which are usually disregarded in theory, are possibly intrinsic and considerable. Thus, as a first step, we consider the effects of point-type impurity potentials, which are randomly distributed on the lattice sites, on the stability of AF. It is found that a moderate impurity potential does not modify the stability of AF, but a sufficiently strong attractive potential V ($V \lesssim -U$) appreciably weakens an antiferromagnetic state owing to an insufficient screening of V.

1. Introduction

Recent theoretical studies using various refined methods [1, 2] have argued that antiferromagnetism (AF) and phase separation widely prevail in areas near half filling of two-dimensional Hubbard (t-t'-U) models, by which the cuprate superconductors (SCs) have been considered to be well described. And $d_{x^2-y^2}$-wave superconductivity (*d*-SC) appears only in the edge of these areas mainly in the overdoped regime. A similar result was reported for the *d*-*p* model [3]. Such a situation apparently contradicts the experiments on cuprates. This discrepancy indicates that some common mistakes exist or some essential factors are missed in common in the above theories.

A factor often disregarded in the theories of cuprates is the effect of disorder owing to the carrier dopants, which are inherent in the cuprate SCs and are usually situated in the so-called block layers, namely, off the conducting CuO_2 layers. They will work as weak scatterers, in contrast to the in-plane substitutions for Cu sites (strong scatterers). Although the effects of disorder in Hubbard models have been studied by many theorists in various contexts—for instance, Anderson-Mott transitions and vortex core problems—, to our knowledge there are a small number of studies from the viewpoint of how the weak scatterers affect an already existing AF order in the CuO_2 plane [4].

In this publication, we discuss preliminary results of a study in this line obtained using a standard variational Monte Carlo (VMC) method [5, 2], which is useful to solve a model with strong correlation in a reliable manner. As a primary step, we introduce a point-type impurity potential V, which acts on the single site close to an impurity, into the standard Hubbard (t-t'-U) model, and check whether or not the impurity potential destabilizes an antiferromagnetic (AF) state by varying V and other model parameters. Here, however, we focus on a few cases of specific parameter settings.

2. Model and method

As a simplified model of cuprate SCs, we study the following Hubbard model ($U \geq 0$) with spin-independent point impurity potentials on extended square lattices:

$$H = H_{\rm kin} + H_U + H_{\rm imp}$$
$$= -\sum_{(i,j),\sigma} t_{ij}\left(c_{i\sigma}^\dagger c_{j\sigma} + {\rm H.c.}\right) + U\sum_j n_{j\uparrow}n_{j\downarrow} + V\sum_{\ell=1}^{N_{\rm imp}}\sum_\sigma n_{\ell\sigma}, \qquad (1)$$

where $n_{j\sigma} = c_{j\sigma}^\dagger c_{j\sigma}$, and (i,j) indicates the sum of pairs on sites i and j. In this work, the hopping integral t_{ij} is t for nearest neighbors (≥ 0), t' for diagonal neighbors, and 0 otherwise ($H_{\rm kin} = H_t + H_{t'}$). We use t and the lattice spacing as the units of energy and length, respectively. Because the value of t'/t is estimated as -0.1–-0.4 for hole-doped cuprate SCs, we use a typical value $t'/t = -0.3$ in the following. Because the value of U/t in hole-doped cuprates is considered larger than the Mott-transition points $U_{\rm c}/t = 6$–8, we discuss the V dependence for $U/t = 12$.

In this work, we assume that the impurity potential V comes from the carrier dopants situated in adjacent block layers (weak scatterers), and acts on respective single sites for simplicity. Therefore, the index ℓ in $H_{\rm imp}$ runs over the "impurity sites" $\mathbf{r}_\ell$ ($\ell = 1, \cdots, N_{\rm imp}$), and the number of impurities is equal to that of doped holes: $N_{\rm imp} = N_{\rm s} - N$ with N and $N_{\rm s}$ being the numbers of electrons and sites, respectively. We also assume that the impurity sites are randomly distributed. Practically, V may be positive and weak, but we widely varies it to grasp overall features. In this publication, we mainly consider cases for $V < 0$, because the calculations have not yet been stable for large positive V/t. Thus, we also discuss U/t dependence for $V/t = -4$. Although the present setting of impurity potential may be simplistic, we believe that fundamental aspects of impurity potentials can be captured. We will treat more realistic settings in coming publications. We impose the periodic-antiperiodic boundary conditions for the consistency with previous studies.

To the Hamiltonian Eq. (1), we apply a variational Monte Carlo (VMC) method [5], which enables us to exactly estimate expectation values with respect to many-body wave functions. As a trial wave function, we use a Jastrow type: $\Psi = P\Phi$, where P indicates a product of many-body projection (Jastrow) factors and Φ is a mean-field-type one-body wave function. For the many-body factor, we use $P = P_Q P_{\rm G}$. In the onsite (Gutzwiller) projection $P_{\rm G}$, we distinguish ordinary and impurity sites as

$$P_{\rm G} = \prod_{j(\neq\ell)}\left[1-(1-g)n_{j\uparrow}n_{j\downarrow}\right]\prod_{\ell=1}^{N_{\rm imp}}\left[1-(1-g_{\rm imp})n_{\ell\uparrow}n_{\ell\downarrow}\right], \qquad (2)$$

where g and $g_{\rm imp}$ are variational parameters. If the optimized value of g (or $g_{\rm imp}$) is smaller [larger] than 1, the effective onsite interaction is repulsive (attractive). P_Q is an asymmetric projection between a nearest-neighbor doubly occupied site (doublon) and an empty site (holon) [6, 7, 8],

$$P_Q = \prod_j\left[1-\zeta_{\rm d}d_j\prod_\tau(1-h_{j+\tau}) - \zeta_{\rm h}h_j\prod_\tau(1-d_{j+\tau})\right], \qquad (3)$$

where $d_j = n_{j\uparrow}n_{j\downarrow}$, $h_j = (1-n_{j\uparrow})(1-n_{j\downarrow})$, and τ runs over the nearest-neighbor sites of site j. Here, $\zeta_{\rm d}$ and $\zeta_{\rm h}$ are variational parameters. As shown before [9], the doublon-holon binding factor is crucial for appropriately treating Mott physics for $U \gtrsim U_{\rm c}$ ($U_{\rm c}$: Mott transition point). For simplicity, we do not distinguish impurity sites from ordinary sites in P_Q in this paper.

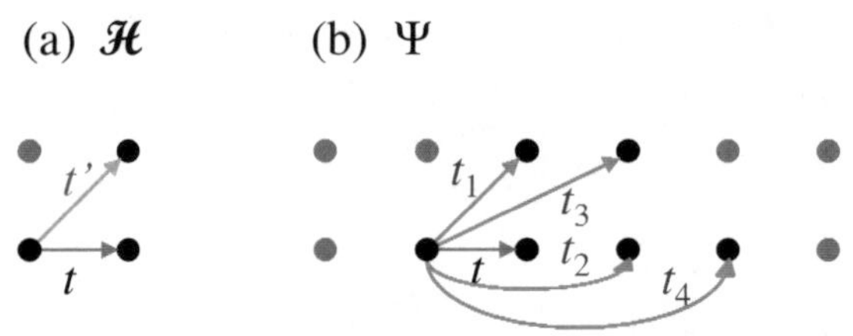

Figure 1. (a) Hopping processes in Hamiltonian [Eq. (1)], and (b) those corresponding to band adjusting parameters t_η in trial wave functions [Eq. (5)]. In both, t is the unit.

For obtaining the one-body part Φ, we diagonalize the following mean-field Hamiltonian,

$$H_{\rm MF} = H_{\rm kin}^{\rm BR} + U\sum_{j,\sigma}\langle n_{j\sigma}\rangle n_{j-\sigma} + V_{\rm eff}\sum_{\ell=1}^{N_{\rm imp}}\sum_{\sigma} n_{\ell\sigma}\,. \tag{4}$$

Following Ref. [2], a band-renormalization effect (BRE) is introduced in the kinetic part as

$$H_{\rm kin}^{\rm BR} = -t\sum_{\langle i,j\rangle,\sigma}\left(c_{i\sigma}^{\dagger}c_{j\sigma} + {\rm H.c.}\right) - \sum_{(i,j),\sigma} t_\eta\left(c_{i\sigma}^{\dagger}c_{j\sigma} + {\rm H.c.}\right) \tag{5}$$

where the first term indicates the nearest-neighbor hopping and the second term virtual hoppings up to fifth neighbors as shown in Fig. 1. Here, t_η's ($\eta = 1$–4) are variational parameters; t_1 is optimized independently of t' in Eq. (1). For $U/t \to 0$, the optimized band parameters are reduced to $t_1 \to t'$ and $t_\eta \to 0$ for $\eta = 2$–4. As shown before [2], BRE up to t_4 should be included for a state with an AF order. In the second term of Eq. (4), we take account of AF order of the nesting vector $\mathbf{Q} = (\pi,\pi)$ as

$$\langle n_{j\sigma}\rangle = \frac{n}{2} + \delta_{\rm AF}\ {\rm sign}(\sigma)\ e^{i\mathbf{Q}\cdot\mathbf{r}_j}, \tag{6}$$

where $n = N/N_{\rm s}$, ${\rm sign}(\sigma) = 1$ or -1 according to $\sigma =\uparrow$ or $\downarrow$, and $\delta_{\rm AF}$ is a variational parameter that controls the magnitude of AF order. In the last term of Eq. (4), a screening of impurity potential owing to electron correlation is considered; $V_{\rm eff}/V$ is an important parameter.

Diagonalizing $H_{\rm MF}$ of an $N_{\rm s} \times N_{\rm s}$ matrix, we obtain eigenenergies ε_n and eigenvectors ϕ_n with corresponding quasiparticle operators $\alpha_{n\sigma}^{\dagger}$, which are linear combinations of $c_{j\sigma}^{\dagger}$. Using $\alpha_{n\sigma}^{\dagger}$, one-body part is written as

$$\Phi = \prod_{n\,(\varepsilon_n \leq \varepsilon_{\rm F})}\prod_{\sigma}\alpha_{n\sigma}^{\dagger}|0\rangle\,, \tag{7}$$

where $\varepsilon_{\rm F}$ indicates the energy of the highest occupied level (HOMO). The trial wave function Ψ ($= P\Phi$) is reduced to the exact noninteracting ground state of Eq. (1) for $U = 0$, and becomes identical to the correlated AF state with BRE used in a preceding study for the pure state [2] for $V = 0$, a large U/t and a small δ. We represent a correlated AF [paramagnetic (PM)] state as $\Psi_{\rm AF}$ [$\Psi_{\rm PM} = \Psi_{\rm AF}(\delta_{\rm AF} = 0)$].

We compute expectation values of Ψ using a VMC method similarly to Ref. [2]. In this paper, we exhibit preliminary data for a system with $N_{\rm s} = 10 \times 10$ sites and 92 electrons, namely, impurity density and doping rate (hole density) are $N_{\rm imp}/N_{\rm s} = \delta = 0.08$. In most cases, we will simultaneously display the data for 4–8 configurations of impurity sites $\{\mathbf{r}_\ell\}$, which are randomly created. We found that $\{\mathbf{r}_\ell\}$ dependence is small except for special cases, where, for instance, plural impurity clusters are accidentally created.

3. Results

3.1 U/t dependence in attractive potential

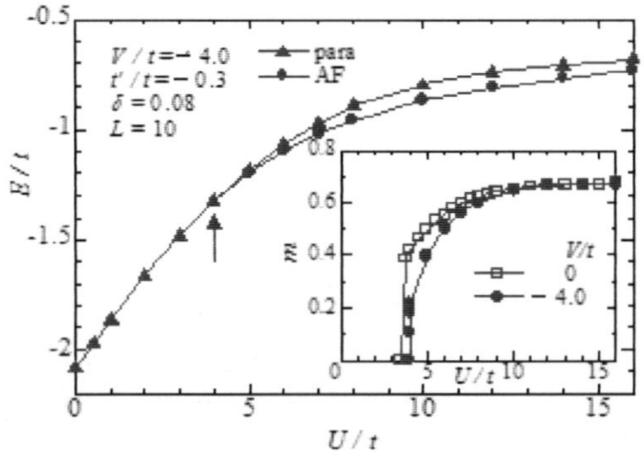

Figure 2. Total energies per site of the AF and PM states are plotted as functions of U/t. In the inset, the sublattice magnetization is compared between the pure case ($V/t = 0$, Ref. [2]) and a case with impurity potential ($V/t = -4$) as functions of U/t.

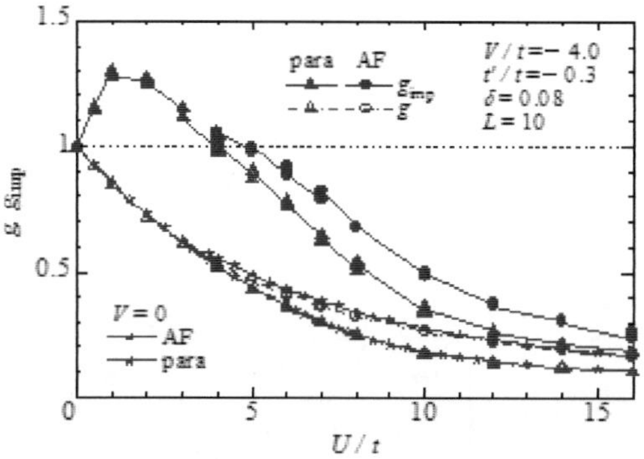

Figure 3. U/t dependence of the optimized onsite correlation parameters are compared between the AF and PM states. For comparison, the result for the pure case ($V/t = 0$, Ref. [2]) are added with thin lines and small stars.

We start with U/t dependence in a moderate attractive potential ($V/t = -4$). In the previous study for the pure model ($V/t = 0$) [2], we found that an AF order is absent for $U < U_{\rm AF} \sim 3.75t$ for $t'/t = -0.3$ and $\delta = 0.08$. In the main panel of Fig. 2, the total energy per site for $V/t = -4$ are compared between the AF and PM states. Corresponding to the case of $V = 0$, the AF state has a lower energy than that of the PM state for $U > U_{\rm AF} \sim 4t$ and the staggered magnetization (an order parameter for AF)

$$m = \frac{2}{N_{\rm s}} \left| \sum_j e^{i\mathbf{Q}\cdot\mathbf{r}_j} \langle S_j^z \rangle \right| , \tag{8}$$

becomes finite as shown in the inset of Fig. 2. For intermediate values of U/t, m in the present case ($V/t = -4$) is somewhat smaller than that in the pure case, but for $U/t \gtrsim 10$, m is almost identical in the two cases. This is because the screening for V becomes effective for strongly correlated cases, as we will discuss shortly.

In Fig. 3, the optimized onsite correlation parameters are plotted. The parameter for the ordinary sites g is almost identical between $V/t = 0$ and -4 for any U/t and irrespective of whether the state is AF or PM; g monotonically decreases as the repulsive interaction U/t increases. On the other hand, the onsite factor for the impurity sites $g_{\rm imp}$ becomes $g_{\rm imp} > 1$ for $U \lesssim |V|$; the effective correlation in the impurity sites is attractive in this regime. For $U > |V|$, however, $g_{\rm imp}$ also monotonically decreases and approaches the value of the pure case. This is again screening effect owing to the electron correlation on the impurity potential. In the main panel of Fig. 4, we show U/t dependence of $V_{\rm eff}/V$ for the two states. As U/t increases, $V_{\rm eff}/V$ rapidly decreases and almost vanishes for $U \gtrsim W$ ($W = 8t$: band width). Thus, the attractive impurity potential V is almost completely screened in the strongly correlated regime. Such a tendency of the screening effect on impurity potentials owing to strong correlations has been known in various contexts [10].

3.2 V/t dependence in strongly correlated regime

Let us consider how the effect of V evolves as V varies in a strongly correlated regime ($U/t = 12$). First, we look at the energy difference between $\Psi_{\rm AF}$ and $\Psi_{\rm PM}$: $\Delta E = E({\rm PM}) - E({\rm AF})$, which is plotted in Fig. 5. The energy gain by the AF order ΔE changes only very slightly for $-W \lesssim V \lesssim t$. Corresponding to this behaviour of ΔE, the gap parameter $\delta_{\rm AF}$ and the order parameter m also vary only very slightly in this range of V/t. Thus, a point impurity potential does not affect the robustness of the AF state realized for the pure state ($V/t = 0$) at least for moderate (especially attractive) values of V/t. This is because the screening of impurity potential is very effective and $V_{\rm eff}/V$ almost vanishes in this regime both for $\Psi_{\rm PM}$ and $\Psi_{\rm AF}$, as shown in the inset of Fig. 4.

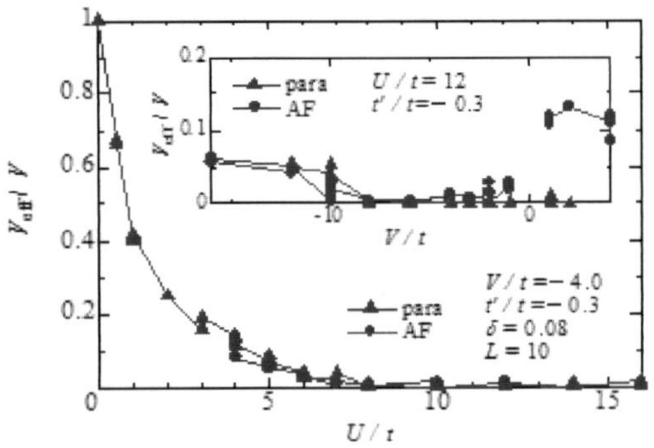

Figure 4. The effective impurity potential ($V_{\rm eff}/V$) is shown as a function of U/t for $V/t = -4$ in the main panel, and as a function of V/t for $U/t = 12$ in the inset. In both, the values of the AF and PM states are compared.

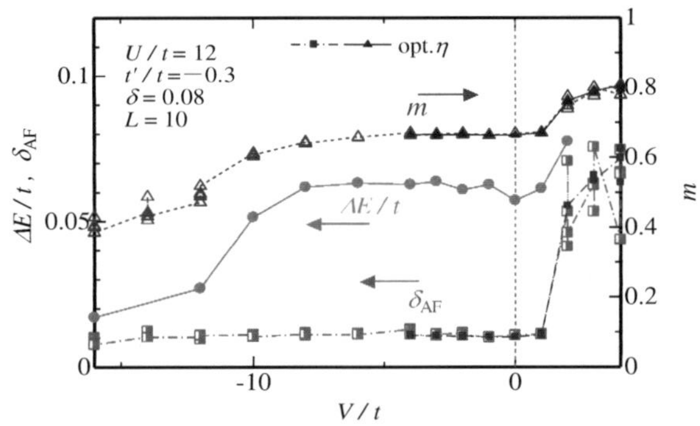

Figure 5. V/t dependence of energy gain by AF order $\Delta E/t$, sublattice magnetization m ($m = 1$ for the full moment), and optimized AF gap parameter $\delta_{\rm AF}$. ΔE is estimated from the averages of plural impurity-site configurations.

On the other hand, in the regime of strong impurity potential of $V \lesssim -U$, m and especially $\Delta E/t$ decrease as $|V|/t$ increases. This is because the screening effect becomes somewhat weak, as seen in the inset of Fig. 4. Consequently, effective onsite correlation in the impurity sites again becomes attractive as shown in Fig. 6, where g and $g_{\rm imp}$ are plotted as functions of V/t. The change in g is very small, whereas $g_{\rm imp}$ monotonically increases as V/t decreases for $V/t < 0$ and becomes larger than 1 for $V \lesssim -U$. The relation $g_{\rm imp}({\rm AF}) > g_{\rm imp}({\rm PM})$ holds for $V < 0$. This feature is better understood by the behaviour of double occupancy d and electron density at impurity sites $n_{\rm imp}$ defined as

$$d = \frac{1}{N}\sum_j \langle n_{j\uparrow} n_{j\downarrow} \rangle, \qquad n_{\rm imp} = \frac{1}{N_{\rm imp}} \sum_{\ell=1}^{N_{\rm imp}} \sum_{\sigma} \langle n_{\ell\sigma} \rangle. \tag{9}$$

V/t dependence of these quantities are shown in Fig. 7. Note that, as V/t decreases, $n_{\rm imp}$ approaches 2, for which all impurity sites are doubly occupied. This is reflected in the marked increases of average double occupancy d for $V \lesssim -U = -12t$. To conclude, in the cases of strong attractive impurity potential ($V \lesssim -U$), stability of the AF state appreciably deteriorates owing to the decrease in number of singly occupied (spin-alive) sites as a result of the increase in double occupancy in the impurity sites.

4. Summary

In this paper, we discussed the effect of a point-type impurity potential on a PM and an AF states in the two-dimensional Hubbard model on the basis of preliminary VMC calculations. The impurity potential is almost completely screened by the electron correlation for $U/t \gtrsim 8$ and moderate potential strength ($-U \lesssim V \lesssim 2t$). In this case, the features of the AF and PM states is not modified by the impurity potential, namely the AF state remains very stable as in the pure model ($V = 0$). However, if the attractive impurity potential is sufficiently strong ($V \lesssim -U$), the AF state is appreciably destabilized. In this case, the screening becomes insufficient and effective onsite interaction in the impurity sites becomes attractive, resulting in a marked increase in double occupancy. This leads to the decrease in number of spin-alive (singly occupied) sites necessary for the AF order. Here, we mainly treated cases of $V/t < 0$. In future publications, we will discuss more realistic cases ($V/t > 0$) for cuprates in detail.

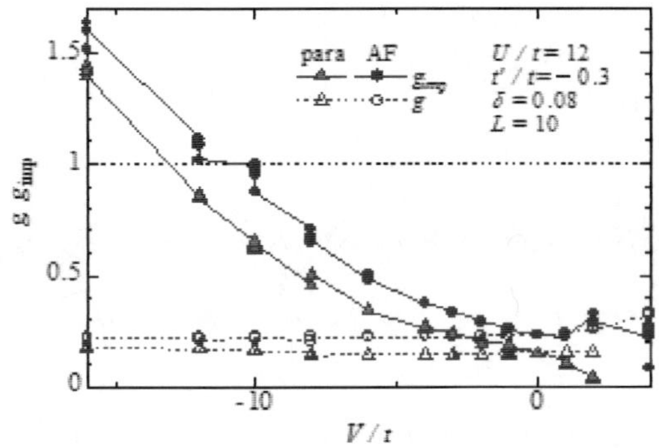

Figure 6. The optimized onsite correlation factors (g, g_{imp}) for $U/t = 12$ are shown as functions of V/t. The data of Ψ_{AF} and Ψ_{PM} are compared.

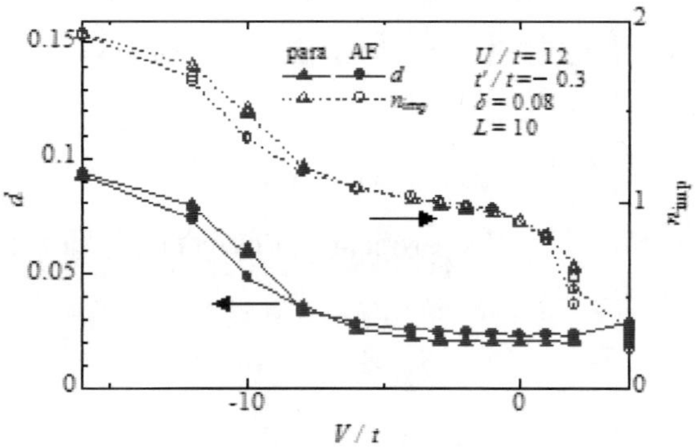

Figure 7. Double occupancy and electron density at the impurity sites are plotted as functions of V/t for AF and PM states. $n_{imp} = 0.92$ for $V = 0$.

References

[1] Otsuki J, Hafermann H and Lichtenstein A I 2014 *Phys. Rev. B* **90** 235132; Misawa T and Imada M 2014 *Phys. Rev. B* **90** 115137; Zheng B-X and Chan G K-L 2016 *Phys. Rev. B* **93** 035126
[2] Sato R and Yokoyama H 2016 *J. Phys. Soc. Jpn.* **85** 074701
[3] Tamura S 2016 Dr. Thesis, Faculty of Science, Tohoku University, Sendai [in Japanese], and in preparation
[4] For instance, Tamaki H and Miyake K 2012 *J. Phys. Soc. Jpn.* **81** 124712
[5] Yokoyama H and Shiba H 1987 *J. Phys. Soc. Jpn.* **56** 1490; Umrigar C J, Wilson K G and Wilkins J W 1988 *Phys. Rev. Lett.* **60** 1719
[6] Kaplan T A, Horsch P and Fulde P 1982 *Phys. Rev. Lett.* **49** 889
[7] Yokoyama H and Shiba H 1990 *J. Phys. Soc. Jpn.* **59** 3669
[8] Yokoyama H, Ogata M, Tanaka Y, Kobayashi K and Tsuchiura H 2013 *J. Phys. Soc. Jpn.* **82** 014707
[9] Yokoyama H, Tanaka Y, Ogata M and Tsuchiura H 2004 *J. Phys. Soc. Jpn.* **73** 1119; Yokoyama H, Ogata M and Tanaka Y 2006 *J. Phys. Soc. Jpn.* **75** 114706
[10] For instance, Fukushima N, Chou C-P and Lee T K 2009 *Phys. Rev. B* **79** 184510; Farhoodfar A, Chen X, Gooding R J and Atkinson W A 2009 *Phys. Rev. B* **80** 045108

IOP Conf. Series: Journal of Physics: Conf. Series **871** (2017) 012033 doi:10.1088/1742-6596/871/1/012033

Theory of high-temperature superconductivity in strongly correlated fermions system

Kazuhisa Nishi

Department of Engineering, University of Hyogo, Himeji, Hyogo 671-2201, Japan

tyus51kzhs@ares.eonet.ne.jp

Abstract. A unified theory to explain generally the properties of high-temperature superconducting cuprates and iron pnictides *etc.* is presented. The theory is based on the extended *d-p* model emphasizing that the electronic state of superconductors can be described by the composed fermions constructed with newly defined operators. The peculiarity in the theory is that the Hamiltonian is so modified by the unitary transformation using these fermion operators as to apply the mean field approximation. It is found that the results can provide a universal explanation about the various superconducting and normal properties of high-temperature superconductors.

1. Introduction

Despite much intensive study about high-temperature superconducting cuprates and iron pnictides etc, its mechanism still remains an unsolved problem [1-8]. Recently, the author proposed the composite fermions theories of these superconductors [9]. These theories are based on the extended *d-p* model emphasizing that the electronic state of superconductors can be described by the composed fermions constructed with newly defined operators. However, it is another problem to unite these theories into a unified theory to explain generally the properties of these superconductors. Here in order to find the unified theory, the high-temperature superconductivity is more generally investigated in strongly correlated fermions system.

2. Effective Hamiltonian

The Hamiltonian is assumed to be an extended *d-p* model for a single layer of square planar

$$H = \sum_{in\sigma} \varepsilon_{dn} d^{+}_{in\sigma} d_{in\sigma} + \sum_{jl\sigma} \varepsilon_{pl} p^{+}_{jl\sigma} p_{jl\sigma} + \sum_{inl\sigma} \sum_{j\in\{inl\}} (\varepsilon_{injl} d^{+}_{in\sigma} p_{jl\sigma} + \text{H.c.}) + \sum_{inm\sigma\sigma'} U_{nm} d^{+}_{in\sigma} d_{im\sigma} d^{+}_{in\sigma'} d_{im\sigma'} + \sum_{injl\sigma\sigma'} V_{injl} d^{+}_{in\sigma} d_{in\sigma} p^{+}_{jl\sigma'} p_{jl\sigma'} \quad (1)$$

where the operator $d^{+}_{in\sigma}$ creates electrons or holes of *n-th* M (Cu or Fe) 3*d* orbital at site *i*, $p^{+}_{jl\sigma}$ creates *l-th* L (oxide or pnictide) *p* electrons or holes at site *j*, $\varepsilon_{dn} = 0$, and ε_{injl} is the nearest-neighbor hopping integral. U_{nn} and $U_{nm} (n \neq m)$ are the Coulomb repulsion and Hund coupling at M site, respectively, and V_{injl} is the interaction between neighboring M and L sites. In (1), the vacuum is defined as M d^{x} and L p^{y} states (x = 10, y = 6 in the cupurate, x = 0, y = 6 in the pnictide). Considering the *d-p* covalency effect, the operator combining the L states around M ion is defined as

IOP Conf. Series: Journal of Physics: Conf. Series **871** (2017) 012033 doi:10.1088/1742-6596/871/1/012033

$$\tilde{p}^{+}_{inl\sigma} = \sum_{j\in\{inl\}} \varepsilon_{injl}\varepsilon^{-1}_{nl} p^{+}_{jl\sigma}, \quad \varepsilon_{nl} = \sqrt{\sum_{j\in\{inl\}} \varepsilon^{2}_{injl}}, \tag{2}$$

where $j = i + \boldsymbol{x}_j$, and $\boldsymbol{x}_j$ presents the directions of M-L bonding. Since the relation of $\sum_j p^{+}_{jl\sigma} p_{jl\sigma} = N_v \sum_i \tilde{p}^{+}_{inl\sigma}\tilde{p}_{inl\sigma}$ (N_v is the valence ratio of d to p) is satisfied in the M-L compounds, Hamiltonian (1) is so changed as

$$H = N_v \sum_{inl\sigma} \varepsilon_{pl}\tilde{p}^{+}_{inl\sigma}\tilde{p}_{inl\sigma} + \sum_{inl\sigma}(\varepsilon_{nl} d^{+}_{in\sigma}\tilde{p}_{inl\sigma} + \text{H.c.}) + \sum_{inm\sigma\sigma'} U_{nm} d^{+}_{in\sigma} d_{im\sigma} d^{+}_{in\sigma'} d_{im\sigma'} + N_v \sum_{inl\sigma\sigma'} V_{inl} d^{+}_{in\sigma} d_{in\sigma}\tilde{p}^{+}_{inl\sigma'}\tilde{p}_{inl\sigma'} \tag{3}$$

Since the operator $\tilde{p}^{+}_{inl\sigma}$ is not orthogonal between the neighboring M sites, it does not exactly satisfy anti-commutation relations. However it is approximated here that $\tilde{p}^{+}_{inl\sigma}$ can be well-defined fermion operator.

To the second order in perturbation theory on the condition of $\varepsilon_{nl} < U_n (= U_{nn})$, let us find out the effective Hamiltonian of Eq. (3). For simplicity, Hund coupling and the interaction between different 3d orbitals are neglected. First the second term of Eq. (3) will be divided into three parts of H_1, H_2 and H_3 as follow as

$$\begin{aligned}
H_1 &= \sum_{inl\sigma}[\varepsilon_{nl}(1-n_{d\sigma'}) d^{+}_{in\sigma}\tilde{p}_{inl\sigma}(1-n_{p\sigma'}) + \text{H.c.}], \\
H_2 &= \sum_{inl\sigma}[\varepsilon_{nl} n_{d\sigma'} d^{+}_{in\sigma}\tilde{p}_{inl\sigma} n_{p\sigma'} + \text{H.c.}] \\
H_3 &= \sum_{inl\sigma}\{\varepsilon_{nl}[n_{d\sigma'} d^{+}_{in\sigma}\tilde{p}_{inl\sigma}(1-n_{p\sigma'}) + (1-n_{d\sigma'}) d^{+}_{in\sigma}\tilde{p}_{inl\sigma} n_{p\sigma'}] + \text{H.c.}\}
\end{aligned} \tag{4}$$

where H_1 (H_2) is the hopping terms in the absence (presence) of fermions with the opposite spin at both d and p sites and H_3 indicates the terms in the presence of fermions with the opposite spin at either d or p sites. Thus, H_3 acts on the interaction U_n as a perturbation, but H_1 and H_2 do not so because the ground state is not assumed here to include the double occupancy states. Further, let us average the occupancy factors such as $(1-n_{d\sigma'})$ by using $\langle 1-n_{d\sigma'}\rangle$. The Coulomb interaction U in the ground state can be effectively eliminated from the starting Hamiltonian because double occupancy states at d-sites are inhibited by Coulomb repulsion. Thus, Hamiltonian (3) is modified by

$$\begin{aligned}
H = {} & N_v \sum_{inl\sigma} \varepsilon_{pl}\tilde{p}^{+}_{inl\sigma}\tilde{p}_{inl\sigma} + \sum_{inl\sigma}\langle 1-n_{d\sigma'}\rangle\langle 1-n_{p\sigma'}\rangle(\varepsilon_{nl} d^{+}_{in\sigma}\tilde{p}_{inl\sigma} + \text{H.c.}) + N_v \sum_{inl\sigma\sigma'} V_{inl} d^{+}_{in\sigma} d_{in\sigma}\tilde{p}^{+}_{inl\sigma'}\tilde{p}_{inl\sigma'} \\
& + \sum_{inl\sigma\neq\sigma'} \left(\langle 1-n_{d\sigma'}\rangle\langle n_{p\sigma'}\rangle + \langle 1-n_{p\sigma'}\rangle\langle n_{p\sigma'}\rangle\right)^2 (-J_1 d^{+}_{in\sigma} d_{in\sigma}\tilde{p}^{+}_{inl\sigma'}\tilde{p}_{inl\sigma'} - J_2 d^{+}_{in\sigma}\tilde{p}_{inl\sigma}\tilde{p}^{+}_{inl\sigma'} d_{in\sigma'})
\end{aligned} \tag{5}$$

where $J_1 = \varepsilon^2_{nl}(U - N_v\varepsilon_{pl} - N_v V_{inl})^{-1}$, $J_2 = \varepsilon^2_{nl}(N_v\varepsilon_{pl} - N_v V_{inl})^{-1}$. Though the latter part of the forth term in (5) corresponds to the kinetic energy due to the hopping between sites, it is neglected here due to the second order kinetic energy. The former part corresponds to the effective anti-ferromagnetic interaction between d-p fermions. Thus, the effective Hamiltonian is finally obtained as

$$H_{eff} = N_v \sum_{inl\sigma} \varepsilon_{pl}\tilde{p}^{+}_{inl\sigma}\tilde{p}_{inl\sigma} + \sum_{inl\sigma}\langle 1-n_{d\sigma'}\rangle\langle 1-n_{p\sigma'}\rangle(\varepsilon_{nl} d^{+}_{in\sigma}\tilde{p}_{inl\sigma} + \text{H.c.}) + N_v \sum_{inl\sigma\sigma'} V d^{+}_{in\sigma} d_{in\sigma}\tilde{p}^{+}_{inl\sigma'}\tilde{p}_{inl\sigma'} \tag{6}$$

where $V = V_{inl} + N_v^{-1} V_{\sigma\sigma'}$ (V_{inl} : interaction between the nearest neighboring sites in (1), $V_{\sigma\sigma'}$: anti-ferromagnetic interaction $V_{\sigma\sigma'} = -\left(\langle 1-n_{d\sigma'}\rangle\langle n_{p\sigma'}\rangle + \langle 1-n_{p\sigma'}\rangle\langle n_{n\sigma'}\rangle\right)^2 J_1 < 0 \quad \sigma \neq \sigma'$).

What's the ground state of this effective Hamiltonian? Here two ground states will be considered depending on the doping conditions. One of these is the states in the neighborhood of the insulator (so-called Mott insulator) and the other is the superconducting state based on the band picture. First let us consider the region in the neighborhood of the insulator. The effective Hamiltonian (6) is appropriate in this case because its representation in real space is very useful in the treatment of this region. In order to enable to apply the mean field approximation, the operators $d^+_{in\sigma}$, $\tilde{p}^+_{inl\sigma}$ are changed by unitary transformations of

$$b^+_{inl\sigma} = \alpha_{nl} d^+_{in\sigma} + \beta_{nl} \tilde{p}^+_{inl\sigma},\ c^+_{inl\sigma} = \beta_{nl} d^+_{in\sigma} - \alpha_{nl} \tilde{p}^+_{inl\sigma},\ \alpha^2_{nl} + \beta^2_{nl} = 1 \tag{7}$$

where b_i^+, c_i^+ are defined as the mixing operators of M and L fermions. For the condition of $\alpha_{nl}\beta_{nl}N_v\varepsilon_{pl} = \langle 1-n_{d\sigma'}\rangle\langle 1-n_{p\sigma'}\rangle\varepsilon_{nl}(\beta^2_{nl}-\alpha^2_{nl})$, the Hamiltonian (6) is transformed into

$$H = N_v \sum_{inl\sigma} \varepsilon_{pl}(\beta^2-\alpha^2)^{-1}(\beta^2 b^+_{i\sigma}b_{i\sigma} - \alpha^2 c^+_{i\sigma}c_{i\sigma}) + N_v \sum_{inl\sigma\sigma'} V(\alpha^2\beta^2 b^+_{i\sigma}b_{i\sigma}b^+_{i\sigma'}b_{i\sigma'} + \alpha^4 b^+_{i\sigma}b_{i\sigma}c^+_{i\sigma'}c_{i\sigma'} + \beta^4 c^+_{l\sigma}c_{i\sigma}b^+_{i\sigma'}b_{i\sigma'} \\ + \alpha^2\beta^2 c^+_{i\sigma}c_{i\sigma}c^+_{i\sigma'}c_{i\sigma'} + \cdots) \tag{8}$$

Notice that the indexes n, l of $\alpha_{nl}, \beta_{nl}, b^+_{nl\sigma}, c^+_{nl\sigma}$ *etc* are omitted to avoid the complex representation. Since there appear many interaction terms in the transformed Hamiltonian (8), this might make the problem somewhat difficult. However, the mixing of d and p fermions can be directly built in this Hamiltonian, which consists of two free fermions and the interaction terms between them. As a result this treatment has an advantage to enable to apply the mean field approximation because of the explicit representation of free fermions and their interaction terms. Next let us consider the case of the superconducting state based on the band picture. Since the representation in momentum space is appropriate in this region, the effective Hamiltonian of (6) is transformed into

$$H = \sum_{\mathbf{k}l\sigma}\varepsilon_{pl}p^+_{\mathbf{k}l\sigma}p_{\mathbf{k}l\sigma} + \sum_{\mathbf{k}nl\sigma}\langle 1-n_{d\sigma'}\rangle\langle 1-n_{p\sigma'}\rangle\varepsilon_{nl}s_{nl}(\mathbf{k})(-id^+_{\mathbf{k}n\sigma}p_{\mathbf{k}l\sigma} + ip^+_{\mathbf{k}l\sigma}d_{\mathbf{k}n\sigma}) + N^{-1}\sum_{\mathbf{kk'k''}nl\sigma\sigma'} V\theta_{nl}(\mathbf{k},\mathbf{k}')d^+_{\mathbf{k}n\sigma}d_{\mathbf{k}'n\sigma}p^+_{\mathbf{k}''l\sigma'}p_{\mathbf{k}-\mathbf{k}'+\mathbf{k}'',l\sigma'} \tag{9}$$

Here $s_n(\mathbf{k}) = \sum_{j\in\{in\}}\varepsilon_{nj}\varepsilon_n^{-1}\exp(i\mathbf{k}\cdot\mathbf{x}_j)$, $\theta_n(\mathbf{k},\mathbf{k}') = \sum_{j\in\{in\}}\varepsilon^2_{nj}\varepsilon_n^{-2}\exp[i(\mathbf{k}-\mathbf{k}')\cdot\mathbf{x}_j]$, x, y refer to the axis for the unit cell, and N is number of M sites in a single layer. The operators $p^+_{\mathbf{k}l\sigma}, d^+_{\mathbf{k}n\sigma}$,which are Fourier transformation of $\tilde{p}^+_{inl\sigma}, d^+_{in\sigma}$, are reconstructed with fermion operators defined by unitary transformation

$$b^+_{\mathbf{k}nl\sigma} = \alpha_{\mathbf{k}nl}d^+_{\mathbf{k}n\sigma} + i\beta_{\mathbf{k}nl}p^+_{\mathbf{k}l\sigma},\quad c^+_{\mathbf{k}n\sigma} = \beta_{\mathbf{k}nl}d^+_{\mathbf{k}n\sigma} - i\alpha_{\mathbf{k}nl}p^+_{\mathbf{k}l\sigma},\quad \alpha^2_{\mathbf{k}nl} + \beta^2_{\mathbf{k}nl} = 1, \tag{10}$$

where $b^+_{\mathbf{k}nl\sigma}, c^+_{\mathbf{k}nl\sigma}$ satisfy anti-commutation relations. Notice that the operators $p^+_{\mathbf{k}l\sigma}$ satisfy anti-commutation relations exactly, but those are approximated operators of $\tilde{p}^+_{inl\sigma}$ in momentum space. For the condition of $\alpha_{\mathbf{k}nl}\beta_{\mathbf{k}nl}\varepsilon_{pl} = \langle 1-n_{d\sigma'}\rangle\langle 1-n_{p\sigma'}\rangle\,\varepsilon_{nl}s_{nl}(\mathrm{k})(\beta^2_{\mathbf{k}nl}-\alpha^2_{\mathbf{k}nl})$, the Hamiltonian (9) is given by

$$H = \sum_{nl}\sum_{\mathbf{k}\sigma}\varepsilon_p(\beta^2_{\mathbf{k}}-\alpha^2_{\mathbf{k}})^{-1}(\beta^2_{\mathbf{k}}b^+_{\mathbf{k}\sigma}b_{\mathbf{k}\sigma} - \alpha^2_{\mathbf{k}}c^+_{\mathbf{k}\sigma}c_{\mathbf{k}\sigma}) + N^{-1}\sum_{nl}\sum_{\mathbf{kk'k''}\sigma\sigma'}V\theta_n(\mathbf{k},\mathbf{k}')(\alpha_{\mathbf{k}}\alpha_{\mathbf{k}'}\beta_{\mathbf{k}''}\beta_{\mathbf{k}-\mathbf{k}'+\mathbf{k}''}b^+_{\mathbf{k}\sigma}b_{\mathbf{k}'\sigma}b^+_{\mathbf{k}''\sigma'}b_{\mathbf{k}-\mathbf{k}'+\mathbf{k}'',\sigma'} \\ + \alpha_{\mathbf{k}}\alpha_{\mathbf{k}'}\alpha_{\mathbf{k}''}\alpha_{\mathbf{k}-\mathbf{k}'+\mathbf{k}''}b^+_{\mathbf{k}\sigma}b_{\mathbf{k}'\sigma}c^+_{\mathbf{k}''\sigma'}c_{\mathbf{k}-\mathbf{k}'+\mathbf{k}'',\sigma'} + \beta_{\mathbf{k}''}\beta_{\mathbf{k}-\mathbf{k}'+\mathbf{k}''}\beta_{\mathbf{k}}\beta_{\mathbf{k}'}c^+_{\mathbf{k}\sigma}c_{\mathbf{k}'\sigma}b^+_{\mathbf{k}''\sigma'}b_{\mathbf{k}-\mathbf{k}'+\mathbf{k}'',\sigma'} + \beta_{\mathbf{k}}\beta_{\mathbf{k}'}\alpha_{\mathbf{k}''}\alpha_{\mathbf{k}-\mathbf{k}'+\mathbf{k}''}c^+_{\mathbf{k}\sigma}c_{\mathbf{k}'\sigma}c^+_{\mathbf{k}''\sigma'}c_{\mathbf{k}-\mathbf{k}'+\mathbf{k}'',\sigma'} + \cdots) \tag{11}$$

where the indexes n, l of $\alpha_{\mathbf{k}nl}, \beta_{\mathbf{k}nl}, b^+_{\mathbf{k}nl\sigma}, c^+_{\mathbf{k}nl\sigma}$ *etc* are omitted to avoid the complex representation. These effective Hamiltonians will be applied for the cuprate and pnictide superconductors in next chapter.

3. Cuprate superconductors

The characteristics of cuprate superconductors is that it is a single d-band (n = 1), superconductive in hole doping, and insulator in non-doping state. Though p-band is multi-orbitals, a single band of l = 1 will be only considered here. First let us consider the case of the nearly non-doped region. Assuming that there is no p-holes in half-filling, the relations of $\langle 1-n_{d\sigma'}\rangle \approx 0.5-(1-\kappa)\delta \approx 0.5$, $\langle 1-n_{p\sigma'}\rangle \approx 1-0.5\kappa\delta \approx 1-0.5\delta, (\kappa\approx 1)$ are derived. Here κ is the doping ratio contributing to p-holes. For the condition of $2\alpha\beta\varepsilon_p = 0.5(1-0.5\delta)\varepsilon(\beta^2-\alpha^2)$, the Hamiltonian (6) is transformed into

$$H = 2\varepsilon_p(\beta^2-\alpha^2)^{-1}\sum_{i\sigma}(\beta^2 b^+_{i\sigma}b_{i\sigma}-\alpha^2 c^+_{i\sigma}c_{i\sigma}) + 2\sum_{i\sigma\sigma'}V(\alpha^2\beta^2 b^+_{i\sigma}b_{i\sigma}b^+_{i\sigma'}b_{i\sigma'}+\alpha^4 b^+_{i\sigma}b_{in\sigma}c^+_{i\sigma'}c_{i\sigma'}+\beta^4 c^+_{i\sigma}c_{i\sigma}b^+_{i\sigma'}b_{i\sigma'} \\ +\alpha^2\beta^2 c^+_{i\sigma}c_{i\sigma}c^+_{i\sigma'}c_{i\sigma'}+\cdots) \quad (12)$$

Here N_v (the valence ratio of d to p) = 2. For the copper oxides it is well known that its ground state shows antiferromagnets (so-called Mott insulator) in the non-doped region [10,11]. Using the composite operators defined here, the non-doped wave function corresponds to

$$|\Psi_{MI}\rangle = \prod_{i\in A}^{N/2}(\alpha b^+_{i\uparrow}+\beta c^+_{i\uparrow})\prod_{j\in B}^{N/2}(\alpha b^+_{j\downarrow}+\beta c^+_{j\downarrow})|0\rangle \quad (13)$$

where A or B shows the sub-space of antiferromagnetic lattice and N is number of Cu sites in a single layer. Since there exists no p-hole in half-filling Mott insulator, it will not be necessary to consider the interaction V in this case. However, in the existence of p-holes, since $V_{\sigma\sigma'}(\sigma\neq\sigma')$ is interpreted to be attractive due to d-p exchange antiferromagnetic interaction, the case of $V(=V_{inl}+N_v^{-1}V_{\sigma\sigma'})<0$ will play an important role for determining the ground state. Considering the experimental facts that the doping can almost supply the p-holes, doping holes are expected to occupy the quantum state corresponding to this situation. For the condition $\alpha^2<\beta^2$ which allows the reliable value of ε_p and ε, the term $b^+_{i\sigma}b_{i\sigma}c^+_{i\sigma'}c_{i\sigma'}$ ($c^+_{i\sigma}c_{i\sigma}b^+_{i\sigma'}b_{i\sigma'}$) can be the most dominant interaction. This means that b-c or c-b pair will mainly contribute to determine the ground state. Thus, for the equivalency of sites, the normalized wave function is assumed to be

$$|\Psi_A\rangle = \prod_i^N(s+tb^+_{i\uparrow}c^+_{i\downarrow})(s+tc^+_{i\uparrow}b^+_{i\downarrow})|\Psi_{MI}\rangle,\quad s^2(s^2+t^2)=1 \quad (14)$$

where the coefficient t indicates the probability of local-antiferro pair state. On the approximation using the relation of $\langle\Psi_A|\sum p^+_{i\sigma}p_{i\sigma}|\Psi_A\rangle\approx 2Ns^2t^2=N\delta$ (δ: the doping ratio relative to half-filling), the ground-state energy is given by

$$E_p=\langle\Psi_A|H|\Psi_A\rangle=N\delta(\varepsilon_p+V),\quad V<0 \quad (15)$$

The energy of the excited states will be evaluated by using the presumed excited state wavefunction as follow as

$$|\Psi_{EX}\rangle = \prod_i^N(s+tb^+_{i\uparrow}c^+_{i\downarrow})(s+tc^+_{i\uparrow}b^+_{i\downarrow})(\beta b^+_{i'\uparrow}-\alpha c^+_{i'\uparrow})|\Psi_{MI}\rangle \quad (16)$$

This is orthogonal to the ground state function and corresponds to breaking up a b-c (c-b) pair in i', the spin-up member going to $b^+_{i'\uparrow}, c^+_{i'\uparrow}$ The excited energy is obtained as $N\delta(\varepsilon_p+V)-2V$. This indicates that there is the energy gap between the ground and the excited states. Thus, the wave function (14) will be identified as the pseudogap state in the nearly non-doped region. The pseudogap energy is then estimated to be $\Delta_p=|V|$ and and the pseudogap temperature T_p is approximated as the $\lambda\Delta_p$ ($\lambda<1$).

Next let us consider the case in the neighborhood of optimally doped region. In this doping region the relation of $\langle 1-n_{d\sigma'}\rangle\approx 0.5-(1-\kappa)\delta\approx 0.5$, $\langle 1-n_{p\sigma'}\rangle\approx 1-0.5\kappa\delta\approx 1-0.5\delta, (\kappa\approx 1)$ will be generally reasonable. On the condition of $\alpha_{\mathbf{k}}\beta_{\mathbf{k}}\varepsilon_p=0.5(1-0.5\delta)\varepsilon s(\mathbf{k})(\beta^2_{\mathbf{k}}-\alpha^2_{\mathbf{k}})$, the Hamiltonian (9) is represented by

$$H=\sum_{\mathbf{k}\sigma}\varepsilon_p(\beta^2_{\mathbf{k}}-\alpha^2_{\mathbf{k}})^{-1}(\beta^2_{\mathbf{k}}b^+_{\mathbf{k}\sigma}b_{\mathbf{k}\sigma}-\alpha^2_{\mathbf{k}}c^+_{\mathbf{k}\sigma}c_{\mathbf{k}\sigma})+N^{-1}\sum_{\mathbf{kk'k''}\sigma\sigma'}V\theta(\mathbf{k},\mathbf{k'})(\alpha_{\mathbf{k}}\alpha_{\mathbf{k'}}\beta_{\mathbf{k''}}\beta_{\mathbf{k-k'+k''}}b^+_{\mathbf{k}\sigma}b_{\mathbf{k'}\sigma}b^+_{\mathbf{k''}\sigma'}b_{\mathbf{k-k'+k''},\sigma'} \\ +\alpha_{\mathbf{k}}\alpha_{\mathbf{k'}}\alpha_{\mathbf{k''}}\alpha_{\mathbf{k-k'+k''}}b^+_{\mathbf{k}\sigma}b_{\mathbf{k'}\sigma}c^+_{\mathbf{k''}\sigma'}c_{\mathbf{k-k'+k''},\sigma'}+\beta_{\mathbf{k''}}\beta_{\mathbf{k-k'+k''}}\beta_{\mathbf{k}}\beta_{\mathbf{k'}}c^+_{\mathbf{k}\sigma}c_{\mathbf{k'}\sigma}b^+_{\mathbf{k''}\sigma'}b_{\mathbf{k-k'+k''},\sigma'}+\beta_{\mathbf{k}}\beta_{\mathbf{k'}}\alpha_{\mathbf{k''}}\alpha_{\mathbf{k-k'+k''}}c^+_{\mathbf{k}\sigma}c_{\mathbf{k'}\sigma}c^+_{\mathbf{k''}\sigma'}c_{\mathbf{k-k'+k''},\sigma'}+\cdots) \quad (17)$$

where $s(\mathbf{k}) = \sin k_x + \sin k_y$, $\theta(\mathbf{k},\mathbf{k}') = \cos(k_x - k_x') + \cos(k_y - k_y')$, and the Cu-O distance is used as the length unit. Here $b^+_{\mathbf{k}\sigma}$ and $c^+_{\mathbf{k}\sigma}$ satisfy anti-commutation relations. Note that $V = V_{ij} + V_{\sigma\sigma'} \approx V_{\sigma\sigma'} < 0$ because in this doping region the Coulomb interaction V_{ij} can be neglected due to the screening effect of carriers. In (17) $b^+_{\mathbf{k}\sigma}$ or $c^+_{\mathbf{k}\sigma}$ fermion has the possibility of creating Cooper pairs, but $b^+_{\mathbf{k}\sigma}$ will contribute to the Cooper formation because the Fermi surface exists in only $b_{\mathbf{k}}^+$ band. Thus the BCS-like wave function is given by

$$|\Psi_S\rangle = \prod_{\mathbf{k}\sigma}(u_{\mathbf{k}} + v_{\mathbf{k}} b^+_{\mathbf{k}\uparrow} b^+_{-\mathbf{k}\downarrow})|0\rangle, \quad u_{\mathbf{k}}^2 + v_{\mathbf{k}}^2 = 1 \tag{18}$$

The ground-state energy is

$$E = \langle \Psi_S | H | \Psi_S \rangle = 2\sum_{\mathbf{k}} \varepsilon_{\mathbf{k}} v_{\mathbf{k}}^2 + N^{-1} \sum_{\mathbf{kk}'} \left[V \alpha_{\mathbf{k}} \alpha_{\mathbf{k}'} \beta_{\mathbf{k}} \beta_{\mathbf{k}'} \theta_{\mathbf{kk}'} u_{\mathbf{k}} v_{\mathbf{k}} u_{\mathbf{k}'} v_{\mathbf{k}'} \right] \tag{19}$$

where the relation $\varepsilon_{\mathbf{k}} = 0.5\{\varepsilon_p + [(1-0.5\delta)^2 \varepsilon^2 s^2(\mathbf{k}) + \varepsilon_p^2]^{1/2}\} - \varepsilon_F$ is defined, measuring the energy relative to the Fermi level ε_F. By minimizing E with respect to $u_{\mathbf{k}}$, $v_{\mathbf{k}}$, the gap equation is obtained as

$$\Delta_{\mathbf{k}} = -0.5 \sum_{\mathbf{k}'} \Delta_{\mathbf{k}'} V_{\mathbf{kk}'} (\Delta_{\mathbf{k}}^2 + \varepsilon_{\mathbf{k}}^2)^{-1/2}, \quad V_{\mathbf{kk}'} = N^{-1} V \alpha_{\mathbf{k}} \alpha_{\mathbf{k}'} \beta_{\mathbf{k}} \beta_{\mathbf{k}'} \theta_{\mathbf{kk}'} \tag{20}$$

Replacing the sum in gap equation by an integral, the solution which is even in **k** is given by $\Delta_{\mathbf{k}} = \Delta_0 \alpha_{\mathbf{k}} \beta_{\mathbf{k}}$ $(\cos k_x \pm \cos k_y)$. The solution decreasing the Coulomb interaction is $\Delta_{\mathbf{k}} \propto (\cos k_x - \cos k_y)$ which agrees with the experimental fact about the anisotropic gap of high-temperature cuprate superconductors [12]. Thus Δ_0 is determined by the relation

$$1 = -0.5\pi^{-2} V \int \alpha_{\mathbf{k}}^2 \beta_{\mathbf{k}}^2 (\cos k_x - \cos k_y) \cos k_x [\Delta_0^2 \alpha_{\mathbf{k}}^2 \beta_{\mathbf{k}}^2 (\cos k_x - \cos k_y)^2 + \varepsilon_{\mathbf{k}}^2]^{-1/2} d\mathbf{k} \tag{21}$$

Using several approximation on the band structure, the superconductive maximum gap energy and the ground-state energy are given by

$$\begin{aligned} \Delta_s &= 2\varepsilon_p \delta [1 + (1-0.5\delta)^2 \varepsilon^2 / \varepsilon_p^2]^{1/2} \exp(1/N_F V), \\ E_s &= N\delta\varepsilon_p - N\varepsilon_p \delta^2 [1 + (1-0.5\delta)^2 \varepsilon^2 / \varepsilon_p^2]^{1/2} \exp(2/N_F V) \end{aligned} \tag{22}$$

where N_F is the density of states at Fermi level. Denote that the energy relative to Cu 3*d* level is measured and the contribution of the $c_{\mathbf{k}\sigma}$ fermions to the total system energy is considered.

In order to evaluate the intermediate state between two doped regions, it is necessary to estimate the doping dependency of the free energy of the pseudogap and the superconductive states. First let us consider the *d-p* interaction V which is the variable of the free energy. The value of $V_{\sigma\sigma'}$ is proportional to ε^2 due to *d-p* anti-ferromagnetic exchange interaction. However, increasing of doping rate δ, ε will be effectively $\varepsilon(1-0.5\delta)$, because the hopping probability from Cu sites to O sites can be proportional to $(1-0.5\delta)$ due to exclusion principle. As a result, V can be approximated as $V = V_0 \varepsilon^2 \varepsilon_p^{-2} (1-0.5\delta)^2$. Thus, the free energy and gap energy of the pseudogap and superconductive states can be obtained as a function of δ. In those states, the interaction term of the free energy initially decreases with increasing doping, and passes a minimum point. Since these minimum points are generally different, it is plausible that there exists the mixed state in the intermediate region. Of course, there may be unknown states in the intermediate doped region. However, if the energy of those ground-states is smaller than the mixed states, they should be evidently detected in various experiments. There is so far no evidence about the existence of those intermediate states. Though those unknown states can contribute to the excited states, they could be at least neglected as the ground-state. This also agrees with the recent research in which the pseudogap state competes with the superconductivity [13]. Thus, an overall wavefunction is assumed to be

$$|\Psi(\delta)\rangle = f^{1/2}|\Psi_p(\delta_p)\rangle + (1-f)^{1/2}|\Psi_s(\delta_s)\rangle \quad (23)$$

where δ_p, δ_s are the doping quantity of PG and superconductive states and the total doping quantity$\delta$is defined to be $\delta = f\,\delta_p + (1-f)\,\delta_s$. The total free energy is then given by $G = f\,E_p(\delta_p) + (1-f)\,E_s(\delta_s)$, $(0 < f < 1)$ and the intermediate region is determined by minimizing G with respect to δ, δ_p, δs. Let us evaluate the doping quantity δ_{pm}, δ_{sm} by minimizing $E_p(\delta_p)$ and $E_s(\delta_s)$ and the minimum point of the total free energy in the condition of δ_p m < δ_{sm}. As the case of $\delta < \delta_{pm}$ means $\delta_p = \delta$, $\delta_s = 0$, the system will show a pure PG state. At $\delta = \delta_{pm}$ superconductivity appears in the PG state and the mixed state will be maintained until the PG state disappear at the doping quantity δ_{sm} satisfying $\delta = \delta_s$. In more increasing of δ there exists a pure superconductor. Thus, the overall wave function will be represented as three regions of $f = 1$ $(\delta < \delta_{pm})$, $f = (\delta_{sm}-\delta)(\delta_{sm}-\delta_{pm})^{-1}$ $(\delta_{pm} < \delta < \delta_{sm})$, and $f = 0$ $(\delta > \delta_{sm})$, respectively. For both regions of $\delta < \delta_{pm}$ and $\delta_{pm} < \delta < \delta_{sm}$, Δ_p and T_p will decrease with increase of δ and vanish at $\delta = \delta_{sm}$. Since the superconductive region appearing at $\delta = \delta_{pm}$ can be identified to be composed of very small superconductive region, the sufficient coherency to detect the net superconductivity will be less formed. Therefore, the superconductive state for $\delta_{pm} < \delta < \delta_{sm}$ could be given by the effective gap $\Delta_{eff} = (\delta-\delta_{pm})\Delta_s(\delta_s)(\delta_s-\delta_{pm})^{-1}$. This also indicates that the superfluid density δ_{sf} changes as $\delta-\delta_{pm}$ in the region of $\delta_{pm} < \delta < \delta_{sm}$. The superconductor will be the optimal doped state at the neighborhood of $\delta = \delta_{sm}$, and in the over-doped region Δ_s and T_c will decrease with δ according to doping dependency. Thus, an overall phase diagram is theoretically obtained by considering three types of electronic states.

4. Pnictide superconductors

The characteristics of pnictide superconductors is that it is multi orbitals, metal in non-doping, and superconductive in electron doping [14]. Since pnictides is metal in non-doping, the momentum representation is appropriate in the neighborhood of this doping region. Then, it seems that it isn't needed to consider PG state on the condition of electron doping. However, if hole doping can enhance the strongly correlation between fermions, PG state may appear in the suitable condition of hole doping. The BCS-like wave function, which is the ground state of (11), is given by

$$|\Psi_S\rangle = \prod_{\mathrm{k}\sigma}(u_{\mathrm{k}} + v_{\mathrm{k}} b^+_{\mathrm{k}\uparrow} b^+_{-\mathrm{k}\downarrow})|0\rangle, \quad u^2_{\mathrm{k}} + v^2_{\mathrm{k}} = 1 \quad (24)$$

Thus, the following gap equation is given by

$$\Delta_{\mathrm{k}n} = -0.5\sum_{\mathrm{k}'}\Delta_{\mathrm{k}'n}V_{\mathrm{kk}'n}(\Delta^2_{\mathrm{k}'n} + \varepsilon^2_{\mathrm{k}'n})^{-1/2}, \quad V_{\mathrm{kk}'n} = N^{-1}V_n\alpha_{\mathrm{k}n}\alpha_{\mathrm{k}'n}\beta_{\mathrm{k}n}\beta_{\mathrm{k}'n}\theta_n(\mathrm{k},\mathrm{k}') \quad (25)$$

Since Fe-3d electrons are multi orbitals, it is needed to select the dominant orbitals of Fe contributing to superconductivity [15]. The orbitals of d_{xz}, d_{yz} that have the strong hopping transition to the nearest Pn sites and intersect the Fermi surface are considered here. First let us consider the symmetry of superconductive paring wavefunctions [16]. The gap equation (25) is then considered in the case of two only orbitals d_{xz}, d_{yz}. For simplicity single p-orbital will be only considered. Considering that the square lattice of Fe^{2+} is tetrahedrally coordinated with Pn anions $s_{xz}(\mathbf{k}), s_{yz}(\mathbf{k}), \theta_{xz}(\mathbf{k},\mathbf{k}')$ and $\theta_{yz}(\mathbf{k},\mathbf{k}')$ are calculated as $s_{xz}(\mathbf{k}) = \cos k_x$, $s_{yz}(\mathbf{k}) = \cos k_y$, $\theta_{xz}(\mathbf{k},\mathbf{k}') = \cos(k_x - k'_x)$, $\theta_{yz}(\mathbf{k},\mathbf{k}') = \cos(k_y - k'_y)$ where the half of Fe-Fe distance is used as the length unit. In the case of d_{xz} $(n = 1)$, replacing the sum in (25) by an integral, the solution which is even in $\mathbf{k}$ is given by $\Delta_{\mathrm{k}1} = \Delta_0\alpha_{\mathrm{k}1}\beta_{\mathrm{k}1}\cos k_x$. Then, Δ_0 is then determined by the relation

$$1=-0.5\pi^{-2}V_1\int\alpha_{\mathrm{k}1}^2\beta_{\mathrm{k}1}^2\cos^2k_x(\Delta_0^2\alpha_{\mathrm{k}1}^2\beta_{\mathrm{k}1}^2\cos^2k_x+\varepsilon_{\mathrm{k}1}^2)^{-1/2})d\mathrm{k} \tag{26}$$

Eq. (26) is also represented as $1=-\pi^{-2}lV_1\int_0^{E_0}\alpha_{\mathrm{k}1}^2\beta_{\mathrm{k}1}^2\cos^2k_x(\Delta_0^2\alpha_{\mathrm{k}1}^2\beta_{\mathrm{k}1}^2\cos^2k_x+\varepsilon_{\mathrm{k}1}^2)^{-1/2}N_F(\varepsilon_{\mathrm{k}1})d\varepsilon_{\mathrm{k}1}$. Here l is the length of Fermi line in Brillouin zone, E_0 is the energy width contributing to the Cooper formation, and N_F is the density of states at Fermi surface. The gap energy $\Delta_{\mathrm{k}1}$ is then obtained by solving the gap equation regarding Δ_0. In the case of d_{yz} (n = 2), using the transformations of $k_x \to k_y$, $V_1 \to V_2$, $\alpha_{\mathrm{k}1} \to \alpha_{\mathrm{k}2}$, $\beta_{\mathrm{k}1} \to \beta_{\mathrm{k}2}$, the similar gap equation is obtained. Since the band of d_{xz}, d_{yz} are almost the same configuration, the value of ε_1 will be equal to that of ε_2. These conditions mean the equivalency of $V_n, \alpha_{\mathrm{k}n}, \beta_{\mathrm{k}n}, b_{\mathrm{k}n\sigma}^+, c_{\mathrm{k}n\sigma}^+$ *etc* and indicate that the total gap energy is presented as $\Delta_{\mathrm{k}}=\Delta_{\mathrm{k}1}+\Delta_{\mathrm{k}2}\propto\cos k_x+\cos k_y$. Notice that the total gap energy is given by $\Delta_{\mathrm{k}}=\sum_n\Delta_{\mathrm{k}n}$ because the gap equations can be separated in the ground state energy. It is thus concluded that the superconductive paring of Fe-Pn is *s*-like state different from the *d*-state of cuprates. Also, it is suggested that d_{xz}, d_{yz} orbitals contribute to superconductivity in a cooperating way. However, if the equivalency of these orbitals is broken, it may lead to the different pair symmetry due to the modification such as the change of the crystal structure. Second let us consider the structural dependency of the critical temperature T_c. It is well known as an empirical rule that the optimal T_c is obtained when the bond angle of Pn-Fe-Pn approaches $109°5'$ of a regular tetrahedron [17,18]. Surprisingly, the superconductivity of Fe-Pn is very sensitive to the change from the regular tetragonal structure of Pn anions. This suggests that Fe-Pn bonding state is strongly correlated with superconductive properties. Since the crystal structure is determined by the state with the maximum atomic bonding energy, the regular tetrahedron indicates that the Fe-Pn bonding energy is in the maximum state. On the other hand, according to this theory, the superconductivity is determined by the state of the Fe-Pn composite system and the antiferromagnetic interaction forming Cooper pairs is proportional to ε_n^2 that is the square of Fe-Pn hopping integral. The Fe-Pn bonding energy will be approximately proportional to the value of Fe-Pn hopping integral. Thus, the modification of the regular tetrahedron will weaken the superconductivity of Fe-Pn. Third let us discuss the doping dependency of the superconductive properties. Though some Fe-Pn materials exhibit low-temperature (T_c~4K) superconductivity without doping, the high T_c iron-based superconductors appear by a partial replacement of F^- or H^- on at the oxygen site [19,20]. In general these iron-based superconductors indicate the phase diagram that the T_c appears when the antiferromagnetism disappear or weaken, and disappear with increasing of doping level. This phase diagram can be explained in a similar way to the case of cuprates. In non-doping state, Fe ions are the superexchange antiferromagnetic state through the sites of pnictides that are non-magnetic materials (closed shell). However, it is recognized that there is the change from the closed shell state related to the mechanism of superexchange antiferromagnetism. This means that the doped electrons will be supplied in Fe-3d sites and the partial empty sites of pnictide. Therefore, in the similar situation with the case of cuprates, the doping will create the antiferromagnetic interaction contributing the superconductive state and reversely will weaken the superexchange antiferromagnetic interaction between Fe ions.

5. Conclusions

In order to find the unified theory to explain generally the properties of high-temperature superconductors such as cuprates and iron pnictides, a theory of high-temperature superconductivity in strongly correlated fermions system is derived from an extended *d-p* model reconstructed with unitary-transformed fermions. It is indicated that the effective interaction between these newly defined fermions can determine the pseudogap and the superconductive states in cuprates and pnictides superconductors. It is also found that the theory can explain the various properties such as pseudogap energy, superconductive gap energy, critical temperature, superconductive anisotropy, and doping dependency and structural dependency of the electronic states in those superconductors. The important conclusion is that the pseudogap state and the Cooper pair formation in high-temperature superconductors could be only caused by the interplay between the *d-p* hybridization and the *d-p* exchange antiferromagnetic

IOP Conf. Series: Journal of Physics: Conf. Series **871** (2017) 012033 doi:10.1088/1742-6596/871/1/012033

interaction. This also suggests that the new selection of composite fermions may open an alternative root to improve the critical temperature of superconductors.

Acknowledgments
The author thanks the Yukawa Institute for Theoretical Physics at Kyoto University, where this work was initiated during the YITP-W-16-10 on "Recent topics and developments on superconductors"

References

[1] Anderson P W 1987 *Science* **235** 1196
[2] Emery V J 1987 *Phys. Rev. Lett.* **58** 2794
[3] Zhang F C and Rice T M 1988 *Phys. Rev.* **B 37** 3759
[4] Massidda J Yu and Freeman A J 1988 *Physica* **C152** 273
[5] Varrma C M, Littlewood P B, Schmit-Rink S, Abrahams E and Ruckenstein A E 1989 *Phys. Rev. Lett.* **63** 1996
[6] Monthoux P, Balatsky A V and Pines D 1991 *Phys. Rev. Lett.* **67** 3448
[7] Alexandrov A S and Mott N F 1993 *Supercond. Sci. Technol.* **6** 215
[8] Leggett A J 2006 *Nat. Phys.* **2** 134
[9] Nishi K 2012 *Phys. Proc.* **27** 80; 2014 *Phys. Proc.* **58** 18; 2016 *Phys. Proc.* **81** 17
[10] Vaknin D *et al.* 1987 *Phys. Rev. Lett.* **58** 2802
[11] Shirane G *et al.* 1987 *Phys. Rev. Lett.* **59** 1613
[12] Tsuei C C and Kirtley J R 2000 *Rev. Mod. Phys.***72** 969
[13] Kondo T, Khasanov R, Takeuchi T, Schmalian J and Kaminski A 2009 *Nature* **457** 296
[14] Hosono H and Kuroki K 2015 *Physica* **C514** 399
[15] Singh D J *et al.* 2008 *Phys. Rev. Lett.* **100** 237003
[16] Mazin I I *et al.* 2008 *Phys. Rev. Lett.* **101** 057003
[17] Lee C H *et al.* 2008 *J. Phys. Soc. Jpn.* **77** 083704
[18] Zhao J *et al.* 2008 *Nat. Mater.* **7** 953
[19] Kamihara Y *et al.* 2006 *J. Am. Chem. Soc.* **128** 10012
[20] Kamihara Y *et al.* 2008 *J. Am. Chem. Soc.* **130** 3296

Carrier Density Dependence of the Critical Temperature in $BaHfO_3$ doped $SmBa_2Cu_3O_y$ Films

S. Sato, Y. Tsuchiya, Y. Ichino and Y. Yoshida

Department of Energy Engineering and Science, Nagoya Unviersity, Furo-cho, Chikusa-ku, Nagoya 464-8601 Japan

E-mail: sato-shun16@ees.nagoya-u.ac.jp

Abstract. $BaHfO_3$ (BHO) doped $SmBa_2Cu_3O_y$ (SmBCO) films show high critical current density (J_c) under magnetic fields. However, excessive addition of BHO causes degradation of superconducting properties. In this study, SmBCO films with different BHO contents were prepared, and the T_c, carrier density and *c*-axis length were measured. As a result, it was clarifiedthat as the BHO content increases, the *c*-axis length of SmBCO extends, and T_c and the carrier density decrease. In order to increase the carrier density, we ptimized the oxygen annealing condition. As a result, the T_c and the carrier density were increased by annealing at low temperature for a long time.

1. Introduction

Superconducting magnetic energy storage (SMES) and magnetic resonance imaging (MRI) are used under magnetic field, and the improvement of these performances inhigher magnetic fields is expected. However, in the superconductors, J_c decreases under magnetic fields. intorducing artificial pinning centers (APCs) in superconductors is one of the promising methods for improving J_c under magnetic field. $BaMO_3$ (BMO: M=Zr, Sn, Hf) forms nanorods in $REBa_2Cu_3O_y$ (REBCO: RE=Rare Earth) films fabricated by vapor phase epitaxy [1]. Doping BMO in REBCO films can suppress decrease of the J_c under magnetic fields because BMO nanorods behave as APCs. The number density of nanorods in REBCO films increase with increasing BHO content,and the irreversible magnetic field is expected to be improved. However, the excess BMO addition in REBCO films cause decrease of the T_c [2]. As the number density of BMO nanorods increases, the lattice strain increases, and the *c*-axis length of REBCO is extended [3]. The extension of *c*-axis length causes the oxygen deficiency [4].

In this study, we aimed to improve superconducting properties of REBCO films with a large amount of BMO at 77 K high field which can be obtained by cooling with liquid nitrogen. At first, wemeasuredthe T_c, *c*-axis length and carrier density of SmBCO films with different BHO contents. Then, in order to improve the T_c of the films, which contains large volume fraction of BHO of 15.3 vol.%, and we optimized the annealing condition such as annealing temperature and holding time.

2. Experimental details

We fabricated SmBCO films on $LaAlO_3(100)$ substrates by pulse laser deposition (PLD) method using a Nd:YAG pulsed laser. The O_2 partial pressure during the deposition, the distance between the substrate and the targets, and the laser energy density were 53 Pa, 42.5 mm and 2.0 J/cm^2, respectively. Alternating targets technique was used to introduce various contents of BHO of 0 ~ 15.2 vol.% into

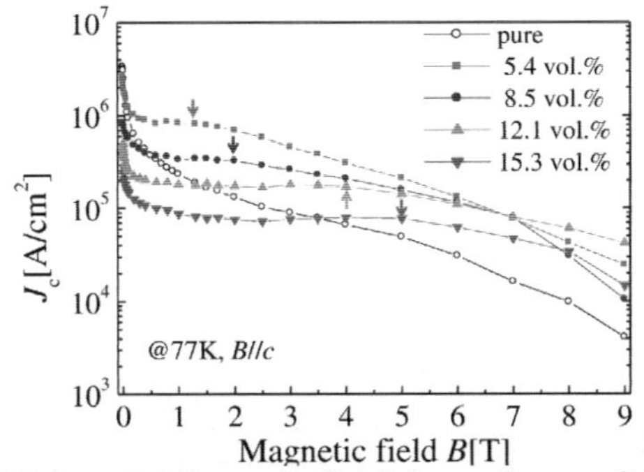

Figure 1. Magnetic field dependence of J_c at 77 K, *B*//*c* for BHO doped SmBCO and non-doped SmBCO films. The arrows indicates matching magnetic fields (B_ϕs).

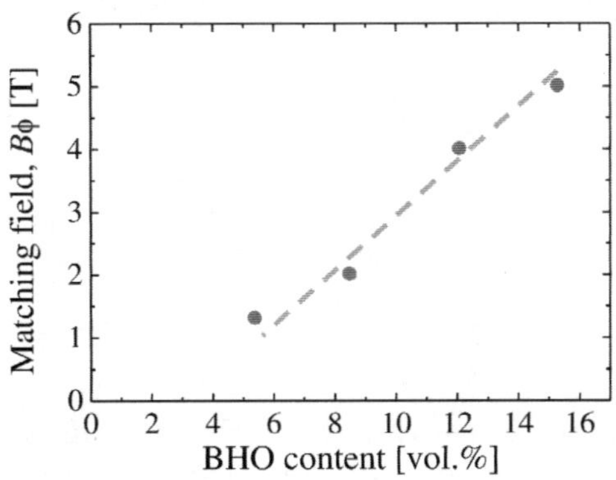

Figure 2. BHO content dependence of matching field.

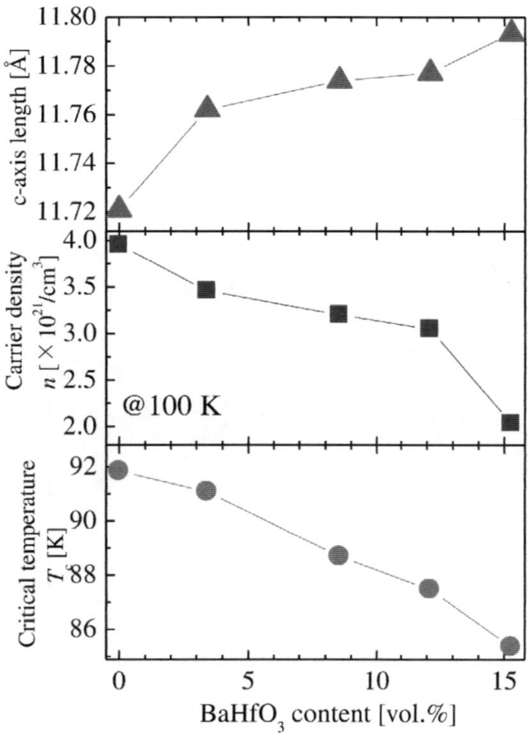

Figure 3. BHO content dependence of *c*-axis length, carrier density and T_c.

SmBCO films [3]. After the deposition, oxygen was introduced into the chamber, and the substrate was rapidly cooled down to room temperature. In order to supress The oxgen deficiency of the REBCO films, post annealing in an oxygen atmosphere was performed. The annealing temperature was varied from 150 to 450°C and the annealing time was examined from 1.5 to 6 hours.

The Crystal orientation of the films were determined by X-ray diffraction (XRD) method, and inductively coupled plasma (ICP) emission spectroscopy was used to determine the film thickness and the BHO content. The J_c and the T_c were measured using DC four-terminal method by physical properties measurement system (PPMS). The carrier density was calculated based on Hall coefficient obtained by Hall effect measurement at 100 K.

3. Results and discussion

3.1. Dependence of T_c and carrier density on BHO content

After deposition, SmBCO films with various BHO contents were post-annealed in an oxygen atmosphere at 350°C for 3 hours, which is an optimum post annealing condition for a pure SmBCO film. Figure 1 shows Magnetic field dependence of J_c at 77 K, *B*//*c* for BHO doped SmBCO and non-doped SmBCO films. The J_c decreased with increasing BHO content. Figure 2 shows BHO content dependence of matching field (B_ϕ). Here, the B_ϕ is defined by the end of the apparent magnetic field of the plateau in the J_c-*B* curve shown in Figure 1. It indicates that the number density of BHO nanorods increases for the higher BHO content because Bphi is propotional to nanorod density in the single vortex pinning state .

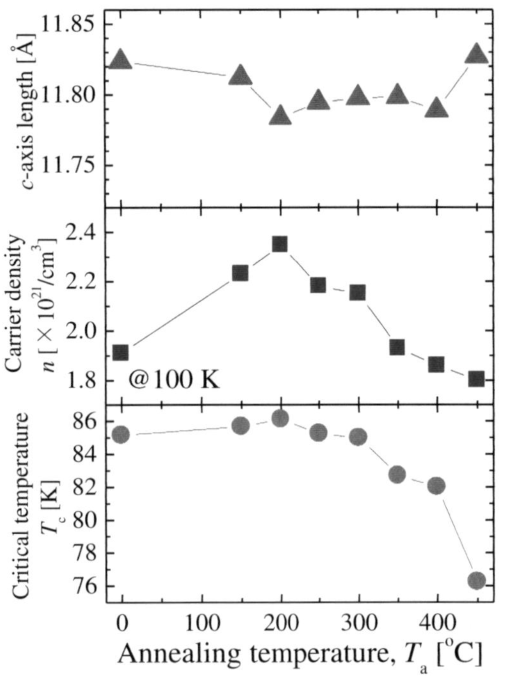

Figure 4. Annealing temperature dependence of *c*-axis length, carrier density and T_c in BHO 15.3 vol.% doped SmBCO films.

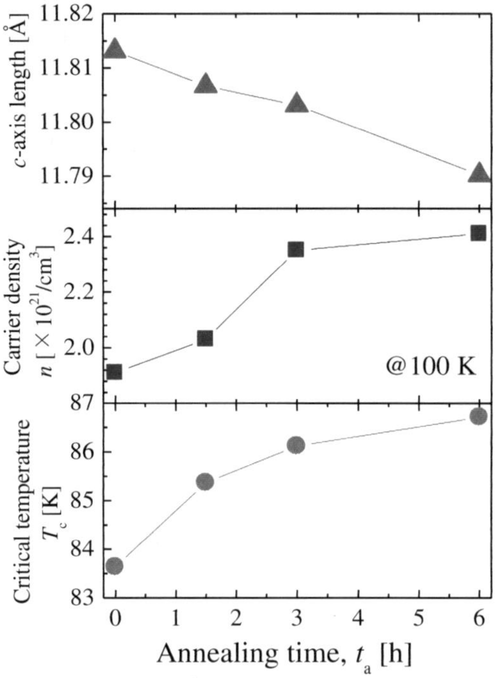

Figure 5. Annealing time dependence of *c*-axis length, carrier density and T_c in BHO 15.3 vol.% doped SmBCO films.

Figure 3 shows BHO content dependence of *c*-axis length, carrier density, and T_c. The carrier density varied in the range of 2.15×10^{21} /cm^3 to 3.95×10^{21} /cm^3, which is equivalent to the reported values of $YBa_2Cu_3O_y$ films [5,6]. Although B_ϕ increased with increasing BHO content, T_c decreased. In Figure 3, the carrier density decreases and the *c*-axis length is extended with decreasing the T_c. These indicate that BHO nanorods extend *c*-axis length of SmBCO by lattice strain, and extention of *c*-axis length form oxygen vacancies in SmBCO. As a result, we clarified that the lower density of the hole carrier suppresses the Tc in the SmBCO films with the higher BHO contents possibly due to interfacial strain between SmBCO and BHO.

3.2. Annealing temperature dependence of T_c and carrier density

The SmBCO films with the higher BHO contents show poor properties due to a large amount of oxygen deficiency. Therefore, in order to improve the superconducting properties, it is necessary to dope sufficient oxygen and increase the carrier density by optimizing post annealing condition in an oxygen atmosphere. Figure 4 shows annealing temperature (T_a) dependence of *c*-axis length, carrier density and T_c. The post annealing time (t_a) is fixed at 3 hours. T_c and carrier density increased with increasing T_a up to 200°C, and then monotonically decreased. The *c*-axis length showed the opposite trend. The optimum annealing condition of pure SmBCO film was 350°C, while the optimum annealing condition of the BHO 15.3 vol.% doped SmBCO film was 200°C. In BHO heavy doped SmBCO films, it is considered that oxygen is easy to desorb from SmBCO matrix at high T_a because the *c*-axis length of 15.3vol.% BHO doped film is elongated by lattice strain.

3.3. Annealing time dependence of T_c and carrier density

It is expected that oxygen annealing at low T_a needs a long time to reach the equilibrium oxygen content. To optimize t_a, the T_a was kept at 200°C, and t_a was varied from 0 to 6 hours. Figure 5 shows t_a dependence of *c*-axis length, carrier density and T_c of SmBCO films with 15.3 vol.% BHO. As the annealing time increases, T_c and carrier density increase and the *c*-axis length become shorter. As a result, the T_c of the SmBCO film with BHO of 15.3 vol.% was improved from 85.4 K to 86.7 K.

4. Summary

In this study, we found T_c decreased because of oxygen deficiency by measurering *c*-axis length, carrier density and T_c, therefore we attempted to increase T_c by optimizing annealing conditions of BHO heavy doped SmBCO films. The optimum annealing condition of BHO heavy doped SmBCO thin films was different from that of pure SmBCO thin films, and annealing at a low temperature for a long time was necessary compared with non-doped SmBCO thin film. In addition, it is suggested that higher T_c can be obtained by performing further hole doping.

Acknowledgement
This work was partly supported by a Grant-in-Aid for Scientific Research (15H04252, 15K14301, 15K14302 and 16H04512), JST-ALCA and NU-AIST alliance project

References

[1] MacManus-Doriscoll J L, Foltyn Q X, Wang H, Serquis A, Civale L, Maiorov B, Hawley M E, Maley M P and Peterson D E 2004 *Nature Materials* **3** 439
[2] Tsuruta A, Yoshida Y, Ichino Y, Miura S, Ichinose A and Matsumoto K 2014 *J. Cryo. Super. Soc. Jpn.* **49** 117
[3] Tsuruta A, Yoshida Y, Ichino Y, Ichinose A, Matsumoto K, and Awaji S 2014 *Supercond. Sci.* **27** 065001
[4] Osuquiguil E, Menhoudt M, Wuyts B and Bruynseraede Y 1992 *Appl. Phys. Lett.* **60** 1627
[5] Abe Y, Segawa K and Ando Y 1999 *Phys. Rev. B* **60** R05155
[6] Carrington A, Walker J C, Mackenzie A P and Cooper J R 1993 *Phys. Rev. B* **48** 13051

IOP Conf. Series: Journal of Physics: Conf. Series **871** (2017) 012035 doi:10.1088/1742-6596/871/1/012035

Effect of a superconducting coil as a fault current limiter on current density distribution in BSCCO tape after an over-current pulse

M Tallouli[1], O. Shyshkin[2], S. Yamaguchi[1]
[1]Center of Applied Superconductivity and Sustainable Energy Research (CASER), Chubu University, Matsumoto-cho 1200, Kasugai, Aichi 487-8501, Japan
[2]V. N. Karazin Kharkiv National University, 4 Svobody Sq., Kharkiv, 610022, Ukraine
E-mail: yamax@isc.chubu.ac.jp

Abstract. The development of power transmission lines based on long-length high temperature superconducting (HTS) tapes is complicated and technically challenging task. A serious problem for transmission line operation could become HTS power cable damage due to over-current pulse conditions. To avoid the cable damage in any urgent case the superconducting coil technology, i.e. superconductor fault current limiter (SFCL) is required. Comprehensive understanding of the current density characteristics of HTS tapes in both cases, either after pure over-current pulse or after over-current pulse limited by SFCL, is needed to restart or to continue the operation of the power transmission line. Moreover, current density distribution along and across the HTS tape provides us with the sufficient information about the quality of the tape performance in different current feeding regimes. In present paper we examine BSCCO HTS tape under two current feeding regimes. The first one is 100A feeding preceded by 900A over-current pulse. In this case none of tape protection was used. The second scenario is similar to the fist one but SFCL is used to limit an over-current value. For both scenarios after the pulse is gone and the current feeding is set up at 100A we scan magnetic field above the tape by means of Hall probe sensor. Then the feeding is turned of and the magnetic field scanning is repeated. Using the inverse problem numerical solver we calculate the corresponding direct and permanent current density distributions during the feeding and after switch off. It is demonstrated that in the absence of SFCL the current distribution is highly peaked at the tape center. At the same time the current distribution in the experiment with SFCL is similar to that observed under normal current feeding condition. The current peaking in the first case is explained by the effect of an opposite electric field induced at the tape edges during the over-current pulse decay, and by degradation of superconductivity at the edges due to penetration of magnetic field in superconducting core during the pulse.

1. Introduction

The next advance in world electric power supply systems is the use of long-length HTS tapes in electric power transmission cables to transport large DC currents. Long-length cables, in this way, are supposed to suffer the extreme environment and hence the fault conditions are expectable [1]. Due to human factor and technical errors during the manufacturing and maintenance of power transmission line the current transport properties of the line could be affected as well. One of the critical events that could happen is the high over-current pulse, as a result of short circuit, with current value exceeding the critical current of HTS tapes [2, 3]. In this case the operation of the line should be either stopped and then restarted or continued if any protection was used. In this case the complete knowledge of current redistribution in HTS tapes is of high importance to undertake the future actions.

In our study we put BSCCO HTS tape in extreme operating regime, i.e. we apply the over-current pulse reaching 900A current value with the duration of 5ms. After that the current feeding is set up at 100A transport DC for 1 hour. We consider two different operating scenarios, one is the over-current pulse acting on the unprotected HTS tape, and the other is the same pulse acting on the tape protected by means of superconducting coil technique SFCL.

After the pulse is gone and the transport current is set up to its constant value of 100A we provide magnetic field scanning above the tape with use of Hall probe. When the scanning is finished we turn of the power supply simulating the urgent shut down of power transmission line and repeat the scanning of the magnetic field again. In order to obtain current distribution in HTS tape we developed MATLAB code that solves inverse problem and calculates the current density distribution from the magnetic field dataset [4-6]. The detailed analysis of the current distribution demonstrates the essential difference of the current flow in unprotected tape and HTS tape protected by SFCL.

2. Experiment

2.1. Experiment set-up

The experimental setup consists of an open cryostat in which a Hall probe and scanning system, HTS tape and fibre reinforced plastic (FRP) plate as shown figure 1(a) are mounted. For the experiment we use the BSCCO HTS tape [6] with I_c=200 A critical current fixed on the FRP plate. The Hall probe (HHP-VP, AREPOC s.r.o) has an active area of 0.05 x 0.05 mm^2 and a sensitivity of 49.7 mV/T at 10 mA with the sampling time of 1 second. It is positioned at 200 μm above the tape surface and a stepping motor with the accuracy 0.05 ±0.0025 mm per step controls the Hall probe position across the tape. Both the HTS tape and the Hall probe sensor are directly cooled using liquid nitrogen. On Fig. 1(b) the geometry of experimental configuration with the Hall probe, HTS tape and the current direction are shown. The current is running in the X direction along the tape, and the Hall probe moves in Y direction across the tape. The Hall probe measures vertical B_z component of the self-magnetic field. Usually during the experiment the Hall probe is positioned at the middle of the HTS tape in X direction [7].

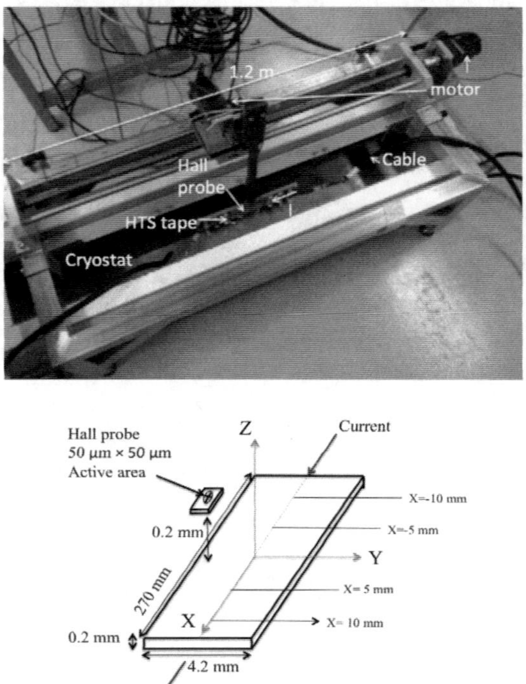

Figure. 1. (a) Photo of Experimental device. (b) Experimental configuration.

The electric circuit of the experiment is shown in figure 2. A DC power supply (DCPS) has its own switch and current controller, and the maximum DC current of the DCPS is 100 A. DCPS is connected to a shunt resistor and current transducer (CT) to measure the current. A mechanical switch is connected to the circuit, and it is a no-fuse-breaker (NFB) switch. The electric circuit can provide current feed operations simulating the over-current pulse operation and we have a possibility to use a superconducting coil to protect the circuit by smoothing the over-current pulse. On figure 3 the SFCL is presented.

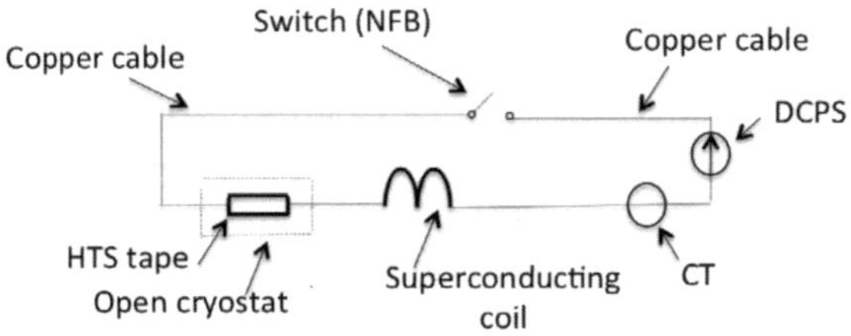

Figure 2. Electric circuit used in the experiment.

The SFCL used in our experiment is made by a 27.6 m long of BSCCO wire with Ic=160 A. SFCL inductance is 0. 67 mH. The coil is fixed inside a cryostat holder filled with liquid nitrogen. We use pomp to pomp out the air inside the holder. This operation changes the nitrogen state from liquid to solid nitrogen. The superconductor coil then can support the heat generated from the short over current pulse and protect the HTS wire.

The over current pulse in very beginning generates an inductance, which by it's turn generates number of oscillations in the electric circuit. If we take into account the inside value of the DCPS we can compare our circuit to a normal LCR circuit. The oscillations period is then proportional to the SFCL inductance combined with the inside value of the DCPS capacitance.

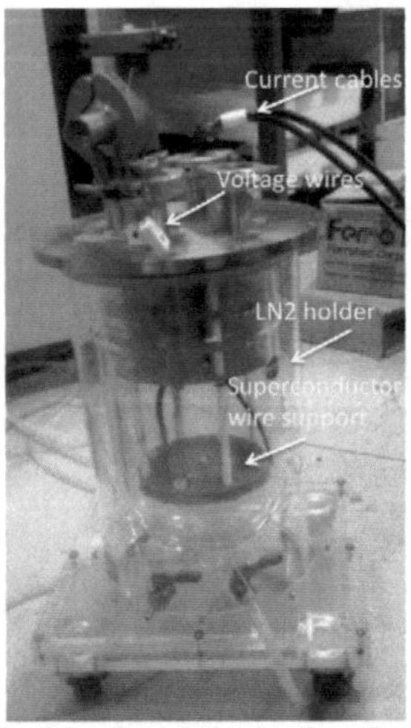

Figure. 3. Photo of the Superconductor Fault Current Limiter (SFCL) device

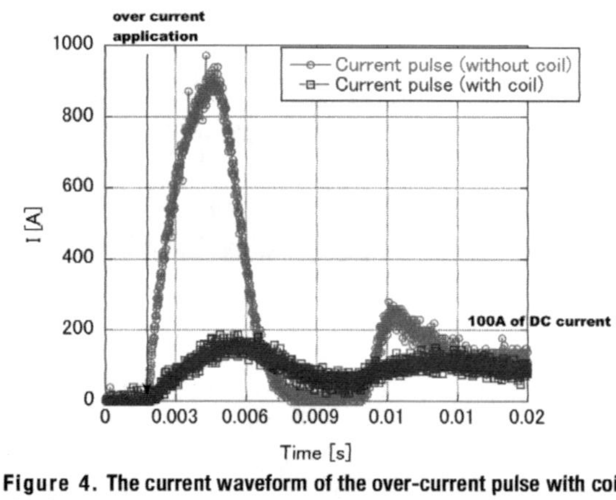

Figure 4. The current waveform of the over-current pulse with coil circuit.

2.2. Over-current pulse operation

This operation simulates the fault current condition. At first the NFB is open, the DCPS is set to 100 A where its voltage is maximum at 5 V according to the power supply controller. When the NFB is switched on, the zero resistivity of the HTS tape allows an over-current pulse to reach 900 A for about 5 ms, then the current is regulated at 100 A by the DCSP after some oscillations within 0.1 s. This evolution of the current is shown on Fig. 4 with red color. The accuracy of the current controller is better than 0.1%. HTS tapes are maintained in liquid nitrogen temperature, important to cool tapes after pulse, by continuously refilling the open cryostat. The peak current is ~900A, whereas the critical current of the BSCCO tape is 200 A. Therefore, the HTS tape matrix, which is copper and silver with resistivity ~$10^{-9}\Omega$m at liquid nitrogen temperature, carries the majority of current during the pulse [8].

2.3. Normal operation

In order to study the effect of SFCL on the experimental circuit and on the HTS tape performance we add to a circuit a superconducting coil of 0.6 mH shown in figure 3. Then we reproduce the same experiment of the over-current pulse. The oscillations of the feeding current during the first milliseconds is this case are shown on figure 4 with blue colour. As one can see instead of high over-current peak the oscillations of the current around 100 A value and further stabilization to a constant value of 100 A take place. Hence adding the superconducting coil to the circuit eliminates the over-current pulse. The current feeding in this case is equivalent to that is a normal current feeding when the transport current is rising to the value 100A by 1A/s steps.

3. Experimental results and Analysis

3.1. Self-magnetic field and permanent magnetic field measurement

The measurements of magnetic field induced by the transport current above the tape are performed for 100A constant current feeding. The Hall probe is moved across the tape in Y direction and along the tape in X direction in order to obtain 2D magnetic field map above the curtain segment of HTS tape. Then we cut the current completely and we repeat measurements for the magnetic field induced by permanent current. The knowledge of the permanent current value gives us the information about the power stored in the transmission line after the shut down procedure.

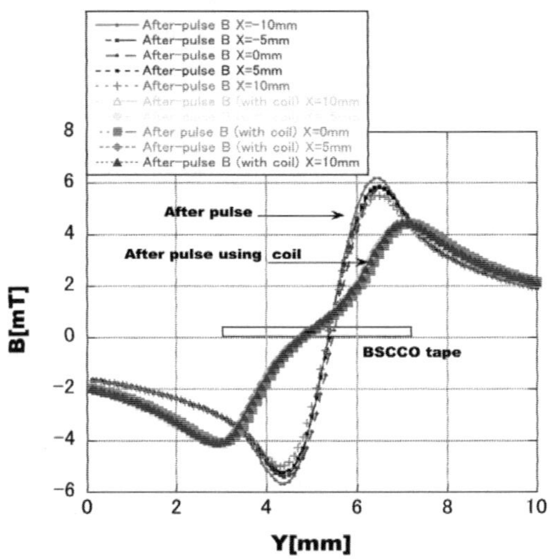

Figure. 5. Self-Magnetic field profiles of the BSCCO tape after over-current pulse (with and without coil).

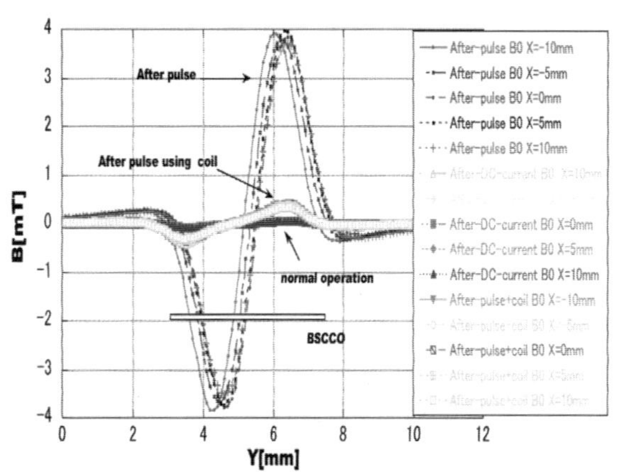

Figure. 6. Permanent magnetic field (B0) profiles of the BSCCO tape of normal (with and without coil) and over-current operations.

On figure 5 the vertical components B_z of self-magnetic field after the over-current pulse passed are demonstrated. Two sets of curves are related to the cases with and without over-current coil application. In case when SFCL is not used the magnetic field maximum and minimum absolute values are higher then in case of application of SFCL. Moreover, the peaks are shifted towards the tape center that is evidence of peaked current density profile.

The magnetic field profile related to the SFCL application reflects the fact that current density profile has a flattening and expanded through the tape width. Also we can see that the magnetic field is less homogeneous along the X direction for the pulse-affected tape then for the case of coil application. We predict the over-current pulse effect on the current density homogeneity inside the BSCCO tape.

The figure 6 shows the permanent magnetic field B_z measurements. Even the over-current pulse duration of 5 ms we can see that it has an effect on the permanent magnetic field after cutting the feeding current. The permanent magnetic field is also almost 8 times higher after the over-current pulse without SFCL protection. The effect of the superconducting coil to eliminate the over-current pulse and it's effect on current flow in HTS tape is obvious that is also confirmed after cutting the feeding current.

3.2. Current density calculation

In order to get the current density profile of the HTS tape from the magnetic field dataset, we have to solve the inverse problem. In order to do that the MATLAB code with several calculation methods was developed. It is described in details in the references [5], [9]. Here we briefly outline that we use the solver for the set of linear equations that are obtained from the Biot-Savart law. And for the solution we use Cramer's rule.

We use 35 data points of the self-magnetic field and solve for 35 points of current density, assuming the self-magnetic field is constant in time. The measurement time of these 35 data points of the self-magnetic field is 2 minutes; therefore the total current changes by 1.65 % during the measurement is acceptable error. If we assumed that the self-magnetic field profile is constant to solve the equation, the calculation result would be within the measurement error. For solving the inverse problem, we also used a constraint that the total current is 100 A.

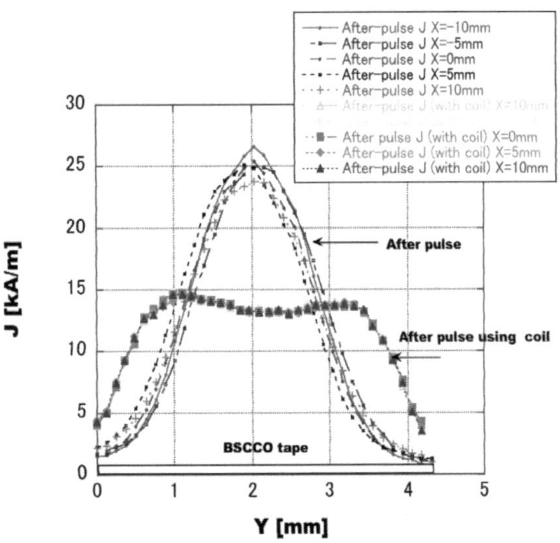

Figure 7. Current density profiles of the BSCCO tape showing the effect of the coil on direct current density.

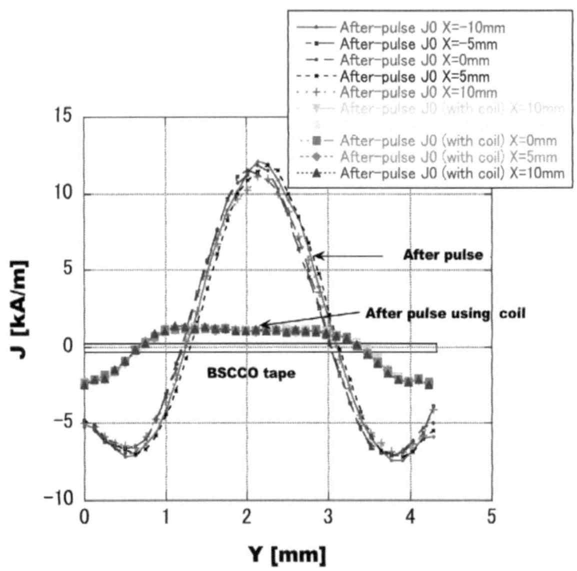

Figure 8. Current density profiles of the BSCCO tape showing the effect of the coil on permanent current density.

The transport current density profiles of the BSCCO tape for both cases with SFCL and without SFCL are shown in figure 7. As shown in the figure the current density after the over-current pulse is peaked at the center. The current is almost conducted by the center of superconducting filaments. The effect of the over-current pulse could be explained by an opposite electric field that is inducted near the tape' edges during the pulse decay phase. Additionally high current causes magnetic field penetration in the superconducting core of the tape that resulting in superconductivity degradation at the edges. That is the explanation for the peaked current profiles [8].

After applying the superconducting coil is to eliminate high over-current pulse the effect of the pulse as opposite electric field and superconductivity degradation on the edges is eliminated as well. The permanent current density profiles are presented in figure 8. As one can see the current stored in HTS tape after switching of the transport current is much larger in case of over-current pulse effect. The center positive and edge negative current values are almost 10 times higher then in case of eliminated over-current pulse. This is important fact that should be taken into account solving the problem of the restarting power transmission line after emergency shut down. The negative current density at the edges is much more important also for the after pulse phase however this considerably eliminated after adding the over-current pulse and the permanent current is then distributed along all the BSCCO tape.

The inhomogeneity of the magnetic field and also the permanent magnetic field are are in agreement with the calculation of the direct and permanent current densities as shown in figure 7 and figure 8. The distribution of the current density in the BSCCO superconducting filament is not equal that leads to different of the current density distribution in the BSCCO tape along the tape's length. However for the normal operation or by using the superconducting coil to eliminate the over-current effect on the tape we have a homogenous current density distribution.

The electric circuit used in this paper simulated also the basic circuit of the current power coverture [10]. The inductor component acts, in such circuits, as a current source, discharging current (which can be relatively high) that flows through the circuit as the magnetic flux collapses and protect then the circuit from over-currents. Current power coverture is characterised then by a high inductance. Our superconductor coil, which has also a high inductance, can be effective in such role.

4. Conclusion

The superconductor coil was used to eliminate the over-current pulse and hence to protect the electric circuit. Even that the critical current of the coil's tape is 160A, our experiment show that this technique of SFCL has the effect to delete the pulse. At the very beginning of the pulse the induction of the coil has an effect eliminating the opposite current, and even that the current shows some oscillations in the electric circuit, current densities show homogeneous and uniform current density distributions. These characteristics have a direct and important effect on the DC current distribution on superconductor cables transporting DC current. Also as we know the permanent current also has a result on different phase of DC current feeding. The use of SFCL as we show in our paper helps eliminating any effect of the over-current pulses on the permanent current distribution and then on the next current feeding phases.

5. References

[1] Yamaguchi S, Ivanov Y, Sun J Watanabe H, Hamabe 2011 *Physica C.* **36** 1131
[2] Kim J H, Park M, Ali M H, Cho J, Kim S 2008 *IEEE. Trans. Appl. Supercond.* **18** 1139
[3] Ishigohka T, and Kurahashi 2002 *Elect. Eng in Japan.* **141** 34
[4] Usak P 1999 *Physica C.* **316** 229

[5] Shyshkin O A, Kazarinov Y G, Tallouli M, Fumakinwa T and Yamaguchi S 2016 *IEEE. Trans. Appl. Supercond.* **26** 9000404
[6] Ayai N, Ymazaki K Kikuchi M Osabe and G, Takaaze 2009 *IEEE. Trans. Appl. Supercond.* **19** 3014
[7] Tallouli M, Sun J, Ninomia A, Hamabe A, Hamabe M, Watanabe H, Chikumot, Charfi S K and Yamaguchi S 2015 *IEEE. Trans. Appl. Supercond.* **25** 8000704
[8] Tallouli M, Sun J, Chikumoto N, Otabe E S, Shyshkin O, Kaddour S C and Yamaguchi S 2016 *Cryogenics.* **77** 53
[9] Tallouli M, Sun J, Shyshkin O, Hamabe M, Watanabe H, Chikumoto N, Kaddour S C, Yamaguchi S 2016 *IEEE. Trans. Appl. Supercond.* **26** 8001505
[10] Wu Bin, Pont J, Rodriguez J, Bernet S and Kourou S 2008 *IEEE. Trans. Appl. Supercond.* **55** 2786

Acknowledgments
The authors acknowledge Mr. K. Yoshimura, Mr. Y. Iwata and Mr. Sh. Ooshima for preparing the experiment. The authors also thank Prof. A. Iiyoshi, the Board Chairperson of Chubu Univ. for his continuous encouragement through the study.

Epitaxial growth of superconducting MgB_2 thin films with a Mg buffer layer at 110 °C

Hiroaki Shishido[1,2]*, Takatoshi Nakagami[1], Takuya Yoshida[1], and Takekazu Ishida[1,2]

[1]Department of Physics and Electronics, Graduate School of Engineering, Osaka Prefecture University, Sakai, Osaka 599-8531, Japan

[2]Institute for Nanofabrication Research, Osaka Prefecture University, Sakai, Osaka 599-8531, Japan

E-mail: shishido@pe.osakafu-u.ac.jp

Abstract. Since the discovery of MgB_2, its application to superconducting electronics has been limited by the absent of proper microfabrication techniques. In this study, we grew crystalline MgB_2 thin films using molecular beam epitaxy at a low substrate temperature of 110 °C under ultra-high vacuum of about 10^{-6} Pa. MgB_2 thin films were deposited with an epitaxial Mg buffer layer on *c*-plane 4H-SiC or sapphire substrates. In spite of the low growth temperature, superior crystallinity and surface flatness were confirmed by in situ reflection high-energy electron diffraction and X-ray diffraction measurements. Moreover, we successfully confirmed the occurrence of a sharp superconducting transition at 27 K.

The present growth temperature was lower than any in prior reports on superconducting MgB_2 thin films, and is lower than the applicable temperature of an organic-based lift-off resist. Our new MgB_2 thin film growth process is promising for the development of an alternative nanofabrication technique for MgB_2 thin films by means of a standard lift-off process with an organic resist.

1. Introduction

MgB_2 has been expected to apply to various superconducting devices because of its remarkably high superconducting critical temperature $T_c = 39\,K$[1], small penetration depth $\lambda \simeq 40\,nm$[2] and extremely short electron-phonon scattering time $\tau_{e-ph} \simeq 3\,ps$[3]. Superconducting devices made of MgB_2 thin films have been limited by the absence of convenient nanofabrication processes partly due to higher growing temperature of high quality MgB_2 thin films. Several research groups have developed techniques for the growth of high quality MgB_2 film, such as the two step method[4], hybrid physical chemical vapor deposition[5] and sputtering[6]. To fabricate a superconducting thin film, these methods require elevated substrate temperatures at above 700 °C during the deposition or a post-deposition annealing process at temperatures as high as 600 °C–800 °C. High temperature procedures are, however, not preferential for achieving a sharp interface between the superconducting MgB_2 and insulator layers in multilayer superconducting devices. Thus, techniques for as-grown synthesis based on low temperature processes have been developed based on the molecular beam epitaxy (MBE)[7, 8, 9, 10], sputtering[11], and co-evaporation[12]. Some studies have attempted to perform a low temperature growth using epitaxial buffer layers with good lattice matching such as AlN (1.9 %)[13], TiZr (3.6 %)[14], and

Ti (4.1 %)[15]. The in-plane axis of MgB_2 tends to align with that of the buffer layer, while the substrate temperature remains low during the entire process. In particular, among these bilayer films, MgB_2/Ti films, that showed a sharp superconducting transition at $T_c = 36$ K, were obtained by the growth at the lowest temperature of 200 °C[15]. This temperature is higher than the applicability limit of the temperature for a standard organic resist. Nanostructures can be fabricated via a conventional lift-off process with an organic resist only at temperatures lower than 150 °C. Therefore, a lift-off-like process using an inorganic material as a resist[16], Ar ion milling[17], and focused ion beam (FIB) milling[18] has been applied to the fabrication of MgB_2 nano-wire structures. However, the physical dry etching process often causes severe damage to the superconducting film.

We have developed the fabrication method of epitaxial MgB_2 thin films at a significantly low substrate temperature of 110 °C by the MBE method[19]. In this study, we describe the fabrication of epitaxial MgB_2 thin films with epitaxial Mg buffer layer on 4H-SiC or sapphire substrates. We note that the substrate temperature was obtained by radiation heating from the Knudsen cells (K cells) in the chamber. Mg has a hexagonal crystal structure with the lattice constants $a = 3.2125$ Å and $c = 5.2132$ Å. The lattice mismatch between Mg and MgB_2 is thus 4.1 %, and is the largest compared with the mismatch values for AlN, TiZr, and Ti. Although good lattice matching is generally effective for promoting epitaxial growth, we note that the wetting at interfaces between the buffer layers and the MgB_2 layer is also important for achieving the epitaxial growth. One can expect good wetting at an interface between Mg and MgB_2 layers. An as-grown film showed a sharp superconducting transition at $T_c = 27$ K.

2. Experimental method

Thin films of MgB_2 were grown by the MBE method. The pressure during the growth process was about 2×10^{-6} Pa. The elementary materials Mg and B were deposited by the K cells. The temperature of each K cell was stabilized at the set point within ±0.1 °C during deposition. This ensured the high stability of each deposition rate for the whole duration of the film deposition. The (001) surface of a 4H-SiC or sapphire was used as a substrate. The substrate temperature was determined to remain at 110°C within the accuracy of 5°C by the several irreversible thermo-labels placed on top surface of the substrate holder. It increased due to the radiation heating from the K cells while we did not use the substrate heater for intentional heating. The deposition rate was monitored by a quartz oscillating monitor located below the substrate. We evaluated a deposition rate by monitoring a resonant frequency change at intervals of 1 minute to recover a very low rate. We found that the deposition rate thus obtained was well stabilized within ±0.014 Å/min with respect to the deposition rate of 0.7 Å/min. The Mg buffer layer was deposited at the typical deposition rate of 3 Å/min to form a 20 to 30-nm-thick film. After the deposition of the Mg buffer layer, the MgB_2 layer was deposited by the co-evaporation of Mg and B. The deposition-rate ratio between Mg and B was 1:2, which is the same as the MgB_2 stoichiometric ratio. This is in marked contrast with the preceding reports on MgB_2 films, where excess Mg was supposed to be necessary to grow MgB_2 films. The typical deposition rate and film thickness of the MgB_2 layer were 0.7 Å/min and 60 to 100 nm, respectively. Surface flatness and crystallinity during the deposition were monitored by the reflection high energy electron diffraction (RHEED). The structure of the films was investigated using the X-ray diffraction (XRD) method and the surface flatness of the films was measured using an atomic-force microscope (AFM). The resistivity measurements were performed using the conventional four-probe method.

3. Result and discussion

Figures 1 (a) and (c) show the RHEED patterns for Mg layer on the 4H-SiC and sapphire substrates with the layer thickness of 30 and 22 nm, respectively. The Mg RHEED image shows

IOP Conf. Series: Journal of Physics: Conf. Series **871** (2017) 012036 doi:10.1088/1742-6596/871/1/012036

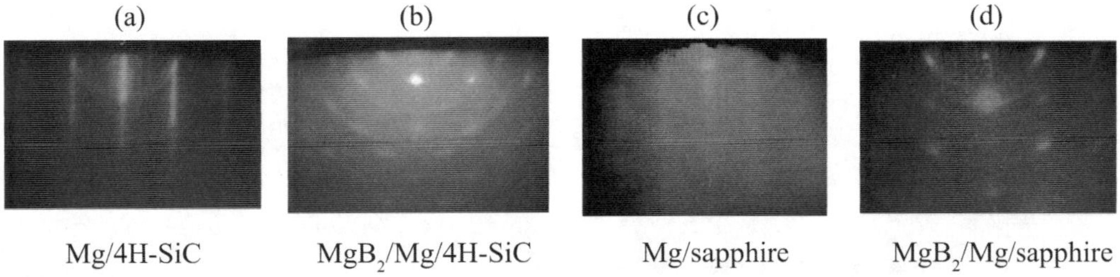

Figure 1. RHEED images for (a) Mg and (b) MgB_2 on a 4H-SiC substrate and (c) Mg and (d) MgB_2 on a sapphire substrate.

a sharp streak pattern, indicating epitaxial growth with good crystallinity and atomic-scale surface flatness of Mg thin films. On the other hand, RHEED images of the MgB_2 with the layer thickness of 100(62) nm on the 4H-SiC(sapphire) substrate are spot-like, implying relatively rough surface(see Figs. 1 (b) and (d)). With the rotation of the incident direction of the electron beam, both patterns changed, ensuring an inplane alignment of Mg and MgB_2 thin films.

The presence of a relatively rough MgB_2 film surface detected by RHEED was directly confirmed by examining the AFM image of MgB_2 films shown in Figs. 2 (a) and (b) on the 4H-SiC and sapphire substrates, respectively. While some protuberances are seen in the images, we estimate the average surface roughnesses to be 3.4 nm and 2.1 nm on the 4H-SiC and sapphire substrates, respectively. Although the origin of these protuberances is not fully understood as of now, it is possible that excess Mg, B, and MgO may have given rise to such protuberances.

Figures 3 (a) and (b) show the interplane XRD pattern for the MgB_2/Mg bilayer films grown on the (001) surface of the 4H-SiC and sapphire substrate, respectively. As shown in Figs. 3 (a) and (b), (001) and (002) peaks of MgB_2 and (002) and (004) peaks of Mg are observed. This is clear evidence of the successful growth of the MgB_2/Mg bilayer thin films. We also observed a contamination peak at around 44° marked by †: the origin of this peak has not been identified yet. Exclusive appearance of the (00l) peaks in MgB_2 and Mg revealed that the film has a distinct c-axis oriented structure. The XRD rocking-curve measurements performed for the (002) MgB_2 peak and for the (002) Mg peak of MgB_2/Mg/sapphire are shown in Figs. 3 (c) and (d), respectively. The 0.55° full width at half maximum (FWHM) of the (002) Mg peak is much smaller than the 1.91° FWHM of the (002) MgB_2 peak. This corresponds to the fact that

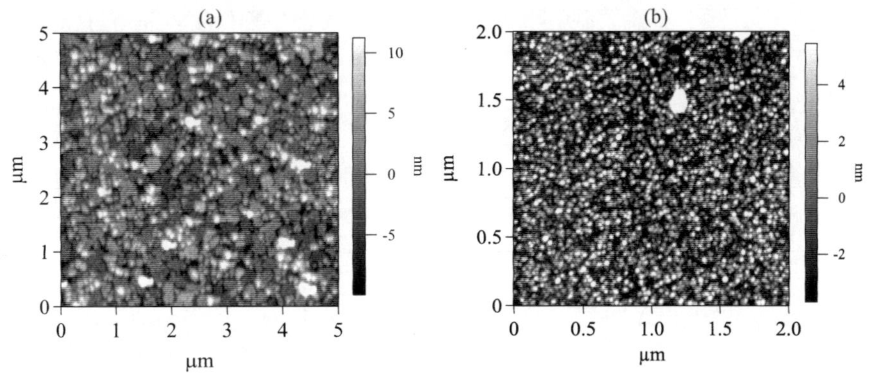

Figure 2. Atomic force microscope image for MgB_2 on a (a) 4H-SiC and (b) sapphire substrate.

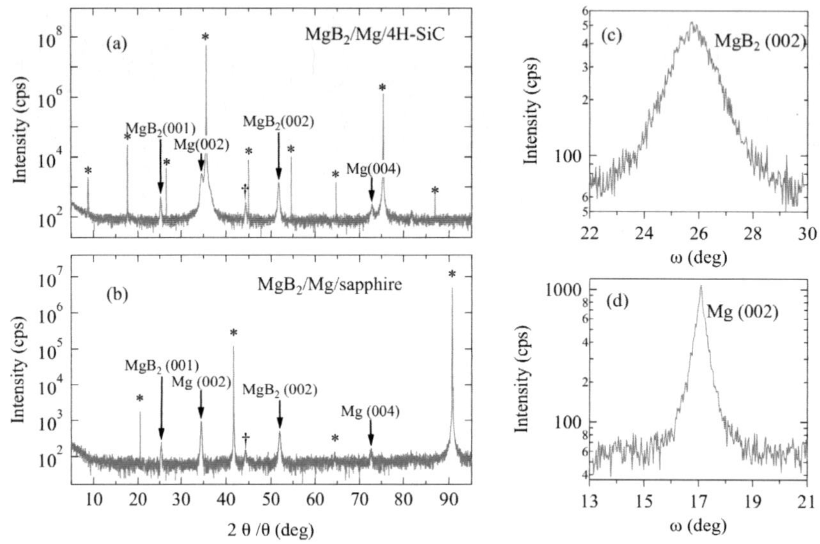

Figure 3. Interplane X-ray diffraction pattern for MgB_2 on a (a) 4H-SiC and (b) sapphire substrate. Substrate peaks are shown by *. Impurity peak appears at 44 °, as shown by †. Rocking curves for (c) MgB_2 (002) peak and (d) Mg (002) peak on a sapphire substrate.

the RHEED pattern of Mg is noticeably sharper than that of MgB_2, as shown in Fig. 1 (c) and (d). We note that almost the same rocking-cures are obtained in MgB_2/Mg/4H-SiC, as 0.25 ° FWHM of the (004) Mg peak and 1.54 ° FWHM of the (002) MgB_2 peak, respectively[19].

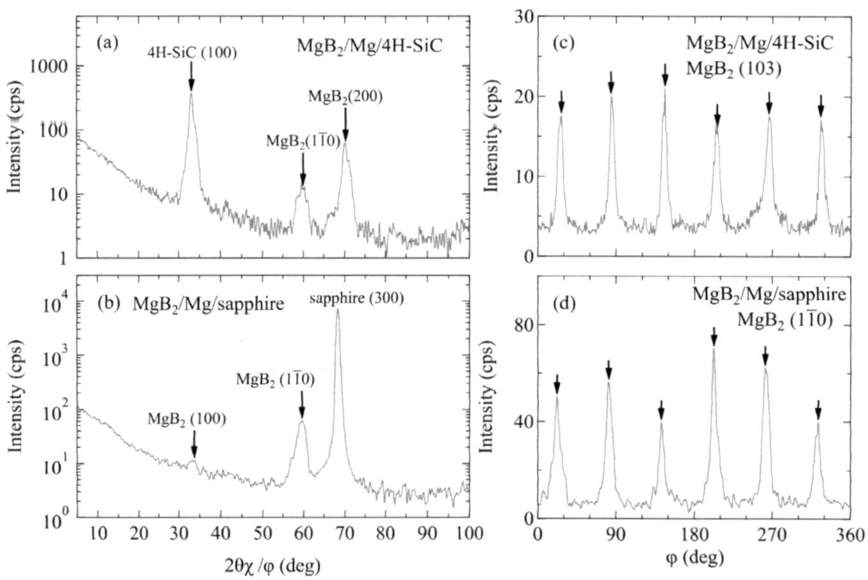

Figure 4. In-plane X-ray diffraction pattern for MgB_2 on a (a) 4H-SiC and (b) sapphire substrate. ϕ-scan of the MgB_2 (c) (103) peak on a 4H-SiC substrate and (d) $(1\bar{1}0)$ peak on a sapphire substrate.

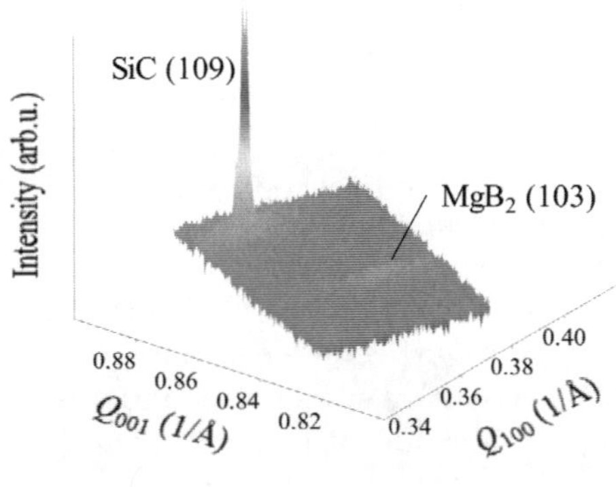

Figure 5. Reciprocal lattice mapping around the MgB_2 (103) peak for a MgB_2/Mg/4H-SiC.

The grazing-incidence XRD profile for a MgB_2/Mg/4H-SiC and MgB_2/Mg/sapphire films are shown in Figs.4 (a) and (b), respectively. The (200)(($1\bar{1}0$)) MgB_2 peak and the (100)((300)) peak of the 4H-SiC(sapphire) substrate are observed, indicating that MgB_2 grains were grown with the MgB_2[100]||4H-SiC[100](MgB_2[100]||sapphire[$1\bar{1}0$]). Observation of the finite ($1\bar{1}0$) ((100)) MgB_2 peak in MgB_2/Mg/4H-SiC (MgB_2/Mg/sapphire) films indicates the existence of unoriented grains. We also determined the degree of epitaxy for Mg buffer layer by employing Mg[100]||4H-SiC[100] and Mg[100]||sapphire[$1\bar{1}0$] in the grazing-incidence XRD measurements. Figures 4 (c) and (d) show the ϕ-scan pattern of the (103) and ($1\bar{1}0$) MgB_2 peak for MgB_2/Mg/4H-SiC and MgB_2/Mg/sapphire films, respectively. Clear six-fold peaks are observed, consistent with the six-fold symmetry of MgB_2. Based on the RHEED and XRD patterns, we conclude that we have successfully obtained partially epitaxial MgB_2 films, while some of the grains are randomly oriented for the in-plane axis.

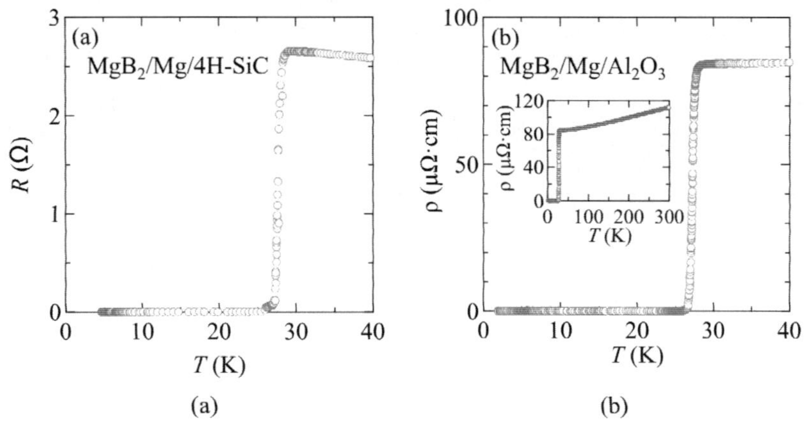

Figure 6. Temperature dependence of the electrical resistance for MgB_2 on a (a) 4H-SiC and of the electrical resistivity for MgB_2 on a (b) sapphire substrate.

Figure 5 shows the X-ray reciprocal lattice mapping for MgB_2/Mg/4H-SiC. The (103) MgB_2 peak is clearly observed; using the peak position, we determined the MgB_2 lattice constants as $a = 3.075$ Å and $c = 3.576$ Å. These results are accurate within 0.4 % and 1.5 % for the bulk a and c values, respectively, thus showing a fairly good agreement with the bulk values. The 0.98 ° FWHM of the (103) MgB_2 peak in the rocking curve was comparable with that of the (002) peak. The average coherent domain size D is estimated as 57 nm by applying the Debye-Scherrer formula $D = 0.9\lambda/\beta\cos\theta$, where λ is the X-ray wave length, β is a FWHM for a diffraction peak in the 2θ scan.

Temperature dependence of the electrical resistance R for MgB_2/Mg/4H-SiC and of the electrical resistivity ρ for MgB_2/Mg/sapphire are shown in Figs. 6 (a) and (b), respectively. Below room temperature, R increases with decreasing temperature because of the influence of a semiconducting resistance by 4H-SiC substrate. As shown in the inset of Fig. 6 (b), ρ decreases with decreasing temperature as expected in a normal metal in MgB_2/Mg/sapphire because sapphire is a good insulator in contrast with semiconducting 4H-SiC. Superconductivity appears below $T_c = 27.2$ K(26.9 K), with a relatively narrow superconducting transition width $\Delta T_c = 0.9$ K(0.5 K) for MgB_2/Mg/4H-SiC(MgB_2/Mg/sapphire).

It is well-known that annealing drastically improves T_c in MgB_2. Shibata *et al.* indicated that the film surface of MgB_2 is deteriorated in air[18]. Therefore, the lower T_c values of our as-grown MgB_2 thin films relative to that of the bulk samples are due to the absence of post-deposition-annealing and to surface oxidization.

4. Summary

We have grown superconducting MgB_2 thin films with a buffer layer of epitaxial Mg on the (001) 4H-SiC or sapphire substrate surface. We have successfully grew MgB_2 thin films with the substrate temperature lower than 110 °C. To the best of our knowledge, this is the lowest deposition temperature for a partial epitaxial growth of MgB_2 thin films. T_c values of the present thin films are comparable to the previously reported results; the observed suppression of thin film T_c is known to be related to the decreased the substrate temperature. The presence of the Mg buffer layer is crucial for reducing the epitaxial temperature. MgB_2 thin films with relatively good crystallinity and superconducting properties were obtained. We believe that this offers a new process for fabricating nanostructured MgB_2 films with the aid of an organic-based lift-off resist. Our new MgB_2 thin film growth process is promising for the development of a convenient nano-fabrication technique for MgB_2 thin films based on a standard lift-off process with a commercial organic resist. We think that our innovative method will have a strong impact on the expansion of utilizing the MgB_2 materials in future.

Acknowledgement

This study is supported by Grant-in-Aid for Scientific Research (C) (No. 1681202500) from Japan Society for the Promotion of Science (JSPS), and a special grant for faculty innovation from Graduate School of Engineering at Osaka Prefecture University.

References

[1] Nagamatsu J, Nakagawa N, Muranaka T, Zenitani Y and Akimitsu J 2001 *Nature* **410** 63
[2] Jin B B, Dahm T, Iniotakis C, Gubin A I, Choi E M, Kim H J, Lee S I, Kang W N, Wang S F, Zhou Y L, Pogrebnyakov A V, Redwing J M, Xi X X and Klein N 2005 *Supercond. Sci. Technol.* **18** L1
[3] Xu Y, Khafizov M, Satrapinsky L, Kus P, Plecenik A and Sobolewski R 2003 *Phys. Rev. Lett.* **91** 197004
[4] Kang W N, Kim H J, Choi E M, Jung C U and Lee S I 2001 *Science* **292** 1521
[5] Zeng X, Pogrebnyakov A V, Kotcharov A, Jones J E, Xi X X, Lysczek E M, Redwing J M, Xu S, Li Q, Lettieri J, Schlom D G, Tian W, Pan X and Liu Z K 2002 *Nat. Mat.* **1** 35
[6] Ermolov S N, Indenbom M V, Rossolenko A N, Bdikin I K, Uspenskaya L S, Stepakov N S and Glebovski V G 2001 *JETP Lett.* **73** 557

[7] Ueda K and Naito M 2001 *Appl. Phys. Lett.* **79** 2046
[8] Erven A J, Kim T H, Muenzenberg M and Moodera J S 2002 *Appl. Phys. Lett.* **81** 4982
[9] Jo W, Huh J -H, Ohnishi T, Marshall A F, Beasley M R and Hammond R H 2002 *Appl. Phys. Lett.* **80** 3563
[10] Harada Y, Takahashi T, Iriuda H, Kuroha M, Nakanishi Y and Yoshizawa M 2005 *Physica C* **426-431** 1453
[11] Saito A, Kawakami A, Shimakage H and Wang Z 2002 *Jpn. J. Appl. Phys.* **41** L127
[12] Shimakage H, Saito A, Kawakami A and Wang Z 2003 *IEEE Trans. Appl. Supercond.* **13** 3309
[13] Ueda K and Makimoto T 2006 *Jpn. J. Appl. Phys.* **45** 5738
[14] Sakata O, Kimura S, Tkata M, Yata S, Kato T, Yamanaka K, Yamada Y, Matsushita A and Kubo S 2004 *J. Appl. Phys.* **96** 3580
[15] Harada Y, Yamaguchi H, Takahashi T, Iriuda H, Oba T and Yoshizawa M 2007 *IEEE Trans. Appl. Supercond.* **17** 2883
[16] Shibata H, Asahi M, Maruyama T, Akazaki T, Takesue H, Honjo H and Tokura Y 2009 *IEEE Trans. Appl. Supercond.* **19** 358
[17] Shibata H, Akazaki T and Tokura Y 2013 *App. Phys. Express* **6** 023101
[18] Shibata H, Akazaki T and Tokura Y 2013 *Supercond. Sci. Technol.* **26** 035005
[19] Shishido H, Yoshida T and Ishida T 2015 *App. Phys. Express* **8** 113101

The microstructure and properties of J-R Nb3Al superconducting wires fabricated by diffusion method at an intermediate temperature

Wenjia Lin[1], Liyuan Xu[1], Wenjie Zhang[1], Liang Zheng[1], Zhou Yu[1]*, Pengyuan Li[2], Xifeng Pan[2], Guo Yan[2], Yong Zhao[1, 3]* and Yong Zhang[1]*

[1] Key Laboratory of Magnetic Levitation Technologies and Maglev Trains, Ministry of Education of China, and Superconductivity and New Energy R&D Center, Southwest Jiaotong University, Chengdu 610031, P.R.China
[2] Western Superconducting Technoligies Co., Ltd., Xi'an 710018, P.R.China
[3] School of Materials Science and Engineering, University of New South Wales, Sydney 2052 NSW, Australia

* Corresponding author. Tel.: +86-28-87600786; fax: +86-28-87600785.
E-mail address: yongzhang@swjtu.cn; yuzhou@swjtu.cn; yzhao@swjtu.cn.

Abstract. The precursor Nb-Al strands were fabricated by the method of Jelly-Roll. The effects of sintering temperature and time on the diffusion degree and superconducting property of the binary Nb-Al system were investigated under isothermal heat treatment at 1100 -1500 °C. The results show that the Tc of superconducting wires increased with the heat-treatment temperature rising at the isothermal region. Excessive heat-treatment time led to the decrease of the properties of Nb3Al wires under a certain heat-treatment temperature. The magnetic analysis results showed that the optimal Tc of J-R Nb3Al wires was about 16.8 K at a teat-treatment temperature of 1500 °C for 3 h.

1. Introduction

Since the low strain-stress property of Nb_3Sn superconductors, the alternative Nb_3Al superconductor has been deeply studied d for the use of D-shapes coils in the International Thermonuclear Experimental Reactor (ITER) plan [1, 2]. Although the Nb_3Al superconductor has a better property than Nb_3Sn, the manufacture is still of a big problem up to now. Unlike the Nb_3Sn superconductor, Nb_3Al superconductor can't be fabricated by the bronze process because of the easy formation of Cu-Nb-Al ternary compound [3]. In addition, according to the Nb-Al binary phase diagram [4], the stable stoichiometric phase of Nb_3Al can only exist at 1940 to 2060 °C, and the ratio of Al in the stable Nb_3Al is far away from the standard stoichiometry at a lower temperature, resulting in a bad superconductivity.

Nowadays, there are generally three kinds of methods for Nb_3Al superconductor fabrications. The first one is prepared via low temperature diffusion [5], such as the high energy ball milling method [6]; The second one is prepared via high temperature diffusion, such as the laser-irradiated method [7]; The third one is prepared via Rapid Heating Quenching and Transformation (RHQT) method [8, 9].

In the low temperature diffusion method, the T_c value of samples is just around 15 K, which is far away from the theoretical value. As for the RHQT method, the precursor Jelly-Roll (J-R) Nb/Al wires were quickly heated by the ohmic heating and then quickly cooled down in a gallium liquid and eventually the nearly stoichiometric Nb_3Al superconducting wires were obtained with fine grains after the subsequent annealing treatment [10, 11]. Although the most popular method of fabricating Nb_3Al superconductor is RHQT, there still are some problems in making long superconducting wires. What's more, the highest T_c was about 18.5 K obtained by the laser-irradiated method, but the J_c value was very low because of the coarse grains for this kind of high temperature diffusion method [12].

Attempts have been made to obtain the Nb_3Al superconducting wires with excellent properties. However, few investigations have been performed to clarify the effect of sintering temperature, sintering time on T_c in total at an intermediate temperature region. Therefore, in order to further understand the detailed relationships, this study focuses on the influence of T_c values in different sintering temperatures and time periods by the method of intermediate temperature diffusion.

2. Experimental

The J-R single-core Cu/Nb_3Al precursor wires, fabricated using Nb layer of 150 μm and Al layer of 50 μm [13, 14], were supplied by the WST Co., Ltd. The corroded Cu shell wires were annealed at a temperature ranging from 1100 to 1500 °C for 1 to 12 h. The diameter of wires was 1.3 mm. Each of the samples was heated by a heating rate of 100 °C/h and then cooled down to room temperature after switching off the furnace. During the annealing process, all samples were protected at the flowing Argon condition.

The surface of annealed wires was sanded and the phase composition was analysed by X-ray diffraction (XRD, Philips X'Pert MRD) using Cu Kα radiation. The microstructure and composition of the Nb_3Al wires were investigated by Field-emission scanning electron microscopy (FESEM, JEOL) and energy dispersive X-ray analysis (EDX), respectively. For all the Nb_3Al wires, Specimen of 1mm long were cut from each sample and the magnetization as a function of temperature (M-T) were measured by using the magnetic properties measurement system (MPMS, Quantum Design) under an applied field of 0.002 T and a temperature range of 9-25 K with a zero-field cooling (ZFC) process.

3. Results and Discussion

Figure 1(a) shows the *M-T* curves of Nb_3Al wires annealed at 1100-1500 °C for 3 h. It can be seen that all the samples have superconducting transition phenomena above the temperature of 12 K. In addition to the sample after annealing at 1200 °C, the diamagnetic signal and transition temperature of samples annealed at a temperature range of 1100-1450 °C gradually increased and their transition width becomes narrowed with the increase of annealing temperature. The results suggest that the diffusion rate of Al atoms is faster and the formation of Nb_3Al phase is increased with increasing the heat-treatment temperature, and that the Nb and Al atomic ratio of Nb_3Al phase is more close to the stoichiometric value of 3:1. The sample after annealing at 1200 °C has a narrow transition width and a low T_c of about 14.3 K. It is indicated that the Nb_3Al phase of significantly deviated from the stoichiometric ratio of 3:1. As shown in Figure 1(a), the sample annealed at 1500 °C has a highest $T_{c\text{-onset}}$ of about 16.8 K, but the diamagnetic signal decreased and the transition width increased. It can be inferred that a small amount of Nb_3Al phase has formed with relatively close to the stoichiometric ratio when the heat-treatment temperature reached 1500 °C. Figure 1(b) shows the curves of the $T_{c\text{-onset}}$ against the annealing temperature. when the annealing time is 3 h, the $T_{c\text{-onset}}$ is improved with the increase of the heat-treatment temperature. At an annealing temperature range of 1100-1500 °C, a liner relationship formula can be drawn:

$$T_c = 3.85704 + 0.00862 \times \mathrm{T}$$

Where T_c and T are the superconducting transition temperature and annealing temperature, respectively.

Figure 2 displays the XRD patterns of Nb_3Al wires annealed at different temperatures for 3 h. All the XRD lines in Figure 2 show diffraction peaks of Nb_3Al and Nb phases. When the heat-treatment temperature is 1100 °C, the XRD diffraction patterns of samples are mainly Nb diffraction peaks. A number of diffraction peaks of impurity phase appear at around $2\theta = 30°$ with only a small amount of Nb3Al diffraction peaks in the XRD line. That is why the magnetic performance of the sample is very bad. With the increase of the annealing temperature, the diffraction peaks of the impurity phase decrease as (Figure 2), so the properties of the samples are gradually improved, indicating consistent performance results as shown in Figure 1(a). The XRD patterns are almost the same for the samples annealed at the temperature range of 1200-1500 °C, and the peaks of Nb_3Al, Nb phase and some impurity phases be detected. It is worth noting that the XRD pattern of the sample annealed at 1500 °C has no obvious peaks of impurity phases, but the transition width is very large. It is suggested that the Nb_3Al phase with different Nb/Al atomic ratios were formed in the sample.

Figure 3 shows the typical SEM images of the Nb_3Al wires annealed at 1100-1500 °C for 3 h. The precursor wire is Nb/Al single-core wire prepared by J-R method. As seen from Figure 3(a-1), (b-1), (c-1) and (d-1), there is still a coiled lamellar structure with Nb rod at the center in all the samples. Figure 3(a-2) shows that there are bright stripes, gray areas and holes: the bright stripes are the unreacted Nb, the gray region is the Nb/Al diffusion layer, and the holes are formed by the melting of the Al layer. There is a large amount of unreacted Nb in the samples annealed at 1100 °C, which may be due to the poor superconducting transition performance. The Nb layer decreases obviously with the increase of the heat-treatment temperature as shown in figure 3(a-2), (b-2), (c-2) and (d-2), suggesting that the Nb and Al react more fully with increasing the heat-treatment temperature. It should be noted that there are two regions of dark gray and grayish white in the images as shown in figure 3 (b-1) and (c-1) with the heat-treatment temperature increasing to 1200 °C and 1450 °C. As indicated in Figure 3(d-2), there are only grayish white region after annealing at 1500 °C, and the grayish white region should be formed by grain growth at a high temperature.

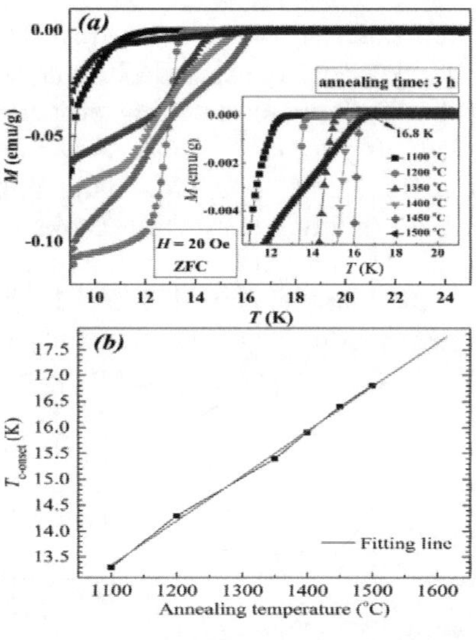

Figure 1. (a) M-T curves of Nb3Al wires annealed at 1100-1500 oC for 3 h and (b) the Tc-onset against the annealing temperature.

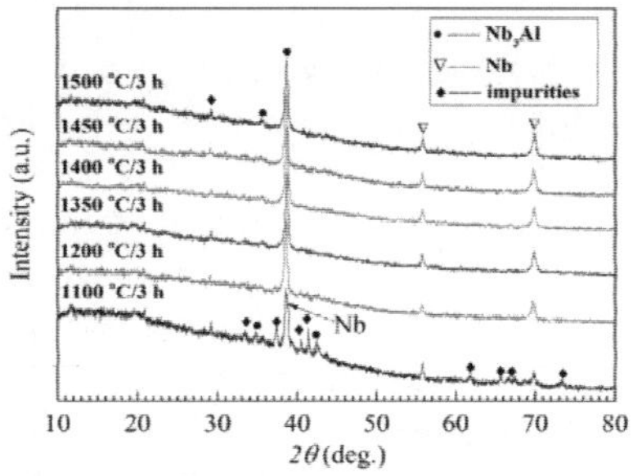

Figure 2. XRD patterns of Nb_3Al wires annealed at 1100-1500 °C for 3 h

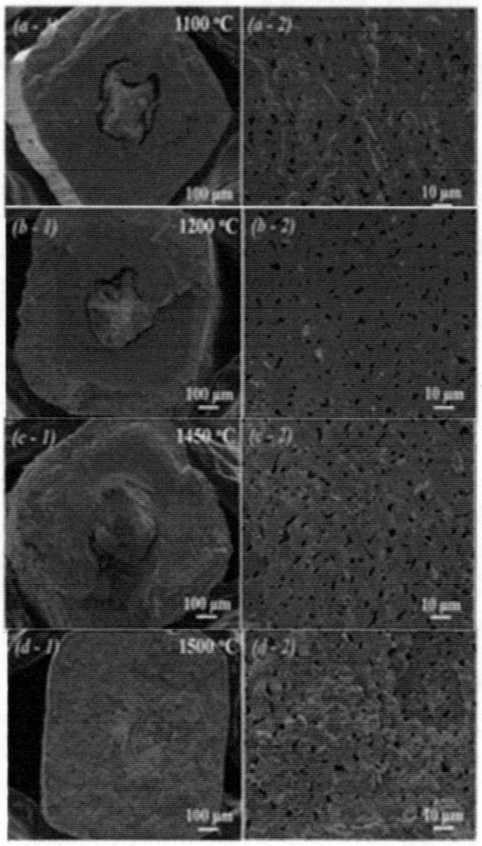

Figure 3. Typical SEM images of samples annealed at different temperatures for 3 h with 100x and 1000x magnifications. (a-1) and (a-2): 1100 °C; (b-1) and (b-2): 1200 °C; (c-1) and (c-2): 1450 °C; (d-1) and (d-2): 1500 °C.

Figure 4 shows the *M-T* curves of Nb_3Al wires annealed at 1400 °C for different time. As shown in Figure 4, the properties of the samples are gradually improve with the increase in the heat-treatment time. As can been seen from the inset of figure 4, the sample annealed for 1 h has poor properties, suggesting that the reaction of Nb and Al is not sufficient, and there is only a very small amount of superconducting phase formed in the sample. The samples annealed at 1400 °C for 3, 6 and 12 h has a same $T_{c\text{-onset}}$ of about 15.8 K, but the superconducting diamagnetic signal increases with the increase in the annealing time, indicating that the formation of Nb_3Al phase increases. This suggest that the diffusion of Al and Nb atoms is more fully with the increase in the heat-treatment time, thus, the Nb_3Al phase content increases and the lattice defects decrease in samples after annealing at 1400 °C.

Figure 5 shows the XRD patterns of the samples annealed at 1400 °C for different time periods. The XRD lines of all samples have clearly peaks of Nb_3Al and Nb phases. The sample annealed for 1 h has formed Nb_3Al phase, but the superconducting property was not so good and the superconducting transition was almost invisible as shown in Figure 4, indicating that the reaction of Nb and Al is not sufficient at 1400 °C for 1 h. With the heat treatment time increasing from 3 to 12 h, there are obvious diffraction peaks of impurity phase at $2\theta = 35.1°$ and 40.6° in the XRD patterns. The Nb_3Al peak increases and the Nb peaks decrease obviously in the XRD line of the sample annealed for 12 h, indicating that the reaction of Al and Nb is more complete as the heat treatment time is prolonged.

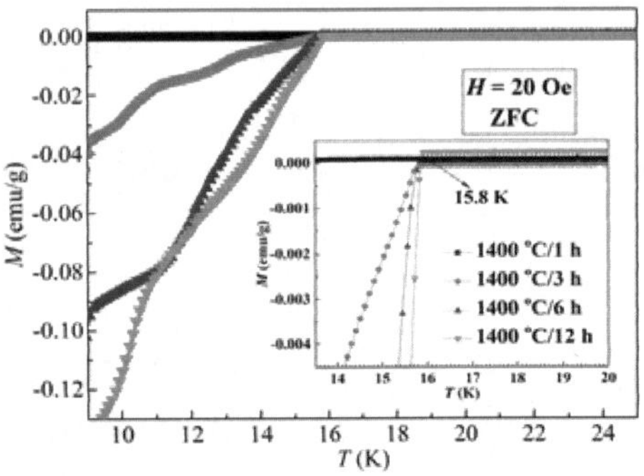

Figure 4 *M-T* curves of Nb_3Al wires annealed at 1400 °C for 1, 3, 6 and 12 h.

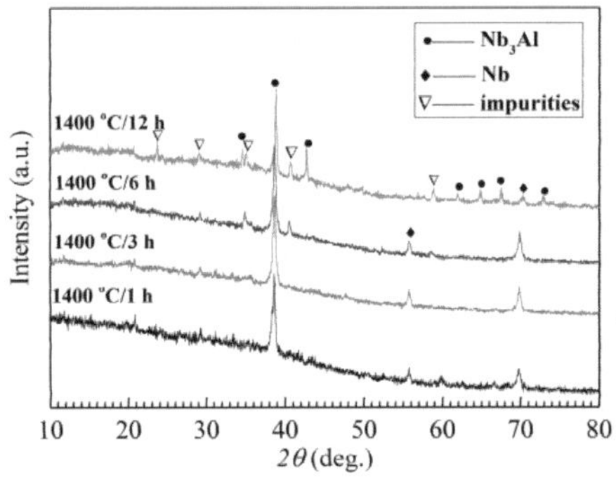

Figure 5 XRD patterns of samples annealed at 1400 °C for 1, 3, 6 and 12 h.

In order to further illustrate the effect of heat-treatment time on the properties of Nb_3Al wires, the magnetization behavior have been investigated for Nb_3Al wires annealed at 1450 °C for different time. As shown in Figure 6(a), the sample annealed at 1450 °C for 3 h has a relatively good property, such as a higher T_c and greater diamagnetic signal, comparing with those annealed for 1, 6 and 12 h. The sample annealed for 3 h has a higher $T_{c\text{-onset}}$ of about 16.4 K suggesting that the sample annealed at 1450 °C for 3 h can formed Nb_3Al phase with a relatively good quality. As seen from Figure 6(b), the properties of the sample increased first and then decreased with the annealing time increasing from 1 to 12 h. It can be inferred by comparing the results of Figure 6(a) and Figure 4 that the optimization properties is obtained in a shorter annealing time with the increase in the heat-treatment temperature, and an excessive annealing deteriorates the properties of the sample.

Figure 7 shows typical EDX images of Nb_3Al wires annealed at 1400 °C for 6 h and 1450 °C for 3 h,. As shown Figure 7(a), the Nb/Al atomic ratio of the sample annealed at 1400 °C for 6 h is about 5:1, much greater than the stoichiometric ratio of 3:1, thus resulting in a severe decrease in magnetic T_c of about 15.8 K (Figure 4). A shown in Figure 7(b), the Nb/Al atomic ratio of the sample annealed at 1450 °C for 3 h is close to 3:1, resulting in higher Tc of about 16.4 K than the sample annealed at 1400 °C for 6 h. It is suggested that the Nb/Al atomic ratio of the formation of Nb_3Al in the samples more close to the stoichiometric ratio of 3:1 when the sample annealed at higher temperature, and that increasing the annealing temperature is more beneficial to improve the properties of Nb_3Al wires than prolonging the annealing temperature.

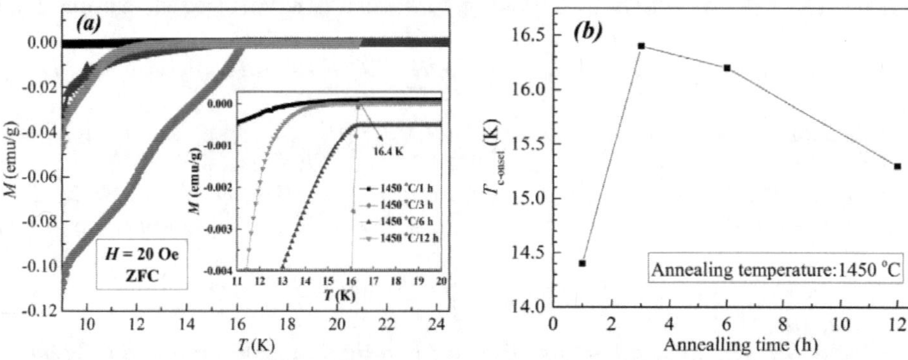

Figure 6 (a) *M-T* curves of Nb_3Al wires annealed at 1450 °C for 1, 3, 6, and 12 h. (b) The $T_{c\text{-onset}}$ versus annealing time at the annealing temperature of 1450 °C.

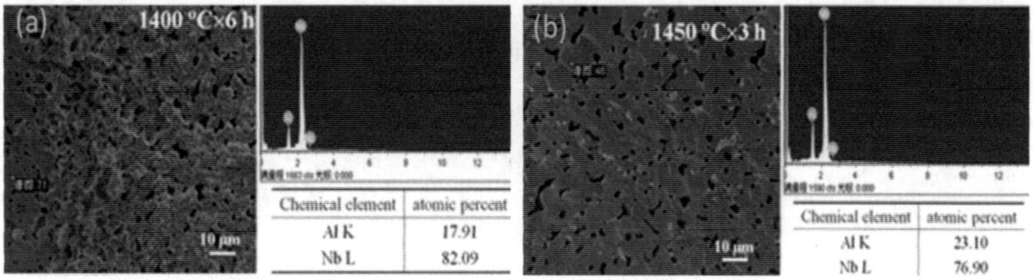

Figure 7 EDX images of Nb_3Al wires (a) annealed at 1400 °C for 6 h and (b) annealed at 1450 °C for 3 h.

4. Conclusions

In this study, a series of J-R single core Nb_3Al wires have been heat-treated at an intermediate temperature of 1100-1500 ℃. The T_c of the superconducting wires increases relevantly with the increase in the sintering temperature. There are some determined T_c values in different sintering time when the sintering temperature is in a certain range. The optimal performance of samples is determined after sintering at 1500 °C for 3 h, and the obtained T_c is about 16.8 K.

Acknowledgements
The authors are grateful for the financial support of the National Natural Science Foundation of China (Grant Nos. 51271155, and 51302224, and 51377138), the Sichuan Province Science Foundation (Grant No. 2017JY0057), the 863 Program 2014AA032701, Program of International S&T Cooperation (Grant No. 2013DFA51050).

References

[1] Takeuchi T, Kuroda T, Itoh K, Kosuge M, Iijima Y, Kiyoshi T, Matsumoto F and Inoue K 1992 *Journal of fusion energy* **11** 7
[2] Kuroda T, Wada H, Bray S L and Ekin J W 1993 *Fusion Engineering and Design* **20** 271
[3] Takeuchi T 2000 *Supercond. Sci. Technol.* **13** R101
[4] Jorda J, Flukiger R, Junod A and Muller J 1981 *IEEE Trans. Magn.* **17** 557
[5] Takeuchi T, Iijima Y, Kosuge M and Kuroda T 1989 *IEEE Trans. Magn.* **25** 2068
[6] Sumption M D, Peng X, Lee E and Buta F 2003 *IEEE Trans. Appl. Supercond.* **13** 3486
[7] Kosuge M, Iijima Y, Takeuchi T and Watanabe K 1993 *IEEE Trans. Appl. Supercond.* **3** 1010
[8] Saito S, Sugawara H, Kikuchi A, Iijima Y and Inonu K 2002 *Physica C* **372** 1373
[9] Saito S, Sugawara H, Yuuki J and Kodaira N 2004 *IEEE Trans. Appl. Supercond.* **14** 1024
[10] Iijima Y, Kogochi M, Takeuchi T and Inoue K 1994 *Adv. Cryogen. Eng.* **40** 899
[11] Takeuchi T 2002 *IEEE Trans. Appl. Supercond.* **12** 1088
[12] Schaper W, Kehlenbeck M, Zimmermann I and Freyhardt H C 1989 *IEEE Trans. Magn.* **25** 1988
[13] Cui L J, Zhang P X, Pan X F, Yan G, Qi M, Wang D Y, Liu X H, Feng Y and Bai Z M 2015 *Materials China* **34** 64
[14] Cui L J, Yan G, Pan X F, Zhang P X, Qi M, Liu X H, Feng Y, Chen Y L and Zhao Y 2015 *Physica C* **513** 24

IOP Conf. Series: Journal of Physics: Conf. Series **871** (2017) 012038 doi:10.1088/1742-6596/871/1/012038

Temperature Distribution and Critical Current of Long HTS Cables Cooled with Subcooled Liquid Nitrogen

V S Vyatkin[1], Y V Ivanov[1,2], H Watanabe[1,2], N Chikumoto[1,2], and S Yamaguchi[1,2]

[1]Chubu University, 1200 Matsumoto-cho, Kasugai, Aichi 487-8501, Japan.
[2]Ishikari Superconducting DC Power Transmission System Research Association, 3-13 Moriya-cho, Kanagawa-ku, Yokohama 221-0022, Japan

(vyatkin@isc.chubu.ac.jp)

Abstract. Cooling of the long HTS power transmission lines performs by pumping of subcooled liquid nitrogen (LN2) along the cable. The temperature of LN2 along the cable increases due to the heat losses of the cryostat and heat generation in the HTS cable. The experiment using test cable line in Ishikari shows that flow rate of 35 L/min retains increasing of LN2 temperature by 1 K per 1 km of length. The technology when the back flow of LN2 cools the radiation shield surrounding the cable pipe is also applied in Ishikari-2 project. In this case the ambient heat flow into cable pipe is 50 times less than that without radiation shield. Back flow of LN2 removes almost all heat coming from the environment. When transport current is close to the critical value the Joule heat of HTS cable is significant. This heat additionally increases the temperature of LN2 flowing along the HTS cable. Near the outlet the temperature of HTS cable is maximal and the local critical current is minimal. The current matching critical current criterion of average electrical field of $E_0 = 10^{-4}$ V/m provides the voltage drop and significant Joule heat at the hot end of the cable. It can lead the damage of the cable. The present work contains analysis of temperature distribution along the cable and the way to achieve the fail-safe operation of long HTS cable cooled by subcooled LN2. We also performed extrapolation of obtained results for several times longer cable lines by decreasing the LN2 flow rate.

1. Introduction

In recent years, the large progress achieved in the field of power transmission using HTS cables. In Chubu University study of the DC HTS cables starts more than 10 years ago. First, it was 20 m HTS DC cable [1]. Next project was 200 m cable installed in 2008 (CASER project) [2]. The continuation of this work is the 500 m HTS DC cable between solar panel plant and data center in Ishikari [3, 4]. The last project is the design of 1000 m HTS DC loop cable in Ishikari [3, 4].

The present paper describes 1 km Ishikari-2 cable from Joule heal losses point of view. The cable contains two straight parts with length of 469 m and 373 m, and U-turn part with length of 125 m. The parts have different structure and different heat properties and give possibility to study the behavior of long cable system for energy transmission. The measurements of critical current of long HTS cable line is not easy because the temperature along the cable varies. The hot part of the cable is dangerous place for burning the cable because the almost all Joule heat generates in the hot place. Our previous paper devoted to the critical current of the long HTS DC cable theoretically discusses 500 m cable [5]. The present paper is the application of these ideas to the 1 km cable in Ishikari-2 project using real experimental data on heat losses in the system. The investigation of heat properties of cable longer than

IOP Conf. Series: Journal of Physics: Conf. Series **871** (2017) 012038 doi:10.1088/1742-6596/871/1/012038

existing now was performed by simulation of 5 L/min LN2 flow rate. This will be future experiment in Ishikari-2 cable.

2. Initial data and results of calculation

The calculations based on the experimental data obtained from the Ishikari-2 project. The cable has 3 sections with length of L_1= 469 m, L_2= 125 m, L_3= 373 m. The structure of sections is shown in Fig. 1. Section 1 made with radiation shield cooled with return LN2 pipe [4]. The heat leak into cable pipe of the Section 1 is minimal. Section 2 is in U-turn area. The section are 2 and 3 without radiation shield. Two joints installed between sections. The joints have own heat leak and temperature of LN2 before and after joints is different.

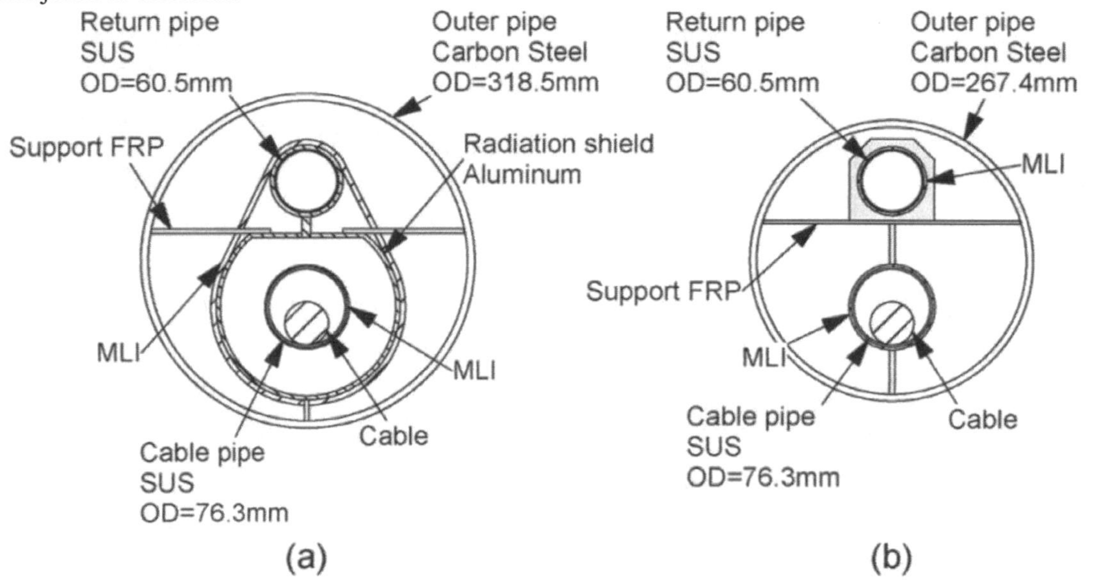

Figure 1. The cross-section of Ishikari-2 cryogenic pipe [4]. (a) Part of line with radiation shield connected to the pipe with return LN2 flow. (b) Part of line without radiation shield.

The result of temperature measurements at 35.17 L/min LN2 flow rate is shown in Fig. 2.

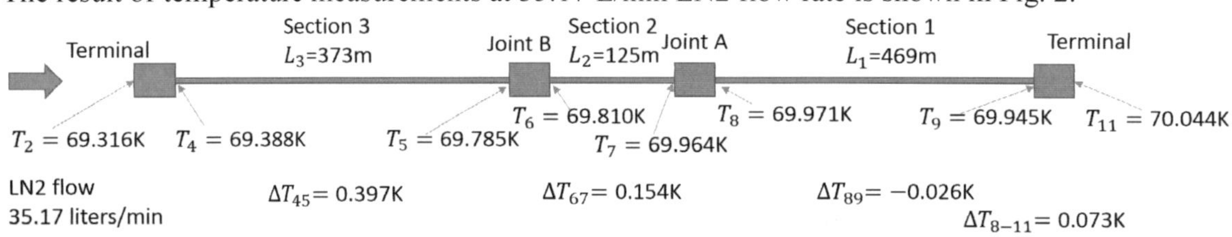

Figure 2. The temperature distribution along the 1 km Ishikari-2 cable line with LN2 flows and without transport current. LN2 starts from section 3 and the temperature increases along the line. In the last part of cable (Section 1) with radiation shield temperature does not increase.

It is possible to see that liquid nitrogen starts from terminal of section 3 with temperature of $T_4 =$ 69.388 K comes to joint B with temperature of $T_5 = 69.785$ K. The temperature drop of $\Delta T_{45} = 0.397$ K corresponds to heat leak

$P_3 = \Delta T_{45} \cdot \rho \cdot F \cdot C_p = 0.397\text{K} \cdot 808\frac{\text{kg}}{\text{m}^3} \cdot 35.17\frac{10^{-3}\text{m}^3}{60\ \text{sec}} \cdot 1.97\frac{10^3\text{J}}{\text{kg}\cdot\text{K}} = 370.4\ \text{W}$

for section 3. It is $\frac{P_3}{L_3} = \frac{370.4W}{373m} = 0.993$ W/m of ambient heat inleak in section 3.

The ambient heat leak into section 2 is $\frac{P_2}{L_2} = 1.15\frac{W}{m}$. It is close to the same value of section 3. The situation is different for section 1. The radiation shield prevents heat leak into the cable pipe of the section 1, and the temperature does not change along the section 1.
The important thing is the critical current of the HTS cable. The temperature dependence of HTS tape critical current shown in Fig. 3 was used to estimate critical current of the cable in whole.

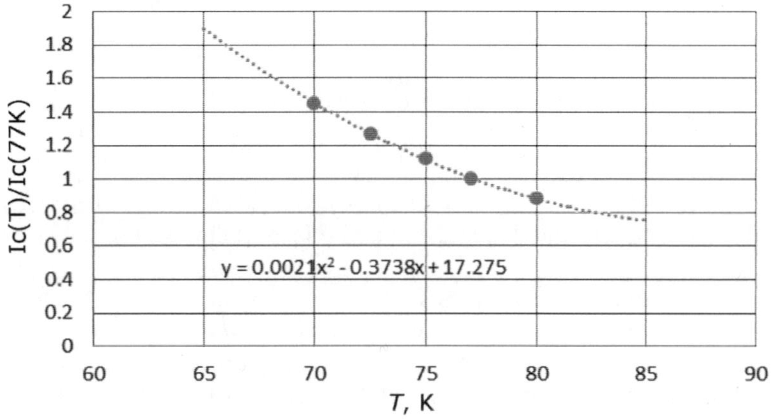

Figure 3. Temperature dependence of relative critical current of the Bi2223 HTS tape used in the Ishikari project.

The local temperature of the cable determines the local critical current. Fig. 4 illustrates the dependence of temperature, voltage (now it is 0 because transport current is 0), and local critical current of the cable on the position.

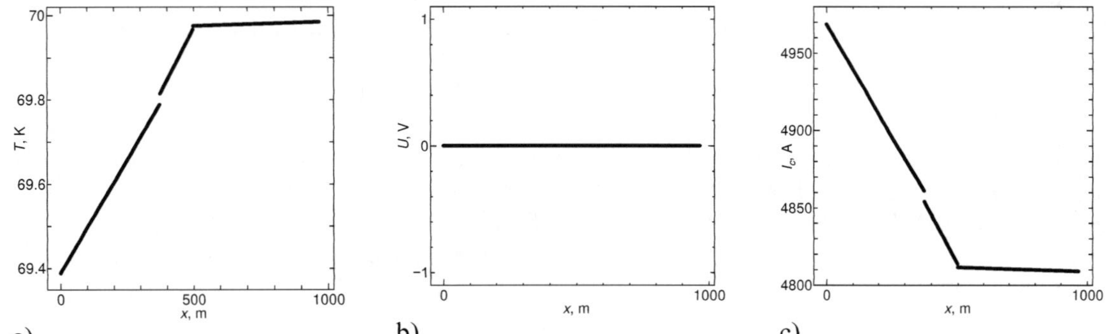

Figure 4. The dependence of temperature (a), voltage (b), and local critical current (c) on the position. The LN2 flow rate is 35.17 L/min and transport current is 0 A.

The obtained heat leak data allow to calculate the influence of flow rate on temperature distribution. The decreasing of flow rate can simulate the system operation when the HTS cable length becomes longer. Fig. 5 shows the same dependences as in Fig. 4 but at the LN2 flow rate of 5 L/min.

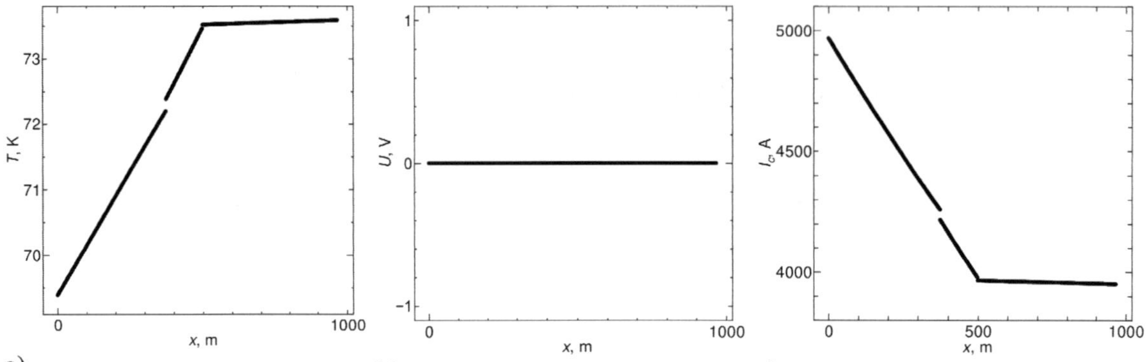

a) b) c)

Figure 5. The dependence of temperature (a), voltage (b), and local critical current (c) on the position. The LN2 flow rate is 5 L/min and transport current is 0 A.

The transport current flowing in the HTS cable heats up the cable. The Joule losses are not homogeneously distributed along the cable. The voltage drop depends on the value of transport current and the proximity of transport current to the critical current according the voltage-current characteristic. Paper [5] describes the method of calculation of voltage drop in the long cable with LN2 flow and transport current. This method includes the splitting of the cable on small parts with size x_i and local temperature T_i. Local critical current $I_c(T_i)$ is obtained using data from Fig. 3, and then we can calculate electrical field in this element E_i.

$$E_i = E_0 \cdot \left(\frac{I}{I_c(T_i)}\right)^n,$$

where $E_0 = 10^{-4}$ V/m – criterion of critical current.

The n-value of the cable was 10 – 20. The length of the cable in the paper [5] was 500 m. In our case the length is 1000 m and step of calculation is $x_i = 1\ m$. The value of transport current takes to obtain the total voltage on the cable corresponding to the criterion of average critical current with average electrical field $E_0 = 10^{-4}$ V/m. Total voltage of the cable is

$$U = \sum_{i=1}^{N} E_i \cdot x_i = U_0 = E_0 \cdot N = 0.1\text{ V}.$$

The result of calculation for 1 km line with ambient heat inleak shown in the Fig. 6d. It is possible to see that the maximal local Joule heat is about 1.5 times larger than average ($E_0 = 10^{-4}$ V/m). It means that hot end of the cable is overloaded by 1.5 times in terms of heat generation.

The situation with LN2 flow rate of 5 L/min is different. The value of electrical field at the end becomes much higher than $E_0 = 10^{-4}$ V/m. Fig. 7 shows the same dependence like in Fig. 6 but for LN2 flow of 5 L/min.

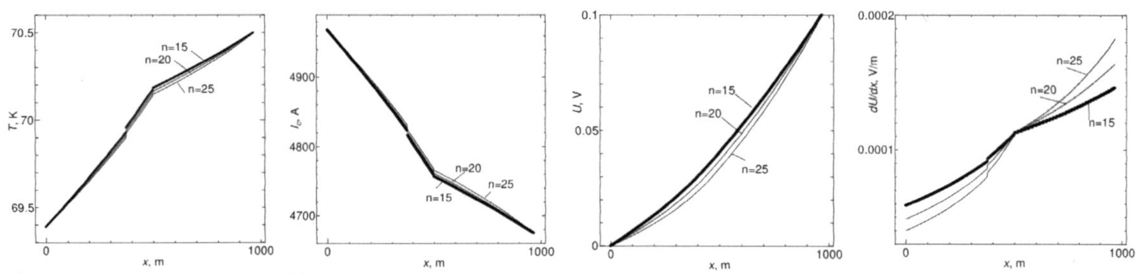

a) b) c) d)

Figure 6. The dependence of temperature (a), local critical current (b), voltage (c) and local electrical field dU/dx (d) on the position in case of LN2 flow rate of 35 L/min and transport current of 4795 A (n=15), 4792 A (n=20), 4789 A (n=25) to obtain the total voltage drop 0.1V.

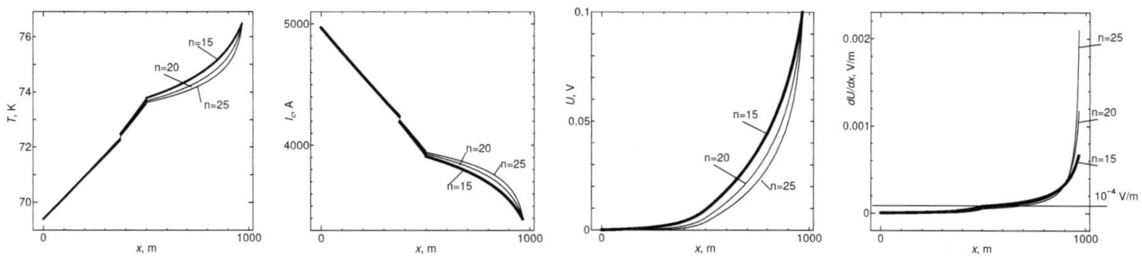

a) b) c) d)

Figure 7. The dependence of temperature (a), local critical current (b), voltage (c) and local electrical field dU/dx (d) on the position in case of LN2 flow rate of 5 L/min and transport current of 3847 A (n=15), 3840 A (n=20), 3837 A (n=25) to obtain the total voltage drop of 0.1 V.

From Fig. 7d it is possible to see that the electrical field at the hot end of the cable is more than 10 times higher than average electrical field. It means that the Joule heat, which is electrical field multiply by current ($P = I \cdot E$) is also 10 times higher at the hot end. The transportation of current with value equals to the critical current obtained with criterion 10^{-4} V/m is dangerous for the cable.

The measurement of critical current is better to perform by pulse current with duration about 2 sec. This estimation comes from the average heat capacity of HTS tape material (mainly stainless steel) and resistivity of the tape at low temperature [6]. The increasing of the temperature in this case even the cable comes to the normal state will be less than 200 K.

In order to estimate the safe value of maximal nominal current the same characteristics were calculated in case of 90% value of the measured critical current. The temperature, critical current, voltage and electrical field distributions along the length are shown in the Fig. 8.

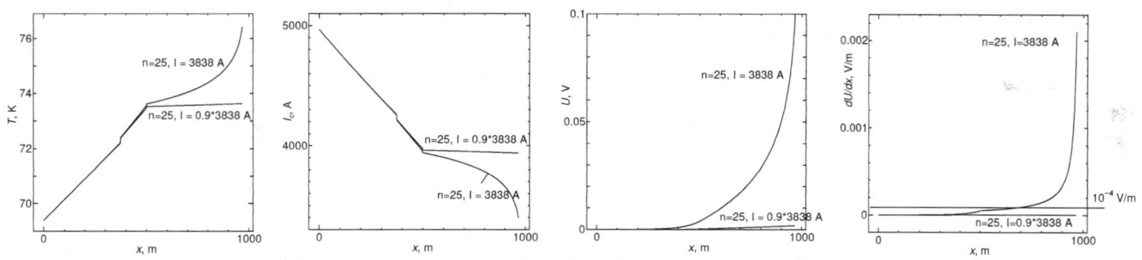

a) b) c) d)

Figure 8. The dependence of temperature (a), local critical current (b), voltage (c) and local electrical field dU/dx (d) on the position in case of LN2 flow rate of 5 L/min and transport current of 3837 A (n=25) to obtain the total voltage drop of 0.1 V.

The value of transport current for different LN2 rates was also calculated. The result is shown in Fig. 9. It can be seen the dependence on LN2 flow rate, but not on *n*-value. The reason is the temperature of hot end is different for different LN2 flow rates. The critical current of the cable decreases for the case of low LN2 flow because the temperature of the hot end of HTS cable becomes higher than in case of high LN2 flow rate.

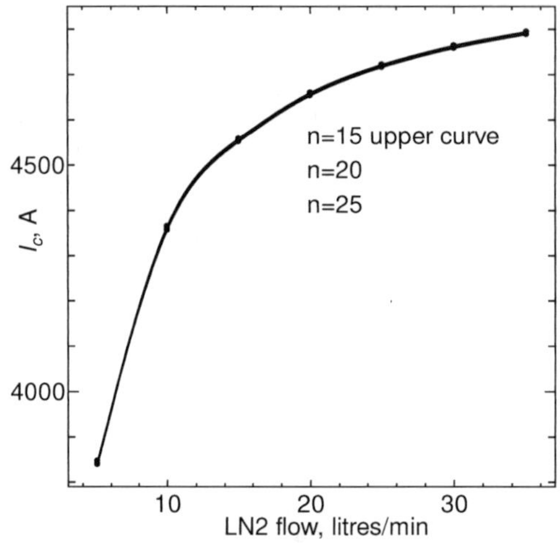

Figure 9. The dependence of critical current of 1 km Ishikari-2 type cable line for different LN2 flow rates for different *n*-values.

3. Conclusion

The present research deals with calculations of heat processes in 1 km cable using experimental data from Ishikari-2 project. It is important for the experiments critical current measurement of HTS cable in Ishikari-2.

The estimation of critical current is only possible by measurement of voltage across the cable. However, the voltage according the criterion of 10^{-4} V/m is dangerous from cable thermal stability point of view because of the nonuniform temperature distribution along the cable. The hot end of the cable will be overloaded in spite of the average voltage across the cable is enough low. To solve this problem the measurements can be performed supplying with current pulses and critical current can be detected with criterion of 10^{-4} V/m and the further account that the transport current should not be more than value 10% less than the value of measures transport current. It makes the safe exploitation of the cable.

References

[1] Yamaguchi S, Hamabe M, Yamamoto I, Famakinwa T, Sasaki A, Iiyoshi A, et al. 2008. Research activities of DC superconducting power transmission line in Chubu University. *J. Phys.: Conf. Ser.* **97** 012290

[2] Ivanov Y V, Watanabe H, Hamabe M, Sun J, Kawahara T and Yamaguchi S 2011. Circulation pump power for 200 m cable experiment. *Physica C: Supercond.* **471** 1308-12

[3] Chikumoto N, Watanabe H, Ivanov Y V, Takano H, Yamaguchi S, Koshizuka H, et al. 2016. Construction and the circulation test of the 500-m and 1000-m DC superconducting power cables in Ishikari. *IEEE Trans. Appl. Supercond.* **26** 5402204

[4] Yamaguchi S, Koshizuka H, Hayashi K and Sawamura T 2015. Concept and design of 500 meter and 1000 meter DC superconducting power cables in Ishikari, Japan. *IEEE Trans. Appl. Supercond.* **25** 5402504

[5] Vyatkin V S, Sun J, Ivanov Y V, Chikumoto N, Watanabe H, Shyshkin O and Yamaguchi S 2016. Dependence of DC HTS cable critical current on the temperature distribution along the cable. *Phys. Procedia* **81** 191-4

[6] Yim S W, Lim S H, Kim H R, Hwang S D and Kishio K 2005. Electrical behavior of Bi-2223/Ag tapes under applied alternating over-currents. *IEEE Trans. Appl. Supercond.* **15** 2482-7

IOP Conf. Series: Journal of Physics: Conf. Series **871** (2017) 012039 doi:10.1088/1742-6596/871/1/012039

Optimal, Nanodefect Configurations via Strain-Mediated Assembly for Optimized Vortex-Pinning in Superconducting Wires from 5K-77K

A. Goyal[1,2] and S. H. Wee[1,3]

[1]Oak Ridge National Laboratory, Oak Ridge, TN 37831

[2]Now at University at Buffalo, Buffalo, NY 14227

[3]Now at Tribogenics Inc., 5440 McConnell Ave, Los Angeles, CA 90066

Email: agoyal@buffalo.edu

Abstract. Engineered nanoscale defects within $REBa_2Cu_3O_{7-\delta}$ based coated conductors are of great interest for enhancing vortex-pinning, especially in high-applied magnetic fields. We have conducted extensive research to optimize vortex-pinning and enhance J_c via controlled introduction of various types of nanoscale defects. This paper provides a summary on how density, morphology, and composition of these engineered nanoscale defects affects vortex-pinning in different field and angular regimes in the temperature regime of 5K-77K. It is shown that certain nanodefect configurations that provide the best performance at high-operating temperatures also result in the optimal properties with record values of the critical current density and the pinning force at low operating temperatures and high-applied magnetic fields.

1. Introduction

High Temperature Superconductors (HTS) wires or coated conductors based on $REBa_2Cu_3O_{7-\delta}$ (REBCO) hold promise to revolutionize the electric grid of the 21st century. The superconducting wire is processed in the form of an epitaxial film deposited by different methods on a near single-crystal flexible template comprising a metal substrate and epitaxial oxide buffer layers. The template is manufactured using two major technologies: the rolling assisted biaxially textured substrate (RABiTS) technology [1-3], and the ion beam assisted deposition (IBAD) technology [4–6]. Although such superconducting wires can carry several orders of magnitude more power than copper wires of the same cross section, further performance improvements are necessary for the superconducting technology to become cost-competitive, especially in the presence of high-applied magnetic fields. This objective can be achieved by introducing and controlling nano-sized defects and non-superconducting phases within the superconducting film's matrix to pin the flux-lattice. The first demonstration of significantly enhanced flux-pinning was via incorporation of a YBCO non-superconducting phase Y_2BaCuO_5 by intermittently depositing YBCO layers and discontinuous 211 layers by PLD using two deposition targets [7]. This approach resulted in pancake-like 211 particles with an average size of 15 nm distributed in the YBCO film, which produced some enhancement in pinning for H//*c*-axis and much higher improvements for H//*ab*-planes. The second demonstration of significantly enhanced flux-pinning was by Driscoll et al. via incorporation of randomly oriented, disk shaped $BaZrO_3$ (BZO) nanoparticles with a modal size of 10 nm [8]. It is well known that nanoscale columnar defects generated by irradiating high-temperature superconducting materials with heavy ions significantly enhance the in-field critical current density. Hence, scientists world-wide had sought means to produce such nanoscale columnar

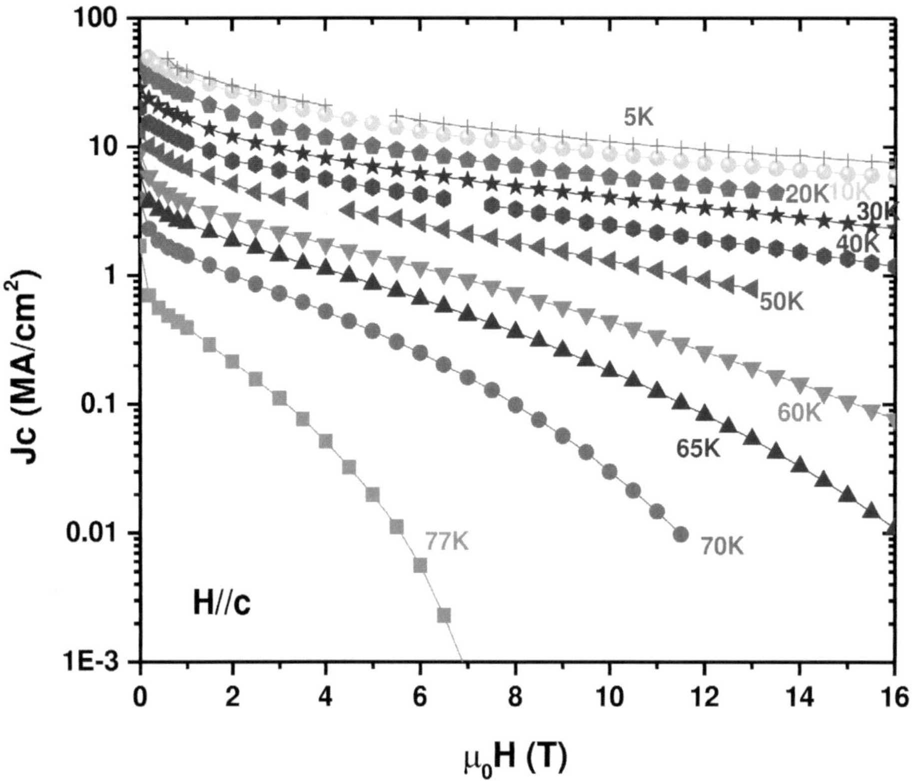

Figure 1. Critical current density, J_c, as a function of applied magnetic field at various measurement temperatures for YBCO film with 2vol% BZO additions. A self-field J_c of ~ 60MA/cm^2 is observed at 10K for 1.4 μm thick films. The extrapolated value of self-field J_c at 5K is in the range of 70-80MA/cm^2.

defects in HTS materials without the expense and complexity of ionizing radiation to realize even better vortex-pinning and superior properties in high-applied magnetic fields. Goyal et al. were the first to demonstrate incorporation of irradiation-free, nanoscale, columnar defects comprised of self-assembled BZO nanodots and nanorods resulting in strongly enhanced flux-pinning in REBCO films [9-11]. These results were subsequently reproduced world-wide [12-14]. Kang et al. were the first to demonstrate that such self-assembled, nanoscale columnar defects can be sustained through the thickness in 3-micrometer-thick YBCO films epitaxially grown via pulsed laser ablation on rolling assisted biaxially textured substrates [15]. Enhancements of the critical current, I_c, in self-field as well as excellent I_c retention in high-applied magnetic fields in the operating temperature range 65-77K were achieved in the thick films via incorporation of a periodic array of BZO columnar defects, extending through the entire thickness of the film. These columnar defects are highly effective in pinning the superconducting vortices or flux lines, thereby resulting in the substantially enhanced performance of this wire. This technology was transferred to industry and the same self-assembly process was also demonstrated for MOCVD deposited HTS wires on IBAD templates [16]. In this paper, we show that the strain-driven, self-assembly process invented by Goyal et al. results in stellar

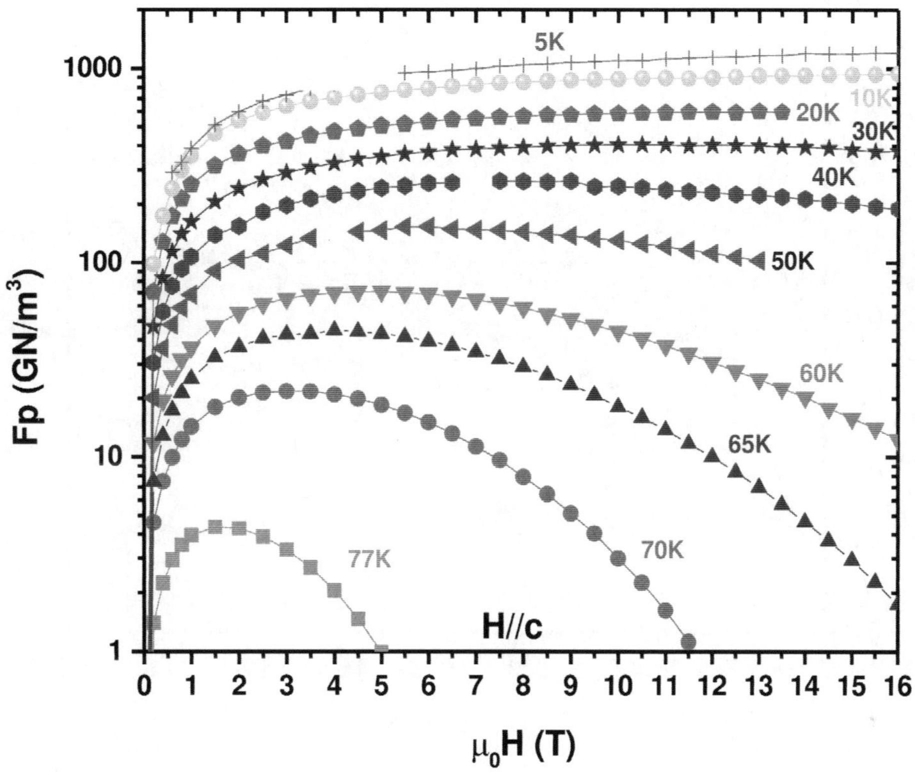

Figure 2. Pinning force, F_p, as a function of applied magnetic field at various measurement temperatures for a film with 2vol% BZO additions. The pinning force approaches $1TN/m^3$ at 10K; 16T and crosses $1TN/m^3$ at 5K, 7T.

vortex-pinning at high operating temperatures (65-77 K) as well as at low operating temperatures (4.2-30 K) in high-applied magnetic fields. At 65-77 K, incorporation of nanoscale BZO columns provides strong correlated pinning as expected. BZO nanocolumn incorporation also significantly enhances critical current density, J_c, and mitigates the J_c anisotropy at lower temperatures (4.2-30 K) and magnetic fields at up to 16 T (with no sign of decreasing at higher fields), via weak uncorrelated pinning created by oxygen point defects around the nanoscale columnar defects. This weak uncorrelated pinning, shows at 5 K, a record high flux pinning force density, $F_p > 1.5$ TN/m^3 in 1.4 μm thick films and a J_c ~20% of depairing current density J_d was achieved. This J_c value is more than 100 times higher than the optimized NbTi superconductors. In addition, J_c keeps rising with BZO concentration up to 4 vol %.

2. Experimental

All YBCO films with and without different vol% of BZO were prepared by pulsed laser deposition (PLD) using a KrF (λ=248 nm) excimer laser at a repetition rate of 10 Hz. Laser energy density and

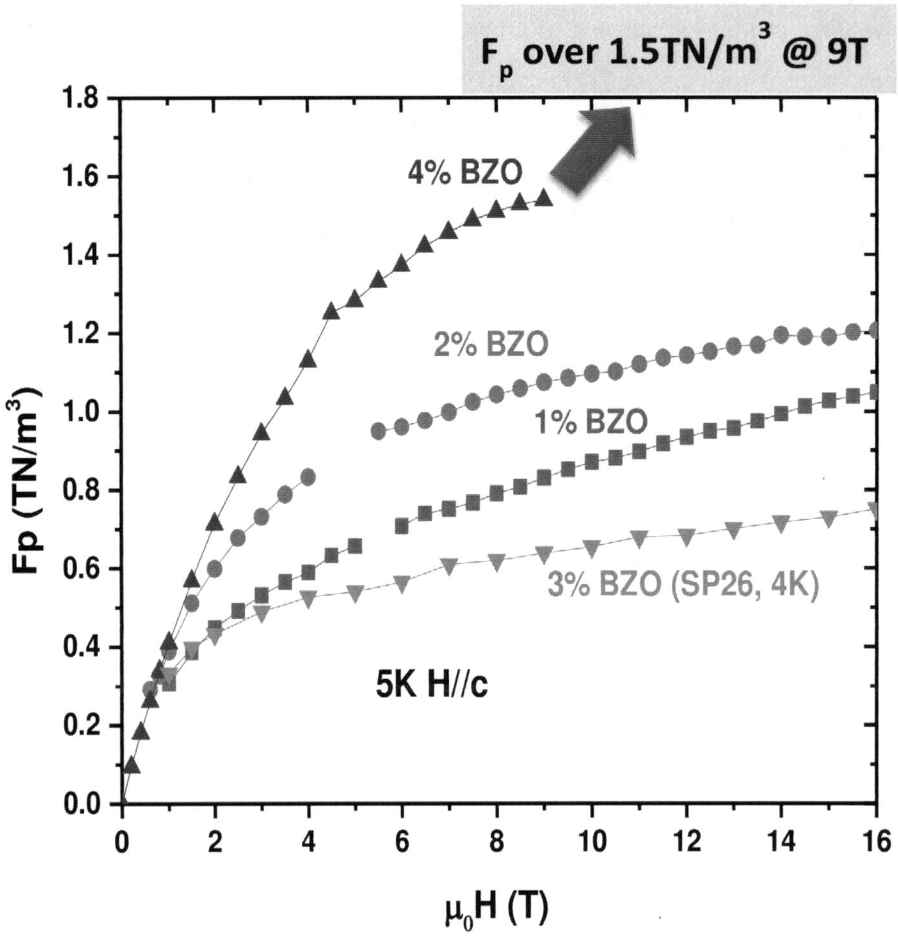

Figure 3. Pinning force, F_p, as a function of applied magnetic field at various measurement temperatures for films with different vol% BZO additions. The pinning force approaches 1TN/m^3 at 5K, 16T for the 1 vol% BZO film; crosses 1TN/m^3 at 5K, 7T for the 2vol% BZO film and exceeds 1.5TN/m^3 at 9T for the 4vol% BZO film. All films were 1.4 μm thick.

substrate to target distance were optimized to be 2 J/cm^2 and 5 cm, respectively. The PLD target (2 inch diameter, 0.25 inch thick) was made by mixing commercially available YBCO powder and $BaZrO_3$ nanopowder using standard ball mixer, followed by densification at 950°C for 2 h. The film growth temperature, T_s, was 790°C and the oxygen partial pressure, P(O_2), was 230 mTorr. All depositions were performed on IBAD-MgO templates with a $LaMnO_3$ cap layer that were supplied from Superpower Inc. All YBCO films deposited were 1.4 μm thick. After deposition, samples were in-situ annealed at T_s = 500°C and P(O_2) = 500 Torr, and ex-situ annealed at 500°C for 1 h in flowing O_2 gas after depositing sputtered Ag electrodes onto the films. The 0.5 cm wide × 2 cm long sample was patterned into a ~0.2 mm wide bridge by laser-scribing, due to the limits on the maximum measuring current in the characterization system. The standard four-point probe method with a voltage criterion of 1 μV cm^{-1} was used for the transport measurements. Microstructures were characterized by transmission electron microscopy using a Hitachi HF-3300 TEM/STEM with a secondary detector and an energy-dispersive x-ray spectrometer at 300 kV and

IOP Conf. Series: Journal of Physics: Conf. Series **871** (2017) 012039 doi:10.1088/1742-6596/871/1/012039

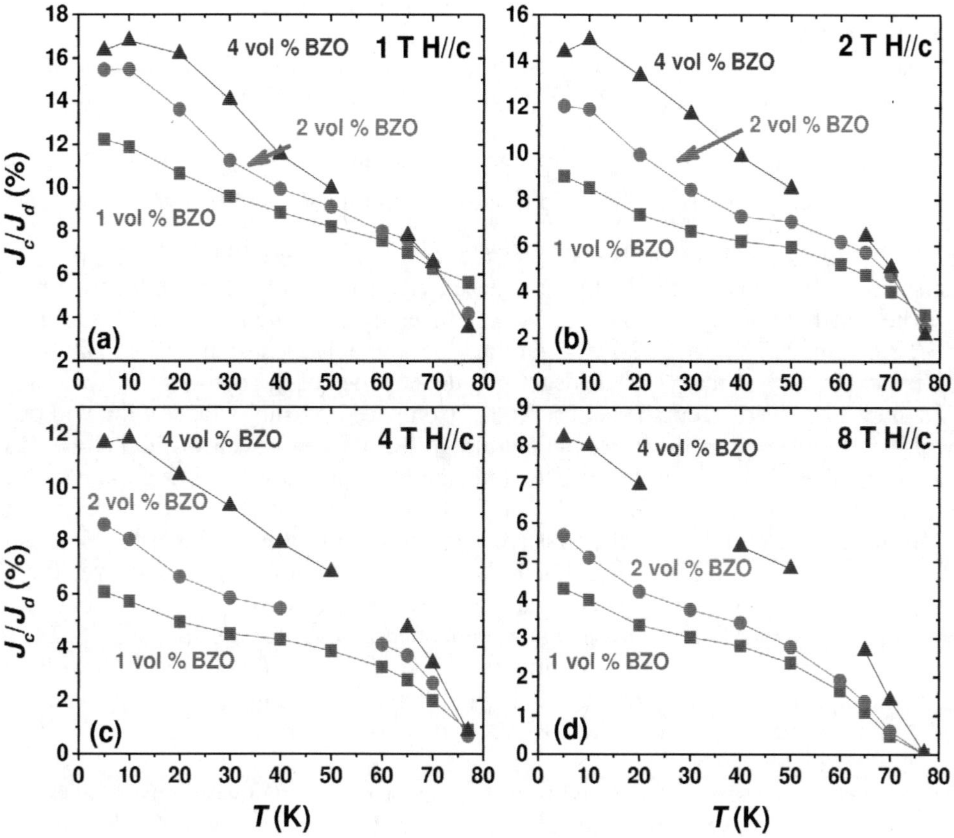

Figure 4. The ratio of transport and depairing current density J_c/J_d as a function of temperature for 1, 2 and 4 vol% BZO additions. J_c/J_d increases with decreasing temperature and vol% BZO additions.

field emission scanning electron microscopy using Hitachi S4800 FEG-SEM. TEM foils were prepared by the focus ion beam (FIB) technique, followed by low voltage ion milling and plasma cleaning. The 4.2 K and high field four-probe J_c measurements were performed in a 52-mm warm bore 31 T Bitter magnet at the National High Magnetic Field Laboratory (NHMFL), fitted with a 38 mm bore liquid He cryostat. The 10–77 K measurements were carried out in a 16 T Physical Property Measurement System (PPMS). Samples were always rotated with respect to the external magnetic field around the axis parallel to the current direction to maintain a maximum Lorentz force configuration. The angle $\theta = 0$ is defined as the applied magnetic field perpendicular to the tape plane whose parallelism to the crystallographic c-axis direction has a typical uncertainty of 1-4° produced by an offset caused by the IBAD process [17]. Due to the high I_c values observed at lower measured temperatures, samples with different geometries were prepared in order to avoid harmful Joule heating and overstressing by the large Lorentz forces ($I_c \times B$) possible in different regimes of H and T. A narrow link ~ 10 μm wide and 200 μm long, were patterned by SEM/FIB so as to restrict I_c to ~ 5 A when the sample was measured in helium gas between 10 and 77 K. Larger bridges about 1 mm wide and 1 cm long were patterned leaving the silver layers present for the high current measurements. Two different home-made I_c probes equipped with rotating sample platforms were used. One had a maximum current-carrying capability of ~ 500 A for high I_c measurement in a liquid-helium bath, while the second had ~ 5 A capability in the gas-cooled PPMS cryostat for studies at temperatures above 10 K.

3. Results and Discussion

We have previously reported results on YBCO films of thickness 0.8 μm deposited on LMO/IBAD-MgO/Hastelloy templates via PLD. Detailed XRD analysis based on BZO peak broadening estimates that the size of BZO nanophase is ~6-7 nm and remains nearly unchanged up to 10 vol% BZO, which is also confirmed by transmission electron microscopy (TEM) analysis [18]. YBCO+BZO films also keep excellent texture quality with small full-width-half-maximum values of omega and phi scans (Δω and Δφ) even up to 10 vol% BZO. Δω and Δφ values of YBCO phase are in the ranges of 1.3°~2° and 3.2°~3.6°, respectively, and those of the BZO nanophase range from 2.1° to 2.8° and from 5.8° to 6.2°, respectively. The self-field J_c at 77 K and 65 K is noticeably improved with BZO additions. The highest J_c of 4.1 MA/cm^2 at 77 K and 12.2 MA/cm^2 at 65 K are achieved with 1 vol% BZO-additions [18]. In the presence of applied magnetic fields, strong dependence of the optimum BZO doping level on the applied magnetic field strength is observed. Compared to the YBCO film without BZO additions, the 1-4 vol% BZO-doped samples show a significantly improved in-field J_c performance with higher pinning forces, F_p particularly for H//c, indicating the efficacy of BZO columnar defects as strong *c*-axis pinning centers. The in-field J_c data at 77 K show that the 1 vol% BZO-doped film has the highest J_c in the self-field and up to the first cross-over field of ~1.6 T. When the field is higher than ~1.6 T, the 2 and 4 vol% BZO doping levels lead to the highest J_c up to and over the second cross-over field of ~7.3 T, respectively. The J_c data at 65 K also show the same trend, but the cross-over fields are shifted to higher fields [18]. Angular dependent J_c data for the samples at 77 K and 65 K show massive enhancement in J_c in all field orientations compared to the YBCO film with no BZO additions. In summary, the YBCO films with BZO additions in the range of 1-4 vol% BZO, show excellent performance in the range of 65-77K [18].

In superconducting films with BZO additions, the incorporated nanocolumns give rise to local strain modulations, which have important effects not only on pinning but also on oxygen stoichiometry and T_c. Differently from (and in addition to) the macroscopic, unidirectional strain that arises in YBCO films due to epitaxial growth on a substrate with slightly different lattice parameter, the array of vertical interfaces between BZO nanocolumns and YBCO matrix produces a much more localized, non-monotone strain field, which varies on the nanoscale and produces measureable broadening of XRD peaks. XRD analysis shows that such a local strain increases with increasing BZO concentration and thus nanocolumns density. In addition to an increase in local strain, increasing BZO additions also results in a significant drop in T_c with a linear decrease in T_c of ~ 0.4 K/ vol % [19]. A mechanism for this reduction in T_c via local strain can be due to locally reduced oxygen concentration or presence of oxygen vacancies. The local strain can be quantified directly through the Z-contrast scanning TEM images by measuring the local variation of the YBCO c-axis parameter by measuring unit cell dimensions along the c-axis. Such an analysis shows that the strain is peaked away from the BZO column - YBCO interface in the YBCO phase [19]. We then measured the local oxygen stoichiometry by electron energy loss spectroscopy (EELS) elemental mapping of film regions incorporating a BZO nanocolumn. The oxygen signal shows a minimum in the YBCO region immediately surrounding the rod, which is highly strained [19]. Oxygen deficiency was also confirmed via local probing of the Cu valence using the Cu-$L_{2,3}$ edge [19]. A picture then emerges in which the strain surrounding the nanorods hinders proper YBCO oxygenation and this oxygen deficiency explains the reduced T_c in YBCO films with self–assembled BZO nanorods. The diameter of the BZO rods is 6 - 8 nm and the average spacing between BZO rods is ~20 nm for a BZO concentration of 4 vol %. The observed spatial extent of oxygen depletion is 3 - 4 times the BZO rod diameter; therefore an overlap of oxygen-depleted YBCO regions surrounding BZO nanocolumns is expected to occur near or above this concentration, leading to the disruption of current percolation through the high T_c phase and to a lower measured T_c. These oxygen vacancies or point defects if present should provide excellent collective pinning at lower temperatures. If this is the case, then there exists the possibility of BZO nanocolumns to not only provide excellent vortex-pinning at higher operating temperatures but also at lower operating temperatures.

Fig. 1 shows the J_c as a function of applied magnetic field at various measurement temperatures for 1.4 μm thick YBCO film with 2 vol% BZO additions. A self-field J_c of ~ 60 MA/cm^2 is observed at 10 K. The extrapolated value of self-field J_c at 5 K is in the range of 70-80 MA/cm^2. The J_c for the film is high throughout the operating temperature range as shown in the figure. These values are the highest reported to date for such thick films at 5 K as opposed to a measurement temperature of 4.2 K (there is a massive increase in J_c from 5 K to 4.2 K). It should be noted that these values cannot be compared to thinner films in the thickness range of 0.2-0.8 μm, as typically for all in-situ fabricated films, there is a strong exponential decrease of J_c with film thickness.

Pinning force, F_p, as a function of applied magnetic field at various measurement temperatures for a film with 2 vol% BZO additions is shown in Fig. 2. The pinning force approaches 1 TN/m^3 at 10 K, 16 T and crosses 1 TN/m^3 at 5 K, 7 T. These values of pinning force are the highest reported to date for a 1.4 μm thick film with only 2 vol% BZO additions. The F_p as a function of applied magnetic field at various measurement temperatures for 1.4 μm thick films with different vol% BZO additions is shown in Fig. 3. The F_p approaches 1 TN/m^3 at 5 K, 16 T for the 1 vol% BZO film; crosses 1 TN/m^3 at 5 K, 7 T for the 2 vol% BZO film and exceeds 1.5 TN/m^3 at 9 T for the 4 vol% BZO film. Once again, these are record performance values and the highest reported to date. These high values of J_c and F_p in YBCO films with BZO additions were first presented at the Electronic Materials and Applications (EMA) meeting in January of 2013 [20].

The temperature dependence of J_c/J_d, the ratio of the measured J_c to the calculated depairing current density, J_d at 4.2K, can provide the pinning evolution with temperature [21]. As shown in Fig. 4, at T > 65 K, due to strong thermal fluctuation, J_c/J_d is dramatically reduced with increasing temperature. Additionally, this decay of J_c/J_d is worsened by increasing of BZO concentration. This fast decline of J_c/J_d partially caused by the reduced T_c, and/or might be related to the high-density of straight BZO nanorods [22]. Between, ~30 K and ~60 K, J_c/J_d shows flattening temperature dependence. With further decreasing temperature, weak pinning takes over and J_c/J_d rapidly increases with decreasing temperature. This dramatically enhancement of J_c/J_d at T < 30 K becomes more pronounced at higher BZO concentration and lower fields. In optimized Nb-Ti thin films, the maximum reached J_c is of order of 10% J_d [23], much lower than □20% J_d achieved in these thick and practical REBCO conductors.

In summary, we have shown possible optimal, nanodefect configurations via strain-mediated self-assembly for optimized vortex-pinning in superconducting wires in broad magnetic field and temperature regimes. At higher operating temperatures, the BZO columnar defects provide strong correlated vortex-pinning resulting in high-J_c and excellent performance in high-applied magnetic fields. At low operating temperatures, oxygen point defects created by local strain at YBCO/BZO interfaces in the films provide excellent weak collective-pinning resulting in very-high J_c (with J_c/J_d more than 100 times higher than the optimized NbTi superconductors) and record F_p values exceeding 1.5 TN/m^3. The transport properties reported in this paper surpass all previously reported values in the temperature regime of 5-77 K for 1.4 μm thick films with only up to 4 vol% BZO additions [24]. The results reported in this paper provide a route for fabrication of commercial HTS wires with outstanding superconducting transport properties over a broad operating temperature regime.

Acknowledgments
Authors wish to acknowledge Aixia Xu and David Larbalestier for transport measurements performed at NHFML.

IOP Conf. Series: Journal of Physics: Conf. Series **871** (2017) 012039 doi:10.1088/1742-6596/871/1/012039

References

[1] Goyal A et al. 1996 *Appl. Phys. Lett.* **69** 1795.
[2] Goyal A, Paranthaman M and Schoop U 2004 *MRS Bulletin* **29** 552.
[3] Goyal A 2005 Epitaxial superconductors on rolling-assisted-biaxially-textured-substrates (RABiTS) *Second Generation HTS Conductor* ed A. Goyal (Kluwer, Norwell, MA, 2005), Chapter 2, pages 29-46.
[4] Iijima Y, Tanabe N, Ikeno Y, and Kohno O 1991 *Physica C* **185** 1959-1960.
[5] Wu X D, Foltyn S R, Arendt P N, Townsend J, Adams C, Tiwari P, Coulter J W, Peterson D E 1994 *Appl. Phys. Lett.* **65** 1961.
[6] Arendt P 2005 IBAD Template for HTS Coated Conductors *Second Generation HTS Conductor* ed A. Goyal (Kluwer, Norwell, MA, 2005), Chapter 1, pages 1-28.
[7] Haugan T, Barnes P N, Wheeler R, Meisenkothen F, and Sumption M 2004 *Nature* **430** 867.
[8] Macmanus-Driscoll J et al. 2004 *Nat. Mater.* **3** 439.
[9] Goyal A et al 2004 Presented at the 2004 Annual Peer Review of the US Superconductivity Program for Electric Systems (July 2004). http://www.energetics.com/ meetings/supercon04/pdfs/presentations/f rabits strategic talk ffy04 final.pdf
[10] Goyal A et al. 2004 CCA 2004: Presented at the Int. Workshop on Coated Conductor for Applications (Oiso Prince Hotel, Kanagawa, Japan, Nov. 2004).
[11] Goyal A et al. 2005 *Supercond. Sci. Technol.* **18** 1533.
[12] Yamada Y et al. 2005 *Appl. Phys. Lett.* **87** 132502.
[13] Varanasi C, Burke J, Wang H, Lee J H, and Barnes P 2008 *Appl. Phys. Lett.* **93** 092501.
[14] Mele P, Matsumoto K, Ichinose A, Mukaida M, Yoshida Y, Horii S and Kita R 2008 *Supercond. Sci. Technol.* **21** 125017.
[15] Kang et al. 2006 *Science* **311** 5769.
[16] Chen et al. 2009 *Appl. Phys. Lett.* **94** 062513.
[17] Chen et al. 2009 *Supercond. Sci. Technol.* **22** 0550131.
[18] Wee et. al. 2013 *Nature Magazine's Scientific Reports* **3** Article number: 2310.
[19] Cantoni et al. 2011 *ACS Nano* **5** 4783–4789.
[20] Goyal et al. 2013 Electronic Materials & Applications in Symposium S12: Recent Developments in High-Temperature Superconductivity, http://ceramics.org/wp-content/uploads/2013/01/ema2013_finalprogram.pdf, page 14.
[21] Blater et al. 1994 *Rev. Mod. Phys.* **66** 1125.
[22] Maiorov et al. 2009 *Nat. Mat.* **8** 398.
[23] Stejic et al. 1994 *Phys. Rev. B* **49** 1274.
[24] Obradors et al., 2014 *Supercond. Sci. Technol.* **27** 044003.

IOP Conf. Series: Journal of Physics: Conf. Series **871** (2017) 012040 doi:10.1088/1742-6596/871/1/012040

Effect of lanthanum addition on the fabrication of $REBa_2Cu_3O_y$ (RE=Gd, Sm and Dy) thin films by metal organic deposition using fluorine-free solutions

S Kato[1], R Kita[1], N Kobayashi[2] and O Miura[2]
[1]Shizuoka University, Johoku 3-5-1, Hamamatsu, Shizuoka 432-8561, Japan
[2]Tokyo Metropolitan University, Minamiosawa 1-1, Hachioji, Tokyo 192-0364, Japan

E-mail: kita.ryusuke@shizuoka.ac.jp

Abstract. The effects of lanthanum (La) additions of 0.5–10 mol% on the superconducting properties and crystal growth of $REBa_2Cu_3O_y$ (REBCO, RE = Gd, Sm, and Dy) films fabricated by the fluorine-free metal-organic deposition method using 2-ethylhexanate solutions were investigated. La-doped REBCO films showed the decrease in the number density of holes on the film surface and the increase in the XRD peak intensity of the REBCO superconducting phases. La-added GdBCO with La contents of 0.5–3 mol%, and SmBCO and DyBCO films with La contents of 0.5–10 mol% showed improved critical current density (J_c) at the self-magnetic field compared to the pure films. For the GdBCO film, the J_c value at the self-magnetic field increased from 1.32 to 2.38 MA/cm^2 upon La addition.

1. Introduction

$REBa_2Cu_3O_y$ (RE = Y and other rare earth elements) superconducting thin films are promising materials for high-temperature superconducting wires because of their high critical current densities (J_c's) under large magnetic fields. Metal-organic deposition (MOD) is a suitable technique for mass production of REBCO thin films because it involves a non-vacuum and cost-effective process. Furthermore, the MOD process using fluorine-free solutions (FF-MOD), i.e., the process employed in the present study, is relatively simple since it requires no water vapor for removing fluorine from the films, unlike MOD using trifluoroacetates (TFA-MOD).

The development of REBCO films with superior high J_c in magnetic fields (B) is essential for applications such as strong magnetic field generating coils and superconducting cables. Recently, the introduction of artificial pinning centers (APC) into REBCO films has been studied intensively, with the goal of enhancing J_c-B performance [1-3]. In particular, REBCO films with Ba oxides or RE oxides were reported to have high pinning forces [3, 4]. The addition of RE oxides into REBCO was also reported to be effective in enhancing the J_c-B performance of pulsed-laser-deposited YBCO films and sintered films [5-7]. Very recently, we have reported the enhanced J_c-B performance of FF-MOD-REBCO films achieved by Ho addition [8]. However, there are few reports on the effects of RE_2O_3 or RE addition on MOD films.

In the present study, we have investigated the effects of lanthanum (La) addition on the superconducting properties and crystal growth of REBCO (RE = Gd, Sm and Dy) films fabricated by the FF-MOD method.

2. Experimental

The starting solutions for REBCO (RE = Gd, Sm and Dy) films were prepared by mixing stoichiometric amounts of Gd-, Sm-, Dy-, Ba-, and Cu-2-ethylhexanate (2-EH) solutions (Nihon Kagaku Sangyo Co., Ltd.). La-added REBCO films were fabricated by mixing REBCO solutions with 0.5, 1.0, 3.0, 5.0, and 10.0 mol% La-2-EH. The pure and La-added REBCO precursor films were fabricated by coating the mixed solutions on $LaAlO_3$ (100) single-crystal substrates, dried at 120°C for 30 min, and then calcined at 600°C for 30 min in air. The coating and calcination procedure was repeated three times. The precursor films were fired at a heating rate of 5°C/min from 840°C to 900°C, kept at 900°C for 30 min, and then cooled to room temperature in flowing N_2 gas with an oxygen content of less than 2 ppm. Finally, the films were annealed at 350°C for 2 h in O_2. The final thickness of the films was approximately 300 nm. The phases in the films were identified by X-ray diffractometry (XRD) with CuKα radiation. The electrical resistivity of the films was measured by the standard four-probe technique to determine the critical temperature (T_c) values. The J_c values of the films under varying magnetic fields ($0 < B < 1$ T) were calculated from B-M curves obtained by a superconducting quantum interference device, by using an extended version of Bean's critical state model [5]. The surface of the films was characterized by scanning electron microscopy (SEM).

3. Results and Discussion

Figure 1 shows the La content dependence of the surface SEM images for the GdBCO, SmBCO and DyBCO films. Many holes were observed for the surface of the pure GdBCO film (figure 1(a)). The GdBCO film with La addition below 3 mol% showed decrease in the average number density of holes on the film surface from 17.3 to 2.8 in the area of 10×10 μm^2. The average size of the hole for the GdBCO film without and with 3mol% La addition was estimated to be 0.36 μm and 0.18 μm, respectively. This indicated that the number density of holes and relatively large holes were decreased by 3mol%-La addition. The GdBCO film with La addition over 3mol% showed cracks on the surface. A large number of fine grains due to the poor grain growth were observed for the pure DyBCO surface(figure 1(g)). The La-added DyBCO film over 3 mol% showed the relatively flat and dense surface. These results suggest that La addition has effects on the improvement of the surface flatness and decrease in the hole density for the MOD-GdBCO, SmBCO and DyBCO films.

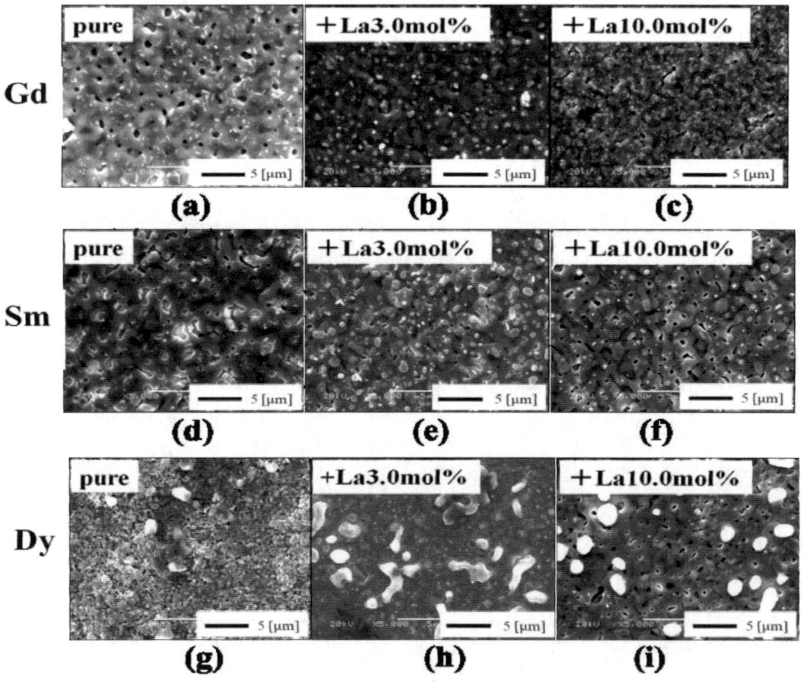

Figure 1. SEM images of pure and La-added REBCO film surfaces.

Figure 2 shows the dependence of the (006) XRD peak intensity on La content for GdBCO, SmBCO and DyBCO films. For the GdBCO film, as the amount of La addition increased, the XRD peak intensity increased up to 3 mol% and decreased thereafter. This indicates that a La addition of up to 3 mol% enhances the crystal growth of the superconducting phase. SmBCO and DyBCO films showed an increase in XRD peak intensity with La addition. In particular, the increase in the XRD peak intensity for the DyBCO films agrees with the results from the SEM analysis in figures 1(g) and 1(h). These results suggest that La addition is effective in promoting the crystallization and/or crystal growth of the superconducting phase. The decrease in the number density of holes on the film surface and the promotion of the film crystallization may relate to the improved lattice matching between REBCO and pseudo-cubic LAO with a lattice constant of a = 3.79 Å, because of the decrease in the lattice constants of a and b by the La substitution for the Ba site in the REBCO structure as reported for $La(Ba_{2-x}La_x)CuO_y$ [9], or the decrease in the b-axis for $Y(Ba_{2-x}La_x)CuO_y$ compound [10].

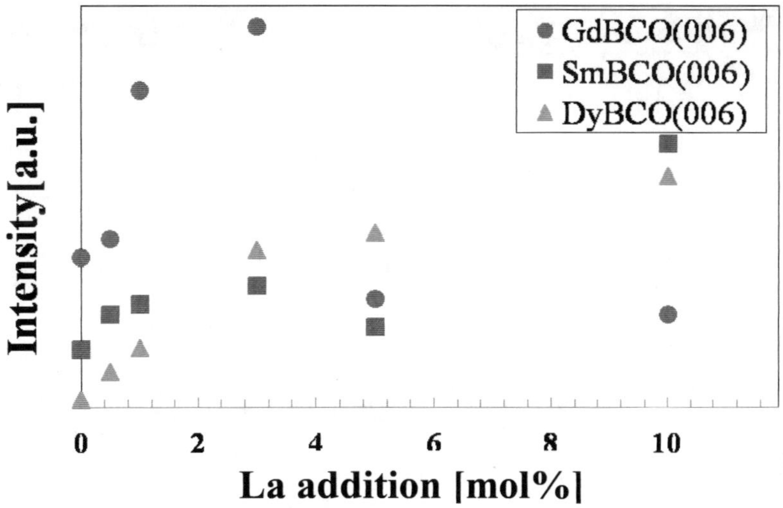

Figure 2. Dependence of (006) XRD peak intensity on La addition for REBCO films.

Figure 3 shows the J_c-B performance for the GdBCO, SmBCO, and DyBCO films. The T_c values for pure and La-added REBCO films were measured to be 89 - 90 K. The La-added GdBCO film showed an improved J_c-B performance compared to the pure GdBCO film below 3 mol% (figure 3(a)). This agrees with the trend revealed by the SEM (figure 1) and XRD analyses (figure 2). The J_c value at the self-magnetic field for the 1 mol%-La-added-GdBCO film (2.38 MA/cm^2) was 1.8 times as high as that of the pure GdBCO film (1.32 MA/cm^2). The J_c-B performance also improved upon La addition in the case of the SmBCO and DyBCO films. The J_c at the self-magnetic field for the 3-mol%-La-added SmBCO film increased by a factor of four compared to the pure SmBCO film. The 3-mol%-La-doped DyBCO film showed a J_c of 0.89 MA/cm^2, even though the pure DyBCO film with T_c of 89 K showed no superconducting current. The enhancement in the self-magnetic field J_c and the J_c-B performance of the La-added films was presumably due to the decrease in the number density of holes on the film surface and improved crystallinity induced by La addition, as discussed in figures 1 and 2.

J_c is known to be proportional to $B^{-\alpha}$ under a relatively low magnetic field ($0 \leq B \leq 1$ T) [11-12]. Here, the power exponent α is a parameter for describing the pinning performance in superconducting thin films under a magnetic field: a smaller α value means higher J_c-B performance. The La content dependence of α for La-added REBCO films is shown in figure 4. The α value of the GdBCO and DyBCO films was almost constant below a La content of 3 mol%, and increased with increasing La content above 3 mol%. By contrast, the α values of the SmBCO films showed a decreasing trend with increasing La content. The lowest α for La-added REBCO films was approximately 0.72, which is comparable to the α values of the Ho-added MOD-GdBCO films in our previous study [8]. A relatively low α value of 0.53 is reported for MOD-REBCO thin films with a pinning center such as $BaZrO_3$ [13]. Therefore, we speculate that the decrease in α is due to the improvement in the number density of holes on the film surface and crystallinity of the films.

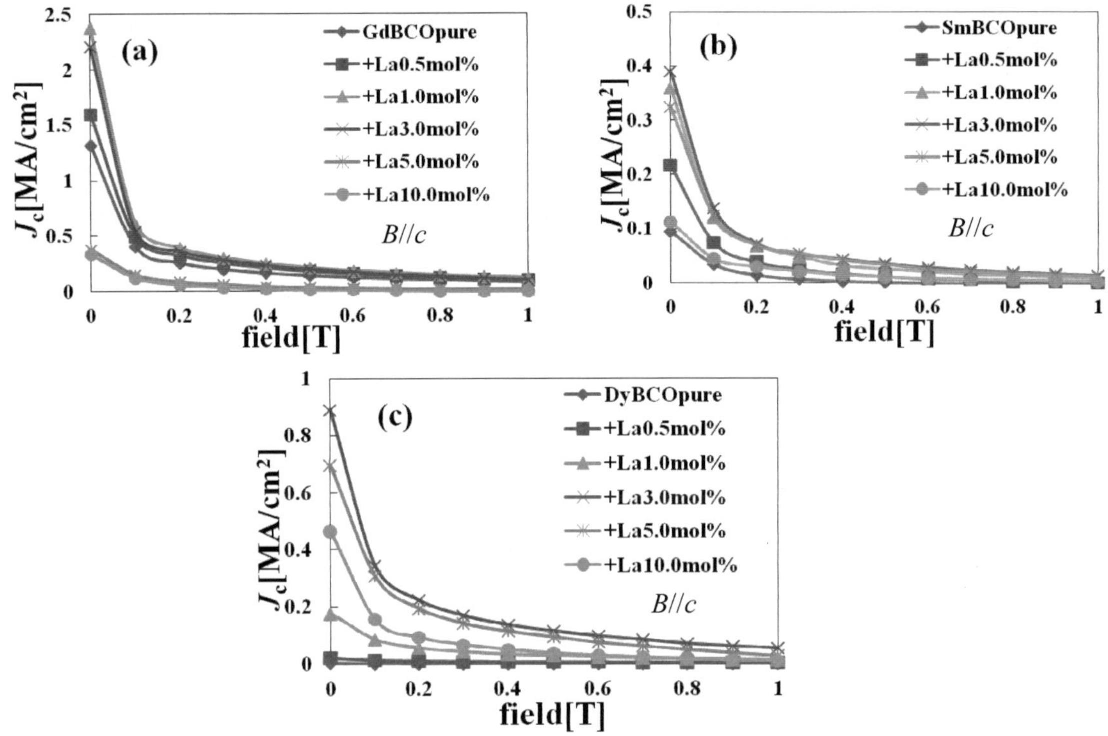

Figure 3. Magnetic field dependence of J_c for pure and La-added REBCO films at 77 K: (a) GdBCO, (b) SmBCO, and (c) DyBCO films.

Figure 5 shows the dependence of c-axis length on La addition for GdBCO, SmBCO, and DyBCO films. All the films expect the SmBCO exhibited the same trend: The c-axis length increased with increasing La content. This increase was presumably caused by the partial substitution of La for the

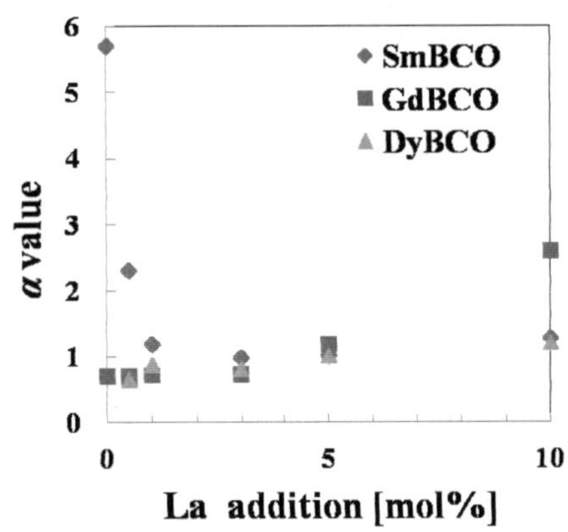

Figure 4. La addition dependence of α values for pure and La-added REBCO films.

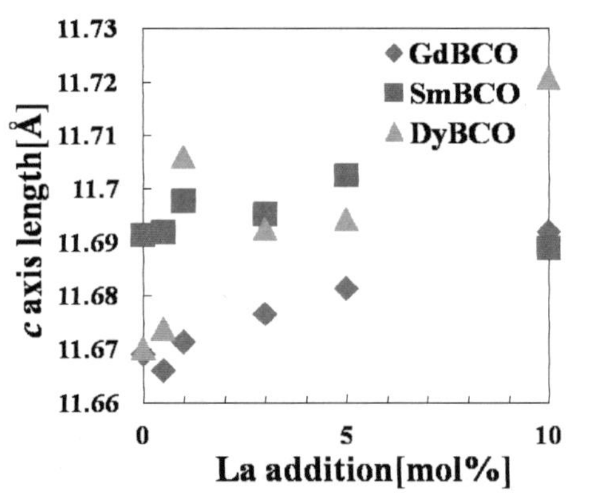

Figure 5. Dependence of c-axis length on La addition for REBCO films.

Gd, Sm, and Dy sites in the REBCO structure, given the larger ionic radius of La^{3+} (0.105 nm) as compared to those of Gd^{3+}(0.094 nm), Sm^{3+}(0.096 nm), and Dy^{3+}(0.091 nm). Furthermore, the change in the *c*-axis length of the DyBCO film was greater than that of the SmBCO film, which is attributable to the difference in ionic radius between Dy^{3+} and Sm^{3+}.

4. Conclusions

We have investigated the effects of La addition on the superconducting properties and crystal growth of REBCO films (RE = Gd, Sm, and Dy) fabricated by the FF-MOD method using 2-ethylhexanate solutions. La-doped REBCO films showed an improvement in surface flatness and a decrease in density of holes on the surface. The SEM and XRD results suggested that La addition was effective in promoting crystallization and crystal growth of the superconducting phase. In particular, for the 1 mol%-La-added-GdBCO film, the J_c value at the self-magnetic field improved 1.8-fold (2.38 MA/cm^2) relative to the corresponding value for the pure GdBCO film (1.32 MA/cm^2). However, the J_c-B performance did not improve upon La addition in the case of the REBCO films.

Acknowledgement

This work was partially supported by JSPS KAKENHI Grant Number 25390054.

References

[1] Matsushita T 2000 *Supercond. Sci. Technol.* **13** 730-737
[2] Larbalestier D C, Gurevich A, Feldmann D M and Polyanskii A 2001 *Nature* **414** 368-377
[3] Matsumoto K and Mele P 2010 *Supercond. Sci. Technol.* **23** 014001
[4] Mele P, Matsumoto K, Horide T, Miura O, Ichinose A, Mukaida M, Yoshida Y and Horii S 2007 *Supercond. Sci. Technol.* **20** 616-620
[5] Song X, Zhijun C, Kim S, Feldmann D, Larbalestier D, Reeves J, Xie Y and Selvamanickam V 2006 *Appl. Phys. Lett.* **88** 212508
[6] Jun B H, Jung S A, Park S D, Park B J, Han Y H and Kim C J 2011 *Physica* C **471** 876-879
[7] Kita R, Hosoya N, Otawa N, Kawabata S, Nakamura T, Miura O, Mukaida M, Yamada K, Ichinose A, Matsumoto K, Horii S and Yoshida Y 2009 *Physica* C **469** 1157-1160
[8] Suzuki H, Kita R, Kubo H, Miura O, Yamada K and Kaneko K 2015 *Physics Procedia* **65** 133
[9] Segre C U, Dabrowski B, Hinks D G, Zhang K, Jorgensen J D, Beno M A, and Schuller I K 1987 *Nature* **329** 227-229
[10] Harada Tand Yoshida K 2003 *Physica* C **391** 1-7
[11] Campbell A M and Evetts J E 1972 *Adv. Phys.* **21** 199-428
[12] Klaassen F C, Doornbos G, Huijbregtse J M, van der Geest R C F, Dam B and Griessen R 2001 *Phys. Rev. B* **64** 184523
[13] Suzuki H, Kita R, Otsu Y, Miura O, Yamada K and Kaneko K 2014 *Physics Procedia* **58** 162-165

Influence of chemical etching and heat-treatment on the structure and superconducting properties of YGdBCO coated conductors

M J Wang[1], W T Wang[1] , L Liu[1], B L Huo[1], X Yang[1], C H Cheng[2] and Y Zhao[1,2]

[1]Key lab of Magnetic Levitation Technologies and Maglev Trains(Ministry of Education of China), Key Laboratory of Advanced Technologies of Materials (Ministry of Education of China), Superconductivity and New Energy R&D Center, Mail stop 165#, Southwest Jiaotong University, Chengdu, Sichuan 610031, China.

[2]School of Materials Science and Engineering,University of New South Wales, Sydney, NSW 2052, Australia.

Corresponding author's e-mail addresses: wtwang@swjtu.edu.cn (W.T. Wang)
yzhao@swjtu.edu.cn (Y. Zhao)

Abstract. The effects of chemical etching (to remove metal stabilizer layers) and novel heat treatment process on the structure and superconducting properties of YGdBCO CCs for preparing a superconducting joint were studied. After removing the Cu stabilizer layer with the $FeCl_3$ alcohol solution, the mixture of $NH_3.H_2O$ and H_2O_2 was used to remove Ag stabilizer layer with various conditions such as etching temperature and time. Due to the decomposition of YGdBCO at high temperature, few secondary phases such as YGd211 and $BaCuO_2$ were detected after partial melting. It is interested to note that these secondary phases were not detected after recrystallization at a relatively lower temperature. According to the pseudo-binary phase diagrams of Lee [1], the peritectic reaction of YGd211 was occurred and the YGd123 particle was aligned again along *c*-axis. Additionally, the oxygenation annealing process was indispensable to restore the degraded superconducting properties of YGdBCO CCs caused by the oxygen diffusion out of itself during heat treatment process. The above results were favorable to prepare the superconducting joint of YGdBCO CCs in our future work.

1. Introduction

Since the discovery of YBCO in 1987, conductor development has advanced to the stage where YBCO superconducting tape in kilometer-length is commercially available [2]. A lot of applications such as transformers, fault current limiters, motors, cables with YBCO CCs have been conducted. The joining technique of the YGdBCO CCs is indispensable for these applications. Various non-superconducting joints such as soldering joints and diffusing joints with Ag stabilizer layers are prevailing recently. However, the resistances of these joints are inevitable due to the existence of stabilizers or solders, which cause energy loss. Therefore, the development of superconducting joint is urgently required.

Recently, Park *et al* [3] and Xinzhe Jin *et al* [4] have reported the successful manufacture of GdBCO superconducting joints. However, the secondary phases such as Y211、$BaCuO_2$ and CuO have been retained within the interface of the joints, which negatively affect the superconductivity [3]. On the other hand, we have reported a promising heat-treatment technique called "melting diffusion and texture merging" for preparing the superconducting joint of YGdBCO CCs [5]. During the texture merging process, these secondary phases have been found to be transformed into YGd123 phase again.

In this work, the effects of jointing parameters such as the chemical etching process, the melting temperature, the oxygen annealing(OA) duration time on the structure and superconducting properties of YGdBCO CCs were discussed in detail.

2. Experimental

Figure 1 shows the configuration of the YGdBCO CCs (SCS4050, SuperPower Inc.) used in this study. It was 4 mm wide and ~100μm thickness. Buffer layers (~0.2μm) were deposited on the hastelloy substrate layer (~50μm) via ion-beam-assisted deposition (IBAD) and the magnetron sputtering system. The YGdBCO superconducting layer (~1μm) was deposited on the buffer layers by metal organic chemical vapor deposition (MOCVD), and the Ag stabilizer layers (~2μm) were then sputtered on the YBCO superconducting layer and the bottom of the substrate layer. The Cu stabilizers layer (~20μm) were finally electroplating on the top and bottom of the Ag layer [6].

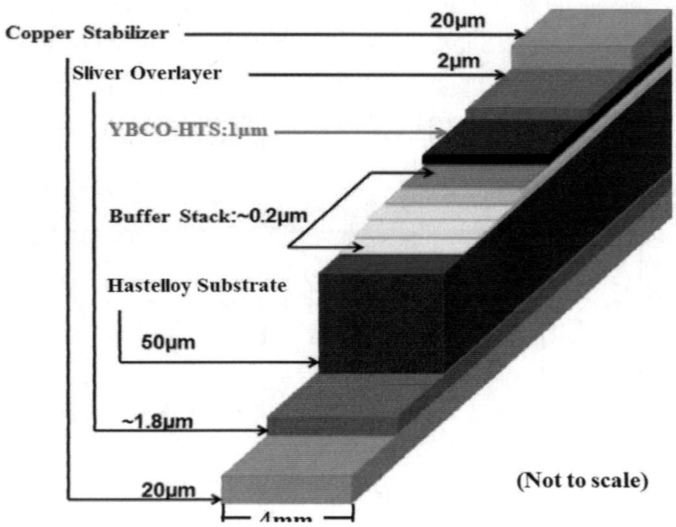

Figure 1. Configuration of the YGdBCO CCs used in this study (SuperPower)

For a superconducting joint of YGdBCO CCs which is in the case of a face-to-face junction, a chemical etching method was used to remove the metal stabilizer layers and the evenly etched surface was achieved without damaging the structure and superconductivity of YGdBCO CCs. Figure 2 shows a schematic drawing of the chemical etching process. Firstly, the $FeCl_3$ alcohol solution was used to remove the Cu stabilizer layer. Secondly, the mixture solution of $NH_3.H_2O$ and H_2O_2 (the ratio was 1:1) was used to remove the Ag stabilizer layer[7] and the etching temperature and the duration was investigated.

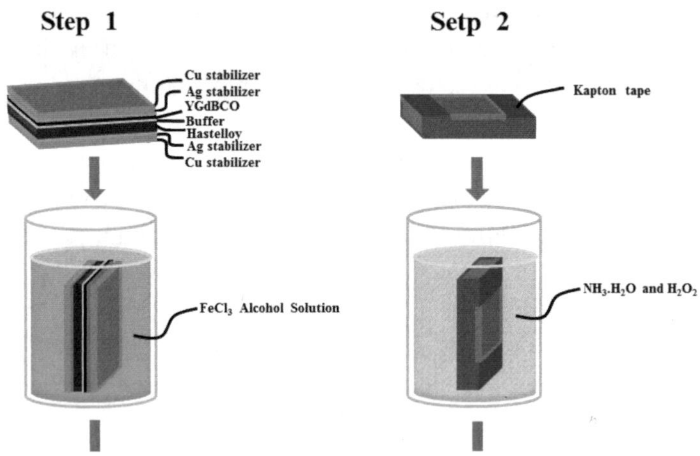

Figure 2. A schematic drawing of the chemical etching process

Figure 3 shows the heat-treatment cycles of partial melting, recrystallization and additional oxygenation processes. Firstly, the samples were held at various melting temperatures for 1min and then were held at 770 ℃ for 60min for recrystallization under a reduced PO_2(0.4Torr). Secondly, the samples were held at 400°C for various times under a flow of pure oxygen (250cm^3 min^{-1}) to restore the superconductivity.

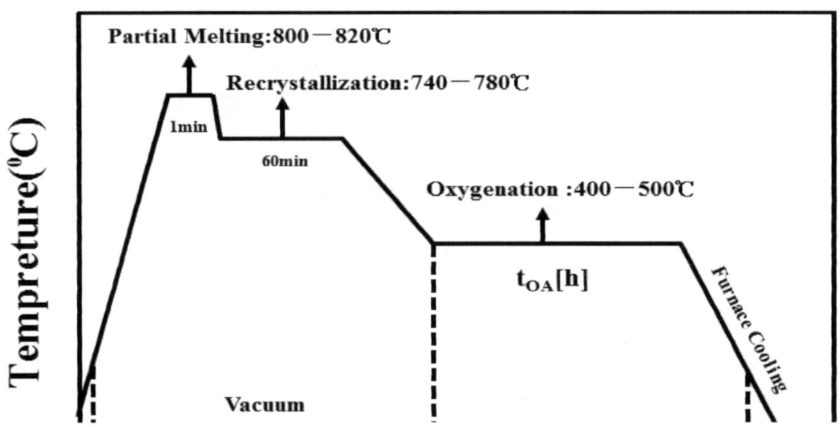

Figure 3. Heat-treatment cycles for the partial melting, recrystallization and additional oxygenation processes.

The self-field critical current(I_c) of the samples were measured using a 0.1μV/cm criterion in a bath of liquid nitrogen (77 K). A Philips X'Pert MRD diffractometer with Cu Kα radiation was used to record the θ–2θ X-ray diffraction patterns, which characterized the change conditions of phase purity of YGdBCO. The microstructural analyses of the YGdBCO layer were performed by the scanning electron microscope (SEM) equipped with an energy dispersive X-ray analysis. The HAS (Hall sensor array) system was used to measure the critical current(I_c) values of the samples after oxygenation [8].

3. Results and discussion

Figure 4 shows *V-I* curves of the samples after etching Cu and Ag stabilizer layers. As shown in Figure 4(a) and (b), the I_c value after etching Cu stabilizer layer was coincident with the virgin one. There was nearly no degradation of I_c value after etching Ag stabilizer layer when the etching temperature and time were 5°C and 10s, respectively. However, when the temperature increased to 25°C, the I_c value has decreased by 23%. Similarly, it has decreased by 36% when the etching time increased to 120s. In conclusion, it is suitable to peel the Cu stabilizer layer with $FeCl_3$ alcohol solution and the Ag stabilizer layer with the mixture solution of $NH_3.H_2O$ and H_2O_2 (the ratio was 1:1).

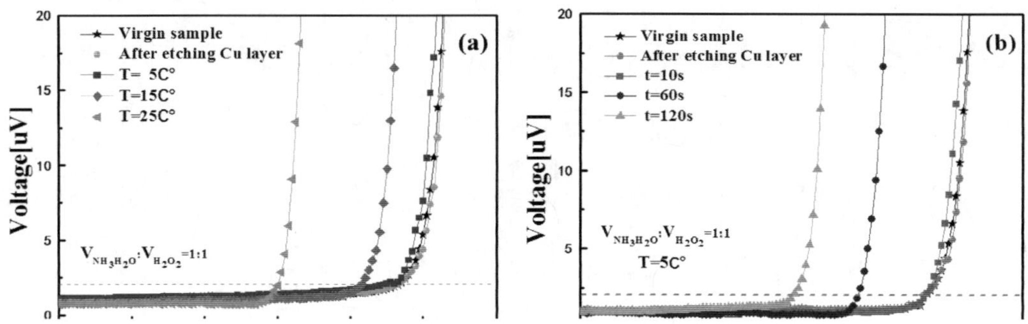

Figure 4. *V–I* curves of the YGdBCO CCs before and after chemical etching with various temperatures (a) and times (b).

As shown in Figure 5, all the samples show strong (00l) peaks demonstrating good c-axis orientations. The MgO and Ni peaks were derived from the buffer and substrate layers of YGdBCO CCs. Although there were some YGd211 peaks for the virgin sample, after partial melting process, the YGd211 peaks became stronger and some more secondary phases such as $BaCuO_2$ were detected. Particularly, when the partial melting temperature increased to 820℃, the peaks of YGd211 and $BaCuO_2$ became more obvious. These results illustrated that the decomposition of YGd123 occurs. As shown in Figure 5(d), after partial melting at 800℃and recrystallization at 770℃, the peaks of YGd211 and $BaCuO_2$ generated in partial melting process disappeared, which demonstrated that the *c*-axis of YGd123 particle was aligned again driven by the undecomposed YGdBCO grains [9]. In conclusion, the liquid phases of $BaCuO_2$ and CuO generated at 800℃ were favorable to the diffusion connection of superconducting joint and the *c*-axis of YGd123 particle was aligned again after recrystallization at 770℃, which could be favorable to form the texture merged interface of joint.

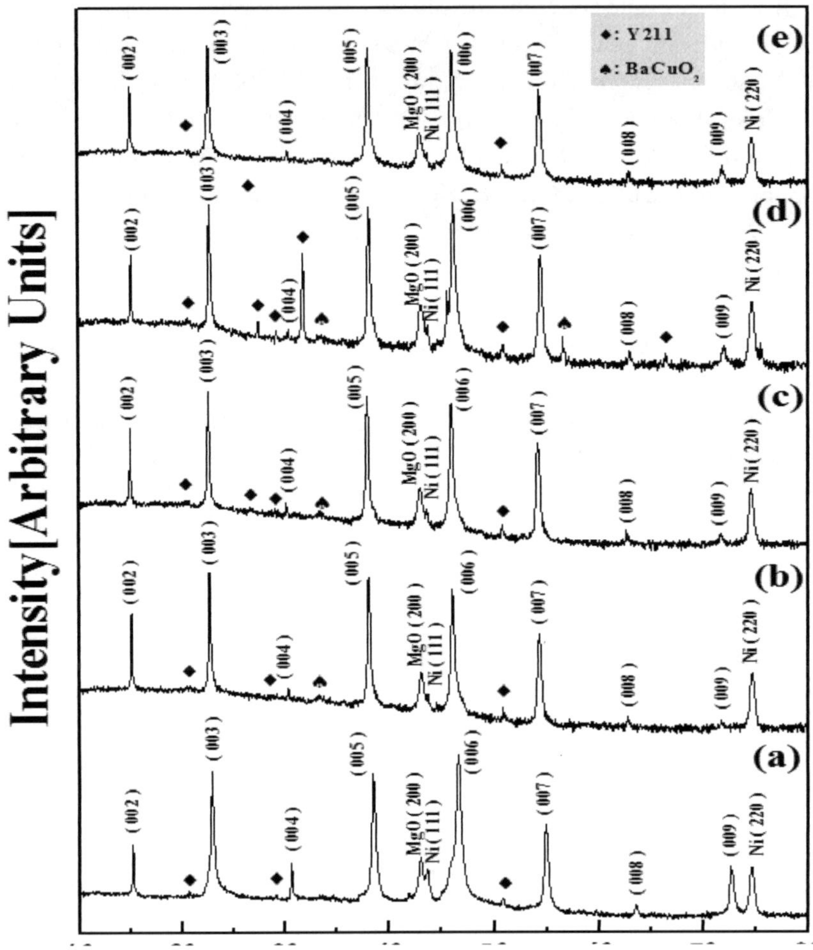

Figure 5. The XRD patterns of the virgin sample (a), the samples after partial melting at 800℃ (b), 810℃(c), 820℃(d) and the sample after partial melting at 800℃, recrystallization at 770℃ (e).

As shown in Figure 6, the surface of YGdBCO was partially melted at 800℃ compared with the virgin tape. Furthermore, the degree of partial melting was further strengthened at 810℃ and 820℃. Besides, the size of Nano-scale fine grain became larger with the increasing of temperature. These results were consistent with the above XRD patterns which demonstrated the decomposition of YGd123. As for Figure 6(e), after partial melting at 800℃ and recrystallization at 770℃, the smooth and dense surface morphology was performed. It illustrated that the texture growth with c-axis orientation was occurred during recrystallization process. In conclusion, the above results also declared that the heat-treatment technique of partial melting and recrystallization was suitable to the texture merging of YGdBCO CCs.

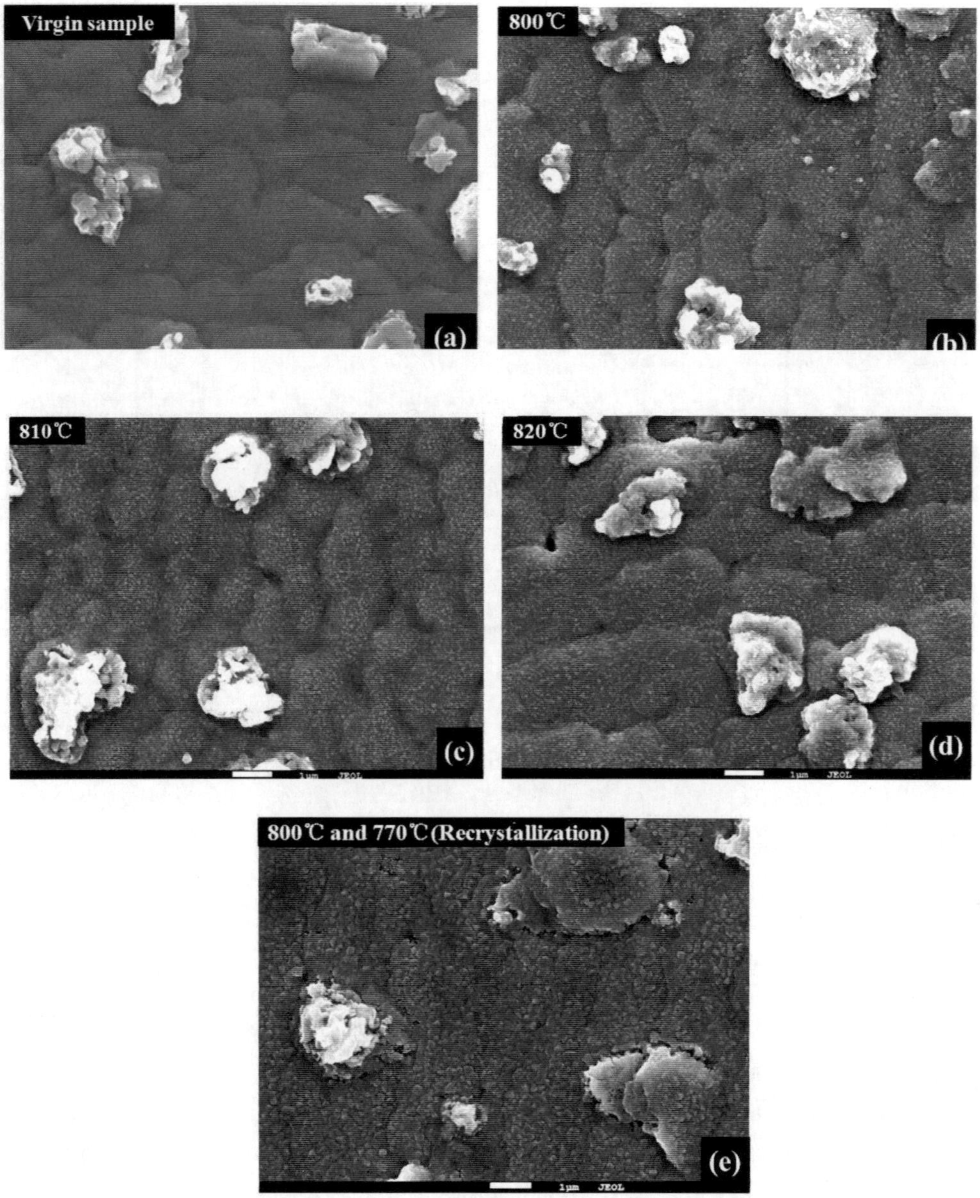

Figure 6. The typical SEM images of the virgin sample (a)， the samples after partial melting at 800 ℃ (b), 810 ℃ (c), 820 ℃ (d) and the sample after partial melting at 800 ℃, recrystallization at 770℃(e).

The oxygen deficiencies of YGdBCO would affect its superconducting properties and cause the phase transition from an orthorhombic phase to a tetragonal phase. This phase transition could become reversible through an oxygenation annealing process. Therefore, we studied the effect of duration times at 400℃ on the recovering of superconducting properties under a flow of pure oxygen. As shown in Figure 7, the critical currents of the virgin tape and samples after oxygenation process were measured by the Hall Sensor Array system. As can be seen, the I_c value of the sample after oxygenation for 50h, 100h, 200h has been restored to 44,

56.2 and 62.5% of the virgin sample, respectively. With the increasing of t_{OA} up to 250h, the I_c value is up to 88% of the virgin sample, which confirms the superconductivity has been almost restored.

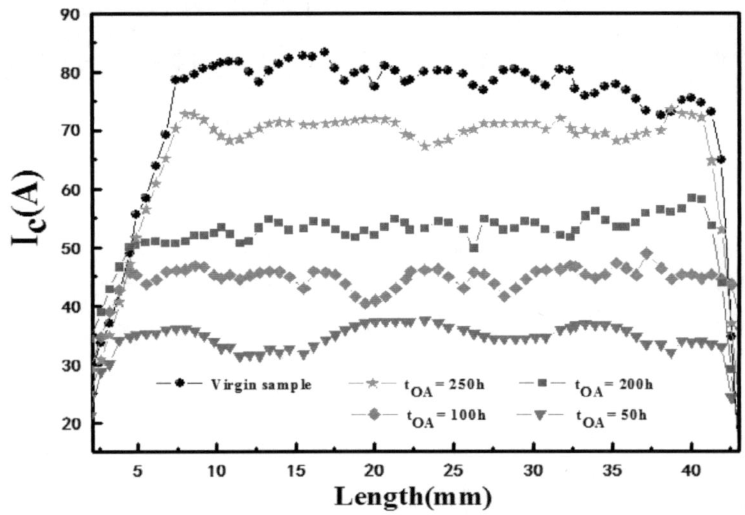

Figure 7.The critical current of virgin sample and samples after oxygenation processes at 400℃ for various duration times.

4. Conclusion

In this work, the $FeCl_3$ alcohol solution and the mixture of $NH_3.H_2O$ and H_2O_2 were used to peel the Cu stabilizer layer and Ag stabilizer layer of the YGdBCO CCs, respectively. With the optimized etching parameters (5℃,10s), the stabilizer can be successfully removed without influence the superconductivity of the CCs. The influences of partial melting temperature on the structure and surface morphology were investigated. The YGd211 and $BaCuO_2$ non-superconducting phases, which were generated during the partial melting at 800℃, were transformed back into YGd123 phase during the recrystallization process at 770℃ under a reduced PO_2(0.75Torr). After oxygenation at 400℃ for 250h under a flow of pure oxygen, the I_c value recovered up to 88% of the virgin sample, which confirms the superconductivity of samples heat-treated at high temperature has been almost restored. Future studies on the superconducting joint of the YBCO CCs will be carried out using the joining parameters obtained in this study.

Acknowledgments

We thank the financial support from the Program of International S&T Cooperation (Grant No. 2013DFA51050), the National Nature Science Foundation of China (grant No. 51271155, 51377138), the Fundamental Research Funds for the Central Universities under the No. 2682016ZDPY10, Science and Technology Project in Sichuan Province (2017JY0058), Science and Technology Project in Sichuan Province (2017JY0058), and the Doctorial Innovation Funds under the No.A0920502051410-3.

References

[1] Byeong J L and Dong N L, 1991, *J. Am. Ceram. Soc*, **74**, 78-84.

[2] Weijun Y, Juan B, Seungyong H and Yukikazu I, 2009, *IEEE Trans. Appl. Supercond*, **19**, 2261-2264.

[3] Yeonjoo P, Myungwhon L, Heesung A, Yoon H C and Haigun Lee, 2014, *NPG Asia Materials*, **6**, 1-5.

[4] Xinzhe J, Yoshinori Y, Hideaki M and Yoshiki T, 2015, *Supercond. Sci. Technol*, **28**, 075010 (6pp).

[5] Wentao W, Mingjiang W, Lian L, Baolei H, Zhou Y, Yong Z, Cuihua C and Yong Z, 2017, *IEEE Trans. Appl. Supercond*, **27**, 6603405(5pp).

[6] Yifei Z, Lehner T F, Fukushima T, Sakamoto H and Hazelton D W, 2013, *IEEE 2013 International Conference on Applied Superconductivity and Electromagnetic Devices(Beijing:China)*, 1-25.

[7] Kato J Y, Sakai N, Tajima S, Miyata S, Konishi M, Yamada Y, Chikumoto N, Nakao K, Izumi T and Shiohara Y, 2006, *Physica C*, **4**, 686-688.

[8] Yunpeng Z, Liyuan Liu, Gang W, Xinsheng Y and Yong Zhao, 2016, *IEEE Trans. Appl. Supercond*, **26**, 8400304 (4 pp).

[9] Yoshio M, Kazuo M, Rikur O and Yoshio K, 1992, *Jpn. JAppl. Phys*, **31**, 2709-2715.

[10] Eatough M O, Ginley D S, Morosin B and Venturini E L, 1987, *Appl. Phys.Lett*, **51**, 367-368.

Strain characteristics of BMO doped REBCO coated conductors fabricated by hot-wall PLD

S. Fujita[1], S. Muto[1], T. Yoshida[1], H. Sato[1], M. Igarashi[1], K. Kakimoto[1], Y. Iijima[1] and K. Naoe[1]

[1] Fujikura Ltd., 1440, Mutsuzaki, Sakura, Chiba 285-8550, Japan

shinji.fujita@jp.fujikura.com

Abstract. REBCO coated conductors (CCs) are expected to show high performance for superconducting applications due to their high tensile strength and high critical current density under magnetic fields. Fujikura Ltd. (Fujikura) has developed REBCO CCs with high performance by ion beam assisted deposition (IBAD) and pulsed laser deposition (PLD) method. It is generally known that the critical current (I_c) of a common superconducting wire varies with the applied longitudinal tensile strain, but that of Fujikura's commercial CC hardly varies due to the peculiarities of the REBCO crystal orientation. Recently, we have developed $BaMO_3$ (BMO, M : Zr or Hf) doped REBCO CCs in order to enhance their in-field critical current density. In this study, we investigated the strain dependence of the BMO doped REBCO CCs. As a result, it was confirmed that the I_c change with strain increased by BMO doping, even though the crystal orientation did not change.

1. Introduction

Introduction of artificial pinning centers (APCs) such as $BaZrO_3$ (BZO), $BaHfO_3$ (BHO) into $REBa_2Cu_3O_X$ (REBCO, RE = rare earth) film has been studied by many researchers in order to improve the in-field critical current (I_c) [1-2]. Especially, excellent properties are reported for BHO doped REBCO film by the pulsed laser deposition (PLD) method [3-5]. Fujikura Ltd. (Fujikura) has developed the production techniques for REBCO coated conductors (CCs) by using large-area ion-beam-assisted deposition (IBAD) and hot-wall type PLD [6-8]. Recently, we have been developing the BMO (M = Zr or Hf) doped REBCO CCs using these techniques. On the other hand, the REBCO CCs have very high tensile strength due to the Hastelloy™ substrates and they are expected for high field magnet applications [9]. It is well known that the I_c of the REBCO CCs varies with applied strain [10-12]. However, it has been reported that the I_c of Fujikura's commercial REBCO CCs hardly varies with applied strain [13]. This phenomenon is considered to depend on the orientation direction of the REBCO crystal due to the structure of the buffer layers. Although the strain effect of the BMO doped REBCO CCs on I_c has also been studied [14], there are few reports. In addition, the mechanical properties of the BMO doped REBCO CCs with our CC structure are completely unknown. In this study, we investigated the axial tensile and the bending strain characteristics of the BMO doped REBCO CCs fabricated by IBAD and hot-wall PLD.

2. Experimental

2.1. Evaluation samples

The structure of the evaluation CC samples is shown in figure 1. The width of the CC was 4 mm or 5 mm, and the stabilization layer was copper-plated with 20 μm thickness. The REBCO layer was deposited by using hot-wall type PLD apparatus. The specifications of the fabricated samples are shown in table 1. Sample A is a standard CC of Fujikura (FYSC-SCH04). In sample B, in order to change the crystal orientation of REBCO, an $LaMnO_3$ (LMO) layer is deposited by PLD between the MgO layer and the CeO_2 layer. The other structures were the same as sample A. Samples C, D and E were doped with BMO in the REBCO layer. The structure of these buffer layers was the same as sample A. As a method of BMO doping, REBCO (RE = Gd or Eu) targets mixed with BZO or BHO were used. In samples A, C, D, and E, the *a*- or *b*-axis of REBCO crystals was rotated by 45° from the longitudinal direction as shown on the right side of figure 2. In sample B, the *a*- or *b*-axis was parallel to the longitudinal direction as shown on the left side of figure 2 due to the presence of LMO.

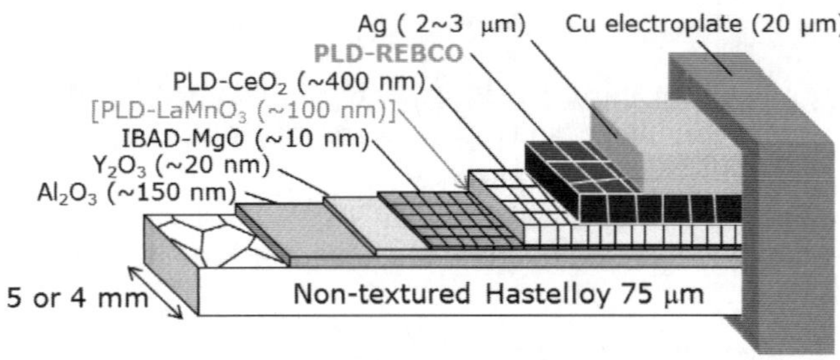

Figure 1. The structure of the evaluation samples.

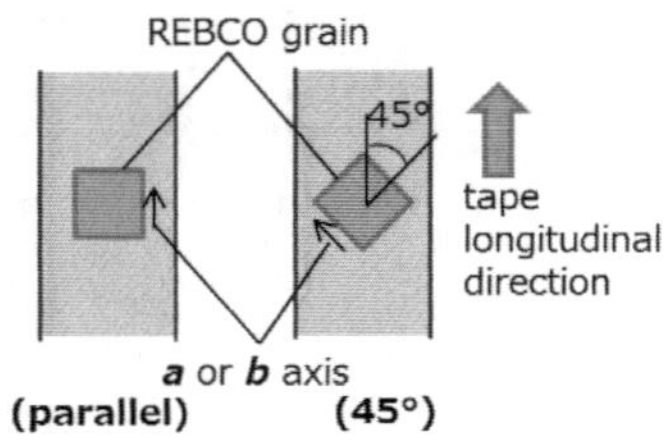

Figure 2. The schematic diagram of the REBCO crystal orientation.

Table 1. Specifications of the evaluation samples.

Sample (CC width)	REBCO layer (thickness)	Buffer layer structure	REBCO crystal orientation	I_c at 77 K, self-field [A]
A (4 mm)	GdBCO Pure (1.9 μm)	$Al_2O_3/Y_2O_3/$ MgO/CeO_2	45°	219
B (5 mm)	GdBCO Pure (2.4 μm)	$Al_2O_3/Y_2O_3/$ $MgO/LMO/CeO_2$	parallel	229
C (5 mm)	GdBCO 5.0 mol% BZO (1.8 μm)	$Al_2O_3/Y_2O_3/$ MgO/CeO_2	45°	154
D (4 mm)	GdBCO 3.5 mol% BHO (1.3 μm)	$Al_2O_3/Y_2O_3/$ MgO/CeO_2	45°	91
E (4 mm)	EuBCO 3.5 mol% BHO (1.1 μm)	$Al_2O_3/Y_2O_3/$ MgO/CeO_2	45°	88

2.2. *Measurement procedure*

Axial tensile tests and bending tests were conducted in liquid nitrogen. Both ends of the sample were clamped by two copper electrodes in order to measure I_c in self-field which was defined at 1 μV/cm of the electric field. For the axial tensile tests, the tensile testing machine and the jig as shown in figure 3 were used. The distance between chuck electrodes was 160 mm and the distance between voltage taps was 100 mm. Axial tensile stresses and axial tensile strains were measured using a load cell (10 kN) and a strain gauge glued onto the surface near the center of the sample respectively. The velocity of the tensile stroke was 4.8 mm/min. For the bending tests, a Goldacker type bending apparatus [15] as shown in figure 4 was used. The minimum bending radius in this apparatus was 5 mm. Bending strain is applied to the sample by turning the handle at the top of the apparatus. Depending on the bending direction, either tensile bending strain or compressive bending strain can be applied. The bending strain ε_b was converted from the bending radius R using the following equation:

$$\varepsilon_b = \frac{d/2}{R} \tag{1}$$

where d is the thickness of the Hastelloy substrate. In these tests, the I_c under the strained state (I_c^{load}) and under the unloaded state after applying strain (I_c^{unload}) were measured, and the applying strain was gradually increased until the I_c^{unload} obviously decreased.

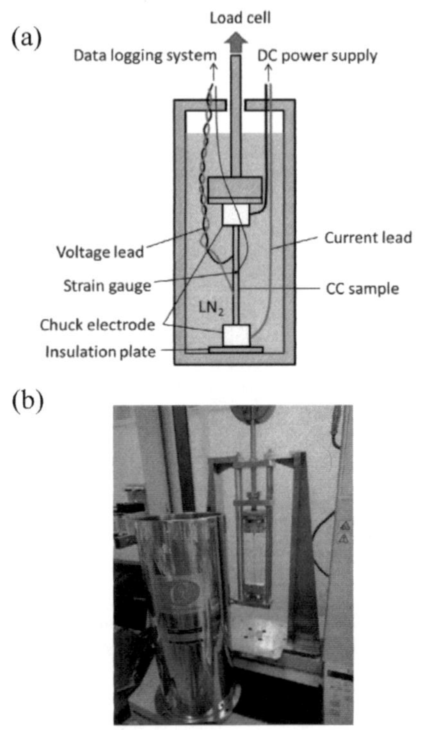

Figure 3. The schematic(a) and the picture(b) of the tensile test apparatus in liquid Nitrogen (LN_2).

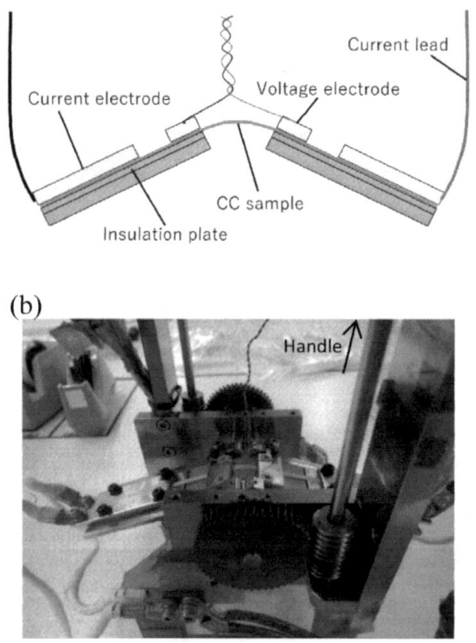

Figure 4. The schematic(a) and the picture(b) of the bending test apparatus in liquid Nitrogen (LN_2).

3. Results and discussion

3.1. Axial tensile tests

The axial tensile strain dependence of I_c at 77 K in self-field are shown in figure 5. The values of the vertical axis are the normalized I_c with the initial I_c (I_{c0}) before loading. The round plots indicate I_c^{load} and the square plots indicate I_c^{unload}. Although sample A had almost no decline in I_c^{load} with applied strain, sample B in which the crystal orientation was changed obviously decreased I_c^{load} with applied strain. These results are attributable to the REBCO crystal orientation as described in [13]. Samples C, D and E doped with BMO in the REBCO layer also decreased I_c^{load} with applied strain, although the REBCO crystal orientation was the same as in sample A. This phenomenon is not fully understood, but it is considered that BMO have some influence on I_c-strain characteristics of REBCO. It cannot be explained by crystal orientation, and there is the possibility of distortion effect of I_c at the grain boundaries in REBCO as described in [14]. The BMO doped REBCO seems to have larger I_c irreversible degraded strain than pure REBCO, but this is considered to be due to the thin film thickness [12].

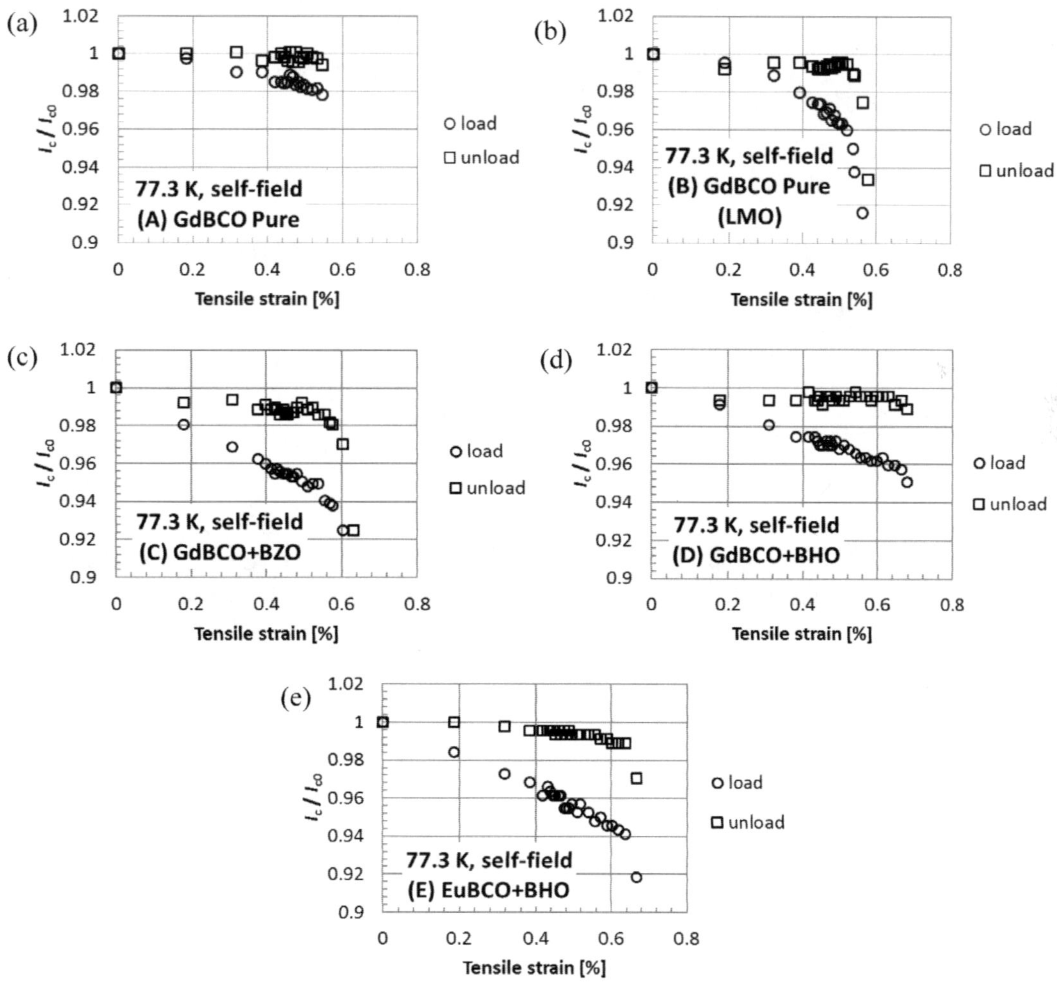

Figure 5. Applied axial tensile strain dependence of I_c for (a) sample A, (b) sample B, (c) sample C, (d) sample D, (e) sample E at 77 K in self-field.

3.2. Bending tests

Figure 6 shows the I_c/I_{c0} at 77 K, self-field as a function of the bending strain. The solid lines indicate quadratic function fitting lines. As with the results of the axial tensile test, the I_c changes due to applied strain increased in the case of the BHO doped REBCO compared to the pure GdBCO. Furthermore, the I_c on the compressive strain region increased in the case of BHO doped REBCO, and the magnitude of the increase was larger in sample E than in sample D. The I_c peak strain (ε_p) also shifted to the compressive strain direction in the order of sample A, sample D and sample E. The graph drawn by normalizing the vertical axis with the maximum I_c ($I_{c\ max}$) and converting the horizontal axis to ε_b-ε_p is shown in figure 7. The solid lines were fitted with the following equation [11, 12]:

$$\frac{I_c}{I_{c\max}} = 1 - a(\varepsilon_b - \varepsilon_p)^2 \tag{2}$$

Here, the coefficient a is called strain sensitivity or a-value. It can be seen that sample D and sample E are located on almost the same curve. This suggests that the strain sensitivity hardly changes depending on the type of rare earth, although the peak strain shifts. The values of ε_p and a-value derived from measurement results are listed in table 2. As for sample A, the a-value almost agrees with the value of Fujikura CC shown in [13], and the ε_p is also close to the value of pure GdBCO in [14]. The a-value increased with the doping of BHO, but it is still much smaller than that of the other studies [14, 16]. This difference is considered to be due to the REBCO crystal orientation.

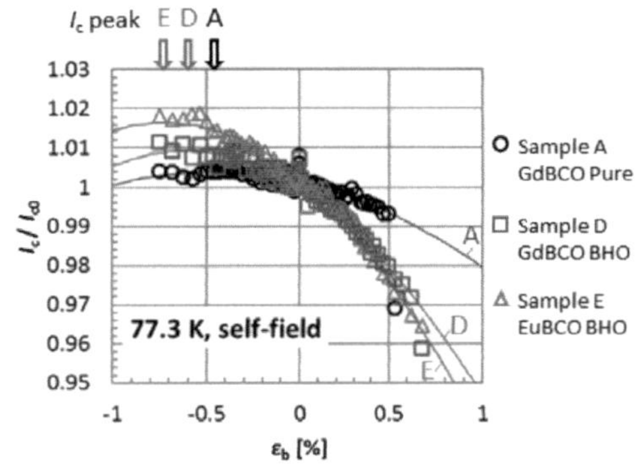

Figure 6. Bending strain dependence of I_c at 77 K in self-field for samples A, D, E.

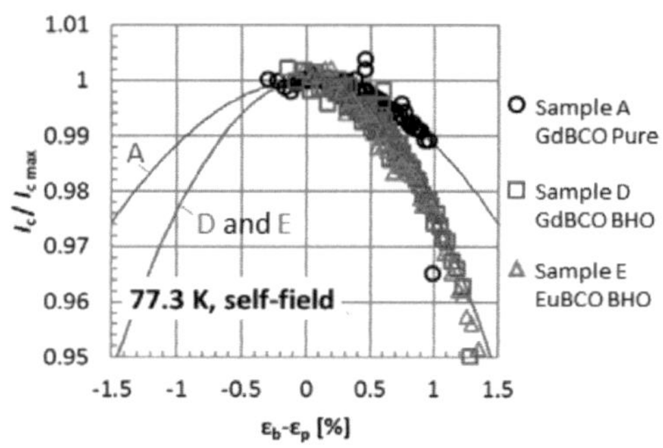

Figure 7. Normalized bending strain with peak strain dependence of the normalized I_c with the maximum I_c at 77 K in self-field.

Table 2. The values of ε_p and a-value derived from measurement results.

Sample	ε_p	a-value
A	-0.457 %	0.012
D	-0.602 %	0.024
E	-0.724 %	0.024

4. Conclusion

The axial tensile strain and the bending strain dependences of I_c for pure GdBCO and BMO (M = Zr, Hf) doped REBCO (RE = Gd, Eu) fabricated by hot-wall PLD were investigated at 77 K in self-field.

It was confirmed that the pure GdBCO film deposited on our IBAD substrate had a significantly low strain sensitivity, while the strain sensitivity increased by changing the crystal orientation by means of adding an LMO buffer layer between the MgO layer and the CeO_2 layer. Furthermore, the strain sensitivity increased similarly by doping of BMO. In the bending tests, it was confirmed that the I_c peak strain shifted in the direction of increasing compressive strain by introducing BHO, but EuBCO was larger in shift than GdBCO. Also, the strain sensitivity was not changed depending on the type of rare earth.

Acknowledgement
A part of this work is based on results obtained from a project commissioned by the New Energy and Industrial Technology Development Organization (NEDO).

References

[1] Matsumoto K and Mele P 2010 *Supercond. Sci. Technol.* **23** 14001.
[2] Goyal A, Kang S, Leonard K J, Martin P M, Gapud A A, Varela M, Paranthaman M, Ijaduola A O, Specht E D, Thompson J R, Christen D K, Pennycook S J and List F A 2005 *Supercond. Sci. Technol.* **18** 1533-1538.
[3] Tobita H, Notoh K, Higashikawa K, Inoue M, Kiss T, Kato T, Hirayama T, Yoshizumi M, Izumi T and Shiohara Y 2012 *Supercond. Sci. Technol.* **25** 062002.
[4] Tsuruta A,Yoshida Y, Ichino Y, Ichinose A, Matsumoto K and Awaji S 2013 *IEEE Trans. Appl. Supercond.* **23** No. 3 8001104.
[5] Yoshida T, Ibi A, Takahashi T, Yoshizumi M, Izumi T and Shiohara Y 2014 *Physica C* **504** 42-46.
[6] Hanyu S, Tashita C, Hanada Y, Hayashida T, Morita K, Sutoh Y, Igarashi, Kakimoto K, Kutami H, Iijima Y and Saitoh T 2010 *Supercond. Sci. Technol.* **23** 014017.
[7] Kakimoto K, Igarashi M, Hanada Y, Hayashida T, Tashita C, Morita K, Hanyu S, Sutoh Y, Kutami H, Iijima Y and Saitoh T 2010 *Supercond. Sci. Technol.* **23** 014016.
[8] Iijima Y, Adachi Y, Fujita S, Igarashi M, Kakimoto K, Ohsugi M, Nakamura N, Hanyu S, Kikutake R, Daibo M, Nagata M, Tateno F and Itoh M 2015 *IEEE Trans. Appl. Supercond.* **25** No, 3 6604104.
[9] Fujita S, Satoh H, Daibo M, Iijima Y, Itoh M, Oguro H, Awaji S and Watanabe K 2015 *IEEE Trans. Appl. Supercond.* **25** No. 3 8400304.
[10] Shin H S, Kim K H, Dizon J R C, Kim T Y, Ko R K and Oh S S 2005 *Supercond. Sci. Technol.* **18** S364-S368.
[11] Laan D C van der and Ekin J W 2007 *Appl. Phys. Lett.* **90** 052506.
[12] Sugano M, Shikimachi K, Hirano N and Nagaya S 2010 *Supercond. Sci. Technol.* **23** 085013.
[13] Sugano M, Machiya S, Oguro H, Sato M, Koganezawa T, Watanabe T, Shikimachi K, Hirano N, Nagaya S, Izumi T and Saitoh T 2012 *Supercond. Sci. Technol.* **25** 054014.
[14] Usami T, Yoshida Y, Ichino Y, Sugano M, Machiya S, Ibi A and Izumi T 2016 *IEEE Trans. Appl. Supercond.* **26** No. 4 8400104.
[15] Goldacker W, Schlachter S I, Nast R, Reiner H, Zimmer S, Kiesel H and Nyilas A 2002 *Adv. Cryog. Eng.(Mater.)* **48** pp.469-476.
[16] Sugano M, Machiya S, Sato M, Koganezawa T, Shikimachi K, Hirano N and Nagaya S 2011 *Supercond. Sci. Technol.* **24** 075019.

IOP Conf. Series: Journal of Physics: Conf. Series **871** (2017) 012043 doi:10.1088/1742-6596/871/1/012043

Critical current density of YBCO films with different configurations of columnar defects in longitudinal magnetic field

T Sueyoshi[1], Y Iwanaga[1], T Kai[1], T Izumi[1], T Fujiyoshi[1] and N Ishikawa[2]

[1] Department of Computer Science and Electrical Engineering, Kumamoto University, 2-39-1 Kurokami, Chuo-ku, Kumamoto 860-8555, Japan
[2] Japan Atomic Energy Agency (JAEA), Tokai-mura, Ibaraki 319-1195, Japan

E-mail: tetsu@cs.kumamoto-u.ac.jp

Abstract. Critical current density (J_c) properties in longitudinal magnetic fields were investigated for $YBa_2Cu_3O_y$ thin films with columnar defects (CDs), where different configurations of CDs were systematically installed into the films by using heavy-ion irradiation: a parallel configuration of CDs aligned along the *c*-axis and two bimodal splay configurations composed of CDs crossing at $\pm\theta_i$ relative to the *c*-axis, where the splay plane defined by the two irradiation angles is perpendicular or parallel to the transport current direction. The unirradiated film under the longitudinal magnetic field shows a J_c peak in the magnetic field dependence of J_c, which is 1.1 times higher than the self-field J_c. For the irradiated films with the parallel CD configuration, on the other hand, the J_c is lower than that for the unirradiated film in all magnetic fields and the value of J_c decreases with increasing CD density. Such degradation effect by CDs under longitudinal magnetic field was observed even for the bimodal splay configurations. These results are attributed to local meandering of current flow induced by CDs extending through the film thickness, which deteriorates the force-free condition.

1. Introduction

In recent years, superconducting applications contributing to high energy efficiency and low-carbon energy have been actively developed, since $REBa_2Cu_3O_y$ (REBCO, RE: Rare Earth element) coated conductors are commercially available now. One of the applications peculiar to superconducting tapes such as the coated conductors is a force-free cable [1, 2]. This utilizes an increment of critical current density J_c in a magnetic field parallel to the transport current, i.e. the so-called longitudinal magnetic field effect, where no Lorentz force works on flux lines, namely the force-free state is induced [3-5]. The flux lines under the force-free state, on the other hand, receive a force-free torque resulting from current-induced rotational shear distortion of flux lines [6]. Hence, flux pinning for the force-free torque (i.e. pinning torque) is one of the key issues to improve J_c even in the longitudinal magnetic field: there is still room for improvement in J_c under the longitudinal magnetic field by additional pinning centers (PCs).

The improvement of J_c in the longitudinal magnetic field has been experimentally observed by introducing additional PCs: neutron irradiation defects in Nb_3Sn tapes [4], $BaHfO_3$ nano-particles in

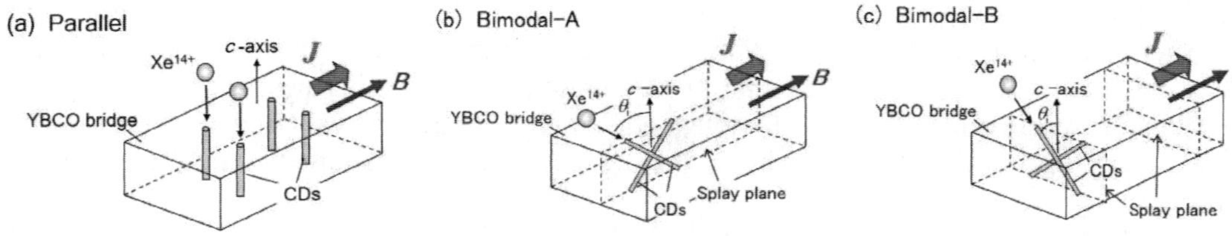

Figure 1. Illustration of various configurations of CDs installed into YBCO films by heavy-ion irradiation: (a) parallel, (b) bimodal-A, and (c) bimodal-B.

SmBCO films [7, 8] and Y_2O_3 nano-particles in YBCO films [9]. Common feature to these PCs effective for the longitudinal magnetic field effect is spherical shape, i.e. random PCs. In a magnetic field perpendicular to transport current (maximum Lorentz force configuration), on the other hand, linear defects such as dislocations [10], nano-rods [11] and columnar defects (CDs) [12], are most effective to immobilize flux lines in high-T_c superconductors. Moreover, the value of J_c in any direction can be customized by the dispersion in direction of linear defects [13]. Despite of such high performance pinning, there has been little or no research for the influence of linear defects on the longitudinal magnetic field effect so far.

In this work, we systematically explored the influence of CDs on J_c properties of YBCO thin films in the longitudinal magnetic field, where various configurations of CDs were installed into the films by using heavy-ion irradiation, as shown in figure 1: a parallel configurations of CDs aligned along the *c*-axis with different densities of CDs, and two bimodal splay configurations composed of CDs crossing at $\pm\theta_i$ relative to the *c*-axis, where the splay plane defined by the two irradiation angles is parallel (bimodal-A) or perpendicular (bimodal-B) to the transport current direction.

2. Experimental details

The samples used in this work were *c*-axis oriented YBCO thin films, which were fabricated by a pulsed laser deposition (PLD) technique on the (100) surface of $SrTiO_3$ substrates. The thickness of the films was about 300 nm. The films were patterned in a bridge geometry with about 40 μm width and 1 mm length before irradiation.

After the patterning process, the samples were irradiated with 200 MeV Xe ions at the tandem accelerator of JAEA in Tokai, Japan. To install various configurations of crossed CDs into the films, the incident angle of the ion beam was tilted off the *c*-axis by $\pm\theta_i$ in two geometries where a splay plane defined by the two irradiation angles is perpendicular (bimodal-A) or parallel (bimodal-B) to the bridge direction (current flow direction), as shown in figure 1. The electronic stopping power S_e for the irradiation is about 28 keV/nm at the surface of the sample. This creates continuous CDs of 5 ~ 9 nm diameter along the incident ion beam directions, which were confirmed by a cross-sectional transmission electron microscopy image in our previous work [14]. Table 1 lists the specifications of the samples in this work. The matching field B_ϕ, which corresponds to the planar density of CDs, is defined with respect to each irradiation direction. For the bimodal configurations, the fluence for each irradiation direction is half the total one.

Table 1 Samples in this work.

Sample	Configuration of CDs	θ_i	B_ϕ [T]	T_c [K]	J_{c0} [MA/cm^2]
Pure	-	-	-	89.3	2.21
PR05	Parallel	0°	0.5	89.5	3.49
PR10	Parallel	0°	1.0	89.1	2.68
BA80	Bimodal-A	±80°	1.0	89.2	3.62
BB45	Bimodal-B	±45°	1.5	89.2	2.79
BB85	Bimodal-B	±85°	1.5	89.2	2.39

Transport properties were measured using a four-probe method, The value of J_c was determined using a criterion of electrical field, 1 μV/cm. For the J_c measurements as a function of applied magnetic field direction in maximum Lorentz force configuration, the transport current was applied always in the direction perpendicular to the magnetic field. In order to evaluate the J_c properties in longitudinal magnetic field, on the other hand, the magnetic field was applied parallel to the bridge direction (i.e. current flow direction).

3. Results and discussion

Firstly, we investigated the angular dependence of J_c under the maximum Lorentz force configuration in order to check the effectiveness of flux pinning by various configurations of CDs. Figure 2a shows the angular dependence of J_c for the unirradiated- and the irradiated-samples with $B_\phi = 1$ T, where θ is the angle between the magnetic field and the *c*-axis of the YBCO films. The unirradiated sample indicates a smooth angular dependence of J_c exhibiting a minimum at $B \parallel c$, which is typical of uncorrelated flux pinning due to naturally occurring growth defects [15]. For the parallel CD configuration, by contrast, a large enhancement of J_c centered at $B \parallel c$ can be observed, although the J_c was slightly reduced around $B \parallel ab$ compared to the unirradiated sample. This means that CDs with $\theta_i = 0°$ effectively work as *c*-axis correlated PCs. The bimodal-A configuration with $\theta_i = \pm 80°$, on the other hand, shows an overall shift upward in J_c compared to the unirradiated sample, where the angular behavior of J_c consists of a broad peak centered at $B \parallel ab$ and an additional peak at $B \parallel ab$. The J_c behavior is typical of the random pinning, as well as the unirradiated sample. For the bimodal-A configuration with the large CD crossing angle, the CDs provide uncorrelated pinning instead of correlated pinning, because flux lines interact with the largely inclined CDs at the intersection points in any direction of magnetic field [16].

In figure 2b, the angular dependence of J_c is represented for the bimodal-B configurations with $\theta_i = \pm 45°$ and $\pm 85°$, in comparison with the unirradiated sample. For $\theta_i = \pm 45°$, a large enhancement of J_c can be observed in a wide angular range, resulting in not so much a peak as a plateau-like behavior. The plateau-structure of the J_c centered at $\theta = 0°$ is attributed to the uncorrelated pinning by crossed CDs with $\theta_i = \pm 45°$, which provides strong pinning around $B \parallel c$ for low magnetic field [14]. The J_c of the bimodal-B configuration with $\theta_i = \pm 85°$, in contrast, falls below that of the unirradiated samples around $B \parallel c$. In the vicinity of $B \parallel ab$, on the other hand, a pronounced J_c peak emerges for $\theta_i = \pm 85°$, whereas the J_c at $B \parallel ab$ deteriorates for most of the other CD configurations: the CDs with $\theta_i = \pm 85°$ provide the extremely effective flux pinning for $B \parallel ab$.

From the above results, it is shown that CDs work as effective PCs in various directions of magnetic field depending on the configuration of CDs under the maximum Lorentz force

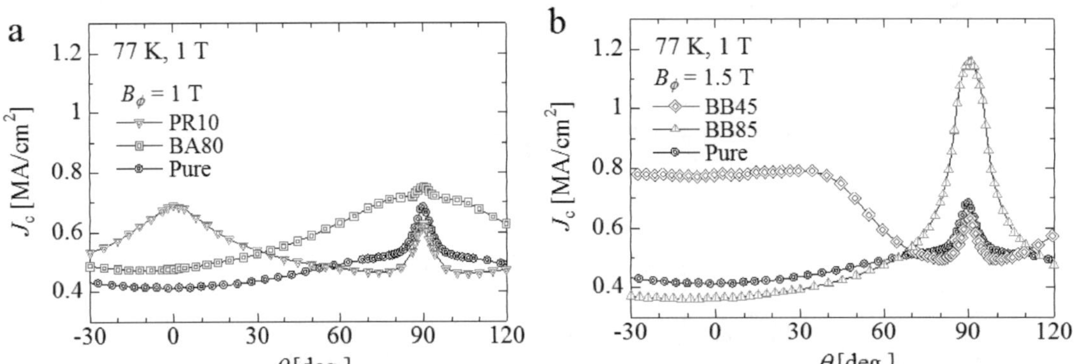

Figure 2. Angular dependence of J_c for YBCO films with different CD configurations under the maximum Lorentz force configuration: (a) parallel and bimodal-A, and (b) bimodal-B configurations, compared with an unirradiated sample.

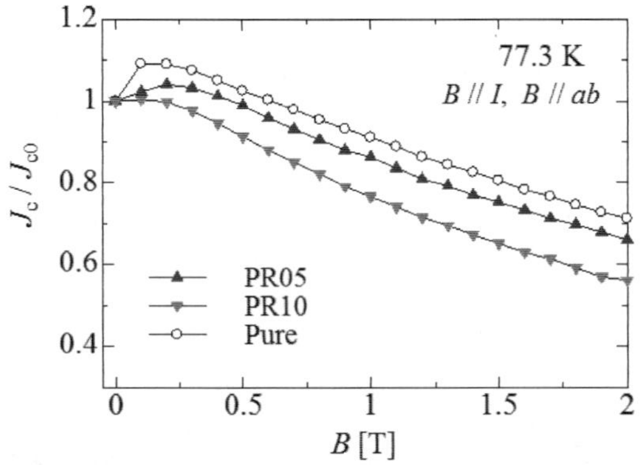

Figure 3. Magnetic field dependence of normalized J_c for YBCO films including a parallel CD configuration in longitudinal magnetic fields.

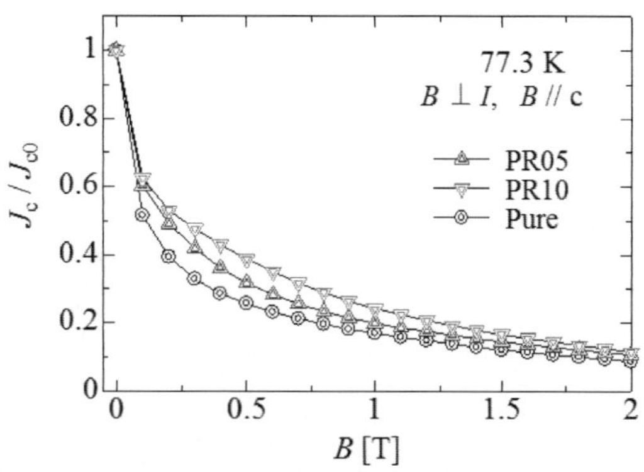

Figure 4. Magnetic field dependence of normalized J_c for YBCO films including a parallel CD configuration in transverse magnetic fields.

configuration. Next, we discuss the influence of these configurations of CDs on J_c under the force-free configuration. Figure 3 shows the magnetic field dependence of normalized J_c by the self-field critical current density J_{c0} for the parallel CD configurations with B_ϕ = 0.5 T and 1 T in the longitudinal magnetic field, in comparison with the unirradiated film. For the unirradiated sample, the J_c becomes higher than the J_{c0} up to 0.7 T, leading to the peak structure. The peak value of J_c near 0.2 T was about 10 % higher than that of J_{c0}, which is about same as the case of a SmBCO film with $BaHfO_3$ nanoparticles [7, 8]. This suggests the naturally occurring growth defects provide effective flux pinning for the unirradiated film. For the irradiated films, on the other hand, the J_c is lower than that for the unirradiated film in all magnetic fields. Furthermore, the value of J_c decreases with increasing CD density, although the J_c increases with increasing CD density at $B \parallel c$ under the maximum Lorentz force configuration, as shown in figure 4. For the increase in J_c by the longitudinal magnetic field effect, the current flow direction needs to be parallel to the magnetic field direction. The CDs, however, would disturb the current flow direction, since the CDs perpendicular to the current direction extend through the film thickness: the parallel relation between the current and the magnetic field is locally loosened by the CDs. The high density of CDs increases the degree of the meandering of the current flow. Therefore, the introduction of CDs with θ_i = 0° reduces the J_c in the longitudinal magnetic field.

Figure 5 shows a comparison of J_c between the parallel and the bimodal-A configurations under the longitudinal magnetic field. The bimodal-A configuration indicates the lowest J_c of the three films, although the J_c of the bimodal-A configuration is the highest at $B \parallel ab$ under the maximum Lorentz force configuration as shown in figure 2(a). This is attributed to larger meandering of current flow due to the inclination of CDs: the diameter of CDs in the *ab*-plane is proportional to $1/\cos\theta_i$, so that the current more meanderingly flows along the *ab*-plane. It should be noted that the J_c of the bimodal-A configuration is the same as or a little higher than that of the parallel one in low magnetic field up to 0.1 T. This suggests that the crossed CDs extending to the magnetic field direction slightly contribute to the flux pinning in a lower magnetic field, because flux lines are subjected not only to the force-free torque but also to the Lorentz force directed to the thickness direction due to the distortion of flux lines by the self-field [8].

Figure 6 represents the magnetic field dependence of J_c in the longitudinal magnetic field for bimodal-B configurations with θ_i = ±45° and ±85°. The value of J_c decreases with increasing CD crossing angle. In particular, the sample with θ_i = ±85° indicates the rapid reduction of J_c without showing a peak of J_c, although the J_c for θ_i = ±85° is the highest of all the films at $B \parallel ab$ under the

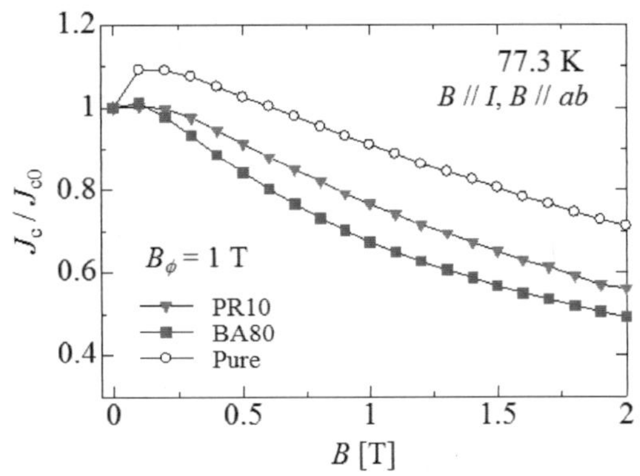

Figure 5. Magnetic field dependence of normalized J_c for YBCO films with the parallel and the bimodal-A configuration in longitudinal magnetic fields.

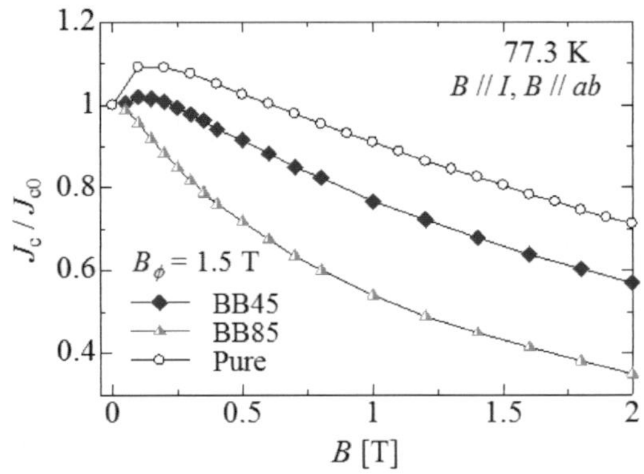

Figure 6. Magnetic field dependence of normalized J_c for YBCO films with the bimodal-B configuration in longitudinal magnetic fields.

maximum Lorentz force configuration as shown in figure 2b. The meandering of the current, which reduces the longitudinal magnetic field effect, is more frequent with increasing CD crossing angle, since the crossed CDs is always perpendicular to the current direction for the bimodal-B configurations and the inclined CDs extending through the sample thickness becomes longer as $1/\cos\theta_i$.

From the above results, it is shown that any configuration of CDs provides the deterioration effect for the flux pinning in longitudinal magnetic fields. One reason for the deterioration effect is that CDs extending through the film thickness induces local meandering of the current flow so as to loosen the parallel relation between the current and the magnetic field. It was reported that weak links of grain boundaries in high-T_c superconductors induce the local meandering of current flow, leading to the reduction of the longitudinal magnetic field effect [17]. Another possible reason is that CDs exhibiting a preferential direction for flux pinning cannot provide the effective pinning torque against the rotational torque, i.e. the force-free torque. In contrast, the uncorrelated PCs such as nano-particles could act as effective PCs in the longitudinal magnetic field, because the uncorrelated PCs provide the isotropic pinning against any orientations of magnetic field. In fact, the effectiveness of nano-particles for the flux pinning in the longitudinal magnetic field has been reported for SmBCO films including $BaHfO_3$ nano-particles [7, 8] and YBCO films with Y_2O_3 nano-particles [9].

4. Conclusions

We systematically investigated the influence of CD configurations on J_c in a longitudinal magnetic field, by using heavy-ion irradiation. All of the CD configurations deteriorated the J_c properties in the longitudinal magnetic field in comparison with the unirradiated sample, although the CDs acted as effective PCs in different directions of magnetic field depending on the configuration of CDs under the maximum Lorentz force configuration: the value of J_c decreased with increasing CD density and/or CD crossing angle in the longitudinal magnetic field. The installation of CDs extending through the film thickness induces local meandering of the current flow over the thickness direction, resulting in larger suppression of the longitudinal magnetic field effect.

Acknowledgements

This work was supported by KAKENHI (25234567 and 16K06269) from the Japan Society for the Promotion of Science and was partly performed under the Common-Use Facility Program of JAEA.

References

[1] Matsushita T, Kiuchi M and Otabe E S, 2012 *Supercond. Sci. Technol.* **25** 125009
[2] Vyatkin V S, Kiuchi M, Otabe E S, Ohya M and Matsushita T, 2015 *Supercond. Sci. Technol.*

IOP Conf. Series: Journal of Physics: Conf. Series **871** (2017) 012043 doi:10.1088/1742-6596/871/1/012043

28 015011

[3] Bychkov Yu F, Vereshchagin V G, Zuev M T, Karasik V R, Kurganov G B and Mal'tsev V A, 1969 *JETP Lett.* **9** 404

[4] Cullen G W and Novak R L, 1964 *Appl. Phys. Lett.* **4** 147

[5] Maiorov B, Jia Q X, Zhou H, Wang H, Li Y, Kursunovic A, MacManus-Driscoll J L, Haugan T J, Barnes P N, Foltyn S R and Civale L, 2007 *IEEE Trans. Appl. Supercond.* **17** 3697

[6] Matsushita T, 1985 *J. Phys. Soc. Jpn.* **54** 1054

[7] Tsuruta A, Watanabe S, Ichino and Y, Yoshida Y, 2014 *Jpn. J. Appl. Phys.* **53** 078003

[8] Sugihara K, Ichino Y and Yoshida Y, 2015 *Supercond. Sci. Technol.* **28** 104004

[9] Kido R, Kiuchi M, Otabe E S, Matsushita T, Jha A K and Matsumoto K, 2016 *Phys. Proc.* **81** 117

[10] Lowndes D H, Christen D K, Klabunde C E, Wang Z L, Kroeger D M, Budai J D, Zhu S and Norton D P, 1995 *Phys. Rev. Lett.* **74** 2355

[11] Macmanus-Driscoll J L, Foltyn S R, Jia Q X, Wang H, Serquis A, Civale L, Maiorov B, Hawley M E, Maley M P and Peterson D E 2004 *Nature Mater.* **3** 439

[12] Civale L, Marwick A D, Worthington T K, Kirk M A, Thompson J R, Krusin-Elbaum L, Sun Y, Clem J R and Holtzberg F 1991 *Phys. Rev. Lett.* **67** 648

[13] Maiorov B, Baily S A, Zhou H, Ugurlu O, Kennison J A, Dowden P C, Holesinger T G, Foltyn S R and Civale L 2009 *Nature Mater.* **8** 398

[14] Sueyoshi T, Sogo T, Nishimura T, Fujiyoshi T, Mitsugi F, Ikegami T, Awaji S, Watanabe K, Ichinose A and Ishikawa N 2016 *Supercond. Sci. Technol.* **29** 065023

[15] Civale L, Maiorov B, Serquis A, Willis J O, Coulter J Y, Wang H, Jia Q X, Arendt P N, MacManus-Driscoll J L, Maley M P and Foltyn S R 2004 *Appl. Phys. Lett.* **84** 2121

[16] Furuki Y, Sueyoshi T, Kai T, Iwanaga Y, Fujiyoshi T and Ishikawa N 2015 *Physica C* **518** 58

[17] Safar H, Coulter J Y, Maley M P, Foltyn S, Arendt P, Wu X D and Willis J O 1995 *Phys. Rev. B* **52** R9875

IOP Conf. Series: Journal of Physics: Conf. Series **871** (2017) 012044 doi:10.1088/1742-6596/871/1/012044

Nanostructural Characterization of Low Resistance Joints Using Ag Pastes for $GdBa_2Cu_3O_{7-x}$ Coated Conductors

Tomohiro Kato[1], Takato Machi[2], Daisaku Yokoe[1], Ryuji Yoshida[1], Takeharu Kato[1], Teruo Izumi[2], Tsukasa Hirayama[1] and Yuh Shiohara[3]

[1] Nanostructures Research Laboratory, Japan Fine Ceramics Center, 2-4-1 Mutsuno, Atsuta-ku, Nagoya, 456-8587, Japan
[2] Research Institute for Energy Conservation, Department of Energy and Environment, National Institute of Advanced Industrial Science and Technology, 1-2-1 Namiki, Tsukuba, 305-8564, Japan
[3] Industrial Superconductivity Technology Research Association, 3-2-1 Sakado, Takatsu-ku, Kawasaki, 213-0012, Japan (former)

e-mail: tkato@jfcc.or.jp

Abstract. $GdBa_2Cu_3O_{7-x}$ coated conductors were splice jointed by a face-to-face manner using a paste containing nano-sized Ag particles under a pressure of about 50 MPa at 150 °C for 1 hr. The low electrical resistance of 6 nΩ at the joint was attained. Nanostructural characterizations of the starting Ag paste and the jointed region of the coated conductors were carried out using scanning electron microscopy and transmission electron microscopy. The size of the Ag particles in the starting pastes were confirmed to be a few tens of nanometers in diameter. The size of Ag particles became larger during the jointing process. Both the surfaces of the stabilizing Ag layers were partially bonded by the Ag particles. No oxides or other elements were detected in the region of the bonding parts.

1. Introduction

Commercial $REBa_2Cu_3O_{7-x}$ (REBCO, RE; Y, Gd etc) coated conductors with high critical current (I_c) values have been produced [1-3]. These coated conductors are promising candidates for industrial applications, such as superconductor cables, magnets and medical instruments. However, it is inevitable to develop a joint technique of the coated conductors for such practical applications [4-7]. Solder jointing [4,5], low resistance jointing studies using silver diffusion [6] and a superconducting joint technique [7] have been reported. However, these joint techniques were performed at higher temperatures (~500 °C) than the oxygen annealing temperature (~500 °C) of the REBCO crystals for carrier doping, so that REBCO crystals had an oxygen deficiency. Recently, a low electric resistance joint technique was developed using a paste containing nano-sized Ag particles at 150 °C [8]. The electric resistance of 6 nΩ at the joint was attained [8]. In this study, we characterized the nanostructures of the starting Ag particles and the jointed part of the coated conductors by scanning electron microscopy (SEM) and transmission electron microscopy (TEM) in detail.

2. Experimental

An Ag paste (L-Agl TeH, ULVAC Inc.) [8] was used as a starting paste. The paste was dispersed on a Cu grid with carbon thin film and dried. The Ag particles on the grid was characterized by TEM. A

Hastelloy™ tape with textured CeO_2/$LaMnO_3$/ion beam assisted deposition-MgO/Y_2O_3/Gd-Zr-O layers was used as a substrate for coated conductors. GdBCO superconductive layers were fabricated on the Hastelloy™ by pulsed laser deposition. Ag stabilizing layers were deposited by sputtering on the GdBCO layers. The surfaces of the Ag stabilizing layers were mechanically polished. These coated conductors were splice jointed by a face to face manner using a paste containing nano-sized Ag particles in air under a pressure of about 50 MPa at 150 °C for 1 hr [8]. The jointed coated conductors were sectioned by an Ar-ion beam at the accelerating voltage of 6 kV and then the cross-sections were further etched using the Ar-ion beam in a Hitachi IM4000 ion milling system. The cross-sections were observed by a Hitachi SU8000 scanning electron microscope (SEM) at the accelerating voltage of 1 kV. Furthermore, the jointed regions were thinned in a Hitachi NB5000 focused ion microscopy (FIB)-SEM system equipped with a micro-sampling system for preparation of a transmission electron microscopy (TEM) specimen, which was pasted on the edge of an Al foil support [9]. In addition, the cross-sectional specimen was further thinned using a Gatan PIPS ion milling system at an accelerating voltage of 1-0.5 kV to remove the FIB damaged layers formed on the foil specimen [9]. Both Ag particles in the staring paste and the specimen were examined in a TOPCON EM-002BF TEM with twin energy dispersive spectroscopy (EDS) detectors at the accelerating voltage of 200 kV.

3. Results and discussion

A TEM image of the Ag particles in the starting paste and a selected area diffraction pattern (SADP) of them are shown in Figs. 1 (a) and (b), respectively. The size of the Ag particles are a few tens of nanometers in diameter. However, these Ag particles are coated by some amorphous materials considered to be polymers included in the starting paste. These reflection rings in Fig. 1 (b) correspond to Ag metal.

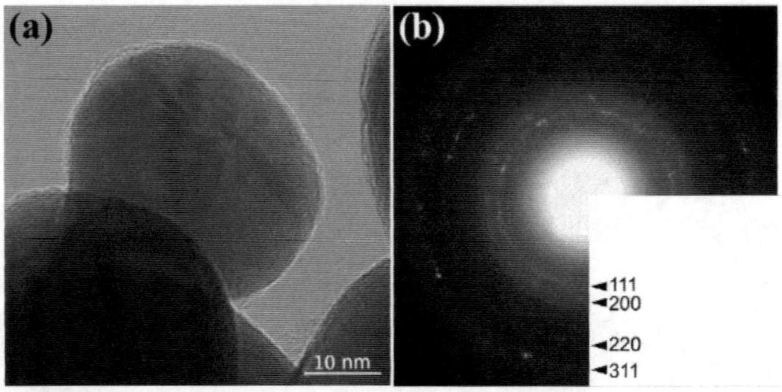

Figure 1. (a) TEM image of Ag particles in a starting Ag paste. (b) SADP corresponding to the Ag particles with plane indexes of Ag.

A cross-sectional SEM image of the jointed coated conductors using back scattering electrons is shown in Fig. 2. Figure. 3 shows a higher magnification SEM image of the jointed region corresponding to the broken square in Fig. 2. The stabilizing Ag layers are partially connected by the Ag particles indicated by arrows in Fig. 3. The size of Ag particles changing from the starting paste is much smaller than that of Ag grains in the stabilizing layers. A cross-sectional TEM image and a SADP of the connected region are shown in Figs. 4 (a) and (b), respectively. The size of Ag particles in the connecting region are about 50 nanometers, so that Ag particles have grown larger during the jointing process. Reflection rings correspond to Ag metal as shown in Fig. 4 (b). A cross-sectional high angle annular dark field (HAADF) image of an unconnected region of the jointed coated conductors is shown in Fig. 5. Figures. 6 (a) - (c) show the results of EDS analysis of the regions A - B in Fig. 5, and Ag particles of the starting pastes. Oxygen cannot be detected in the region A and the starting Ag particles, but can be detected in the region B. (Since the TEM specimen was pasted on the Al foil support [9], the Al peak of around 1.5 keV in Figs. 6 (a), (b) was caused from the Al supports. Therefore, neither the Ag particles nor the stabilizing

Ag layers dose not include Al. The Ag particles was put on a Cu grid with carbon thin film. The Cu peak of around 0.9 keV in Fig. 6 (c) was from the Cu grid.) Figure. 7 shows a SADP acquired from such the region containing oxygen. Reflection spots in Fig. 7 mainly correspond to Ag metal, but some spots, indicated by arrows, correspond to the (110) of Ag_2O. These oxides are consider to be formed on the unconnected regions during the jointing process. With increasing the splice jointed regions, nucleation of such the oxides on the unconnected regions could be suppressed. In addition, the electrical joint resistance between the coated conductors should be lower.

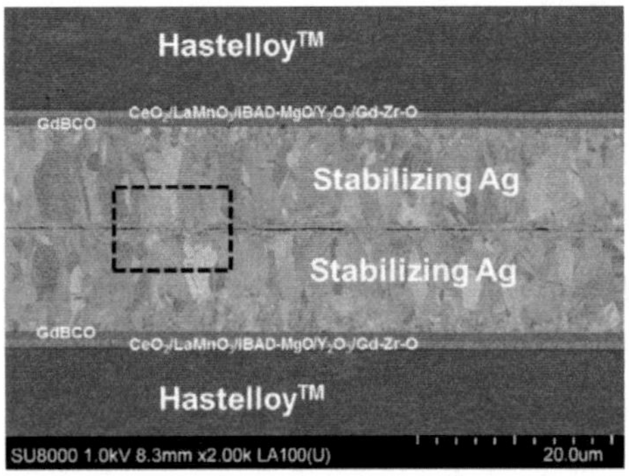

Figure 2. Cross-sectional SEM image of jointed GdBCO coated conductors using back scattering electrons.

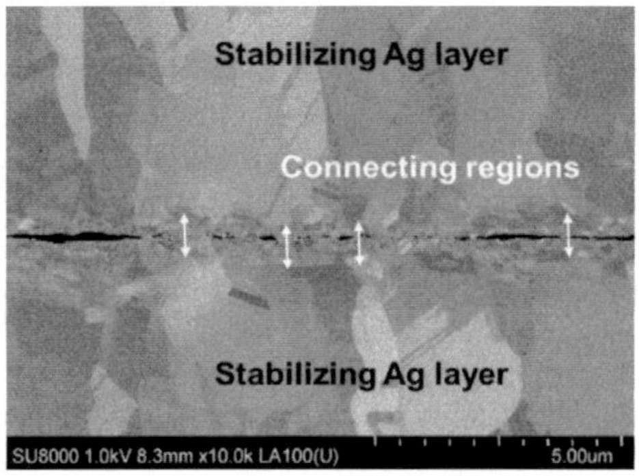

Figure 3. Cross-sectional SEM image of bonding regions corresponding to the broken square in Fig. 2.

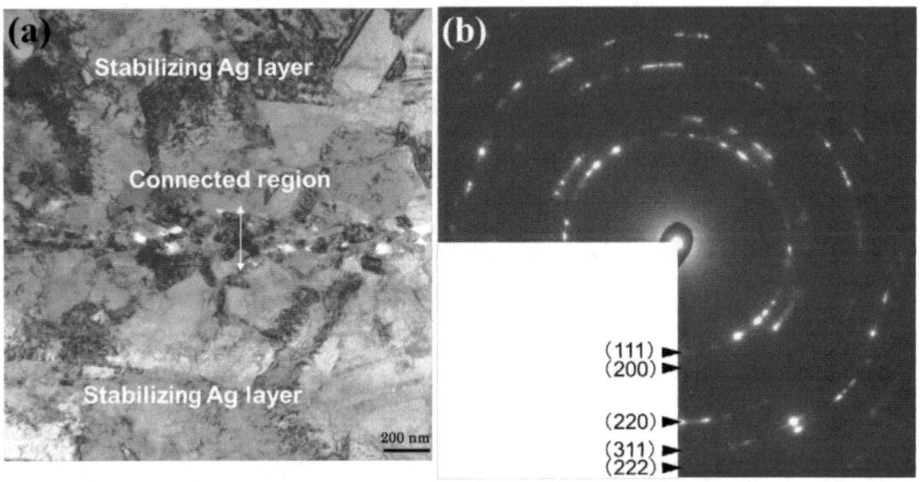

Figure 4. (a) Cross-sectional TEM image of a connected region and (b) SADP corresponding to the region of (a) with the plane indexes of Ag.

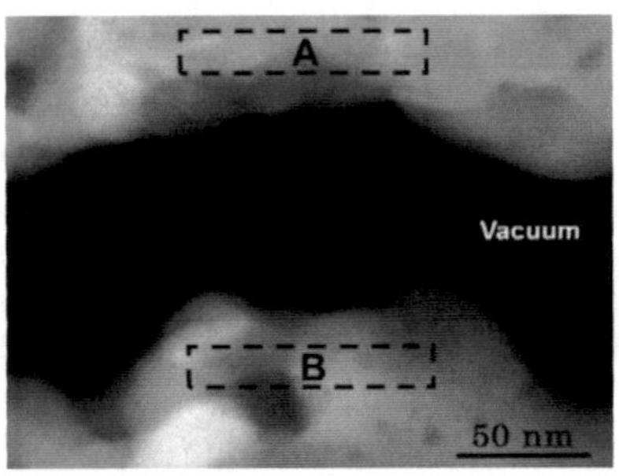

Figure 5. HAADF image of an unconnected region of the jointed coated conductors.

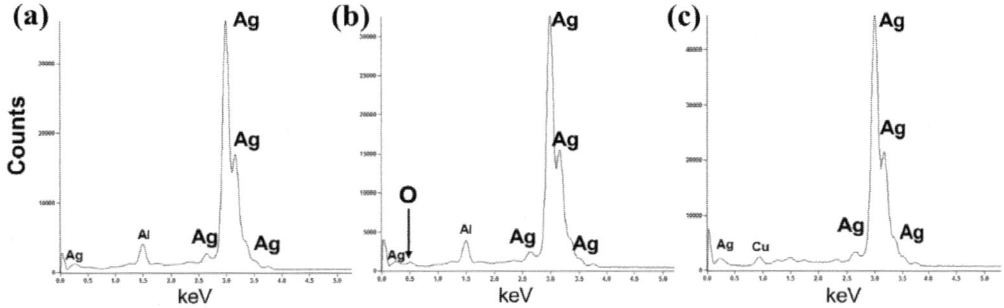

Figure 6. Results of EDS analysis of Ag particles. (a) and (b) corresponds to the regions A and B in Fig. 5, respectively. (c) Ag particles of the starting pastes.

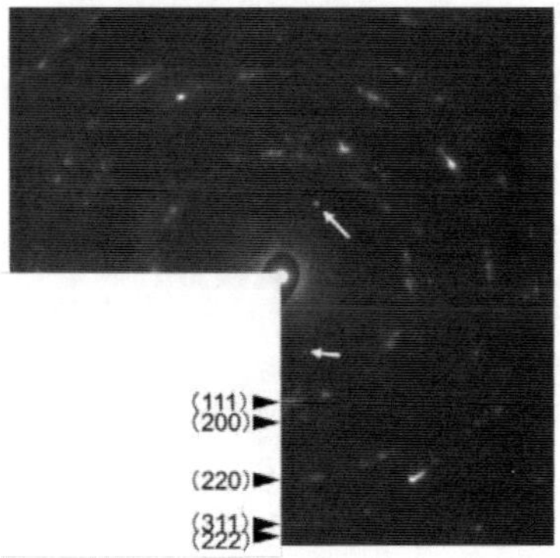

Figure 7. SADP corresponding to Ag particles containing oxygen with the plane indexes of Ag. Arrows indicate reflection spots of the (110) of Ag_2O.

4. Conclusions

We observed the nanostructures of the Ag particles in the starting pastes and jointed regions of GdBCO coated conductors using the Ag pastes in which the low electrical resistance of 6 nΩ was attained. The size of Ag particles in the starting paste was a few tens of nanometers in diameter. According to the selected area diffraction patterns, no Ag oxides were found in the starting paste. Both the surfaces of the stabilizing Ag layers were partially bonded by the Ag particles. The larger the splice jointed area, the lower the electrical joint resistance.

Acknowledgements

This work was supported by the Ministry of Economy, Trade and Industry (METI) and Japan Agency for Medical Research and Development (AMED) as 'Development of Fundamental Technologies for HTS Coils'.

References

[1] Kutami H, Hayashida T, Hanyu S, Tashita C, Igarashi M, Fuji H, Hanada Y, Kakimoto K, Iijima Y and Saitoh T 2009 *Physica C* **469** 1290

[2] Selvamanickam V, Chen Y, Xiong X, Xie Y Y, Martchevski M, Rar A, Qiao Y, Schmidt R M, Knoll A, Lenseth K P and Weber C S 2009 *IEEE Trans. Appl. Supercond.* **19** 3225

[3] Rupich M W, Li X P, Thieme C, Sathyamurthy S, Fleshler S, Tucker D, Thompson E, Schreiber J, Lynch J, Buczek D, DeMoranville K, Inch J, Cedrone P and Slack J 2010 *Supercond. Sci. Technol.* **23** 14015

[4] Onabe K, Nagaya S, Iijima Y, Sadakata N, Saitoh T and Kohno O 1998 *Adv. in Supercond.* **10**

603

[5] Sugano M, Nakamura T, Shikimachi K, Hirano N and Nagaya S 2007 *IEEE Trans. Appl. Supercond.*, **17** 3067

[6] Kato J Y, Sakai N, Tajima S, Miyata S, Konishi M, Yamada Y, Chikumoto N, Nakao K, Izumi T and Shiohara Y 2006 *Physica C* **445-448** 686

[7] Park Y, Lee M, Ann H, Choi Y H and Lee H 2014 *NPG Asia Mater.* **6** e98

[8] Nakanishi T, Machi T, Izumi T, Teranishi R., Kato T, Kato T and Hirayama T 2016 *Physica Procedia* **81** 105

[9] Sasaki H, Matsuda T, Kato T, Muroga T, Iijima Y, Saitoh T, Iwase F, Yamada Y, Izumi T, Shiohara Y and Hirayama T 2004 *J. Electron Microsc.* **53** 497

IOP Conf. Series: Journal of Physics: Conf. Series **871** (2017) 012045 doi:10.1088/1742-6596/871/1/012045

A Study on Critical Current of Twisted Soldered-Stacked-Square (3S) HTS Wire with 1mm Width

M Hu[1], Z L Xie[1], D M Xi[2], Y K Zhou[1], Z Y Li[1], Z Y Hong[1], Z J Jin[1], Z Yin[3], X F Wang[3] and Y L Zhang[3]

[1] Department of Electrical Engineering, Shanghai Jiao Tong University, No.800 Dongchuan Road, Minhang District, Shanghai, 200240, China

[2] Electric power college, Inner Mongolia University of technology, No. 49 Aimin Street, Hohhot City, Inner Mongolia, 010080, China

[3]State Grid Shanghai Electric Power Company, Jiading Supply Company, Shanghai, 201800, China

E-mail: lizhuyong@sjtu.edu.cn

***Abstract*--**In high temperature superconducting (HTS) applications, the external magnetic field affects the performance of the critical current and the AC loss. To improve the performance, a novel soldered-stacked-square (3S) HTS wire is developed through narrowing, stacking and soldering processes using 2G narrow wires with 1 mm width. In this paper, the narrowing process of the wire with 4 mm width was showed and the stacking and soldering processes were also presented. The 3S wire was made of three narrow HTS tapes and four brass strips. For the 3S HTS samples, the typical electrical and mechanical properties have been tested. Based on that, single 3S wire and double 3S wires were twisted. Then the critical currents of the samples were tested at different twist pitches. The test results showed that the measured critical currents of 3S wire samples were between 120 A and 150 A. The critical current of the single 3S wire at average twist pitch of 50~70 mm was about 8% lower than that of the original wire. And the critical current of double 3S wires was about 10% lower than the original critical current at a twist pitch of 65 mm.

1. Introduction

The second generation high temperature superconductor tapes with excellent high current performance have been applied to different areas. However, the external magnetic field could reduce the performance of critical current and AC loss. Many different structures such as assembled, stacked, or twisted of tapes were explored in order to reduce influence of the external magnetic field on the critical current and AC loss. The results showed that magnetization losses were apparently reduced, however the transport current losses increased slightly [1]. In this paper [2], the fabrication process and the characteristics of Twisted Stacked-Tape Cable with width of 4 mm was recommended. In the paper [3], the AC loss of the twisted and non-twisted BSCCO tapes were measured at different frequency. The critical current was measured according to different torsion of YBCO and BSCCO conductors with width of 4.5 mm, and the results showed that when the twist pitch was close to 60 mm, the critical current was reduced to 70% in the paper [4]. Though various structures have been developed, they are mostly based on 4 mm or over 4 mm width HTS wires.

However, in this paper, a novel wire was suggested and successfully developed through narrowing, stacking and soldering processes using 2G narrow wires with 1 mm width, which is named Soldered-Stacked-Square (3S) HTS wire. The content of this paper was described as follows. Firstly, the narrowing process of the wire with 4 mm width and the stacking and soldering processes were presented in detail. Secondly, several samples were prepared and the critical currents were tested. Then the samples were twisted with single 3S wire and double 3S wires. And the critical currents of twisted samples were measured and analyzed at different pitches.

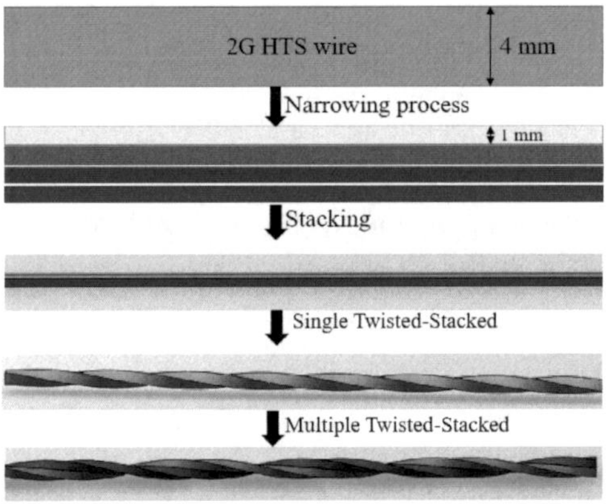

Figure 1. Schematic illustration for the manufacture process of novel 3S HTS narrow wires.

2. Manufacture process of 3S wires

Figure 1 shows the schematic illustration for fabrication processes of the soldered-stacked-square (3S) HTS wire including twisted wires. Firstly, an original 2G tape with 4 mm width is incised into several narrow wires with the same width 1 mm by a cutting device which is the commercial one frequently used to fabricate flat copper wires for conventional transformers. The 1 mm HTS wire is the narrowest width of the HTS tape, and it is firstly reported by our research group [1]. Then the narrow wires are immersed together into a solder bath, and undergone stacking and soldering processes simultaneously. As shown in figure 2 (a), all of the fabrication processes are completed in a reel-to-reel machine, which is mainly composed of several unwinding reels, a solder bath and one winding reel. In this study, the fabricated 3S wires were made of three narrow HTS wires and four brass strips. Thus the cross section of the 3S wire was near to square. The parameters of the narrow HTS wire and brass strip were listed in table 1. The arrangement of the narrow HTS wires and brass strips are shown in figure 2 (b).

Table 1. Parameters of the HTS wire and brass strip.

Material	Width	Thickness
2G HTS tape	1 mm	0.1 mm
Brass	1 mm	0.2 mm

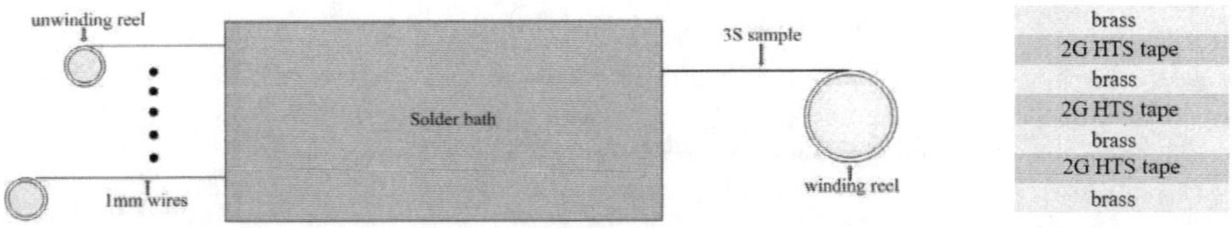

(a) (b)

Figure 2. Schematic illustration: (a) stacking process of narrow wires. (b) the cross section of the 3S wire.

Figure 3 (a) and (b) show the photos of the twisted 3S wire samples with single wire and double wires, respectively. The full length of the single wire and the double wires were 420 mm and 400 mm. Double 3S wires should be twisted around each other as tightly as possible. Their cross-sections were square shape with 1 mm width, just as figure 2 (b) shows. In order to evaluate the effect of twist pitch on the critical current of 3S wires, several 3S wire samples have been prepared. The samples were divided into two groups: *A*) Single 3S wire, such as No.1 No.2 and No.3 samples. *B*) Double 3S wires, such as No.4 and No.5 samples. The specifications of the samples were listed in table 2. The cross-section of the 3S wire is approximately 1×1 mm^2.

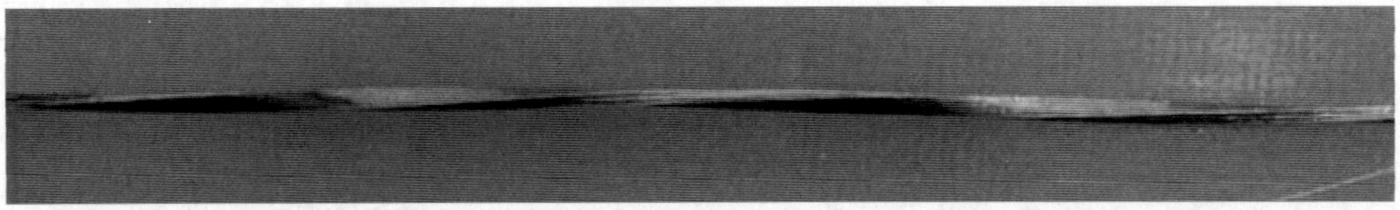

(a)

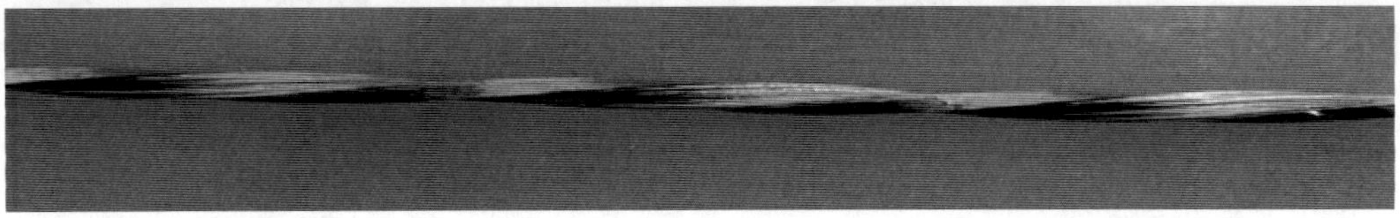

(b)

Figure 3. Schematic diagram of twisted 3S wires: (a) with single wire, (b) with double wires.

Table 2. Specifications of 3S wire samples.

3S wire samples	Cross-section area	Length	Number of 3S wires
No.1	1×1 mm^2	420 mm	1
No.2	1×1 mm^2	420 mm	1
No.3	1×1 mm^2	420 mm	1
No.4	1×1 mm^2	400 mm	2
No.5	1×1 mm^2	400 mm	2

3. Results and analysis

3.1 Critical currents of the 1 mm HTS wires

The original 2G tape with 4 mm width was incised into four narrow wires with the same width of 1 mm. In this paper the four narrow wires were named as piece 1~4. Firstly, the critical currents of the narrow wires were tested in liquid nitrogen of 77 K by the four-terminal method. As shown in figure 4, the critical current distributions in length direction show good uniformity. Meanwhile, except for pieces 1~4, many other narrow wires were also cut and tested. The critical current distributions in length direction of them also showed good uniformity. The feasibility of the narrowing process was verified. However, although they are cut from the same original tape, the critical currents among narrow wires are different. The non-uniformity of critical current in width direction might accounts for the difference. The critical current of 3S wire could be designed by using narrow wires with different number and critical current.

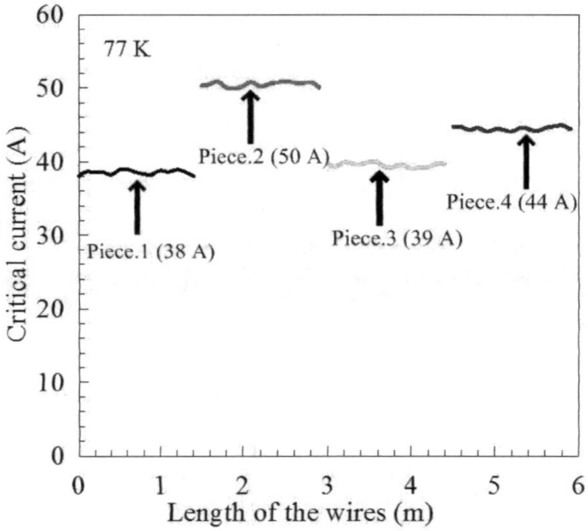

Figure 4. Critical currents of the narrow HTS wires.

3.2 Measurement of twist pitch of 3S wire

Firstly, all samples were fabricated using the narrow wires of piece 2. Before being twisted, the critical currents of all samples were tested. Figure 5 indicates the critical currents of the group *A*. The lengths of the three single samples were the same with 420 mm. The length between the two voltage leads was 250mm. The critical currents of sample 1~3 are 143 A, 138 A and 138 A, respectively. The critical current of group *A* without any degradation is 150 A. Therefore, the critical current of the sample 1 decline 5% and the critical currents of the sample 2 and sample 3 decline 8% after stacking and soldering process. For group *B*, the lengths of the sample 4 and sample 5 are the same with 400 mm and the length of between the two voltage leads is 200 mm. The critical currents of sample 4 and sample 5 are 260 A and 252 A, respectively. The critical current of group *B* without any degradation is 300 A. Therefore, the critical current of the sample 4 and sample 5 decline 13% and 16% after stacking and soldering process. Then these samples were twisted by a device whose one side was fixed and the other side could be rotated. The details of the test machine and method were described in the paper [5]. The critical current was tested at different twist pitches.

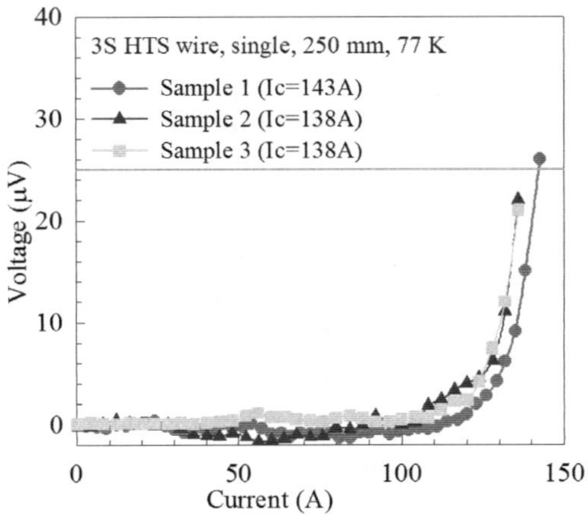

Figure 5. Measured critical currents of samples with single 3S

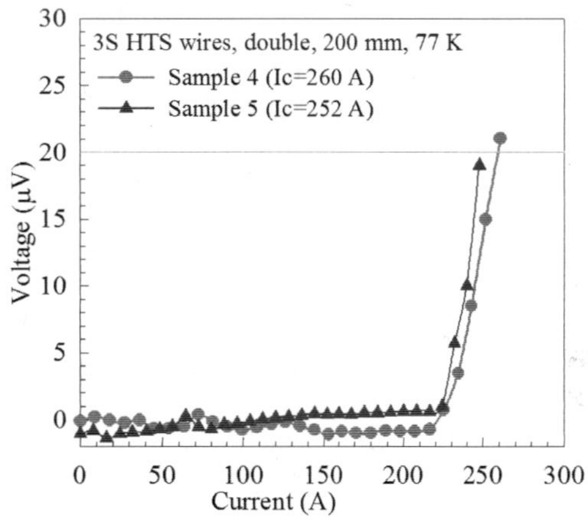

Figure 6. Measured critical currents of samples with two 3S

The normalized critical current means that the measured critical current after being twisted is divided by the original critical current before being twisted. The tolerable twist pitch is defined as the pitch when the original critical current remains 90%. In figure 7, we could observe that the critical current will decline as the twist pitch decreases. The average tolerable twist pitch of sample 1 is approximately 50 mm and the average twist pitch of sample 2 and sample 3 are both 70 mm. The uncertainty and nonuniformity of the twist process might account for the difference of twist pitch. Thus, we could obtain that the tolerable twist pitch of the single sample is between 50 mm and 70 mm. This value could meet the requirement of actual superconducting applications. The influence of twist pitch on the critical currents of double 3S wires is shown in figure 8. As can be seen in figure 8, the critical current increase a little when the twist pitch is bigger than 200 mm. However, when the twist pitch is smaller than 200 mm, the critical current declines as the twist pitch decreases, just the same as that of single 3S wires. The critical current is smaller than the original critical current when the twist pitch is smaller than 80 mm. The tolerable twist pitch of double 3S wires is about 65 mm. The tolerable twist pitch of double 3S wire is near to that of the single 3S wire.

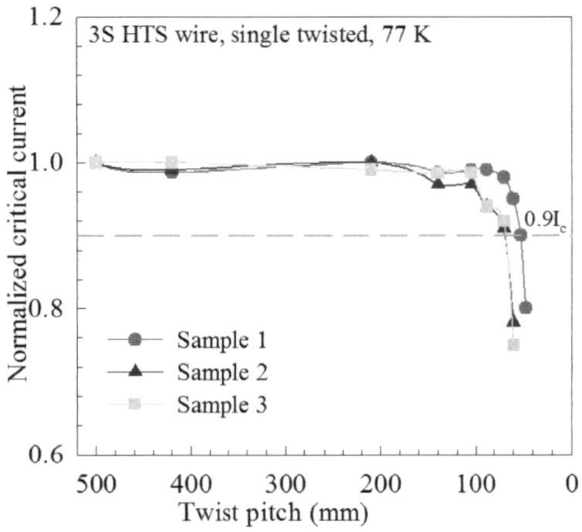

Figure 7. Dependence of twist pitch on critical current for 3S wire samples with group A, sample 1~3.

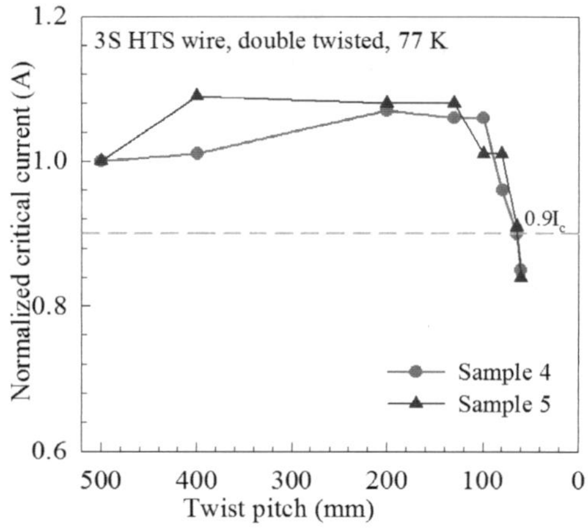

Figure 8. Dependence of twist pitch on critical current for 3S wire samples with group B, sample 4~5.

4. Conclusion

In this paper, a novel soldered-stacked-square (3S) HTS wire is developed through narrowing, stacking and soldering processes using 2G narrow wires with 1 mm width. The critical currents of single and two 3S HTS wires were tested in liquid nitrogen of 77 K before and after being twisted. The following conclusions could be obtained based on the experiment results.

The critical currents of the 3S wires approximately decline 5%~8% after the stacking and soldering process. The tolerable twist pitch of the single 3S wire is approximately between 50 mm and 70 mm. And the tolerable twist pitch of double 3S wires is approximately 65 mm. Therefore, the 3S HTS wires could be applied to several actual superconducting applications such as large HTS coil, HTS cable, HTS motor and so on. Next we will study the influence of the number of 1 mm HTS wires on the performance of 3S HTS wires.

Acknowledgement

This work was supported in part by the National Natural Science Foundation of China (Project 51577119).

References

[1] Li Z, Li J, Wang Y, Yao Z, Kang Z, Yuan B, Yang Z, Jin Z and Hong Z 2016 *IEEE Trans. Appl. Supercond.* **26** 8201104

[2] Takayasu M, Mangiarotti F J, Chiesa L, Bromberg L and Minervini J V 2013 *IEEE Trans. Appl. Supercond.* **23** 4800104

[3] Jiang Z, Amemiya N, Ayai N and Hayashi K 2003 *IEEE Trans. Appl. Supercond.* **13** 3557-3560

[4] Fleiter J, Sitko M and Ballarino A 2013 *IEEE Trans. Appl. Supercond.* **23** 8000204.

[5] Badcock R A, Long N J, Mulholland M, Hellmann S, Wright A and Hamilton K A 2009 *IEEE Trans. Appl. Supercond.* **19** 3244–3247

[6] Wang Y, Baasansuren S, Xue C and Hasegawa T 2016 *IEEE Trans. Appl. Supercond.* **26** 4804406

Determination of Pinning Parameters in Flux Creep-Flow Model for *E-J* characteristics of High Temperature Superconductors by using Differential Evolution

E S Otabe[1], T Taguchi[1], Y Tsuruda[1] and R Funaki[2]

[1] Department of Computer Science and Electronics, Kyushu Institute of Technology, 680 – 4 Kawazu, Iizuka, Fukuoka, 820 – 8502, Japan
[2] Department of Electrical Engineering, Kyushu University, 744 Motooka Nishi-ku, Fukuoka, 819 – 0395, Japan

otabe@cse.kyutech.ac.jp

Abstract. The pinning parameters such as strength of pinning force, temperature dependence of pinning force and so on using in flux creep-flow model to explain electric field vs current density (*E-J*) characteristics were determined by Differential Evolution (DE). DE is one of the methods in Evolutionary Computation (EC) to find an optimization of a problem. First, a model data of *E-J* characteristics in which the pinning parameters were given was prepared, and it was confirmed that DE can find the given pinning parameters from the model data. Then, DE and mesh method were used to determine the pinning parameters in experimental *E-J* characteristics of $GdBa_2CuO_{7-\delta}$ high temperature superconductor. In mesh method, the all combinations of pinning parameters with constant interval for each parameter are calculated, and best set of pinning parameters is selected. It was found that DE shows better performance than mesh method in terms of calculation time and accuracy for determining pinning parameters.

1. Introduction

Electric field vs current density (*E-J*) characteristics of high temperature superconductor (HTS) are important for many kind of applications using superconductor, since other characteristics such as magnetic field dependence of critical current density, irreversibility field are obtained from *E-J* characteristics. It is well known that flux creep-flow model can explain *E-J* characteristics of HTS by using several pinning parameters such as the strength of pinning force, the temperature dependence of pinning force and so on, since the flux creep phenomenon due to thermal agitation is severe at high temperatures [1]. However, it is difficult to determine the pinning parameter to fit the experimental and calculated results, since the number of fitting parameters is 4 to 5. In our previous work, Genetic Algorithm (GA) was used to determine the pining parameters of the flux creep-flow model [2]. It was found that GA was effective to determine the parameters. However, since GA has complicated procedure for calculation, the calculation time tends to long and the computer program is difficult. In the present study, we use Differential Evolution (DE) to determine the pinning parameters. DE is one of the methods in Evolutionary Computation (EC) to find an optimization of a problem, and it was proposed by R. Storn and K. Price in 1995 [3, 4]. In Ref. 4, DE is compared with several optimization-search programs including GA and it is found that DE shows better performance in many kinds of benchmark tests. DE is known to be able to be applied many kinds of problem for searching optimum condition. Recently, it is shown that the modified DE is effective for solving combinatorial optimization problems [5, 6].

In this study, several data sets of E-J characteristics are used, and the pinning parameters of the flux creep-flow model are deduced by DE. We discuss the advantage of using DE by comparing with other solving method.

2. Calculation

It is well known that the electric field vs current density (E-J) characteristics of high temperature superconductors can be explained by the flux creep-flow model [1]. In this model, several pinning parameters have to be determined. The strength of flux pinning, A, temperature and magnetic field dependences of critical current density for the case of virtual flux-creep free, m and γ are assumed in the equation of virtual critical current density, J_{c0}, as

$$J_{c0} = A\left(1-\frac{T}{T_c}\right)^m B^{\gamma-1}\left(1-\frac{B}{B_{c2}}\right)^2. \tag{1}$$

In practical superconductors, the flux pinning strength is known to be widely distributed and is expressed as

$$f(A) = K\exp\left[-\frac{(\log A - \log A_m)^2}{2\sigma^2}\right], \tag{2}$$

where A_m is the most probable value of A, σ^2 is a parameter representing a distribution width and K is a normalization constant. In this study, the pinning parameters, A_m, σ^2, m and γ are determined in terms of fitting of the theoretical and experimental results by using Differential Evolution (DE).

In DE, after initialize individuals, for each individual $\boldsymbol{x}$, evolution is determined by so called mutant vector $\boldsymbol{M}$ calculated from randomly picked up other individuals $\boldsymbol{x}_1$, $\boldsymbol{x}_2$, $\boldsymbol{x}_3$ given by

$$\boldsymbol{M} = \boldsymbol{x}_1 + F(\boldsymbol{x}_2 - \boldsymbol{x}_3), \tag{3}$$

where $0 < F < 1$ is a constant representing the scale factor of vector length. Then trial vector $\boldsymbol{y}$ is generated by crossing of $\boldsymbol{x}$ and $\boldsymbol{M}$. Finally, $\boldsymbol{x}$ is compared with $\boldsymbol{y}$ and if random number is smaller than CR and $\boldsymbol{y}$ is better than $\boldsymbol{x}$, $\boldsymbol{x}$ is replaced and evolved as $\boldsymbol{y}$, where CR is crossover probability. The evaluation value d in the present study is given by

$$d = \frac{1}{N}\sum\left[\log(E_{\text{exp}}) - \log(E_{\text{theo}})\right]^2 \tag{4}$$

where E_{exp} and E_{theo} are experimental and theoretical results of electric field, N is the number of sampling. Further information is in Refs. [1, 2]. In the above procedure, it is found that the all individuals move to better positions than one generation before. In present study, $\boldsymbol{x}$ is the vector consisted from 4 pinning parameters, A_m, σ^2, m and γ. The number of individuals is 30, and the number of generations is up to 20000.

First, DE is confirmed with using a model data of E-J in which the pinning parameters are given. Then an experimental data of E-J of $GdBa_2CuO_{7-\delta}$ (Gd-123) coated conductor of 1.56 μm thickness at 20 and 40 K is used and pinning parameters are determined by DE. For comparison of DE, the mesh method is used. In mesh method, the all combinations of pinning parameters with constant interval for each parameter are calculated, and best set of pinning parameters is selected. In present study, $11^4 = 14641$ sets of parameters are calculated, i.e., each parameter is divided in 11 cases.

3. Results and discussion

First, DE is used to deduce for the model data of E-J and tries to find the given pinning parameters. It is confirmed that DE can deduce the given parameters. Then, DE is used to deduce the pinning parameters for experimental data of Gd-123 superconductor. Figure 1 shows the comparison of E-J of

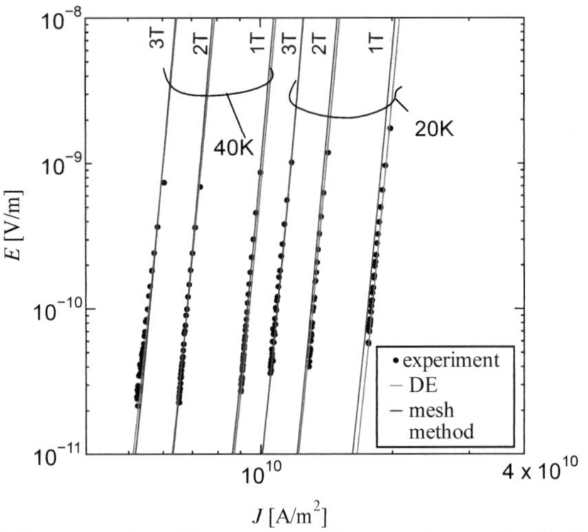

Figure 1. Experimental result of E-J characteristics and theoretical results by pinning parameters determined by DE and mesh method.

Table 1. Comparison of pinning parameters, evaluation value d and calculation time for two methods.

	mesh method	Differential Evolution
$\log_{10}(A_m)$	11.00	11.15
σ	1.00×10^{-2}	1.12×10^{-3}
γ	6.40×10^{-1}	6.31×10^{-1}
m	3.60	3.70
d	4.18×10^{-2}	9.19×10^{-3}
Time [s]	232	79

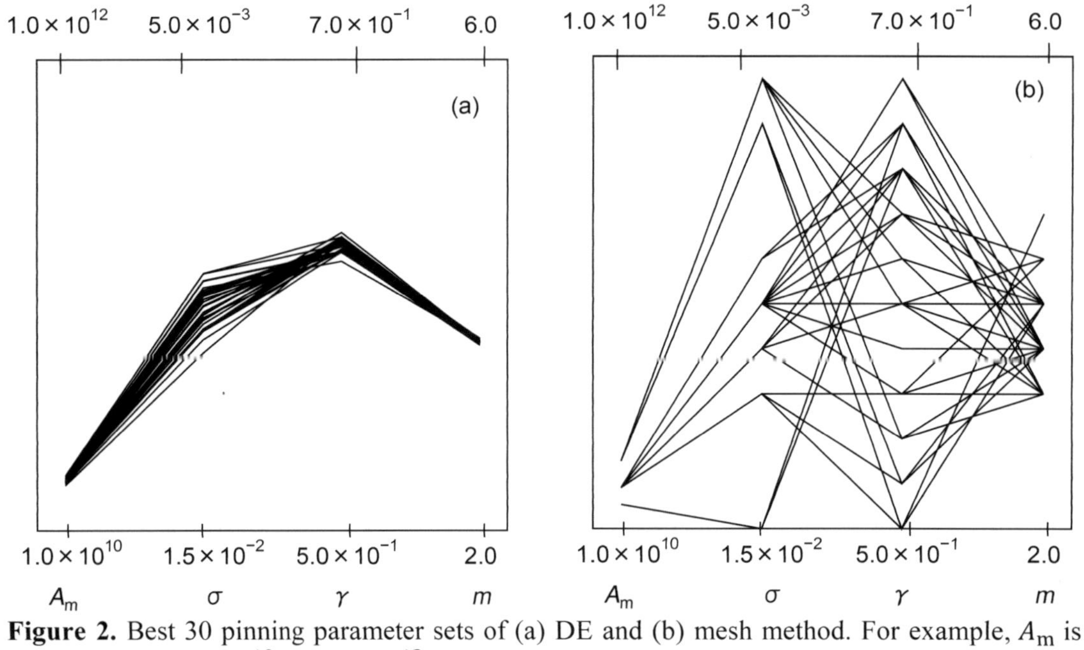

Figure 2. Best 30 pinning parameter sets of (a) DE and (b) mesh method. For example, A_m is in the range of 1.0×10^{10} – 1.0×10^{12} from lower abscissa to upper abscissa. Solid line connects the set of pinning parameters.

experimental result, prediction by flux-creep flow model using pining parameters determined by DE and mesh method. Table 1 lists the pinning parameter determined by two methods and the evaluation value d and time for calculation. The scale factor of vector length $F = 1$ and crossover probability $CR = 0.5$ are used in the calculation of DE. It is found that determination of parameters by DE is better than that of the mesh method. Since pinning parameters can be set more flexible in DE than the mesh method, d is smaller in DE.

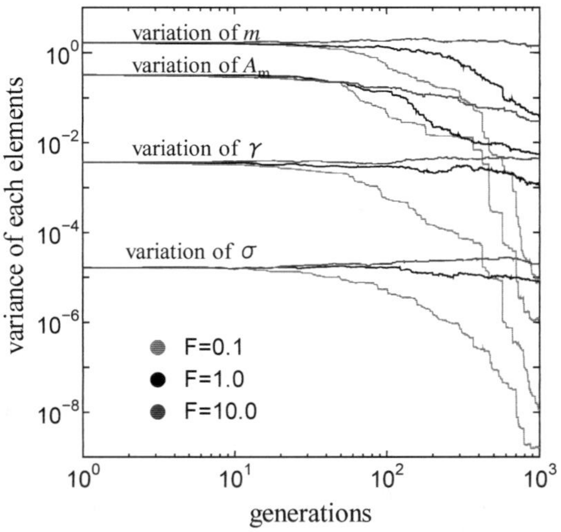

Figure 3. Generation dependence of variance of each pinning parameter for various scale factor of vector length F.

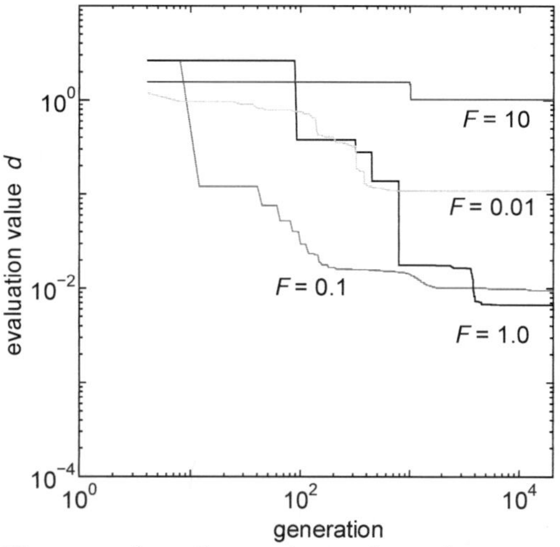

Figure 4. Generation dependence of evaluation value d for various scale factor of vector length F.

In figure 2 (a) and (b), the best 30 sets of pinning parameters by DE and mesh method are shown. The sets are almost same in DE, i.e., after evolution all sets move toward almost same area where d is enough small. On the other hand, the sets of parameters in the mesh method are distributed widely. This means that there are many local minimum of d in addition of optimum condition. Therefore, it is expected to use DE to solve optimization problems when using superconductors such as designs of superconducting motor and magnet.

Figure 3 shows variance of each parameter as a function of generation. It is found that the variance of each parameter decreases as increasing generation, as well as decreasing the scale factor of vector length F. It is also confirmed that the variance of each parameter is smallest for the case of $F = 0.01$. However, this does not mean that the obtained pinning parameters are best. That is, the individuals does not move from the local minimum area due to small value of $F = 0.01$, and the best pinning parameters are obtained for the case of $F = 1$.

Finally, generation dependence of evaluation value d for various value of F is shown in figure 4. The evaluation value monotonically decreases as increasing generation for each value of F. It is found that the case of $F = 1$ reaches the best value of d.

4. Conclusion

In the present study, the four pinning parameters of the flux creep-flow model to explain E-J characteristics in superconductors were determined by Differential Evolution (DE), which is one of the method of Evolutionary Computation. And the results were compared by the mesh method. In mesh method, the all combinations of pinning parameters with constant interval for each parameter are calculated, and best set of pinning parameters is selected. It is found that the evaluation value of DE is better than the mesh method, and the calculation time is shorter in DE than the mesh method. Therefore, it is useful to use DE for determining the pinning parameters of the flux creep-flow model for E-J characteristics in superconductors.

References

[1] Matsushita T 2006 *Flux Pinning in Superconductors* (Berlin: Springer) p. 333.
[2] Minami J, Edamoto T, Uryu K, Ishibashi T, Otabe E S and Kiuchi M 2012 *Physics Procedia* **27**

IOP Conf. Series: Journal of Physics: Conf. Series **871** (2017) 012046 doi:10.1088/1742-6596/871/1/012046

244
[3] Storn R and Price K 1996 *the International Conference on Evolutionary Computation* (ICEC1996) 842
[4] Storn R and Price K 1997 *Journal of Global Optimization* **11** 341
[5] Funaki R, Takano H and Murata J 2013 *SICE Annual Conference* (SICE) 1550
[6] Funaki R, Takano H, Murata J 2016 *SICE Journal of Control, Measurement, and System Integration* **9** 18

Numerical reproduction of screening-current-induced fields in HTS tape windings using finite element method

Yuma Okabe, Tomokazu Honda and Kazuhiro Kajikawa

Department of Electrical and Electronic Engineering, Kyushu University, 744 Motooka, Nishi-ku, Fukuoka 819-0395, Japan

E-mail: okabe@sc.kyushu-u.ac.jp

Abstract. The screening-current-induced fields in one of the high temperature superconducting (HTS) coils fabricated previously with coated conductors are evaluated numerically by using a one-dimensional finite element method, in which only the perpendicular component of a current vector potential is considered due to a very thin superconductor layer in the coated conductor. It is assumed that the voltage-current characteristics in the superconductor layer can be expressed by the critical state or n-value model, in which the field-dependent critical current density is also taken into account. The numerically calculated results of the screening-current-induced fields are compared with the experimental results carried out previously.

1. Introduction

Superconducting magnets for magnetic resonance imaging and nuclear magnetic resonance require a highly uniform magnetic field in their central part, so that these magnets have been usually wound with multifilamentary wires composed of low temperature superconductors. On the other hand, high temperature superconducting (HTS) wires have also been developed and become commercially available, and these wires are in the form of a flat tape with very large cross-sectional aspect ratio. Therefore, if this type of wire is used as a winding for superconducting magnet and a transport current is applied to it, the central magnetic field has very low uniformity because of a screening current induced in the HTS tape [1]. In order to eliminate the screening-current-induced fields (SCFs), the methods based on the abnormal transverse-field effect [2–5] or vortex shaking effect [6–8] have been proposed and validated experimentally [9–11]. Although the theoretical expressions for the decays of magnetizations in a single superconductor (SC) strip have been derived [6–8], there has been no useful method to estimate the decay of SCF in an HTS coil during the application of a cyclic external magnetic field.

In this study, the SCFs in simplified models of an HTS coil are numerically calculated using a finite element method [1, 12]. The decays of SCFs due to applications of cyclic fields are also evaluated on the basis of the theoretical expression for single SC strip [7, 8]. Furthermore, the obtained numerical results are compared with experimental results carried out previously [11].

2. Modeling and calculation method

The specifications of the HTS tape and coil used in the previous experiments are listed in Tables 1 and 2, respectively [11]. The width w of tape and the thickness d of SC layer are 5.03 mm and 2.6 μm, respectively. The critical current I_{c0} at 77 K in self-field is 257 A. The fabricated HTS coil is composed of four double pancakes without any joint between them. The inner and outer diameters of coil are 66.0 mm

Table 1. Specifications of HTS tape [11].

Parameter	Value
Tape width (without insulator), w	5.03 mm
Tape thickness (without insulator)	0.159 mm
Thickness of SC layer, d	2.6 μm
Critical current at 77 K in self-field, I_{c0}	257 A

Table 2. Specifications of HTS coil [11].

Parameter	Value
Inner diameter	66.0 mm
Outer diameter	70.7 mm
Height	42.0 mm
Number of turns	84

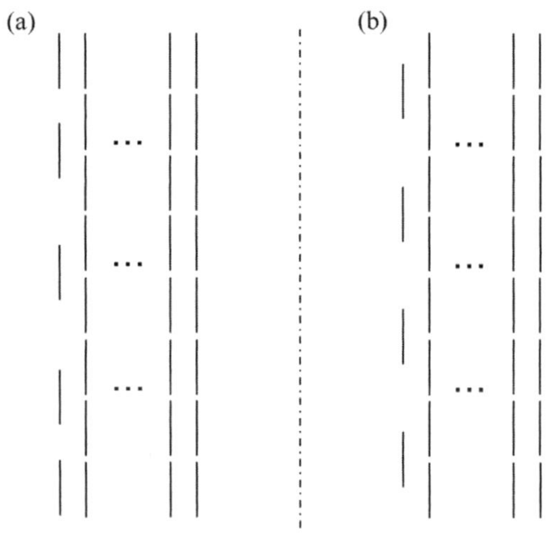

Figure 1. Cross-sectional views of axisymmetrical models for numerical calculations. (a) 85 strip loops (model 1) and (b) 84 strip loops (model 2)

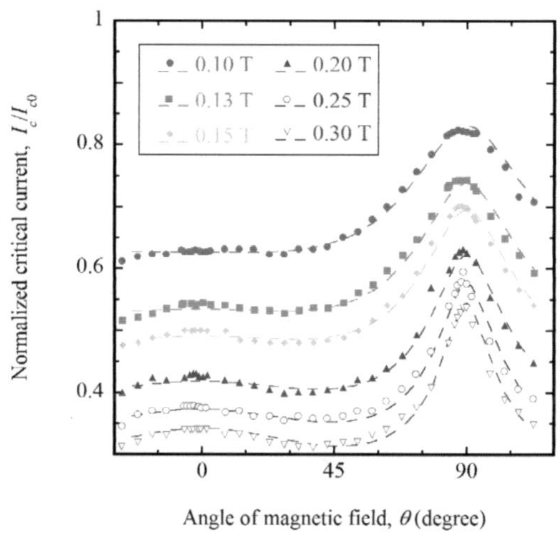

Figure 2. Comparison between experimental results [14] and approximate curves of critical currents in HTS tape.

and 70.7 mm, respectively. The height and turn number of coil are 42.0 mm and 84, respectively. Figure 1 shows the cross-sectional views of axisymmetrical models for numerical calculations, where two typical cross-sections of the HTS coil are considered. One is the cross-section drawn in figure 1(a) at which the current terminals are attached and there are 85 cylindrical SC strips, whereas the other is figure 1(b) in which we begin to wind the double pancakes and the number of strips is 84. The former and latter are called as models 1 and 2, respectively, in this study.

When we consider the current vector potential $\boldsymbol{T}$ defined by $\boldsymbol{J} = \nabla \times \boldsymbol{T}$ with the current density $\boldsymbol{J}$, the vector potential and current density in the cylindrical coordinate system (r, φ, z) have only the radial and azimuthal components as $\boldsymbol{T} = (T_r, 0, 0)$ and $\boldsymbol{J} = \left(0, J_\varphi, 0\right)$, respectively, which have a relationship given by $J_\varphi = \partial T_r/\partial z$, because the thickness d of the strip loop is much smaller than the width w. In this case, we can obtain the governing equation from Faraday's low as [1]

$$\frac{\partial}{\partial z}\left(\rho \frac{\partial T_r}{\partial z}\right) = \mu_0 d \frac{\partial}{\partial t} \sum_{j=1}^{N_e} F(r, z; r', z') \frac{\partial T_r}{\partial z'} \Delta z' + \frac{\partial B_{er}}{\partial t}, \tag{1}$$

where ρ is the resistivity of strip, N_e the total number of elements, which have the width $\Delta z'$ equal to one another, and B_{er} the radial component of external magnetic field. We consider that the self-field in the radial direction at the observation point (r, z) is generated by the current $\partial T_r/\partial z'$ flowing in the element, whose central coordinate is represented by (r', z'). Since the structure of equation (1) in the cylindrical coordinate system is identical with that for the Cartesian coordinate system, we can use a program code for numerical calculations developed previously [12], except for the factor F determined by the geometrical arrangement of strips. The factor F in the cylindrical coordinate system can be expressed by [1]

$$F(r,z;r',z') = \frac{1}{4\pi}\frac{z-z'}{r}\frac{\sqrt{(r+r')^2+(z-z')^2}}{(r-r')^2+(z-z')^2}\left\{(2-k^2)E(k)-2(1-k^2)K(k)\right\}, \tag{2}$$

where $K(k)$ and $E(k)$ are the complete elliptic integrals of the first and second kinds, respectively, with the elliptic modulus k given by

$$k = \sqrt{\frac{4rr'}{(r+r')^2+(z-z')^2}}. \tag{3}$$

In the numerical calculations carried out in this study, the external field is always set to zero as $B_{er} = 0$. The governing equation (1) is discretized by means of the Galerkin method and backward difference method for space and time, respectively.

We use the critical state model (CSM) or n-value model (NVM) for the electric field vs. current density property in the SC strip. The resistivity ρ as a function of the local current density $J_\varphi = \partial T_r/\partial z$ for the CSM and NVM can be expressed by [13]

$$\rho(J_\varphi) = \begin{cases} 0, & |J_\varphi| \le J_c, \\ \rho_f\left(1-\dfrac{J_c}{|J_\varphi|}\right), & |J_\varphi| > J_c, \end{cases} \tag{4}$$

$$\rho(J_\varphi) = E_c\frac{|J_\varphi|^{n-1}}{J_c^n}, \tag{5}$$

respectively, where J_c is the critical current density and n is the n-value. The flux-flow resistivity ρ_f and the electric-field criterion E_c are set as $\rho_f = 1\times10^{-7}$ Ωm and $E_c = 1\times10^{-4}$ V/m, respectively. In order to take into account the dependence of critical current I_c at 77 K on the magnitude B and angle θ of applied magnetic field, the corresponding experimental results for a similar HTS tape [14] are approximated using the equation

$$\frac{I_c(B,\theta)}{I_{c0}} = \frac{\alpha_1 B^{-\Gamma_1}}{\sqrt{\cos^2\theta+\gamma_1^{-2}\sin^2\theta}} + \frac{\alpha_2 B^{-\Gamma_2}}{\sqrt{\gamma_2^{-2}\cos^2\theta+\sin^2\theta}}, \tag{6}$$

where the mathematical symbols, α_1, α_2, Γ_1, Γ_2, γ_1 and γ_2, are the fitting parameters given by $\alpha_1 = 0.0159$, $\alpha_2 = 0.0988$, $\Gamma_1 = 0.92$, $\Gamma_2 = 0.65$, $\gamma_1 = 1 + 19.4B$ and $\gamma_2 = 1 + 1.20B$ on the basis of the least squares method if the unit of field B is tesla. Figure 2 shows the comparison between the experimental results and approximate curves. The critical current density J_c in the SC strip as a function of field magnitude B and angle θ is expressed by $J_c(B,\theta) = I_c(B,\theta)/(wd)$. In the case of the Bean model [15], in which the critical current density is independent of the applied magnetic field, we use J_{c0} $(= I_{c0}/\{wd\})$ as a constant critical current density. Although the n-value also depends on the field magnitude and angle, we have no reliable information about the n-value in the present HTS tape. Therefore, we assume that the n-value has a property similar to the critical current as

$$\frac{n(B,\theta)-1}{n_0-1} = \frac{I_c(B,\theta)}{I_{c0}}, \tag{7}$$

where n_0 is the n-value at 77 K in self-field given by $n_0 = 35$ [14]. On the other hand, the n-value is fixed at n_0 for the case of constant n in the NVM. In the numerical calculations, it is assumed that all the discretized elements in an SC strip under consideration at each time step have an identical critical current density and n-value represented typically at the center of the strip. Since the resistivity in the CSM or NVM has a strong non-linearity as can be seen in equation (4) or (5), the equation (1) is solved iteratively at each time step by means of the Newton–Raphson method.

The sweep rates of currents applied to the HTS coil are fixed at 1 A/s. When the transport currents become equal to the pre-determined values, the transport currents are held maximum. The SCFs B_{SCF} in the axial direction at the center of HTS coil, which is set as the origin of cylindrical coordinates, are estimated by two different methods as [1, 16]

$$B_{SCF} = \mu_0 \sum_{i=1}^{M} \frac{3m_i r_i^2 z_i}{2\left(r_i^2 + z_i^2\right)^{5/2}}, \tag{8}$$

$$B_{SCF} = \mu_0 d \sum_{j=1}^{N_e} G(r', z') \frac{\partial T_r}{\partial z'} - B_0, \tag{9}$$

where M is the number of strips, m_i the radial magnetic moment per unit length in the i-th strip, (r_i, z_i) the coordinate at the center of the i-th strip, B_0 the ideal central field for the case where the currents uniformly flow over the strips and G the geometrical factor determined by the strip arrangement given by [1]

$$G(r', z') = \frac{1}{2}\left\{ \frac{z' + \Delta z'/2}{\sqrt{(r')^2 + (z' + \Delta z'/2)^2}} - \frac{z' - \Delta z'/2}{\sqrt{(r')^2 + (z' - \Delta z'/2)^2}} \right\}. \tag{10}$$

Equations (8) and (9) give us the numerical results identical to each other.

3. Calculated results

Figure 3 shows the comparison between the calculated results of SCFs for models 1 and 2 using the CSMs with constant J_c and $J_c(B, \theta)$. The HTS coils are monotonically energized from 0 A at the constant sweep rate of 1 A/s. In the case of the CSMs, the calculated electromagnetic fields are always in quasi-equilibrium, so that the instantaneous SCFs during the excitations for all the cases can be equal to the final values after the excitations are suddenly stopped. The experimental results of SCFs are also plotted with the symbols [11]. It can be seen in figure 3 that all the results including the experiments have a good agreement with one another in the cases of small applied currents. On the other hand, the discrepancies arize when the applied currents become larger, but the shapes of curves are almost similar to one another. This means that all the curves have a peak and the SCFs decrease with increasing the applied currents in their large range. It is also found that the differences between models 1 and 2 are very small and less than 7.4%. Therefore, the following calculated results are obtained only for model 1.

Figure 4 shows the time evolutions of SCFs during excitations up to 100 A and subsequent current holdings for three different transport properties in the SC strips. In the case of the CSM with $J_c(B, \theta)$, the SCFs during the excitation directly give us the steady-state calculated results as mentioned above. On the other hand, the calculated results of SCFs for the NVMs after the excitations ($t > 100$ s) gradually decrease with time due to the flux creep as shown in figure 4, and should become zero in the infinite time. In the experiments carried out previously, the readings of outputs from a Hall probe using a voltmeter to observe central magnetic fields in the order of 100 mT were carried out after first 4 digits of the output values became unchanged, which took roughly 1 minute [11]. This means that the measurement errors in the experimental results of SCFs shown in figure 3 were less than 0.1 mT. Therefore, we pick up the SCFs at three minutes after the excitations as the typical values of SCFs for the cases with the NVMs, whose magnitudes of time evolutions are less than 0.1 mT/min. The calculated results of SCFs after the excitations are plotted in figure 5, where three different transport properties in the SC strips are used. By using the NVM with $J_c(B, \theta)$ and $n(B, \theta)$, the obtained SCFs almost agree with the experimental results as can be seen in figure 5.

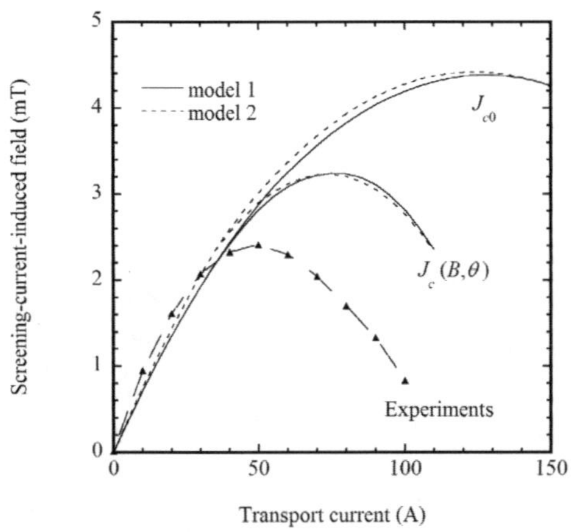

Figure 3. Comparison between calculated results of SCFs for models 1 and 2 using CSM. The symbols are the experimental results [11], and the broken line is a guide to the eye.

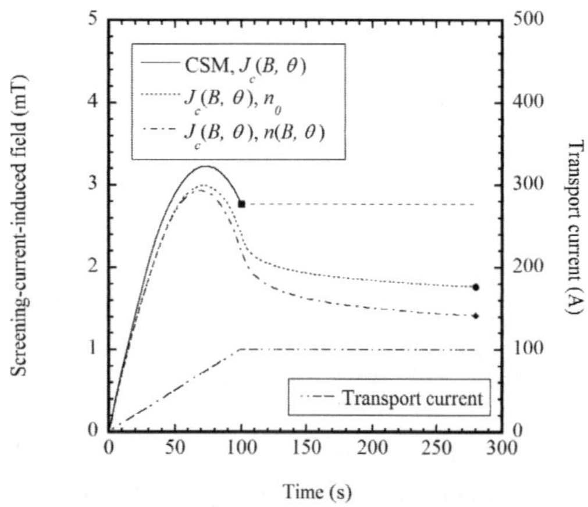

Figure 4. Comparison among time evolutions of SCFs during excitations up to 100 A and subsequent current holdings for three different transport properties in SC strips.

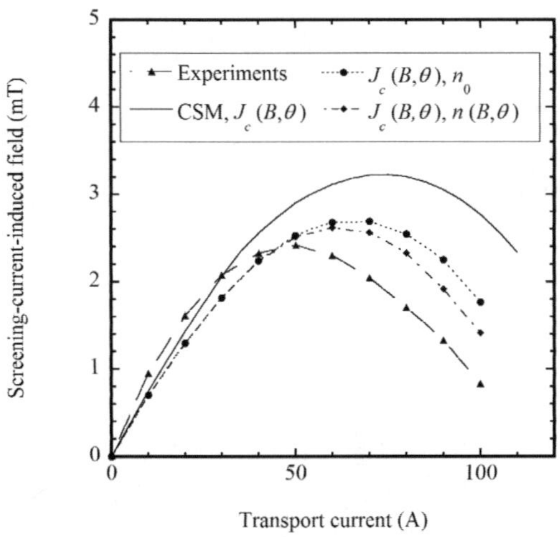

Figure 5. Comparison among calculated results of SCFs for three different transport properties in SC strips. The experimental results [11] are also plotted for comparison. The lines except for the solid line are a guide to the eye.

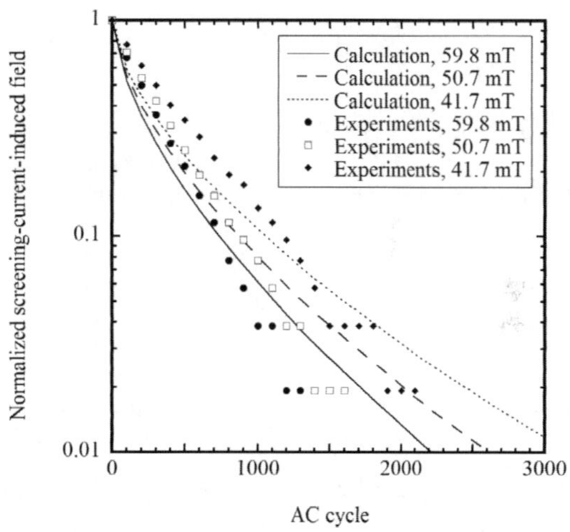

Figure 6. The decays of SCFs for applications of AC magnetic fields in azimuthal direction with three different amplitudes larger than full penetration field. The symbols show the experimental results [11], and the lines are the calculated results.

Figure 6 shows the decays of SCFs for the applications of AC magnetic fields in the azimuthal direction with three different amplitudes larger than the full penetration field of 32.1 mT. The symbols obtained in the experiments [11] are compared with the calculated results drawn using the curves, which are estimated in this study as follows. The critical current $I_c(B_i, \theta_i)$ in every strip after the excitation up to 100 A is obtained using equation (6), where B_i and θ_i are the magnitude and angle, respectively, of magnetic field applied at the center of the i-th strip. Then, the SCF B_{SCF} as a function of AC cycle N is

IOP Conf. Series: Journal of Physics: Conf. Series **871** (2017) 012047 doi:10.1088/1742-6596/871/1/012047

estimated using equation (8) with the magnetic moment m_i given by [7,8]

$$m_i = m_0 \exp\left[-1.68\left(\frac{4B_m d}{\mu_0 I_c(B_i,\theta_i)}N\right)^{0.64}\right], \tag{11}$$

where m_0 is the initial radial magnetic moment per unit length in an SC strip loop under consideration after the excitation of the HTS coil and B_m the amplitude of AC field. The initial moment m_0 in every strip obtained from the finite element analysis using the NVM with $J_c(B,\theta)$ and $n(B,\theta)$ is used here. It can be seen in figure 6 that the experimental results are almost reproduced using the above-mentioned procedure for calculation.

4. Conclusion

The SCFs at the center of the HTS coil after initial excitations were numerically evaluated by means of the one-dimensional finite element method, where the CSM or NVM was used for the voltage-current characteristics of the SC strip loops. The NVM with the dependence of the critical current density and n-value on the magnitude and angle of applied magnetic field most closely reproduced the experimental results of SCFs carried out in the previous work. The applications of the existing theoretical expression for the single strip to all the turns with the separate critical currents and initial magnetic moments obtained from the numerical calculations almost agreed with the experimental results on the decays of SCFs in the HTS coil exposed to the azimuthal AC magnetic fields.

References

[1] Amemiya N and Akachi K 2008 *Supercond. Sci. Technol.* **21** 095001
[2] Funaki K and Yamafuji K 1982 *Japan. J. Appl. Phys.* **21** 299
[3] Funaki K, Niidome T and Yamafuji K 1982 *Japan. J. Appl. Phys.* **21** 1121
[4] Funaki K, Noda M and Yamafuji K 1982 *Japan. J. Appl. Phys.* **21** 1580
[5] Funaki K, Niidome T and Yamafuji K 1983 *Tech. Rep. Kyushu Univ.* **56** 45
[6] Brandt E H and Mikitik G P 2002 *Phys. Rev. Lett.* **89** 027002
[7] Mikitik G P and Brandt E H 2003 *Phys. Rev. B* **67** 104511
[8] Brandt E H and Mikitik G P 2004 *Supercond. Sci. Technol.* **17** S1
[9] Kajikawa K and Funaki K 2011 *Supercond. Sci. Technol.* **24** 125005
[10] Kajikawa K and Funaki K 2012 *IEEE Trans. Appl. Supercond.* **22** 4400404
[11] Kajikawa K and Okabe Y 2016 *IEEE Trans. Appl. Supercond.* **26** 4400504
[12] Kajikawa K, Mawatari Y, Iiyama Y, Hayashi T, Enpuku K, Funaki K, Furuse M and Fuchino S 2006 *Physica C* **445–448** 1058
[13] Kajikawa K, Hayashi T, Yoshida R, Iwakuma M and Funaki K 2003 *IEEE Trans. Appl. Supercond.* **13** 3630
[14] Daibo M, Fujita S, Haraguchi M, Iijima Y, Itoh M and Saitoh T 2012 *IEEE Trans. Appl. Supercond.* **22** 3900204
[15] Bean C P 1962 *Phys. Rev. Lett.* **8** 250
[16] Kajikawa K, Gettliffe G V, Chu Y, Miyagi D, Lécrevisse T P, Hahn S, Bascuñán J, and Iwasa Y 2015 *IEEE Trans. Appl. Supercond.* **25** 4300305

Evaluation of Magnetic Cutting and Polishing with Superconducting Bulks

Y Hiramatsu[1], Y Takahashi[1], E S Otabe[1], K Suzuki[2], Y Tanaka[2] and M Kiuchi[1]

[1] Department of Computer Science and Electronics, Kyushu Institute of Technology, Kawazu 680 – 4 , Iizuka, Fukuoka 820 – 8502, Japan
[2] Department of Mechanical Information Science and Technology, Kyushu Institute of Technology, Kawazu 680 – 4, Iizuka, Fukuoka 820 – 8502, Japan

E-mail: otabe@cse.kyutech.ac.jp

Abstract. In this paper, magnetic levitation tool with superconducting bulks is introduced as a new hollow machining technology. Magnetic levitation tool is the machine that magnet levitates above superconducting bulks and driving force of rotating magnet shaves the object. This tool is expected to use for a grinding machine and machining device because of hollow machining and micromachining by strong fixing. For using magnetic levitation tool, the attractive force, the repulsive force and rotating torque are important for grinding machine, machining outer surface and both, respectively. These forces are calculated by FEM, and compared with experimental results. The experimental results are agreed well with calculated results. However, the attractive force is one order smaller than that required in chemical mechanical polishing.

1. Introduction

Processing technologies are important for creating various metal parts, for example, automobiles, aircrafts and medical applications. To create metal parts, there are various hollow machining technologies, including multi-axis processing machines. Magnetic processing is one of those techniques and is actually used [1 – 3]. In this magnetic processing, magnetic abrasive grains inside the object polishes by attractive force of permanent magnet outside the object. This magnetic polishing has a disadvantage that it is hard to deal with complicated processing, since adjusting the magnetic force is difficult. In other hollow machining process, there is lathe machining. This is a method of processing the object polished by rotating lathe. In this method, although more delicate processing is possible, processing is limited by arm. In order to deal with the disadvantages of these machining methods, a magnetic levitation tool using flux pinning of superconductor is proposed. The magnetic levitation tool is the tool that fixes a permanent magnet in the air and shaves the object by rotating the permanent magnet. By stabilizing the permanent magnet, it is possible to take stability on the left and right balance, and it is expected to process detailed compared with conventional magnetic processing. Here, this magnetic levitation tool is named SUAM (SUperconductive Assisted Machine).

SUAM is expected to apply to cutting and polishing. For cutting, a strong force is required to shave the object roughly, whereas for polishing, it is required a stable force for finer processing. In the case of cutting, a stainless steel blade is attached to the levitated magnet. In the case of polishing, an abrasive or a slurry is used.

In this research, in order to clarify the behavior of the superconductor in SUAM, attractive force, repulsive force and torque were measured by basic method, and experimental results are compared with the calculated result by Finite Element Method (FEM). Furthermore, the influence on the attractive force and the repulsive force due to the initial magnetization distance of the magnet was calculated. Discussion is given for the possibility of SUAM for hollow machining process.

2. Experiment

Figure 1 shows experimental setup of SUAM. In this study, we used four quadrangular superconducting bulks and ring magnet for SUAM. $GdBa_2Cu_3O_{7-\delta}$ was used for a quadrangular superconducting bulk, which side is 35 mm and thickness is 10 mm. In experiment, the single-sided, four-pole type magnet was used in which N and S poles are alternately arranged in the azimuthal direction as shown in figure 2. This is because the one-sided, single-pole type magnet cannot prevent movement in the azimuth direction due to the azimuthal isotropy and cannot rotate itself following the rotation of the bulks. The surface magnetic flux density of this quadrangular ring magnet was 450 mT, the inner diameter was 20 mm, the outer diameter was 58 mm, and weight was 186 g. In this experiment, first, the magnet was placed at appropriate height above the bulks. Then, the superconducting bulks were cooled by liquid nitrogen and the magnetic field was trapped into the bulk. The magnet levitated above the superconductor. After trapped, the magnet moved to vertical direction and rotational direction, and the attractive force, repulsive force and rotational torque were measured by spring scale. The results of this experiment are average of experiment values performed three times.

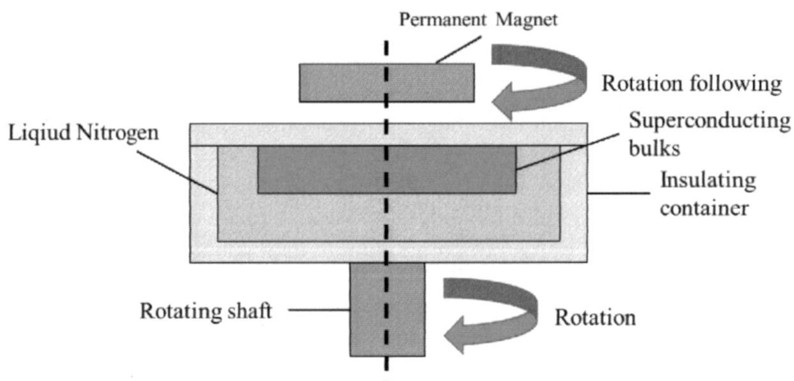

Figure 1. Schematic illustration of SUAM (SUperconductive Assisted Machine).

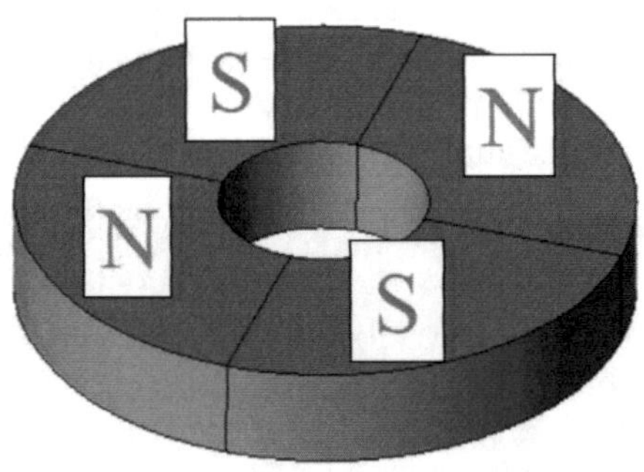

Figure 2. 4-pole magnet used in the SUAM.

3. Calculation Method

In the present study, JMAG-Designer 15.0 was used as Finite Element Method (FEM) simulation software. In this calculation, the permanent magnet was separated from the superconducting bulk by a certain distance, and the superconducting bulk was cooled to trap the magnetic field. For simplified calculation, calculation of temperature dependence of superconductivity was omitted. First, the conductor was set as normal conductor in the initial state, and a permanent magnet was placed over the superconducting bulk with distance of 10 mm. Hence, the magnetic field was penetrated into the bulk. Then, the conductor changed from normal conductor to superconductor in order to reproduce cooling magnetization in the magnetic field. In this way, the magnetic field was trapped in the bulk. Experimental values were used for J_c-B characteristics and n-value models were used for E-J characteristics [4].

4. Results and Discussion

Figure 3 shows experimental results of attractive and repulsive forces when the initial magnetization distance is 10 mm, and calculation results by FEM. The maximum repulsive force is 120 N at 1.0 mm. Here, since the gravitational force is 1.8 N (0.19 kg), the levitation distance is 9 mm which is 1 mm lower than the initial magnetization distance. The maximum attractive force due to the Lorentz force is about 1.8 N. In this study, experimental results and calculation results compare with Chemical Mechanical Polishing (CMP). CMP is the polishing technology, which is currently used as a standard technique for polishing of semiconductor [5 – 7]. Considering CMP, pressure is required 10 – 30 kPa [5 – 7]. Since the contact area of this tool is 2300 mm^2, attractive force should be 23 – 70 N. Therefore, attractive force by present apparatus of 1.8 N does not satisfy this value. In practical case, the initial magnetization distance must be optimized. Therefore, the attractive force was calculated when the initial magnetization distance is changed to 5 – 15 mm.

Figure 4 shows the maximum attractive force when the initial magnetization distance is changed. Since the magnetic field trapped by the superconductor increases, the maximum attractive force increases as decreasing initial magnetization distance. When the initial magnetization distance is 5 mm, the maximum attractive force is 5.9 N. Although it is not beyond the attraction force required in CMP as above mentioned, it is necessary to improve the apparatus such as using a stronger magnet. However, it is difficult to increase the magnetic field of permanent magnet. Hence, using a weight on the magnet seems to be useful for increasing attractive force, since the repulsive force is enough large. For example, when the weight of 1.0 kg is put on magnet, it can generate 9.8 N. Then, the levitation

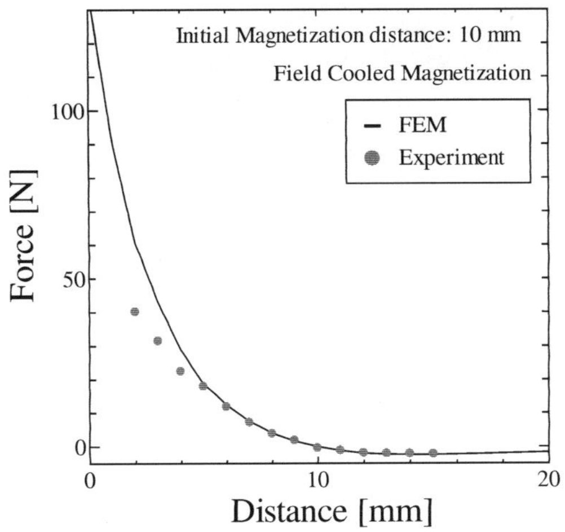

Figure 3. Repulsive force at distance below 10 mm and attractive force at distance above 10 mm.

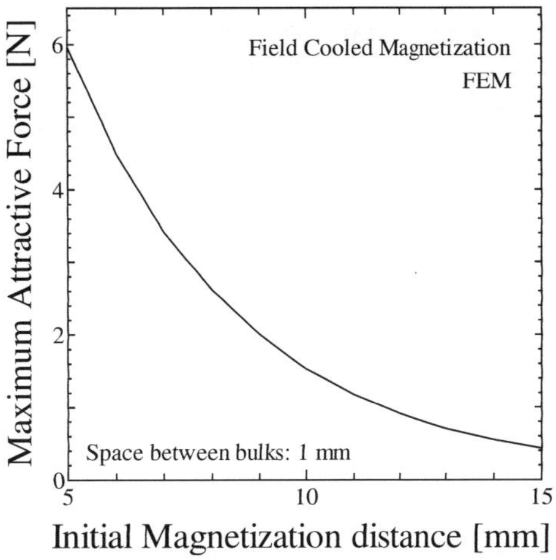

Figure 4. Maximum attractive force as a function of initial magnetization distance.

height becomes 6 mm. However, this value is not enough to use SUAM as a hollow machining process. Then, the weight is put with considering the levitation distance. It is expected that the pressure 10 kPa of a conventional CMP will be achieved by adding the weight of 2.3 kg on the permanent magnet. Furthermore, the effect of varying the distance between the bulks is considered as the factor of the change in the maximum attractive force.

Figure 5 shows the maximum attractive force as changing the space between quadrangular superconducting bulks. The initial magnetization distance is 10 mm. As shown in figure 5, the maximum attractive force decreases as increasing the space between bulks. This occurs because the trapped magnetic field decreases due to larger space between the bulks. This effect is significant, and a difference in attractive force between 1 mm and 10 mm is about 0.45 N. Hence, it is necessary to make the space between the bulks as narrow as possible. Since the superconducting bulks is subjected to resin processing in practical situation, it is necessary to make as close contact as possible in consideration of its thickness. It is found that the current bulk space of 1 mm is reasonable, since attractive force does not change largely from 0.1 to 2 mm.

Figure 6 shows experimental result and calculated results by FEM of the rotational torque when the initial magnetization distance is 10 mm. Here, φ represents the angle of movement of the magnet and the superconductor from the initial magnetization position. It is found that FEM results agree well with experimental result. Since the displacement in the superconductor becomes large, as the angle increases, the torque of the magnet becomes larger. When the angle reaches 45 degrees, the maximum torque is obtained. In the experimental results, the magnet got out at a point exceeding 45 degrees, and measurement could not be done. In FEM, the torque becomes 0 at 90 degrees, and then the torque works in the reverse direction. Torque draws a curve like a sine wave. When a torque greater than the maximum torque is worked, the position of the magnet is moved by the greater force. Therefore, it can be said that the torque at 45 degrees is the torque limit of SUAM.

Figure 7 shows calculation results by FEM for rotational torque in the case of changing initial magnetization distance from 5 – 15 mm. The maximum torque is obtained at 45 degree, regardless of initial magnetization distance. Maximum torque increases as decreasing the initial magnetization distance, and maximum torque at 5 mm is 0.48 N・m. It is found that the phase of rotational torque is not affected by initial magnetization distance.

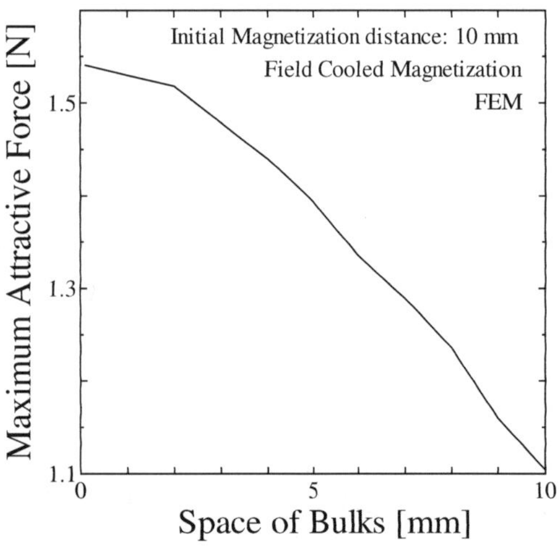

Figure 5. Maximum attractive force as a function of space between bulks.

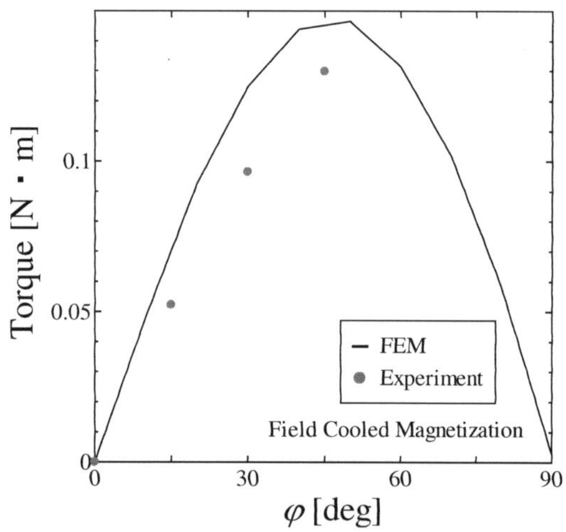

Figure 6. Rotational torque as a function of angle when initial magnetization distance is 10 mm.

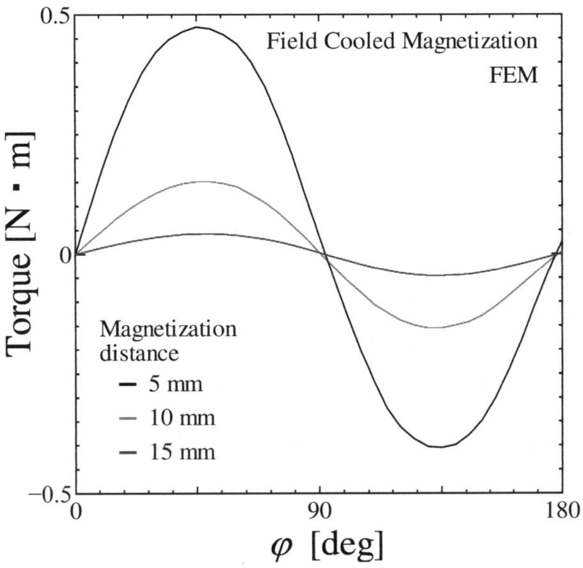

Figure 7. Rotational torque as a function of angle with various initial magnetization distance.

5. Conclusion

In this study, attractive force, repulsive force and rotational torque of SUAM were experimentally measured. The FEM results of repulsive force and attractive force agreed well with the experimental results. We also compared FEM result and experimental result of the rotational torque. The FEM result agreed well with the experimental result. The maximum attractive force in various setting is 6.0N which is one order lower than that expected in Chemical Mechanical Polishing (CMP). From these results, it is concerned that SUAM at present stage cannot provide the same polishing performance as CMP. Therefore, it is necessary to use strong magnet in order to obtain stronger pressure and put the weight on the magnet.

References

[1] Yoshioka M, Hira S and Takeuchi H 1999 *Proceeding of the Third International Conference on Abraisive Technology* 469.
[2] Wang Y, Hu D 2005 International *Journal of Machine Tools & Manufacture* **45** pp 43–49
[3] Boggs T, Carroll R, Tran-Son-Tay R, Yamaguchi H, Al-Mousily F and DeGroff C 2013 *Jounal of Medical Devices* **8** 011005 pp 1–8
[4] Zhai W, Shi Y, Durrell J H, Dennis R Anthony, Zhang Z and Cardwell D A 2015 *Crystal Growth Design* **15** pp 907–914
[5] Krishnan M, Nalaskowski J K and Cook L M 2010 *Chemical Reviews* **110** 178–204
[6] Kasai T and Bhushan B 2008 *Jounal of Physics: Condensed Matte*r **20** 225011
[7] Marinescu I D, Uhlmann E and Doi T K 2007 *Handbook of Lapping and Polishing*

IOP Conf. Series: Journal of Physics: Conf. Series **871** (2017) 012049 doi:10.1088/1742-6596/871/1/012049

Magnetic flux invasion in REBCO bulk magnets with varying pre-magnetized flux distributions in multiple-PFM processes

T Oka[1], K Hara[1], A Takeda[1], J Ogawa[1], S Fukui[1], T Sato[1], K Yokoyama[2] and A Murakami[3]

[1] Niigata University, 8050 Ikarashi-Ninocho, Nishi-Ward, Niigata 950-2181, Japan
[2] Ashikaga Institute of Technology, 268-1 Ohmae-cho, Ashikaga, Tochigi 326-8558, Japan
[3] Ichinoseki National College of Technology, Takanashi, Hagisho, Ichinoseki, Iwate 021-8511, Japan

okat@eng.niigata-u.ac.jp

Abstract. The motion of magnetic flux invading into the HTS bulk magnets were experimentally studied in their pulsed field magnetization processes. The authors paid attention to the effects of the shapes of the pre-magnetized trapped flux distributions before the successive field applications by means of varying the M-shaped distribution. We estimated the differences of the magnetic flux motions between the Dy123 and Gd123 systems, which might have different J_c properties of each sample. As for the Dy123 system, the increase of remaining flux in the periphery region of the M-shaped distribution resulted in the decrease of flux-trapping according to the promotion of flux invasion. On the other hand, trapped flux density has been raised to 3.4 T owing to the effective suppression of flux invasion for the Gd123 bulk magnet. The experiments showed that the peak heights and the positions of the formerly trapped M-shaped fluxes precisely affect the heat generation and the trapped field performance.

1. Introduction

It is well-known that the largely-grown RE–Ba–Cu–O (RE = Y, Sm, Dy, Gd, as RE123) high temperature superconductors (abbreviated as HTS), which are fabricated by the so-called melt-process behave as the trapped field magnets or the bulk magnets [1], [2]. Recently, the world high trapped field of 17.6 T has been reported by Durrell *et al.* [3] by means of the field cooling (FC) process accompanied with a large scale superconducting solenoid magnet and under the special reinforcement to promote mechanical strength of the bulk magnet against the fracture stress due to the magnetic force caused by the strong pinning effect. Although the performances of the trapped fields in FC clearly express the potential abilities of their magnetic field emission, the usage of large scale superconducting magnets would have to limit the application areas for the bulk magnets. While the field trapping ability is usually evaluated in the FC process, various kinds of pulsed field magnetization methods (abbreviated as PFM) have been developed as more compact and practical magnetization methods than FC.

On the course of PFM process, the flux motion in the sample causes the local heat, raises the temperature, lowers the critical current density, and resultantly degrades the trapped field ability [4]. Mizutani *et al.* [5] showed that the iterative application of pulsed fields applied at low temperature regions beneath 77 K is effective for enhancing the trapped field in the bulk magnets. In the past, some other studies have clarified that the existing magnetic flux beforehand in the bulk magnet substantially affects the magnetic flux invasion in the following pulsed field application. Sander *et al.* [6] reported that iterative pulse field application with lowering the sample temperature is effective to promote the trapped field. In 2005, Fujishiro *et al.* [7] reported the strongest field trapping of 5.2 T for the first time in the world by MMPSC method they call, applying iterative pulsed fields at two different temperature stages at 47 K and 28 K.

In the study, the authors take a couple of approaches to promote the field trapping ability of the bulk magnets. One is the approach on the material fabrication. The authors have reported the fabrication of the twin-seeded samples and the anisotropic flux motion in the field application processes [8]. Since the applied magnetic fluxes invade into the bulk magnets from outside to the sample centre, we must take the shape of samples [9] or pulse coils [10], macroscopic grain structure [11], angle dependence [12], the macroscopic defects in the samples [13, 14] into account when we design the activation systems.

The other approach is on the field application method, considering such as the intensity, iterative number, temperature to control the flux motion, heat generation [15]. In the paper, we focus on the shapes of the existent magnetic flux which was trapped beforehand the successive field applications. With use of densely-fabricated Dy123 and Gd123 bulk magnets, we estimated the flux motion and field trapping ability during and after various iterative PFM modes. We aim to clarify the magnetic flux behavior in the PFM operations to improve the field trapping property.

2. Experimental procedure

2.1. Bulk magnets

Figure 1(a) shows a photograph of the melt-processed Dy123 bulk sample manufactured by Nippon Steel and Sumitomo Metals Co. LTD with the dimensions of 30 mm in diameter and 15 mm in thickness. As employed was the unique heat treatment which was operated in the oxygen atmosphere, the mechanical strength of thus densely-made sample was substantially promoted by reducing internal voids [16]. A top view of the sample shows us that the crystal growth begins from the seed crystal attached at the centre toward the periphery of the sample. One sees the obvious traces of crystal growth and the cross-shaped lines showing like the boundaries, which we call growth sector boundaries GSB and growth sector regions GSR. The Gd123 bulk magnet with the same shape in dimensions was employed as a comparison with Dy123 system. The systems have different properties with each other mainly with respect to their J_c values. In Figure 1(b), the positions of the Hall sensor are indicated at four positions (r = 0, 3, 6, 9 mm) to detect the flux motion during and after the PFM processes. They were changed in order in the same manner on each PFM process.

2.2. Experimental setup

Figure 1(c) schematically shows the whole equipment which was employed in this study. The experimental procedure is shown elsewhere [15, 17]. The bulk magnet is cooled to the lowest temperature of 30 K by a Gifford-McMahon refrigerator (AISIN SEIKI Co., GR-103). The successive magnetic fields around 5-7 T were generated and applied to the Dy123/Gd123 bulk magnets by feeding currents to the pulse coil from the condenser bank (SBV-10124, Nihon Denji Sokki Co). The changes

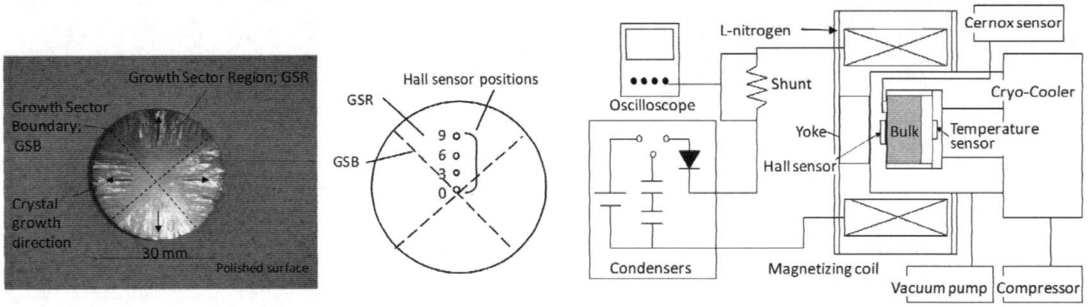

(a) (b) (c)

Figure 1. A surface view of Dy123-based bulk magnet (a), the Hall sensor positions on it (b), and the schematic illustration of the whole experimental setup for pulsed field magnetization (c).

of the magnetic flux density were measured at various points on the bulk magnet surface during and after every magnetizing procedure, as shown in Figure 1(b). The magnetizing electrical circuit is composed of a condenser bank, a shunt resistor, and a 112-turn pulse coil. The cross section of copper wire is 3.0×1.24 mm^2, which is dipped in the liquid nitrogen vessel to reduce the resistance by cooling it to 77 K. The coil constant is 0.925 mT/A. An iron yoke is attached to the bulk magnet to attract the magnetic flux during the pulsed field application.

3. Results and Discussion

3.1. Successive-PFM processes in various pre-shaped magnetic flux distributions

It is reported that the remnant trapped field distributions affects the successive magnetic flux application and their field-trapping behaviours [18, 19]. The authors paid attention to the pre-shaped magnetic flux distributions in the successive PFM operations. Various pre-shaped distributions were formed at the centre of the bulk magnet. Figure 2 shows the trapped flux distributions before (a) and after (b) 7-T application in the successive-PFM processes of 5(40K)-5-7, 5-5-7, and 6-5-7T. The data were measured at r=0-9 mm on the bulk surface, and were plotted on the x-axis which was reflected at the y-axis in each plot area. The legends show the order of the pulsed field applications.

As shown in Figure 2, we obtained various pre-shaped flux distributions by various successive field applications at the center of the sample, while the same distribution was observed at the periphery of the sample. Although the final flux trapping was insufficient when we took the concave distribution at the centre portion, the steep and fine conical flux distributions were obtained when we formed the pre-activated trapezoid shapes. The pre-shaped flux distribution is found to affect the final trapped field distribution. As shown in Figure 3, we exhibit the same flux distributions as shown in Figure 2, and compare it with those of the sample which has weak flux distribution on the periphery. The trapped field apparently degraded at the centre portion in comparison with Figure 2. On the other hand, although the same flux distributions were formed at the periphery, insufficient flux trapping there resulted in the insufficient flux invasion into the centre portion of the bulk magnet.

Figure 4 shows the data of increasing speed (T/s) of the magnetic flux penetration, which corresponds to Figures 2 and 3. The data were derived from the inclination in each time-evolutional profile measured at the centre surface during the 7-T PFM operation. Among the group of 50-70 T/s, the data of 5-5-7 T field application only shows the extremely rapid motion of around 140 T/s. This suggests that the changes of the magnetic flux at the centre portion related to the heat generation and degradation of J_c values. The pre-shaped flux distribution apparently affects to the successive field application and resultant flux trapping ability.

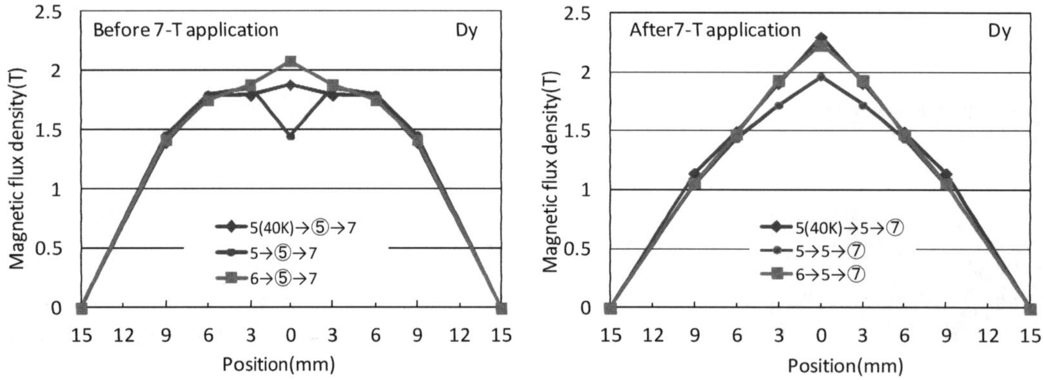

(a) (b)

Figure 2. Trapped field distributions before 7-T application (a) and after (b) for the various field distribution shapes formed at the centre before the applications.

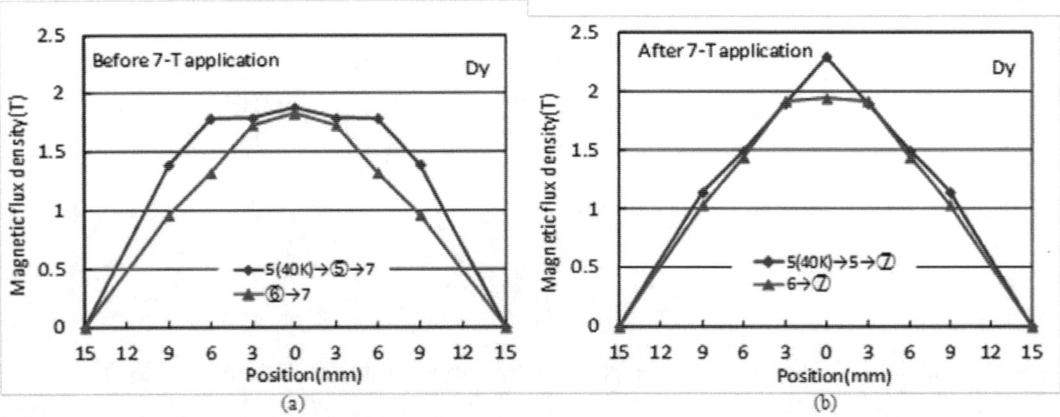

Figure 3. Trapped field distributions before 7-T application (a) and after (b) for the various field distribution shapes formed at the periphery of the sample before the applications.

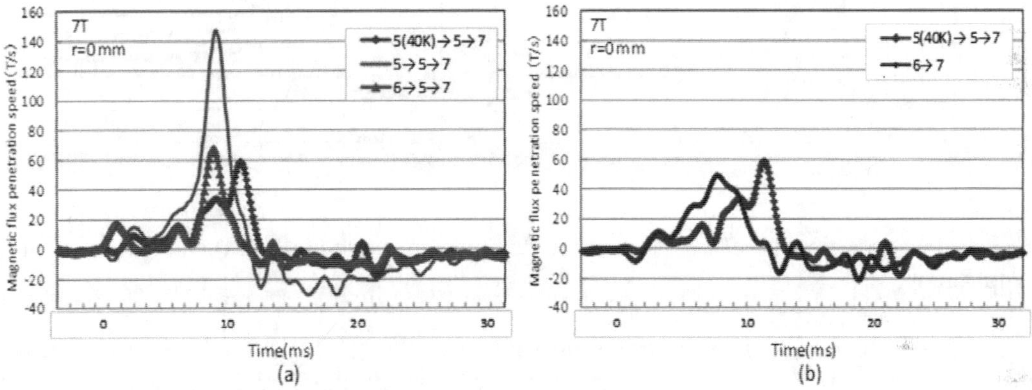

Figure 4. Time evolution profiles of magnetic flux penetration speeds for the distribution-shape varieties shown at the and during 7-T application for the various field distribution shapes formed at the centre (a) and at the periphery (b) before the applications.

3.2. Successive-PFM in various M-shaped flux trapping

According to the Fujishiro's report [20], the pre-activated "M-shaped" flux distribution was proved to be effective for high-field trapping. We investigated the flux trapping performance and the magnetic flux motion which were affected by various M-shaped flux distribution on the sample periphery. The Gd123-based sample was employed in addition to Dy123 bulk magnet. Figure 5 shows four types of the pre-activated M-shaped flux distributions. Various pre-shaped flux distributions were prepared on their periphery with the same magnetic flux trapping of 1.5 T at the centre portion on the sample surface.

The changes of the flux invasion into the sample during the final 7-T application are plotted in Figure 6. In Dy system, the flux shows the sluggish invasion on its periphery, and the applied flux lately invades in comparison to other areas. This means that the existent magnetic flux apparently restricts the flux invasion.

On the other hand, the flux invasion was stopped at the periphery in Gd system, as shown in Fig. 5b, whereas the field invasion in the case of 5-7 application was enhanced, as shown in Fig. 6b. Therefore, the existent M-shaped distribution promotes the flux invasion in the case of 5-7 application, and lowers the trapped field in comparison with single 7-T application. We observe the promotion of flux invasion and restriction of heat generation when we apply the 5-5-7 T with the substantial flux trapping on its periphery.

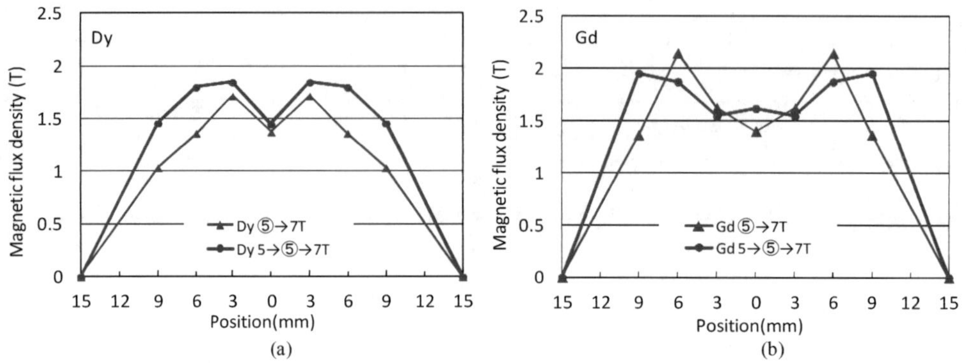

Figure 5. Trapped field distributions before 7-T application for Dy123 (a) and Gd123 (b).

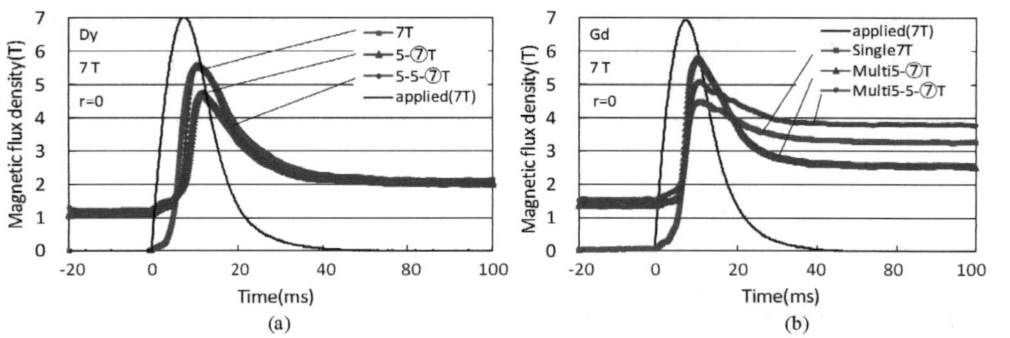

Figure 6. Time evolutional profiles of magnetic flux invasions into Dy123 (a) and Gd123 (b) during 7-T application.

Figure 7 shows the resultant trapped flux distributions after final 7-T application to various pre-activated M-shaped Gd123 system. The highest trapped flux of 3.43 T was obtained when the 5-5-7 successive fields were applied, which showed the fine conical distribution. Although we see no differences at each peripheral flux trapping, we observe various trapped flux distributions which correspond to the respective pre-activated M-shape at the centre portion.

Figure 8 shows the magnetic flux penetration speeds measured at the centre. In the Dy system, the flux invasion for 5-7 and 5-5-7 data occurs late in comparison with single 7-T application, which relates the weak and sluggish penetration into the sample. In case of Gd, the slight excess speed for 5-7 application might result in the higher heat generation and lower flux trapping at the centre part. The discussion of heat generation would have to wait the results of the direct temperature measurement on each position of the sample surface. The trapped flux performances are listed in Table 1. The final

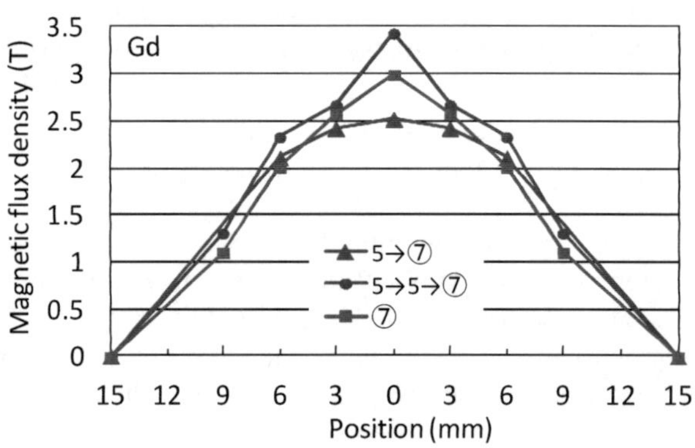

Figure 7. Trapped field distributions after 7-T application for Gd123.

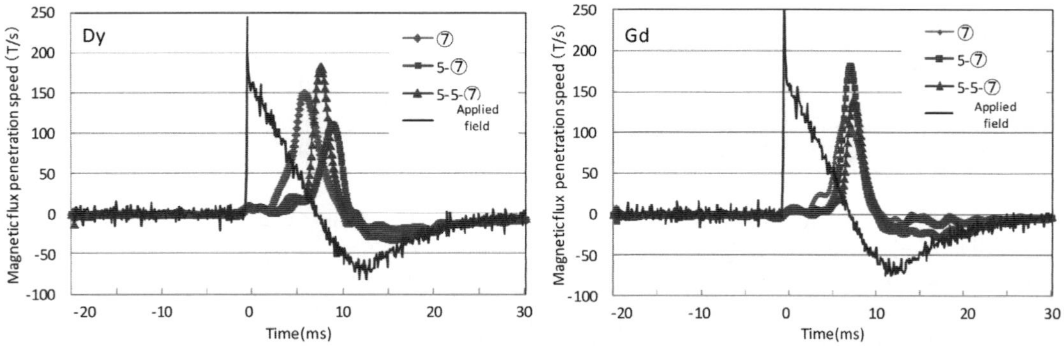

Figure 8. Magnetic flux penetration speeds for Dy123 (a) and Gd123 (b) during 7-T application to the sample showing different M-shaped distributions.

Table 1. Final trapped fields after various activation modes.

System	7	5-7	5-5-7
Dy	1.87	2.07	2.06
Gd	2.99	2.52	3.43

The data were measured at the center of the bulk surface [T].

trapped flux is finely found to correspond to the pre-shaped flux distribution. One can see distinctive difference in the flux trapping ability between Dy and Gd systems, which clearly reflects the J_c performances of the samples chosen.

4. Conclusion

We precisely investigated the effects of applied magnetic flux invasion to the pre-activated samples of Dy and Gd123 systems having various M-shaped flux distributions. The bulk magnets trap the substantial remnant flux on the periphery areas of the samples, and the distributions relax the field gradient when the successive pulsed-field was applied. This phenomenon is capable of suppress the flux motion and heat generation. The final flux invasion tends to be promoted when the steep M-shaped distribution is formed. When we form the weak magnetic field area showing M-shape at the centre portion, the successive applied flux invades toward the centre portion without any blocking by the pre-activated magnetic flux.

Acknowledgments

The bulk sample was supplied by Dr. Teshima in Nippon Steel and Sumitomo Metal Co. The authors give him great thanks for the courtesy.

References

[1] Wipf S and Laquer H 1989 *IEEE Trans. Magn.* **25** 1877
[2] Tomita M and Murakami M 2003 *Nature* **421** 517
[3] Durrell J, Dennis A, Jaroszynski J, Ainslie M, Palmer K, Shi Y-H, Campbell A, Hull J, Strasik M, Hellstrom E and Cardwell D 2014 *Supercond. Sci. Technol.* **27** 082001
[4] Itoh Y and Mizutani U 1996 *Jpn. J. Appl. Phys.* **35** 2114
[5] Mizutani U, Ikuta H, Hosokawa T, Ishihara H, Tazoe K, Oka T, Itoh Y, Yanagi Y and Yoshikawa M 2000 *Supercond. Sci. Technol.* **13** 836
[6] Sander M, Sutter U, Koch R and M. Klaeser M 2000 *Supercond. Sci. Technol.* **13** 841
[7] Fujishiro H, Tateiwa T, Fujiwara A, Oka T and Hayashi H 2006 *Phys. C.* **445-448** 334

[8] Oka T, Miyazaki T, Ogawa J, Fukui S, Sato T, Yokoyama K and Langer M 2015 *Supercond. Sci. Technol.* **29** 024003
[9] Tateiwa T, Sazuka Y, Fujishiro H, Hayashi H, Nagafuchi T and Oka T 2007 *Phys. C.* **463-465** 398
[10] Ida T, Matsuzaki H, Akita Y, Izumi M, Sugimoto H, Hondou H, Kimura Y, Sakai N, Nariki S, Hirabayashi I, Miki M, Murakami M and Kitano M 2004 *Phys. C* **412-414** 638
[11] Fujishiro H, Arayashiki T, Tamura T, Naito T, Teshima H and Morita M 2012 *Jpn. J. Appl. Phys.* **51** 093005
[12] Ainslie M, Zou J, Mochizuki H, Fujishiro H, Shi Y-H, Dennis A and Cardwell D 2015 *Supercond. Sci. Technol.* **28** 125002
[13] Yokoyama K, Oka K, Kondo N and Hosaka S 2013 *Phys. C* **484** 343
[14] Fujishiro H, Naito T and Oyama M 2012 *Phys. Procedia* **36** 687
[15] Oka T, Ota H, Shimoda T, Ogawa J, Fukui S, Sato T, Yokoyama K, Murakami A and Langer M 2014 *IEEE Trans. on Appl. Supercond.* **25** 6800904
[16] Murakami A, Miyata H, Hashimoto R, Otaka K, Katagiri K and Iwamoto A 2009 *Phys. C* **469** 1207
[17] Yanagi Y, Itoh Y, Yoshikawa M, Oka T, Ikuta H and Mizutani U 2005 *Supercond. Sci. Tech.* **18** 839
[18] Oka T, Seki H, Ogawa J, Fukui S, Sato T and Yokoyama K 2011 *IEEE Trans. Appl. Supercond.* **21** 3356
[19] Oka T, Ishiduka D, Otsuka T, Ogawa J, Fukui S, Sato T, Yokoyama K and Murakami A 2013 *J. Supercond. Nov. Magn.* 26 1301
[20] Fujishiro H, Hiyama T, Tateiwa T, Yanagi Y and Oka T 2007 *Phys. C* **463–465** 394

Influence of Artificial Defects on Trapped Field Performance in a Superconducting Bulk Magnet

K Yokoyama[1], K Eranda[2], Y Zhao[2], A Katsuki[2], A Miura[2] and T Oka[3]

[1] Division of Mechanical and Electrical Engineering, Ashikaga Institute of Technology, 268-1 Omae-cho, Ashikaga, Tochigi 326-8558, Japan
[2] Department of Electrical and Electronic Engineering, Graduate School of Engineering, Ashikaga Institute of Technology, 268-1 Omae-cho, Ashikaga, Tochigi 326-8558, Japan
[3] Education Center for Engineering and Technology Faculty of Engineering. Niigata University, 2-8050, Ikarashi, Nishi-ku, Niigata 950-2181, Japan

E-mail: k-yokoyama@ashitech.ac.jp

Abstract. In this study, we attempt to improve the trapped magnetic field by pulsed-field magnetization using REBCO bulk with small holes processed artificially. In our previous study, a hole was opened in the growth sector region (GSR) in accordance with the idea that the performance of the growth sector boundary (GSB) was not reduced because the critical current density of the GSB was higher than that of the GSR. In the next study, a small hole was processed on the GSB to decrease the magnetic shield toward the magnetic flux that entered through the GSB. The experimental results suggested that placing a hole on the GSB was advantageous to trapped field performance as compared with placing a hole at the GSR. In this paper, three more holes were processed in the bulk with a single hole on the GSB, and trapped field performance was investigated. Although the performance was decreased overall in the bulk with four holes, we confirmed that the peak value of total magnetic flux shifted to the low applied magnetic field, although the maximum value was almost the same as that of the bulk with a single hole.

1. Introduction

We attempted to improve the magnetic field of REBCO bulk magnet by pulsed-field magnetization (PFM) and to develop an efficient magnetization method. PFM is a useful method for industrial applications because the bulk can be activated by a conventional copper coil and a condenser bank without an expensive superconducting magnet. However, there is a problem that a trapped field is small due to a decrease of J_c caused by large heat generation when a magnetic field is applied, because the magnetic flux penetrates by overcoming the magnetic shield. Then, several advanced PFM methods [1, 2], such as IMRA method [3, 4], were developed to improve the trapped field. In this method, the inhomogeneous characteristic of the bulk material was utilized. The magnetic flux was introduced from the low J_c part of the bulk at the first pulsed-field application; thereafter, that part was filled by applying low fields. In REBCO bulk superconductors, growth sector boundaries (GSBs) and growth sector regions (GSRs) are formed during the manufacturing process, and the J_c characteristic of the GSB is higher than that of the GSR because many RE211 particles, which are the pinning center, exist on the GSB. However, the quality of the bulk material has become uniform with advances in material fabrication techniques, and there is no difference in the J_c at both areas. Therefore, it becomes difficult to trap the high magnetic field by PFM due to the increase of the magnetic shield with the increase of J_c, although the trapped field activated by the field-cooling method is improved [5]. If a

IOP Conf. Series: Journal of Physics: Conf. Series **871** (2017) 012050 doi:10.1088/1742-6596/871/1/012050

high magnetic field is applied to overcome the shield, a substantial amount of heat is generated, and, consequently, the trapped field is reduced. Therefore, it is necessary to supply the magnetic field to the bulk efficiently while restraining the generation of heat.

We have proposed artificially processing small holes in a part of the bulk material for the purpose of supplying the magnetic field to the bulk easily and estimated the magnetizing performance [6, 7]. A hole is not a superconductor; thus, it becomes easy for the magnetic flux to penetrate from the portion with the hole due to a decrease of the magnetic shield. In the first part of our study, a hole was opened at the GSR based on the idea that the magnetic flux was introduced from the GSR with the low J_c characteristic, and it was trapped on a GSB with a high J_c [6]. In the next step, the hole was processed on the GSB based on the concept that the magnetic flux penetrated from the GSB as well as the GSR [7]. When comparing the magnetizing performance of both bulks, it was confirmed that the magnetic flux could penetrate in a lower applied field in the case in which the hole existed on the GSB, although the amplitude of the trapped field was almost the same. Although the hole with a 2-mm diameter was drilled at one of four places on the GSB in our previous study, a hole of the same size was processed at all places on the GSB completely, as an extreme example in this paper, and the magnetizing performance was investigated by applying a single pulsed field while varying the amplitude and the temperature.

2. Experimental

The highly *c*-axis-oriented GdBCO bulk material with dimensions of 60 mm in diameter and 20 mm thick consisted of 75.0 wt.% Gd123, 24.5 wt.% Gd211, and 0.5 wt.% Pt powder (Nippon Steel & Sumitomo Metal Corporation), and it was reinforced mechanically with a 2-mm thick stainless steel ring (SUS316L). Figure 1(a) shows the bulk sample with a single hole 2 mm in diameter at the rim of the sample on the GSB that was used in our previous study. In this paper, each single hole was processed in three other places of the GSB in the same sample and filled with solder, as shown in Figure 1(b). Hereinafter, they are referred to as "single-hole bulk" and "four-hole bulk," respectively. The sample was installed to the second stage of a two-stage GM refrigerator using a sample holder. The refrigerator had an ultimate temperature of 13 K and a cooling capacity of 5 W at 20 K. Four Hall sensors (BHT-921, F.W.BELL) were adhered on the sample surface, as shown in the inset of Figure 2, with Kapton tapes to investigate the behaviour of the magnetic flux during magnetization. After a vacuum vessel was attached and evacuated by a diffusion pump, the bulk was cooled to 20, 30, 40, and 50 K using a temperature controller. A single pulsed field was applied while varying the amplitudes of 3.1, 3.9, 4.6, 5.4, 6.2, and 7.0 T for each temperature, where the rising time of the pulse was 10 ms. The magnetic flux density was monitored with a sampling rate of 100 μs during magnetization, and the trapped field distribution at 7 mm above the bulk surface was measured by scanning the three-dimensional Hall sensor (BH-703, F.B.BELL) with a 2-mm pitch after magnetization. Moreover, the total magnetic flux was calculated using the distribution data.

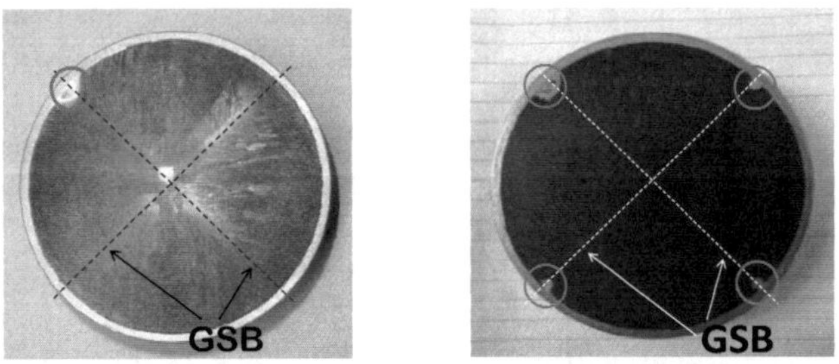

(a) single-hole bulk (b) four-hole bulk

Figure 1. Photographs of bulk superconductors with a single hole and four holes

IOP Conf. Series: Journal of Physics: Conf. Series **871** (2017) 012050 doi:10.1088/1742-6596/871/1/012050

3. Results and discussion

3.1. Time responses of magnetic flux on the bulk surface

Figure 2 shows the time responses of the magnetic flux density on the bulk surface in (a) the single-hole bulk and (b) the four-hole bulk at 20 K. For the applied field $\mu_0 H$ = 3.9 T, only the value of H1 became large, indicating that the magnetic flux penetrated preferentially from the portion with the hole in the single-hole bulk. On the other hand, in the four-hole bulk, the value of H2 increased as compared with that of the single-hole bulk, indicating that the magnetic flux entered from the portion where other holes existed. There was no difference in the values of H3 and H4 in both bulks. For $\mu_0 H$ = 5.4 T, the value of H2 of the four-hole bulk decreased as compared with that of the single-hole bulk. It was considered that the magnetic flux was trapped on the GSB in the single-hole bulk, while the pinning center decreased by processing small holes in the four-hole bulk. The result of $\mu_0 H$ = 7.0 T had the same tendency as that of $\mu_0 H$ = 5.4 T.

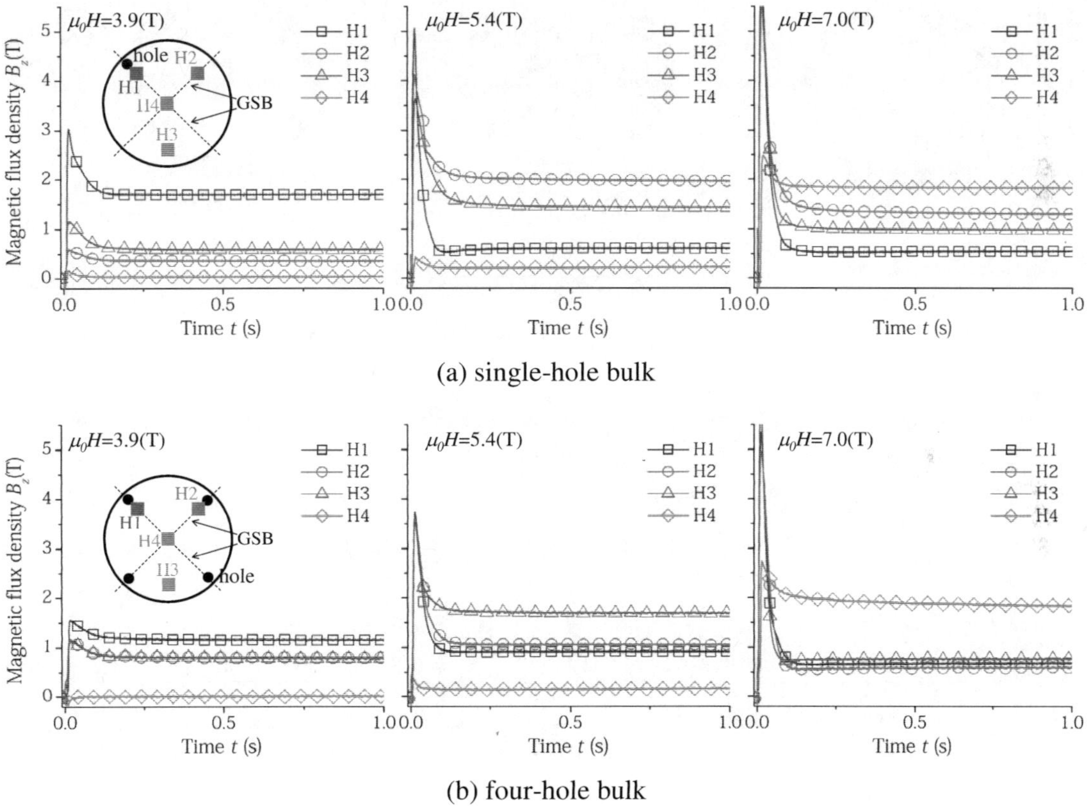

Figure 2. Time responses of magnetic flux density on the bulk surface at 20 K.

3.2. Trapped field distributions

Figure 3 shows trapped field distributions in (a) the single-hole bulk and (b) the four-hole bulk at 20 K. For $\mu_0 H$ = 3.9 T, small peaks existed in the portions with holes in both bulks, indicating that the magnetic flux penetrated from that portion, and they were trapped there. In the single-hole bulk, the distribution of $\mu_0 H$ = 5.4 T was mostly concentric; moreover, that of $\mu_0 H$ = 7.0 T changed to a square shape, indicating that the magnetic flux was trapped on the GSB. However, there was a small distortion in the portion with the hole (H1), namely, the influence of the small hole was somewhat apparent. In the four-hole bulk, the distributions of $\mu_0 H$ = 5.4 and 7.0 T were diamond shape, indicating that the magnetic flux on GSBs decreased conspicuously near the small hole.

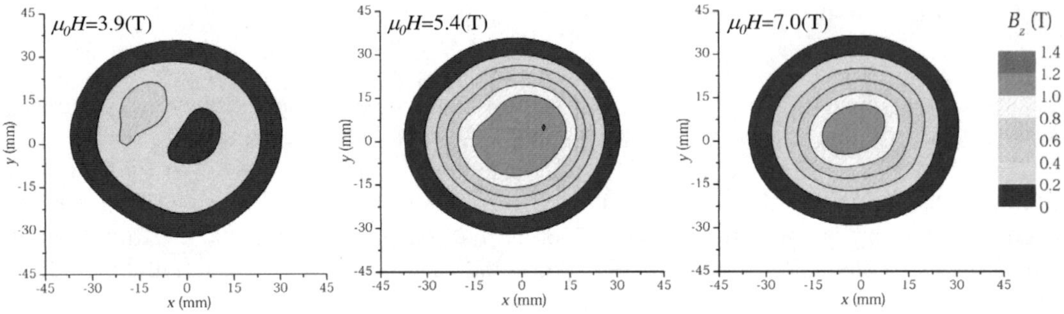

(a) single-hole bulk

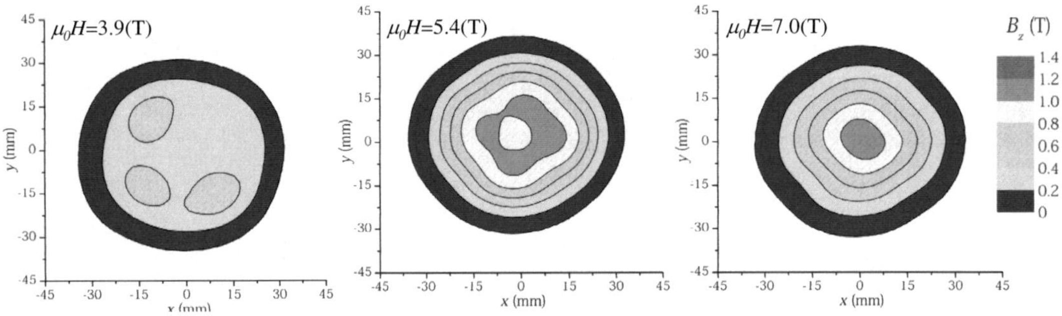

(b) four-hole bulk

Figure 3. Trapped field distributions at 20 K.

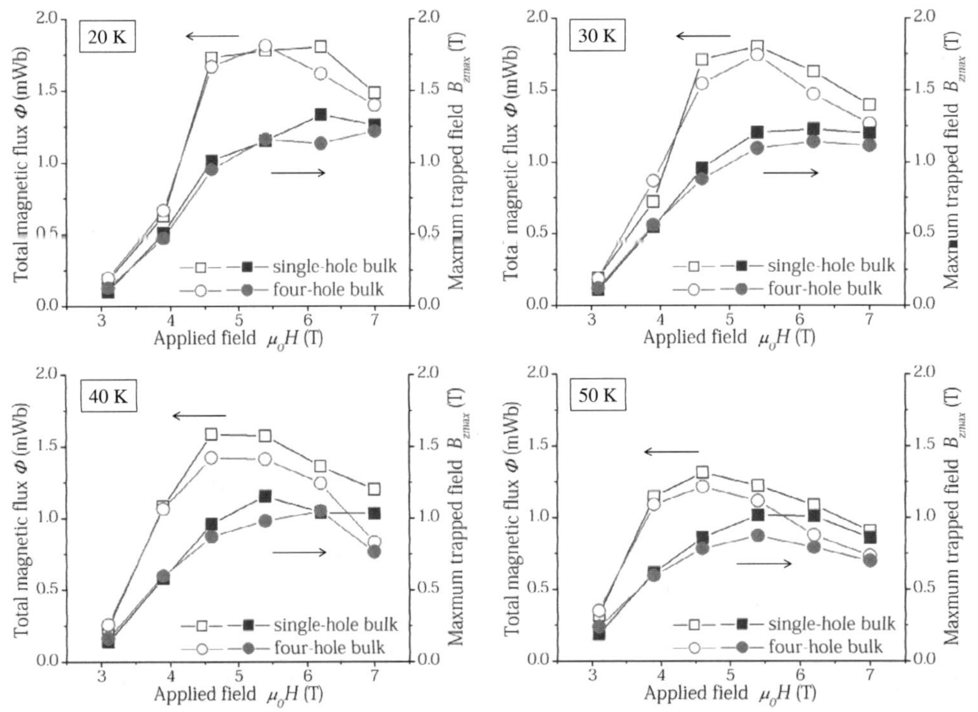

Figure 4. Comparisons of total magnetic flux (open symbols, left axis) and maximum trapped flux density (solid symbols, right axis) as a function of applied field between the single-hole bulk (blue and square symbols) and the four-hole bulk (red and circle symbols) at 20-50 K.

3.3. Total magnetic flux and maximum trapped field density

Figure 4 compares the total magnetic flux, Φ (open symbols, left axis), and the maximum trapped flux density, B_{zmax} (solid symbols, right axis), as functions of the applied field of the single-hole bulk (blue and square symbols) and the four-hole bulk (red and circle symbols) at 20-50 K, where the B_{zmax} was the maximum value in the trapped field distribution, as shown in Figure 3. Although there was almost no difference between the bulks in low applied fields, both the Φ and the B_{zmax} of the single-hole bulk were larger than those of the four-hole bulk, suggesting that four holes are too much because the flux flow was increased by decreasing the number of pinning centers. However, we can see that the peak value of the total magnetic flux shifted to the low applied magnetic field, although the maximum value did not change at 20 K. We will consider the number and the position of small holes aiming at a magnetizing method that can trap a high magnetic field with a low applied field.

4. Conclusions

We attempt to improve the trapped magnetic field by PFM using a REBCO bulk with small holes processed artificially. Since the trapped field performance when the hole was opened on the GSB was superior to the performance when the hole was existed at the GSR in our previous experiments, in this paper, we increased the number of holes opened on the GSB and estimated the trapped field property. As determined from the time responses of flux density during magnetization, the magnetic flux penetrated easily for low applied fields in the four-hole bulk, while the trapped field was decreased for high applied fields. A similar tendency was also seen in the magnetic field distribution results, and the magnetic flux was not trapped on the GSB, due to the flux flow in high applied fields. When comparing the total magnetic flux and the maximum flux density between the singe-hole bulk and the four-hole bulk, both values of the single-hole bulk were higher than those of the four-hole bulk in high applied fields. Surplus holes had a harmful influence on the trapped field characteristic; therefore, it is necessary to investigate the number and position of small holes most suited to trapping a high magnetic field by low applied field.

Acknowledgments
This work was supported by JSPS KAKENHI Grant Number JP15K05951.

References

[1] Sander M, Shutter U, Koch R and Läser M 2000 *Supercond. Sci. Technol.* **13** 841
[2] Fujishiro H, Tateiwa T, Fujiwara A, Oka T and Hayashi H 2006 *Physica C* **445-448** 334
[3] Mizutani U, Oka T, Itoh Y, Yanagi Y, Yoshikawa M and Ikuta H 1998 *Appl. Supercond.* **6** 235
[4] Ikuta H, Hosokawa T, Ishihara H, Yoshikawa M, Yanagi Y, Itoh Y, Oka T and Mizutani U 2001 *IEEE Trans. Appl. Supercond.* **11** 3716
[5] Durrell J H, Dennis A R, Jaroszynski J, Ainslie M D, Palmer K G B, Shi Y-H, Campbell A M, Hull J, Strasik M, Hellstrom E E and Cardwell D A 2014 *Supercond. Sci. Technol.* **27** 082001
[6] Yokoyama K, Oka T, Kondo N and Hosaka S 2013 *Physica C* **484** 343
[7] Yokoyama K, Kulawansha E, Yuanding Z and Oka T 2016 *IEEE Trans. Appl. Supercond.* **26** 8201704

IOP Conf. Series: Journal of Physics: Conf. Series **871** (2017) 012051 doi:10.1088/1742-6596/871/1/012051

Effects of carbon nanotube addition on superconductivity in Y-Ba-Cu-O bulk superconductors

K Inoue, Y Miyake, M Miryala and M Murakami

Department of Materials Science and Engineering, Shibaura Institute of Technology, Toyosu 3-7-5, Koto-ku, Tokyo, 135-8548, Japan

E-mail: k-inoue@sic.shibaura-it.ac.jp

Abstract. Bulk Y-Ba-Cu-O superconductors have significant potential for engineering applications due to high critical current density, which is attributed to the presence of pinning centers such as Y_2BaCuO_5. The introduction of nano-sized secondary phase is known to act as more effective pinning center than those in micron sizes. The diameter of carbon nanotube (CNT) is close to that of the coherence length of high-temperature superconductors, which is expected to improve the flux pinning performance. We have investigated the effects of CNT addition on the microstructure, superconducting transition temperature (T_c), and critical current density (J_c) of $YBa_2Cu_3O_x$ (Y123) based bulk superconductors. SEM observation showed the distribution of needle-like particles around 100 nm in length in the Y123 matrix for the CNT added samples. The highly porous texture was also observed for the excess addition of CNT. T_c was enhanced from 90.5 K to 91.8 K with increasing CNT addition. It is probable that carbon originated from CNT suppressed oxidation and carrier doping. J_c exhibited the highest value for 0.25 wt% CNT added sample. This suggests that nano-sized needle-like particles act as effective pinning centers. However, a further increase of CNT led to the decline of J_c, which suggests that there is an optimum amount of CNT for the improvement of J_c. The secondary peak was observed for the sample with 1 wt% CNT addition, where CO_3 substitutions with Cu site at the Cu-O chain might induce oxygen vacancies leading to the field induced pinning.

1. Introduction

Melt-processed Y-Ba-Cu-O superconductors have significant potential for engineering applications due to high critical current density (J_c) and high trapped magnetic fields, which is attributed to the presence of the pinning centers such as the micro-sized secondary phase Y_2BaCuO_5 (Y211) in the $YBa_2Cu_3O_x$ (Y123) matrix. It is known that J_c values are inversely proportional to the size of pinning centers. Nano-sized secondary particles are known to be more effective as the pinning center than those in micron sizes. It has been reported that J_c values were enhanced with the addition of nano-sized secondary phases such as ZrO_2 [1], Nb_2O_5 [2], TiO_2 [3], and $Y_2Ba_4CuMO_y$ (Y-2411) (M = Nb, W, Ag, Bi, etc.) [4] as well as Y211 [5] into the Y123 materials.

We focus on carbon nanotube (CNT) as another candidate of nano-sized pinning centers. The diameter of CNT is close to the coherence length of the high-temperature superconductors, which may influence the flux pinning behavior. It was reported that in the Y123 polycrystals CNTs are segregated along grain boundaries, which leads to the increase in J_c [6].

IOP Conf. Series: Journal of Physics: Conf. Series **871** (2017) 012051 doi:10.1088/1742-6596/871/1/012051

We have fabricated the melt-processed Y-Ba-Cu-O bulk samples with the addition of CNT and investigated the superconductivity and microstructure in order to elucidate the effects of CNT on the pinning performance.

2. Experimental

Y123 and Y211 powders were mixed in a molar ratio of Y123:Y211 = 10:4 with the addition of 1 wt% CeO_2. The multi-wall CNT was added to the mixture and ultrasonically dispersed with an organic solvent. The amount of CNT was varied from 0 to 1 wt%. The resulting liquid mixture was then dried and uniaxially pressed into pellets. We have fabricated bulk Y-Ba-Cu-O samples with the cold seeding method. Nd-Ba-Cu-O seed crystal was placed on the top of a precursor pellet at room temperature. The typical heat schedule for growing bulk superconductors is as follows. At first, the precursor sample was heated to the maximum temperature 1040°C and held for 1 hour. After the sample was decomposed, it was cooled to the growth temperature 995°C and held for 25 hours. The melt grown samples were annealed at 400°C for 100 hours in flowing oxygen.

Microstructures were observed with a scanning electron microscope (SEM). J_c was estimated from magnetization hysteresis loops measured with a Quantum Design MPMS-XL SQUID magnetometer. The hysteresis curves were measured at 77 K. The superconducting transition temperature (T_c) was also estimated with the same SQUID magnetometer. The magnetic field was applied parallel to the *c*-axis.

3. Results and discussion

Figures 1 show the top surfaces of the melt-grown Y-Ba-Cu-O samples with 0, 0.25, 0.5, and 1 wt% CNT additions. A single domain was formed without spontaneous nucleation of other grains at 0, 0.25, and 0.5 wt% CNT samples. In the case of 1.0 wt% CNT-added sample, the nucleation of other grains was observed. The size of single domain becomes slightly smaller with increasing the content of CNT. This suggests that the crystal growth rate is reduced by adding CNT.

Figure 2 shows SEM photographs of the cleavage plane for bulk Y-Ba-Cu-O samples fabricated with the addition of 0.25 and 1 wt% CNT. One can observe needle-like particles with around 100 nm in length finely dispersed in the Y123 matrix. These fine particles seem to be originated from CNT since they were not observed in undoped sample. CNT raw material originally has the length of several microns. Presumably, high temperature melt process caused thermal decomposition and the fragmentation of CNT.

Figure 1. Top surfaces of melt-grown Y-Ba-Cu-O samples with 0, 0.25, 0.5, and 1 wt% CNT additions.

Figure 2. SEM photographs of the cleavage plane for the bulk Y-Ba-Cu-O samples fabricated with the addition of 0.25wt% (left) and 1 wt% CNT (right).

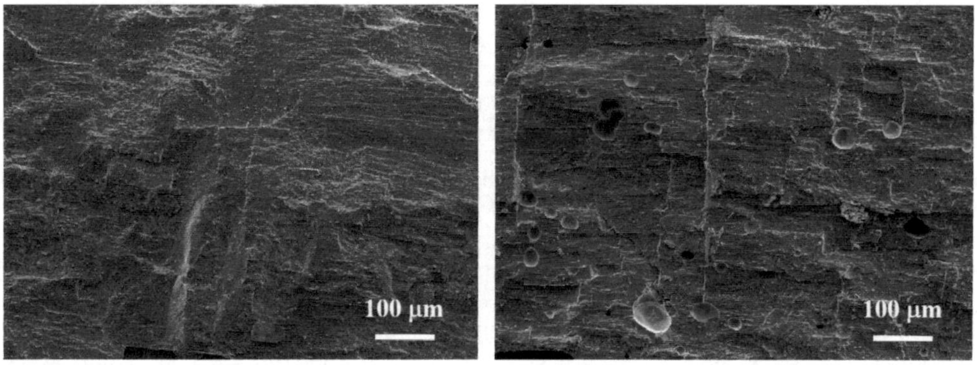

Figure 3. SEM photographs at a low magnification for the bulk samples fabricated with the addition of 0.25wt% (left) and 1 wt% CNT (right).

The fine particles are densely distributed in the central region, but their density is low in the peripheral part for the sample with 0.25 wt% CNT. On the other hand, they are distributed in the whole region for the sample with 1 wt% CNT. In local areas one can find fine particles overlapped each other and in disorder. The sample with 1 wt% CNT contains a number of particles longer than 200 nm unlike 0.25 wt% CNT sample. These results suggest that the addition of highly concentrated CNT do not inhibit the grain growth of Y123 phase and remain as CNT even after the melt-process.

Figure 3 shows SEM photographs at a low magnification for bulk samples fabricated with the additions of 0.25 and 1 wt% CNT. The voids and cracks were observed for the sample with 1 wt% CNT in contrast to 0.25 wt% CNT with relatively dense texture. Sofie *et al.* [7] reported that a bulk sample has the highly porous texture due to CO_2 gas released from $BaCO_3$ used as a raw material. In the present Y-Ba-Cu-O samples, similar effect might occur due to CO_2 originated from thermally decomposed CNT.

Figure 4 shows the temperature dependence of magnetization under 10 Oe for the samples with the addition from 0 to 1 wt% CNT. The sharp transitions were observed even for the CNT-added samples. T_c was enhanced from 90.5 K to 91.8 K as the addition of CNT was increased up to 1 wt%.

IOP Conf. Series: Journal of Physics: Conf. Series **871** (2017) 012051 doi:10.1088/1742-6596/871/1/012051

It is known that the oxygen annealing at 400 °C results in the overdoped state in the Y123 system [8]. Our pristine bulk samples are also considered to be overdoped. It is reported that carbon suppresses oxidation or carrier doping in the polycrystalline [9] and melt-processed samples [7, 10]. There are two possible effects of carbon on oxidation. One is that the diffusion of oxygen is prevented by carbon at grain boundaries. The other is that the substitution of CO_3 group with Cu site at the Cu-O chain reduces the oxygen content. Such substitution is revealed by the structure analysis of the neutron powder diffraction for carbon-doped $YBa_2Cu_{2.85}(CO_3)_{0.15}O_{6.73}$ phase [11]. Although the correlation between oxygen (or carrier) and carbon has not been clarified, the thermal equilibrium state of oxygen content might be affected by the modification of crystal structure due to the incorporated carbon. Nakashima *et al.* [10] have reported that the carrier concentration decreases with increasing carbon content from the results of thermoelectric power measurements for several samples annealed at the same temperature of 400°C. Yamamoto *et al.* [9] have performed the quantitative analysis of oxygen and carbon, and found that an increase in carbon concentration leads to a decrease of oxygen content. They also found that a degree of oxygen deficiency increased for the sample containing carbon from the measurements of the zero-field Cu nuclear quadrupole resonance (NQR) spectra, and discussed that a part of chain site oxygen is pushed out together with Cu, since CO_3 group is located at the Cu chain site.

Reducing carbon by the optimization of calcination and synthesis condition or the use of BaO_2 instead of $BaCO_3$ as a raw material leads to the promotion of oxidation and thereby overdoping [9, 10]. In the present case, the carrier concentration might be decreased from the overdoped to the optimum doped region with the suppression of oxidation due to partial carbon decomposition from CNT.

The magnetic field dependence of J_c at 77 K are shown in figure 5 for the samples with 0 - 1 wt% CNT additions. J_c is enhanced with the addition of CNT in relatively low magnetic fields. The sample with 0.25 wt% CNT addition exhibited the highest J_c value of about 50000 A/cm^2 in self field. This suggests that nano-sized needle-like particles are active as pinning centers. However, a further increase of CNT led to the decline of J_c.

It is reported that the highly porous texture due to CO_2 gas released from $BaCO_3$ results in the decease of J_c [7]. In our highly CNT-concentrated sample, the porous texture is observed as shown in figure 3. This can be considered as one of the causes for degraded J_c. Another possible cause is the reduction of the superconducting Y123 matrix with increasing the amount of second phase impurities. The area of Y123 matrix is reduced with increasing CNT as shown in figure 2, which might lead to the decline of J_c.

The secondary peak was observed for the sample with 1 wt% CNT addition. Sofie *et al.* [7] have observed the peak effect for the samples with more than 0.8 wt% carbon, but not with 0.475 wt% carbon, and discussed this phenomenon as follows. $Y_nBa_{2n}Cu_{3n-1}CO_3O_{7n-1}$ type oxycarbonate is formed due to CO_3 substitutions with Cu site at the Cu-O chain. This induces oxygen vacancies, which cause the field induced pinning and thus the secondary peak effect [12]. A similar effect might occur for our samples, although a further study will be necessary for the clarification.

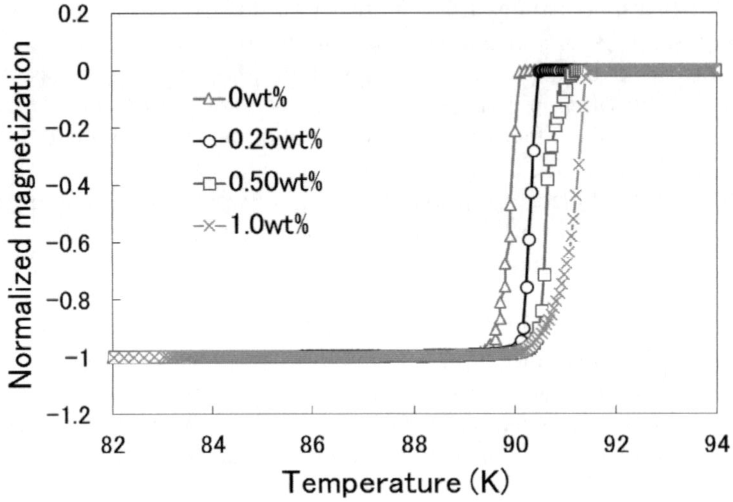

Figure 4. Temperature dependence of magnetization for Y-Ba-Cu-O samples with 0 - 1 wt% CNT additions in the presence of magnetic field of 10 Oe.

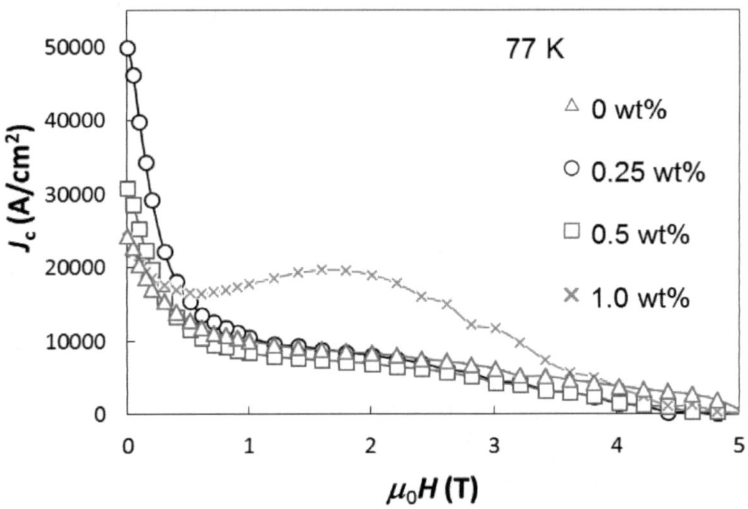

Figure 5. Magnetic field dependence of J_c at 77 K for the samples with the addition between 0 and 1 wt% CNT.

4. Summary

We have investigated the effects of CNT addition on the microstructure and pinning performance in bulk Y-Ba-Cu-O superconductors. It was observed that needle-like particles with around 100 nm in length were distributed in the Y123 matrix. These needle-like particles seem to be originated from CNT. The bulk samples have the highly porous texture for the excess addition of CNT. T_c was increased by 1.3 K with adding CNT. The carrier concentration might be decreased from the overdoped to the optimum doped region by the suppression of oxidation due to carbon decomposition

IOP Conf. Series: Journal of Physics: Conf. Series **871** (2017) 012051 doi:10.1088/1742-6596/871/1/012051

from CNT. The sample with 0.25 wt% CNT addition exhibited the highest J_c value of about 50000 A/cm^2 in self field, which is twice as large as that for the undoped sample. This suggests that nano-sized needle-like particles are effective pinning centers. However, a further increase of CNT led to the decrease of J_c because of the increase of the defects and the decrease of Y123 matrix area. The optimum amount of CNT is considered to be around 0.25 wt% for the improvement of J_c.

References

[1] Iida K, Hari Babu N, Reddy E S, Shi Y-H, Cardwell D A 2004 *Supercond. Sci. Technol.* **18** 249-54
[2] Muralidhar M, Sakai N, Jirsa M, Murakami M, Hirabayashi I 2003 *Appl. Phys. Lett.* **92** 162512
[3] Muralidhar M, Jirsa M, Tomita M 2010 *Physica* C **470** 592-7
[4] Hari Babu N, Reddy E S, Cardwell D A, Campbell A M, Tarrant C D, Schneider K R 2003 *Appl. Phys. Lett.* **83** 4806-8
[5] Nakazato K, Muralidhar M, Koblischka M R, Murakami M 2014 *Cryogenics* **63** 129-32
[6] Dadras S, Liu Y, Chai Y S, Daadmehr V, Kim K H 2009 *Physica* C **469** 55-9
[7] Sofie S W, Dogan F 2002 *Supercond. Sci Technol.* **15** 735-40
[8] Misochko O V, Tajima S, Miyamoto S, Kobayashi H, Kagiya S, Watanabe N, Koshizuka N, Tanaka S 1995 *Phys. Rev.* B **51** 1346-9
[9] Yamamoto A, Hirose K, Itoh Y, Kakeshita T, Tajima S 2005 *Physica* C **421** 1-9
[10] Nakashima T, Tazaki Y, Ishii Y, Horii S, Shimoyama J, Kishio K 2007 *IEEE Trans. Appl. Supercond.* **17** 2996-9
[11] Karen P, Kjekshus A, Huang Q, Karen V, Mighell A D, Santoro A, 2000 *Physica* C **336** 279-86
[12] Matano T, Koblischka M R, Veal B W, Murakami M, Takizawa T 2000 *Supercond. Sci. Technol.* **13** 805-6

IOP Conf. Series: Journal of Physics: Conf. Series **871** (2017) 012052 doi:10.1088/1742-6596/871/1/012052

Effect of growth temperature on properties of bulk $GdBa_2Cu_3O_y$ superconductors grown by IG process

Y. Nakanishi, S. Pavan Kumar Naik, M. Muralidhar* and M. Murakami

[1]Department of Materials Science and Engineering, Shibaura Institute of Technology,

3-7-5 Toyosu, Koto-ku, Tokyo 135-8548, Japan

*miryala1@shibaura-it.ac.jp

Abstract. Recently, the flux pinning performance of Y-123 material was dramatically improved by optimizing the processing conditions during the Infiltration-Growth (IG) process. In the present work, we adapted the top- seeded IG technique and produced several 20 wt.% AgO_2 doped bulk $GdBa_2Cu_3O_y$ (GdBCO or Gd-123) samples at different growth temperatures of 983, 985, 986, 987, 989, 990, and 992 °**C** with the same holding time of 25 h. Our results indicate that the bulk GdBCO samples dwelled at 986 °**C** is the optimum temperature to produce large Gd-123 growth without spontaneous nucleation. Magnetization measurements on 986 °**C** dwelled composite showed a sharp superconducting transition with $T_{c,\ onset}$ at 93.6 K. The critical current density at 77 K and self-field was 16,500 A/cm^2. The critical current density values are correlated with the microstructural changes in the bulk Gd-123 materials.

1. Introduction

Recently, the use of bulk $LREBa_2Cu_3O_{7-\delta}$ (LREBCO or LRE-123, with light rare-earth elements such as Y, Nd, Sm, Gd, etc.,) superconductors in various scientific and technological applications is increased due to their high trapping magnetic fields as compared to the conventional best permanent magnets available in the market [1, 2]. Producing single grained $LREBa_2Cu_3O_{7-\delta}$ (GdBCO or Gd-123) bulks is, therefore, important in various scientific and technological applications. Compared to $YBa_2Cu_3O_{7-\delta}$ (YBCO or Y-123), there is a chance to form a small fraction of solid solutions in GdBCO which contributes to the flux pinning. These solid solutions are of nanometre size and can effectively pin the flux at higher applied fields [4-7].

Over the past few decades, several groups have put efforts to introduce Ag into the LREBCO superconductors. In melt-growth (MG)-processed LREBCO superconductors a considerable work is done on the addition of Ag [8-10]. The introduction of Ag in LREBCO can modify the microstructures of the LREBCO superconductors; hence, the critical current density (J_c) and the mechanical properties are improved without reducing the T_c. A careful review of the literature clearly shows that the doped Ag fuses the gaps between the platelets and thus reduces the density of micro-cracks in the final composites. Furthermore, the non-superconducting 211 phase particles in the 123 matrix are refined [11-13] with the addition of Ag. In addition, the peritectic reaction temperature (T_p) at which the peritectic reaction of the LREBCO superconductors starts is lowered by ~ 15 °**C** to 30 °**C** depending on the content of Ag added to the precursor [14-17]. Thus, the bulk LREBCO superconductors can be processed at very low temperatures and it is possible to control the 211 grain growth. Another important result of the Ag addition is, due to the reduction in T_p, various 123 seeds with lower melting points can also be utilized for the fabrication of single grain bulks.

In our previous work [18] on the melt processed single grained mixed LREBCO sample, a systematic study of addition of Ag (5 wt.% to 30 wt.%) was investigated. The addition of 20 wt. % Ag_2O resulted in a large critical current density of 10 kA cm^{-2}, which was maintained even up to a field of 7 T at 77 K. Compared to the melt-growth process, the IG process provides many advantages such as shape forming, finer particles in the final products without using any grain refiners such as Pt, CeO etc., reduced macro defects, etc [18-20]. Recently, we investigated the effect of the addition of nanometer-sized inclusions on bulk YBCO superconductors [21-22]. The difficulties encountered in IG processing is detailed these papers. The long term goal is to produce Ag-doped single grained bulk GdBCO superconductors. In the present work, we investigated the effect of different time dwells on the growth sectors of Ag-doped GdBCO bulk composites. the utilization of bulk $LREBa_2Cu_3O_{7-\delta}$ (LREBCO or LRE-123, light rare earth elements such as Y, Nd, Sm, Gd, etc.,) superconductors in various scientific and technological applications are increased due to their high trapping magnetic fields compared to conventional best magnets available in market. Producing single grained $Ba_2Cu_3O_{7-\delta}$ (GdBCO or Gd-123) bulks are important in various scientific and technological applications. Compared to $YBa_2Cu_3O_{7-\delta}$ (YBCO or Y-123) there is a chance to form a small fraction of solid solutions in GdBCO which helps in flux pinning. These solid solutions are of the order of nanometric size and effectively pins the flux at higher fields [1-6].

2. Experimental details

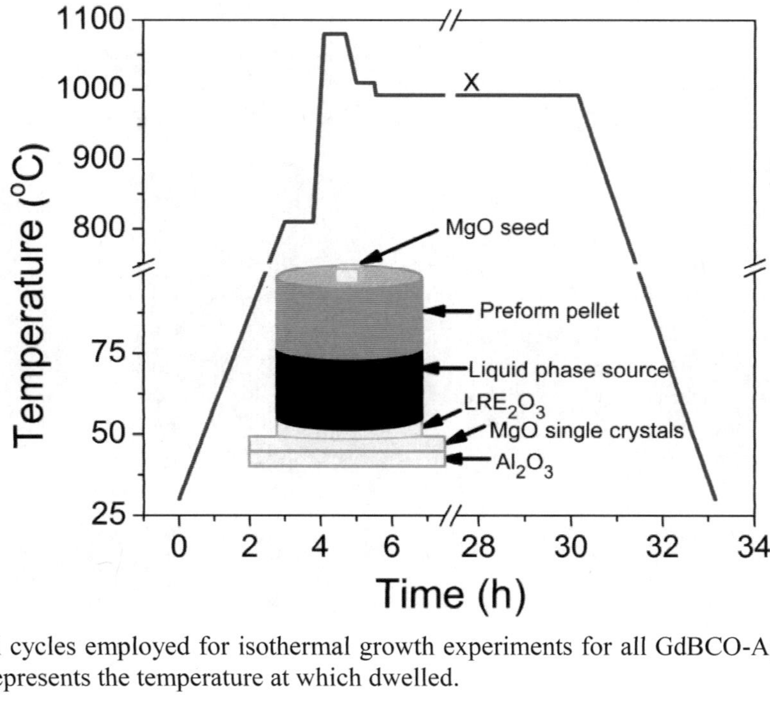

Figure 1. Thermal cycles employed for isothermal growth experiments for all GdBCO-Ag composites. In thermal cycle, X represents the temperature at which dwelled.

The precursor powders of Gd-211 and liquid phases are synthesized using the conventional solid state route. For this, raw materials Gd_2O_3, BaO and CuO are mixed for several hours and the mixture is then sintered at different temperatures, i.e., at 840 °**C**, 860 °**C** and 880 °**C**, respectively. After each sintering, the powders are grinded mechanically for thorough mixing. The phase purity of all the precursor powders is confirmed with X-ray diffractometer. 20 wt.% of Ag_2O is added to the Gd-211 precursor powder and mixed into intimate mixture using a mill with electric motor. The 123 and mixture of

211+Ag_2O phase powders are pressed into pellets using a uniaxial press. As discussed in the earlier section, the addition of 20 wt.% of Ag_2O reduces the peritectic temperature of GdBCO bulk superconductors and enables to process the material at ~30 °**C** lower than the actual peritectic temperature, i.e., 1030 °**C**. Therefore, we aimed to perform the growth at a temperature window of 983 °**C** to 993 °**C**. The heat treatment schedules utilized for processing all the bulks are described in fig. 1. The schematic diagram of the sample assembly employed for IG processing is also displayed in fig. 1. For all the samples, the temperature dwell of 25 h is employed. After processing the Gd-986 bulk composite, oxygenation was carried out at 450 °**C** to 400 °**C** for 175 h to assure the superconductivity. During the oxygenation, the bulk LREBCO samples changes the non-superconducting tetragonal phase to the orthorhombic superconducting phase. The measurement of the superconducting critical temperature of the Gd-986 composite is performed employing a Superconducting Quantum Interference Device magnetometer (Quantum Design, Model MPMS7).

3. Results and discussion

The cold seeding method is employed for all the samples in the present study; single crystalline MgO is used as seed. Single grains can be produced by employing two different ways (i) thermal gradient method and (ii) an isothermal method. In the thermal gradient method, the temperature changes constantly throughout the growth process. The crystal growth of the LREBCO bulk composite depends on the temperature window allowed for slow cooling during the growth. In the isothermal growth method, below the peritectic temperature of LREBCO bulk, a constant temperature will be employed. In the present study, isothermal growth method is used for understanding the heating cycles to be exploit the IG processing of GdBCO-Ag system. The photographs of the as processed bulk GdBCO composites by different dwells are presented in figs. 2 (a)-(f) for Gd-983, Gd-985, Gd-986, Gd-990, and Gd-992, respectively. MgO seed is employed for all the composites for epitaxial growth, but self-nucleation can be observed in all the samples. Clear growth sectors can be realised in Gd-986 composite. Tuning the heating parameters can produce single grains in Ag doped GdBCO bulks.

Figure 2. The photographs of top surface of the as processed GdBCO-Ag bulk composites for which isothermal dwell of 25 h is employed. The photographs clearly demonstrate the formation of growth sector in Gd-986 sample.

The isothermal dwell time of 25 h is employed at different temperatures to explore at which temperature the growth is starting. It is well known that the growth rate of the LREBCO is about ~ 0.2 mm/h at the growth parameters considered here. The growth sector is small due to the small temperature window, through which full single-grain GdBCO composites cannot be processed. The present work confirms that the slow cooling through 986 °**C** would produce single grains of Ag-doped GdBCO bulk superconductors. For that reason for further superconducting investigations, we have chosen Gd-986 composite.

The critical temperature of the Gd-986 composite is measured through the temperature dependence of the ac susceptibility (χ(T)) and the plot is shown in fig. 3. The onset of T_c is found to be 93.6 K with a transition width of 4 K in χ(T) curve which represents the formation of a small fraction of solid solutions of the type $Gd_{1-x}Ba_{2+x}Cu_3O_{7-\delta}$. The field dependence of the critical current density of Gd-986 superconductor is calculated using the extended Bean critical state model and the curve is displayed in fig 3 (b). The self-field J_c of 16.5 kA/cm^2 is achieved for bulk GdBCO superconductor which is doped with 20 wt.% Ag. Extensive microstructural and magnetic studies are underway to investigate the reasons behind the suppression of the J_c curve.

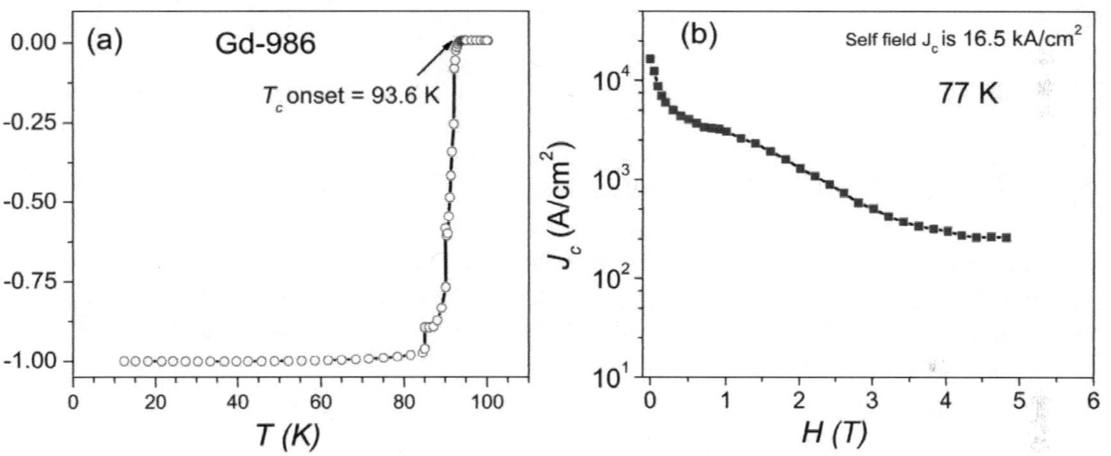

Figure 3. Panel (a) show the ac susceptibility measurement on Gd-986 composite and panel (b) represents the $J_c(H)$ variation at 77 K of the same composite.

The formation of a large fraction of solid solutions in Gd-986 sample can be attributed to the employment of less time for the peritectic reaction. It is noteworthy to mention here that allowing enough time i.e., a constant thermal gradient for peritectic reaction may result a high quality bulk GdBCO-Ag superconducting system. With these inputs, the mass production of GdBCO-Ag bulks can be carried out. The mass production of bulk LREBCO superconductors is a hot topic in recent applications due to the dramatic increase of the use of these materials in various fields.

4. Conclusions

The isothermal growth of the bulk GdBCO superconductors were investigated in the temperature window of 983 °**C** to 992 °**C** with the addition of 20 wt.% of Ag. For all the samples, 25 h dwell time is employed. Large epitaxial growth is observed in GdBCO sample dwelled at a temperature of 986 °**C**. The employment of the single crystalline MgO seed resulted in multi-nucleation in all the composites. The magnetic measurements on oxygenated GdBCO bulk composite show superconducting nature with an onset of T_c at 93.5 K and a self-field J_c of 16.5 kA/cm^2 at 77 K.

Acknowledgments

The paper was supported by Grant-in-Aid FD research budget code: 112261, SIT (Shibaura Institute of Technology). S. Pavan Kumar Naik would like to acknowledge gratefully SIT, Tokyo, Japan.

IOP Conf. Series: Journal of Physics: Conf. Series **871** (2017) 012052 doi:10.1088/1742-6596/871/1/012052

References

[1]. Tomita M, Murakami M 2003 Nature, **421** 517.
[2]. Muralidhar M, Tomita M, Suzuki K, Jirsa M, Fukumoto Y and Ishihara A 2010 *Supercond. Sci. Technol.* **23** 045033.
[3]. Li B, Xu K, HaraS , Zhou D, Zhang Y and Izumi M 2012 *Physica C* **475** 51–56.
[4]. Muralidhar M, Suzuki K., Fukumoto Y, Ishihara A and Tomita M 2011 *IEEE Transactions on Applied Superconductivity* **21** 3.
[5]. Shi Y, Hari Babu N, Iida K and Cardwell D A 2007 *Supercond. Sci. Technol.* **20** 38–43.
[6]. Muralidhar M, Sakai N, Chikumoto N, Jirsa M, Machi T, Wu Y, Murakami M 2002 Phys. Rev. Lett. **89** 237001.
[7]. Yang W M, Zhi X, Chen S L, Wang M, Li J W, Ma J and Chao X X 2014 Physica C **496** 1–4.
[8]. Diko P, Krabbes G and Wende C 2001 Supercond. Sci. Technol. **14** 486
[9]. Matsui M, Sakai N, Murakami M and Osamura 2003 *Supercond. Sci. Technol.* **16** 1086.
[10]. Cheng L, Guo L S, Wu Y S, Yao X and Cardwell D A 2013 Journal of Crystal Growth **366** 1–7.
[11]. Muralidhar M, Kenta N, Zeng X L, Koblischka M R, Diko P, Murakami 2016 *Phys. stat. solidi (a)* **213** 443.
[12]. Sudhakar Reddy E and T Rajasekharan 1998 *Supercond. Sci. Technol.* **11** 523–534.
[13]. Pavan Kumar Naik S, Missak Swarup Raju P, Rajasekharan T and Seshubai V 2016 *IEEE transactions on Applied Superconductivity* **26** 8002206.
[14]. Maeda J, Izumi T and Shiohara Y 1999 *Supercond. Sci. Technol.* **12** 232
[15]. Yoo S I, Takebayashi S, Hayashi N, Nagashima K, Sakai N and Murakami M 1997 Advances in Superconductivity, X, Proc. ISS'97 (Springer-Verlag, Tokyo) p. 697.
[16]. Ikuta H, Hosokawa T, Yoshikawa M and Mizutani U 2000 *Supercond. Sci. Technol.* **13** 1559.
[17]. Muralidhar M, Jirsa M, Sakai N and Murakami M 2001 *Jpn. J. Appl. Phys.* **40** 6329–6334
[18]. Cloots R, Koutzarova T, Mathieu J P, Ausloos M 2005 *Supercond. Sci. Technol.* **18** R9.
[19]. Pavan Kumar Naik S, Devendra Kumar N, Missak Swarup Raju P, Rajasekharan T and Seshubai V 2013 *Physica C* **487** 72-76.
[20]. Hari Babu N, Iida K, Shi Y and Cardwell D A 2006 Physica C **445–448** 286–290.
[21]. Pavan Kumar Naik S, Missak Swarup Raju P, Seshubai V and Rajasekharan T 2016 *Supercond. Sci. Technol.* 29 (2016) 115001.
[22]. Pavan Kumar Naik S and Seshubai V 2017 *Journal of Physics and Chemistry of Solids* **101** 65–73.

IOP Conf. Series: Journal of Physics: Conf. Series **871** (2017) 012053 doi:10.1088/1742-6596/871/1/012053

Superconducting Performance, Microstructure and SEM by EDX Analysis of IG Processed $YBa_2Cu_3O_y$ Bulk Superconductors by Top and Interior Seeding Methods

N. Ide[1], M. Muralidhar[1*], M. Radusovska[2], P. Diko[2], M. Jirsa[3], M. Murakami[1]

[1] Superconducting Material Laboratory, Graduate School of Science and Engineering, Shibaura Institute of Technology, 3-7-5 Toyosu, Koto-ku, Tokyo 135-8546, Japan
[2] Institute of Experimental Physics SAS, Watsonova 49, SK-04353 Kosice, Slovak Republic
[3]EInstitute of Physics CAS, Na Slovance 2, CZ-18221 Praha 8, Czech Republic

[*]miryala1@shibaura-it.ac.jp

Abstract. The top-seeded and interior seeded methods, together with infiltration growth (IG) technique were used to produce $YBa_2Cu_3O_y$ (Y-123) samples with Y_2BaCuO_5 (Y-211) secondary phase particles. T_c (onset) was around 91.5. K. When interior seeding process was used, a complete growth of Y-123 single grain starting at the lower part of the bulk was observed by optical microscopy. The Y-211 particles dispersion was quite uniform in lower and upper parts of the samples, both in the *a*- and *c*-axis growth sectors, both at the beginning and end of the grain growth. In the sample produced by infiltration growth and interior seeding the critical current density at 77 K with *H*//*c*-axis was 44,000 A/cm^2 and 7,750 A/cm^2 in self-field and 2 T, respectively. It is a good basis for optimization of processing conditions, which can further improve the superconductor's performance and enable to grow large-size Y-123 bulks.

1. Introduction

Bulk $REBa_2Cu_3O_y$ (RE: Y, Nd, Sm, Gd, NEG) superconducting magnets can trap magnetic fields by order of magnitude higher than the best classical hard magnets and are therefore promising as permanent magnets for use in magnetic drug delivery system (MDDS), for construction of small mobile diagnostic devices, for water cleaning technologies etc. [1-6]. The drug delivery system could be controlled by the superconducting magnetic force in the body. As a result, a high drug concentration can be delivered in a controlled way to the target diseased tissue [2]. For all these applications good quality samples are needed, with a uniform performance and a high critical current density [7,8]. For this purpose, single domain bulk samples are produced among other techniques by infiltration-growth (IG) [9]. It allows for fabrication of near-net-shape products, uniform microstructure and performance [10,11]. Furthermore, the density of pores is quite low in IG samples, which results in a uniform distribution of the RE-211 particles and a reduced amount of cracking. However, due to limitations in the processing, one cannot produce thick samples. To overcome this drawback, we tried to produce the bulk $YBa_2Cu_3O_y$ material using the interior seeded method, together with infiltration growth (IG) technique and compared the results of top-seeded and IG + internal seed processed samples. The results are presented in this paper. They indicate that high quality single-domain material can scale up from laboratory to industrial production by using the interior-seeded infiltration growth process. A further processing optimization is needed to prepare a high performance Y-123 bulk materials by means of this method.

2. Experimental details

High-purity commercial powders of Y_2O_3, $BaCO_3$ and CuO were mixed in a nominal composition of Y_2BaCuO_5 and calcined at 900° C for 24 h. In parallel, the powders of BaO_2 and CuO were mixed in a nominal composition of $Ba_3Cu_5O_8$ (Y-035). Eventually, the pre-forms of Y-211 were placed on Ba_3Cu_5O and then an Nd-123 melt-textured seed was placed at the centre of the top surface of the Y-211 pre-forms for the growth of single Y-123 grains. We call it a cold top-seeded method combined with the infiltration growth (IG) (see Fig.1, left). In the case of the internal seeded infiltration growth (IG), the seed is placed at the top centre of the RE-211 preform, too, but covered by another RE-211 preform (see Fig.1, right). Yb_2O_3 and MgO plates were used to suppress liquid loss and spontaneous Y-123 nucleation. The samples 20 mm in diameter and 8 mm in thickness were produced by the IG process. The sintering temperature profile for large-grain Y-123 production was then designed based on the isothermal results reported elsewhere. The sample was heated at a rate of 100 °C/h to 880 °C and held there for 15 min, and further heated in 5 h to the temperature of 1040 °C and held there for 50 min. Then, the temperature was lowered in 60 min to 1000 °C, and slowly cooled to 925 °C at a cooling rate of 0.3 °C/h. Finally, the temperature was lowered with a cooling rate of 100 °C/h to 100 °C, followed by furnace cooling to room temperature.

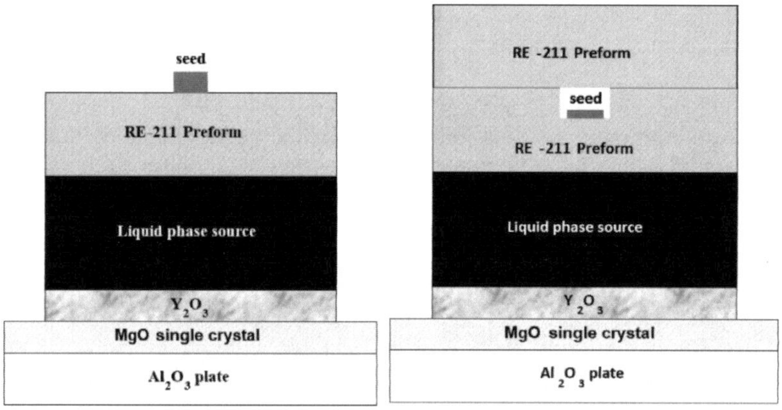

Figure 1. A typical arrangement used in the IG experiment. A source of liquid phase ($Ba_3Cu_5O_y$) is supported on a chemically inert, impervious substrate. The pre-form fabricated of 211 powder is placed on top of it. A *c*-axis Nd-123 seed crystal is placed on top (left) or interior (right) of the pre-form, at the center.

The top surface of the as-grown sample clearly indicated that sample was fully processed without multi-nucleation (see Fig. 2, right). The melt-textured samples were annealed at 450 °C for 250 h in flowing pure O_2 gas. The microstructure of these samples was studied with an optical micrograph. Small test specimens with a size of $1.5 \times 1.5 \times 0.5$ mm^3 were cut from the bulk samples. Measurements of the critical temperature (T_c) and magnetization hysteresis (M-H_a) loops around 77 K in fields from −1 to +5 T were done using a commercial SQUID magnetometer (Quantum Design, model MPMS5). Critical currents were estimated using the extended Bean's critical state model formula for a rectangular sample given by

$$J_c = 20\,\Delta m/[a^2 d\,(b-a/3)] \quad (1)$$

where d is the sample thickness, a, b are transversal dimensions, $b \geq a$, and Δm is the difference of magnetic moments during increasing and decreasing field in the M-H loop [12].

3. Results and discussion

Figure 1 shows the top surfaces of the top-seeded (left) and interior seeded (middle) Y-123 materials produced from Y-211 pre-form sintered at 925 °C, with the final melt growth performed during a slow cooling process, as described in the experimental section. All samples were 19 mm in diameter. For the seed in the interior of the sample in Fig.2 (middle), 2.5 mm bed hole was made, seen clearly after the melt growth process (Fig. 2., right).

Figure 2. Top seeded IG processed Y-123 superconductor prepared in air by IG process using Y-211 pre-form (left). Interior seeded Y-123 superconductor produced by IG technique, top view (middle), and side cross-section view (right).

Note that the four-fold growth facet lines were visible on the top surface of all Y-123 samples prepared by both seeding methods, which clearly indicated that the samples were grown into single-grains, typical for the samples processed by top-seeded melt-growth process [13-15]. Due to the latter fact, bulk single-grain Y-123 or LRE-123 materials can be routinely produced by batch production [16]. The IG-process combined with the interior seed method is new and we would like to see the performance of this new material processing in comparison with the top-seeded method.

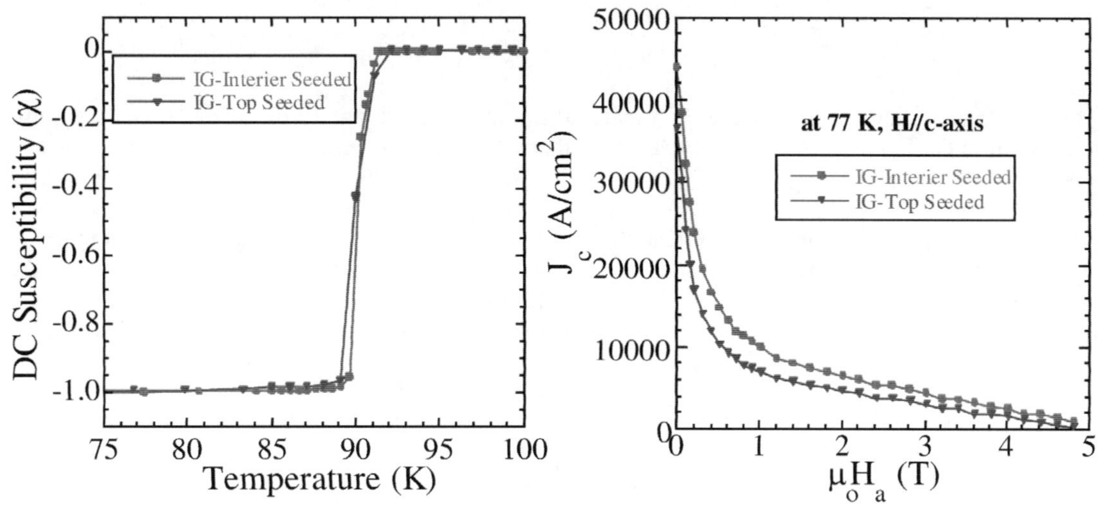

Figure 3. Superconducting transitions temperatures (left) and field dependence of the super-current density (right) at 77 K for specimens cut 2 mm below the seed of the single-domain Y-123 bulk superconductors produced by top seeding and interior seeding of Y-211 pre-form, by the infiltration growth and slow cooling in air method.

Figure 3 (left) presents the example of the temperature dependence of the dc susceptibility after zero-field-cooling (ZFC) in magnetic field of 1mT for both types of samples. A sharp superconducting transition with onset around 92.1 K and transition width of less than 1.9 K was observed, which indicated that the top-seeded IG processed sample was of an excellent performance. The other sample, produced by interior seeding, showed superconducting transition onset slightly lower, at around 91.5 K, but the transition was narrower, ≈ 1.5 K, as compared with the samples produced by top-seeded IG process. *M-H* curves were measured on the same samples at 77 K with the field applied parallel to the *c*-axis and critical current density, J_c, was calculated from the M-H curves using the extended Bean model formula, as shown in Fig. 3 (right). The interior seeded sample showed J_c at 77 K, in self-field and 2 T of about 44,000 A/cm^2 and 7,750 A/cm^2, respectively (Fig.3, right), slightly higher than that in the Y-123 sample produced by top seeding and lower than the Y-123 material produced Y-211 pre-form sintered from 900 °C to 1100 °C [17]. There, critical current density of 37,000 A/cm^2 and 6,000 A/cm^2 at self-field and 2 T, respectively, were detected at 77 K and *H*//*c*-axis. Anyway, a refinement of secondary phase particles and a processing optimization can further significantly improve the already quite good performance of the interior seeded Y-123 material.

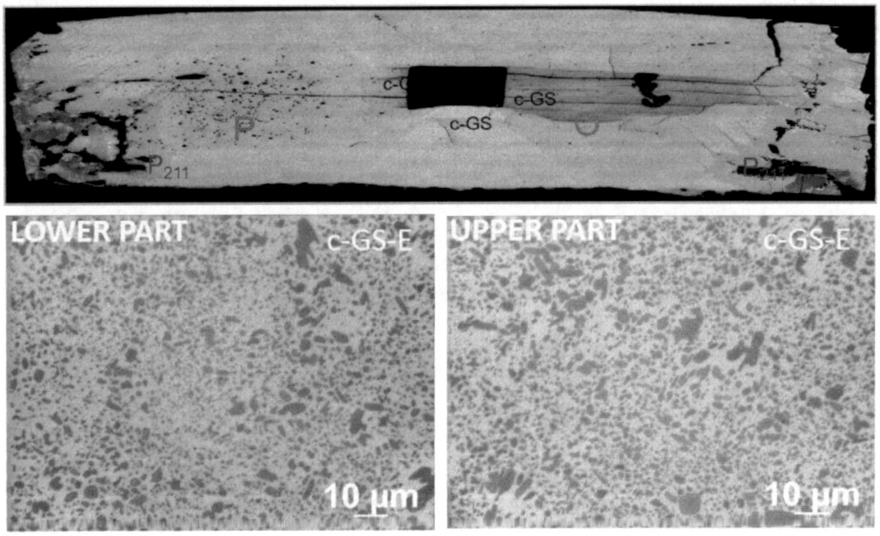

Figure 4. Optical micrograph of the Y-123 material cut along a/c- plane (top), secondary phase particles dispersion in c-GS-E at the upper and lower part of the Y-123 bulk material prepared by interior seeded method.

The structure of the interior-seeded Y-123 material is seen on the sample's cut in the a/c–plane polished up to optical smooth. Figure 4 shows the optical micrograph in polarized light of the Y-123 material produced by the interior seeding method combined with the IG technique. Figure shows the growth starting from the seed. A complicated growth of Y-123 started at the lower part of the bulk. The c-growth sectors (c-GS) are bright due to a low Y-211 content. The 211 particles were pushed away from the sample rim (P_{211}). The dark-yellow part of the sample is oxygenated (O) due to the extended cracks along a/b-planes. The black spherical objects are pores (P). The high-magnification micro-graphs of the top and bottom part of the sample clearly indicate the uniform dispersion of the secondary 211 particles (Fig. 4., bottom). Beside the fine precipitates some larger particles appear in some parts. It shows a space for a further refinement of the secondary phase, which is known to improve critical current density (J_c) in any superconductor.

4. Conclusions

We produced single-grain bulk Y-123 superconductors by infiltration growth process combined with the top and interior seeding methods. Magnetization results confirmed that samples were of good quality with superconducting transition starting at ≈ 91.5 K (interior seeding) and at ≈ 92.1 K (top seeding). The transition was 1.5 K and 1.9 K wide, respectively. The respective J_c-values at 77 K, *H*//*c*-axis, were 44,000 A/cm^2 and 37,000 A/cm^2 (self-field) and 7,750 A/cm^2 and 6,000 A/cm^2 (2 T). Although the electromagnetic and superconducting performance of the interior-seeded material was somewhat better than that of the top-seeded one, the microstructural observation by optical microscopy indicated that a further refinement of Y-211 is possible that can further improve the performance of the Y-123 material prepared in this promising way. The present results indicate that interior seeding can help to prepare thicker high-quality IG-processed Y-123 bulks, with a higher growth speed along *c*-axis direction.

Acknowledgments
The paper was supported by Grant-in-Aid FD research budget code: 112261, SIT (Shibaura Institute of Technology) and by APVV No. 0330-12, VEGA 2/0121/16.

References

[1]. Takeda S-I and Nishijima S 2007 *IEEE Trans. Appl. Supercond.* **17** 2178.
[2]. Muralidhar M, Koblischka M R, Tomita M 2012 *Microscopy Book Series* **5** 1468.
[3]. Tomita M, Fukumoto Y, Suzuki K, Ishihara A and Muralidhar M 2011 *J. of Appl. Phys*, **109** 023912.
[4]. Muralidhar M 2015 *Superconductivity: Today and Tomorrows Applications*, Muralidhar M (Ed.), Nova Science Publishers **1**.
[5]. Muralidhar M, Koblischka M R and Murakami M 2000 S*tudies of high temperature superconductors,* Narlikar A V (Ed.,) Nova Science Publishers **31** 89.
[6]. Murakami M 1992 *Melt processed high temperature superconductors,* Murakami M (Ed.), World Scientific Publisher Co. Singapore **1**.
[7]. Muralidhar M, Nariki S, Jirsa M, Wu Y and Murakami M 2002 *Appl. Phys. Lett.* **80** 1016.
[8]. Muralidhar M, Sakai N, Chikumoto N, Jirsa M, Machi T, Wu Y, Murakami M 2002 *Phys. Rev. Lett.* **89** 237001.
[9]. Mahmood A, Jun B-H, Han Y H, Kim C-J 2010 *Supercond. Sci. Technol.* **23** 065005.
[10]. Nakazato K, Muralidhar M, Koblischka MR, Murakami M 2014 *Cryogenics* **63** 125.
[11]. Muralidhar M, Nakazato K, Zeng X L, Koblischka M R, Diko P, Murakami M 2015 *Phys. Stat. Sol. A*, DOI 10.1002/pssa.201532632.
[12]. Chen D X and Goldfarb R B 1989 *J. Appl. Phys.* **66** 2489.
[13]. Cardwell D A 1998 *Mater. Sci. Eng. B.* **53** 1.
[14]. Salama K and Lee D F 1994 *Supercond. Sci. Technol.* **7** 177.
[15]. Kim C J and Hong G W 1999 Supercond. Sci. Technol. **12** R27.
[16]. Muralidhar M, Tomita M, Suzuki K, Jirsa M, Fukumoto Y and Ishihara A 2010 *Supercond. Sci. Technol.* **23** 045033.
[17]. Muralidhar M, Ide N, Koblischka M R, Diko P, Inoue K and Murakami M 2016 *Supercond. Sci. Technol.* **29** 054003

Trapped Field Performance of the $Gd_{1+x}Ba_{2-x}Cu_3O_{7-\delta}$ Bulk Superconductors Fabricated by Cooling-Rate-Control-Melt-Growth

R Matsuumi[1], M Izumi[1] and X Yao[2]

[1] Laboratory of Applied Physics, Tokyo University of Marine Science and Technology, 2-1-6 Etchu-jima, Koto-ku, Tokyo 135-8533 Japan
[2] State Key Lab for Metal Matrix Composites, Key Lab. of Artificial Structures & Quantum Control (Ministry of Education), Dept. of Physics and Astronomy, Shanghai Jiao Tong University, 800 Dongchuan Road, Shanghai 200240, China

Email:izumi@kaiyodai.ac.jp

Abstract. For $LRE_{1+x}Ba_{2-x}Cu_3O_{7-\delta}$ (LRE: Light rare-earth superconductors, e.g., Nd, Sm and Gd), cation off-stoichiometry caused by the substitution of LRE^{3+}/Ba^{2+} degrades their superconducting properties. Cooling-Rate-Control-Melt-Growth (CRCMG) method is an effective approach to suppress the substitution of RE element on Ba sites. To investigate the relationship between the cooling rate and trapped field performance in $GdBa_2Cu_3O_{7-\delta}$ bulk, we fabricated some full-growth $Gd_{1+x}Ba_{2-x}Cu_3O_{7-\delta}$ bulks with different second-stage cooling rates (0.1, 0.3, 0.5, 0.8, and 1.0 °C/h) by the CRCMG method, and studied the correlation of the cooling rate and the trapped flux of these bulks. From the results, we found that the maximum trapped flux density and the trapped flux of all bulks have few difference at liquid nitrogen temperature, which indicates that the cooling rate seems to affect insignificantly the trapped flux property of $Gd_{1+x}Ba_{2-x}Cu_3O_{7-\delta}$ bulks.

1. Introduction

Due to its high flux-trapping capability, $RE_{1+x}Ba_{2-x}Cu_3O_{7-\delta}$ bulk superconductors are attractive in industrial power applications such as motors [1, 2]. $RE_{1+x}Ba_{2-x}Cu_3O_{7-\delta}$ superconductors have high superconducting transition temperature (T_c) and high flux-trapping capability. Particularly, $LRE_{1+x}Ba_{2-x}Cu_3O_{7-\delta}$ superconductors could exhibit higher T_c and larger critical current density (J_c) than the other $RE_{1+x}Ba_{2-x}Cu_3O_{7-\delta}$ superconductors. For example, $Gd_{1+x}Ba_{2-x}Cu_3O_{7-\delta}$ have been reported to trap a magnetic field of 3 T at 77 K [3]. However, due to the LRE^{3+} and Ba^{2+} ionic radius are similar, the substitution of LRE^{3+} element on Ba^{2+} sites easily occurs. And cation off-stoichiometry caused by this substitution might lead to lower T_c and degrades the performance of bulk, which is common in air-processed $LRE_{1+x}Ba_{2-x}Cu_3O_{7-\delta}$ bulk. Therefore, we need to control the LRE^{3+}/Ba^{2+} substitution for bulk applications.

Previous studies have shown that when the $LRE_{1+x}Ba_{2-x}Cu_3O_{7-\delta}$ bulk are prepared in a reduced oxygen atmosphere, the substitution of Ba^{2+} site by LRE^{3+} ion was significantly reduced (the so-called Oxygen-Control-Melt-Growth OCMG) [4, 5]. On the other hand, Composition-Control-Melt-Growth (CCMG) method also can suppress LRE^{3+}/Ba^{2+} substitution [6, 7]. Recently, it has been reported that the Cooling-Rate-Control-Melt-Growth (CRCMG) method could effectively control the cation stoichiometry for air-processed $Sm_{1+x}Ba_{2-x}Cu_3O_{7-\delta}$ bulk, gaining a comparable T_c to those from OCMG and CCMG [8]. To further study on the advantages of the CRCMG method, we fabricated several $Gd_{1+x}Ba_{2-x}Cu_3O_{7-\delta}$ bulks by the CRCMG method, using a variety of second-stage cooling rate (SSCR) during the growth of these bulks as 0.1, 0.3, 0.5, 0.8, and 1.0 °C/h. Finally, we conducted field-cooled magnetization of these bulks to clarify the relationship between the cooling rate and the trapped field.

2. Experimental

We prepared $GdBa_2Cu_3O_{7-\delta}$ (Gd123) starting powder by calcining commercial Gd_2O_3, BaO_2 and CuO at 920 °C for 24 h and repeated for two times. Commercially available Gd_2BaCuO_5 (Gd211) powders, Ag_2O and Pt were used as starting materials. The powders of the $GdBa_2Cu_3O_{7-\delta}$ and the Gd_2BaCuO_5 were mixed in a molar ratio of $GdBa_2Cu_3O_{7-\delta}$: Gd_2BaCuO_5 = 10:3. 0.5 wt% Pt and 10 wt% Ag_2O were also added to the mixture in order to improve the mechanical property and coarsen Gd_2BaCuO_5. The powders were thoroughly mixed and then uniaxially pressed into a pellet with diameter of 20 mm as a precursor. The precursor used for this process is shown in figure 1(a). The precursor was placed on $ErBa_2Cu_3O_{7-\delta}$ pellet used as a liquid source [9]. The commercial Nd123/Y123/MgO thin films, a 600 nm Nd123 films with a 25 nm Y123 buffer layer on the MgO substrate, grown by thermal co-evaporation, were provided by the German Company of Ceraco ceramic coating GmbH. A 2 mm×2 mm thin films were placed on the top of the pre-form at the beginning. The assembly was conducted above a thin Yb_2O_3 pellet to support the liquid phase, then onto Y-stabilized ZrO_2 supporting rods inside a conventional box furnace.

The bulk were first heated to 900 °C and kept for 4 h, further heated to 1090 °C and kept for 1 h. Then, the temperature was decreased to 1011 °C, which was the starting temperature of slow cooling stage. All bulks were cooled at a rate of 0.3 °C/h from 1011 °C to 1008 °C; then the bulks were cooled at different second-stage cooling rates of 0.1, 0.3, 0.5, 0.8, and 1.0 °C/h until crystal growth was completed, as shown in figure 1(b) and Table 1. Annealing process was carried out in flowing oxygen at 450 °C to 300 °C for 200 h. After annealing process, height of the bulks was adjusted to 6 mm.

We measured the trapped field of these bulks, which were magnetized by the field-cooled magnetization with a 1 T external magnetic field paralleled to the c-axis at LN2 temperature. Trapped field of

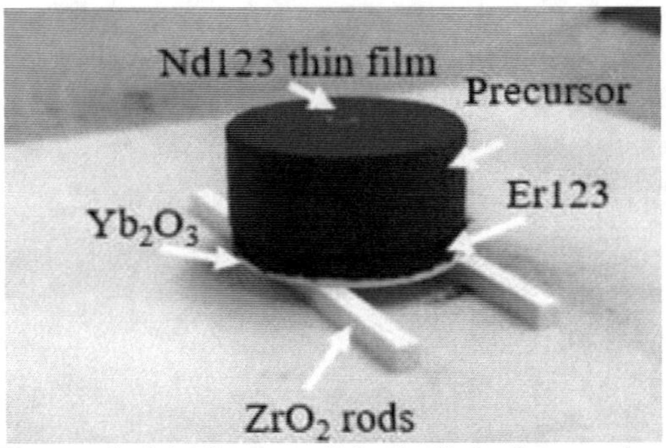

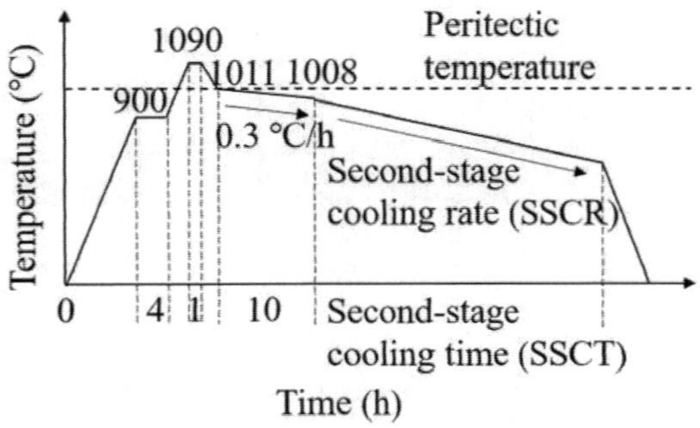

Figure 1. (a) Picture of the bulk before melt growth. (b) Heat treatment process of the fabrication of $Gd_{1+x}Ba_{2-x}Cu_3O_{7-\delta}$ bulks by CRCMG.

Table 1. Growth parameters of $Gd_{1+x}Ba_{2-x}Cu_3O_{7-\delta}$ bulks.

	SSCR (°C/h)	SSCT (h)
B1	0.1	50
B2	0.3	30
B3	0.5	30
B4	0.8	25
B5	1.0	20

each bulk was measured 15 minutes later after the magnetization by an automatic Hall probe scanning system (sensor: F. W. Bell, BHT-921) [1-3]. The sensor was located 0.5 mm above the surface of the bulks. The sampling interval is 1 mm during trapped field measurement.

3. Results and discussion

The top surface view of $Gd_{1+x}Ba_{2-x}Cu_3O_{7-\delta}$ bulks with diameter of 16 mm is shown in Figure 2. From these pictures, we found that all the bulks have no self-nucleation. Full-growth $Gd_{1+x}Ba_{2-x}Cu_3O_{7-\delta}$ bulks were successfully prepared with the different second-stage cooling rates by the CRCMG method. The trapped flux density distributions of each bulk are displayed in Figure 3. We observed that all the bulks have single growth domains, which indicates the bulks have no remarkable weak links.

Figure 2. Top view of $Gd_{1+x}Ba_{2-x}Cu_3O_{7-\delta}$ bulks B1, B2, B3, B4 and B5 grown with different second-stage cooling rates 0.1, 0.3, 0.5, 0.8 and 1.0 °C/h, respectively.

From Table 2, we observed the maximum values of trapped flux density at liquid nitrogen temperature is 0.390 T, 0.406 T, 0.398 T, 0.390 T, and 0.394 T for B1, B2, B3, B4, and B5, respectively. And the maximum trapped flux density reaches 0.406 T for B2. The lowest maximum trapped flux density is 0.390 T for B4.

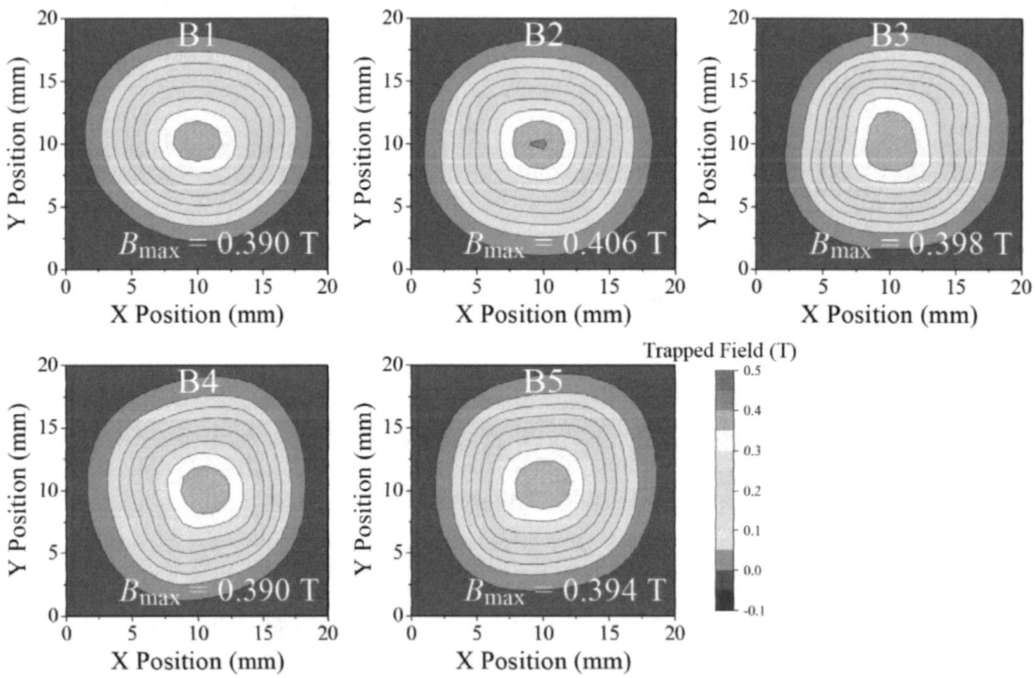

Figure 3. The corresponding trapped flux density distribution of $Gd_{1+x}Ba_{2-x}Cu_3O_{7-\delta}$ bulks B1, B2, B3, B4 and B5 with different cooling rates 0.1, 0.3, 0.5, 0.8, and 1.0 °C/h at liquid nitrogen temperature.

The maximum trapped field of B2 increases by 4% compared with that of B4. These results, demonstrate that the maximum trapped flux density of the $Gd_{1+x}Ba_{2-x}Cu_3O_{7-\delta}$ bulks prepared with different second-stage cooling rates from 0.1 °C/h to 1.0 °C/h does not evidently change. The maximum integrated trapped flux reaches 35.7 μWb for B3. The minimum integrated trapped flux is 32.2 μWb for B1. The maximum integrated trapped flux increases by 11 % compared with the minimum one.

Table 2. The maximum trapped field density and integrated trapped field of $Gd_{1+x}Ba_{2-x}Cu_3O_{7-\delta}$ bulks with different second-stage cooling rates (SSCR).

	SSCR (°C/h)	Maximum trapped flux density (T)	Integrated trapped flux (μWb)
B1	0.1	0.390	32.2
B2	0.3	0.406	33.7
B3	0.5	0.398	35.7
B4	0.8	0.390	32.2
B5	1.0	0.394	35.0

It indicates that the different second-stage cooling rates do not provide obvious influence on the integrated trapped flux of the bulks, which is consistent with the results of the trapped flux density. As illustrated above, the different second-stage cooling rates from 0.1 °C/h to 1.0 °C/h do not affect the trapped field performance of the $Gd_{1+x}Ba_{2-x}Cu_3O_{7-\delta}$ bulks.

Almost the same trapped flux property was observed regardless of different fabrication process with the different second-stage cooling rates. We think that considerable homogeneity has been achieved according to the sample preparation in which $ErBa_2Cu_3O_{7-\delta}$ pellet employed as a liquid source [9]. One of the reasons may come from a smaller range of Gd^{3+}/Ba^{2+} substitution in $Gd_{1+x}Ba_{2-x}Cu_3O_{7-\delta}$ than in $Sm_{1+x}Ba_{2-x}Cu_3O_{7-\delta}$ since the Gd^{3+} ion is smaller than Sm^{3+} or Nd^{3+} ions [3]. The maximum substitution content in the $Gd_{1+x}Ba_{2-x}Cu_3O_{7-\delta}$ is $x = 0.1$ [10]. This value is small compared with that of other $LRE_{1+x}Ba_{2-x}Cu_3O_{7-\delta}$ superconductors. As a result, the trapped field performance of $Gd_{1+x}Ba_{2-x}Cu_3O_{7-\delta}$ bulk is not remarkably changed by alternating the second-stage cooling rate.

4. Conclusion

We succeeded in preparing five full-growth $Gd_{1+x}Ba_{2-x}Cu_3O_{7-\delta}$ bulks with different second-stage cooling rates 0.1, 0.3, 0.5, 0.8, and 1.0 °C/h using the CRCMG method. Almost the same trapped field performance was observed regardless of different fabrication process with the different second-stage cooling rates. As a result, it is suggested that second-stage cooling rates from 0.1 °C/h to 1.0 °C/h do not influence the trapped flux of $Gd_{1+x}Ba_{2-x}Cu_3O_{7-\delta}$ bulks. Further study will be carried out to clarify J_c-B characteristics, T_c, and lattice parameters on the samples with approval of homogeneity.

References

[1] Miki M, Felder B, Tsuzuki K, Xu Y, Deng Z, Izumi M, Hayakawa H, Morita M and Teshima H 2010 *Supercond. Sci. Technol.* **23** 124001
[2] Zhang Y, Zhou D, Ida T, Miki M and Izumi M 2016 *Supercond. Sci. Technol.* **29** 044005
[3] Nariki S, Sakai N and Murakami M 2005 *Supercond. Sci. Technol.* **18** S126
[4] Sakai N, Yoo S.I, Watahiki M, Higuchi and Murakami M 1995 *Adv. Supercond.* **VII** 345
[5] Oda M, Yao X, Yoshida Y and Ikuta H 2009 *Supercond. Sci. Technol.* **22** 075012
[6] Cardwell D.A, Shi Y, Hari Babu N and Iida K 2009 *Phys. C supercond.* **469** 1146
[7] Dai J.Q, Zhao Z.Q and Xiong J.W 2003 *Supercond. Sci. Technol.* **16** 815
[8] Cui X.X, Wang W, Zhuang Y.F, Guo L.S, Xiang H.Yao X, Pan B, Ikuta H and Zou Z.Q 2016 *J. Alloys Comp.* **663** 10
[9] Zhou D, Hara S, Li B, Noudem J and Izumi M 2014 *Supercond. Sci. Technol.* **27** 7
[10] Xu C, Hu A, Sakai N, Izumi M and Hirabayashi I 2005 *Phys. C supercond.* **417** 77

IOP Conf. Series: Journal of Physics: Conf. Series **871** (2017) 012055 doi:10.1088/1742-6596/871/1/012055

Mechanical properties of EuBaCuO superconducting bulk material at liquid nitrogen temperature

A Murakami[1] **and A Iwamoto**[2]

[1] National Institute of Technology, Ichinoseki College, Takanashi, Hagisho, Ichinoseki, Iwate 021-8511, Japan
[2] National Institute for Fusion Science, 322-6 Oroshi-cho, Toki, Gifu 509-5292, Japan

E-mail: amura@ichinoseki.ac.jp

Abstract. In order to understand the mechanical properties of $EuBa_2Cu_3O_y$ "EuBaCuO" superconducting bulk material, 3- and 4-point bending tests were carried out at room and liquid nitrogen temperature (77.3 K) for specimens cut from a EuBaCuO single-grain bulk material. The Young's modulus evaluated through the 3-point bending test and that through the 4-point bending test were similar to each other. On the other hand, the 3-point bending strength was higher than the 4-point bending strength, which is due to the smaller maximum bending moment area of the specimen in the 3-point bending test. The Young's modulus at liquid nitrogen temperature was higher than that at room temperature (RT), which is due to the decrease of the interatomic distance by cooling. The bending strength at 77 K was also higher than that at RT. These mechanical property data of the EuBaCuO bulk material were comparable to those of other $REBa_2Cu_3O_y$ "REBaCuO" bulk materials (RE: rare-earth elements).

1. Introduction

It is well-known that REBaCuO (RE: Y or rare-earth elements) single-grain bulk materials have excellent and unique features such as large current transport capacity, automatically stable levitation and extremely high trapped field ability. Various superconducting devices using such excellent and unique features of REBaCuO bulk materials are being developed. Improvement of the superconducting properties and enlargement of the single-grain size of REBaCuO bulk materials are useful for the development of high-performance superconducting devices. REBaCuO bulk materials are subjected to the electromagnetic force and thermal stress in the superconducting devices [1-4]. Since improvement of the superconducting properties and enlargement of the single-grain size cause the increases of the electromagnetic force and thermal stress, improvements of the mechanical properties are also useful for the development of superconducting devices. Also, understanding of the mechanical properties at cryogenic temperature is informative for designing the superconducting devices. NMR application using EuBaCuO bulk material has been suggested [5-7]. However, the mechanical properties of EuBaCuO bulk material have not been understood extensively in comparison with those of other REBaCuO bulk materials such as YBaCuO, GdBaCuO and DyBaCuO. In this study, mechanical properties of a EuBaCuO single-grain bulk material were evaluated through 3- and

4-point bending tests at room and liquid nitrogen temperatures for specimens cut from the bulk material. Effects of loading method and testing temperature on the mechanical property data are discussed.

2. Experimental procedure

One $EuBa_2Cu_3O_y$ single-grain bulk sample fabricated by Nippon Steel & Sumitomo Metal Corporation was used for 3- and 4-point bending tests at room and liquid nitrogen temperatures. The precursor for the bulk sample was heated up to 1443 K, kept at that temperature for 1 h and then cooled down to 1310 K. After that, one $NdBa_2Cu_3O_y$ seed crystal was placed on the top of it in air. After the crystal growth, oxygen annealing was conducted at 673 K for 100 h. The diameter of the bulk sample was 60 mm. 25 mol% Eu_2BaCuO_5 and 10 wt% Ag were added to the bulk sample. Bending test specimens with the dimensions of 2.8 x 2.1 x 24 mm^3 were cut from the bulk sample (Figure 1). Schematic illustrations of the 3- and 4-point bending tests, together with the bending moment diagrams (BMD), are shown in Figure 2 (a) and (b), respectively. Loading was carried out under the crosshead speed of 0.2 mm/min. Strain was measured by using a strain gauge glued to the centre of the tensile side surface of the specimens. In the bending tests at 77 K, specimen was placed on the bending test jig and immersed into the liquid nitrogen bath, together with the bending test jig. After the bending tests, side surfaces of the fractured specimens were polished by using lapping sheets. The polished surfaces were observed by using an optical microscope. Porosities (area fraction of pores) on the optical micrographs were evaluated through image analysis. Stress caused by 3-poing bending loading σ_3 and that by 4-point bending loading σ_4 were calculated by the following equations, respectively. Where P is applied load, L is outer supporting span, l is inner loading span, w and t are width and thickness of the specimen, respectively.

$$\sigma_3 = \frac{3PL}{2wt^2} \tag{1}$$

$$\sigma_4 = \frac{3P(L-l)}{2wt^2} \tag{2}$$

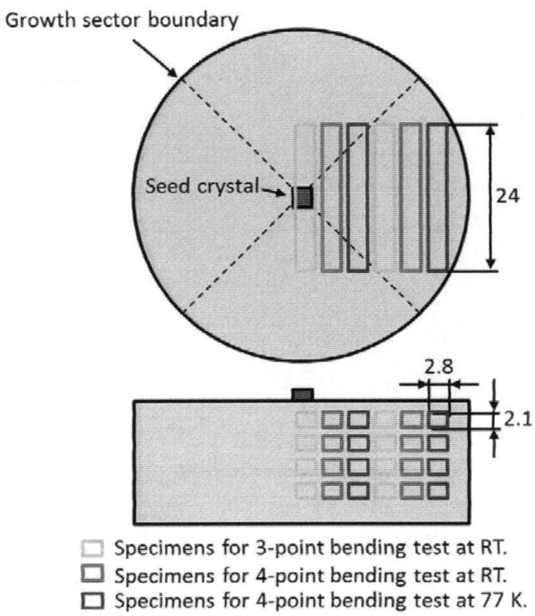

Figure 1. Schematic illustration of cutting out of specimens for 3- and 4-point bending tests at room and liquid nitrogen temperatures.

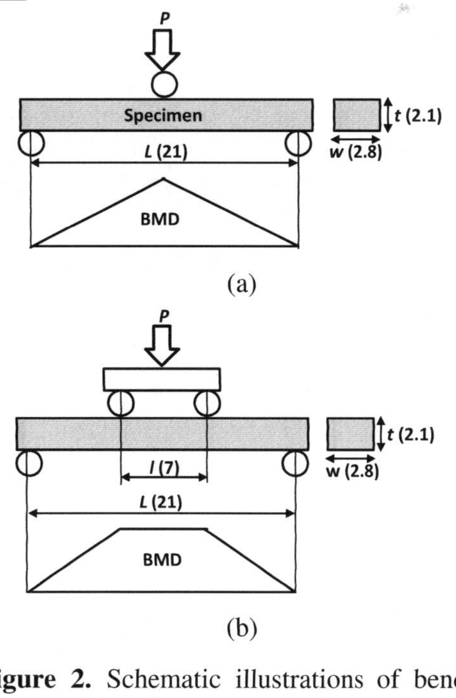

Figure 2. Schematic illustrations of bending tests and bending moment diagrams. (a) 3-point bending test and (b) 4-point bending test.

3. Results and discussion

3.1. Effects of loading method on mechanical property data

Figure 3 shows stress-strain curves obtained through the 3- and 4-point bending tests for specimens at room temperature. Stress-strain behaviours are almost linear until the fracture. Such stress-strain behaviour is typical of brittle materials. There is no significant difference in the slope of the stress-strain curves between the 3- and the 4-point bending tests. Young's modulus data evaluated from the slope of the initial part of the stress-strain curves are shown in Figure 4. Since the Young's modulus is intrinsic to the material, the Young's modulus evaluated through the 3-point bending test and that through the 4-point bending test were similar to each other. On the other hand, the 3-point bending strength was higher than the 4-point bending strength (Figure 5). It is attributable to that the maximum bending moment area of the 3-point bending test specimen is smaller than that of the 4-point bending test specimen (see Figure 2). Both the 3- and 4-point bending tests are commonly carried out to evaluate the mechanical properties of brittle materials. However, the bending strength evaluated through the 3-point bending test is overestimated in comparison with that through the 4-point bending test. Thus, the mechanical properties at 77 K mentioned in the followings were evaluated through the 4-point bending test.

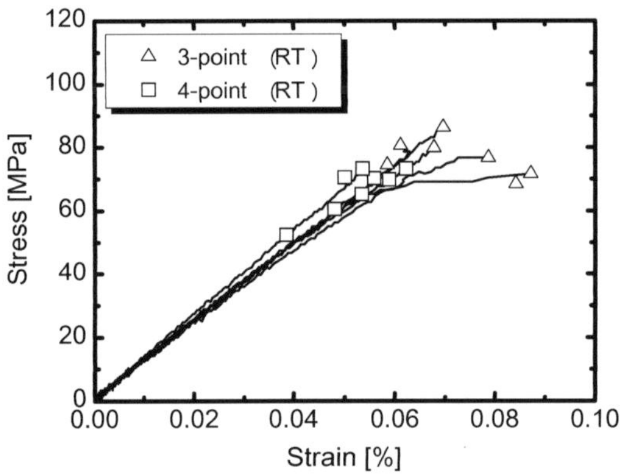

Figure 3. Stress-strain curves obtained through 3- and 4-point bending tests at room temperature (RT). Data points represent fracture point of specimens.

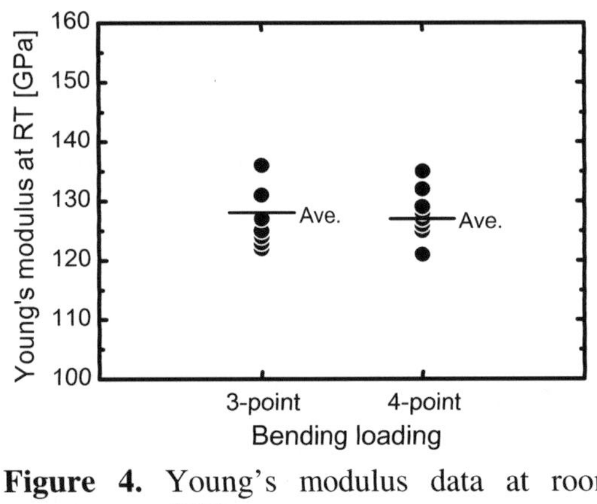

Figure 4. Young's modulus data at room temperature (RT).

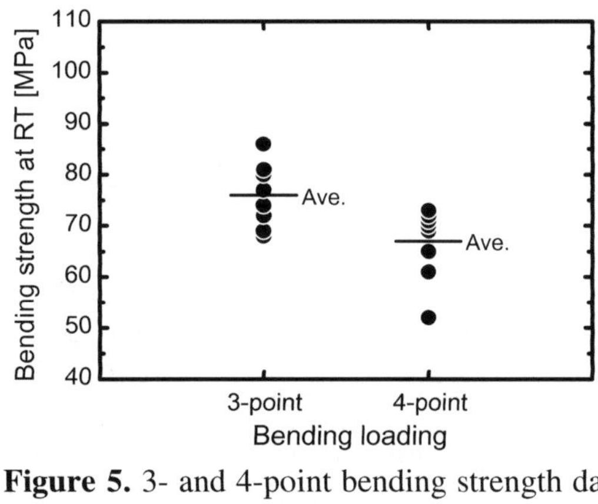

Figure 5. 3- and 4-point bending strength data at room temperature (RT).

3.2. Effects of testing temperature on mechanical property data

Figure 6 shows stress-strain curves obtained through the 4-point bending tests for specimens at liquid nitrogen temperature (77 K). Stress-strain curves at room temperature (RT) are also shown for reference. Slopes of the stress-strain curves at 77 K are steep in comparison with those at RT. Young's modulus data at 77 K and RT are shown in Figure 7. The Young's modulus at 77 K was higher than that at RT. The bending strength at 77 K was also higher than that at RT (Figure 8). These temperature dependences are mainly due to the decrease of the interatomic distance by cooling. Figure 9 shows Weibull plots of the 4-point bending strength data. 3-point bending strength data are also shown for reference. Weibull plot is commonly carried out to evaluate the scatter of the strength data. The slope obtained through the linear fitting of the data represents the Weibull coefficient m; larger Weibull coefficient value means smaller scatter of the strength data. Since the Weibull coefficient of the 3-point bending strength data at RT and that of the 4-point bending strength data at 77 K are similar to each other, it is deduced that there is no significant difference in the Weibull coefficient between the bending strength data at RT and 77 K although the Weibull coefficient of the 4-point bending strength at RT is smaller than that at 77 K.

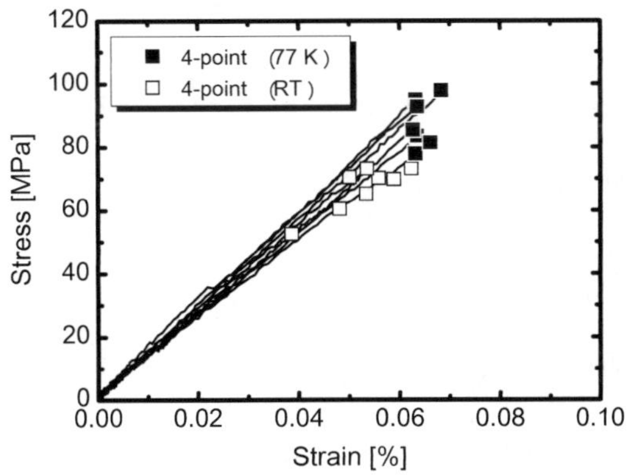

Figure 6. Stress-strain curves obtained through 4-point bending tests at room and liquid nitrogen temperatures (RT and 77 K). Data points represent fracture point of specimens.

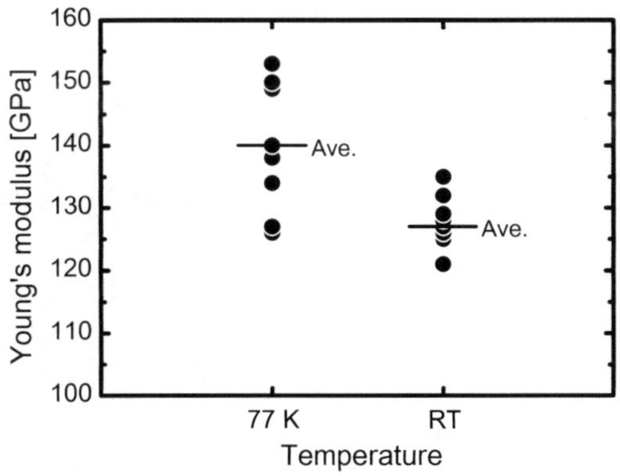

Figure 7. Young's modulus data evaluated through 4-point bending tests at room and liquid nitrogen temperatures (RT and 77 K).

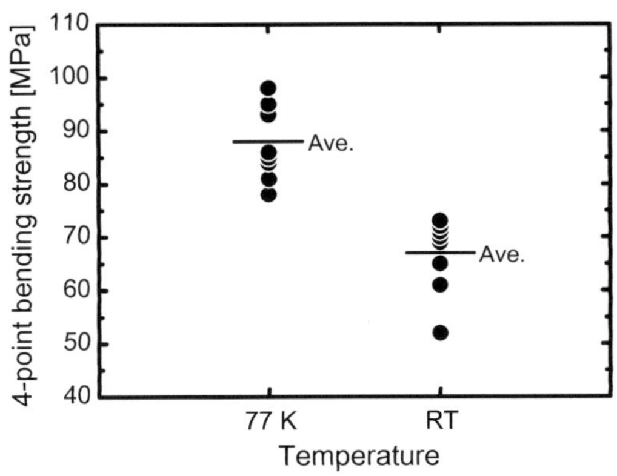

Figure 8. 4-point bending strength data at room and liquid nitrogen temperatures (RT and 77 K).

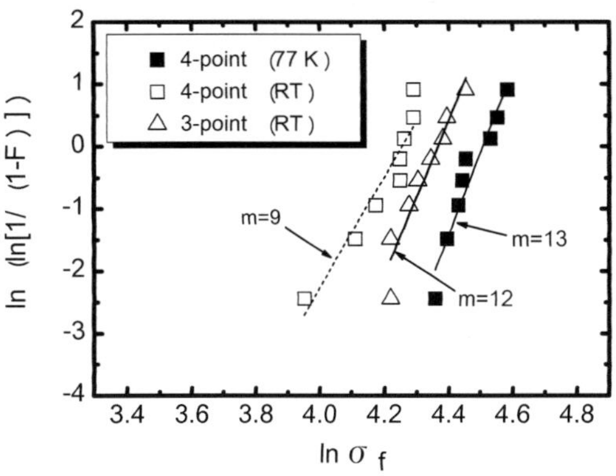

Figure 9. Weibull plots of 3- and 4-point bending strength data at room and liquid nitrogen temperatures (RT and 77 K). F and m represent cumulative fracture probability and Weibull coefficient, respectively.

IOP Conf. Series: Journal of Physics: Conf. Series **871** (2017) 012055 doi:10.1088/1742-6596/871/1/012055

Figure 10 shows optical micrograph of the polished surface of the fractured bending test specimen. Black and white areas are pores and Ag particles, respectively. In order to evaluate the porosity, pores on the optical micrographs were extracted and the area fractions of pores were measured by using image analysis software. Relationship between the bending strength and the porosity is shown in Figure 11. Approximation curves of the 3-point bending strength data at 77 K of a DyBaCuO bulk material reported in [8] and the 3-point bending strength data at room temperature of a DyBaCuO reported in [9] are also shown for reference. Since low porosity specimens cut from the region just adjacent to the top surface of the DyBaCuO bulk materials were used for the bending tests in the previous studies [8,9], approximation curves (solid and broken lines in Figure 11) were obtained. On the other hand, such approximation curves were not obtained for the EuBaCuO bulk material because the low porosity specimens were not used for the bending tests in this study. It can be seen that the 3-point bending strength data at RT of the EuBaCuO bulk material are comparable to the approximation curve of the 3-point bending strength data at RT of the DyBaCuO bulk material (Figure 11). Also, it can be seen that the 4-point bending strength data at 77 K of some specimens of the EuBaCuO bulk material are comparable to the approximation curve of the 3-point bending strength data at 77 K of the DyBaCuO bulk material. Since 3-point bending strength is commonly higher than the 4-point bending strength, it is deduced that the bending strength at 77 K of the EuBaCuO bulk material is also comparable to that of the DyBaCuO bulk material.

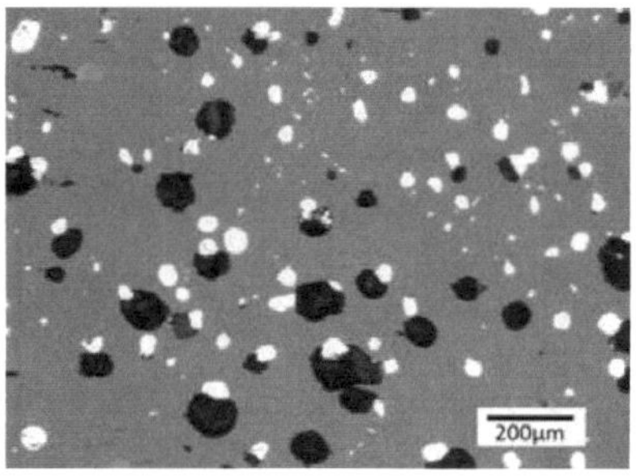

Figure 10. Optical micrograph of polished surface of bending test specimen.

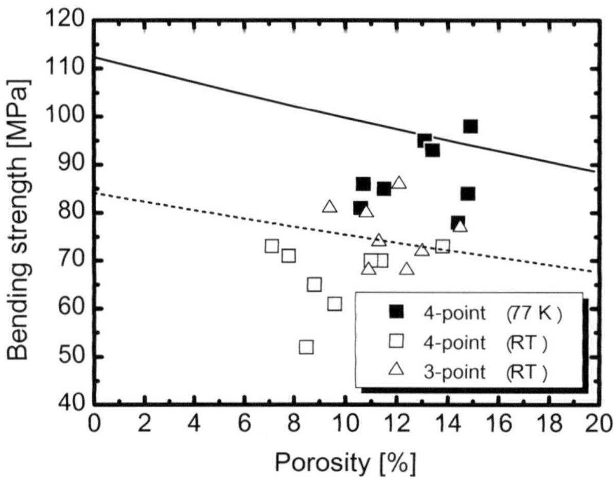

Figure 11. Relationship between bending strength and porosity of the EuBaCuO specimens. Solid and broken lines represent approximation curves of 3-point bending strength for DyBaCuO at 77 K reported in [8] and RT reported in [9], respectively.

IOP Conf. Series: Journal of Physics: Conf. Series **871** (2017) 012055 doi:10.1088/1742-6596/871/1/012055

4. Summary

Mechanical properties of a EuBaCuO single-grain bulk material were evaluated at room and liquid nitrogen temperatures through 3- and 4-point bending tests for specimens cut from the bulk material. While the Young's modulus evaluated through the 3-point bending test and that through the 4-point bending test were similar to each other, the 3-point bending strength was higher than the 4-point bending strength. It is due to the smaller maximum bending moment area of the specimen in the 3-point bending test. Both the Young's modulus and the bending strength at liquid nitrogen temperature were higher than those at room temperature, which are mainly due to the decrease of the interatomic distance by cooling. Taking into account the porosity, these mechanical property data of the EuBaCuO bulk material were comparable to those of DyBaCuO bulk materials.

Acknowledgements

Authors appreciate experimental supports by Dr. H. Teshima of Nippon Steel & Sumitomo Metal Corporation and students of National Institute of Technology, Ichinoseki College. This work was supported in part by the NIFS Collaboration Research Program (NIFS15KECA033).

References

[1] Ren Y, Weinstein R, Liu J, Sawh R P and Foster C 1995 *Physica C* **251** 15
[2] Miyamoto T, Nagashima K, Sakai N and Murakami M 2000 *Physica C* **340** 41
[3] Miyamoto T, Nagashima K, Sakai N and Murakami M 2001 *Physica C* **349** 69
[4] Tsuchimoto M and Murata K 2003 *Physica C* **392-396** 575
[5] Nariki S, Sakai N, Hirabayashi I, Yoshikawa M, Itoh Y, Nakamura T and Utumi H 2008 *Physica C* **468** 1451
[6] Sakai N, Nariki S, Teshima H, Morita M, Hirabayashi I, Nakamura T, Itoh Y, Yoshikawa M and Utsumi H 2011 *J. Cryo. Soc. Japan* **46** 131
[7] Nakamura T, Itoh Y, Yoshikawa M, Sakai N, Nariki S, Hirabayashi I and Utsumi H 2011 *J. Cryo. Soc. Japan* **46** 139
[8] Murakami A, Otaka K, Miura T and Iwamoto A 2011 *IEEE Trans. Appl. Supercond.* **21** 3107
[9] Murakami A, Miyata H, Hashimoto R, Katgiri K and Iwamoto A 2008 *Physica C* **468** 1395

Record critical current density in bulk MgB_2 using carbon-coated amorphous boron with optimum sintering conditions

M. Muralidhar[1*], M. Higuchi[1], P. Diko[2], M. Jirsa[3], M. Murakami[1]

[1] Superconducting Material Laboratory, Graduate School of Science and Engineering, Shibaura Institute of Technology, 3-7-5 Toyosu, Koto-ku, Tokyo 135-8546, Japan
[2] Institute of Experimental Physics SAS, Watsonova 49, SK-04353 Kosice, Slovak Republic
[3]EInstitute of Physics CAS, Na Slovance 2, CZ-18221 Praha 8, Czech Republic

[*]miryala1@shibaura-it.ac.jp

Abstract. We report on the synthesis and characterization of a sintered bulk MgB_2 material produced at an *optimized sintering temperature* with a *varying content of carbon-encapsulated* amorphous boron. A series of MgB_2 bulks was prepared with 0%, 1.5%, 2.8%, 7.3%, 12% and 16.5% of carbon-encapsulated boron. In the samples with 12% of carbon-encapsulated boron, Mg and MgB_2C_2 formation was observed. T_c was around 38.4 K for the pure MgB_2 and decreased with increasing carbon content up to 25 K for 16.5 % of carbon-encapsulated boron. The highest J_c values of 470 kA/cm^2 and 310 kA/cm^2, in the self-field and 1 T, respectively, were achieved at 20 K, in the MgB_2 sample with 1.5% of carbon-encapsulated boron. It proved that the optimized sintering conditions together with the appropriate amount of the carbon-coated boron were able to bring critical current performance of bulk MgB_2 material up to the level necessary for real technical applications.

1. Introduction

A significant effort has been made in the development of MgB_2 materials processing, characterization, and applications [1, 4]. The superconducting transition temperature of MgB_2 is significantly lower than that of $YBa_2Cu_3O_y$ "Y-123", however, MgB_2 benefits from some BCS-like superconducting features, e.g. a large coherence length and weak magnetic relaxation. It allows for a better Josephson junction fabrication. Within the group of intermetallic superconductors, the critical temperature is still rather high, which with use of a refrigerator allows for a rather low cooling costs. The cheaper technology and the quite high critical current density (J_c) in the polycrystalline state make this material promising for applications in power industry, magnetic resonance imaging (MRI), fault current limiters (FCL), Josephson junctions and SQUIDS, energy storage, levitation devices and powerful super-magnets operating at around 20 K [4, 5].

For superconducting super-magnet applications, it is necessary to produce a good quality and high performance bulk MgB_2 material with high J_c and low production cost. To improve the critical current density of the MgB_2 material, a variety of processing techniques have been developed including chemical doping [6], refining of the initial particle size by ball milling [7], and irradiation [8]. As a result, the transport J_c values at liquid helium temperature and 10 T reached the level of 10^5 A/cm^2 for Si-doped MgB_2 samples. Similar improvements were also observed with additions of carbon, boron carbide, carbon nanotubes, carbohydrates or hydrocarbons, graphene oxide, Ni-Co-B nanoparticles [9,10]. However, the critical current density values of MgB_2 materials still need a further improvement for high magnetic field applications. Recent microstructural observations by AFM and SEM indicated that a proper sintering time is very important to control the grain size of the MgB_2 material [11]. X-ray diffraction, DC magnetization and scanning electron microscopy (SEM) analyses proved that single-phase and homogenous MgB_2 bulks were produced by sintering at 775 °C and 800 °C, which exhibited the highest critical current density recorded so far. The samples sintered at 775 °C for 1 h showed J_c of 270 kA/cm^2 at 20 K [12]. A trapped field value of 1.51 T at 20 K was achieved in such MgB_2 samples of 20 mm diameter and 7 mm thickness. The simple sintering route is attractive for mass production of bulk MgB_2 materials. However, a further improvement in J_c is needed for such a simple technology. Utilizing the carbon-encapsulated boron might help, since carbon doping of MgB_2 superconducting wires had very positive results, record critical current densities being reported in such tapes and wires [13]. Recent reports on bulk MgB_2 samples doped by carbon encapsulated boron showed superior critical current density as compared to pure samples [14,15]. Processing optimization is still needed for bulk MgB_2 material to reach a uniform carbon dispersion in the bulk MgB_2 material. One of the points is to find an optimum content of the carbon-encapsulated boron in the compound.

In the present paper, we report on superconducting properties, X-ray diffraction, and chemical analysis of the carbon-coated boron bulk MgB_2 materials fabricated by a simple sintering method, exhibiting a high J_c and a homogenous carbon distribution in the whole sample.

2. Experimental details

The bulk MgB_2 samples were fabricated by using in-situ solid state reaction. The commercial powders (Furu-uchi Chemical Corporation, Japan) of Mg metal (99.9% purity, 200 meshes) and amorphous B powder (99% purity, 300 meshes) were mixed in the nominal ratio of Mg: B = 1: 2. To test the effect of carbon-coated boron on flux pinning and the associated J_c value, five kinds of carbon-encapsulated B powder (98% purity) were utilized in production of the bulk MgB_2 material, namely PVZ-C15 (1.5% C), PVZ-C28 (2.8% C), PVZ-C73 (7.3% C), PVZ-C120 (12% C) and PVZ-C165 (16.5% C). All the carbon-encapsulated B powders were produced at PAVEZYUM, Advanced Chemicals, Turkey. More details of the carbon-encapsulated B powder production can be found elsewhere [14]. The starting powders were thoroughly ground in a glove box in argon atmosphere. The powder mixture was pressed into pellets of 20 mm in diameter and 7 mm thick using a uniaxial press of 200 MPa. The consolidated pellets were then wrapped in tantalum foils and subjected to the heat treatment in Ar atmosphere in a tube furnace. All samples were heated to the target sintering temperature of 805 °C and kept there for 3 h in flowing argon gas. Finally, the temperature was lowered to room temperature at a cooling rate of 100 °C/h. The constituent phases of the samples were identified with a high-resolution automated X-ray powder diffractometer (RINT2200), using Cu-K$^\alpha$ radiation generated at 40 kV and 40 mA. The microstructure of these samples was studied with a scanning electron microscope (SEM) and chemical composition was estimated by SEM by EDX analysis.

Small specimens with dimensions of 1.5×1.5×0.5 mm^3 were cut from bulk MgB_2 samples and subjected to the measurements of critical temperature (T_c) and magnetization hysteresis loops (*M-H* loops) in applied magnetic fields from −1 to +5 T at 20 K using a SQUID magnetometer (Quantum Design, model MPMS5). The magnetic J_c values (in MA/cm^2) were estimated based on the extended Bean critical state model using the relation

$$J_c = 20\, \Delta m / [a^2 d\, (b - a/3)] \tag{1}$$

where d is the sample thickness, a, b are cross sectional dimensions, $b \geq a$, (a, b, d in mm) and Δm (in emu units, 1 emu≡10^{-3} Am^2) is the difference of magnetic moments during increasing and decreasing field in the *M-H* loop [16].

3. Results and discussion

X-ray diffraction patterns of bulk MgB_2 materials produced using various carbon-encapsulated boron powders showed that lower contents of C always gave a single-phase bulk MgB_2 material. Fig.1 shows the X-ray diffraction patterns of bulk MgB_2 materials produced utilizing a low and a high content of C in the boron powder. Both samples were sintered at 805 °C in argon atmosphere. From the figure it follows that the main phase was MgB_2 in both samples, similar with our earlier report [3].

However, with the C content in the B powders around 12 wt% and 16.5wt% the minor impurity phases i.e. MgB_2C_2 and Mg phases appeared (see Fig. 1, right). Several other authors also found the minor phase of MgB_2C_2 in bulk MgB_2 samples with carbon-doping [9, 17]. In our case, we did not observe any minor phase in the samples with a lower carbon content in the carbon-encapsulated boron powder than 7.5%. Thus, less content of carbon than 7.5% in the carbon-encapsulated boron is needed to produce a MgB_2 material without impurity phases.

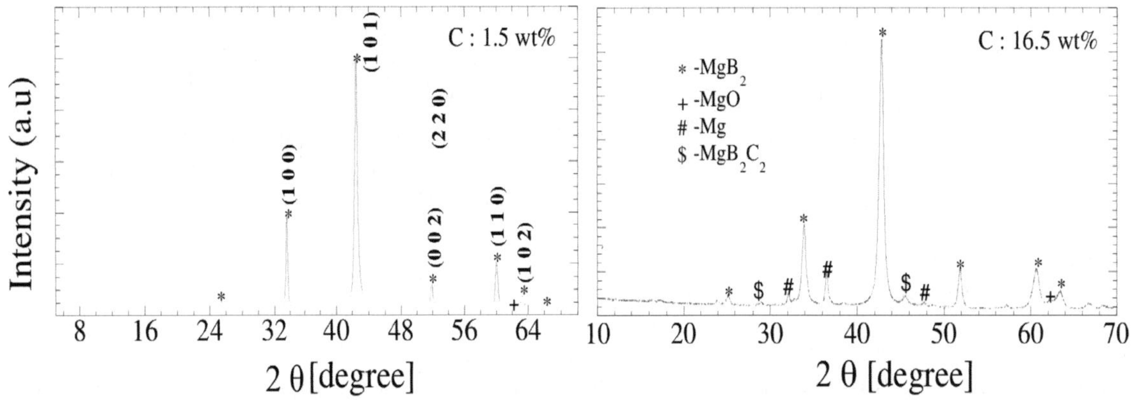

Figure 1. X-ray diffraction patterns of bulk MgB_2 superconductor produced using the carbon-encapsulated boron powders and sintered at 805 °C in argon atmosphere. Low C content in B powders (left figure); high C content in B powders (right figure).

Figure 2 shows the DC magnetic susceptibility in magnetic field of 1 mT as a function of temperature for the MgB_2 materials produced various contents of carbon in the carbon-encapsulated boron. All samples exhibited a sharp superconducting transition. A high onset T_c close to 38.5 K was found for the pure MgB_2 sample, with increasing carbon content in the carbon encapsulated boron T_c decreased.

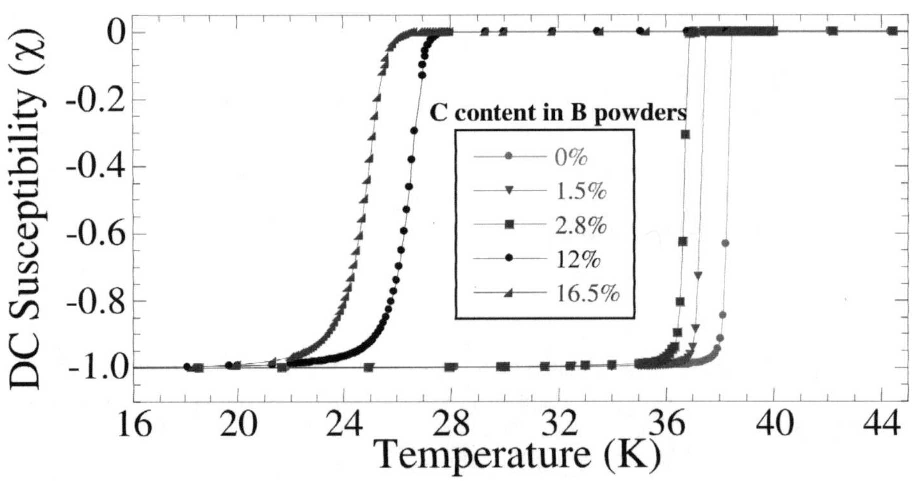

Figure 2. Superconducting transition in the bulk MgB_2 materials produced with various carbon contents in the carbon-encapsulated B powders, all sintered at 805 °C for 3h in argon.

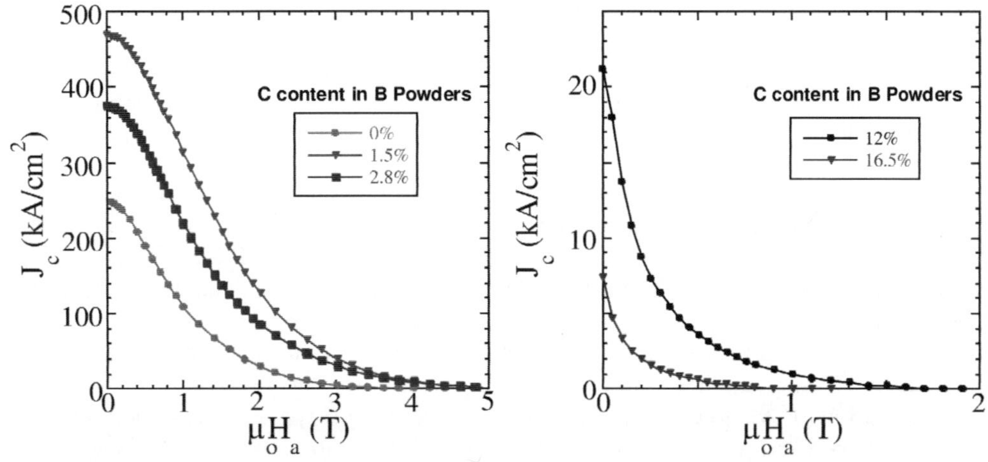

Figure 3. Field dependence of the critical current densities (T=20K) for MgB_2 superconductor produced using the carbon-encapsulated boron powders and sintered at 805 °C in argon atmosphere. Lower C content in B powders (left figure); Higher C content in B powders (right figure).

The samples with low C contents, 1.5wt% and 2.8 wt%, showed T_c (onset) 37.6 K and 37 K, only slightly lower than for the pure sample [18]. On the other hand, the samples with high carbon contents in carbon-encapsulated boron showed rather lower superconducting transitions, close to 25 K (see Fig. 2). In MgB_2 tapes and wires doped by carbon it is believed that C substitutes for boron, which stiffens the optical E_{g2} phonon mode, strongly linked to the anisotropic σ band and hence to a lower transition temperature [9]. The results indicate that the carbon-encapsulated boron powders with a low content of C (<2.8%) will help to optimize production of high quality bulk MgB_2 materials.

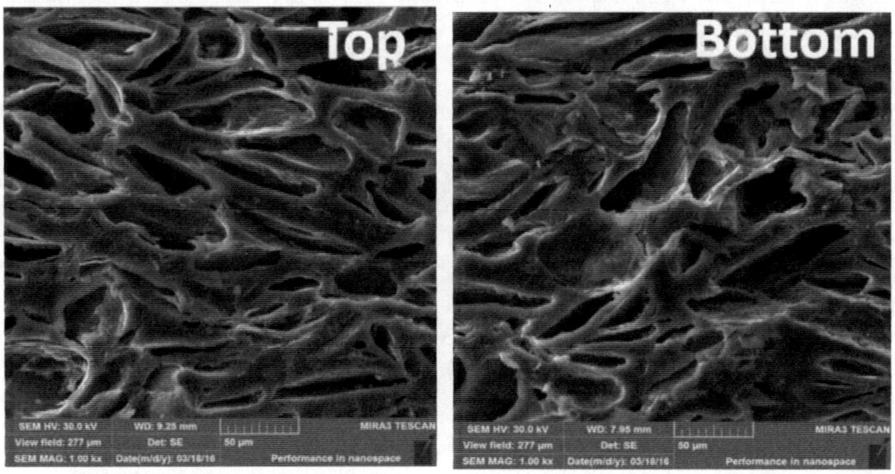

Figure 4. Low magnification scanning electron micrographs of the MgB_2 superconductor produced with the carbon-encapsulated boron powder with 2.8 wt% content of carbon and sintered at 805 °C in argon atmosphere.

The critical current density at 20 K as a function of applied magnetic field is shown in Figure 3. for bulk MgB_2 materials with various contents of carbon in the carbon-encapsulated boron powder, all sintered at 805 °C for 3h in argon atmosphere. High J_c values were achieved in bulk MgB_2 material produced with low contents of carbon in the carbon-encapsulated boron (Fig.3, left). In the sample with 1.5% of carbon in the carbon-encapsulated boron the self-field critical current density of 470 kA/cm^2 was observed. The high-field J_c and the irreversibility field, H_{irr}, values increased in the samples with the carbon content in the carbon-encapsulated boron up to 2.8wt% and decreased in samples with a higher carbon contents. Especially the samples with 12wt% and 16.5% of carbon showed a dramatic decrement in the high-field J_c and H_{irr} values (see Fig.3, right). It is consistent with the shift of the experimental temperature to the rather low T_c in these materials.

To elucidate microstructure, SEM by EDX analysis was performed the MgB_2 sample with 2.8 wt% of carbon in the carbon-encapsulated boron powder. The SEM image taken at a lower magnification shows that the MgB_2 matrix contains numerous voids of various shapes and sizes (Fig. 4). The porosity presents a serious problem in the sintered bulk MgB_2 material. The microstructure and chemical analysis were tested at several positions of the same pellet, on its top and bottom as presented in Fig. 4 and Table 2. The carbon content was measured by EDAX without calibration, therefore the measured values are only semi-quantitative. In any case, the obtained carbon contents in various positions are quite uniform (Table 1). These results prove that the carbon content over the MgB_2 pellet is quite homogenous, which indicates that the carbon-encapsulated boron with a low optimum amount of carbon is an effective way how to produce a high quality MgB_2 with effective pinning structure and electromagnetic performance at a reasonably high operating temperature.

Table 1. Carbon content estimated by SEM-EDAX analysis for MgB_2 superconductor produced using the 2.8 wt% carbon-encapsulated boron powder.

Position	**1**	**2**	**3**	**4**	**average**
Bottom	3.2	3.0	2.6		2.86
Top	3.0	3.0	2.8	2.8	2.90

4. Conclusions

High performance bulk MgB_2 materials were prepared utilizing the boron powders pre-encapsulated by carbon in various contents and sintered at optimum sintering conditions. The magnetization results confirmed that 1.5wt% of carbon in the carbon-encapsulated boron powder did not affect much the superconducting transition temperature, in contrast to the MgB_2 materials using the carbon-encapsulated boron powders with 12wt% or 16.5% carbon. The MgB_2 material produced utilizing the 1.5 wt% carbon-encapsulated boron powder and sintered at 805 °C for 3 h showed a high critical current density of 470 kA/cm^2 and 310 kA/cm^2 at 20 K in self–field and 1 T, respectively. This indicates that the carbon-encapsulated boron powder has a high potential for use in batch production of large single grain MgB_2 bulks, reducing the production time and cost, just this being the key to transfer the technology from laboratory to industrial production.

Acknowledgments

The paper was supported by Grant-in-Aid FD research budget code: 112261, SIT (Shibaura Institute of Technology) and by APVV No. 0330-12, VEGA 2/0121/16.

References

[1]. Kim J H, Zhou S, Hossain M S A, Pan A V, Dou S X 2000 *Appl. Phys. Lett.,* **89** 142505.

[2]. Durrell J H, Dancer C E J, Dennis A, Shi Y, Xu Z, Campbell A M, Hari Babu N, Todd R I, Grovenor C R M, Cardwell D A 2012 *Supercond. Sci. Technol.* **25** 112002.
[3]. Muralidhar M, Inoue K, Koblischka MR, Tomita M, Murakami M 2014 *J. Alloys Compd.* **608** 102.
[4]. Vinod K, Abhilash Kumar R G, Syamaprasad U 2007 *Supercond. Sci. Technol.* **20** R1.
[5]. Perini E, Ginuchi G 2009 *Supercond. Sci. Technol.* **22** 045021.
[6]. Dou S X, Soltanian S, Horvat, Wang X L, Zhou S H, Ionescu M, Liu H K, Munroe P, Tomsic M, *Appl. Phys. Lett.* 2002, **81**, 3419-3421.
[7]. Gumbel A G, Eckert J, Fuchs G, Nenkov K, Müller K H, Schultz L 2002 *Appl. Phys. Lett.*, **80** 2725.
[8]. Bugoslavsky Y, Cohen L F, Perkins G K, Polichetti M, Tata T J, Gwilliam R, Caplin A D 2001 *Nature* **411** 561.
[9]. Mickelson W, Cumings J, Han W Q, Zettl A 2002 *Phys. Rev. B* **65** 052505.
[10]. Sudesh, Kumar N, Das S, Bernhard, Varma G D 2013 *Supercond. Sci. Technol.* **26** 095008.
[11]. Muralidhar M, Nozaki K, Kobayashi H, Zeng X L, Koblischka-Veneva A, Koblischka MR, Inoue K, Murakami M 2015 *J. Alloys Compd.* **649** 833.
[12]. Muralidhar M, Kenta N, Koblishka MR, Murakami M 2015 *Physica Status Solidi a*, **212**, 2141.
[13]. Gao Z S, Ma Y W, Zhang X P, Wang D L, Yu Z G, Yang H, Wen H H, Mossang E 2007 *J. Appl. Phys.* **102** 013914.
[14]. Barua S, Hossain Md S Al, Ma Z, Patel D, Mustapic M, Somer M, Aelcuk S, Kokal I, Morawski A, Cetner T, Gajda D, Dou S X 2015 *Scripta Materialia* **104** 37.
[15]. Muralidhar M, Higuchi M, Jirsa M, Diko P, Koka I, Murakami M 2016 IEEE transactions on Applied Superconductivity at press.
[16]. Kim J H, Zhou S, Hossain M S A, Pan A V, Dou S X 2000 *Appl. Phys. Lett.,* **89** 142505.
[17]. Saengdeejing A, Saal J E, Wang Y, Liu Z K 2007 Appl. Phys. Lett. **90** 151920.
[18]. Muralidhar M, Inoue K, Koblischka MR, Murakami M,. Murakami M 2015 *Adv. Eng. Mater.*, **17** 831.

IOP Conf. Series: Journal of Physics: Conf. Series **871** (2017) 012057 doi:10.1088/1742-6596/871/1/012057

The performance improvement of MgB_2 prepared by the Mg diffusion method with the MgB_4 addition

H Zhang[1], L Li [1], Y Zhao[1,2], Y Zhang[1]

[1]Key Laboratory of Maglev Train and Maglev Technology of Ministry of Education, Superconductivity and New Energy R&D Center, Southwest Jiaotong University, Chengdu 610031, China,

[2]School of Physical Science and Technology, Southwest Jiaotong University, Chengdu 610031, China

zhanghong@home.swjtu.edu.cn

Abstract. The Mg diffusion method with the MgB_4 addition is carried out to improve the MgB_2 application performance. The precursor B powders and MgB_4 powders are uniformly mixed as the ratio of （1- 4x）: x mol %, (x = 0, 1, 2, 5, 10 and 20). The mixed powders are pressed into bulks and the excess of Mg powders are added into the iron pipe. Then the heat treatment is carried out under the pure argon atmosphere. Finally, the high-quality MgB_2 bulks are achieved with high density. The superconducting critical temperature of the MgB_2 bulks decreases slightly with the increase of MgB_4 addition content. However, the critical current density increases significantly at the high-fields. It is perhaps due to the small-sized MgB_4 impurities in MgB_2 bulk function as effective pinning centers and generates significant lattice distortion in the MgB_2 lattice to yield improved J_c performance in the applied field. It suggests that further improvement in the superconducting properties of the MgB_2 added with the MgB_4 will be possible via a better control of the composition and processing parameters. This also provides a reference for improving the preparation of MgB_2 wire.

1. Introduction

High critical current density and high irreversible field of the MgB_2 is one of the keys to the large-scale application of MgB_2. Because these properties are very sensitive to preparation conditions, a large amount of research about MgB_2 preparation conditions were reports[1~4], but there are still no optimal results. Although the commonly used method of solid state reaction is simple, high porosity is one of the important factors that limit the current transport capacity of MgB_2 materials[5]. On the other hand, Mg is a kind of material which can be very easy to be oxidized and will inevitably bring oxygen into the raw material during the preparation of MgB_2 by solid state reaction. Accumulation of MgO in the MgB_2 grain boundary leads to a decrease in the J_c value of MgB_2[6]. Several reports indicating that the diffusion method[7~9] can avoid void formation in the MgB_2 and is effective in forming high-density MgB_2. By a theoretical calculation, the density of MgB_2 materials prepared by diffusion method can reach 100% of the theoretical density of MgB_2. At present, the density of MgB_2 wire prepared by diffusion method has reached more than 95% of MgB_2 theoretical density in the laboratory[10]. At the same time, Mg diffusion into B materials and forming a dense MgB_2 at the interface can effectively prevent MgO from getting into the MgB_2 superconductor and the purity of MgB_2 sample is improved. Therefore, the MgB_2 current transport capacity is effectively improve by the diffusion method preparation.

However, the critical current density of the wire prepared by the diffusion method is still decreasing very fast at high magnetic fields, and the methods of element doping are needed to improve the critical current density at high magnetic fields[11,12]. Other impurity elements are usually introduced into the MgB_2 to enhance the flux pinning but reduce the superconductivity of MgB_2. So the doping amount is limited.

MgB_4 addition is introduced in this paper, which can avoid the introduction of other elements in to the MgB_2. Also, MgB_4 as the precursor of the MgB_2 forming reaction can promote the formation of MgB_2. At the same time, it is reported that the small self-generated coherent MgB_4 particles trapped in the MgB_2 can be used as an effective pinning center to improve the critical current density of the MgB_2 [13]. Therefore, in this paper, a certain amount of MgB_4 is added in to the precursor to prepare the MgB_2 by the diffusion method, which can artificially control the amount of MgB_4 in the MgB_2 and provide references to the actual production process.

2. Experiments

MgB_4 are prepared by *in-situ* solid-state reaction between Mg (99.99%) and B (99.9%) in the nominal stoichiometry of Mg:B = 1:4. They are mixed and ground, followed by pelletization. A high temperature is required to synthesize MgB_4 and an attempt has been done in the temperature 1050°C. Due to the high sintering temperature, a large amount of MgO is formed in those samples. In order to remove the MgO, acid washed is performed and a "cleaner" sample of MgB_4 is obtained.

MgB_2 bulks samples are prepared by an Mg diffusion method. B powder (purity >99.9% amorphous) and a amount of MgB_4 are mixed with the ration（1-4x）: x,（ x = 0,1mol%, 2mol%, 5mol%, 10mol% and 20mol%, respectively）. The well-mixed powders are pressed into bulks with the diameter of 10 mm and the thickness of about 1 mm under a pressure of ~20MPa. The bulks are sealed in the stainless steel tube which filling with amount of Mg powder and put into furnace for heating to 700 ℃ for 5 hours. Then the furnace cools to room temperature.

The sample microstructure, crystal structure and phase composition are analyzed by the scanning electron microscopy (SEM) and *X*-ray diffraction (XRD). The superconducting transition and magnetic hysteretic loop are measured by Quantum-Design SQUID XL. The J_c values of the MgB_2 slices are determined using Bean critical state model: $J_c = 20 \times (\Delta M)/[a(1-a/3b)]$ with a, b ($a \le b$) being the edge lengths of a thin sheet in the plane. T_c is defined as the onset temperature at which diamagnetic properties are observed.

3. Results and discussion

Figure 1 shows the typical X-ray diffraction θ-2θ patterns of MgB_4 powders which before and after acid leaching. MgB_4 appears as a primary phase with MgO as the secondary phase. The high sintering temperature in prolonged time produced a relatively larger amount of MgO and this was inevitable due to the presence of residual O_2 during the sintering[14]. The MgO impurity was then removed by immersing the samples in HNO_3 solution for 5 minutes. As shown in figure 1, the peaks of MgO were reduced to a much lower intensity. Thus, a "cleaner" sample of MgB_4 was obtained and can be used later for synthesizing MgB_2.

Figure 2 shows the typical θ-2θ X-ray diffraction patterns of MgB_2 with the various MgB_4 addition contents of 0 to 20 mol%. In addition to the diffraction peaks of MgB_2, the diffraction peaks of MgB_4,

MgO and pure Mg are also can be detected as shown in figure 2. The pure Mg presents because of the magnesium residues in the bulk during the liquid magnesium diffusion into the bulk.

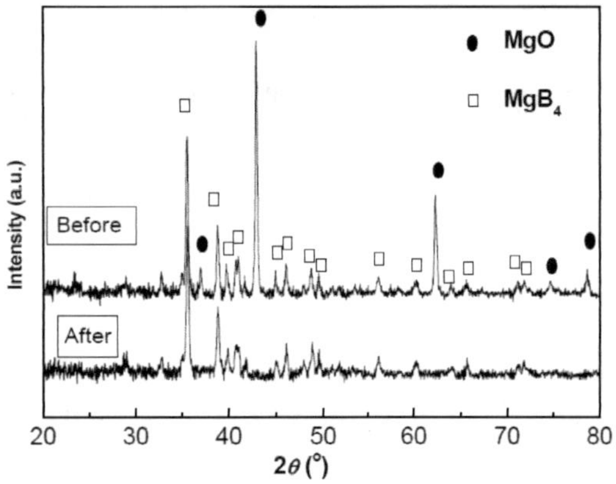

Figure 1. Typical X-ray diffraction *θ-2θ* patterns of MgB_4 powders which before and after acid leaching.

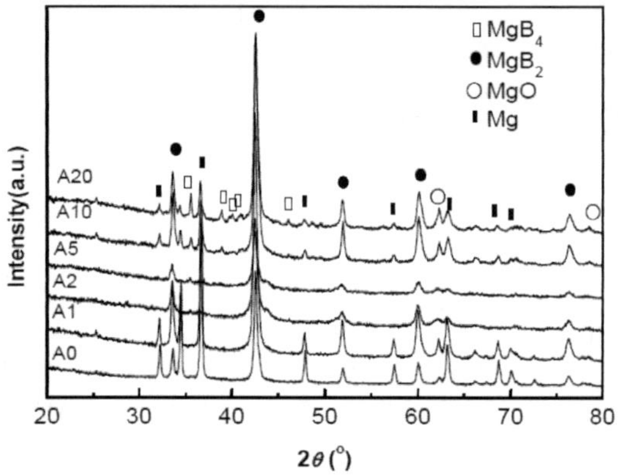

Figure 2. Typical X-ray diffraction *θ-2θ* patterns of MgB_2 samples with the various MgB_4 addition of: (A0) 0; (A1) 1 mol %; (A2) 2 mol %; (A5) 5 mol %; (A10) 10 mol % ; (A20) 20 mol %.

Table 1. The relative content of MgB_4 and MgO in the MgB_2 bulk

the relative content (mol%)	A0	A1	A2	A5	A10	A20
$\frac{I_{MgB_{4(121)}}}{I_{MgB_{4(121)}}+I_{MgB_{2(101)}}} * 100\%$	0	4.28	10.10	14.11	7.58	8.02
$\frac{I_{MgO_{(220)}}}{I_{MgO_{(220)}}+I_{MgB_{2(101)}}} * 100\%$	4.75	5.36	11.47	10.38	5.69	7.16

The relative contents of MgB_4 and MgO in the MgB_2 bulk are calculated from the intensity value of M (M are MgB_4(*121*) and MgO(*220*), respectively) and MgB_2(*101*) diffraction peaks by the expression: $I_M/(I_{MgB2(101)}+I_M)$. Table 1 show the relative contents of MgB_4 and MgO in the MgB_2 bulk which prepared with the various MgB_4 addition of 0 to 20 mol%. As shown in table 1, the relative MgO contents in the MgB_2 samples prepared with the MgB_4 addition are more than that of the samples prepared without the MgB_4 addition. It seems that the addition of MgB_4 introduced the MgO into the MgB_2 bulk. This could be due to a small portion of MgO which the MgB_4 precursor contains and a little amount of MgO comes from the outside of the precursor bulk. At the same time, the presence of MgB_4 in the MgB_2 bulk suggests that MgB_4 does not fully participate in the reaction of the MgB_2 formation.

Figure 3 shows scanning electron microscope (SEM) images of the MgB_2 samples with the various MgB_4 addition of 0 to 20 mol%. All the samples show the high density and uniform characteristic. It is means that the diffusion process is effective in achieving higher-density MgB_2.

IOP Conf. Series: Journal of Physics: Conf. Series **871** (2017) 012057 doi:10.1088/1742-6596/871/1/012057

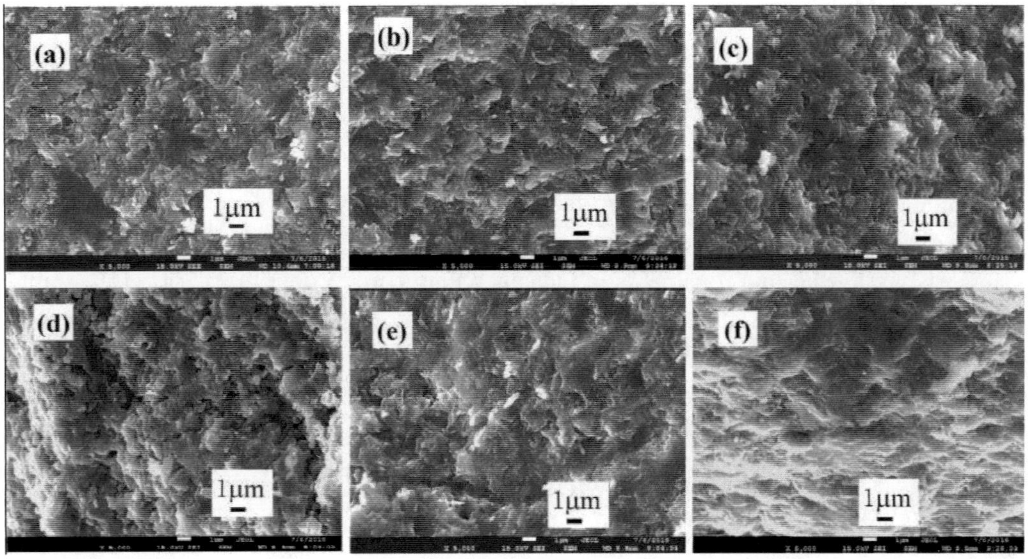

Figure 3. Scanning electron microscope (SEM) images of MgB_2 samples with various MgB_4 addition of: (a) 0 ; (b) 1 mol % ; (c) 2 mol %; (d)5 mol % ; (e) 10 mol % ; (f) 20 mol %.

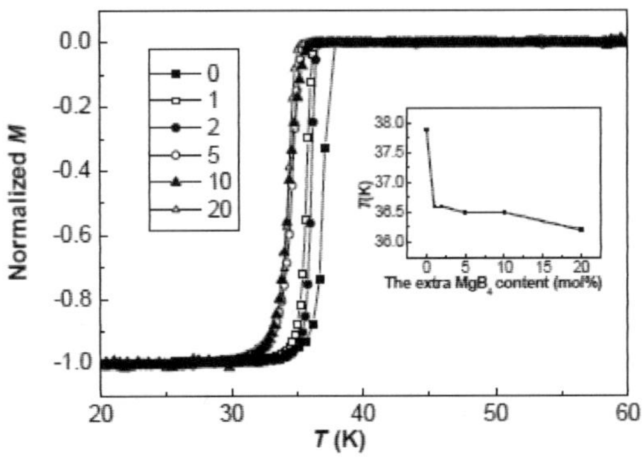

Figure 4. Normalized magnetization versus temperature curve of the MgB_2 samples with various MgB_4 addition of: (0) 0; (1) 1 mol %; (2) 2 mol %; (5)5 mol %; (10) 10 mol %; (20) 20 mol %. Insert is the superconducting critical temperature as a function of MgB_4 addition content.

Figure 4 shows normalized magnetization versus temperature curve of the MgB_2 samples with various MgB_4 addition of 0 to 20 mol%. Inserted is the superconducting critical temperature as a function of MgB_4 addition content. It can be seen that the T_c decreases slightly with the precursor MgB_4 addition content increasing. The T_c decreasing in this work may be due to the impurity phase existences of MgB_4 and MgO in the MgB_2 samples as shown in the figure 2.

Figure 5 shows the magnetic critical current density (J_c) versus the magnetic field of the MgB_2 samples with various MgB_4 additions at different temperatures. The various MgB_4 additions are 0 to 20 mol % and the temperature is 10 K to 20 K. It is noted that the J_c of the MgB_2 decreases with the precursor MgB_4 addition at low-fields compared with that of the MgB_2 sample without the precursor MgB_4 addition. However, the J_c increases with MgB_4 addition at high-fields and the 5 mol% MgB_4 addition sample has the highest J_c value. The enhancement of J_c value at the high-fields is perhaps due to the small-sized MgB_4 impurities in MgB_2 bulk prepared by Mg diffusion method. The MgB_4 impurities has been observed to function as effective pinning centers and generate significant lattice distortion in the MgB_2 lattice to yield improved J_c performance in the applied field[13]. When the MgB_4 addition content more than 5 mol%, the J_c value decrease perhaps due to more precursor MgB_4 contribute into the formation reaction of the MgB_2 which leads to the reduction of the final MgB_4 remaining amount in the MgB_2 as shown in the table 1. To verify the improved flux pinning in the MgB_2 material added by MgB_4, micro-structural experiments are necessary. This study is under the way. It suggests that further improvements in the superconducting properties of the MgB_2 bulk added with the MgB_4 will be possible via a better control of the composition and processing parameters.

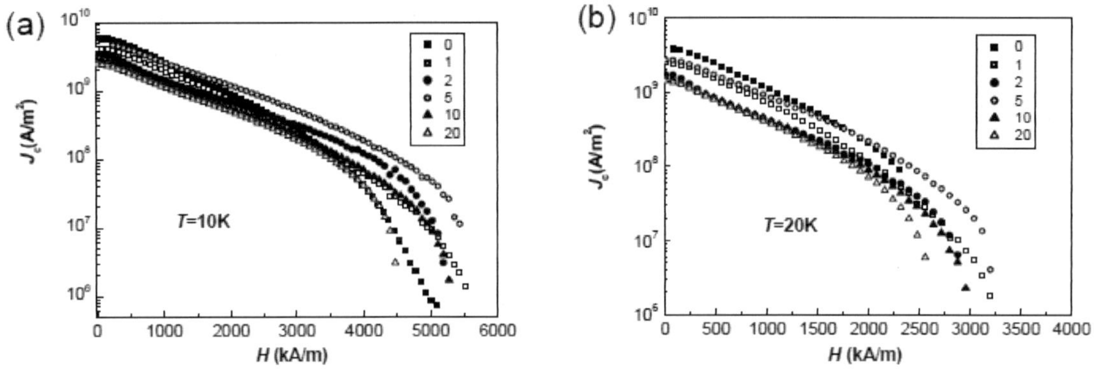

Figure 5. The magnetic critical current density (J_c) versus the magnetic field of the MgB_2 samples with various MgB_4 addition of: (0) 0; (1) 1 mol %; (2) 2 mol %; (5)5 mol %; (10) 10 mol %; (20) 20 mol %. (a) T = 10K; (b) T = 20K.

4. Summary

The high-quality MgB_2 bulks are achieved with high density by the Mg diffusion method. The superconducting critical temperature decreases slightly with the increase of precursor MgB_4 addition. However, the critical current density increases significantly at the high-fields. It suggest that further improvement in the superconducting properties of the MgB_2 bulk added with the MgB_4 will be possible via a better control of the composition and processing parameters.

Acknowledgments

In this paper, the authors wish to thank the fund for the project support, including the Fundamental Research Funds for the Central Universities (A0920502051619-75), the National Natural Science Foundation of China (51377138 and 51271155), the National Science and Technology Projects Foundation（2013DFA51050, 2013GB114003-1, 2014AA032701）and the Sichuan Foundation (2017JY0058).

References

[1] Ye S, Matsumoto A, Togano K, Zhang Y C, Ohmura T, and Kumakura H 2014 *Supercond. Sci. Technol.* **27** 055017

[2] Nardelli D, Matera D, Vignolo M, Bovone G, Palenzona A, Siri A S, and Grasso G 2013 *Supercond. Sci. Technol.* **26** 075010

[3] Parizh M, Lvovsky Y and Sumption M 2016 *Journal of Alloys and Compounds* **687** 616

[4] Hu r J M, Togano K, Matsumoto A, Kumakura H, Wada H and Kimura K 2008 *Supercond. Sci. Technol.* **21** 032001

[5] Feng Y, Zhao Y, Pradhan A K, Zhou L, Zhang P X, Liu X H, Ji P, Du S J, Liu C F, Wu Y and Koshizuka N 2002 *Supercond. Sci. Technol.* **15** 12

[6] Kováč P, Kulich M, Haessler W, Hermann M, Melišek T and Reissner M 2012 *Physica C: Superconductivity* **477** 20

[7] Wang D, Zhang X P, Tang S P, Xu D , Yao C , Dong C H, Xu Z T, Ma Y W, Oguro H, Awaji S and Watanabe K, 2015 *Supercond. Sci. Technol.* **28** 105013

[8] Togano K, Nakane T, Fujii H, Takeya H, and Kumakura H 2006 *Supercond. Sci. Technol.* **19** L17

[9] Li G Z, Sumption M D, Susner M A, Yang Y, Reddy K M, Rindfleisch M A, Tomsic M J, Thong C J, and Collings E W 2012 *Supercond. Sci. Technol.* **25** 5023

[10] Ueda S, Shimoyama J-i, Iwayama I, Yamamoto A, Katsura Y, Horii S and Kishio K 2005 *Appl. Phys. Lett.* **86** 2502

[11] De Silva K S B, Aboutalebi S H, Xu X, Wan g X L, Li W X, Konstantinov K and Dou S X 2013*Scripta Materialia* **69** 437

[12] Wong D C K, Yeoh W K, De Silva K S B, Kondyurin A, Bao P, Li W X, Xud X, Peleckis G, Dou S X, Ringer S PA and Zheng R K 2015 *Journal of Alloys and Compounds* **644** 900

[13] Cai Q, Liu Y C, Ma Z Q, Li H J, and Yu L M, 2013 *Appl. Phys. Lett.* **103** 132601

[14] Tan K L, Tan K Y, Lim K P, Halim S A and Chen S K 2011 *Solid State Science and Technology* **19** 15

IOP Conf. Series: Journal of Physics: Conf. Series **871** (2017) 012058 doi:10.1088/1742-6596/871/1/012058

Production and Characterization of Bulk MgB_2 Material made by the Combination of Crystalline and Carbon Coated Amorphous Boron Powders

K. Hiroki[1], M. Muralidhar[1*], M.R. Koblischka[2], M. Murakami[1]

[1] Superconducting Material Laboratory, Graduate School of Science and Engineering, Shibaura Institute of Technology, 3-7-5 Toyosu, Koto-ku, Tokyo 135-8546, Japan
[2]Institure of Experimental Physics, Saarland University, P.O. Box 151150, 66041Saarbrucken, Germany

*miryala1@shibaura-it.ac.jp

Abstract. The object of this investigation is to reduce the cost of bulk production and in the same time to increase the critical current performance of bulk MgB_2 material. High-purity commercial powders of Mg metal (99.9% purity) and two types of crystalline (99% purity) and 16.5 wt% carbon-coated, nanometer-sized amorphous boron powders (98.5% purity) were mixed in a nominal composition of MgB_2 to reduce the boron cost and to see the effect on the superconducting and magnetic properties. Several samples were produced mixing the crystalline boron and carbon-coated, nanometer-sized amorphous boron powders in varying ratios (50:50, 60:40, 70:30, 80:20, 90:10) and synthesized using a single-step process using the solid state reaction around 800 °C for 3 h in pure argon atmosphere. The magnetization measurements exhibited a sharp superconducting transition temperature with $T_{c,onset}$ around 38.6 K to 37.2 K for the bulk samples prepared utilizing the mixture of crystalline boron and 16.5% carbon-coated amorphous boron. The critical current density at higher magnetic field was improved with addition of carbon-coated boron to crystalline boron in a ratio of 80:20. The highest self-field J_c around 215,000 A/cm^2 and 37,000 A/cm^2 were recorded at 20 K, self-field and 2 T for the sample with a ratio of 80:10. The present results clearly demonstrate that the bulk MgB_2 performance can be improved by adding carbon-coated nano boron to crystalline boron, which will be attractive to reduce the cost of bulk MgB_2 material for several industrial applications.

1. Introduction

Since 2001, the discovery of superconductivity in MgB_2 material has led to notable progress concerning especially the understanding of the origin of the large T_c, processing, characterization, and industrial applications [1-3]. The intermetallic material is more attractive for the next generation of superconducting applications because of the lack of weak-links at the grain boundaries and a high critical transition temperature of 39 K as compared to the conventional NbTi. The superconducting transition temperature of MgB_2 is significantly lower than that of $YBa_2Cu_3O_y$ "Y-123", instead, MgB_2 benefits of a high critical current density (J_c) in the polycrystalline state, which makes these materials promising candidates for several industrial applications including the next generation of super-magnets for medical devices, electrical power system, transportation and powerful super-magnets operating at around 20 K [3-9]. For superconducting super-magnet applications, it is required to produce good quality, low cost MgB_2 material with high J_c and an acceptable mechanical performance. To improve the critical current density of the MgB_2 material, a variety of processing techniques have been developed and studied in terms of commercially usefulness [9,10]. On the other hand, to improve the critical current density of these materials a variety of dopants [11-14] were adopted and very successful results were observed, especially using carbon-based dopants, e.g., carbon, boron carbide, carbon nanotubes, carbohydrates or hydrocarbons, graphene oxide, carbon-coated boron powders, etc. [15-19]. As a result, several researchers observed a dramatic improvement of the superconducting properties such as the irreversibility field, H_{irr}, the upper critical field, H_{c2}, or the critical current density, J_c, under high magnetic fields. Further, significant improvements in self-field performance were reached by optimizing the microstructure [10]. Our recent results also clarified that atomic force microscopy and EBSD observations clearly indicated that the observed grains exhibit grain sizes in the nanometer range and their density was high in the samples sintered for 3 h at 800 °C and 805 °C. Further, the samples sintered at 805 °C for 3 h showed a high critical current density of 245 kA/cm^2 at 20 K [10]. Moreover, the samples sintered at 775 °C for 1 h showed a high critical current density of 270 kA/cm^2 at 20 K [20]. The latest reports also suggested that the bulk MgB_2 materials' performance can be further improved by tuning both the optimum sintering temperature and the optimal sintering time. To utilize the bulk MgB_2 material for several industrial applications, one needs batch production, cheaper processing, and a high performance. In the present work, we study the effect of mixture of crystalline and 16.5 wt% carbon-coated, nanometer-sized amorphous boron powders were mixed in a nominal composition of MgB_2 to reduce the boron cost and studied the X-ray diffraction and superconducting performance especially the T_c and J_c.

2. Experimental details

The bulk polycrystalline MgB_2 samples were fabricated by using in-situ solid state reaction. High-purity commercial powders (Furu-uchi Chemical Corporation, Japan) of Mg metal (99.9% purity, 325 meshes) and 16.5 wt% carbon-encapsulated amorphous nano-boron powders (85% purity) were mixed in a nominal ratio of Mg: B = 1: 2. For any industrial applications, low cost raw materials are crucial to reduce the products final costs. Amorphous boron powders always yield high performance as compared to the crystalline boron powders. However, the cost of the amorphous or carbon coated amorphous boron powders is three to four times higher as compared to the crystalline boron powders. Therefore, to reduce the cost of the boron and to optimize the high critical current density, several samples were produced mixing the crystalline boron and the 16.5% carbon-coated, nanometer-sized amorphous boron powders in varying ratios, i.e., 50:50, 60:40, 70:30, 80:20, 90:10. More details for the 16.5 wt% carbon-encapsulated nano-boron powders (provided by PAVEZYUM, Advanced Chemicals, Turkey) can be found elsewhere [21]. The powder mixture was pressed into pellets of 20 mm in diameter and 4 mm in thickness using an uniaxial pressing machine. The consolidated pellets were then wrapped in tantalum foils and subjected to the heat treatment in Ar atmosphere in a tube furnace. The samples were heated to the target sintering temperature of 800 °C and kept there for 3 h in flowing argon gas. Finally, the temperature was lowered to room temperature at a cooling rate of 100 °C/h.

The constituent phases of the samples were identified with a high-resolution automated X-ray powder diffractometer (RINT2200), using Cu-K_α radiation generated at 40 kV and 40 mA. Small specimens with dimensions of 1.5 × 1.5 × 0.5 mm^3 were cut from bulk MgB_2 samples and subjected to the measurements of the critical temperature (T_c) and magnetization hysteresis loops (*M-H* loops) in applied magnetic fields from –1 to +5 T at temperatures of 20 K using a SQUID magnetometer (Quantum Design, model MPMS5). The magnetic J_c values were estimated based on the extended Bean critical state model using the relation

$$J_c = 20\ \Delta m/[a^2 d\ (b-a/3)] \quad , \qquad (1)$$

where d is the sample thickness, a, b are cross sectional dimensions, $b \geq a$, and Δm is the difference of magnetic moments during increasing and decreasing field in the *M-H* loop [22].

IOP Conf. Series: Journal of Physics: Conf. Series **871** (2017) 012058 doi:10.1088/1742-6596/871/1/012058

3. Results and discussion

X-ray diffraction patterns observed on the bulk MgB_2 produced using the mixture of the crystalline boron and 16.5% carbon-coated, nanometer-sized amorphous boron powders in varying ratios i.e. 50:50, 60:40, 70:30, 80:20, 90:10 are presented in Fig. 1. From the figure it is clear that the main phase was MgB_2 for all the samples, which is similar to the earlier reports [9]. There was a very small trace of MgO. Thus is in agreement with the fact that oxygen could trap the material during sample pressing into pellets or wrapping into tantalum foils before loading the furnace [20]. The magnified view of the region around 56 degrees, which shows the peaks (0 0 2) and (1 1 0) in detail (see Fig.1, right). From the figure it is clear that the (1 1 0) peak for the 16.5% carbon-encapsulated Boron added samples is slightly shifted to higher angles as compared to the sample produced using the crystalline boron. The (0 0 2) peak position stood unchanged for all samples studied. It indicates that carbon partly substitutes for boron in the lattice of MgB_2, which is similar to earlier reports [14,21].

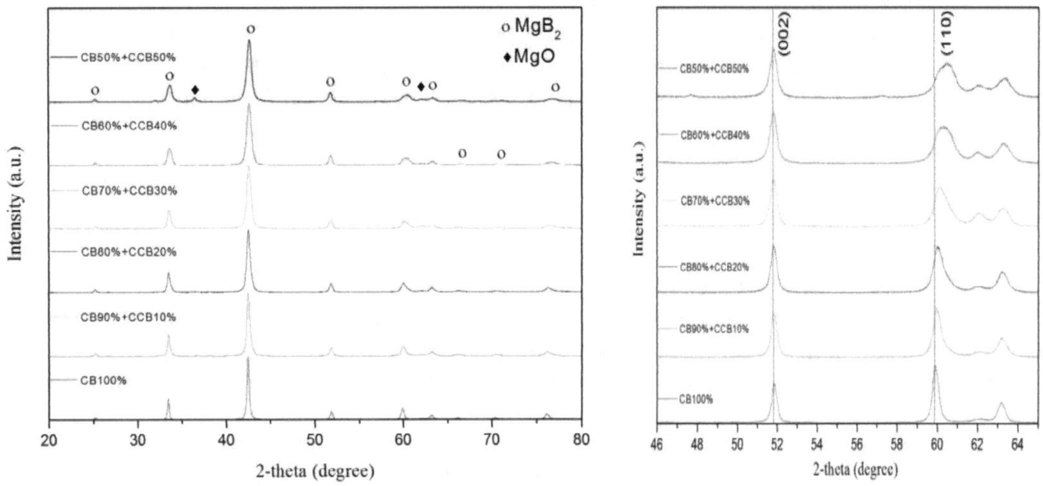

Figure 1. X-ray diffraction patterns of bulk MgB_2 samples produced using the mixture of the crystalline boron and 16.5% carbon-coated, nanometer-sized amorphous boron powders in varying ratios (50:50, 60:40, 70:30, 80:20, 90:10) and sintered at 800 °C for 3h in argon atmosphere (left); the enlarged view of (0 0 2) and (1 1 0) peaks for the same material (right).

Figure 2 shows temperature dependence of magnetic susceptibility curves of MgB_2 samples produced using the mixture of the crystalline boron and 16.5% carbon-coated, nanometer-sized amorphous boron powders in varying ratios i.e. 50:50, 60:40, 70:30, 80:20, 90:10 in a magnetic field of 1 mT. It is evident that superconducting transition temperature ($T_{c,onset}$) is decreased with increasing the 16.5% carbon-coated, nanometer-sized amorphous boron in the crystalline boron (see Table 1). The high T_c of 38.65 K was observed for sample produced utilizing the crystalline boron. On the other hand, MgB_2 sample produced mixture of the crystalline boron and 16.5% carbon-coated, nanometer-sized amorphous boron powders, i.e., 50:50 is 32.99 K. The good T_c indicate that all processed bulk MgB_2 samples were good in quality (see Table 1).

The observed superconducting transition temperatures are similar to recently reported bulk MgB_2 material fabricated with a sintering process [19]. These results suggest that superconducting transition temperature are not effected much when we add the carbon-coated, nanometer-sized amorphous boron powders in crystalline boron.

In order to evaluative the critical current performance of the MgB_2 material produced by the mixture of the crystalline boron and 16.5% carbon-coated, nanometer-sized amorphous boron powders, we measured the magnetization hysteresis loop (*M-H* loop) in fields from –1 to +5 T at 20 K using a SQUID magnetometer (Quantum Design, model MPMS5) and the critical current density (J_c) was estimated using Bean's critical state formula and it was presented in Figure 3.

Table1. Superconducting transitions temperatures and critical current density at 20K for MgB_2 materials

Boron content in MgB_2	T_c (K)	J_c at 20 K (A/cm^2)			
		Self-field	2 T	3 T	4 T
CB 100%	38.66	211083	24813	4386	158
CB 90%+CCB 10%	38.10	214755	37029	10413	1997
CB 80%+CCB 20%	36.80	197490	33399	11638	2947
CB 70%+CCB 30%	35.76	152899	17894	6291	851
CB 60%+CCB 40%	34.99	73880	5245	1410	11
CB 50%+CCB 50%	32.99	25151	621	41	

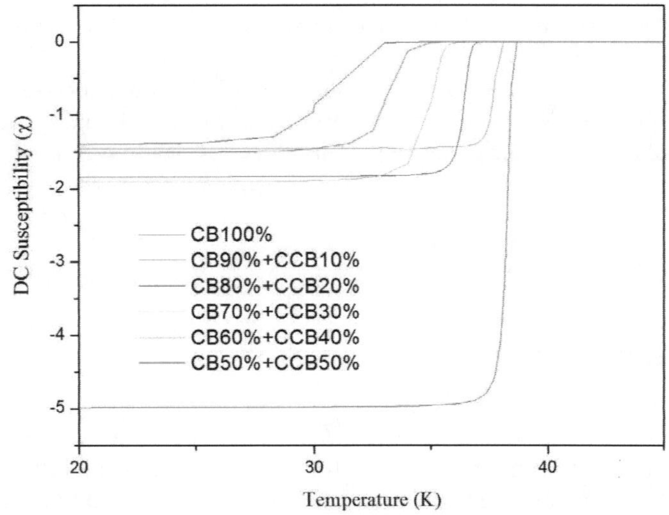

Figure 2. Superconducting transition in the bulk MgB_2 materials produced using the mixture of the crystalline boron and 16.5% carbon-coated, nanometer-sized amorphous boron powders and sintered at 800 °C for 3 h in argon atmosphere and measured in a magnetic field of 1 mT.

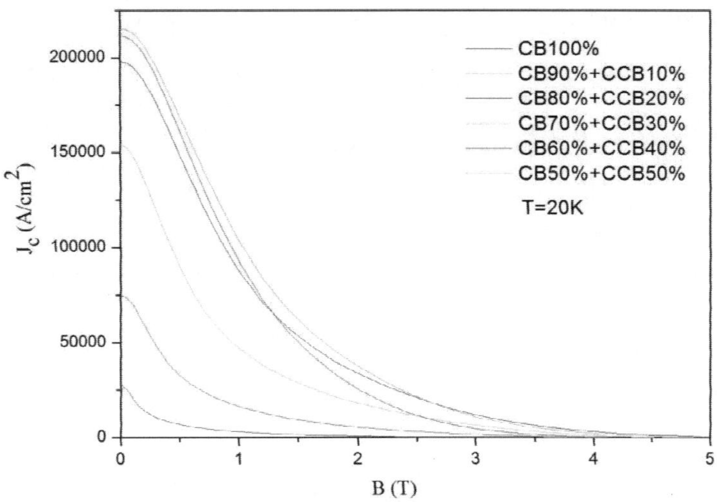

Figure 3. Field dependence of the critical current densities (T = 20 K) for MgB_2 superconductor produced using the mixture of the crystalline boron and 16.5% carbon-coated, nanometer-sized amorphous boron powders and sintered at 800 °C for 3 h in argon atmosphere.

IOP Conf. Series: Journal of Physics: Conf. Series **871** (2017) 012058 doi:10.1088/1742-6596/871/1/012058

It was notable that critical current density was increased especially at higher magmatic field the samples produced utilizing the mixture of 16.5% carbon-coated, nanometer-sized amorphous boron powders (see Fig. 3). Further, the highest self-field J_c was recorded to be 215,000 A/cm^2 at 20 K in MgB_2 samples produced mixture of crystalline boron and the carbon-coated amorphous boron powders ratio is 90:10. These values are still high as compared to samples produced from 100% amorphous boron powders [20]. The summary of the J_c performance for all produced samples is given in Table 1. It is clear that high field critical current performance was more than double in samples the carbon-coated amorphous boron powders ratio is 10:90 and 20:80 as compared to samples made by crystalline boron powders. Further, it is important to note that the irreversibility field was improved with increasing the carbon-coated amorphous boron and crystalline boron powders ratio from 10:90, 20:80, and 30:70; higher ratios lead to a decrease of the irreversibility fields. The carbon-coated amorphous boron and crystalline boron powders ratio of 80:20 exhibited the highest irreversibility among all the samples. It has been reported by several authors that carbon substituted at boron sites is effective in improving the critical current performance, especially at high magnetic fields and decreases the superconducting transition temperature. Further, it was also clarified in bulks and tapes that carbon doping introduced defects into the sample, which can act as strong pinning centers, leading to the improved performance at higher fields, and as a result, to an improvement in irreversibility. Moreover, the present results indicate that the 16.5 wt% carbon-coated amorphous boron powders addition to the crystalline boron is crucial to improve the quality of bulk MgB_2 materials made by crystalline boron powders, which will be defiantly reduce the prize of the MgB_2 simples and eventually improve the high field performance for the industrial applications.

4. Conclusions

Several bulk MgB_2 samples were produced mixing the crystalline boron and carbon-coated, nanometer-sized amorphous boron and synthesized using a single-step process using the solid state reaction at 800 °C for 3 h in pure argon atmosphere. X-ray diffraction results indicated that all samples are single phase MgB_2 along with small quantity of MgO. DC susceptibility vs. temperature measurements showed a sharp superconducting transition around 38 K. The bulk MgB_2 samples made by crystalline and carbon-coated, amorphous boron ratio 90:10 and 80:20 exhibited enhanced J_c value at 2 to 4T at 20K. These results imply that high flux pinning performance of the sintered bulk MgB_2 material can be produced using the small quantity of 16.5 wt% carbon-coated boron powders addition to the crystalline boron powders, which will be very useful to reduce the boron cost by means of production of cheap superconducting super-magnets.

Acknowledgments

The paper was supported by Grant-in-Aid FD research budget code: 112261, Shibaura Institute of Technology (SIT).

References

[1]. Vinod K, Abhilash Kumar R G, Syamaprasad U 2007 *Supercond. Sci. Technol.* **20** R1.
[2]. Perini E, Ginuchi G 2009 *Supercond. Sci. Technol.* **22** 045021.
[3]. Kajikawa K, Nakamura T 2009 *IEEE Trans. Appl. Supercond.* **19** 1669.
[4]. Gunchi G, Ripamonti G, Cavallin T, Bassani E 2014 *Cryogenics* **46** 23.
[5]. Kim J H, Oh S, Heo Y-U, Hata S, Kumakura H, Matsumoto A, Mitsuhara M, Choi S, Shimada, Maeda Y M, MacManus-Driscoll J L, Dou S X, 2012 *NPG Asia Materials*, **4** doi:10.1038/am.2012.3.
[6]. Noudem N J, Aburras M, Bernstein P, Chaud X, Muralidhar M, Murakami M 2014 *J. Appl. Phys.* **116** 163916.
[7]. Naito T, Sasaki T, Fujishiro H 2012 *Supercond. Sci. Technol.* **25** 2012 095012.

[8]. Durrell J H, Dancer C E J, Dennis A, Shi Y, Xu Z, Campbell A M, Hari Babu N, Todd R I, Grovenor C R M, Cardwell D A 2012 *Supercond. Sci. Technol.* **25** 112002.
[9]. Muralidhar M, Inoue K, Koblischka MR, Tomita M, Murakami M 2014 *J. Alloys Compd.* **608** 102.
[10]. Muralidhar M, Nozaki K, Kobayashi H, Zeng X L, Koblischka-Veneva A, Koblischka MR, Inoue K, Murakami M 2015 *J. Alloys Compd.* **649** 833.
[11]. Kumar D, Pennycook SJ, Narayan J, Wang H, Tiwari A 2003 Supercond. Sci. Technol. **16** 455.
[12]. Muralidhar M, Inoue K, Koblischka MR, Murakami M,. Murakami M 2015 *Adv. Eng. Mater.*, **17** 831.
[13]. Feng W J, Xia T D, Liu T Z, Zhao W J, Wei Z Q 2005 Physica C **425** 144.
[14]. Wozniak M, Juda K L, Hopkins S C, Gada D, Glowacki B A 2008 *Supercond. Sci. Technol.* **21** 035006.
[15]. Kim J H, Zhou S, Hossain M S A, Pan A V, Dou S X 2000 *Appl. Phys. Lett.,* **89** 142505.
[16]. Kim J H, Zhou S, Hossain M S A, Pan A V, Dou S X 2006 *Appl. Phys. Lett.* **89** 142505.
[17]. Gao Z S, Ma Y W, Zhang X P, Wang D L, Yu Z H, Yang H, Wen H H, Mossang E 2007 *J. Appl. Phys.* **102** 013914.
[18]. Vajpayee A, Jha R, Srivastava A K, Kishan H, Tropeano M, Ferdeghini C, Awana V P S 2011 *Supercond. Sci. Technol.* **24** 45013.
[19]. Muralidhar M, Higuchi M, Jirsa M, Diko P, Koka I, Murakami M 2016 *IEEE transactions on Applied Superconductivity* at press.
[20]. Muralidhar M, Kenta N, Koblischka MR, Murakami M 2015 *Physica Status Solidi a*, **212**, 2141.
[21]. Barua S, Hossain Md S Al, Ma Z, Patel D, Mustapic M, Somer M, Aelcuk S, Kokal I, Morawski A, Cetner T, Gajda D, Dou S X 2015 *Scripta Materialia* **104** 37.
[22]. Chen D, Goldfarb R B 1989 *J. Appl. Phys.* **66** 2489.

IOP Conf. Series: Journal of Physics: Conf. Series **871** (2017) 012059 doi:10.1088/1742-6596/871/1/012059

Microstructure and critical current density in MgB_2 bulk made of 4.5 wt% carbon-coated boron

M. Higuchi[1], M. Muralidhar[1*], M. Jirsa[2], M. Murakami[1]

[1] Superconducting Material Laboratory, Graduate School of Science and Engineering, Shibaura Institute of Technology, 3-7-5 Toyosu, Koto-ku, Tokyo 135-8546, Japan
[2]EInstitute of Physics CAS, Na Slovance 2, CZ-18221 Praha 8, Czech Republic

*miryala1@shibaura-it.ac.jp

Abstract. Superconducting performance and its uniformity was studied in the single-step sintered MgB_2 bulk prepared with 4.5 wt% of carbon in the carbon-encapsulated boron. The 20 mm in diameter MgB_2 pellet was cut into several pieces from bottom to top and the microstructure, superconducting transition temperature (T_c onset), and critical current density at 20 K were studied. DC magnetization measurements showed a sharp superconducting transition with onset T_c at around 35.5 K in all positions. SEM analysis indicated a dispersion of grains between 200 and 300 nm in size, as the main pinning medium in this MgB_2 superconductors. The critical current density at 20 K was quite uniform, around 330 kA/cm^2 and 200 kA/cm^2 at self-field and 1 T, respectively, for all measured positions. The results indicate that the *carbon-encapsulated* boron is very promising for production of high quality bulk MgB_2 material for various industrial applications.

1. Introduction

Practical applications of bulk MgB_2 magnets are similar to those of melt-textured Y-123, namely magnets for nuclear magnetic resonance (NMR), magnetic resonance imaging (MRI), fault current limiters, non-contact bearings for liquid pumping, and for magnetic shielding screens [1-3]. To fully utilize the bulk MgB_2 material, especially for superconducting super-magnets, the optimization of the flux pinning and further improvement in critical current density (J_c) are very important. To improve the critical current density of bulk MgB_2 material a variety of processing techniques has been developed [4-8]. As for all functional materials, the microstructure control is crucial. In MgB_2 it concerns mainly the grain size, governing the low field critical current density [9]. In our recent report we showed that J_c and the trapped magnetic field in bulk MgB_2 materials were dramatically improved when the material was processed at around 800 °C for an optimized sintering time [9]. Under optimized conditions uniformly distributed nanometer sized grains formed and a coarsening of the MgB_2 grains was prevented, resulting in an enhanced grain boundary pinning [10]. The critical current density was dramatically improved, especially with doping by metals like Ag, Bi, Sn, Si, Te, Cu etc. [11-14]. On the other hand, the high-field critical current performance was enhanced utilizing carbon-based dopants [15-19]. These dopants improve the high-field flux pinning by substituting carbon into boron lattice sites [20]. The resulting point-like arbitrary pinning disorder improves superconducting properties such as J_c, H_{irr}, and H_{c2} [20]. A recent report clarified that the optimized sintering temperature combined with carbon-encapsulated boron powders dramatically improved flux pinning and critical current density in bulk MgB_2 materials [19]. As a result, the highest J_c values at 20 K, of 375 kA/cm^2 and 220 kA/cm^2, in the self-field and 1 T, respectively, were achieved in the MgB_2 sample with 2.8% of carbon-encapsulated boron.

These results are very promising. However, for trapped-field applications of bulk MgB_2 materials it is very important to achieve such properties uniformly in the whole superconductor pellet. In the presented study we synthesized a bulk MgB_2 material utilizing the carbon-encapsulated boron powder with 4.5 wt% carbon and checked the effect of carbon distribution over a large bulk MgB_2 material on the superconducting performance. The original pellet was cut into several pieces and x-ray diffraction, superconducting transition temperature, and critical current density at 20K were studied on the small samples from different positions.

2. Experimental details

Commercial powders (Furu-uchi Chemical Corporation, Japan) of Mg (99.9% purity, 200 meshes) and carbon-encapsulated amorphous B powder with 4.5% carbon (98% purity, 300 meshes) were mixed in the nominal ratio of Mg: B = 1: 2. The carbon-encapsulated B powders were produced at PAVEZYUM, Advanced Chemicals, Turkey. More details of the carbon-encapsulated B powder production can be found in Ref. [21]. The starting powders were thoroughly ground in a glove box in argon atmosphere. The powder mixture was pressed into pellets of 20 mm in diameter and 7 mm thick using a uniaxial press of 200 MPa. The consolidated pellets were then wrapped in titanium foils and subjected to the heat treatment in Ar atmosphere in a tube furnace. All samples were heated to the target sintering temperature of 805 °C and kept there for 3 h in flowing argon gas. Finally, the temperature was lowered to room temperature at a cooling rate of 100 °C/h. The constituent phases of the samples were identified with a high-resolution automated X-ray powder diffractometer (RINT2200), using Cu-Kα radiation generated at 40 kV and 40 mA. The microstructure of these samples was studied with a scanning electron microscope (SEM).

Small specimens with dimensions of $1.5 \times 1.5 \times 0.5$ mm^3 were cut from MgB_2 pellets and subjected to the measurements of critical temperature (T_c) and magnetization hysteresis loops (*M-H* loops) in applied magnetic fields from –1 to +5 T at 20 K using a SQUID magnetometer (Quantum Design, model MPMS5). The magnetic J_c values (in MA/cm^2) were estimated based on the extended Bean critical state model using the relation

$$J_c = 20\ \Delta m/[a^2 d(b-a/3)] \qquad (1)$$

where d is the sample thickness, a, b are cross sectional dimensions, $b \geq a$, (a, b, d in mm) and Δm (in emu units, 1 emu$\equiv 10^{-3}$ Am^2) is the difference of magnetic moments during increasing and decreasing field in the *M-H* loop [22].

3. Results and discussion

To study the bulk MgB_2 performance at various positions, small samples were cut from the pellet as shown in Fig. 1 (left). 3 pieces were selected in top and 3 pieces in bottom of the pellet (see Fig. 1, left). Fig. 1 (right) shows the X-ray diffraction pattern for the sample from the top centre. It can be seen that the sample mainly consists of the MgB_2 phase and a small quantity of the MgO phase. The latter phase was found to arise during the samples pressing into pellets after the samples being wrapped into tantalum foils before loading them into the tube furnace [23]. Several pieces selected from various positions on the pellet exhibited the same structure, with the main phase of MgB_2 and a small trace of MgO. It follows from the XRD results that the 4.5 wt% of carbon in the carbon-encapsulated boron does not destroy structure of the bulk MgB_2 material. To support this statement, we measured transition temperature and critical current density at 20 K on the samples selected at top and bottom of the pellet. The results are presented in following section.

Temperature dependence of magnetization of the samples selected from top (T1, T2, T3) and bottom (B1, B2, B3) of the pellet is presented in Fig.2. All samples exhibited a sharp superconducting transition with the onset T_c around 35K. The T_c (onset) values for the T1, T2, T3 positions were 35.5 K, 35.8 K, and 36 K, respectively (Fig.2., left). T_c of 35 K, 35.5 K, and 36.1 K was observed at B1, B2, and B3 positions, respectively (Fig.2., right). Compared to a pure MgB_2 material, there is only a slight depression in T_c (onset) in the present samples, in accord with our earlier reports [10]. The results indicate that a uniform high quality MgB_2 material can be produced utilizing the carbon-encapsulated B powder with 4.5 wt% of carbon.

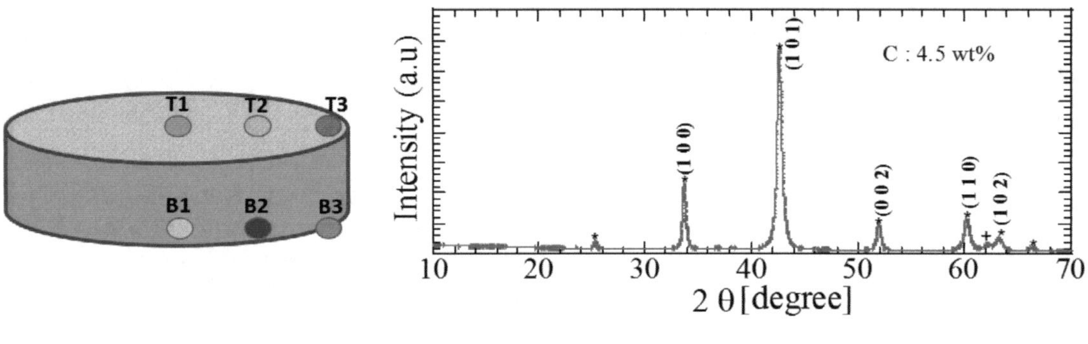

Figure 1. Scheme of the bulk MgB_2 pellet showing the sample positions selected for magnetization measurements (left). X-ray diffraction pattern of the sample from the top centre (right, *MgB_2 phase and + MgO phase).

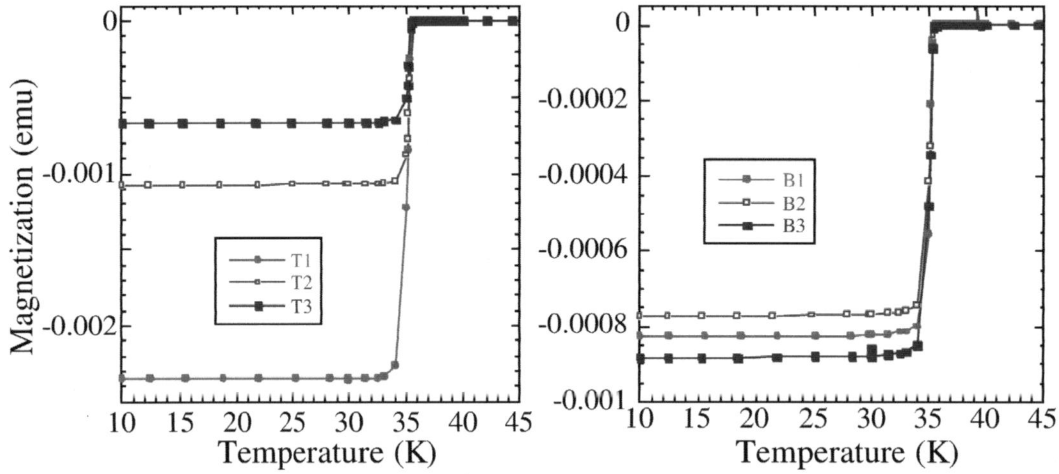

Figure 2. Superconducting transitions in the samples selected from top (left) and bottom (right) of the MgB_2 pellet. The MgB_2 was produced using carbon-encapsulated B powder with 4.5 wt% carbon and sintered at 805 °C for 3h in argon.

To see the effect of carbon distribution on critical current density, the magnetic J_c values were determined from magnetization hysteresis loops (*M-H* loops) detected by a SQUID magnetometer in applied magnetic fields from –1 to +5 T at 20 K. Figure 3 shows J_c values at 20 K as a function of applied magnetic field for the samples selected from the top (Fig. 3., left) and bottom (Fig. 3., right) of the MgB_2 pellet. All the curves show almost same critical current density values. The high-field J_c values and the irreversibility field were same in all positions. The self-field critical current density between 300 kA/cm^2 and 330 kA/cm^2 was observed in all the samples. At 1 T and 2 T the critical current density reached the values of around 200 kA/cm^2 and 75 kA/cm^2, respectively. These results again indicate that the performance of the bulk MgB_2 material is nearly same in various positions of the large bulk. It is a good message for mass production of the MgB_2 super-magnets.

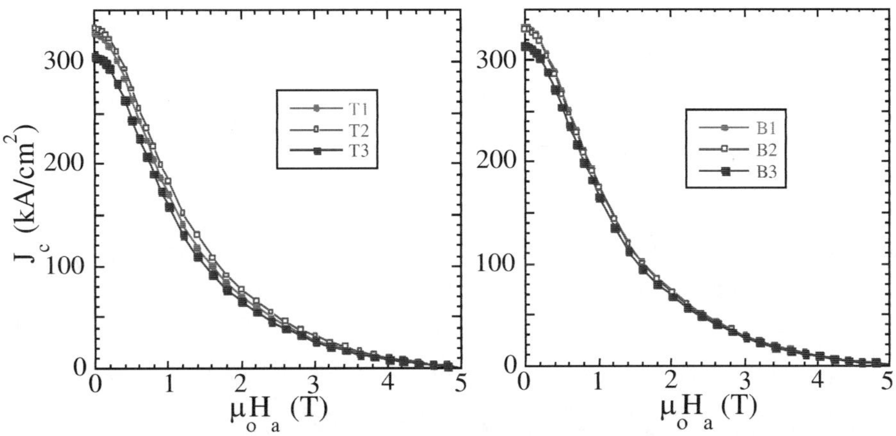

Figure 3. Field dependence of critical current density (T = 20 K) of the MgB_2 samples from different positions in top (left) and bottom (right) of the pellet.

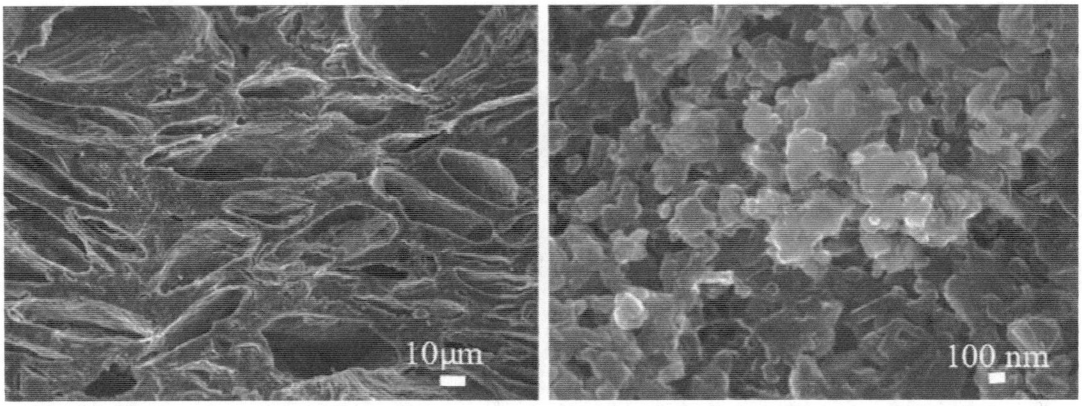

Figure 4. High and low magnification SEM images of the bulk MgB_2 sample produced utilizing the carbon-encapsulated B powder with 4.5 wt% carbon and sintered at 805 °C for 3h in argon atmosphere.

The microstructural features in the MgB_2 material were tested by scanning electron microscopy (SEM) as shown in Fig. 4. The SEM image taken at a lower magnification shows that the MgB_2 matrix contains numerous voids of various shapes and sizes (left). The porosity presents a serious problem in MgB_2 materials. To overcome this problem, numerous measures have been tested, like sintering time optimization, silver addition, use of carbon-based dopants etc. On the other hand, it is evident from the SEM micrograph (Fig. 4., right) that the size of the MgB_2 grains is very small, in nanometer size. One can see a rather dense structure of 200 to 300 nm grains on the high magnification image that are most probably responsible for the high J_c in this material. The smaller the grains, the more effectively they can act as strong flux pinning centres. These results again imply that high pinning performance of the sintered MgB_2 material can be enhanced by means of carbon-encapsulated boron along with optimum sintering temperature.

4. Conclusions

In the present work, a high quality superconducting bulk MgB_2 was produced utilizing the carbon encapsulated B powder with 4.5 wt% carbon and sintered at 805 °C for 3h in argon atmosphere. The effect of carbon encapsulated B powder distribution in the MgB_2 bulk on electromagnetic and superconducting properties was studied. Only a negligible variation was observed in the superconducting transition temperature measured in various spots at top and bottom parts of the pellet. The critical current density and irreversibility field values were found almost same at top and bottom of the large bulk MgB_2. The best J_c value at 20 K was around 330 kA/cm^2 in self-field and 200 kA/cm^2 at 1 T. The observed material homogeneity is by far better than in cuprate melt-textured bulks [24]. These results suggest that carbon coating of the boron used by fabrication of bulk MgB_2 is very promising for production of superconducting super-magnets for several industrial applications.

Acknowledgments
The paper was partly supported by Shibaura Institute of Technology (SIT) Research Centre for Green Innovation and Grant-in-Aid FD research budget code: 112261.

References

[1]. Vinod K, Abhilash Kumar R G, Syamaprasad U 2007 *Supercond. Sci. Technol.* **20** R1.
[2]. Perini E, Ginuchi G 2009 *Supercond. Sci. Technol.* **22** 045021.
[3]. Kajikawa K, Nakamura T 2009 *IEEE Trans. Appl. Supercond.* **19** 1669.
[4]. Gunchi G, Ripamonti G, Cavallin T, Bassani E 2014 *Cryogenics* **46** 23.
[5]. Muralidhar M, Inoue K, Koblischka MR, Tomita M, Murakami M 2014 *J. Alloys Compd.* **608** 102.
[6]. Noudem N J, Aburras M, Bernstein P, Chaud X, Muralidhar M, Murakami M 2014 *J. Appl. Phys.* **116** 163916.
[7]. Naito T, Sasaki T, Fujishiro H 2012 *Supercond. Sci. Technol.* **25** 2012 095012.
[8]. Durrell J H, Dancer C E J, Dennis A, Shi Y, Xu Z, Campbell A M, Hari Babu N, Todd R I, Grovenor C R M, Cardwell D A 2012 *Supercond. Sci. Technol.* **25** 112002.
[9]. Muralidhar M, Inoue K, Koblischka MR, Tomita M, Murakami M 2014 *J. Alloys Compd.* **608** 102.
[10]. Muralidhar M, Nozaki K, Kobayashi H, Zeng X L, Koblischka-Veneva A, Koblischka MR, Inoue K, Murakami M 2015 *J. Alloys Compd.* **649** 833.
[11]. Kumar D, Pennycook SJ, Narayan J, Wang H, Tiwari A 2003 *Supercond. Sci. Technol.* **16** 455.
[12]. Muralidhar M, Inoue K, Koblischka MR, Murakami M,. Murakami M 2015 *Adv. Eng. Mater.*, **17** 831.
[13]. Feng W J, Xia T D, Liu T Z, Zhao W J, Wei Z Q 2005 *Physica C* **425** 144.
[14]. Wozniak M, Juda K L, Hopkins S C, Gada D, Glowacki B A 2008 *Supercond. Sci. Technol.* **21** 035006.
[15]. Kim J H, Zhou S, Hossain M S A, Pan A V, Dou S X 2000 *Appl. Phys. Lett.,* **89** 142505.
[16]. Kim J H, Zhou S, Hossain M S A, Pan A V, Dou S X 2006 *Appl. Phys. Lett.* **89** 142505.
[17]. Gao Z S, Ma Y W, Zhang X P, Wang D L, Yu Z H, Yang H, Wen H H, Mossang E 2007 *J. Appl. Phys.* **102** 013914.
[18]. Vajpayee A, Jha R, Srivastava A K, Kishan H, Tropeano M, Ferdeghini C, Awana V P S 2011 *Supercond. Sci. Technol.* **24** 45013.
[19]. Muralidhar M, Higuchi M, Jirsa M, Diko P, Koka I, Murakami M 2017 *IEEE Trans. Appl. Supercond.* **27** 62021104 pp.4.
[20]. Kim J H, Oh S, Heo Y-U, Hata S, Kumakura H, Matsumoto A, Mitsuhara M, Choi S, Shimada, Maeda Y M, MacManus-Driscoll J L, Dou S X, 2012 NPG Asia Materials, **4** doi:10.1038/am.2012.3.
[21]. Barua S, Hossain Md S Al, Ma Z, Patel D, Mustapic M, Somer M, Aelcuk S, Kokal I, Morawski A, Cetner T, Gajda D, Dou S X 2015 *Scripta Materialia* **104** 37.
[22]. Chen D, Goldfarb R B 1989 *J. Appl. Phys.* **66** 2489.
[23]. Muralidhar M, Kenta N, Koblishka MR, Murakami M 2015 *Physica Status Solidi a*, **212**, 2141.
[24]. Jirsa M, Muralidhar M, Murakami M, Noto K, Nishizaki T, Kobayashi N, 2001 *Supercond. Sci. Technol* **14** 50.

Carrier-Doping Dependence of Critical Current Density in $Ba_{1-x}K_xFe_2As_2$ Single Crystals and Superconducting Wires

S Pyon, T Suwa, and T Tamegai

Department of Applied Physics, The University of Tokyo, Hongo, Bunkyo-ku, Tokyo113-8656, Japan

E-mail: pyon@ap.t.u-tokyo.ac.jp

Abstract. The carrier-doping dependence of J_c in $Ba_{1-x}K_xFe_2As_2$ single crystals and superconducting wires were investigated. $Ba_{1-x}K_xFe_2As_2$ single crystals ($0.24 < x < 0.4$) are synthesized by commonly used FeAs self-flux method, while $Ba_{1-x}K_xFe_2As_2$ wires are fabricated by powder-in-tube (PIT) method and hot isostatic pressing (HIP) technique, using polycrystalline $Ba_{1-x}K_xFe_2As_2$ ($0.25 < x < 0.4$) prepared by solid state reaction and silver and copper tubes. Their magnetic J_c was characterized by magnetization measurements. We found that the x-dependence of J_c in single crystals shows the peak in under-doped region around $x \sim 0.28$ similar to previous study. In the HIP wire, although significant peak is not observed, J_c shows a broad plateau in a wide doping region of $0.3 < x < 0.4$. These results indicate that J_c in the under-doped $Ba_{1-x}K_xFe_2As_2$ HIP wire is controlled by interglanular J_c.

1. Introduction

Iron-based superconductors (IBSs) are promising candidates for high-field applications of superconductors because they exhibit high critical temperature T_c, large upper critical field H_{c2}, and relatively low anisotropy compared with cuprate superconductors. Among them, $A_{1-x}K_xFe_2As_2$ (A = Ba, Sr) compounds (122-type) are considered to be most attractive candidates. They have small anisotropies of 2-3, moderate T_c's, large H_{c2}'s, and large J_c's [1-2]. Actually high critical current density is realized in superconducting wires or tapes using $A_{1-x}K_xFe_2As_2$ [3-10]. In these report, K content x in the staring materials for superconducting wires and tapes are fixed to 0.4, where T_c is optimized [11,12]. Recently, however, Song *et al.* reported distinct doping dependence of J_c and T_c in $Ba_{1-x}K_xFe_2As_2$ single crystals, which were synthesized by KAs self-flux method [13]. The x-dependence of J_c is characterized by a peak at $x = 0.3$, which corresponds to the under-doped region. The origin of the characteristic doping dependence of J_c may be connected with the pinning forces related to the orthorhombic phase domain boundary or to the chemical inhomogeneity introduced by the dopant substitutions [13]. This result implies that higher J_c may also be achieved not at the optimal-doping level of $x \sim 0.4$ for T_c but at under-doped region in 122-type superconducting wires and tapes. However, to our knowledge, there have been no systematic studies on the x-dependence of J_c in $Ba_{1-x}K_xFe_2As_2$ wires and tapes. Furthermore, we were also interested in the investigation of x-dependence of J_c in $Ba_{1-x}K_xFe_2As_2$ single crystals synthesized by commonly used FeAs self-flux method, since the process of crystal growth should affect chemical homogeneities which generate pinning forces.

In this work, we focused on carrier-doping dependence of J_c in $Ba_{1-x}K_xFe_2As_2$ single crystals and superconducting wires. Single crystals are synthesized by FeAs flux method. And superconducting wires were fabricated by powder-in-tube (PIT) method using well characterized $Ba_{1-x}K_xFe_2As_2$

polycrystalline powders. J_c and T_c in prepared samples were evaluated from magnetization measurements.

2. Experiment

Single crystals of (Ba,K)Fe_2As_2 were synthesized by self-flux method using FeAs flux. We used Ba pieces, K ingots and FeAs powder as starting materials. FeAs was prepared by placing stoichiometric amounts of As pieces and Fe powder in an evacuated quartz tube and reacting them at 700°C for 40 h after heating them at 500°C for 10 h. A mixture with a ratio of Ba:K:FeAs = 1-x:1.1x:4 was placed in an alumina crucible and sealed in a stainless steel containers [14] in a nitrogen-filled glove box. In order to compensate the loss of elements, 10% excess K was added. They were heated for 1 h at 1200~1250°C followed by cooling down to 900°C at a rate of 5°C/h. Obtained single crystals were separated from flux and cut into rectangular shape as shown in fig.1(a). In order to characterize the compositions of pieces of single crystals, a scanning electron microscope equipped with an energy-dispersive x-ray spectroscopy (SEM-EDX) was used (S-4300, Hitachi High-Technologies equipped with EMAX x-act, HORIBA). Bulk magnetization was measured by a superconducting quantum interference device (SQUID) magnetometer (MPMS-5XL, Quantum Design).

(Ba,K)Fe_2As_2 superconducting wires were fabricated by ex-situ PIT method. Polycrystalline powders of $Ba_{1-x}K_xFe_2As_2$ were prepared by the solid-state reaction. Ba pieces, K ingots, Fe powder, and As pieces were used as starting materials. In order to compensate the loss of elements, the starting mixture contained 15% excess K and 5% excess As. They were mixed in a nitrogen atmosphere more than 10 hour using a ball-milling machine and densely packed into a niobium tube. The niobium tube was then put into a stainless steel tube and sealed in nitrogen filled glove box for heat treatment at 900°C for 30 h. The surface of the obtained polycrystalline ingot was polished to remove small impurities. It was then ground into powder using an agate mortar in nitrogen filled glove box. The phase identification of the sample was carried out by means of powder x-ray diffraction (XRD) with Cu-Kα radiation (Smartlab, Rigaku). Ground powders were filled in a silver tube with an outer diameter of 4.5 mm and an inner diameter 3 mm, then cold drawn into a square shape with a diagonal dimension of ~1.2 mm. After cutting it into short pieces, one of the pieces was put into 1/8 inch copper tube and redrawn down to the diagonal dimension of 1.2 mm. After that, both ends of the wires were sealed by using an arc furnace. The sealed wires were sintered using the hot isostatic pressing (HIP) technique. Wires and tapes were heated for 0.5 h at 700°C in argon atmosphere under the pressure of 9 MPa. An optical micrograph of the cross section of the HIP wire is shown in fig. 1(b). Bulk magnetization was measured by the same method and intergranular J_c was evaluated using the extended Bean model.

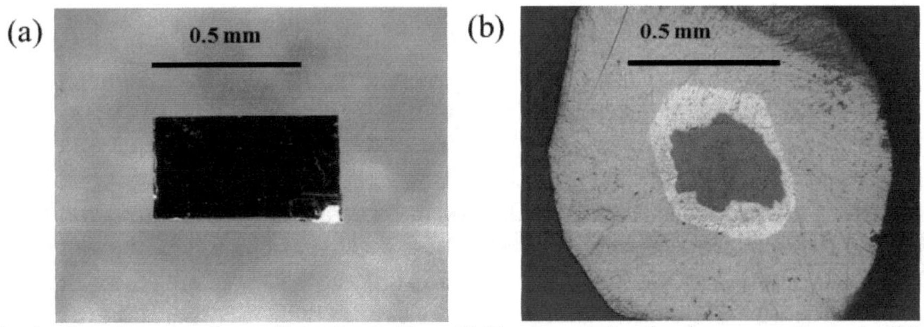

Figure 1. Optical micrographs of typical $Ba_{1-x}K_xFe_2As_2$ (a) single crystal and (b) HIP wire, respectively.

3. Results and Discussion

Before the fabrication of wires, polycrystalline powders of $Ba_{1-x}K_xFe_2As_2$ were characterized by magnetization and powder x-ray diffraction measurements. Fig. 2(a) shows the temperature dependence of magnetization of several $Ba_{1-x}K_xFe_2As_2$ polycrystalline powders with x from 0.25 to 0.40. All the samples show bulk superconductivity. While diamagnetism starts at ~38 K in all the sample, the magnetization curves of under-doped samples show broad drop, especially in $x \sim 0.25$. These broadenings indicate the distribution of carrier concentration in polycrystalline powders. Considering the superconducting dome in $Ba_{1-x}K_xFe_2As_2$ phase diagram [11], broadening of the superconducting transition caused by the distribution of carrier contents should become wider at under-doped region. It should be noted that no impurity phases are detected in the x-ray diffraction pattern of these powders. As shown in fig. 2(b), (0 0 2) powder x-ray diffraction peaks of several $Ba_{1-x}K_xFe_2As_2$ powders shows the systematic shift. These results suggest that obtained polycrystalline powders are single phases, respectively, and carrier content of x is systematically controlled. Using these powders, HIP wires were fabricated.

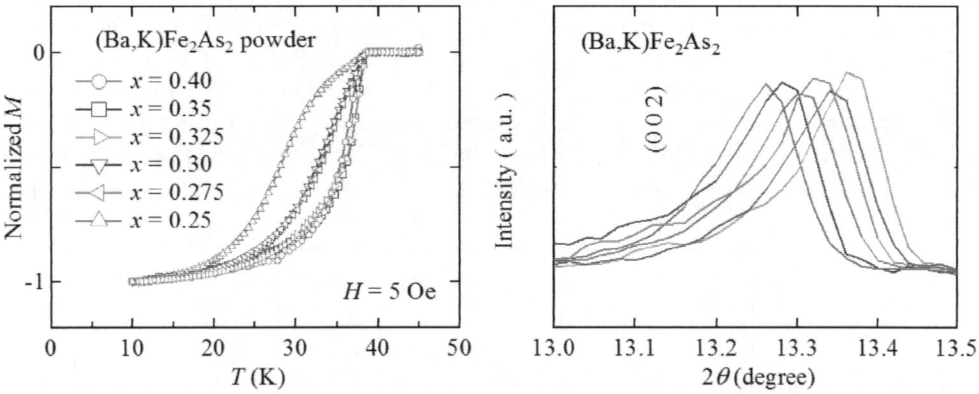

Figure 2. (a) Temperature dependence of magnetization of several $Ba_{1-x}K_xFe_2As_2$ polycrystalline powders. (b) Normalized (0 0 2) powder x-ray diffraction peak of several $Ba_{1-x}K_xFe_2As_2$ powders.

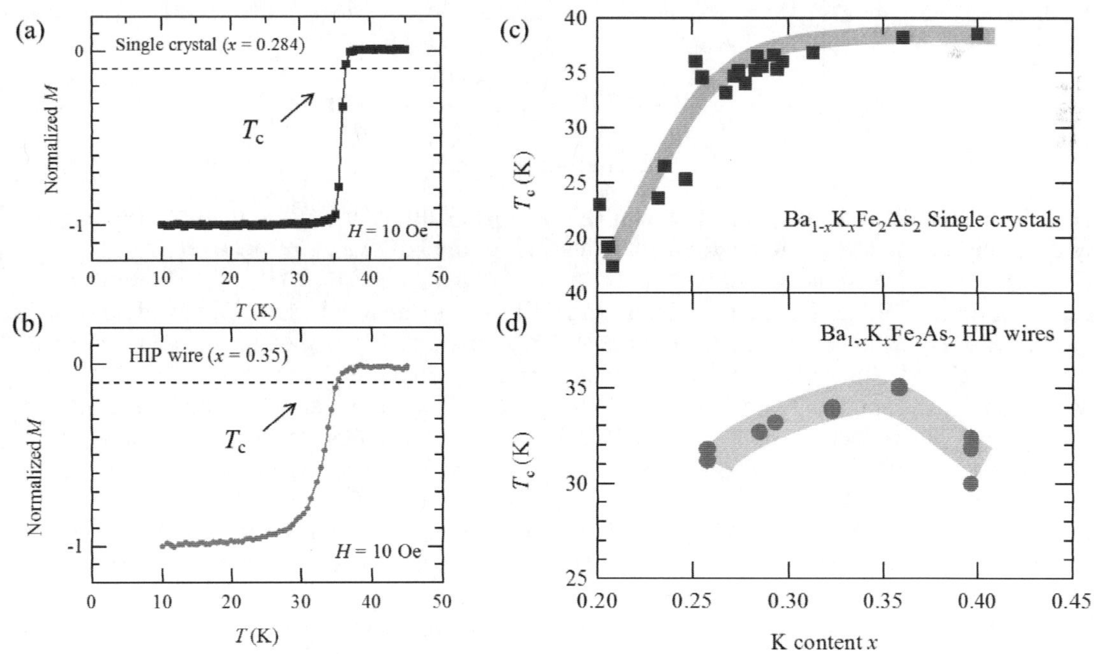

Figure 3. Typical temperature dependence of magnetizations of several $Ba_{1-x}K_xFe_2As_2$ (a) single crystal and (b) HIP wire, respectively. K content dependence of T_c in several $Ba_{1-x}K_xFe_2As_2$ (c) single crystals and (d) HIP wires, respectively. T_c is defined at the temperature where normalized magnetization reaches at -0.1 as shown in (a) and (b).

To evaluate the carrier-concentration dependence of superconducting properties, such as T_c and J_c, K content of samples should be estimated. From SEM-EDX analyses, the chemical compositions in single crystal and the core of HIP wires were evaluated. In the case of HIP wire, nominal and analyzed carrier contents are almost identical. In the case of single crystals, analyzed K content x is quite different from the nominal value in some pieces. For example, analyzed carrier content in single crystals with nominal carrier content $x \sim 0.28$ are distributed in the range of $0.20 < x < 0.30$. So we perform EDX analysis for each single crystal before evaluating its T_c and J_c.

Next, we evaluate T_c in both single crystals and HIP wires from temperature dependence of magnetization. Typical temperature dependence of magnetization are shown in fig. 3 (a) and (b). Similar to magnetization of polycrystalline powders in fig.2, drop of magnetization tends to be broad in under-doped samples. In this study, T_c is defined at the temperature where normalized magnetization reaches at -0.1 as shown in fig. 3 (a) and (b). Evaluated T_c of single crystals and HIP wires are summarized in fig. 3(c) and (d), respectively. T_c in single crystals is the highest at $x \sim 0.4$. This is consistent with the previous reports. T_c is reduced with decreasing x moderately above $x \sim 0.3$, and rapidly below $x \sim 0.3$. In the HIP wires, x-dependence of T_c shows a broad peak.

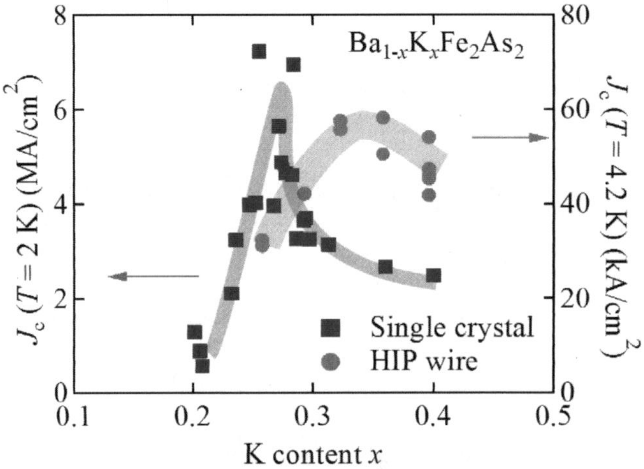

Figure 4. K content dependence of J_c in several $Ba_{1-x}K_xFe_2As_2$ single crystals and HIP wires. J_c in self field is evaluated at 2 K and 4.2 K for single crystals and for the HIP wires, respectively.

Magnetic J_c in all of these single crystals and HIP wires are evaluated. Obtained results are summarized in fig.4. J_c in self field is evaluated at 2 K and 4.2 K for single crystals and for the HIP wires, respectively. Maximum of J_c in single crystals reaches around 5-7 MA/cm^2 at $x \sim 0.28$. This is relatively high value compared with the J_c at $x \sim 0.4$ where T_c is optimized as shown in fig. 3. In our sample grown by FeAs flux method, J_c takes its maximum in the under-doped region. This result is roughly consistent with previous report by Song *et al.*, where KAs flux was used [13], although the peak position is not $x \sim 0.30$. This difference may be caused by the difference in the starting materials for synthesis. In the case of x dependence of J_c in the HIP wire, significant peak is not observed at $x \sim 0.28$. However, J_c shows a broad plateau in a wide doping region of $0.3 < x < 0.4$. J_c is reduced only below $x \sim 0.30$. We can safely conclude that the shift of maximum of J_c from optimal region to under-doped region is the common feature in 122-type single crystals and wires.

In fig. 5, K content dependences of both T_c and J_c in several $Ba_{1-x}K_xFe_2As_2$ HIP wires are shown. Both dependences show broad peaks around $x \sim 0.35$. In contrast to the doping dependence of T_c and J_c in single crystals, maximum values are realized in almost the same doping level, although the peak is shifted to under-doped region. This fact suggests that intergranular J_c is dominant compared with intragranular J_c in the core of the HIP wire. Furthermore, sintering conditions for wires and tapes using

IOP Conf. Series: Journal of Physics: Conf. Series **871** (2017) 012060 doi:10.1088/1742-6596/871/1/012060

under-doped polycrystalline powders have not been well studied yet. In this sense, J_c in 122-type superconducting wires and tapes may further be enhanced by controlling the carrier content and the optimization of sintering process. Obtained results will expand the doping level of starting materials for IBS high-J_c superconducting wires and tapes.

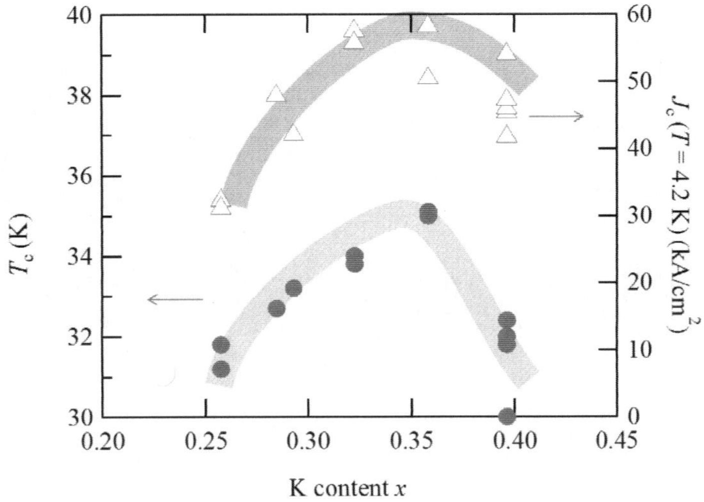

Figure 5. K content dependence of both T_c and J_c in several $Ba_{1-x}K_xFe_2As_2$ HIP wires.

4. Summary

The K content dependence of J_c in $Ba_{1-x}K_xFe_2As_2$ single crystals and HIP wires were investigated. $Ba_{1-x}K_xFe_2As_2$ single crystals are synthesized by FeAs flux method. $Ba_{1-x}K_xFe_2As_2$ HIP wires are fabricated using polycrystalline $Ba_{1-x}K_xFe_2As_2$. We found that the x-dependence of magnetic J_c in our single crystals shows the peak in under-doped region around $x \sim 0.28$. In the HIP wire, although significant peak is not observed, J_c shows a broad plateau in a wide doping region of $0.3 < x < 0.4$. These results indicate that J_c in the under-doped $Ba_{1-x}K_xFe_2As_2$ HIP wire is controlled by intergranular J_c.

Acknowledgments

This work was partially supported by a Grant-in-Aid for Young Scientists (B) (No.16K17745), the Japan-China Bilateral Joint Research Project by the Japan Society for the Promotion of Science (JSPS).

References

[1] Ma Y 2012 *Supercond. Sci. Technol.* **25** 113001

[2] Pallecchi I, Eisterer M. Malagoli A and Putti M 2015 *Supercond. Sci. Technol.* **28** 114005

[3] Weiss J D, Tarantini C, Jiang J, Kametani F, Polyanskii A A, Larbalestier D C and Hellstrom E E 2012 *Nat. Mat.* **11** 682

[4] Hecher J, Baumgartner T, Weiss J D, Tarantini C, Yamamoto A, Jiang J, Hellstrom E E, Larbalestier D C and Eisterer M 2016 *Supercond. Sci. Technol.* **29** 025004

[5] Pyon S, Suwa T, Park A, Kajitani H, Koizumi N, Tsuchiya Y, Awaji S, Watanabe K and Tamegai T 2016 *Supercond. Sci. Technol.* **29** 115002

[6] Pyon S, Tsuchiya Y, Inoue H, Kajitani H, Koizumi N, Awaji S, Watanabe K and Tamegai T 2014 *Supercond. Sci. Technol.* **27** 095002

[7] Gao Z, Togano K, Matsumoto A and Kumakura H 2014 *Sci. Rep.* **4** 4065

[8] Lin H, Yao C, Zhang X, Dong C, Zhang H, Wang D, Zhang Q, Ma Y, Awaji S, Watanabe K *et al.* 2014 *Sci. Rep.* **4** 6944
[9] Pyon S, Yamasaki Y, Kajitani H, Koizumi N, Tsuchiya Y, Awaji S, Watanabe K and Tamegai T 2015 *Supercond. Sci. Technol.* **28** 125014
[10] Pyon S, Taen T, Ohtake F, Tsuchiya Y, Inoue H, Akiyama H, Kajitani H, Koizumi N, Okayasu S and Tamegai T 2013 *Appl. Phys. Express* **6** 123101
[11] Rotter M, Tegel M and Johrendt D 2008 *Phys. Rev. Lett.* **101** 107006
[12] Sasmal K, Lv B, Lorenz B, Guloy A M, Chen F, Xue Y-Y and Chu C-W 2008 *Phys. Rev. Lett.* **101** 107007
[13] Song D, Ishida S, Iyo A, Nakajima M, Shimoyama J, Eisterer M and Eisaki H 2016 *Sci. Rep.* **6** 26671
[14] Kihou K, Saito T, Ishida S, Nakajima M, Tomioka Y, Fukazawa H, Kohori Y, Ito T, Uchida S, Iyo A *et al.* 2010 *J. Phys. Soc. Jpn.* **79** 124713

IOP Conf. Series: Journal of Physics: Conf. Series **871** (2017) 012061 doi:10.1088/1742-6596/871/1/012061

Development of Fe-based superconducting wires for liquid-hydrogen level sensors

S Ishida, Y Tsuchiya, Y Mawatari, H Eisaki, A Nakano and Y Yoshida

National Institute of Advanced Industrial Science and Technology (AIST), Tsukuba, Ibaraki 305-8568, Japan

E-mail: s.ishida@aist.go.jp

Abstract. We developed liquid-hydrogen (LH_2) level sensors with $Ba(Fe_{1-x}Co_x)_2As_2$ superconducting wires (Co-Ba122 wires) as their detection elements. We fabricated Co-Ba122 wires with different Co concentrations x by using the powder-in-tube method. The superconducting transition temperatures of the wires were successfully controlled in the range of 20–25 K by changing x from 0.06 to 0.10. The resistance–temperature curves of the wires exhibited sharp superconducting transitions with widths of 0.5–1.0 K. In addition, we performed an operation test of the Co-Ba122 level sensors with LH_2. Close correspondence between the output resistance and the actual LH_2 level was observed for a sensor equipped with $x = 0.09$ wire, demonstrating that this sensor can accurately measure LH_2 levels.

1. Introduction

Hydrogen is a potential source of clean energy whose liquid form is advantageous for transportation and storage due to its high density. To ensure the safety of these processes, accurate monitoring of the liquid hydrogen (LH_2) levels in tanks is required. Superconducting wires can be used as the detection elements of LH_2 level sensors because their output resistances are sensitive to the LH_2 level. For superconducting wires composed of a given material to be usable in LH_2 level sensors, the superconducting transition temperature (T_c) must be higher than the boiling point of LH_2 ($T_{LH2} \sim 20.3$ K at atmospheric pressure). In the LH_2 level sensors developed so far, MgB_2 superconducting wires with $T_c = 30$–35 K have been used as the detection elements [1–3]. However, those LH_2 level sensors require heating systems to improve their level detection accuracies because the value of T_c for MgB_2 is rather high compared with T_{LH2}; thus, MgB_2 wires become superconducting even at temperatures at which hydrogen is in its gaseous form (GH_2). Accordingly, superconducting materials with T_c values close to T_{LH2} are considered to be good candidates for the materials of the wires used as the detection elements of LH_2 level sensors.

Iron-based superconductors [4] have attracted significant attention due to their high T_c values and high upper critical fields, which are suitable for high-field applications. To evaluate the practical application potential of such superconductors, iron-based superconducting wires and tapes have been developed to achieve high critical currents in high magnetic fields [5–7]. Iron-based superconductors are also advantageous for use in sensor applications due to their wide variety of T_c values. Here, we focus on an iron-based superconductor, $Ba(Fe_{1-x}Co_x)_2As_2$ (Co-Ba122) [8], as a candidate material for the detection elements of LH_2 level sensors. Co-Ba122 possesses a maximum T_c of 25 K, and T_c can be tuned by varying the Co concentration x [9,10]. Since T_c can be adjusted until it is close to T_{LH2}, it is interesting to examine the feasibility of using Co-Ba122 wires as the detection elements of LH_2 level sensors.

In this study, we fabricated Co-Ba122 superconducting wires with different x by using the powder-in-tube method. We successfully controlled the T_c values of the wires in the range of 20–25 K by varying x from 0.06 to 0.10. The superconducting transitions in the resistance–temperature curves were quite sharp, with widths of 0.5–1.0 K. We then developed LH_2 level sensors based on the Co-Ba122 wires and performed an operation test with LH_2. The output resistances and actual LH_2 levels corresponded closely with one another, demonstrating that the level measurements were highly accurate [11].

2. Experimental

2.1. Fabrication of $Ba(Fe_{1-x}Co_x)_2As_2$ superconducting wires

Polycrystalline powders of Co-Ba122 ($x = 0.06$–0.10) were synthesized via a solid-state reaction. Before synthesizing the polycrystalline samples, BaAs, Fe_2As, and Co_2As were prepared as precursors. The starting materials (Ba (3N, chunks), Fe (4N, 150 μm mesh), Co (4N, 150 μm mesh), and As (6N, 1-5 mm grains)) were weighed and heated for 10 h at 650 °C (BaAs) and 850 °C (Fe_2As and Co_2As). The precursors were weighed and mixed with a ratio of BaAs : Fe_2As : Co_2As = 1 : 1 - x : x. The mixture was sealed in a quartz tube and heated at 900 ℃ for 48 h. All of the materials were handled in a nitrogen-filled glove box.

The Co-Ba122 superconducting wires were fabricated by using the *ex-situ* powder-in-tube method. The polycrystalline bulks were ground into powders with an agate mortar in a glove box. The powders were filled into a silver tube with outer and inner diameters of 6 mm and 4.3 mm, respectively. The silver tube was groove-rolled into a square wire with a $1.2 \times 1.2\ mm^2$ cross-section. The wires were cut into 4-cm- and 20-cm-long pieces, sealed into a quartz tube, and heated at 880 °C for 20 h. An example scanning electron microscope (SEM, Hitachi High-Technologies TM3000) image of the transverse cross-section is presented in Figure 1(a). The resistances of the short wires were measured by using a four-probe method, and the contacts were made of solder.

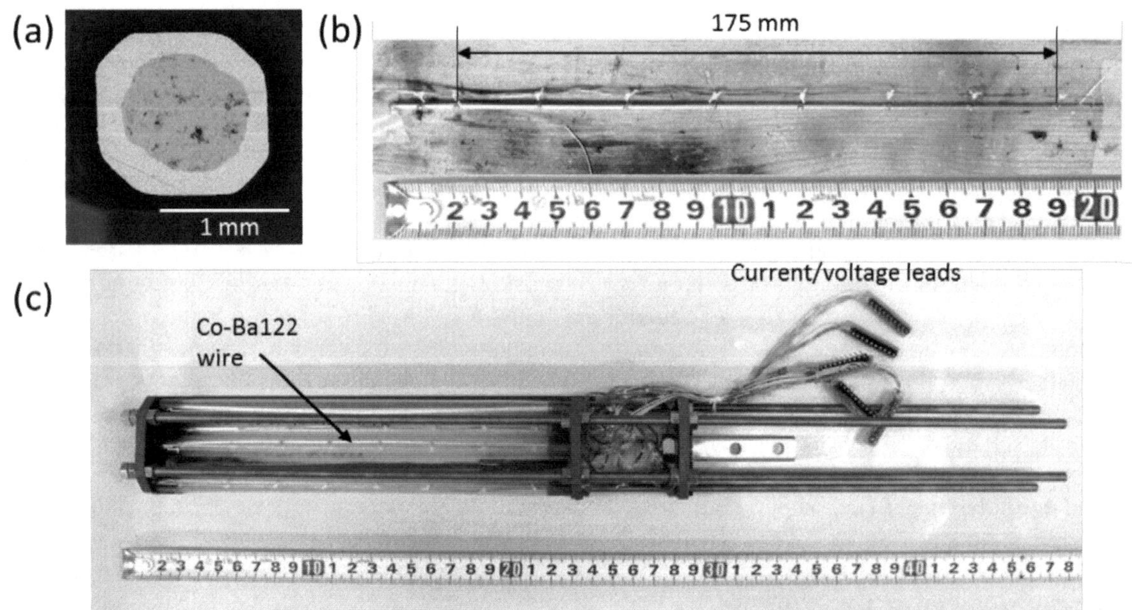

Figure 1. (a) SEM image of the transverse cross-section of a Co-Ba122 wire. The gray and silver regions correspond to the superconducting core and silver sheath, respectively. Photographs of (b) a 20-cm-long Co-Ba122 wire and (c) an assembled level sensor.

2.2. Development of the liquid hydrogen level sensors and operation test

Figures 1(b) and 1(c) present photographs of a 20-cm-long Co-Ba122 wire used for the LH_2 level sensor and an assembled level sensor, respectively. The contacts were formed by soldering and were separated by 175 mm. The wires were placed into straws, which were used as protecting tubes.

Figure 2(a) shows a photograph of the setup used in the operation test. The LH_2 level sensor was installed in the cryostat with observation slits. The LH_2 was prepared using the hydrogen liquefier [12]. It was transferred from the liquefier to the cryostat through a transfer tube. Figure 2(b) is a schematic of the experimental setup, which consisted of a Co-Ba122 level sensor, current source (denoted as I in Figure 2(b)), voltmeter (V), and personal computer (PC). The output resistance of each sensor was measured by using a four-probe method. The actual LH_2 level was checked by reading a scale attached to the level sensor. The output resistance and actual LH_2 level were recorded every 5 min during the natural evaporation of the LH_2.

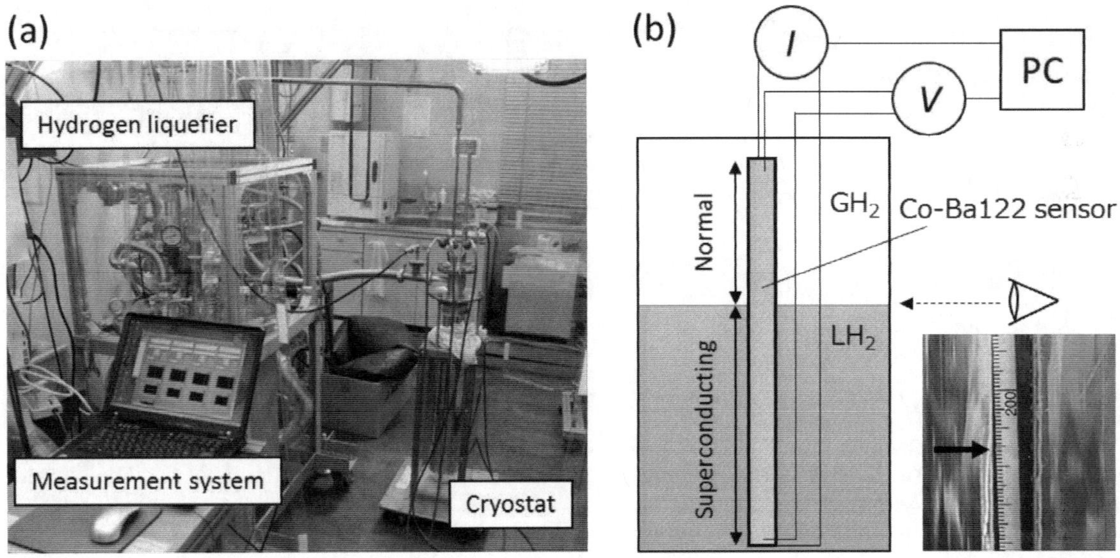

Figure 2. (a) Photograph and (b) schematic of the experimental setup used in the operation test.

3. Results and discussion

3.1. Characterization of $Ba(Fe_{1-x}Co_x)_2As_2$ superconducting wires

Figure 3(a) depicts the temperature (T) dependences of the resistances (R) of 4-cm-long Co-Ba122 wires with $x = 0.06$, 0.07, 0.08, 0.09, and 0.10. A relatively small current of 10 mA was applied to perform the measurements. For each of the wires, a large drop in R corresponding to the superconducting transition is observable. This large drop is necessary to obtain a high signal-to-noise ratio without applying a large current. The normal-state R of each wire exhibits a weak T dependence characterized by a ratio of $R(300\text{ K})/R(30\text{ K}) = 1.3$–$1.5$. The weak T dependence is advantageous because it results in the output R of the wires not being affected significantly by the gradient or variations of T in GH_2. Figure 3(b) shows a magnified section of the R–T curves around T_c. The superconducting transitions of the wires are very sharp with the widths ΔT_c of 0.5–1.0 K, except in the case of $x = 0.06$ (ΔT_c was determined by the difference between the values of T corresponding to R values equal to 5% and 95% of the normal-state R). A sharp transition is also essential to achieve high-resolution liquid level detection. The broad transition exhibited by the $x = 0.06$ wire is a result of the existence of the antiferromagnetic–orthorhombic phase on the underdoping side of the superconducting dome, where slight inhomogeneities in the Co atom distribution cause significant spatial variation of T_c. Figure 3(b) confirms that T_c varies systematically with x for the Co-Ba122 wires; the T_c values determined based on the values of T at which $R = 0$ mΩ were found to be 16.3 K, 24.1 K, 23.4 K, 21.8 K, and 20.1 K for x = 0.06, 0.07, 0.08, 0.09, and 0.10, respectively. The x dependence of T_c is summarized in the inset of Figure 3(b). These results illustrate that T_c can be controlled in Co-Ba122 by adjusting x. The x = 0.06 and 0.10 wires with T_c lower than T_{LH2} are not available because $R > 0$ at $T = T_{LH2}$. Among the investigated wires, the x = 0.09 wire, for which T_c is 1.5 K higher than T_{LH2}, is expected to be most suitable for use in LH_2 level sensors.

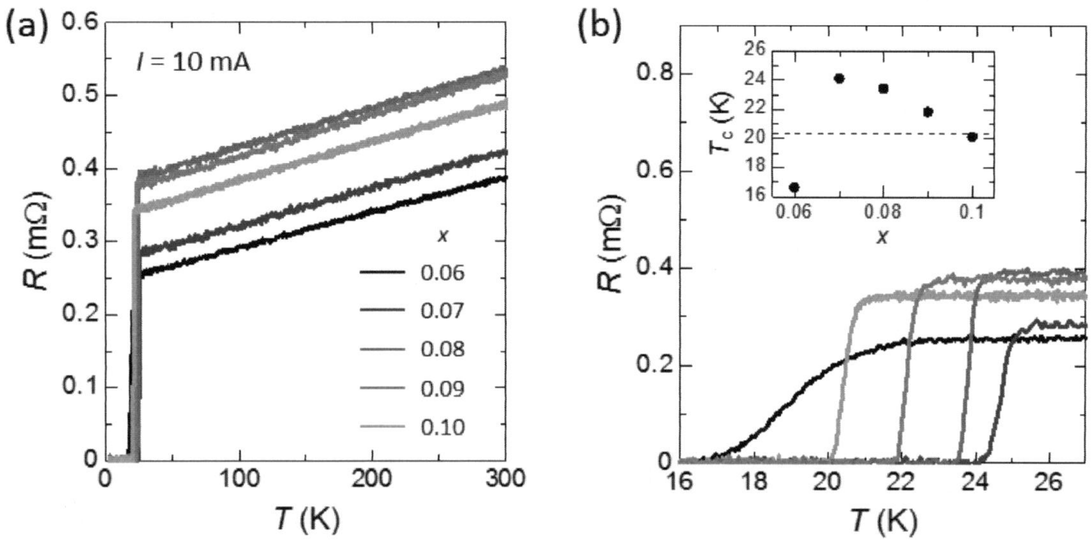

Figure 3. (a) T dependence of R for Co-Ba122 wires with x = 0.06, 0.07, 0.08, 0.09, and 0.10. (b) Magnified section of the R–T curves around T_c. The inset shows the x dependence of T_c. The dashed line indicates T_{LH2} (= 20.3 K).

3.2. Operation test of level sensors with liquid hydrogen

We assembled level sensors using the Co-Ba122 wires discussed in Section 3.1 and performed an operation test with LH_2. The sensors equipped with x = 0.07 and 0.09 wires (T_c = 24.1 K and 21.8 K) were labelled as Sensors A and B, respectively. Figure 4(a) shows the time (t) dependences of the LH_2 level (blue) and output R of Sensors A (black) and B (red). For an LH_2 level of 180 mm (at which point we defined t to be 0 min), which was 5 mm above the upper voltage contacts of the sensors, both sensors exhibit $R = 0$ mΩ. Sensor A shows $R = 0$ mΩ until $t \sim 30$ min, when the LH_2 level reaches 130 mm, i.e., 45 mm below the upper voltage contact. This finding indicates that the upper part of Sensor A, which was already in GH_2, remained superconducting due to the high T_c of the wire used in Sensor A. For $t >$ 30 min, R is finite and increases with t in response to the decrease of the LH_2 level. However, R continues to increase even after the LH_2 level is below the lower voltage contact. These results indicate that LH_2 levels determined based on the output R of Sensor A will be overestimates. Thus, as has been reported previously [1–3], Sensor A requires a heating system to be capable of detecting LH_2 levels accurately.

On the other hand, Sensor B, for which the T_c of the wire is closer to T_{LH2} than that of the wire in Sensor A, exhibits a finite R at $t \sim 5$ min, which corresponds to an LH_2 level of around 170 mm, i.e., 5 mm below the upper voltage contact. This finding indicates that Sensor B responds to LH_2 level decreases more accurately than Sensor A does. Furthermore, R is almost constant for Sensor B after the LH_2 level effectively reaches 0 mm. Based on the t dependences of the LH_2 level and R, we plotted the LH_2 level against R for Sensor B; the resulting graph is presented in Figure 4(b). The two quantities exhibit a strong linear correlation described by LH_2 level (mm) = 169.7 - 1.674R (mΩ) with a linear

IOP Conf. Series: Journal of Physics: Conf. Series **871** (2017) 012061 doi:10.1088/1742-6596/871/1/012061

correlation coefficient $|r|^2$ of greater than 0.99. Although Sensor B slightly overestimates the LH_2 level, its measurements are reasonably accurate, and it does not require a heating system. In addition, more accurate measurements could be achieved if T_c were made closer to T_{LH2} by refining x.

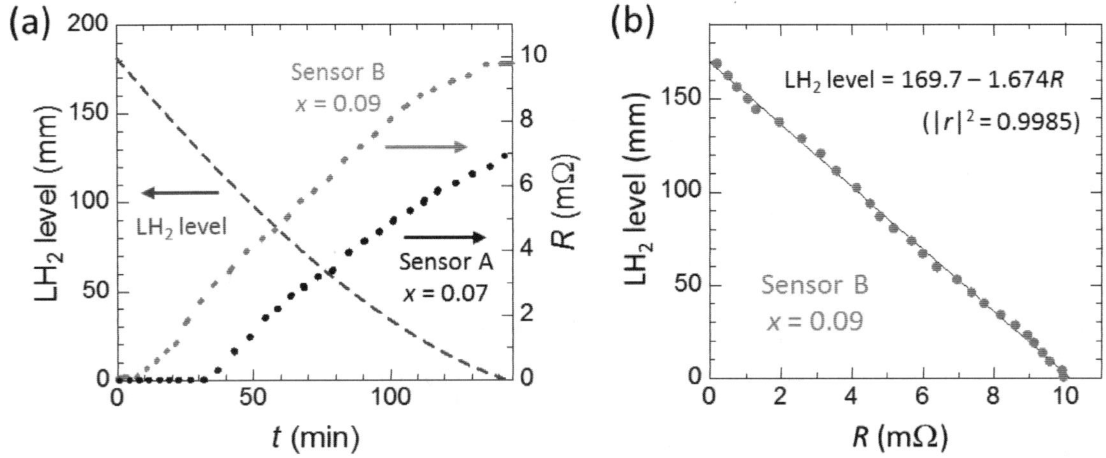

Figure 4. (a) t dependences of the LH_2 level (blue) and output R of Sensor A (x = 0.07 wire, black) and Sensor B (x = 0.09 wire, red). (b) Relationship between R of Sensor B and the LH_2 levels derived from (a). The black solid line indicates the linear fit determined using the least-squares method. The fitting function is also provided in the panel.

4. Summary

We developed LH_2 level sensors with Co-Ba122 superconducting wires as their detection elements. The T_c values of the wires were successfully controlled in the range of 20–25 K by varying x. In addition, the Co-Ba122 wires exhibited sharp superconducting transitions, and their normal-state R values showed weak T dependences; both of these characteristics make these wires advantageous for use in level sensors. In addition, we performed an operation test of the Co-Ba122 level sensors with LH_2. We demonstrated that the Co-Ba122 level sensor whose wires had T_c close to T_{LH2} yielded reasonably accurate level measurements without requiring a heating system. The accuracy will be further improved by fine-tuning T_c.

References

[1] Haberstroh C, Dehn G and Kirsten G 2007 *Proc. ICEC21-ICMC2006* 357
[2] Kajikawa K, Tomachi K, Tanaka K, Funaki K, Kamiya T, Okada M and Kumakura H 2009 *Proc. ICEC22-ICMC2008* 425
[3] Takeda M, Matsuno Y, Kodama I, Kumakura H and Kazama C 2009 *IEEE Trans. Appl. Supercond.* **19** 764
[4] Kamihara Y, Watanabe T, Hirano M and Hosono H 2008 *J. Am. Chem. Soc.* **130** 3269
[5] Ma Y 2012 *Supercond. Sci. Technol.* **25** 113001
[6] Hosono H, Tanabe K, Takayama-Muromachi E, Kageyama H, Yamanaka S, Kumakura H, Nohara M, Hiramatsu H and Fujitsu S 2015 *Sci. Technol. Adv. Mater.* **16** 033503
[7] Pallecchi I, Eisterer M, Malagoli A and Putti M 2015 *Supercond. Sci. Technol.* **28** 114005
[8] Sefat A S, Jin R, McGuire M A, Sales B C, Singh D J and Mandrus D 2008 *Phys. Rev. Lett.* **101** 117004
[9] Chu J-H, Analytis J G, Kucharczyk C and Fisher I R 2009 *Phys. Rev. B* **79** 014506
[10] Ishida S, Nakajima M, Liang T, Kihou K, Lee C H, Iyo A, Eisaki H, Kakeshita T, Tomioka Y, Ito T and Uchida S 2013 *J. Am. Chem. Soc.* **135** 3158
[11] Ishida S, Tsuchiya Y, Mawatari Y, Nakano A, Eisaki H and Yoshida Y 2016 JP-A-2016-081394

IOP Conf. Series: Journal of Physics: Conf. Series **871** (2017) 012061 doi:10.1088/1742-6596/871/1/012061

[12] Nakano A, Maeda T, Ito H, Masuda M, Kawakami Y, Kato A, Tange M, Takahashi T and Matsuo M 2010 *Intt. J. Hydrogen Energy* **35** 9088

IOP Conf. Series: Journal of Physics: Conf. Series **871** (2017) 012062 doi:10.1088/1742-6596/871/1/012062

Large and field-insensitive critical current densities in $(Sr,Na)Fe_2As_2$ superconducting tapes

T Suwa[1], S Pyon[1], A Park[1], T Tamegai[1], Y Tsuchiya[2], S Awaji[2], and K Watanabe[2]

1. Department of Applied Physics, The University of Tokyo, Hongo, Bunkyo-ku, Tokyo 113-8656, Japan
2. High Field Laboratory for Superconducting Materials, Institute for Materials Research, Tohoku University, Sendai 980-8577, Japan

Email : tamegai@ap.t.u-tokyo.ac.jp

Abstract. We have fabricated $(Sr,Na)Fe_2As_2$ tapes by optimizing the sintering temperature and time, and characterized their transport J_c at 4.2 K up to 14 T. The transport J_c for the $(Sr,Na)Fe_2As_2$ tape sintered at 875°C reaches 4.9 x 10^4 A/cm² under self-field, and it is suppressed only down to 1.9 x 10^4 A/cm² even at 14 T. We succeeded in fabricating very field-insensitive tapes, which are very promising for practical applications. We also report compositional distributions and magneto-optical images of current distribution in tapes. Magneto-optical imaging of the core of the tape clarified inhomogeneous intergranular current distributions.

1. Introduction

The iron-based superconductor was first discovered by the Hosono group in 2008 [1]. Soon after the discovery, several types of high temperature superconductivity in iron pnictides were discovered. Among them, superconducting wires and tapes have been fabricated mainly by using $(AE,K)Fe_2As_2$ materials (AE = Ba or Sr), so called '122-type' [2-6], because of their high upper critical field $H_{c2}(0)$ [7, 8] and low anisotropy $\gamma(\sim 2)$ [8, 9], as well as their high transition temperature T_c (~36-38 K) [10, 11]. At present, the transport J_c for these tapes has reached 10^5 A/cm^2 [12, 13], which is the level for practical applications, at 4.2 K under high magnetic fields.

In addition to these K-doped materials, there are analogous Na-doped ones: $(AE',Na)Fe_2As_2$ (AE' = Ba, Sr, or Ca). Among them, polycrystalline samples of $(Sr_{1-x}Na_x)Fe_2As_2$ with the maximum T_c of 36.5 K for x = 0.55 [14], and large transport J_c at 20 K in a $(Sr,Na)Fe_2As_2$ tape were reported [15]. Although these Na-doped materials are very promising, superconducting wires and tapes using Na-doped ones have not been studied extensively.

In this paper, we report the relationship between transport J_c of $(Sr,Na)Fe_2As_2$ tape and sintering temperature. The value of transport J_c was the largest when the tape was sintered at 875°C, although the core of tape was reacted with silver sheath. In the tape which was sintered at 850°C for 2.5 h, the value of J_c was almost as high as that of the tape which was sintered at 875°C, and the surface of silver sheath has not been reacted with the core. We performed a scanning electron microscope observations, compositional analyses with an energy dispersive x-ray spectroscopy (SEM-EDX), and magneto-optical (MO) observations for the tape sintered at 850°C.

2. Experiment

In fabricating superconducting tapes of $(Sr_{1-x}Na_x)Fe_2As_2$, we basically followed the method reported in Ref. [15]. First, we synthesized polycrystalline samples of $(Sr_{1-x}Na_x)Fe_2As_2$. Since it is reported that a sample with Na concentration $x = 0.55$ exhibited the highest T_c (36.5 K), we selected x as 0.55 [14]. Polycrystalline samples were synthesized at 720°C for 18 h using the stainless steel pipe, where pellets of starting materials (SrAs, NaAs, Fe_2As) were placed. Then, the reacted materials were ground into powder, pelletized, and reacted at 720°C 18 h again. The prepared polycrystalline materials were ground into powder with an agate mortar in a nitrogen filled glove box. The ground powder was tightly packed into a silver tube with outer and inner diameters of 4.6 and 3 mm, respectively. The tube was groove-rolled into a wire with 1.3 x 1.3 mm^2 cross section. Then, the wire was flat-rolled into a tape form with a thickness and width of 0.25 and 2.2 mm, respectively. The tape was cold-pressed under a pressure of 0.54 GPa. Finally, the tape was heated for sintering. Sintering temperature was 800°C, 825°C, 850°C, 860°C, and 875°C, and the sintering temperature dependence of J_c was studied.

The J_c was measured at 4.2 K by the standard four-probe method in magnetic fields up to 14 T. Powder x-ray diffraction (XRD) measurements were conducted using Cu*Kα* radiation. Magnetic measurements were conducted using a commercial SQUID magnetometer (MPMS-XL5, Quantum Design). Compositional analyses and visualization of magnetic induction in the core were conducted using SEM-EDX (S-4300, Hitachi High-Technologies equipped with EMAX x-act, HORIBA) and MO imaging, respectively.

3. Result and discussion

3.1. Transport J_c measurements at 4.2 K in magnetic fields up to 14 T

Figure 1(a) shows J_c versus magnetic field at 4.2 K for the $(Sr,Na)Fe_2As_2$ tapes sintered at various temperatures. The J_c in $(Sr,Na)Fe_2As_2$ tape sintered at 875°C reaches 4.9 x 10^4 A/cm^2 under self field, and it is suppressed only down to 1.9 x 10^4 A/cm^2 even at 14 T. Although J_c in 122 type superconducting tapes usually goes down most sharply in the range between 0 T and 2 T [16], J_c in our tapes sintered at 875 °C was hardly suppressed in this range. J_c(14 T) / J_c(0 T) was calculated from figure 1(a), and plotted in figure 1(b) for each sintering temperature. For the tape sintered at 875°C, J_c(14 T) / J_c(0 T) = 0.39. For $(Ba,K)Fe_2As_2$ tape, J_c(0 T) = 2.5 x 10^5 A/cm^2, and J_c(14 T) = 1.9 x 10^4 A/cm^2, then J_c(14 T) / J_c(0 T) = 0.08 [16]. For Nb_3Sn tape, J_c(0 T) = 8.2 x 10^5 A/cm^2, and J_c(14 T) = 1.0 x 10^5 A/cm^2, then J_c(14 T) / J_c(0 T) = 0.13 [17]. This indicates that suppression of J_c by magnetic field in $(Sr,Na)Fe_2As_2$ tape is very weak.

From figures 1(a) and (b), it is suggested that the tapes sintered at higher temperature exhibit higher and more field-insensitive value of J_c. However, if the tape is sintered at temperatures over 875°C, severe reaction between the core and Ag sheath was observed. So, a range between 850 and 875°C is a critical region where the core begins to melt and reacts with the silver sheath. Probably, this partial melting of the core leads to improvement of connections between grains and higher inter-granular J_c.

3.2. X-ray diffraction analyses

Figure 2 shows XRD patterns for the core surface of the $(Sr,Na)Fe_2As_2$ tapes sintered at various temperatures. Compared with the reference peak of polycrystalline samples, the intensity of (002) peak is higher. This result indicates that the core is well textured with its *c*-axis perpendicular to the tape plane. We believe that this is due to the uniaxial cold press and it helps to achieve higher value of J_c. Also, we can identify the FeAs impurity peaks in tapes sintered at 800°C and 825°C, while this peak vanishes in the tape sintered at 850°C. The presence of impurity peaks at 800°C and 825°C partially explains why J_c of the tapes sintered at temperatures over 850°C is higher. In addition, we can identify the Fe_2As impurity peaks in all tapes. Fe_2As may have been produced by the reaction between

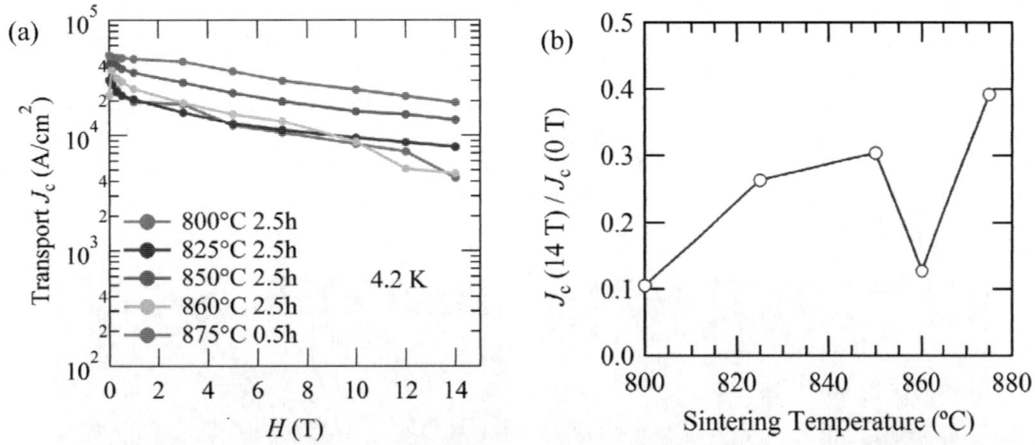

Figure 1. (a) J_c versus magnetic field at 4.2 K for the $(Sr,Na)Fe_2As_2$ tapes sintered at various temperatures. (b) J_c(14 T) / J_c(0 T) as a function of sintering temperature calculated from the data in figure 1(a). For the tape sintered at 875°C, J_c(14 T) / J_c(0 T) = 0.39. This means that critical current densities are very field-insensitive.

Ag sheath and As as will be shown in section 3.3, which decreased the amount of As in the core and helped the formation of Fe_2As. Alternatively, when the tape was sintered, As in the core may have vaporized and escaped from the core. Whichever the case, addition of As to the starting powders may reduce the Fe_2As impurity phase in the core, and contribute to enhance the transport J_c.

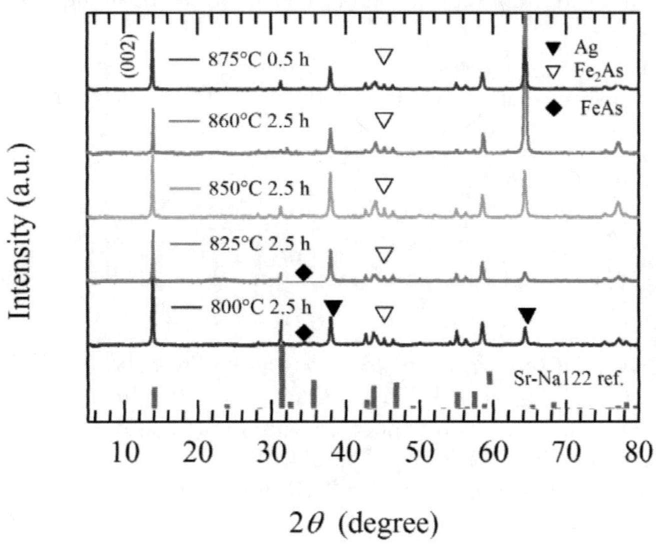

Figure 2. XRD patterns for the core surface of the $(Sr,Na)Fe_2As_2$ tapes sintered at various temperatures.

3.3. SEM images and compositional analyses

Figures 3(a) and (b) show SEM images of the surface of the core of the tape sintered at 850°C. Atomic ratios in four different regions on the surface of the tape marked in figures 3(a) and (b) are summarized in Table 1. The ideal atomic ratio is Sr : Na : Fe : As = 9 : 11 : 40 : 40 for $x = 0.55$. In region 1, there are impurities which seem like Fe_2As. In region 2, compounds of Fe and As and Ag are identified. This is consistent with SEM images reported in Ref. [15]. According to their report, there is a possibility that Ag sheath reacts with the core and penetrates into the core, and that these compounds of Fe and As and Ag partially complement impurity regions. These compounds are identified only in tapes sintered at temperatures over 850°C and can contribute to the large and field-insensitive J_c. In regions 3 and 4, atomic ratios are approximately equal to those of $Sr_{0.45}Na_{0.55}Fe_2As_2$, consistent with the starting composition.

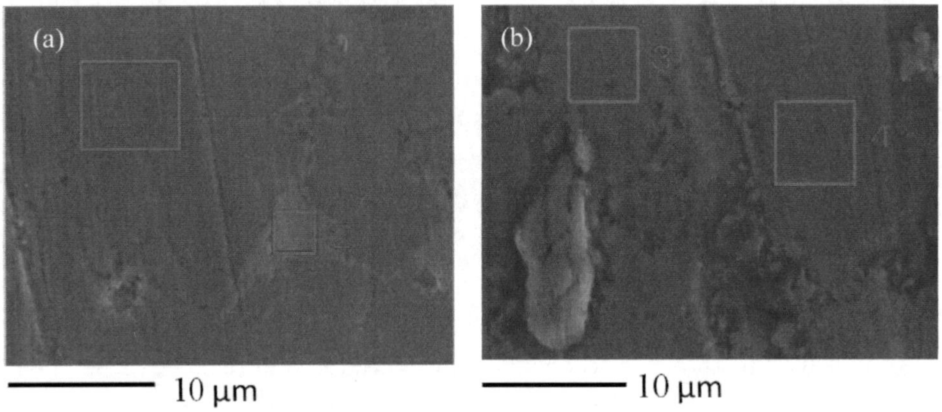

Figure 3. (a) (b) SEM images of the surface of the tape sintered at 850°C.

Table 1. Atomic ratios in four different regions on the surface of the tape marked in figures 3(a) and (b).

	Region 1	2	3	4
atom	ratio (%)	ratio (%)	ratio (%)	ratio (%)
Sr	4.2	2.7	10.1	7
Na	5.9	4.5	12.4	11
Fe	50.7	12.8	37.1	44
As	39.2	12.5	40.1	37.3
Ag	0.1	67.5	0.3	0.7

3.4. MO imaging

Figure 4(a) shows MO images of the surface of the tape sintered at 850°C. This image was taken at 5 K after cycling the magnetic field up to 0.1 T and back to zero. Bright regions trap higher magnetic field, and correspond to grains that are well connected. Figure 4(b) shows temperature dependence of residual magnetic induction along the red line in figure 4(a). Thickness of the core is roughly 100 μm, and the highest trapped field is approximately 60 mT. From this trapped field value, we can roughly evaluate $J_c \sim 6 \times 10^4$ A/cm². Figures 4(c) and (d) show optical images of the surface of the tape sintered at 850°C. In figure 4(c), the bright region on the left side shows an unpolished Ag sheath. In figures 4(c) and (d), slightly bright spots are high-density regions of (Sr,Na)Fe_2As_2 grains corresponding to bright regions in figure 4(a). These images indicate that high-density regions are localized, and that there are some regions where grains are weakly connected each other and inter-granular J_c is small. To improve the connectivity between grains, the hard and homogeneous core in the tape should be realized, and this will lead to further enhancement of J_c.

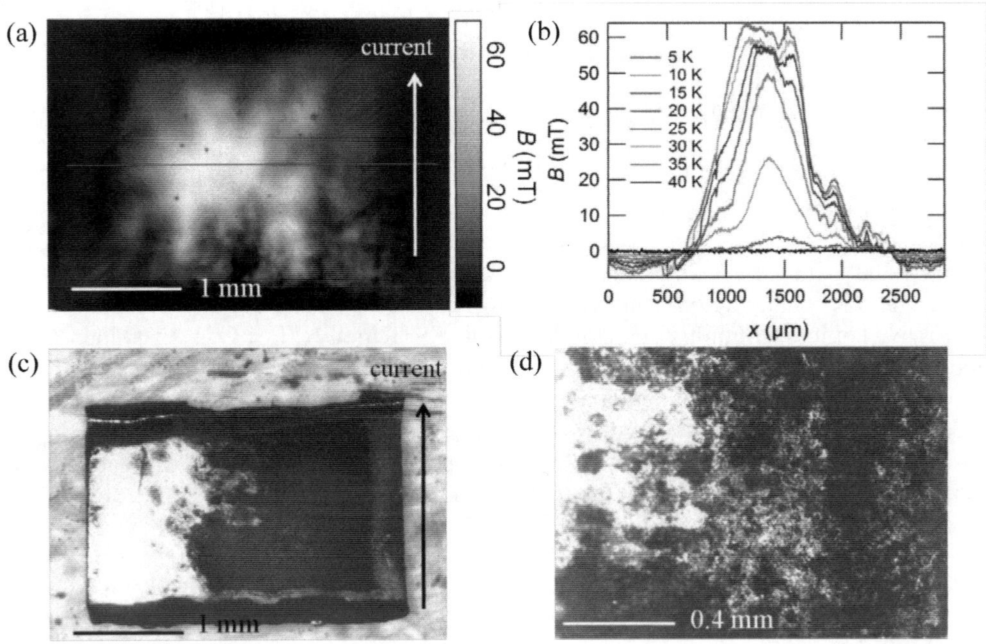

Figure 4. (a) MO images of the surface of the tape sintered at 850°C. Bright regions are trapping higher magnetic field. (b) Residual magnetic induction profile along the red line in (a). (c)-(d) Optical images of the surface of the tape sintered at 850°C. In (c), bright region on the left side shows an unpolished Ag sheath. In (c) and (d), slightly bright spots are high-density regions of (Sr,Na)Fe_2As_2 grains and correspond to bright regions in (a).

4. Summary

We have reported the fabrication and characterization of (Sr,Na)Fe_2As_2 superconducting tapes prepared under different conditions including transport J_c up to 14 T and magneto-optical observation of the trapped field. In the tape sintered at 875°C, the value of J_c reached 4.9 x 10^4 A/cm^2 under self-field, and it is suppressed only down to 1.9 x 10^4 A/cm^2 even at 14 T. We succeeded in fabricating very field-insensitive tapes, which are very promising for practical applications. Magneto-optical imaging of the core of the tape clarified inhomogeneous intergranular current distributions.

References

[1] Kamihara Y, Watanabe T, Hirano M and Hosono H 2008 *J. Am. Chem. Soc.* **130** 3296
[2] Ma Y W, Gao Z S, Qi Y P, Zhang X P, Wang L, Zhang Z Y and Wang D L 2009 *Physica* C **469** 651
[3] Togano K, Matsumoto A and Kumakura H 2011 *Appl. Phys. Express* **4** 043101
[4] Pyon S, Tsuchiya Y, Inoue H, Kajitani H, Koizumi N, Awaji S, Watanabe K and Tamegai T 2014 *Supercond. Sci. Technol.* **27** 095002
[5] Pyon S, Yamasaki Y, Kajitani H, Koizumi N, Tsuchiya Y, Awaji S, Watanabe K and Tamegai T 2015 *Supercond. Sci. Technol.* **27** 125014
[6] Weiss J D, Tarantini C, Jiang J, Kametani F, Polyanskii A A, Larbalestier D C and Hellstrom EE 2012 *Nat. Mater.* **11** 682
[7] Putti M, Pallecchi I, Bellingeri E, Cimberle M R, Tropeano M, Ferdeghini C, Palenzona A, Tarantini C, Yamamoto A, Jiang J, Jaroszynski J, Kametani F, Abraimov D, Polyanskii A, Weiss JD, Hellstrom EE, Gurevich A, Larbalestier DC, Jin R, Sales BC, Sefat AS, McGuire MA, Mandrus D, Cheng P, Jia Y, Wen HH, Lee S and Eom CB 2010 *Supercond. Sci. Technol.* **23** 034003

IOP Conf. Series: Journal of Physics: Conf. Series **871** (2017) 012062 doi:10.1088/1742-6596/871/1/012062

[8] Wang Z S, Luo H Q, Ren C and Wen H H 2008 *Phys. Rev.* B **78** 140501R
[9] Yuan H Q, Singleton J, Balakirev F F, Baily S A, Chen G F, Luo J L and Wang N L 2009 *Nature* **457** 565
[10] Rotter M M, Tegel M, Johrendt D, Schellenberg I, Hermes W and Pöttgen R 2008 *Phys. Rev.* B **78** 020503
[11] Sasmal K, Lv B, Lorenz B, Guloy A M, Chen F, Xue Y Y and Chu C W 2008 *Phys. Rev. Lett.* **101** 107007
[12] Zhang X, Yao C, Lin H, Cai Y, Chen Z, Li J, Dong C, Zhang Q, Wang D, Ma Y, Oguro H, Awaji S and Watanabe K 2014 *Appl. Phys. Lett.* **104** 202601
[13] Gao Z, Togano K, Matsumoto A and Kumakura H 2014 *Sci. Rep.* **4** 4065
[14] Shinohara N, Tokiwa K, Fujihisa H, Gotoh Y, Ishida S, Kihou K, Lee C H, Eisaki H, Yoshida Y and Iyo A 2015 *Supercond. Sci. Technol.* **29** 115002
[15] Iyo A, Shinohara N, Tokiwa K, Ishida S, Tsuchiya Y, Ishii A, Asou T, Nishio T, Matsuzaki K, Takeshita N, Eisaki H and Yoshida Y 2015 *Supercond. Sci. Technol.* **28** 105007
[16] Pyon S, Suwa T, Park A, Kajitani H, Koizumi N, Tsuchiya Y, Awaji S, Watanabe K and Tamegai T 2016 *Supercond. Sci. Technol.* **28** 062001
[17] Togano K, Matsumoto A and Kumakura H 2012 *Solid State Commun.* **152** 740

IOP Conf. Series: Journal of Physics: Conf. Series **871** (2017) 012063 doi:10.1088/1742-6596/871/1/012063

Fabrication of FeSe superconducting films with chemical transport deposition process

J Q Feng[1], S N Zhang[1,3], J X Liu[1,2], Q B Hao[1], C S Li[1] and P X Zhang[1]

[1]Northwest Institute for Non-Ferrous Metal Research, 710016, Xi'an, China
[2]School of Materials and Metallurgical, Northeast University, Shenyang, 110016, China

Abstract. FeSe Superconducting films were fabricated with a chemical transport deposition process. During the fabrication process, Fe foils were adopted as substrates and Se powders were put at one end of the tube furnace. During the heating process, Se powders were vaporized, and vaporized atoms were carried by Ar flow and deposited on the Fe substrates. With a heat treatment process under proper temperature, superconducting tetragonal β-FeSe phase can be obtained. The effects of key parameters, including the sintering temperatures and the distances between Fe substrates and Se source on the phase composition and morphology of the obtained films were systematically investigated. The superconducting transition temperature of 7.8 K was obtained on the optimized film. By further optimization of the heat treatment process, it is promising to fabricate FeSe films with higher superconducting phase content and better superconducting properties.

1. Introduction

In 2008, the superconducting transition with critical temperature T_c of ~8 K was discovered in tetragonal β-FeSe system by Hsu *et. al.* [1]. Then in the following decades, it attracted more and more attentions from the standpoints of both fundamental researches and practical applications. FeSe based superconductors, known as "11" systems, exhibit the simplest lattice structures among all the superconductors, which is only composed of -Fe-Se- layers as superconducting layers with no blocking layers. Based on the similar Fe_2Se_2 layer structures, $FeSe_{1-x}Te_x$ and $FeTe_{1-x}S_x$ also register their distinct presence in this iron chalcogenide system, with both T_c values of ~15 K. It is interesting that the T_c values of FeSe system is quite tunable. It can be greatly enhanced up to 37 K under high pressure [2] and to 65 K by preparing unit cell FeSe ultrathin films [3-6]. At the same time, the advantages such as high upper critical field, H_{c2} of ~47 T [7], low cost and low toxicity of starting materials compared to FeAs-based superconductors all suggest their potentials in practical applications.

Recently, aiming at practical applications, many researchers paid attentions to the fabrication of FeSe wires or tapes [8]. However, although the current carrying density of FeSe (or Fe(Se,Te)) single crystals and films can all reach as high as 10^5 Acm^{-2} at 4.2 K [9-11], that of FeSe wires still remains only about 100 Acm^{-2} [12-16]. The main factor which limits the current transport is the intergrain weak link. Both the low superconducting filaments density and the existence of secondary phase are responsible for the intergrain weak links. The fabrication of thin films is a possible way to realize practical applications of superconductors, such as YBCO coated conductors [17]. Recently, the developments of FeSe thin film fabrication techniques are mostly focused on either pulsed laser

[3] Corresponding author. Email Address: snzhang@c-nin.com;

deposition (PLD) [11, 16, 18, 19] or Molecular Beam Epitaxy (MBE) method for the fabrication of single layer thin films [3, 6, 20-22], which are both very costly. Electrochemical deposition is a potential low-cost method. However, the crystallization of obtained films is not good enough [23-25]. Thus it is necessary to develop a new method for the fabrication of β-FeSe thin films with high transport properties.

In this study, chemical transport deposition (CTD) method is designed and optimized for the preparation of β-FeSe superconducting films. The effects of key processing parameters, including the sintering temperatures and the distances between Se source and Fe substrates on the phase composition and morphology of obtained films are investigated. β-FeSe superconducting film with the critical temperature of 7.8 K is obtained, which suggests the potential of this method on the fabrication of β-FeSe films in an industrial scale.

2. Experimental

Fe foils with the purity of 99.9% were carefully washed by ethanol and deionized water and punched into small rectangular pieces with the dimension of 1.5×1.0 cm^2 as substrates. During the fabrication process, high purity elemental Se (99%) powders (from Alfa Aesar) were adopted as starting materials and put on one end of the tube furnace, where the temperature is ~100 °C lower than that at the center of the furnace. And Fe substrates were put at the center of the tube furnace with the distance carefully measured as *d*. The change of distances between Fe and Se was realized by changing the positions of Fe. Ar flow with the flowing rate of 1 L/min was adopted as the carrying gas. The sintering process was performed by heating these substrates up to T_{max} with T_{max} = 500 and 600 °C and kept for 12 h, respectively, then slowly cooling to room temperature with the cooling rate of ~25 °C/h.

Phase composition of all the obtained thin films was determined by X-ray diffractometer (XRD, Bruker D8 Advance) with CuKα (λ=1.5406 Å) radiation. And a phase purity parameter, $F_{\beta\text{-FeSe}}$ related to the tetragonal phase content ratio was calculated as,

$$F_{\beta-FeSe} = \frac{\sum I_{\beta-FeSe}}{\sum I_{\beta-FeSe} + \sum I_{\delta-FeSe}} \times 100\% \qquad \text{Eq.1}$$

where $I_{\beta\text{-FeSe}}$ and $I_{\delta\text{-FeSe}}$ represent for the diffraction peak intensities of β-FeSe and δ-FeSe phase, respectively. The morphologies of thin films were characterized by scanning electron microscopes (SEM, JEOL-6390A). And the superconducting critical temperature was measured by magnetization method, which was carried out on Superconducting Quantum Interference Device (SQUID, MPMS-XL-7) with the applied field of 10 Oe from 4.2 K to 30 K.

3. Results and Discussions

The XRD patterns of obtained films with the sintering temperatures of 500 and 600 °C are plotted in Figure 1 (a) and (b), respectively. The films are named 500-1#~500-4# and 600-1#~600-4# based on the different distances from Fe substrates to Se source of *d*=30, 35, 40, and 45 cm, respectively. As shown in both figures, these thin films are all composed of tetragonal β-FeSe phase and hexagonal δ-FeSe phase. The small content of residual Se is due to the Se deposition during the cooling process. It is known that the tetragonal β-FeSe phase is the superconducting phase. Thus the contents of β-FeSe phase in all the films are estimated with the phase purity parameter $F_{\beta\text{-FeSe}}$ and plotted in each inset. It can be noticed that with the sintering temperature of 500 °C, the β-FeSe phase purity parameters are all smaller than 60%, suggesting that the sintering temperature of 500 °C is not high enough for the formation of β-FeSe phase. With the increase of sintering temperature to 600 °C, the β-FeSe phase content greatly increases, and the maximum phase purity parameter of 91% is achieved.

The distances between Se source and Fe substrates, *d*, also plays an important role for the film preparation. In both sintering process, with the increase of *d*, the β-FeSe phase purity parameter increases then decreases, and the optimized value is *d*=40 cm under both conditions. Considering that all the substrates are in the flat-temperature zone, with the temperature variation within ±1 °C, the different phase purity parameters should be attributed to the different Fe:Se ratios on each film. Based on our previous study, the increase of Fe:Se ratio can lead to the increase of β-FeSe phase content [26]. The amount of Se that reaches Fe substrate mostly depends on the distances between the Se source and Fe substrates. Therefore, more Se can deposit on sample #1 with the shortest *d* value. And under the same sintering condition, more hexagonal δ-FeSe phase can be formed. The decrease of β-FeSe phase content in film 4# may be attributed to the larger cooling rate, considering that it is at the edge of our flat-temperature zone.

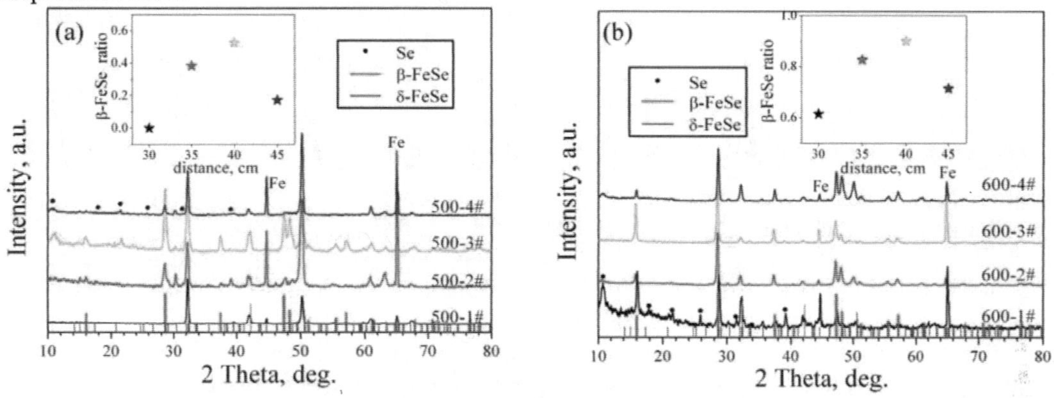

Figure 1. X-ray diffraction patterns of FeSe thin films sintered at different temperature of (a) 500 and (b) 600 °C, respectively. The red-purple and dark-green lines are standard diffraction lines of β-FeSe and δ-FeSe from the JCPDS card of 85-0735 and 75-0608, respectively.

The obtained films are peeled off the substrates and the cross section images of the obtained films sintered under 600 °C at the position of 2, 3, and 4# are obtained as shown in Figure 2 (a~c). It can be observed that the thickness of each film is uniform with the variation within 3.0 μm. However, the thicknesses of films changes obviously with different *d* values, which decrease from ~200 μm to ~165 μm. As discussed above, the different thicknesses should be attributed to the Se deposition content, that with the increase of *d*, less Se powders can be deposited on the Fe films. The top view morphology of film surface of film 600-3# is shown in Figure 2(d). The observed ripples and topographical irregularities are due to the unevenly deposition of Se. In Figure 2 (a~c), the formation of columnar grain can be observed on the top part of the films. And the grain size obviously decreases closer to the substrate. It is because the nucleation rate of FeSe grains is larger on the Fe substrates. And on the top of the film, more stress can be released, thus the grains can grow bigger.

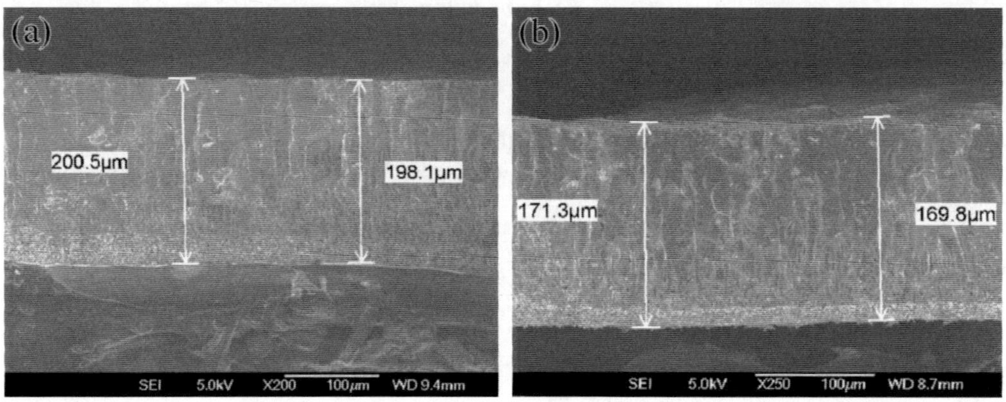

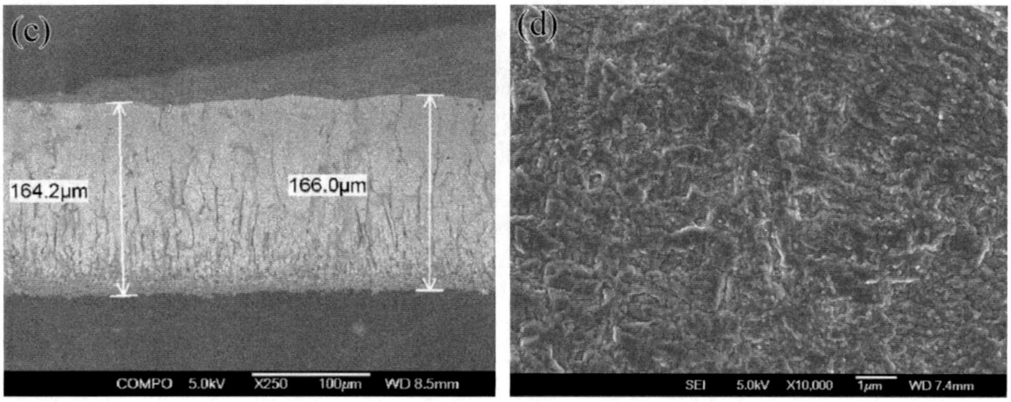

Figure 2. (a~c) Cross section images of FeSe thin films sintered at 600 °C at position 2~4#, respectively. (d) surface morphology of 600-3# film.

The magnetization of obtained film 600-3# is measured after peeling off the film from Fe substrate as shown in Figure 3. The positive magnetization over the entire temperature range should be related to the residual Fe substrates on the film. The field cool and zero field cool (FC-ZFC) curves present a superconducting transition at 7.8 K, which should be the critical temperature of obtained β-FeSe phase. However, the small superconducting phase content can be attributed to both the existence of hexagonal δ-FeSe phase and the chemical composition off-stoichiometry of β-FeSe phase. Thus with further optimizations of both the deposition and heat treatment process, better superconducting property can be expected.

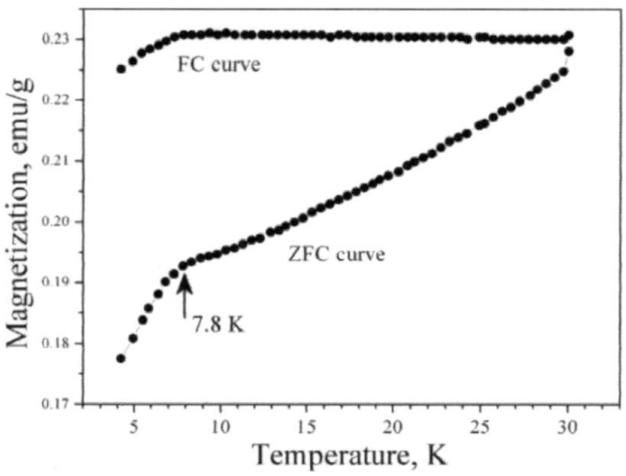

Figure 3. FC-ZFC curves of 600-3# film.

4. Conclusions

Chemical transport deposition technique for the fabrication of FeSe superconducting films has been successfully developed. The optimal key parameters, including the sintering temperatures and the distances between Fe substrates and Se source were obtained. It is concluded that within the flat-temperature zone, the increase of *d* could lead to the decrease of Se deposition content. Thus the thickness and phase composition of the films were all changed. The optimized film with critical temperature, T_c of 7.8 K was obtained with the sintering temperature of 600 °C and *d* of 40 cm. Further optimization of the heat treatment process is on the way in order to obtain the films with higher superconducting phase content and better superconducting properties.

References

[1] Hsu F C, Luo J Y, Yeh K W, Chen T K, Huang T W, Wu P M, Lee Y C, Huang Y L, Chu Y Y, Yan D C and WU M K 2008 Proc. Natl. Acad Sci USA **105** 14262.

[2] Bendele M, Pomjakushina E, Conder K, Khasanov R and Keller H 2014 *J. Supercond. Nov. Magn.* **27** 965.
[3] He S L, He J F, Zhang W H, Zhao L, Liu D F, Liu X, Mou D X, Ou Y B, Wang Q Y, Li Z, Wang L L, Peng Y Y, Liu Y, Chen C Y, Yu L, Liu G D, Dong X L, Zhang J, Chen C T, Xu Z Y, Chen X, Ma X C, Xue Q K and Zhou X J 2013 *Nat. Mater.* **12** 605.
[4] Li W, Ding H, Zhang P, Deng P, Chang K, He K, Ji S, Wang L, Ma X, Wu J, Hu J P, Xue Q K and Chen X 2013 *Physical Review B* **88** 140506.
[5] Zhang W H, Li Z, Li F S, Zhang H M, Peng J P, Tang C J, Wang Q Y, He K, Chen X, Wang L L, Ma X C and Xue Q K 2014 *Phys. Rev. B* **89** 060506(R).
[6] Liu D F, Zhang W, Mou D, He J, Ou Y B, Wang Q Y, Li Z, Wang L, Zhao L, He S, Peng Y, Liu X, Chen C, Yu L, Liu G, Dong X, Zhang J, Chen C, Xu Z, Hu J, Chen X, Ma X, Xue Q K and Zhou X J 2012 *Nature Commun.* **3** 931.
[7] Fang M H, Yang J H, Balakirev F F, Kohama Y, Singleton J, Qian B, Mao Z Q, Wang H D and Yuan H Q 2010 *Phys. Rev. B* **81** 020509.
[8] Ma Y W 2012 *Supercond. Sci. Technol.* **25** 113001.
[9] Yadav C S and Paulose P L 2011 *Solid State Commun.* **151** 216.
[10] Zhang W H, Sun Y, Zhang J S, Li F S, Guo M H, Zhao Y F, Zhang J M, Peng J P, Xing Y, Wang H C, Fujita T, Hirata A, Li Z, Ding H, Tang C J, Wang M, Wang Q Y, He L, Ji S H, Chen X, Wang J F, Xia Z C, Li L, Wang Y Y, Wang J, Wang L L, Chen M W, Xue Q K and Ma X C 2014 *Chin. Phys. Lett.* **31** 017401.
[11] Si W D, Han S J, Shi X Y, Ehrlich S N, Jaroszynski J, Goyal A and Li Q 2013 *Nature Commun.* **4** 1347.
[12] Mizuguchi Y, Izawa H, Ozaki T, Takano Y and Miura O 2011 *Supercond. Sci. Technol.* **24** 125003.
[13] Gao Z S, Qi Y P, Wang L, Wang D L, Zhang X P, Yao C and Ma Y W 2011 *Supercond. Sci. Technol.* **24** 065022.
[14] Mizuguchi Y, Deguchi K, Tsuda S, Yamaguchi T, Takeya H, Kumakura H and Takano Y 2009 *Appl. Phys. Express* **2** 083004.
[15] Ozaki T, Deguchi K, Mizuguchi Y, Kawasaki Y, Tanaka T, Yamaguchi T, Tsuda S, Kumakura H and Takano Y 2011 *Supercond. Sci. Technol.* **24** 105002.
[16] Tsukada A, Luna K E, Hammond R H, Beasley M R, Zhao J F and Risbud S H 2011 *App. Phys. A* **104** 311.
[17] Grilli F and Pardo E 2010 *Supercond. Sci. Technol.* **23** 115018.
[18] Peng R, Xu H C, Tan S Y, Cao H Y, Xia M, Shen X P, Huang Z C, Wen C H P, Song Q, Zhang T, Xie B P, Gong X G and Feng D L 2014 *Nature Commun.* **5** 5044.
[19] Yoshida K, Yoshida Y, Ichino Y, Takai Y, Ichinose A, Matsumoto K, Kiss T and Mukaida M 2011 *Physica C* **471** 1185.
[20] Wang L L, Ma X C, Chen X and Xue Q K 2013 *Chinese Phys. B* **22** 086801.
[21] Li Z, Peng J P, Zhang H M, Zhang W H, Ding H, Deng P, Chang K, Song C L, Ji S H, Wang L L, He K, Chen X, Xue Q K and Ma X C 2014 *J. Phys.: Condens. Matter* **26** 265002.
[22] Berlijn T, Cheng H P, Hirschfeld P J and Ku W 2014 *Physical Review B* **89** 020501.
[23] Chen P Y, Hu S F, Liu R S and Huang C Y 2011 *Thin Solid Films* **519** 8397.
[24] Demura S, Okazaki H, Ozaki T, Hara H, Kawasaki Y, Deguchi K, Watanabe T, Denholme S J, Mizuguchi Y, Yamaguchi T, Takeya H and Takano Y 2013 *Solid State Commun.* **154** 40.
[25] Demura S, Ozaki T, Okazaki H, Mizuguchi Y, Hara H, Kawasaki Y, Deguchi K, Watanabe T, Yamaguchi T, Takeya H and Takano Y 2012 *J. Phys. Soc. Jpn.* **81** 043702.
[26] Zhang S N, Ma X B, Liu J X, Feng J Q, Li C S and Zhang P X 2015 *Mater. Chem. Phys.* **163** 587.

IOP Conf. Series: Journal of Physics: Conf. Series **871** (2017) 012064 doi:10.1088/1742-6596/871/1/012064

$FeTe_{0.6}Se_{0.4}$ bulk single crystals with high critical current densities under magnetic fields

Y Tanaka[1]*, Y Mizuguchi[1] and O Miura[1]

[1] Dept. of Electrical and Electronic Engineering, Tokyo Metropolitan University, Hachioji 192-0397, Japan

e-mail: tanaka-yuuji1@ed.tmu.ac.jp

Abstract. We have fabricated $FeTe_{0.6}Se_{0.4}$ large-size bulk single crystals with high critical current densities (J_c) under magnetic fields. $FeTe_{0.6}Se_{0.4}$ single crystals were prepared by the melting method with two stage heat treatments. The X-ray diffraction (XRD) analysis and the remanent magnetization method showed that the fabricated bulk crystals are intrinsically single crystals from viewpoint of the critical current flow. Temperature dependence of magnetization revealed that low-T_c region exists inside the crystals. The magnetization curves indicated the typical fishtail type, and the magnetic J_c under the magnetic field parallel to the c-axis at 4.2 K achieved 0.36 and 0.2 MA / cm^2 at 0 T and 5 T respectively. From the temperature scaling behavior of flux pinning properties we speculated that low-T_c regions near excess Fe moderately distributed inside the crystals are dominant pinning centers in high fields at low temperatures.

1. Introduction

The discovery of iron-based superconductors (IBSs) at 26 K in LaFeAsO (1111 system) doped with fluoride was a surprise since most Fe compounds are magnetic [1]. Soon after the first report, the superconducting transition temperature (T_c) was raised to as high as 55 K by replacing La with other rare earths such as Ce [2], Pr [3], Nd [4], Sm [5], or Gd [6]. In addition, other crystal structure types of layered Fe compounds were found to become superconductors; for example, $Ba_{1-x}K_xFe_2As_2$ (122 system) [7], $Li_{1-x}FeAs$ (111 system) [8], and FeSe (11 system) [9]. FeSe has the simplest crystal structure among the IBSs and less toxicity as compared to the FeAs systems, which are advantageous for applications. Partial substitution of Se by Te can effectively enhance T_c to around 14 K [10]. Furthermore, upper critical field (H_{c2}) is as high as around 50 T [11]. Due to their high superconductivity performance in high magnetic fields, $FeTe_{1-x}Se_x$ wire fabrication has been studied by several researchers [12-14]. We have fabricated $FeTe_{0.4}Se_{0.6}$ superconducting tapes by a newly developed chemical-transformation powder-in-tube method [15]. However, the critical current densities (J_c) was as low as 3.0×10^3 A / cm^2 at 4.2 K, which was not enough for practical use. This is due to the formation of non-superconducting phase at the grain surface of crushed crystal resulting in the low electrical connectivity between the grains [16]. Therefore, to improve J_c properties for $FeTe_{1-x}Se_x$ system the fabrication of large-size bulk single crystals and studies for the flux pinning mechanisms are needed.

In this paper, we have fabricated $FeTe_{0.6}Se_{0.4}$ large-size bulk single crystals with high J_c by the melting method with two stage heat treatments. J_c in magnetic fields and the flux pinning properties were investigated. We also studied the effective pinning centers in the crystals.

2. Experimental

$FeTe_{0.6}Se_{0.4}$ was prepared by the melting method. Iron powder (99.9%), selenium chips (99.999%) and tellurium chips (99.999%) were used as starting materials. The stoichiometric quantities of 2 g were put into small double quartz tubes, evacuated at less than 5 Pa, and sealed. Then it was heated up to 850 °C and kept for 10 h, followed by slow cooling down to 600 °C at a rate of 4 °C / h. After that, the temperature was rapidly cooled down to 400 °C and kept for 200 h for crystal homogenization. Finally it was cooled down to room temperature by shutting down the furnace. The as-grown crystal is a columnar shape single crystal with 8 mm in diameter and 10 mm in length. The crystal structure was confirmed by X-ray diffraction (XRD) with Cu Kα radiation (λ = 1.541841 Å).

A photograph of the as-grown single crystal is shown in figure 1. The crystal is easily cleaved perpendicular to the c-axis. The perpendicular direction to the center axis of the as-grown columnar sample is the c-axis direction. Magnetization was measured by a superconducting quantum interference device (SQUID) magnetometer and T_c was determined from the temperature dependence of magnetization when a magnetic field of 10 Oe was applied to the c-axis direction. We confirmed that a large shielding current flows at all over the crystal by the remanent magnetization method [17, 18]. Hence, J_c was estimated from the hysteresis width of magnetization curves by Bean's critical state model [19].

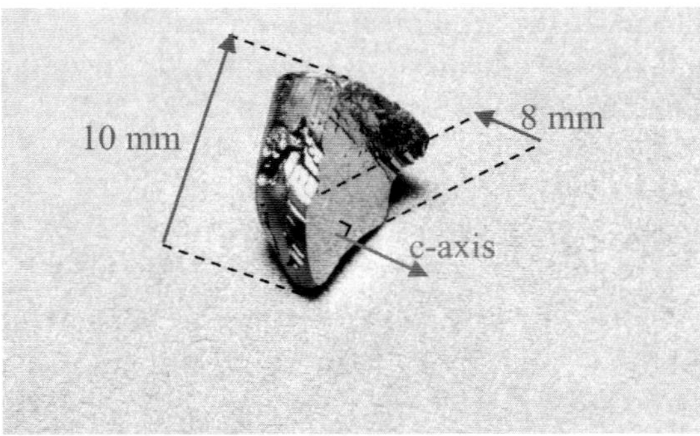

Figure 1. A photograph of the as-grown single crystal with a cleavage plane.

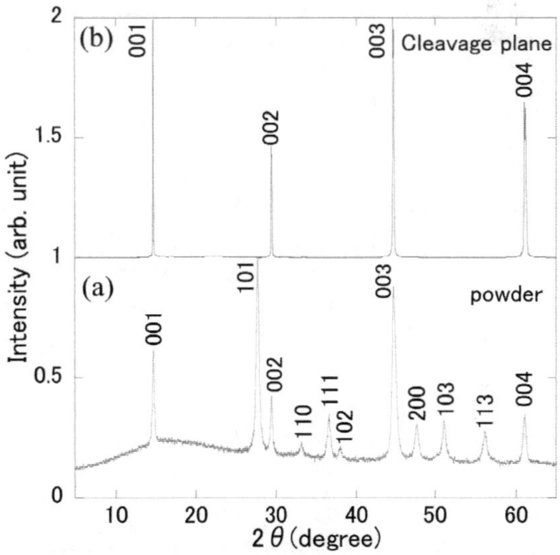

Figure 2. XRD patterns of the fabricated crystal for (a) powder and (b) a cleavage plane.

3. Result and discussion

Figures 2 show the XRD patterns for (a) powder and (b) a cleavage plane of the fabricated $FeTe_{0.6}Se_{0.4}$ crystal. Figure 2 (a) shows that there are no impurity peaks [10]. Figure 2 (b) shows only the peaks of the (0 0 l) planes. Therefore, we confirmed that it is a highly pure and c-axis oriented crystal.

Figure 3 shows the temperature dependence of magnetization when applying a magnetic field of 10 Oe to the c-axis direction under ZFC and FC conditions. It shows the sharp transition with T_c onset of 14 K. This indicates that the fabricated crystal is a relatively uniform single crystal. On the other hand, the positive magnetization due to excess Fe was observed. The excess Fe concentration at the interlayer sites between Fe_2Te_2 layers can be turned by initial amount of Fe and annealing conditions [20]. A slight decrease in diamagnetism was also observed between 5 to 8 K in the ZFC conditions. It probably comes from low-T_c regions with a relatively high Fe concentration moderately distributed inside the crystal. They have the potential to act as effective flux pinning centres as well as dislocations in the crystal.

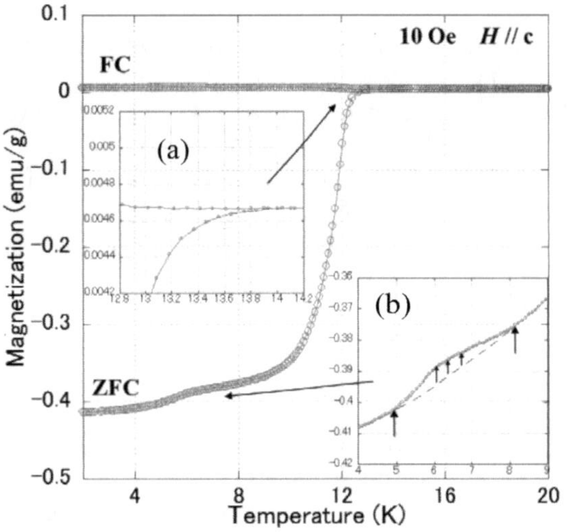

Figure 3. Temperature dependence of magnetization applying 10 Oe to the c-axis direction. (a) of insets shows enlargement near Tc. (b) of insets shows enlargement near 5 to 8 K in ZFC.

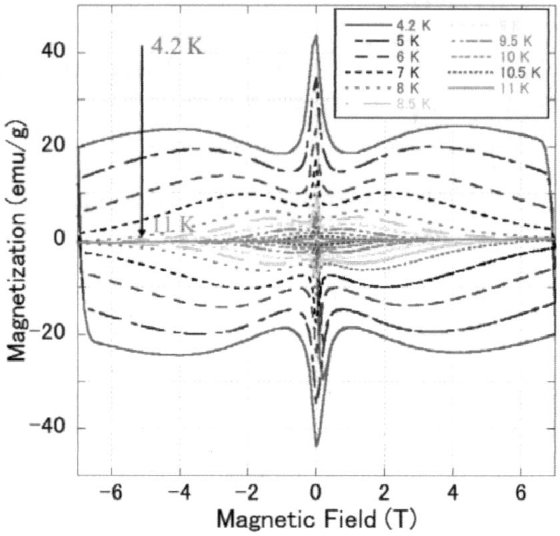

Figure 4. Magnetic field dependence of magnetization applying magnetic field to the c-axis direction for each temperature.

Figure 4 shows the magnetic field dependence of magnetization for each temperature when a magnetic field is applied to the c-axis direction. Fishtail shape magnetization curves appear from 4.2 K to 10 K. The fishtail shape indicates that the pinning force densities (F_p) in the magnetic fields is strong. As the temperature increases, the magnetization and the fishtail shape decreases.

J_c was estimated from the Bean model using the magnetization width (ΔM) of M-B curves. Temperature dependence of J_c-B properties under magnetic fields is shown in figure 5 (a). J_c at 4.2 K achieved 0.359 and 0.195 MA / cm^2 at 0 T and 5 T respectively. In addition, it keeps over 0.10 MA / cm^2 in higher magnetic fields. Figure 5 (b) shows temperature dependence of F_p-B properties. F_p achieved 115 GN / m^3 at 4.2 K under 6.5 T. As the rise of temperature, F_p decreased and the peak F_p shifted to the low magnetic field side.

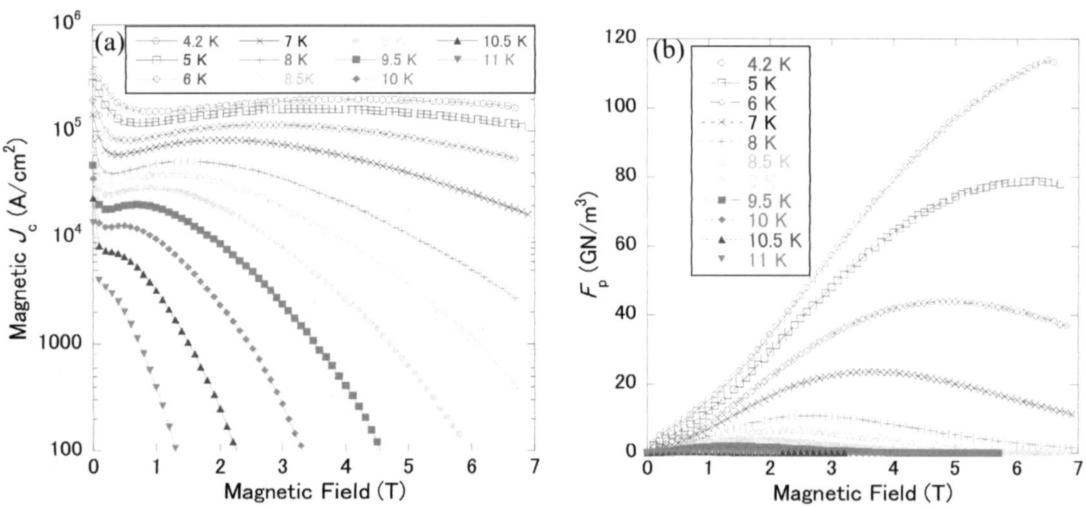

Figure 5. Temperature dependence of (a) J_c-B properties (b) F_p-B properties.

Figure 6 shows the irreversible fields (B_{irr}) for the crystal where the magnetization hysteresis disappears. When the irreversible temperature (T_{irr}) is defined as the temperature where the magnetization hysteresis derived from the pinning current disappears, the fittig was estimated from the formula of $B_{irr}(T) = B_{irr}(0)[1-(T / T_{irr})^2]$ [21] with $B_{irr}(0) = 16.6$ T, $T_{irr} = 11.6$ K.

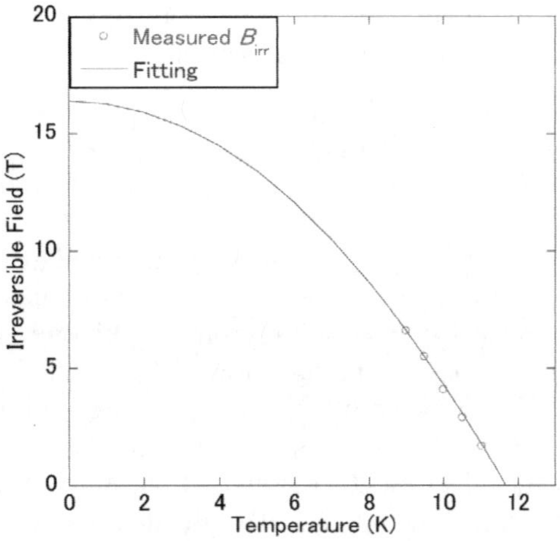

Figure 6. The fitting result of B_{irr} by $B_{irr}(T) = B_{irr}(0)[1-(T / T_{irr})^2]$.

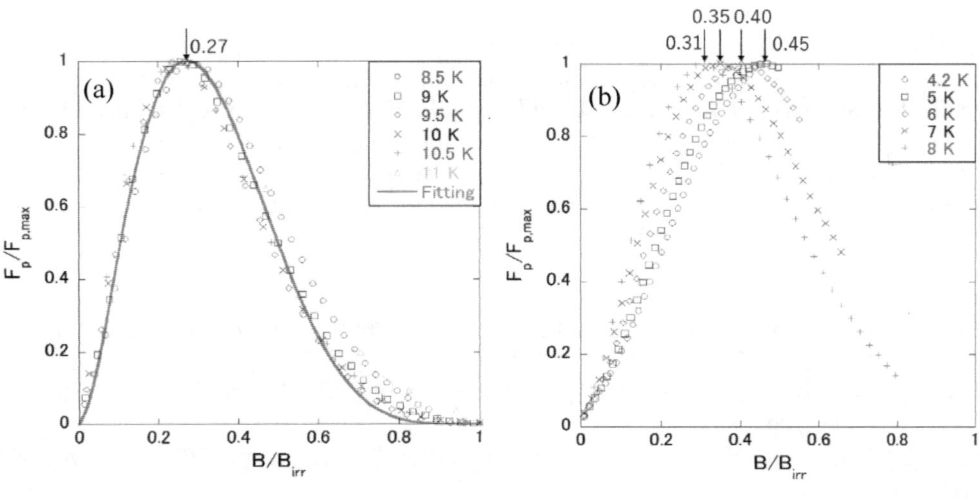

Figure 7. The normalized F_p-B properties for each temperature (a) scaling (b) not scaling.

Figure 7 shows normalized F_p-B properties using estimated B_{irr} for the $FeTe_{0.6}Se_{0.4}$ bulk crystal. In high temperatures above 8.5 K the temperature scaling behaviour of the flux pinning properties was clearly observed with the fitting parameters $p = 1.7$, $q = 4.6$ and $b_{max} = 0.27$ using the scaling formula of $F_p / F_{pmax} = Cb^p(1-b)^q$, where $b = B / B_{irr}$. Here p and q are just fitting parameters which are not based on a specific pinning model. This is nearly consistent with previous studies for $FeTe_{0.6}Se_{0.4}$ and $FeTe_{0.7}Se_{0.3}$ by different fabrication methods reported by Yadav C S et al. [22] and Bonura M et al. [23] respectively. On the other hand, in low temperatures under 8 K the temperature scaling behavior was not observed. The peak F_p / F_{pmax} shifted to high field side with decreasing temperature and reached 0.45 at 4.2 K. This suggests that the different pinning mechanisms are working between low and high temperatures. The likely candidates for pinning centres are both low-T_c regions and dislocations. In several papers, the coherence length for Fe(Te,Se) has been reported [20, 24]. Since the precise determination of coherence length of Fe(Te,Se) is difficult, the estimated value could vary depending on the method of analysis. However, we previously evaluated the coherence length of $FeTe_{0.6}Se_{0.4}$ as around several nm to tens nm by the effective mass model [20]. Those values may correspond to that for low-T_c regions, which would be expected to work as effective pinning centers. The elementary pinning force f_p for very thin low-T_c superconducting pins in which the spatial variation of the order parameter of superconducting matrix is very small can be expressed by the sum of potential energy interaction and kinetic energy interaction,

$$f_p = 0.479\pi \frac{d_p \xi}{\mu_0}(B_{cp}^{\ 2} - B_c^{\ 2}) + 0.296\pi \frac{d_p}{\mu_0 \xi}(B_{cp}^{\ 2}\xi_p^{\ 2} - B_c^{\ 2}\xi^2) \cdot \quad (1)$$

Where μ_0 is the permeability in vacuum, $d_p$is the pin size, ξ, ξ_p and B_c, B_{cp} are the coherence lengths and thermodynamic critical fields of superconducting matrix and superconducting pins, respectively [25, 26]. At low temperatures under T_c for low-T_c regions the kinetic energy interaction is dominant pinning energy, and generates repulsive pinning force. This behavior was first observed on the temperature dependence of F_p for NbTi multifilamentary wires with different kinds of artificial superconducting pins [27-29]. From the observed temperature scaling behavior for $FeTe_{0.6}Se_{0.4}$ with low-T_c regions it is speculated that the low-T_c regions act as temperature-induced repulsive pinning centres at low temperatures. On the other hand, at high temperatures the normal conductive pins changed from low-T_c regions and dislocations are considered to work as attractive normal pins, and hence the temperature scaling is expected. Further investigation is needed to identify actual pinning centres.

4. Conclusion

We have successfully fabricated $FeTe_{0.6}Se_{0.4}$ large-size bulk single crystals by the melting method with two stage heat treatments. Temperature dependence of magnetization revealed that low-T_c region exists inside the crystals. J_c under magnetic fields parallel to the c-axis at 4.2 K achieved 0.36 and 0.2 MA/cm^2 at 0 T and 5 T respectively. From the temperature scaling behavior of flux pinning properties we speculated that low-T_c regions near excess Fe moderately distributed in the crystals are dominant pinning centers in high fields at low temperatures.

References

[1] Kamihara Y, Watanabe T, Hirano M and Hosono H 2008 *Journal of the American Chemical Society* **130** 3296–97.
[2] Chen G F, Li Z, Wu D, Li G, Hu W Z, Dong J, Zheng P, Luo J L, and Wang N L 2008 *Physical review letters* **100** 247002.
[3] Ren Z A, Yang J, Lu W, Yi W, Che G, Dong X, Sun L, and Zhao Z 2008 *Materials Research Innovations* **12** 56-57.
[4] Ren Z A, Yang J, Lu W, Yi W, Shen X L, Li Z C, Che G C, Dong X L, Sun L L, Zhou F, and Zhao Z X 2008 *Europhysics Letters* **82** 57002.
[5] Z. Ren, W. Lu, J. Yang, W. Yi, X. Shen, Z. Li, G. Che, X. Dong, L. Sun, F. Zhou, and Z. Zhao 2008 *Chinese Physics Letters* **25** 2215.
[6] Cheng P, Fang L, Yang H, Zhu X, Mu G, Luo H, Wang Z, and Wen H 2008 *Science in China Series G: Physics, Mechanics and Astronomy* **51** 719-22.
[7] Rotter M, Tegel M, and Johrendt D 2008 *Physical Review Letters* **101** 107006.
[8] Wang X C, Kiu Q Q, Lv Y X, Gao W B, Yang L X, Yu R C, Li F Y and Jin C Q 2008 *Solid State Communications* **148** 538-40.
[9] Hsu F C, et al. 2008 *Proceedings of the National Academy of Sciences* **105** 14262-64.
[10] Sales B C, Sefat A S, McGuire M A, Jin R Y, Mandrus D and Mozharivskyj Y 2009 *Physical Review B* **79** 094521.
[11] Kida T, Matsunaga T, Hagiwara M, Mizuguchi Y, Takano Y and Kindo K 2009 *Journal of the*

Physical Society of Japan **78** 113701.

[12] Ozaki T, Deguchi K, Mizuguchi Y, Kumakura H and Takano Y 2011 *IEEE Transactions on Applied Superconductivity* **21** 2858-61.

[13] Ozaki T, Deguchi K, Mizuguchi Y, Kawasaki Y, Tanaka T, Yamaguchi T, Kumakura H and Takano Y 2012 *Journal of Applied Physics* **111** 112620.

[14] Izawa H, Mizuguchi Y, Ozaki T, Takano Y, and Miura O 2011 *Japanese Journal of Applied Physics* **51** 010101.

[15] Izawa H, Mizugchi Y, Miura O 2014 *Physica C: Superconductivity* **504** 77-80.

[16] Izawa H, Tanaka Y, Mizugchi Y, Miura O 2016 *Japanese Journal of Applied Physics* **55** 053101.

[17] Yamamoto A, et al. 2008 *Superconductor Science and Technology* **21** 095008.

[18] Otabe E S, et al. 2012 *Progress in Superconductivity and Cryogenics* **14** 1-7.

[19] Bean C P 1964 *Reviews of Modern Physics* **36** 31.

[20] Hamada K, Izawa H, Mizuguchi Y and Miura O 2013 *Physics Procedia* **45** 81-84.

[21] Matsushita T 2007 *Flux Pinning in Superconductors* (Berlin: Springer Berlin Heidelberg) p4.

[22] Yadav C S and Paulose P L 2011 *Solid State Communications* **151** 216-18.

[23] Bonura M, Giannini E, Viennois R and Senatore C 2012 *Physical Review B* **85** 134532.

[24] Klein T, et al. 2010 *Physical Review B* **82** 184506.

[25] Matsushita T 1996 *Proceedings of the 8th International Workshop on Critical Currents in Superconductors World Scientific* 63-68.

[26] Matsushita T 2000 *Superconductor Science and Technology* **13** 730.

[27] Zhu Y, Miura O and Ito D 2002 *IEEE transactions on applied superconductivity* **12** 1117-20.

[28] Zhu Y, Miura O, Okubo T, Ito D and Endo S 2000 *IEEE transactions on applied superconductivity* **10** 1046-49.

[29] Zhu Y, Miura O, Hayakawa K and Ito D 2001 *IEEE transactions on applied superconductivity* **11** 3808-11.

Run-to-Run Yield Evaluation of Improved Nb 9-layer Advanced Process using Single Flux Quantum Shift Register Chip with 68,990 Josephson Junctions

Shuichi Nagasawa and Mutsuo Hidaka

National Institute of Advanced Industrial Science and Technology (AIST), 1-1-1 Umezono, Tsukuba 305-8568, Japan

E-mail: s-nagasawa@aist.go.jp

Abstract. We have improved an Nb 9-layer advanced process (ADP2) using a plasma-enhanced chemical vapor deposition (PECVD) SiO_2 insulator. We could solve the problems due to the surface profile of the PECVD SiO_2 by adding an ion-milling process. The reliability of the ADP2 was evaluated by the operating yield of the SFQ shift registers (SRs). The SRs with six different bit sizes from 16 to 2560 were designed to evaluate the circuit size dependency of their yields. The total number of SRs is 16, and the total number of junctions per chip is 68,990. By introducing the PECVD SiO_2 insulator, we could obtain a high operating yield of more than 80% even for large-scale 2560-bit shift registers with more than 10,000 junctions. Moreover, we could obtain nine perfect chips (in which all the SRs in the chip functioned) out of 26 measured chips fabricated using the PECVD SiO_2. From these results, we believe that the fabrication yield of the ADP2 is at a level of 100,000 Josephson junction circuits.

1. Introduction

Superconducting digital devices are superior because they have lower power dissipation and higher speed operation compared to those of conventional semiconductor devices. However, the high performance of superconducting digital devices has not been achieved sufficiently since their integration level is still low. Fabrication technology that enables large-scale integration is required to realize the high performance of superconducting digital devices [1, 2].

We have been developing an Nb 9-layer advanced process (ADP2) for large-scale superconducting single flux quantum (SFQ) circuits [3]. In the ADP2, the bias-sputtered SiO_2 had been used for an inter-layer insulator, however we replaced the bias-sputtered SiO_2 with a plasma-enhanced chemical vapor deposition (PECVD) SiO_2, which has the advantages of good step coverage, low film stress, lower temperature process, and fewer particles. However, at first, there was a problem due to the surface profile of the PECVD film when applying it to the actual device structure.

Therefore, to improve the surface profile of the PECVD film, we introduced an ion-milling process. In this paper, we describe the inter-layer insulator forming methods using PECVD SiO_2 deposition and ion milling. Then, we describe the yield evaluation of the ADP2 by using shift registers (SRs).

2. Initial problem using PECVD and improved method

We introduced a PECVD system (Samco PD-270STL) in March 2014. The current process conditions of the PECVD SiO_2 are as follows: wafer temperature of 80°C, TEOS/O_2 gas pressure of 30Pa, flow rate of 2.5/75 sccm, and RF power of 125W. Although PECVD SiO_2 films have the aforementioned advantages, there was an initial problem due to the surface profile. Figure 1 shows the difference of the surface profile between bias-sputtered SiO_2 and PECVD SiO_2 films. The bias-sputtered SiO_2 film has an appropriate angular slope on the step edges of the underlying Nb patterns (figure 1 (a)). On the other hand, the PECVD SiO_2 film has a surface profile with sharp step edges and narrow concaves (figure 1 (b)). After the upper Nb layer is formed, etching residues tend to be left in these regions. This was a serious problem since these etching residues led to intra-layer leaks of the upper Nb patterns.

Therefore, we improved the surface profile of the PECVD SiO_2 film by adding an ion-milling process after PECVD SiO_2 deposition. Figure 2 shows the process flow of the inter-layer insulator forming methods using the PECVD SiO_2 deposition and the ion milling. First, 400 nm SiO_2 film is deposited by PECVD over the Nb patterns with 400 nm thickness. Then, about half the thickness of the SiO_2 film is etched by Ar ion milling. The ion beam is accelerated with a voltage of 400 eV at an Ar pressure of 5×10^{-2} Pa. The wafer is mounted on the plate that rotates during the milling operation, and the incident angle of the ion beam is 20 degrees to the rotation axis of the plate. By using this ion milling, we can eliminate sharp edges, and obtain a smooth surface profile. Then, additional PECVD SiO_2 is deposited to obtain the desired thickness. We could solve the problem of the intra-layer leaks by using this method.

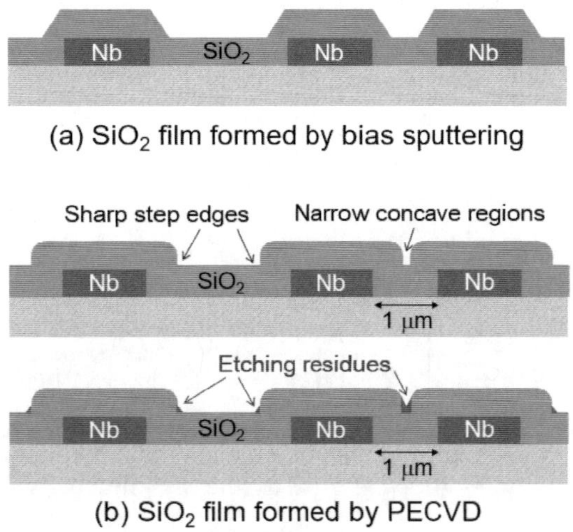

Figure 1. Difference of surface profile between bias-sputtered SiO_2 and PECVD SiO_2 films.

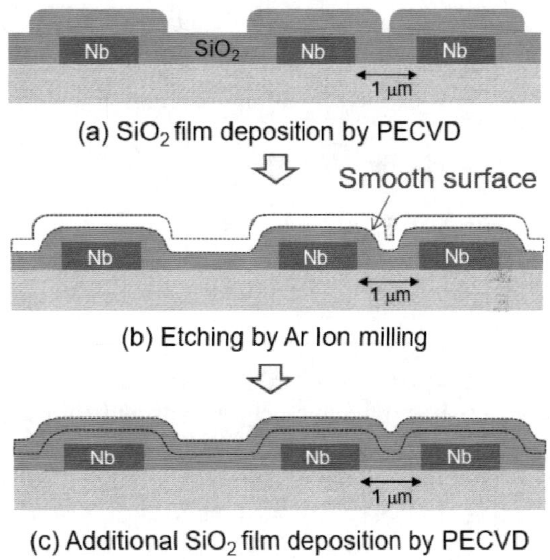

Figure 2. Inter-layer insulator forming methods using PECVD SiO_2 deposition and ion milling.

3. Nb 9-layer advanced process using PECVD SiO_2 insulator

Figure 3 shows the device structure of the Nb 9-layer advanced process (ADP2) [3]. Every Nb layer for M1-M7 is planarized by the caldera planarization technology except for the upper 2-Nb layers. We simply replaced bias-sputtered SiO_2 with PECVD SiO_2 for planarized insulator layers between M1 (DCP) and M8 (BAS) since the above mentioned sharp step edges and narrow concaves of the PECVD SiO_2 are eliminated by the planarization before forming the upper Nb layer. However, for the insulator layer between M8 (BAS) and M9 (COU) which is not planarized, we introduced the technology described in section 2 to eliminate the problem of the intra-layer leaks. Moreover, to increase reliability, we added a 2nd ion milling process to the process flow of figure 2, as shown in Table 1. This procedure, which uses PECVDs and ion millings is also used for our unplanarized standard processes (STP2, HSTP).

Table 1. Process flow for interlayer insulation between M8 (BAS) and M9 (COU).

Step	Process	Thickness
1	1st PECVD	400 nm
2	1st Ar Milling	- 200 nm
3	2nd PECVD	400 nm
4	2nd Ar Milling	- 200 nm

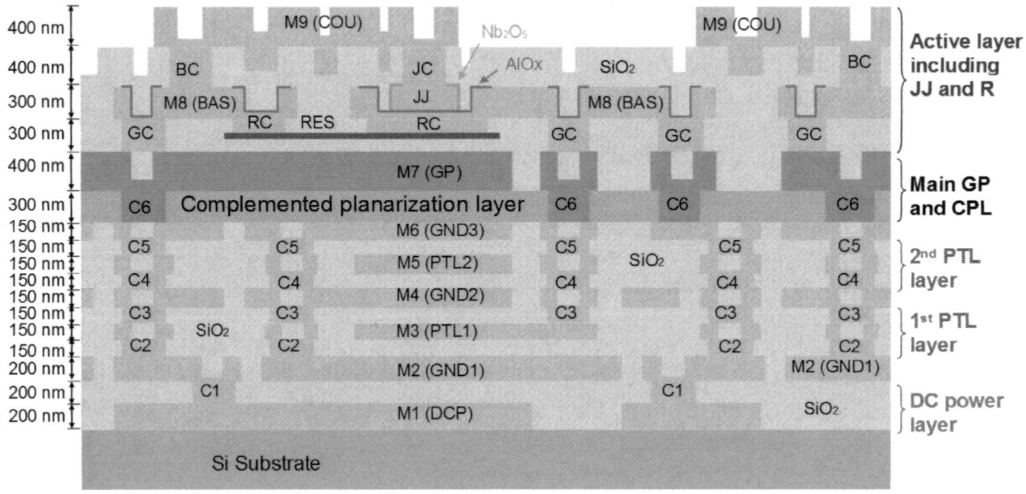

Figure 3. Device structure of Nb 9-layer advanced process (ADP2).
JJ: Nb/AlOx/Nb junction, M1-M9: Nb layers, M2, M4, M6, M7: Ground planes, RES: Mo resistor (thickness: 45 nm), C1-C6, GC, RC, BC, JC: Contacts between metal layers, SiO_2: Interlayer insulator.

4. Yield evaluations of ADP2 using shift registers

Process evaluations have been performed for every fabrication by using both the diagnostic chips and shift register (SR) chips [3]. In this paper, we describe the yield evaluation using shift registers.

Figure 4 is a schematic of the shift register cell, and figure 5 is its device structure and layout design. The cell is composed of four junctions, inductors, and resistors. Since this shift register was designed to evaluate the reliability of the fabrication process (ADP2), most components were composed with minimum feature sizes (JJ: 1 μm square, line width: 1 μm, contact: 0.7 μm square) for ADP2. Since the DC bias is fed from the bottom DCP layer to the upper junction layer through the DCVIA, all layers are used. However, since this SR does not need PTL interconnections, dummy patterns composed of PTL1 and PTL2 layers are included in one of the two cells as shown in figure 5 (b).

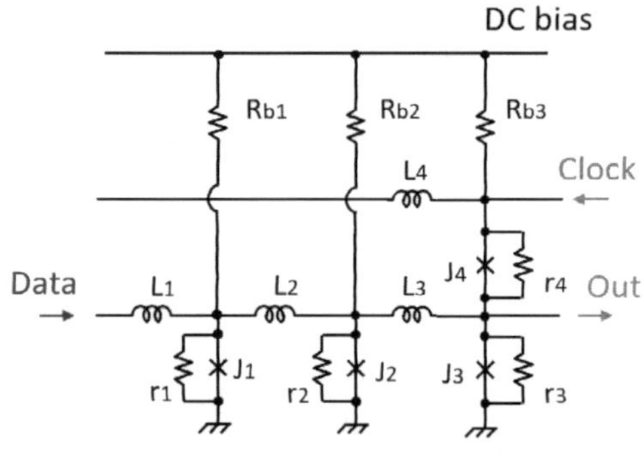

Figure 4. Schematic of shift register cell.
J_1 = 0.1mA, J_2 = 0.12mA, J_3 = 0.11mA, J_4 = 0.1mA, L_1 = 12 pH, L_2 = 10 pH, L_3 = 20 pH, L_4 = 3.9 pH, DC bias = 0.172mA

Figure 6 (a) shows the layout of a shift register chip. SRs with six different bit sizes from 16 to

2560 were designed to evaluate the circuit size dependency of their yields. The specifications (the number of Josephson junctions, bias current values, and number of circuits) are listed in Table 2. The details of the SR design have been described in other papers [4, 5].

Figure 6 (b) shows the shot map of the ADP2. The shift register chips (C9) are included in these positions on the 3-inch wafers for every fabrication run of the ADP2.

Table 2. Shift register chip specifications. Total junctions per chip: 68,990.

Bit capacity of SR	No. of JJ	SR_bias (mA)	No. of SRs
16-bit	90	2.75	2
64-bit	282	11	2
160-bit	666	28	2
640-bit	2589	110	2
1280-bit	5153	220	4
2560-bit	10281	440	4

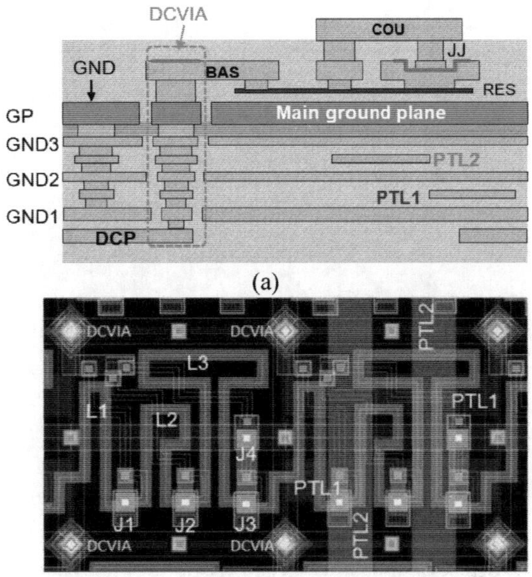

Figure 5. (a) Device structure of SR cell, (b) Layout for SR cell (2 bits).

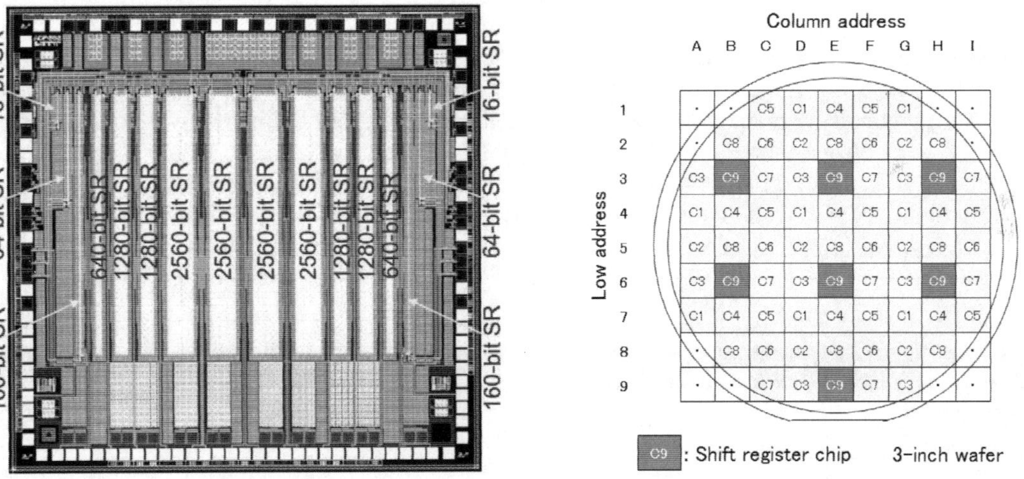

Figure 6. (a) Layout of shift register chip (chip size: 7.1mm × 7.1mm), (b) Shot map of ADP2.

Figure 7 shows the run-to-run variations of the yields for the SRs. In this figure, to clarify differences in the circuit size dependency of the yield, only two kinds of SRs are shown: the smallest 16-bit SR in the blue bar graphs and the largest 2560-bit SR in the red bar graphs. For each fabrication run, since the number of measured circuits is different, the number of measured circuits and the number of correct operations for 2560-bit SRs are shown on each yield bar graph. For example, for ADP644 No.2, sixteen 2560-bit SRs were measured, and correct operations were obtained for fifteen circuits of 2560-bit SRs, giving a yield of 93.8%. We replaced the bias-sputtered SiO_2 with the PECVD SiO_2 from ADP643 No.1. Although the yields of the small-scale 16-bit SRs were so far relatively high, the yields of the large-scale 2560-bit SRs had a large run-to-run variation and were often very small or zero. However, we could obtain very high yields with a good run-to-run reproducibility for the fabrication runs after the implementation of the PECVD SiO_2.

No. of meas.	8	8	8	8	8	8	8	8	8	8
No. of oper.	5	1	7	1	0	0	3	1	1	5

No. of meas.	12	12	12	12	12	12	12	12	12	12	12	12	12	12	12	12	12	12
No. of oper.	1	1	3	0	0	0	0	0	1	0	0	0	1	3	3	0	2	8

No. of meas.	12	12	8	8	12	16	24	12	for 2560-bit SR
No. of oper.	7	10	7	7	10	15	21	8	

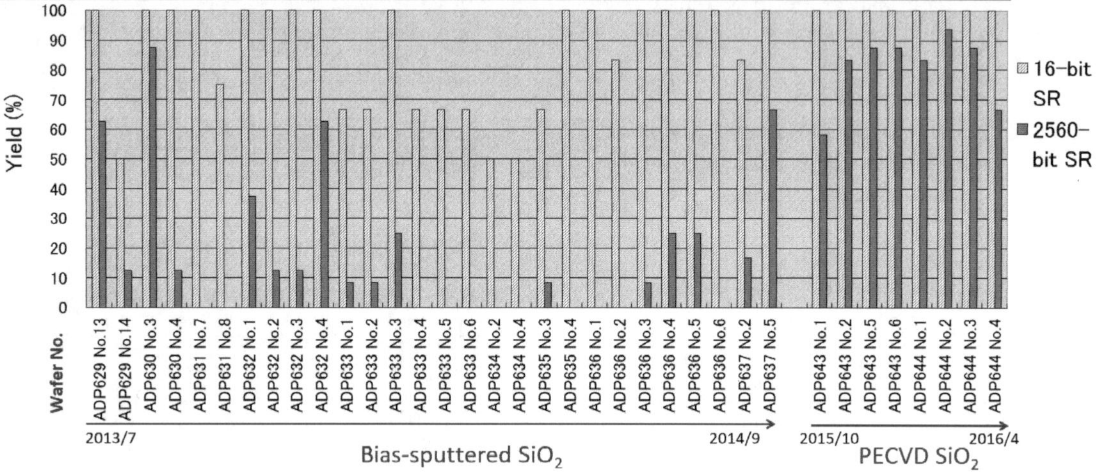

Figure 7. Run-to-run variation of yields for 16-bit and 2560-bit SRs. Here, yield means ratio of number of correct operation circuits to number of measured circuits.

Figure 8 shows a summary of the measured operating yield depending on the circuit size for the fabrication runs before and after using the PECVD. The number of measured circuits, number of correct operating circuits, and operating yields depending on the circuit size are shown on the bar graph. These tables are summaries of 74 measured chips of the fabrications before using the PECVD, and 26 measured chips of the fabrications after using the PECVD. Before using the PECVD, the operating yields decrease drastically with increasing bit capacity, as shown in figure 8 (a). We obtained only 47 correct operation circuits out of the 296 measured circuits for the largest 2560-bit SRs, that is, a poor operating yield of 15.9%. On the other hand, after using the PECVD, the reduction of the operating yields is very small, as shown in figure 8 (b). We could obtain 85 correctly operating circuits out of the 104 measured circuits, that is, a high operating yield of 81.7% even for the large-scale 2560-bit SRs with more than 10,000 junctions.

Bit Capacity	16	64	160	640	1280	2560
No. of meas.	148	148	148	148	296	296
No. of oper.	125	110	93	61	94	47
Yield (%)	84.5	74.3	62.8	41.2	31.8	15.9

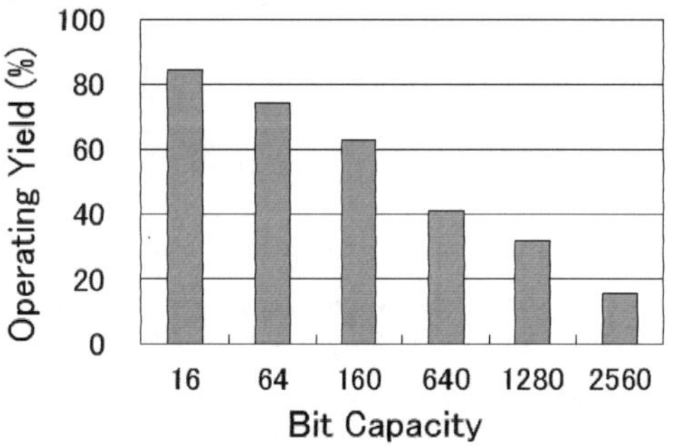

(a)

Bit Capacity	16	64	160	640	1280	2560
No. of meas.	52	52	52	52	104	104
No. of oper.	52	51	50	45	93	85
Yield (%)	100.0	98.1	96.2	86.5	89.4	81.7

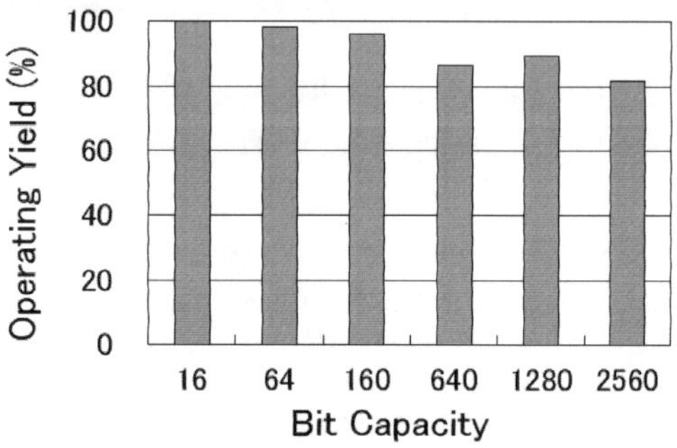

(b)

Figure 8. Measured operating yield depending on circuit size of 16- to 2560-bit SRs.
(a) Summary for fabrication runs (ADP629 No. 13 – ADP637 No. 5) using bias-sputtered SiO_2.
(b) Summary for fabrication runs (ADP643 No. 1 – ADP644 No. 4) using PECVD SiO_2.

Figure 9 shows an example of measured bias margins for SRs in a perfect chip (ADP644 No.2 Chip E6). The perfect chip means that all SRs in the chip were correctly operated. In this figure, the bias margins of top-to-bottom SRs correspond to those of left-to-right SRs in the chip shown in figure 6 (a). The wide operating margins of more than ±20% were obtained for all SRs. The reduction of the bias margin is small for increasing the circuit size from 16bit to 2560-bit. This means that circuit parameter variations due to the fabrication process are very small. In this chip, all SR circuits were correctly operated, which means that the SFQ circuits including a total of 68,990 Josephson junctions were successfully operated.

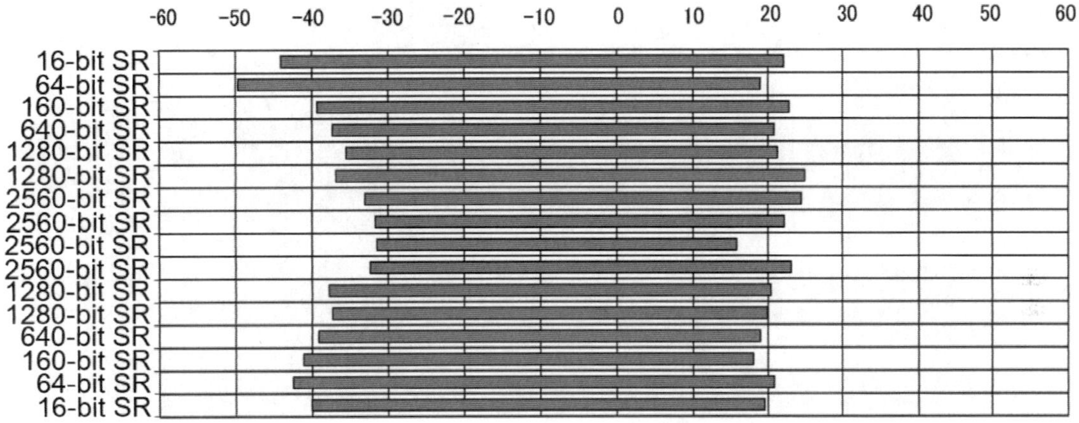

Figure 9. Measured bias margins for SRs in perfect chip (ADP644 No.2 Chip E6). Horizontal axis indicates ratio (%) for designed SR bias currents, as shown in Table 2.

Figure 10 shows the yields of the SRs on individual chips for the two wafers before and after using the PECVD. In these figures, the yields are represented as percentages of the correctly operated SRs for all the 16 measured SRs, regardless of the differences in their bit capacities and measured bias margins. These are examples of good fabrication runs including perfect chips. Before using the PECVD SiO_2, there was a large difference of the yields as shown in figure 10 (a) even in the good fabrication run. However, after using the PECVD SiO_2, the uniformity of the yields in the wafer has substantially improved, as shown in figure 10 (b). We could obtained three perfect chips within the measured 6 chips in this wafer. Other three chips were also very high yields. In this wafer, the number of the malfunctioning SRs was just 4. The number of defects is also considered to be just 4 or so in the SRs including in total 413,940 Josephson junctions for the 6 chips since these malfunctions are considered to correspond to defects on a one-to-one basis. This defect number shows that the fabrication yield of our ADP2 is at a level of 100,000 Josephson junction circuits. Moreover, we could successfully obtain nine perfect chips out of the 26 measured chips fabricated recently using the PECVD SiO_2.

5. Discussion

From the above mentioned results, we consider that the reliability of the ADP2 has improved substantially by the interlayer insulator forming methods using the PECVD SiO_2 deposition and ion milling.

The run-to-run reproducibility of the fabrication using the bias-sputtered SiO_2 was not good. We think that this was mainly due to the number of adherent particles on the bias-sputtered SiO_2 films. Many particles often occurred during bias-sputtering depositions since SiO_2 films adhered inside the bias-sputtering chamber were easy to peel due to their high film stress. Most causes for the defects such as open or short circuits and degradation of the junction characteristics are considered to be in these particles [6].

On the other hand, since our PECVD system is kept in a good condition by scheduled cleanings for both the PECVD chamber using C_2F_6 gas plasma and the SiC substrate tray using buffered hydrofluoric acid (NH_4F and HF mixture), it is considered that the PECVD SiO_2 films with very few particles were formed with an excellent run-to-run reproducibility. We could replace the bias-sputtered SiO_2 with the PECVD SiO_2 to the actual device structure of the ADP2 by adding the ion milling process. We believe that this led to the substantial improvement of the yields and run-to-run reproducibility for the SRs.

Chip No.	B3	B6	D3	D6	F3	F6
No. of Meas.	16	16	16	16	16	16
No. of oper.	8	14	6	6	15	16
Yield (%)	50.0	87.5	37.5	37.5	93.8	100.0

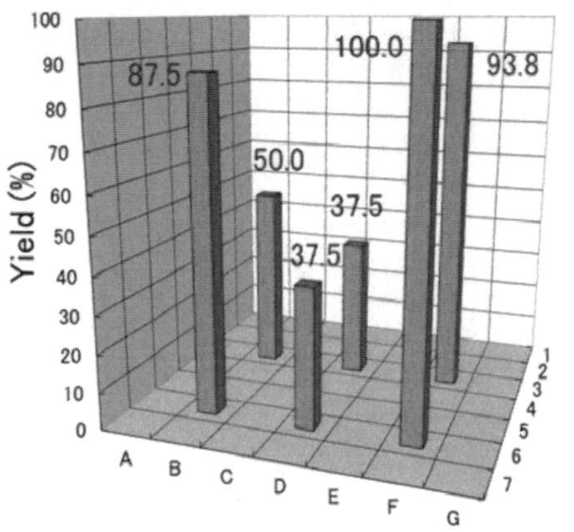

Chip No.	B3	B6	E3	E6	H3	H6
No. of Meas.	16	16	16	16	16	16
No. of oper.	15	16	15	16	14	16
Yield (%)	93.8	100.0	93.8	100.0	87.5	100.0

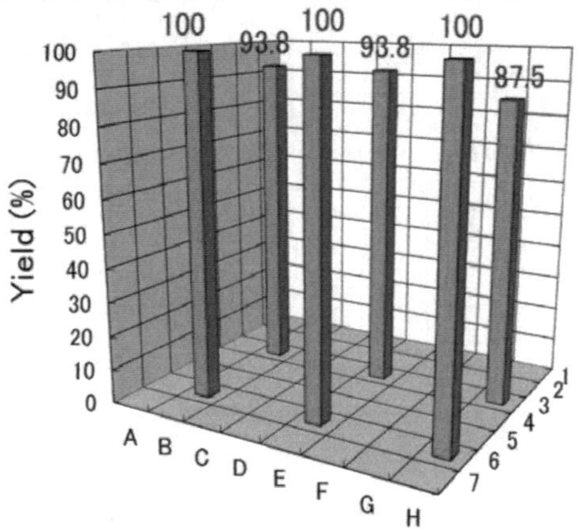

(a) (b)

Figure 10. Examples of yields of shift registers depending on chip positions of wafer fabricated using (a) bias-sputtered SiO_2 (ADP623 No. 2) and (b) PECVD SiO_2 (ADP644 No. 3).

6. Conclusion

We replaced bias-sputtered SiO_2 with PECVD SiO_2 for the Nb 9-layer advanced process (ADP2). The surface profile of the PECVD SiO_2 film was improved by adding an ion-milling process after the PECVD SiO_2 deposition. We solved the problems of the PECVD by using the improved method. For the recent fabrication runs using PECVD SiO_2, we could obtain a high operating yield of more than 80% with an excellent run-to-run reproducibility even for the large scale 2560-bit SRs with more than 10,000 junctions. Moreover, we could successfully obtain nine perfect chips out of the 26 measured chips fabricated using the PECVD SiO_2. From these results, we believe that the reliability of the ADP2 has improved substantially by the inter-layer insulator forming methods using the PECVD SiO_2 deposition and ion milling, and the fabrication yield of our ADP2 is at a level of 100,000 Josephson junction circuits.

Acknowledgments

This work is supported by the ALCA-JST project, and the fabrications were performed in the clean room for analog-digital superconductivity (CRAVITY) of AIST.

References

[1] Tolpygo S K, Bolkhovsky V, Weir T J, Johnson L M, Gouker M A and Oliver W D 2015 Fabrication process and properties of fully-planarized deep-submicron Nb/Al-AlOx/Nb Josephson junctions for VLSI circuits *IEEE Trans. Appl. Supercond.* **25** 1101312

[2] Tolpygo S K, Bolkhovsky V, Weir T J, Wynn A, Oates D E, Johnson L M and Gouker M A 2016 Advanced fabrication processes for superconducting very large-scale integrated circuits *IEEE Trans. Appl. Supercond.* **26** 1100110

[3] Nagasawa S, Hinode K, Satoh T, Hidaka M, Akaike H, Fujimaki A, Yoshikawa N, Takagi K and Takagi N 2014 Nb 9-layer fabrication process for superconducting large-scale SFQ circuits and its process evaluation *IEICE Trans. Electron.* **E97-C** 132-140

[4] Nagasawa S, Satoh T, Hinode K, Kitagawa Y, Hidaka M, Akaike H, Fujimaki A, Takagi K, Takagi N and Yoshikawa N 2009 New Nb multi-layer fabrication process for large-scale SFQ circuits *Physica C* **469** 1578-84

[5] Nagasawa S, Hinode K, Satoh T and Hidaka M 2011 Evaluation of advanced fabrication process using both diagnostic chips and shift register chips *Proceedings of Superconducting SFQ VLSI Workshop* 77-82

[6] Hidaka M, Nagasawa S, Satoh T and Hinode K 2015 Defects of Nb/AlOx/Nb Josephson junctions caused by underneath fine particles *15th International Superconductive Electronics Conference (ISEC)* DOI: 10.1109/ISEC.2015.7383437

Thermally-fluctuated single-flux-quantum pulse intervals reflected in input-output characteristics of a double-flux-quantum amplifier

Yoshinao Mizugaki, Yoshiaki Urai and Hiroshi Shimada

Department of Engineering Science, The University of Electro-Communications (UEC Tokyo), Chofu, Tokyo 182-8585, Japan

E-mail: y.mizugaki@uec.ac.jp

Abstract. A double-flux-quantum amplifier (DFQA) is a voltage multiplier of quantum accuracy, which we have employed at the final stage of a single-flux-quantum (SFQ) digital-to-analog converter (DAC). We recently found that experimental input–output (IO) characteristics of DFQAs were always slightly different from numerical results assuming ideally-periodic SFQ pulse trains. That is, experimental IO characteristics obtained using an over-biasing method were gradually deteriorated near their maximum operation voltages. Numerical simulation including the over-biasing method at a finite temperature suggested that the difference was likely to be attributed to thermally-fluctuated intervals of input SFQ pulses.

1. Introduction

Over-biasing a Josephson junction beyond its critical current is a simple method for feeding SFQ pulse trains into Josephson circuits, which we refer to as an "over-biasing method." We have employed the over-biasing method for evaluation of double-flux-quantum amplifiers (DFQAs) [1], especially for determination of their maximum input voltage $V_{\mathrm{IN-MAX}}$ and the corresponding Josephson frequency $f_{\mathrm{IN-MAX}} = V_{\mathrm{IN-MAX}}/\Phi_0$, where Φ_0 is a flux quantum [2, 3]. We have recently found that there is always slight difference between the experimental and numerical input-output (IO) characteristics near $V_{\mathrm{IN-MAX}}$ [4], for which the input SFQ pulses are fed by the over-biasing method in experiments and by periodic current pulses in simulation.

In this paper, we present experimental IO characteristics of a 100-fold DFQA and compare them with numerical results assuming different input methods of SFQ pulses. Numerical simulation including the over-biasing method at a finite temperature suggests that thermally-fluctuated pulse intervals gradually deteriorate IO characteristics of the DFQA near $V_{\mathrm{IN-MAX}}$ when the over-biasing method is employed for feeding SFQ pulses.

2. Experiments

We designed a 100-fold DFQA including 99-stacked three-junction loops (3JLs). The circuit configuration with the measurement set-up is shown in figure 1(a). In experiments, the input SFQ pulse train was generated by over-biasing a Josephson junction in front of a 10-stage Josephson transmission line (JTL). In other words, the input voltage V_{IN} was generated by

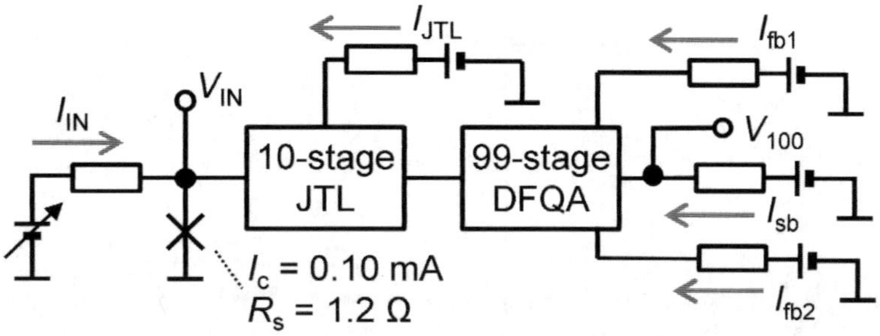

(a)

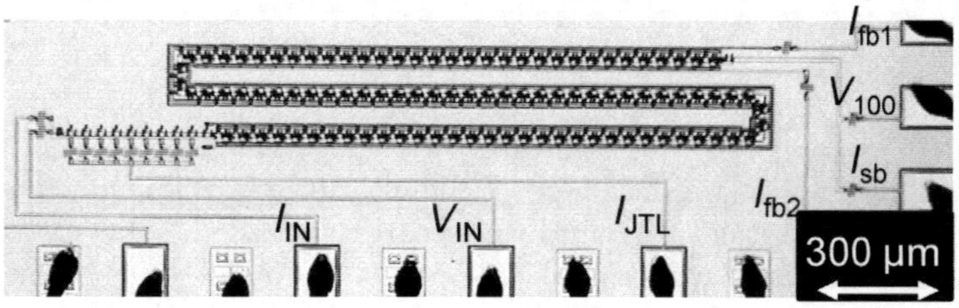

(b)

Figure 1. (a) Schematic configuration of a tested 100-fold DFQA comprising 99-stage 3JLs with the measurement set-up. (b) Photomicrograph of a test circuit fabricated using the AIST Nb STP2.

applying excess input current $I_{\rm IN}$ to the Josephson junction. The input and output voltages ($V_{\rm IN}$ and V_{100}) were measured using digital multi-meters.

Test chips were fabricated using the AIST 25-μA/μm^2 Nb standard process 2 (STP2). A photomicrograph of a fabricated circuit is shown in figure 1(b). In measurement, a test chip was cooled down in liquid helium.

3. Results

The experimental IO ($V_{\rm IN}$–V_{100}) relationships are presented in figure 2 as cross marks, where the ideal 100-fold voltage multiplication ($V_{100} = 100V_{\rm IN}$) is indicated by a solid straight line. It should be noted again that the over-biasing method was used for feeding SFQ input pulses. $V_{\rm IN-MAX}$ and $f_{\rm IN-MAX}$ are determined as 69.2 μV and 33.4 GHz, respectively, with multiplication errors less than 1%. These are typical values for our 100-fold DFQAs. It is confirmed that the experimental results gradually separated from the ideal $V_{100} = 100V_{\rm IN}$ line as $V_{\rm IN}$ increases beyond $V_{\rm IN-MAX}$.

Numerical simulation was executed assuming an ideally-periodic pulse current source for SFQ feeding. That is, periodic SFQ pulses were fed to the DFQA. The numerical IO characteristics are plotted as crosses combined with plus marks in figure 2. It is confirmed that numerical V_{100} follows the ideal relationship up to 77 μV, of which the corresponding frequency is 37 GHz, and that the numerical V_{100} values drop rapidly for $V_{\rm IN} > 77$ μV. Such rapid drop is not observed in the experimental results.

4. Discussion

To figure out the origin of such gradual degradation observed in the experimental IO characteristics, we employed the second simulation model where the over-biasing method was used for feeding the input SFQ pulses. Thermally-induced currents in resistors at 4.2 K were also included [5].

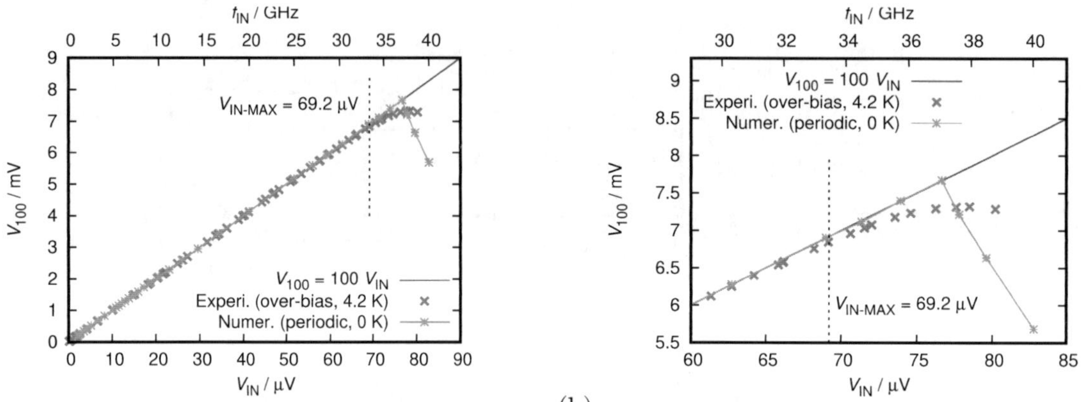

(a) (b)

Figure 2. Experimental and numerical V_{IN}–V_{100} characteristics. The lower and upper horizontal axis represent the input voltage V_{IN} and the corresponding SFQ repetition frequency f_{IN}, respectively. The vertical axis represents the output voltage V_{100}. The over-biasing method was used in experiments for feeding SFQ input pulses. In simulation, periodic SFQ pulses were fed into the 100-fold DFQA. (a) Wide range. (b) Narrow range near V_{IN-MAX}.

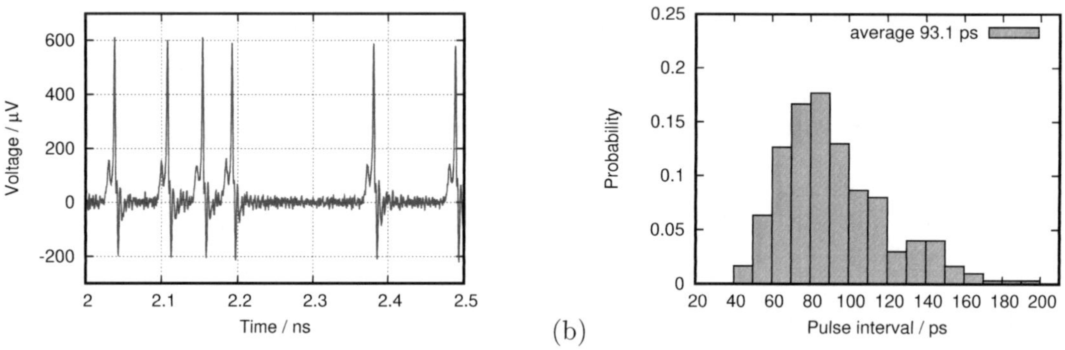

(a) (b)

Figure 3. (a) Numerical input voltage waveform generated using the over-biasing method at 4.2 K. I_{IN} was adjusted to realize SFQ pulse intervals of 93.1 ps on average. SFQ pusles are not exactly periodic. (b) Histogram of SFQ pulse intervals for $V_{IN} = 22.2\ \mu$V, for which the corresponding f_{IN} and $1/f_{IN}$ are 10.7 GHz and 93.1 ps. The pulse intervals deviate between 40 and 200 ps.

An example of input voltage waveforms generated using the over-biasing method is shown in figure 3(a). I_{IN} was adjusted to realize SFQ pulse intervals of 93.1 ps on average, for which V_{IN} was calculated to be 22.2 μV. It is confirmed that pulse intervals are fluctuated. Figure 3(b) shows a histogram of 300 SFQ pulse intervals. (We have confirmed that the number of pulse intervals should be more than 100 for the statistical processing in this work.) The pulse intervals deviate between 40 and 200 ps. That is, SFQ pulses generated using the over-biasing method at 4.2 K are not exactly periodic.

Such deviated pulse intervals should be reflected in the IO characteristics of DFQAs near V_{IN-MAX}. Open squares in figure 4(a) represent the IO characteristics obtained using the second simulation model, that is, the circuit model including the over-biasing method at 4.2 K. Although there are still quantitative difference for the voltage range of $V_{IN} > 77\ \mu$V, gradual

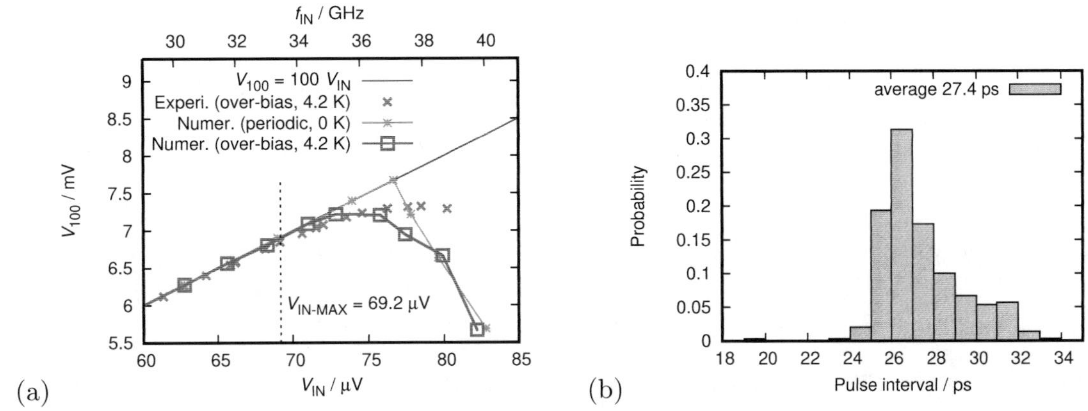

Figure 4. (a) Numerical V_{IN}–V_{100} characteristics of the second simulation model shown with other data on figure 2(b). The over-biasing method at 4.2 K was included in simulation. (b) Histogram of SFQ pulse intervals for $V_{IN} = 75.6$ μV ($f_{IN} = 36.5$ GHz, $1/f_{IN} = 27.4$ ps). Pulse intervals shorter than 27 ps are out of the correct operation range of the DFQA.

degradation similar to the experimental V_{IN}–V_{100} characteristics are reproduced.

Figure 4(b) shows a histogram of 300 intervals in the input SFQ pulse train for $V_{IN} = 75.6$ μV. The corresponding f_{IN} and $1/f_{IN}$ are 36.5 GHz and 27.4 ps, respectively. It is found that the pulse intervals deviate between 19 and 34 ps, of which the corresponding frequencies are 52 and 29 GHz. As described in the previous section, the numerical simulation demonstrates that the DFQA operates for periodic SFQ pulses up to 37 GHz, and thus, pulse intervals shorter than 27 ps (= 1/37 GHz) are out of the correct operation range. The percentage of intervals shorter than 27 ps is as much as 22%, which cannot be ignored from the viewpoint of DFQA operation. That is, thermally-fluctuated SFQ pulse intervals should be a reason for the gradual degradation observed in experiments.

5. Conclusion

We presented experimental IO (V_{IN}–V_{100}) characteristics of a 100-fold DFQA fabricated using the Nb integration technology. The voltage multiplication was gradually deteriorated near the maximum input voltage. Numerical simulation assuming periodic input SFQ pulses did not exhibit such gradual degradation, whereas another numerical simulation assuming the over-biasing method with thermal noise current demonstrated gradual degradation similar to the experimental results. The numerical results suggested that SFQ pulse trains generated by the over-biasing method contained non-negligible deviation of SFQ pulse intervals.

Acknowledgments

The authors are grateful to M. Maezawa for initiation of the present work. They also thank M. Moriya, Y. Takahashi, T. Watanabe, K. Sawada, and other lab members for fruitful discussion and technical supports. This work was partially supported by JSPS KAKENHI 15K13999, and by VLSI Design and Education Center (VDEC), the University of Tokyo in collaboration with Cadence Design Systems, Inc. The circuits were fabricated in the clean room for analog-digital superconduc-tivity (CRAVITY) of National Institute of Advanced Industrial Science and Technology (AIST) with the standard process 2 (STP2). The stable supply of liquid helium from the Coordinated Center for UEC Research Facilities is also acknowledged.

Appendix A. Distribution of SFQ pulse intervals generated using the over-biasing method

Although timing jitters induced by thermal noise have been investigated for two decades in the field of SFQ digital electronics [6], over-biased conditions are outside the scope of digital circuit operations. In this appendix, we describe the distributions of SFQ pulse intervals generated using the over-biasing method at 4.2 K.

Fitting results for the SFQ pulse intervals shown in figures 4(b) and 3(b) are respectively presented in figures A1(a) and A1(b). The normal distribution $N(\mu_{\mathrm{N}}, \sigma_{\mathrm{N}}^2)$ and the logarithmic normal distribution $LN(\mu_{\mathrm{LN}}, \sigma_{\mathrm{LN}}^2)$ are employed for fitting. Their probability density functions are expressed as $P_{\mathrm{N}}(x) = \left(\sqrt{2\pi}\sigma_{\mathrm{N}}\right)^{-1} \exp\left\{-\left(x-\mu_{\mathrm{N}}\right)^2 / \left(2\sigma_{\mathrm{N}}^2\right)\right\}$ and $P_{\mathrm{LN}}(x) = \left(\sqrt{2\pi}\sigma_{\mathrm{LN}}x\right)^{-1} \exp\left\{-\left(\ln x-\mu_{\mathrm{LN}}\right)^2 / \left(2\sigma_{\mathrm{LN}}^2\right)\right\}$, respectively.

In figure A1(a), there is no significant difference in the both fitting results for the average interval of 27.4 ps. On the other hand, the pulse intervals for the average interval of 93.1 ps are fitted better by using the logarithmic normal distribution, as shown in figure A1(b). Another fitting for the average interval of 215 ps (figure A1(c)) clearly demonstrates the adequacy of the logarithmic normal distribution.

In addition, the variances σ_{N}^2 and σ_{LN}^2 increase rapidly as the average intervals increases. This feature should be related to nonlinear characteristics of a Josephson junction. The I–V curve of a resistively-shunted-junction (RSJ) model is presented in figure A1(d), where the

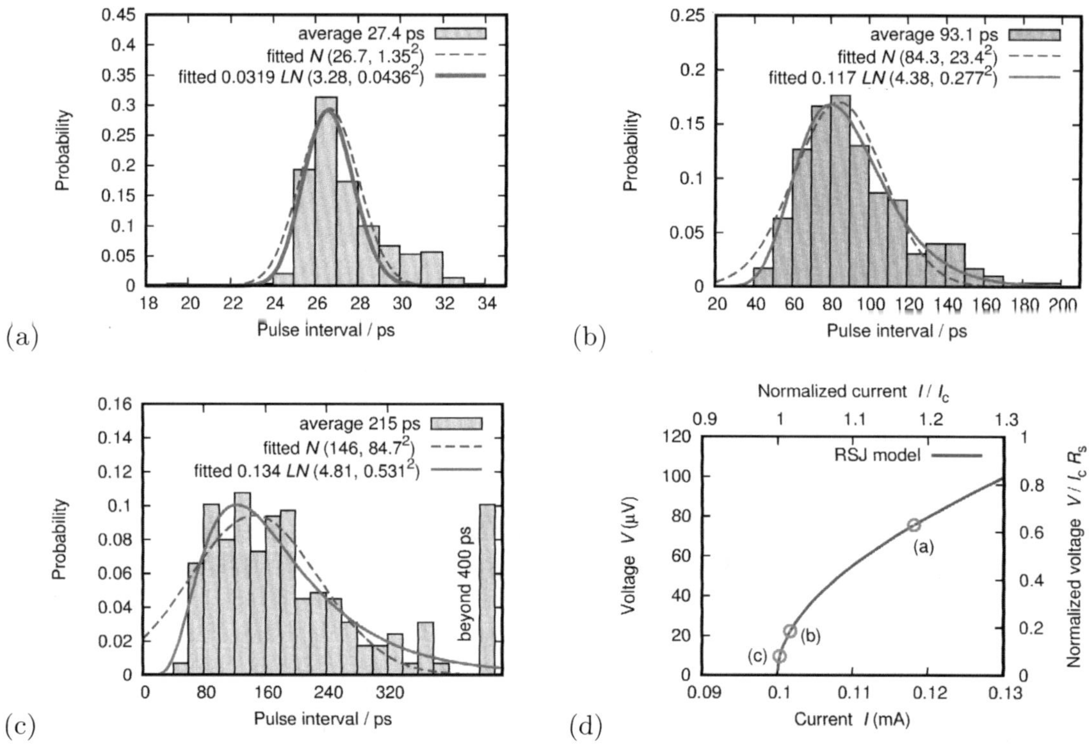

Figure A1. Probability distributions fitted to the SFQ pulse intervals of (a) 27.4, (b) 93.1, and (c) 215 ps on average. Both the normal distribution and the logarithmic normal distribution are employed for fitting. (d) I–V curve of an RSJ model. The average voltages corresponding to the SFQ pulse intervals of (a), (b), and (c) are indicated by open circles.

critical current and the shunting resistance are assumed to be 0.10 mA and 1.2 Ω. The average voltages corresponding to the average intervals for figures A1(a), A1(b), and A1(c) are indicated by open circles. The I–V curve is expressed as $V = R\sqrt{I^2 - I_c^2}$, whereas its derivative dV/dI is given by $RI/\sqrt{I^2 - I_c^2}$. (Since I_c is assumed to be much larger than the thermal current $2\pi k_B T/\Phi_0$ of 0.18 μA, the rounding effect around $I = I_c$ can be neglected [7].) dV/dI increases as I decreases down to I_c, which means that the voltage fluctuation is enhanced as the voltage decreases to zero. The voltage fluctuation is proportional to the frequency fluctuation, resulting in the enhancement of the variances σ_N^2 and σ_{LN}^2 at low voltages.

It should be noted that the large fluctuations at low voltages (long pulse intervals) do not modify the IO charcteristics of DFQAs unless they exceed the operation margins of DFQAs.

References

[1] Herr Q P 2005 Stacked double-flux-quantum output amplifier *IEEE Trans. Appl. Supercond.* **15** 259–62
[2] Sato Y, Moriya M, Shimada H, Mizugaki Y and Maezawa M 2013 Design and operation of 1000-fold voltage multiplier based on double-flux-quantum amplifier *Physics Procedia* **45** 221–4
[3] Mizugaki Y, Sato Y, Shimada H and Maezawa M 2014 Input-output characteristics of a 999-stage double-flux-quantum amplifier designed for 1000-fold voltage multiplication *Japan. J. Appl. Phys.* **53** 053101-1–4
[4] Urai Y, Shimada H and Mizugaki Y 2015 Feeding methods of SFQ pulse trains in evaluation of DFQ amplifiers *IEICE Technical Report* SCE2015-11 (in Japanese).
[5] Satchell J 1997 Stochastic simulation of SFQ logic *IEEE Trans. Appl. Supercond.* **7** 3315–8
[6] e.g., Ortlepp T and Uhlmann F H 2005 Noise induced timing jitter: a general restriction for high speed RSFQ devices *IEEE Trans. Appl. Supercond.* **15** 344–7
[7] Amegaokar V and Halperin B I 1969 Voltage due to thermal noise in the dc Josephson effect *Phys. Rev. Lett.* **22** 1364–6

Experimental and simulation results of a symmetrical pad to reduce a stray ground current in superconducting integrated circuits

H Suzuki, T Ono and N Yoshikawa

Yokohama National University, 79–1, Tokiwadai, Hodogaya-ku, Yokohama , 240–8501, Japan

suzuh@ynu.ac.jp

Abstract. The method of extracting bias current is widely used in large scale SFQ circuits to avoid harmful effect by spreading the bias return current on a ground plane, which causes malfunction or degrading the operating margin by the stray magnetic field. Although extracting the bias current is effective, we found that the extraction using a normal pad structure with a bias pad and a current extracting pad still produces a large stray ground current (magnetic field) for its asymmetric structure. We proposed a bilaterally symmetrical pad structure, which can reduce the stray ground current significantly in superconducting circuits. The bias current is extracted from a nearby ground pad of a bias pad in usual method, whereas the bias current is extracted from two of the symmetrical ground pads which are arranged both sides of the bias pad in the symmetrical structure. Then, the bias current is extracted on halves from each of the two pads symmetrically, reducing the stray ground current. First, we have simulated the ground current by using FastHenry program. The results showed that the symmetrical pad structure outside of ground plane reduced spread of the ground return current significantly. Next, we have designed a test chip and confirmed the effectiveness of the symmetrical structure experimentally.

1. Introduction

Integration level of superconducting circuits has been increased step by step, especially in single flux quantum (SFQ) circuits [1]–[4]. In the SFQ LSI circuits with several to ten thousand junctions, the bias currents exceed 1 A. The bias currents are applied to circuit elements through bias lines from bias supply pads and returned through a ground plane in SFQ circuits. The return current induces undesirable magnetic field, which causes harmful effect in the circuit operation [5]–[7]. Several methods are used for reducing the effect [8]–[12]. One of the methods is extracting the bias current from the neighboring pad of the bias current supply pad. This improved the stability and the bias current margins in the circuit operations. However, the harmful effect of the ground current seems to be still existed, especially in large-scale circuits.

We have analyzed the current distribution on the ground plane by using 3D inductance calculation program, FastHenry [13]. Then, we have found that relatively small difference of the position of the pads for supplying and extracting bias current produces large spread of the ground current. To avoid the spread, we proposed a symmetric pad structure outside of the ground plane for extracting bias current symmetrically [14]. According to the simulation using FastHenry program, the current on the ground plane in this symmetrical structure was reduced more than one order of magnitude than that of the usual asymmetric pad structure with bias extraction in practically usable circuit area.

To verify the results of the simulation, we have designed a test chip and confirmed the effectiveness of the symmetric pad structure experimentally.

2. Basic analysis of bias current extraction using FastHenry program

2.1. *Model for FastHenry analysis*

Usually, bonding pads are arranged peripherally in a chip and bias current is applied from a bias pad through a bias line to circuit elements. We made a simple analysis model with extracting the bias return current as shown in figure 1. The bias current is applied from Port1 and is flowed to the ground plane (GP) at Port2 through the bias line. Then, the current is extracted from Port3 connecting to the GP. The ground plane was 5 mm × 5 mm in size which is usually same as the die size, and the bias line was 50 μm in width. Port2 was located in the center of the chip. The number of filaments and the ratio of adjacent filaments along the thickness, Z, were set to five and two, respectively, as shown in figure 1. The number of segments for X and Y direction in ground plane was set to 200 or 400 in the following analysis. The London penetration depth of the bias line and ground plane was set to 87 nm.

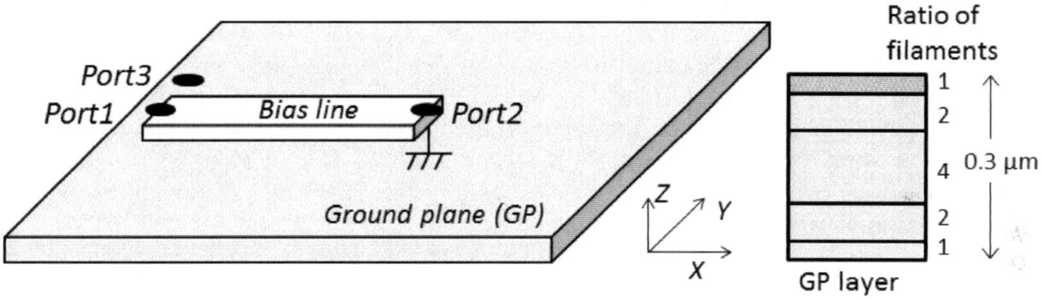

Figure 1. Basic analysis model of bias current and ground return current.

2.2. *Results in Bias Current Extraction*

First, we have analyzed for the simple structure with bias current extraction. Figure 2 shows the color contour map of the return current in the top filament on the ground plane, in which the number of the segment is 400 × 400. The bias current was normalized with 1 A, which was flowed on the 50 μm-wide bias line. The numerical number in the legend such as 2.940×10^7 means current density parallel to X and Y plane of the top filament with thickness of 0.03 μm on GP in units of A/m^2. Note that the ground current of three or four orders less current should be considerable because the critical current and operating current is several hundred micro amperes in SFQ circuits.

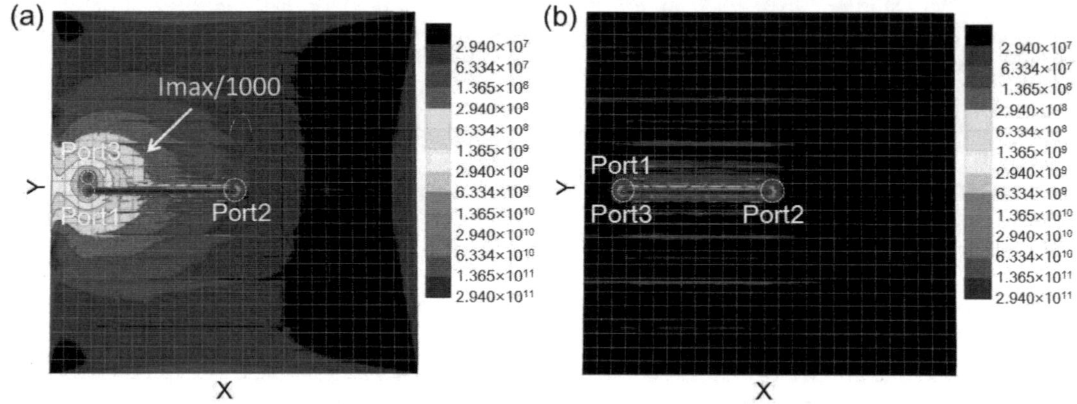

Figure 2. Colour contour maps of simulated ground current in the case of bias extraction, (a) when the bias current supply port, Port1, and the bias current extraction port, Port3, are in different position, (b) when Port1 and Port3 are in the same position.

The ground return current is color contour mapped with 12 levels in four orders of log scale. The return current initially flows back under the bias line and then spreads over the ground plane in near Port1 and Port3. The area flowing larger than 1/1000 of the maximum current which is the current under the bias line was spread to around 1mm far from Port1 and Port3. Moreover, the area larger than 1/10000 of the maximum current extend to more than half of the chip. In figure 2(a), the bias current extracting position, Port3, was 200 µm apart from the bias current supplying position, Port1, which seemed to be relatively small. However, the small difference produced widely spread ground current. In figure 2(b), Port3 and Port1 were placed in the same position. This ideal case showed no current spread on a ground plane in the Ports. Note that the current spread is large even with relatively small difference of 200 µm between current supplying and extracting points.

3. Analysis of ground return current for four pad structures

The distribution of the ground current for the four structures shown in figure 3 was simulated by FastHenry. In figure 3, (a) and (c) have asymmetrical pad structure, whereas (b) and (d) symmetric structure. In addition, pads are place on the ground plane in (a) and (b), whereas (c) and (d) outside of the ground plane. The simulation condition was the same as that described in Section 2. The number of the segment was set to 200 × 200 to save the calculation time, since there was no obvious difference for that of 400 × 400 segments. Figure 4(a)–(d) show color contour maps of the ground current, which correspond to the structure shown in figure 3(a)–(d). All the pad size is 100 × 100 µm and the distance between the bias pad and the extraction pad is also 100 µm. The width of the bias line is 50 µm in width. When compared asymmetric pad structures, (a) and (c), with symmetric ones, (b) and (d), the spread of the ground current was reduced obviously for the later ones. Moreover, when compared the pads on GP, (a) and (b), to the pads outside of GP, (c) and (d), the latter ones can reduce the current spread.

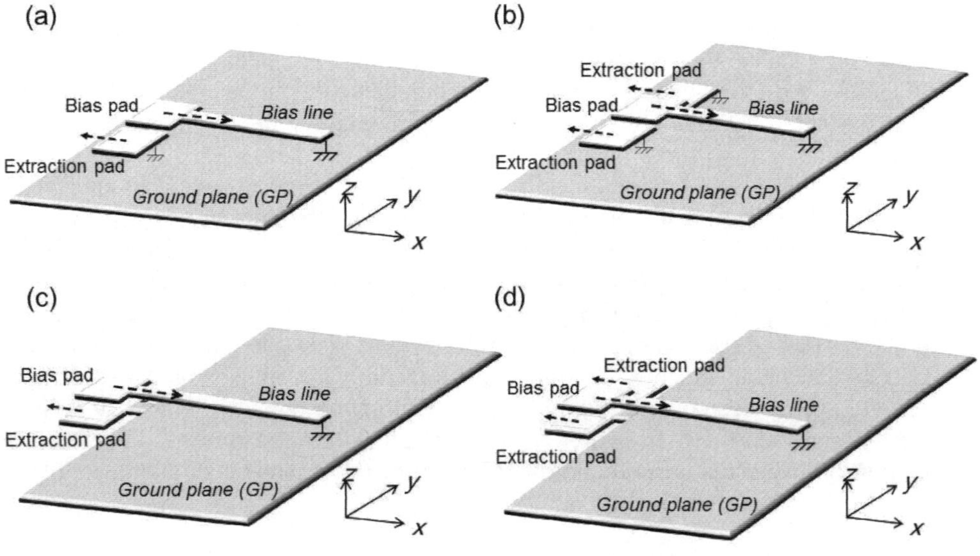

Figure 3. Four pad structures for simulation; (a) normal asymmetrical pad structure on a ground plane, (b) symmetrical pad structure on a ground plane, (c) asymmetrical pad structure outside of a ground plane, and (d) symmetrical pad structure outside of a ground plane.

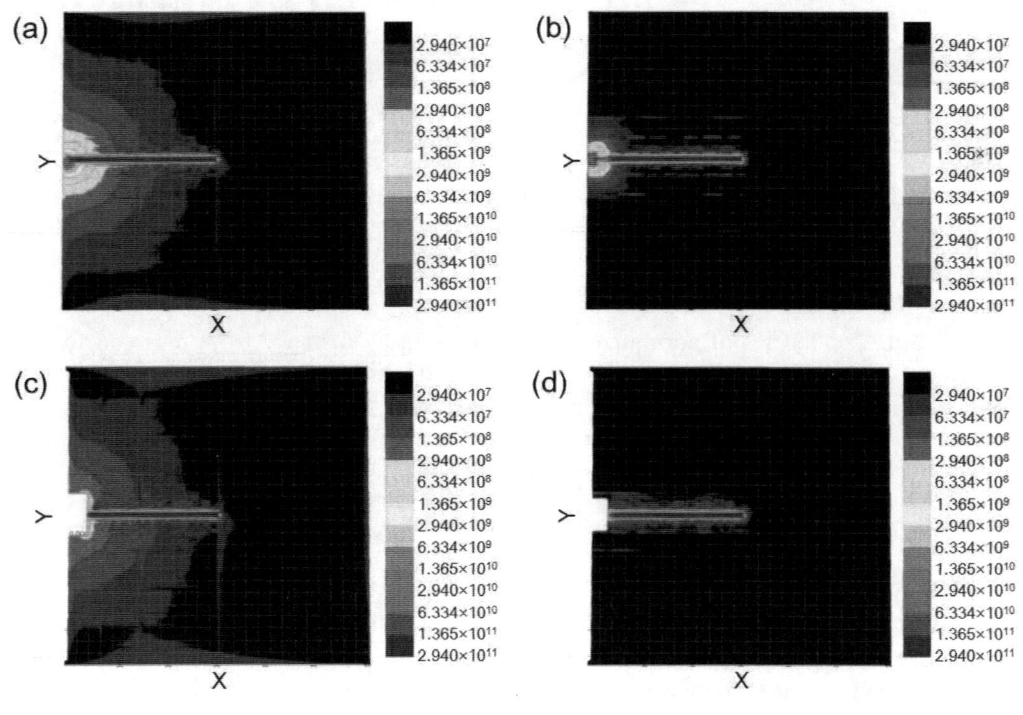

Figure 4. Simulated colour contour maps in (a) normal asymmetrical pad structure on a ground plane, (b) symmetrical pad structure on a ground plane, (c) asymmetrical pad structure outside of a ground plane, and (d) symmetrical pad structure outside of a ground plane.

We have quantitatively compared the ground current distribution in the four pad structures. The calculations were performed without a bias line to eliminate the effect from the bias line and clarify the contour map, since the contour map without the bias line had no special difference from that with the bias line. Figure 5 shows the dependences of the ground current on the distance from the center of bias current supply pad. The data were obtained with drawing contour lines on the ground current color map, which fitted well with circle lines. The ground current was normalized with 1 A as with other results, so a current of 10^{-3} corresponds to flowing 1 mA in 50 μm-wide ground plane. In the normal pad structure, circuits must be placed 0.5 mm away from the bias supply pad for reducing the ground current to less than 10^{-3}. When considering a 2 μm-wide inductive line in SFQ circuit, the ground current under the line corresponds to be as large as 40 μA in 10^{-3}. It seems to affect the operation of SFQ circuits. Although the ground current should be reduced less than 10^{-4}, it is impractical in the normal pad structure. The bilaterally symmetrical pad structures on a ground plane and outside of a ground plane structure can reduce well the spread of the ground current as shown in figure 5. Especially, one can see that the symmetrical pad structure outside of GP is superior to reduce the ground current than that on GP. In particular, the normalized ground current can be reduced to 10^{-4} at d ~ 0.5 mm. Furthermore, the ground current can be reduced to more than one order of magnitude less than that of the normal structure at d > 0.5 mm.

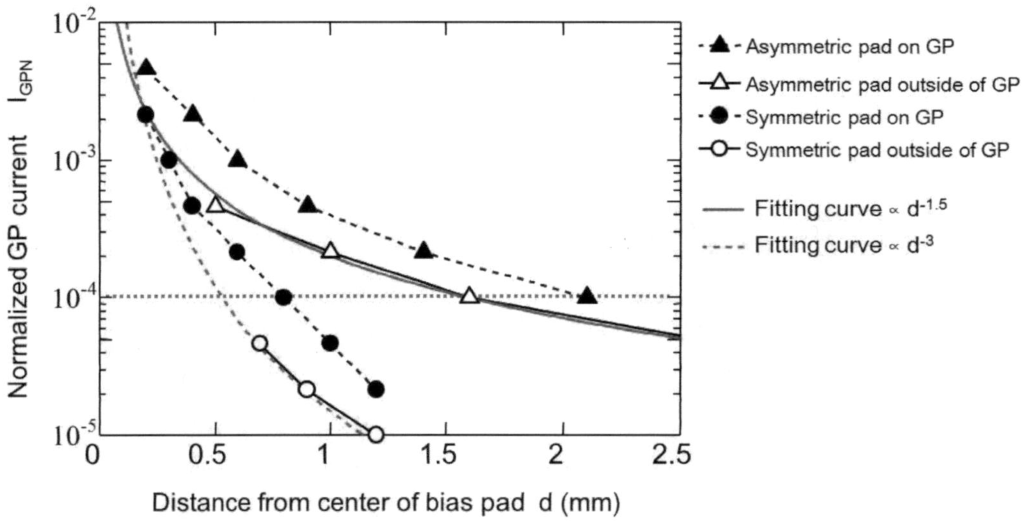

Figure 5. Dependencies of normalized ground current on distance from bias pad for four pad structures, where distance between bias pad and extraction pad is 100 μm. Fitting curves are also drawn for the asymmetric and symmetric pad structures outside of ground plane.

Fitting curves for the asymmetric and symmetric pad structures outside of GP are also drawn in figure 5. According to approximate expression for current elements from the bias pad to the extraction pads by using Biot-Savart law, the magnetic field was decayed with proportional to d^{-2} in the asymmetric structure and d^{-4} in the symmetric structure. On the other hand, the simulated results are well fitted with $d^{-1.5}$ and d^{-3} in respectively. In the approximation, we assumed that the distance from the bias pad, d, is large enough compared to the pad-to-pad distance, S. The difference seems that the assumption d >> S is not enough for the simulated results. It is also possible that the difference is due to Meissner effect on ground plane, which cannot be included in the approximate expression. However, the fitting curves are well represents a feature of the difference of the symmetric and asymmetric pads.

As shown in figure 2, the distance between the bias pad and the bias extraction pad is important for reducing the spread of the GP return current. We simulated GP current for the four structures with several pad-to-pad distances as shown in figure 6. Figure 6(a) shows the results for pads on GP and

IOP Conf. Series: Journal of Physics: Conf. Series **871** (2017) 012067 doi:10.1088/1742-6596/871/1/012067

(b) shows that for outside of GP. It is obvious that the smaller the pad-to-pad distance, the smaller the GP current spread in all of the structure. In addition, the GP current is reduced by shortening pad-to-pad distance more effectively in the symmetrical pad structure than that in asymmetrical one.

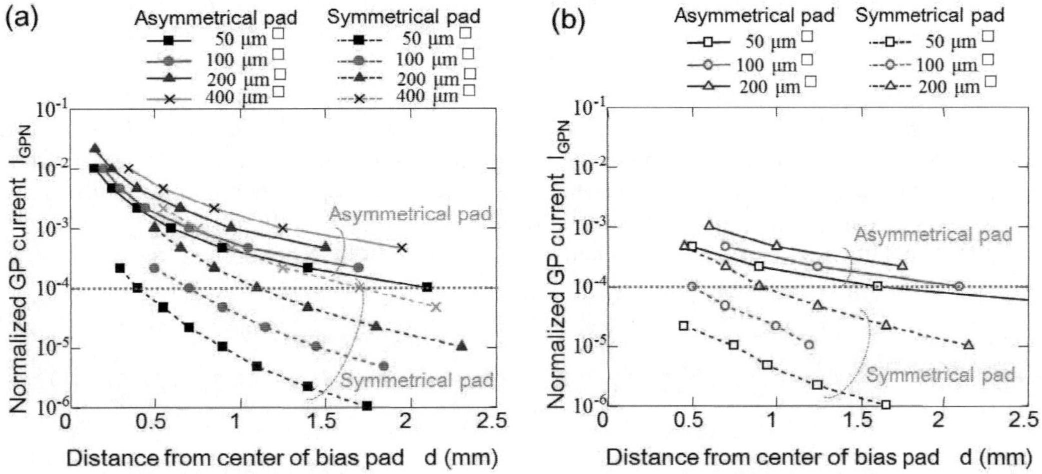

Figure 6. Influence of pad size, distance between bias pads and bias extraction pads, on ground current; (a) pads are located on GP and (b) pads are located outside of GP.

4. Experimental results

To confirm the validity of the simulated results, we designed a test chip using AIST advanced process 2 (ADP2). Figure 7 shows the layout of the test circuit, where the symmetrical pad structure outside of GP is arranged on the left half and that on GP is arranged on the right half. The test circuit of 2.2 mm × 6.5 mm in size was placed near the edge on a die of 7 mm × 7 mm in size. To separate electrically and magnetically to other circuit area, 100 μm around of the ground plane was removed in the test circuit as shown in figure 7. X and Y direction SQUID pairs are placed on three different positions for measuring GP current on both structures. Each positon to the pads are the same for both structures. The SQUIDs are shunted with resistors, where McCumber parameter, β_c, was set to 1. The width of the inductance line of the SQUID is 3 μm. The pad size is 100 × 150 μm and the distance between the bias pad and the extraction pad is 120 μm. The bias current is extracted equally from both of PAD A and PAD B for evaluating the symmetrical pad structure, while the bias current is extracted from either of them for evaluating the asymmetrical pad structure.

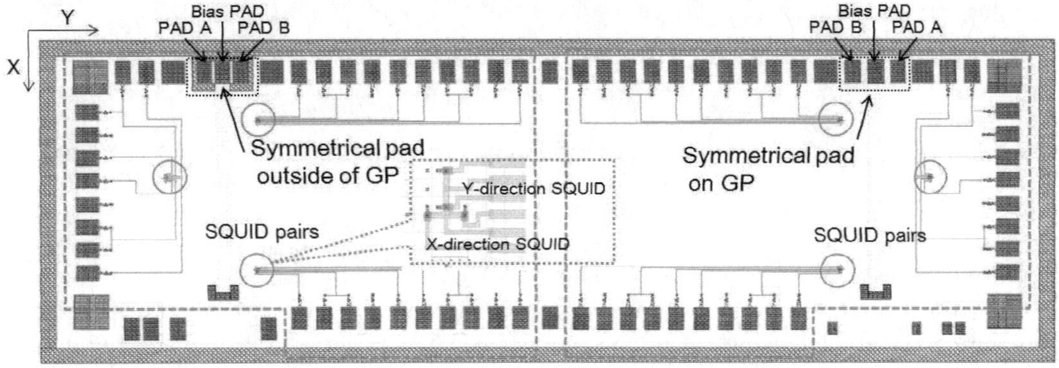

Figure 7. Layouts of a test chip for evaluating symmetrical and asymmetrical pad structures. Pads are located outside of ground plane in left half, whereas they are located on ground plane on right half.

The ground current was measured using a shift of well-known V–Φ_{ext} (I_{ctl}) characteristic of the SQUID, where V is the voltage of the SQUID, Φ_{ext} is the external magnetic flux induced by the control current, I_{ctl}. Bias current of 0.4 A was applied to the bias pad and extracted from PAD A and/or PAD B. Then, we measured the shift in V–Φ_{ext} (I_{ctl}) characteristic, I_{shift}, which was induced by the GP current. Square-root of sum of squares of the measured I_{shift} for X and Y direction was calculated and was normalized to accorded with the simulated results. Figure 8 shows the experimental results, where the normalized GP current was divided by 2.5 so as to be fitted to the simulated value. The measured dependencies on the distance from the pads are well agreed with the simulated results except for the symmetrical pad structure on GP. We have not solved the discrepancy between the measured and the simulated results for the symmetrical pad structure on GP at present. Since another chip showed the same dependence, it seems that the experimental results are correct and the simulated result has lack of some factors. In either case, the symmetrical pad outside of GP structure is the best solution for reducing the spread of GP current.

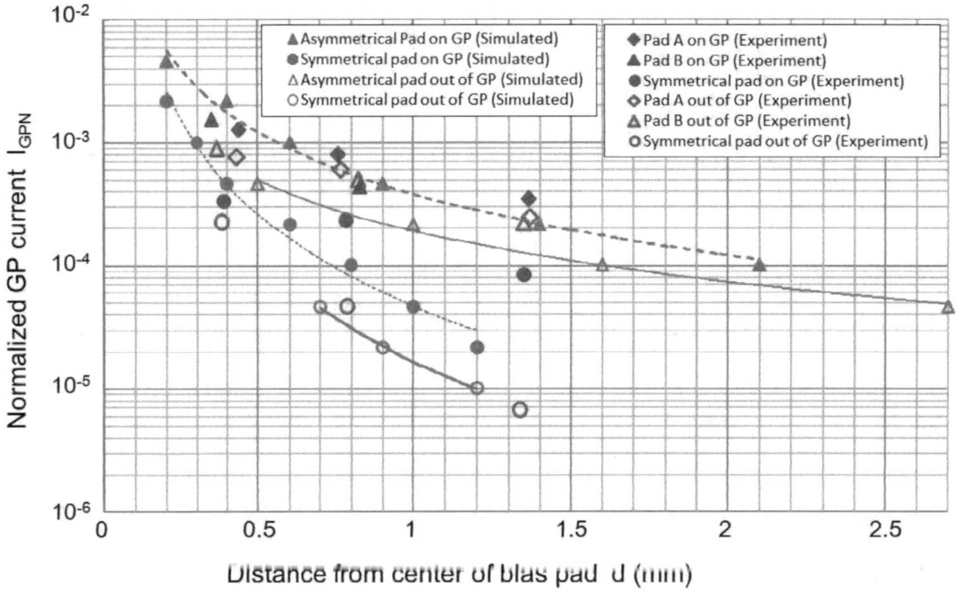

Figure 8. Experimental results of ground current dependencies on the distance from the pads for the four pad structure. Simulated results are also shown for comparison.

5. Conclusion

We have investigated bias return current on a ground plane with simulation using FastHenry program and with experiment. Even extracting the bias current from the adjacent ground pad of the bias supply pad, the ground current spread extensively. We have found that the cause was due to the relatively small difference between the bias supply point and the extraction point. To solve the problem, we proposed the symmetrical pad structure, which has bilaterally symmetric bias current extraction pads to a bias pad. We simulated the ground current using FastHenry program and experimentally measured the ground current for the symmetrical and the asymmetric pad structures. Then, we confirmed the bias current extraction using the symmetric pad structure was able to reduce the spread of the ground current significantly. Moreover, we have found out that arranging the pad outside of ground plane was more effective than that on ground plane to reduce the ground current. The symmetrical pad structure outside of ground plane structure would improve the operation in large scale superconducting circuits.

Acknowledgments

This work was supported by a Grant-in-Aid for Scientific Research (S) (No. 26220904) from the Japan Society for the Promotion of Science (JSPS). The devices were fabricated in the clean room for analog-digital superconductivity (CRAVITY) in National Institute of Advanced Industrial Science and Technology (AIST). The authors thank Prof. Y. Mizugaki for his help for using FastHenry program and useful comments.

References

[1] Likharev K and Semenov K V 1991 RSFQlogic/memory family: A new Josephson-junction technology for sub-terahertz-clock-frequency digital systems *IEEE Trans. Appl. Supercond.* **1** 3–28

[2] Tanaka M, Kondo T, Nakajima N, Kawamoto T, Yamanashi Y, Kamiya Y, Akimoto A, Fujimaki A, Hayakawa H, Yoshikawa N, Terai H, Hashimoto Y and Yorozu S 2005 Demonstration of a single-flux-quantum microprocessor using passive transmission lines *IEEE Trans. Appl. Supercond.* **15** 400–404

[3] Kato T, Yamanashi Y, Yoshikawa N, Fujimaki A, Takagi N, Takagi K and Nagasawa S 2013 60-GHz demonstration of an SFQ half-precision bit-serial floatin-point adder using 10 kA/cm^2 Nb Process *14th International supercond. Electronics coference*

[4] Ono T, Suzuki H, Yamanashi Y and Yoshikawa N 2016 *Design and implementation of an SFQ-based single-chip FFT processo*r *Apped. Superconductivity Conference*; to be published in *Tran. Appl. Supercond.* (2017)

[5] Terai H, Kameda Y, Fujimaki A and Wang Z 2003 The effect of dc bias current in large-scale SFQ circuits *IEEE Trans. Appl. Supercond.* **13** 502–506

[6] Suzuki M, Maezawa M and Hirayama F 2004 Effects of magnetic fields induced by bias currents on operation of RSFQ circuits *Physica C Supercond.* **412-414** 1576–1579

[7] Kadin M A, Webber J R, Sarwana and S 2005 Effects of superconducting return currents on RSFQ circuit performance *IEEE Trans. Appl. Supercond.* **15** 280–283

[8] Akaike H, Fujimaki A, Satoh T, Hinode K, Nagasawa S, Kitagawa Y and Hidaka M 2007 Effects of a dc-power layer under a ground plane in SFQ circuits *IEEE Trans. Appl. Supercond.* **17** 466–469

[9] Tanaka M, Akaike H and Fujimaki A 2009 Control of return current flows in the single-flux-quantum circuits using positive/negative bias supply lines *Physica C Supercond.* **469** 1666–1669

[10] Kuroiwa K, Moriya M, Kobayashi T and Mizugaki Y 2011 Evaluation of two methods for supressing ground current in the superconducting integrated circuits *IEICE Trans. Electron.* **E94–C** 296–300

[11] Terai H, Yorozu S, Fujimaki A, Yoshikawa N and Wang Z 2006 Signal integrity in large-scale single-flux-quantum circuit *Physica C Supercond.* **445-448** 1003–1007

[12] Suzuki H, Oikawa M, Nishii K and Hidaka M 2013 Investigation of a 5-bit Flash-Type SFQ A/D Converter Using 10 kA/cm^2 Niobium Process and Locally Isolated Grounds *IEEE Trans. Appl. Supercond.* **23** 1400805

[13] http://www.wrcad.com/freestuff.html

[14] Suzuki H, Peng X and Yoshikawa N 2015 Investigation of Reducing Harmful Effect of Bias Return Current on Ground Plane in Superconducting Integrated Circuits *15th International supercond. Electronics coference*

IOP Conf. Series: Journal of Physics: Conf. Series **871** (2017) 012068 doi:10.1088/1742-6596/871/1/012068

High-Speed Superconductive Decimation Filter for Sigma-Delta Analog to Digital Converter

Tomu Wakamatsu[1], Yuki Yamanashi[1], and Nobuyuki Yoshikawa[1]
[1]Department of Electrical and Computer Engineering, Yokohama National University, 79-5 Tokiwadai, Hodogaya, Yokohama, Japan

Corresponding author: yamanashi-yuki-kr@ynu.ac.jp

Abstract. A superconducting decimation filter is required to convert high-speed output data from a superconducting sigma-delta analog to digital (A/D) modulator to low-speed data for data acquisition by room-temperature electronics. Because the operating frequency of the conventional superconducting decimation filter is lower than that of the maximum operation frequency of A/D modulator, the system performance of the superconducting A/D converter is limited by the decimation filter. We propose a decimation filter that can operate at the sampling frequency of the A/D modulator by hybridizing a shift-register-based and a counter-based decimation filters. The investigated decimation filter can be implemented with a practical circuit area. We designed and tested the investigated decimation filter. The simulation result indicates that the maximum operation frequency of the designed decimation filter is 39.8 GHz assuming the 2.5 kA/cm^2 Nb fabrication process. We experimentally confirmed the low-speed operation of the designed decimation filter with the bias margin of 93.8%–110.8%.

1. Introduction

A sigma-delta analog to digital converter (ADC) composed of a single-flux-quantum (SFQ) circuit is superior to a semiconductor ADC in terms of the signal-to-noise-ratio, dynamic range, and linearity [1, 2]. Moreover, the power consumption of the SFQ ADC is extremely lower than that of semiconductor ADC even though it can operate at the sampling rate of several tens GHz. In order to build the sigma-delta ADC system based on the SFQ circuit, a decimation filter that converts ultra-high-speed output data from the analog to digital (A/D) modulator to the low-frequency parallel data is required after the A/D modulator to transmit the data to room-temperature instruments. The decimation filter has to operate at the same frequency as the sampling frequency of the sigma-delta ADC. The shift-register-based decimation filter can operate at the frequency of the A/D modulator [3]. However, the shift-register-based decimation filter is not used in the practical SFQ ADC systems because the large circuit area is needed to implement the shift-register-based decimation filter with the high decimation rate [4]. The counter-based decimation filter [5] has been used because of its small circuit area. However, the operating frequency of the counter-based decimation filter is low compared to the A/D modulator and the decimation filter limits the performance of the SFQ ADC systems.

In this paper, we propose an SFQ decimation filter that can operate at the high clock rate and can be implemented on a small circuit area by hybridising the shift-register-based and the counter-based decimation filters.

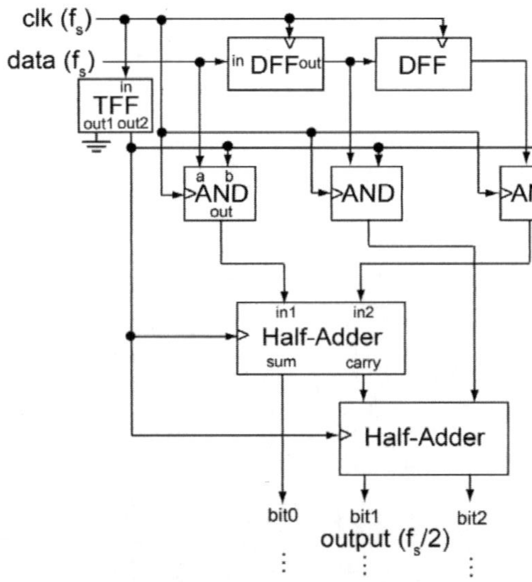

Figure 1. Block diagram of 2nd-order shift-register-based decimation filter with the decimation rate of 2.

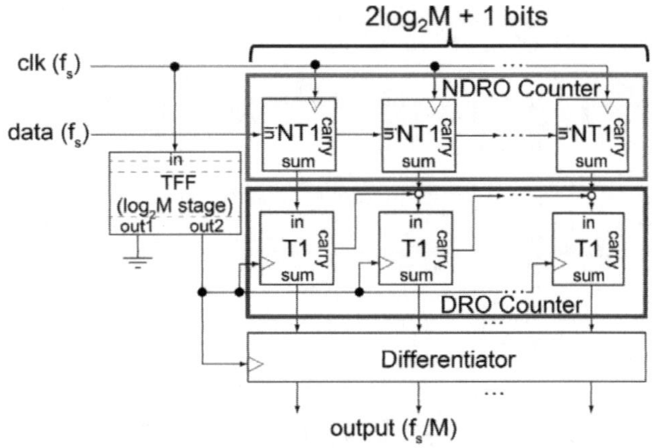

Figure 2. Block diagram of 2nd-order counter-based decimation filter with the decimation rate of *M*.

2. High-speed hybrid decimation filter

The transfer function of the decimation filter is expressed as

$$\mathrm{H(z)} = (1 + z^{-1} + z^{-2} + \cdots + z^{-M})^N, \tag{1}$$

where *M* and *N* are the decimation rate and order of the filter, respectively. To implement the SFQ decimation filter that has the transfer function (1), two implementation methods has been investigated.

Figure 1 shows a block diagram of the 2nd-order shift-register-based SFQ decimation filter with the decimate rate *M* = 2. The shift-register-based decimation filter is composed of a delay-flip-flop(DFF)-based Shift register, AND gates, half-adders [6] and a toggle-flip-flop (TFF). The transfer function of the decimation filter shown as figure 1 is expressed as

$$\mathrm{H(z)} = (1 + z^{-1})^2. \tag{2}$$

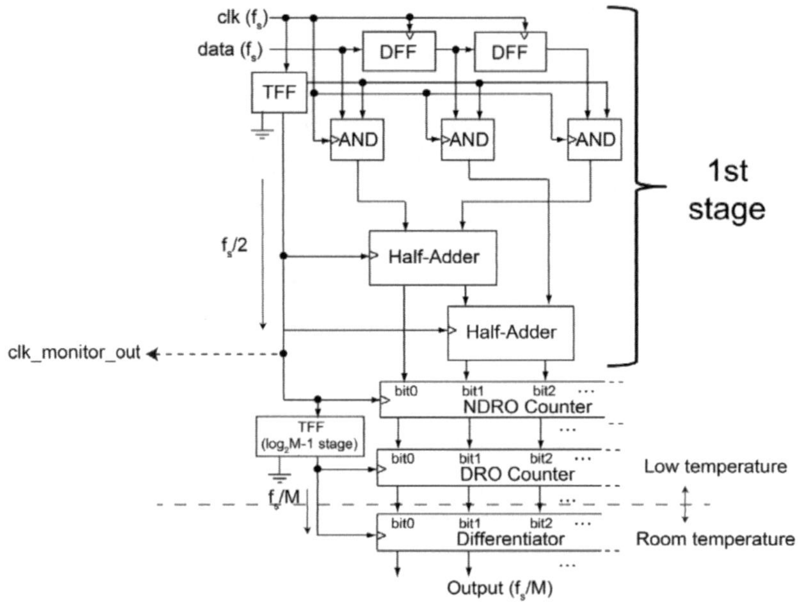

Figure 3. Block diagram of the 2nd-order hybrid decimation filter with the decimation rate $M = 2$. The clk_monitor_out is used only in the measurement.

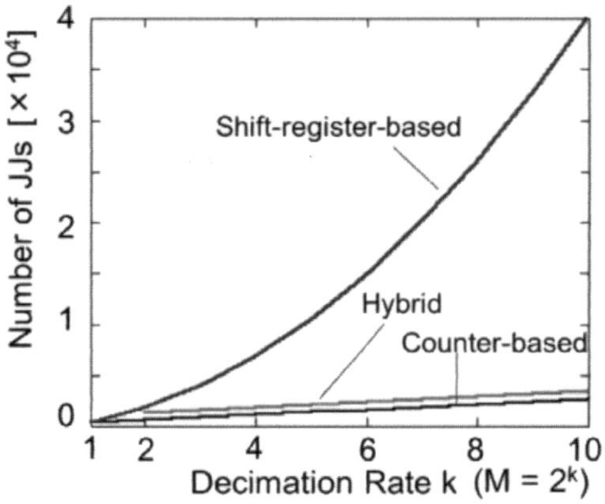

Figure 4. Comparison of dependences of the number of JJs on decimate rate.

By cascading the decimation filters shown in figure 1, the decimation filter with the higher decimation rate can be implemented. Though the operation speed of the shift-register-based decimation filter is high, the circuit area of the decimation filter with the decimation rate of M increases with order of $3^\wedge(\log_2 M)$. This results in drastic increase in the circuit area with increase in M.

Figure 2 shows a block diagram of the 2nd-order counter-based decimation filter with the decimate rate of M. The counter-based decimation filter is composed of a non-destructive-readout (NDRO) binary counter composed of non-destructive T1 flip-flops (NT1s), a destructive-readout (DRO) binary counter composed of T1 flip-flops, TFFs and a differentiator. Because M can be increased by increasing the bits of the NDRO and DRO counters, the counter-type decimation filter with the high decimation rate is implemented with small the circuit area compared to the shift-register-based decimation filter [5]. The circuit area of the counter-based decimation filter increases with order of

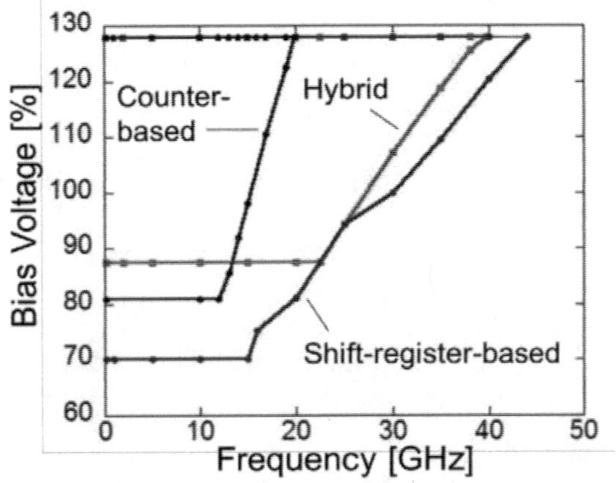

Figure 5. Comparison of dependence of the bias margin on the operation frequency. The bias voltage is normalized by the standard value of the CONNECT cell library [7].

$\log_2 M$. However, the operating frequency of the counter-based decimation filter is low compared to the shift-register-based filter because of the non-destructive operation.

Figure 3 shows the block diagram of the proposed hybrid decimation filter that hybridizes the shift-register-based and the counter-based decimation filters. In this hybrid decimation filter, the shift-register-based decimation filter is used as the first stage of the decimation filter because the first stage of the filter processes the high-speed data output from the A/D modulator at the sampling frequency of the SFQ ADC. Because the output data rate from the first stage decimation filter is half of that of the sampling frequency, the following stage can be composed of the counter-based decimation filter to reduce the circuit area. The differentiator is implemented by the room-temperature circuit because it does not need high-speed operation.

We estimated the number of Josephson junctions (JJs) required to implement the hybrid decimation filter and the maximum operation frequency assuming using the AIST 2.5 kA/cm^2 Nb standard 2 process (AIST-STP2) [8]. Figure 4 shows comparison of dependences of the number of JJs to implement the three types of decimation filters on the decimation rate. This estimation indicates that the circuit area of the hybrid decimation filter is much smaller than that of the shift-register-based filter and almost the same as the counter-based filter. Assuming the decimation rate of $M = 2^{10}$, the hybrid decimation filter can be implemented by approximately 3500 JJs. Therefore, the hybrid decimation filter, which has the practical decimation rate, can be implemented on a single chip.

Figure 5 shows comparison of dependences of the bias margins of decimation filters on the operating frequency. The operation frequency of the hybrid decimation filter is almost the same as the shift-register-based decimation filter. The maximum operating frequency of the hybrid decimation filter is estimated to be 39.8 GHz.

3. Circuit implementation and test results

We designed and implemented the 2nd-order hybrid decimation filter with the decimation rate $M = 4$ using the CONNECT cell library [8] and the AIST-STP2 [7]. The number of bits in NDRO and DRO counters is 5 and the counters can count the data input up to 31 (= 2^5-1). Figure 6 shows the microphotograph of the implemented hybrid decimation filter. The number of JJs is 1204. Figure 7 shows the test result of the low-speed test when the input test pattern is "1111111111111111". By substituting the input pattern into the transfer function (1), 16 is obtained every for inputs. The obtained data output are 10, 26, 10 (overflow of 42), and 26 as shown in Figure 7. Because the output data are incremented by 16 every output cycle, the obtained output pattern shows the correct operation of the decimation filter. The measured normalized dc bias margin was 93.8%–110.8%. The measured

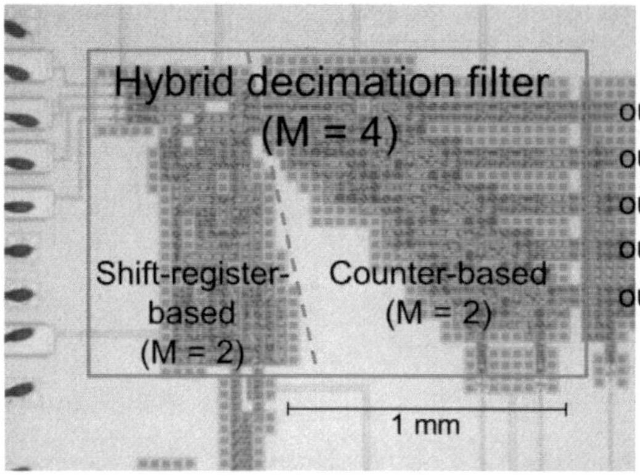

Figure 6. Microphotograph of the test circuit of the hybridized decimation filter.

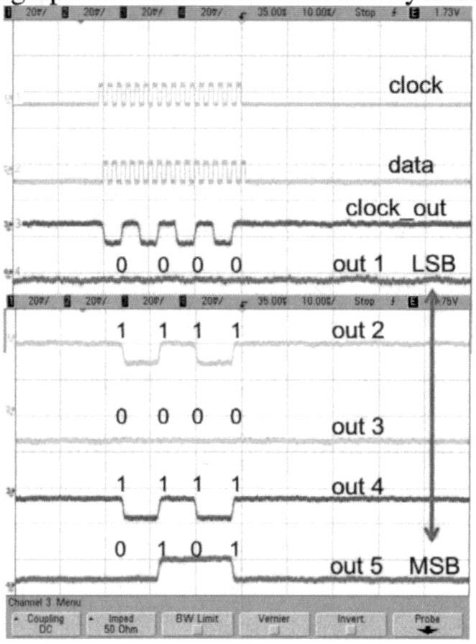

Figure 7. Low-speed test result of the hybrid decimation filter. Input test pattern is '1111111111111111'. Output data of 01010 (= 10), 11010 (= 26), 01010 (= 10 = 42-2^5), and 11010 (= 26) are obtained every four clock inputs. "clk_monitor_out" is used to check the operation of dc/sfq converters and the TFF.

bias margin was narrower than the simulation result. We think this is because of the discrepancy of the circuit parameters between designed and fabricated circuits. According to the data sheet of the fabricated wafer, the critical current density of Josephson junctions and resistivity of the resistance layer are 95% and 91 % of the designed values, respectively. We think reduction in the critical current of Josephson junctions and resistance of on-chip bias resistors is the main reason of the narrow and shifted margin.

4. Conclusion

We proposed the new high-speed decimation filter that realizes both high-speed operation and the small circuit area. Our estimation indicates that the total number of the Josephson junctions required to implement the proposed decimation filter with the high decimation rate can be drastically reduced compared to the conventional shift-register-based decimation filter. We implemented the decimation filter with the decimation rate of 4 using the 2.5 kA/cm^2 Nb fabrication process and confirmed the correct operation by a low-speed function test.

Acknowledgments
The authors thank Dr. H. Suzuki for fruitful discussions. This work was supported by JSPS KAKENHI Grant Number JP26220904. The circuits were fabricated in the clean room for analog-digital superconductivity (CRAVITY) of National Institute of Advanced Industrial Science and Technology (AIST) with the standard process 2 (STP2). The AIST-STP2 is based on the Nb circuit fabrication process developed in International Superconductivity Technology Center (ISTEC).

References

[1] Mukhanov O A, Gupta D, Kadin A M and Semenov V K 2004 *Proc. IEEE* **92** 1564
[2] Worsham A H, Miller D L, Dresselhaus P D, Miklich A H and Przybysz J X 1999 *IEEE Trans. Appl. Supercond.* **9** 3157
[3] Hasegawa H, Hashimoto T, Nagasawa S, Suzuki H, Miyahara K, and Enomoto Y 2001 *IEEE Trans. Appl. Supercond.* **11** 517
[4] Vernik I V, Kirichenko D E, Filippov T V, Talalaevskii A, Sahu A, Inamadr A, Kirichenko A F, Gupta D, and Mukhanov O A 2007 *IEEE Trans. Appl. Supercond.* **17** 442
[5] Filippov T V, Pflyuk S V, Semenov V K and Wiborg E B 2001 *IEEE Trans. Appl. Supercond.* **11** 545
[6] Tanaka M, Yamanashi Y, Kamiya Y, Akimoto A, Irie N, Park H, Fujimaki A, Yoshikawa N, Terai H, and Yorozu S 2007 *IEEE Trans. Appl. Supercond.* **17** 516
[7] Yorozu S, Kameda Y, Terai H, Fujimaki A, Yamada T and Tahara S 2002 *Physica C* **378** 1471
[8] Hidaka M, Nagasawa S, Satoh T, Hinode K and Kitagawa Y 2006 *Supercond. Sci. Technol.* **19** S138

Improving the spatial resolution of superconducting tunnel junction THz wave detectors using a modified $LiNbO_3$ absorber

M Sone[1], N Igarashi[1], M Naruse[1], H Myoren[1], Y Sasaki[2], C Otani[2], and T Taino[1]

[1] Graduate School of Science and Engineering Saitama University, 225 Shimo-Ohkubo, Sakura-ku, Saitama 338-8570, Japan
[2] RIKEN, 519-1399 Aramaki Aoba, Aoba-ku, Sendai, Miyagi 980-0845, Japan

taino@mail.saitama-u.ac.jp

Abstract. We proposed a superconducting tunnel junction (STJ) detector for terahertz (THz) waves with a modified substrate absorber. The absorber of each STJ was isolated by trenches in order to restrict the diffusion of phonons generated by THz waves, which are made with a dicing process. We evaluated the diffusion characteristics of phonons in the $LiNbO_3$ substrate by detecting THz photons. We found that the phonon diffusion length in a 500-μm-thick $LiNbO_3$ substrate was over 1 mm. No degradation of the current – voltage (*I–V*) characteristics of the STJ after performing the dicing process.

1. Introduction

Terahertz (THz) waves are categorized between millimeter radio waves and infrared radiation. Since soft materials exhibit a high transmittance and many materials show specific absorption spectra in the THz range, THz waves have a great potential for use in a nondestructive imaging tool [1-3]. In fact, the detection of illegal drugs within envelopes has been done by THz waves [2]. To perform THz imaging and measurement of an absorption spectrum, a THz detector equipped with a fast response time, a broad band, and a large detection area is required. A superconducting tunnel junction (STJ) is one of the candidates for the THz detector. The STJ is attractive as a detector because of its high speed, high sensitivity, and wide band. We previously developed a STJ with a substrate absorber for THz photons [4]. The STJ detects phonons generated in the substrate when THz waves are illuminated from the back. These phonons are isotropically diffused in the substrate. Therefore, the spatial resolution of its detector is poor.

In this study, we fabricated a new STJ with a modified substrate absorber for the THz detector. The absorber size of each STJ was isolated by trenches in order to restrict the diffusion of phonons generated by THz waves, which are made with a dicing process. The trenches with a mesh pattern are formed from the back. Each STJ is fabricated on the front of the absorber and set in the center of each absorber divided by trenches. The absorber size can be modified by changing the trench pitch and affects the spatial resolution. The fabrication method and the evaluated results are reported.

2. Improvement of spatial resolution

The STJ consists of three layers, the top superconducting layer, thin insulation layer, and base superconducting layer. Usually, the STJ absorbs photons by the top layer. On the other hand, when a substrate is used as the photon absorber, STJs detect phonons by the base layer. Our STJ, of which the fabrication process was described previously, was on an $LiNbO_3$ (LN) substrate [4]. The LN is the best material for detecting the THz waves.

Figure 1(a) shows a cross-sectional view of the STJs with a conventional LN absorber. The THz waves are illuminated from the back. The phonons diffuse isotropically in the substrate. The STJ detects phonons generated at different positions of the LN, the STJ cannot identify the illuminated positions of individual THz waves. Therefore, we proposed the modification of the absorber to improve the spatial resolution, as shown in Figure 1(b). Trenches formed on the opposite side of the STJs restrict the diffusion of phonons in the substrate. The THz waves were also absorbed at the bottom of the trenches. In order to eliminate generated phonon at the bottom of the trenches, the THz waves must be reflected at those areas. Therefore, an Al layer is deposited in the trenches. In order to decide the appropriate trench pitch, the evaluation of the phonon diffusion length in the substrate is required.

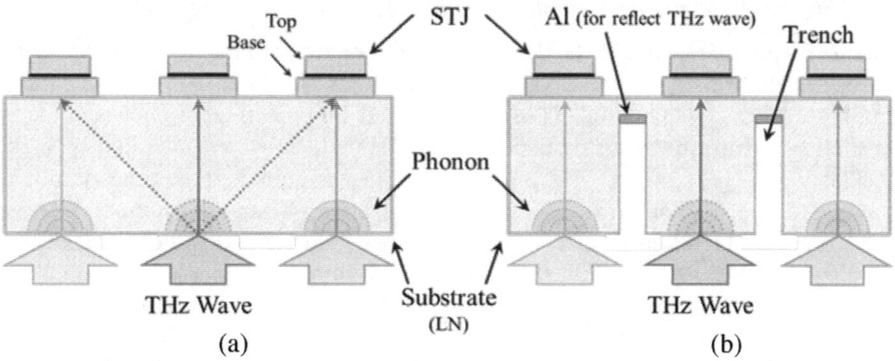

Figure 1. (a) STJ with a conventional substrate absorber (b) STJ with a modified substrate absorber.

3. Formation of trenches and Al deposition

There are two alternative processes to form the trenches on the LN substrate. One is the dry etching method, the other is the dicing method. It is difficult to etch the substrate by the dry etching method because of its large thickness of 500 μm. Although an etching yield of 10 μm was realized by using a nickel mask exhibiting a high selectivity ratio for the dry etching process [5], the nickel mask cannot be used in the process of STJs because of its magnetism. Therefore, we adopted the dicing method to form the trenches. It is important to determine the optimal conditions to prevent chipping at the edge of the trench. We succeeded in forming the trenches under the following conditions: speed of blade movement of 1 mm/s and rotation speed of 20000 rpm. The time of the making one trench was 25 second. Figure 2 shows a side view and an oblique view of the trenches. The trenches showed no chipping and the depth of the trenches was 320 μm, which is about 10% larger than the design value. The fabrication process to deposit the Al layer on the bottom of the trenches is as follows. First, the photoresist was coated on the substrate. Then the substrate with the photoresist was cut for forming the trenches. Without removing the photoresist on the LN, the Al was deposited by the sputtering from the back. After deposition of the Al layer on the substrate, the Al layer on the photoresist was removed using a lift-off process.

4. Phonon diffusion in the substrate

First of all, we fabricated the STJ with a substrate absorber for the THz photons. The STJ array with 25 (5 × 5) pixels was fabricated on the LN substrate of 500 μm thickness. The structure of the STJ was Nb (150 nm)/Al (20 nm)/AlO_x/Al (20 nm)/Nb (150 nm) with the Josephson current density of about

IOP Conf. Series: Journal of Physics: Conf. Series **871** (2017) 012069 doi:10.1088/1742-6596/871/1/012069

100 A/cm^2. The size of each STJ was 100 µm square. STJs were arranged with a pitch of 360 µm on the LN substrate of 500 µm thickness.

Figure 2. (a) A side view and (b) the oblique view of the trenches.

It is important, for our proposed THz detector, to evaluate the phonon diffusion in the substrate when THz waves are illuminated to the LN substrate, in order to reveal the optimal parameters of the trench pattern. To confirm the phonon diffusion, we used STJs without trenches. THz waves were illuminated to the LN substrate through a Si lens. The THz waves were generated by a ring terahertz parametric oscillator (R-TPO) as the THz pulse source [6]. Figure 3 shows the output pulse signal corresponding to THz pulse input from one STJ in the array. Then the output pulse signals were observed using the other 11 STJs. Figure 4 shows the dependence of the pulse height of the output signal on the STJ position. In this figure, the STJ situated with a distance more than 1 mm from the irradiated STJ can detected the generated phonons. When the intensity decay of the maximum output signal is 1/e, the phonon diffusion length of the LN substrate is estimated to be 1 mm. Therefore, if the device needs the spatial resolution of less than 1 mm, the LN has to cut into the desired size. Moreover, this result is the first demonstration of detection by the STJ array with a substrate absorber in the THz range.

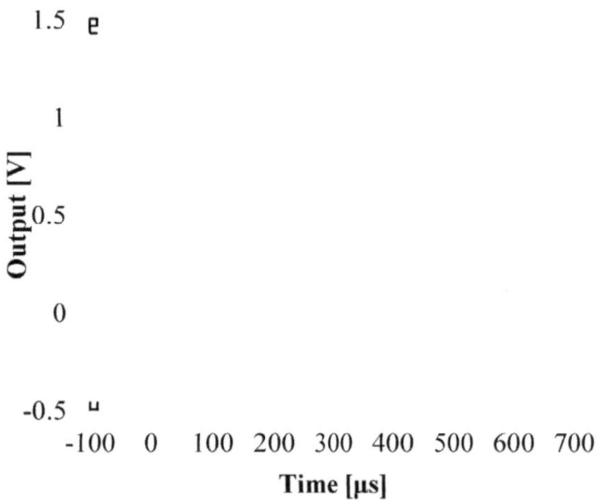

Figure 3. The output pulse from one STJ in the array.

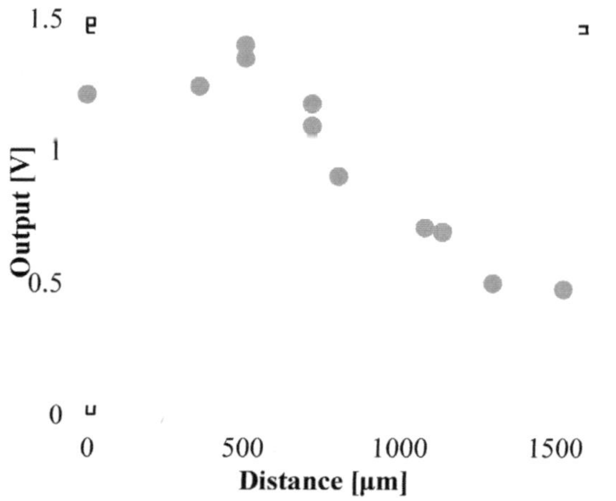

Figure 4. The dependence of the pulse height of the output signal on the STJ position.

5. *I–V* characteristics of STJ with trenches

To evaluate the effect of the dicing process on the STJ performance, the electric characteristics of the STJ were measured. The *I–V* characteristic of the STJ on the LN without trenches and with trenches at

4.2 K is shown in Figure 5(a) and (b), respectively. Each STJ of 100 μm square was fabricated on the front of the absorber and set in the center of each absorber divided by the trenches.

Figure 5(c) shows the temperature dependence of the subgap leakage current of the STJs without the trenches and with trenches. In these figures, similar results were observed in the temperature range from 4.2 to 0.3 K. Therefore, the formation of the trenches and the deposition of the Al layer do not affect the quality of the STJ.

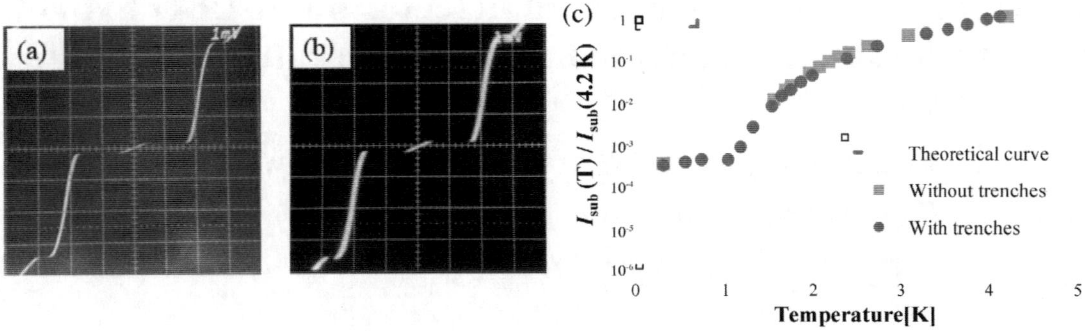

Figure 5. Current–voltage characteristics of the STJ (a) before the formation of trenches, (b) after the formation of trenches and (c) the temperature dependence of the subgap leakage current at the temperature range from 4.2 to 0.3 K.

6. Summary

We proposed and fabricated a STJ with a modified substrate absorber for THz photons. Our proposed STJ has trenches to restrict the diffusion of phonons. We obtained the diffusion of phonons in the THz range using the STJ array without trenches. The effect of phonon diffusion ranged over 1 mm with a 500 μm thickness of the LN substrate. Then, we succeeded in forming trenches on the LN substrate by the dicing method and the Al deposition for the reflect of the THz wave in the trenches. STJs on the LN without trenches and with trenches showed the same *I–V* characteristics at the temperature range from 4.2 to 0.3 K. Our future aim is to demonstrate for detecting performance of the STJ with trenches.

Acknowledgements

The authors would like to express thanks N. Furukawa, and S. Mima for their kind support with the fabrication process. This work was partially supported by the Takahashi Industrial and Economic Research Foundation.

References

[1] Dobroiu A, Yamashita M, Ohshima Y, Morita Y, Otani C, Kawase K 2004 *Appl. Opt.* **43** 5637
[2] Kawase K, Ogawa Y, Watanabe Y 2003 *Opt. Exp.* **11** 2549
[3] Hoshina H, Hayashi A, Miyoshi N, Miyamaru F, Otani C 2009 *Appl. Phys. Lett.* **94**, 123901
[4] Otani C, Taino T, Nakano R, Hoshino K, Shibuya T, Myoren H, Ariyoshi S, Sato H, Shimizu H, Takada S, Kawase K 2005 *IEEE Trans. Appl. Supercond.* **15** 591
[5] Benchabane B, Robert L, Bauch J, Khelif A, Laude V 2009 *J. Appl. Phys.* **105** 094109
[6] Minamide H, Kawase K, Imai K, Sato A, Ito H 2001 *Rev. Laser Eng.* **29** 744 (in Japanese)

Development of STJ for neutron detector on Si-LBO hybrid substrate by surface-activated room-temperature bonding

S Endo[1,2], G Fujii[2], M Ukibe[2], H Takagi[2], M Ohkubo[2], M Naruse[1], H Myoren[1], C Otani[3], and T Taino[1]

[1] Graduate School of Science and Engineering Saitama University, 225 Shimo-Ohkubo, Sakura-ku, Saitama, 338-8570, Japan
[2] AIST, 1-1-1 Umezono, Tsukuba-shi, Ibaraki 305-8568, Japan
[3] RIKEN, 519-1399 Aramaki-Aoba, Aoba-ku, Sendai, Miyagi 980-0845, Japan

taino@mail.saitama-u.ac.jp

Abstract. We have been developing superconducting tunnel junctions (STJs) for neutron detector on a Si-$Li_2B_4O_7$ (LBO) hybrid substrate. An active area of each STJ was limited by dicing in order to obtain high spatial resolution. The Si-LBO hybrid substrate was bonded by a surface-activated room-temperature bonding. In this bonding method, the Si and LBO surfaces were sputter-etched by high energy Ar ion beam and bonded in a vacuum. We fabricated STJs on the Si-LBO hybrid substrate and measured their current-voltage characteristic. Their leakage currents (I_{leak}) were < 10 nA, and so we succeeded in fabricating STJs on the Si-LBO hybrid substrate, which can be expected to operate as neutron detectors with high performance.

1. Introduction

Neutron has a low cross section to metals, although its cross section to light elements, such as hydrogen and lithium is high. For this reason, neutron has been expected as a probe in order to search the state of hydrogen and lithium surrounded within metal frame non-destructively. For example, neutron diffraction is well known to be a useful technique for analyzing a bulk texture of metallic materials because of a large penetration depth of the neutron beam, and utilized in many industrial fields such as an automotive industry [1]. In those research fields, it is necessary for a neutron detector to exhibit high detection efficiency, high spatial resolution, and high counting rate capability, simultaneously. Superconducting tunnel junctions (STJs) on a single crystal $Li_2B_4O_7$ (LBO) substrate is one of the promising neutron detector [2]. This detector has the excellent performance that a detection efficiency for thermal neutrons (1.8 Å) is nearly 100 % when enriched $^{6}Li^{10}BO$ with a thickness of 300 μm is used as the substrate. On the other hand, it is difficult to directly identify the irradiating position because the detector utilizes the whole substrate as an absorber for neutrons and generated quasiparticles are diffused in the whole substrate [3].

In this paper, we have proposed and fabricated a new STJ with a high spatial resolution for neutrons. The STJ was fabricated on the Si-LBO hybrid substrate by surface-activated room-temperature bonding. The fabrication process and the measured results are reported.

2. Our proposed STJ for neutron detector

Figure 1 (a) shows the cross-sectional view of the STJ on the LBO substrate. Incoming neutron into the substrate can be detected by utilizing nuclear reactions of neutron with ^{6}Li and ^{10}B, which lead to the excitation of a large number of phonons in the substrate. After these reactions, phonons propagate in the substrate, and are measured by STJs. It is difficult to directly identify the irradiating position without taking the coincidence of the pulse height from neighboring STJs because the detector utilizes the whole substrate as an absorber for neutrons. In order to solve that difficulty, we have proposed a new STJ detector as follow. The new STJ detector is fabricated on the Si-LBO hybrid substrate, which is made by the bonding of Si and LBO substrate. The Si-LBO hybrid substrate is diced to limit the active area for the improvement of the spatial resolution while keeping high detection efficiency, as shown in Figure 1 (b), (c) [4]. The STJs are located on the Si substrate and trenches between STJs are formed from the LBO substrate. The diffusion of phonons generated in the LBO due to the capture of neutrons is restricted because the connected areas of the Si between each STJ are quite small. Phonons generated in the LBO are propagated to Si without reflection because LBO and Si have very good acoustic impedance matching (Z_{LBO} = 17.7, and Z_{Si} = 17.4). In addition, neutron penetrate the trenches because the nuclear reaction between neutron and Si is very low.

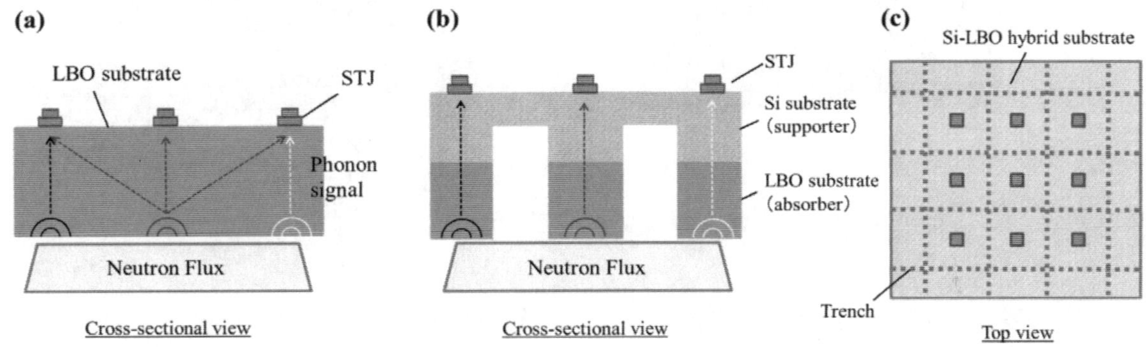

Figure 1. Cross-sectional view of **(a)** the conventional STJ for neutron detector on LBO substrate and **(b)** our proposed STJ detector for neutron detector on Si-LBO hybrid substrate and **(c)** Top view of our proposed STJ detector for neutron detector on Si-LBO hybrid substrate.

3. Si-LBO hybrid substrate by surface-activated room-temperature bonding

3.1 Surface-activated room-temperature bonding

As mentioned above, the Si-LBO hybrid substrate is key technology for our proposed detector. The Si-LBO hybrid substrate was bonded by surface-activated room-temperature bonding [5]. In this bonding method, the contact surfaces are sputter-etched by ion or neutralized atom beam of inert gas, such as Ar and are bonded in vacuum. The sputter-etching removes contaminant layers and absorbed molecules on the surface. Then, surface atoms have activity to form chemical bonds between other atoms even at room temperature. Therefore, it is expected that the phonon signal attenuation at Si-LBO bonding interface can be ignored because no intermediate layer exists at the Si-LBO bonding interface.

3.2 Forming of Si-LBO hybrid substrate by surface-activated bonding

3inch Si and LBO wafers were used to make hybrid substrates. Table 1 shows bonding conditions of Si-LBO substrate. Figure 2 shows the cross-sectional view optical micrograph of Si/LBO interface by surface-activated bonding. In this figure, it cannot see any intermediate layer in the interface. Figure 3 shows the top view photograph of the Si-LBO hybrid substrate from the LBO side. The Newton's

IOP Conf. Series: Journal of Physics: Conf. Series **871** (2017) 012070 doi:10.1088/1742-6596/871/1/012070

rings were observed in the figure. In this region, the Si and LBO are not bonded well. The surface smoothness of the substrate is very important for the bonding. The rms roughness of the LBO has 0.7 nm using an atomic force microscope. To achieve good bonding result, the rms roughness of the substrates needs to be less than 0.3 nm [6]. Although Newton's rings existed in small area of the substrate, we have achieved the formation of the Si-LBO hybrid substrate by surface-activated bonding.

Table 1. Bonding conditions.

Si/LBO Substrate Size	3 inch / 3 inch
Si/LBO Substrate Thickness	400 μm / 250 μm
Etching Gas	Ar fast atom beam
Degree of vacuum	1×10^{-5} Pa
Load	10000 N
Bonding Time	60 sec

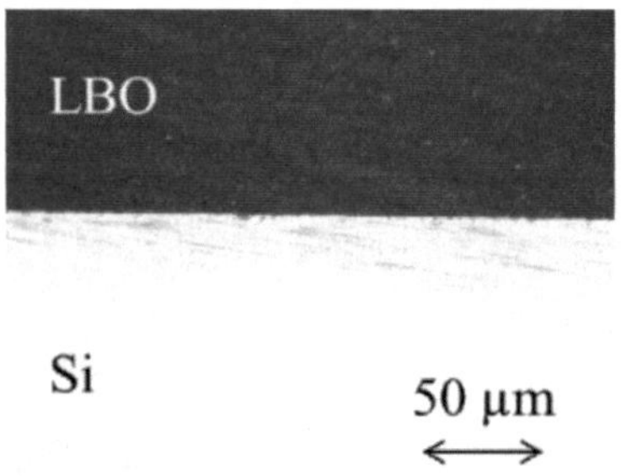

Figure 2. The cross-sectional view optical micrograph of Si/LBO interface by surface-activated bonding.

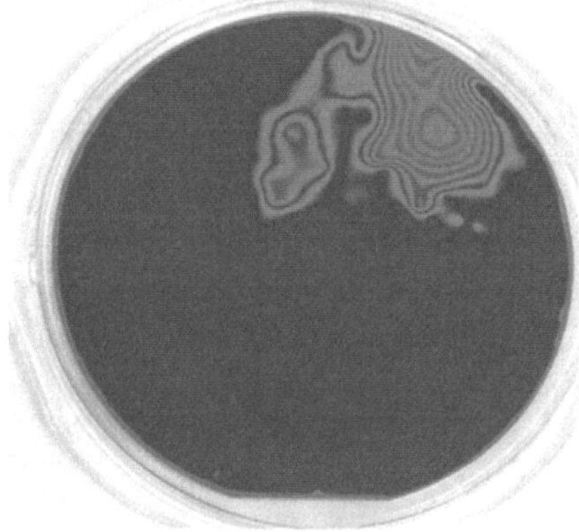

Figure 3. The top view photograph of Si-LBO hybrid substrate from the side of the transparent(LBO) substrate.

4. Fabrication and evaluation of STJ on Si-LBO hybrid substrate

4.1. Fabrication process

We fabricated the prototype of STJs on the Si-LBO hybrid substrate with a size of 10 × 10 mm square. Figure 4 (a) shows a cross-section view of prototype STJ neutron detectors and Figure 4 (b) shows the top view photograph of the fabricated STJ. Prototype STJs had a Nb/Al-AlO_x/Al/Nb layer structures of 100/70/70/300 nm in this study. The Nb/Al multilayer was fabricated in the Clean Room for Analog & digital superconductiVITY (CRAVITY) [7]. The tunnel barrier was formed by conventional O_2 oxidation to achieve a critical current density (J_C) of about 200 A/cm^2. SiO_2 layer with a thickness of about 100 nm was deposited on the LBO side of the hybrid substrate by tetraethyl orthosilicate (TEOS) chemical vapor deposition (CVD) for preventing LBO from deliquescent.

25-pixel STJ arrays fabricated on the chip and the size of all the STJs was 100 × 100 μm. The STJs were fabricated via conventional photolithography, DC magnetron sputtering, a lift-off technique, reactive-ion etching, and wet etching.

4.2. Current-voltage (I-V) characteristic

The STJ properties were measured with a cryogen-free ^{3}He cryostat with a base temperature of 0.31 K. The *I-V* characteristics were measured to evaluate leakage currents (I_{leak}). The I_{leak} was defined by a sub gap current value around ~0.4 mV in a magnetic field of about 10 mT. Figure 5 shows an *I-V* characteristic of the STJ on the Si-LBO hybrid substrate. The STJ shows an excellent *I-V* characteristic having a low sub gap current of less than 10 nA. A dynamic resistance (R_d) was 96 kΩ. The numbers of short and open (N_{open}) junctions were only two and one, respectively. From these results, our proposed STJ on Si-LBO hybrid substrate can be expected to operate as a neutron detector. In the near future, it is irradiated with the neutron in order to confirm the operation as the neutron detector.

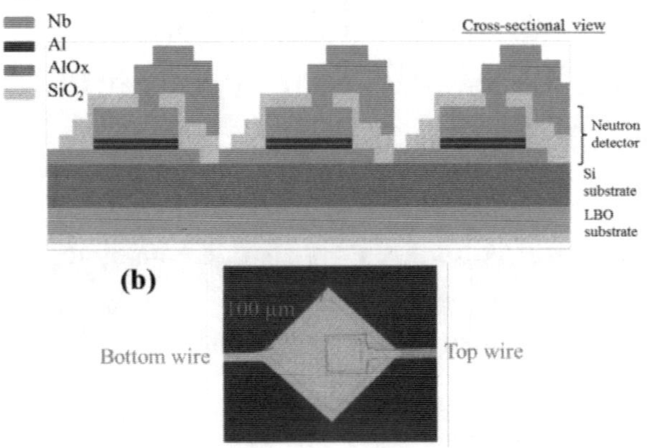

Figure 4. (a) Schematic of prototype STJ for neutron detector and **(b)** top view optical micrograph of fabricated STJ for neutron detector.

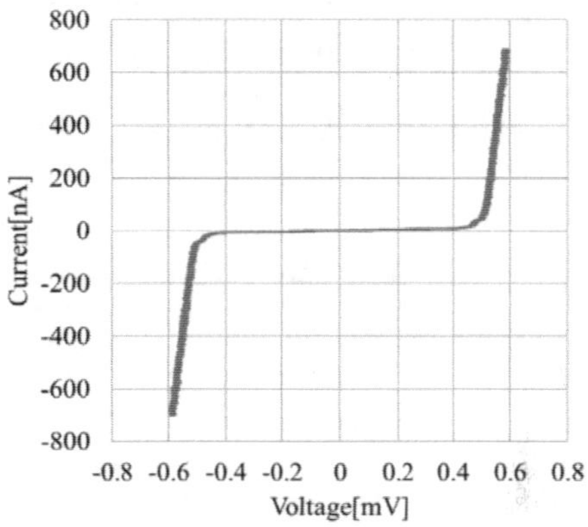

Figure 5. *I-V* characteristics of a prototype of STJ on Si-LBO hybrid substrate at 0.31 K.

5. Conclusion

We developed new STJ neutron detector utilizing a Si-LBO hybrid substrate in order to improve its spatial resolution. The Si-LBO hybrid substrate was formed by surface-activated bonding method. We fabricated STJs on the Si-LBO hybrid substrate with a 10 × 10 mm square. The STJ showed excellent *I-V* characteristics having a low sub gap current of less than 10 nA. As the results, our proposed STJs on the Si-LBO hybrid substrate can be expected to operate as neutron detectors.

Acknowledgements

This work was supported by the Takahashi Industrial and Economic Research Foundation. The devices were fabricated in the clean room for analog-digital superconductivity (CRAVITY) in National Institute of Advanced Industrial Science and Technology (AIST).

References

[1] Ikeda Y, Taketani A, Takamura M, Sunaga H, Kumagai M, Oba Y, Otake Y, Suzuki H 2016 *Nucl. Instr. and Meth.* **A530** 61
[2] Nakamura T, Katagiri M, Ukibe M, Ikeuchi T, Ohkubo 2004 *Nucl. Instr. and Meth.* **A 520** 67
[3] Kurakado M, Kamihirata S, Kagamihara A, Hirota K, Hashimoto H, Sato H, Hotchi H, Shimizu H, Taniguchi K 2003 *Nucl. Instr. and Meth.* **A 506** 134
[4] Endo S, Naruse M, Myoren H, Otani C, Taino T 2015 *76th JSAP Autumn Meeting* 14p-4A-21 (in Japanese)
[5] Takagi H, Maeda R, Hosoda N, Suga T 1999 *Appl. Phys. Lett.* **74** 2387
[6] Takagi H, Maeda R, Chung R, Hosoda N, Suga T 1998 *Jpn. J. Appl. Phys.* **37** 4197
[7] Web site: https://unit.aist.go.jp/riif/openi/cravity/en/index.html

Stress control of reactively sputtered thick NbN film on Si wafer changing the location of the substrate Si wafer against the Nb target on a magnetron cathode

Y Suzuki[1], N Iguchi[1], K Adachi[1], T Hioki[1], A Ichiki[1], C-W Hsu[2], S Kumagai[2], M Sasaki[2] and T Motohiro[1, 2]

[1] Institutes of Innovation for Future Society/Graduate School of Engineering, Nagoya University, Furo-cho, Chikusaku, Nagoya, 464-8603, Japan
[2] Graduate School of Engineering, Toyota Technical Institute, Hisakata 2-12-1, Tempaku-ku, Nagoya 468-8511, Japan

motohiro@gvm.nagoya-u.ac.jp

Abstract. We have been developing a superconducting NbN thin film coil in a spiral trench on a Si-wafer using MEMS technology. Connecting the coils on the different wafers using wafer-bonding process, a cylindrical wafer stack is to be formed as a unit of a compact SMES. The critical current density of our NbN film was measured to be around 1100 A/mm^2. We measured critical current I_c of 47 mA for the previously fabricated coil of the film thickness t_f = 0.5 μm. I_c in the spiral coil increases with t_f. However, if we make the NbN film thicker, the film is apt to have higher lateral force caused by tensile or compressive stress which can cause peeling of the film from the Si substrate. It is well known that the stress of the sputtered thin films can be controlled from tensile to compressive stress by controlling the bombardment of high energy particles including argon atoms backscattered from the target surface. Based on this knowledge, a specially designed sputter-deposition apparatus was fabricated in which the substrate can be located not only at the different target-to-substrate distances but also at several different lateral distances from the central axis of the target (off-axis lateral shift). Using this apparatus, various stress conditions could be realized which contributed to fabrication of thick NbN film spiral coil in the trench. The film stress was calculated from bending analysis of the substrate Si wafer by stylus method using Stoney's formula. The maximum compressive stress of 2.5 GPa was measured. By an off-axis lateral shift, t_f could be increased from 0.5 to 1 μm. By increasing sputtering gas pressure from 0.7 to 2 Pa, the compressive stress could be mitigated and t_f could be further increased from 1.0 to 3 μm. Up to now, we measured I_c of 220 mA for a NbN spiral coil at t_f around 3 μm. More detailed adjustment of the deposition condition will bring further increase in t_f, and hence I_c into sight.

1. Introduction

We have been developing a superconducting NbN thin film coil in a spiral trench on a Si-wafer using MEMS technology as shown in figures.1(a) and (b).[1]. Connecting the coils on the different wafers in series using wafer-bonding process, a cylindrical wafer stack is to be formed as shown in in figure.1(c). The superconductive contacts between the neighboring wafers via through-holes are to be attained, following the established conventional multi-wafer interconnection technology as described in detail in our previous paper [1]. A compact SMES system is to be composed of required numbers of the stacks and cryogenic refrigerators as well as other indispensable appendage systems such as thermal and magnetic shields as schematically shown in figure 1(d), for example. The critical current density of our NbN film was measured to be around 1100 A/mm^2 [2]. We measured critical current I_c of 47 mA for the previously fabricated coil of the film thickness t_f = 0.5 μm. figures 2(a) and (b) shows schematics of the cross-sectional structure of the reactively sputter-deposited NbN thin film spiral coil lined with Cu plated layer in the trench. I_c in the spiral coil increases with t_f, and hence the amount of energy storage. Therefore, we would like to increase t_f from the present status as shown in figure 2(a) to figure 2(b) in future. Figure 2(c) and (d) are cross-sectional SEM images of the trench after NbN deposition in the present status. However, if we make the NbN film thicker, the film is apt to have higher lateral force caused by tensile or compressive stress which can cause peeling of the film from the Si substrate as shown in figures 3 (a)-(c). It is well known that the stress of the sputtered thin films can be controlled from tensile to compressive stress by controlling the bombardment of high energy particles including argon atoms backscattered from the target surface [3]-[13]. In this paper, conditions of reactive sputter-deposition of NbN superconductive film are studied to mitigate film stress to realize thick NbN films showing high I_c. Here, it should be noted that elongation of lattice parameters from that of the bulk value of NbN has been known to cause increase in T_c in reactively sputtered thin NbN films for the studies on Josephson tunnel junctions [14]. Therefore, a certain amount of film stress may improve superconducting properties of NbN thin films. However, what we experienced was that thicker NbN films deposited at the corresponding high T_c condition easily break up and peel off from the substrate. Based on this experience, stress free is the best condition as a priority target in this paper. Utilization of any possible effects of the film stress on the superconducting properties can be targets on the next stage.

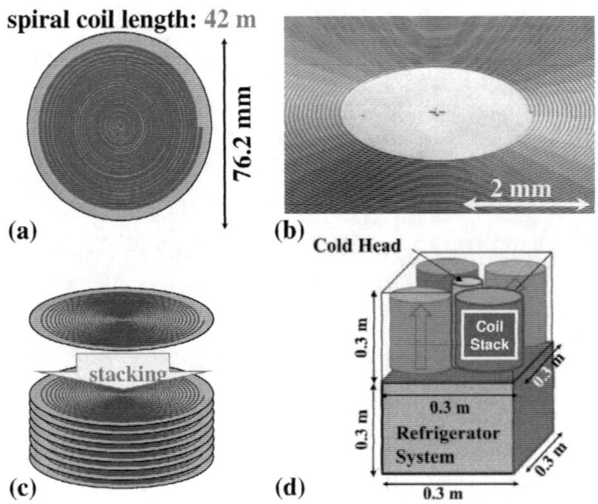

Figure 1. A compact SMES to be formed. (a) a design of a spiral trench on a Si wafer, (b) a SEM image of the central part of the spiral trench formed by MEMS process. (c) a schematics of a stack of Si wafers with spiral superconducting coils connected in series between the wafers, (d) a schematic of a typical example of a compact SMES system.

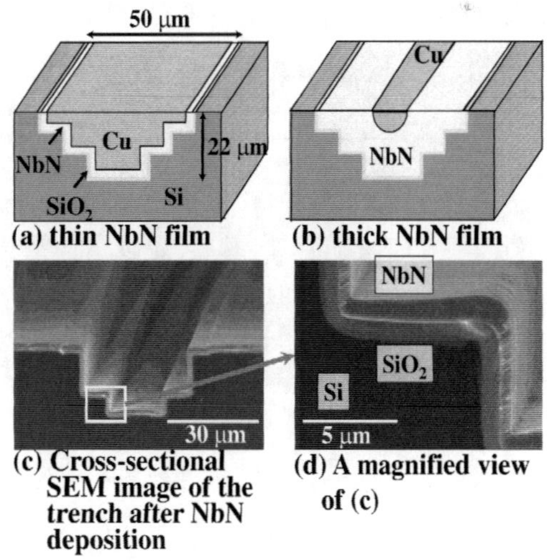

Figure 2. Schematics of the cross-sectional structure of the reactively sputter-deposited NbN thin film spiral coil lined with Cu plated layer in the trench: (a) and (b). Cross-sectional SEM image of the trench after NbN deposition: (c) and (b). The trench geometries are different between (a)(b) and (c)(d).

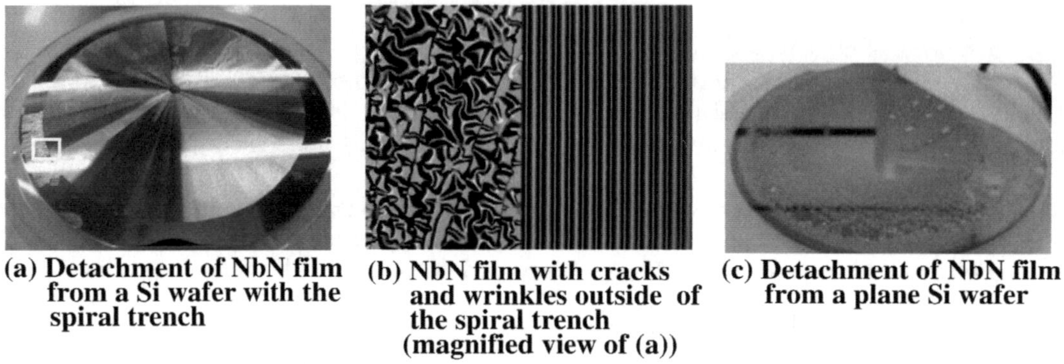

Figure 3. Wrinkles and cracks in the NbN film in the outer side of Si wafer with a spiral trench (a), (b), Detach-ment of NbN films from a plane Si wafer without a trench(c).

2. Experimental

2.1. Instrumentation

Based on the knowledge extensively studied on the effect of bombardment of high energy particles in sputter deposition[3]-[13], a specially designed sputter-deposition apparatus was fabricated in which the substrate can be located not only at the different target-to-substrate distances but also at several different lateral distances from the central axis of the target (off-axis lateral shift)[15]. Figure 4(a) shows a top view of the sputtering chamber showing a ceiling circular flange on which a rotatable substrate holder is mounted. Since the circular flange is fixed to the chamber with 24 sets of bolts and nuts, the position of the rotating axis of the substrate holder can be laterally shifted from the central axis of the target mounted on the bottom flange of the chamber by rotating and fixing the ceiling flange by every 15 degree of angle around the center of the ceiling flange. Figures 4(c) and (d) show

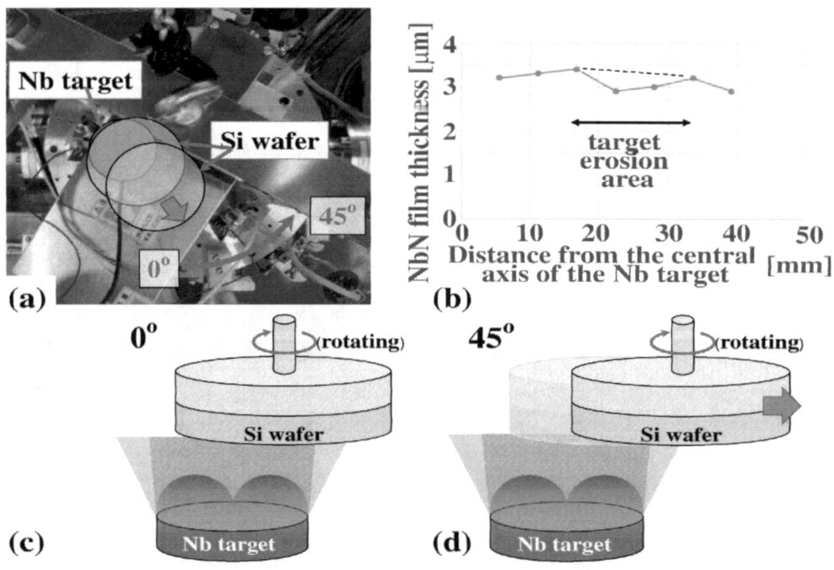

Figure 4. Lateral shift of the rotating axis of the substrate holder from the central axis of the target.

schematics to show this lateral shift of the rotating substrate holder in the case of 0 and 45 degrees of angle, respectively. Figure 4(b) shows a typical examples of NbN film thickness distribution as a function of the distance from the central axis of the target when the ceiling flange was rotated and fixed to the chamber by 45 degrees of angle. The target used was of metallic Nb of 3N grade and 76.2 mm in diameter. The perpendicular distance between the target surface and substrate surface could be changed between 50 and 100 mm, but was fixed to 50 mm in this study. The temperature of the substrate holder could be increased and held up to 873 K by radiation heating. The chamber can be pumped down to 1 x 10^{-4} Pa using turbo-molecular pump. The sputtering gas flow rate was 44 sccm for Ar and 6 sccm for N_2, respectively. The substrate holder was rotated at 10 rpm.

2.2. *Measurement of the stress in the NbN film*

Reactive sputter-deposition of NbN film on Si wafers of 101.6 mm in diameter was performed in various conditions. The substrate Si wafer had not gone through the process for removal of the surface oxide layer before the deposition. In general, metallic thin film sputter-deposited at lower sputtering gas pressure than a certain critical value shows compressive stress which would likely extend its area after deposition resulting in formation of wrinkles by local detachment from the substrate as shown in figure 3(b) or bending of the substrate Si wafer convexly as shown in figure 5 (a) if the adhesive strength is too high to cause the detachment [8]. At higher sputtering gas pressure than the critical value, the film is to show tensile stress and would likely dwindle its surface area resulting in bending of Si wafer in a concave shape as shown in figure 5 (c). Since the thermal expansion coefficient is not so different between Si and NbN, the stress causing the substrate bending can be considered as the intrinsic stress of the film. By measuring this concavity or convexity using stylus method as shown in figures 5(a)-(d), the intrinsic film stress was obtained by the following Stoney's formula [16].

$$\sigma_f = \frac{E_s t_s^2}{(1 - v_s) 6 R t_f}$$

,where σ_f stands for the intrinsic stress, E_S for Young's modulus of Si substrate, t_s for thickness of Si substrate, v_S for Poisson's ratio of Si substrate, R for curvature radius of bending, t_f for thickness of NbN film, respectively.

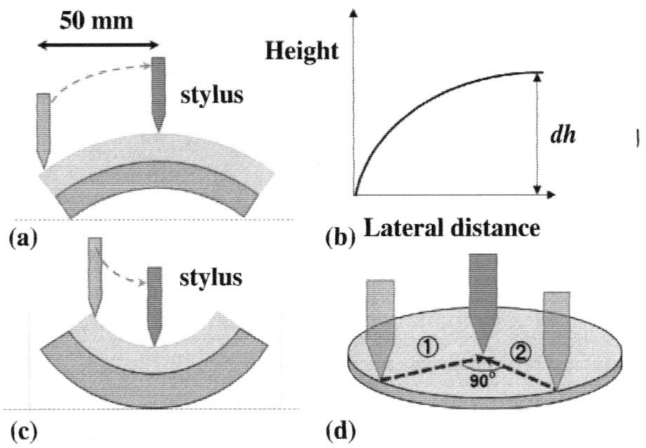

Figure 5. The measurement of intrinsic film stress by convexity or concavity of the sample wafer using stylus method.

3. Results

3.1. Effect of off-axis lateral shift of location of rotating substrate holder

Figure 6 (a) shows a photograph of a Si wafer of 76.2mm in diameter with a spiral groove after deposition of NbN as thick as 0.3 μm at the rotating substrate location as shown in figure 4(c). Although the NbN film in the area of the spiral groove showed no detachment, some detachment could be found in the peripheral region outside of the spiral groove. Therefore, t_f=0.3 μm was considered to be the upper limit of NbN film thickness without remarkable detachment. Figure 6 (b) shows the

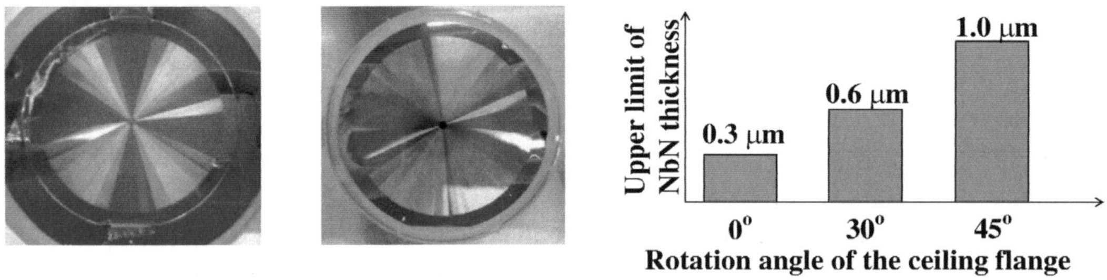

(a) Rotation angle : 0° with film detachment, t_f=0.3 μm **(b) Rotation angle : 45° without film detachment, t_f=1 μm** **(c) Upper limit of NbN thickness without film detachment at different rotation angles**

Figure 6. Increase in the upper limit of NbN thickness without detachment from the substrate with increase in the rotation angle of the ceiling flange

similar photograph to figure 6(a) but at the rotating substrate location as shown in figure 4(d). The upper limit of NbN film thickness without remarkable detachment increased to 1 μm. Figure 6 (c) summarizes the increase of the upper limit of t_f with increase in the rotation angle of the ceiling flange around the center of the flange.

3.2. Effect of sputtering gas pressure on the NbN film stress

Figure 7 (a) shows *dh* shown in figure 5 (b) as a function of lateral distance for different deposition times. The deposition time was proportional to film thickness measured by SEM observation. The

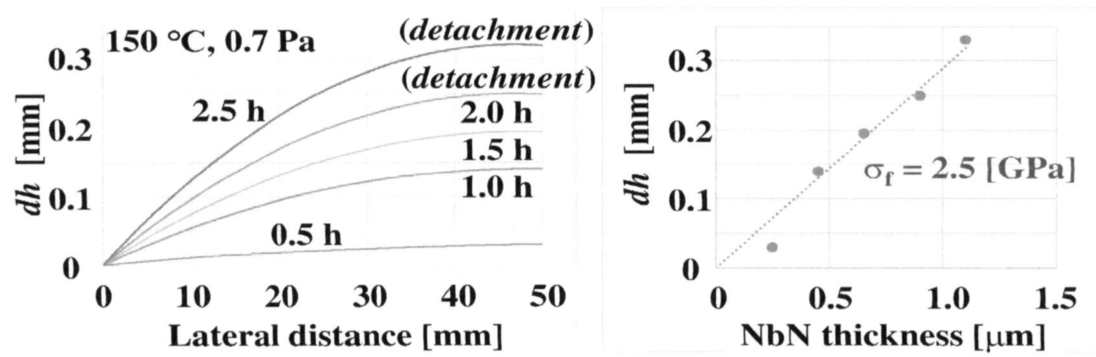

(a) *dh* as a function of the lateral distance from the edge of the Si substrate **(b) *dh* at the center of the Si substrate as a function of the film thickness**

Figure 7. Increase of bending of the sample by compressive stress of the NbN film with increase in film thickness and obtained value for the compressive stress using Stoney's equation.

lateral distance 0 mm and 50 mm correspond to the edge and the center of the Si wafer, respectively. The substrate temperature was 150 degrees of centigrade and sputtering gas pressure was 0.7 Pa. In this condition, deposition rate was 0.6 μm/h and the film stress was compressive. Detachment of the film from Si substrate took place for the deposition time longer than 2.0 h. The value of *dh* was almost proportional to the NbN film thickness as shown in figure 7 (b). From the proportional constant, the film stress σ_f was obtained to be 2.5 GPa using the Stoney's formula. Figures 8(a) and (b) show variations of *dh* with the lateral distance from the edge of the substrate for different sputtering gas pressure. It was shown that intrinsic compressive stress in NbN films were considerably mitigated and even changed to the tensile stress by increasing sputtering gas pressure as in the case of metallic films studied by D. W. Hoffman [8]. By the adjustment of sputtering gas pressure, the compressive stress could be mitigated and NbN film as thick as 3 μm could be deposited without detachment.

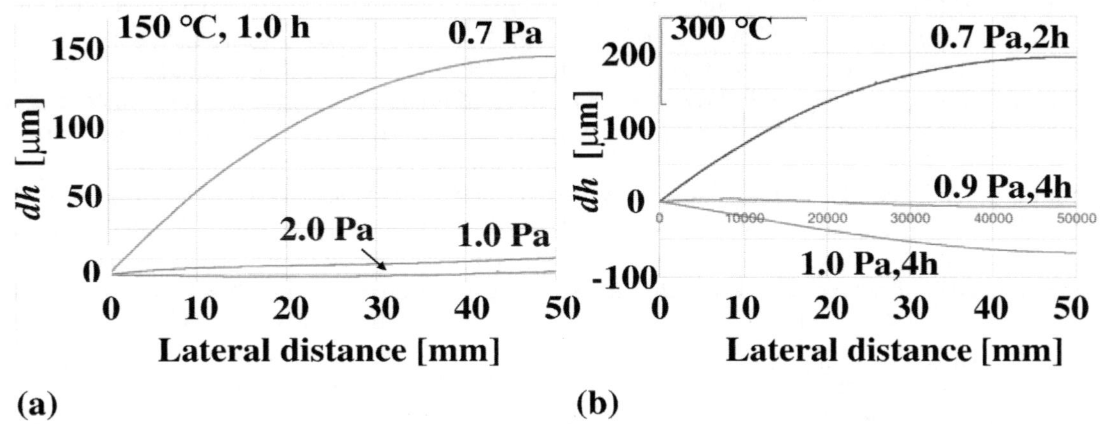

Figure 8. Reduction of the compressive film stress by increasing sputtering gas pressure for the case of the substrate temperature of (a) 150 degrees of centigrade and (b) 300 degrees of centigrade.

4. Discussion

The intrinsic compressive stress in the reactively sputtered NbN films was found to be changed to tensile stress by increasing sputtering gas pressure. This phenomenon is apparently similar to the well-known phenomena in metallic films [8]. In these metallic films, it was considered that energetic particles coming from the target bombarded growing film on the substrate and caused compressive stress. This phenomenon was called as atomic peening effect. By increasing the sputtering gas pressure, those energetic particles were considered to be effectively scattered by the sputtering gas atoms and the atomic peening effect was mitigated. Reduction of the atomic peening effect was considered to cause growth of thin film of lower density than the bulk materials, and cause tensile stress during the densification procedure. As for the energetic particles, neutralized sputtering gas ions once accelerated toward the target and backscattered at the target surface were considered to be one of the candidates. It was shown experimentally that target of elements heavier than sputtering gas atom tended to cause intense atomic peening effect and hence remarkable compressive stress of the film. Since Nb has heavier atomic mass than argon and nitrogen, it is likely for the same phenomenon to take place. On the other hand, there found an unusual dip in film thickness distribution as shown in figure 4(b). The position of the dip seems to correspond the erosion ring of the target which is specific to the planar magnetron type cathode. The similar phenomena were reported by D. J. Kester and R. Messier in the case of four perovskite targets: $PbTiO_3$, $CaTiO_3$ and $SrTiO_3$ and $BaTiO_3$ [12]. In these cases, the dips were more remarkably formed in the target materials which yield oxygen negative ions more easily. Therefore, the energetic particles in this case was considered to be oxygen negative ions accelerated in the cathode fall region towards the substrate. K. Tominaga et al. reported the similar dips of film thickness distribution in the case of ZnO target [13]. These oxygen negative ions accelerated towards the substrate can be also effectively scattered by sputtering gas atoms, and hence the increase in the scattering gas pressure may mitigate the atomic peening effect. In the present case, it is the point of discussion whether nitrogen negative ions may take the similar role as oxygen or not. However, phenomenologically, the similar phenomenon seems to be going on apparently.

5. Conclusion

By adjusting the sputtering gas pressure and mitigating the compressive film stress at the rotation angle of the ceiling flange of 45 degree of angle, the NbN films of the thickness more than 3 μm could be formed after continuous 12 hours' deposition without detachment as shown in figure 9. After the deposition of NbN, Cu plating and CMP were performed to complete the spiral coil as having been reported in our previous report [1]. The input forward RF power was 200 W while the reflected power was almost zero W. The substrate temperature was 300 degree of centigrade. Figure 10 shows temperature dependence of resistivity of a spiral coil in the trench as shown in figure 2(a) on the Si wafer of 76.2 cm in diameter measured by two-terminal method in the cryostat. Clear change of resistivity from that of the Cu coil to that of the superconducting NbN coil only with the contact resistance around 2 ohm was observed at the transition temperature around 15.5 K. The maximum critical current was 220 mA at 4 K. More detailed adjustment of the deposition condition will bring further increase in the film thickness to fill the trench as shown in figure 2 (b), and hence further increase in the critical current into sight.

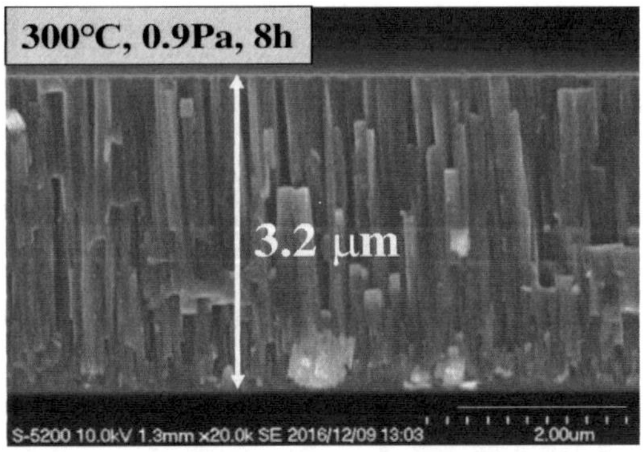

Figure 9. SEM image of a cross-sectional view of a thick NbN film

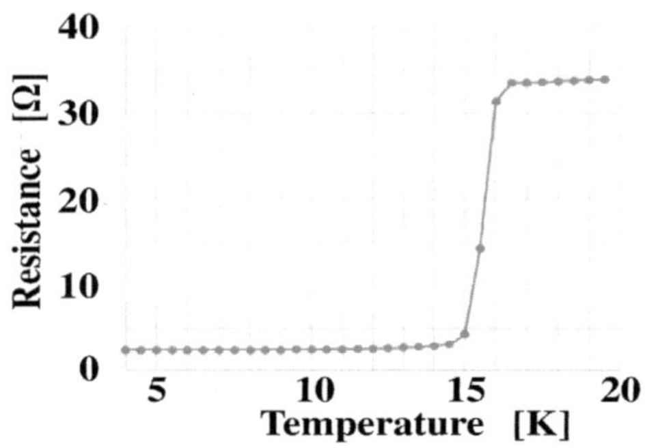

Figure 10. Temperature dependence of resistivity of a spiral coil in the trench measured by two terminal method.

Acknowledgement
This study is based on results from a project commissioned by NEDO (15101789-0). Authors would like to express their thanks to Mr. Tetsukazu Onoue for his contribution to fabricate the newly designed sputter-deposition apparatus. We also thank Mr. Y. Ichiki for his help in the present experiments. Authors are also grateful to TOYOTA Central R&D Labs., Inc. and Mr. N. Sugimoto for their cooperation.

References

[1] Sugimoto N, Iguchi N, Kusano Y, Fukano T, Hioki T, Ichiki A,Bessho T and Motohiro T, 2017, *Supercond. Sci. Technol.* **30,** 015014

[2] Sugimoto N and Motohiro T 2012 *Vacuum* **93,** 13

[3] Schwaltz D C and Jones R E, 1970 *IBM J. Res. Develop.* **14**, 52

[4] Lee W W Y and Oblas D,197, *J. Appl. Phys.* **46**, 1728

[5] Winters H F and Kay E, 1967, *J. Appl. Phys.* **38**, 3928

[6] Brodie I, Lamont L T and Jepsen R. L. 1968 *Phys. Rev. Lett.* **21**, 1224

[7] Kay E, Parmigiani F and Parrish W 1987, *J. Vac. Sci. Technol.* A**5**, 44

[8] Hoffman D W 1982 Film Stress Diagnostics in the Sputter Deposition of Metals *Proc. 7th Int. Conf. Vacuum Metallurgy (Nov. 26th-30th, Tokyo)* , The Iron and Steel Institute of Japan/The Japan Institute of Metals/The Vacuum Society of Japan, Tokyo, pp.145-156

[9] Thornton J A, Tabock J and Hoffman D W 1979 *Thin Solid Films* **64**, 111

[10] Cuomo J J, Gambino R J, Harper J M E and Kuptsis J D 1977 *IBM J. Res. Develop.* **21**, 580

[11] Cuomo J J, Gambino R J, Harper J M E and Kuptsis J D 1978 *J. Vac. Sci. Technol.* **15**, 281

[12] Kester D J and Messier R 1986 *J. Vac. Sci. Technol.* A**4**, 496

[13] Tominaga T, Ueshiba N, Shintani Y and Tada O 1981 *Jpn. J. Appl. Phys.* **20**, 519

[14] Wang Z, Kawakami A, Uzawa Y and Komiyama B 1996 *J. Appl. Phys.* **79**, 7837

[15] Shiosaki Y, Yasumoto Y and Kawabata A 1974 *Proc. 6th Conf. Solid Stae Devices*, Tokyo; 1975 *Suppliment to the Journal of the Japan Society of Applied Phys* **44**, 115

[16] Hoffman R W 1966 The Mechanical Properties of Thin Condensed Films *Physics of Thin Films –Advances on Research and Development–* vol 3 ed G Hass G and R E Thun (New York and London: Academic Press) pp.211-273

IOP Conf. Series: Journal of Physics: Conf. Series **871** (2017) 012072 doi:10.1088/1742-6596/871/1/012072

HTS-SQUID NDE Technique for Pipes based on Ultrasonic Guided Wave

Y Hatsukade[1], N Masutani[1], S Teranishi[1], K Masamoto[1], S Kanenaga[1], S Adachi[2] and K Tanabe[2]

[1]Kindai University, 1 Umenobe, Takaya, Higashi-Hiroshima, Hiroshima 739-2116, Japan

[2]SUSTERA, 2-11-19 Minowa-cho, MinatoKita-ku, Yokohama, Kanagawa 223-0051, Japan

hatsukade@hiro.kindai.ac.jp

Abstract. This article describes research on the novel high-temperature superconductor (HTS) superconducting quantum interference device (SQUID) non-destructive evaluation (NDE) technique for metallic pipes based on ultrasonic guided waves. We constructed HTS-SQUID NDE system for pipes based on ultrasonic guided waves, which were generated and received by means of the magnetostrictive effects. Using the system, we measured magnetic signals due to T (0, 1) mode ultrasonic guided waves that transmitted on aluminium pipe, and investigated influences of measurement parameters to the magnetic signals, such as direction of a HTS-SQUID gradiometer, lift-off distance, and intensity and frequency of input current fed to a magnetostrictive transmitter. With the gradiometer oriented parallel to the pipe axis, more than 10 times larger signals were measured compared with that oriented perpendicular to the pipe axis. Magnetic signals measured by the gradiometer were inverse proportional to the power of the list-off distance, and proportional to the intensity of the input current up to 1 A_{pp}. Relation between the frequency of the input current and the measured signal was shown and discussed.

1. Introduction

Ultrasonic guided waves refer to elastic waves in ultrasonic frequencies (10 – 100 kHz) that propagate in a bounded medium such as pipes, plates, etc., parallel to the plane of its boundary [1-4]. Properties of the guided waves are very complex. In pipes, the guided waves exit in three different wave modes: longitudinal (L), torsional (T), and flexural (F). Furthermore, each mode has parameters on circumferential direction n and wall thickness direction m, i.e., L (0, m), T (n, m), and F (n, m). In spite of such complexities, with adequate selection and proper control of wave mode and frequency, the ultrasonic guided waves are excellent for globally inspecting a large area of a structure from sensor location [4-8]. In long-range ultrasonic guided wave inspection, defects in pipes are inspected by detecting waves that reflected from the defects. Generally, piezoelectric and magnetostrictive transducers are used in such inspection. In recent years, T (0, 1) and L (0, 2) mode ultrasonic guided waves are mainly used, and especially T (0, 1) mode is easy to handle because it has no velocity dispersion and its group velocity is constant in a material, e.g., about 3.1 km/s in aluminium. Magnetostrictive methods, which utilize magnetostrictive and inverse magnetostrictive effects, has advantage over piezoelectric transducers in that no direct physical contact to pipe surface is required in case that the pipe is ferromagnetic material, and its conversion efficiency from electrical energy to mechanical energy and vice versa is sufficiently good for practical use.

To generate T (0, 1) mode ultrasonic guided waves in ferromagnetic and non-ferromagnetic pipes, magnetostrictive sensors (MsSs) combining a magnetized nickel plate and an induction coil are used [8]. In this case, the thin nickel plate is pre-magnetized toward its longitudinal direction and adhered firmly on a circumference of the pipe surface. The nickel generates ultrasonic guided waves when it is exposed in a magnetic field in the frequency range of 10-100 kHz from the coil, and the waves propagate in the pipe. The MsS can also receive the guided waves, because magnetization in the nickel plate changes when the nickel is deformed (or strained) by the propagated guided waves, leading to generate a magnetic field. The MsS picks up the magnetic field by the induction coil. We have been developing a novel non-destructive evaluation (NDE) technique using a high-temperature superconductor (HTS) superconducting quantum interference device (SQUID) to measure the magnetic field from the magnetized nickel in place of an induction coil [9]. Since the HTS-SQUIDs are superior to induction coil with regard to magnetic field sensitivity, it is expected that smaller defects in pipes will be detected by using the HTS-SQUIDs. In this study, we investigated the novel SQUID NDE technique for pipes while changing parameters in measurements of magnetic signals due to T (0, 1) mode ultrasonic guided waves. Direction of the HTS-SQUID gradiometer, lift-off distance between the SQUID and the pipe surface, and intensity and frequency of input current fed to an MsS on an aluminium pipe were changed in the measurements.

2. SQUID NDE system for pipe inspection

We constructed the HTS-SQUID NDE system for pipes utilizing the MsS as ultrasonic guided wave transmitter and receiver. The schematic diagram of the system including the aluminium sample pipe with two MsSs is shown in Figure 1. Details of the system are described in ref. [9]. A HTS-SQUID gradiometer based on ramp-edge Josephson junctions with two 1mm x 1mm differentially-connected pickup coils was used. The gradiometer was cooled at 69 K by a coaxial pulse tube cryocooler in a cryostat. The flux white noise level of the gradiometer was about 10-15 $\mu\Phi_0/Hz^{1/2}$ from 10 to 30 kHz. Due to the specification of a commercial SQUID electronics, the flux sensitivity above 30 kHz slightly decreased with increase in the frequency. The single sine wave voltage at a few tens kHz from the function generator is amplified and converted into single sine wave current to generate an excitation field at one MsS, which transmits T (0, 1) mode ultrasonic guided waves on the pipe toward $+x$ and $-x$

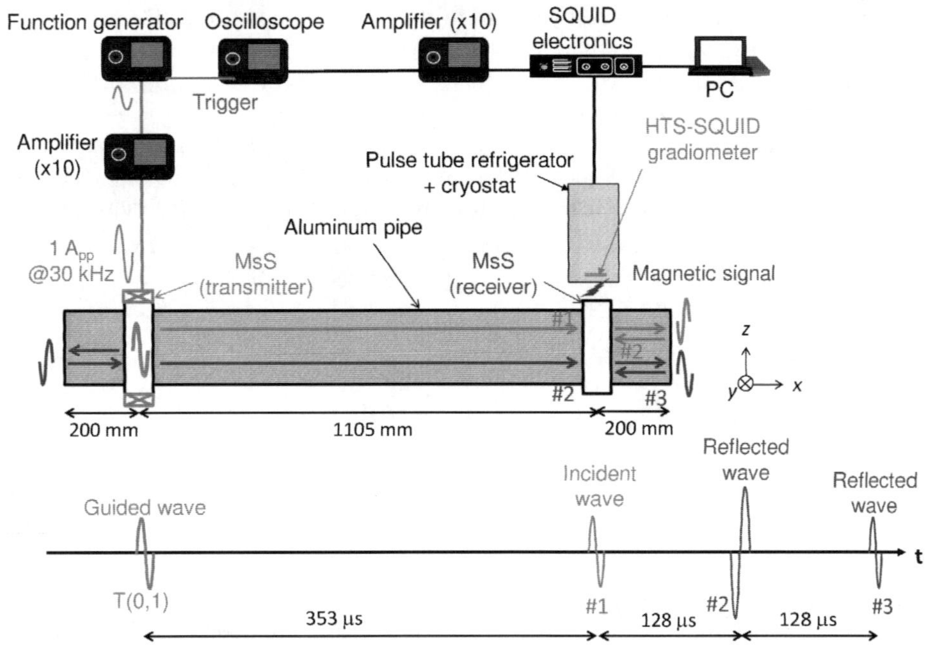

Figure 1. Schematic diagram of HTS-SQUID NDE system for pipes. Length of pipe, positions of MsSs, estimated pathways and time of flight of T (0, 1) mode guided waves are also shown.

directions. The wound coil around the nickel thin plate had 20 turns and the diameter of 50 mm. Under the gradiometer, the other MsS without an induction coil receives and converts the guided waves into magnetic field, which is measured by the gradiometer. Basically, it was set to measure dB_z/dx with a lift-off of about 2 mm and above the pipe axis, although the direction of the gradiometer was occasionally changed in the following measurements. The dimension of the sample aluminium pipe is 3.9 mm in thickness, 60 mm in outer diameter, and 2000 mm in length. Two previously magnetized thin nickel plates of 0.2 mm in thickness, 20 mm in width and 204 mm in length were wound around the circumferences of the pipe and firmly adhered as parts of the MsSs. The positions of the MsS on the pipe are depicted in figure 1. Since the group velocity of T (0, 1) mode guided wave in aluminium is about 3130 mm/ms, time of flight of the transmitted guided waves from transmitter to the receiver can be calculated from the distance between them. The estimated pathways and time of flight of the transmitted waves toward $+x$ and $-x$ directions are also illustrated in figure 1. It is known that the guided waves reflect both at pipe's ends and at defects, including the change of the thickness of the pipe, with reversed phase [10].

3. Experiments and results

In the first experiment, T (0, 1) mode guided waves were generated with the current input of 1 A_{pp} at 30 kHz to the MsS transmitter, and magnetic signals, whose component was dB_z/dx, due to the guided waves at the MsS receiver were measured by the SQUID gradiometer. With the current input of 1 A_{pp}, the applied field was about 0.5 mT_{pp} near the transmitter. Figure 2 shows the experimental result. At 0 μs, the guided waves were generated and started to propagate toward $+x$ and $-x$ directions. Measured single sine signals at around 350 μs, 480 μs, and 600 μs, which are labelled as #1, #2 and #3, are the incident wave, the reflected wave, in which reflected wave at the left end and reflected wave at the right end were superposed on each other, and the twice reflected wave at the left and the right ends in turn.

Next, we changed the measurement parameters. At first, the direction of the gradiometer was changed to measure dB_z/dy with the same lift-off of 2 mm. The same measurement as above was done and the measurement result is also shown in figure 2, with offset of about -12 mΦ_0 for visibility. Apparently, in the waveform of dB_z/dy, there are the same kinds of signals due to the guided waves, however, the amplitudes of the signals are 20-30 times smaller than those in the waveform of dB_z/dx. The results indicates that the magnetic field generated by means of the inverse magnetostrictive effect of the wound nickel thin plate produces the magnetic field oscillating parallel to the x direction. It can be said that the field change should be axisymmetric. Therefore, the flux coupled to the gradiometer to measure dB_z/dy was cancelled by the differential pickup coil of the gradiometer.

Secondary, the lift-off was changed from 2mm to 20 mm while adopting the same measurement procedure. Relationship between signal peak-peak amplitude of #1 and lift-off distance is shown in figure 3. The amplitude decreased inversely proportional to square of the lift-off distance.

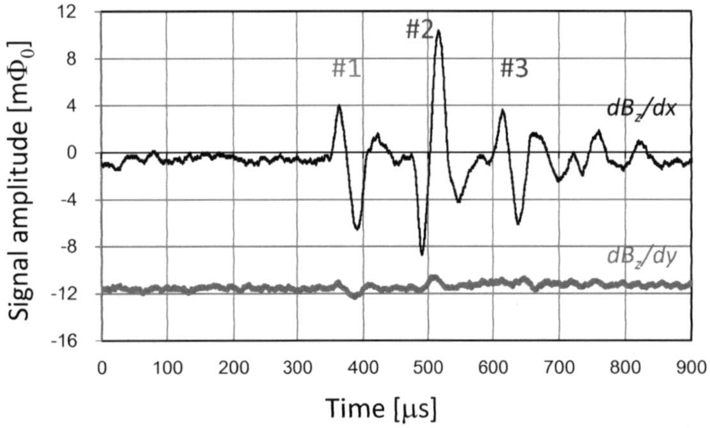

Figure 2. Measurement results of dB_z/dx (upper) and dB_z/dy (lower) above MsS receiver (magnetized nickel thin plate) by HTS-SQUID gradiometer. Offset of about -12 mΦ_0 was added on waveform of dB_z/dy for visibility.

IOP Conf. Series: Journal of Physics: Conf. Series **871** (2017) 012072 doi:10.1088/1742-6596/871/1/012072

Thirdly, the current amplitude into the MsS transmitter was changed from 0 to 1 A_{pp} at 30 kHz. Relationship between signal peak-peak amplitude of #1 and current peak-peak amplitude is shown in figure 4. As expected, the signal amplitude increased with the current amplitude. However, it is thought that if the excitation field is too strong to disturb the pre-magnetization of the nickel thin plate, the generated guided waves should be also disturbed. Now we are studying this phenomenon while increasing the current amplitude more than 1 A_{pp} up to 12 A_{pp}.

Finally, the frequency of the current into the MsS transmitter was changed from 10 to 100 kHz at 1 A_{pp}. Relationship between signal peak-peak amplitude of #1 and frequency of the input current is shown in figure 5. We corrected the amplitudes according to the sensitivity of the HTS-SQUID gradiometer, which slightly decreases with increase of the frequency [9]. As shown in figure 5, the signal amplitude slightly increased with frequency from 10 to 40 kHz, and it saturated over around 50 kHz. We briefly discuss the results in figure 5. The wave length of T (0, 1) mode guided wave at 40 kHz in aluminium at the group velocity of 3130 mm/ms is calculated to be about 78 mm, and a quarter of the wave length ($\lambda/4$) roughly matches to the width of the nickel thin plate 20 mm, as shown schematically in Fig. 6. We think the relation between the wave length and the width of the nickel thin plate may determine the frequency dependence of the magnetic signals. More profound research is now being studied.

4. Conclusion

In this paper, we investigated the novel SQUID NDE technique for pipes while changing parameters in the measurements of the ultrasonic guided waves on the aluminium pipe. The experimental results showed that the magnetic field change in the MsS receiver occurred toward the x (axis) direction. The magnetic signals due to the guided waves were inversely proportional to the lift-off, and the gradiometer could measure the magnetic signal with even the lift-off of 20 mm. The magnetic signals

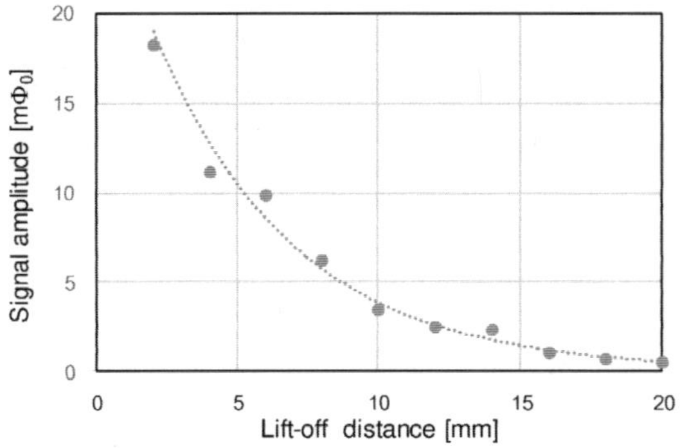

Figure 3. Signal amplitude of guided wave #1 shown in figure 2 vs. lift-off distance.

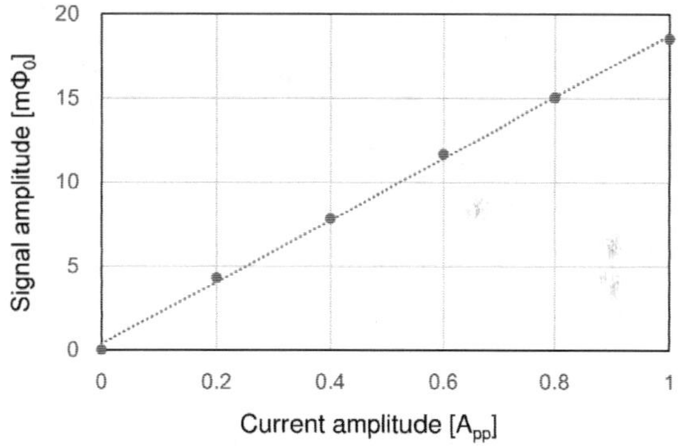

Figure 4. Signal amplitude of guided wave #1 shown in figure 2 vs. input current amplitude.

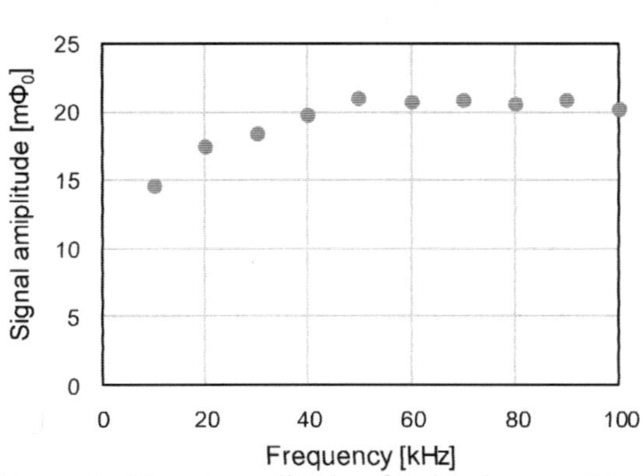

Figure 5. Signal amplitude of guided wave #1 shown in figure 2 vs. frequency of input current.

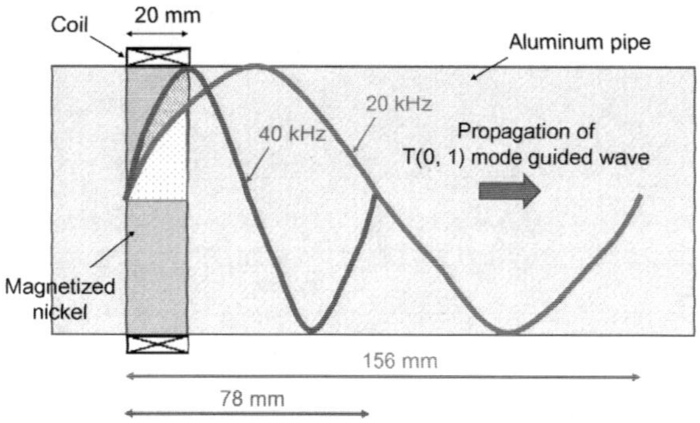

Figure 6. Relationship of width of magnetized nickel of MsS and wave length of guided waves.

due to the guided waves were proportional to the amplitude of the input current up to 1 A_{pp}. The magnetic signal increased with frequency of the input current, but it saturated around 40 -50 kHz, where a quarter of the wave length of the guided wave was roughly the same as the width of the nickel plate of the MsS.

Acknowledgments

Authors would like to acknowledge great support and encouragement from Mr. Takashi Ikeda of CXR Corporation and Dr. Yoshikazu Tanaka of Hiroshima University for this work. This work was supported partly by Suzuki Foundation, Japan, and is now supported partly by MEXT/JSPS KAKENHI Grant Number 16K06395 and Electric Technology Research Foundation of Chugoku, Japan.

References

[1] Graff K F 1975 *Wave Motion in Elastic Solids* (Oxford: Clarendon Press)
[2] Achenbach J D 1985 *Wave Propagation in Elastic Solids* (Amsterdam: Elsevier Science Publishers B. V.)
[3] Rose J L 1999 *Ultrasonic Waves in Solid Media* (Cambridge: Cambridge University Press)
[4] Rose J L 2014 *Ultrasonic Guided Waves in Solid Media* (New York: Cambridge University Press)
[5] Kwun H and Teller C M 1994 *Mat. Eval.* **52** 503
[6] Kwun H and Bartels K A 1998 *Mat. Ultrasonics* **36** 171
[7] Alleyne D N, Pavlakovic B, Lowe M J S and Cawley P 2001 *AIP Conf. Proc.* **557** 180
[8] TanakaY, Tamato E and Fujimoto Y 2010 *International Journal of Applied Electromagnetics and Mechanics* **33** 1237
[9] Hatsukade Y, KobayashiT, Nakaie S, Masutani N and Tanaka Y 2017 *IEEE Trans. Appl. Supercond.* **27** 1600104
[10] Tanaka Y, Ikeda T, Toiyama K, Kuwako S and Fujimoto Y 2006 *Transactions of the Japan Society of Mechanical Engineers* **72** 951 (in Japanese)

IOP Conf. Series: Journal of Physics: Conf. Series **871** (2017) 012073 doi:10.1088/1742-6596/871/1/012073

Quantitative and high-resolution magnetic images obtained by STM-SQUID microscope with distance modulation technique

T Yokocho, H S Akaba and Y Miyato

Department of Engineering Science, Osaka University, Toyonaka, Osaka, 560-8531, Japan

E-mail: miyato@sup.ee.es.osaka-u.ac.jp

Abstract. We have developed an STM-SQUID microscope, in which a scanning tunnelling microscope (STM) is combined with an rf-SQUID. The issue in our STM-SQUID microscope was that the obtained magnetic image was not the same as the sample's ideal magnetic field distribution. This was because the magnetic image could be affected not only by the sample's local magnetic field distribution, but also by the magnetic field distribution above the sample, namely "background field". In this work, we applied the distance modulation technique to our microscope. In this technique, the change of the SQUID output signal was detected while the distance between the sample and the probe tip was modulated in a constant amplitude. As a result, the influence of the background field was cancelled out, and the magnetic information near the sample surface was largely extracted. We successfully obtained the quantitative magnetic images in ~100 nm spatial resolution by using the distance modulation technique.

1. Introduction

Recently, many types of SQUID microscopes have been developed. In typical SQUID microscopes, small-sized dc-SQUIDs are applied to get high sensitivity and high spatial resolution[1]–[3]. However, our SQUID microscope, namely an STM-SQUID microscope, is completely different from typical SQUID microscopes[4],[5]. In our STM-SQUID microscope, an rf-SQUID is equipped because of easy fabrication and wirings[6]. Remarkably, a high permeability probe is also used as a flux guide from the sample to the SQUID. The high permeability probe plays some important roles to equivalently shorten the gap between the sample surface and the SQUID, and to equivalently miniaturize the size of SQUID[7]. Owing to the probe's roles, we successfully observed the sample's fine magnetic domain structures in sub-100 nm spatial resolution, as shown in figure 1(b). In this paper, the magnetic images obtained by directly mapping the output signal of the SQUID electronics are called "normal images".

However, in the normal images obtained by our STM-SQUID microscope, image-processings were required to obtain the clear images. In the raw normal images, magnetic field distributions tended to be largely tilted, and the fine magnetic domain structures were easily hidden in the large contrast change, as shown in figure 1(c). Therefore, the STM-SQUID microscope had the issues that magnetic field distribution might be distorted by image-processings, and that the normal image was not perfectly the same as the sample's local magnetic field distribution. It was required to resolve these issues to improve the quantitative performance of the STM-SQUID microscope.

In this work, we firstly analized the cause and solution of the issues by means of the electromagnetic simulations. From the analysis results, we focused on the distance modulation technique[8] for the solution, and we investigated its effect for the issues. Finally, we demonstrated the measurements of typical samples by the STM-SQUID microscope using distance modulation technique.

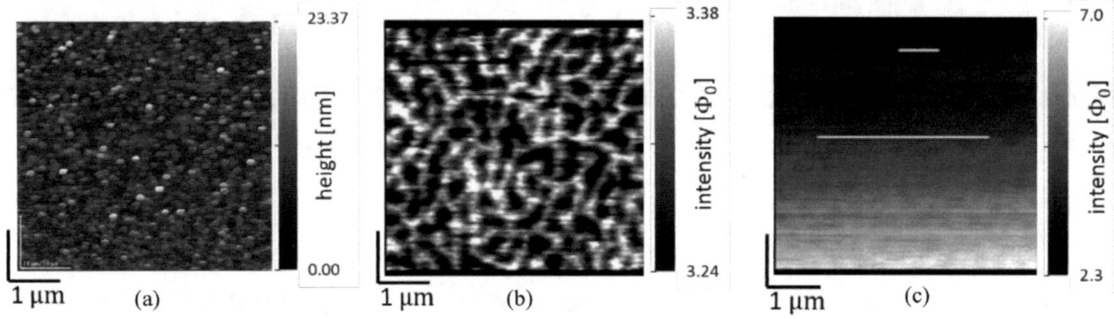

Figure 1. Example of measurement results obtained by the STM-SQUID microscope. (a) Topographic and (b) normal magnetic images of a nickel thin film sample after image-processings including flattening. (c) Its raw magnetic image.

2. Simulation analysis

2.1. Comparison of original local flux density with flux distribution measured by STM-SQUID microscope

To consider the cause of the issues, we simulated the local magnetic field distribution over a sample and its magnetic flux distribution measured by the STM-SQUID microscope. Figure 2 shows these simulation models. A meander line, through which constant current flows, was applied to the sample model. Because the meander line generated the alternative magnetic field distribution as shown in figure 2(a), it can imitate the magnetic material having periodic domain structure. In this paper, the static distance between the probe tip and sample is defined by "h". In the simulation, we calculated the magnetic flux density distribution over the sample. Figure 3(a) and (b): the results of figure 2(a) show the magnetic field distribution along the z-axis direction (B_z) at the plane of h =50 nm above the sample surface, and its line profile along the A-B line, respectively. On the other hand, in the STM-SQUID microscope measurement model, we employed the typical sizes and relative permeability of the actual rf-SQUID and probe. We also adopted the same sample model as shown in figure 2(a). In the simulation, we calculated the magnetic flux inside the SQUID hole at each position. Obtaining the distribution by mapping the magnetic flux inside the SQUID hole is equivalent to obtaining the normal magnetic image by the actual STM-SQUID microscope. Similarly, figure 4(a) and (b): corresponding to the results of figure 2(b) show the magnetic flux distribution mapped by the STM-SQUID microscope at h = 50 nm and its line profile along A-B line, respectively.

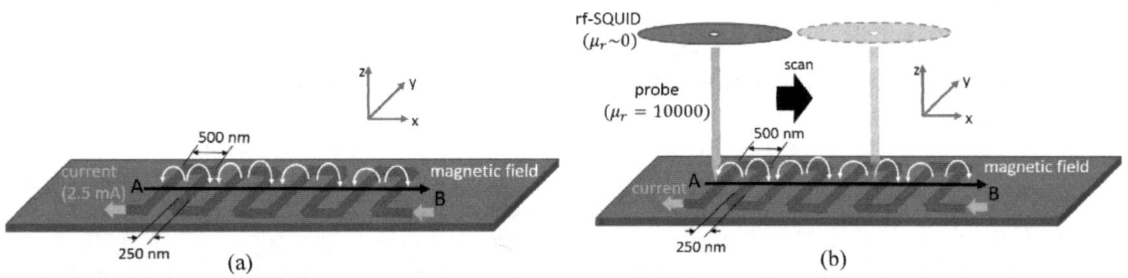

Figure 2. Schematic simulation models of (a) a meander line sample representing a magnetic domain structure and (b) measurement executed by the STM-SQUID microscope.

In figure 3(b), the flux density distribution oscillates around at 0 mT. Compared with figure 3(b), figure 3(d), illustrating the magnetic flux profile taken by the STM-SQUID microscope has large signal offset of ~20 aWb, superimposed on the variation of the local magnetic field. Besides, the variation of figure 3(d) is deformed from the shape of figure 3(b). It suggests that the real normal image can be affected by the same phenomena.

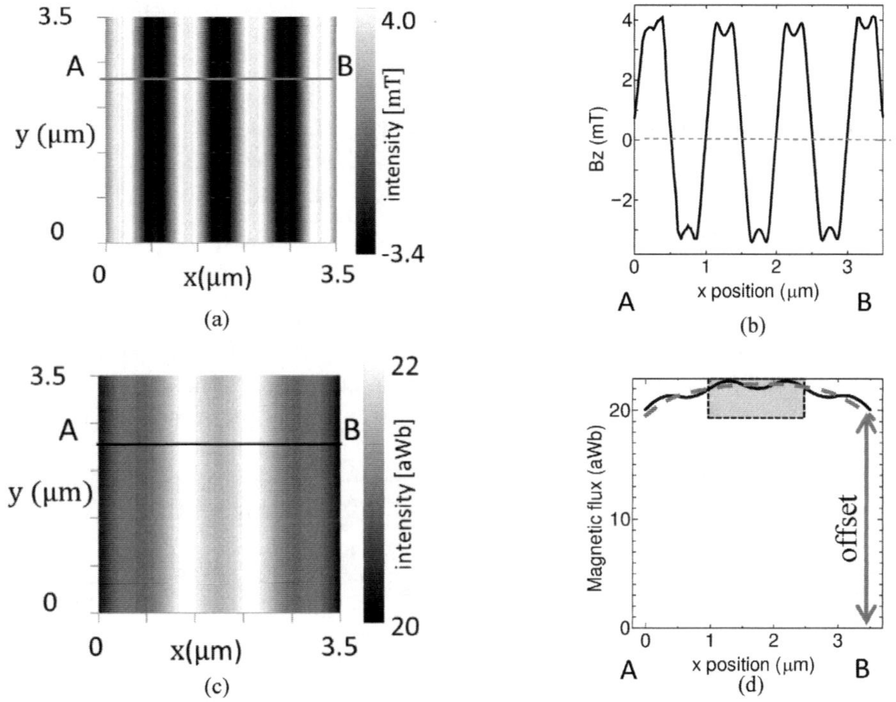

Figure 3. (a) Local magnetic flux density distribution over the meander line sample, and (b) its line profile along A-B line. (c) Magnetic flux distribution measured by the STM-SQUID microscope, and (d) its line profile along A-B line.

2.2. Discussion about the cause of issues

To consider the cause of the normal image's issues, we analized the magnetic field distribution around the probe tip. Figure 4(a) shows the magnetic flux density distribution in the case of $h = 150$ nm. Each triangle marker indicates the vector direction of the magnetic flux density. The magnetic field is distributed as the probe concentrates the magnetic flux from the side surface. The field intensity becomes smaller with the distance from the sample surface longer. However, the probe length is relatively so long that the influence of the magnetic flux coming from the probe side cannot be disregarded.

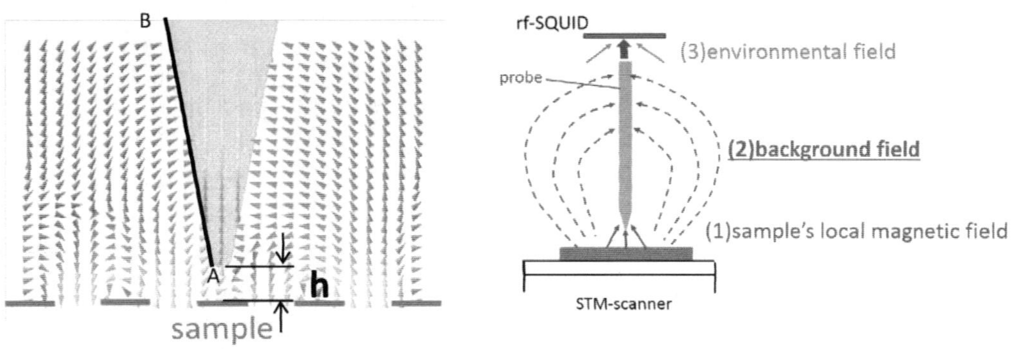

Figure 4. (a) Magnetic flux density distribution around the probe tip ($h = 150$ nm). (b) Main contributions to the normal image taken by the STM-SQUID microscope.

This result indicated that the normal image was mainly affected by three contributions from the magnetic field: sample's local field, background field, and environmental field. First, the sample's local magnetic field distribution originates from the fine magnetic structure distributed on the sample's surface. This first contribution corresponds to the flux coming from the probe tip. Next, the background field originates from magnetic field distributed over and away from the sample. This second contribution corresponds to the averaged flux coming from the upper area of the probe. Third, the environmental field originates from the external magnetic field coming to the STM-SQUID microscope instrument. This third contribution corresponds to the external magnetic flux directly coming to the instrument. The contribution of background and the environmental fields can hide the local magnetic field distribution because they are sometimes too large to ignore against the local magnetic field contribution.

2.3. Distance modulation technique

To consider the solution for cancellation of the background field influence, we analized the magnetic field distribution around the probe tip by changing the static distance between the probe tip and the sample: h. We obtained the line profiles from simulations, and subtracted each other. One example of the results is shown in figure 5. The result shows that contribution of the magnetic field at region around the probe tip (i) is dominant, but that at region around the upper area of the probe (ii) is almost zero, as indicated in figure 5(b). Consequently, we concluded that a magnetic image obtained by detecting the signal change at different distances between the probe tip and the sample surface could be less affected by influence of the background field. It can be realized by the distance modulation technique.

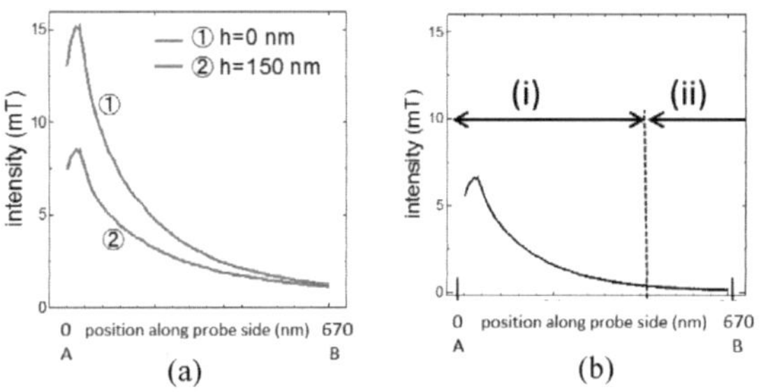

Figure 5. (a) Line profiles of the magnetic flux density along A-B line as shown in figure 4(a). (b) Line profile obtained by subtraction of ①-②.

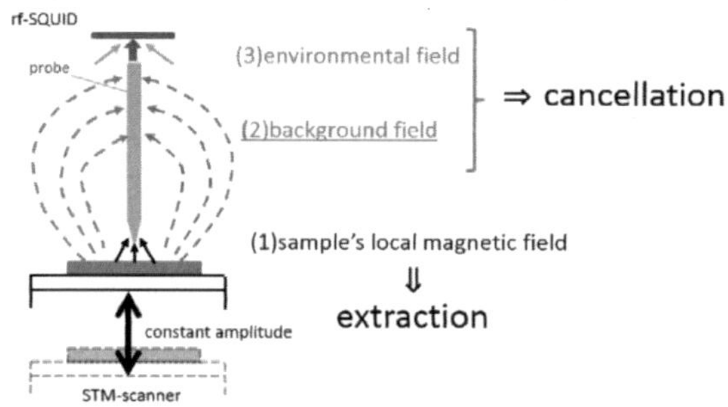

Figure 6. Schematic illustration of the distance modulation technique.

In the distance modulation technique, the change of SQUID signal synchronized to the modulation frequency was detected while the distance between the probe tip and the sample surface was modulated in a constant amplitude, as shown in figure 6. It is expected that the sample's local magnetic field is largely extracted and the background field is cancelled out. We simulated the STM-SQUID microscope using the distance modulation technique. For this simulation, we calculated the profiles of the flux change by subtracting the magnetic flux obtained at two different h conditions at each position. Figure 7(a) shows an example of the simulation result, where the modulation amplitude and h were set at 40 $\mathrm{nm_{p-p}}$ and 50 nm, respectively. The simulation result indicates that the signal offsets obtained by the distance modulation was reduced to 0.16 aWb, which was about 100 times smaller than the normal result as shown in figure 7(b). In addition, distribution of the magnetic flux obtained by the distance modulation becomes similar to the sample's local magnetic field distribution, as shown in figure 3(b). Thus, the distance modulation technique could reduce the background field influence.

IOP Conf. Series: Journal of Physics: Conf. Series **871** (2017) 012073 doi:10.1088/1742-6596/871/1/012073

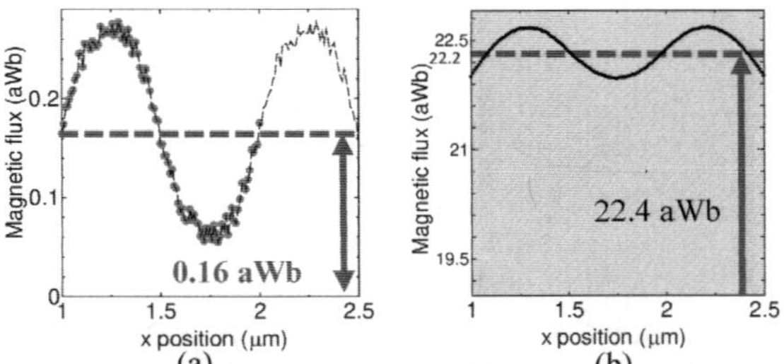

Figure 7. Profile comparison between the distance modulation and the normal results. (a) Magnetic flux distribution obtained by the distance modulation technique, assuming the modulation amplitude of 40 nm_{p-p}. (b) Magnetic flux distribution taken from the meshed area in Figure 3(d).

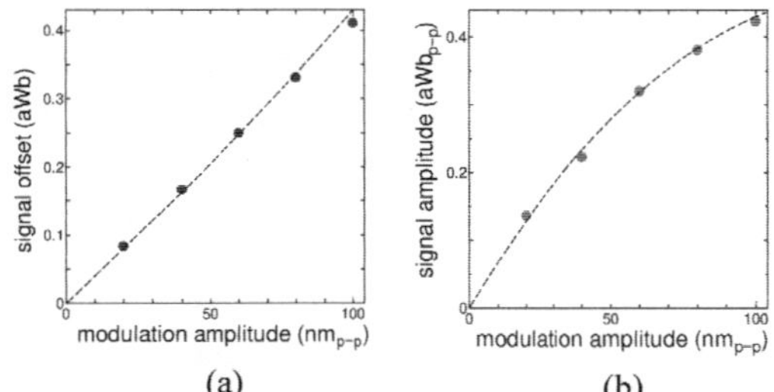

Figure 8. Dependence of (a) the signal offset and (b) the signal amplitude on modulation amplitude (h=50 nm).

We also investigated the dependence of the signal offsets and amplitudes on the modulation amplitude. We calculated the signal offsets and amplitudes while changing the modulation amplitude, as shown in figure 8(a) and (b), respectively. The result shows that the signal offset becomes smaller as the modulation amplitude is decreased, as shown in figure 8(a). From this result, we concluded the modulation amplitude should be small to cancel out the background field effect ideally. However, the signal amplitude becomes smaller as the modulation amplitude is decreased, as shown in figure 8(b). In other words, the signal amplitude may become so small to be hidden by noises, if the modulation amplitude is too small.

3. Experimental results

3.1. Setup of STM-SQUID microscope

Our STM-SQUID microscope was based on a commercially available STM-microscope system (JSPM-5410, JEOL). Figure 9 shows the basic setup of the STM-SQUID microscope. A laboratory-made cryostat head was mounted instead of the default STM head. An rf-SQUID made of YBCO film with 120 nm thickness, as shown in figure 9(b), was fixed inside the cryostat. The high-permeability probe, as shown in figure 9(c), was also fixed just under the the SQUID hole of the rf-SQUID. The probe was made of a permalloy (alloy of iron and nickel) wire with 90 μm diameter and ~5 mm length. The probe tip was sharpened with its curvature radius from 50 to 150 nm. The STM feedback-system controlled the probe tip-sample distance at ~1 nm by measuring the tunnelling current between the probe tip and sample surface. From the STM-feedback signal, the sample's topographic image was obtained. We used a commercially available SQUID electronics, supporting the flux-locked loop (FLL) control to detect the magnetic field intensity quantitatively. From the FLL output, normal magnetic images were obtained.

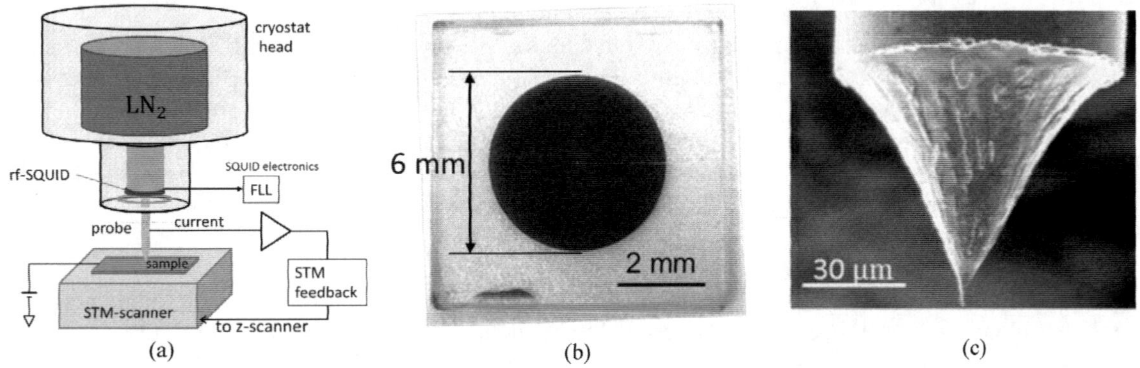

Figure 9. (a) Setup of the STM-SQUID microscope, consisting of (b) an rf-SQUID and (c) a high-permeability probe.

3.2. Signal flows of for distance modulation technique

Figure 10 shows the signal flows for obtaining the magnetic images by the STM-SQUID microscope using the distance modulation technique. We adopted the two-pass method for obtaining the normal image and the modulation image in the same condition. In this method, the probe scanned every line twice. In the first pass, the STM feedback control was executed to keep the probe tip-sample distance and obtain the topographic image. At this timing, the normal image was obtained by directly mapping the FLL output as shown in figure 10(a)′. The image-processings were performed after the measurement, as shown in figure 10(a). In the second pass, the distance modulation technique was executed. For the application of the distance modulation technique, we used a lock-in amplifier to detect the change of the SQUID output signal while the distance between the probe tip and the sample was modulated at the frequency of ~8 kHz. From the lock-in amplifier, we obtained two images: the amplitude image corresponding to the intensity of the magnetic field (figure 10(b)) and the phase image corresponding to the polarity of the magnetic field (figure 10(c)). After the measurement, the modulation image (figure 10(d)) was reconstructed by combining the amplitude and phase images.

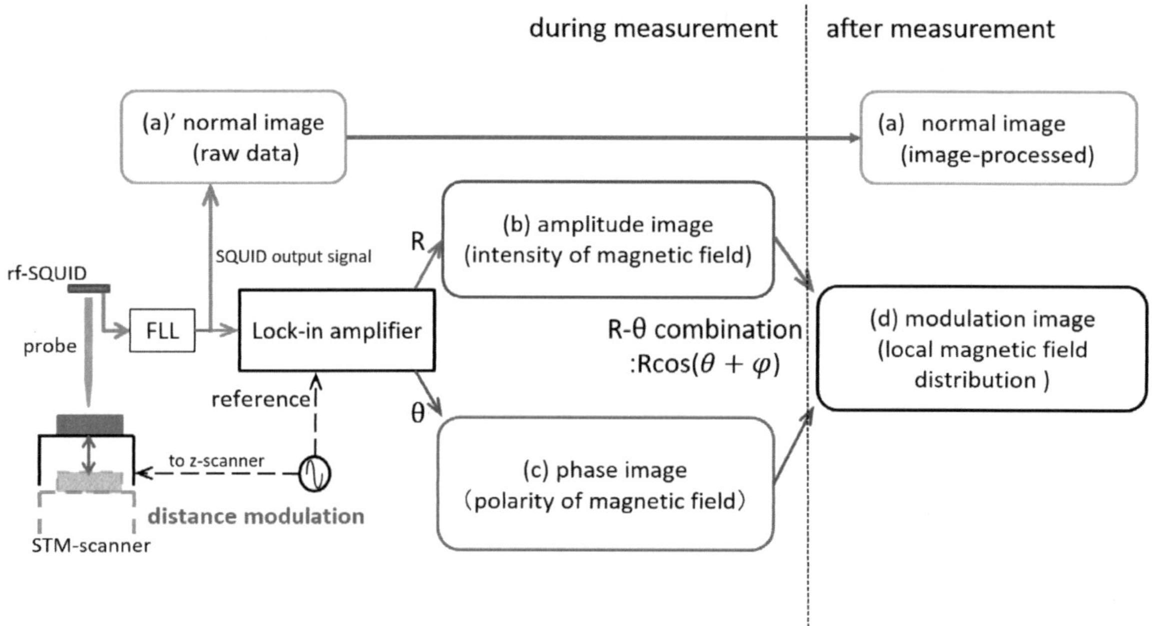

Figure 10. Signal acquisition flows of the magnetic images taken by the STM-SQUID microscope using the distance modulation technique.

3.3. Demonstration of measurements using distance modulation technique

We measured an iron garnet film and a nickel thin film samples by the STM-SQUID microscope using the distance modulation technique. Magnetic domain structures of the both samples show the maze patterns[9],[10]. The magnetic domain sizes of the garnet and the nickel samples are tens of μm and a few hundreds of nm, respectively. We measured the garnet and the nickel samples to confirm the effect of distance modulation technique and to confirm the spatial resolution of our microscope, respectively. The iron garnet surface was coated by gold thin films in order to detect the tunnelling current because it was an insulator.

Figure 11 shows the magnetic images of the garnet sample. In figure 11(a), the normal image has large offset and the sample's magnetic domain structure is blurred. In figure 11(b), the magnetic field intensity is almost constant, but is close to zero at the boundary of magnetic domain. In figure 11(c), the phase of the magnetic field changes by 180 degrees, domain by domain. Compared with figure 11(a) illustrating the normal image, figure 11(d), illustrating the modulation image shows the magnetic domain structure clearly even though any image-processings were not carried out. Judging from the amplitude and phase images, we supposed that the sample had a perpendicular domain structure. Moreover, judging from the modulation image, we conclude that distance modulation technique could reduce the background field influence.

On the other hand, figure 12 shows the magnetic images of the nickel sample. Figure 12(a), illustrating the normal image is distributed by noises even after the image-processings carried out. In figure 12(b), the magnetic field intensity is weak at the boundary. In figure 12(c), the phase of magnetic field is almost constant. In the same manner of the garnet result, the modulation image reveals the fine magnetic domain structure clearly without any image-processings, as shown in figure 12(d). Fewer noises are observed in the modulation image than the normal image. Judging from the amplitude and phase images, we supposed that the sample had a horizontal domain structure. Moreover, according to the modulation image, the domain size is approximately 200 nm. Thus, the spatial resolution of the STM-SQUID microscope is estimated at ~100 nm.

4. Summary

The background field over the sample strongly affects the normal magnetic image obtained by the STM-SQUID microscope. It causes the issue that the normal image is not completely same as the ideal magnetic field distribution of the sample because the high permeability probe transfers the magnetic flux not only from the probe tip but also from the upper side of the probe. The distance modulation technique is very effective to cancel out the background field influence. We successfully obtained the local magnetic field distribution of the nickel sample in ~100 nm spatial resolution by the STM-SQUID microscope by using the distance modulation technique.

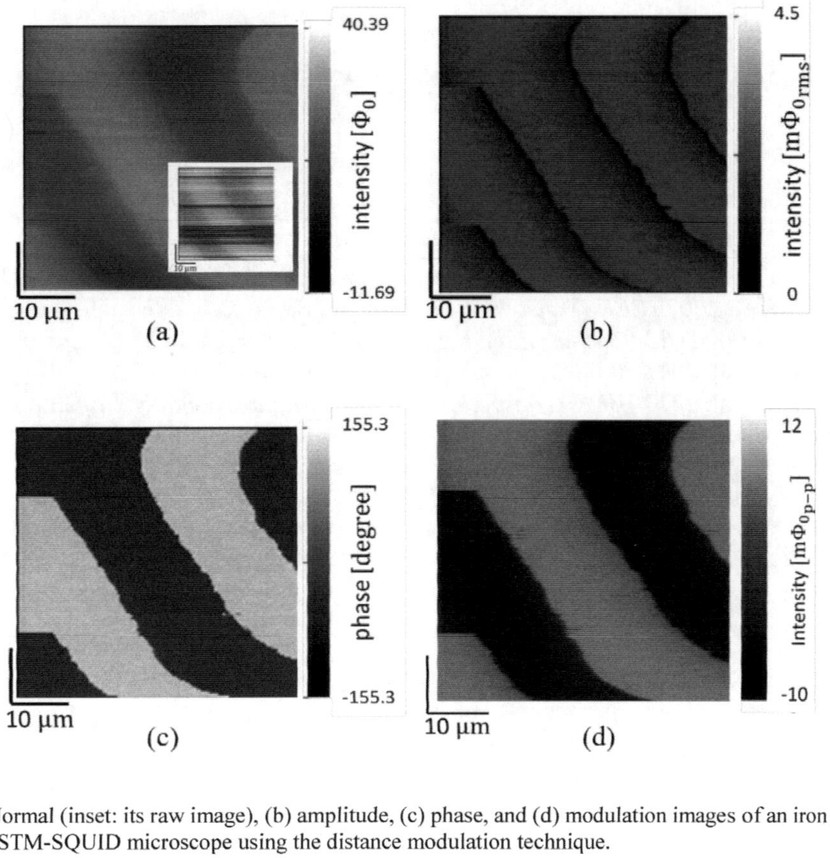

Figure 11. (a) Normal (inset: its raw image), (b) amplitude, (c) phase, and (d) modulation images of an iron garnet sample obtained by the STM-SQUID microscope using the distance modulation technique.

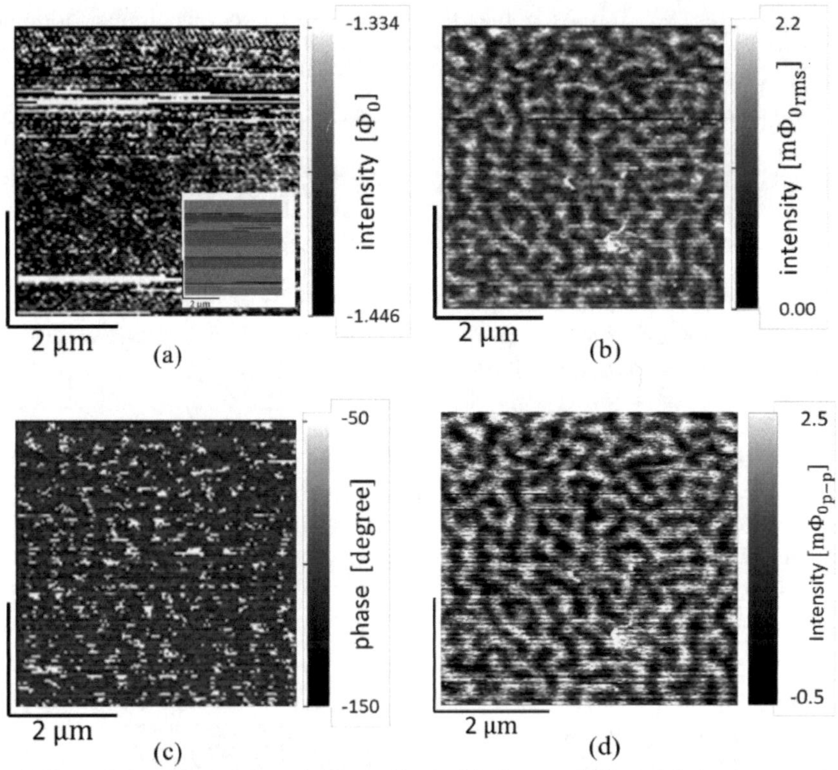

Figure 12. (a) Normal (inset: its raw image), (b) amplitude, (c) phase, and (d) modulation images of a Ni thin film sample obtained by the STM-SQUID microscope using the distance modulation technique.

References

[1] Finkler A et al. 2010 *Nano Letters* **10** pp.1046–9
[2] Anahory Y et al. 2014 *Nano Letters* **14** pp.6481–7
[3] Shibata Y et al. 2016 *Superconductor Science and Technology*IOP Publishing **29** pp.104004
[4] Hayashi T et al. 2007 *IEEE Transactions on Applied Superconductivity* **17** pp.792–5
[5] Watanabe N et al. 2013 *Journal of the Magnetics Society of Japan* **37** pp.235–8
[6] Hisayama K et al. 2015 *Physics Procedia*Elsevier B.V. **65** pp.189–92
[7] Watanabe N et al. 2013 *IEEE Transactions on Applied Superconductivity* **23** pp.3–6
[8] Hayashi T et al. 2007 *Superconductor Science and Technology* **20** pp.S374–9
[9] Xia WX et al. 2010 *Journal of Applied Physics* **108** pp.3–7
[10] Mishina H et al. 2009 *Journal of Applied Physics* **105** pp.093911-1–093911-5

Defect Detection of Pipes using Guided Wave and HTS-SQUID

N Masutani[1], S Teranishi[1], K Masamoto[1], S Kanenaga[1], Y Hatsukade[1], S Adachi[2] and K Tanabe[2]

[1]Kindai University, 1Umenobe, Takaya, Higashi-Hiroshima, Hiroshima 739-2116, Japan

[2]SUSTERA, 2-11-19 Minowa-cho, MinatoKita-ku, Yokohama, Kanagawa 223-0051, Japan

1633850010h@hiro.kindai.ac.jp

Abstract. In this paper, we investigated ultrasonic guided wave measurements on aluminium pipes with defects using high temperature superconducting (HTS) SQUID gradiometer and magnetostrictive sensors (MsS), which utilized pre-magnetized nickel thin plates. One of the pipes was provided with an axial defect, and the other was provided with a circumferential defect. The MsSs were used by adhering them on circumferences of the pipes to generate ultrasonic guided waves and to receive the waves by converting them into the magnetic signals, which were measured by the HTS-SQUID gradiometer. Guided wave measurements using the MsSs and the gradiometer demonstrated that magnetic signals of about 3.3 mΦ_0 due to reflected waves from both the defects were successfully detected. Reflected wave signals of about 1.4 mΦ_0 from thin nickel plates of the MsS on the pipes were also detected.

1. Introduction

In recent years, in order to maintain the health of pipes in social infrastructures, importance of non-destructive inspection (NDI) is increasing. Ultrasonic guided wave testing is mainly used for NDI of long pipelines, for example, in chemical plants [1-5]. Ultrasonic guided waves used for this inspection are mainly T (0, 1) mode and L (0, 2) mode guided waves. The T (0, 1) mode guided wave is easy to handle because it has no velocity dispersion and its group velocity is constant in a material. Therefore, the T (0, 1) mode guided waves are used as transmission/reception signals. Magnetostrictive method and piezoelectric method are often used in the ultrasonic guided wave testing. In such case, it is necessary to bring the ultrasonic transducers into contact with target pipe. When the pipe is wrapped with insulating materials, it is necessary to peel off the materials at a measuring part. It leads to cost and time.

From these backgrounds, we have introduced high temperature superconductor (HTS) superconducting quantum interference device (SQUID) as a receiver in the magnetostrictive method. HTS-SQUIDs have extremely high sensitivity in a frequency range from DC to a few MHz [6], which includes the frequency range from 10 to 100 kHz of the ultrasonic guided wave. Furthermore, it has the very low intrinsic magnetic noise of about 10 $\mu\Phi_0/Hz^{1/2}$ (Φ_0 is the flux quantum). From these characteristics, we thought that a vibration due to the guided wave in a pipe can be measured without contact by a HTS SQUID while converting the vibration into a magnetic signal based on the inverse magnetostrictive effect of the receiving part of the magnetostrictive method. Therefore, we have studied the novel NDI technique to measure magnetic signals due to ultrasonic guided waves on an aluminium pipe combining a HTS-SQUID gradiometer and the magnetostrictive method, which utilizes magnetized nickel to convert vibration into magnetic signal and vice versa. In the previous study [7, 8], ultrasonic guided waves generated in a target pipe by the magnetostrictive method were successfully detected by the HTS-SQUID gradiometer above a magnetized nickel wound and adhered around the circumference of the pipe. In this study, we demonstrated detection of reflected wave signals due to defects in axial and circumferential directions in sample aluminium pipes using the HTS-SQUID gradiometer and the magnetostrictive method.

2. Samples and measurement system

2.1. Samples

As samples, aluminium pipes of 3.9 mm in thickness, 60 mm in outer diameter, and 2000 mm in length were used. At each position of 200 mm, 750 mm and 1300 mm away from the left end of each pipe, a thin nickel plate of 0.2 mm in thickness, 20 mm in width and 157 mm in length and previously magnetized toward its longitudinal direction by a strong permanent magnet, was wound around the circumference of the pipe and firmly adhered by resin bond. The thin nickel plate at 200 mm was wound with a coil of 25 turn and 60 mm in diameter and was used as a transmitter of T (0, 1) mode guided wave. When ac current of 1 A_{pp} was fed to the coil, we measured a generated magnetic field of about 0.5 mT_{pp} just below the coil using a Hall sensor. Another magnetized thin nickel plate of the same dimension at 750 mm was used to receive an ultrasonic guided wave and convert it to a magnetic signal. One of the pipes (labelled as #1) was provided with a penetrating defect of 100 mm in length in the axial direction and 2 mm in width, which was made at 975 mm from the left end of the pipe. The other pipe (labelled as #2) was provided with a penetrating defect of 50 mm in width in the circumferential direction and 2 mm in length, which was made at 1030 mm from the left end. Schematic figures of the defects and pipes are shown in Figure 3 (b) and Figure 4 (b). We noticed that residual magnetization leaked from the magnetized nickels at each face-to-face part of them. When the SQUID gradiometer was brought close to the parts, the SQUID gradiometer could not be locked properly with a commercial flux locked loop circuit. We defined the parts of the nickels was at 0 degree in the circumference of each pipe (see Figures 2-4). In this definition, the centres of the defects were set at 180 degrees.

2.2. SQUID NDI system

We constructed the SQUID NDI system for pipe based on the magnetostrictive effects using the HTS-SQUID gradiometer. Figure 1 shows the schematic diagram of the system. Details of the system are described in refs. [7, 8]. As a sensor, the HTS-SQUID gradiometer with ramp-edge Josephson junctions, which has a planar differential pickup coil that consists of two 1 mm x 1 mm square coils, was used. In the measurement system, the gradiometer was cooled at about 69 K by a coaxial pulse-tube cryocooler [9]. Flux noise level of the HTS-SQUID gradiometer is about 10-15 $\mu\Phi_0/Hz^{1/2}$ in a range from 10 Hz to 30 kHz. One cycle of sine wave voltage at 30 kHz from a function generator was amplified into current of 1 A_{pp} by a power amplifier, and the current was repeatedly applied to the coil of the transmitter with a period of 0.33 s to generate an excitation magnetic field. The T (0, 1) mode ultrasonic guided waves were generated by magnetostrictive effect in the thin nickel plate, and they propagated the opposites directions (+x and –x) through the pipe's wall. When the guided wave reached the thin nickel plate of the receiver, the vibration due to the guided wave was converted into a magnetic signal by the inverse magnetostrictive effect. The magnetic signal was measured by the HTS-SQUID gradiometer, which was placed above the center of the receiver. The HTS-SQUID gradiometer was set to measure dB_z/dx above the magnetized nickel thin plate on the sample pipe with a lift-off distance of about 2 mm. Output voltage of the SQUID electronics passed through a high-pass filter (HPF) with cut-off frequency of about 1 kHz and then was amplified using a low-noise amplifier with a gain of 20 dB and noise level of 4 $nV/Hz^{1/2}$. The voltage from the SQUID electronics were measured by a digital oscilloscope and the data were averaged 256 times on the oscilloscope.

3. Experiments and results

3.1. Experiment on aluminum pipe without defect

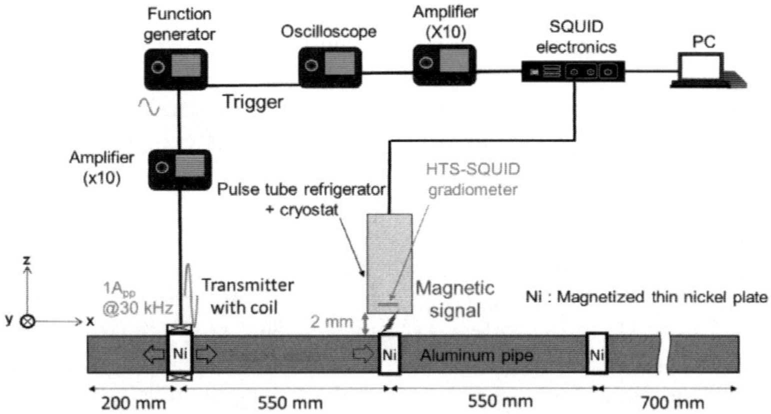

Figure 1. Schematic diagram of SQUID NDI system for pipes.

We investigated effects of defects on magnetic signals due to ultrasonic guided waves using the aluminium pipes #1 and #2. Using the pipe #1 *without defect*, we measured magnetic signals due to the T (0, 1) mode ultrasonic guided wave with the HTS-SQUID gradiometer, which was set above the pipe at 90 degrees (see figure 2 (b)). Figure 2 (a) shows the measurement result. At about 180 μs, a magnetic signal of about 9.1 mΦ_0 in peak-peak amplitude labelled as "a" was measured. At about 305 μs, a magnetic signal of about 8.3 mΦ_0 labelled as "b" with reversed phase to "a" was measured. After the signal "b" was generated, a series of smaller signals including "c" than the signals "a" and "b" was measured at 440 μs, 560 μs, and 660 μs. The group velocity of a T (0, 1) mode torsional wave propagating in aluminium pipe is known to be about 3100 mm/ms. From the distance between the transmitter and the receiver and the guided wave velocity, start timing of the magnetic signal due to the incident wave measured by the SQUID gradiometer is calculated to be 177 μs, which agrees well with the timing of "a". The transmitter generated the guided waves propagating to both sides of the transmitter. One wave propagated toward the $+x$ direction and generated the magnetic signal "a" at 177 μs. Meanwhile, the other wave propagated toward the $-x$ direction, and then reflected at the left end of the pipe with the reversed phase, then propagated toward the $+x$ direction to generate a magnetic signal at the thin nickel plate. Start timing of this signal measured by the gradiometer is calculated to be 306 μs, which agrees well with the timing of "b". After the signal "b" was generated, the small signal "c" was measured. In the previous research, it was reported that the guided waves were partly reflected at a nickel thin plate adhered around the circumference of the pipe, and the reflected waves were detected by means of a magnetostrictive sensor [10]. It means that thickening of the pipe's wall due to the adhesion of the nickel plate, as well as defects such as thinning of a pipe's wall, makes a guided wave reflect. Some portion of the wave reflected firstly at the left end of the pipe was reflected again at the nickel plate, and it was again reflected at the left end and some portion of this wave transmitted through the nickel plate to arrive at the nickel plate of the receiver, as shown in figure 2 (b). Start timing of this transmitted wave measured by the gradiometer is calculated to be 435 μs, which agrees well with the timing of "c". It is considered that the small signals after "c" at 560 μs and 660 μs were originated in the guided waves repeatedly reflected at the thin nickel plates and also at the ends of the pipe and transmitted through the plates. From these results, in the guided wave measurement using the HTS-SQUID gradiometer, it was found that this method can not only measure the reflected waves from the ends of the pipe but also the reflected waves from the thin nickel plate.

3.2. Experiments on aluminum pipes with defects

3.2.1 In case of axial defect

The penetrating defect toward the axial direction was produced at 975 mm in the x direction and at 180 degrees of the pipe #1, and guided wave measurement was performed under the same conditions as in the experiment in *3.1*. Figure 3 (a) shows the measurement result and figure 3 (b) shows the estimated propagation paths of signals "d", "e", "*e'*", and "f". As shown in figure 3 (a) , the magnetic signal "d" at 169 μs and the signal "e" at 302 μs with reversed phase to "d" were the same incident and reflected waves as "a" and "b" in figure 2 (a). Comparing the signal "e" in figure 3 (a) with the signal "b" in figure 2 (a), the signal "e" was a slightly wider in time. Comparing the amplitude ratio of the signals "a" over "b" with that of "d" over "e", the latter is a bit larger. We considered this reason as follows: the incident wave that generated "d" propagated toward the +*x* direction, and reflect at the axial defect, and the phase of the wave was reversed. This reflected wave is labelled as "*e'* " in figure 3 (b). From the propagation path of signal "e" and "*e'* ", they should superpose at around 314 μs. Because the timings of "e" and "*e'* " are slightly different and some parts of the signals cancelled each other, the signal ratio "d"/"e" become larger than that of "a"/"b". The signal of about 3.3 mΦ_0 labelled as "f", which was not observed on the same pipe without the defect as shown in figure 2 (a), was measured at 440 μs. The signal "f" was considered to be originated in the wave that generated the signal "e", which propagated toward the +*x* direction through the receiver after first reflection at the left end, and reflected by the defect with the reversed phase and returned to the receiver. The start timing of the signal "f" measured by the gradiometer was calculated to be 446 μs, which agrees well with the timing of "f".

3.2.2 In case of circumferential defect

The penetrating defect toward the circumferential direction was produced at 1030 mm in the x direction and at 180 degrees of the pipe #2, and guided wave measurement was performed under the same conditions as *3.1* and *3.2.1*. Figure 4 (a) shows the measurement result and figure 4 (b) shows the estimated propagation paths of the signals "g", "h", "i" and "j". Regarding the signals "g" and "h", the results were similar to those in figure 2 (a) and figure 3(a). However, just after the signal "h" was measured, a signal of about 4.2 mΦ_0 labelled as "i" was measured by the gradiometer. The propagation path of the wave that generated this signal was the same as that of "*e'* " in figure 3 (a). However, compared with the position of the axial defect, the position of the circumference defect was 35 mm further away from the nickel plate of the receiver. The start timing of the signal "i" is calculated to be 347 μs. Therefore, the signal "i" was measured separately from the signal "h". The propagation path of the wave that generated the signal "j" was almost the same as the signal "f". The start timing of "j" is calculated to be 379 μs, which agrees well with the timing of "j", so "j" was also originated in the wave reflected at the defect. As shown in figure 4 (a), continuous signal was generated after the signal "i". As shown in ref. [11], it is supposed that standing waves was probably generated due to the relation that the wavelength of the guide wave (about 100 mm at 30 kHz) was twice as long as the length of the defect (about 50 mm in width).

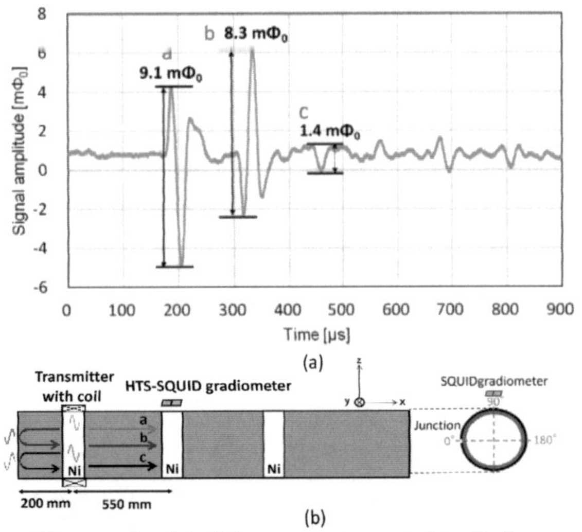

Figure 2. (a) Measurement result of the sample #1without defect. (b) Propagation paths of the signals "a", "b" and "c".

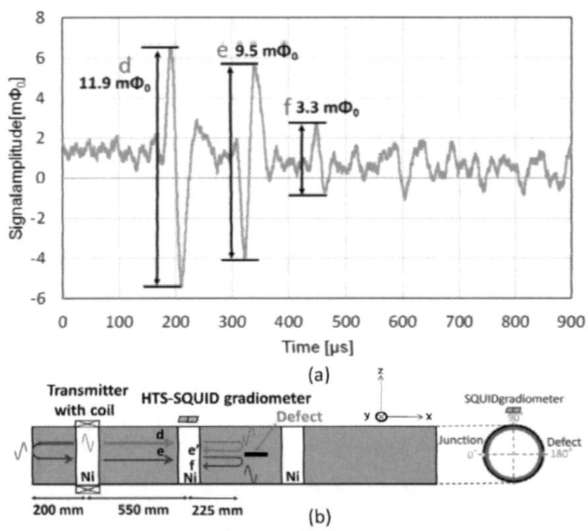

Figure 3. (a) Measurement result of the sample #1 with axial defect. (b) Propagation paths of signals "d", "e", "*e'* " and "f".

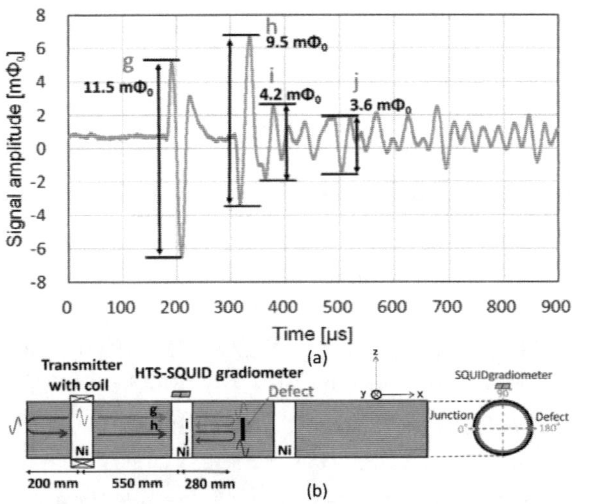

Figure 4. (a) Measurement result of the sample #2 with circumference defect. (b) Propagation paths of signals "g", "h", "i" and "j".

4. Conclusion

In this paper, the SQUID NDI system for pipes based on ultrasonic guided wave testing was firstly applied to the aluminium pipe #1 without defect. The magnetic signals due to the T (0, 1) mode guided waves at about 3100 mm/ms, which included the incident wave and the reflected waves from the sample end and the thin nickel plate of the transmitter, were measured. In the measurements using the aluminium pipe #1 and #2 with the axial and the circumferential defects respectively, the magnetic signals due to the reflected waves from the defects were successfully measured. The standing wave was observed in the latter case and the occurrence of the wave is probably due to the relationship that the wavelength of the guided waves was twice longer than the length of the defect. At the present time, we are developing a new method to generate T (0, 1) mode guided waves without contact to a sample pipe, which does not require magnetized thin nickel plates.

Acknowledgments

Authors would like to acknowledge great support and encouragement from Mr. Takashi Ikeda of CXR Corporation and Dr. Yoshikazu Tanaka of Hiroshima University for this work. This work was supported partly by Suzuki Foundation, Japan, and is supported partly by MEXT/JSPS KAKENHI Grant Number 16K06395 and Electric Technology Research Foundation of Chugoku, Japan.

References

[1] Lowe M J S, Alleyne D N and Cawley P 1998 *Ultrasonics* **36** 147
[2] Nagai T, Hyodo M, and Takamura K 2002 *Journal of JSNDI* **51** 622 (in Japanese)
[3] Nishino H 2003 *Journal of JSNDI* **52** 654 (in Japanese)
[4] TanakaY, Ikeda T, Toiyama K, Kuwako S and Fujimoto Y 2006 *Transactions of the Japan Society of Mechanical Engineers* **72** 951 (in Japanese).
[5] Tanaka Y, Tamato E and Fujimoto Y 2010 *International Journal of Applied Electromagnetics and Mechanics* **E33** 1237
[6] Clarke J and Braginski A I 2004 *The SQUID Handbook* (Weinheim: Wiley-VCH Verlag GmbH & Co. KGaA)
[7] Masutani N, Teranishi S, Masamoto K, Kanenega S and Hatsukade Y 2016 *IEICE Tech. Rep.* **116** 75 (in Japanese)
[8] Hatsukade Y, Kobayashi T, Nakaie S, Masutani N and Tanaka Y 2017 *IEEE Trans. Appl. Supercond.* **27** 1600104.
[9] Hatsukade Y, Inaba T, Maruno Y and Tanaka S 2005 *IEEE Trans. Appl. Supercond.* **15** 723
[10] Kondo H, Yoshida K and Nishino H 2008 *JSME annual meeting* **1** 335 (in Japanese).
[11] Furukawa K and Furumura I 2011 *NDEC Technical Review* **7** 5 (in Japanese).

IOP Conf. Series: Journal of Physics: Conf. Series **871** (2017) 012075 doi:10.1088/1742-6596/871/1/012075

Vector sensor for scanning SQUID microscopy

Vu The Dang[1,2,*], Masaki Toji[1], Ho Thanh Huy[1,2], Shigeyuki Miyajima[3], Hiroaki Shishido[1,4], Mutsuo Hidaka[5], Masahiko Hayashi[6] and Takekazu Ishida[1,4]

[1] Department of Physics and Electronics, Osaka Prefecture University, Sakai 599-8531, Japan

[2] University of Sciences, Vietnam National University HCMC, Ho Chi Minh, Viet Nam

[3] National Institute of Information and Communications Technology, Kobe 651-2492, Japan

[4] Institute for Nanofabrication Research, Osaka Prefecture University, Sakai 599-8531, Japan

[5] National Institute of Advanced Industrial Science and Technology, Tsukuba 305-8568, Japan

[6] Faculty of Education and Human Studies, Akita University, Akita 010-8502, Japan

*E-mail: vu-dang@pe.osakafu-u.ac.jp

Abstract We plan to build a novel 3-dimensional (3D) scanning SQUID microscope with high sensitivity and high spatial resolution. In the system, a vector sensor consists of three SQUID sensors and three pick-up coils realized on a single chip. Three pick-up coils are configured in orthogonal with each other to measure the magnetic field vector of X, Y, Z components. We fabricated some SQUID chips with one uniaxial pick-up coil or three vector pick-up coils and carried out fundamental measurements to reveal the basic characteristics. Josephson junctions (JJs) of sensors are designed to have the critical current density J_c of 320 A/cm^2, and the critical current I_c becomes 12.5 μA for the 2.2μm x 2.2μm JJ. We carefully positioned the three pickup coils so as to keep them at the same height at the centers of all three X, Y and Z coils. This can be done by arranging them along single line parallel to a sample surface. With the aid of multi-layer technology of Nb-based fabrication, we attempted to reduce an inner diameter of the pick-up coils to enhance both sensitivity and spatial resolution. The method for improving a spatial resolution of a local magnetic field image is to employ an XYZ piezo-driven scanner for controlling the positions of the pick-up coils. The fundamental characteristics of our SQUID sensors confirmed the proper operation of our SQUID sensors and found a good agreement with our design parameters.

1. Introduction

A DC Superconducting quantum interference device (SQUID) consists of a superconductor ring and two Josephson junctions (JJs). It was invented by Jaklevic *et al.* of Ford Research Labs [1] after theoretical prediction of the effect in 1962 by Josephson [2]. The SQUID devices are useful in various fields such as magnetics, electronics, biology, spintronics, security analysis, agriculture, information technology, communication, material science, and physics [3-9]. In those applications, the DC SQUID is very sensitive to the magnetic field because it utilizes an interference effect of two JJs [10,11]. Therefore, it has been an essential component in constructing scanning SQUID microscopy for mapping a local magnetic field. A conventional scanning SQUID microscope has a single scanning pick-up coil

IOP Conf. Series: Journal of Physics: Conf. Series **871** (2017) 012075 doi:10.1088/1742-6596/871/1/012075

and its SQUID readout circuit to obtain the distribution of local magnetic field component only along the *Z* direction.

In order to obtain a vector magnetic field, we proposed a vector scanning SQUID microscope with high sensitivity of magnetic field and high spatial resolution of local field mapping [12]. Ketchen *et al.* also studied a vector pick-up system using three SQUIDs with a vector pick-up coil [13]. We intend to realize a scanning SQUID microscope with high spatial resolution without losing the merit of high sensitivity in magnetic field. This possibly comes from reduced size in the pick-up coil. It has been known for a long time that a legacy SQUID sensor has a trade-off relationship between sensitivity and spatial resolution. Since an inductance of pick-up coil is a good measure of identifying the sensitivity of sensor, the increased diameter of the pick-up coil would enhance the sensitivity of the SQUID sensor. However, it also yields to lose a spatial resolution, directly. In addition, a SQUID microscope requires a lift-off distance between the sample surface and the SQUID pick-up coil. Therefore, the actual observation of the local magnetic field has a deteriorated spatial resolution than that of the original magnetic field distribution at the sample surface. If a lift-off distance is different among three pick-up coils, it gives a difference in the broadening effect on each coil. Regarding this annoying problem, we employ a multiple winding pick-up coil with the aid of the Nb multi-layer fabrication process of the CRAVITY facility of Advanced Industrial Science and Technology (AIST). Our system consists of three SQUID sensors with the three pick-up coils, which are located along the single line and are orthogonal with each other for sensing *X*, *Y* and *Z* components.

2. Design to fabricate SQUID Sensor.

Figure 1 shows the schematic equivalent circuit of the sensor of one-coil SQUID designed by us. A pick-up coil is connected directly to a SQUID washer, where two JJs of a DC SQUID were designed to have the critical currents I_c of 12.5 μA. We employ a feedback coil in a gradiometer structure, which couples with SQUID magnetically in order to reduce the effect of external magnetic noise.

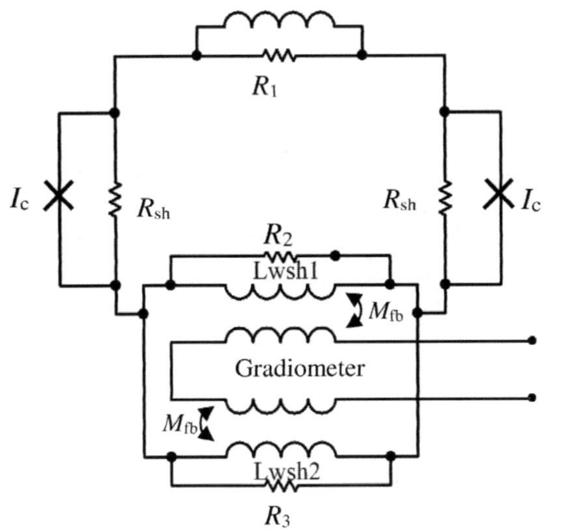

Figure 1. Equivalent circuit of a SQUID sensor with one pick-up coil for sensing a magnetic field. R_1, R_2, and R_3 are damping resistors, and R_{sh} is a shunt resistor for each JJ. The circuit parameters are designed as $R_1 = R_2 = R_3 = 1\ \Omega$, $R_{sh} = 7.8\ \Omega$, $I_c = 12.5\ \mu\text{A}$, $L_{coil} = 10$ pH, $L_{wsh} = 50$ pH, and $M_{fb} \sim 50$ pH.

The SQUID sensors were designed by using a CADENCE layout editor software in the standard AIST-STP2 for the multi-layer Nb thin film. We consider that the sensitivity and the resolution of a sensor are primarily dominated by the inductance of pick-up coil or equivalently by the size of coil. The geometry of the pick-up coil affects the procedure how to scan the coil during measurements and how to process observed data to recover the magnetic field at a certain point. Therefore, we designed to fabricate three different type pick-up coils as shown in Figure 2. Figure 2 (a) shows a single-turn pick-up coil in a circular form with an inner diameter 8 μm. Figure 2 (b) shows a double-turn pick-up coil in a circular form with an inner diameter 6 μm. We intend that the effective area of the coil (a) is similar to that of coil (b). We expect that both sensors have a similar sensitivity and a similar inductance while the number of windings is different from each other. Due to the smaller diameter, the coil (b) is supposed to have a higher spatial resolution than the coil (a). Figure 2 (c) shows the pick-up coil in a square form with an outer size 10 μm. We consider that the square coil (c) is suitable for achieving an enhanced spatial resolution when the scanning is done in the 2D Cartesian coordinate in a step, which is appreciably smaller than the coil diameter. The coil (c) is also preferential in treating the data procession in our numerical method of the inverse Biot-Savart law for recovering a true local magnetic field at the sample surface [14]. We can easily define a pixel with the size smaller than the pick-up size by comparing a difference of magnetic flux between data at (i,j) position, the $(i+1,j)$ position, and the $(i, j+1)$ position, where indices i, j are row and column of scanning mesh, respectively. Our scanning system is able to scan at a step of 10 nm at 4K by using the piezo-driven XYZ stages.

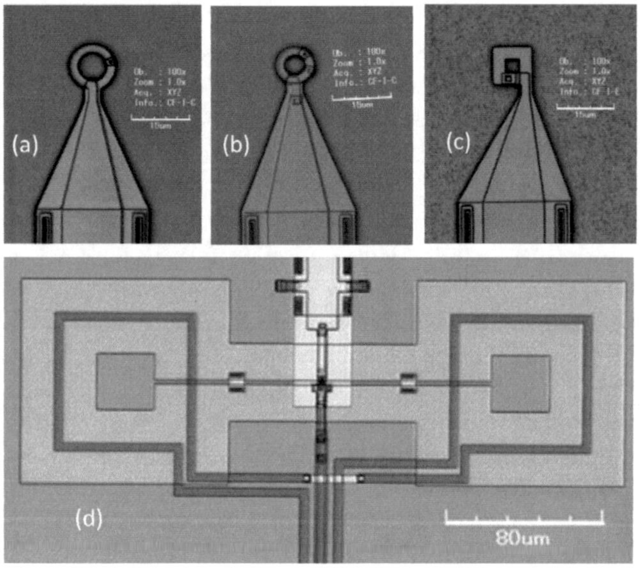

Figure 2. The uniaxial pick-up coil of the sensor was fabricated in three different types. The one-turn pick-up coil (a) is in a circular form with an inner diameter of 8 μm. The two-turn pick-up coil (b) is in a circular form with an inner diameter of 6 μm. The single-turn pick-up coil (c) is in a square form of the outer edge 10 μm. The SQUID circuit (d) of the sensor is coupled with the pick-up coil.

The SQUID sensors were designed by us and were fabricated with the aid of the CRAVITY process, where the number of Nb layers is kept at four and the critical current density J_c of a JJ is 320 kA/cm^2 [15-18]. Figure 3 shows a photograph of a typical vector SQUID sensor, and shows that the sensor was fabricated successfully. The design in the present work is an improved version of that employed in our preceding reports [12,19]. We revised the positions of three pick-up coils on the chip. They are not only orthogonal with each other, but the centers of pick-up coils are also located along the same line. This makes it possible to keep the same distance between three pick-up coils and the sample surface (see an enlarged photo of the three coils in Figure 3). This means that the vector pick-up coil catches the three components of a magnetic field vector at the same height from the sample surface. We suppose that the distance between the Z-coil center and the X-coil center can be chosen as an integer multiple of the scanning step. The distance between the Z-coil center and the Y-coil center is also chosen in a similar fashion. We plan to develop the measurement software so as to synthesize a magnetic field vector at the same point from the XYZ scanned data of the SQUID microscope to assist data acquisition. The sensitivity of the vector pick-up coil is designed to be improved by a factor of two by using a double-winding pick-up coils. Our vector SQUID sensor was fabricated on a chip size of 2.9 mm x 2.9 mm, and the vector pick-up coil is located at a corner of the chip. Our DC SQUID sensor consists of two JJs, which were designed to have a junction area of each JJ of 2.2 μm x 2.2 μm and the critical current I_c of

12.5 μA. It was reported that the typical variation in JJ critical currents is better than 2% [20]. In each single sensor, the pick-up coil is connected to a SQUID washer directly, which was covered fully by superconducting Nb film. In order to reduce the external electric noise in the readout SQUID signal, we used a gradiometer structure with a feedback coil, which is fully coupled with a SQUID washer magnetically (Figure 2 (d)).

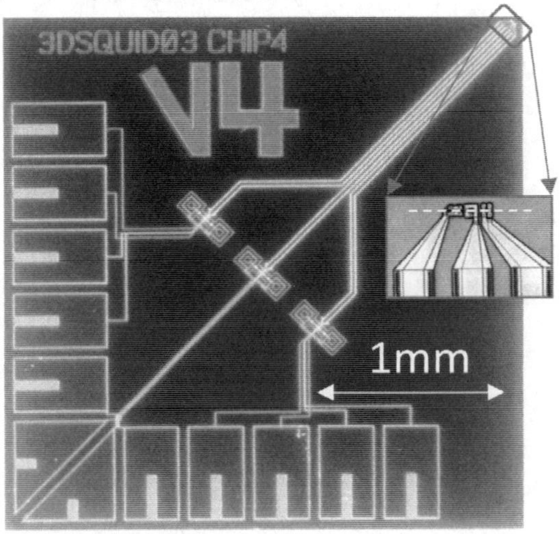

Figure 3. Photograph of a typical vector SQUID sensor fabricated by us. Inset is an expanded photo of the vector three XYZ pick-up coils with two-turn windings. They are orthogonal to each other and their center positions are located along a single line at the same height from the sample surface.

3. Measurement Results

In order to measure the characteristics of a vector SQUID sensor, we constructed a measurement system of the *I-V* curve as described in Figure 4. We use a Gifford–McMahon cryocooler to reduce the temperature of a SQUID sensor down to 4 K. This refrigerator is able to measure the *I-V* curve for long-time data acquisition in liquid helium at 4 K, and the temperature can be cooled down to 1.5 K for 60 minutes by pumping our liquefied helium in the sample space. The cryostat system was fully suitable for measurements at high frequencies because we use the semi-rigid coaxial cables and SMA connectors for leads from room temperature down to cryogenic temperatures. A digital-analog converter DAC (National Instrument Inc., NI9263 module) has a resolution 16 bit to feed a voltage in the range of ± 10V with a 0.3-mV resolution. The operation speed reaches 100 k/s/ch, the output impendence is lower than 0.1 Ω and a current driver of 1 mA which are fine enough to cover a necessary current to apply a SQUID sensor. An analog-digital converter ADC (National Instrum. Inc., NI9239 module) has a resolution 24 bit, an input noise lower 70 μVrms, and an input impendence as large as 1 MΩ. This means our system has a resolution of 2 μV and the full-scale ranges ±10V in a single range. In order to improve the performance of the ADC module and to reduce the noise of signal, we used a low noise preamplifier (Ithaco Inc., Model 1201) with a high input impedance (100 MΩ) and a variable gain from 10 to 1000. A current source (Keithley Inc., Model 2400) can feed the desired current with a computer via the GP-IB interface. A temperature controller (Lakeshore Inc., Model 340) was used to monitor and control the temperature of a SQUID sensor. We also installed the several low pass filters (LPF) consisting of R, C components to reduce the noise in the signal produced by the DAC module, the current source, and the commercial power line. All of the equipment were connected to the computer by the USB interface or the GPIB interface, and were controlled by a LabVIEW (National Instrum. Inc.) program coded by ourselves. The SQUID sensor is shielded both by the double-wall permalloy cylinders at room temperature and the single-wall cryogenic permalloy cylinder at 4K.

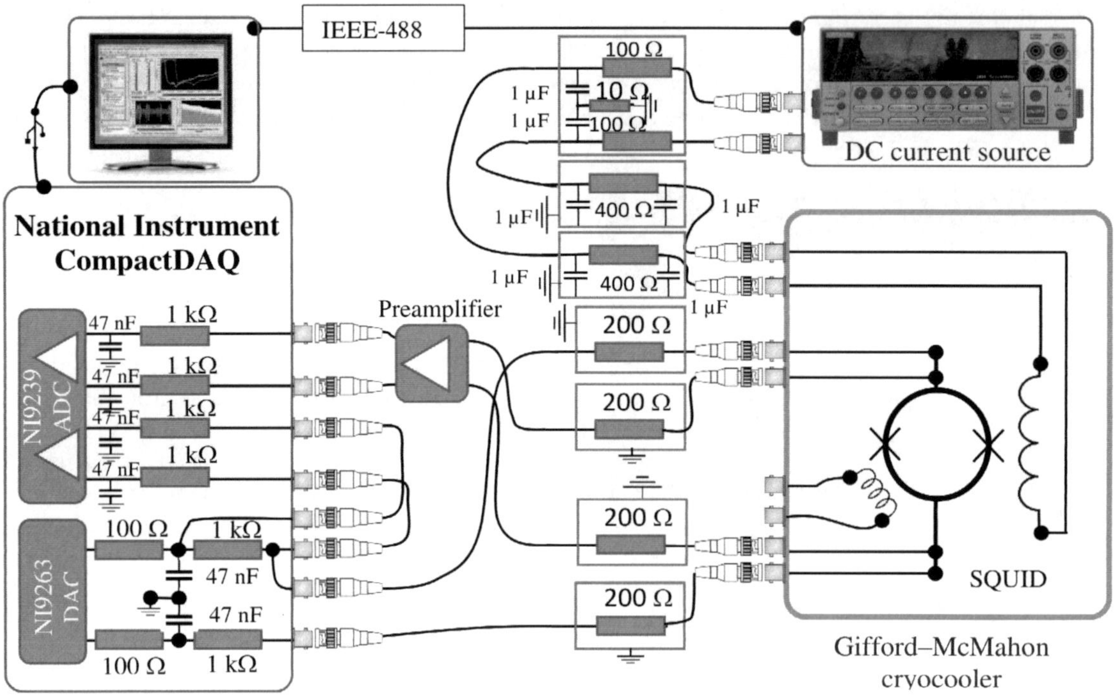

Figure 4. The *I-V* curve measurement system for testing the SQUID characteristics. The system is controlled by a LabVIEW program. The cryogenic temperature is achieved by a Gifford-McMahon cryocooler. We inserted several passive filters to reduce the noise in measurement signals.

Figure 5 shows the typical *I-V* curves of three XYZ channels of the SQUID sensor on the same chip (a vector SQUID sensor as in Figure 3). The measurement results were obtained as the three-different *I-V* curves, of which the characteristics are almost the same with each other (see Figure 5). This confirms that our design to fabricate the three XYZ sensors is precise enough to work properly under the same condition. The critical current I_c of each sensor is 27 μA, which is in good agreement with our design value of 25 μA (= 2 x 12.5 μA). We installed an additional coil of 20 turns in our system to produce an external magnetic field at the position of the sensor, where the pick-up coils were placed at the center of the 20-turn coil. We applied an external magnetic field so as to adjust a current fed into the 20-turn coil. The inset of Figure 5 shows that there is an oscillation in the *I-V* curves when the external magnetic field is changed systematically. All of the measurement results confirm that we fabricated the vector SQUID sensors successfully as designed by us.

We also measured an inductance of pick-up coil by using a test chip, which was fabricated simultaneously on the same wafer with the vector SQUID sensors with an additional lead to feed an additional current to the coil to know an inductance. By applying a controlled current directly through two terminals of pick-up coil, we found the periodic oscillation of the *I-V* curve. From the period of the oscillation, we identified the inductance of *Z* pick-up coil as 10.35 pH and the inductance of XY pick-up coils as 9.86 pH. Those results give us sound evidence that our fabrication is good enough to make sure a proper operation as intended in our design. We note that the inductance of the X pick-up coil is almost the same with those of the Y and Z coils. This again indicates that the sensitivity of three XYZ sensors is supposed to be approximately the same with each other.

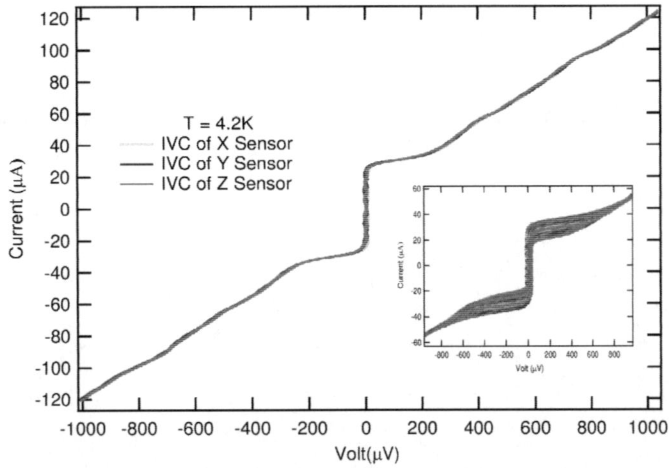

Figure 5. The three *I-V* curves of the vector SQUID sensor for the XYZ coils with the critical current of 27 μA (typical). The SQUID sensor is cooled down to 4.2 K using a Gifford–McMahon cryocooler. Inset shows the *I-V* curve under the various different magnetic field used for the SQUID modulation.

Figure 6 shows the *V-Φ* characteristics of the sensor S1 (see Figure. 2(a)) and the sensor S2 (see Figure. 2(b)) when an external magnetic field is changed systematically under the SQUID bias current I_b of 30 μA. We find that the sensor S2 shows the four periods of the Φ_0 oscillation while the sensor S1 shows the three periods of the Φ_0 oscillation in 60 μT. They can be interpreted that the areas of sensor with 2-turn pick-up coil (226μm^2) is slightly larger than that of the single-turn pick-up coil (201 μm^2). Therefore, the inductance of the 2-turn coil is higher than that of the single-turn coil, and it is responsible for the higher sensitivity of the sensor S2 than the sensor S1. This confirms that our idea to improve the inductance of pick-up coil works well. In addition, the magnetic field penetration into the coil wire should be taken into account in estimating the effective areas of the pick-up coils. We are going to investigate further in this respect in our future work.

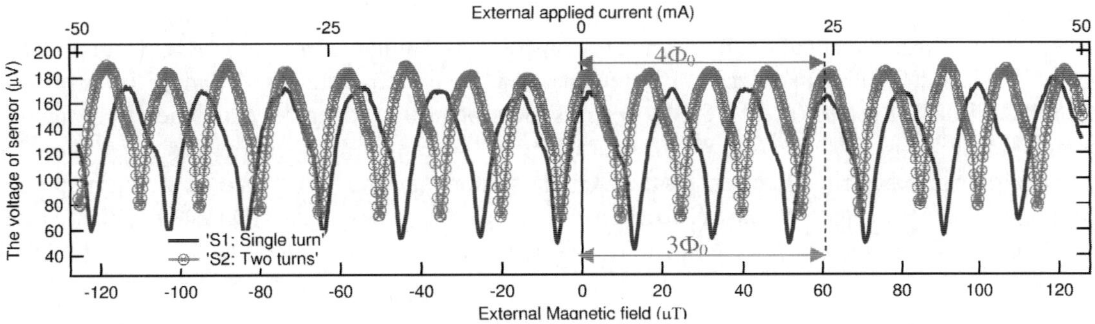

Figure 6. The *V-Φ* characteristics of the dc SQUID sensors. The sensor S1 has a single-turn pick-up coil while the sensor S2 has a two-turn pick-up coil. The sensor voltage was measured as a function of the external applied magnetic field.

To understand the proper operation of our SQUID sensor or to simulate the operation of flux locked loop (FLL), we feed a current into the feedback coil when the SQUID sensor is biased under a certain current. When an external magnetic field is applied to the coil, the voltage of the sensor changes systematically. We confirmed that this shift in the voltage could be compensated by feeding a proper current to the feedback coil. In Figure 7, we show the current applied to feedback coil of the SQUID sensor as a function of the external magnetic field when the bias voltage of the SQUID sensor is kept at 190 μV. The first operation point, the second operation point, *etc.* respectively show the working points in the different periodic branches of the *V-Φ* characteristics with the step Φ_0 of each branch. We consider that this is sound evidence to convince that our SQUID sensor works properly even if it is used in an FLL mode as designed by us.

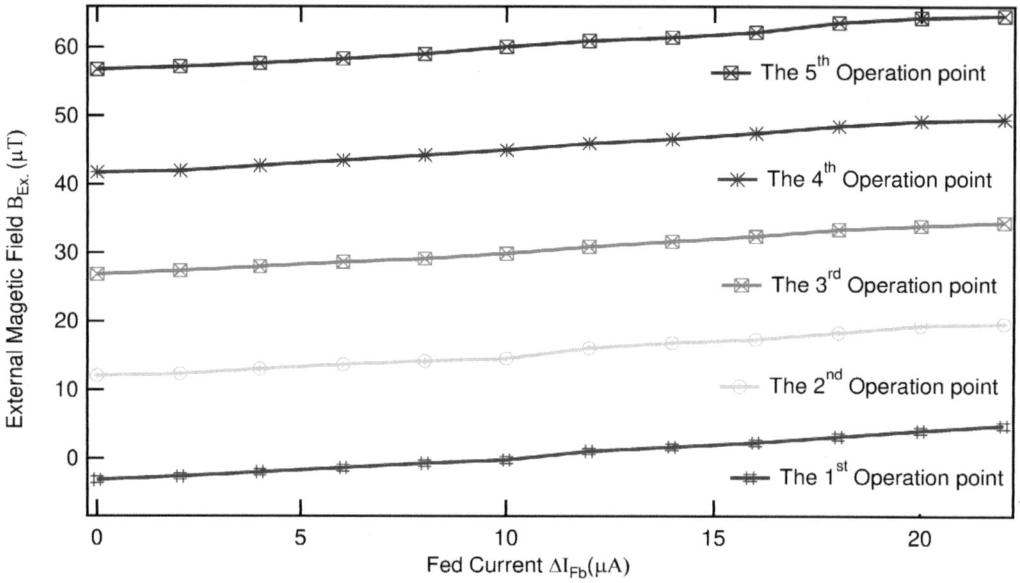

Figure 7. The external magnetic field as a function of current fed into the feedback coil. We find a good linearity of the external magnetic field as a function of feeding current in the periodic *V-Φ* characteristics.

In the next step of our research, we are going to install our sensor with a commercial SQUID readout circuit (Magnicon Inc., Model XXF-1) to achieve an automatic measurement system for the XYZ three channels.

4. Summary

We successfully fabricated the SQUID sensors to construct a scanning XYZ vector SQUID microscope. A vector scanning SQUID sensor was fabricated on the single chip by using the Nb multi-layer process. The characteristics of the different SQUID sensors have been investigated systematically by using a laboratory-made measurement system. Our measurements such as the *I-V* curves and the *V-Φ* curves were used to obtain the critical current of the SQUID sensors and the inductance of the pick-up coils. We confirmed that our sensor worked properly as we designed by using the CAD system. We are under construction of the cryogenic system with an *XYZ* piezo-driven scanner to realize a vector SQUID microscope.

Acknowledgment

This work was supported by Grant-in-Aid for Challenging Exploratory Research (No. 25600018, No.15K13979) and Grant-in-Aid for Young Scientists (B) (No. 26820130 and No. 26800192) from Japan Society for the Promotion of Science (JSPS). This work is partially supported by a special grant for faculty innovation from Graduate School of Engineering at Osaka Prefecture University. This work is benefitted by the use of VLSI Design and Education Center (VDEC), the University of Tokyo with the collaboration with Cadence Corporation. The SQUID sensors in this work were fabricated in CRAVITY facilities at AIST. We thank M. Maezawa (AIST) for his comprehensive supports and his help in design and S. Nagasawa (AIST) for his kind suggestion.

References

[1] Jaklevic R C, Lambe J, Silver A H, and Mercereau J E, *Phys Rev Lett* 1964 **12** 159-160
[2] Josephson B D, *Phys Lett* 1962 **1** 251–253
[3] Kirtley J R, Ketchen M B, Stawiasz K G, Sun J Z, Gallagher W J, Blanton S H, and Wind S J, *Appl Phys Lett* 1995 **66** 1138–1140
[4] Ho T H, Kato M, and Ishida T, *Supercond Sci Technol* 2013 **26** 065001
[5] Fong L E, Holzer J R, McBride K K, Lima E A, Baudenbacher F, and. Radparvar M, *Rev Sci Instrum* 2005 **76** 053703
[6] Zutic I, Fabian J, and Sarma S D, *Rev Mod Phys* 2004 **76** 323–410
[7] Tanaka S, Fujita H, Hatsukade Y, Otani T, Suzuki S, and Nagaishi T, *Supercond Sci Technol* 2007 **20** S385–S388
[8] Tsukada K and Kiwa T, *IEEE Trans Magn* 2007 **43** 2683–2685
[9] Vu T D, Matsumoto H, Miyoshi H, Ho T H, Shishido H, Kato M, and Ishida T, *Physica C: Superconductivity and its Applications* 2017 **533** 44–48
[10] Lenz J E, *Proc IEEE* 1990 **78** 973-989
[11] Chemla Y R, Grossman H L, Poon Y, McDermott R, Stevens R, Alper M D, and Clarke J, *Proc Natl Acad Sci* 2000 **97** 14268-14272
[12] Miyajima S, Okamoto T, Matsumoto H, Ho T H, Hayashi M, Maezawa M, Hidaka M, and Ishida T, *IEEE Trans Appl Supercond* 2015 **25** 1600704
[13] Ketchen M B, Kirtley J R, and Bhushan M, *IEEE Trans Appl Supercond* 1997 3139–3142
[14] Hayashi M, Ebisawa H, Ho T H, and Ishida T, *Appl Phys Lett* 2012 **100** 182601
[15] Tanaka M, Yamanashi Y, Irie N, Park H-J, Iwasaki S, Takagi K, Taketomi K, Fujimaki A, and Yoshikawa N, Terai H, *IEEE Trans Appl Supercond* 2007 **17** 474–477
[16] Iwasaki S, Tanaka M, Yamanashi Y, Park H, Akaike H, Fujimaki A, Yoshikawa N, Takagi N,Murakami K, and Honda H, *Supercond Sci Technol* 2007 **20** S328
[17] Miyajima S, Kusumoto T, Ito K, Akita Y, Yagi I, Yoshioka N, Ishida T, Miki S, Wang Z, and Fujimaki A, *IEEE Trans Appl Supercond* 2013 **23** 1800505
[18] Sano K, Yamanashi Y, and Yoshikawa N, *IEICE Trans Electron* 2014 **E97-C** 182–187
[19] Vu T D, Ho T H, Miyajima S, Matsumoto H, Miyoshi H, Okamoto T, Maezawa M, Hidaka M, and Ishida T, *IEEE: Superconductive Electronics Conference (ISEC)* 2015, **SQ-P30** 15721011
[20] Nagasawa S, Hinode K, Satoh T, Hidaka M, Akaike M. Fujimaki A, Yoshikawa N, and Takag K, *IEICE Trans Electron* 2014 **E97-C** 132-140

IOP Conf. Series: Journal of Physics: Conf. Series **871** (2017) 012076 doi:10.1088/1742-6596/871/1/012076

Ultra-Low Field SQUID-NMR using LN_2 Cooled Cu Polarizing Field coil

K Demachi, S Kawagoe, S Ariyoshi and S Tanaka

Toyohashi University of Technology, 1-1 Hibarigaoka, Tempaku-cho, Toyohashi, Aichi 441-8580, Japan

E-mail address: tanakas@ens.tut.ac.jp

Abstract. We are developing an Ultra-Low Field (ULF) Magnetic Resonance Imaging (MRI) system using a High-Temperature Superconductor superconducting quantum interference device (HTS rf-SQUID) for food inspection. The advantages of the ULF-NMR (Nuclear Magnetic Resonance) / MRI as compared with a conventional high field MRI are that they are compact and of low cost. In this study, we developed a ULF SQUID-NMR system using a polarizing coil to measure fat of which relaxation time T_1 is shorter. The handmade polarizing coil was cooled by liquid nitrogen to reduce the resistance and accordingly increase the allowable current. The measured decay time of the polarizing field was 40 ms. The measurement system consisted of the liquid nitrogen cooled polarizing coil, a SQUID, a Cu wound flux transformer, a measurement field coil for the field of 47 μT, and an AC pulse coil for a 90°pulse field. The NMR measurements were performed in a magnetically shielded room to reduce the environmental magnetic field. The size of the sample was ϕ35 mm × L80 mm. After applying a polarizing field and a 90°pulse, an NMR signal was detected by the SQUID through the flux transformer. As a result, the NMR spectra of fat samples were obtained at 2.0 kHz corresponding to the measurement field B_m of 47 μT. The T_1 relaxation time of the mineral oil measured in B_m was 45 ms. These results suggested that the ULF-NMR/MRI system has potential for food inspection.

1. Introduction

One of the advantages of Ultra-low field NMR/MRI (ULF NMR/MRI) is that the restriction of the magnetic field homogeneity is moderated because the magnetic field is much lower than that of a traditional MRI. Therefore the size of the ULF MRI system becomes smaller and it can be made at a lower cost [1-2]. Moreover, using a modified inversion recovery technique, enhancement of the contrast with T_1 dispersion at a specific part of a tissue in a wide range of fields is possible [3].

Since our previous ULF MRI food inspection experiment system used a permanent magnet, which provided a polarizing field, the magnet was placed well away from the measurement position preventing the leakage field, which degrades the measurement field homogeneity; in this case, a sample transportation mechanism was equipped [4-5]. Therefore, it is difficult to measure the NMR signal of a sample with short relaxation time, such as fat, because of the longer transportation time.

In the paper, we demonstrate a ULF SQUID-NMR system using a 77 K cooled Cu wound polarizing coil to measure the fat, which has a short relaxation time without transportation.

IOP Conf. Series: Journal of Physics: Conf. Series **871** (2017) 012076 doi:10.1088/1742-6596/871/1/012076

2. Experimental setup and Measurement Method

2.1. ULF SQUID-NMR system using LN_2 cooled Cu polarizing field coil

Figure 1(a) shows a schematic of the ULF SQUID-NMR system. The main system consists of the following units: Helmholtz type measurement field (B_m) coil, an AC pulse (B_{AC}) coil, a polarizing field (B_P) coil, a pick up coil and an NMR spectrometer (Kea^2, *Magritek*, New Zealand). A specially designed LN_2 Dewar with a room temperature bore dimension of ϕ50 mm × L 230 mm was prepared. The polarizing coil was installed in the Dewar; the pickup coil was placed at the center of the room temperature bore. A SQUID was separately cooled by liquid nitrogen in an aluminium Dewar (8 ℓ), which was surrounded by a three-layer mu-metal cylinder with a 2-mm thickness in order to attenuate environmental magnetic noise. The detected signal was transferred to the SQUID via a flux transformer. The NMR measurements were performed in a magnetically shielded room to attenuate the Earth's magnetic field.

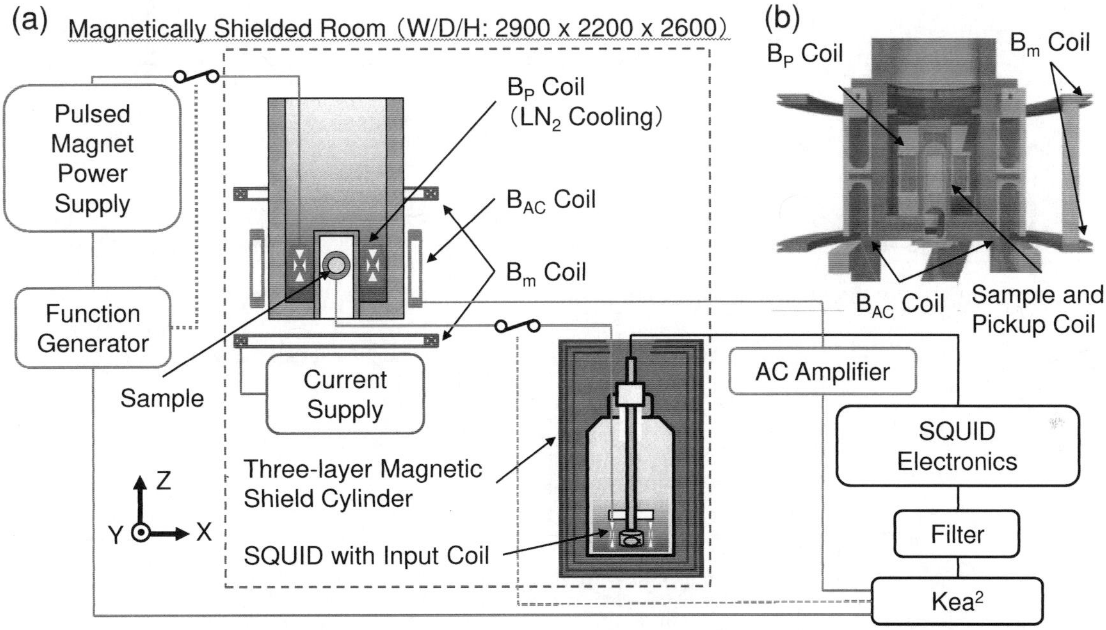

Figure 1. Experimental setup. (a) Schematic drawing of the ULF SQUID-NMR using Cu polarizing coil. (b) Cross sectional view of the measurement system.

Figure 1(b) shows the cross sectional view of the measurement system, which involves B_P coil, B_m coil and B_{AC} coil. The B_P coil and the B_m coil were coaxially aligned; thus the B_P // B_m configuration can avoid any signal losses during a nonadiabatic switching event [6]. The pickup coil was arranged orthogonally to the B_{AC} coil and the B_m coil. The HTS rf-SQUID and driving electronics used in the experiment are made by *Jülicher SQUID GmbH* in Germany. The B_{AC} coil and the B_P coil were connected to an AC power amplifier (*R&K*, A1.5K030-4040R) and a pulsed magnet power supply (*Kudo-denki*, Japan, 30 V × 50 A), respectively.

The signal was filtered (*Stanford Research System*, SR650, HPF: 1.5 kHz, LPF: 3.8 kHz) and recorded by Kea^2 with amplifier (Kea^2 internal receiver amp, gain: +40 dB). The Kea^2 also controlled the sequence for B_P, B_{AC} and the reset pulse of the SQUID.

2.2. Cu wound flux transformer

Figure 2(a) shows the detail of the flux transformer, which consists of a pickup coil and an input coil. The NMR signal from a sample was detected by the pickup coil. Then, the signal was transferred to the HTS rf-SQUID via the input coil, with which the number of turns was set to 950. The flux transformer circuit was electrically opened by a mechanical relay during the polarization so that the SQUID was not exposed to the strong magnetic field.

Figure 2(b) shows the schematic drawings of the two different pickup coils. One was a round shape solenoid coil and was prepared for a "Small" plastic canister with a capacity of 5.6 mℓ; the other was a racetrack shape solenoid coil and was prepared for a "Large" glass bottle with a capacity of 50 mℓ. The number of turns of each coil was 205 and 200, respectively.

The measured mutual inductance between the SQUID and the input coil was 15.5 nH at 77 K. The noise profiles of the SQUID were measured in both cases of the SQUID itself and with the flux transformers. As shown in Figure 2(c) the noise increased by a factor of two to three when the flux transformer was coupled. The white noise of the SQUID itself was approximately 100 fT/Hz$^{1/2}$.

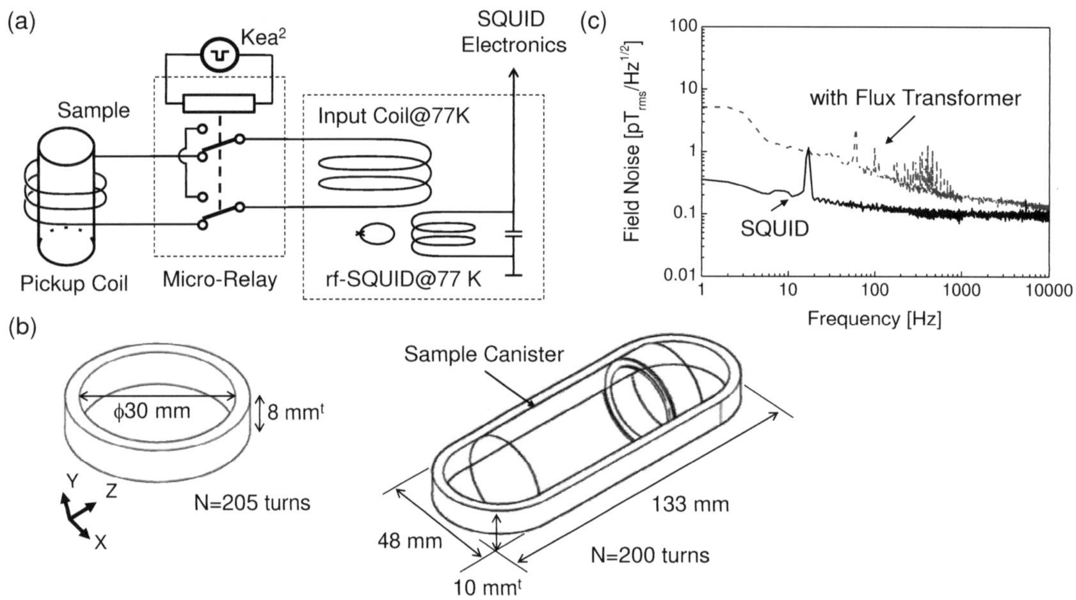

Figure 2. (a)Schematic of flax transformer. (b) Round shape and racetrack shape pick up coil. (c) Noise spectrum density of the SQUID itself and with flux transformer measured in the magnetically shielded room.

2.3. LN_2 cooled Cu polarizing field coil

Although the Cu wound polarizing coil has a drawback, producing joule heat due to an electric resistance, there are some advantages, such as having no residual magnetic field which may become the origin of measurement field inhomogeneity or a quench event when changing the coil current. Another advantage is that the resistance of the coil wire can be reduced at 77 K by a factor of 1/6 compared with that at room temperature. As a result, it is possible to increase the allowable current with low loss comparing with water-cooled Cu coils. Specifications of the B_P coil are listed in Table 1.

The B_P coil was installed inside the specially designed LN_2 Dewar so that it can effectively polarize the sample in the room temperature bore. The wire diameter of the coil was set as ϕ1.8 mm to harmonize the impedance of the coil with the specification of the power supply.

Table 1. Parameters of polarizing coil.

	Design	Experimental
Inner Diameter	80 mm	80 mm
Outer Diameter	134 mm	130 mm
Length	80 mm	80 mm
Wire Diameter	ϕ1.8 mm	ϕ1.8 mm
Number of Turns	660	687
Resistance [R.T.]	1.5 Ω	1.7 Ω
Resistance [77 K]	0.259 Ω	0.278 Ω
Inductance	29.9 mH	32.5 mH
Central Magnetic Field	313mT@50 A	313 mT @50 A

Figure 3 shows the measured field distribution in the X-Z plane at the sample position. It was found that 300 mT can be obtained in a range of 30 mm × 30 mm when a current of 50 A is applied; a current-magnetic field conversion coefficient at the sample center can be calculated as 6.26 mT/A. When turning off the current, no large ringing was observed in the magnetic field, and the measured decay time of the magnetic field was 40 ms.

Figure 4 shows the change of the voltage across the B_P coil while keeping a constant current of 50 A. The voltage increased up to 30 V in 90 sec associated with the increase of coil resistance due to the Joule heat. This voltage is in the upper limit of the rating of the power supply. Therefore, when the polarization time is supposed to be 5 sec each, a total of 18 sets of continuous measurements are possible. When the current was reduced to 35 A, 200 times of the measurement could be performed because of less Joules heat. Therefore, regarding the T_1B_m measurement, a current of 35 A passed to reduce the heat generation of the B_P coil.

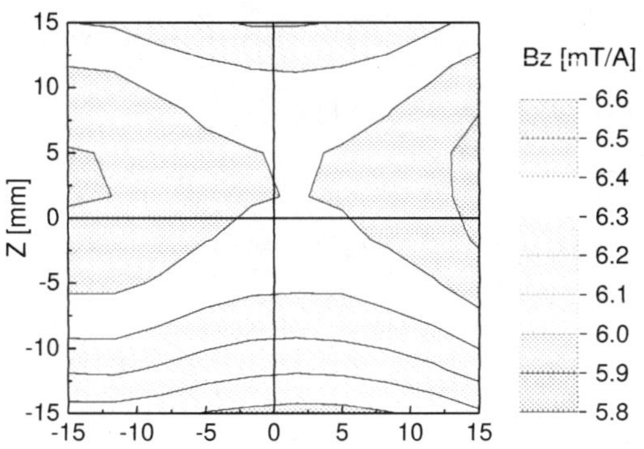

Figure 3. Magnetic field distribution in X-Z plane.

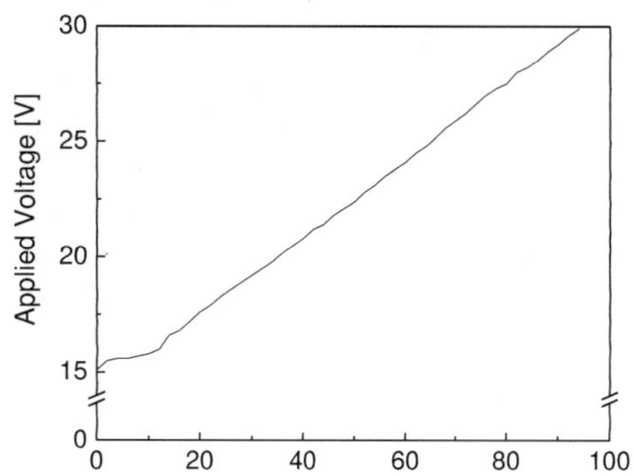

Figure 4. Time trace of the applied voltage.

IOP Conf. Series: Journal of Physics: Conf. Series **871** (2017) 012076 doi:10.1088/1742-6596/871/1/012076

3. Result and Discussions

3.1. NMR measurement of water and fat

The pulse sequence for the NMR signal measurement is shown in Figure 5. Firstly, the measurement field B_m of 47 μT was applied and kept constant until the end of the measurement. Then the sample was polarized by the B_P for 5 s. In this period the flux transformer circuit was electrically opened so that the SQUID was not exposed to the strong polarizing field. After the B_P current had dropped sufficiently, the B_P circuit was disconnected by a mechanical relay. Subsequently, a 90°pulse was applied to rotate the magnetization of the sample. We note that B_P delay time is the time between the turn-off edge of the B_P field and the turn-on edge of 90° pulse. The B_P delay time of 50 ms was selected in the NMR measurements, while it was varied in the following T_1 measurements. After closing the flux transformer circuit and setting the SQUID, the NMR signal was then detected by the SQUID via the flux transformer. The output signal of SQUID was acquired by Kea[2] through the filter for 512 ms.

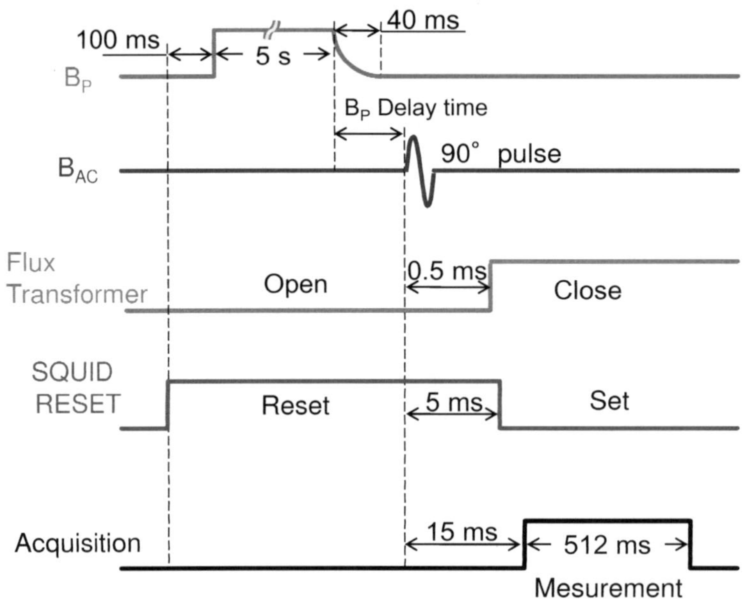

Figure 5. Pulse sequence for NMR signal measurement.

Figure 6 (a) shows the FID trace of a "Small" sample containing water of 5.6 mℓ and a "Large" sample containing water of 50 mℓ. Since the signal was down converted with a local frequency of 1.99 kHz, the period of the oscillation was 0.1 sec, which corresponds to the Larmor frequency of 2.0 kHz. The signal intensity of the large sample was three times larger than that of the smaller one. However the T_2* relaxation time of the large sample is much smaller than that of the smaller one. This may attribute to the inhomogeneity of the measurement field. The spectrum of each small and large sample is shown in Figure 6(b). The NMR spectrum of each water sample shows the peak at 2.0 kHz corresponding to the measurement field strength of 47 μT. The half-value width of large sample was wider than that of small sample. Additionally, both samples had an asymmetric peak spread, which showed the tail on the lower frequency side. The reason for the asymmetry was considered to be a fluctuation of the B_m current due to the voltage induced by of the polarizing field switching.

Mineral oil (*Johnson's*, baby oil) and lard were prepared as samples. As for the lard, both a solid state and a liquid state, melted by heat were used. They were filled in the large sample container, which was easier than in the small one to exchange specimens. The spectra of those samples are shown in Figure 7. There were peaks for the mineral oil and liquid lard but no peak for the solid lard. The intensities of the peaks were smaller than that of the water sample as shown in Figure 6(b) because of the difference of the material.

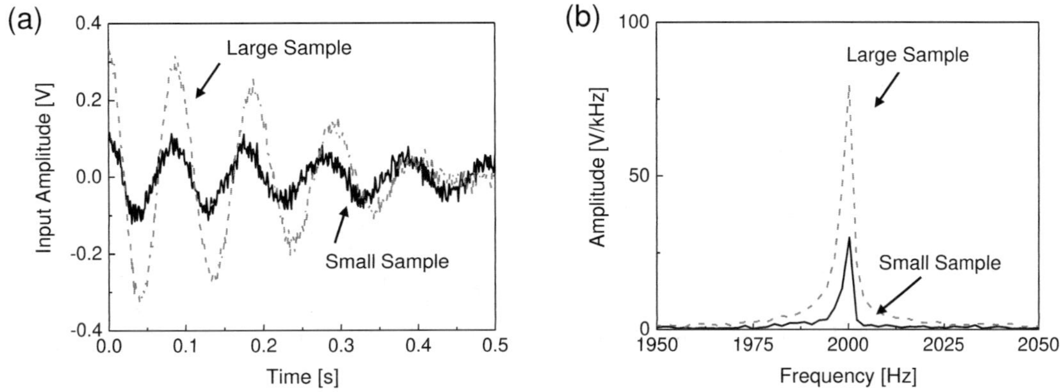

Figure 6. NMR signal of the water samples. (a) FID signal of each large and small sample. (b) NMR spectra.

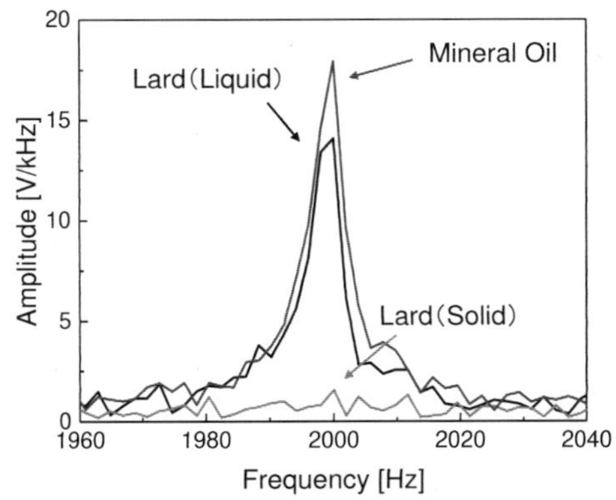

Figure 7. NMR spectra of the fat samples.

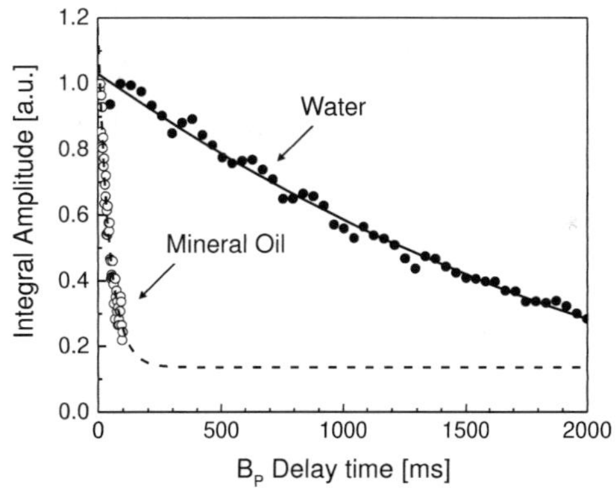

Figure 8. Dependence of the amplitude on B_P delay time.

3.2. T_1 measurement in B_m

It was reported that images of gel phantoms with T_1 weighted contrast demonstrated dramatic enhancement in low field [3]. To investigate whether the enhancement of T_1 contrast could be applied to fat with short relaxation time, T_1 relaxation time in the measurement field B_m was measured. The NMR signal was measured by changing the B_P delay time, which was the time between the turn-off edge of the B_P field and the turn-on edge of the 90° pulse as shown in Figure 5. When the B_P delay time was longer than the relaxation time T_1, it was hard to distinguish the signal from the noise since the signal intensity became extremely low. Therefore, the signal of the mineral oil can be observed until only a B_p delay time of 100 ms, while that of the water sample can be taken until 2000 ms as shown in Figure 8. The values were normalized after integration in the bandwidth of 10 Hz at the peak frequency since the shape of each spectrum is different and the spectrum has some area. The T_1 relaxation time was calculated from the time constant of the exponential decay by fitting the plots of the NMR observed signal. As shown in the Figure, it was found that the T_1 relaxation time of the mineral oil sample was 45 ms, which was much shorter than T_1 of water sample, 2.6 s.

4. Conclusions

In this study, we developed a ULF SQUID-NMR system using an LN_2 cooled Cu polarizing field coil aiming for food inspection. In the system, the NMR signal was detected by the HTS rf-SQUID with the Cu wound flux transformer. The handmade polarizing coil was able to continuously apply a polarizing field of 300 mT at maximum. It was demonstrated that the NMR signal of fat was successfully measured in the system. The results suggested that the ULF SQUID-NMR/MRI system has potential for food inspection.

References

[1] Kawagoe S, Toyota H, Hatta J, Ariyoshi S and Tanaka S 2016 Ultra-low field MRI food inspection system prototype. In: *Physica C: Superconductivity and its Applications*,

[2] Hatsukade Y, Abe T, Tsunaki S, Yamamoto M, Murata H and Tanaka S 2013 Application of ultra-low field HTS-SQUID NMR/MRI to contaminant detection in food *IEEE Transactions on Applied Superconductivity* **23** 1602204-

[3] Lee S K, Möβle M, Myers W, Kelso N, Trabesinger A H, Pines A and Clarke J 2005 SQUID-detected MRI at 132 μT with T_1-weighted contrast established at 10 μT–300 mT *Magnetic resonance in medicine* **53** 9-14

[4] Fukumoto S, Tsunaki S, Chigasaki T, Hatsukade Y and Tanaka S 2013 ULF-NMR system using HTS-SQUID and permanent magnet *Physica C: Superconductivity* **484** 202-5

[5] Yamamoto M, Toyota H, Kawagoe S, Hatta J and Tanaka S 2015 Development of Ultra-low Field SQUID-MRI System with an LC Resonator *Physics Procedia* **65** 197-200

[6] Qiu L, Zhang Y, Krause H-J, Braginski A I and Offenhäusser A 2009 Low-field NMR measurement procedure when SQUID detection is used *Journal of magnetic resonance* **196** 101-4

IOP Conf. Series: Journal of Physics: Conf. Series **871** (2017) 012077 doi:10.1088/1742-6596/871/1/012077

2D-MPI System using second harmonic with HTS-SQUID

K Kobayashi, Y Sanada, S Ariyoshi and S Tanaka

Toyohashi University of Technology, 1-1 Hibarigaoka, Tempaku-cho, Toyohashi, Aichi 441-8580, Japan

E-mail address: tanakas@ens.tut.ac.jp

Abstract. Magnetic Particle Imaging (MPI) is an imaging technique, with high sensitivity and high speed imaging, utilizing non-linear magnetization response for the detection of super-paramagnetic iron oxide nanoparticles (MNP). MPI measures the magnetization change in the MNP under the AC magnetic field. Since the signal of the magnetization change is much smaller than the signal generated by the AC magnetic field, the signal response at the fundamental frequency cannot be used. Accordingly, it is the third harmonic of the response that is generally measured. However this method has disadvantages that the power of the AC magnetic field is large and as a result the system increased in size. In this study, we investigated the 2D imaging using a second harmonic, which can be theoretically obtained with a greater signal than in the case of the third harmonic. Moreover, an HTS-SQUID (Superconducting Quantum Interference Device) magnetometer was employed in the study to enhance the peak signal to noise ratio (PSNR). A 2D-MPI system, which enables an imaging by scanning a DC magnetic field under a gradient magnetic field, was constructed. As a result, PSNR was increased by 1.5 times using SQUID with a Flux Transformer.

1. Introduction

MPI (Magnetic Particle Imaging) is an imaging technique, utilizing non-linear magnetization response for detection of magnetic nanoparticles [1]. Since it has high time resolution, the position and quantity of magnetic nanoparticles injected into the body are possible to grasp in real time. In the signal measurement of MPI, the most commonly employed method is detection of the third harmonic of the magnetic response, but it has disadvantages that power of the AC magnetic field is large and as a result the system increased in size. Therefore, we investigated a method to detect the second harmonic of the magnetization response, externally applying an AC magnetic field with the application of a DC magnetic field [2-6]. In this study, we improved the peak signal-to-noise ratio (PSNR) by applying an HTS-SQUID to a 2D imaging system, which detects the second harmonic of the magnetization response from magnetic nanoparticles.

IOP Conf. Series: Journal of Physics: Conf. Series **871** (2017) 012077 doi:10.1088/1742-6596/871/1/012077

2. Principle of MPI using second harmonic

In order to perform MPI using the second harmonic, it is necessary to apply a gradient magnetic field, a DC magnetic field, and an AC modulation magnetic field to the imaging region. Figure 1 shows the principle of MPI using the second harmonic. First, we describe 1D-imaging of the Z direction. By generating a gradient magnetic field in the imaging region, as shown in Figure 1 (a), a point FKP (Field Knee Point), indicated as a small square mark in the Figure, at which the magnetic nanoparticle is saturated, is generated as positive and negative points (solid line as shown in Figure 1 (b)). The dotted arrows shown in Figure 1 (a) indicate magnetic field lines generated by a pair of NdFeB permanent magnets. By applying a DC magnetic field under this gradient magnetic field, the FKP is moved (dotted line as shown in Figure 1 (b)). Applying an AC magnetic field while scanning, the second harmonic signal, which is twice the frequency of the applied AC magnetic field frequency, becomes its maximum value when a magnetic field corresponding to FFP is applied to the position of the sample. Since + FKP and - FKP differ in phase by 180°, peaks with different polarity are detected. When the DC magnetic field is increased, even after applying an AC magnetic field, the second harmonic cannot be detected because the magnetic nanoparticles are magnetically saturated. For example, when the sample is placed in the center of the imaging region, a signal as shown in the left of Figure 1 (c) is obtained. However, as it is difficult to identify the position, the signal is differentiated as shown in the right of Figure 1 (c), so that a peak appears at the position of the sample. In the MPI using the second harmonic, this differentiated value is used for imaging. The DC field in the Z direction is scanned to obtain a 1D image. After scanning along the Z direction, the DC field in the X direction is changed with a step. By repeating the procedure, a 2D image is obtained.

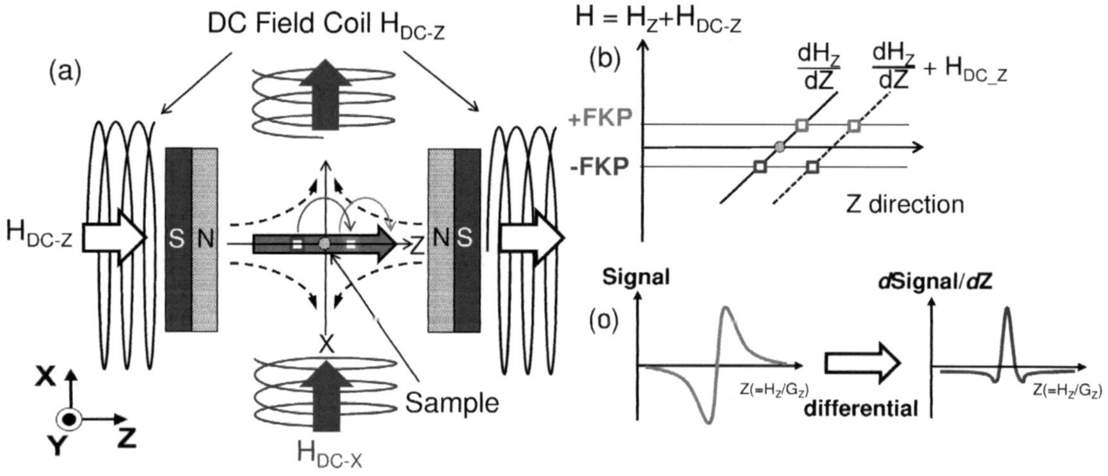

Figure 1. Principle of the MPI using second harmonic. (a) Schematic drawing of 2D-MPI system, (b) Gradient magnetic field in the imaging region, (c) Signal processing.

2.1. *Experimental system*

In a 2D-MPI system, an AC magnetic field coil and a DC magnetic field coil were installed to apply DC and AC magnetic fields in the X and Z directions. The experimental system is shown in Figure 2. By controlling the output of the current source with a PC, the DC magnetic field was adjusted so that the scanning step became 0.5 mm. A signal of MPI was measured under the conditions of an AC magnetic field H_{AC-Z} = 1.59 mT_{p-p}/μ_0 at 5.4 kHz, H_{ac-X} = 2.32 mT_{p-p}/μ_0 at 5.4 kHz, a DC magnetic field H_{DC-Z} = ±32 mT/μ_0 and H_{dc-X} = ±12.6 mT/μ_0. The gradient magnetic fields, which were generated by a permanent magnet, were G_Z = 3.2 mT/mm and G_Z =1.3 mT/mm in the Z and X directions, respectively. The imaging region was set in the range of ± 10 mm from the center in both the Z and X directions. As a sample,

Resovist 20 μℓ, in which magnetic nanoparticles with iron oxide (γ-Fe_2O_3) as a core were dispersed in water, was used. It was then put in a sample holder with a dimension of ϕ3 x 5 mm and placed in the center of the imaging region. Applying the DC and the AC magnetic field, the magnetization response from Resovist was detected by the differential detection coil. Then the signal from the magnetically coupled HTS-SQUID was connected to the lock-in-amplifier and the second harmonic component was measured. To reduce the noise, the signals were averaged by five times. The SQUID and the input coil were cooled with liquid nitrogen and they were placed in a three-layer permalloy magnetic shield. The measured signal was differentiated and imaged.

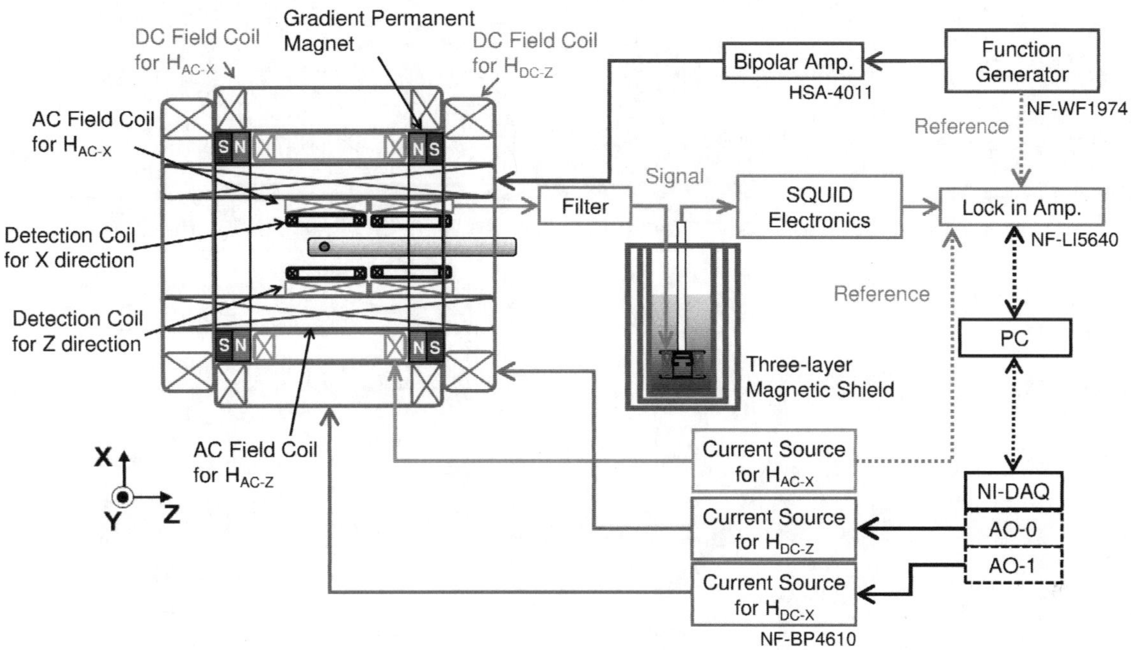

Figure 2. MPI system.

2.2. *Flux Transformer*

The flux transformer consists of a differential detection coil, a filter, an input coil and a HTS rf-SQUID.The HTS rf-SQUID and driving electronics used in the experiment are made by *Jülicher SQUID GmbH* in Germany. The circuit diagram is shown in Figure 3. When the SQUID was not used, the detection coil was directly connected to the lock-in-amplifier as shown in Figure 3 (a). When the SQUID was applied, the detection coil was connected to the input coil through a filter as shown in Figure 3 (b). The input coil and the HTS-SQUID were cooled at 77 K and magnetically coupled with a mutual inductance of 16 nH. Regarding the filter set, a band elimination filter (BEF) with a center frequency of 5.3 kHz and low pass filter were designed. Figure 4 shows the frequency characteristic of the flux transformer. Figure 4 (a) shows the spectra of the signal detected by the SQUID when the white noise of 1 $V_{p\text{-}p}$ was applied to the filter by the function generator (NF-WF 1974). In this case, the detection coil was removed. Although the center frequency of the BEF was designed as 5.3 kHz, the measured frequency was 5.4 kHz; as can be seen in Figure 4 (a). For this reason, the fundamental frequency of the applied AC magnetic field was set as 5.4 kHz. Since the resonant frequency was 10.8 kHz, the detection frequency of the lock-in-amplifier was set at 10.8 kHz as well. The shift of the frequency by 0.1 kHz may arise from an error of the capacitor. Figure 4 (b) shows the ambient magnetic field noise of the flux transformer without water sample, which was measured by the SQUID. The peak spectrum at the fundamental frequency of 5.4 kHz was 135 $fT/Hz^{1/2}$ as small as the white noise level, while the commercial frequency of 60 Hz was as large as 8.06 pT/Hz$^{1/2}$. Concerning the detection frequency of 10.8 kHz, which is the second harmonic, the peak spectrum increased as 452 fT/Hz$^{1/2}$ by the LC resonant. This means that the fundamental frequency was sufficiently supressed and that the filter set worked well. Thus, measurement was carried out using the flux transformer with the filter set.

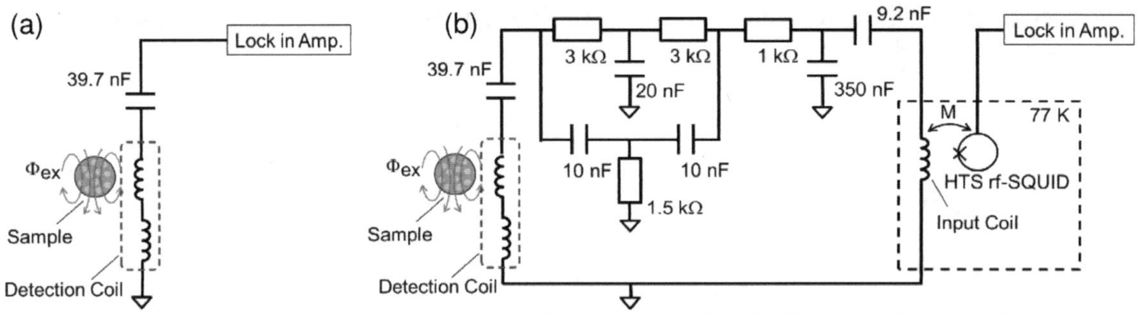

Figure 3. Circuit diagram. (a) Direct amplification, (b) Using SQUID and flux transformer.

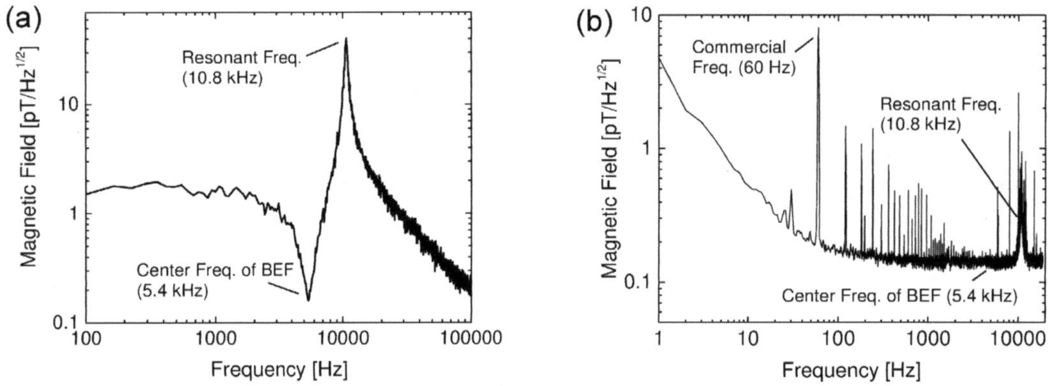

Figure 4. Signal intensity dependent on frequency. (a) Spectra of filter, (b) Spectra of ambient magnetic noise of flux transformer.

3. Results and Discussion

Using the 2D-MPI system, measurements were carried out in the both cases, using the SQUID with the flux transformer or not. First, the signal of the detection coil was directly connected to the lock-in-amplifier. The sample Resovist 20μℓ was placed at the center of the imaging region. Then the sequence was started and the signal was recorded. For comparison, both of the signals with and without the sample were obtained. The phase was automatically adjusted by the lock-in-amplifier so that dispersion of the second harmonic component became a minimum. Figure 5 shows the result of the differentiated value of the measured signal in the Z direction. Figure 5 (a) shows 1D image at X = 0. As shown in the figure, the position of the peak was the same as the sample position. This indicates that the sample position was successfully identified by the system. Figure 5 (b) shows 2D contour images, which 1D image in the Z direction was repeatedly taken by changing the position in the X direction from X = -10 mm to 10 mm by a step of 0.5 mm. In this study, in order to evaluate the quality of the detected signal, the peak signal-to-noise ratio PSNR was defined and introduced. The PSNR was defined as the value, which the peak value of the signal was subtracted by the averaged noise and followed by division by standard deviation of the noise when the sample was placed in the center of the imaging region. From the measurement results shown in Figure 5 (b), by calculation using the following values: the peal value of the signal 6.4 x 10^{-5} V/mm, the average value of noise -2.3 x 10^{-7} V/mm, the standard deviation of noise 1.5 x 10^{-6} V/mm, the PSNR of 42 can be obtained.

Subsequently, measurement was performed using the SQUID and the flux transformer. The output signal of the SQUID was lock-in amplified and differentiated. Figure 6 (a) shows a 1D image obtained at X = 0; the position of the peak was the same as the sample position as well as in the case of measurement without the SQUID. Figure 6 (b) shows a 2D contour image made by repeating the 1D scanning along the Z direction at the position of X = -10 mm to 10 mm. By calculation using the following values: the peal value of the signal 1.4 x 10^{-3} V/mm, the average value of noise -3.8 x 10^{-6} V/mm, the standard deviation of noise 2.2 x 10^{-6} V/mm, the PSNR of 64 was obtained.

Comparing the images between the direct lock-in amplification and via the SQUID amplification, the PSNR of the image obtained using the SQUID is 1.5 times larger than that obtained by direct amplification. We note that the contour images shown in Figure 5 (b) and Figure 6 (b) were elongated in the X direction. This is because the gradient magnetic field in the X direction was smaller than that in the Z direction, as a result, the resolution in the X direction may be poor.

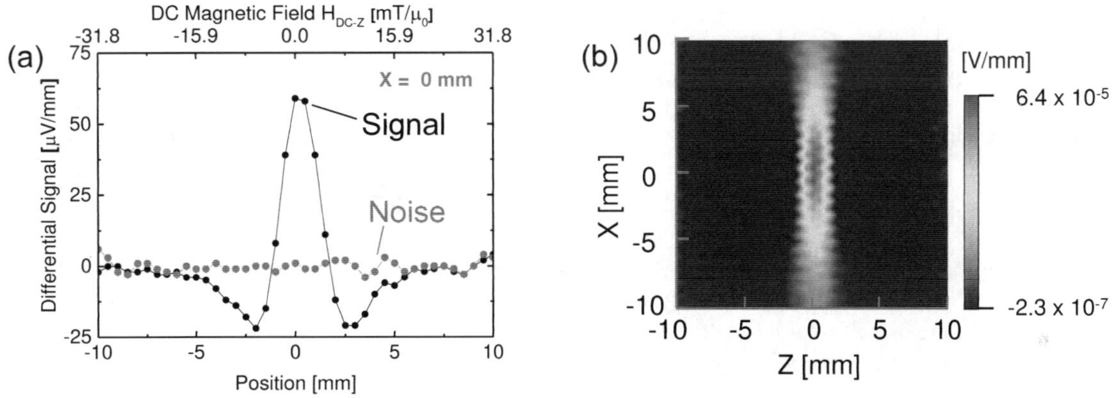

Figure 5. Measurement results by direct amplification. (a) 1D image with X = 0, (b) 2D contour image map.

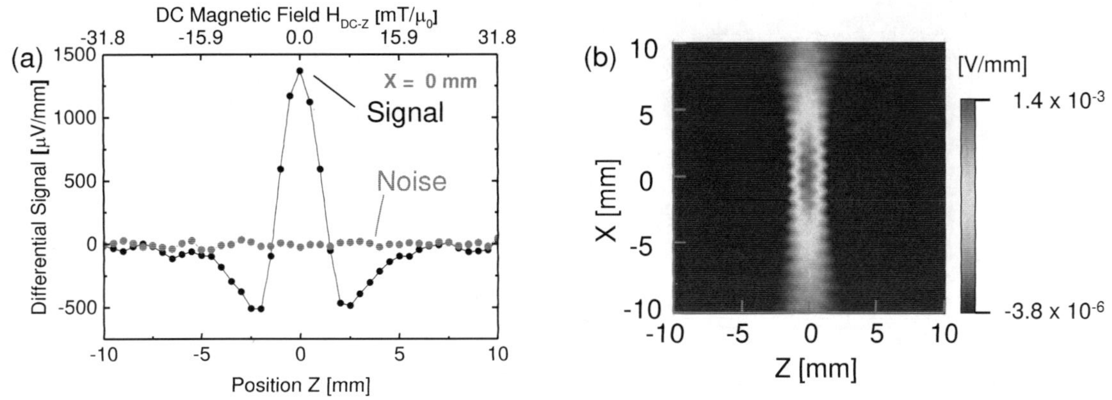

Figure 6. Measurement results is using SQUID. (a) 1D image with X = 0, (b) 2D contour image map.

4. Conclusion

In this study, we examined the second harmonic 2D-MPI system applying the flux transformer and HTS-SQUID to enhance the peak signal-to-noise ratio. The second harmonic component with a frequency of 10.8 kHz was amplified via the LC resonator with a resonant frequency of 10.8 kHz connected with the band elimination filter with a center frequency of 5.4 kHz.

From the measured results, the position of the peak was the same as the sample position. This indicates that the sample position was successfully identified by the system. Utilizing the SQUID and the flux transformer, the PSNR measured in the Z direction could be improved by 1.5 times; the advantage of using the SQUID amplification could be demonstrated. In the future, by applying a proper system function and a correction coefficient, the quality of the image will be improved.

References

[1] Gleich B and Weizenecker J 2005 Tomographic imaging using the nonlinear response of magnetic particles *Nature* **435** 1214-7
[2] Zhang Y, Murata H, Hatsukade Y and Tanaka S 2013 Superparamagnetic nanoparticle detection using second harmonic of magnetization response *Review of scientific instruments* **84** 094702
[3] Tanaka S, Murata H, Oishi T, Suzuki T and Zhang Y 2015 2D magnetic nanoparticle imaging using magnetization response second harmonic *Journal of Magnetism and Magnetic Materials* **383** 170-4
[4] Oishi T, Suzuki T, Yi Z and Tanaka S 2015 Magnetic Particle Imaging Using Second-Harmonic Response of Magnetization *SENSORS AND MATERIALS* **27** 1045-52
[5] Tanaka S, Murata H, Oishi T, Hatsukade Y, Zhang Y, Horng H-E, Liao S-H and Yang H-C 2015 Imaging of Magnetic Nanoparticles Using a Second Harmonic of Magnetization With DC Bias Field *IEEE Transactions on Magnetics* **51** 1-4
[6] Tanaka S, Oishi T, Suzuki T, Ohtani T and Ariyoshi S 2015 Imaging of Magnetic Nanoparticles Using Second-Harmonic Signals *IEEE Transactions on Magnetics* **51** 1-4

IOP Conf. Series: Journal of Physics: Conf. Series **871** (2017) 012078 doi:10.1088/1742-6596/871/1/012078

Investigation on readout coil design for fluxed locked loop control of HTS rf-SQUID

Y Miyato and H Itozaki

Department of Engineering Science, Osaka University,
1-3 Machikaneyama, Toyonaka, Osaka 560-8531, Japan

E-mail: miyato@sup.ee.es.osaka-u.ac.jp

Abstract. We investigated the readout coil, electro-magnetically coupled with a HTS rf-SQUID for the flux-locked loop control. The design and size of the readout coil affected the SQUID performances. Among the tested combinations of different readout coils with the rf-SQUID, the rectangular coil, which just surrounded the slit in the rf-SQUID, was advantageous for the better performance. We also demonstrated the rf-SQUID operation with the rectangular coils made of the thin flexible print circuit board, which could be put on the rf-SQUID stably.

1. Introduction

Superconducting quantum interference devices (SQUIDs) are well known as highly sensitive magnetic sensor for various applications. We have developed HTS rf-SQUIDs for the application of SQUID microscopy [1]–[3]. An rf-SQUID measures the extremely weak magnetic field by detecting its resonance change caused by the external quasi-static magnetic field. The SQUID performance depends on the designs of not only the rf-SQUID itself, especially the slit inside it [3], but also the readout coil combined with it for fluxed locked loop (FLL) control. In this study, we focused on the readout coil to improve the SQUID performance. Figure 1 shows the schematic illustration of the total system for the HTS rf-SQUID magnetometer. In this system, one readout coil can serve as three functions: the transmission coil for the rf-magnetic field to excite the rf-SQUID at the resonant frequency, the reception coil for the rf-magnetic field returned back from the resonated rf-SQUID, and the feedback coil for FLL control to measure the intensity of the external magnetic field quantitatively. For the transmission and receipt of the rf-magnetic field, the readout coil should be coupled to the rf-SQUID in the proper coupling condition. In contrast, for the FLL feedback, the weak coupling should

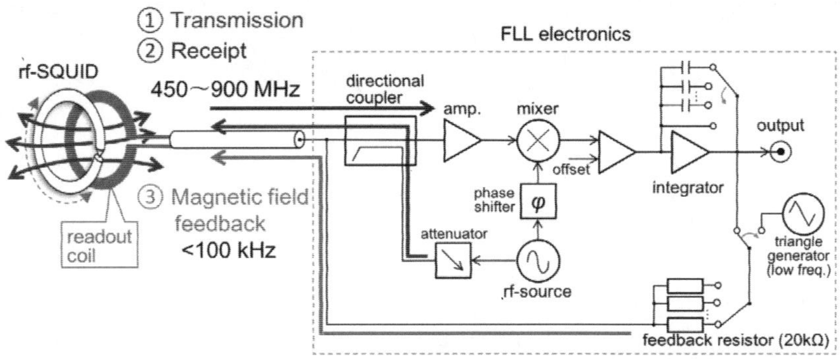

Figure1. Total measurement system of an rf-SQUID magnetometer.

be better for the sensor sensitivity as long as the noise is not increased. It was supposed that the coil design should affect the SQUID performance. We investigated the coil influences to the conventional rf-SQUID operated with a substrate resonator [4], and to our modified rf-SQUID having a meandered slit, which can be operated without a resonator [3]. The SQUID performances were investigated by changing the combinations of the rf-SQUIDs with various coils with the different sizes and designs.

2. Setup for measurements of the rf-SQUIDs

The rf-SQUIDs and the substrate resonator were made of the YBCO thin films on bi-crystal and mono-crystal $SrTiO_3$ (STO) substrates, respectively. Figures 2(a) and (b) respectively show our conventional rf-SQUID and the substrate resonator fabricated in reference to a commercially available rf-SQUID set (Jülicher SQUID GmbH). On the other hand, figure 3(a) shows our modified rf-SQUID having a meandered slit, which plays the resonator role to adjust the resonant frequency. Each outer diameter of both rf-SQUIDs was set at 3.5 mm.

Readout coils having different designs were also made with the printed circuit boards (PCBs) having a 35 μm copper layer by a CNC milling machine. We firstly prepared two types of the coils: square and circular coils as shown in figures 4 and 5, respectively. Based on the results of two type coils, we also prepared the rectangular type coils as shown in figure 6. Each wire width of the readout coils was set at 0.5 mm in the PCB drawing, but the actual width was 0.5±0.2 mm due to the machining accuracy of the milling machine. For the square type (figure 4), we fabricated four coils having the different inner side length: L (L=5, 4, 3 and 2 mm). For the circular type (figure 5), we made three coils having the different inner diameter: R (R=5, 4 and 3 mm). For the rectangular type (figure 6), we made two coils having the different inner side width: W (W=0.8 and 0.4 mm), keeping the inner side length L=1.8 mm, in which the coil wire just surrounded the slit and hole in the rf-SQUID.

The rf-SQUID was operated using the commercially available FLL electronics (HTSL-RF-SQUID-ELEKTRONIK ver.5, Jülicher SQUID GmbH) [5]. The V-Φ characteristics and the noise density spectra were obtained by the open-loop (FLL off) and the closed-loop (FLL on), respectively. For the measurements of the rf-SQUID, the rf-SQUID was mounted with the PCB coils using the

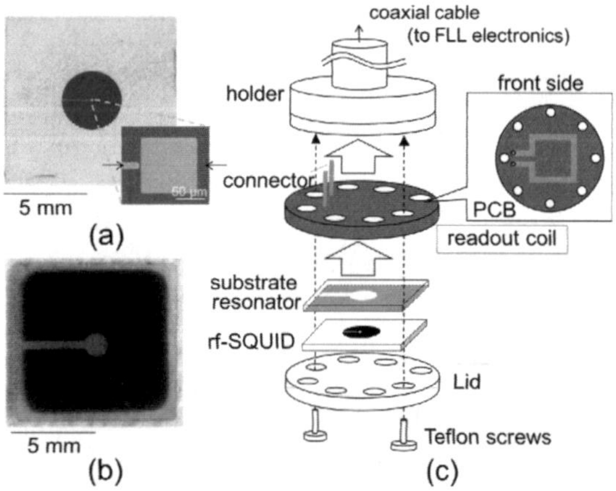

Figure 2. Optical microscope images of (a) our conventional rf-SQUID and (b) substrate resonator. Inset shows the magnified image around the center. The square bright region corresponds to the SQUID hole, and arrows indicate the grain boundary. (c) Assembly of the rf-SQUID and resonator inside the holder. The rf-SQUID was combined with the substrate resonator in a flip-chip configuration.

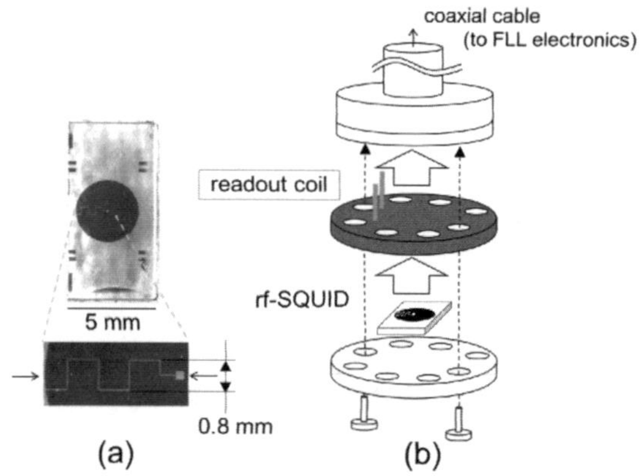

Figure 3. Optical microscope images of (a) our modified rf-SQUID. Inset shows the magnified image around the slit. Arrows indicate the grain boundary. (b) Assembly of the rf-SQUID inside the holder. The substrate resonator was omitted differently from figure 2(c).

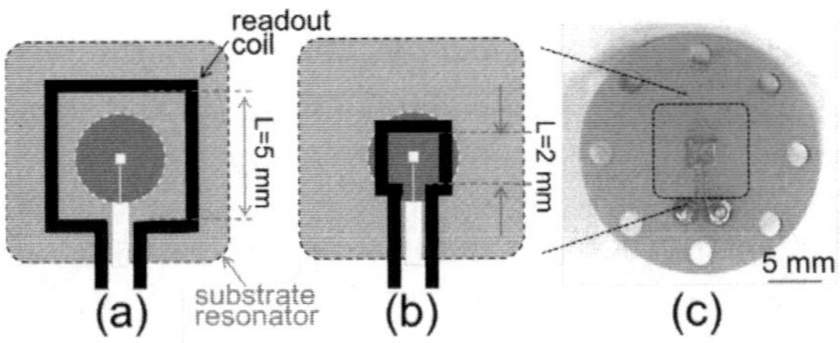

Figure 4. Square type of readout coils. Schematics of the coil shapes with (a) L=5 mm and (b) L=2 mm. Outlines of our conventional rf-SQUID and resonator were superimposed. (c) Photograph of a fabricated coil.

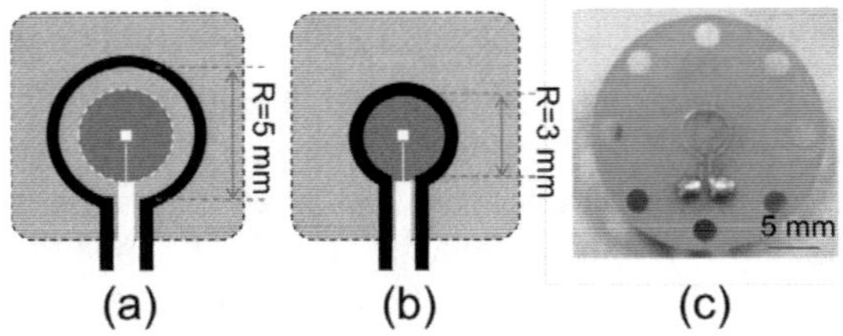

Figure 5. Circular type of readout coils. Schematics of the coil shapes with (a) R=5 mm and (b) R=3 mm. Outlines of our conventional rf-SQUID and resonator were superimposed. (c) Photograph of a fabricated coil.

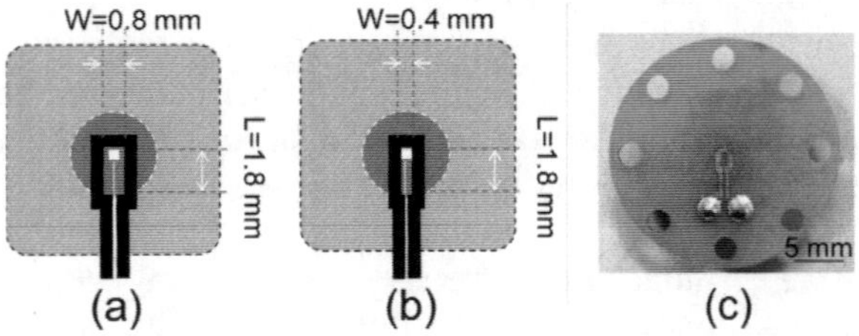

Figure 6. Rectangular type of readout coils. Schematics of the coil shapes with (a) W=0.8 mm and (b) W=0.4 mm, while keeping the side length L=1.8 mm. Outlines of our conventional rf-SQUID and resonator were superimposed. (c) Photograph of a fabricated coil.

holder of the FLL electronics accessory, as shown in figure 2(c) or 3(b). The rf-SQUID measurements were carried out in the liquid nitrogen Dewar flask, around which a Helmholtz coil was wound (diameter: 13 cm, 5 turns in each side), inside a μ-metal magnetic and a copper EM shields. The V-Φ curve and the flux conversion coefficient, defined as α, for the FLL-output voltage signal were measured by sweeping the static magnetic field from the readout coil, and by applying the external magnetic field from the Helmholtz coil, respectively.

3. Experimental results

For the comparison among the combinations of the different readout coils, we extracted characteristic features from the measurement results, related with the transmission, the feedback and the output signal. In the transmission measurement, the frequency and the power of the rf-source must be optimized to get the largest amplitude appearing in the V-Φ curve, as shown in the example in figure

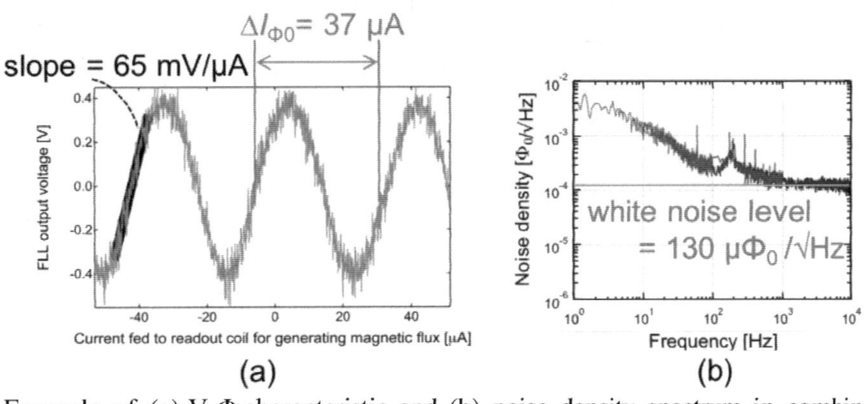

Figure 7. Example of (a) V-Φ characteristic and (b) noise density spectrum in combination of our conventional rf-SQUID with the square coil (the case of L=5 mm).

IOP Conf. Series: Journal of Physics: Conf. Series **871** (2017) 012078 doi:10.1088/1742-6596/871/1/012078

7(a). These optimized values can make the rf-SQUID resonated. From the optimized condition, the period and the maximum slope values of the V-Φ curve were obtained, as shown in figure 7(a). Note that every V-Φ curve's horizontal axis in this paper means the current fed to the readout coil, related with magnetic flux for the modulation or FLL feedback. The period, defined as $\Delta I_{\Phi 0}$, means the application current for generating 1 Φ_0 flux quanta of the magnetic flux from the readout coil to the rf-SQUID, corresponding to the magnetic coupling efficiency between the rf-SQUID and the coil. The smaller $\Delta I_{\Phi 0}$, the bigger coupling efficiency. The slope mainly affects the slew rate and the bandwidth of the FLL control. In the feedback measurement, the effective area and the flux conversion coefficients, defined as A_{eff} and α, respectively, were obtained by using the Helmholtz coil. In the output measurement, the white noise levels were obtained from the noise density spectra, in which the noise level was settled to be constant at the high frequency range, as shown in figure 7(b).

3.1. Coil combinations with our conventional rf-SQUID

We firstly measured the combinations of the square and circular coils, as shown in figures 4 and 5, with our conventional rf-SQUID as shown in figure 2. As described in the following section 3.1.2, we assumed that it would be preferable for the readout coil to surround the slit in the rf-SQUID. Then, we prepared the rectangular coils as shown in figure 6, and also measured their combinations.

3.1.1. Result Comparison. The total measurement results are listed up in this section. The transmission frequencies and A_{eff} values for the combinations with the different readout coils were almost constant at 620±2 MHz and 0.42±0.04 mm^2, respectively. Figures 8, 9, 10, 11 and 12 respectively show dependence of the transmission power, $\Delta I_{\Phi 0}$ value of each V-Φ curve, slope of each V-Φ curve, coefficient α, and white noise level, on the coil feature size (L, R and W).

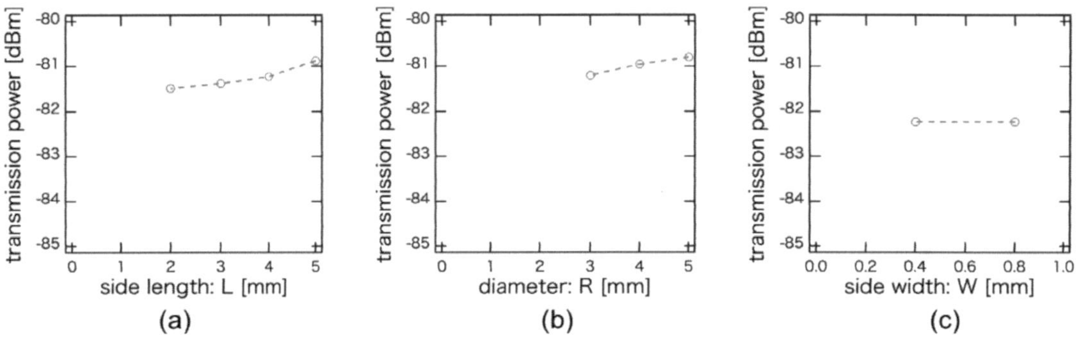

Figure 8. Transmission powers for the combinations with (a) square, (b) circular and (c) rectangular coils as functions of the inner side length: L, diameter: R and side width: W, respectively.

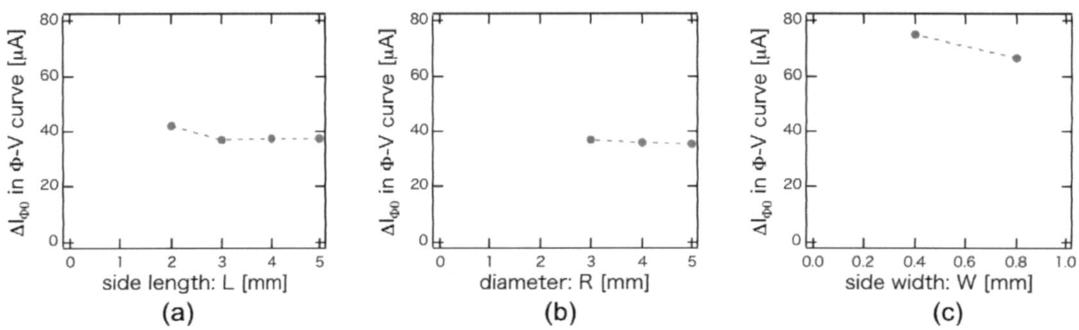

Figure 9. $\Delta I_{\Phi 0}$ values of each V-Φ curve for the combinations with (a) square, (b) circular and (c) rectangular coils as functions of the inner side length: L, diameter: R and side width: W, respectively.

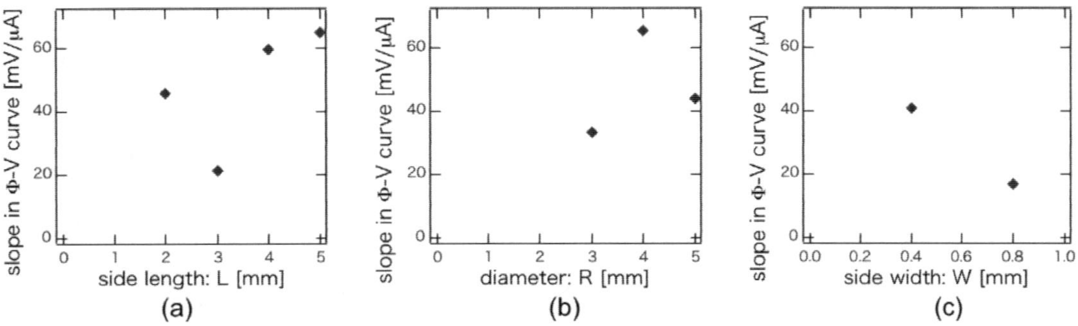

Figure 10. Slopes of each V-Φ curve for the combinations with (a) square, (b) circular and (c) rectangular coils as functions of the inner side length: L, diameter: R and side width: W,

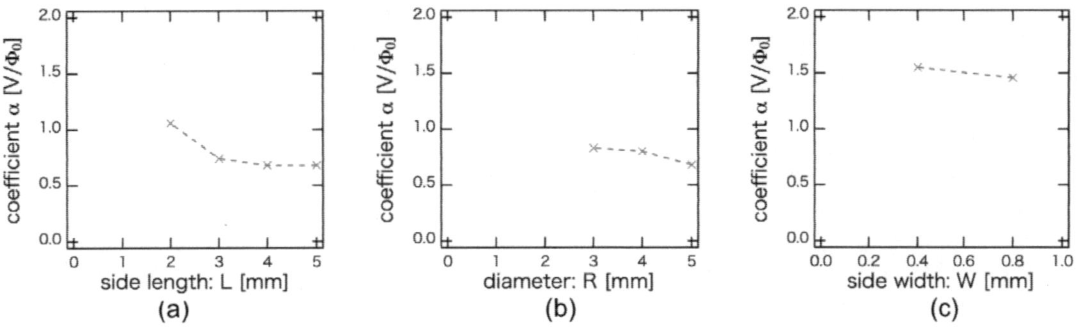

Figure 11. Coefficients α for the combinations with (a) square, (b) circular and (c) rectangular coils as functions of the inner side length: L, diameter: R and side width: W, respectively.

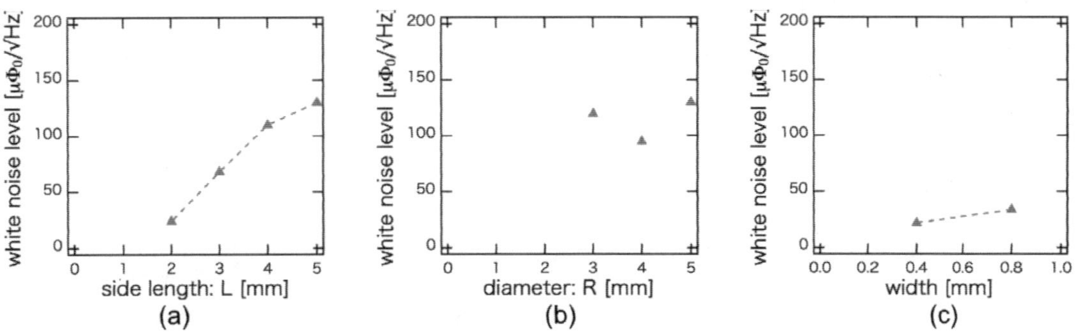

Figure 12. White noise levels for the combinations with (a) square, (b) circular and (c) rectangular coils as functions of the inner side length: L, diameter: R and side width: W, respectively.

3.1.2. Discussion. From the measurement results, the characteristic features except the transmission frequency and A_{eff} value depended on the readout coil. The reason why the transmission frequency and A_{eff} value showed the independence of the coil design or size is because the resonant frequency and the effective area are little influenced by the coil, but strongly depend on the rf-SQUID and the resonator, themselves. In contrast, the reason why the other features depended on the coil must be because magnetic coupling between the rf-SQUID set and the coil played an important role for the operation. In figure 8, the transmission power is slightly decreased with decreasing the coil size. It indicates that the magnetic coupling for the rf-magnetic field to excite the rf-SQUID set was mildly improved due to reduction of the energy required to reach the same resonant condition. Figure 13 shows the simulation result of the rf-magnetic field over the rf-SQUID, excited in resonance. The field is mainly distributed around the slit and hole. As the result, it is advantageous for the coil to surround the slit of the rf-

SQUID for receipt of the rf-magnetic field. This was why we reached concept of the rectangular type and prepared the rectangular coils. Conversely, in figure 9, $\Delta I_{\Phi 0}$ values seems to be increased with decreasing the coil size. In contrast to the rf-magnetic coupling, it is supposed that the static magnetic coupling between the rf-SQUID set and the coil became weaker due to increase of the application voltage required to cause the same flux change against the rf-SQUID. These results suggest that the magnetic coupling behaved differently between the rf-range for transmission/receipt and the low frequency range for the feedback. In figure 10, the slopes of each V-Φ curve are somewhat fluctuated, but the reason is unclear at present. In addition, in figure 11, the coefficient α is also increased with decreasing the coil size. The α is related with mutual inductance, or the static magnetic coupling between the rf-SQUID and the coil for the FLL feedback control. It is consistent with $\Delta I_{\Phi 0}$. If the voltage noise from the FLL electronics is dominant, the larger α is desirable to reduce the flux noise density. Consequently, the flux noise density was successfully reduced with decreasing the coil size, especially in the rectangular coils, as shown in figure 12. For the comparison, we specially pick up the V-Φ characteristic and noise density spectrum in combination with the rectangular coil (W=0.4 mm), as shown in figure 14, which showed the lowest white noise level among the tested combinations. The V-Φ curve shape is largely changed, and the white noise level is greatly reduced from figure 7.

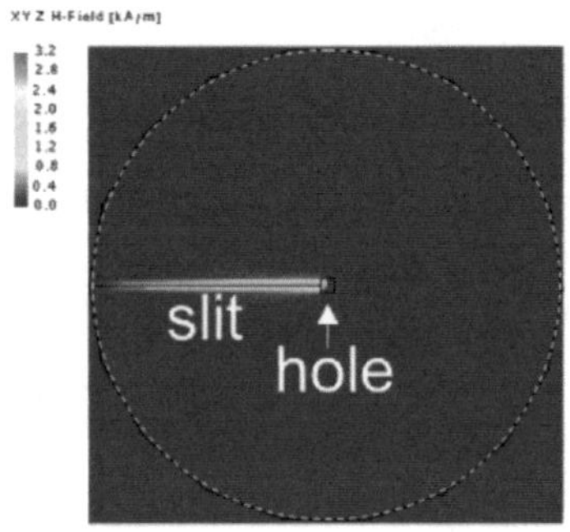

Figure 13. Simulation result of rf-magnetic field distribution in the resonated rf-SQUID.

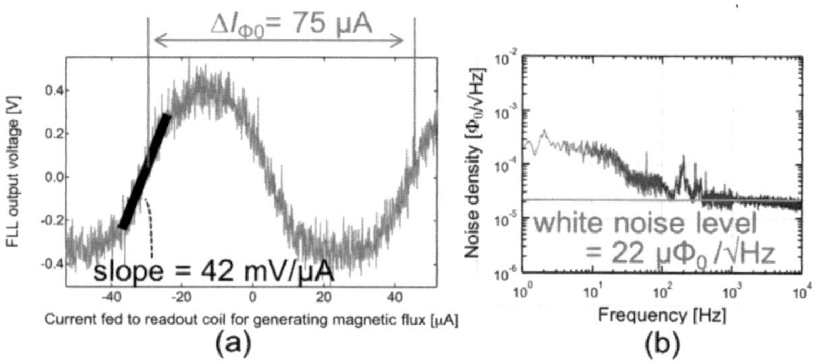

Figure 14. (a) V-Φ characteristic and (b) noise density spectrum in combination of our conventional rf-SQUID with the rectangular coil (W=0.4 mm).

3.2. Coil combinations with our modified rf-SQUID

We also investigated combinations of our modified rf-SQUID, as shown in figure 3(a), with the square, circular and rectangular coils, as shown in figures 4, 5 and 6. In each combination, the transmission frequency and effective area were 543±2 MHz and 0.31±0.02 mm^2, respectively. In this frequency range, the rf-SQUID could be operated without a resonator by our electronics. The coil combinations of our modified rf-SQUID showed the similar dependence to our conventional rf-SQUID results, as mentioned previous section. The smaller coil size, especially in the rectangular coils, the lower white noise level the combination showed. Here, we specially pick up the V-Φ characteristics and noise density spectra for the combinations of the square coil (L=5 mm) and two rectangular coil (W=0.8 and 0.4 mm), as shown in figures 15 and 16, respectively. Compare to the conventional rf-SQUID, the V-Φ characteristics were deformed probably because the optimum rf-power was out of the minimum power limit of the FLL electronics (the limit: −82.2 dBm); nevertheless, we measured the noise spectra as far as we could adjust the transmission condition to get the largest amplitude within the limit in each V-Φ curve (the optimized transmission power: ~−82 dBm). Among the coil combinations, the rectangular coil with W=0.8 mm showed the lowest white noise density. Our modified rf-SQUID had the meandered slit with the meander-width of 0.8 mm, as shown in figure 3(a). In contrast, for the combination of the conventional rf-SQUID having the straight slit, the rectangular coil with the narrowest width of W=0.4 mm showed the best performance. It suggests that it should be more advantageous for the readout coil width to just fit into the slit width of the rf-SQUID.

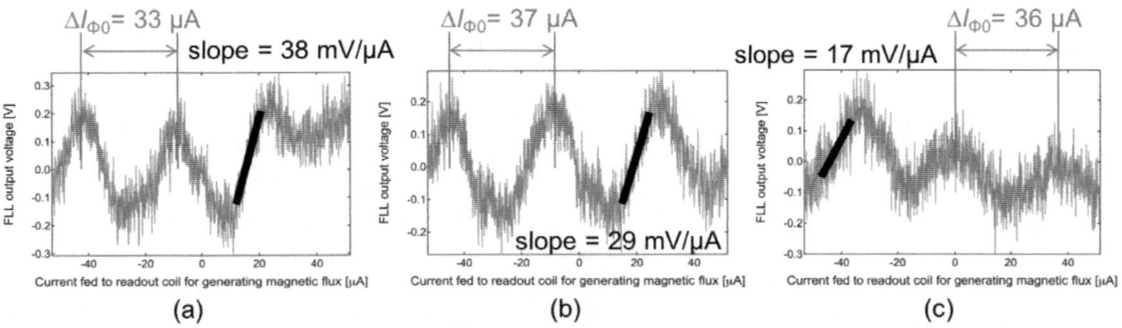

Figure 15. V-Φ characteristics in combinations of our modified rf-SQUID with (a) the square coil (L=5 mm), (b) the rectangular coil (W=0.8 mm) and (c) the rectangular coil (W=0.4 mm).

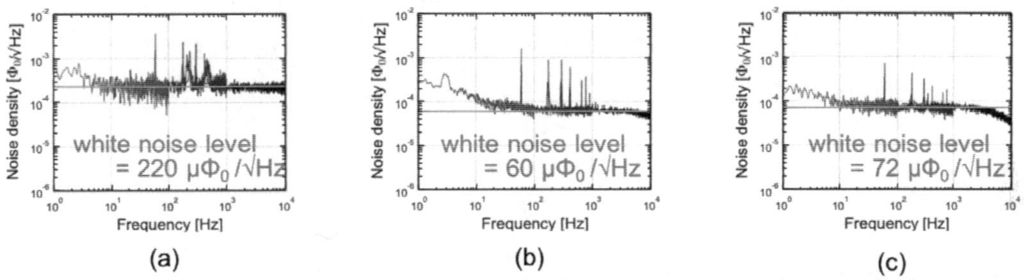

Figure 16. Noise density spectra in combinations of our modified rf-SQUID with (a) the square coil (L=5 mm), (b) the rectangular coil (W=0.8 mm) and (c) the rectangular coil (W=0.4 mm).

4. Rectangular coil on flexible PCB

One of the applications of our rf-SQUID is for an STM SQUID microscope, in which the rf-SQUID is combined with STM microscopy [1]. In this microscope, the high-permeability probe transfers the magnetic flux from the sample surface to the rf-SQUID. It is important to make the gap distance between the probe end and the rf-SQUID as short as possible, so as not to be brought in contact for the thermal insulation. The rf-SQUID should be mounted on the sapphire rod cooled by liquid N_2. The readout coil must be put on the surface of the rf-SQUID. However, it is necessary for the readout coil not to interfere this gap. Therefore, we applied thin flexible PCBs to the readout coils. The flexible PCB coils were made of a 35 μm thick copper layer on a polyimide substrate, of which total thickness was 90 μm including a polyimide cover and glue layers. For our conventional and modified rf-SQUIDs, the same rectangular designs with W=0.4 mm and W=0.8 mm as figures 6(a) and (b) were employed to the flexible PCB coils, as shown in figures 17(a) and (b), respectively.

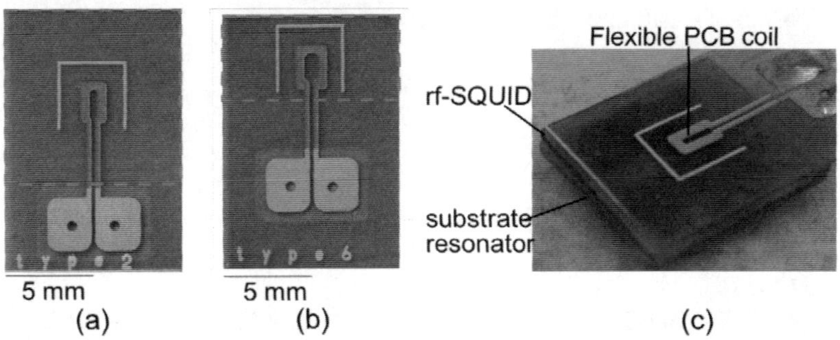

Figure 17. Flexible PCB coils for (a) our conventional and (b) modified rf-SQUIDs. (c) Photograph of example in combination of the conventional rf-SQUID with the flexible PCB coil.

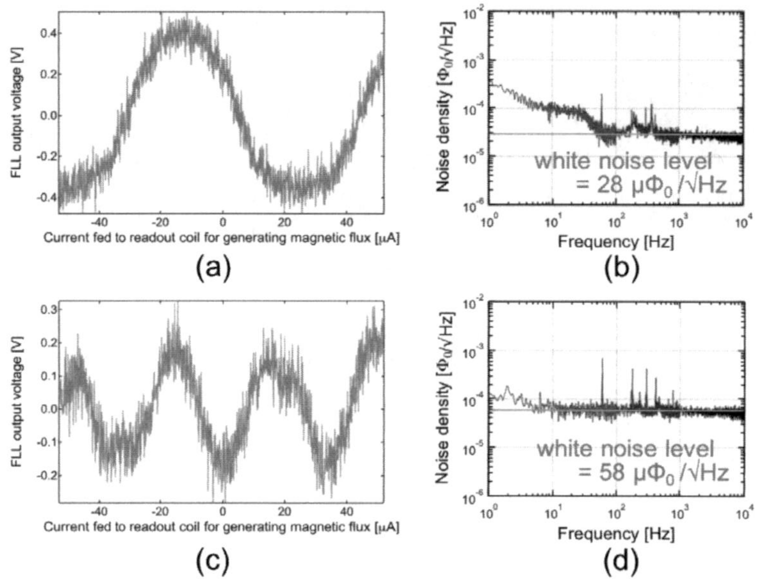

Figure 18. (a), (c) V-Φ characteristics and (b), (d) noise density spectra of the combinations with flexible PCB coils. (a), (b) and (c), (d) correspond to combinations of our conventional and modified rf-SQUIDs with the rectangular coils with W=0.4 mm and 0.8 mm, respectively.

The flexible PCB coils were fixed on the top surface of the rf-SQUIDs by using the vacuum grease, and were connected to the FLL electronics from its contact pads, soldered to a coaxial cable. Figure 18 shows the measurement results in combinations of our conventional and modified rf-SQUIDs with each flexible PCB coil. The results were almost the same as results in the rigid PCB coils, as mentioned in the previous section (figures 14, 15(b) and 16(b)). The flexible PCB coils improved the SQUID performance, similarly to the rigid PCB coils. Thus, these flexible PCB coils can be also applicable to the STM-SQUID microscope to improve its performance.

5. Summary

The SQUID performances were compared by combining the different readout coils with the rf-SQUIDs. The coil significantly affected the white noise level of the rf-SQUID. When the rectangular coil, which just surrounded the slit and hole of the rf-SQUID, was employed, the rf-SQUID showed the lowest white noise level among the tested coils. The results suggested that the performance could be affected by the magnetic coupling between the rf-SQUID and the coil. In addition, the flexible PCB coils were very useful for stable operation. Thus, it is important to optimize the readout coil as well as the rf-SQUID itself for improvement of the SQUID performance.

References

[1] Tachiki M, Hayashi T and Itozaki H 2009 *IEEE Trans. Appl. Supercond.* **19** 874
[2] Miyato Y, Ashizuka T, Matsui Y and Itozaki H 2015 *IEEE Trans. Appl. Supercond.* **25** 1600804
[3] Sakai A, Ashizuka T, Miyato Y and Itozaki H 2015 *Phys. Procedia* **65** 181
[4] Zhang Y, Schubert J, Wolters N, Banzet M, Zander W and Krause H J 2002 *Phys. C Supercond.* **372–376** 282
[5] Krause H J, Wolters N and Zhang Y 2012 *Phys. Procedia* **36** 306

Simulations of chaos generation from Josephson junctions with various junction parameters

R Hiwatashi[1], **Y Tamura**[1], **H Shimakage**[1]

[1] Graduate School of Science and Engineering, Ibaraki University, Hitachi, Ibaraki 316-8511, Japan

15nm616r@vc.ibaraki.ac.jp

Abstract. It is well known that voltage waveforms between electrodes of Josephson junctions under irradiation of a microwave behave chaos characteristics under appropriate conditions. In order to apply the chaos to a random number generator, we have been studying Josephson chaos by simulations. In the simulation, the Josephson junction is assumed to fabricate with YBCO materials. We used a RCSJ model in order to present an equivalent circuit of the Josephson junction, and derived a derivative equation. Lyapunov exponents, which determined if the state of the Josephson junction was chaotic or not, were calculated from time evolutions of voltages obtained from the equation. In the simulation, junction parameters were assigned feasible values for an actual YBCO Josephson junctions. As a result, we found that chaos can be generated by adjusting element parameters. Moreover, we found that there were lower limits in the resistance values for generation of chaos. In addition, we found that frequency margins, at which the chaos was obtained, were broadened by decrease of the resistance and increase of capacitance.

1. Introduction

A random number generator is used for cryptographic communication [1], Monte Carlo simulation [2], a probability model [3], and so on. In particular, high-quality random numbers are required for security applications. In general, there exists two ways to generate the random numbers; one is a pseudo random number, which is generated by a computer algorithm, and the other is a physical random number, which is generated from a physical random phenomenon. The pseudo-random number generator can be easily implemented on any platform, and a random number generation speed depends only on the processing hardware. But, since its random number series was always determined by an initial condition of the algorithm, it is reproducible. On the other hand, the physical random number generator was affected by probabilistic physical phenomena such as thermal noise [4] in resistors or diodes, radiations [5], photons [6-8], and so on. Therefore, the random number output from the physical random number generator has an unreproducible characteristic, and is suitable for security application. However, generally, the generation speed of the physical random number generator is relatively slow, and the size of the generator itself is relatively large. For example, the generation speed of random numbers using photons is reported of about 16 Mbit/s [9]. Some approaches to solve these problems have been reported. The physical random number generator using semiconductor laser chaos as chaotic phenomena have been studied, and random number generation at 12.5 Gbit/s has been successfully achieved [10]. However, there is a periodicity due to return light of the laser, and it can't be completely removed.

On the other hand, the chaos phenomenon is also observed in a voltage oscillation from a Josephson junction under irradiated with microwaves. Up to now, the studies of chaos in the Josephson junction have been reported so far [11-13]. Main purposes of these studies were to study chaos theory and to suppress occurrence of chaos in a Josephson voltage standard. But, it is expected that chaotic oscillation by Josephson junction can be applied as a high-quality random number source. In fact, it has been reported that high-quality white noise can be generated by chaotic oscillation from Josephson junction by a numerical simulation [14]. Therefore, it has been shown that chaos from the Josephson junction behaves highly randomly [15]. Moreover, as the voltage in the Josephson junctions are oscillates at the frequency of several hundreds of GHz, it has the potential to extract the high-speed voltage output [16].

Our research objective is to realize a physical random number generator using chaos oscillation by the Josephson junction. In order to realize that, it is necessary to know whether chaos can be generated by using actual junction parameters of the Josephson junctions. In this paper, we report on the chaos simulations for various junction parameters, which are realized in actual devices.

2. Simulation

2.1. Simulation method

Josephson junction is represented by an equivalent circuit, in which a term responsible for a current flowing through a resistance R and a capacitance C is connected in parallel to a term responsible for Josephson current according to $I_c \sin\varphi$. This circuit is called RCSJ model and is often used for analysis of Josephson junctions. An external microwave irradiation $I_{rf} \sin \omega t$ and current source I_0 are added to its equivalent circuit, and a model of a circuit for generating chaos is shown in Fig. 1. A circuit equation is written from the circuit of Fig. 1, and Eq. (1) is obtained by normalization of the circuit equation.

$$i_0 + i_{rf} \sin \Omega\tau = \sin\varphi + \frac{d\varphi}{d\tau} + \beta \frac{d^2\varphi}{d\tau^2} \tag{1}$$

Where $i_0 = I_0 / I_c$, $i_{rf} = I_{rf} / I_c$, $\Omega = \hbar\omega / 2eI_cR$, $\beta = 2eI_cCR^2 / \hbar$, $\tau = 2eI_cR / \hbar$. A simulation of the voltage behavior was carried out using by Eq. (1). A time evolution of the voltage was calculated by a 4th-order Runge-Kutta method, a Lyapunov exponent λ was derived based on it, and a chaotic state was judged. The parameters of the junctions used in the simulation were set to values that can be realized in actual Josephson junction parameters. Since an I_cR of a general YBCO Josephson junction is about 2 mV, the resistance value and the critical current value were set to I_c = 1 mA, R = 2 Ω in this simulation, respectively. The capacitance value was set as C = 82.24 fF so as to set a McCamber parameter to be 1. Since we focused on simple operation of chaos generation as a first trial, the bias current was set to 0 mA. Using these parameters, we investigated the chaotic state in range of the irradiation frequency from 100 GHz to 1 THz.

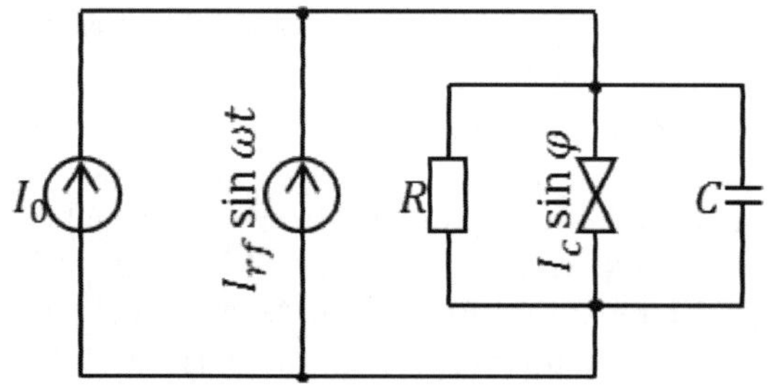

Figure 1. Resitively shunted junction model.

2.2. *Adjustment of resistance and capacitance value*

First, the Lyapunov exponent was calculated using the initial parameters shown in the Table. 1, but the Lyapunov exponent was negative within the examined frequency. Therefore, we predicted that the frequency, at which the Lyapunov exponent becomes positive, would appear by changing the junction parameters. We tried to change the resistance, since it can be easily changed by resistive shunt on the Josephson junctions. Simulations were performed with resistance values of 2, 1.5, and 1 Ω. As the I_c value was unchanged, the I_cR_ns changed with the R_ns. The value of the Lyapunov exponent also decreased with decrease of the resistance value, and the frequency, at which the Lyapunov exponent was positive, did not appear. Next, we tried to change the capacitance value. As we supposed to use the shunt capacitance to the Josephson junction, the capacitance value would be increase because of parallel configuration. Therefore, simulation was performed by changing the capacitance to 164.4, 246.7, 328.9, and 411.2 fF, which are integer number times of initial capacitance value. Although we surveyed if the Josephson junctions in the chaos state at the capacitance value of 164.4 fF, the frequency, at which the

Table 1. The initial parameters used to calculate the Lyapunov exponent

I_0 [mA]	I_{rf} [mA]	R [Ω]	I_c [mA]	C [fF]
0	1	2	1	82.24

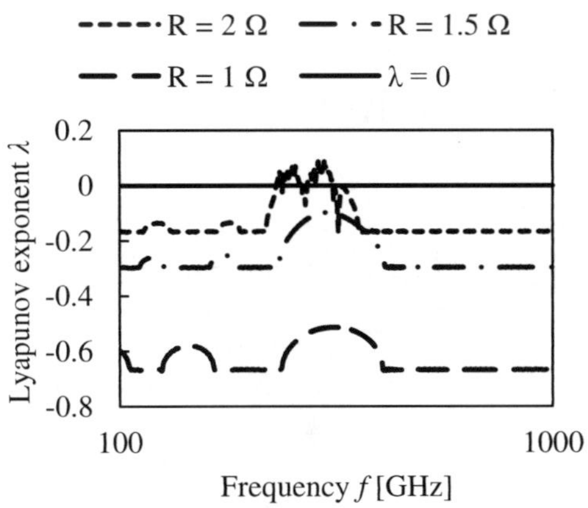

C = 246.7 fF

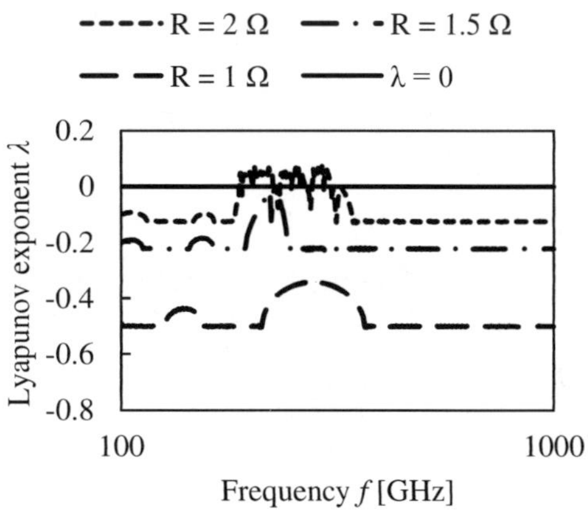

C = 328.9 fF

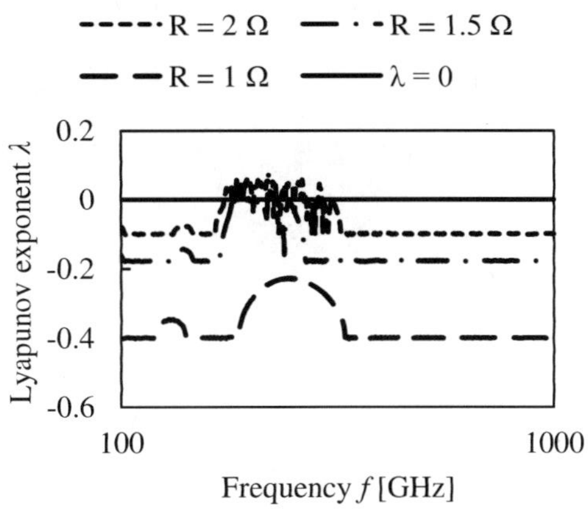

C = 411.2 fF

Figure 2. Frequency dependences of Lyapunov exponent and frequency when resistance values are 2, 1.5, and 1 Ω.

IOP Conf. Series: Journal of Physics: Conf. Series **871** (2017) 012079 doi:10.1088/1742-6596/871/1/012079

Lyapunov exponent was positive, did not appear. On the other hand, when we surveyed it as the same way at the capacitance values of the other values, we found the chaos state. The result is shown in Fig.

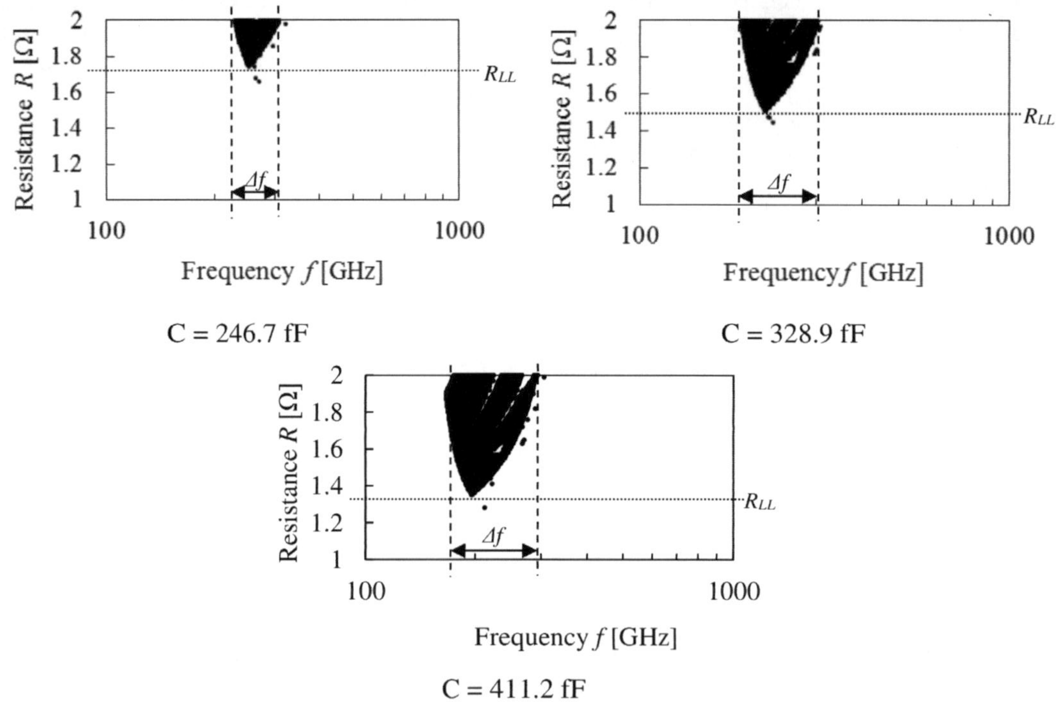

Figure 3. Results of mapping of the regions of chaos generation.

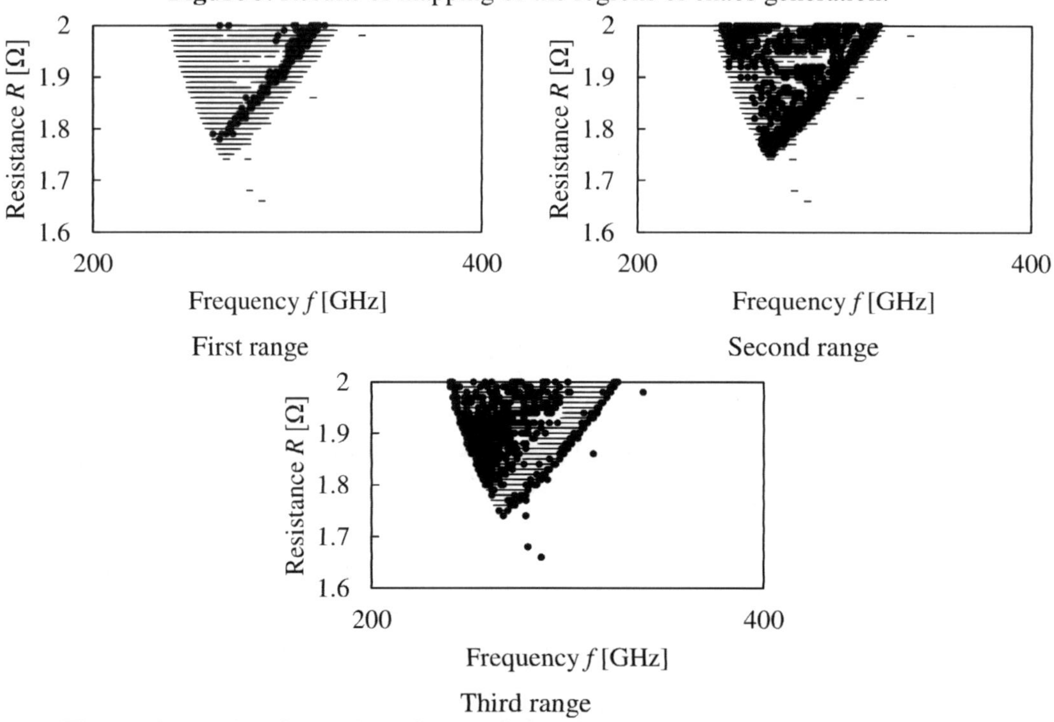

Figure 4. Results of mapping of the regions of chaos generation. The points in each map were plotted for the first range (0.063795<λ≤0.095696), the second range (0.031898<λ≤0.063795), and the third range (0.000003<λ≤0.031898).

IOP Conf. Series: Journal of Physics: Conf. Series **871** (2017) 012079 doi:10.1088/1742-6596/871/1/012079

2. The positive Lyapunov exponents were obtained intermittently for any resistance values. To clarify the details of the region, at which the Josephson junctions were in the chaos state, we mapped the points of the positive Lyapunov exponents in Fig. 3. Here, in Fig.3 a vertical and horizontal axes are the frequency and the resistance, respectively. From the data of first range, second range, third range in Fig.3, the lower limit of the resistance values (R_{LL}s) were 1.74, 1.51, and 1.35 Ω, respectively. As the

Table 2. The parameters to simulate the voltage waveforms

λ	I_0 [mA]	I_{rf} [mA]	f [GHz]	R [Ω]	I_c [mA]	C [fF]
0.095696	0	1	265	1.85	1	246.7
0.063205	0	1	265	1.83	1	246.7
0.031892	0	1	267	1.96	1	246.7

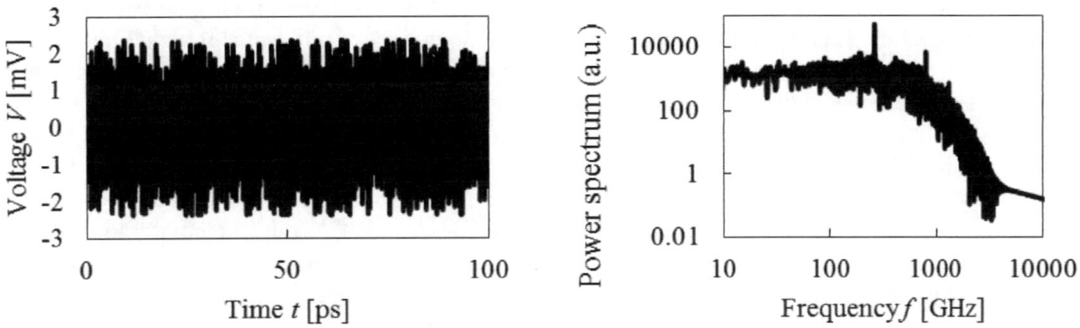

First range $\lambda = 0.095696$

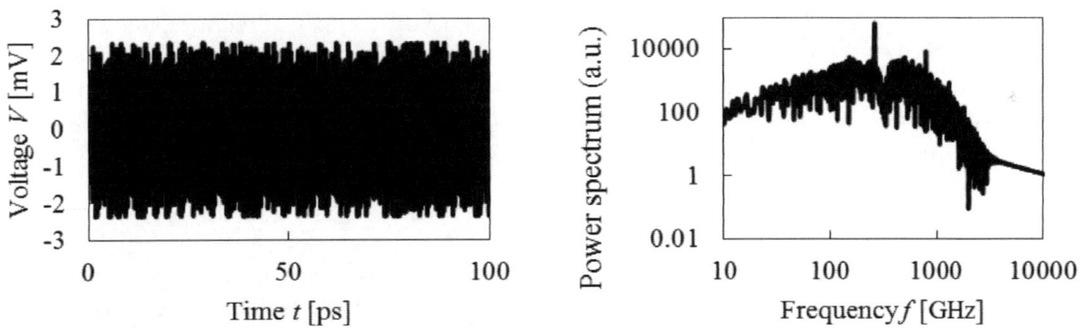

Second range $\lambda = 0.063205$

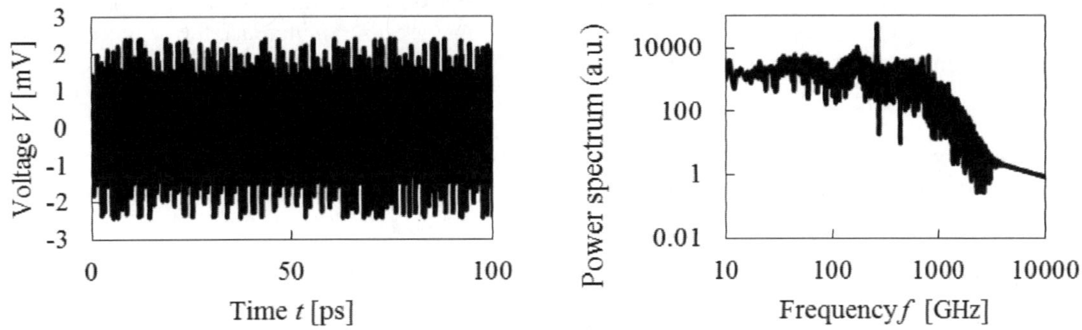

Third range $\lambda = 0.031892$

Figure 5. Voltage waveforms and power spectra.

clear chaos regions were found in Fig. 3, we defined Δf as a frequency bandwidth at the resistance of 2 Ω. From the data of first range, second range, third range in Fig.3, Δf were 230-309, 188-303, and 174-293 GHz, respectively. As larger the capacitance value was, as broader the frequency bandwidth was, which indicated that a broad frequency margin would be utilized for a real random number generation. In addition, it was also found that increase of the capacity value made the chaos occurrence region shift to a lower frequency region, which indicated that we can use electromagnetic waves of relatively available GHz order.

2.3. Relationship between a magnitude of Lyapunov exponent and white noise
In general, if a Lyapunov exponent is positive, that is, the system is in the chaos state, the signal should be irregular. However, in some case, power spectra of waveforms may not be white-noise like. As the magnitude of the Lyapunov exponent might affect the power spectrum, we investigated an influence of the Lyapunov exponent for the voltage waveform and the power spectrum. Since it is difficult to set the capacitance to extremely large values for the actual Josephson junction, we investigated the influence only when the capacitance value is 246.7 fF. In order to clarify the influence of the Lyapunov exponent, a range from the maximum value 0.095696 to the minimum value 0.000003 of the Lyapunov exponent, which are obtained in our simulation, was equally divided into three. A first range is $0.063795 < \lambda \leq 0.095696$, a second range is $0.031898 < \lambda \leq 0.063795$, and a third range is $0.000003 < \lambda \leq 0.031898$. Figure 4 represents mappings of the regions of the chaos generation at the capacitance of 246.7 fF. Points in each map were plotted for the first, second, and third ranges. It was found that the region having a large Lyapunov exponent exists on the high frequency side as compared with the region having a small one. In addition, the voltage waveforms and the power spectra at the largest Lyapunov exponent λ = 0.095696(first range), 0.063205(second range), 0.031892(third range) in each range were calculated with the parameters shown in the Table 2., and are shown in Fig. 5. Although there were no clear significant differences in appearance in the waveforms due to differences of the Lyapunov exponents, clear differences below the irradiation frequency of 265, 265, 267 GHz was found in the power spectra. In the power spectrum of the second range, the power was not flat below the irradiation frequency. In the power spectrum of the third range, a dip around 100 GHz was observed. On the other hand, it was considered that the power spectrum of the first range was white noise. Therefore, we found that chaos can be generated with the available parameters in actual elements by adjusting element parameters

3. Conclusion and discussion
We described the simulations of chaos generation by using superconducting Josephson junctions in order to create a circuit to generate chaos. It was found that the range of irradiation frequency, in which the chaos generate, can be broaden by increase of the capacitance of the Josephson junction. However, chaos that can be used as white noise is a region with a large Lyapunov exponent. Therefore, when creating an actual element, it is necessary to operate in the region where the Lyapunov exponent having a large value concentrates. As a future schedule, we will consider how to increase the capacitance value in the actual element. Moreover, we plan to create the chaos generation circuit for the actual element.

Acknowledgement
This work is supported by the Telecommunications Advancement Foundation.

References
[1] Stinson D R 1995 *Cryptography: Theory and Practice* (CRC Press, Boca Raton, FL).
[2] Metropolis N and Ulam S 1949 *J. Am. Stat. Assoc.* **44** 335
[3] Asmussen S and Glynn P W 2007 *Stochastic Simulation:Algorithms and Analysis* (Springer-Verlag, New York)
[4] Holman W T, Connelly J A and Dowlatabadi A B 1997 *IEEE Trans. Circuits Syst. I* **44** 521

IOP Conf. Series: Journal of Physics: Conf. Series **871** (2017) 012079 doi:10.1088/1742-6596/871/1/012079

[5] Isida M and Ikeda Y 1956 *Ann. Inst. Stat. Math.* **8** 119
[6] Stefanov A, Gisin N, Guinnard O, Guinnard L and Zbinden H J 2000 *J. Mod. Opt.* **47** 595
[7] Jennewein T, Achleitner U, Weihs G, Weinfurter H and Zeilinger A 2000 *Rev. Sci. Instrum.* **71** 1675
[8] Dynes J F, Yuan Z L, Sharpe A W and Shields A J 2008 *Appl. Phys. Lett.* **93** 031109
[9] http://www.idquantique.com/
[10] Reidler I, Aviad Y, Rosenbluh M and Kanter I 2009 *Phys. Rev. Lett.* **103** 024102
[11] Braiman Y and Goldhirsch I 1991 *Phys. Rev. Lett.* **66** 2545
[12] Bartuccelli M, Christiansen P L, Pedersen N F and Soerensen M P 1986 *Phys. Rev. B* **33** 4686
[13] Abraham, Atkin I L and Wilson A 1999 *IEEE Trans. Appl. Supercond.* **9** 4166
[14] Kautz R L 1999 *J. Appl. Phys.* **86** 5794
[15] González J A, Reyes L I, Suarez J J, Guerrero L E and Gutiérrez G 2002 *Phy. Lett. A* **295(1)** 25-34
[16] Shimakage H and Tamura Y 2015 *IEEE Trans. Appl. Supercond.* Vol. 25 No. 3 1800604

IOP Conf. Series: Journal of Physics: Conf. Series **871** (2017) 012080 doi:10.1088/1742-6596/871/1/012080

Evaluation of detectable angle of mid-infrared slot antennas

R Obara[1], J Horikawa[2], H Shimakage[1] and A Kawakami[3]

[1] Graduate School of Science and Engineering, Ibaraki University, Hitachi, Ibaraki 316-8511, Japan
[2] National Institute of technology, Fukui College, Sabae, Fukui 916-0064, Japan
[3] National Institute of Information and Communications Technology, Kobe, Hyogo 651-2492, Japan

E-mail: 15nm605g@vc.ibaraki.ac.jp

Abstract. For evaluations of a mid-infrared (MIR) detectors with antenna, we constructed an angular dependence measurement system of the antenna properties. The fabricated MIR detector consisted of twin slot antennas and a bolometer. The area of the slot antennas was designed to be 2.6 × 0.2 μm^2 as to resonate at 61 THz, and they were located parallel and separated 1.6 μm each other. The bolometer was fabricated using by a 7.0-nm thick NbN thin film, and located at the center of the twin antennas. We measured polarization angle dependence and directivity, and showed that the MIR antennas have polarization dependence and directivity like radiofrequency antennas.

1. Introduction

Applications using infrared lights are useful for science and engineering technologies. For example, an infrared spectroscopy is important application to analyze any kinds of molecules. An infrared light is emitted from all of materials so that it has been used for a thermography and a night vision camera. They are the reasons why high sensitive mid-infrared (MIR) detectors are required for a security device, a mobility radar, and an astronomy. In particular, explosives such as a triacetone triperoxide (TATP) have strong absorption features in the MIR region. As the TATP is easy to synthesize, it would be very dangerous material so that MIR detectors detecting TATPs are useful for a security system at the airport [1]. Also, a MIR radar will be able to improve resolution of detectors because the MIR wavelength is shorter than millimeter and sub-millimeter wavelength. For example, a millimeter wave (94 GHz) and MIR (28.3 THz) dual band detector has been reported, and each detector play roles in the permeability of smoke and the high resolution, respectively [2]. Thus, MIR detectors are required to have fast and high-sensitive characteristics.

Detectors such as a photomultiplier tube (PMT), a HgCdTe photoconductor (MCT), an InSb photovoltaic detector, and so on, are utilized as the MIR detectors with fast response speed, and their detection principles are based on the particle nature as photon. In general, such detectors usually suffer from a tradeoff between sensitivity and speed because the sensitivity depends on the detection area, and their electrostatic capacitance tends to increase with the area. Thus, it is difficult to realize the MIR detectors with both of fast and high-sensitive characteristics. In order to realize fast detectors without sacrificing detection area, we have proposed to use antennas for detecting MIR waves [3]. It means that the MIR power was collected by using the wave property of MIR instead of particle property. Our group has already reported hot electron bolometer (HEB) mixers with high sensitivity and fast responsiveness in far infrared (FIR) region [4]. Although our objective is to apply the antenna technology to MIR region, MIR wavelength is several microns so that it is more difficult to fabricate MIR antennas compared with far-infrared antennas. However, recent rapid progress of e-beam lithography technique has made it possible to fabricate extremely small pattern, which means that it can be applied to MIR antenna fabrications. Thus, we have been tried to study HEB mixers [5].

Up to now, studies to apply MIR antennas to detectors has been reported; such as a bolometer coupled with 10×10 dipole array antenna [6], bowtie antennas [7], square-spiral antennas [8], and so on. In our previous work, we fabricated MIR detectors with nano-slot antenna, and have measured its antenna properties using FTIR [3]. Here, in our experiment, we used the HEB as a detector. Our results indicated that the nano-slot antenna has been properly functioned as the ordinary antenna. Furthermore, we evaluated its response speed of 0.3 ns by the experiment of MIR pulses irradiation with approximately 150 photons per 1 ns [9]. Although we confirmed the antenna function as described above, we haven't reported the antenna directionality yet. As the antenna's fundamental property, the measurement of the directionality is requisite. In this paper, we report on the evaluation of detectable angle of MIR twin slot antennas coupled with a superconducting bolometer.

2. Structure and Setup

2.1. Structure of MIR detector

We fabricated MIR detectors, which consist of twin slot antennas and bolometers. Here, we will describe a fabrication process of the MIR detector. A NbN-MgO-NbN (3.5-0.55-7.0 nm) film was evaporated on the MgO substrate and patterned by using e-beam lithography. In order to separate the functions for receiving MIR waves and detecting MIR power, twin slot antennas and a bolometer were fabricated, respectively. The slot antennas were designed 2.6 × 0.2 μm^2 as to resonate at 61 THz. Here, the frequency of the resonation was calculated using the dielectric constant of a MgO substrate is approximately 2.7 at 61 THz [10]. The slot antennas were located parallel and separated at intervals of 1.6 μm each other (Fig. 1). The bolometers, each of which was fabricated as a 7.0-nm thick NbN thin film (Fig. 2), were located at the center of the twin slot antennas. The bolometer was designed 0.2 × 0.2 μm^2 which is smaller than a diffraction limit of MIR. The critical temperature (T_C) of the NbN bolometer is approximately 13 K so that the detector has to be operated at cryogenic temperature. The antennas and the bolometer were connected using by a coplanar waveguide (CPW). The CPW has 0.2 μm line width and 0.1 μm gap width. In addition, we fabricated a choke filter on the CPW to filter out the MIR waves to electrodes, and the dc bias cuts between both ends of the bolometer.

2.2. Experimental Setup

For evaluating the angle dependence of its sensitivities, we set up directionality evaluation system (Fig. 3). The detector was cooled to 4 K in a Gifford-McMahon (GM) refrigerator, cooling ability of which was 0.1 W. The refrigerator was on a rotatable stage and the detector was mounted at the center of the rotation. MIR laser was introduced through a CaF_2 window, which transmitted MIR laser power over 90 %. A 4.89-μm quantum cascade laser (QCL) was used as a MIR irradiation source. It was mounted on the X, Y, Z position adjustable stage, and inclined 45° with respect to the X axis. Two polarizers in front of QCL were used. One of the polarizers was positioned near the refrigerator so as to adjust the laser polarization, and another polarizer was used for modulating laser power. For measuring I-V curves, DC source was operated with the ± 0.2 mA range. At first, we measured critical current (I_C) of the bolometer and laser power dependence of I_C so as to check the alignment of the laser spot position. Secondly, we measured the I_C and the angle of polarized plane dependence so that we have confirmed whether the detector with antenna detect wave property of MIR. Finally, a directivity of the detector was measured when it mounted horizontal and vertical to X axis (Fig. 4). We measured angle dependences of the detector sensitivity by rotating the refrigerator. We assumed that the directivity would be symmetry because the antenna and bolometer configurations are also symmetry [11]. Thus, we measured antenna radiation pattern from 0° to 80° . We will show the results in next section.

IOP Conf. Series: Journal of Physics: Conf. Series **871** (2017) 012080 doi:10.1088/1742-6596/871/1/012080

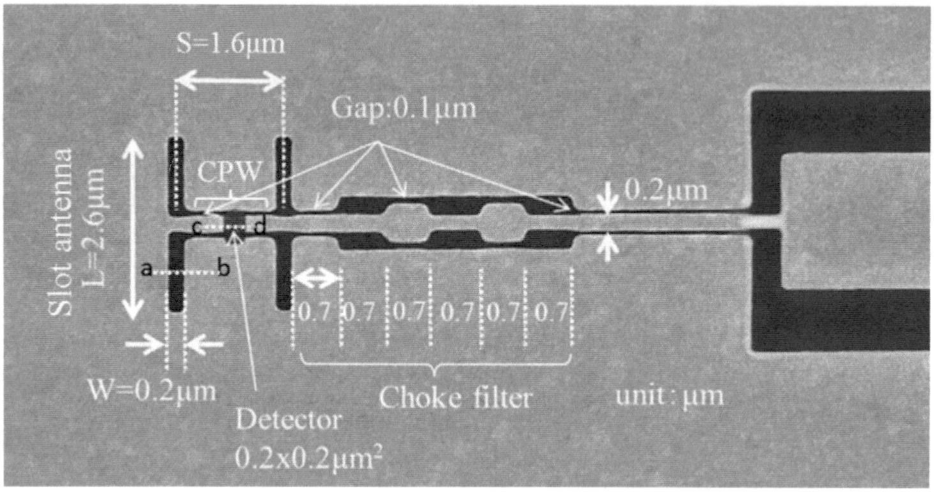

Figure 1. SEM image of MIR detector.

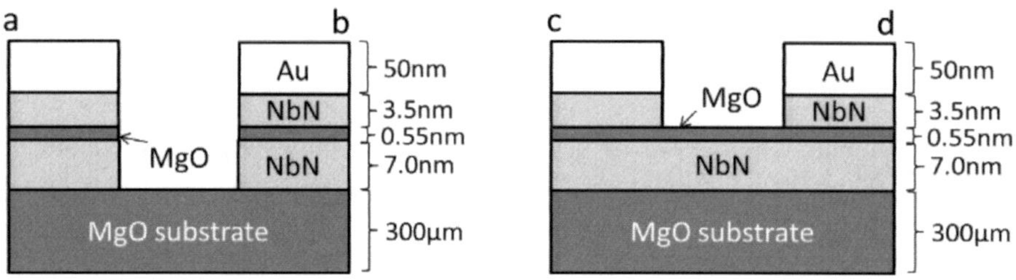

Figure 2. Cross section of the MIR antenna (Left) and bolometer (Right).

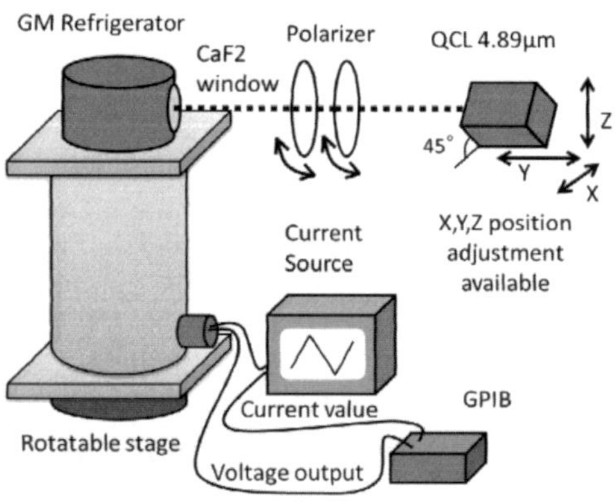

Figure 3. Evaluation system of the MIR detector.

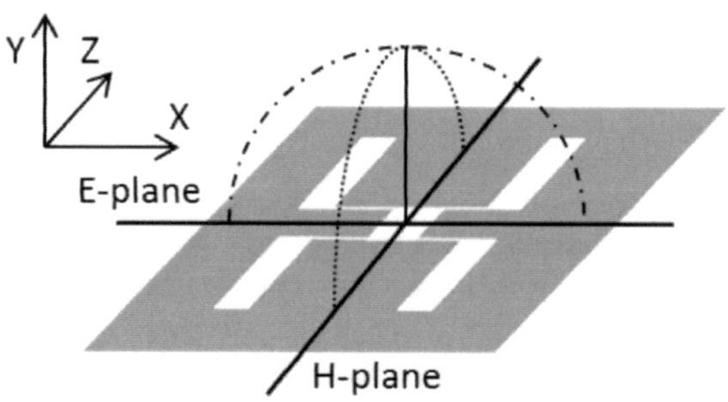

Figure 4. E-plane and H-plane of the twin slot antenna.

3. Results

Figure 5 shows the I-V characteristics of the NbN bolometer without the laser radiation, and the I_C was observed of 0.135 mA. Although the measurement temperature of the bolometer was not monitored directly, the T_C of the bare NbN film was checked to be 12.6 K by another measurement. Figure 6 shows the MIR laser power dependence of NbN-bolometer's I_C. Here, the laser was operated at CW mode, and the power was scanned from 0 to 50 mW. We observed the I_C approximately decreased as the laser power was increased, and the I_C was decreased to 0.098 mA at the maximum laser power. The dependence indicated that the laser was properly focused on the bolometer. Here, although the I_C was not decreased monotonous, we guess that the coupling of the MIR antenna and the bolometer changed by laser power.

Next, we observed the polarization dependence of the I_C to confirm the function of the antenna. The laser power was kept at 50 mW by adjusting two polarized filters in front of the CaF_2 window. The result of the polarization dependence was shown in Fig. 7. The angle relationships of the polarization and the antennas were shown in insets of Fig. 7. That is, when the degree of polarization was 0°, the electrical fields of the MIR were parallel to the slot antennas. When the degree of polarization was 90°, the electrical fields of the MIR were vertical to the antennas and the resonance was occurred. We designed the antenna configuration to resonate at 61 THz. The result showed the clear suppression of the I_C at the degree of polarization of 90°. The result indicated that the detection mechanism was not by the nature of the particle, but by the nature of MIR waves, which means the MIR slot antenna was properly functioned as the antenna.

Finally, we measured the angle dependence of the I_C depression ratio $\left(1-\frac{I_C}{I_{Cmax}}\right)$. The results with the polarization angle 0° and 90° are shown in Fig. 8. The clear angular dependences were observed, and they were very like the dependence of that of a radiofrequency antenna. When the detector was positioned at the parallel to the E-plane, the I_C depression ratio was around 15 % at 0°, which was like a dip characteristic. Each antenna of the twin antennas received signals, and the signals transmitted through the transmission line to the detector. The result showed that the signals from twin antenna were interfered and cancelled. We think that the dip characteristic of the E-plane data in Fig. 8 was attributed to the interference. In the measurement, the CPW and bolometer impedances didn't appropriately match, so it is expected that the signals didn't absorbed in the bolometer. To estimate the exact behaviour, the dielectric constant of the MgO substrate must be known accurately. As the dielectric constant of MgO was reported of around 1.65 at the wavelength of 4.89 μm [10], the length of the transmission line was almost coincident with the half wavelength of the waveguide. When we measured the angular dependence at the parallel to the H-plane, we observed the strong directivity, but no dips were observed. It was because the phase of the MIR waves incident on the two antennas are the same regardless of the incident angle. We thought that the signal in the E-plane data in Fig. 8 have unnecessary signal from the reflection of the back side of the MgO substrate. Thus, to remove the reflection of the back side, we set up an absorber, which has 98.6 % absorption ratio at 61 THz, behind the substrate. The result of the angle dependence was shown in Fig. 9. This result was roughly in agreement with the result shown in Fig. 8. Because of the existence of the impedance mismatching between CPW and bolometer, the signal received by one antenna will go through the bolometer and interfere with the signal received at the other antenna. When the incident angle is 0°, if the transmission line is in an ideal state, the signal received by the bolometer will be twice that of a single antenna. However, we observed the dip at 0° and we thought the result is due to impedance mismatching.

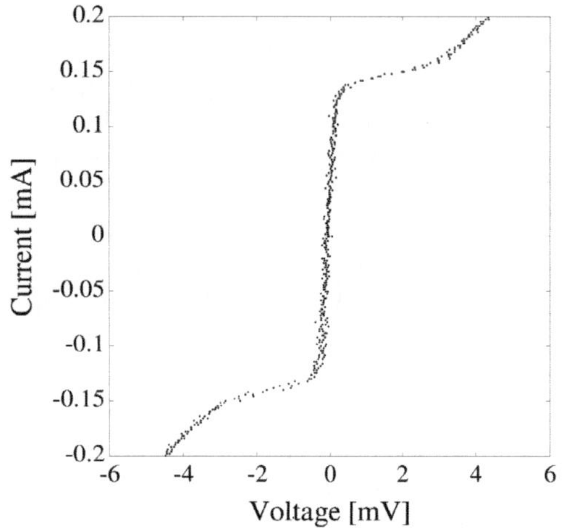

Figure 5. I-V characteristic of the bolometer without laser radiation

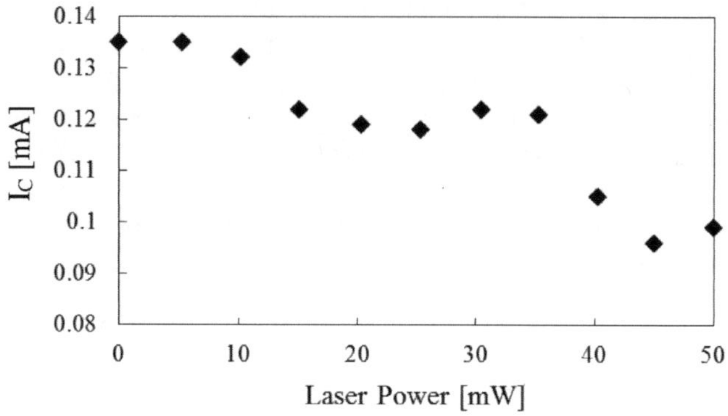

Figure 6. Laser power dependence of I_C.

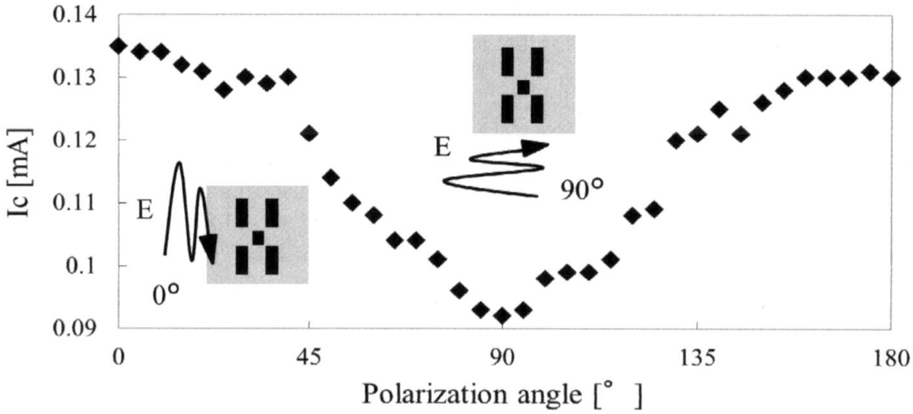

Figure 7. Polarization angle dependence of I_C

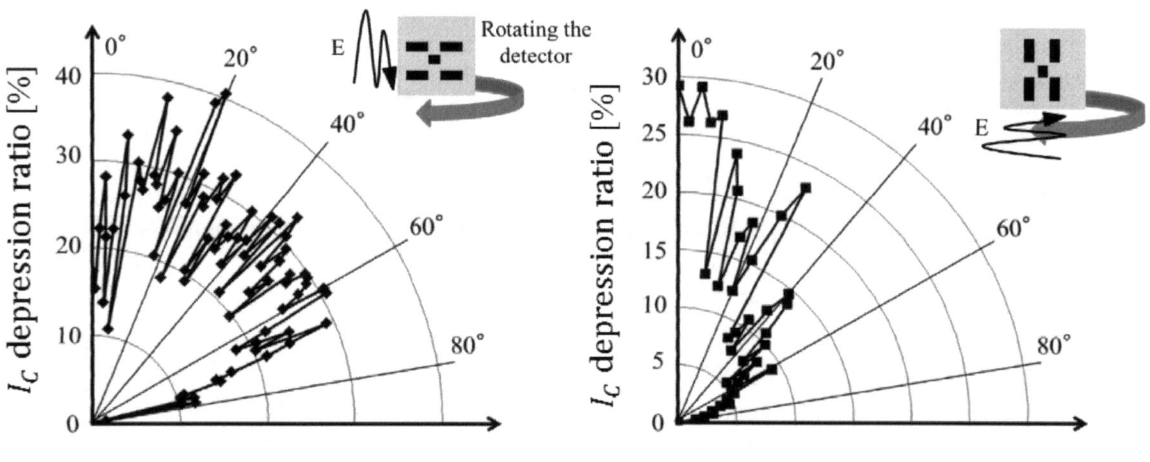

Figure 8. E-plane (Left) and H-plane (Right) antenna pattern of the twin slot antennas

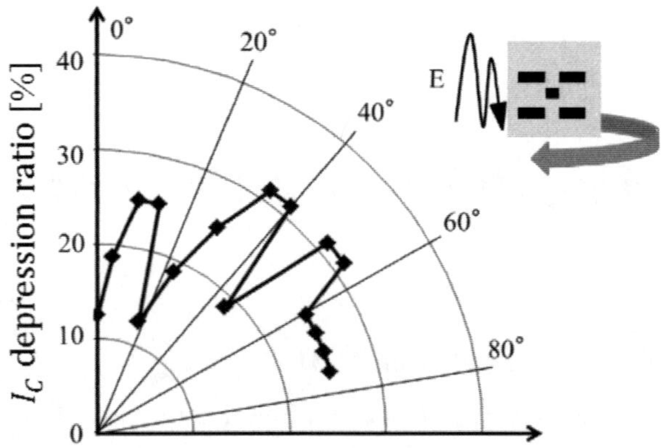

Figure 9. E-plane antenna pattern of the twin slot antennas without reflectance

4. Conclusion

We constructed the directivity evaluation system for the MIR detectors with nano-antenna. We measured characteristics of the MIR detector with twin slot antenna and NbN bolometer. The results showed that the very small MIR antennas have polarization dependence and directivity. Thus, it operated like a radiofrequency antenna. The results enable us to fabricate phased-array antenna by reference to the results and improve its sensitivity. The MIR phased-array antenna will be established the large detection area and high sensitivity by gathering many antennas signal. In the future, we adapt this technique to development both of high-speed and sensitive MIR detector for security system and advanced sensing technique.

References

[1] Bauer C, Willer U, Lewicki R, Pohlkotter A, Kosterev A, Kosynkin D, Tittel F K and Schade W 2009 *3rd International Workshop on Infrared Plasma Spectroscopy, Journal of Physics: Conference Series* **157** 012002

[2] Abdel-Rahman M R, Monacelli B, Weeks A R, Zummo G and Boreman G D 2005 *Optical Engineering* vol. 44(6) 066401

[3] Horikawa J, Kawakami A, Hyodo M, Tanaka S, Takada M and Shimakage H 2014 *Infrared Physics & Technology* vol. 67 pp. 21-24

[4] Irimajiri Y, Kawakami A, Morohashi I, Kumagai M, Sekine N, Nagano S, Ochiai S, Tanaka S, Hanado Y, Uzawa Y and Hosako I 2015 *IEEE transactions on terahertz science and technology* vol. 5 No. 6 pp. 1154-1159

[5] Kawakami A, Shimakage H, Horikawa J, Hyodo M, Saito S, Tanaka S and Uzawa Y 2016 *IEEE Transactions on Applied Superconductivity* vol. 27 Issue. 4 2300105

[6] González F J, Gritz M A, Fumeaux C and Boreman G D 2002 *International Journal of Infrared and Millimeter Waves* vol. 23 Issue. 5 pp.785-797

[7] Blanchard R, Boriskina S V, Genevet P, Kats M A, Tetienne J P, Yu N, Scully M O, Negro L D and Capasso F 2011 *Optics Express* vol. 19 No. 22 pp. 22113-22124

[8] González F J, Ilic B, Alda J and Boreman G D 2005 *IEEE Journal of Selected topics in Quantum Electronics* vol. 11 No. 1 pp. 117-120

[9] Kawakami A, Shimakage H, Horikawa J, Hyodo M, Saito S, Tanaka S and Uzawa Y 2016 *1st Asia ICMC and CSSJ 50th Anniversary Conference* 1D-p05

[10] Stephens R E and Malitson I H 1952 *Journal of Research of the National Bureau of Standards* vol. 49, No. 4 pp. 249-252

[11] Button K J 1983 *Academic Press, Inc.* vol. 10 pp. 8

IOP Conf. Series: Journal of Physics: Conf. Series **871** (2017) 012081 doi:10.1088/1742-6596/871/1/012081

Calculations of superconducting parametric amplifiers performances

T Goto[1], M Takeda[2], S Saito[3], H Shimakage[1]

[1] Graduate School of Science and Engineering, Ibaraki University, Hitachi, Ibaraki 316-8511, Japan
[2] Graduate School of Integrated Science and Technology, Shizuoka University, Hamamatsu, Shizuoka 432-8011, Japan
[3] Advanced ICT Research Institute, NICT, Koganei, Tokyo 184-8795, Japan

e-mail : 16nm612y@vc.ibaraki.ac.jp

Abstract. A superconducting parametric amplifier is an electromagnetic wave amplifier with high-quality characteristics such as a wide bandwidth, an extremely low noise, and a high dynamic range. In this paper, we report on the estimations of a YBCO superconducting parametric amplifier characteristic. The YBCO thin films were deposited on an MgO substrate by a pulsed laser deposition method. Based on the measured YBCO thin film parameters, theoretical calculations were implemented for evaluations of kinetic inductance nonlinearities and parametric gains. The nonlinearity of the YBCO thin film was estimated to be stronger than a single crystal NbTiN thin film. It is indicated that the YBCO parametric amplifier has a potential to be realized the amplifier with the high parametric gain. It is also expected that it could be operated in the range of the high frequency band, at the high temperature, and low applied current.

1. Introduction

In the fields of cryogenic electronics, there are some applications, such as superconducting qubit experiments for the realization of a quantum computer [1,2], a microwave kinetic inductance detectors (MKIDs) for radio astronomical observation [3,4], and so on. For such applications, a realization of a microwave amplifier with quantum noise limit, low power and broadband operations is desired [5]. However, in general, it is difficult to simultaneously satisfy low noise and broadband characteristics. For example, the HEMT amplifiers have a wideband characteristic, but have relatively high noise temperatures. The superconducting Josephson amplifiers have noise performances approaching the quantum limit, but the bandwidth is narrow [6]. On the other hand, a superconducting parametric amplifier was recently reported [7], and it has been shown that it has high-quality characteristics, such as a broad bandwidth, an extremely low-noise property, and a high dynamic range.

The parametric amplification process in the superconducting parametric amplifier uses a kinetic inductance (L_k), which has a nonlinear property. Since a total inductance (L_T) of the superconducting thin films is the sum of a magnetic inductance (L_m) and L_k, it is essential that L_k is dominant compared with L_m. L_k nonlinearly changes according to the applied current I as follows:

$$L_k(I) \approx L_k(0)\left[1 + \left(\frac{I}{I^*}\right)^2\right] \tag{1}$$

where I^* is same order as the critical current (I_C) of the superconductors. Its amplification process is a degenerate four-wave mixing, which is also utilized in an optical fiber parametric amplification process, but the phase matching method is different from that of the optical fiber. In the parametric amplifier, phase matching is realized by periodically changing the line impedance of a superconducting transmission line [7].

As the kinetic inductance nonlinearity per unit length is not so much, the length of the transmission line has to be long to obtain a high parametric gain. When a 2.2 m long coplanar waveguide (CPW) by using a polycrystalline NbTiN thin film was fabricated, the parametric gain was 15 dB in the microwave band [8]. We also have been studying the kinetic inductance nonlinearity of a single-crystal NbTiN and the polycrystalline NbTiN thin films to realize the superconducting parametric amplifier [9]. We have already reported that the nonlinearity of the single-crystal NbTiN is three times as much as that of the polycrystalline NbTiN. We are continuing the further study by using high temperature superconductors in order to operate it at the higher temperature. In this paper, we report on calculation results of the kinetic inductance nonlinearity for the YBCO thin films.

2. YBCO thin film fabrication and superconducting properties

The YBCO thin films were deposited on the MgO substrate by a pulsed laser deposition (PLD) method and treated by annealing in oxygen atmosphere. In the deposition process, the distance between the target and the substrate was 3.5 cm, a YAG laser energy density was 1.26 J/cm^2, a repetition frequency was 10 Hz, an oxygen pressure in the chamber was 400 mTorr, and a substrate temperature was 820 °C. The annealing treatment after the deposition was carried out for 60 minutes in the oxygen pressure of 475 Torr and the temperature of 430 °C. The YBCO thin film thickness was 180 nm. The thin film was processed into narrow lines with a line width of 6 μm and a line length of 1 mm by a photolithography and a dry etching. The superconducting critical temperature (T_C) and the critical current density (J_C) with a 6 μm wide YBCO line were measured by a four-terminal measurement. Figure 1(a) shows the temperature-resistivity characteristics of the YBCO thin film with bias currents of 0.1 to 15.0 mA, and T_C was obtained of 78 K for 0.1mA bias current. Next, we extracted the J_C values from the data of Fig. 1(a). Figure 1(b) shows the J_C characteristics of the YBCO thin film, and J_C at 77 K was obtained of 1.0×10^5 A/cm^2. The value of J_C is smaller than that of the general one. However, low J_C is suitable for parametric amplifier operations because of low drive current operation. From the next section, we will simulate the performance of the YBCO parametric amplifiers using the obtained parameters.

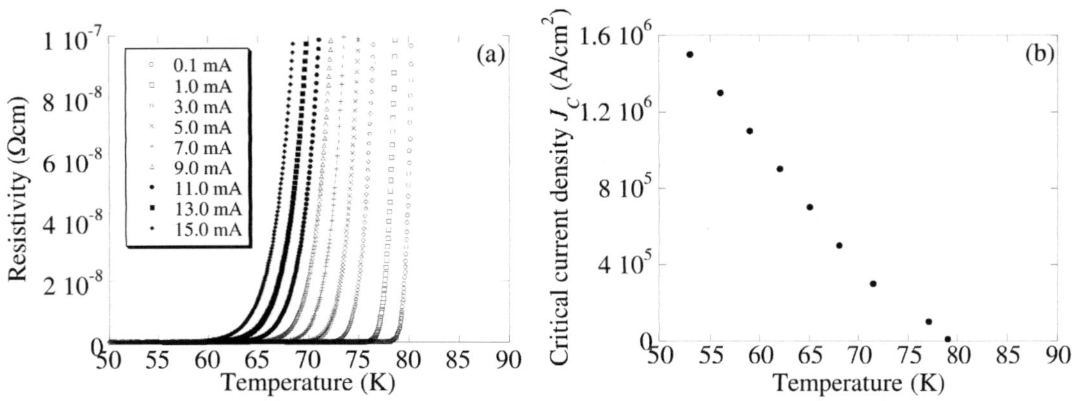

Figure 1. (a) Temperature-resistivity characteristics of the patterned YBCO thin film with 6 μm width. (b) Temperature-critical current density characteristics of the YBCO thin film.

3. Superconducting parametric amplifier using YBCO thin film CPW

3.1. Kinetic inductance nonlinearity

In this section, we calculate the kinetic inductance nonlinearity in the YBCO thin film CPW. The L_T per unit length in superconducting CPW is expressed as

$$L_T = L_m + L_k. \tag{2}$$

When the film thickness d is thinner than the penetration depth λ ($d < 2\lambda$), both of the inductances are expressed by conformal mapping by TEM mode approximation [10].

$$L_m = \frac{\mu_0}{4}\frac{K(k')}{K(k)} \tag{3}$$

$$L_k = \mu_0 \frac{\lambda^2}{dw} g(s,w,d) \tag{4}$$

Where μ_0 is the permeability of vacuum, d is the film thickness, w is the width of signal line, s is the gap width between the signal line and the ground line, $K(k)$ is the complete elliptic integral of first kind and parameters k and k' is

$$k = \frac{w}{w + 2s},\ k' = \sqrt{1 - k^2}. \tag{5}$$

Moreover, $g(s,w,d)$ is a geometrical factor of CPW given as:

$$g(s,w,d) = \frac{1}{2k^2K(k)^2}\left[-\ln\frac{d}{4w} - \frac{w}{w+2s}\ln\frac{d}{4(w+2s)} + \frac{2(w+s)}{w+2s}\ln\frac{s}{w+s}\right] \tag{6}$$

From the above equations, the magnetic inductance depends only on the line pattern, whereas the kinetic inductance depends on both of the line pattern and the penetration depth. From Eq. (2), in order for the nonlinearity of the kinetic inductance to appear conspicuously in the superconducting transmission line, the kinetic inductance have to be dominant in the total inductance. Therefore, we examined the line pattern dependences of the kinetic inductance. Figure 2(a) and 2(b) show the film thickness and the line pattern dependences of the kinetic inductance, respectively. The kinetic inductance was normalized with the total inductance. From these results, the kinetic inductance becomes dominant by thinning the film thickness and narrowing the line width.

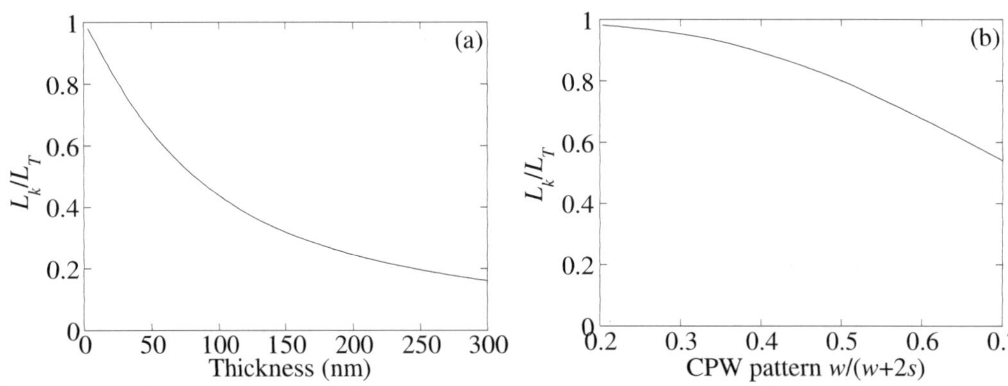

Figure 2. Kinetic inductance in superconducting CPW. The kinetic inductance was normalized with the total inductance. The vertical axes are (a) the film thickness d dependences (w = 4 μm, s = 2 μm), and (b) the line pattern $w/(w+s)$ dependences (d = 40 nm).

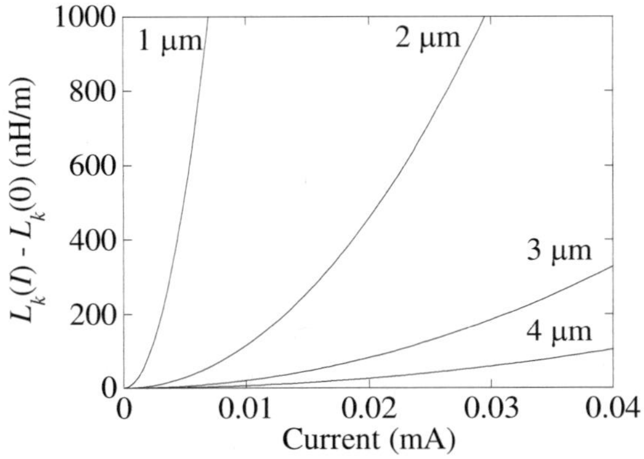

Figure 3. Kinetic inductance variations with respect to applied current.

Next, we calculated the kinetic inductance nonlinearity when the current was applied. In the calculation, the dimensions of the transmission line were set to the film thickness d = 40 nm, the width of signal line w = 1~4 μm, and the gap width between the signal line and the ground line s = 2 μm. In any case, since $L_k/L_T > 0.7$ is satisfied, the kinetic inductance is dominant in the total inductance. When the external current is applied to the superconducting transmission line, the kinetic inductance would change as shown in Eq. (1). Figure 3 shows the calculation results of the kinetic inductance variation $L_k(I) - L_k(0)$ in the existence of the applied current. Here, since $I^* \approx I_C$, calculations were implemented with assuming that I^* in Eq. (1) was I_C of the YBCO thin film. The I_C was calculated from J_C obtained in Section 2. The YBCO penetration depth λ at zero-temperature was set to 135 nm from the data of the reference [11], and an operating temperature was set to 77 K. Here, the longer the penetration depth, the stronger the kinetic inductance nonlinearity. From the Fig. 3, it is indicated that the reduction of the width of the signal line made, it possible to obtain a large change in the kinetic inductance with respect to the applied current. We fitted the curves in Fig. 3 to a polynomial, and derived their quadratic coefficient, which was defined as χ. In the case of the line width of 4 μm, the quadratic coefficient χ was 6.504×10^{-2} H/A^{2}m. In the previous study, the quadratic coefficients of the single crystal NbTiN thin film and the polycrystalline NbTiN thin film of the same pattern were 3.436 $\times 10^{-4}$ H/A^{2}m and 1.202×10^{-4} H/A^{2}m, respectively. Therefore, the nonlinearity of the YBCO thin film was stronger than the single crystal NbTiN thin film. That is, if the YBCO thin film is used, the high parametric gain can be obtained at the high temperature and a small applied current.

3.2. Calculation of amplification

The amplification process of the superconducting parametric amplifier is the degenerated four-wave mixing. Therefore, the amplification gain can be calculated by solving a coupled wave equation of the optical fiber parametric amplifier [12]. We assumed $2f_p = f_s + f_i$ where f_p, f_s and f_i, are a pump frequency, a signal frequency and an idler frequency, respectively. In the superconducting transmission line,

$$\Delta\beta = \beta(f_s) + \beta(f_i) - 2\beta(f_p) \tag{7}$$

is established in the linear dispersion relation, where $\beta(f)$ is a propagation constant. The parametric amplification gain derived from the coupled wave equation is

$$G_s = 1 + (\Delta\theta)^2. \tag{8}$$

At this point, $\Delta\theta$ is the amount of a nonlinear phase shift caused by the pump wave current.

$$\begin{aligned}\Delta\theta &= [\beta(0) - \beta(I)]l \\ &= \omega\left[\sqrt{L_T(0)C} - \sqrt{L_T(I)C}\right]l\end{aligned} \tag{9}$$

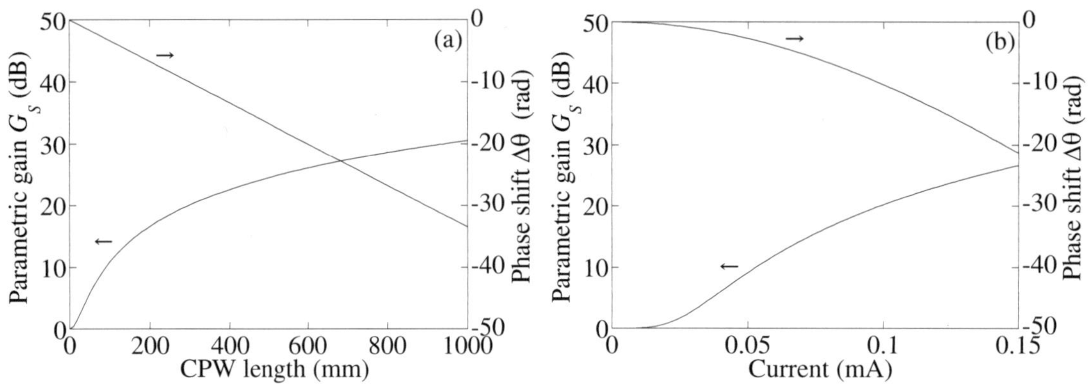

Figure 4. (a) Parametric gain and nonlinear phase shift as a function of the line length. Applied current was set to be 0.08 mA. (b) Parametric gain and nonlinear phase shift as a function of pump current. Line length was set to be 200 mm.

where l is the line length of the CPW. We calculated the amplification gain using the above equations. The CPW patterns were set to w = 4 μm, s = 2 μm and d = 40 nm. We assumed f_p = 4 GHz and $f_s \approx f_p$. Figure 4(a) is the parametric gain G_s and the nonlinear phase shift $\Delta\theta$ as a function of the line length l when the pump wave current is 0.08 mA. The longer the line length is, the larger the phase shift is, which indicated that the parametric gain also increases. In addition, we described that the YBCO thin film CPW has strong nonlinearity of the kinetic inductance in section 3.1, high parametric gain can be expected even with a short line length. Figure 4(b) shows the parametric gain as a function of pump current. The CPW length was set to be 200 mm. It is possible to realize high parametric gain even with a small pump current. As a result, a large parametric gain was expected for the YBCO, even if their configurations were with shorter line lengths and smaller pump current.

4. Conclusion

In order to realize the superconducting parametric amplifier, we calculated the kinetic inductance nonlinearity and the parametric gain in the YBCO thin film CPW. The kinetic inductance nonlinearity of YBCO was stronger than the single crystal NbTiN having strong nonlinearity. Therefore, the YBCO has a potential to be realized the high parametric gain, and it can be expected to operate in the range of the high frequency band, at the high temperature, and the small applied current. Moreover, from the fact that YBCO thin film CPW's kinetic inductance nonlinearity is strong, the use of the YBCO thin film CPW is considered to be useful for the realization of a other kinetic inductance devices.

Although some calculations were performed assuming that the phase matching condition was satisfied, the transmission line configurations have to be properly designed to realize the phase matching in the real operation. We are also studying for the realization of the phase matching for YBCO transmission lines, the results will be described elsewhere in the future. Moreover, we are planning to examine the nonlinearity and amplification gain in CPW more accurately for the parametric amplifier, and proceed the fabrication of superconducting CPW element.

Acknowledgment
This work was supported by JSPS KAKENHI Grant Number 26420305.

References

[1] Clarke J and Wilhelm F K 2008 *Nature* **453** 817.
[2] Lupascu A, Saito S, Picot T, de Groot P C, Harmans C J P M and Mooij J E 2007 *Nature. Phys* **3** 1031.
[3] Day P K, LeDuc H G, Mazin B A, Vayonakis A and Zmuidzinas J 2003 *Nature* **425** 817.

IOP Conf. Series: Journal of Physics: Conf. Series **871** (2017) 012081 doi:10.1088/1742-6596/871/1/012081

[4] Baselmans J 2012 *J. Low. Temp. Phys* **167** 292.
[5] Clerk A A, Devoret M H, Girvin S M, Marquardt F and Schoelkopf R J 2010 *Rev. Mod. Phys* **82** 1155.
[6] Muck M and Welzel C 2003 *Appl. Phys. Lett* **82** 3266.
[7] Eom B H, Day P K, LeDuc H G and Zmuidzinas J 2012 *Nature. Phys* **8** 623.
[8] Bockstiegel C, Gao J, Vissere M R, Sandberg M, Chaudhuri S, Sanders A, Vale L R, Irwin K D and Pappas D P 2014 *J. Low. Temp. Phys* **176** 476.
[9] Takeda M, Kojima T, Saito A, Makise K and Shimakage H 2016 *JJAP. Conf. Proc* **4** 011502.
[10] Watanabe K, Yoshida K, Aoki T and Kohjiro S 1994 *Jpn. J. Appl. Phys* **33** 5708.
[11] Pond J M, Carroll K R, Horwitz J H, Chrisey D B, Osofsky M S and Cestone V C 1991 *Appl. Phys. Lett* **59** 3033.
[12] Hansryd J, Andrekson P A, Westlund M, Li J and Hedekvist P O 2002 *IEEE J. Selected Topics in Quantum Electron* **8** 506.

IOP Conf. Series: Journal of Physics: Conf. Series **871** (2017) 012082 doi:10.1088/1742-6596/871/1/012082

$I - V$ characteristics and THz radiation properties of Bi2212 mesas

Akinobu Irie[1], Takashi Watanabe[1], Yoshihiko Yajima[1] and Kazuhiro Yamaki[1]

[1] Department of Electrical and Electronic Systems Engineering, Utsunomiya University, 7-1-2 Yoto, Utsunomiya 321-8585, Japan

E-mail: `iriea@cc.utsunomiya-u.ac.jp`

Abstract. Large mesa structures composed of intrinsic Josephson junctions (IJJs) are expected to be one of a promising source of THz electromagnetic wave. In order to study the collective dynamics in large mesas with and without the influence of self-heating effect, two types of mesa structures are fabricated. One is a double-mesa structure which enables the non-uniform bias current distribution to examine the heat-flow in large mesas. We find that the current lead geometry strongly affects the collective switching of IJJs into resistive state even if the number of IJJs in the mesas is the same. Another is a standard rectangular mesa structure used in the emission experiments. In this mesa, we observe THz radiation from IJJs biased at the low bias region, where the self-heating effect is almost negligible, on inner branches of multiple $I-V$ structures and find that all of observed intense inner-branch emissions correspond to the same cavity resonance mode.

1. Introduction

Intrinsic Josephson-junction (IJJ) stacks in $Bi_2Sr_2CaCu_2O_y$ (Bi2212) are a promising source to emit the intense, continuous, and monochromatic terahertz waves [1, 2, 3]. It is considered that the emission power is characterized by the number N of synchronizing IJJs in the stack and thus straight way to increase the power is to make mesa consisting of a larger number of IJJs. However, the increase in N is known to cause the significant heating problem which results in reduction of voltage across the stack. In addition, one of the reasons that a variety of $I - V$ curves observed for large IJJ stacks may also be due to peculiar heat-generation and heat-flow processes in individual samples because it is clear that such a self-heating induces the spatial variation of critical current and bias current densities. Indeed, at high bias where the $I - V$ curve shows back bending a hot spot has been observed to be formed inside the mesa and it is considered that the hot spot plays a role for synchronization [4, 5, 6, 7]. Furthermore, a IJJ stack with a large number of junctions have complex dynamical behavior due to coupling effect between adjacent IJJs in the stack, so that in a large IJJ stack switching process of individual IJJs into voltage state is also affected by heating generated from them in the resistive state because of their large critical currents.

In this study, we fabricated the IJJ mesas with different configurations of current injection to produce the spatial distribution of bias current density and found that the $I - V$ curves of the mesas were strongly influenced by the current lead structure even for the same number of junctions because of different thermal distribution. Furthermore, we present the observation

IOP Conf. Series: Journal of Physics: Conf. Series **871** (2017) 012082 doi:10.1088/1742-6596/871/1/012082

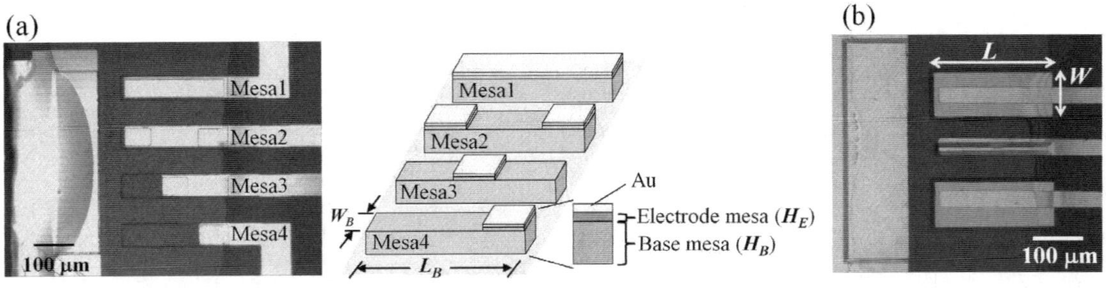

Figure 1. (a) A photograph of four double-mesa structures and schematic view of the sample. Dimensions of each mesa are listed in Table 1. (b) A photograph of the sample used in emission experiments. Lateral dimensions of the mesa are $L = 275$ μm, $W = 111$ μm and its thickness is $H = 0.84$ μm corresponding to the number of junctions of 560.

Table 1. Dimensions of double-mesa structures used in this work. N_B and N_E are the number of IJJs in the base and electrode mesas, respectively.

	Base mesa				Electrode mesa				
Mesa	L_B (μm)	W_B (μm)	H_B (μm)	N_B	L_E (μm)	W_E (μm)	H_E (μm)	N_E	
1	270	60	0.9	600	263	47	0.12	80	
2	270	60	0.9	600	68	47	0.12	80	$\times 2$
3	270	60	0.9	600	68	47	0.12	80	
4	270	60	0.9	600	68	47	0.12	80	

result of THz radiation from the IJJ stack biased at the low bias region, where the self-heating effect is almost negligible, on inner branches of multi-branched $I - V$ curves.

2. Experimental

The Bi2212 single crystals with critical temperatures T_c of $\sim$ 85 K were grown by a conventional melting method [8] and two types of mesa structures were fabricated from them using standard photolithography and Ar ion milling. One is a double-mesa structure which consisting of electrode and base mesas. As shown in Fig. 1(a), four double-mesa structures were formed on the same crystal. Dimensions of each mesa are listed in Table 1. In this case, the bias current is supplied to the base mesa through the electrode mesas. Another is a conventional mesa structure as shown in Fig. 1(b), which was used in emission experiments. Lateral dimensions of the mesa are $L = 275$ μm, $W = 111$ μm and its thickness is $H = 0.84$ μm corresponding to the number of junctions of 560. Detail of the fabrication process were described elsewhere [9].

For both samples, the $I - V$ characteristics were measured by two probe method and in emission experiments the radiation from IJJ stacks was detected in situ by using high sensitive IJJ detector [9]. The contact resistance is subtracted in all $I - V$ characteristics presented here.

3. Results and discussion

3.1. Influence of heat-transfer on $I - V$ characteristics

Figure 2 shows the $I - V$ characteristics at 4.2 and 77 K for four mesas of the double-mesa structures as shown in Fig. 1(a). Due to two probe configuration these $I - V$ curves are a sum

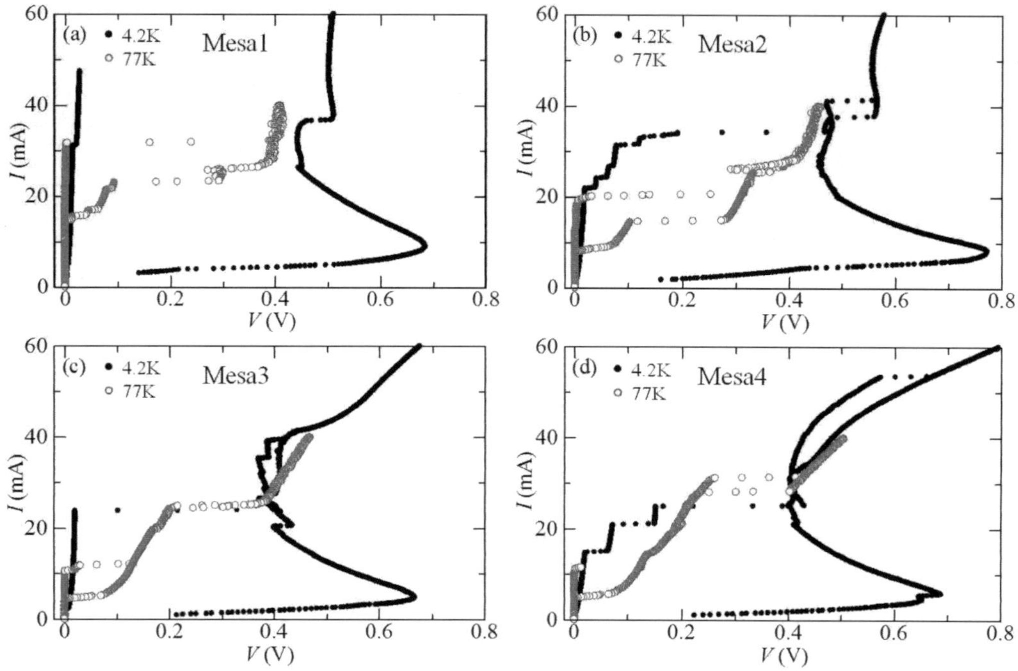

Figure 2. $I-V$ characteristics of double-mesa structures as shown in Fig. 1 at 4.2 and 77 K.

of $I-V$ curves of the base and electrode mesas. At 4.2 K the $I-V$ characteristics of Mesa1 and Mesa3 show that all IJJs in the stack switch into the voltage state simultaneously when the bias currents reach to 47.4 mA for Mesa1 and 23.8 mA for Mesa3, respectively, which correspond to the critical current $I_c^{(E)}$ of the electrode mesa. This collective voltage jump indicates that the temperature of all IJJs in the mesa increased suddenly although for Mesa3 the bias current injection into base mesa is non-uniform. On the other hand, in the case of Mesa2 and Mesa4 there are several stepwise voltage jumps and the simultaneous switching has occurred only partially in the mesas. For these mesas heat-generation starts from both ends or one end of base mesa at which the bias current was applied and self-heating region may increase with increasing bias current. Therefore, from their $I-V$ curves we can estimate the critical current $I_c^{(B)}$ of the base mesa as well as $I_c^{(E)}$, whose values are listed in Table 2. However, at 77 K the features of $I-V$ curves for Mesa2, Mesa3 and Mesa4 are different from those at 4.2 K as mentioned above. As seen from Figs. 2(b)-(d), the $I-V$ curves are characterized by two critical currents determined by the ratio of areas of electrode and base mesas but the ratio of the first and second voltage jumps is not consistent with $N_B/N_E = 7.5$ in all cases. This suggests that the IJJs in the top portion of the base mesa switch into the voltage state simultaneously due to heat-transfer from the electrode mesa when IJJs in electrode mesa switch into the voltage state. Hypothetical temperature distribution inside the mesas presumed from the observed $I-V$ curves is shown in Fig. 3. Furthermore, such a contribution of Joule heating from the electrode mesa to base mesa may increase with increasing N_E because the Joule heat generated in the electrode mesa is proportional to the resistance R_E of the electrode mesa, i.e., N_E, in the voltage state. Indeed, we observed that the $I-V$ curves systematically vary depending on N_E and will discuss in detail elsewhere.

Table 2. Critical currents of the electrode mesa, $I_c^{(E)}$ and the base mesa, $I_c^{(B)}$ at 4.2 K and 77 K. The data were obtaind from Fig. 2.

	T=4.2 K		T=77 K	
Mesa	$I_c^{(E)}$ (mA)	$I_c^{(B)}$ (mA)	$I_c^{(E)}$ (mA)	$I_c^{(B)}$ (mA)
1	47.4	unspecified	31.6	unspecified
2	21.8	33.0	19.3	25.9
3	23.8	unspecified	10.4	24.6
4	14.9	24.9	11.3	31.2

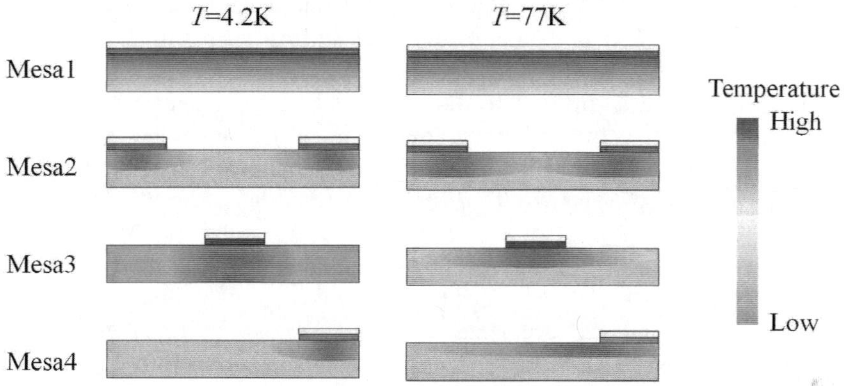

Figure 3. Hypothetical temperature distribution inside the base mesas immediately after the electrode mesa switches into the voltage state at 4.2 and 77 K.

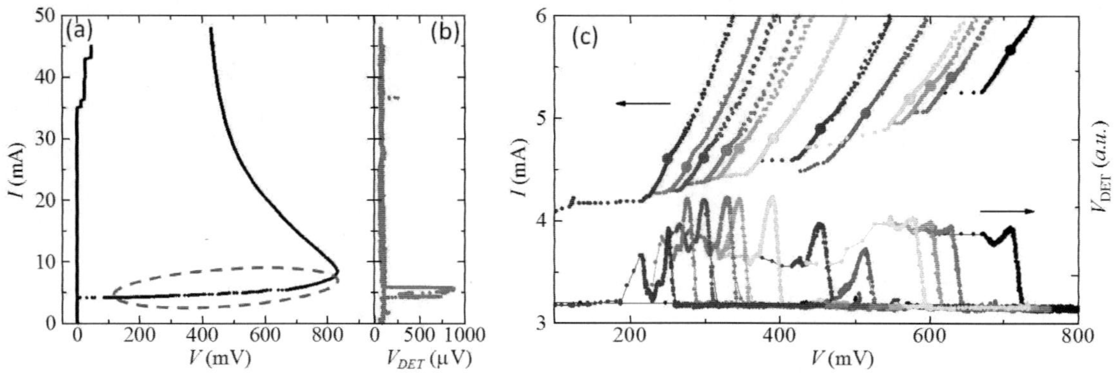

Figure 4. (a) $I-V$ characteristic of the emitter sample at 4.2 K as shown in Fig. 1. (b) Response of IJJ detector as a function of bias current of the emitter. (c) Multi-branching $I-V$ characteristics in the low bias region denoted by the dashed circle in Fig. 4 (a) and the emission properties of the emitter biased at inner branches.

3.2. Emission properties

A large mesa structure has been used in most studies of the THz emission from IJJs and it is known that two types of emissions have been observed. One is the emission in low current region and another is in high current region. The difference between these emission is the magnitude of contribution of the self-heating. The high-bias emission have been observed under the situation where a hot spot is formed while in the case of the low-bias emission the influence of self-heating is almost negligible. In this paper we focus on the low-bias emission because the self-heating effect makes the analysis of the emission properties complicated.

Figures 4(a) and 4(b) show $I-V$ characteristic and the response of IJJ detector as a function of bias current, respectively, for the sample ($N = 560$) as shown in Fig. 1(b) at 4.2 K. The critical current is 45 mA and the maximum voltage of 830 mV is appeared at $I = 8.3$ mA. It is found that from the response of IJJ detector the sample emitted the electromagnetic wave at the current range between 4.3 and 5.7 mA in the low bias region. The largest response of the detector was observed at the bias point of $I = 5.64$ mA and $V = 700$ mV. Although the emission frequency could not be measured directly in our experimental system, it is roughly estimated to be 0.6 THz from the Josephson voltage-frequency relation $f = V/(N\Phi_0)$ using the values of $V = 700$ mV and $N = 560$, where Φ_0 is the flux quantum. Here, we assumed that all IJJs in the mesa participate in the emission because the corresponding bias point is on the outermost branch. Although it is difficult to distinctly determine the cavity mode, the estimated emission frequency of 0.6 THz is qualitatively consistent with the cavity resonance frequency of $f(2,0) = 0.64$ THz or $f(2,1) = 0.66$ THz calculated by a general expression of the geometrical resonance for rectangular mesa, given by [10]

$$f(m,n) = \frac{c_0}{2n_I}\sqrt{\left(\frac{m}{W}\right)^2 + \left(\frac{n}{L}\right)^2}, \tag{1}$$

where c_0 is the speed of light in vacuum, $n_I = 4.2$ is the refractive index of Bi2212 [2], m and n are integers. On the other hand, multiple-branch structure can be observed in the $I-V$ characteristics by sweeping up and down the bias current in the low bias region. Figure 4(c) shows the observed inner branches and corresponding detector responses displayed with the same color. The filled circles on the inner branches indicate the bias point at which the detector signal becomes maximum.

We roughly estimated the number of active IJJs for each inner branches from the ratio of

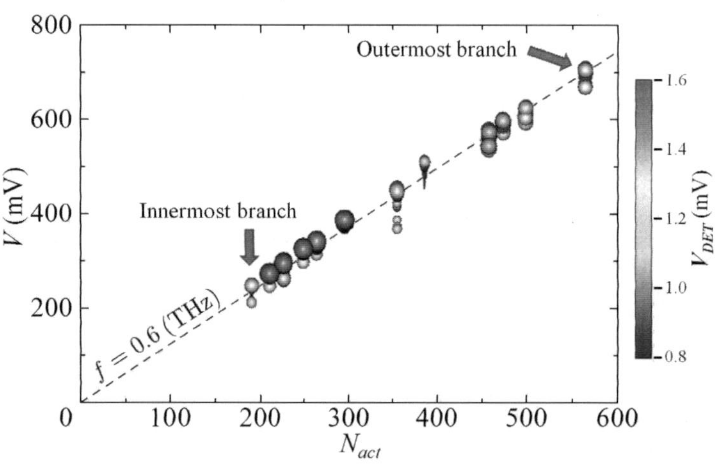

Figure 5. Voltage ranges emitting on the inner branches taken from Fig. 4(c) as a function of the number of active IJJs N_{act}. N_{act} was estimated from the ratio of the minimum voltage of the outermost branch and inner ones shown in Fig. 4(c). The color and the size of the symbols corresponds to the voltage of the detector V_{DET}. The dashed line indicates the ac Josephson relation $V = N_{act} f \Phi_0$ for $f = 0.6$ THz.

the minimum voltage of the outermost branch and inner ones shown in Fig. 4(c). Figure 5 shows the emitter voltage ranges on each inner branches, at which the intense emission was observed, as a function of estimated the number of active IJJs. From this, we found that all of observed intense emission on inner branches also correspond to the cavity resonance mode of $(m, n) = (2, 0)$ or (2,1) and the smallest number of IJJs contributing to the emission is estimated to be 189, which corresponds to 34% IJJs in the mesa. This may be not strange because only voltage corresponding to such a mode is suitable for the voltage range of each inner branch and the expected bias points for other modes deviate from the low bias region of each inner branch.

4. Conclusion

We have fabricated large IJJ mesas with different configurations of current injection and have discussed heat-flow in them. We observe that $I-V$ characteristics are strongly influenced by the current lead geometry even for the mesas with the same number of junctions because of different thermal distribution. Furthermore, we observe the THz radiation from the IJJ emitter mesa biased at inner branches of multiple-branched $I-V$ characteristics. We find that all of observed intense emissions on inner branches correspond to the same cavity resonance mode and in the case of the observed innermost branch 34% IJJs in the mesa contribute to such emission.

References

[1] Ozyuzer L, Koshelev A E, Kurtur C, Gopalsami N, Li Q, Tachiki M, Kadowaki K, Yamamoto T, Minami H, Yamaguchi H, Tachiki T, Gray KE, Kwok W -K and Welp U, 2007 *Science* **318** 1291

[2] Minami H, Kakeya I, Yamaguchi H, Yamamoto T and Kadowaki K 2009 *Appl. Phys. Lett.* **95** 232511

[3] Wang H B, Guénon S, Yuan J, Iishi A, Arisawa S, Hatano T, Yamashita T, Koelle D and Kleiner R 2009 *Phys. Rev. Lett.* **102** 017006

[4] Wang H B, Guénon S, Gross B, Yuan J, Jiang Z G, Zhong Y Y, Grünzweig M, Iishi A, Wu P H, Hatano T, Koelle D and Kleiner R 2010 *Phys. Rev. Lett.* **105** 057002

[5] Guénon S, Grünzweig M, Gross B, Yuan J, Jiang Z G, Zhong Y Y, Li M Y, Iishi A, Wu P H, Hatano T, Mints R G, Goldobin E, Koelle D, Wang H B, Kleiner R 2010 *Phys. Rev. B* **82** 214506

[6] Minami H, Watanabe C, Sato K, Sekimoto S, Yamamoto T, Kashiwagi T, Klemm R and Kadowaki K 2014 *Phys. Rev. B* **89** 054503

[7] Benseman T M, Koshelev A E, Vlasko-Vlasov V, Hao Y, Kwok W -K, Welp U, Keiser C, Gross B, Lange M, Kölle D, Kleiner R, Minami H, Watanabe C, Kadowaki K 2015 *Phys. Rev. Applied* **3** 044017

[8] Irie A, Sakakibara M and Oya G, 1994 *IEICE Trans. Electron.* **E77-C** 1191

[9] Oikawa D, Irie A and Yamaki K 2013 *IEEE Trans. on Appl. Supercond.* **23** 1500604

[10] Klemm R. A and Kadowaki K 2010 *J. Phys.: Condens. Matter.* **22** 375701

IOP Conf. Series: Journal of Physics: Conf. Series **871** (2017) 012083 doi:10.1088/1742-6596/871/1/012083

Design and Test Results of Superconducting Magnet for Heavy-Ion Rotating Gantry

S Takayama[1], K Koyanagi[1], H Miyazaki[1], S Takami[1], T Orikasa[1], Y Ishii[1], T Kurusu[1], Y Iwata[2], K Noda[2], T Obana[3], K Suzuki[4], T Ogitsu[4] and N Amemiya[5]

[1] Toshiba corporation, 2-4 Tsurumi, Yokohama 230-0045, Japan
[2] National Institute of Research Science, 4-9-1 Anagawa, Inage, Chiba 263-8555, Japan
[3] National Institute for Fusion Science, 322-6 Oroshi-cho, Toki-city, Gifu 509-5292, Japan
[4] High Energy Accelerator Research Organization, 2-4 Shirane, Shirakata, Tokai 319-1195, Japan
[5] Kyoto University, Kyoto-Daigaku-Katsura, Nishikyo, Kyoto 615-8510, Japan

Shigeki2.takayama@toshiba.co.jp

Abstract. Heavy-ion radiotherapy has a high curative effect in cancer treatment and also can reduce the burden on patients. These advantages have been generally recognized. Furthermore, a rotating gantry can irradiate a tumor with ions from any direction without changing the position of the patient. This can reduce the physical dose on normal cells, and is thus commonly used in proton radiotherapy. However, because of the high magnetic rigidity of carbon ions, the weight of the rotating gantry for heavy-ion therapy is about three-times heavier than those used for proton cancer therapy, according to our estimation. To overcome this issue, we developed a small and lightweight rotating gantry in collaboration with the National Institute of Radiological Sciences (NIRS). The compact rotating gantry was composed of ten low-temperature superconducting (LTS) magnets that were designed from the viewpoint of beam optics. These LTS magnets have a surface-winding coil-structure and provide both dipole and quadrupole fields. The maximum dipole and quadrupole magnetic field of the magnets were 2.88 T and 9.3 T/m, respectively. The rotating gantry was installed at NIRS, and beam commissioning is in progress to achieve the required beam quality. In the three years since 2013, in a project supported by the Ministry of Economy, Trade and Industry (METI) and the Japan Agency for Medical Research and Development (AMED), we have been developing high-temperature superconducting (HTS) magnets with the aim of a further size reduction of the rotating gantry. To develop fundamental technologies for designing and fabricating HTS magnets, a model magnet was manufactured. The model magnet was composed of 24 saddle-shaped HTS coils and generated a magnetic field of 1.2 T. In the presentation, recent progress in this research will be reported.

1. Introduction

Because of its high curative effects and low burden on patients, heavy-ion radiotherapy is becoming more widespread, and the total number of patients treated with heavy-ion radiotherapy has been increasing. In order to reduce the physical dose on normal cells, it is preferable to irradiate the tumor with the heavy-ion beam from various directions. A rotating gantry is a suitable apparatus for meeting this requirement. Rotating gantries are already used in proton radiotherapy; however, the rotating gantries required for heavy-ion radiotherapy will be much larger because the magnetic rigidity of the heavy ions is about three-times larger than that of protons. There is only one heavy-ion gantry in the world, located in Heidelberg, which had been operating until 2014. The total weight of the gantry was reported to be 600 tons [1], whereas the typical weight of existing proton gantries is between 100 and 200 tons. To reduce the size of the heavy-ion gantry, we developed LTS magnets in collaboration with NIRS [2], and their installation in a superconducting rotating gantry was completed successfully in 2015. Beam commissioning is in progress to achieve the required beam quality [3]. In this paper, manufacturing and test result of the LTS magnets are reported.

For the purpose of further size reduction of the rotating gantry, we have been developing fundamental technologies for an HTS gantry in the three years since 2013, in a project supported by METI and AMED. In this project, a model magnet composed of 24 saddle-shaped HTS coils was fabricated and generated a magnetic field of 1.2 T [4]. In this paper, the fabrication and test result of the HTS magnet are reported.

2. LTS rotating gantry

2.1. Layout of LTS gantry

The layout of the LTS rotating gantry which was developed in collaboration with NIRS is shown in Figure 1. It is an isocentric gantry with an axial length of 5.5 m and a radius of 13 m. Ten LTS magnets of five different types were mounted on it. A pair of scanning magnets was arranged on the top of the gantry, and three pairs of steering magnets and beam monitors were placed in the straight section. A dose monitor, a beam position monitor, and a ridge filter were arranged downstream of the beam line. These devices were mounted on a cylindrical structure, allowing the gantry to irradiate patients with ions at angles of over ±180 degree by rotating this structure.

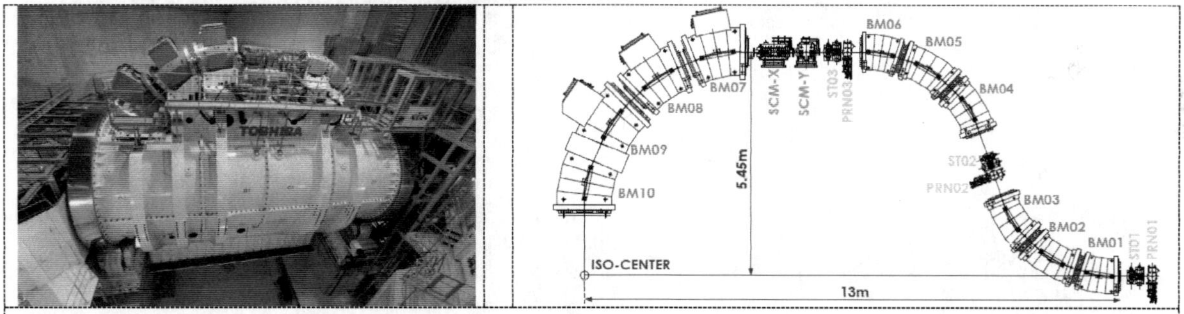

Figure 1. Left: Photograph of the rotating gantry at NIRS. Right: Layout of the gantry. The gantry consists of ten superconducting magnets (BM01-10), a pair of scanning magnets (SCM-X and SCM-Y), and three pairs of steering magnets and beam profile-monitors (ST01-03 and PRN01-03).

2.2. Design and test result of LTS magnets

The specifications of the LTS magnets are shown in Table 1. The maximum dipole field was 2.88 T, and the maximum quadrupole field was 9.3 T/m, and these values were determined by the beam optics [2]. Photographs of one of the LTS magnets and a coil are shown in Figure 2. The coils were saddle-shaped, and the positions of the superconductors followed a cos-theta distribution. The LTS magnets, except BM7 and BM8, have a dipole coil and quadrupole coil in a coaxial arrangement for shortening the gantry. The arrangement of the superconductors was determined by three-dimensional magnetic field analysis so that the integral dipole and quadrupole field were uniform [2]. Using an NC winding machine, conductors having diameters of 0.9 mm for BM1-6 and 1.3 mm for BM7-10 were wound so as to adhere to the outer surface of a bent cylinder with an accuracy of 0.1 mm. To reduce AC losses,

NbTi wire with a filament diameter of 10 μm and CuNi barriers were employed. Each coil can be excited independently, and the LTS magnets were connected to power supplies placed on the non-rotating area with power cables through a cable spool located at the end of the gantry. The return yoke was arranged on the outside of the coil. It enhanced the magnetic field intensity and supported the magnetic force. To suppress eddy currents, the return yoke was constructed from laminated electromagnetic steel sheets. The magnetic field at the mid-plane of the return yoke was set to be less than 2 T in order to reduce the influence of saturation. Moreover, the vacuum vessel was made of iron for reducing magnetic field leakage. The magnetic field distributions of all manufactured LTS magnets were measured and their performance was checked [5]. The alignment reference point was determined from the measurement results, and the magnets were installed with an accuracy of ±0.3 mm.

Table 1. The specifications of the LTS magnets.

parameter	BM1 and BM6	BM2-BM5	BM7	BM8	BM9-10
Bending angle (deg.)	18	26	22.5	22.5	22.5
Bending radius (m)	2.3	2.3	2.8	2.8	2.8
Bore radius (mm)	30	30	85	120	145
Reference area (mm)	ϕ40	ϕ40	60×60	80×80	100×100
Maximum dipole field (T)	2.88	2.88	2.37	2.37	2.37
Maximum quadrupole field (T/m)	9.3	9.3	-	-	1.3
Field uniformity (dipole)	$\leq 1\times10^{-3}$			$\leq 1\times10^{-4}$	
Field uniformity (quadrupole)			$\leq 1\times10^{-3}$		

Figure 2. Photographs of the LTS magnet (left:BM4) and the saddle-shaped LTS coil (right).

A conduction cooling system with no coolant was adopted because the magnets were rotated. Small Gifford-McMahon (GM) cryocoolers having a cooling capacity of 1.5 W at 4.2 K were used, and the LTS coils and yokes were cooled by the cryocoolers through pure aluminum plates with a thickness of 1 mm. Three sets of cryocoolers were mounted on each of BM1-6, and four sets of cryocoolers were mounted on each of BM7-10, giving a total of 34 small cryocoolers in the LTS gantry. The compressors were arranged in the non-rotating area and were connected to the cryocooler with flexible helium gas lines with a length of 30 m via the cable spool. When cooling the magnets with only the cryocoolers, it took about 8 days for the initial cooling at BM4. To evaluate the influence of rotation on the cooling capacity, a continuous rotation test was carried out. The prototype LTS magnet was fabricated, and the coil temperature was measured while rotating the magnet repeatedly from –180 degree to +180 degree. The rotating speed was 1.5 rpm, so that the acceleration acting on the cryocooler mounted on the prototype magnet was the same value (0.002G) as the acceleration that acts on the cryocooler mounted on the gantry. As a result of the test, there was no change in the coil temperature even after changing the position 33,000 times. This number of position changes is equivalent to 1.5 years of operation. Moreover, it was checked that there were no faults, such as leaks or bad insulation, in the flexible helium gas line and power cables at the time of rotation. In order to control the irradiation depth by the beam energy, the operating currents of the LTS magnets were also changed in a short time according to the energy. At the time of the magnetic field change, the coils generated heat due to AC loss. Then, the coil temperature was measured using an operating pattern in which the coil was excited from 430 MeV/u to 80 MeV/u in 200 steps. The test results are shown in Figure 3. The average coil temperature was initially 3.7 K, and with the excitation pattern, the average coil temperature was stabilized at about 4 K, and a transient temperature rise of 0.5 K was observed [6,7]. On the other hand, there was no quenching. To reduce the treatment time, the magnets were rotated in the excited state. Even in such a case, there was no remarkable coil temperature rise or quenching. The cooling curve when training quench occurred in BM6 is shown in Figure 4. Re-cooling was completed in about 2 hour.

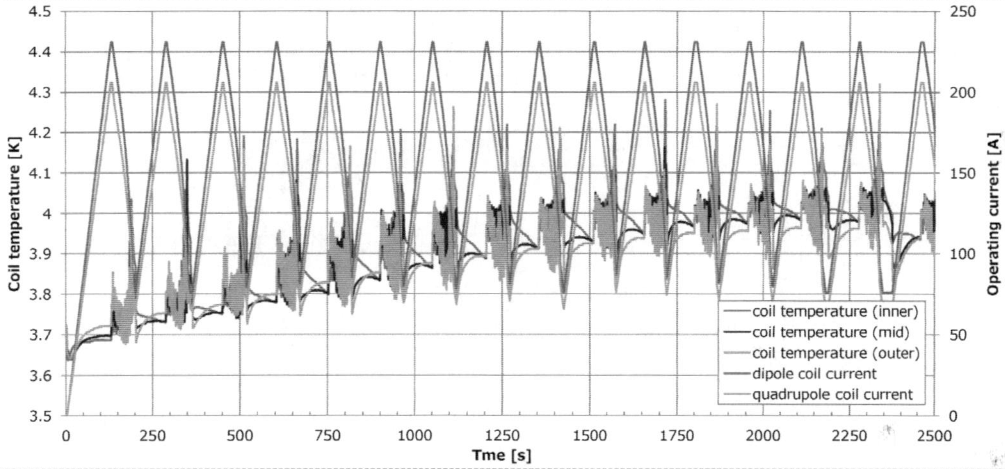

Figure 3. Test results of operating pattern in which coil was excited from 430 MeV/u to 80 MeV/u in 200 steps (BM10).

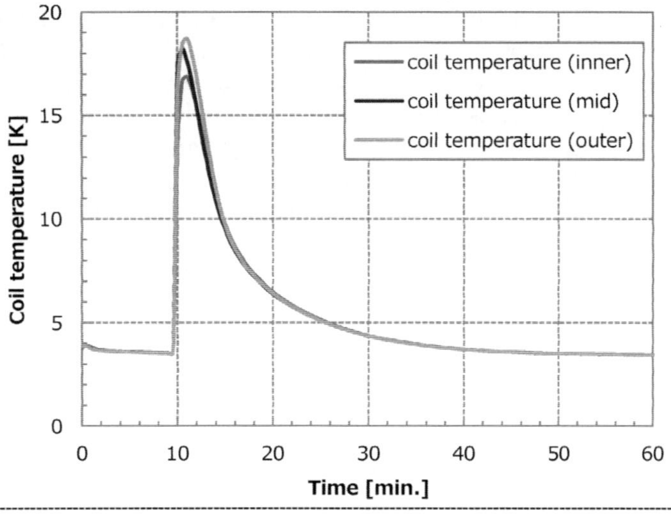

Figure 4. The re-cooling curve after quenching was occurred (BM6).

3. HTS rotating gantry

3.1. Layout of HTS gantry

For the purpose of further reducing the size of the rotating gantry, an R&D project to develop fundamental technologies for HTS gantries started in 2013, supported by METI and AMED. An HTS magnet can generate a high magnetic field and can be cooled efficiently because of the high operating temperature. On the other hand, there are some issues that should be considered, for example: the influence of tape magnetization and manufacturing accuracy on the field quality [8], the thermal stability of the conduction-cooled HTS coils under an alternating magnetic field [9], and methods to protect the coils from thermal runaway caused by an anomalous thermal input such as that due to beam loss. In order to clarify the required specifications, conceptual design of an HTS gantry was carried out. The layout of this HTS gantry is shown in Figure 5 [10]. This design was based on the LTS gantry, and it has an axial length of 9.2 m, a radius of 3.9 m, and a total weight of 177 tons. Ten HTS magnets of four different types were mounted on the HTS gantry.

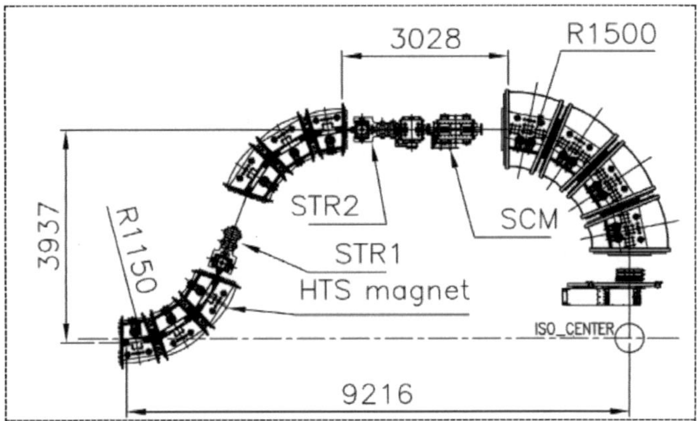

Figure 5. Layout of HTS rotating gantry.

3.2. Design and test result of HTS magnets

The specifications of the HTS magnets are shown in Table 2. The maximum dipole field of the HTS magnet should be approximately 6 T, and the maximum quadrupole field should be 33 T/m. These values were estimated from the beam optics [11]. The HTS coils composing the magnets of the rotating gantry were designed based on the required intensity, quality, and spatial size of the magnetic field. Also, in this study, REBCO conductor having a width of 4 mm and a thickness of 0.1 mm was considered because of its performance advantages in high magnetic fields and at high temperatures. To generate the magnetic field efficiently, the HTS coils were formed in a saddle shape. The return yoke was arranged on the outside of the coil for enhancement of the magnetic field and supporting the magnetic force. Conduction cooling was adopted because the magnets were rotated. To evaluate this design, a model magnet was fabricated.

Table 2. The specifications of the HTS magnets.

parameter	A type	B type	C type	D type
Bending angle (deg.)	18	26	22.5	22.5
Bending radius (m)	1.15	1.15	1.5	1.5
Maximum dipole field (T)	5.8	5.8	4.5	4.5
Maximum quadrupole field (T/m)	15.5	33	-	1.7
Field uniformity (dipole)	$\leq 1\times10^{-3}$		$\leq 1\times10^{-4}$	
Field uniformity (quadrupole)	$\leq 1\times10^{-3}$			

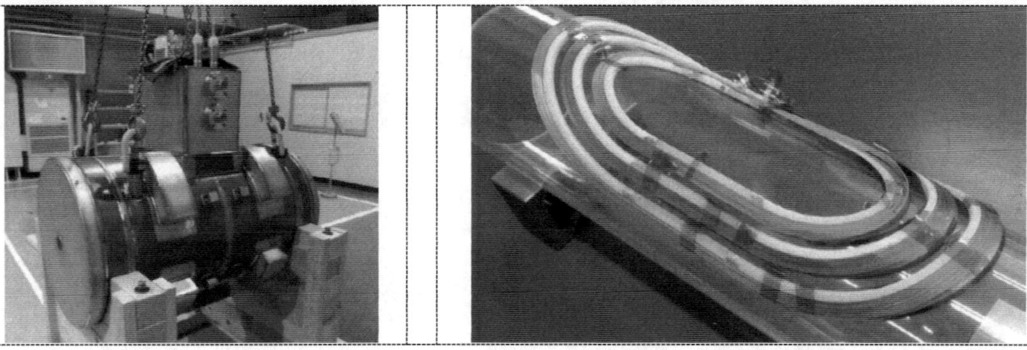

Figure 6. Photographs of the HTS model magnet (left) and the saddle-shaped HTS coil (right).

Photographs of the model magnet and the HTS coil are shown in Figure 6. This model magnet was a short model, and the specifications are shown in Table 3. Twenty-four saddle-shaped REBCO coils and iron yokes were applied, and these were cooled down to approximately 4 K by a GM cryocooler. To reduce stress in the windings, these saddle-shaped coils were fabricated by an automatic winding machine with an eight-axis synchronous control function [12]. To confirm the positional displacement of each turn in the HTS coils quantitatively, the roughness of the fabricated HTS coil surface was measured by a laser displacement meter, and the winding accuracy was measured to be 0.2 mm in the tape edge direction [8]. The superconducting properties of all saddle-shaped coils constructing the model magnet were confirmed under a liquid nitrogen environment, and the HTS coils were stacked in four layers and attached to the iron yoke. After assembly of the model magnet, initial cooling was carried out, which took about 13 days. The operating current was increased with a sweep rate of 5-10 A/minute, and the behaviors of the coil temperature and coil voltage generated in the saddle-shaped coils were measured. The temperature profiles of the operating current are shown in Figure 7. An anomalous temperature rise was observed at an operating current of 153 A, and the generated magnetic field at the center of the beam duct was 1.2 T at that time. Temperature rises also can be found at 1.14 hour, 1.34 hour, and 1.54 hour in Figure 7. This phenomenon was assumed to be explained by flux jump. In this event, the dipole field strength changed coincidentally less than 100 ppm level [4]. Since there was no anomalous temperature rise even after re-excitation under 153 A, field quality measurement was conducted with a harmonic coil system at an operating current of 120 A [4]. Large quadrupole and sextupole coefficients were observed compared with the values predicted in the simulation, and these were considered to be due to partial misalignment of the coils. Further investigation of the model magnet is on-going.

IOP Conf. Series: Journal of Physics: Conf. Series **871** (2017) 012083 doi:10.1088/1742-6596/871/1/012083

Table 3. The specifications of the model magnet.

parameter	value
Outer dimension (mm)	680×1200
Magnetic field at center of beam duct (T)	1.2
Rated current (A)	153
Radius of beam duct (mm)	30
Inner radius of REBCO coils (mm)	60
Outer radius of REBCO coils (mm)	80
Inner radius of iron yoke (mm)	95
Outer radius of iron yoke (mm)	205
Coil length (mm)	340
Number of REBCO coils	24
Number of turn for each coil	50
Total length of REBCO conductor (m)	820
Coil inductance (mH)	288.9

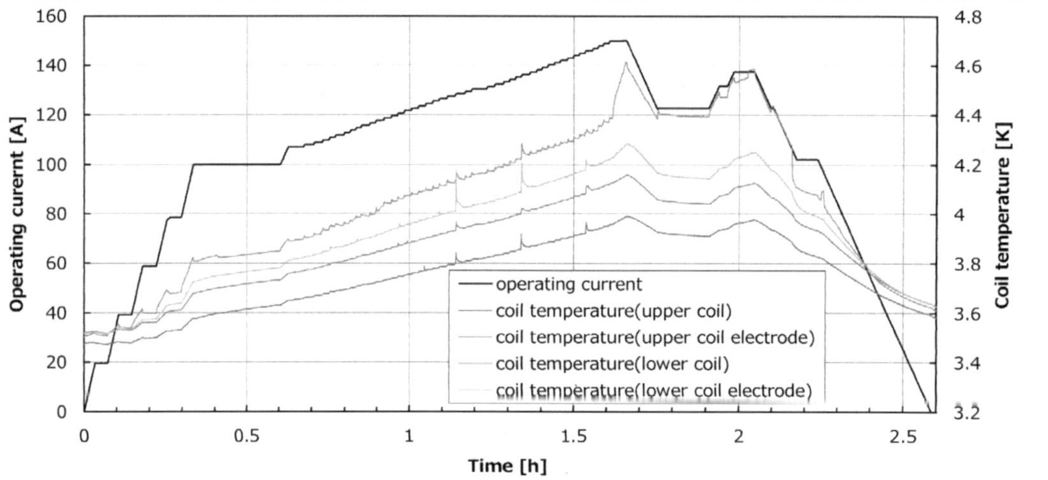

Figure 7. Excitation test result of model magnet.

4. Conclusion

To reduce the size of the rotating gantry for heavy-ion radiotherapy, we have been developing LTS magnets and HTS magnets. For the LTS magnets, the results of temperature and magnetic field measurements were positive, and manufacturing and installation of an LTS gantry were successfully completed. Beam commissioning to achieve the required beam quality is on-going. For the HTS magnets, we have been developing fundamental technologies for REBCO coils, and a model magnet composed of saddle-shaped REBCO coils was fabricated. The model magnet could generate a magnetic field of 1.2 T at the center of the beam duct. The field quality was measured using a harmonic coil system, and large multipole coefficients were observed. These were considered to be due to misalignment of the HTS coils. Further investigation of the model magnet is on-going.

Acknowledgments
This work is supported by the Japanese Ministry of Economy, Trade and Industry (METI) and the Japan Agency for Medical Research and Development (AMED).

References

[1] Eickhoff H 2002 *Proc. of the 8th European Particle Accelerator Conf.* p.2730.
[2] Iwata Y, Noda K, Shirai T, Murakami T, Furukawa T, Mori S, Fujita T, Itano A, Shouda K, Mizushima K, Fujimoto T, Ogitsu T, Obana T, Amemiya N, Orikasa T, Takami S, Takayama S and Watanabe I 2012 *Phys. Rev. ST Accel. Beams*, **15** 044701.
[3] Iwata Y, Fujimoto T, Matsuba S, Fujita T, Sato S, Furukawa T, Hara Y, Mizushima K, Saraya Y, Tansho R, Saotome N, Shirai T and Noda K 2016 *Nucl. Instrum. and Meth. in Phys. Res. A* **834** 71-80.
[4] Suzuki K, Ogitsu T, Ishii Y, Koyanagi K, Takayama S, Tasaki K and Amemiya N 2017 *IEEE Trans. Appl. Supercond.*, **27** 4600405.
[5] Iwata Y, Suzuki S, Noda K, Shirai T, Murakami T, Furukawa T, Fujita T, Hara Y, Mizushima K, Sato S, Mori S, Shouda K, Arai H, Fujimoto T, Orikasa T, Takayama S, Ogitsu T, Obana T and Amemiya N 2014 *IEEE Trans. Appl. Supercond.,* **24** 4400505.
[6] Iwata Y, Noda K, Murakami T, Shirai T, Furukawa T, Fujita T, Mori S, Mizushima K, Shouda K, Fujimoto T, Ogitsu T, Obana T, Amemiya N, Orikasa T, Takami S and Takayama S 2013 *Nucl. Instrum. and Meth. in Phys. Res*. B **317** 793-797.
[7] Iwata Y, Noda K, Shirai T, Murakami T, Fujita T, Furukawa T, Mizushima K, Hara Y, Suzuki S, Sato S, Shouda K, Fujimoto T, Arai H, Ogitsu T, Obana T, Amemiya N, Orikasa T, Takayama S, Nagamoto Y and Yazawa T, *Proc. of PAC2013, Pasadena, CA USA,* FRXB1.
[8] Takayama S, Koyanagi K, Tosaka T, Tasaki K, Kurusu T, Ishii Y, Amemiya N, Suzuki K, Ogitsu T, Iwata Y and Noda K 2016 *IEEE Trans. Appl. Supercond.,* **26** 4402304.
[9] Takayama S, Koyanagi K, Tosaka T, Tasaki K, Kurusu T, Ishii Y, Amemiya N, Suzuki K, Ogitsu T, Iwata Y and Noda K 2016 *IEEE Trans. Appl. Supercond.*, **26** 4402404.
[10] Tasaki K, Koyanagi K, Takayama S, Ishii Y, Kurusu T, Amemiya N, Ogitsu T, Iwata Y and Noda K 2016 *Phys. Proc.* **81** 162-165.
[11] Takayama S, Koyanagi K, Yamaguchi A, Tasaki K, Kurusu T, Ishii Y, Amemiya N, Ogitsu T and Noda K 2015 *Phys. Proc.* **67** 879-884.
[12] Koyanagi K, Takayama S, Miyazaki H, Tosaka T, Tasaki K, Kurusu T and Ishii Y 2015 *IEEE Trans. Appl. Supercond.* **25** 4003104.

IOP Conf. Series: Journal of Physics: Conf. Series **871** (2017) 012084 doi:10.1088/1742-6596/871/1/012084

Locating of normal transitions in a Bi2223 high temperature superconducting coil by using capacitor type voltage terminals and the active power method

N Nanato, K Okura, H Kumagai, H Aoyama

Graduate School of Natural Science and Technology, Okayama University, 3-1-1, Tsushima Naka, Kita-ku, Okayama, 700-8530, Japan

nanato@okayama-u.ac.jp

Abstract. It is important to locate positions of normal transitions in a high temperature superconducting (HTS) coil for identifying its design and fabrication weakness. In this paper, the authors propose a locating method by using capacitor type voltage terminals and the active power method. The former is a method to measure voltage in a HTS coil without electric contact and the latter is a method to detect the normal transitions by measuring active power dissipated in the coil. Combination of the two methods can achieve precise location of the normal transitions. The authors show usefulness of the method through experimental results for a Bi2223 HTS coil.

1. Introduction

AC HTS power applications have been developed in recent years such as motors and transformers [1]. For identifying their safety design and fabrication weakness, it is important to understand behaviors of HTS coils after occurrence of superconducting to normal transitions. Locating the normal transitions is very useful to understand the behaviors. Generally, many voltage taps are placed directly on the coil for the purpose to locate the normal transitions. However the placement of the taps may cause serious problems such as high voltage sparks because insulations of the coil has to be removed to solder the taps on the coil. The authors have presented capacitor type voltage terminals as a method to measure voltages in the coil without electric contact [2]. Reference [3] shows that this method is useful for locating the positions of the normal transitions. In this paper, the authors propose a more precise location method using the capacitor type voltage terminals and the active power method [4]. The latter method can recognize polarity of the measured voltage by the former method and the polarity contributes to make the location method more precise. Also, temperature in the normal areas has to be kept under allowable temperature of the coil windings to avoid their deterioration due to excessive heating in the occurrence of the normal transitions. The active power method is originally a quench detection / protection method for a superconducting coil [4] and therefore the proposed method can safely locate the positions of the normal transitions. As experimental results for a Bi2223 HTS coil, it was found that the proposed method enabled to locate the positions of the normal transitions without exceeding the allowable temperature more precisely than the conventional method.

2. Principle of location method

As shown in Figure 1, a pair of electrically conductive sheets is attached on the surface of an HTS coil

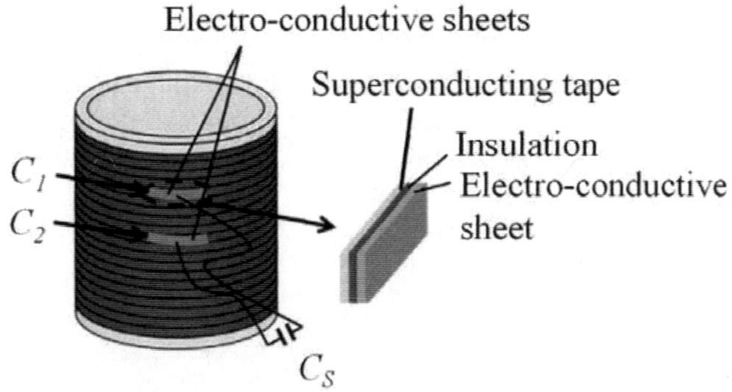

Figure 1. Capacitor type voltage terminals.

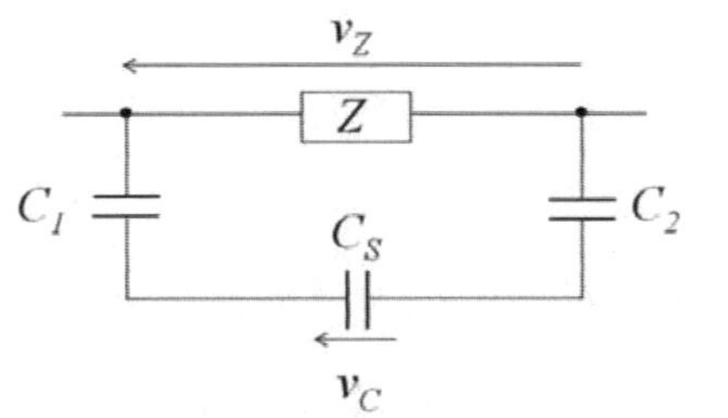

Figure 2. An equivalent circuit.

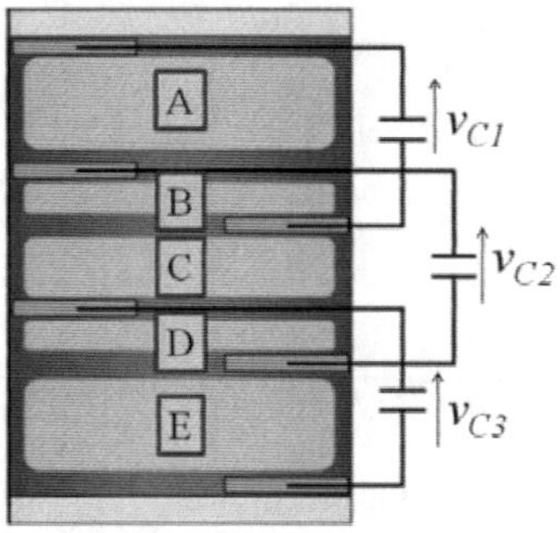

Figure 3. An outline of an HTS coil with three voltage terminals.

and is connected with an external capacitor. Then insulation exists between an HTS wire and the each sheet as shown in right hand side of Figure 1 and therefore a capacitor is formed at the each sheet. Its equivalent circuit is shown as Figure 2. Z and v_z are respectively an impedance and a voltage between the two sheets, C_1 and C_2 are capacitances at the two sheets, and C_s and v_c are respectively a capacitance and a voltage of the external capacitor. Then following equation is derived.

$$v_z = \frac{C_1C_2 + C_1C_s + C_2C_s}{C_1C_2} v_c = \alpha v_c. \tag{1}$$

The equation shows a voltage v_z in the HTS coil can be measured without electric contact by measuring v_c. Also, the electro-conductive sheets are adhesive aluminum tapes. Each sheet has thickness of 0.05 mm and its adhesive agent is also electrically conductive. The external capacitor was connected to the electro-conductive sheet by soldering.

Figure 3 shows an outline of an HTS coil which three pairs of electrically conductive sheets are attached on. Then, principle of the location method is explained as follows. v_{c1}, v_{c2} and v_{c3} are shown as following equations.

$$\alpha_1 v_{c1} = L_1 \frac{di}{dt}, \tag{2}$$

$$\alpha_2 v_{c2} = L_2 \frac{di}{dt}, \tag{3}$$

$$\alpha_3 v_{c3} = L_3 \frac{di}{dt}, \tag{4}$$

where i is a transport current through the coil and L_1, L_2 and L_3 are self-inductances between the two sheets for measuring v_{c1}, v_{c2} and v_{c3}, respectively. In actual measurement, all the sheets are attached so that all of α_1, α_2 and α_3 are the same α. In order to detect the normal transitions, the inductive voltages have to be canceled. Following equations achieve the cancellation.

$$v_{12} = \alpha(v_{c1} - k_{12}v_{c2}), \tag{5}$$
$$v_{23} = \alpha(v_{c2} - k_{23}v_{c3}), \tag{6}$$
$$v_{31} = \alpha(v_{c3} - k_{31}v_{c1}), \tag{7}$$

where

$$k_{12} = \frac{L_1}{L_2}, \tag{8}$$
$$k_{23} = \frac{L_2}{L_3}, \tag{9}$$
$$k_{31} = \frac{L_3}{L_1}. \tag{10}$$

In actual measurement, all the sheets are attached so that all of k_{12}, k_{23} and k_{31} are set to 1. This means $L_1 = L_2 = L_3$. Then $v_{12} = v_{23} = v_{31} = 0$ in a superconducting state. On the other hand, for example, when a normal transition occurs in area "A" shown in Figure 3, only v_{c1} has a resistive voltage and then following equations are derived.

$$v_{12} = \alpha(v_{c1} - k_{12}v_{c2}) = \alpha(ri + L_1\frac{di}{dt} - k_{12}L_2\frac{di}{dt}) = \alpha ri, \tag{11}$$
$$v_{23} = \alpha(v_{c2} - k_{23}v_{c3}) = \alpha(L_2\frac{di}{dt} - k_{23}L_3\frac{di}{dt}) = 0, \tag{12}$$
$$v_{31} = \alpha(v_{c3} - k_{31}v_{c1}) = \alpha\left\{L_3\frac{di}{dt} - k_{31}\left(ri + L_1\frac{di}{dt}\right)\right\} = -\alpha ri, \tag{13}$$

where r is a resistance of the normal area. v_{12} is a resistive voltage with same polarity as the current i and v_{31} has a reversed polarity to the polarity of v_{12}. The polarity can be recognized by transforming the voltages into active powers shown as following equations.

$$P_{12} = \alpha(v_{c1} - k_{12}v_{c2})i = \alpha ri^2 \rightarrow \text{Averaging} \rightarrow P'_{12} = \alpha rI^2 > 0, \tag{14}$$
$$P_{23} = \alpha(v_{c2} - k_{23}v_{c3})i = 0 \rightarrow \text{Averaging} \rightarrow P'_{23} = 0, \tag{15}$$
$$P_{31} = \alpha(v_{c3} - k_{31}v_{c1})i = -\alpha ri^2 \rightarrow \text{Averaging} \rightarrow P'_{31} = -\alpha rI^2 < 0, \tag{16}$$

where I is a root mean square value of i. According to the above results, when a normal transition occurs in area "A" shown in Figure. 3, $P'_{12} > 0$, $P'_{23} = 0$ and $P'_{31} < 0$.

When a normal transition occurs at "B", v_{c1} and v_{c2} have same resistive voltage and v_{c3} has no resistive one. Then

$$P_{12} = \alpha(v_{c1} - k_{12}v_{c2})i = \alpha(ri^2 - ri^2) = 0 \rightarrow \text{Averaging} \rightarrow P'_{12} = 0, \tag{17}$$
$$P_{23} = \alpha(v_{c2} - k_{23}v_{c3})i = \alpha ri^2 \rightarrow \text{Averaging} \rightarrow P'_{23} = \alpha rI^2 > 0, \tag{18}$$

IOP Conf. Series: Journal of Physics: Conf. Series **871** (2017) 012084 doi:10.1088/1742-6596/871/1/012084

$$P_{31} = \alpha(v_{c3} - k_{31}v_{c1})i = -\alpha r i^2 \;\rightarrow \text{Averaging} \rightarrow P'_{31} = -\alpha r I^2 < 0. \tag{19}$$

Table 1. Logic of locating the positions of the normal transitions.

The place of the occurrence of normal transitions	A	B	C	D	E
P'_{12}	+	0	–	–	0
P'_{23}	0	+	+	0	–
P'_{31}	–	–	0	+	–

All cases of the normal transitions are summarized in Table 1.
The areas of the normal transitions can be located by the logic. Higher resolution of the location method can be achieved by increasing the number of pairs of electrically conductive sheets.

3. Locating test for a Bi2223 HTS coil

3.1 Experimental setup
Figure 4 shows a photograph of a test Bi2223 HTS coil. Table 2 shows specifications of the HTS coil, which is a solenoidal coil wound with $Bi_2Sr_2Ca_2Cu_3O_x$ (Bi2223) tape on Glass Fiber Reinforced Plastic bobbin. Table 3 shows specifications of capacitors. Three voltage terminals were attached on the HTS coil so that v_{c1} was a voltage across the 1st to 4th turn, v_{c2} was 3rd to 7th turn and v_{c3} was 6th to 9th turn. Figure.5 shows an experimental circuit. AC current of 60 Hz and 110 A_{peak} was supplied to the HTS coil which was cooled in liquid nitrogen. The normal transitions were caused by heaters mounted on five areas A – E of the HTS coil shown in Figure 3. Each heater has height of 4.2mm and width of 12mm. The thyristor switch is turned off to protect the HTS coil from excessive heating in the normal zone when any of positive active powers such as equations (14) and (18) become larger than a specified positive threshold or any of negative ones such as equations (16) and (19) become smaller than a specified negative one. The specified threshold is decided by coupling of 1D heat equation and electric circuit equation [5].

3.2 Experimental results
Figure 6 shows experimental results for the occurrence of the normal transition at "A". Figures 6 (a), (b) and (c) show the active power signals. P'_{12} increased from about 13 s (positive), P'_{23} was almost constant (zero) and P'_{31} decreased from about 13 s (negative). These results agreed well with the logic shown in Table 1.

Figure 4. A photograph of the HTS coil.

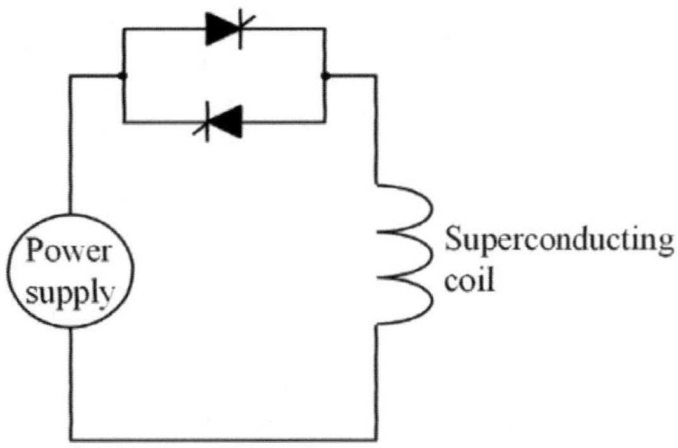

Figure 5. An experimental circuit.

Table 2. Specifications of an HTS coil.

Inner diameter (mm)	147
Outer diameter (mm)	150
Height (mm)	105
Number of turns	9.5
Inductance (μH)	11.9
I_C (at 77 K, self-field)(A)	193.5

Table 3. Specifications of capacitors.

External capacitors : C_{s1}, C_{s2}, C_{s3} (nF)	100
Electro-conductive sheet capacitors : C_1 (nF)	0.275
: C_2 (nF)	0.270
: C_3 (nF)	0.282
: C_4 (nF)	0.275
: C_5 (nF)	0.267
: C_6 (nF)	0.272

Figure 6 (d) shows the current through the circuit. The current was shut off at 21.29 s because the active power P'_{12} reached a specified threshold of "5 W". Figure 6 (e) shows temperature of the normal zone. The temperature increased after the normal transition and then it decreased after the shutdown of the current. Its maximum value is 167 K which is allowable temperature for the test coil [6].

Figure 7 shows experimental results for the occurrence of the normal transition at "B". Figures 7 (a), (b) and (c) show the active power signals. P'_{12} was almost constant (zero), P'_{23} increased from about 10 s (positive) and P'_{31} decreased from about 10 s (negative). These results also agreed well with the logic shown in Table 1. Figure 7 (d) shows the current through the circuit and Figure 7 (e) shows temperature of the normal zone. As well as Figure 6 (e), maximum temperature 134 K of the normal zone was under the allowable temperature.

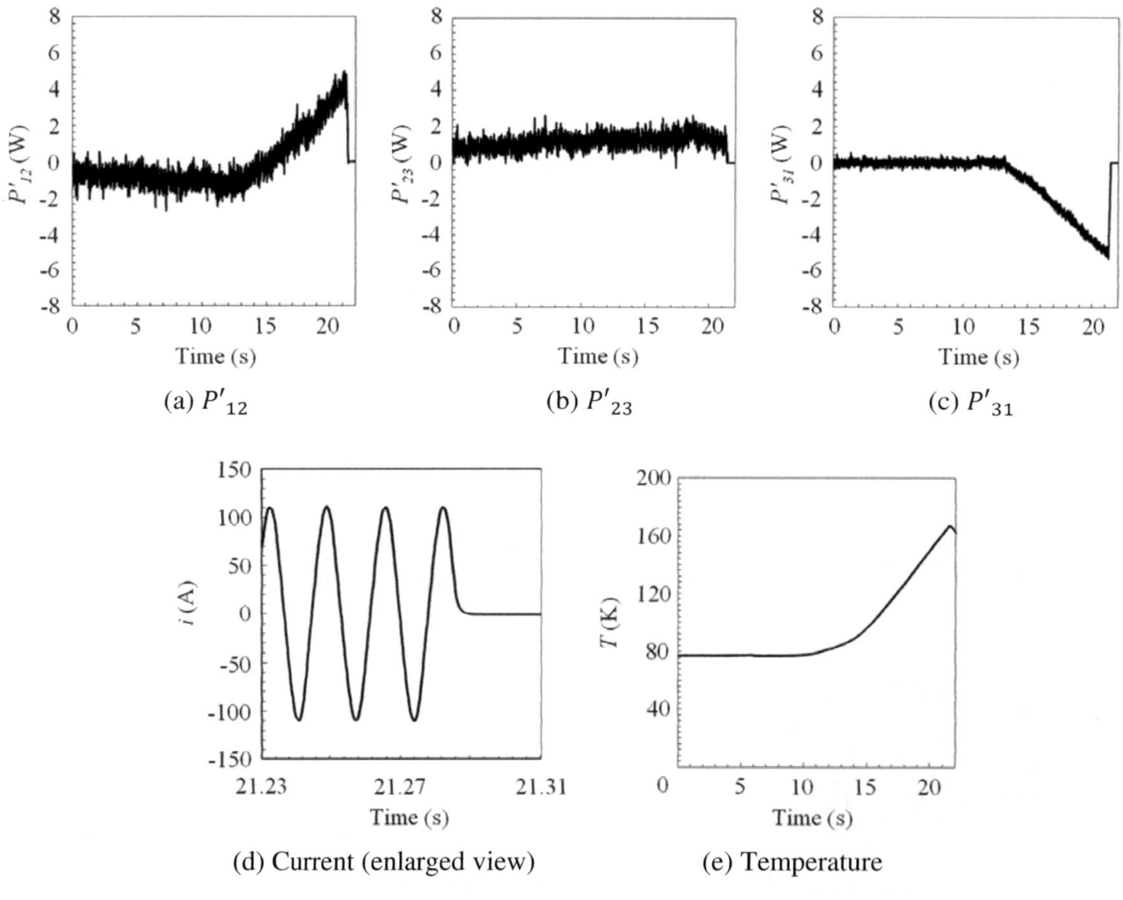

Figure 6. Experimental results for the normal transition at "A".

From the test results, it was verified that the proposed method could locate the positions of the normal transitions so as to suppress the temperature of the normal zone under the allowable temperature.

4. Conclusions

In this paper, the authors proposed a method to locate normal transitions in HTS coils by using capacitor type voltage terminals and the active power method. The proposed method can locate the positions of the normal transitions without electric contact so as to suppress the temperature of the normal zone under the allowable temperature. The experimental results showed validity of the method for a Bi2223 HTS coil. Hereafter the authors will study higher resolution of the method.

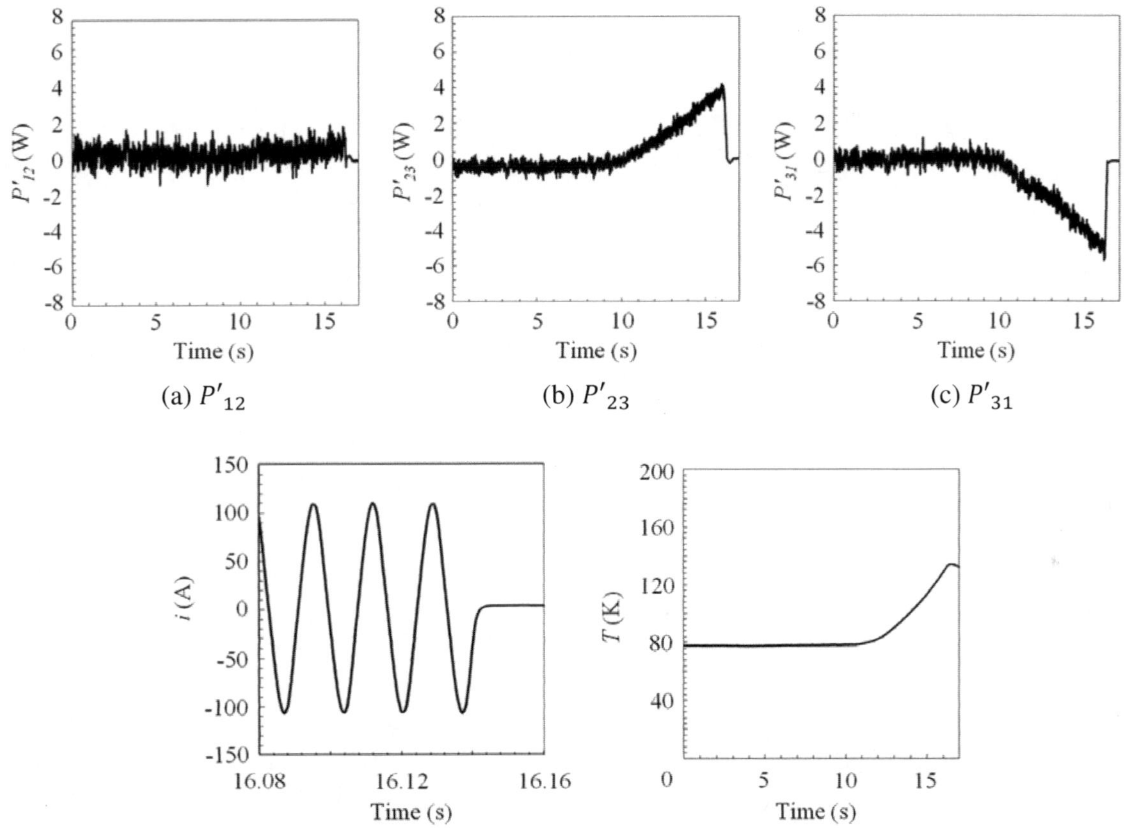

(a) P'_{12} (b) P'_{23} (c) P'_{31}

(d) Current (enlarged view) (e) Temperature

Figure 7. Experimental results for the normal transition at "B".

References

[1] Shiohara Y, et, al. Future prospects of high Tc superconductors-coated conductors and their applications 2013 *Physica C* **484** 1

[2] Nanato N and Nishiyama K Non-destructive Detection of Normal Transitions in High Temperature Superconducting Coil 2014 *Physics Procedia* **58** 260

[3] Nanato N and Nishiyama K Locating of normal transitions in a Bi2223 high temperature superconducting coil by non-contact voltage measurement method 2015 *Cryogenics* **72** 53

[4] Nanato N and Kobayashi Y Quench Detection and Protection for High Temperature Superconducting Transformers by Using the Active Power Method 2014 *Physics Procedia* **58** 264

IOP Conf. Series: Journal of Physics: Conf. Series **871** (2017) 012084 doi:10.1088/1742-6596/871/1/012084

[5] Nanato N, Asai W and Murase S Study on criterion for quench detection/protection of superconducting magnet based on active power method 2012 *Physics Procedia* **27** 416
[6] Iwasa Y 2009 *Case Studies in Superconducting Magnets, Design and Operational Issues Second Edition, Springer* 505

Detection of normal transitions in a hybrid single-phase Bi2223 high temperature superconducting transformer by using the active power method and a magnetic flux detection coil

N Nanato, S Nakamura, S Tanaka

Graduate School of Natural Science and Technology, Okayama University, 3-1-1 Tsushima Naka, Kita-ku, Okayama, 700-8530, Japan

nanato@okayama-u.ac.jp

Abstract. The authors have been developing a hybrid single-phase Bi2223 high temperature superconducting (HTS) transformer used in the AC current source with rated current of over 500A. Its primary coil is a copper coil and secondary coil is a Bi2223 HTS coil. In this paper, the authors propose a new detection method of normal transitions in the secondary coil by using the active power method and a magnetic flux detection coil attached on the inside of the secondary coil. In the proposed method, the normal transitions are detected by measuring active power dissipated in the secondary coil, and induced voltage of the magnetic flux detection coil by a primary and leakage flux of the secondary coil enables to calculate the active power dissipated in only the secondary coil. As experimental results for a hybrid single-phase Bi2223 HTS transformer, it was found that the proposed method enabled to detect the normal transitions in its secondary superconducting coil.

1. Introduction

The authors have been developing a small and light AC power source with a hybrid single-phase Bi2223 HTS transformer for supplying a large current. The transformer consists of a primary copper coil and a secondary Bi2223 HTS coil. It is important to detect normal transitions in the secondary coil to protect from excessive heating in the normal area. The authors have presented the active power method as a detection method of the normal transitions in a superconducting coil [1]. In the active power method, the normal transitions are detected as active power signal in the superconducting coil. The hybrid transformer has a primary copper coil and an iron core and therefore no-load losses are always consumed in the transformer. Then the conventional active power method cannot detect accurately the normal transitions due to the no-load losses [1].

The authors propose a new detection method by using the active power method and a magnetic flux detection coil attached on the inside of the secondary coil. The magnetic flux detection coil contributes to measure active power dissipated in only the secondary coil. This paper shows usefulness of the proposed method for detection of the normal transitions in the hybrid HTS transformer.

2. Protection tests for a hybrid single-phase Bi2223 HTS transformer

2.1. *Principle of detection of normal transitions by using a magnetic flux detection coil*

In the Hybrid single-phase Bi2223 HTS transformer, copper loss in the primary coil and iron loss are always consumed as no-load losses in the transformer. In conventional active power method, no-load losses are regarded as active power due to the normal transitions, therefore false detections may happen. In the proposed method, the voltage of the magnetic flux detection coil enables to calculate the active power dissipated in only the secondary coil. Mounting positon of the magnetic flux detection coil and an equivalent circuit converted to the primary side of the hybrid single-phase Bi2223 HTS transformer are shown in Figures 1 and 2, respectively. In Figure 2,

v_1 : primary voltage, i_1 : primary current, r_1 : primary resistance,
x_1 : leakage primary reactance, i_0 : excitation current, g_0 : excitation conductance,
b_0 : excitation susceptance, i_2 : secondary current, x_2 : leakage secondary reactance,
r_2 : secondary resistance generated after the normal transitions,
v_{SC2} : secondary voltage, Z : secondary load impedance,
a : turn ratio of the primary coil and secondary coil,
a' : turn ratio of the magnetic flux detection coil and the secondary coil ($a = a'$),
v_f : voltage of the magnetic flux detection coil.

As shown in Figure 1, the magnetic flux detection coil is mounted according to the secondary coil on the inside of the bobbin and therefore it generates an induced voltage by a primary flux and leakage flux of the secondary coil. Then $a'v_f$, which is a conversion voltage on primary side of v_f, is shown in Figure 2. Also the turn ratio a equals to a'. A secondary resistive voltage is detected by difference between $a'v_f$ and av_{sc2} (a conversion voltage on primary side of v_{sc2}) and active power dissipated in the secondary coil is calculated as follows.

$$P_2 = (a'v_f - av_{sc2})\frac{i_2}{a} = (a^2r_2\frac{i_2}{a})\frac{i_2}{a} = r_2{i_2}^2. \qquad (1)$$

In this method, the active power P_2' is not affected by the signal of primary winding resistance r_1 and excitation conductance g_0. Moreover, the protection system becomes more simple than the conventional one because the detection can be achieved by measuring only three signals of the secondary voltage, the secondary current and the voltage of the magnetic flux detection coil.

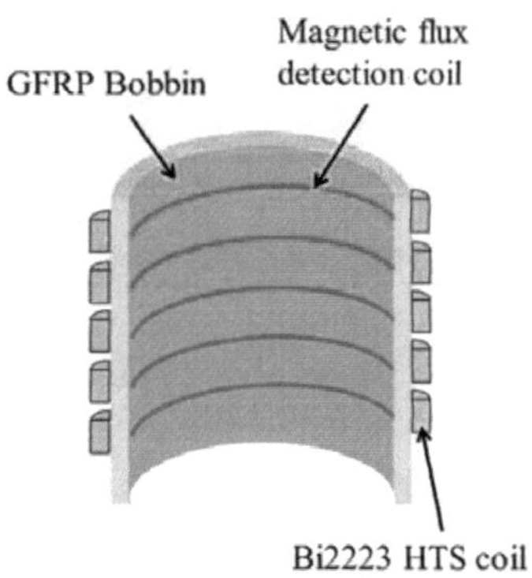

Figure 1. Mounting position of the magnetic flux detection coil.

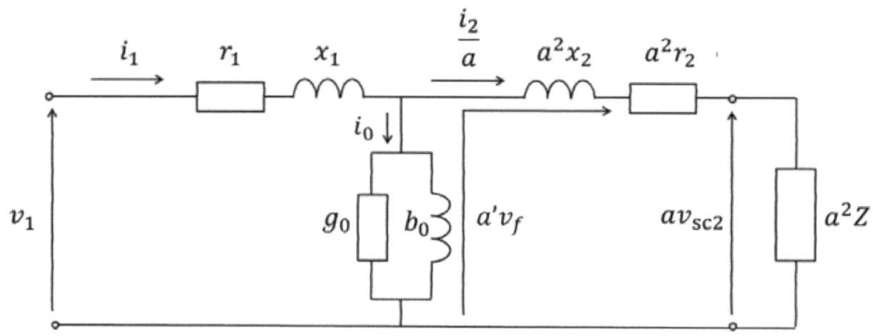

Figure 2. An equivalent circuit converted to the primary side of the hybrid single-phase Bi2223 HTS transformer.

2.2. *Protection tests of the hybrid HTS transformer*

In order to verify the above method, detection and protection tests for a hybrid single-phase Bi2223 HTS transformer were carried out. The configuration and specifications of a hybrid single-phase Bi2223 HTS transformer are shown in Figure 3 and Table 1. The primary copper coil is a square shape solenoid coil to supress a leakage flux and the secondary HTS coil is a cylindrical solenoid coil to supress bending stress. The cooling container is made of styrene foam and has a cylindrical bore for the iron core and then only the secondary HTS coil is cooled in liquid nitrogen. A protection circuit is shown in Figure 4. In order to reduce an electromagnetic noise in calculated P_2, it is filtered by a low pass filter. The resultant active power " P_2' " reaches a specified threshold P_{th}', the thyristor switches are turned off and transport current is shut off.

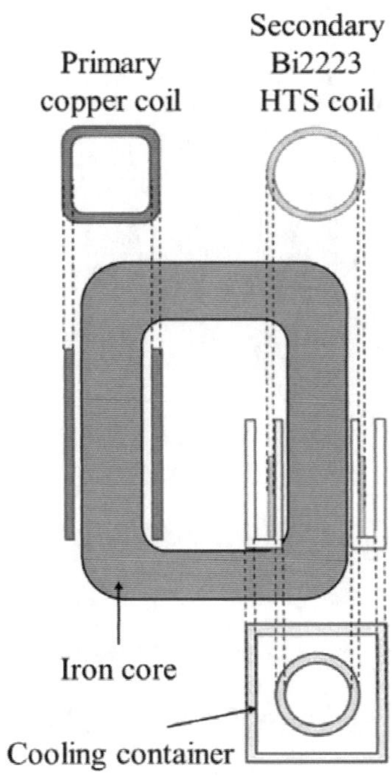

Figure 3. Configuration of the hybrid single-phase Bi2223 HTS transformer.

Table 1. Specifications of a hybrid single-phase Bi2223 HTS transformer.

	Primary coil (Copper)	Secondary coil (Bi-2223)
	53 mm (One side length)	83 mm (Inner diameter) 89 mm (Outer diameter)
Height	220 mm	60 mm
Number of Turns	120	5
I_C (at 77 K, self-field)	-	760 A

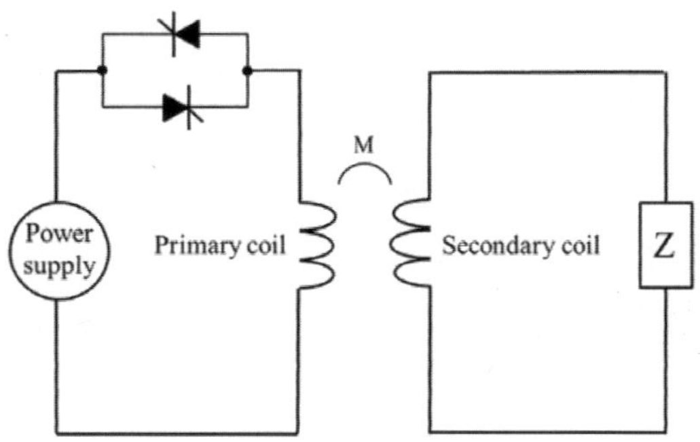

Figure 4. A protection circuit.

In the protection tests of the transformer, transport current of 21 A_{peak}, 60 Hz was supplied to the primary coil and then a current of 500 A_{peak} was supplied to the secondary coil. The secondary side circuit of the transformer was connected to a resistor of 1.1 mΩ (=Z). The normal transitions were occurred by a heater mounted on the secondary one. The tests for normal transitions in the secondary coil was carried out and the results are shown in Figure 5. Figures 5 (a)-(d) show expanded waveforms for time axes in a superconducting state. Figure 5 (a) shows the primary current of 21 Apeak, 60 Hz and (b) shows the secondary current of 500 A_{peak}, 60 Hz, which was transformed according to the turn ratio of the transformer as mentioned above. Figure 5 (c) show the voltage across the secondary coil and it equals to the voltage of the connected resistor. Figure (d) shows the voltage across the magnetic flux detection coil and it is the induced voltage by a primary flux and leakage one of the secondary coil. Figure 5 (e) shows the active power signal P_2' in the secondary coil before and after the normal transition. In a superconducting state until about 150 s, P_2' was constant with a small value. Then it increased drastically because the secondary resistance generated by the normal transition. It does not have no-load losses and therefore the normal transition can be detected with a higher SN ratio than that of the conventional method [1]. The thyristor switches were turned off when P_2' reached a specified threshold P_{th}'=25 W [2]. Figure 5 (f) shows temperature of the normal zone near the heater on the secondary coil. The maximum temperature of the secondary coil was suppressed to 178K which was lower than permissive temperature [3].

From the test results, it was verified that the protection system based on the proposed method worked successfully for the hybrid HTS transformer.

The small value shown in Figure 5 (e) in the superconducting state is supposed to be a signal due to AC loss in the secondary coil. The signal is much smaller than the threshold P_{th}' and therefore it causes no false detection.

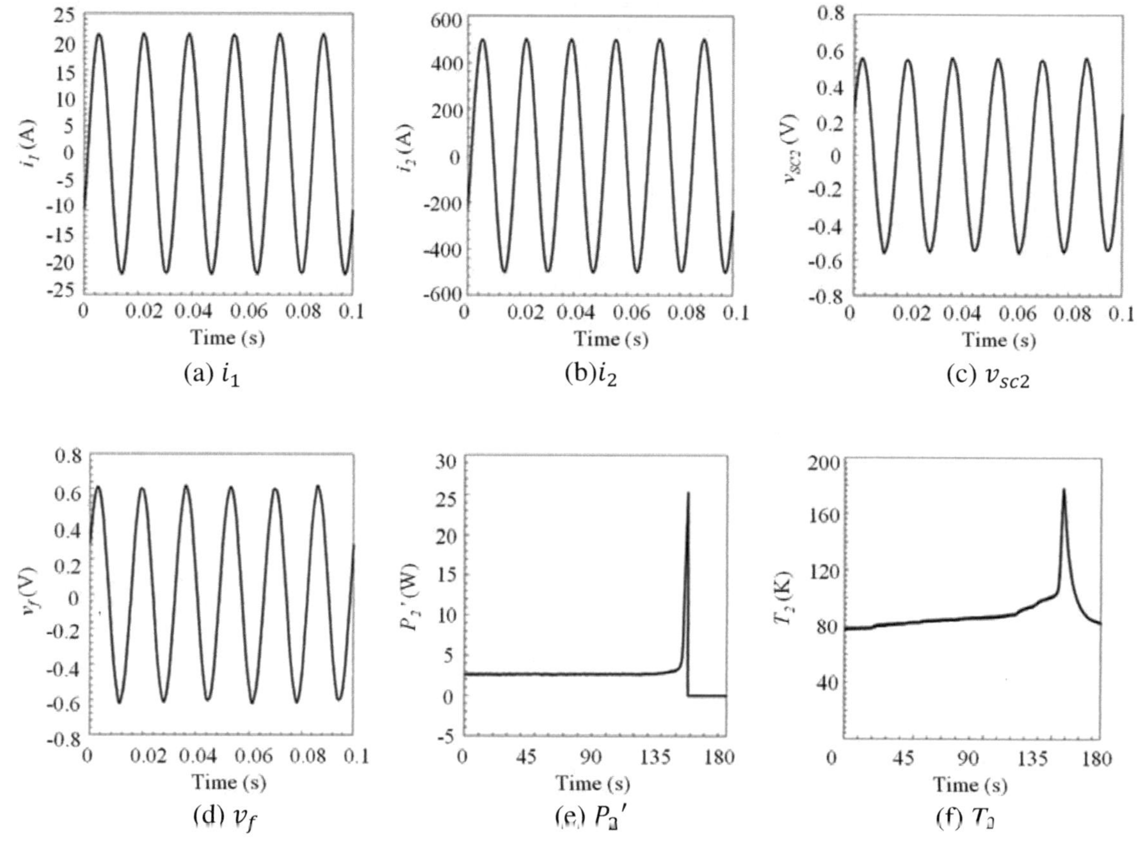

Figure 5. Test results for the hybrid single-phase Bi2223 HTS transformer.

3. Conclusions

The authors proposed a detection and protection system which consists of the active power method using the magnetic flux detection coil for normal transitions in a hybrid single-phase HTS transformer. The system can detect only the normal transitions in the secondary HTS coil regardless of the resistance of the primary copper coil and iron loss. They verified usefulness of the system for the hybrid HTS transformer through the experimental results.

Acknowledgements

This work was supported by JSPS KAKENHI Grant Number 15K18020.

References

[1] Nanato N, Kobayashi Y 2014 *Physics Procedia* **58** 264
[2] Nanato N, Asai W and Murase S 2012 *Physics Procedia* **27** 416
[3] Iwasa Y 2009 *Case Studies in Superconducting Magnets, Design and Operational Issues Second Edition, Springer* 468

IOP Conf. Series: Journal of Physics: Conf. Series **871** (2017) 012086 doi:10.1088/1742-6596/871/1/012086

Performance Test of a Metal Insulated HTS Magnet with Conduction Cooling

J Choi[1,2], C S Hwang[3], C K Lee[2] , S K Kim[3] , M Park[2] and I K Yu[2]
[1]Supercoil Co., Ltd, Changwon, GN, 51140, Korea
[2]Chanwon National University, Changwon, GN, 51140, Korea (corresponding author)
[3]Korea Electrotechnology Research Institute, Changwon, GN, 51541, Korea

yuik@changwon.ac.kr

Abstract. In the fabrication and operation of an HTS magnet, ensuring thermal stability against uneven quench is the most important factor. A sample HTS magnet was designed and fabricated with the metal insulation (MI) method and its fundamental characteristic analysis was conducted under the liquid nitrogen cooling system. On the basis of the electromagnetic analysis results, the thermal and mechanical structural design and detailed experimental analysis of the MI HTS magnet under the conduction cooling condition were performed in this paper. The conduction cooling condition was achieved using the 1st stage GM cryo-cooler. The characteristic resistances and the charging and discharging times of the magnet were measured according to the operating temperature of 32 K of the HTS magnet. The long-term current flowing test was conducted by monitoring the coil temperatures. In addition, the thermal stability of the HTS magnet was analyzed when the over current flowed into the magnet. The test results will be applied to the large size HTS MI magnet.

1. Introduction

In the fabrication and operation of an HTS magnet, ensuring thermal stability against uneven quench is the most important factor. The no insulation (NI) winding method allows the HTS magnet to maintain the best thermal stability by distributing the quench energy evenly [1-4]. The NI HTS magnet has the characteristic resistance connected to the inductance of the magnet in parallel. The charging and discharging time are dependent on the characteristic resistance. In the case of, however, the characteristic resistance of the NI HTS magnet is too small, the charging and discharging time takes too long to generate the desired magnetic field.

In 2013, our research team had developed and tested a NI HTS magnet for a 10kW-class prototype DC induction heater. In the previous research, the characteristic resistance was measured to be 98 μΩ, and it required 2,300 s to charge the current of the magnet of 230 mH with the maximum magnetic flux density [5-7]. Thus, we employed the winding method of metal insulation (MI) using stainless steel to reduce the long charging and discharging time. A sample HTS magnet was designed and fabricated with the MI method and its fundamental characteristic analysis was conducted under the liquid nitrogen cooling system. The characteristic resistance was measured to be 2.9 mΩ and the charging and time constant of the discharging mode was calculated to be 2.5 s. For the higher than critical current flowed to the sample magnet, detailed investigations through voltage and temperature characteristic analysis were conducted [8].

On the basis of the electromagnetic analysis results, the thermal and mechanical structural design of the MI HTS magnet under the conduction cooling condition were performed and the detailed experimental analysis results were presented in this paper. The minimum cooling temperature of the 1st

stage of the conduction cooling system was 14.6 K, while the total heat invasion was calculated to be approximately 25 W. The temperature of the HTS magnet was saturated to be 21.4 K after 6 hr. The inductance of the MI HTS magnet including iron core was calculated to be 52.5 mH. The characteristic resistance was calculated to be 20.6 mΩ. This characteristic resistance is higher than that of the MI HTS magnet without the stycast painted. At that time, the time constant was measured to be 2.5 s. The input current was hold on the 600 A, and the quench occurred. However, the HTS tape in the coil was not damaged, because the current bypassed into the different layers of the magnet.
Through the characteristic analysis results of the MI HTS magnet, the large scale implementation of 2G HTS technology for commercial application has been being researched. For various designs of the HTS commercial applications, their optimal designs will be presented.

2. Design of a MI HTS magnet under the conduction cooling system

Prior to the real fabrication and experiment, the cooling system with cryostat for the HTS magnet was designed. In this section, the heat transfer and mechanics analysis using a FEM tool was performed and the results were presented. When the conduction cooling system design is performed, the heat invasion components including conductive and radiative parts should be considered above all [9-12]. A MI HTS magnet with a cryostat was fabricated and tested using the conduction cooling method. As shown in Figure 1, the first stage GM cryo-cooler was installed on the cryostat and the MI HTS magnet was equipped with the cryostat using GFRP supporters. In addition, the metal current leads were connected to the current terminals of the magnet. The height of the cryostat was 709 mm. The cryostat radius was fabricated to be 240 mm. The detailed specifications for the cryostat with the conduction cooling system will be presented in this section.

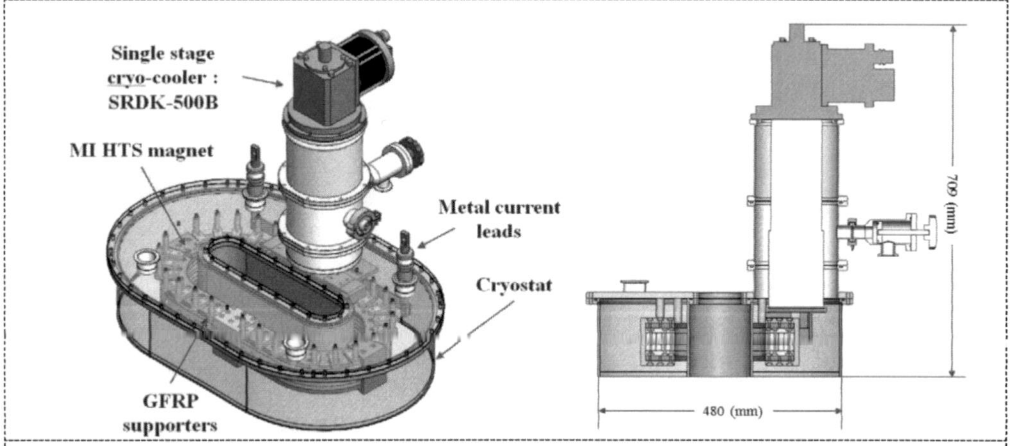

Figure 1. 3D design model of the MI HTS magnet with conduction cooling system.

2.1. Specifications of the MI HTS magnet and the cryostat

The experimental specifications of the MI HTS magnet and the cryostat were presented in detail, as indicated in Table 1. The HTS tape from the SuNAM in Korea was used for the MI HTS magnet. The minimum critical current of the HTS tape was measured to be 365 A under the LN_2. The total length of the HTS tape was 134 m. The inductance of the magnet was measured to be 4.5 mH. The critical current of the MI HTS magnet under the 77.4 K and 32 K were measured to be 196 A and 600 A, respectively. The single stage GM cryo-cooler donated by Sumitomo Korea was used on the fabricated MI HTS magnet. The saturated temperature of the 1st stage of the conduction cooling system was measured to be 14.6 K with no current condition, and the total heat invasion was 25 W as shown in the cooling curve of the cyro-cooler in Figure 2. The critical current at the target cooling temperature of 40 K of the MI HTS magnet was estimated to be 430 A, considering only the perpendicular magnetic flux density as shown in Figure 3. The perpendicular magnetic flux density was 1.9 mT at the current of 1 A from the FEM analysis results.

Table 1. The experimental specifications for the MI HTS magnet fabricated with the cryostat using a conduction cooling.

Parameter	Value
HTS tape maker	SuNAM, Korea
HTS tape (Width × Thickness)	W12.1 (±0.1) mm × T100 (±15) μm
SUS tape (Width × Thickness)	W12 mm × T100 μm
Minimum critical current (77 K)	≥365 A (copper laminated)
Substrate material	Hastelloy
HTS magnet type	MI, racetrack, double pancake
Size (Radius × Length)	R115 mm × L300 mm
Number of turns in a SPC	50
Total length of the HTS tape	134 m
Inductance of the magnet without iron core	4.5 mH
Critical current at 77.4 K	196 A
Critical current at 32 K (ramping rate 1 A/s)	600 A
Target operating current of the magnet	350 A at 40 K
Position of the magnetic field sensor	(x, y, z) = (27.5 mm,0,0)
B_{norm} at the center of the magnet	3.43E-4 (T/A)
Maximum perpendicular magnetic flux density from FEM analysis results	1.90E-3 (T/A)
Cryo-cooler	Sumitomo 1st GM cryo-cooler SRDK-500B
Total heat invasion	25 W under no load condition
Minimum cooling temperature	14.6 K at the 1st stage

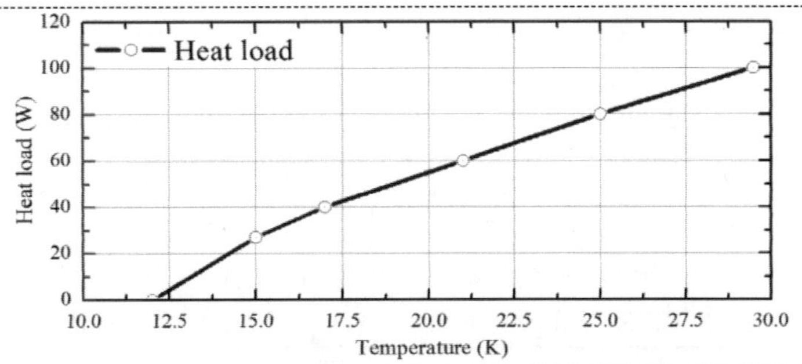

Figure 2. Cooling curve according the temperature of the single stage cryo-cooler (ref. by the Sumitomo Korea).

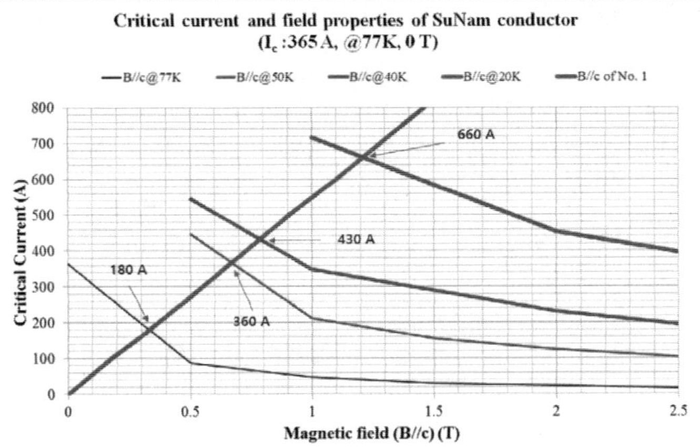

Figure 3. Critical current estimation curves of the MI HTS magnet according to operational temperatures.

2.2. Thermal analysis of the MI HTS magnet with conduction cooling system

The heat transfer analysis of the HTS magnet with GFRP supporters was performed and the results were presented in Figure 4. The 8 GFRP supporters were installed to fix the HTS magnet. Two brass current leads including the copper terminals on the room temperature of 300 K were connected to the 1st stage of the conduction cooling system. The total heat loads of 8 GFRP supporters were calculated to be 1.93 W, as described in Table 2. The two metal current leads have the heat loads of 42.7 W generated by the target current of 350 A under the temperature varying from 300 K to 20 K. The radiation loss of the magnet was calculated to be 3.73 W. After the cooling was completed, the saturated temperature of the 1st stage was 19.2 K and the highest temperature of the magnet was 24.4 K. The total heat load of the 1st stage was calculated to be 48.8 W.

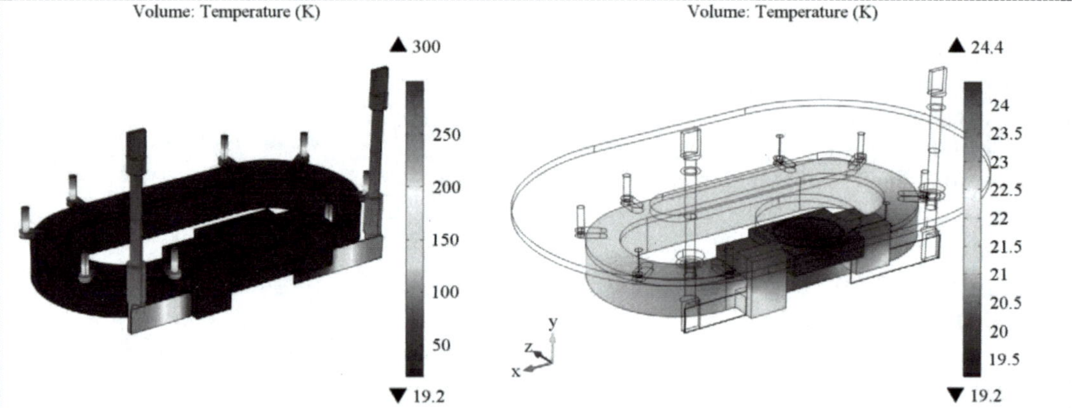

Figure 4. Heat transfer analysis results of the MI HTS magnet with the operating current of 350 A.

Table 2. Heat load calculation of the MI HTS magnet with the operating current of 350 A.

Stages	Heat transfer methods	Parameters	Qty.	FEM results
1st stage	Conduction	Metal current leads with the current of 350 A	2 ea	42.7 (W)
	Conduction	GFRP supporters (50 mm)	8 ea	1.93 (W)
	Conduction	Copper current leads at HTS magnets	2 ea	0.42 (W)
	Radiation	Radiative heat loads (MLI): emissivity:0.03	1 ea	3.73 (W)
Total heat load of the 1st stage				48.8 (W)

2.3. Thermal analysis of the MI HTS magnet with conduction cooling system

As shown in Figure 5, the body load condition caused by Lorentz forces from the HTS magnet was applied to the FEM analysis model and the fixed constraint at the bottom surface of the cryostat was applied for the stress analysis. The maximum stress of the MI HTS magnet was 13.5 MPa at the supporters arranged in the straight section of the cryostat. The HTS magnet was assumed that it was under solid state with the painted stycast, a kind of cryogenic epoxy. The coil maximum stress was calculated below 1 MPa. These results were acceptable according to the material properties of the GFRP [13].

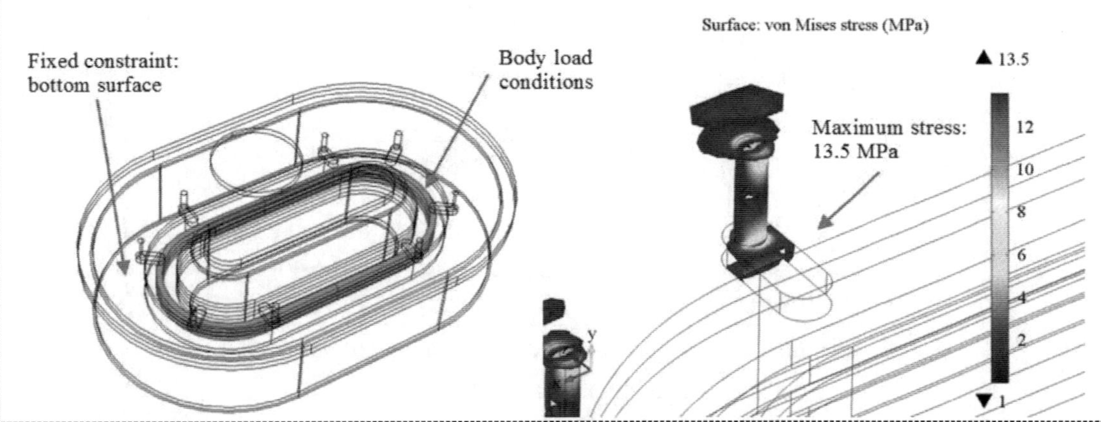

Figure 5. Mechanical structural analysis results of the MI HTS magnet.

3. Fabrication and the experimental results of the MI HTS magnet with conduction cooling system

3.1. Numbering

The fabrication process of the MI HTS magnet with conduction cooling system was described in Figure 6. First, the MI HTS magnet was fabricated and assembled with each part. The stycast was painted on the side of the magnet to improve the thermal contact between the bobbin cover and the HTS coil. The HTS magnet was installed on the cryostat and the two metal current leads were connected on the copper terminals of the HTS magnet. The multi-layer insulation (MLI) was wound around the HTS magnet to shield the radiation heat invasion.

Figure 6. Fabrication process of the MI HTS magnet with the conduction cooling system.

3.2. Conduction cooling test results

The temperature curves of the MI HTS magnet were presented in Figure 7 when the magnet cooled down. The minimum cooling temperature of the 1st stage of the conduction cooling system was 14.6 K, while the total heat invasion was calculated to be approximately 25 W. The temperature of the HTS magnet was saturated to be 21.4 K after 6 hr. In comparison with the thermal FEM analysis results in Figure 8, the difference is about 1.4 K as represented in Table 3. In the real conduction system, the radiation loss was minimized by over 25 turns of the MLI and heat load in the contact parts of the metal current leads decreased.

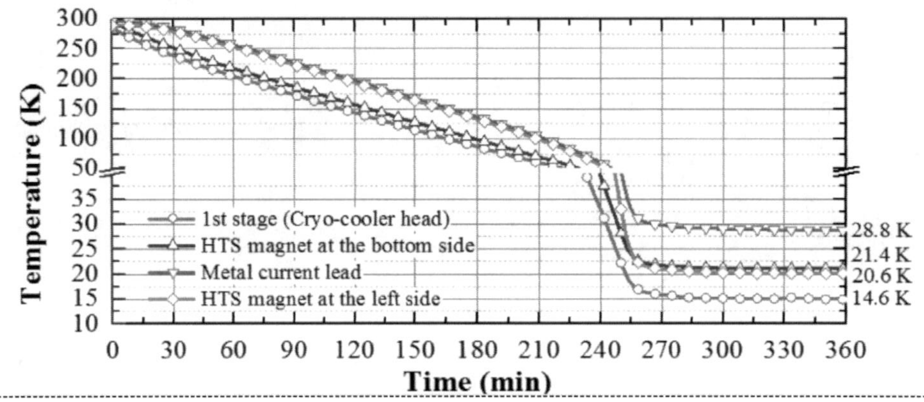

Figure 7. The cooling temperatures of the MI HTS magnet under the conduction cooling operation.

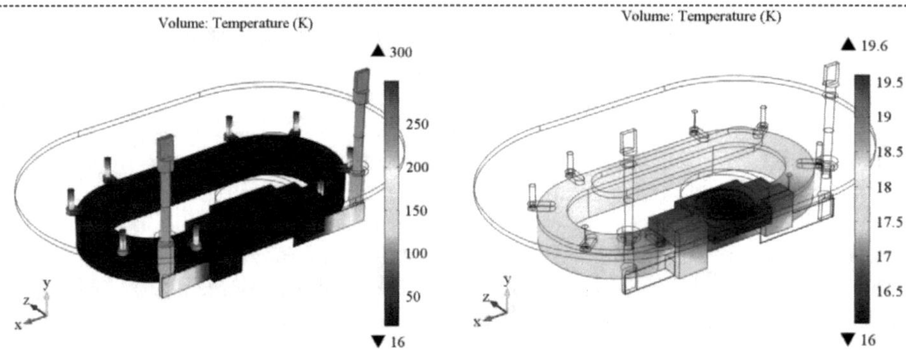

Figure 8. Heat transfer analysis results of the MI HTS magnet with no load condition.

Table 3. Comparison with heat loads of the MI HTS magnet between in FEM analysis and in the real experiment with no load condition.

Stages	Heat transfer methods	Parameters	Qty.	FEM results	Experimental results
1st stage	Conduction	Metal current leads with no load condition	2 ea	25.3 (W)	
	Conduction	GFRP supporters (50 mm)	8 ea	1.94 (W)	
	Radiation	Radiative heat loads (MLI): emissivity:0.03	1 ea	4.75 (W)	
Total heat load of the 1st stage				31.99 (W)	25 (W)
The saturated temperature of the 1st stage				15	14.6 (W)

3.3. Charging and discharging test results of the MI HTS magnet

The input current and the terminal voltage curves of the MI HTS magnet were presented in Figure 9. The inductance of the MI HTS magnet including iron core was calculated to be 52.5 mH. The series resistance of the MI HTS magnet was calculated to be 17.5 μΩ. In the case of the charging operation, the characteristic resistance was calculated to be 20.6 mΩ. This characteristic resistance is higher than that of the MI HTS magnet not to paint the stycast. At that time, the time constant was measured to be 2.5 s as shown in Figure 10. The temperature of the MI HTS magnet didn't increase at that point, the temperatures at the 1st stage and the metal current lead started increasing.

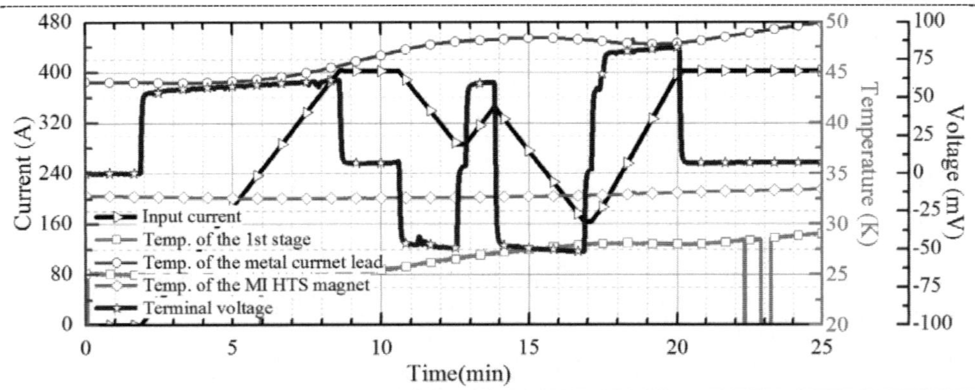

Figure 9. Charging and discharging curves of the MI HTS magnet at the specific temperature.

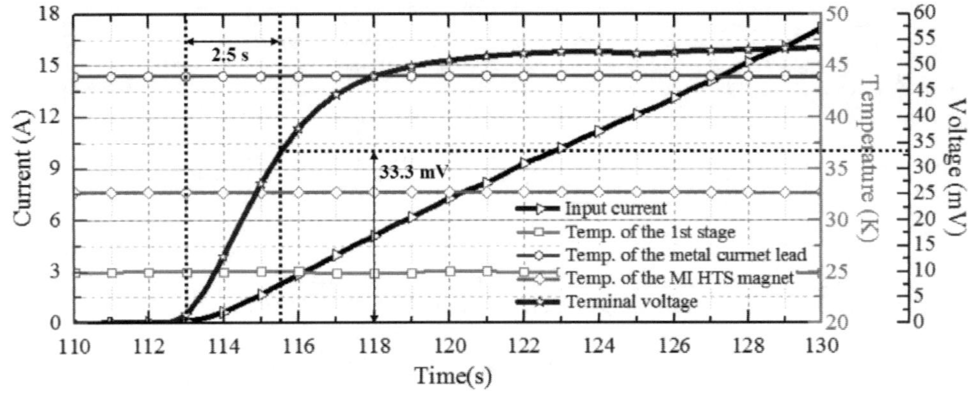

Figure 10. Charging curve of the MI HTS magnet.

3.4. *Over current test results*

When the input current flowed into the magnet until 600 A with the ramping rate of 1 A/s, the terminal voltage and temperatures were presented in Figure 11. The AC loss in the charging sequence was generated and it made the coil temperature increase slightly. The input current was hold on the 600 A, and then the quench occurred. However, the HTS tape in the coil was not damaged, because the current bypassed into the different layers of the magnet on the basis of the R//L circuit. More details about the magnet protection scheme of MI HTS magnet were referred to [8].

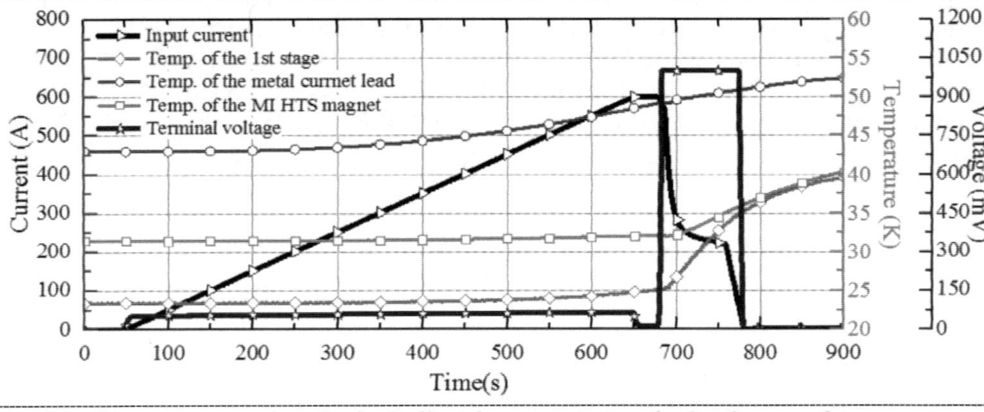

Figure 11. Overcurrent test results including the current, terminal voltage and temperature of the MI HTS magnet.

3.5. Thermal stability of the conduction cooling system during long-term current flowing test

The current of 350 A was supplied to the magnet for 8 hours and the temperatures at the 1st stage and the MI HTS magnet was saturated to be 36 K and 37.5 K, respectively, as presented in Figure 12. The results demonstrate that the long-term operation of the MI HTS magnet is stable at the target excitation current.

The final temperature of the HTS magnet is a very important factor to consider thermal stability for the HTS magnet to be re-excited within a reasonable timeframe after the quench. The total absorbable heat energy of the HTS magnet bobbin and other structures is calculated according to the temperature. The minimum quench duration is determined by considering the limitations of the temperature increases of the HTS magnets. The total absorbable heat energy of the bobbin and other structures in the cooling system was calculated as 300 kJ at the limiting temperature of 50 K. The quench duration should be below 1 s, after which time the HTS magnet is excited again.

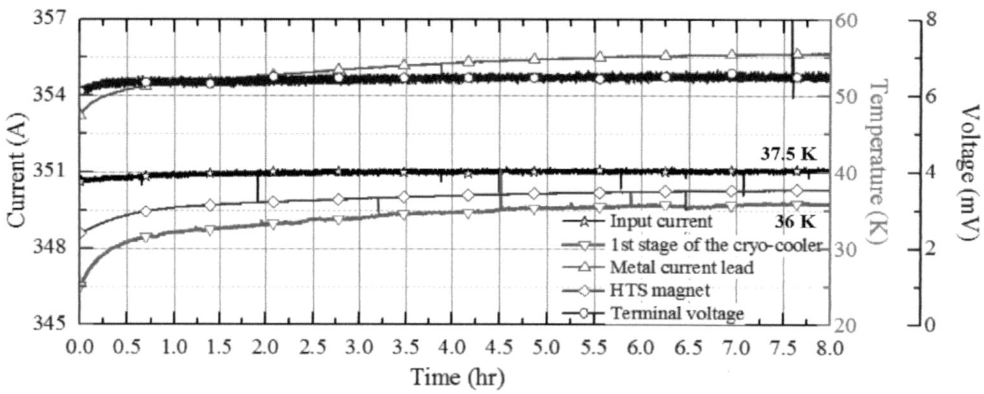

Figure 12. The temperature characteristics during the long-term current flowing test.

4. Conclusions

The thermal and mechanical structural designs of the MI HTS magnet under the conduction cooling condition were performed and the detailed experimental analysis results were presented in this paper. The saturated temperature of the 1st stage of the conduction cooling condition achieved at 14.6 K using the 1st stage GM cryo-cooler. The characteristic resistance was calculated to be 20.6 mΩ. The time constant of the charging and discharging sequences of the magnet were measured to be 2.5 s at the operating temperature of 24.5 K. The current of 350 A was supplied to the magnet for 8 hr and the temperatures at the 1st stage and the MI HTS magnet was saturated to be 36 K and 37.5 K, respectively. In addition, the thermal stability of the HTS magnet was tested, when the current of 600 A flowed into the magnet, continuously. The temperature of the magnet reached 40 K after about 200 s. It was demonstrated that the HTS magnet was thermally stable, even when the quench had occurred. The large scale implementation of 2G HTS technology for commercial application has been being researched through the characteristic analysis results of the MI HTS magnet in this study. The optimal designs for the superconducting application will be presented in the near future.

Acknowledgments

This work was supported by the Power Generation & Electricity Delivery Core Technology Program of the Korea Institute of Energy Technology Evaluation and Planning (KETEP), granted financial resource from the Ministry of Trade, Industry & Energy, Republic of Korea. (No. 20142020103940)

References

[1] Hahn S, Park D K, Bascuñán J and Iwasa 2011 *IEEE trans. on appl. Supercon.* **21**(3) 1592-1595
[2] Wang X, Hahn S, Kim Y, Bascuñán J, Voccio J, Lee H and Iwasa 2013 *Supercon. Sci. and Tech.* **26**(3) 035012

[3] Hahn S, Kim Y, Keun Park D, Kim K, Voccio J P, Bascuñán, J and Iwasa 2013 *Appl. Phys. letters* **103**(17) 173511
[4] Hahn S, Kim Y, Ling J, Voccio J, Park D K, Bascuñán J and Iwasa 2013 *IEEE Trans. on Appl. Supercon.* **23**(3) 4601705-4601705
[5] Choi J, Kim K, Park M, Yu I K, Kim S, Sim K and Kim H J 2014 *Physica C: Supercon. & its Appl.* **504** 120-126
[6] Choi J, Kim K, Park M, Yu I K, Kim S and Sim K 2015 *J. of Supercon. & Novel Magn.* **28**(2) 657-661
[7] Choi J, Kim K, Kim S K, Kim S, Sim K, Park M and Yu I K 2016 *Physics Procedia* **81** 149-153
[8] Choi J, Kim S K, Kim S, Sim K, Park M and Yu I K 2016 *IEEE Trans. on Appl. Supercon.* **26**(3) 1-5
[9] Yeom H K 2008 *Progress in Superconductivity and Cryogenics*, **10**(1) 62-66
[10] Han P, Wu Y, Liu H, Li L and Yang H 2015 *Physics Procedia* **67** 360-366
[11] Kim S, Sim K, Kim H J, Bae J H, Lee E Y, Seong K C and Jeong S 2009 *Cryogenics* **49**(6) 294-298
[12] Yeom H K, Hong Y J, Park S J, Seo T B, Seong K C and Kim H 2007 *IEEE Trans. on Appl. Supercon.* **17**(2) 1955-1958
[13] Kim K M, Kim A R, Park H Y, Kim J G, Park M, Yu I K and Won Y J 2010 *IEEE Trans. on Appl. Supercon.* **20**(3) 1900-1903

IOP Conf. Series: Journal of Physics: Conf. Series **871** (2017) 012087 doi:10.1088/1742-6596/871/1/012087

Study on the Persistent Current Switch in HTS Coils Wound with 2G Wire for Compact NMR Magnets

K Sugo, S B Kim, R Nomura and H Ueda

Graduate School of Natural Science and Technology, Okayama University, 3-1-1, Tsushimanaka, Kita-ku, Okayama 700-8530, Japan.

kim@ec.okayama-u.ac.jp

Abstract. High temperature superconducting (HTS) magnets wound with REBCO wires are used in many applications, and the superconducting magnet which is operated in persistent current mode has many advantages. Therefore, we have been developing compact NMR relaxometry devices using HTS coils operated at liquid nitrogen temperature. The required strength and homogeneity of magnetic field of proposed NMR relaxometry devices are 1.5 T and 150 ppm/cm^3 respectively. The proposed HTS magnet for NMR relaxometry device consists with stacked HTS double pancake coils wound with REBCO wires and operated by persistent current mode (PCM) using superconducting joint between REBCO wires. In PCM operation, the ability of persistent current switch (PCS) is very important, so, in this study, the thermal properties of YBCO wire against the various thermal inputs by heater were investigated experimentally to design the PCS for PCM HTS magnets. The thermal behaviors of the YBCO wires were measured as functions of amount of heat input using two types of epoxy resin. The current bypassing properties on the YBCO loop coil with developed PCS were studied experimentally.

1. Introduction

We have been developing the compact NMR/MRI device using HTS bulk magnets [1], [2], and the compact NMR relaxometry device using HTS wires with the persistent current mode (PCM) operation [3]. In PCM operation, the superconducting joints between stacked HTS coils are required. So, the superconducting joint techniques between HTS wires are very important issue and many researches has been studying. Recently, the K·JOINS, Inc. has developed successfully the high performance superconducting joints between REBCO tape wires by partial melting diffusion and oxygenation annealing process [4]. The development of high performance persistent current switch (PCS) is required to develop the compact NMR relaxometry operated in PCM. In this study, thin strain gauges were used as thermal heater and the thermal characteristics of two types of epoxy resin (putty and liquid) were compared to develop the compact PCS operated at liquid nitrogen temperature. Therefore, the thermal behaviors of YBCO wires by the heaters were studied experimentally to design the thermal type PCS for NMR relaxometry. In order to investigate the PCM using the developed PCS, the loop-shaped YBCO wire fabricated by wind-and-flip winding method [5]-[7] was prepared and the performances of the developed PCS operating in liquid nitrogen bath were estimated.

IOP Conf. Series: Journal of Physics: Conf. Series **871** (2017) 012087 doi:10.1088/1742-6596/871/1/012087

2. Thermal diffusion properties of 2G wires to develop the PCS

2.1. Experimental details

In order to design the PCS for NMR relaxometry, the thermal behaviors of YBCO wires by heater were investigated experimentally. Figure 1 shows the schematic draws of front views of YBCO wires impregnated by putty and liquid epoxy resins used in experiments. The 100 μm thickness YBCO wires are used, and the thickness of Cu stabilizer is 20 μm, thickness of YBCO layer is 1 μm, and the critical current at 77.4 K is 100 A. The four strain gauges with 1 kΩ were used as a heater and they were attached to the front and back sides at center region of sample wire. The transport current of heater was 50 mA and the thickness, length and width of heater are 50 μm, 11 mm and 5 mm, respectively. Four thermocouples were attached on the front side with 1 cm interval as shown in figure 1. The thermal properties of YBCO wires against the various input heat were measured using two types of epoxy resin (putty and liquid), and the samples were cooled by liquid nitrogen.

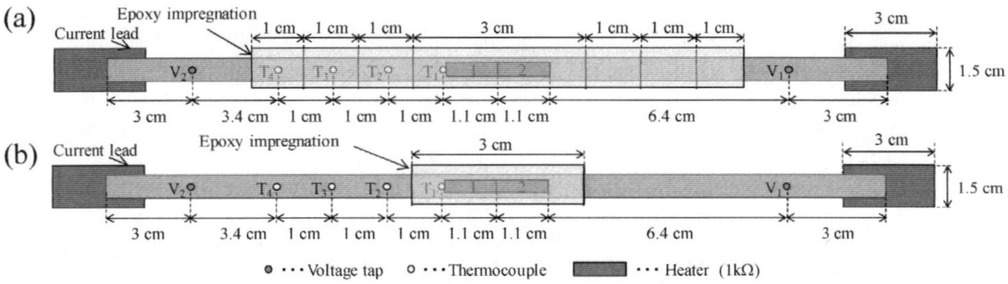

Figure 1. The schematic draws of front views of YBCO wire impregnated by (a) putty and (b) liquid epoxy resins

2.2. Measured the thermal properties of YBCO by heater

The thermal behaviors of YBCO wires with/without transport currents (0 A and 20 A) against the different input heat (10 W×30 s and 10 W×3 s) were measured using two types of epoxy resin (putty and liquid) in the liquid nitrogen bath. The impregnation lengths of putty and liquid epoxy are 9 cm and 3 cm, respectively, and the thickness of both epoxy resins is 1 cm. Figures 2 and 3 show the measured temperature properties and distributions in the longitudinal direction of YBCO wire impregnated by putty and liquid epoxy resin in the liquid nitrogen bath. They were measured at T1-T4 when the transport currents are 0 A (input heat:10 W×30 s) and 20 A (input heat:10 W×3 s). From figures 2 and 3, the maximum temperatures of both epoxy resins were measured at T1, and the maximum temperature on putty and liquid epoxy resins with 0A are 139.4 K and 202.1 K, and 122.3 K and 149.1 K when the 20 A transported. The temperature gradient in the longitudinal direction of YBCO wire impregnated by the liquid epoxy resin is larger than the putty epoxy resin even the length of putty epoxy is longer than that of liquid epoxy when the 0 A and 20 A were transported. The results of subtracting the temperature profile of the putty epoxy resin (Fig.2(a)) from the temperature profile of the liquid epoxy resin (Fig.2(b)) after thermal inputting by the heater are shown in figure 2 (c). On inputting the heat, the temperature at T1 of the liquid epoxy resin rises several tens of kelvin higher than that of the putty epoxy resin, and the temperature distribution from T1 to T3 also higher. When the heat input by the heater was finished, the required time of liquid epoxy resin (27 s) for the temperature at T1 to drop from 140 K to the liquid nitrogen temperature (77 K) is also longer than that of the putty epoxy resin (10 s). Therefore, it is clarified that the heat insulating ability of the liquid epoxy resin is much better than the putty epoxy resin. The cooling condition of the epoxy resin is determined by thermal conductivity, heat capacity and surface contact condition between epoxy resin and sample wire. We know the values of thermal conductivity and heat capacity of liquid epoxy resin, but these values of putty epoxy resin are still unknown. In this experiment, we were not able to measure the surface contact condition between epoxy resin and sample wire. However, it was proved that the thermal conductivity of the liquid epoxy resin is lower than that of the putty epoxy resin, and the liquid epoxy resin is suitable for making an adiabatic conduction. When the same amount of heat has been inputted, the thermal insulation material which can make a high temperature gradient in the short area is very effective for the compact thermal type PCS. From these results, it is suitable to use a liquid epoxy resin as a thermal insulation material for the thermal type PCS.

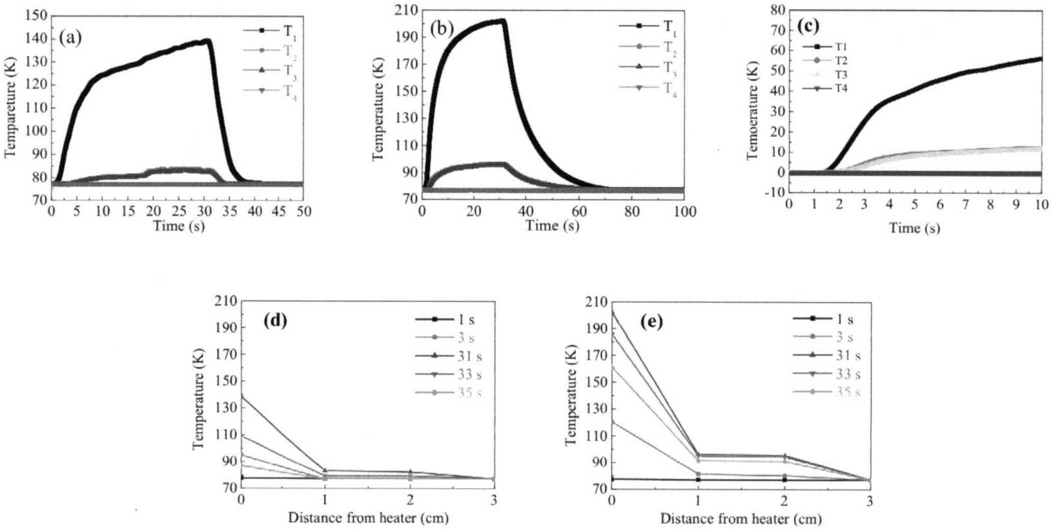

Figure 2. The measured temperature profiles and temperature distributions in the longitudinal direction of YBCO wire impregnated by (a), (d) putty and (b), (e) liquid epoxy resins at T1-T4 when the transport current is 0 A (input heat:10 W×30s). Figure (c) shows the temperature profile obtained by subtracting figures (b) to (a).

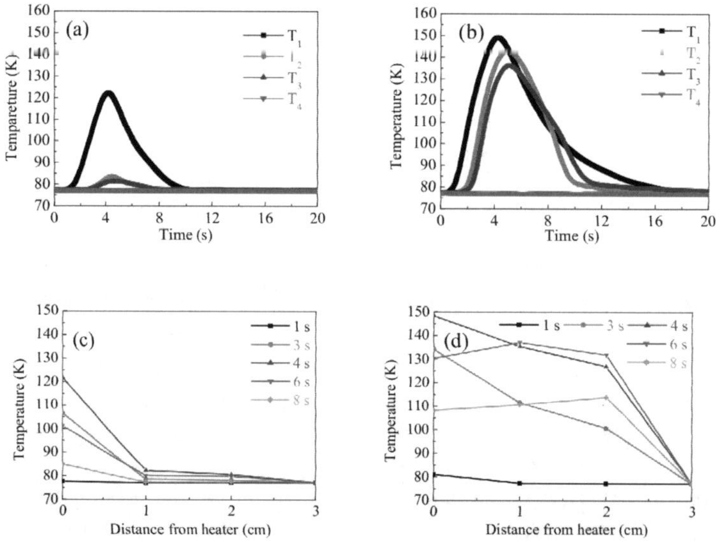

Figure 3. The measured temperature profiles and temperature distributions in the longitudinal direction of YBCO wire impregnated by (a), (c) putty and (b), (d) liquid epoxy resins at T1-T4 when the transport current is 20 A (input heat:10 W×3 s).

3. Division of Current in the Loop-shaped YBCO Wires by Thermal PCS

In order to confirm the performance of the developed PCS, the experiments using the stacked HTS DP coils with superconducting joints are necessary. However, in this study, we performed the experiments using a simple loop-shaped YBCO wire instead of the HTS magnet.

3.1. Experimental details and the current division by different inductance

The loop-shaped HTS coil made by wind-and-flip winding method [5] - [7] was prepared and provided by W.S. Kim in Korea Polytechnic University. Figure 4 (a) shows the schematic top view of the loop-shaped YBCO wire. YBCO wire (SuperPower Inc.) with 12 mm width was cut longitudinally at its center except both of end parts, and a joint-less superconducting closed loop circuit consists of sliced YBCO wire with 6 mm width was prepared as shown in figure 4 (b). Figure 4 (c) shows the photograph of experimental setup to estimate the performances of the developed PCS. This coil is cooled down at 77 K in the liquid nitrogen bath. The four thermocouples were attached at near the heaters with 1 cm interval to measure the thermal properties during the PCS operating. Figure 5 shows the measured temperature distribution in the longitudinal direction when the thermal inputs are delayed with 2 and 3 seconds from the current transporting. In order to surely realize a PCM, it is necessary to create a normal region of over 90 K which is a critical temperature of the YBCO wire as long as 2.2 cm in PCS at least. The two hall sensors (B_1, B_2) were attached on the surface of YBCO wire to measure the amount of both currents (I_1, I_2). In the superconducting closed loop circuit as shown in figure 4, the current flows to the two paths proportional to their inductances because both paths have no electrical resistance. In this case, the inductances of both paths are proportional to the length of their path, and the length of shorter path with PCS is 280 mm and the other one is 580 mm. Figure 6 shows the measured magnetic flux density (B_1, B_2) by Hall sensor and the calculated current (I_1, I_2) by (B_1, B_2) when the transport current is 20 A. The current flows to the PCS (I_1) is two times larger than I_2 because the current flows to the two paths in the superconducting closed loop is inversely proportional to their inductances.

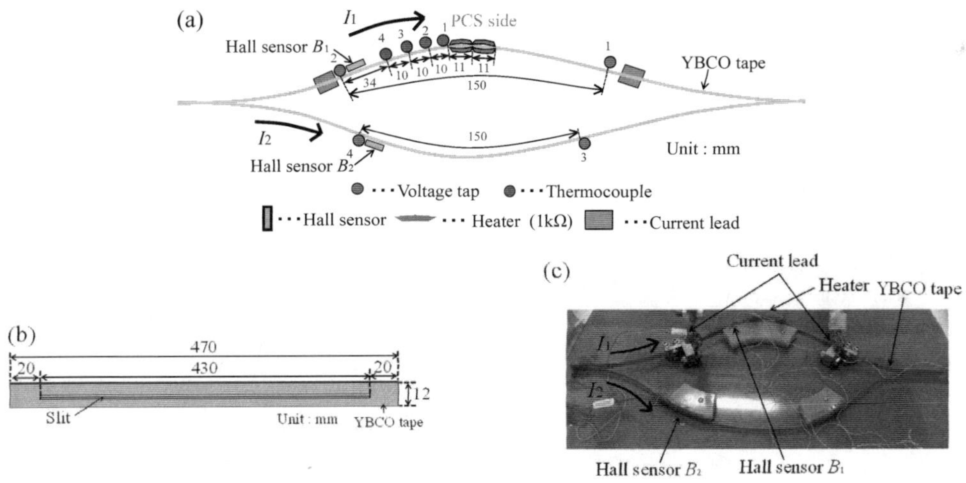

Figure 4. (a) the schematic top view of the loop-shaped YBCO wire, (b) the schematic draw of YBCO wire with 12 mm width was cut longitudinally at its center except both of end parts and (c) the photograph of experimental setup.

IOP Conf. Series: Journal of Physics: Conf. Series **871** (2017) 012087 doi:10.1088/1742-6596/871/1/012087

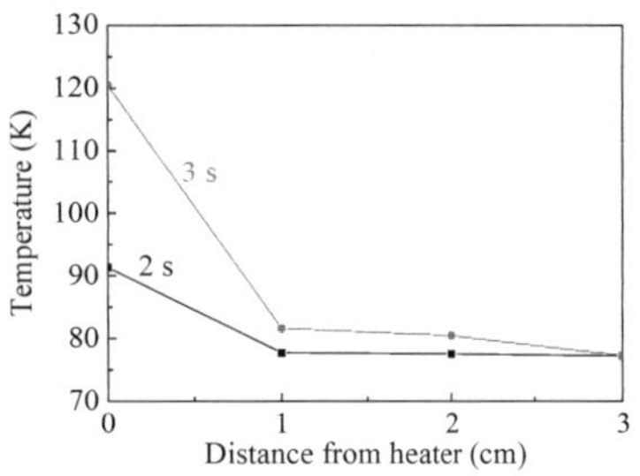

Figure 5. The measured temperature distribution in the longitudinal direction.

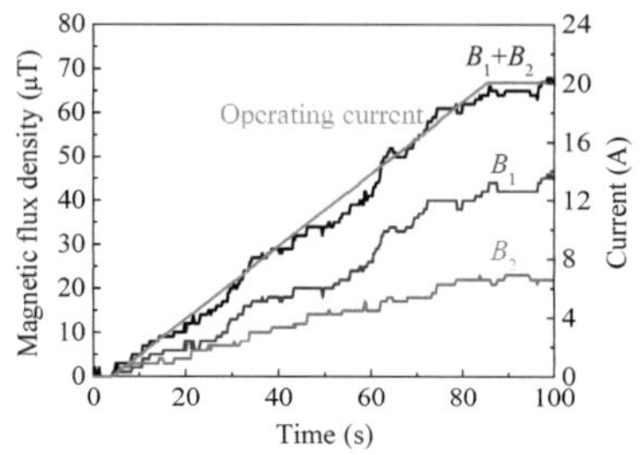

Figure 6. The measured magnetic flux density (B_1, B_2) by Hall sensors and the calculated current (I_1, I_2) by B_1 and B_2 when the 20 A was transported.

3.2. Measured the turn on/off properties of PCS

The stability of generated magnetic field operated in the persistent current mode is better than driven mode superconducting magnets used the achievable best power supplies, and no more energy is needed. In order to go to PCM, the supply current has been adjusting until the desired magnetic field was obtained, and then PCS was turned off. It is necessary to determine the time to stop the current transporting in closed loop circuit from the time of stopping the thermal input in PCS. Therefore, in this study, the sufficient cooling time to recovery the normal stated PCS to the superconducting state was studied experimentally, and the cooling time of PCS was investigated until the 13 s with 1 s interval.

Figures 7 (a) and (b) show the measured path currents and thermal input, and transported current was started to reduce to 0 A with 10 and 11 seconds delayed from stopping the thermal input of PCS. In Figure 7 (a), PCS was not sufficiently cooled, the current I_2 was reduced to 6 A because there were still remained the electrical resistance and operated as RL circuit. However, when the cooling time was longer than 11 seconds, PCS was fully recovered to the superconducting state, the persistent current mode was obtained as shown in figure 7 (b). So, we know that the required cooling time in order to fully return from the normal state to the superconducting state of the developed thermal PCS is required over 10 seconds.

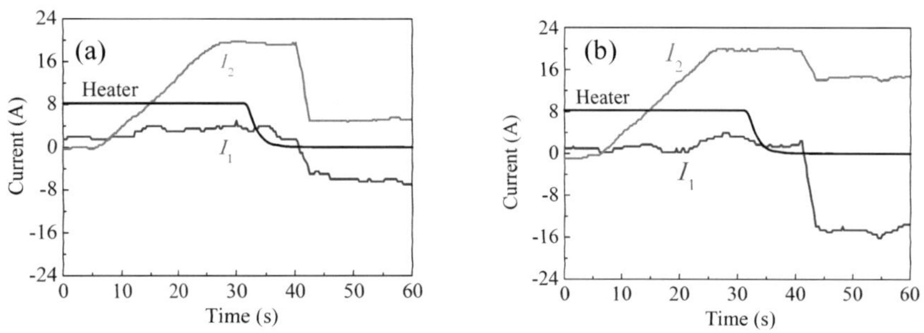

Figure 7. The measured currents I_1 and I_2 by 20 A, and the thermal inputs by heater are delayed with (a) 2 and (b) 3 seconds from the current transporting.

4. Conclusion

We have been developing the thermal type PCS operating in PCM for compact NMR devices. In order to design the thermal insulation of PCS for NMR relaxometry, the thermal properties of YBCO wires by input heat were investigated experimentally. From the results of thermal behaviors using the two different epoxy resins, the liquid epoxy resin is better than putty epoxy resin for using thermal insulation, even though putty epoxy resin is able to be handled easily. From the simple experimental test using the liquid epoxy resin and a loop-shaped YBCO wire which has two different inductances, the PCM operating was confirmed by developed thermal type PCS.

References

[1] Kim S B, Kitamura H, Ishizuka D and Miyazawa D 2015 The study to improve the field homogeneity of the NMR relaxometry magnets using the packaged HTS bulks with various iron rings *IEEE Trans. Appl. Supercond.* **25** no.3 4602204.

[2] Kim S B, Hayashi N, Kitamura H and Eritate I 2014 Study on the remagnetization process to improve the field homogeneity of the stacked HTS bulk magnets *IEEE Trans. Appl. Supercond.* **24** no. 3 4301405.

[3] Kim S B, Takahashi M, Saito R, Park Y J, Lee M W, Oh Y K and Ann H S 2015 Current bypassing properties by thermal switch for PCS application on NMR/MRI HTS magnets *Physics Procedia* **65** 149-152.

[4] Park Y J, Shin H J, Kim Y G, Oh Y K and Lee H 2013 Effects of melting diffusion and annealing in oxygen on superconducting characteristics of GdBCO coated conductors: preliminary results *IEEE Trans. Appl. Supercond.* **23** no.3 6600804.

[5] Kim W S, Park C, Park S H, Lee J, Song J B, Lee H, Lee H G, Hong G W and Choi K 2009 Magnetic field stability of a small YBCO magnet in persistent current mode *IEEE Trans. Appl. Supercond.* **19** no. 3 2194-2197.

[6] Lee S, Kim W S, Kim Y, Park S H, Lee J K, Hahn J H, Hong G W, Park I H, Park C and Choi K 2013 Characteristics of an HTS pancake coil in persistent current mode using wind-and-flip winding method *IEEE Trans. Appl. Supercond.* **23** no. 3 4601305.

[7] Kim W S, Lee S, Kim Y, Lee J Y, Park S H, Lee J K, Hong G W, Han J and Choi K 2015 Persistent current mode of a 1-T-Class HTS pancake coil for NMR/MRI applications *IEEE Trans. Appl. Supercond.* **25** no.3 4301504.

Numerical study to obtain the improved field homogeneity of HTS bulk magnet with enlarged inner diameter for compact NMR

S Fukada, S B Kim, T Nakagawa, R Nomura and H Ueda

Graduate School of Natural Science and Technology, Okayama University, 3-1-1, Tsushimanaka, Kita-ku, Okayama 700-8530, Japan

kim@ec.okayama-u.ac.jp

Abstract. The strength and homogeneity of the magnetic field required for NMR relaxometry device are 1.5 T and 150 ppm/cm^3 respectively. It is relatively easy to generate the trapped magnetic field over 1.5 T at 77.4 K using the stacked HTS bulks, and 150 ppm/cm^3 field homogeneity was obtained using the fabricated field compensation methods on HTS bulks magnet with 20 mm inner diameter. However, it is still hard to obtain 150 ppm/cm^3 field homogeneity without field compensation methods. In this paper, in order to improve the trapped magnetic field homogeneity, the HTS bulk magnet with enlarged inner diameter was proposed and studied as the functions of size and shape of HTS bulk. The analytical study based on FEM was carried out to optimize the shape of HTS bulk magnet with enlarged inner diameter.

1. Introduction

Nuclear magnetic resonance (NMR) device is used for analyzing tool of organic compound and has been paid attention in food, medicine and drag industries. The NMR signal resolution is improved by increasing the magnetic field strength, and 1 GHz-class (23.5 T) NMR magnet wound with low temperature superconducting conductors such as NbTi and Nb_3Sn has been already developed. On the other hand, a new type NMR magnet consisted of the stacked HTS bulk annuli (ring-shaped) where magnetic fields were trapped by field cooling method had been suggested and fabricated by Nakamura at RIKEN in Japan [1], [2]. The magnetically charged HTS bulk magnet for NMR device does not need a power supply and additional coolant supply system. Therefore, this is not only economical but also compact [3], [4]. Furthermore, the NMR relaxrometry device was produced by Stelar s.r.l., and its field homogeneity is 150 ppm/cm^3 [5]. We have been developing a new compact NMR magnet using stacked HTS bulk magnets. The strength and homogeneity of the magnetic field required for NMR relaxometry device are 1.5 T and 150 ppm/cm^3 respectively, these values are higher than the permanent magnet NMR and lower than the conventional NMR device. In a previous study, we have already achieved the target values of 1.5 T and 150 ppm/cm^3 by using passive and active field compensation methods [6], [7] and remagnetization method [8]. However, it is still hard to obtain 150 ppm/cm^3 field homogeneity using the conventional superconducting magnet (SCM) without field compensation methods. The magnetic field distributions and intensities of the HTS bulk depend on the magnetic field of SCM. The field homogeneity of our SCM is more than 600 ppm/cm^3, and it is impossible to achieve 150 ppm/cm^3 without any shimming methods. In general, the NMR magnet with a large sample space is required to provide the space for shimming coils. In this study, the shape design of HTS bulk magnet was studied using FEM based analysis to obtain the field homogeneity of 150 ppm/cm^3 without field compensation methods and to provide the enlarged inner space.

2. HTS bulk magnet for NMR relaxometry

In this study, $GdBa_2CuO_x$ (GdBCO) HTS bulks and SCM were used. Each annulus HTS bulk have 5 mm thickness, 20 mm inner diameter and 60 mm outer diameter. The SCM (JASTEC.Inc.) which have been used in our laboratory have a 100 mm room temperature bore size and 10 T magnetic field strength. The field homogeneity of SCM is 610 ppm/cm^3 in the center region along the ±5 mm axial direction, and the almost same scale SCM was used in analysis. In this study, the trapped magnetic fields of HTS bulk magnet were obtained by field cooling (FC) method with 1.6 T magnetization field at 77.4 K. The current flowed in the HTS bulk during the FC process were induced by the Bean's critical state model and n-value model. The trapped magnetic fields of HTS bulks were calculated by their currents. In this study, the properties of the enlarged inner diameter (ID) HTS bulk magnet were investigated. Figure 1(a) shows the scaled cross-sectional drawing of the "Basic model" of HTS bulk magnet with various ID when the outer diameter (OD) was fixed at 60 mm. Figure 1(b) shows the "Hollow model", the inside of a center part of HTS bulk magnet having an inner diameter of 20 mm is hollowed out in the radial direction. The height (5 and 10 mm) and diameter of enlarged hollow space were used as parameters when the OD was fixed at 60 mm. Figure 1(c) shows the "Bugle model" with the various height (5 and 10 mm) and diameter of the enlarged hollow space while maintaining the radial thickness of HTS bulk at 20 mm. Therefore, the diameter of the enlarged hollow space in figures 1(b) and 1(c) is defined as ID, and the total volume of HTS bulk of each model was different. So, in this study, the three models were examined with various inner diameter and height of hollow space to investigate the optimal shape that enhances the field homogeneity of HTS bulk magnets.

As a result of analysis, the calculated field distributions along z-axis and r-axis are shown in figure 2. In figure 2, ID of Basic model is 20 mm, ID of Hollow and Bugle models is 30 mm and the height of hollow space is 5 mm. From figure 2, the magnetic field strengths of the Hollow model and the Bulge model were lower than the Basic model. This result is due to the difference of volume of the HTS bulk in the radial direction of the measurement space. Figure 3 shows the calculated field homogeneity at the center region along ±5 mm in the z-axis and r-axis directions of the three models as shown in figure 1. From figure 3, the magnetic field homogeneity of the Basic model and the Hollow model was deteriorated because the presence area of the magnetic flux was enlarged by expanding inner diameter. In addition, the field homogeneity of r-axis is usually superior to z-axis, as shown in figure 3, so we mainly discuss the field distribution and homogeneity of z-axis. Figure 4 shows the calculated field homogeneity along ±5 mm in the z-axis direction of Hollow and Bulge models as a function of ID value. From figure 4, the Bulge model has better uniformity than the Hollow model because the part protruding outermost of the Bulge model plays the role of a notch coil as shown in figure 5. Figure 5 shows the calculated current density distributions in the r-z plane of the Hollow and Bulge model magnets. From figure 5, supercurrent of the Bulge model locally flows to bulging part. Therefore, degradation of uniformity due to the expansion of inner diameter was suppressed in the Bulge model. From the above results, the Bulge model should be suggested to enlarge the ID of HTS bulk magnet with high field homogeneity.

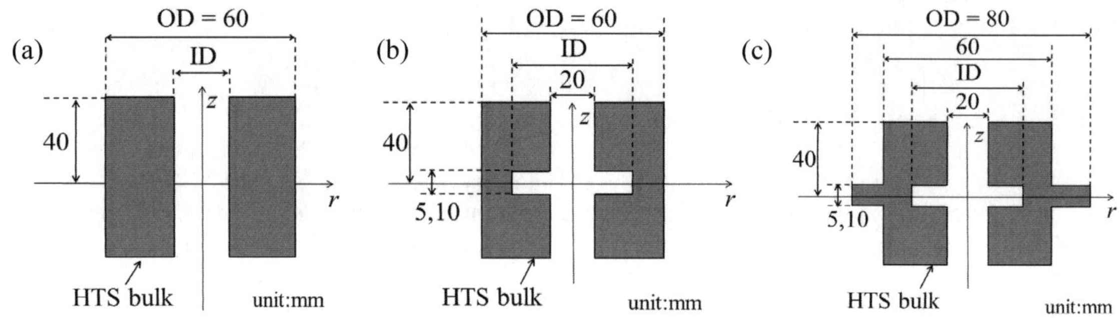

Figure 1. Scaled cross-sectional drawing of the (a) Basic model, (b) Hollow model and (c) Bulge model.

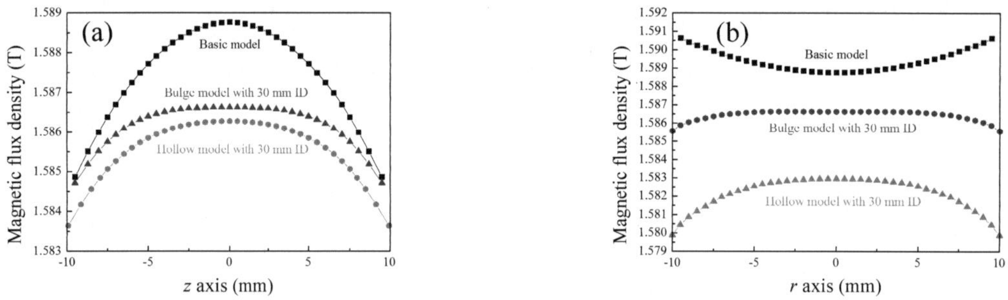

Figure 2. Calculated field distributions on the (a) z-axis and (b) r-axis when the height of enlarged hollow space in Hollow and Bulge models is 10 mm.

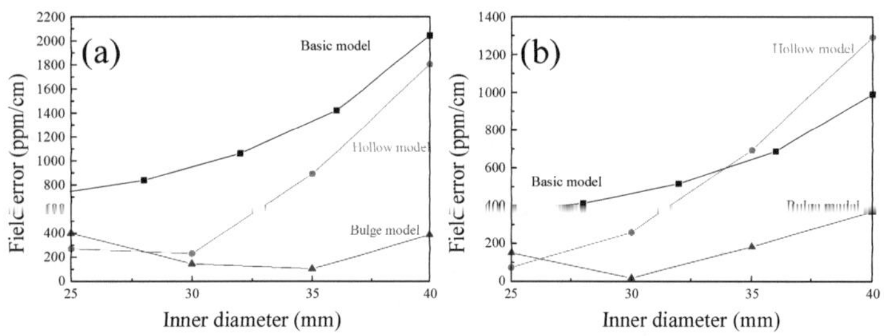

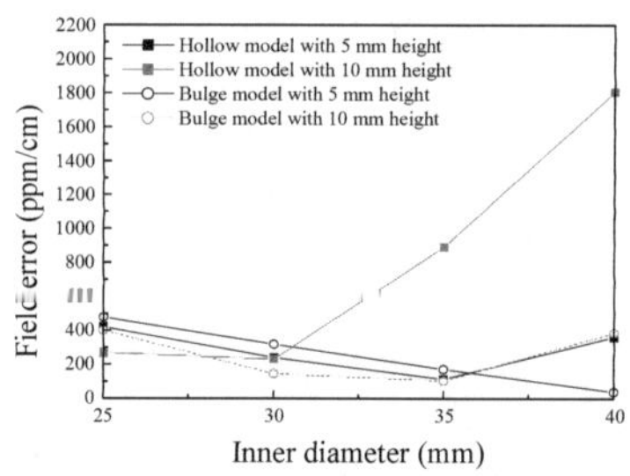

Figure 3. Calculated field homogeneity on the (a) z-axis and (b) r-axis directions as a function of ID value when the height of enlarged hollow space of Hollow and Bugle models is 10 mm.

Figure 4. Calculated field homogeneity on the z-axis direction as the functions of ID value and height of enlarged hollow space.

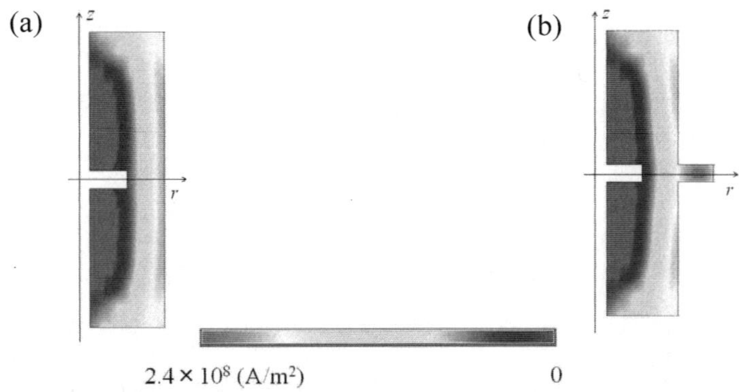

Figure 5. Calculated the current density distributions in the *r-z* plane of (a) Hollow model and (b) Bulge model.

3. Field distribution and homogeneity of Bulge model

It is clear that the Bulge model is suitable for enlarging the inner diameter and improving the magnetic field homogeneity. Therefore, the possibility of enlarging the inner diameter was investigated as functions of ID and OD values when the heights of the hollow space are 4 and 5 mm.

3.1. The Bulge model with 5 mm and 10 mm height of enlarged hollow space

In order to investigate the influence by the size of the ID of the Bulge model, an analytical model with a fixed OD are prepared firstly. Figure 6 shows the calculated field distributions in *z*-axis as a function of ID value when the heights of hollow space are 5 mm and 10 mm and the OD of HTS bulk is fixed at 80 mm. Figure 7 shows the calculated field homogeneity along ±5 mm in the *z*-axis and *r*-axis directions as a function of ID value when the heights of hollow space are 5 mm and 10 mm. From figure 6, the magnetic field strength at center part was reduced with increasing ID because the total volume of HTS bulk was decreased with increasing the ID. From figure 7, the magnetic field homogeneity was deteriorated because the presence area of the magnetic flux was enlarged by expanding ID. In the Bulge model with fixed 80 mm OD, there was no ID value to achieve the target magnetic field homogeneity of 150 ppm/cm in the height and the radial directions. Therefore, we examined the Bulge model with various height and diameter of the hollow space while maintaining the radial thickness of HTS bulk at 20 mm. Figure 8 shows the calculated field homogeneity along ±5 mm in the z-axis and *r*-axis directions when the heights of hollow space are 5 mm and 10 mm. As shown in figure 8, the target field homogeneities below 150 ppm/cm were obtained when the ID 40 mm, OD 80 mm and 5 mm height of hollow space, and ID 30 mm, OD 70 mm and 10 mm height.

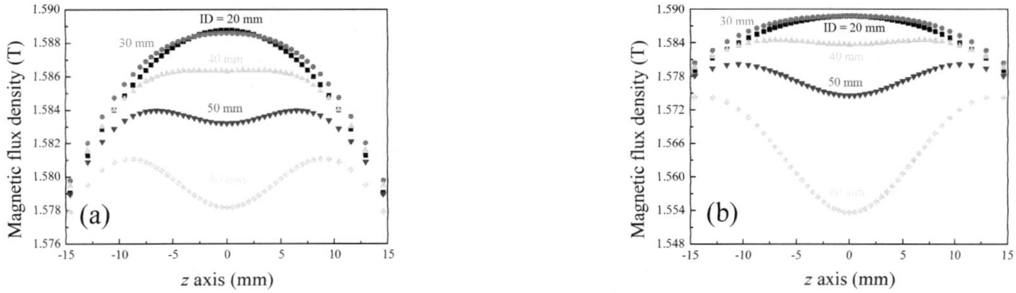

Figure 6. Calculated the field distributions along *z*-axis direction of Bulge model with fixed 80 mm OD as a function of ID value and the heights of enlarged hollow space are (a) 5 mm and (b) 10 mm.

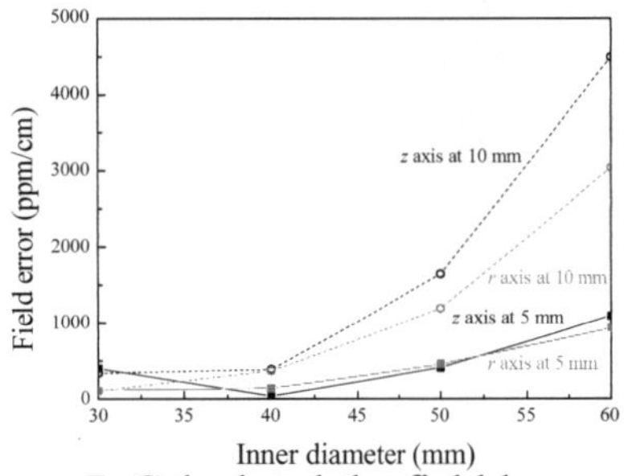

Figure 7. Calculated the field homogeneity on the z-axis and r-axis directions of Bulge model with fixed 80 mm OD as a function of ID value and the heights of enlarged hollow space are 5 mm and 10 mm.

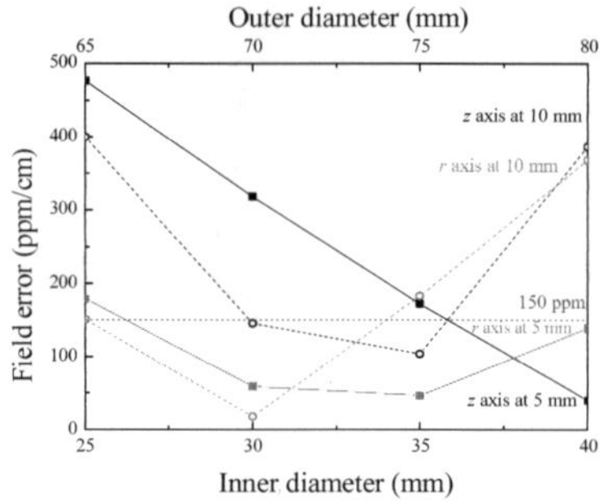

Figure 8. Calculated the field homogeneity on the z-axis and r-axis directions of Bulge model maintaining the radial thickness of HTS bulk at 20 mm as a function of ID value and the heights of enlarged hollow space are 5 mm and 10 mm.

3.2. *Reducing of the Number of HTS Bulk*

Reducing the number of HTS bulk is directly linked to reducing the cost and size of HTS bulk magnets. So, the analytical study to reduce the number of HTS bulk without decreasing the field homogeneity was carried out using modified HTS bulk magnets used for real applications as shown in figure 9. In previous section, the target field homogeneities below 150 ppm/cm were obtained in the two Bulge models with ID 40 mm, OD 80 mm, 5 mm height and ID 30 mm, OD 70 mm, 10 mm height, respectively. Therefore, we considered how much the number of HTS bulk could be reduced while maintaining the target field homogeneity using two Bulge models. Figure 9(a) shows the model which has 5 mm height of hollow space and HTS bulks were stacked from 11 to 17. Figure 9(b) shows the model which has 10 mm height of hollow space and HTS bulks were stacked from 8 to 16. Figure 10 shows the calculated field homogeneity along ±5 mm in the z-axis and r-axis directions as a function of the number of bulks when the gap length between HTS bulks except center region were 1 mm and 3 mm. We succeeded to reduce the total number of HTS bulk from 16 up to 10 while keeping the target field homogeneity of 150 ppm/cm.

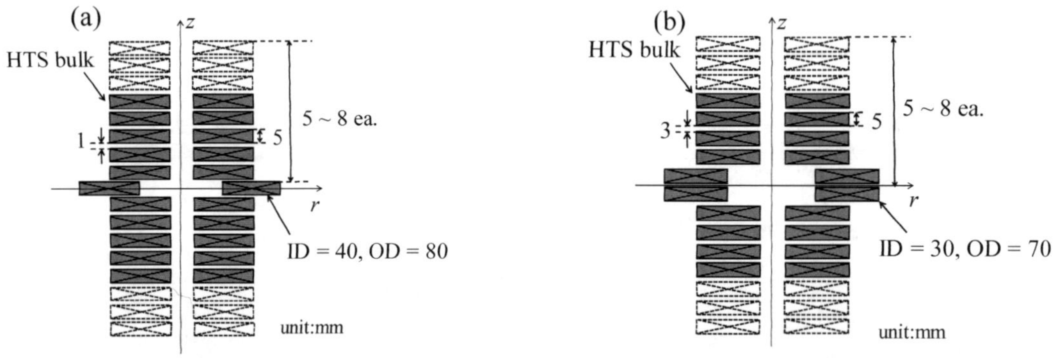

Figure 9. Scaled cross-sectional drawing of Bulge model magnets with different enlarged hollow space to study the reducing the number of HTS.

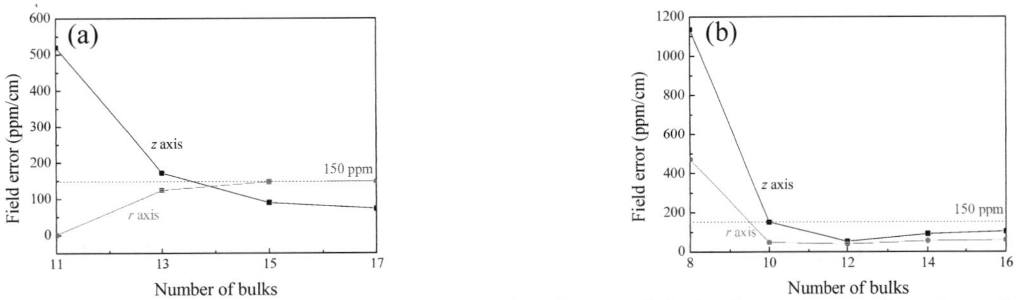

Figure 10. Calculated the field homogeneity of Bulge models in figure 9 as a function of the number of HTS bulks.

4. Conclusions

We have been developing NMR relaxometry magnet using stacked HTS bulk with inner diameter of 20 mm. NMR relaxometry magnet requires the field strength of 1.5 T and field homogeneity of 150 ppm/cm^3. It is easy to obtain the field strength. However, it is still hard to obtain 150 ppm/cm^3 field homogeneity. In this study, FEM based analysis was carried out to optimize the HTS bulk magnet for NMR relaxometry with enlarged inner diameter. The bulge model was mainly studied analytically because it has better uniformity than the single model and the hollow model. The field homogeneities at sample space were improved with enlarged ID and OD HTS bulks located at centre part in the bulge model. We obtained the high field homogeneity using the proposed 16-stacked bulge model, and succeeded to reduce the number of HTS bulk to 10-stacked while keeping the target field homogeneity of 150 ppm/cm^3.

References

[1] Nakamura T, Itoh Y, Yoshikawa M, Oka T and Uzawa J 2006 Development of a superconducting magnet for nuclear magnetic resonance using bulk High-Temperature superconducting materials *Concepts in Magnetic Resonance Part B* p 65.

[2] Iwasa Y 2006 HTS and NMR/MRI magnets: Unique features, opportunities, and challenges *Physica C* **445-448** pp 1088-1094.

[3] Hahn S, Kim Y, Voccio J, Park D K, Gagnon B, Bascuñán J, Tomita M and Iwasa Y 2013 Bulk and Plate Annulus Stacks for Compact NMR Magnets: Trapped Field Characteristics and Active Shimming Performance *IEEE Trans. Appl. Supercond* **23** no. 3, 4300504.

[4] Ogawa K, Nakamura T, Terada Y, Kose K and Haishi T 2011 Development of a magnetic resonance microscope using a high Tc bulk superconducting magnet *Appl. Phys.* L98 234101.

[5] http://www.ffcrelax.com/schoolNMR/school2009/download/pdf/Ferrante/SchoolNMR.pdf

[6] Kitamura H, Kim S B, Hayashi N, Ishizuka D and Miyazawa D 2014 The Study to Improve the Field Homogeneity of the HTS Bulk Magnets for NMR Relaxometry Device by Passive Compensation Methods *Physics Procedia* **58** 298-301.

[7] Kim S B, Ishizuka D, Kitamura H and Miyazawa D 2015 Study on the Shape Optimization of the HTS Bulk Magnets With Active Compensation for Compact NMR Relaxometry Magnets *IEEE Trans. Appl. Supercond* **25** no. 3 4602404.

[8] Kim S B, Hayashi N, Kitamura H and Eritate I 2014 Study on the Remagnetization Process to Improve the Field Homogeneity of the Stacked HTS Bulk Magnets *IEEE Trans. Appl. Supercond* **24** no. 3 4301405.

IOP Conf. Series: Journal of Physics: Conf. Series **871** (2017) 012089 doi:10.1088/1742-6596/871/1/012089

Design of conduction cooling system for a high current HTS DC reactor

Van Quan Dao, Taekue Kim, Thang Le Tat, Haejin Sung, Jongho Choi, Kwangmin Kim, Chul-Sang Hwang, Minwon Park, and In-Keun Yu

Changwon National University, 20 Changwondaehak-ro, Changwon 51140, Rep. of Korea

yuik@cwnu.ac.kr

Abstract. A DC reactor using a high temperature superconducting (HTS) magnet reduces the reactor's size, weight, flux leakage, and electrical losses. An HTS magnet needs cryogenic cooling to achieve and maintain its superconducting state. There are two methods for doing this: one is pool boiling and the other is conduction cooling. The conduction cooling method is more effective than the pool boiling method in terms of smaller size and lighter weight. This paper discusses a design of conduction cooling system for a high current, high temperature superconducting DC reactor. Dimensions of the conduction cooling system parts including HTS magnets, bobbin structures, current leads, support bars, and thermal exchangers were calculated and drawn using a 3D CAD program. A finite element method model was built for determining the optimal design parameters and analyzing the thermo-mechanical characteristics. The operating current and inductance of the reactor magnet were 1,500 A, 400 mH, respectively. The thermal load of the HTS DC reactor was analyzed for determining the cooling capacity of the cryo-cooler. The study results can be effectively utilized for the design and fabrication of a commercial HTS DC reactor.

1. Introduction

Large electric power systems, such as high voltage direct current (HVDC) transmission systems, need a DC reactor with high inductance and high transport current. However, systems typically experience a lot of electrical losses due to the resistance of the metal winding. HTS magnet has zero electric resistance under DC current conditions and it is possible to increase the capacity of the transport current, hence, reduce the reactor's size, weight, and electrical losses [1]-[4].

An HTS DC reactor magnet needs cryogenic cooling to achieve and maintain its superconducting state. There are two cooling methods; the one is pool boiling the HTS magnet in cryogen, and the other is conduction cooling by connecting the HTS magnet to a cryo-cooler directly or indirectly in a vacuum chamber. The conduction cooling system is more effective, smaller, and lighter than the pool boiling method [5]-[7]. An optimal design of the conduction cooling system is an important factor to operate the reactor stably and effectively.

This paper discusses a design of conduction cooling system for a high-current HTS DC reactor. The inductance and the operating current of the HTS DC reactor were 400 mH and 1500 A, respectively. Dimensions of conduction cooling system parts including HTS magnets, bobbin structures, current leads, support bars, and thermal exchangers were calculated and drawn using a 3D CAD program. A finite element method model was developed for determining the optimal design parameters and analyzing the thermal characteristics. For the heat load calculation, the temperatures of the 1st stage, 2nd stage, and radiation shield were 38 K, 7.4 K and 77 K, respectively. The simulation results were analyzed and are described in detail. This study results can be effectively utilized for the design and fabrication of a commercial HTS DC reactor.

2. Design of a 1,500 A, 400 mH class HTS DC reactor

2.1. Structure design of the HTS DC reactor magnet

Figure 1 shows the structure and size of a toroid-type HTS DC reactor magnet. The D-shape HTS double package coil (DPC) was applied to the reactor magnet. The toroid-type magnet consists of 30 DPCs. The inner diameter and outer diameter of the reactor magnet are 439.24 mm and 933.81 mm, respectively. The D-shape DPCs are arranged at an angle of 12° from each other based on the central axis of the toroid-type magnet. That is, all coils constituting the toroid-type magnet are arranged at identical intervals. The wire length of one D-shape DPC is 101.8 m, and the total wire length of the toroid-type HTS DC reactor magnet is about 3,054 m. The basic structure was referred to the previous research [2], and the major dimensions of the HTS DC reactor were modified to minimize the cooling capacity of the cryo-coolers in this paper.

Two layered GdBCO HTS wires were used for the reactor magnet. The critical current () was 1,200 A at a temperature of 77 K. The wire width and thickness are 12 mm and 0.61 mm. The calculated inductance of the reactor magnet was 404.59 mH. The detailed specifications of the 1,500 A, 400 mH class toroid-type HTS DC reactor magnet are shown in table 1.

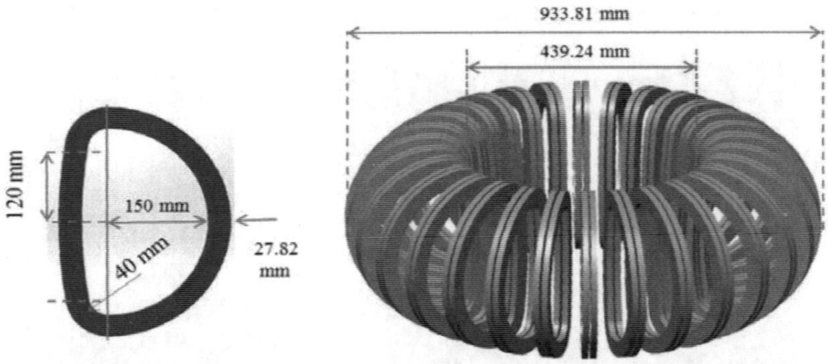

Figure 1. Detail size of the D-shape DPC and magnet structure

Table 1. Heat load of 1,500 A, 400 mH HTS DC reactor

Parameter	Value
Wire type	2G HTS wire
Thickness of wire	0.61 mm
With of wire	12 mm
Operating current	1,500 A
Magnet inductance	404.59 mH
Length of wire (SPC/DPC)	50.9/101.8 m
Number of DPC	30 ea.
Total length of wire	3,054 m
Number of turns	57 turns
Width of magnet	933.81 mm
Height of magnet	370.76 mm

2.2. Electromagnetic analysis result of HTS DC reactor magnet

A 1/10 numerical model of the toroid-type DPC HTS DC reactor magnet was built in the Matlab program as shown in figure 2. The flux density of HTS DC reactor magnet was calculated by a numerical method based on Biot-Sawart law. The highest magnetic flux density area of the D-shape coil needs to be considered for determining the critical current of HTS coil. The magnetic flux density results were shown in figure 3. The maximum perpendicular and the parallel flux density were 1.57 T and 4.37 T, respectively. The values obtained by the numerical calculation were applied to all of the DPCs in order to determine the parameters of the magnet.

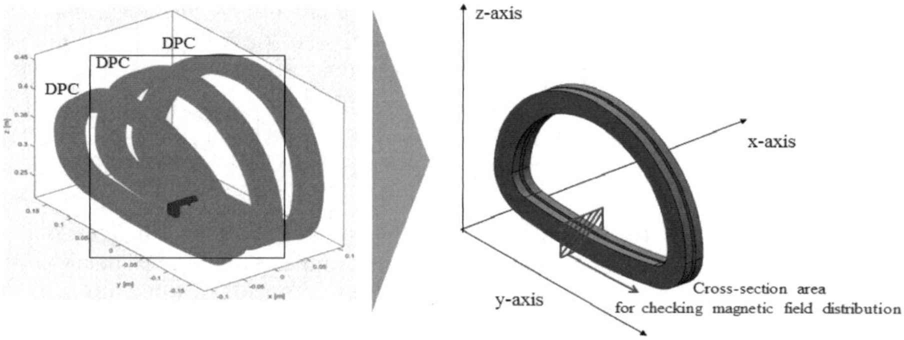

Figure 2. 1/10 numerical model of the toroid-type HTS DC reactor magnet

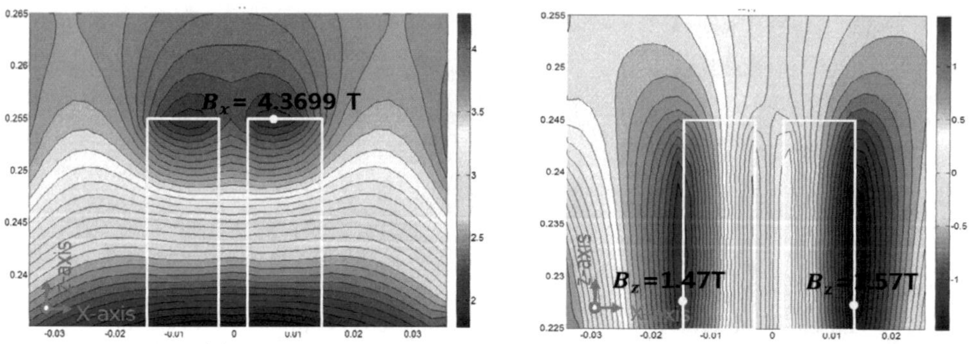

(a) (b)

Figure 3. Magnetic flux density in D-shape DPC: (a) Parallel flux density, (b) Perpendicular flux density

2.3. Conduction cooling system design

In this study, a conduction cooling system was used for cooling the toroid-type HTS DC reactor magnet. In this cooling method, the single-stage (RDK-400) and two-stage (RDK-415D) type Gifford-McMahon (GM) cryo-cooler (Sumitomo Corp.) were adopted. The conduction cooling system of HTS DC reactor mainly consists of the 1st and 2nd stage areas as shown in figure 4.

The 1st stage area is structured to carry currents via the DC reactor magnet in the room temperature section. A current supplied from the outside into the inside of the cryostat through two 800 A class current feedthroughs and then a current is supplied to the reactor magnet through brass current lead, copper stick, and HTS current lead. The 2nd stage area cools HTS DC reactor magnet. 30 DPCs were arranged in a toroid shape through the heat exchanger located in the lower and upper side of the DPC.

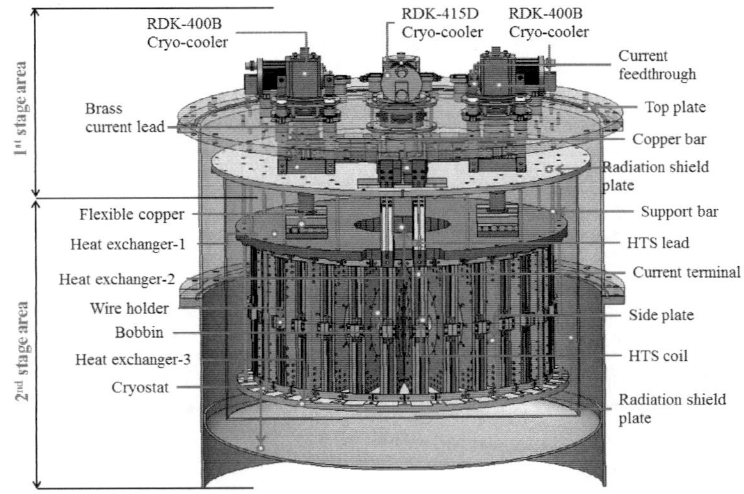

Figure 4. Configuration of the conduction cooling system of HTS DC reactor

2.3.1. Design of the 1st stage area. In the 1st stage area of the conduction cooling system, three cold heads of the GM cryo-coolers (single and two-stage) are responsible for the cooling of current leads and radiation shields.

Several important factors must be determined in order to design current leads, including the choice of materials and the lead geometry (length, cross section area, cooling surface area) [8, 9]. Figure 5 shows schematic layout of a conduction current lead with a length of L, a cross sectional area of A, and a carrying current of I. and represent the temperature of the warm and cold ends, respectively. Assume the lead has a uniform temperature distribution on the cross section, such that k(T) and (T) are the thermal conductivity and the electric resistivity of the material. The minimum heat load can be written as (1), and the optimal length of the current lead can be determined by (2):

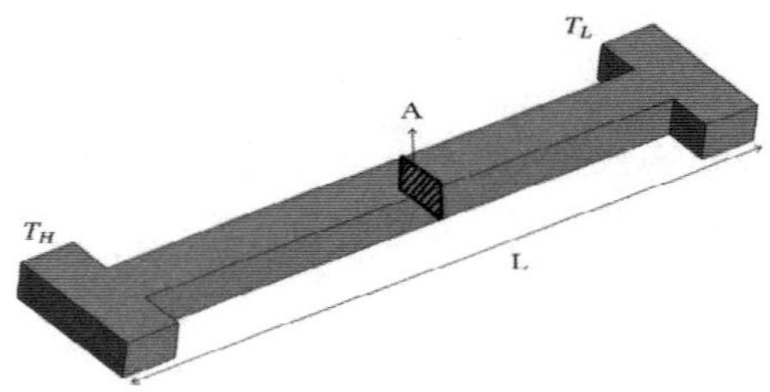

Figure 5. Schematic layout of a conduction current lead

$$Q_{op} = I\left(2 \int_{T_L}^{T_H} \rho(T)k(T)dT \right)^{1/2} \quad (1)$$

$$\frac{L}{A} = \frac{1}{I} \int_{T_L}^{T_H} k(T) \left(2 \int_{T_L}^{T_H} \rho(T)k(T)dT \right)^{-1/2} dT \quad (2)$$

For pure metals, k(T) and (T) are inversely related, according to the Wiedemann-Franz law:

$$k(T)\rho(T) = L_0 T \tag{3}$$

Here, $L_0 = 2.45 \times 10^{-8}$ WΩK^{-2}. Hence, the equation (1) and (2) can be rewritten as:

$$Q_{op} = I\sqrt{L_0(T_H^2 - T_L^2)} \tag{4}$$

$$\frac{L}{A} = \frac{1}{I}\int_{T_L}^{T_H} k(T)\left(L_0(T_H^2 - T_L^2)\right)^{-1/2} dT \tag{5}$$

The current lead system, including current feedthroughs, brass loop current leads, copper bars and HTS leads was designed using a 3D CAD program using calculation results. Figure 6 shows the configuration the current lead system of the 1st stage area in the 1,500 A, 400 mH class HTS DC reactor.

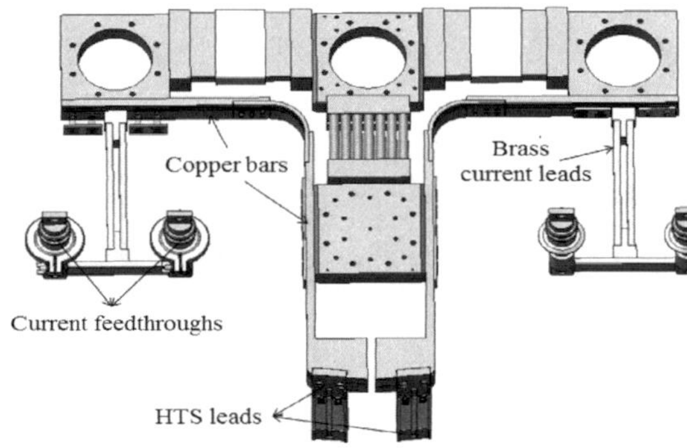

Figure 6. Configuration of the current lead system of the 1st stage area in HTS DC

2.3.2. Design of the 2nd stage area. The 2nd stage cold heads of two GM cryo-coolers (RDK-415D) are responsible for cooling of the toroid-type magnet. Figure 7 shows the detail configuration of the 2nd stage area. The coils are cooled by aluminum conducting bars assembled on the top and bottom of the coils. In addition, between each coil and the cryo-coolers, the heat exchangers made of oxygen free copper (OFCu), aluminum and a flexible copper blade are installed in order to support and cool the 30 double pancake coils simultaneously.

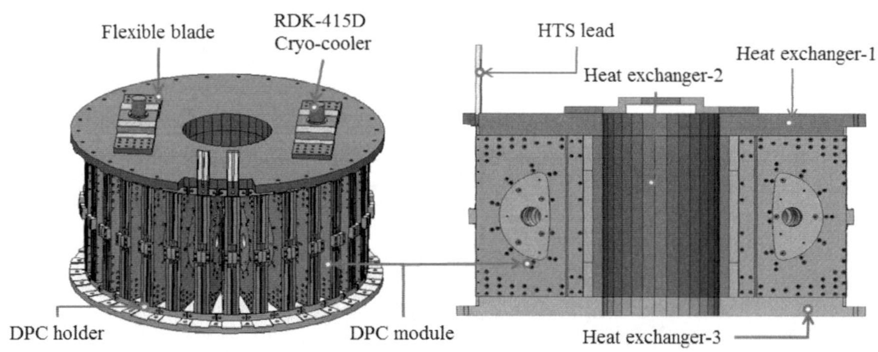

Figure 7. Detail configuration of the 2nd stage area

The bobbin structures and the shape of DPC module are described in detail in figure 8. The DPC module for an HTS DC reactor magnet is composed of two D-shape coils, two bobbins, two side plates and joint parts (joint plate, wire holder).

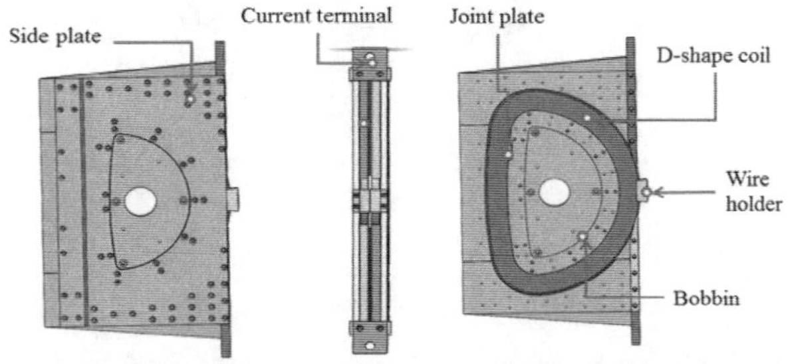

Figure 8. Bobbin structures and the shape of DPC module

3. Thermal analysis and optimal design of the conduction cooling system

3.1. FEM simulation

From the 3D CAD design, the thermal analysis for the 1,500 A, 400 mH HTS DC reactor was implemented. For basic heat load, the 1st stage, 2nd stage, radiation shield, and room temperatures were 40 K, 7 K, 77 K, and 300 K, respectively.

In the 1st stage area, the operating currents in each current feedthroughs were 750 A in the analysis of Joule heating. In order to reduce the Joule heat, the HTS current leads were connected to copper bars. The total heat load of the current leads includes Joule heat and heat invasion at room temperature. Here, the heat invasion at room temperature is conduction heat from outside into the current leads.

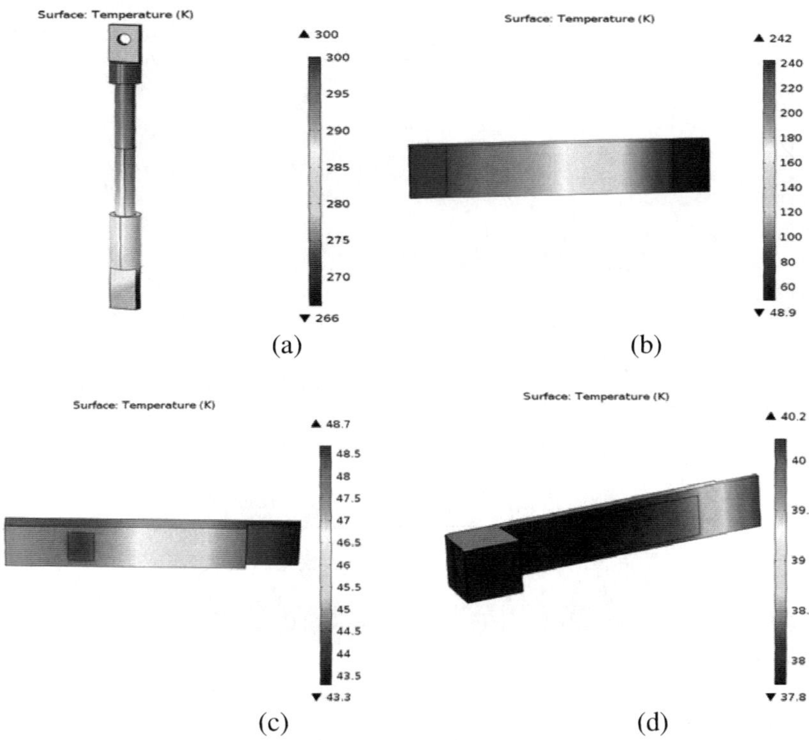

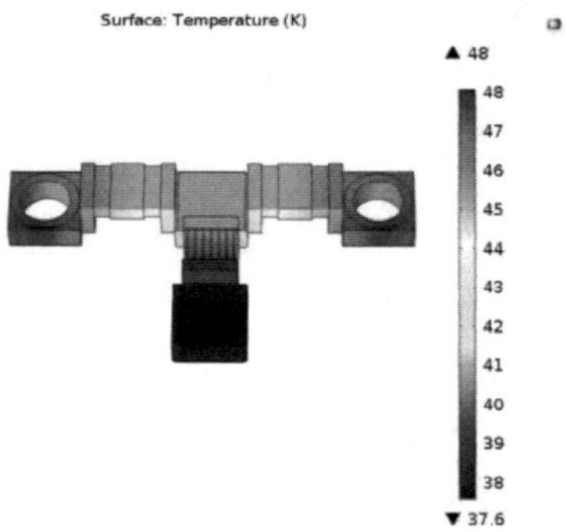

(e)

Figure 9. Thermal analysis results in the 1st stage area: (a) Current feedthrough, (b) Brass current lead, (c)-(d) Copper bars, (e) Cooling blocks

Figure 9 shows the temperature distribution analysis result of the optimal design for the 1st stage area. The temperatures in the copper bars connected to HTS leads and HTS wires were 38 K and 48 K, respectively.

In this paper, the minimized heat load of the current lead was 76 W. The total heat load includes the Joule heat and conduction heat load. To prevent heat transfer from outside to inside of the HTS DC reactor, the conduction heat load of the current lead should be 0 W which was obtained by changing the thickness of the current lead as depicted in figure 10 with major parameters of the current lead.

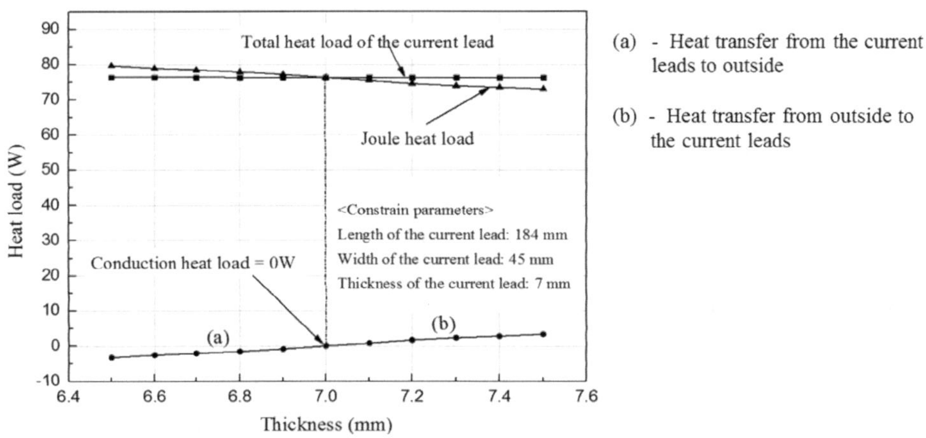

Figure 10. Conduction heat load depending on the thickness of the current lead

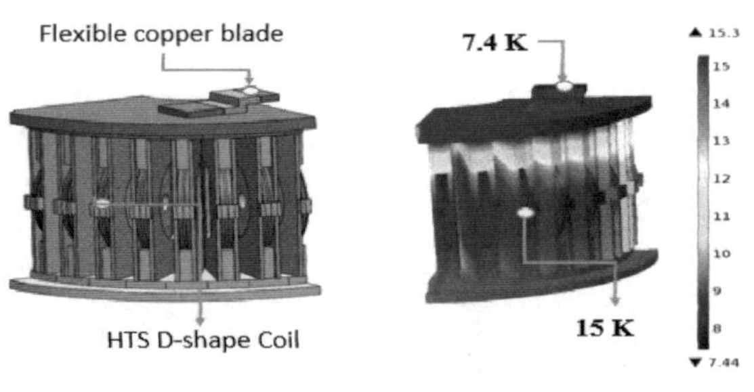

Figure 11. Temperature distribution of the 2nd stage area of HTS DC

In the 2nd stage area, the HTS magnet stably operates at the temperature of under 20 K. Figure 11 shows the temperature distribution analysis result of the 2nd stage area. The temperature in cold head of two-stage cryo-cooler (RDK-415D) is 7.4 K. In the D-shape coil, the temperature is 15 K.

3.2. Conduction cooling system design

In order to analyze the thermal characteristics of the HTS DC reactor, the basic heat load of the system was considered. Basic heat load can be divided into conduction heat and radiation heat. Conduction heat is transferred by the current lead and steel use stainless (SUS) support bars and cooler port. Multi-Layer Insulation (MLI) reduces the radiation heat transferred from outside to inside of the HTS DC reactor. The MLI was wrapped at the surface of the radiation shield. The thermal heat load of the HTS DC reactor in the 1st stage and 2nd stage is shown in detail in table 2.

Table 2. Heat load of 1,500 A, 400 mH HTS DC reactor

1st stage heat load (W)		2nd stage heat load (W)	
Current feedthroughs (4ea)	48.84 W	Conduction heat load from support	1.2 W
Brass current leads (4ea)	104 W	Conduction heat load from HTS lead	0.016 W
Radiation Shield	12.4 W	Magnet joint	6 W
Support bars	6 W	Current terminals	0.018 W
Total heat load	171.24 W	Total heat load	7.234 W

4. Conclusions

The authors designed a conduction cooling system for a toroid-type HTS DC reactor magnet and analyzed its thermal characteristics. The operating current and inductance of the reactor magnet were 1,500 A, 400 mH, respectively. The detailed structure design of conduction cooling system in the 1st stage area and 2nd stage area were described. The optimal heat load of HTS DC reactor was calculated and analyzed by FEM simulation. The minimized heat load of the current lead was 76 W. The conduction heat load of the current lead was 0 W with thickness of 7 mm of the current lead. The temperature at the 1st stage area for the operation of HTS leads and the 2nd stage area for the magnet were 42 K, and 7 K. The heat load of the HTS DC reactor can determine the optimal cooling capacity of the cryo-coolers. The study results can be effectively utilized for the design and fabrication of a commercial HTS DC reactor.

Acknowledgement

This work was supported by the Power Generation & Electricity Delivery Core Technology Program of the Korea Institute of Energy Technology Evaluation and Planning (KETEP), granted financial resource from the Ministry of Trade, Industry & Energy, Republic of Korea. (No.2013T100200094)

References

[1] B. S. Go, K. Kim, M. Park, S. H. Kim, I. K. Yu, S. Lee 2014 Detailed design of a 2,000 A, 400 mH toroid-type HTS DC reactor for an HVDC system *J. Supercond. Novel Magn* **28** no. 2 pp. 629-632

[2] K. Kim, S. K. Kim, M. Park, H.S. Ha, M. H. Sohn, K. Kim, H. Lee, I. K. Yu 2016 Design and Operating Characteristic Analysis of D-Shape HTS Coil for 1500-A 400-mH Class Toroid-Type HTS DC Reactor *IEEE Transactions on Applied Superconductivity* **26** no. 3 pp. 1-5

[3] S. M. Schoenung, W. R. Meier and W. V. Hassenzahl, 1991 A comparison of large-scale toroidal and solenoidal SMES systems *IEEE Transactions on Magnetics* **27** no. 2 pp. 2324-2328

[4] K. Kim, J. G. Kim, H. Jung, S. Kim, S. Lee, M. Park, I. K. Yu 2013 Design of a 400 mH 400 A Toroid-Type HTS DC Reactor Magnet *IEEE Transactions on Applied Superconductivity* **23** no. 3 pp. 4601104-4601104

[5] H. K. Yeom, Y. J. Hong, S. J. Park, T. B. Seo, K. C. Seong and H. J. Kim 2007 Study of Cryogenic Conduction Cooling Systems for an HTS SMES *IEEE Transactions on Applied Superconductivity* **17** no. 2 pp. 1955-1958

[6] K. M. Kim, A. R. Kim, H. Y. Park, J. G. Kim, M. Park, I. K. Yu, S. H. Kim, K. Sim, K. C. Seong, Y. J. Won 2010 Design and Mechanical Stress Analysis of a Toroidal-Type SMES Magnet *IEEE Transactions on Applied Superconductivity* **20** no. 3 pp. 1900-1903

[7] Y. J. Hong, H. K. Yeom, S. J. Park, S. H. Kim and Y. D. Choi 2008 Temperature Distribution of Cryogenic Conduction Cooling System for a HTS SMES *IEEE Transactions on Applied Superconductivity* **18** no. 2 pp. 745-749

[8] Y J. R. Hull 1989 High temperature superconducting current leads for cryogenic apparatus *Energy Conversion Engineering Conference, IECEC-89., Proceedings of the 24th Intersociety,* (Washington, DC) **1** pp.459-464

[9] Ho-Myung Chang, Min Jee Kim 2009 Optimization of conduction-cooled current leads with unsteady operating current *Cryogenics* **49** pp. 210

Load test of Superconducting Magnetic Bearing for MW-class Flywheel Energy Storage System

S Mukoyama[1], K Nakao[1], H Sakamoto[1], K Nagashima[2], M Ogata[2], T Yamashita[2], K Miyazaki[3], H Shimizu[4] and H Sawamura[4]

[1] Furukawa Electric Co, Ltd., 6, yawatakaidan-dore, Ichihara, Chiba 290-8555, Japan
[2] Railway Technical Research Institute, 2-8-38 Hikari-cho Kokubunji, Tokyo, 185-8540, Japan
[3] Public Enterprise Bureau of Yamanashi Prefecture, 1-6-1 Marunouchi Kofu, Yamanashi, 400-8501, Japan
[4] Mirapro Co. Ltd, 1100 Anadaira Sutama-cho Hokuto, Yamanashi, 408-0111, Japan

shinichi.mukoyama@furukawaelectric.com

Abstract. A flywheel energy storage system (FESS) stores electrical power as kinetic energy of a rotating flywheel rotor. Since the storage energy of the FESS is proportional to the weight of the rotor and the square of the rotating speed, the heavy weight and high speed rotor leads a FESS to a high power and a high capacity. However a conventional FESS limits in both the rotor weight and the rotating speed because of using mechanical bearings. A superconducting FESS (SFESS) utilizes a superconducting magnetic bearing (SMB) to levitate and rotate the flywheel rotor that has ton class weight and high speed rotation without mechanical contact. As the SFESS with 300 kW demonstrated at Mt. Komekura in Yamanashi prefecture, the SMB in the SFESS levitated the 4-ton rotor. The SMB consisted of a high temperature superconducting magnet (HTS magnet) and a HTS bulk, and utilized a repulsive force between the HTS magnet and the HTS bulk. The demonstration of the SFESS has been carried out successfully at Mt. Komekura. Now the next step development was started to aim a MW-class SFESS. The MW-class SFESS needs the SMB levitated and withstood a 10 ton-class load. This paper describes a design of the 10 ton-class SMB and the result of the load test of the developed SMB

1. Introduction

A FESS stores an electrical power as a kinetic energy in a rotating flywheel rotor and discharges the power by transforming the kinetic energy into electrical power via an electrical motor and generator. The FESS, which is different from a secondary battery that uses a chemical reaction, performs the operation of long lifetime because it has no chemical degradation like the secondary batteries. Particularly, a high temperature superconducting FESS (SFESS) uses a superconducting magnetic bearing to support a high speed and heavy weight rotor without a mechanical contact, in order that the operation of the long lifetime is realized. The Railway Technical Research Institute and Furukawa Electric have developed the SFESS in the NEDO project with Yamanashi prefecture government, Kubotek and Mirapro [1]-[3]. The SFESS that was developed and was demonstrated in the project had a rotor of 4 tons (2 m in diameter) and provided an electrical power of 300 kW.

A superconducting magnetic bearing (SMB) was consisted of a high temperature superconducting magnet (HTS magnet) in the stator side, and a HTS bulk in the rotor side. The both are made from rare-earth barium copper oxide (REBCO) superconductor. The HTS bulk was levitated by a shielding effect, which was generated with the magnetic field of the HTS magnet. The HTS magnet and the HTS bulk were cooled to cold temperature of several-ten K. The SMB was made of the HTS magnet, the HTS bulk, and a heat insulation chamber, in which the temperature kept at less than 50 K with vacuum and a multi-layer insulation. The 300-kW SFESS that levitated the 4-ton rotor stabilized and averaged output power of the photovoltaic power (PV) plant provided by Public Enterprise Bureau of Yamanashi Prefecture (Yamanashi Prefecture) as figure 1. The Yamanashi Prefecture constructed a one-MW PV plant for the demonstration of the SFESS. The stable electrical power was transmitted to a commercial line of Tokyo Electrical Power Company (TEPCO).

The demonstration system for the field test of the SFESS was consisted of the SFESS, the control system of the network stabilization, and two-way converter. The SFESS charged electric power from the PV plant and discharged the power to stabilize the output power that fluctuated for short period. In the field test, the SFESS was performed to stabilize the output power at the connection point to the electrical power grid of TEPCO. When the output of the PV plant was 500 kW, fluctuation of the power was within +/- 100 kW. The output of the PV plant is shown with a orange color line in graph of figure 1. The SFESS charged and discharge electrical power depending on fluctuation of the output of the PV plants as shown with a green color line. The output at the connection point totally became constant and stable by control of the short period fluctuation, as shown with a red color line.

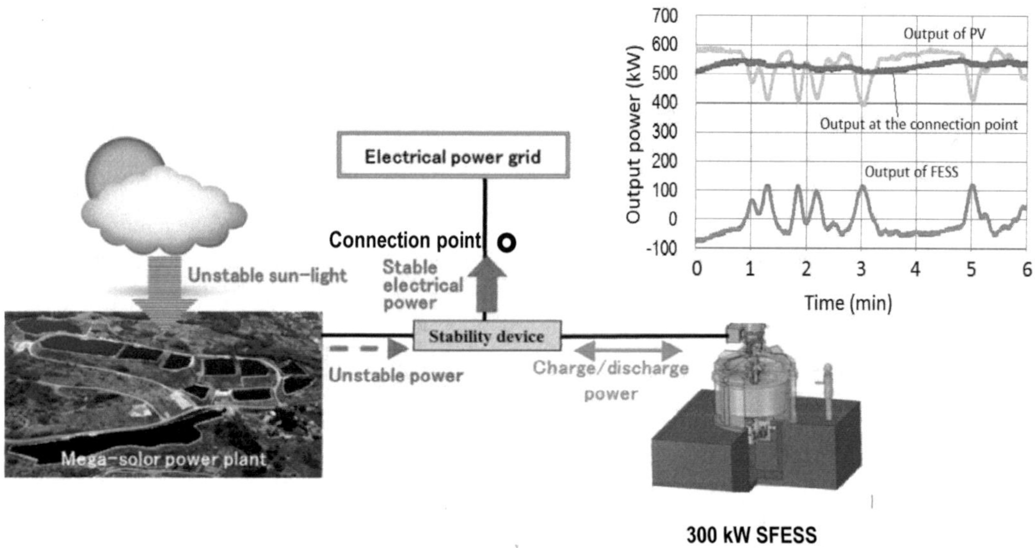

Figure 1. Configuration of the SFESS demonstrated at Mt. Komekura in Yamanashi. The SFESS was charged and discharged by the one-MW PV plant. The graph shows output power of the SFESS, the PV plant, and the connection point to the grid of TEPCO.

2. Configuration of superconducting flywheel energy storage system (SFESS).

Configuration of the SFESS is shown in figure 2. The flywheel rotor of 300-kW SFESS was composed of a carbon fiber glass reinforced plastic (CFRP), which was the world largest volume of 2 m in diameter, 1 m in height, and 4 tons in weight. The motor/generator was a synchronous machine of a commercial permanent magnet type, and was set on top of the FESS. The vacuum seal along the rotor shaft between air and vacuum area in the outer vacuum vessel was applied a magnetic fluid seal. The CFRP rotor and the SMB were in an inner vacuum vessel in which a thin helium gas with 100 Pa was filled. The pressure of the helium gas was chosen to avoid an air friction of the CFRP rotor rotation and to cool the HTS bulk of the SMB. The SMB only supported the radial direction force of the rotor, so the thrust direction force was supported by two active magnetic bearings (AMB) set at above and below of the rotor shaft. Figure 4 shows an overview of the FESS, and the SMB was located at the bottom of the SFESS. The specifications of the SFESS are shown in Table 1.

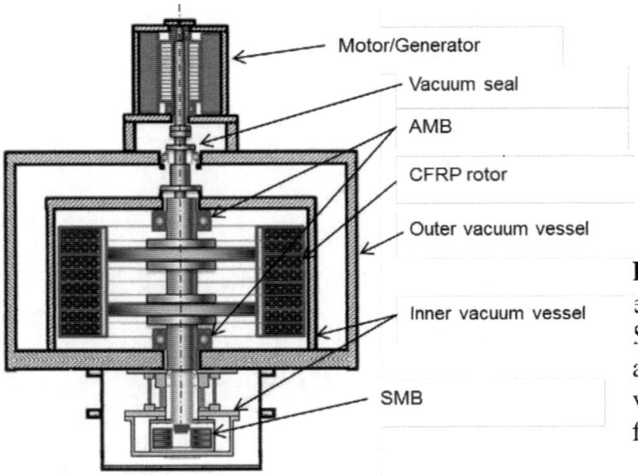

Figure 2. 300-kW HTS flywheel energy storage system (SFESS). The SMB consisted of the HTS magnet and the HTS bulk in the inner vacuum vessel, in which thin He gas filled.

Table 1. Specification of the SFESS.

Item	Specification	Remark
Outer diameter	2,000 mmϕ	consisting of nine layers with a thickness of 100 m
Inner diameter	1,400 mmϕ	
Height	900 mm	
Weight	4,000 kg	excluding the rotating rod
Rotor material	CFRP	
Rotating speed	6,000 rpm	rotating speed at the outermost rim: 628 m/s
Storage capacity	100 kWh	

3. Configuration of superconducting magnetic bearing (SMB).

The SMB developed for the levitation of a 4-ton rotor consisted of the HTS magnet and the HTS bulk. The HTS magnet was composed of 5 double pancake coils, which were made of REBCO wires. The REBCO wires were manufactured by Superpower (type SCS6050AP). The HTS bulk was composed of 3 HTS plates, which Nippon Steel & Sumitomo Metal developed by a quench melt growth (QMG) process [4]. The schematic configuration of the SMB is shown in figure 3. The rotor axis included the HTS bulk was levitated by the magnetic field generated by the HTS magnet as the stator. The specifications of the superconducting wires and the HTS magnet are shown in Table 2.

Table 2. Specifications of the HTS wires and the HTS coil.

Item	Specification
Wire	Product of Superpower (type SCS6050AP)
Width of tape	6 mm
Thickness of tape	0.1 mm
Inner diameter of HTS coil	120 mm
Outer diameter of HTS coil	260 mm
Height of the HTS coil	17.6 mm

IOP Conf. Series: Journal of Physics: Conf. Series **871** (2017) 012090 doi:10.1088/1742-6596/871/1/012090

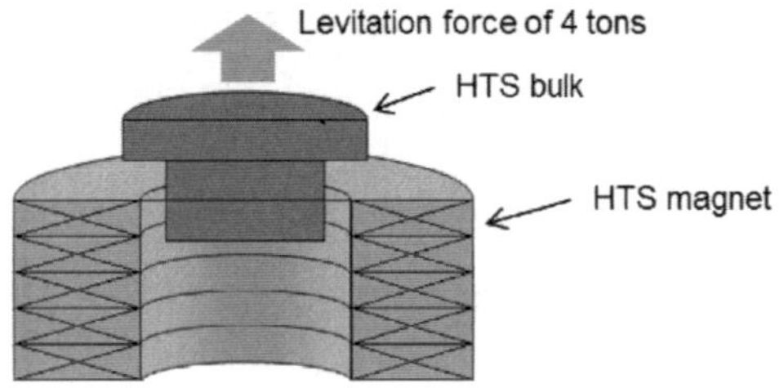

Figure 3. Schematic drawings of the HTS magnet and the HTS bulk in the SMB. The HTS magnet consisted of 5 HTS double pancake coils, and the HTS bulk consisted of 3 HTS plates

4. Test of SMB in SFESS.

The HTS magnet stacked five double pancake coils was cooled by conducting cooling with a GM refrigerator, which had a cooling capacity of 50 kW at 20 K. The HTS bulk was cooled by heat transfer of thin helium gas from cold surface of the HTS magnet.

The SMB was settled in the SFESS prototype (figure 4) and was tested to confirm the levitation properties while supporting the 4-ton rotor. The procedure of the SMB cooling is following; 1) Exhausting air in the vacuum chamber and the inner chamber including the HTS magnets and the HTS bulks for vacuum thermal insulation, 2) Operating the GM refrigerator to cool the magnets, 3) after the coil cooling, introducing helium gas until about 100 Pa to cool the HTS bulks. After cooling to less than 50 K, the levitation height of the rotor was measured under 4-ton load while increasing the magnet operation current. As the result of the current loading test of the SMB, levitation height of the rotor axis is shown in figure 5. The rotor axis was set at 19 mm in height from the original zero position. The 19 mm was height of a touch-down mechanical bearing. The axis was settled on the touch-down bearing with no current load. The rotor was levitated when the coil current exceeded 76 A, and the rotor gained the height by 1 mm per 1 A.

Figure 4. 300-kW SFESS was constructed in a Mirapro factory. The SMB was set on the bottom of SFEE and levitated the FW rotor.

Figure 6 shows the levitation height that depends on the operating coil current in the levitation test. After the magnet engaging, the levitation height was dropping gradually. The coil current was increased with 0.2 A more to keep the constant height of the rotor. We think the height was decreased due to the flux creep in the HTS magnet and the HTS bulk. The long term change of the height is shown in figure 7. The levitation height was changed only in first several hours. Since the magnetic fluxes were finally settled in the superconductors, the SMB provided a stable levitation over 1,000 hours.

IOP Conf. Series: Journal of Physics: Conf. Series **871** (2017) 012090 doi:10.1088/1742-6596/871/1/012090

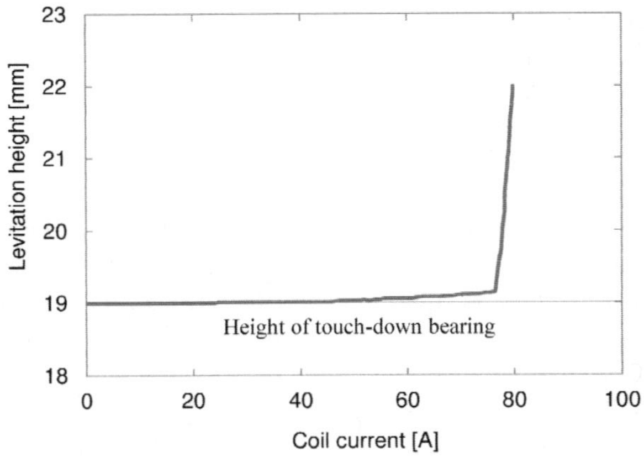

Figure 5. The levitation height of the rotor axis included HTS bulks was measured during the coil current feeding to the SMB magnet. Start height with 19 mm was height of a touch-down bearing, on which the rotor located without the current.

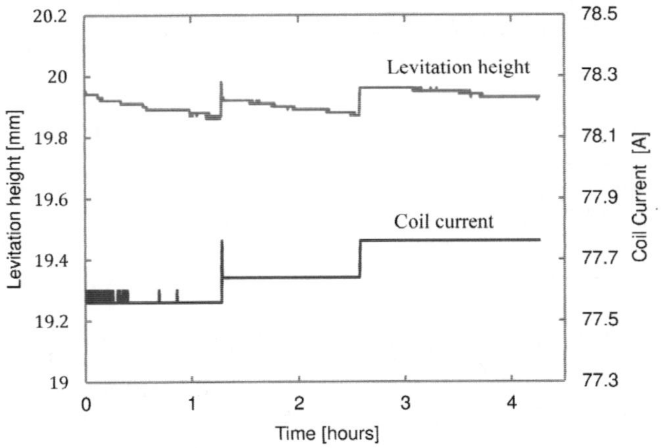

Figure 6. Levitation height depending on time for short time. The coil current increased to keep a constant lever of the height.

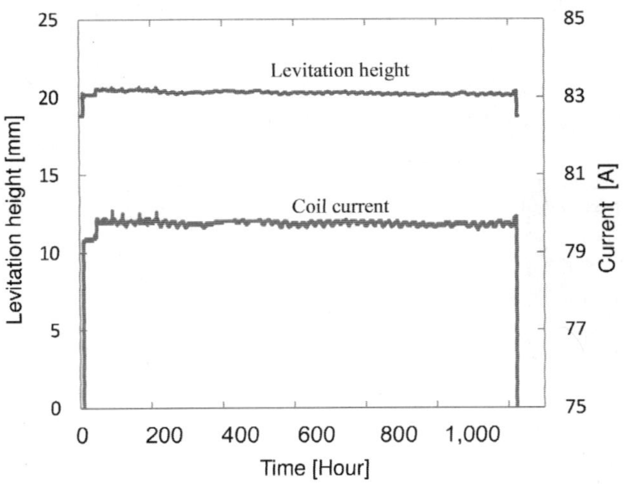

Figure 7. Levitation height depending on time for long time period. After initial adjustment of the current, the current was fixed at a constant value. Fluctuation of the current in the graph was occurred due to temperature changing.

5. Levitation load test of 10 tons in weight.

The capacity target of the SFESS is 1 MW power for stabilization of a network that connects renewable energy sources for short period. The 1 MW class SFESS has the rotor of 10 tons in weight, and the SMB needs levitate force of 10 tons. At first, we designed a 10-ton SMB by using critical current performance of the conventional HTS wire and by the simulation method used on the design of the 4-ton SMB[2]. According to result of the first design, the 10-ton SMB consisted of a HTS magnet of 7 double pancake coils and a HTS bulk of 4 HTS plates. The SMB also needed to be cooled at less than 40 K. However, since the cost percentage of SMB was large in the whole flywheel cost, we tried to reduce number of the coils and quantity of the HTS wires substantially by using the high-Ic wire, which were manufactured by Superpower[5]. The high-Ic wire improved to 1.5 times higher Ic than the conventional HTS wire in magnetic field at 30 K to 50 K as shown in figure 8.

Figure 9 shows result of the FEM analysis of the SMB configuration that consists of 5 double pancake coils and 3 HTS plates. Inner side of the second upper coil was applied the highest magnetic field. Table 3 indicates the maximum magnetic fields of the coil in 4 tons, 10 tons and 15 tons loading respectively.

Table 3. The maximum magnetic field and coil current in the case of 4, 10 and 15 tons loading.

	4 tons	10 tons	15 tons
B_{max} [T]	3.38	5.35	6.55
I_{coil} [A]	74	117	143

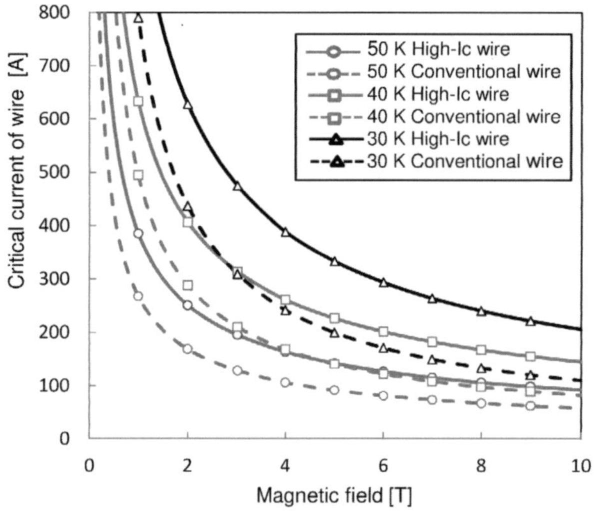

Figure 8. Critical currents of conventional REBCO wire (dotted lines) and high-Ic REBCO wire (solid lines), which were produced by Superpower.

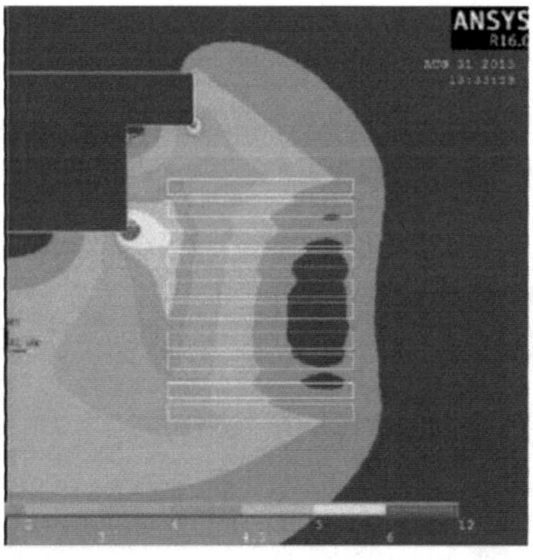

Figure 9. Electromagnetic analysis result for 5 double pancake coils and 3 HTS plates.

IOP Conf. Series: Journal of Physics: Conf. Series **871** (2017) 012090 doi:10.1088/1742-6596/871/1/012090

The load line in the second upper coil is indicated in figure 10. In the case of the SMB that provides 4-ton levitation, the conventional wire has enough Ic at 50 K, which we aim as the operation temperature. However, to realize 10-ton levitation, the conventional wire is cooled at 40 K. Moreover 15-ton levitation needs to cool the conventional wire at less than 30 K. On the other hand, high-Ic wire enables to lift the 10-ton rotor at 50 K. The 15-ton levitation is realized by cooling at 40 K in the high-Ic wire. In figure 10, The SMB that fabricated with the high-Ic wires is more stable and lower cost for cooling than the SMB with the conventional wires.

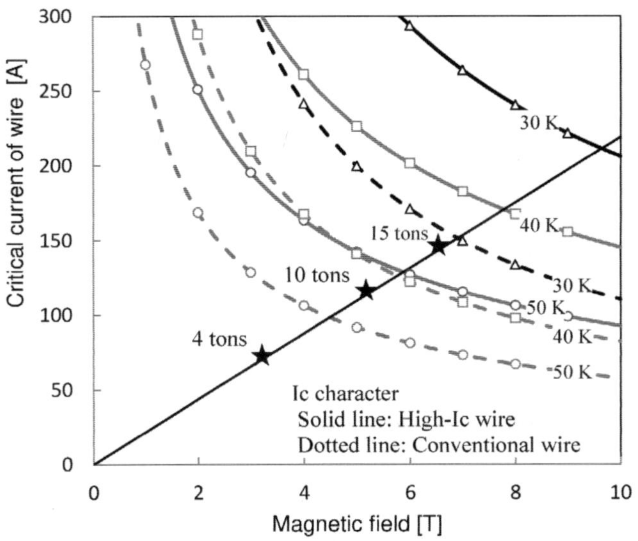

Figure 10. The load line of the 2nd upper coil and Ic characters of the conventional HTS wire and the high-Ic wire produced by Superpower.

Figure 11 shows the test equipment to measure levitation force with a load cell. A SMB for 10-ton levitation test was fabricated by using high-Ic wire to the 2 coils as shown in figure 11. Figure 12 shows test result of levitation force depending of the feeding current into the SMB magnet. The magnets were cooled at 25 K by the GM refrigerator. When the current of 126 A was applied to the magnets, the levitation force of 10 tons (98 kN) was achieved. The HTS bulk was assumed a perfect diamagnetism without magnetic field penetration in the magnetic analysis. Actually, when external magnetic field was increased, magnetic field was entered into surface of the HTS bulk. The difference between the analysis and the test result was caused by the magnetic penetration in the HTS bulk.

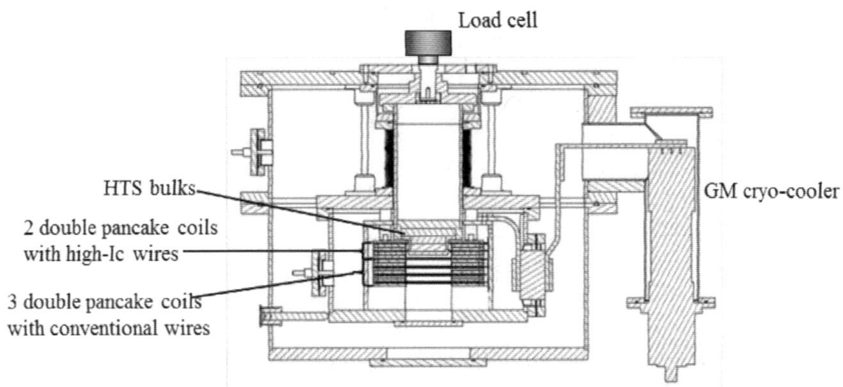

Figure 11. Configuration of 10-ton levitation test equipment.

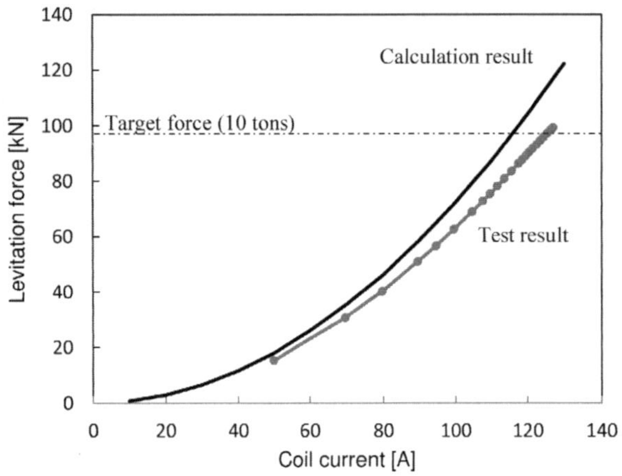

Figure 12. The levitation force depends on the coil current.

6. Summary.

The demonstration of the high temperature superconducting flywheel energy storage system (SFESS) using the superconducting magnetic bearing (SMB) was performed at Mt. Komekura in Yamanashi prefecture. The SMB composed of the HTS magnet and the HTS bulk levitate the 4-ton rotor in stable and safe.

The SMB with 10-ton levitation force was developed for the practical applications. In the first design, we needed 7 pieces of the HTS coils and 4 pieces of the HTS plates if we followed the 4-ton SMB design. In the final design, since the high-Ic wires were applied for the 10-ton SMB, the SMB that was composed of 5 pieces of the HTS coils achieved 10 tons in weight. This result leaded to reduce a cost of the SMB. Moreover, it introduces the other application such as a regeneration storage device of a railway. The regenerative device of the railway is needed with repetition of charge and discharge to more than 400,000 times and 1 MW class in power. The SFESS is the most suitable device against the requirements in energy storage devices. We started to study on the MW-class SFESS as the regeneration energy storage device while cooperating with related departments.

Acknowledgement

This development was supported by NEDO in a part of 'Development of next generation flywheel energy storage system'.

References

[1] Mukoyama S, Matsuoka T, Furukawa M, Nakao K, Nagashima K, Ogata M, Yamashita T, H, Hasegawa Yoshizawa K, Arai Y, Miyazaki K, Horiuchi S, Maeda T and Shimizu H 2015 *Physics Procedia.* **65** p 253

[2] Nakao K, Kasahara H, Hatakeyama H, Matsuoka T and Mukoyama S 2016 *Furukawa Review* **47** p 13

[3] Ogata M, Matsue H, Yamashita T, Hasegawa H, Nagashima K, Maeda T, Matsuoka T, Mukoyama S, Shimizu H and Horiuchi S 2016 *Superconductor Science and Technology* **29** 5

[4] Teshima H and Morita M 2014 *Journal of Physics: Conference Series* **507** 012048

[5] Nakasaki R, Zhang Y, Brownsey P, Sundaram A, Hazelton D, Sakamoto H and Fukushima T 2015 Continuous improvements in performance and quality of 2G HTS wires produced by IBAD-MOCVD for coil applications *MT-24* 4OrBA_03 http://superpower-inc.com/system/files/2015_1018+MT24+4OrBA_03_0.pdf

Vibrational characteristics of a superconducting magnetic bearing employed for a prototype polarization modulator

Yuki Sakurai[1], Tomotake Matsumura[1], Hajime Sugai[1], Nobuhiko Katayama[1], Hiroyuki Ohsaki[2], Yutaka Terao[2], Yusuke Terachi[2], Hirokazu Kataza[3], Shin Utsunomiya[1] and Ryo Yamamoto[3]

[1]Kavli Institute for the Physics and Mathematics of the Universe (WPI), The University of Tokyo Institutes for Advanced Study, The University of Tokyo, 5-1-5 Kashiwanoha, Kashiwa, Chiba 277-8583, Japan
[2]Graduate School of Frontier Sciences, The University of Tokyo, 5-1-5 Kashiwanoha, Kashiwa, Chiba 277-8561, Japan
[3]Japan Aerospace Exploration Agency, Institute of Space and Astronautical Science (ISAS), 3-1-1 Yoshinodai, Chuo-ku, Sagamihara, Kanagawa 252-5210, Japan

yuki.sakurai@ipmu.jp, tomotake.matsumura@ipmu.jp

Abstract. We present the vibrational characteristics of a levitating rotor in a superconducting magnetic bearing (SMB) system operating at below 10 K. We develop a polarization modulator that requires a continuously rotating optical element, called half-wave plate (HWP), for a cosmic microwave background polarization experiment. The HWP has to operate at the temperature below 10 K, and thus an SMB provides a smooth rotation of the HWP at the cryogenic temperature of about 10 K with minimal heat dissipation. In order to understand the potential interference to the cosmological observations due to the vibration of the HWP, it is essential to characterize the vibrational properties of the levitating rotor of the SMB. We constructed a prototype model that consists of an SMB with an array of high temperature superconductors, YBCO, and a permanent magnet ring, NdFeB. The rotor position is monitored by a laser displacement gauge, and a cryogenic Hall sensor via the magnetic field. In this presentation, we present the measurement results of the vibration characteristics using our prototype SMB system. We characterize the vibrational properties as the spring constant and the damping, and discuss the projected performance of this technology toward the use in future space missions.

1. Introduction

A cosmic microwave background (CMB) is electromagnetic microwave radiation from the big bang. It is still observable isotropically from the whole sky today. One of most important research topics in the current cosmology and high-energy physics is to study the cosmic inflation theory, which predicts a rapid expansion of the universe after $\sim 10^{-38}$ seconds from the beginning of the universe [1, 2]. The inflation solves several mysteries of the standard cosmology and it can provide a foothold for new physics in high energy physics. The theory predicts that the inflation left the divergence free pattern in the CMB polarization, called B-mode. Therefore, the experimental discovery of the cosmic inflation is possible by observing the CMB B-mode polarization. A world-wide hunting race is ongoing among CMB polarization experiments. Correspondingly, the instrumental development has progressed rapidly.

One of the critical instruments for a precise measurement of the CMB polarization is a polarization modulator. It consists of an optical element, a half-wave plate (HWP), and a rotational mechanism. The continuously rotating HWP modulates the CMB polarization signal synchronously at the four times of the rotational frequency of the HWP, a few Hz. The continuously rotating HWP has to be maintained at cryogenic temperature (~10 K) in order to reduce the detector noise originating from the excess thermal emission of the HWP itself to the detector. It is therefore necessary to use a rotational mechanism which allows to rotate with minimal heat dissipation at cryogenic temperature.

A superconducting magnetic bearing (SMB) is a contactless bearing [3]. It employs an array of high temperature superconductor tiles as a stator and a permanent magnet as a rotor. The rotor levitates above a stator and spins without contact. There is no friction due to the physical contact, and thus this technology is well matched for use in the polarization modulator, which is required to operate at the cryogenic temperature with minimal heat dissipation. This unique application is first investigated for use in a balloon-borne CMB polarization experiment, EBEX [4]. The successful implementation accelerates the interest for further development toward the next generation experiment from a ground telescope to a satellite mission [5].

While we gain the benefit of low heat dissipation by using the SMB as compare to a mechanical bearing the stiffness of the rotor over the stator is expected to be less rigid as compare to the mechanical bearing. Therefore, the levitation based bearing has an inherent trade-off between the heat dissipation and the stiffness. In this paper, we discuss the design trade-off for use of this SMB technology for a future CMB polarization experiment. We present our prototype SMB system, the rotor diameter of 384 mm, and the measurement results of a spring constant and a damping coefficient to characterize the mechanical properties. In order to characterize the mechanical properties, we use a laser displacement gauge, which we can use in a lab but cannot use during observations. As an alternative solution, we also introduce a Hall sensor to monitor the mechanical properties simultaneously. We discuss the consistency between the two independent measurements and propose the concise passive technic to monitor the vibrational motion of the rotor magnet at the cryogenic temperature. Finally, we project the results toward the design of the SMB system for the polarization modulator.

2. Experimental Setup

We conduct vibration measurements using a Styrofoam bucket with liquid nitrogen as shown in Figure 1 (a). Hull et al. reported the vibrational properties of the SMB system at both the liquid nitrogen and the liquid helium temperature. The results show no significant difference between the two temperature ranges. Furthermore, the rotor diameter of about 400 mm requires a large diameter cryostat for testing, and thus we start characterizing the vibrational properties at a liquid nitrogen temperature although the SMB system is operated inside the 4 K cryostat in an actual CMB observation.

We prepare the SMB system with inner diameter of 384 mm. Due to a large diameter, a ring shaped rotor magnet consists of 16 segmented NdFeB magnets. Each segmented magnet is magnetized axially with a magnetic remnance of 1.24 T. A $Y_{1.65}Ba_2Cu_3O_x$ (YBCO) superconductor is used as a stator. The critical temperature of this high temperature superconductor is ~95 K [7]. The YBCO array is formed in a ring shape using three-seeded YBCO tiles.

The YBCO array is submerged by the liquid nitrogen but the rotor magnet is exposed to the room environment. We set a gap between the rotor magnet and YBCO array with aluminum plates at room temperature, and then a liquid nitrogen is poured with a presence of the magnetic field, i.e. field-cooling process. After the temperature of the YBCO array becomes below its transition temperature, the plates are removed for magnetic levitation. The levitation height, defined as the distance between the bottom of the rotor magnet and the top of the YBCO array, is controlled by the thickness of the aluminum plate. We measure the rotor vibration in the axial direction with four different levitation heights (3, 6, 9 and 12 mm). In order to enhance the vibrational amplitude, an external impulse force is applied by hand in the axial direction.

The displacement of the rotor is monitored by a laser displacement gauge, which is mounted on an aluminum frame at the distance of 260 mm from the rotor magnet as shown in Figure 1 (a). We also monitor the vibration via magnetic field using a cryogenic Hall sensor installed between the rotor magnet and the YBCO array.

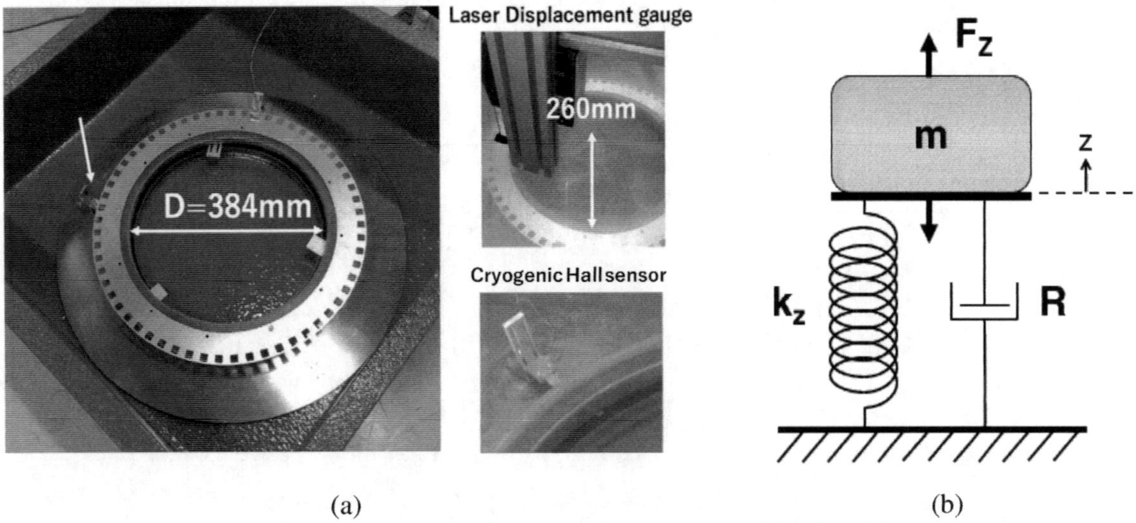

Figure 1. (a) The experimental setup for the vibration measurements for $D = 384$ mm. (b) The Mass-spring-damper model.

3. Result and Discussion

An axial vibration of the rotor can be described by a damping motion of a mass-spring-damper model [8] shown in Figure 1 (b). The motion of this model is expressed by

$$m\frac{\partial^2 x}{\partial^2 t} + R\frac{\partial x}{\partial t} + k_z x = 0, \tag{1}$$

where m is the mass of the rotor (3.0 kg), R is the damping coefficient and k_Z is the spring constant. Under the condition of the damping motion, the solution of the equation (1) is

$$x = Ae^{-\zeta\omega_0 t}\cos(\sqrt{1-\zeta^2}\omega_0 t + \phi), \tag{2}$$

where $\zeta = \mathrm{R}/2\sqrt{\mathrm{mk_z}}$ is the damping ratio and $\omega_0 = \sqrt{\mathrm{k_z}/\mathrm{m}}$ is the undamped angular frequency. The parameters of A and ϕ are the amplitude and the phase determined by the magnitude and the timing of the external impulse force. Thus, these values can be any values.

Figure 2 (a) shows representive outputs of the laser displacement gauge and the Hall sensor as a function of time at the levitation height of 9 mm. The laser displacement gauge output represents the position displacement of the rotor. The Hall sensor measures the magnetic field corresponding to the position of the rotor magnet.

We fit the data with Equation (2) and derive the fitting parameters of ζ and $\omega 0$ for each levitation height. The fitting result of the Hall sensor output with the levitation height of 12 mm is shown in Figure 2 (b). The fitted parameters are summarized with their statistical error values in Table 1. The parameters from the laser displacement gauge and the Hall sensor are consistent. The detailed comparison between the two results is beyond the scope of this paper. In order for the two methods to be identical, the change of the rotor magnetic field and the spatial displacement of the rotor have to be linearly related, but this is not necessary guaranteed. In an actual application, the measurement by using a laser displacement gauge is not ideal because entire SMB system have to be kept below 10 K. This result indicates that the Hall sensor can be an alternative tool against to the laser displacement gauge for the vibration measurement possibly.

Table 1. The summary of the fitted parameters from the vibration measurements for each levitation height (h). The parameters of ζ and ω_0 represent a damping ratio and an undamped angular frequency. The variables of R and k_Z represent a damping coefficient and a spring constant.

h [mm]	laser displacement gauge			
	ζ	ω_0 [rad/s]	R [N·s/m]	k_Z [N/m]
3	$2.1 \pm 0.3 \times 10^{-2}$	331.8 ± 0.8	42.0 ± 5.6	$3.3 \pm 0.02 \times 10^5$
6	$2.7 \pm 0.1 \times 10^{-2}$	194.0 ± 0.2	31.0 ± 2.5	$1.1 \pm 0.01 \times 10^5$
9	$2.3 \pm 0.1 \times 10^{-2}$	138.2 ± 0.2	19.1 ± 2.5	$5.7 \pm 0.02 \times 10^4$
12	$1.8 \pm 0.1 \times 10^{-2}$	111.7 ± 0.1	11.8 ± 1.2	$3.7 \pm 0.01 \times 10^4$
	Hall sensor			
h [mm]	ζ	ω_0 [rad/s]	R [N·s/m]	k_Z [N/m]
3	$2.4 \pm 0.1 \times 10^{-2}$	299.3 ± 0.2	43.5 ± 2.5	$2.7 \pm 0.03 \times 10^5$
6	$2.3 \pm 0.1 \times 10^{-2}$	195.7 ± 0.1	27.1 ± 1.3	$1.2 \pm 0.01 \times 10^5$
9	$2.4 \pm 0.1 \times 10^{-2}$	136.4 ± 0.2	20.2 ± 2.8	$5.6 \pm 0.01 \times 10^4$
12	$2.4 \pm 0.1 \times 10^{-2}$	108.0 ± 0.1	15.7 ± 0.8	$3.5 \pm 0.01 \times 10^4$

We also perform a Fourier transformation to the data from the Hall sensor. Figure 3 (a) shows the result of Fourier transformation for each levitation height. The distinctive peak in the plot corresponds to the natural frequency of the SMB system. The modulation frequency of the signal by the polarization modulator is about 20 Hz. Thus, the natural frequency does not resonate if the levitation height is higher than 12 mm. The spring constant is also derived by

$$k_z = m(2\pi f_0)^2, \quad (3)$$

where f_0 is the natural frequency of the SMB. The spring constant from the Fourier transformation is compared with the fitting result in the figure 3 (b). There is no significant difference depending on the analysis method. For the polarization modulator, the typical requirement of the spring constant is 10^5 Nm. This SMB system can satisfy the requirement if the levitation height is less than 6 mm. From this measurement, we demonstrate that the spring constant and the damping coefficient are inversely

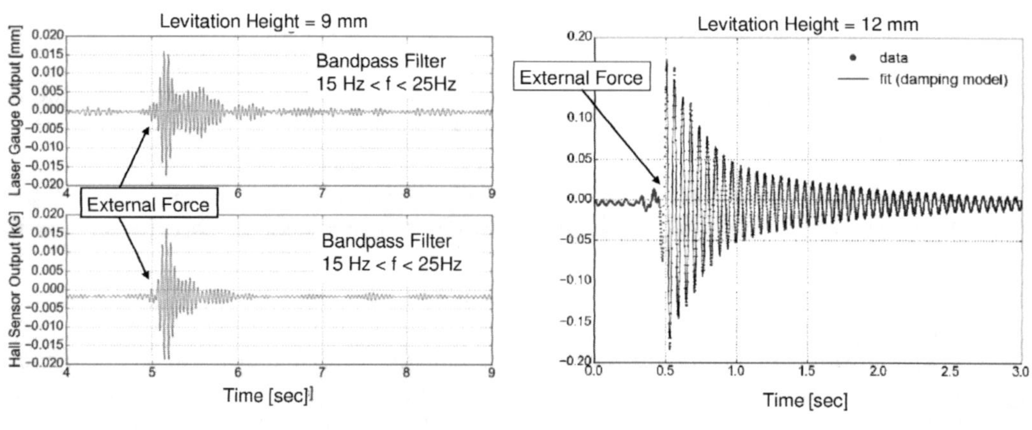

Figure 2. (a) The laser displacement gauge output and the Hall sensor output as a function of time with the levitation height of 9 mm in top and bottom panels, respectively. (b) The fitting result of the Hall sensor output with the levitation height of 12 mm.

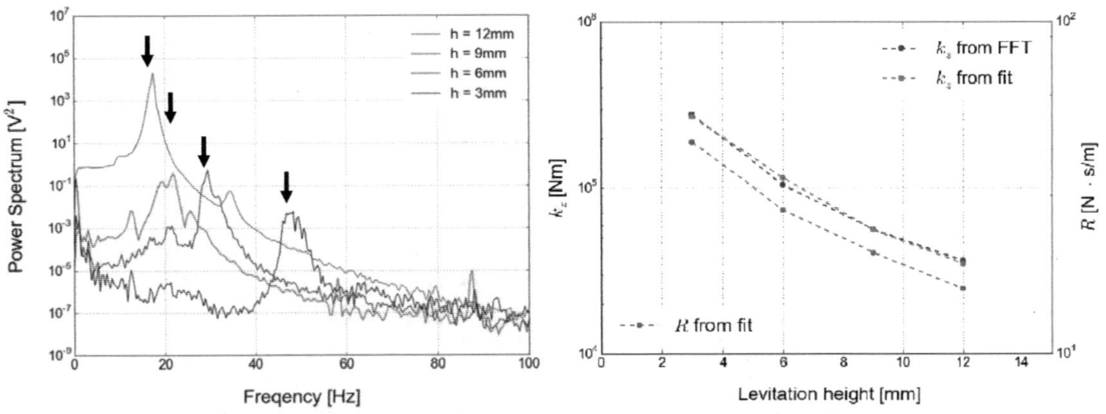

(a) (b)

Figure 3. (a) The power spectrum of the Hall sensor output as a function of frequency for levitation height of 3 mm, 6 mm, 9 mm and 12 mm. (b) The spring constant and the damping coefficient as a function of levitation heights. The red and green points are derived from the fit with damping model, and the blue points are derived from the Fourier transformation.

proportional to the levitation height at the first order. The damping coefficient corresponds to an energy loss associated with magnetic hysteresis. From the viewpoint of the stiffness, it is efficient to set the levitation height as low as possible. However, the increase of the energy loss leads to the noise from the heat dissipation of the HWP. In addition, the heat dissipation from a magnetic friction during the rotation also increases by reducing the levitation height. Thus, the SMB system has a trade-off between the stiffness and the heat dissipation. For the polarization modulator, the levitation height of the SMB is an important parameter to be determined by the experimental requirement. In this measurement, we use the segmented ring magnet with axial magnetization. The further effort of designing an optimal magnetic circuit can be improved the total performance of the SMB system. The measurement and evaluation of the magnetic circuit will be discussed in future papers.

4. Conclusion

We have conducted the vibration measurements of the prototype SMB system with the diameter of ϕ=384 mm for the polarization modulator used in CMB polarization experiments. From the measurement, we derive the spring constant and the damping coefficient using both the laser displacement gauge and the cryogenic Hall sensor. The spring constant is in the order of 10^4 ~ 10^5 N/m and the damping coefficient is 15 ~45 depending on levitation heights.

The result has consistency between the laser gauge and the Hall sensor. We conclude that the Hall sensor can be an alternative method to monitor the vibration of the SMB system. We demonstrate the spring constant and the damping coefficient are inversely proportional to the levitation height. We discuss the trade-off of the SMB system between the stiffness and the heat dissipation. The further effort of designing an optimal magnetic circuit can be improve the total performance of the SMB system.

Acknowledgments
The author would like to thank to Dr. H. Imada at ISAS/JAXA. This work was supported by MEXT KAKENHI Grant Numbers JP15H05441 and JSPS Core-to-Core Program, A. Advanced Research Networks. This work was also supported by World Premier International Research Center Initiative (WPI), MEXT, Japan.

References

[1] Sato K 1981 First-order phase transition of a vacuum and the expansion of the Universe,

IOP Conf. Series: Journal of Physics: Conf. Series **871** (2017) 012091 doi:10.1088/1742-6596/871/1/012091

Monthly Notices of Royal Astronomical Society **195** 46
[2] Guth Alan H, 1981 The Inflationary Universe: A Possible Solution to the Horizon and Flatness Problems *Phys. Rev. D* **23** 347
[3] Hull J 2000 TOPICAL REVIEW: Superconducting bearings, *Superconductor Science and Technology* **13** 2 pp. R1-R15
[4] Klein J et al. 2011 A cryogenic half-wave plate polarimeter using a superconducting magnetic bearing *Proceedings of the SPIE, Cryogenic Optical Systems and Instruments XIII* **8150** 815004
[5] Masumura T et al. 2014 LiteBIRD: mission overview and design tradeoffs, *Proceedings of the SPIE, Space Telescopes and Instrumentation 2014* **9143** 91431F
[6] Matsumura T, Kataza H, Utsunomiya S, Yamamoto R, Hazumi M and Katayama N 2016 Design and Performance of a Prototype Polarization Modulator Rotational System for Use in Space Using a Superconducting Magnetic Bearing *IEEE Transactions on Applied Superconductivity* **26** 3
[7] Hull J and Cansiz A 1999 Vertical and lateral forces between a permanent magnet and a high-temperature superconductor *Journal of Applied Physics* **86** 11
[8] Takahata R and Ueyama H 1992 Characterization of Superconducting Magnetic Bearings (Vibration Damping and Hysteresis of Magnetic Force in Superconductor) *Advances in Superconductivity IV* pp 1089-1092

Side-suspended High-T_c Superconducting Maglev Prototype Vehicle Running at a High Speed in an Evacuated Circular Test Track

Dajin Zhou[1], Lifeng Zhao[1], Chenyu Cui[1], Yong Zhang[1], Jianqiang Guo[1] and Yong Zhao[1,2,*]

[1]Key Laboratory of Maglev Train and Maglev Technology of Ministry of Education, Superconductivity and New Energy R&D Center, Southwest Jiaotong University, Chengdu 610031, China,
[2]School of Physical Science and Technology, Southwest Jiaotong University, Chengdu 610031, China

*Corresponding author: Prof. Yong Zhao, yzhao@home.swjtu.edu.cn

Abstract. High-T_c superconductor (HTS) and permanent magnetic guideway (PMG) based maglev train is intensively studied in China, Japan, Germany and Brazil, mainly through static or vibration test. Amongst these studies, only a few of reports are available for the direct and effective assessment on the dynamic performance of the HTS maglev vehicle by running on a straight or circular PMG track. The highest running speed of these experiments is lower than 50 km/h. In this paper, a side-suspended HTS permanent magnetic guideway maglev system was proposed and constructed in order to increase the running speed in a circular track. By optimizing the arrangement of YBCO bulks besides the PMG, the side-suspended HTS maglev prototype vehicle was successfully running stably at a speed as high as 150 km/h in a circular test track with 6.5 m in diameter, and in an evacuated tube environment, in which the pressure is 5×10^3 Pa.

1. Introduction

Taking advantage of pinning force generated under the interaction between high-T_c superconductor (HTS) bulks and permanent magnetic guideway (PMG), HTS maglev vehicle has excellent autostability [1]. In comparison with Transrapid series electromagnetic suspension (EMS) maglev vehicle of Germany and low-temperature superconducting electrodynamic suspension (EDS) maglev vehicle of Japan, the HTS maglev vehicle dispenses with complex control system. No matter the HTS maglev vehicle under a certain load is static or dynamic, a large suspension gap can be maintained. Further more, the HTS maglev system realizes suspension in any direction, e.g. side suspension, in addition to conventional horizontal suspension.

Currently, research on dynamic operation of HTS maglev vehicle is made mainly through test on a short-distance linear or small-radius circular track. The first HTS maglev circular track co-designed by Chinese and German researchers in 1997 adopted air-core linear synchronous motor with track diameter of 3.5m and the highest speed of 6.5 km/h [2]. At the end of 2000, the research team led by Wang et al from Southwest Jiaotong University (hereinafter referred to as SWJTU) of China designed a manned HTS maglev prototype [3] with a 15.5 m linear test track. In 2004 and 2011, German IFW team designed linear and circular test track systems respectively for manned HTS maglev vehicle [4-5], the latter with length of 80m, allowing for vehicle speed of up to 20 km/h. In 2014, Deng et al established an HTS maglev prototype as well as a 45 m circular track with diameter of 12 m where the manned maglev prototype was driven by a linear motor with the highest speed of 50 km/h [6]. In 2015, a research team from Federal University of Rio de Janeiro of Brazil designed a full-size manned HTS maglev vehicle driven by on-board motor with improved structure as well as a 200m linear test track allowing for test speed up to 7 km/h [7].

(a) (b)

Figure 1. (a) The HTS maglev loop track system in an evacuated tube with a up-down levitation configuration; (b) The SS-HTS maglev loop track system in an evacuated tube.

In order to reduce the air resistance for HTS maglev vehicle, evacuated tube technology was cooperated with the HTS maglev system (see Fig.1(a)) [8]. The first circular track system for evacuated tube HTS maglev vehicle (driven by linear motor) with track diameter of 3 m allowing for vehicle speed up to 21 km/h. The test speed of HTS maglev vehicle, from model vehicle, prototype vehicle, full-size test vehicle to evacuated tube vehicle, has been low due to limited length or radius of linear or circular test track. Though it is theoretically predicated that the highest speed of HTS maglev vehicle is nearly 2,900 km/h, there is no specific dynamic test for verification. In order to demonstrate the potential of higher running speed for HTS maglev system, it is necessary to make new design for the HTS maglev system if the experiments has to be performed in a limited laboratory space. In this paper, we report the design and construction of a side-suspended-HTS maglev circular track system (see Fig. 1(b)), in which the PMG track was laid on the wall to allow the vehicle to operate at a high speed. Since the strong centrifugal force during a high speed running in the circular track was overcome by the repulsive force between the HTS bulks and PMG, a running speed up to 150 km/h was achieved for the testing maglev vehicle in the circular test track with track diameter of 6.5 m and track length of 20.4 m, and in an evacuated tube environment of a pressure 5×10^3 Pa.

2. Structure of HTS maglev loop track in an evacuated tube

The evacuated tube SS-HTS maglev circular track system is mainly composed of SS-HTS maglev system, evacuated tube system and driving system, as shown in Fig. 2. The outer circle diameter and inner circle diameter of the system are 6.77 m and 5.8 m respectively; the system base is 1.33 m high; and the diameter and perimeter of circular PMG are 6.5 m and 20.4 m respectively. As shown at the top left corner of Fig. 2, the SS-HTS maglev system consists of a wall-attached track and a SS-HTS maglev vehicle prototype. The prototype vehicle is placed beside the track and driven by a linear motor after it reaches the superconducting state. Given energy consumption and efficiency of the driving system, segmented power supply control technology is adopted for the linear motor. As shown on the right of Fig. 2, 96 linear units are laid along the entire circular track and powered by a 30 kw three-phase frequency converter while power-on/off of the liner motor unit is controlled by the corresponding solid state relay. Meanwhile, 96 optoelectronic switches are also laid along the circular track to detect the position of the maglev vehicle and a segmented power supply controller is used to control power supply of the motor unit near the maglev vehicle. A 7.5 kw vehicle-mounted vacuum pump is used to control the air pressure in the tube and, in combination with a 1.5 kw small-power mechanical vacuum pump to maintain the set pressure therein.

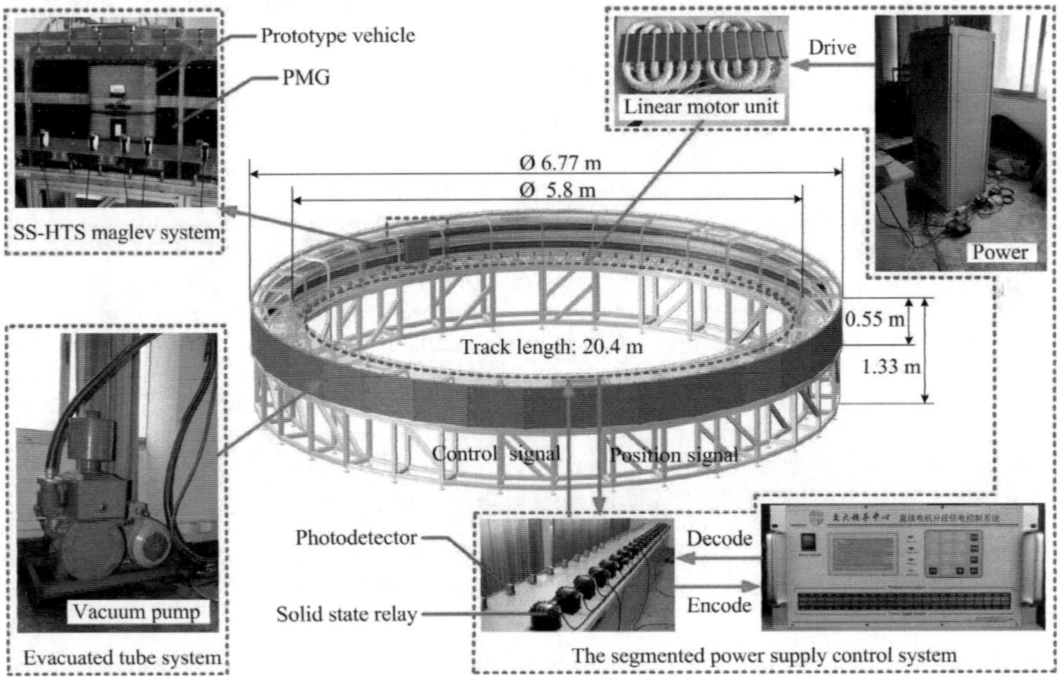

Figure 2. Structure of SS-HTS maglev test circular track system with an evacuated tube, which is consist of SS-HTS maglev system, Evacuated tube system, and the segmented power supply control system.

3. Side-suspended HTS maglev system

SS-HTS maglev system mainly consists of side-suspended track and side-suspended superconducting vehicle, the value of maglev vehicle mass is 5.67kg filled with liquid nitrogen, as shown in Fig. 3(a), 2D cross-section diagram of the system. Different from conventional up-down configuration, superconducting bulks are laid beside the track and through the field cooling process, enable free suspension of the maglev vehicle beside the track. The centrifugal force generated during operation of the maglev vehicle is overtaken by suspension force and gravity by guidance force, as shown in Fig.

3(b) below. When the huge centrifugal force is overtaken by suspension force, a greater centripetal force will be obtained to make the maglev vehicle reach a higher speed in the circular test track, thus preventing the maglev vehicle running off the track along the tangent direction and providing a solution for the maglev vehicle to run across a small-radius curve.

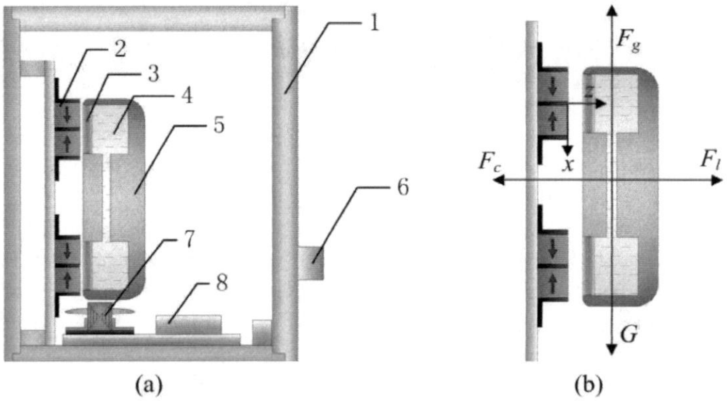

Figure 3. (a) The cross-sectional view of SS-HTS maglev test circular track, 1-evacuated tube, 2-PMG, 3-multiple YBCO bulks, 4-liquid nitrogen, 5-prototype vehicle, 6-photodetector, 7-linear motor, 8-solid state relay; (b) the force of prototype vehicle under the stable operation, F_c is centrifugal force, F_l is levitation force, F_g is guidance force, G is gravity.

As shown in Fig. 4(a), PMG is assembled by two N52 permanent magnet units with opposite polarity and an iron plate (technically pure iron) in between. A higher magnetic intensity can be achieved above the track surface due to aggregation of magnetic field through the iron plate. The permanent magnet unit is 40 × 40 mm in size, the iron plate is 6mm in thickness, and the center-to-center spacing between upper and lower tracks is 220 mm. Fig. 4(b) shows the intensity distribution along the x-axis of z-axis and x-axis components (B_z and B_x) of magnetic fields obtained with the measurement 5 mm and 20 mm above the PMG respectively. Symmetric distribution of B_z and antisymmetric distribution of B_x are seen around the center of PMG. B_z forms a primary magnetic hill above the track center and two secondary magnetic hills with opposite polarity at two sides of the track. The intensity of B_z above the track center is up to 1.5 T and that 5 mm and 20 mm above the track surface, 0.9 T and 0.3 T respectively. The support wall of the track is precisely assembled with high-strength metal components, as shown in Fig. 4(c). The maglev vehicle will get closer to the track under the action of centrifugal force during acceleration. It is found through static simulation test that guidance force will change due to continuous reduction of suspension gap. As shown in Fig. 4(d), triangle array of YBCO bulks can effectively keep constant the guidance force, ensuring high suspension stability of the maglev vehicle during dynamic operation, single YBCO bulk is 30 mm in diameter and 20 mm in thickness.

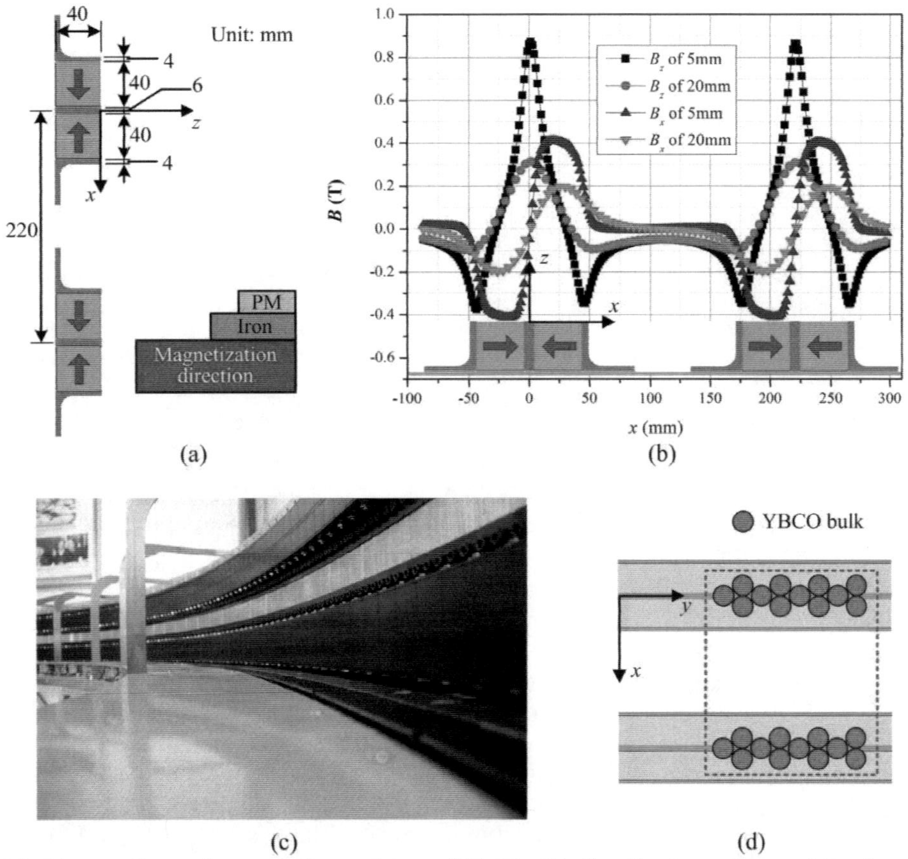

Figure 4. (a) The configuration and dimensions of PMG; (b) distribution of B_z and B_x along the x-axis at positions 5 mm and 20 mm above the top surface of PMG; (c) the local feature photo of side-suspended PMG; (d) triangle arrangement of YBCO bulks.

4. Evacuated tube system

In order to reduce the influence of air flow upon high-speed operation of the maglev vehicle, an evacuated tube is established along the circular track, as shown in item 1 of Fig. 3(a). The evacuated tube, with sectional dimension of 48.5 × 55 cm, is composed of metal base plate and outer wall and acrylic top plate and inner wall. The tube is divided into multiple components for assembly to facilitate transportation and installation and the joint between adjacent components is subject to vacuum sealing. Reinforcing ribs are provided along the inner side of the evacuated tube to prevent excessive deformation of the tube wall under low pressure. Air pressure in the evacuated tube is adjusted and stabilized by primary pump and holding pump, with the specific control principle shown in Fig. 5(a). Assembled by multiple components, the tube may be subject to leakage on a certain degree, leading to pressure drop in the tube. When the pressure is adjusted to the specified value by the primary pump, the power supply of the primary pump will be cut off by air pressure sensor built in the tube and the holding pump will be started to maintain the pressure in the tube. Finally, the air pressure in the tube will be reduced to 650 Pa by the 7.5 kw mechanical vacuum pump within 19 minutes, as shown in Fig. 5(b).

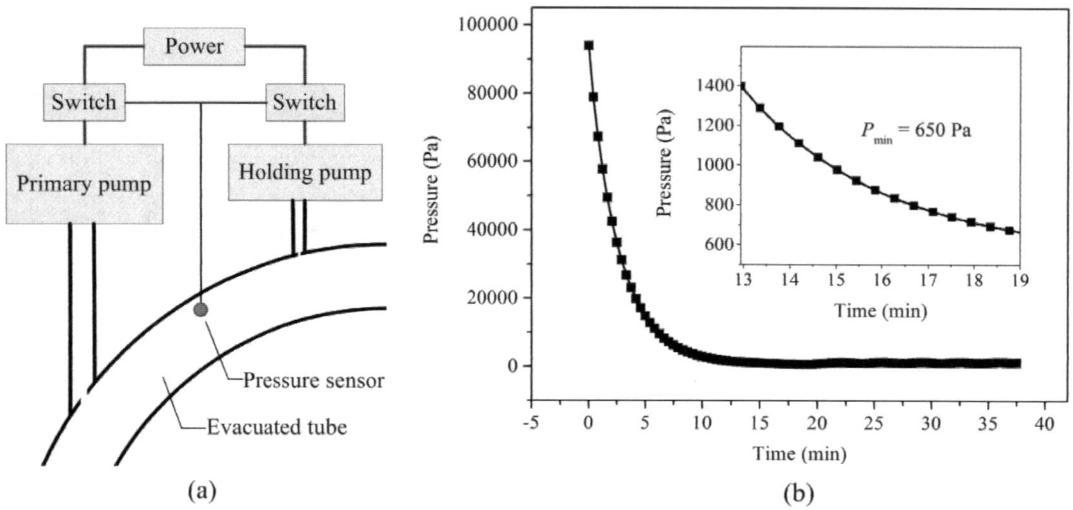

Figure 5. (a) The schematic diagram of evacuated tube system; (b) the curve of pressure in evacuated tube from local atmosphere down to 650 Pa during 19 minutes.

5. Drive system

The driving system is mainly composed of weak current controller, linear motor and three-phase variable frequency power supply. The operating principle of the system is as shown in Fig. 6(a): segmented power supply control of linear motor is realized through FPGA and MCU. MCU refers to the microprogrammed control unit, which is mainly used for receiving address signal of maglev vehicle and sending control signal of linear motor. High-speed encoding of 96 parallel binary address signals (0: maglev vehicle not at current position; 1: maglev vehicle at current position) detected by 96 photoelectric detectors is carried out based on high-speed parallel processing through multiple peripheral ports of FPGA, and then single-character signals are sent to the MCU. MCU will send driving control signals to FPGA according to set power supply parameters and FPGA will convert these signals into 96 parallel binary control signals (0: close solid state relay; 1: open solid state relay) through high-speed decoding and control the power supply of 96 linear motors. Parameters for segmented power supply of linear motor mainly include the number of front, middle and back motors and can be set through operation panel as shown in Fig. 6(b). During operation of the maglev vehicle, it is only necessary to supply power for front, middle and back motors, thus saving the power capacity and effectively improving the driving efficiency of the linear motor, as shown in Fig. 6(c).

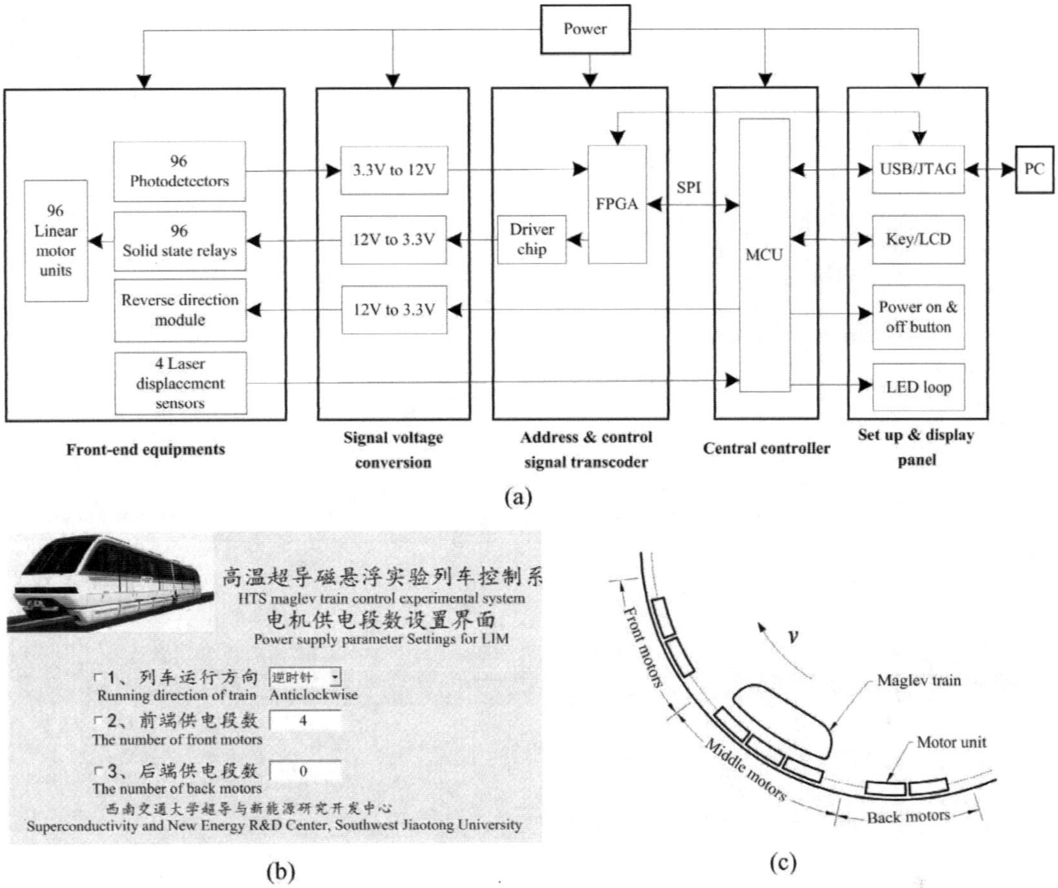

Figure 6. (a) Block diagram of drive system, which is mainly consist of traction motor and weak current controller; (b) power supply parameter setting for linear motor; (c) the schematic of sectional power supply for linear motor, where is covered and closed to maglev train.

Dynamic operation of the evacuated tube SS-HTS maglev circular track system is successfully realized through integration of SS-HTS maglev system, evacuated tube system and segmented linear motor driving system. As is shown in Fig. 7(a), a vehicle-mounted three-axis gyroscope is used to test the instantaneous speed of the maglev vehicle during high-speed dynamic operation. Under tube pressure of 5×10^3 Pa, the maglev vehicle can operate stably at an average speed of 150 km/h, the highest ever recorded in tests of HTS maglev vehicle, as shown in Fig. 7(b). The photo of maglev vehicle during static suspension and that during high-speed dynamic operation are respectively shown in Fig. 7(c) and 7(d).

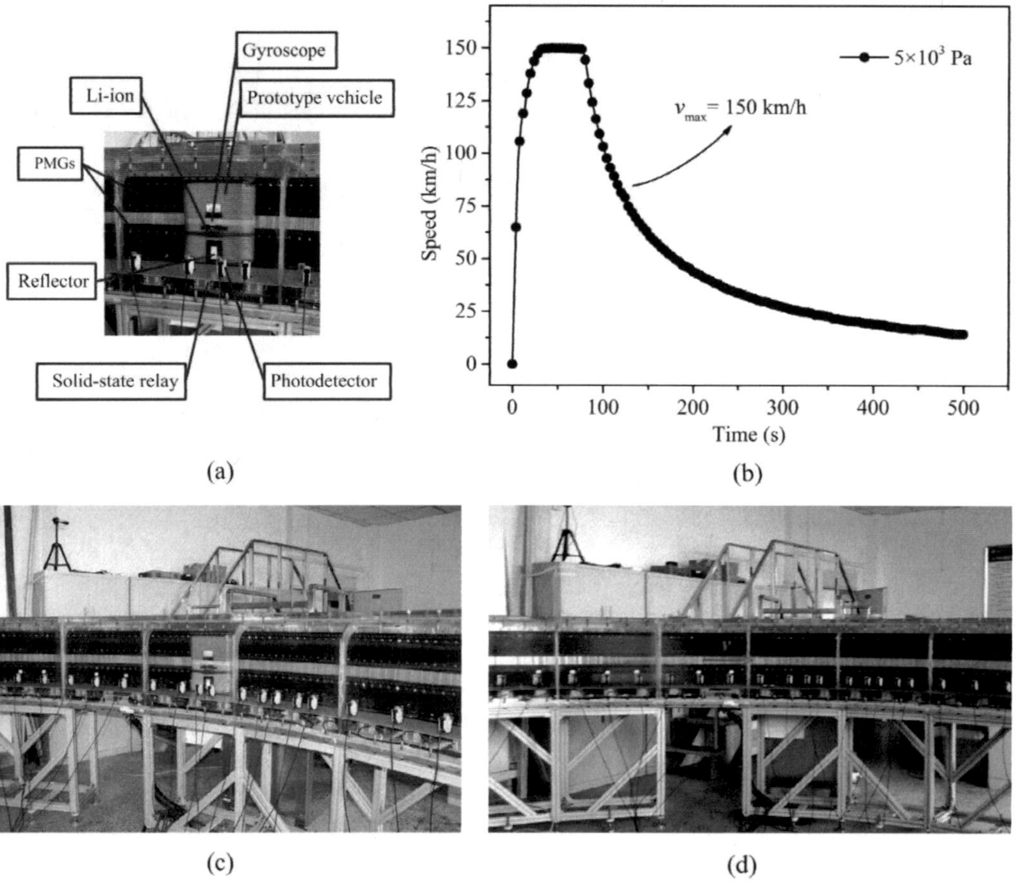

Figure 7. (a) The prototype vehicle with triangle arrangement of YBCO bulks; (b) the maximum speed curve of 150 km/h under 5×10^3 Pa; (c) the local feature photo of prototype vehicle under static state; (d) a snapshot for prototype vehicle running at a high-speed of 150 km/h in SS-HTS maglev test circular track.

Commissioning and test in the latest two months show that the maglev vehicle can operate stably at high speed with vertical displacement change not exceeding 1mm, as confirmed through inspection of air gap of linear motor during operation of the maglev vehicle. So far, the test vehicle has run approximately 1,000 km in the evacuated tube SS-HTS maglev circular track system with zero accident, even though the superconducting bulks has been used for 10 years, directly verifying the long-term stability and safety of HTS maglev vehicle running at high speed.

6. Summary

As is demonstrated through high-speed dynamic test in the evacuated tube SS-HTS maglev circular track system, the maglev vehicle adopting triangle array of YBCO bulks realizes stable operation of up to 150 km/h, a remarkable improvement in the speed of free suspension operation and the highest speed ever recorded in current reports on HTS maglev system. And the operation stability and reliability at 150 km/h of HTS maglev system is proved through long-term commissioning and high-speed operation test.

Acknowledgments

We thank the financial support from the Program of International S&T Cooperation (Grant No. 2013DFA51050), the National Nature Science Foundation of China (grant No. 51271155, 51377138), the Fundamental Research Funds for the Central Universities under the No. 2682016ZDPY10, Science and Technology Project in Sichuan Province (2017JY0058), and the Doctorial Innovation Funds under the No.A0920502051410-3.

References

[1] E. H. Brandt, "Levitation in physics," *Science*, **vol. 243**, no. 4889, Jan. 1989, pp. 349–355.

[2] J. R. Wang et al., "High-T_c superconductive vehicle for maglev model," *Rare Metal Materials and Engineering*, **vol. 27**, no. 4, Aug. 1998, pp. 240–243.

[3] J. Wang et al., "The first man-loading high temperature superconducting maglev test vehicle in the world," *Physica C*, **vol. 378–381**, Oct. 2002, pp. 809–814.

[4] L. Schultz et al., "Superconductively levitated transport system—the SupraTrans project," *IEEE Trans. Appl. Supercond.*, **vol. 15**, no. 2, Jun. 2005, pp. 2301–2305.

[5] L. Schultz et al., "Static and dynamic behavior of a superconducting magnetic bearing using YBCO bulk material," *IEEE Trans. Appl. Supercond.*, **vol. 17**, no. 2, Jun. 2007, pp. 2079–2082.

[6] Z. Deng et al., "A high-temperature superconducting maglev ring test line developed in chengdu, Chian," *IEEE Trans. Appl. Supercond.*, **vol. 26**, no. 6, Jun. 2006, Art. ID. 3602408.

[7] L. S. Mattos et al., "MagLev-cobra operational tests," *IEEE Trans. Appl. Supercond.*, **vol. 26**, no. 3, Apr. 2016, Art. ID. 3600704.

[8] J. Ma, D. Zhou, L. Zhao, Y. Zhang, and Y. Zhao, "The approach to calculate the aerodynamic drag of maglev train in the evacuated tube," *J. Mod. Transport.*, **vol. 21**, no. 3, Sep. 2013, pp. 200–208.

Study of Running Stability in Side-Suspended HTS-PMG Maglev Circular Line System

Dajin Zhou[1], Lifeng Zhao[1], Linbo Li[1], Chenyu Cui[1], Chang-Chun Hsieh[1], Yong Zhang[1], Jianqiang Guo[1] and Yong Zhao[1,2,*]

[1]Key Laboratory of Maglev Train and Maglev Technology of Ministry of Education, Superconductivity and New Energy R&D Center, Southwest Jiaotong University, Chengdu 610031, China,

[2]School of Physical Science and Technology, Southwest Jiaotong University, Chengdu 610031, China

*Corresponding author: Prof. Yong Zhao, e-mail: yzhao@home.swjtu.edu.cn

Abstract. A research on stability of the side-suspended HTS-PMG maglev circular line system is carried out through simulation experiment. The results show that the maglev vehicle will gradually get close to the track surface during acceleration under the action of centrifugal force, leading to decay of guidance force and occurrence of vertical eccentric motion. In case of linear array of $YBa_2Cu_3O_{7-x}$ (YBCO) bulks, the guidance force will be changed with the decreasing of the levitation gap. It can be suppressed through the complex arrangement of YBCO bulks. Fortunately, triangle array of YBCO bulks can effectively keep the guidance force constant and realize stable running during accelerating process of the prototype vehicle. Based on the research on stability of side-suspended maglev vehicle, a side-suspended PMG circular test track with diameter of 6.5 m and circumference of 20.4 m is successfully designed and established, enabling the prototype vehicle to run stably at up to 82.5 km/h under open atmosphere (9.6×10^4 Pa).

1. Introduction

Taking advantage of flux pinning characteristic of YBCO bulk, HTS maglev system has excellent self-stability [1]. It is widely used in magnetic bearing [2, 3], maglev vehicle [4-6], maglev launch [7, 8] and other fields. Compared with conventional wheel rail train, maglev vehicle operates more stably without noise, and has smaller turning radius. It is an important solution for high-speed ground transportation in the future.

Currently, mature maglev train technologies mainly include TR train of Germany [9] and electric dynamic suspension (EDS) train of Japan [10]. It has already been commercialized in TR train and achieved a high speed of 603 km/h in the test of EDS train. Although 29 years has passed since YBCO high-temperature superconductor was discovered, research and development of HTS maglev vehicle is still under laboratory study, mainly in China, Germany, Brazil and Japan [11-14]. In 2016, the team from Federal University of Rio de Janeiro of Brazil designed a full-size manned HTS maglev vehicle and a 200 m-long straight track to carry out the test with a speed of only 7 km/h. Though it is theoretically predicated that the highest speed of HTS maglev vehicle is up to 2900 km/h, there is no actual dynamic test for verification.

Continuous acceleration of maglev vehicle can be realized in a finite-length circular track. Huge centrifugal force, however, will be generated during high-speed operation of maglev vehicle, even causing the maglev vehicle to run off the track. For locomotive, the track is designed with a proper inclination to increase the centripetal force of the train when making a turn. Similarly, the permanent magnetic guideway (PMG) is inclined by 90 ° along the horizontal plane and the HTS maglev vehicle is placed beside the track. The guidance force counteracts the weight and the levitation force counteracts the centrifugal force. It will greatly improve the speed of maglev vehicle by this method. The proposal of side-suspended HTS (SS-HTS) maglev system effectively solves the problem of insufficient centripetal force and off-track of the maglev vehicle along the tangential direction of the track when the maglev vehicle runs above circular track. Currently, there are only a few research reports on SS-HTS maglev system due to a lack of theory and experimental study. When running "sticking to wall" at a high speed, the SS-HTS maglev vehicle will get closer to the track surface under the action of centrifugal force. The dynamic process of maglev vehicle running above a circular track can be simulated by using a servo motor to control the change in levitation gap between the maglev vehicle and the track. In the SS-HTS maglev circular line system, a simulation test for the guidance force is carried out to analyze the stability of guidance force under different arrays of YBCO bulks.

2. Experiments

2.1. *Magnetic track*

In order to provide a high-density magnetic field for the HTS maglev vehicle, the magnetic track is usually assembled with permanent magnet (PM) and iron plate to press PMs with the same pole closer to each other and gather magnetic field through iron plate, thus obtaining a magnetic field stronger than that of a conventional PM. Figure 1(a) shows cross-section structure of the PMG which consists of two PMs with opposite magnetization direction and an iron plate in between. The PM is made of NdFeB (grade: N52), 40 × 40 ×52.5 mm^3 in size, and the iron plate is 6 mm in thickness. As shown in figure 1(c), the characteristics of magnetic field distribution above the PMG are that *z*-component of B (B_z) forms a primary magnetic peak above the center of the PMG and two secondary magnetic peaks located at both sides edges of the PMG respectively, and presents an symmetrical distribution on both sides of the PMG while *x*-component of B (B_x) presents an anti-symmetric distribution. Actually, the distribution of B_z along *x*-axis and *y*-axis direction is asymmetric and inhomogeneous respectively due to the difference in PM performance and gaps between PM and PM, PM and iron plate.

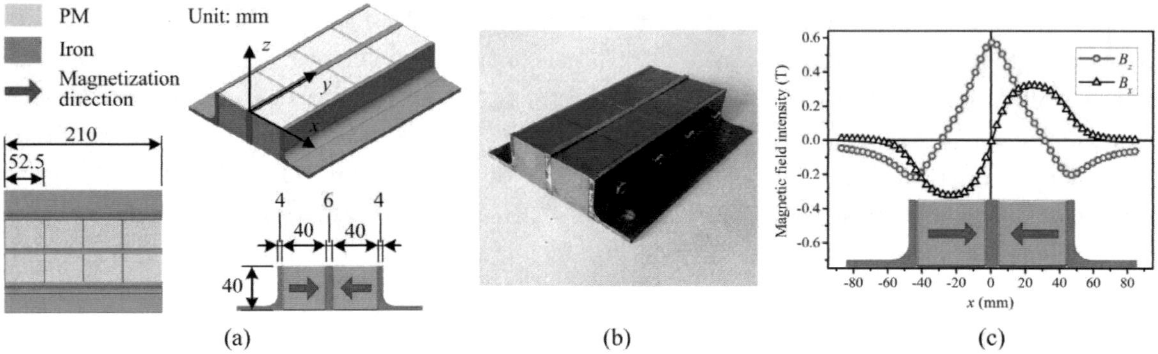

Figure 1. (a) The structure of PMG segment; (b) Photograph of the PMG segment; (c) Distribution of magnetic field intensity at position 10 mm above the PMG segment.

2.2. Experimental principle and equipment

The operating principle of SS-HTS maglev circular line system shown in figure 2 mainly involves the following three processes: (1) Initial field cooling (FC) process: where the relative position of YBCO bulk and PMG is fixed by the aid of a fixture, then liquid nitrogen is injected into the cryogenic container for cooling YBCO bulk. The corresponding levitation gap l is called field cooling height (FCH); (2) static suspension process: after YBCO bulk fully gets into the superconducting state, the fixture is removed. YBCO bulk moves downwards under the action of gravity until a balance between guidance force and gravity is achieved, a static vertical displacement d is generated; and (3) dynamic operation process: during acceleration in circular track, YBCO bulk gets close to the PMG under the action of centrifugal force. The guidance force is decreased with the reduction of levitation gap (as shown in figure 4). YBCO bulk moves downwards until a new balance between guidance force and gravity is achieved, a dynamic vertical displacement s is generated.

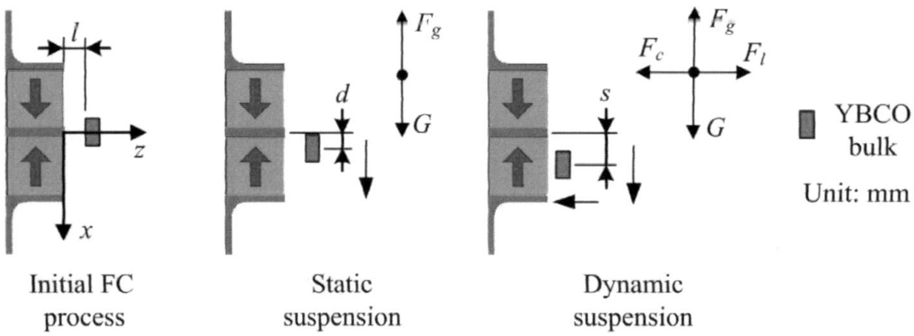

Figure 2. The static and dynamic suspension process of SS-HTS maglev circular line system, l is the levitation gap, d is static vertical displacement and s is dynamic vertical displacement, F_c is centrifugal force, F_l is levitation force, F_g is guidance force, G is gravity.

A simulation test platform as shown in figure 3(c) is designed and established according to the operating principle of SS-HTS maglev circular line system. Columnar single-seed crystal melt-processed YBCO bulk with 30 mm in diameter and 18 mm in thickness is used in the test. Multiple

YBCO bulks are fixed on the inside wall of the cryogenic container. The levitation gap l and the displacement d, s are controlled by servo motors in horizontal and vertical direction respectively. The simulation test mainly involves two processes: (1) the servo motor in vertical direction is actuated after YBCO bulks get into the superconducting state to cause a vertical displacement of the PMG along x-axis direction, to simulate that YBCO bulks move downwards under the action of gravity until a balance between gravity and guidance force is reached; and (2) the servo motor in horizontal direction is actuated to make the cryogenic container get close to the PMG along z-axis direction, to simulate that YBCO bulks get close to the PMG under the action of centrifugal force. During the entire test process, the guidance force is measured by a pull-pressure sensor.

For the purpose of dynamic test in SS-HTS maglev vehicle, a SS-HTS maglev circular track with diameter of 6.5 m and perimeter of 20.4 m is established and an evacuated tube is added to decrease aerodynamic drag, as shown in figure 3(a). A superconducting prototype vehicle adopting triangular array of YBCO bulks is also designed for high-speed dynamic test, as shown in figure 3(b). Linear motor is used for continuous driving during operation of the prototype vehicle and a vehicle-mounted gyroscope is used to measure the speed of prototype vehicle.

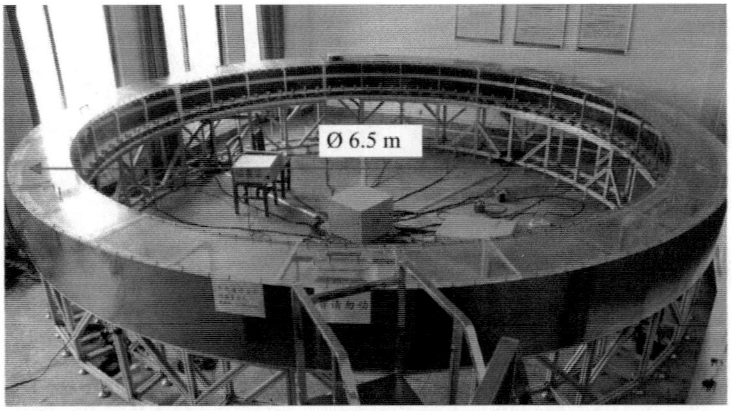

(a)

(b)

(c)

Figure 3. (a) Panoramic photograph of SS-HTS maglev circular track; (b) local features of PMG and prototype vehicle, 1-prototype vehicle, 2-PMG; (c) the experimental setup for measuring the guidance force change with levitation gap, 3-PMG, 4-cryogenic container.

3. Results and discussion

3.1. Linear array of YBCO bulks

The guidance force of YBCO bulks in linear array (see Array 1 in figure 6(a)) is tested. The detail test process is: (1) Make axis of YBCO bulk coincide with symmetry axis of PMG (see Position 3 in figure 5(a)) and inject liquid nitrogen into cryogenic container until all YBCO bulks get into the superconducting state; (2) Start the servo motor in vertical direction to drive the PMG upward by 2 mm (equivalent downward movement of YBCO bulk by 2 mm); (3) Keep the relative position between PMG and container unchanged in vertical direction, start the servo motor in horizontal direction to drive container to get close to PMG until the rated levitation gap of 5 mm is reached. At FCH of 10 mm, 15 mm and 20 mm, the change in guidance force with reduction of levitation gap are obtained respectively, as shown in figure 4. The guidance force decreases with reduction of levitation gap, and becomes negative value at FCH of 15 mm and 20 mm when levitation gap is less than 6 mm and 10 mm respectively. These results show that the instability of guidance force exists in the dynamic operation of SS-HTS maglev circular line system. Thus further research on maintaining the stability of guidance force is required for safe operation of SS-HTS maglev circular line system.

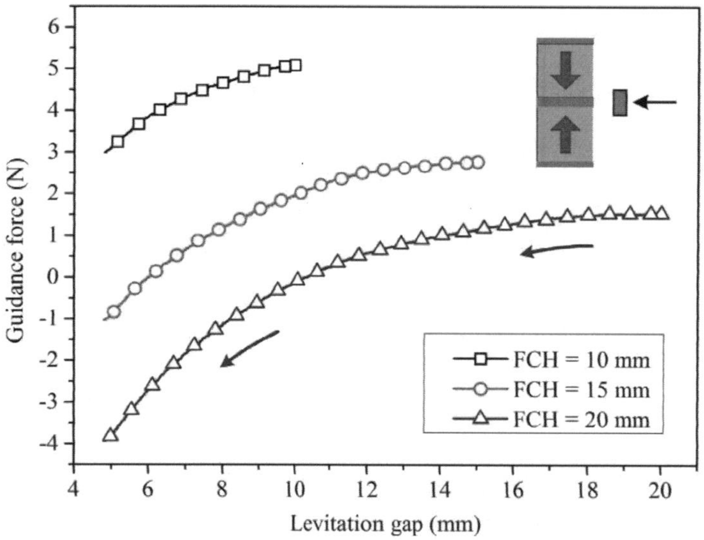

Figure 4. The change in guidance force with the decrease of levitation gap under different conditions of FCH, where the axis of YBCO bulks overlapping with the centerline of PMG (e.g. Position 3 in figure 5(a)).

Through changing the initial field cooling (FC) position between YBCO bulk and PMG in vertical direction, five different linear arrays are obtained, as shown in figure 5(a). According to the foregoing test procedure, the change in guidance force under different initial FC positions is acquired at FCH of 20 mm, as shown in figure 5(b). When the axis of YBCO bulk located in the area of $x < 0$ during static suspension (the location is not the initial FC position, because YBCO bulk drops downward by 2 mm from the initial FC position, as shown in figure 2), the guidance force increases with reduction of levitation gap, corresponding to Position 1 and Position 2. On the contrary, when the axis of YBCO bulk located in the area of $x > 0$ during static suspension, the guidance force decreases and gets negative with reduction of levitation gap, corresponding to Position 3 and Position 4. But the guidance force of Position 5 presents different changing law compared with Position 3 and Position 4. Possibly because the YBCO bulk located at Position 5 is more closer to the lower edge of PMG during side suspension, the guidance force is affected by magnetic field nearby the secondary magnetic peak of B_z, as shown in figure 1(c). These results indicate that the relative position in vertical direction between YBCO bulk and PMG plays a key role in the changing law of guidance force with decreasing the levitation gap. The instability of guidance force still exists in the system with linear array of YBCO bulks. According to the results from figure 5(b), choose proper YBCO bulks located upper half area (x < 0) and lower half area (x > 0) of PMG to combination, the positive guidance force can be overlaid with the negative guidance force to decrease the change of guidance force. It is a feasible method to realize the stabilization in guidance force. Thus, the fluctuation in guidance force during dynamic process of SS-HTS maglev circular line system can be optimized through complex array of YBCO bulks.

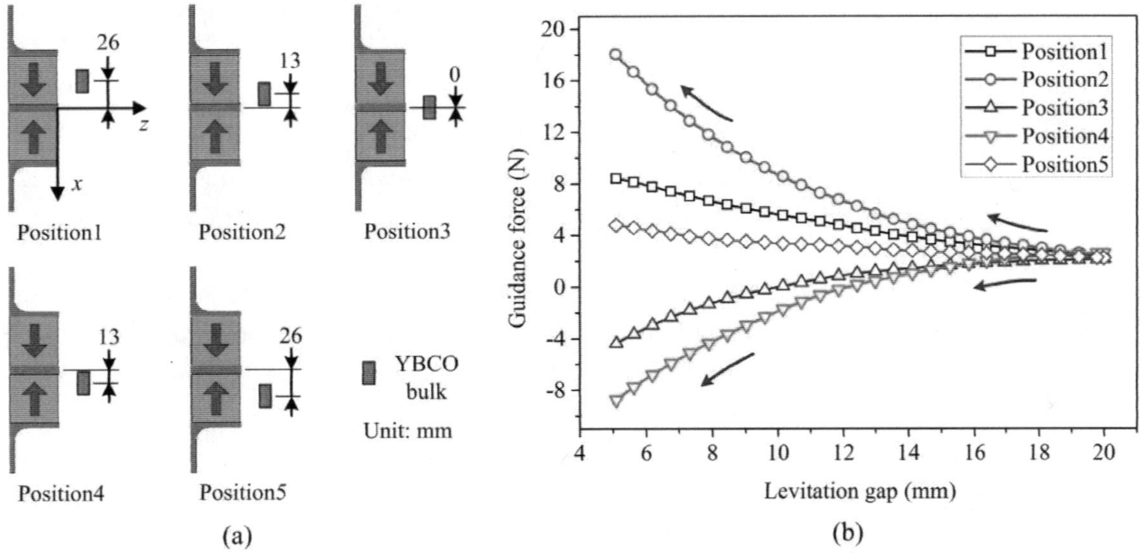

Figure 5. (a) Five different types of linear arrays (e.g. Array 1 in figure 6(a)) of YBCO bulks with the condition of FCH = 20 mm; (b) the change of average guidance force reduced to one bulk with the levitation gap from 20 mm to 5 mm for five types of linear array of YBCO bulks in static experiment.

3.2. Complex array of YBCO bulks

Figure 6(a) gives five different complex arrays of YBCO bulks. By reference to the test procedures of guidance force in linear array of YBCO bulks, the guidance force under different complex arrays of YBCO bulks is tested with levitation gap decreasing from 20 mm to 5 mm. As the quantity of YBCO bulks varies in different complex arrays, the average guidance force reduced to one YBCO bulk is taken for discussion to facilitate comparison and analysis, as shown in figure 6(b). Compared with the linear array (Array 1) of YBCO bulks, it can be found that all other complex arrays (Arrays 2–5) restrain the variation of guidance force in different extent. Complex arrays of YBCO bulks can improve the dynamic stability of SS-HTS maglev circular line system.

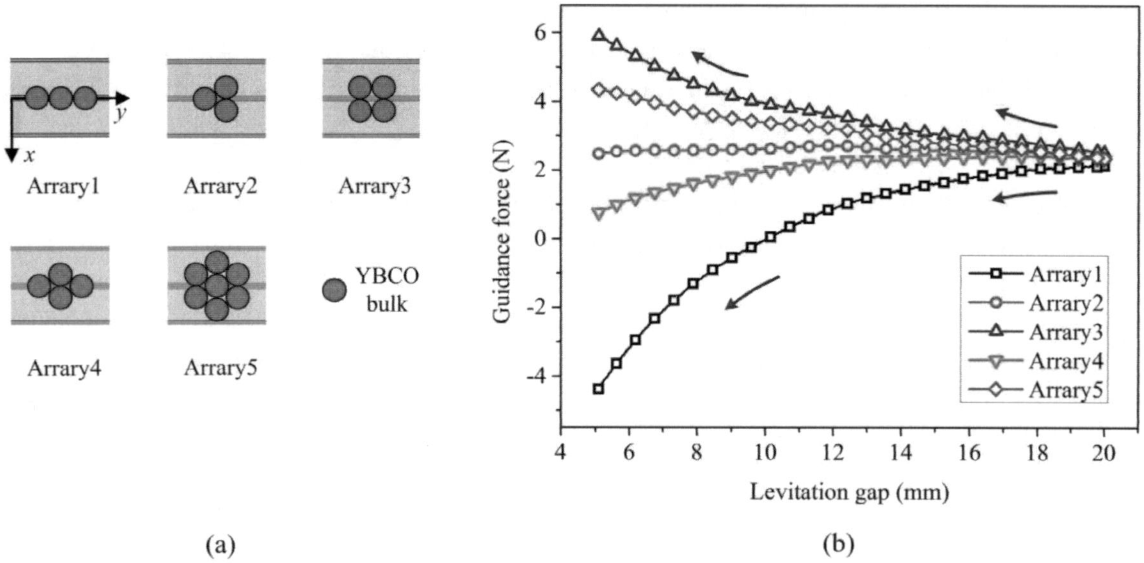

Figure 6. (a) Five different types of complex arrays of YBCO bulks with the condition of FCH = 20 mm; (b) the change of average guidance force reduced to one bulk with the levitation gap decreasing from 20 mm to 5 mm, among which Array 2 (triangular array) presented the minimum change.

According to figure 5(a, b), when the axis of YBCO bulk located in the area of $x < 0$ (the upper half area of PMG) during static suspension, the guidance force increases and keeps positive value with the levitation gap decreasing from 20 mm to 5 mm. Apart from the case of Position 5 (possible reason was given in the previous section), the guidance force decreases and becomes negative value under the axis of YBCO bulk located in the area of $x > 0$ (the lower half area of PMG) during static suspension. When the upper YBCO bulks and lower YBCO bulks are combined together, the variation in complex guidance force will be reduced due to the superposition of positive force and negative force. In fact, when the initial FC positions of two YBCO bulks are symmetrical at both sides of PMG, the absolute values in guidance forces of two YBCO bulks are not equal. Due to the YBCO bulks drop downward by 2 mm during static suspension, result in the upper YBCO bulk is more closer to z-axis (symmetry axis of PMG) compared with the lower YBCO bulk. The absolute value in guidance force of upper YBCO bulk is greater than lower YBCO bulk in the symmetrical initial FC position.

During static suspension, Array 1 only consists of three lower YBCO bulks, the average guidance force presents negative force and maximal variation. Array 3 consists of two upper YBCO bulks and two lower YBCO bulks, positive force dominates in the average guidance force. Array 4 consists of one upper YBCO bulk and three lower YBCO bulks, negative force dominates in the average guidance force. Array 5 consists of three upper YBCO bulks and four lower YBCO bulks, but one lower YBCO bulk closed to the lower edge of PMG presents positive force, result in positive force dominates in the average guidance force. Finally, it can be found that almost constant in average guidance force occurs in triangular array (Array 2), which consists of one upper YBCO bulk and two lower YBCO bulks. It indicates that test vehicle can run in SS-HTS maglev circular line system with the highest dynamic stability under Array 2. For the test results of figure 6(b), in this paper, we only make a qualitative discussion. In the further research, we will give the quantitative interpretation through theoretical calculation and simulation.

A SS-HTS maglev prototype vehicle with triangular array of YBCO bulks is designed, as shown in the illustration of figure 7. The prototype vehicle consists of two cryogenic containers which are respectively placed above the upper PMG and the lower PMG. They are connected by two metal conduits, one for liquid nitrogen injection and the other for gas exhaust. A vehicle-mounted three-axis gyroscope is used to measure the running path and obtain the speed-time curve (include acceleration, constant speed and deceleration process) of the prototype vehicle, as shown in figure 7. Under open-air atmospheric pressure (9.6×10^4 Pa), actual test speed of the prototype vehicle reaches to 82.5 km/h, which is the highest ever recorded without vacuuming.

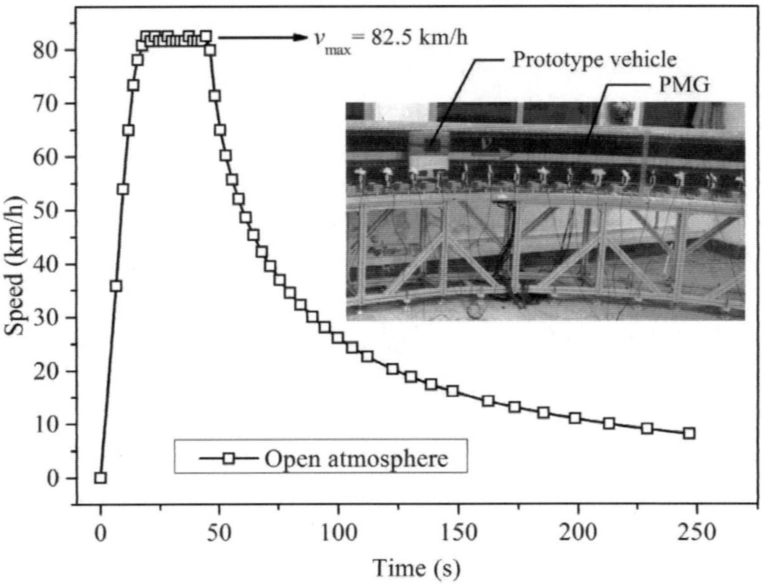

Figure 7. Speed-time curve in dynamic test under ambient pressure (9.6×10^4 Pa), the inset shows the photo of field test.

4. Summary

SS-HTS maglev system helps maglev vehicle break through the restriction of centrifugal force in circular line, greatly improves the running speed of maglev vehicle above circular track and prevents maglev vehicle running off the track along tangent direction. An effective research on dynamic operation process of the SS-HTS maglev circular line system is made through simulation test. The results show that the prototype vehicle gets close to the track surface during acceleration, the guidance force will vary with the reduction of levitation gap, leading to instability of the guidance force. Researches on guidance force under linear and complex arrays of YBCO bulks show that variation of guidance force can be reduced under complex array of YBCO bulks, especially triangular array where the guidance force almost remains unchanged with the reduction of levitation gap, thus ensuring stable high-speed operation of the prototype vehicle above the circular track. Finally, the prototype vehicle based on triangular array of YBCO bulks realizes to run stably above SS-HTS maglev circular track, with the highest speed of 82.5 km/h under open-air atmospheric pressure (9.6×10^4 Pa).

Acknowledgments

We thank the financial support from the Program of International S&T Cooperation (Grant No. 2013DFA51050), the National Nature Science Foundation of China (grant No. 51271155, 51377138), the Fundamental Research Funds for the Central Universities under the No. 2682016ZDPY10, Science and Technology Project in Sichuan Province (2017JY0057), and the Doctorial Innovation Funds under the No.A0920502051410-3.

References

[1] E. H. Brandt, "Levitation in physics," *Science*, **vol. 243**, no. 4889, Jan. 1989, pp. 349–355.

[2] W. Liu et al., "Design of a high-T_c superconductive maglev flywheel system at 100-kW level," *IEEE Trans. Appl. Supercond.*, **vol. 26**, no. 4, Jun. 2016, Art. ID. 5700805.

[3] F. N. Werfel et al., "Towards high-capacity HTS flywheel systems," *IEEE Trans. Appl. Supercond.*, **vol. 20**, no. 4, Aug. 2010, pp. 2272–2275.

[4] F. N. Werfel et al., "Superconductor bearings, flywheels and transportation," *Supercond. Sci. Technol.*, **vol. 25**, no. 1, Dec. 2011, Art. ID. 014007.

[5] F. N. Werfel et al., "Experiments of superconducting maglev ground transportation," *IEEE Trans. Appl. Supercond.*, **vol. 26**, no. 3, Apr. 2016, Art. ID. 3602105.

[6] S. Y. Wang et al., "High temperature superconducting Maglev equipment on vehicle," *Physica C*, **vol. 378–381**, Oct. 2002, pp. 809–814.

[7] W. J. Yang et al., "Construction and performance of HTS maglev launch assist test vehicle," *IEEE Trans. Appl. Supercond.*, **vol. 16**, no. 2, Jun. 2006, pp. 1108–1111.

[8] W. J. Yang et al., "Levitation characteristics in an HTS maglev launch assist test vehicle," *Supercond. Sci. Technol.*, **vol. 20**, Feb. 2007, pp. 281–286.

[9] E. Gottzein, K. -H. Brock, E. Schneider and J. Pfefferl, "Control aspects of a tracked magnetic levitation high speed test vehicle," *Automatica,*, **vol. 13**, no. 3, May 1977, pp. 205–223.

[10] S. Kusada et al., "The project overview of the HTS magnet for superconducting maglev," *IEEE Trans. Appl. Supercond.*, **vol. 17**, no. 2, Jun. 2007, pp. 2111–2116.

[11] J. Wang et al., "The first man-loading high temperature superconducting maglev test vehicle in the world," *Physica C*, **vol. 378–381**, Oct. 2002, pp. 809–814.

[12] L. S. Mattos et al., "MagLev-cobra operational tests," *IEEE Trans. Appl. Supercond.*, **vol. 26**, no. 3, Apr. 2016, Art. ID. 3600704.

[13] L. Schultz et al., "Superconductively levitated transport system—the SupraTrans project," *IEEE Trans. Appl. Supercond.*, vol. 15, no. 2, pp. 2301–2305, Jun. 2005.

[14] M. Okano, T. Iwamoto, M. Furuse, S. Fuchino, and I. Ishii, "Running performance of a pinning-type superconducting magnetic levitation guide," *J. Phys. Conf. Ser.*, **vol. 43**, 2006, pp. 999–1002.

IOP Conf. Series: Journal of Physics: Conf. Series **871** (2017) 012094 doi:10.1088/1742-6596/871/1/012094

Numerical analysis of fundamental characteristics of superconducting magnetic bearings for a polarization modulator

Yusuke Terachi[1], Yutaka Terao[1], Hiroyuki Ohsaki[1], Yuki Sakurai[2], Tomotake Matsumura[2], Hajime Sugai[2], Shin Utsunomiya[2], Hirokazu Kataza[3] and Ryo Yamamoto[3]

[1]Graduate School of Frontier Sciences, The University of Tokyo, 5-1-5 Kashiwanoha, Kashiwa, Chiba 277-8561, Japan
[2]Kavli Institute for the Physics and Mathematics of the Universe (WPI), The University of Tokyo Institutes for Advanced Study, The University of Tokyo, 5-1-5 Kashiwanoha, Kashiwa, Chiba 277-8583, Japan
[3]Japan Aerospace Exploration Agency, Institute of Space and Astronautical Science (ISAS), 3-1-1 Yoshinodai, Chuo-ku, Sagamihara, Kanagawa 252-5210, Japan

E-mail: ohsaki@k.u-tokyo.ac.jp

Abstract. We have carried out numerical analysis of mechanical properties of a superconducting magnetic bearing (SMB). A contactless bearing operating at below 10 K with low rotational energy loss is an attractive feature to be used as a rotational mechanism of a polarization modulator for a cosmic microwave background experiment. In such application, a rotor diameter of about 400 mm forces us to employ a segmented magnet. As a result, there is inevitable spatial gap between the segments. In order to understand the path towards the design optimizations, 2D and 3D FEM analyses were carried out to examine fundamental characteristics of the SMBs for a polarization modulator. Two axial flux type SMBs were dealt with in the analysis: (a) the SMB with axially magnetized permanent magnets (PMs), and (b) the SMB with radially magnetized PMs and steel components for magnetic flux paths. Magnetic flux lines and density distributions, electromagnetic force characteristics, spring constants, etc. were compared among some variations of the SMBs. From the numerical analysis results, it is discussed what type, configuration and design of SMBs are more suitable for a polarization modulator.

1. Introduction

The measurement of the cosmic microwave background (CMB) has provided rich information about the physics of early universe [1, 2]. With a rapid progress in various cosmological observations, the experimental probe for the cosmic inflation is identified as one of the highest priorities in the field of cosmology. The corresponding technological improvements toward the high sensitivity measurements are in high demand. One of the key instruments to achieve this science goal is to employ a polarization modulator. The polarization modulator is an instrument to modulate incident linearly-polarized radiation. It consists of a continuously-rotating optical element, called a half-wave plate (HWP), and a rotational mechanism with a bearing. One requirement that makes this observational instrument challenging to implement is to cool an HWP to below 10 K during observations. Thus, the entire continuously-rotating system must operate at this temperature range. Thus, a conventional mechanical bearing is unsuitable due to the mechanical friction and the resultant heat dissipation.

A superconducting magnetic bearing (SMB) is a contactless bearing, which consists of an array of high temperature superconducting (HTS) bulk tiles and a permanent magnet (PM). Due to its nature of a levitation-based bearing, we identify this bearing to be an attractive candidate for a cryogenically continuously-rotating bearing for the CMB polarization modulator.

An SMB provides nearly but not perfectly frictionless rotation. This is due to a magnetic interaction between a rotor and a stator. The source of friction is originated from the magnetic field inhomogeneity of a rotor magnet [3, 4]. As the rotor spins, this spatial magnetic field inhomogeneity causes a time varying magnetic field for the stator, and a hysteresis and eddy current loss results in friction [5]. The energy loss is eventually dissipated as heat. The diameter of a rotor magnet is determined from the size of HWP, and thus the expected diameter is about 400 mm for CMB observations [6]. This forces us to use a segmented magnet to form a large ring, and the imperfection of a segment rotor magnetization and a fabrication error become a source of a magnetic field inhomogeneity. In order to minimize the heat dissipation to a cryogenic temperature stage from this rotational mechanism, one can explore the design parameters of the SMB, such as the SMB configuration, the rotor magnet geometry, and the levitation height. The magnetic field inhomogeneity is one of the key figure-of-merits, but we also study the stiffness as a figure-of-merit since the expected rotor mass including an optical element amounts about 30 kg.

In this paper, we present the numerical study for design parameter tradeoffs of SMB system for use in a CMB polarization modulator. We show a comparison of the magnetic field inhomogeneities for two segmented-rotor configurations: an axially-magnetized rotor magnet and a radially-magnetized rotor magnet. We also compute the spring constant for the axially-magnetized rotor magnet. Finally, we discuss the applicability of this technology for the polarization modulator as well as room for the optimization in an SMB design.

2. Superconducting magnetic bearing

Figure 1 shows a configuration of our SMB composed of a PM ring and bulk HTS tiles. There are two types of PM arrangements in the rotor, axially magnetized PMs and radially magnetized PMs as shown in figure 2 and 3, respectively. In both PM configurations, an assembled magnet has a gap between the segmented magnets inevitably. This gap results the magnetic field inhomogeneity. In addition to this effect, the ideal magnetization for the radial configuration is to magnetize the PM piece radially. In reality this is difficult to fabricate and we are forced to employ an arc-shaped PM piece that is magnetized in parallel instead of radially. This becomes another source of magnetic field inhomogeneity. The flux density variation needs to be reduced because it becomes a source of energy loss during rotation. We have considered following options to suppress the magnetic inhomogeneity of a segmented rotor magnet. For axially magnetized PMs, the insert of a single-piece of steel plate just below the PMs is effective to reduce the flux density variation [7], but it reduces also the amplitude of flux density. In this analysis, no addition of steel components was considered. As a part of effort to minimize the magnetic field inhomogeneity, the PMs are divided into two vertical stacks. Each are placed with a shift angle of $180/n$ between the upper and lower PM arrangements, where n is the number of PM segments. For a radially magnetized PM, steel pieces are placed on both the inner and outer sides of the PMs as shown in figure 3. The steel component is effective to reduce the flux density variation along the rotational direction. We address the quantitative assessment of these options numerically.

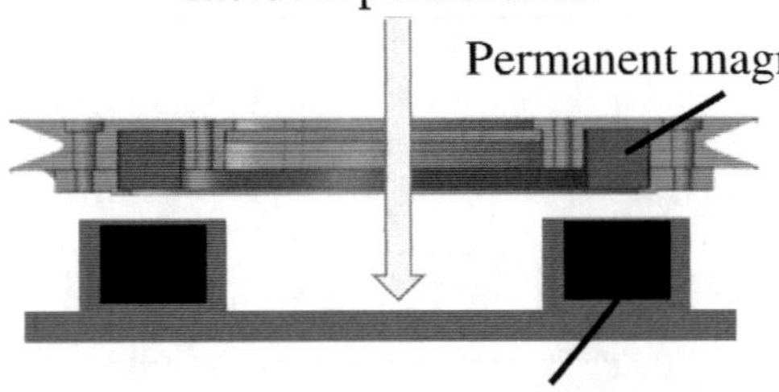

Figure 1. Superconducting magnetic bearing composed of PMs and bulk superconductors.

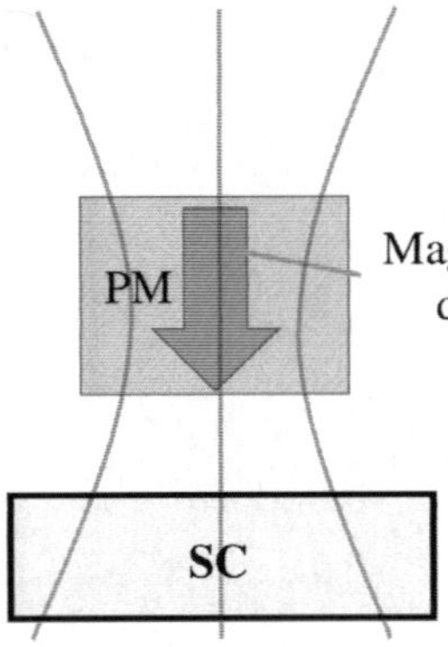

Figure 2. SMB with an axially magnetized PM ring.

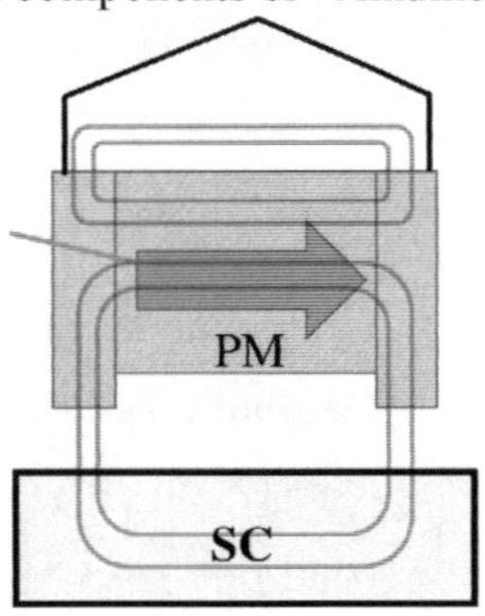

Figure 3. SMB with a radially magnetized PM ring and steel components.

3. Analysis models and conditions

3.1. Fundamental parameters of the SMB

Finite element analysis (FEA) was carried out using a commercial software, COMSOL Multiphysics, to examine the magnetic flux density distribution and the spring constant of the SMBs. Figure 4 shows a SMB model and its basic geometric parameters used in the FEA. Table 1 gives characteristic and geometric parameters of the PMs. The magnetic field of PMs can vary as a function of temperature [8], and we did not take into account this effect in the analysis.

A PM ring is divided into arc-shaped PM components. In the actual fabrication of the PM ring, a small mechanical gap cannot be avoided between adjacent PM components. Here the PM gap of 1 mm is assumed. In the analysis of SMBs, which have radially magnetized PMs with steel magnetic flux paths, the steel pieces with h_m ($> h_{PM}$) in height and w_m in width are placed on both the inner and outer side of the PMs. Here, h_m is set to 18 mm and w_m is 1 mm in all cases. The top surfaces of the PMs and the steel pieces are arranged to become a flat surface as shown in figure 3.

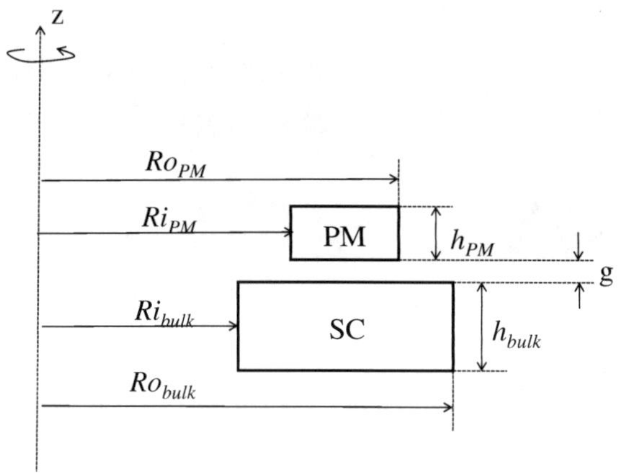

Figure 4. Arrangement of the permanent magnets and the bulk superconductors for FEM analysis.

Table 1. Parameters of permanent magnets.

Parameters	Values
Residual magnetic flux density B_r (T)	1.1
Inner radius Ri_{PM} (mm)	200
Outer radius Ro_{PM} (mm)	216
Height h_{PM} (mm)	16

3.2. Analysis of static spring constants

An axisymmetric FEA of magnetic fields was carried out to examine the magnetic spring constant of the SMB shown in figure 2 and 3. A gap between adjacent PMs was not considered and a radial magnetization of the PMs was assumed. A precise estimation of the magnetic spring constant would require the investigation of dynamic characteristics of the SMB with its superconducting properties such as an *E-J* relation taken into account. However, an approximate analysis based on a simple SMB model was adopted as the first step of the study. Namely, static spring constants were calculated using a magnetic flux frozen model, where an infinite critical current density was assumed. The analysis procedure is as follows:

i) Calculate the magnetic field in the analysis region including the SMB model without the bulk superconductor for a given initial levitation gap; g_0 between the bulk superconductor and the PM. The region of the bulk superconductor is set to air. Magnetic vector potentials in the analysis region are obtained.
ii) Obtain the magnetic vector potentials on the surface of the bulk superconductor region. They are used as one of boundary conditions for the next analysis. This boundary condition corresponds to the magnetic flux frozen state of the bulk superconductor with an infinite critical current density approximation.
iii) The above boundary condition is applied to the surface of the bulk superconductor. A small vertical displacement; Δg is given to the levitated body by changing the gap between the bulk superconductor and the PM. Moreover, calculate the magnetic field in the analysis region.
iv) Calculate the magnetic force acting on the levitated body by the nodal force method.
v) From the obtained magnetic forces for the gaps, g_0 and $g_0+\Delta g$, the spring constant can be estimated.

By changing the initial levitation gap, the relation between the static spring constant and the gap can be obtained. The FEAs were carried out for the SMBs with the axially magnetized PMs and with the radially magnetized PMs. The gap was changed in the range from 2 to 5 mm.

3.3. Analysis of magnetic field inhomogeneity

Three-dimensional FEM analysis was carried out to investigate the distribution of the axial magnetic flux density along a coaxial circular path on the upper surface of bulk superconductor. It was assumed here that the axial flux density distribution could be used as an evaluation parameter of the rotational loss of the SMBs. In the SMB with the axially magnetized PMs the circular path was placed in the middle of the top surface of bulk superconductor as shown in figure 5 while in the SMB with the radially magnetized PMs and the steel magnetic circuit the circular paths are placed on the bulk superconductor surface right below the steel pieces, as shown in figure 6. The radius of the circular path is 208 mm in the case of axial magnetization type. And the radii of the inner and outer circular paths are 199.5 mm and 216.5 mm, respectively, in the case of radial magnetization type. The levitation gap, g, is defined as the minimum distance between the bulk superconductor and the levitated component (the axially magnetized PMs or the steel pieces). The levitation gap was set to 5 mm in the analysis. The distribution of circumferential B_z of the axial magnetization type (n=8) is shown in figure 5. In addition, the distribution of circumferential B_z of the radial magnetization type (16 segments) is shown in Fig. 6. A magnetic flux density variation; ΔB is defined as $B_{max} - B_{min}$ along the circular path. We employ $\Delta B/B$ as a figure-of-merit of the magnetic field inhomogeneity.

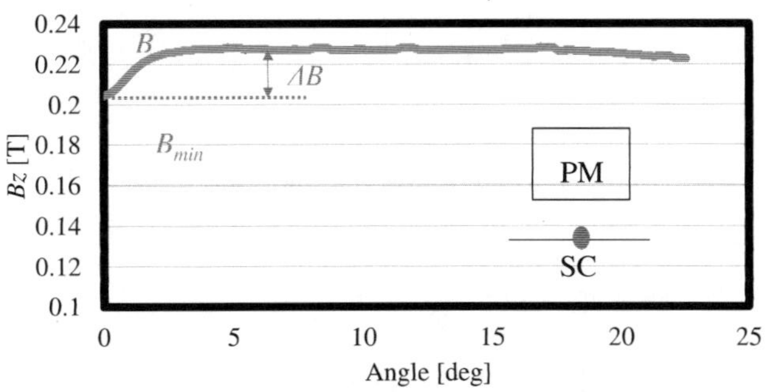

Figure 5. Magnetic flux density distribution generated by the axially magnetized PMs.

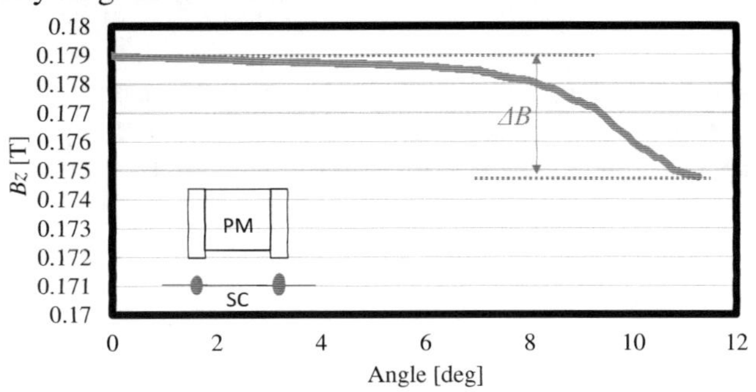

Figure 6. Magnetic flux density distribution generated by the radially magnetized PMs with steel components.

4. Analysis results

4.1. Spring characteristics

Static spring constants of the two types of SMBs were analyzed using the magnetic flux frozen approximation for the bulk superconductor. Figure 7 shows the results of the analysis for the levitation gap range from 2 to 5 mm. The spring constant is on the order of 10^5 N/m and decreases with an increase in the gap. The SMB with PMs having the axial magnetization has a larger spring constant than that with PMs having the radial magnetization and the steel magnetic circuit. An SMB for a smooth and ultra-low loss rotation of the HWP used in CMB polarization experiments is required to have a spring constant of at least a few 10^5 N/m and, if possible, as large as 10^6 N/m. The results shown in figure 7 satisfy the minimum requirement, but a design improvement will be needed for a better SMB performance. The approximation of the analysis has to be carefully considered again. The results were obtained on the approximation of the static analysis and the magnetic flux frozen model. A dynamic motion of the levitated body would generate larger currents in the superconductor than a static case if we consider the *E-J* characteristics of the superconductors with critical state models [9, 10]. Therefore, the static analysis gives us a smaller spring constant. On the other hand, the magnetic flux frozen model assumes an infinite critical current of the superconductor, and gives us a larger spring constant. Therefore, we consider that the static analysis and the magnetic flux frozen model would be an acceptable approximation for the estimation of the spring constant of the SMB in the first step of the study.

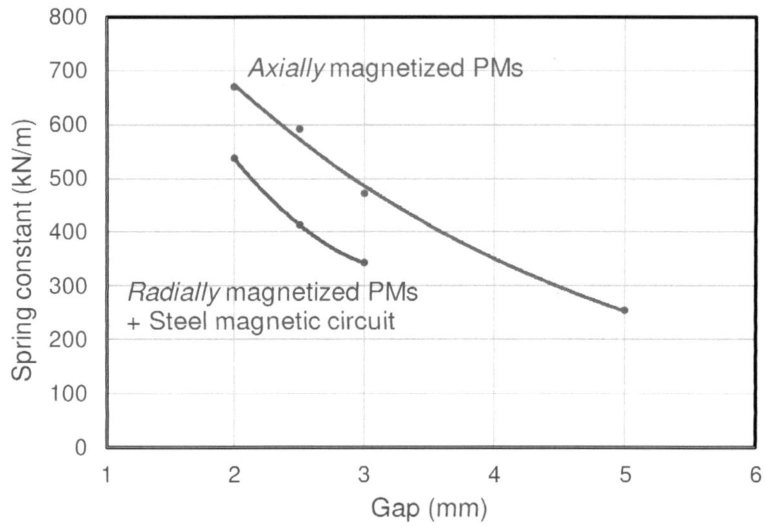

Figure 7. Dependence of the static spring constant on the levitation gap for the two types of SMBs. The SMB with axially magnetize PMs, and the SMB with radially magnetized PMs and the steel magnetic circuit.

4.2. Evaluation of $\Delta B/B$

In the case of the radial magnetization type, the magnetic field, *B*, vs the number of divisions, *n*, is shown in figure 8. The magnetic field, *B*, reaches the peak when *n* is between 8 and 16 due to a predominant effect of the increase in the radial component of magnetic flux density. However, when *n* increased continuously, *B* decreases because of the decrease in the volume of PMs fractionally with respect to the gap volume.

IOP Conf. Series: Journal of Physics: Conf. Series **871** (2017) 012094 doi:10.1088/1742-6596/871/1/012094

The magnetic field inhomogeneity, $\Delta B/B$, keeps decreasing as n increases as shown in figure 8. The design goal of the magnetic field inhomogeneity is $\Delta B/B < 0.01$. The magnetic field inhomogeneity, $\Delta B/B$, is found to be small enough to meet the requirement when n is over about 50 divisions as shown in figure 9.

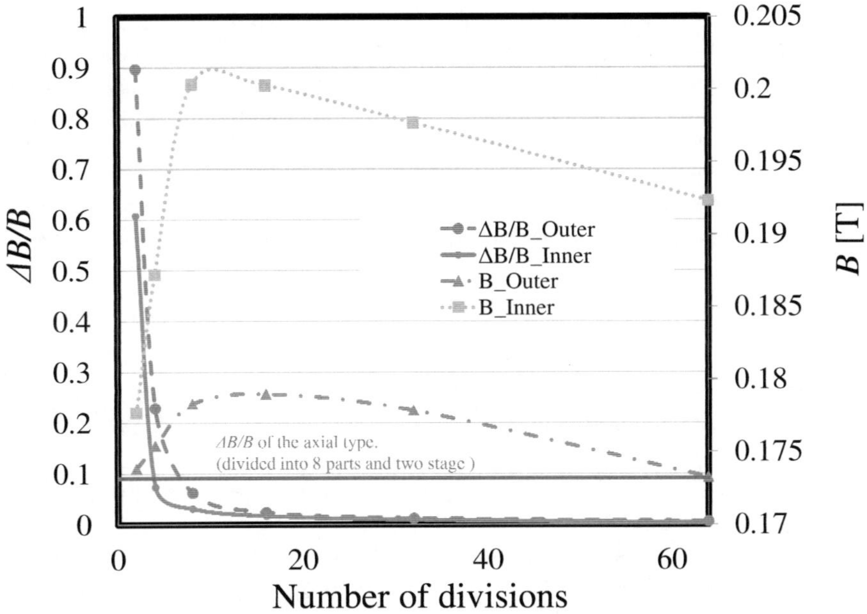

Figure 8. $\Delta B/B$ and B vs number of divisions.

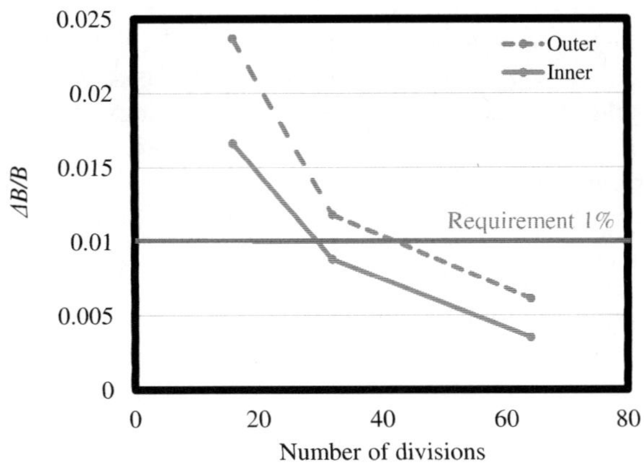

Figure 9. Vertically enlarged version of $\Delta B/B$ vs number of divisions.

5. Conclusions

We conducted the numerical studies to explore the design parameters of SMB using the FEA. We conclude that we did not find no-go result for the SMB to use for the polarization modulator in the CMB experiment even where a rotor diameter is about 400 mm and the rotor PM has to be assembled by a collection of segment PMs. We showed the results of the spring constant for two different rotor PM configurations, the segmented rotor magnet in the axially magnetized and the pseudo-radially magnetized. By employing a magnetic frozen flux approximation, we obtained the spring constants in range of 10^5 N/m. This is well within the range of need for the polarization modulator, and thus this result drive us to move to the detailed implementation of the physical model, including the properties of the superconductor and the dynamical effect, to derive the spring constant. The magnetic field inhomogeneity, $\Delta B/B$, in the axial magnetization of PM was about 10%. However, in the radial magnetization of PM, we demonstrated to suppress the $\Delta B/B$ as low as 1 %, that is the requirement value. This study guides us to construct a rotor magnet as well as investigate the further analyses, including computing the rotational loss with superconducting characteristics.

Acknowledgement

This work was supported by MEXT KAKENHI Grant Numbers JP15H05441 and JSPS Core-to-Core Program, A. Advanced Research Networks. This work was also supported by World Premier International Research Center Initiative (WPI), MEXT, Japan.

References

[1] Sato K 1981 First-order phase transition of a vacuum and the expansion of the Universe *Monthly Notices of the Royal Astronomical Society Vol* 195 no 3 p 467
[2] Guth A H 1981 Inflationary universe: A possible solution to the horizon and flatness problems *Physical Review D Vol 23 no* 2 p 347
[3] Kaneko H, Miyagawa Y, Takahata R and Ueyama, H 1999 Rotation loss characteristics of superconducting magnetic bearings *TEION KOGAKU (Journal of Cryogenics and Superconductivity Society of Japan) Vol* 34 *no* 11 p 678.
[4] Shimizu R., Demachi K and Miya K 1999 Development evaluation method of AC loss in high temperature superconductor *Presented in 11th Symposium on Electromagnetics and Dynamics*
[5] Hull J R, Mulcahy T M, Uherka K L and Abboud R G 1995 Low rotational drag in high-temperature superconducting bearings *IEEE Trans. on Appl. Supercond. Vol* 5 *no 2* p 626
[6] Sakurai Y, Matsumura T, Sugai H, Katayama N, Hiroyuki Ohsaki H, Terao Y, Terachi Y, Kataza H, Utsunomiya S, Yamamoto R 2016 Vibrational characteristics of a superconducting magnetic bearing employed for a prototype polarization modulator *Preprint* ISS2016/AP2-5
[7] Higasa H, Ishikawa F, Kawauchi N, Yokoyama S, Nakamura S, Imaizumi H and Ito N 1994 Experiment of a 100Wh-Class Power Storage System using High-Tc Superconducting Magnetic Bearing *Advances in Superconductivity VI* (Springer Japan) p 1249
[8] Hara T, Tanaka T, Kitamura H, Bizen T, Maréchal X, Seike T, Kohda T and Matsuura Y 2004 Cryogenic permanent magnet undulators *Physical Review Special Topics-Accelerators and Beams Vol* 7 *no* 5 050702.
[9] Bean C P 1964 Magnetization of high-field superconductors *Reviews of Modern Physics Vol* 36 *no* 1 p 31
[10] Kim Y B, Hempstead C F and Strnad A R 1962 Critical persistent currents in hard superconductors *Physical Review Lett. Vol* 9 *no* 7 p 306.

IOP Conf. Series: Journal of Physics: Conf. Series **871** (2017) 012095 doi:10.1088/1742-6596/871/1/012095

Transfer Efficiency and Cooling Cost by Thermal Loss based on Nitrogen Evaporation Method for Superconducting MAGLEV System

Y D Chung[1], D W Kim[2], C Y Lee[3]

[1]Dept. of electrical engineering, Suwon Science College, Kyounggi-do,18516, Korea
[2]Dept. of electrical & electronic engineering, Yonsei University, Seoul, 03722, Korea
[3]Korea Railroad Institute (KRRI) Kyounggi-do, 16105, Korea

Corresponding E-mail: ydchung@ssc.ac.kr

Abstract. This paper presents the feasibility of technical fusion between wireless power transfer (WPT) and superconducting technology to improve the transfer efficiency and evaluate operating costs such as refrigerant consumption. Generally, in WPT technology, the various copper wires have been adopted. From this reason, the transfer efficiency is limited since the copper wires of Q value are intrinsically critical point. On the other hand, as superconducting wires keep larger current density and relatively higher Q value, the superconducting resonance coil can be expected as a reasonable option to deliver large transfer power as well as improve the transfer ratio since it exchanges energy at a much higher rate and keeps stronger magnetic fields out. However, since superconducting wires should be cooled indispensably, the cooling cost of consumed refrigerant for resonance HTS wires should be estimated. In this study, the transmission ratios using HTS resonance receiver (Rx) coil and various cooled and non-cooled copper resonance Rx coils were presented under non cooled copper antenna within input power of 200 W of 370 kHz respectively. In addition, authors evaluated cooling cost of liquid nitrogen for HTS resonance coil and various cooled copper resonance coils based on nitrogen evaporation method.

1. Introduction

Wireless power transfer technologies are spreading in application fields from transferring a few milli-watts of power in biomedical applications [1] up to hundreds of watts of power in automotive applications [2]. Currently, WPT systems with several kilowatts output, intended to wireless power charging devices for transportations such as electric vehicles, trains and underwater ships. Even though it takes an advantage of large air gap, it has not shown the brightness for commercial grade products in the power applications due to efficiency and cost [3], [4].

Generally, as the WPT system adopted normal conducting wire, the size of antenna is too large to be equipped to deliver the large power promptly due to the intrinsically property of the normal conducting wire. As well as, as the copper resonance keep relative low Q factor, the exchanges energy rate is limited. The high Q factor resonators exchange energy at a much higher rate than they lose energy due to low damping ratio for each resonance coil as well as they are possible to keep much stronger magnetic fields out in the peripheral regions. From this reason, the WPT technology has been required for diffusion of various wires to maximize Q value of resonance coils with stability [5]-[7]. As a reasonable approach, the superconducting wire is a noble option in order to improve transfer efficiency and extend the transfer distance with high stability under high power operation. Fortunately, since the superconducting wires keep enough current density and higher Q value, it enables to exchange a massively electric power in spite of a small scale coil as well as to improve the efficiency. From this reason, superconducting resonance coils are promisingly highlighted in the electric vehicle and superconducting MAGLEV as an antenna (Tx) and receiver (Rx) coils, respectively [8]-[12]. As the superconducting wires should be cooled indispensably, the cooling cost of resonance coils should be considered to commercialize. Generally, as the cooling cost of superconducting wires is caused by AC losses, which are the sum of magnetization losses and transport current losses during AC charging. Total power dissipation is indirectly and relatively easily estimated by evaporation of liquid nitrogen compared with voltage terminal or pick-up coils [13].

In this study, operating characteristics for various copper types (solid, litz, tape wires) and HTS Rx coils, which are cooled by liquid nitrogen, are evaluated with RF generator of 370 kHz within 200 W. In addition, power dissipations of HTS, copper cable, solid, litz and tape resonance wires by AC loss are estimated using the nitrogen evaporation method. Based on the total evaporation rate, relatively consumed comparison of liquid nitrogen for various wires is presented.

2. Mechanism and properties

2.1. Structure and mechanism

Figure 1 shows the conceptual design of wireless power charging (WPS) system in MAGLEV train of EDS technology with superconducting magnet. The antenna (Tx) at MAGLEV system, magnetically-coupled multi resonance antennas, which consist of normal conductor and are used in as transmitter via optimizing resonance capacitances according to the positions of receiver (Rx) coil. Phased-multi Tx coils are also implemented that can localize and track the receiver with high power transmission efficiency even when Tx and Rx coils are misaligned. The structure of resonance wireless charging unit of SUWPT system consists of four components as follows; RF source (V_s), impedance matching (IM) circuit, antenna (Tx) coil, and receiver (Rx) coil including load (R_L), as shown in Figure 2 (a). The transferred power is allowed by selected resonance condition for each coil, which contains creating an LC resonance. The resonant degree of coupled coils can be easily variable corresponding to different distance and external factors. The inserted IM circuit plays a major part in the needed band of frequencies to be passed, while it rejects in the all others band. Once resonance coupling condition between Tx and Rx coils is broken, the transmitting waves are reflected back into Tx, and then those cause thermal heating problem. Thus, the IM circuit, which consists of LC resonant circuit, comes into play when electromagnetic wave cannot be sufficiently reduced using capacitors and inductors; it suppresses unnecessary frequency, effectively.

In the WPT technology, the transfer efficiency is generally determined by quality factor Q, which is defined by the ratio of apparent power to the power losses in a device. The antenna coil forms a series RL circuit and the Q factor is expressed as

$$Q = \frac{\omega L}{R} = \frac{2\pi f L}{R} = 2\pi f \frac{Energy\ stored}{Power loss} \quad (1)$$

where f means resonant frequency. The maximized Q value of superconducting wires can be realized based on the optimized joint technique. Stored energy is $LI^2/2$ and power dissipation is RI^2. The Q factor is proportional to stored energy but inversely proportional to power dissipation.

Figure 2 (b) shows the equivalent circuit of the resonance type WPT system. Variable inductances L_{X1}

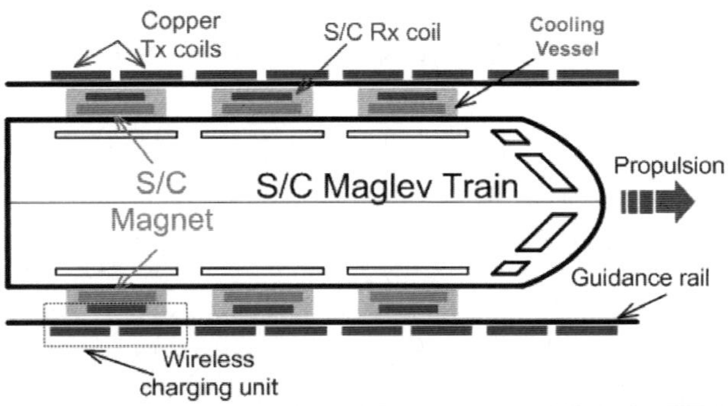

Figure 1. Illustration of superconducting wireless power charging (WPC) system for electrodynamic suspension (EDS)-based superconducting MAGLEV train

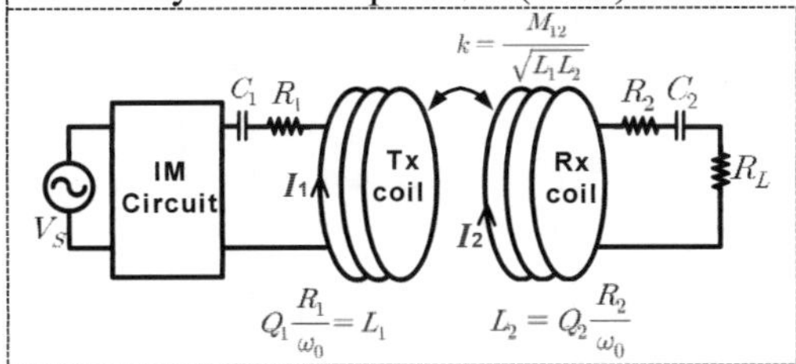

Figure 2 (a). Illustration of simplified circuit model of resonant type superconducting WPT (SUWPT) system

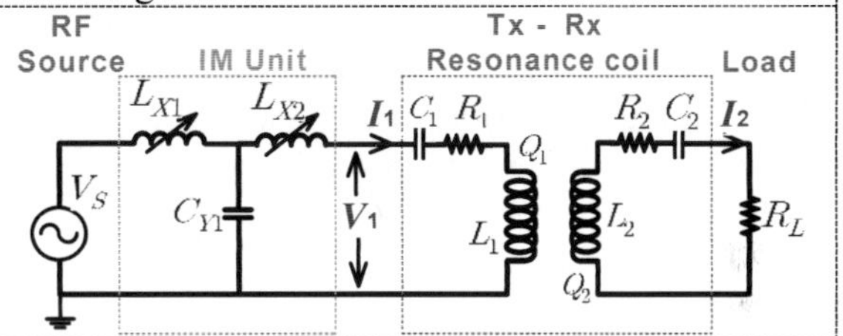

Figure 2 (b). Equivalent circuit diagram of resonance type SUWPT system

and L_{X2} are changed to select the resonance frequency of input source corresponding to LC coupled degree and distance of Tx and Rx coils. Using element circuit analysis, currents of primary and secondary coil are expressed as

$$Z_1 = R_1 \quad Z_2 = R_2 + R_L \quad Z_{12} = j\omega k_{12}\sqrt{L_1 L_2} \quad \omega = \frac{1}{\sqrt{LC}} \tag{2}$$

$$I_1 = \frac{(1+k_{12}^2 Q_2)}{(1+k_{12}^2 Q_1 Q_2)}\frac{V_s}{R_S} \qquad I_2 = \frac{k_{12}\sqrt{Q_1 Q_2}}{1+k_{12}^2 Q_1 Q_2}\frac{jV_s}{\sqrt{R_s R_L}} \tag{3}$$

2.2. Thermal loss of resonance coil

The cooling cost of superconducting wires is caused by AC loss, which is the sum of magnetization losses and transport current losses during AC charging. In addition, in the SUWPT system, the resonance coupling coil also can cause thermal loss depended on the LC resonance condition. The LC resonance of Tx and Rx coils constitute a source resonance circuit to generate an alternative non- radiative magnetic field in the resonance coupling coils. In theory, when the oscillation of source resonance circuit is tuned with receiver resonance circuit, the resonance current is highest. That means thermal loss is minimized in the Tx and the transfer efficiency is maximized. On the other hand, if LC resonance coupling is weakened or enlarged air gap with Rx resonance circuit, resonance current is reflected back into Tx. And then, the thermal loss of Tx coil is maximized. Thus, in this study, the consumed evaporation rates of various Rx coils are examined by liquid nitrogen evaporation method in the next section.

2.3. Nitrogen evaporation method

The measurement device using the nitrogen evaporation method was developed as a practical way to measure AC losses in HTS coils, and is being verified for its measurement accuracy. Compared to conventional electrical measurements, the nitrogen evaporation method has the following beneficial features: Firstly, the amount of AC loss, equivalent to the amount of released gas, can be directly detected. Secondly, there is no specific limitation on shape and size of coils. Thirdly, any calibration is made possible directly by using a heater. The power dissipation by nitrogen evaporation method can be expressed by [14]

$$Q_l = \frac{P}{h_L} \quad (4)$$

Where liquid nitrogen boil-off rate Q_l, power dissipation P, the latent heat of vaporization of liquid nitrogen h_L.

3. Experimental setup and results

In this experiment, authors focused on power transfer characteristics, thermal distributions, transfer ratio and evaporation rate of LN2 for cooled Rx resonance coils. In Figure 3, the Q value of various resonance coils at 77 and 300 K, respectively, is presented under different frequencies. The cooled copper solid, litz, and tape resonance coils are apparently improved over two times compared with non-cooled circumstance since the resonance coils are stabilized by lower resistance As the high Q and transfer efficiency are correlated, the cooling condition has a direct influence on transfer ratio. In Figure 4, the strong resonance circumstance between Tx and Rx coils for solid, litz and tape wires is

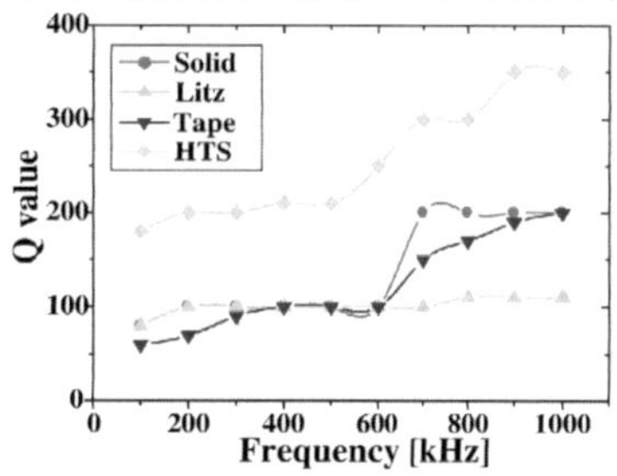

(a)

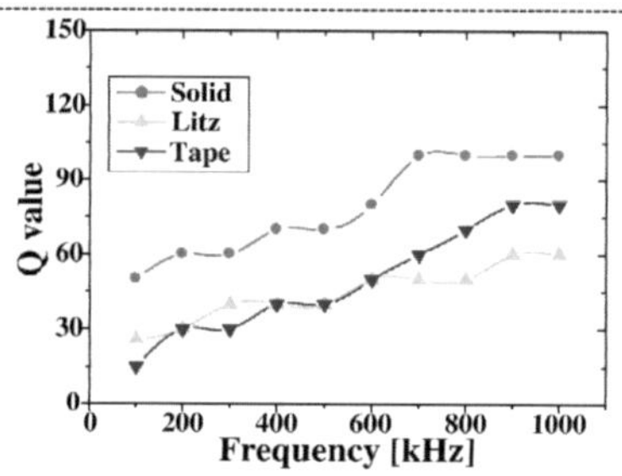

(b)

Figure 3. Measured results of Q value for copper cable, solid, litz, tape and HTS wires within 1 MHz: (a) with 77 K, (b) with 300 K.

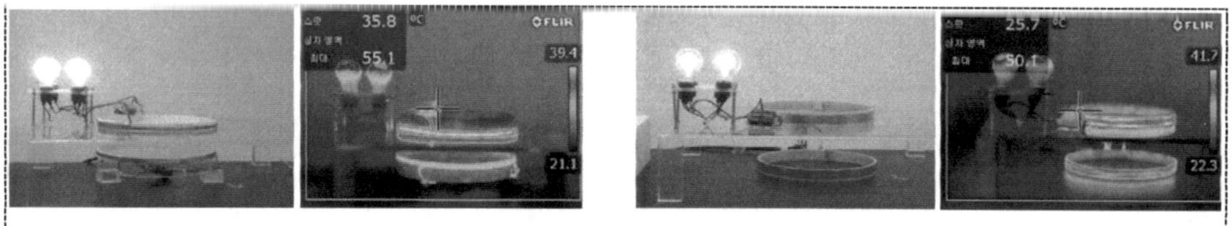

(a) (b)

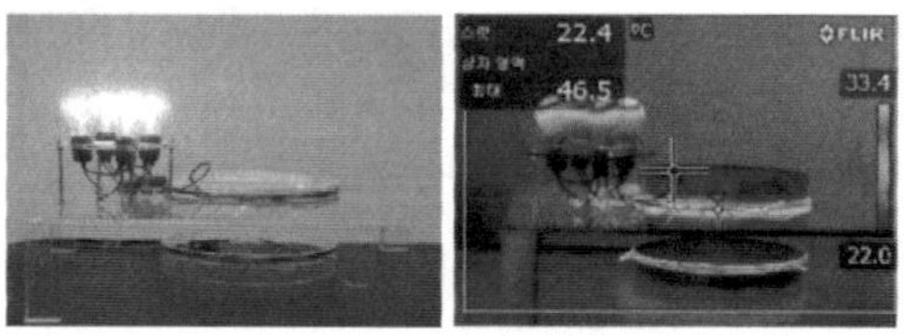

(c)

Figure 4. Photograph of experimental performance and measured thermal distributions for coupled Tx, and Rx coils by thermo-graphic camera at 300 K. The input power is 200 W of 370 kHz RF generator: (a) copper solid wire, (b) litz wire (c) tape wire.

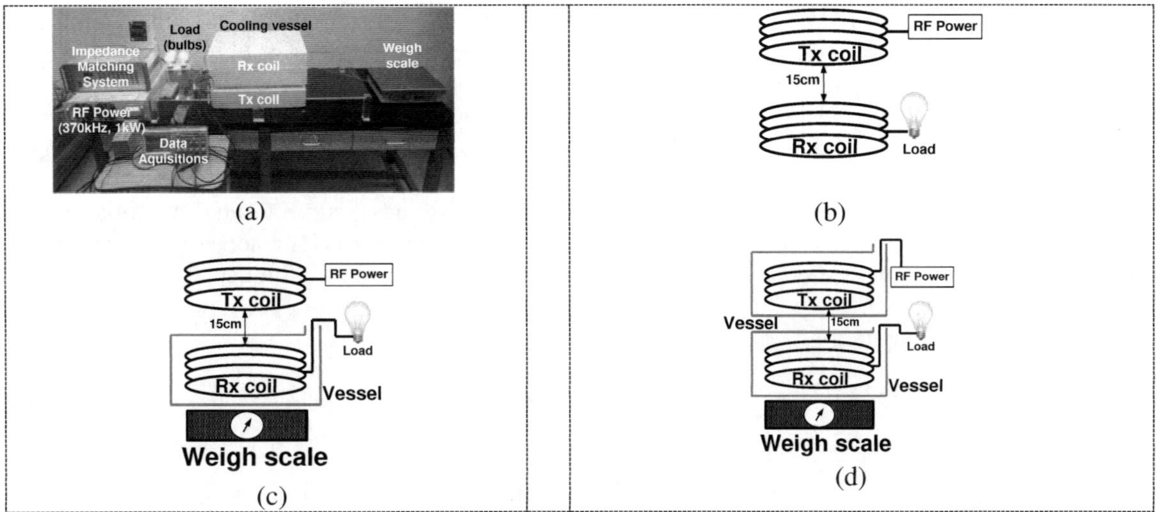

Figure 5. Photograph of experimental performance and setup: (a) experimentally structural elements, (b) experimental setup of non-cooled performance (c) experimental setup of cooled Rx for solid, litz and tape coils (d) experimental setup of HTS coil.

Table 1. Material properties and specifications of HTS and copper resonance coil with 77 K at 370 kHz.

Parameters	**Impedance** [Ω]	**Inductance** [μH]	**Q**	**Dimensions** (thickness *t*, width *w*, diameter *d*)
HTS (YBCO copper clad)	12.28	10.94	200	(*t*: 0.2mm, *w*: 4 mm)
Solid	23.3	10.41	90	(*d*: 1.2mm)
Tape	25.6	10.24	80	(*t*: 0.35mm, *w*:5mm)
Litz	22.8	10.22	95	(no. of wires:150 strand)

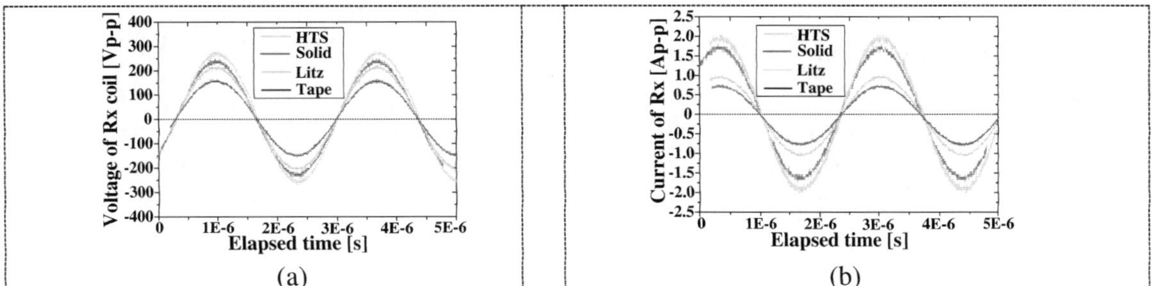

Figure 6. Experimental results of voltage and current distributions at Rx coil in the different resonance Rx coils with input power of 200 W at 77 K:. (a) voltage wave forms (b) current wave forms.

shown under 300 K. The reflected power below 1 W is kept corresponding to input of 200 W. It is confirmed that the relatively higher heat at Rx is caused compared with Tx due to load impedance. From this investigations, authors installed experimental setup and data acquisition elements as shown in Figures 5 (a) and (b). The operating characteristics of copper soild, litz and tape resonance coils are investigated under non-cooled circumstance and only cooled Rx, respectively, as shown in Figure 6 while HTS resonance characteristics are investigated at 77 K under conditions of Figures 5 (c) and (d). The evaporation rate of liquid nitrogen, which is preserved using Styrofoam vessel, is measured by weigh scale. Table I show design specifications and measured properties HTS and various copper resonance coils. Figures 6 (a) and (b) show measured results of voltage and current wave distributions for various resonance coils at 77 K. The advanced order of high transfer ratio for voltage and current is HTS wire, copper solid wire, litz wire and tape wire. In the amplitude of current, the differential gap is eminently widened compared with voltage wave forms. That means the HTS resonance coil can keep strong magnetically resonance coupling. In addition, its transfer intensity is higher than even cooled copper resonance coil. Especially, it is confirmed that it is advantageous for copper solid wire to transfer current wave forms compared

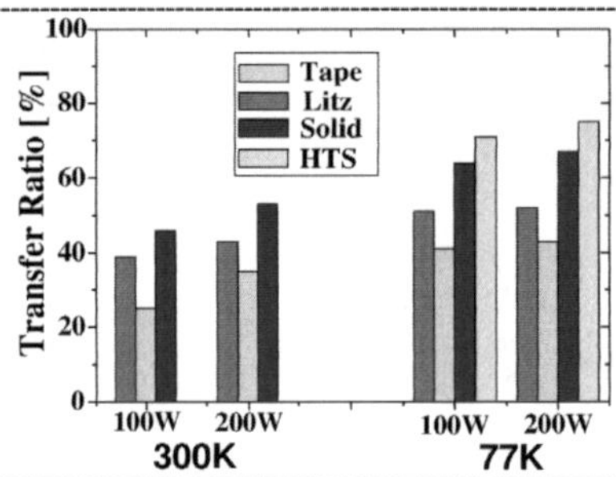

Figure 7. Experimental results of transfer ratio from IM circuit to Rx coil for different resonance coils with input of 100 and 200 W under 300 K and 77 K, respectively.

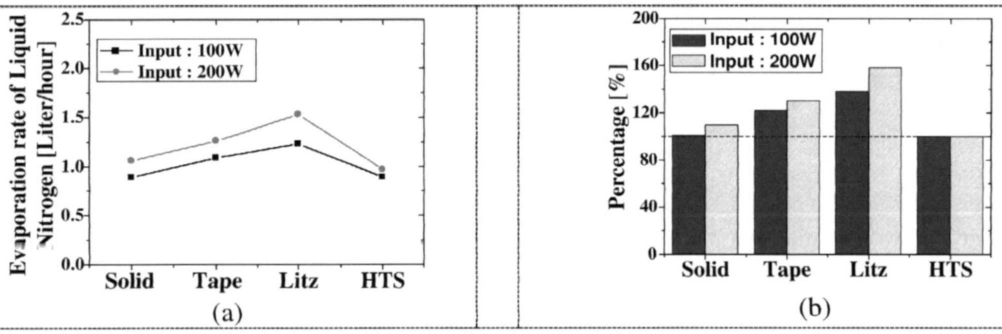

Figure 8. Experimental results of evaporation rate of liquid nitrogen of Rx coils and relative consumption percentage compared with fiducial point of nitrogen consumption for HTS coil with input of 100 W and 200 W: (a) evaporation rate of liquid nitrogen (b) relative comparison for HTS consumption.

with other copper resonance coils. In Figure 7, from a transfer efficiency point of view, the transfer ratio of resonance Rx coils between non-cooled copper and cooled copper including HTS coil are presented. Definitely, the transfer ratio of cooled resonance coil is enhanced over 15 % compared with non-cooled resonance coils. Based on the results, it is certified that the transfer efficiency of copper resonance coils can be effectively promoted by cooling circumstance.

In Figure 8, as a viewpoint of cost and consumption, the evaporation rate of LN2 per one hour for various resonance coils and relative consumed comparison are presented with input power of 100 W and 200 W, respectively. The consumption rate of HTS resonance coil from 100 W to 200 W is increased about 5 %, while the solid, tape and litz resonance coils are increased 10, 12 and 21 %, respectively as shown in Figure 8 (a). As well as, compared with HTS coil even under 77 K, the relative consumptions of LN2 per one hour for copper solid, tape, litz resonance coils are higher about 11, 38 and 63 %, respectively, as shown in Figure 8 (b). The LN2 consumption rates of solid, tape, litz wires are relatively increased compared with HTS wire under increasing input power since the HTS wire keeps higher current critical point. That means thermal loss of HTS wire relatively low under critical current point due to intrinsically characteristics compared with copper material wires. The consumed LN2 of HTS, solid, tape and litz wires are 0.8, 0.9, 1.1 and 1.2 liters with input power of 100 W during one hour. On the other hand, the HTS, solid, tape and litz wires consumed 0.9, 1.1, 1.3 and 1.6 liters of LN2 with input power of 200 W, respectively. The cost of evaporated LN2 for one liter is about 0.5 dollar.

4. Conclusion

Authors successfully achieved the relative comparison for transfer ratio and consumption of liquid nitrogen between HTS and various copper resonance coils at 77 K and 300 K, respectively. It is confirmed that the HTS resonance coil has advantages of enhanced transfer ratio with stability and relatively low cooling cost of LN2 compared with cooled copper resonance coils. Furthermore, the cooling cost can be improved through the optimal design of cooling vessel in the next study. That the copper resonance coil with cooling circumstance can improve for transfer ratio is informative and interesting points. Based on the results, authors expect to apply the HTS resonance coil for superconducting MAGLEV train. In the case of wireless power charging system for superconducting MAGLEV train, the Rx resonance coil is easily installed in the train since the cooling vessel is installed in the train to cool superconducting magnet. Thus, our research target is to evaluate validity and stability of HTS resonance coils to accomplish wireless power charging system for the superconducting MAGLEV train.

Acknowledgments
This work was supported by the Basic Science Research Program through the National Research Foundation of Korea (NRF) funded by Korean Ministry of Education, Science & Technology (MEST) (Grant: 2015R1D1A1A01058286).

References

[1] Strassner, B. and K. Chang, "Microwave power transmission: Historical milestones and system components," *Proceedings of the IEEE*, Vol. 101, No. 6, 1379-1396, June 2013.
[2] Sun, Tianjia, Xie, Xing, Wang Zhihua , "*Wireless Power Transfer for Medical Mocrosystems* (New York: Springer). 2013 pp. 34
[3] Haipeng Dai, Yunhuai Liu, Guihai Chen, Xiaobing Wu, Tian He, "Safe charging for wireless power transfer *Proc. IEEE Conf. on Computer Communications* pp. 1105-1113 April 2014
[4] Ke Wu, Debabani Choudhury, Hiroshi Matsumoto, "Wireless power transmission, technology, and applications ", Proceedings of the IEEE, vol 101, issue 6, pp. 1271-1275, May 2013
[5] J. Casanova, Z. N. Low, J. Lin, "A Loosely Coupled Planar Wireless Power System for Multiple Receivers," *IEEE Transactions on Industrial Electronics*, vol. 56, no. 8, pp. 3060-3068, August 2009.
[6] Kim JW, Son H-C, Kim K-H, Park Y-J., "Efficiency analysis of magnetic resonance wireless power transfer with intermediate resonant coil," *IEEE Antennas and Wireless Propagation Letters*. vol. 10, pp. 389–392, 2011
[7] Dae Wook Kim, Yoon Do Chung, Hyoung Ku Kang, Yong Soo Yoon, Tae Kuk Ko, "Characteristics of contactless power transfer for HTS coil based on electromagnetic resonance coupling," *IEEE Transactions on Applied Superconductivity*, vol. 22, issue 3, pp. 5400604, 2012

[8] Guo Min Zhang, Hui Yu, Liwei Jing, Jincheng Li, Qi Liu, Xianghua Feng, "Wireless power transfer using high temperature superconducting pancake coils," *IEEE Transactions on Applied Superconductivity*, vol. 24, issue 3, pp. 4600505, 2014.
[9] Jun Ogawa, Satoshi Fukui, Tetsuo Oka, Takao Sato, Makoto Shibayama, Masayasu Egawa, Ayumi Sato, "Experimental investigation of loss characteristics with AC and DC offset transport current in HTS double pancake coil," *IEEE Transactions on Applied Superconductivity*, vol. 22, issue 3, pp. 4705004, 2012.
[10] Hui Yu, GuoMin Zhang, Liwei Jing, Qi Liu, Weijia Yuan, Zhongjing Liu, Xianghua Feng, "Wireless power transfer with HTS transmitting and relaying coils," *IEEE Transactions on Applied Superconductivity*, vol. 25, issue 3, pp. 5000405, 2015.
[11] W. Zuo, S. Wang, Y. Liao, Y. Xu, "Investigation of efficiency and load characteristics of superconducting wireless power transfer system," *IEEE Transactions on Applied Superconductivity*, vol. 25, issue 3, pp. 5000106, 2015.
[12] Yoon do Chung, Chang Young Lee, Hyun Chul Jo, Young Gun Park, Seong Woo Yim, "Operating characteristics of contactless power transfer for electric vehicle from HTS antenna to normal conducting receiver," *Physica C*, no. 504, pp. 115-119, September 2014.
[13] H. Okamoto, F. Sumiyoshi, K. Miyoshi, Y. Suzuki, "The nitrogen boil-off method for measuring AC losses in HTS coils," *IEEE Transactions on Applied Superconductivity*, vol. 16, issue 2, pp. 105-107, 2006.
[14] Iwasa Y , "Case studies in superconducting magnets," Springer, pp. 579, 2009.

Study on the Rotation Properties and the Design Issue of Non-Contact Rotating System Using HTS Bulks and Permanent Magnets

R Okamura, S B Kim, Y Ozaki and H Ueda
Graduate School of Natural Science and Technology, Okayama University, 3-1-1, Tsushimanaka, Kita-ku, Okayama 700-8530, Japan

kim@ec.okayama-u.ac.jp

Abstract. In previous study, non-contact rotating system consisting of the ring-shaped high temperature superconducting (HTS) bulks, ring-shaped permanent magnets (PMs) and stator coil was proposed. In this system, HTS bulks were magnetized by PMs and PMs were levitated with strong restoring force. In our previous study, we have constructed the rotating system with the ring-shaped HTS bulks with ID 20 mm, OD 60 mm, and 15-mm thickness. However, since these bulks costs too much, we switched to use HTS bulks with ID 20 mm, OD 60 mm, and 5-mm-thickness to miniaturize the system. However, this system have potential to fail the radial stability of rotating shaft. Therefore, we focused on the rotating and the radial restoring force in terms of the stability of the rotating shaft in the rotating system with 5-mm thickness HTS bulks.

1. Introduction

The application of non-contact levitation technology has advantages such as no energy loss and easy maintenance because of no mechanical friction. For these reasons, the non-contact levitation technologies using electrostatic force, ultrasonic wave, air pressure, electromagnetic force have been studying. Particularly, the technology using electromagnetic force has the advantages such as a large levitation force and easy handling.

The HTS bulks magnetized by field cooling (FC) method[1][2] show both diamagnetic behavior and pinning effect. By using these characteristics, the PMs can be levitated with stability[3]-[6]. From our previous study, it is known that our system using ring-shaped PMs and ring-shaped HTS bulks is very useful to apply to the non-contact rotating system. The maximum rotating speed of 840 rpm was obtained in the non-contact rotating system by using the ring-shaped HTS bulks with ID 20 mm, OD 60 mm, and 15-mm thickness[7]. However, since it costs too much, we switched to use HTS bulks with ID 20 mm, OD 60 mm, and 5-mm-thickness to miniaturize the system. However, it is predicted that the radial stability of rotating shaft will be weakened. Therefore, we studied about the rotating and the radial restoring force in terms of the stability of the rotating shaft using 5-mm-thickness HTS bulks.

2. Rotating system and experimental details

2.1. Rotating system and experiment

The proposed non-contact rotating system is shown in Figure 1. Table 1 shows the specifications of components used in the experiments. Figure 2 (a) and (b) show the measurement results of magnetic field distribution for surface of PM(1) and 12 mm above the surface of PM(1) respectively. PM (1) magnetized in the axial direction was used to increase the axial restoring force of the rotating shaft in the system. Furthermore, two PM (1)s are faced and arranged on the opposite side of HTS bulk in order to increase the radial restoring force. PM (2) for the rotor of motor is the permanent magnet attached to iron shaft and PM (2) is multi-poles magnet which generates a magnetic field in the radial direction. Figure 3 (a) and (b) show measured area of magnetic field distribution around PM(2) and measurement result of magnetic field distribution for PM(2) respectively.

The HTS bulks were magnetized by PM (1)s with FC method with liquid nitrogen. Therefore, PM (1)s were levitated because of the trapped HTS bulks with pinning force. The iron shaft with PM (1) and PM (2) is rotated by applied current to the stator coil using three bipolar amplifiers.

Table 1. The specifications of components used in the experiments

	Inner diameter (mm)	Outer diameter (mm)	Height (mm)
HTS Bulk	20	60	5
PM (1)	10.8	64	13
PM (2)	10.8	16	6
Stator coil	30	90	7

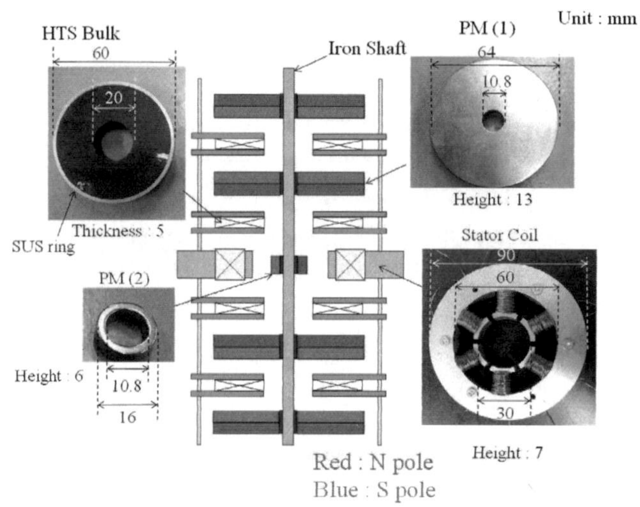

Figure 1. Schematic drawing of proposed the non-contact rotating system and photos of the ring-shaped HTS bulks, 2 types PMs and stator coil.

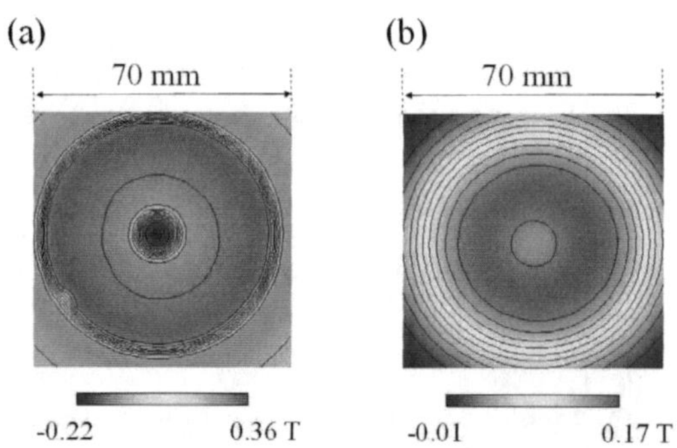

Figure 2. The measurement results of magnetic field distribution, (a) surface of PM(1) and (b) 12 mm above the surface of PM(1).

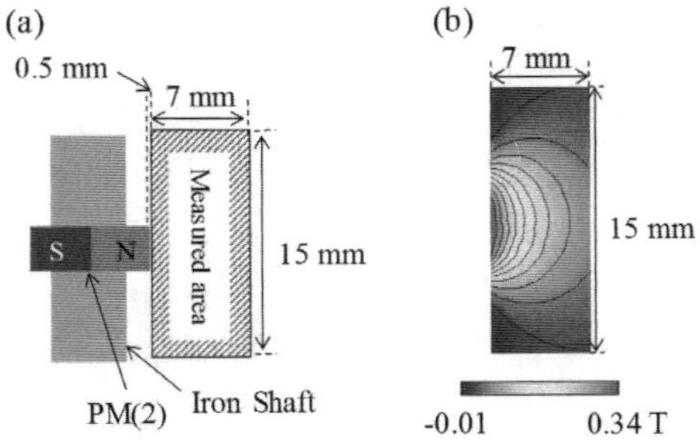

Figure 3. (a) Measured area of magnetic field distribution around PM(2) and (b) Measurement result of magnetic field distribution for PM(2).

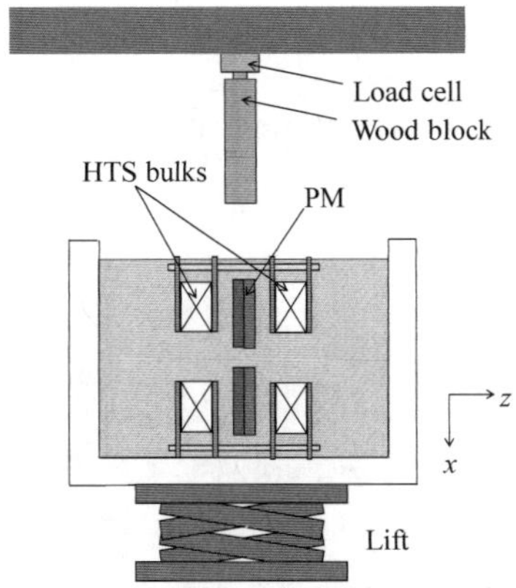

Figure 4. The measuring method of radial restoring forces in the radial directions.

2.2 Experiment of radial restoring force

Figure 4 shows the measurement method of a radial restoring force. The HTS bulks magnetized by field cooling (FC) method show both diamagnetic behavior and pinning effect. The HTS bulks and permanent magnet are magnetically coupled. Then, the PM has the force to stay in the initial position. The HTS bulks were magnetized by PM with FC method, and PM levitated between HTS bulks. Then, we measured the radial restoring force against the displacement from the levitating position by using load cell. And we also measured the radial restoring force when the thickness of HTS bulks are 5 mm and 15 mm, and the several levitation systems were designed and tested experimentally.

3. Results and discussion

In general, the restoring force of proposed rotating system is increased according to increasing the thickness of HTS bulk. So, the thick HTS bulks are very effective for improving the restoring force, and 15-mm thickness HTS bulks were used in our previous study [7]. However, the cost of HTS bulk increases in proportion to its thickness. Therefore, we tried to reduce the thickness of HTS bulks (from 15 mm to 5 mm) used for our rotating system. Firstly, rotating test with the system as shown in Figure 1 were carried out. However, when the rotating system was levitated with FC method, PM(2) as the rotor was adsorbed by the stator coil and could not rotate because the radial stability of the rotating shaft was weakened by changing the HTS bulks with thickness of 15 mm to those with thickness of 5 mm in the rotating system. Therefore, we estimated the radial stability of the rotating shaft by measuring the radial restoring force using 1 unit levitation system with 15-mm and 5-mm thickness HTS bulks as shown in figure 5.

3.1. Radial restoring forces

3.1.1. The comparison by thickness of HTS bulks HTS bulks with a thickness of 15 mm have been used in the previous study. However, in this study, HTS bulks with a thickness of 5 mm have been used. Hence, we compared radial restoring force using 5-mm-thickness HTS bulks and 15-mm-thickness HTS bulks. Figure 5 shows the schematic drawings of the levitation systems (here in after referred to as "1 unit system") without shaft in this experiment.

Figure 6 shows the measurement results of radial restoring forces for 1 unit system without shaft. As seen in figure 4, the restoring force for the system with 15-mm-thickness HTS bulks is larger than that with 5-mm-thickness HTS bulks. This difference of the restoring force is related to the HTS bulks thickness. The amount of trapped magnetic field is larger in a wider range with the larger thickness of the bulk, and the larger restoring force works on the HTS bulk. Hence, we need to consider new structure of the levitation system using 5-mm-thickness HTS bulks.

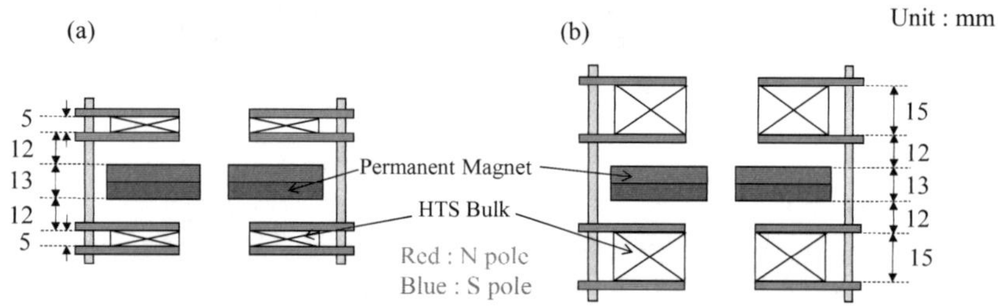

Figure 5. The schematic drawings of the 1 unit levitation system for the measuring radial restoring forces. (a) Using 5-mm-thickness HTS bulks and (b) Using 15-mm-thickness HTS bulks.

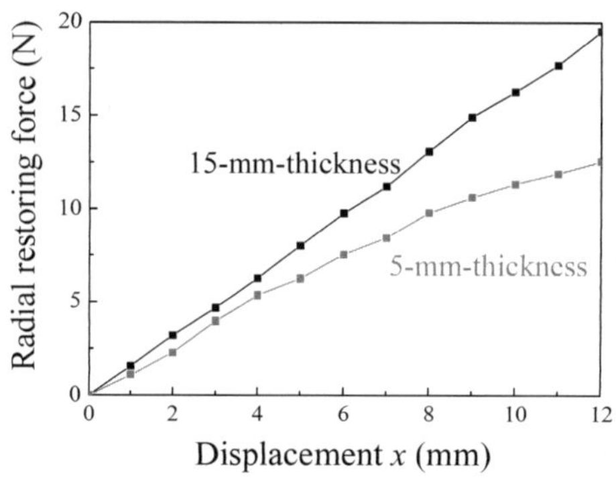

Figure 6. Measured radial restoring forces in 1 unit system using HTS bulks with thicknesses of 5 mm and 15 mm).

3.1.2. The comparison by design of the levitation system To improve the restoring force in the radial direction, the high magnetic field strength and the large magnetic gradient in the radial direction are necessary, and thick HTS bulks are very effective for improving the restoring force. Therefore, we investigated the system structure which can obtain a high restoring force even though using a thin HTS bulk.

Figure 7 shows the schematic drawings of the cross-sectional structure of several levitation systems in the experiments. Figure 7(a) shows a levitating system used in measuring the rotation characteristics in our previous study using 15-mm thickness HTS bulks, and the stator coil was removed to measure the restoring force. Figure 7(b) is the same structure shown in figure 7(a) except that the thickness of HTS bulks is changed from 15 mm to 5 mm. Figure 7(c) basically has the same structure as figure 7(b), but the distance between PM(1) and HTS bulk is expanded from 12 mm to 17 mm in order to reduce the influence by PM(1) against PM(3). In figure 7(d), PM(1)s are arranged face to face with the same polarity at both sides of HTS bulk in order to improve the strength of magnetic field applied in radial direction of HTS bulk (here in after referred to as “4 PMs + 4 Bulks system”). Then, in order to obtain the levitation force in the axial direction, two HTS bulks were placed at the center part of the rotating shaft. In order to confirm the influence by the thickness of HTS bulk, two HTS bulks with 5 mm thickness were stacked as shown in figure 7(e) (here in after referred to as "2 PMs + 2 Bulks system").

Figure 8 shows the measuring area at near the PM(3) and measured the magnetic field distribution around PM(3). In order to increase the restoring force in the radial direction, the PM(3) has a structure which two ring-shaped PMs were combined with opposite direction, and the iron ring was sandwiched by PMs for increasing the magnetic flux in the radial direction. These PM(3)s are directly attached to the rotating shaft and placed in the inner part of the HTS bulks.

Figure 9 shows the measurement results of radial restoring force for the various levitation systems. 4PMs + 4Bulks system and 2 unit system using 5-mm-thickness HTS bulks did not obtain enough radial restoring force to be measured. These systems fell down and contacted inside HTS bulks trying to levitate them horizontally. This is because by using 5-mm-thickness bulks, the range in which the magnetic field generated by PM (3) as shown in figure 8 can be captured is smaller than in the case of using 15-mm-thickness bulks. Therefore, the radial restoring force became smaller by using 5-mm-thickness HTS bulks. 2 PMs + 2 Bulks system was able to measure the radial restoring force. Due to using stacked two 5-mm-thickness HTS bulks, the radial restoring force was larger than that using 5-mm-thickness bulks. However, the shaft contacted inside HTS bulks with 1 mm displacement, and the radial restoring force could not be measured any further. Comparing the radial restoring force of 2 PMs + 2 Bulks system and 2 unit system (using 15-mm-thickness HTS bulks), it can be seen that the radial restoring force of the system using stacked two 5-mm-thickness HTS bulks is smaller than that using 15-mm-thickness HTS bulks. As these results, it was found that radial restoring force is insufficient in any case using 5-mm-thickness HTS bulks. Also, the radial restoring force obtained in the model using stacked two 5-mm-thickness HTS bulks is larger than the model using non-stacked bulks. However, the radial restoring force is still smaller than that in the case with 15-mm-thickness HTS bulks, therefore larger radial restoring force is required for a stable rotation. Therefore, we need to consider the design using stacked two 5-mm-thickness HTS bulks.

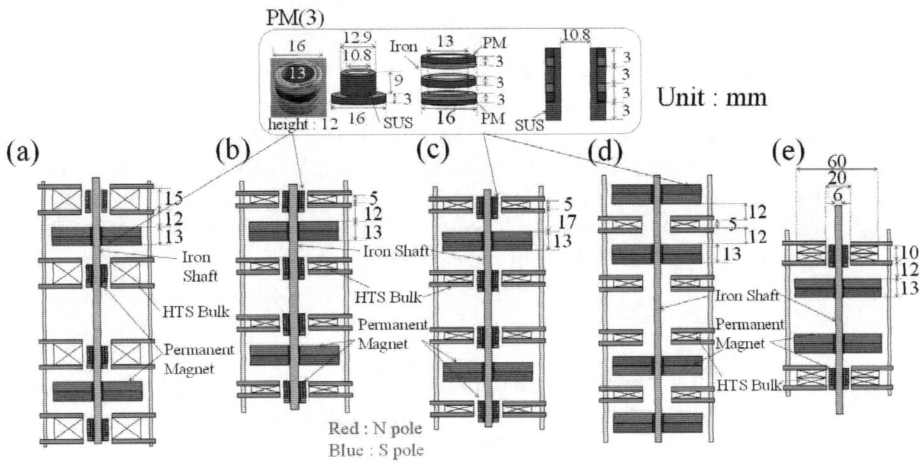

Figure 7. The schematic drawings of the levitation system to measure the radial restoring forces and the photo of the PM(3) as bearing. (a) 2 unit system (using 15-mm-thickness HTS bulks), (b) 2 units system (using 5-mm-thickness HTS bulks, Gap = 12 mm), (c) 2 unit system (using 5-mm-thickness HTS bulks, Gap = 17 mm), (d) 4 PMs + 4 Bulks system and (e) 2 PMs + 2Bulks system.

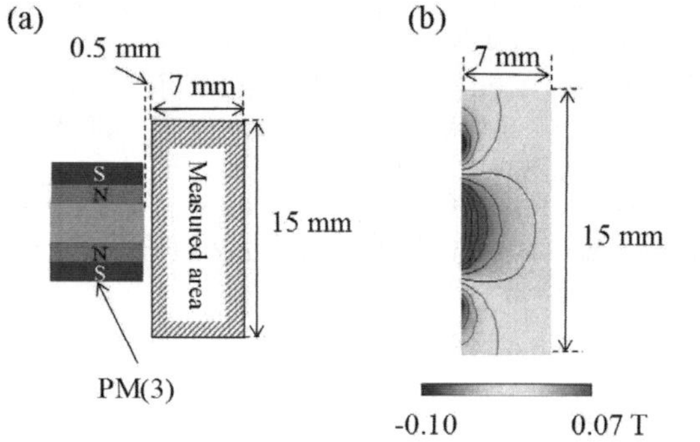

Figure 8. (a) Measuring area at near the PM(3) and (b) measured the magnetic field distribution around the PM(3).

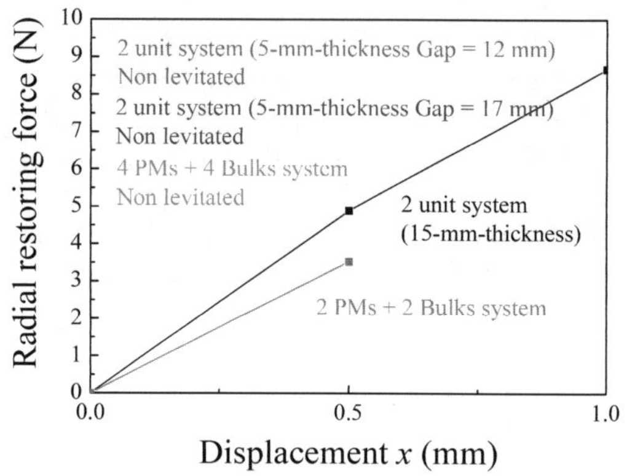

Figure 9. Measured radial restoring force of the various levitation systems.

4. Conclusion

In previous study, we proposed non-contact rotating system composed of HTS bulks, PMs and stator coil. The proposed non-contact levitation system can be used a rotating machine to apply the medical apparatus such as medical mixer. In order to use the developed levitation system as a rotating machine, the stability of the rotating shaft is very important, and it is necessary to improve the restoring force of the rotating shaft. To improve the restoring force in the radial direction, the high magnetic field strength and the large magnetic gradient in the radial direction of HTS bulk are necessary, and thick HTS bulks are very effective for improving the restoring force. However, the cost of HTS bulk increases in proportion to its thickness. So, we studied to improve the strength of the magnetic field in radial direction of HTS bulk and tried to reduce the thickness of HTS bulks for economic reasons. The measured restoring force in the radial direction using 1 unit levitation system with 5-mm thickness HTS bulk was 30% smaller than that of system using the 15-mm thickness HTS bulk. The several levitation systems for improving the restoring force using 5-mm thickness HTS bulks were designed and tested experimentally. However, the designed levitation models using 5-mm thickness HTS bulks did not levitated, and measured the restoring force of 2 PMs+2 Bulks system was also lower than that of the model using the 15-mm thickness HTS bulk. Therefore, it was confirmed that the thickness of the HTS bulk is very effective in our proposed system. However, we need to continue research for reducing the thickness of the HTS bulk to solve the economic problem due to expensive HTS bulks.

References

[1] Kim S B, Matsunaga J, Doi A, Ikegami T and Onodera H 2013 Study on the characteristics of magnetic levitation for permanent magnets and ferromagnetic materials with various sizes using stacked HTS annuli *Physica C* **484** 316-320

[2] Kim S B, Ikegami T, Matsunaga J, Fujii Y and Onodera H June 2014 Development of the non-contact rotating system using combined ring-shaped HTS bulks and permanent magnets *IEEE Trans. Appl. Supercond.* **24** 6800105

[3] Fukusawa Y and Ohsaki H June 1999 Three-Dimensional Structure of Magnetic Field in the Mixed-μ Levitation System using Bulk Superconductors *IEEE Trans. Appl. Supercond.* **9** 980-983

[4] Takao T, Kameyama S, Doi T, Tanoue N and Kamijo H June 2011 Increase of Levitation Properties on Magnetic Shielding Effect of GdBCO Bulk Superconductor *IEEE Trans. Appl. Supercond.* **21** 1543-1546

[5] Ghodsi M, Ueno T, Teshima H, Hirano H and Higuchi T 2006 The characteristics of trapped magnetic flux inside bulk HTS in the Mixed-μ levitation system *Physica C* **445-448** 343-346
[6] Kashiwagi T, Kubota M, Suzuki E, Matsuda T, Hirakawa M, Nakashima H, Ohsaki H and Mizuno T 2003 Levitation characteristics of the superconducting mixed-μ system *Physica C* **392-396** 654-658
[7] Kim S B and Nakamura K June 2016 Development of Non-contact Levitated Rotating Machine Using HTS Bulks and Permanent Magnets *IEEE Trans. Appl. Supercond.* **26** 5202204

IOP Conf. Series: Journal of Physics: Conf. Series **871** (2017) 012097 doi:10.1088/1742-6596/871/1/012097

Electromagnetic Performance Calculation of HTS Linear Induction Motor for Rail Systems

Liu Bin, Fang Jin*, Cao Junci, Chen Jie, Shu Hang, Sheng Long

School of Electrical Engineering, Beijing Jiaotong University, Beijing, China

14117396@bjtu.edu.cn

Abstract. According to a high temperature superconducting (HTS) linear induction motor (LIM) designed for rail systems, the influence of electromagnetic parameters and mechanical structure parameters on the electromagnetic horizontal thrust, vertical force of HTS LIM and the maximum vertical magnetic field of HTS windings are analyzed. Through the research on the vertical field of HTS windings, the development regularity of the HTS LIM maximum input current with different stator frequency and different thickness value of the secondary conductive plate is obtained. The theoretical results are of great significance to analyze the stability of HTS LIM. Finally, based on theory analysis, HTS LIM test platform was built and the experiment was carried out with load. The experimental results show that the theoretical analysis is correct and reasonable.

1. Introduction

In recent years, with the urgent needs of the novel motors, as well as the progress of linear motor and HTS technology, the research of HTS LIM has become a hot spot. Compared with the traditional rotary induction motor, linear induction motor has the advantages of low noise, strong gradeability, non adhesion drive and small passing curve radius [1]. However, the traditional linear induction motor also has lots of defects, such as low power factor and efficiency, unbalanced primary current, distorted end magnetic field [2], which greatly restrict its development.

The combination of HTS technology and the traditional liner motor with HTS materials will produce more superior operating performance than the general motor. In addition, HTS LIM not only has a higher power density and reliability, but also effectively decreases the size of the motor, saving the raw materials of manufacturing motor, reducing the noise generated by motor operation [3-5]. Therefore, the research on LIM HTS has important theoretical significance and commercial value.

In this paper, a mathematical model is established for the HTS LIM that is applied in rail systems, which is based on the electromagnetic parameters and mechanical structural parameters. Then the electromagnetic horizontal thrust, vertical force and the maximum vertical magnetic field of HTS windings are analyzed to verify the theory through the experimental test platform.

2. HTS LIM physical model

2.1. Electromagnetic theory of HTS LIM

According to force characteristic analysis of HTS LIM, the horizontal thrust and vertical force formula of HTS LIM can be obtained in one cycle by the FEM method [6-7]:

$$F_x = \frac{1}{T_e}\int_t^{t+T_e} L_{ef} \cdot \sum_1^n B_{yi} \cdot J_{zi} \cdot S_i dt \tag{1}$$

$$F_y = F_{y1} + F_{y2} \tag{2}$$

$$F_{y1} = \frac{1}{T_e}\int_t^{t+T_e} L_{ef} \cdot \sum_1^n B_{xi} \cdot J_{zi} \cdot S_i dt \tag{3}$$

$$F_{y2} = \frac{\partial W_m}{\partial g_e} = \frac{1}{T_e}\int_t^{t+T_e} \frac{\partial}{\partial g_e} \frac{L_{ef}}{2\mu_0} \sum_1^n B_{yi}^2 . S_i \, dt \tag{4}$$

In the formula, F_x is the HTS LIM horizontal thrust, F_y is the HTS LIM vertical force, in which, F_{y1} is the repulsive force as a result of the interaction between the primary magnetic field and the secondary eddy current and F_{y2} is the attractive force produced between the primary magnetic field and the secondary ferromagnetic material. B_{xi} and B_{yi} are the magnetic flux density components of the *x* and *y* directions within the grid subdivision of the FEM, J_{zi} is the eddy current density in the secondary grid cell, S_i is the area of the grid cell *i*, T_e is a period of time, L_{ef} is the effective length of the primary core, W_m is the magnetic field energy, g_e is the electromagnetic air gap length, u_0 is the vacuum permeability, *n* is the total number of mesh generation in the calculation region.

Table1. Main Parameters of the HTS LIM

Items	Value	Items	Value
Rated power, (kW)	3.5	Number of HTS coils	12
Rated current, (A)	20	Mechanical gap, δ (mm)	12
Rated frequency, *f* (Hz)	10	Width of motor, *D*(mm)	250
Rated speed, (m/s)	3.5	Number of primary slots, N_s	26
Rated thrust, (N)	1000	Height of reaction plate, h_c (mm)	3
HTS material	Bi-2223/Ag	Height of back iron, h_b (mm)	8

2.2. Main parameters of HTS LIM

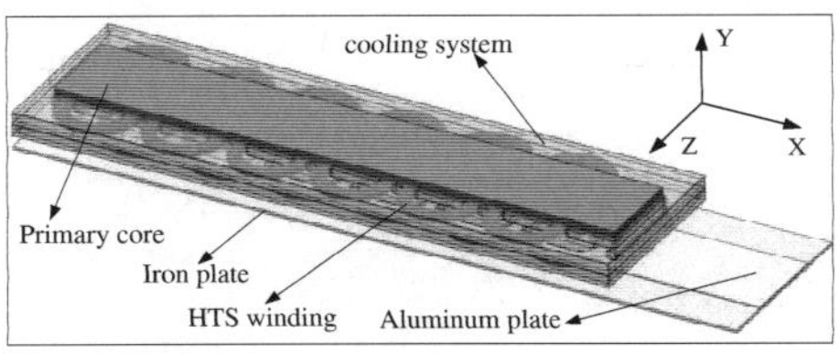

Figure 1. Model of HTS LIM

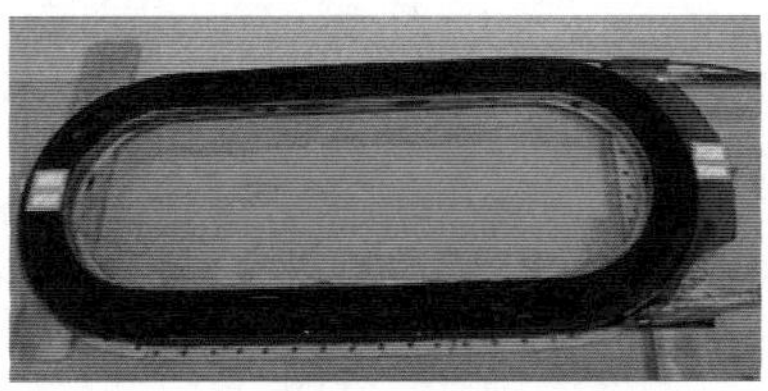

Figure 2. Prototype of HTS coil

In this paper, the main structural parameters of the HTS LIM designed, as shown in Table 1, of which the rated power is 3.5kW and the rated frequency is 10Hz. The designed motor model adopts the structure with unilateral type, short primary and compound secondary made of aluminum and iron, of which the coils are made of Bi-2223/Ag HTS tapes instead of the traditional copper wires as shown in Fig.1.

Table2. Typical Specifications of HTS coil

Items	Value	Items	Value
Critical current of tapes(A)	97(Self field,77K)	Width of tapes(mm)	4.3
Thickness of tapes(mm)	0.24	Turns of coil	240
Length of tapes(m)	254	Resistance(77K)(Ω)	1.2E-5
Inductance(mH)	29	Insulation resistance(MΩ)	1000

The typical specifications of the HTS coils are shown in Table 2. The HTS coil is the compound of HTS tapes and epoxy resin. Because the HTS coil is limited by the curvature radius of tapes and can't be twisted [8], HTS winding is only designed for the double pancake type structure, of which the upper pancake coil and the lower pancake coil are separated by corona resistant polyimide films. The prototype of superconductive winding, which is wound with 240 turns of superconducting tapes, is as shown in Fig.2. Due to the special nature of HTS materials and the particularity on environmental requirements of HTS LIM, HTS windings cooling system must be made with non magnetic stainless steel materials.

3. Research on electromagnetic performance of HTS LIM

3.1. Effect of stator current on electromagnetic characteristics of HTS LIM

HTS LIM can effectively improve the load capacity with HTS windings instead of the traditional copper windings as result of strong current flow ability of superconducting tapes. To study the influence of current size on the electromagnetic characteristics of HTS LIM, the frequency of the power supply is kept constant for 10Hz to observe the variances of the electromagnetic force, vertical force and the maximum vertical field of superconducting windings with the change of the current size.

Fig. 3 indicates the relationship curves of the motor's electromagnetic thrust and vertical force with the primary current respectively. Visible from the graph, with the increase of the current from 4A to 40A, the electromagnetic thrust of HTS LIM increases from 50N to 4697N, and vertical force increases from 44N to 4807N.

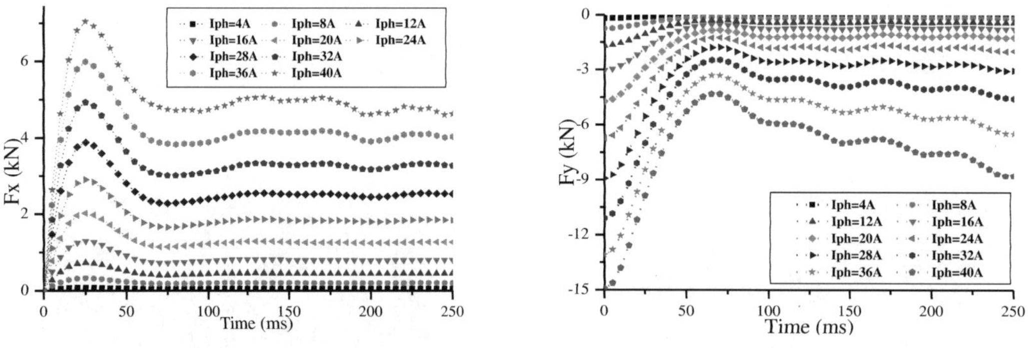

(a) Transient curve of the thrust force (b) Transient curve of the vertical force

Figure 3. Relationship of the force with different exciting current

IOP Conf. Series: Journal of Physics: Conf. Series **871** (2017) 012097 doi:10.1088/1742-6596/871/1/012097

Fig.4 shows the 3D transient waveform of HTS winding subjected to maximum vertical field with the change of the excitation current. It can be seen from the figure that the vertical field shows a linear increasing trend as the current is gradually increasing. When the current is increased from 4A to 40A, the maximum vertical field of the superconducting winding is increased from 0.032T to 0.311T.

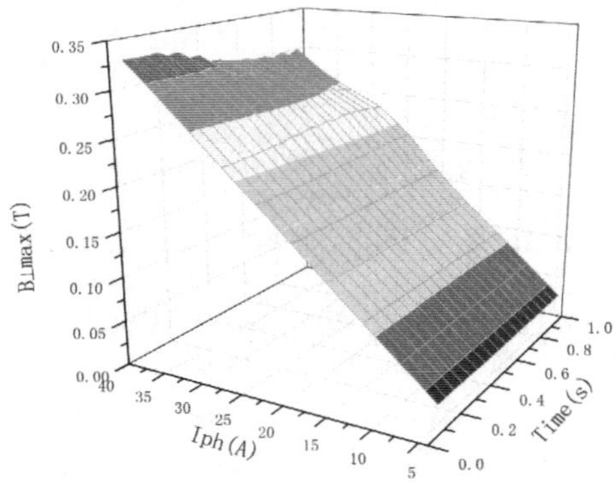

Figure 4. 3D transient $B_{\perp max}$ vs current.

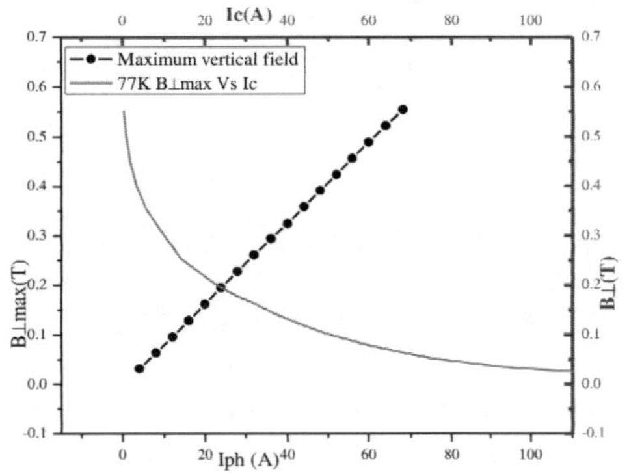

Figure 5. Relationship of maximum current vs maximum B⊥max.

The maximum vertical field curve of HTS winding under different excitation current is compared with that of HTS strip with the critical current in 77K low temperature environment. The comparison curves are shown in Fig. 5. The intersection of the two curves in Fig. 5 is the maximum current that can be applied to the motor designed, which the maximum current is 24A, and the maximum vertical field of the superconducting coil is 0.195T. Therefore, setting the maximum current peak value to 20A is sufficient to ensure safe and stable operation of the HTS LIM.

3.2. Effect of excitation frequency on electromagnetic characteristics of HTS LIM

Figure 6 shows the transient waveforms of the electromagnetic force and the vertical force by changing frequency. The figure shows, with the increase of frequency, the electromagnetic thrust increases first and then decreases, and the vertical force always trends to reduce. The lower the frequency is, the more slowly the stability of the electromagnetic force and the vertical force is at the start time of HTS LIM. When the frequency is increased from 2Hz to 18Hz, the maximum value of the electromagnetic force is 1433N at 6Hz. But the vertical force is reduced from 25.2kN to 371N.

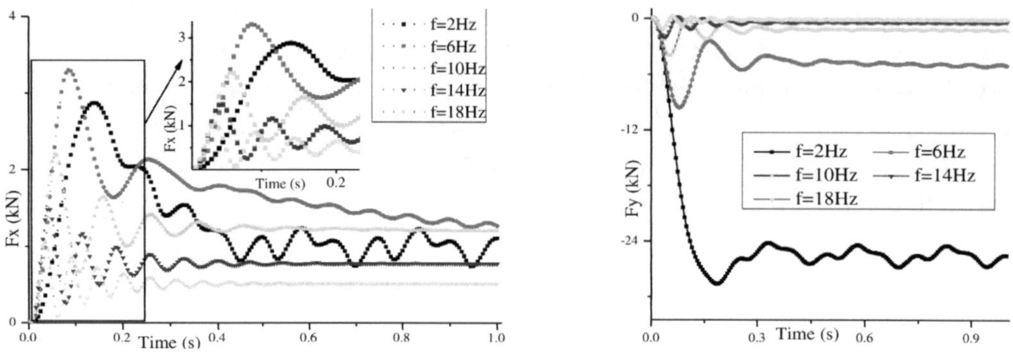

(a) Thrust force vs frequency (b) Vertical force vs frequency

Figure 6. Transient curve of the thrust and vertical force with different frequency

In order to study the effect of frequency on the current capacity of superconducting motor, the frequency of the power supply is chosen for 4Hz to 14Hz to analyze the variation of the maximum vertical field size of the superconducting windings with different current conditions. Fig.7 shows the change curve of the maximum vertical field of the superconducting windings with the maximum input current of HTS LIM at different frequencies, in which the solid line represents the surface vertical field with critical current of Bi-2223/Ag superconducting tapes in 77K low temperature environment. According to the intersection point of the curve in Fig.7, the maximum current curve of the HTS LIM can be obtained with different frequencies, as shown in Fig. 8. It can be seen from Fig. 7 and Fig. 8 that the maximum vertical field of the superconducting winding is increasing with the decrease of the frequency of the power supply under a certain voltage. When the power frequency is increased from 4Hz to 14Hz, the maximum input current of HTS LIM is increased from 22.82A to 24.52A, and the change is not very obvious.

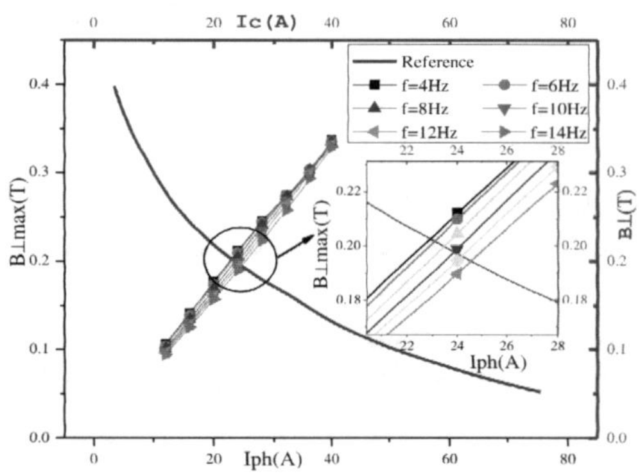

Figure 7. Maximum current vs different frequency

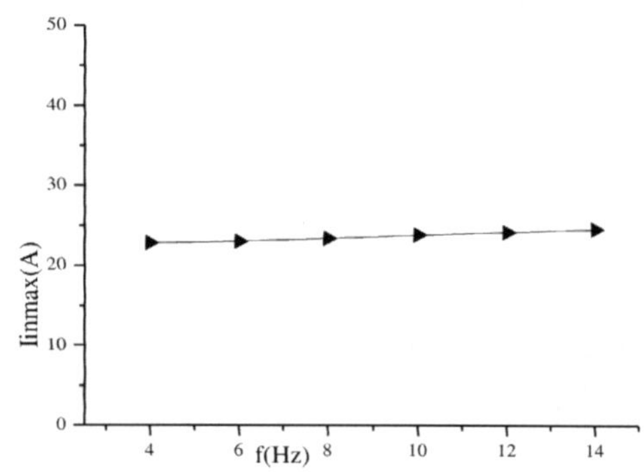

Figure 8. Maximum current vs different frequency

3.3. Effect of thickness of the secondary conductive plate on electromagnetic properties

The secondary plate of HTS LIM is a compound structure made of aluminum and iron. Fig.9 shows the horizontal thrust and the vertical force change curve of the motor with different aluminum thickness. It can be known from the graph that the electromagnetic force increases at first and then decreases with the increase of the thickness of the secondary aluminum plate, but the vertical force continually reduces at the start time of the motor. When the aluminum thickness size is between 2-3mm, the electromagnetic force reaches the maximum value of about 1290N; then the electromagnetic thrust decreases with the increase of the thickness of the aluminum plate, and the electromagnetic force is reduced to 900N when the thickness size increases to 6mm. However, the vertical force of HTS LIM is always shown as attractive between primary and secondary when the thickness size of the aluminum plate is increased from 1mm to 6mm, and its value is gradually reduced from 3188N to 0; when the thickness of the aluminum plate is larger than 6mm, the vertical force is almost 0—that is to say roughly equal attraction and repulsion. If the thickness of aluminum plate is increased continually, the vertical motors will behave as a repulsive force.

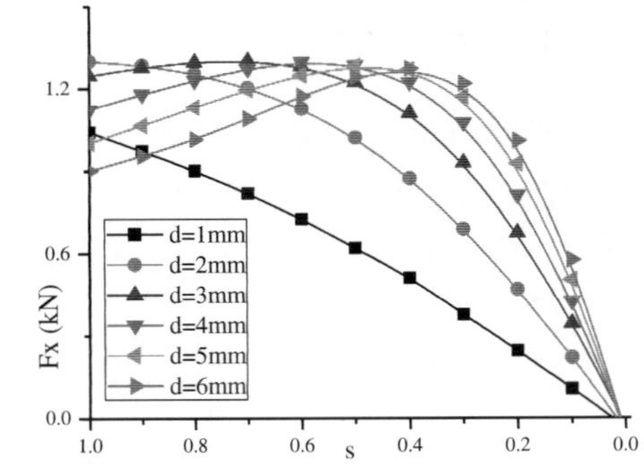

(a) The curve of the electromagnetic force

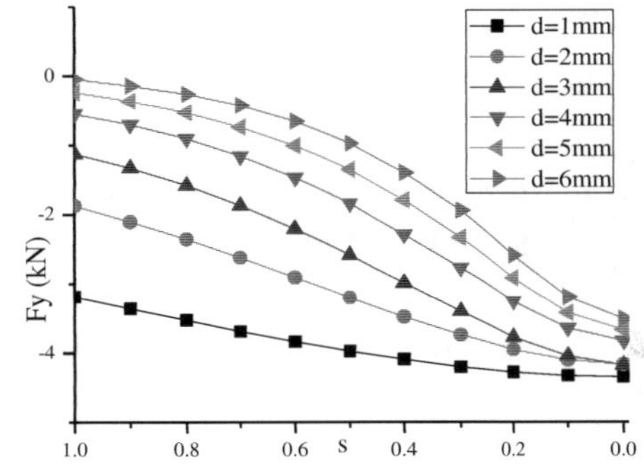

(b) The curve of the vertical force

Figure 9. The curve of F_x and F_y vs s with different aluminum thickness

In practical applications, due to the different expansion coefficient of aluminum and steel materials and the attraction of primary and secondary in the vertical direction when the motor is working(about 4 ~ 6 times of the thrust), the aluminum plate is easy to break off affecting the motor's normal operation. Therefore, the aluminum plate thickness value of 2mm, 3mm, 4mm, 5mm, 6mm is respective chosen to research the current capacity of HTS LIM. Fig. 10 shows the variation curve of the maximum vertical field of the superconducting winding with different aluminum thickness value, in which the solid line represents the surface vertical field with the critical current of the Bi-2223/Ag superconducting tapes in 77K low temperature environment. According to the intersection point of curve in Fig.10, the maximum current curve of motor can be obtained with different aluminum thickness value, as shown in Fig.11.

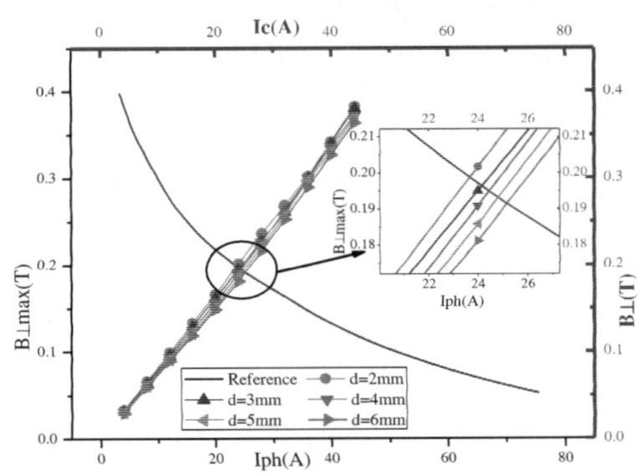

Figure 10. B⊥max vs maximum current with aluminum thickness

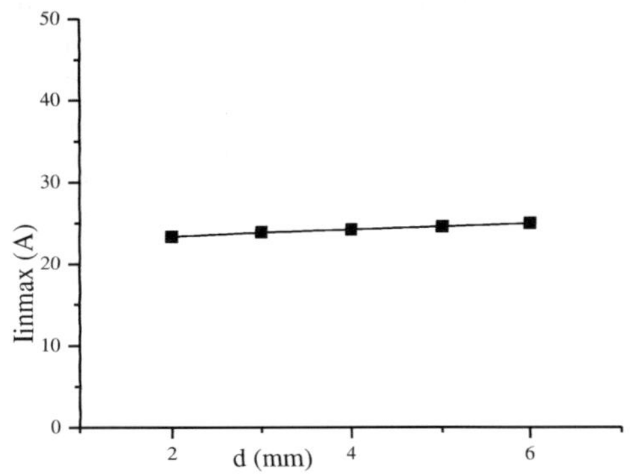

Figure 11. Maximum current vs aluminum thickness value

Fig.10 and Fig.11 show that the maximum input current of HTS LIM is increased from 23.33A to 24.97A when the thickness value of the aluminum plate is increased from 2mm to 6mm. This is because the increase of thickness of the secondary aluminum plate results in a gradual increase of the eddy current induced by the secondary conductive plate and weakening air-gap flux density, which will enlarge the repulsion and reduce the vertical force between the primary and secondary. Similarly, the weakening of the air-gap flux density will also decrease the maximum vertical field of HTS winding and increase the input current HTS LIM.

4. Test

In order to verify the correctness and feasibility of the theoretical calculation, the load test is carried out on the experimental platform which has been built, and the load is a linear motor with a copper winding. Fig.12 shows the experimental platform and control devices. Two sets of completely independent inverter control system are respectively connected with the power line of HTS LIM and the prototype with a copper winding, and the signal line of speed sensor and pull pressure sensor is tied together with the motor power line to move through the pulley with the movement of the car. According to the test platform built, HTS LIM' own friction is measured as 70N.

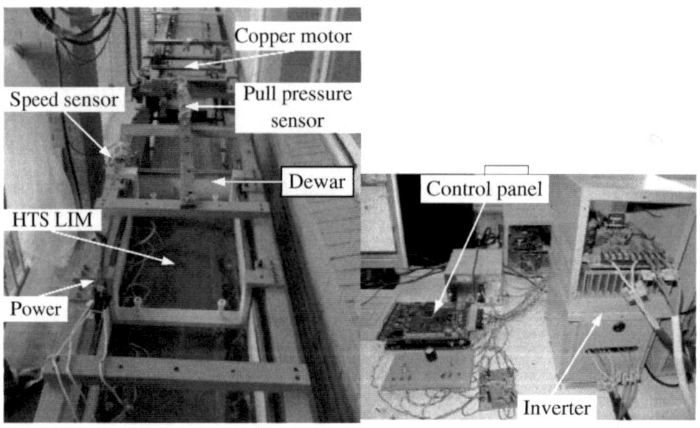

Figure 12. Test platform and control devices of HTS LIM

Due to the limit of railway track length, HTS LIM cannot reach the rated speed 3.5m/s (rated frequency is 10Hz), so the operating power frequency is selected as 4Hz. The horizontal thrust waveform diagram of HTS LIM with load by the test was obtained, as shown in Fig.13. The thrust measured by the sensor 493N plus its own friction 70N, the total thrust of the motor is 563N. Under the same conditions, the simulation thrust waveforms obtained by FEM are shown in Fig.14. The simulation result shows that the thrust of the motor is 499N, which is 0.886 times of the result measured by test. From the above analysis, it can be known that the motor performance requirements can be satisfied.

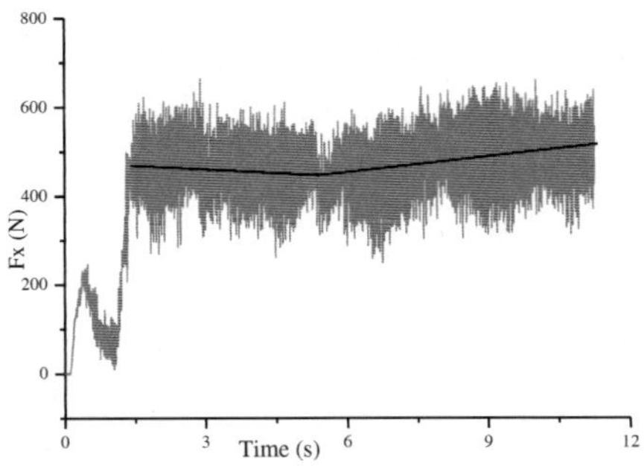

Figure 13. Measuring curve of thrust with load

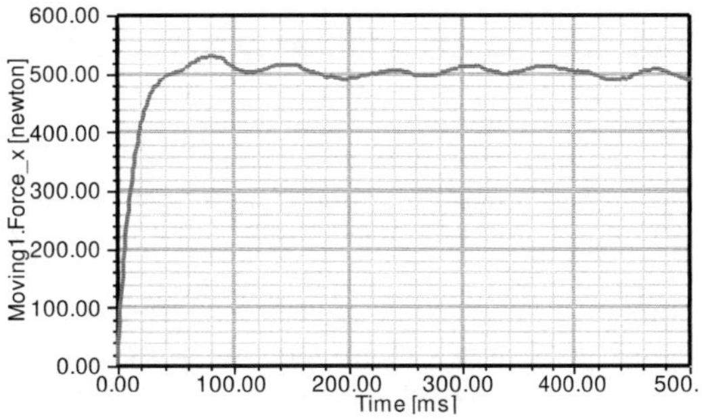

Figure 14. Simulation curve of FEM forthrust

5. Conclusion

In this paper, the effects of electromagnetic parameters and main structural parameters on the electromagnetic force, vertical force and the maximum vertical field of HTS windings are analyzed. Through the research of the vertical field of HTS winding, the rule of the maximum input current of HTS LIM with the stator frequency and the thickness value of the secondary conductive plate is obtained, which is of great significance to the analysis of the stability of HTS LIM. In addition, the load test is carried out on an experimental platform, which has been built with the load that is a linear motor with a copper winding, to compare the test results with the FEM results with the same operation conditions. The comparison shows the reliability of the FEM in the design process of the motor.

Acknowledgement

This research was supported by a grant from the National High Technology Research and Development of China 863 Program (2008AA03A203).

References

[1] Fang J, Sheng L, Li D, Zhao J, Li S, Qin W, Fan Y, Zheng Q L, Zhang W 2012 *Physics Procedia* **27** 408-411

[2] Lu J and Ma W 2011 *IEEE Transactions on Plasma Science* **39** 116-120

[3] Pina J M, Neves M V and Rodrigues A L 2007 *POWERENG 2007. International Conference on. IEEE* 185-190

[4] Jin J X, Zheng L H, Guo Y G, Zhu J G, Grantham C, Sorrell C C and Xu W 2012 *IEEE Transactions on Applied Superconductivity* **225** 5202617-5202617

[5] Pinatti D G, Souza T M and Shigue C Y 1995 *IEEE Transactions on Applied Superconductivity* **5** 453-456

[6] Pinatti D G, Souza T M and Shigue C Y 1995 *IEEE Transactions on Applied Superconductivity* **5** 453-456

[7] Nakamura T, Miyake H, Ogama Y, Morita G, Muta I and Hoshino T 2006 *IEEE transactions on applied superconductivity* **16** 1469-1472

[8] Chen Q, Liu G, Liu Z and Li X 2014 *IEEE Transactions on Applied Superconductivity* **24** 1-5

IOP Conf. Series: Journal of Physics: Conf. Series **871** (2017) 012098 doi:10.1088/1742-6596/871/1/012098

Status of MgB_2 wire and cable applications in Europe

A. Ballarino, R. Flükiger
CERN, 1211 Geneva 23, Switzerland
amalia.ballarino@cern.ch

Abstract. Since its discovery in 2001, MgB_2 has generated interest for practical applications. Its availability in the form of multifilamentary round wire makes it suitable for production of cables. Together with relatively high critical temperature and potential low-cost, this renders it appealing for use in superconducting devices where its limited in-field performance can be tolerated. The state-of-the-art properties of commercially available wire and the potential of MgB_2 conductors for use in superconducting systems are discussed. An overview of high-current electrical transmission projects where MgB_2 has been proposed as an alternative to conventional Nb-Ti or High Temperature Superconductors is presented.

1. Introduction

The discovery in 2001 of the superconductor MgB_2 raised high expectations in the field of fundamental and applied research. The T_c value of this compound, 39 K, is lower than those for High Temperature Superconducting (HTS) materials, but considerably above the value of Nb_3Sn (18 K). Since both Mg and B are abundant and not intrinsically expensive, it was expected that MgB_2 wires fabricated using classical deformation methods would be competitive for superconducting applications above 4.2 K in devices cooled by closed-circuit refrigerators or even in devices cooled by liquid hydrogen. A considerable effort has been made worldwide, both theoretically and experimentally, to understand this exciting material and to develop industrial conductors. In the following, research and development in the field of MgB_2 will be briefly discussed, with a focus on the activities in Europe.

2. Fundamental properties, single crystals

The simple binary, hexagonal compound MgB_2 presents in reality an unexpected complexity. As for other Low Temperature Superconductors (LTS), e.g. Nb_3Sn, MgB_2 was found to be of the BCS type, but superconductivity in MgB_2 was found to be correlated with two distinct energy bands, with energy gaps Δ_σ and $\Delta\pi$, at 7.5 and 2.3 meV [1]. The interplay of the quasi two-dimensional σ-band and the almost isotropic π-band results in an anomalous and complex behaviour of the characteristic lengths and anisotropic critical fields, as described by Gurevich [2, 3]. The existence of two bands was experimentally confirmed calorimetrically, by tunneling and by point contact spectroscopy.

MgB_2 single crystals [4, 5] were the basis of numerous fundamental properties of this compound. Most properties are anisotropic, e.g. the coherence length ξ, the penetration depth λ, the upper critical field B_{c2} and the electrical resistivity ρ. Binary and ternary $MgB_{1-x}C_x$ crystals of a size up to 1.5 x 1 x 0.1 mm^3 and of weight up to 230 μg were grown from flux at pressures as high as of 3 GPa by Karpinski [5] at the ETH in Zurich. Rowell [6] proposed a simple correlation between electric resistivity ρ_0 and grain connectivity, which has been verified by numerous authors. The importance of the mixed state parameters in defining the limiting conditions for loss-free currents and thus J_c in MgB_2 has been discussed in detail in a review by Eisterer *et al* [7] at the ATI in Vienna. The effect of high energy radiation on the electronic properties of MgB_2 has been described by Putti *et al* [8].

3. The effect of Carbon substitution on the properties of MgB_2

A sizeable and reproducible enhancement of J_c of MgB_2 was observed after partial substitution of B by Carbon, after the reaction with C based compounds, in particular SiC [9], malic acid ($C_4H_6O_5$) [10], coronene ($C_{24}H_{12}$) [11] or thin C layers at the surface of B particles [12]. T_c in MgB_2 decreases almost linearly with the increase of the C content: for x = 0.18 in the formula $MgB_{2-x}C_x$ a T_c value of 30 K is

IOP Conf. Series: Journal of Physics: Conf. Series **871** (2017) 012098 doi:10.1088/1742-6596/871/1/012098

obtained. The anisotropy ratio γ (defined by the ratio between $B_{c2}//ab$ and $B_{c2}//c$) decreases, too, from $\gamma = 5.3$ in binary single crystals to ~ 2 in ternary $MgB_{0.87}C_{0.13}$ single crystals [5]. Both gaps Δ_σ and Δ_π approach each other with increasing C content and merge at 13.5% C, only one gap being observed at higher C contents. The enhancement of the residual resistivity ρ_o with lattice disorder in MgB_2 wires with increasing C content was clearly recognized as the dominant effect enhancing the critical current density J_c at high magnetic fields. The upper critical fields $B_{c2}//c$ and $B_{c2}//ab$ as well as the irreversibility fields $B_{irr}//ab$ and $B_{irr}//c$ were found to increase, the value of $B_{c2}//ab$ (T=0 K) approaching 35 T. The vortex pinning behavior of MgB_2 is only little affected by the presence of C on the B sites, which was confirmed by relaxation measurements on bulk samples, which show that the pinning energy U_o of SiC alloyed MgB_2 is unchanged with respect to that of the binary compound [13].

A review of the effect of a large number of elemental additives has been published by Collings [14], showing that C is the only substitute leading to a marked enhancement of J_c in wires at high fields. The list of elements in [14] does not contain oxygen, which plays a particular role: indeed, the filaments in all wires produced by powder metallurgical procedures contain an important part of oxygen, generally attributed to the presence of MgO inclusions. Recently, Prikhna *et al* [15] reported that in reality, around 7 - 10 at. % oxygen is substituting B in the MgB_2 structure. This result, which has yet to be confirmed, would explain the strong difference between the B_{c2} values in thin films and wires mentioned later in this article.

4. Fabrication methods for MgB_2 wires

Soon after the discovery of superconductivity in MgB_2, a strong activity started at the University of Genova on the fundamental properties, but also on methods for wire production. Their results were the basis for the creation of a new company, Columbus Superconductors (Genova), based on the *ex situ* processing method. Today, industrial MgB_2 wires are also produced by Hypertech. Wire developments in view of a possible industrial production are also carried out in Japan at NIMS in Tsukuba, in collaboration with Hitachi. The main requirements for optimized transport properties in MgB_2 wires are elemental B and Mg powders with a high purity and a small particle size. The properties in the high field range are improved by C based additives. The desired final MgB_2 grain sizes being of the order of 10-50 nm, the appropriate initial boron particle size should be ideally in the nanometer range, too. The commonly used matrix materials for industrial MgB_2 wires are Ni and Monel, with a Nb barrier for avoiding diffusion of Mg into the stabilizing Cu or reaction with the Ni in the matrix. The variation of J_c vs. the applied uniaxial strain ε for multi-filamentary wires shows an almost linear increase up to ε ~ 0.4%, after which degradation occurs. Due to the large total surface of the nanopowders, all MgB_2 filaments contain several at. % of oxygen, which gives rise to MgO inclusions and to oxygen substitution of B [15]. The main processing routes for the MgB_2 wires are: a) *ex situ*, b) *in situ* and c) IMD (Internal Mg diffusion); they are briefly described in the following.

4.1. The "ex situ" technique

Pre-reacted MgB_2 powders with a molar ratio of 1:2 are inserted into a Nb tube, surrounded by a $Cu_{.70}Cu_{.30}$ tube and drawn to a wire of about 3.5 mm diameter. Several pieces of wires are then inserted in a Nickel or Monel tube, together with OFHC Cu for stabilization. This composite is groove rolled and drawn to a round wire with a diameter of 2 mm, which is twisted before to be rolled to a tape (typically 3.6 x 0.65 mm^2) or to a wire of about 1 mm diameter, followed by a recrystallization heat treatment of typically 4 minutes at 965°C. Industrial *ex situ* wires produced by Columbus Superconductors exhibit J_c(4.2K) ~ 3 x 10^3 A/mm^2 and 5 x 10^2 A/mm^2 at 3 and 5 T, respectively, for binary wires, and ~ 4 x 10^2 A/mm^2 at 7 T for C alloyed wires (Malagoli *et al* [16], Kario *et al* [17]), as shown in figure 1. These J_c values are lower than those for the other two processing route alternatives, but *ex situ* wires exhibit the highest homogeneity over long lengths, in addition to availability in km lengths produced with industrial and quality-controlled methods.

4.2. The "in situ" technique

For *in situ* MgB_2 wires, the initial MgB_2 mixture consists of Mg and B powders. In Europe, *in situ* wires were developed at the University of Genova, the University of Geneva, and the ESS in Bratislava, while industrial wires using this process are developed at Ohio State University and fabricated by Hyper Tech Research. With respect to the *ex situ* route, this alternative has the advantage that the reaction to MgB_2 occurs at considerably lower temperatures (around 650 °C), thus reducing the reaction layer at the metallic sheath. Even at these temperatures, C can be substituted into the B plane of the MgB_2 lattice. The highest J_c reported so far for *in situ* monofilamentary MgB_2 wires was obtained by Li *et al* [12] using B powder of 100 nm size, pre-doped with various C contents by plasma spray synthesis and 99 % Mg powders of 25 μm size. Multi-filamentary wires of 0.83 mm diameter with a Nb barrier and a Monel outer sheath (MgB_2/Nb/Monel) were manufactured and the reaction was performed at 700°C for 40 min, the final composition in the final MgB_2 filament being 2.54 mol% C. In the MgB_2 layer this wire exhibited J_c(4.2 K) = 1 x 10^2 A/mm^2 at 13.2 T (see figure 1).

4.3. The Internal Mg diffusion (IMD) technique

The internal Mg diffusion process (IMD) is based on the infiltration process developed by Giunchi (Edison, Italy) [18]. The density of the MgB_2 phase in these wires is close to 100%, i.e. much higher than in *in situ* and *ex situ* wires (~ 45 and ~ 70 %, respectively). Hypertech and NIMS/Hitachi use similar processing routes, known as AIMI and IMD, respectively. The higher connectivity leads to considerably higher J_c values in the MgB_2 layer (see figure 1). A pure Mg rod with a diameter of typically 2.0 mm is placed at the center of a Ta or Nb tube, and the hollow space between the metal sheath inner wall and the Mg rod is filled with a B and C, the latter as a thin carbon layer on the surface of the B particles [12] or as $C_{14}H_{12}$ powders [11]. A multifilamentary configuration is obtained by bundling the rods, inserting them into a Cu-Ni tube, cold deforming and reacting. During the heat treatment at 640 °C, Mg diffuses into the B and reacts to form MgB_2 layers of 10-30 μm thickness along the inner wall of the Ta tube.

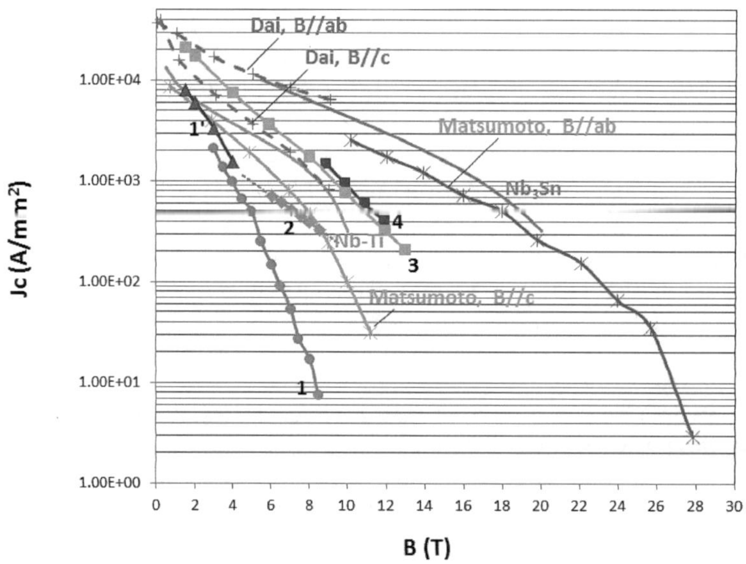

Figure 1. J_c vs. B at 4.2 K for MgB_2 wires, tapes and thin films. 1: J_c for *ex situ* wire, binary [16], 2: J_c for *ex situ* tape, C alloyed [17], 1′: J_c for *ex situ* wire, binary, measured at CERN on a total production of 80 km [34], 3: J_c for *in situ* wire, C alloyed [10], 4: J_c(layer) for AIMI wire, C alloyed [12].
J_c in MgB_2 thin films: binary: C alloyed (12.3 at % C) [19], *B//ab* and *B//c* [21].
Data of Nb-Ti and of state-of-the-art Nb_3Sn wires have been added for comparison.

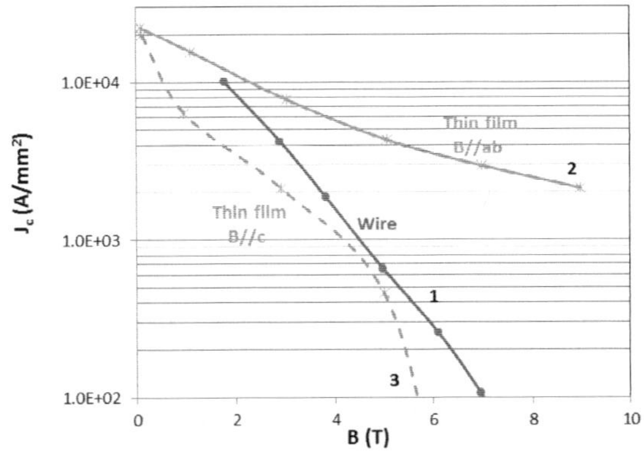

Figure 2. J_c vs. B at 20 K for C alloyed MgB_2 wires and thin films. 1: J_c(layer) for IMD processed wire with SiC additions [11]; 2 and 3: HPCVD processed thin films 12.3 at % C [21] for the orientations *B//ab* and *B//c*, respectively.

In both AIMI and IMD wires, holes are formed at the center of each filament, where the Mg core was located before reaction. During reaction, the C based additives decompose and carbon substitutes the boron in the MgB_2 phase. The values of J_c(layer) are slightly above 1 x 10^3 A/mm^2 at 4.2 K/10 T [11,12]. At 20 K/4 T, a J_c(layer) value close to 3 x 10^3 A/mm^2 was reported by Ye *et al* [12] (see figure 2). The corresponding values of Li *et al* [12] (not shown here) are very similar. The layer J_c of the AIMI wires [12] at 4.2 K exceeds the J_c values reported so far for multifilamentary Nb-Ti wires (see figure 1).

5. Comparison of J_c in MgB_2 wires and in thins films

The J_c vs. B values at 4.2 K for the three routes are shown in figure 1 (J_c is normalized to the superconducting cross section, with the exception of the AIMI process for which the layer J_c is reported). For comparison, the J_c vs. B values for C added MgB_2 thin films, as well those for Nb-Ti and Nb_3Sn wires are also shown in the same figure. At all magnetic fields, the J_c values for *in situ* and AIMI wires are higher than those of Nb-Ti. Due to the very low values of J_c, even for the best thin films in the unfavorable *B*//c orientation, MgB_2 is not competitive with Nb_3Sn at 4.2 K. It must be recognized that the present MgB_2 wires do not take full advantage of the inherent properties of this material. When comparing the J_c values of MgB_2 wires with those of thin films (figure 1), it is seen that the highest J_c values in AIME wires just achieve the same level as the lower J_c(*H*//c) values for thin films [20]. Since the values of B_{c2} for *B//ab* are at least 2 times higher (the highest known value of B_{c2}// is 74 T [20]), it follows that a considerable enhancement of J_c in wires would be expected if its B_{c2} (and B_{irr}) could be raised in wires too. This is also true for the properties of MgB_2 at temperatures in the range between 10 and 25 K, as shown by the variation of J_c vs. B both for AIMI wires[12] and thin films [20], see figure 2. In this temperature range, MgB_2 has still a series of advantages. The potential at 20 K could even be raised if the reasons for the strong discrepancy between critical fields of thin films and wires could be found. The recently reported presence of sizeable quantities of oxygen inside the MgB_2 structure [15] could possibly provide a source of new ideas for tacking this problem.

6. Applications of MgB_2

6.1. General considerations

The promised low price of MgB_2 conductor, in principle enabled by the low cost and wide availability of raw materials, motivated several groups to explore possibilities for use in electrical applications. The price (Euro/kA m) is not yet competitive with that of Nb-Ti at 4.2 K. Processing of *ex situ* and *in situ* wires is based on the well-known Powder-In-Tube (PIT) technology, and while the manufacture of round wire involves the mechanical deformation of alloys or metals (typically Monel® or iron) that are harder and more difficult to draw than copper, it is reasonably expected that the supply of large quantities of conductor will lead to optimization and cost reduction. In addition, margins in state-of-art industrial conductor (filling factor, raw material purity, and density of the superconducting phase) point in the direction of a potential future enhancement of performance and lower cost per kA-m.

To date the use of MgB_2 is limited to demonstrators or niche applications in low (< 5 T) magnetic fields and at temperatures of up to 15 - 20 K. Operation at these higher temperatures is appealing, but MgB_2 is not always the natural choice when compared to HTS BSCCO 2223 and REBCO conductors – in particular when liquid nitrogen cooling is an option. If electrical performance of industrially-produced MgB_2 wires at 15 K - 20 K and medium and low fields, and cost, were to become comparable to those of Nb-Ti at 4.2 K, then MgB_2 could emerge as replacement technology for applications benefiting from higher temperature margin, and/or relying on cryogen-free, helium gas or liquid hydrogen cooling.

Winding and cabling of reacted MgB_2 wire must accommodate the mechanical properties of the conductor. Development wires indicate a potential for critical tensile strains of up to 0.55 % [21] and 0.67 % [22] at 4.2 K for both PIT and IMD technologies. State-of-the-art industrial *ex situ* and *in situ* PIT multi-filamentary wires exhibit critical tensile strain of ~ 0.25 % at room temperature and 0.3 % at 4.2 K, with a minimum bending radius of ~ 100 mm [23-25]. More demanding applications call for Wind & React (W&R) technology with typical reaction temperatures of about 650 °C. Both W&R and React & Wind (R&W) technologies were demonstrated by several groups with the successful assembly and test of prototype pancake and solenoid coils generating magnetic fields of up to about 5 T [26-28].

MRI cryogen-free magnets have been the first commercial application of MgB_2 conductor. Operation at temperatures of 15 – 20 K offers the advantage of a lower operational cost and an enhanced stability compared to Nb-Ti, and a reduction of the system, i.e. superconductor, cost if compared to HTS materials. Already in 2006, only five years after the discovery of the superconducting properties of MgB_2, ASG Superconductors, Columbus Superconductors and Paramed designed and commissioned an open MRI system generating a central field of 0.5 T in a free patient gap of about 60 cm. The system, that today counts 28 units operational in hospitals and clinics world-wide, uses 18 km of MgB_2 multi-filamentary *ex situ* tape operated at 90 A and about 20 K [29].

A racetrack coil made from reacted MgB_2 tape has also designed and tested in the framework of the SR2S European space project [30]. The aim was to explore the potential of MgB_2 in an active magnetic shield to protect astronauts from cosmic radiation during long space missions. For this application, a lightweight multifilamentary tape, with Ti matrix and Al stabilizer, was developed and tested [31].

6.2. High current cables and systems based on the use of MgB_2

Transport of high current requires cabling of MgB_2 tape or wire. First demonstrations of MgB_2 cables were done in the field of DC superconducting power transmission. The higher critical temperature and potential low-cost of MgB_2 triggered proposals of smart superconducting grids combining delivery of liquid hydrogen and electricity. In this context, the industrially available *ex situ* tape produced by Columbus Superconductors was assembled at the JSC VNIIKP institute, in Russia, into initially 10 m and then and 30 m long prototype cables [32]. The cables, which consisted of MgB_2 tapes helically wound around a central former, were measured in cryogenic transfer lines where cooling was provided by forced flow of liquid hydrogen. They reached DC currents of up to ~ 3000 A at 20 K and successfully underwent 50 kV DC electrical insulation tests. This work first demonstrated the possibility of transporting an electrical power of up to ~ 150 MW via MgB_2 cables based on tape superconductor.

For transmission of 1 kA range DC currents, the concept of twisted-pair cables made from tape conductor has been proposed and developed [33]. Two stacks of MgB_2 tapes, each electrically insulated with wrapped polyimide tape, are twisted together to form the two polarities of an electrical circuit (see figure 3). Compact twisted-pair MgB_2 cables were successfully measured at currents of above 4000 A at 4.2 K [33]. A cabling machine able to produce kilometer lengths of this cable geometry with controlled cabling parameters has also been developed and operated [34].

An important step in the field of high-current MgB_2 superconducting transmission systems was the successfully development and commissioning, at CERN, of a helium gas-cooled transfer line incorporating 2 × 20 m long cables, made for the first time from MgB_2 round wire, and operated at a record DC current of up 20 kA at 24 K [35] (see figure 3). This work, which was performed in the framework of the LHC High Luminosity (HL-LHC) upgrade [36], has been the first demonstrator of a novel MgB_2 wire-based electrical system that is being developed for feeding the LHC HL-LHC Nb_3Sn and Nb-Ti magnets [34]. In total eight superconducting transfer lines, called superconducting links, each about 100 m long, will be used to transfer the current from the power converters, located in newly excavated underground galleries, to the LHC main tunnel. A superconducting link consists of a semi-flexible cryostat containing up to about forty MgB_2 cables rated at currents ranging from 0.12 kA to 18 kA, for a total current capacity of about 150 kA at 20 K. The cooling is provided by forced flow of helium gas. At the time when the HL-LHC superconducting link project started, only MgB_2 tape was industrially available, and the choice was made to develop round wires, more suitable for cable assembly. Development of MgB_2 round wire, with electrical performance and mechanical properties enabling use in high-current cables, was launched in a collaboration between CERN and Columbus Superconductors. Passing from the existing industrial tape to a round wire took a significant time and effort, and required production and measurement of several prototype wires with different layouts and composition. Today a round wire for use in cables is available in long lengths. The *ex situ* wire has a diameter of 1 mm. It consists of 37 superconducting filaments, each with an equivalent diameter of about 60 μm, twisted with a pitch of 100 mm. The MgB_2 filaments, made from high purity boron powder, are surrounded by a Niobium barrier and embedded in a Nickel core. The Monel matrix, around the Nickel core, is copper plated, and the copper surface is tinned. Eighty kilometer of wire have been produced by Columbus Superconductors and delivered to CERN in unit lengths of greater than 500 m. Hundreds of measurements performed at CERN and at Columbus on short wire samples confirmed homogeneity and conformity of the electrical performance, which requires a minimum critical current at 25 K and 0.9 T of 186 A. In-depth work was done at CERN to qualify the mechanical behavior of wires and cables [23, 24, 25, 37] (the cables are made from reacted wires), as well as to develop low resistance joints [38], quantify quench behavior of cables and develop associated protection strategies [39, 40]. The superconducting links project has been approved for integration in the LHC machine in 2024, when all hardware associated with the HL-LHC project will be installed in the LHC underground areas. The total quantity of MgB_2 wire required for the series production of the superconducting links is about 1000 km.

Further to the CERN initiative, MgB_2 superconducting technology was also proposed by Prof. Carlo Rubbia, scientific director of the Institute for Advanced Sustainability Studies (IASS) in Potsdam, for an innovative transmission line for long-distance transport of green power [42]. This work has generated a European initiative, the FP7 BESTPATHS (acronym for "Beyond State-of-the-art Technologies For RePowering Ac corridors and Multi-Terminal HVDC Systems") project, presently working on the design of a full-scale 320 kV MgB_2 monopole cable system designed to transfer a current of 10 kA at 20 K for a power of up to 3.2 GW [42, 43]. The cable under study has the same geometry (18 MgB_2 strands helically wound around a copper core, see figure 3) and uses the same wire layout as those developed for the CERN HL-LHC project. In addition, it incorporates all features required for future operation in the grid. The laboratory test of the system is planned by end 2018.

Among future applications that could benefit from the on-going developments, it is worth mentioning the potential use of MgB_2 for the future IGNITOR tokamak [44]. IGNITOR is based on copper coils operated at cryogenic temperatures of about 30 K via forced flow of helium gas. The challenging aspects of the IGNITOR project are the creation of: magnetic systems with a strong field of great volume with the use of cryo-resistive conductor, a unique multipurpose power complex, a system of physical and technological diagnostics of dense thermonuclear plasma, and a system of

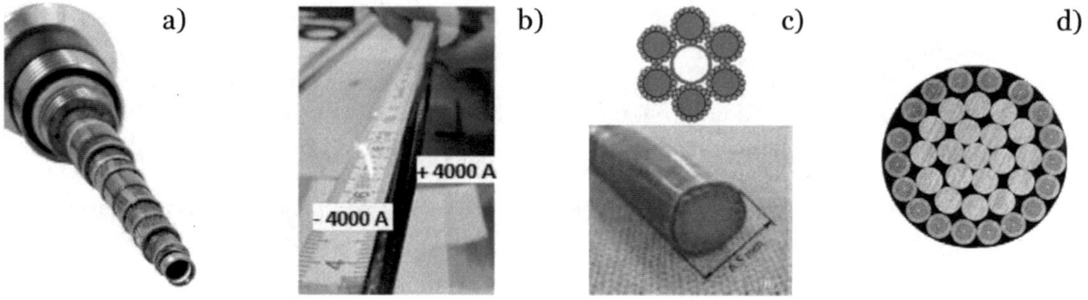

Figure 3. MgB_2 cables. From left: a) prototype power transmission cable made from MgB_2 tape [32] (~ 3000 A DC at 20 K, $\varnothing_{ext}$ ~ 30 mm); b) twisted-pair cable made from MgB_2 tape[33] (~ 4000 A DC at 4.2 K, Φext~5 mm); c) top: prototype cable (consisting of six sub-cables) made from round MgB_2 wire [34,35] (20000 A DC at 24 K, $\varnothing_{ext}$ ~ 20 mm); bottom: one of the six sub-cables each made from 18 MgB_2 wires, of 1 mm diameter, helically wound around a copper core; d) prototype cable made from 18 MgB_2 round wires, of 1.5 mm diameter, helically wound around a copper core (10000 DC at 20 K, Φext~10 mm) [42, 43]. All quoted critical currents are in self-field conditions (≤ 1 T).

intelligent control of processes in plasma. For the poloidal field coils, IGNITOR considers the use of MgB_2 conductor helium gas cooled at about 30 K [45]. This project could open the door to fusion technology for MgB_2.

7. Conclusions

Significant progress has been made in the direction of developing MgB_2 wire with enhanced electrical performance. R&D activity indicates promising superior performance of IDM and AIMI processed wires. Round *ex situ* PIT wire, with homogeneous and quality controlled properties, has been successfully developed and is being produced, in large quantity, for electrical transmission in accelerator technology (HL-LHC superconducting link project). Within the same project, significant effort is on-going for the development of high-current MgB_2 cables and of the associated electrical system. The potential of MgB_2 for future power transmission in the electrical network is also being studied in the framework of a European project (BESTPATHS) gathering together industry, laboratories, and the French transmission system operator (RTE). The first industrial application of MgB_2, i.e. MRI systems based on MgB_2 tape, boasts 28 units operational world-wide. Several demonstrators of R&W and W&R small coils have been made. R&D development is still needed to understand the strong discrepancy between critical fields of thin films and wires and potentials of MgB_2 as high field conductor, as well as to boost performance of MgB_2 at higher (~ 20 K) temperature. Successful developments in these directions could bridge the gap with Nb-Ti conductor.

References

[1] Choi H J, Roundy D, Sun H, Cohen M L and Louie S G 2002 *Nature* **418** 758
[2] Gurevich A 2003 *Phys. Rev. B* **67** 184515
[3] Gurevich A *et al* 2004 *Supercond. Sci. Technol.* **17** 278
[4] Lee S, Masui T, Yamamoto A, Uchiyama H, Tajima S 2003 *Physica C* **397 7**
[5] Karpinski J *et al* 2003 *Supercond. Sci. Technol.* **16** 221
[6] Rowell J M 2003 *Supercond. Sci. Technol.* **16** R17
[7] Eisterer M 2005 *Phys. Stat. Sol. C* 2 1606 14 1606-14
[8] Putti M, Vaglio R and Rowell J M 2008 *Supercond. Sci. Technol.* **21** 043001 (25 pp)
[9] Dou S X *et al* 2001 *Physica C* **361** 79
[10] Susner M A, Sumption M D, Bhatia M, Peng X, Tomsic M J, Rindfleisch M A, Collings E W 2007 P*hysica C* **456** 180
[11] Ye S J, Matsumoto A, Zhang Y C and Kumakura H 2014 *Supercond. Sci. Technol.* **27** 085012;

Kumakura H 2016 in *MgB_2 Superconducting Wires, Basics and Applications* ed R Flükiger (World Scientific) p. 315
[12] Li G Z, Yang Y, Susner M A, Sumption M D, Collings E W 2012 *Supercond. Sci. Technol.* 25 02500
[13] Senatore C, Lezza P, Flükiger R 2006 *Adv. Cryo. Eng.* **52** 654
[14] Collings E W, Sumption M , Bhatia M, Susner M A and Bohnenstiehl S D 2008 *Supercond. Sci. Technol.* **21** 103001
[15] Prikhna T et al. 2016 presented at ASC, accepted for publication in *IEEE Trans. Appl. Supercond.* 2017
[16] Malagoli A, Braccini V 2016 in *MgB_2 Superconducting Wire, Basics and Applications* ed R Flükiger (World Scientific) pp 239-270
[17] Kario A et al. 2012 Physica C 483 pp. 222-224
[18] Giunchi G 2016 in *MgB_2 Superconducting Wire, Basics and Applications* ed R. Flükiger (World Scientific), pp. 159-208
[19] Matsumoto A, Kobayashi Y, Takahashi K-I, Kumakura H, Kitaguchi H 2008 *Appl. Phys. Express* **1** 021702
[20] Dai W *et al* 2011 *Supercond. Sci. Technol.* **24** 125014
[21] Kovac P, Melisek T, Kopera L, Husek I, Polak M, Kulich M 2009 *Supercond. Sci. Technol.* **22**
[22] Nishijima G, Ye S J, Matsumoto A, Togano K, Kumakura H, Kitaguchi H and Oguro H 2012 *Supercond. Sci. Technol.* **25** 054012
[23] Konstantopoulou K, Ballarino A, Gharib A, Stimac A, Gonzalez Garcia M, Perez Fontenla A Sugano M 2016 *Supercond. Sci. Technol.* **29** 084005
[24] Alknes P, Hagner M, Bjoerstad R, Scheuerlein C, Bordini B, Sugano M, Hudspeth J and Ballarino A 2016 *IEEE Trans. Appl. Supercond.* **26** No. 3 8401205
[25] Sugano M, Ballarino A, Bartova B, Bjoerstad R, Scheuerlein C, Grasso G 2015 *IEEE Trans. Appl. Supercond.* **25** No. 3 4801004
[26] Tanaka K, Kitaguchi H, Kumakura H, Yamada H, Hirakawa M, Okada M 2005 *Supercond. Sci. Technol.* **18**
[27] Sumption M D *et al* 2007 *IEEE Trans. Appl. Supercond.* **17** No. 2
[28] Young E A, Falorio I, Pelegrin M J, Yang Y 2015 *IEEE Trans. Appl. Supercond.* **25** No. 3 4600105
[28] http://www.paramed.it/medical-systems/technology-0000184.html
[30] http://www.sr2s.eu/2013-08-01-15-34-14
[31] Musenich R *et al* 2016 *IEEE Trans. Appl. Supercond* **26** No. 4 6200204
[32] Vysotsky V S *et al* 2015 *Physics Procedia* 67 189 – 194
[33] Ballarino A, Fleiter J, Hurte J, Sitko M and Willering G 2012 *Physics Procedia* **36** 855 – 858
[34] Ballarino A 2014 *Supercond. Sci. Technol.* **27** 044024
[35] https://cds.cern.ch/journal/CERNBulletin/2014/16/News%20Articles/1693853?ln=en
[36] Rossi L 2016 The High Luminosity LHC Project. See: http://slideplayer.com/slide/10789990/
[37] Sugano M, Ballarino A, Bartova B, Bjoerstad R, Gerardin A, Scheuerlein C 2016 *Supercond. Sci. Technol.* **29** 025009
[38] Giannelli S, Montenero G, Ballarino A Proceedings of ASC 2016 Denver, Colorado
[39] Giannelli S, Montenero G and Ballarino A 2016 IEEE Trans. Appl. Supercond. **26** 3 5400705
[40] Spurrell J, Young E A, Falorio I, Pelegrin M J, Ballarino A, Yang Y 2015 *IEEE Trans. Appl Supercond.* **25** No. 3 9700201
[41] Rubbia 2011 http://www.iasspotsdam.de/sites/default/files/files/rubbia_presentation.pdf
[42] Ballarino A *et al* 2016 *IEEE Trans. Appl. Supercond.* **26** No. 3 5401705
[43] Bruzek C E, Ballarino A, Escamez G, Giannelli S, Grilli F, Lesur F, Marian A, Tropeano M Proceedings of ASC 2016 Denver, Colorado
[44] Coppi B et al 2013 *Nuclear Fusion* **53** 104013
[45] Spigo G. CERN Private Communication

IOP Conf. Series: Journal of Physics: Conf. Series **871** (2017) 012099 doi:10.1088/1742-6596/871/1/012099

Current-carrying capability of the 100 kA-class HTS STARS conductor for the helical fusion reactor FFHR-d1

Y Terazaki[1], N Yanagi[1,2], S Ito[3], S Hamaguchi[2], H Tamura[2], T Mito[1,2], H Hashizume[3] and A Sagara[1,2]

[1] SOKENDAI (The Graduate University for Advanced Studies), 322-6 Oroshi-cho, Toki 509-5292, Japan
[2] National Institute for Fusion Science, 322-6 Oroshi-cho, Toki 509-5292, Japan
[3] Department of quantum Science and Energy Engineering, Graduate School of Engineering, Tohoku University, 6-6-01-2 Aramaki-Aza-Aoba, Aoba-ku, Sendai, Miyagi 980-8579, Japan

E-mail: terazaki@nifs.ac.jp

Abstract. A 100-kA-class conductor using REBCO high-temperature superconductor (HTS) is proposed as one of the conductor options for the helical coils of the LHD-type helical fusion reactor FFHR-d1. We named it STARS (Stacked Tapes Assembled in Rigid Structure) conductor. Current carrying capability of such a simply-stacked tape conductor is an important issue. We fabricated and tested a 100-kA class STARS prototype conductor sample to verify whether a premature quench occurs. The conductor sample is a one-turn short-circuit coil with a race-track shape having a bridge-type mechanical lap joint. The transport current of the sample was induced by changing the external magnetic field, and the critical current of the sample was measured. A numerical analysis of the critical current is being performed by self-consistently solving the spatial distributions of the current density and magnetic field among the tapes to verify the measured critical current of the samples.

1. Introduction

The conceptual design activities of the LHD-type helical fusion reactor FFHR-d1 are progressing at National Institute for Fusion Science (NIFS). The helical coils of FFHR-d1 have the major radius of 15.6 m and the toroidal magnetic field of 4.7 T [1]. To produce this field, an operating current of 94 kA is required for a conductor of the helical coils at the maximum magnetic field of 11.8 T. We proposed a conductor using the high-temperature superconductor (HTS) as one of the options of the helical coil conductor. This conductor consists of simply-stacked REBCO HTS tapes, a copper stabilizer and a rigid stainless-steel structure. We named it STARS (Stacked Tapes Assembled in Rigid Structure) conductor. The STARS conductor is mechanically rigid because of having no voids due to the fact that there is no twisting nor transposition among tapes. The conductor has high cryogenic stability, and low refrigeration power is supposed to be required with elevated temperature operations at >20 K in comparison to selecting low-temperature superconducting (LTS) conductor options [2, 3]. Furthermore, it is proposed that the large-diameter complex-shape helical coils are constructed by connecting conductor segments [4] (or coil segments [5]), exploiting the advantage of HTS. The conductor without including any twisting and transposition, however, may experience current carrying instability. The purpose of the present study is to examine this feature. On the other hand, it should be noted that the simply-stacked tape configuration does not give any problem to the required accuracy of the magnetic field of the helical fusion reactor, since the inhomogeneity of the conductor current due to magnetic shielding is practically canceled with a toroidal symmetry along the helical coils [4].

The feasibility studies of the HTS option for FFHR have been progressing since 2005 [6], [7], then in our previous study in 2012-2013 [8]-[10], we fabricated and tested 30-kA-class HTS conductor samples using 20 GdBCO tapes of 10-mm width, and measured the critical currents at 45 kA with the temperature of 20 K and bias magnetic field of 6.1 T and at 69 kA with 4.2 K, 1.2 T. In the series of these experiments, the validity of the conductor fabrication and the experimentation methods have been verified to proceed to a 100-kA-class HTS conductor testing.

In this paper, the fabrication and test of a 100-kA-class HTS STARS conductor sample are reviewed and a numerical analysis of the critical current has been improved from the previous study to verify whether a premature quench occurs in the 100-kA test.

2. Experiment

2.1 100 kA-class HTS conductor sample

Figure 1 shows cross-sections of the 100-kA-class HTS conductor designed for FFHR-d1 and a sample conductor that we fabricated and tested. The REBCO-type GdBCO tapes were employed in this experiment produced by Fujikura Ltd. (FYSC-SC10). The tape has the following layered structure: Hastelloy substrate (100 μm), buffer layers (0.5 μm), GdBCO (2.3 μm), silver (8 μm), tin (2-4 μm) and copper (100 μm). The critical current of this tape having a width of 10 mm was ~650 A at 77 K, self-field. Fifty four tapes (3 rows and 18 layers) were simply stacked in a stabilizing oxygen-free copper (OFC) jacket without any impregnation or soldering. The copper jacket was then installed in a stainless-steel jacket, which was assembled by bolts. As is shown in figure 2, the conductor sample formed a one-turn short circuit coil with a racetrack shape having no current leads. For measuring the current of the sample, Rogowski coils and Hall probes were employed [9]. Voltage taps, temperature sensors and heaters were attached to the stainless-steel jacket as indicated in this figure. The straight sections of the sample were surrounded by glass-fiber reinforced plastic (GFRP) jackets for thermal insulation. One side of the straight sections has a bridge-type mechanical lap joint (the "joint part") developed in Tohoku University [11], [12]; while the other side has no joint (the "continuous part"). The tapes were overlapped in a step-wise pattern at the joint part and then the copper side (GdBCO side) of the tapes was mechanically connected face-to-face through inserted indium sheets. The details of the joint part and its experimental results are described in [13].

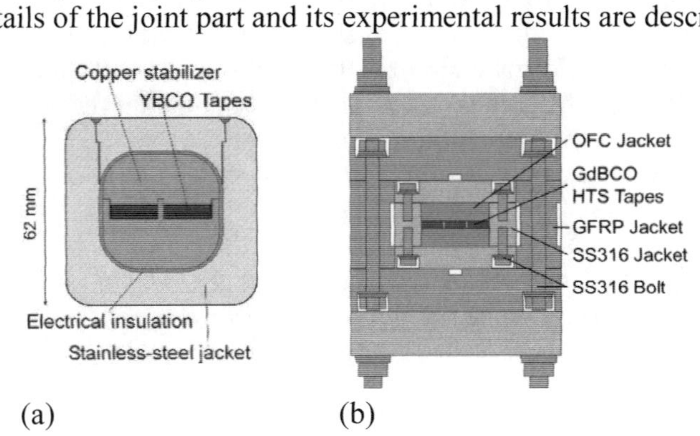

Figure 1. Cross-sectional images of (a) the 100-kA class HTS STARS conductor designed for the helical coils of FFHR-d1 fusion reactor and (b) the prototype conductor sample fabricated in this experiment.

IOP Conf. Series: Journal of Physics: Conf. Series **871** (2017) 012099 doi:10.1088/1742-6596/871/1/012099

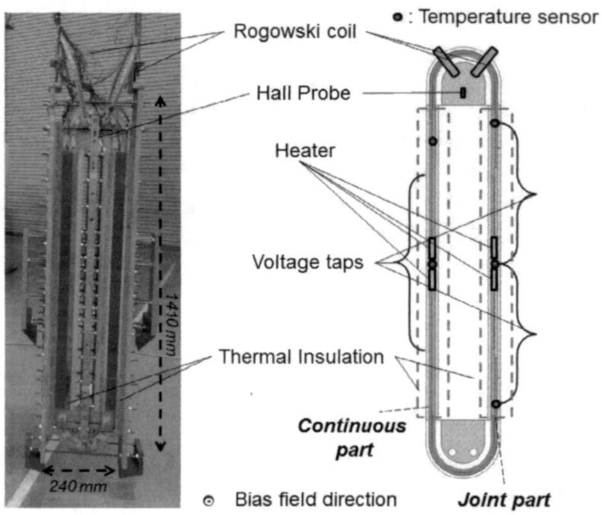

(a) (b)

Figure 2. (a) Photograph and (b) Schematic view of the overall conductor sample. Locations of voltage taps, temperature sensors, heaters and thermal insulations are indicated in (b).

2.2 Experimental Setup

The sample was immersed in liquid helium in the cryostat of the large-scale superconductor testing facility at NIFS. The current was induced in the sample by changing the background magnetic field, which was generated by a 9-T split coil with a 1-m diameter. The sample temperature was controlled by heaters attached on the stainless-steel jackets.

2.3 Experimental Result

Figure 3 shows one of the measurement results of the critical current of the 100-kA HTS conductor sample. The continuous part of the sample was heated up to 45 K (by the sensor reading) using heaters. The current was induced in the sample by changing the bias magnetic field from 3.9 to 2.4 T. The sample quenched after reaching the maximum current of 73 kA. We note that the sample current was evaluated by the Rogowski coils in the present analysis since it was confirmed that the Hall probe signals contained the magnetic field generated by the shielding currents induced in the HTS tapes, whereas the Rogowski coils did not count this. For precisely evaluating the sample current, calibration tests of the Rogowski coils were performed at room temperature [9] and in liquid nitrogen. We also note that the difference between the currents measured by two Rogowski coils was within 1%; hence, only the current measured by one of the Rogowski coils is shown in figure 3.

The voltage waveform of the continuous part of the sample is shown in figure 4 as a function of the sample current. The waveform is obtained by removing the inductive component from the raw data of the voltage. The inductive voltage is estimated by the temporal derivation of the bias magnetic field multiplied by a proper coefficient. The dashed line in the figure is obtained by a regression analysis using the following equation:

$$V = V_C \left(\frac{I_S}{I_C}\right)^n, \quad (1)$$

where V and V_C are the measured voltage at the continuous part and the voltage criterion of 20 μV corresponding to the electric field criterion of 1 μV/cm (here the voltage tap length is 20 cm), respectively. The parameters I_S, I_C and n are the sample current, the critical current of the sample determined by the criterion of 1 μV/cm and the n-value (resistive transition index), respectively. The critical current of the sample was evaluated to be 72.6 kA in this case. We also observed a critical current of 67.4 kA at 45 K, 4.3 T. We note that the sample temperature of 45 K shown here is the temperature read by Cernox temperature sensors attached on the stainless-steel jackets. In this experiment, one of the two heaters attached at the sides of the stainless steel was lost and hence the actual temperature of the HTS tapes might have been lower than this value. The detail of the temperature analysis is described below.

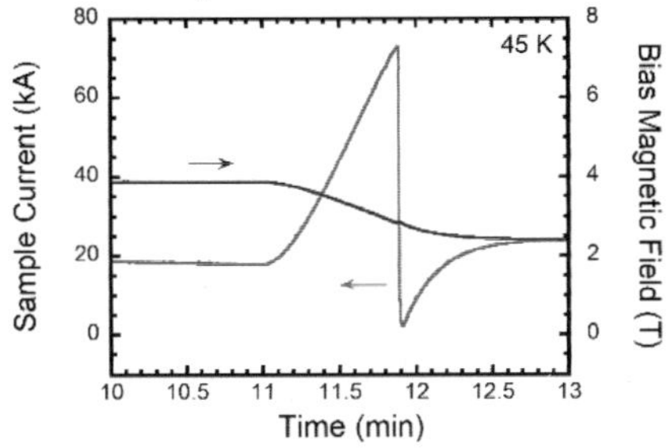

Figure 3. Waveforms of the sample current and the bias magnetic field at 45 K.

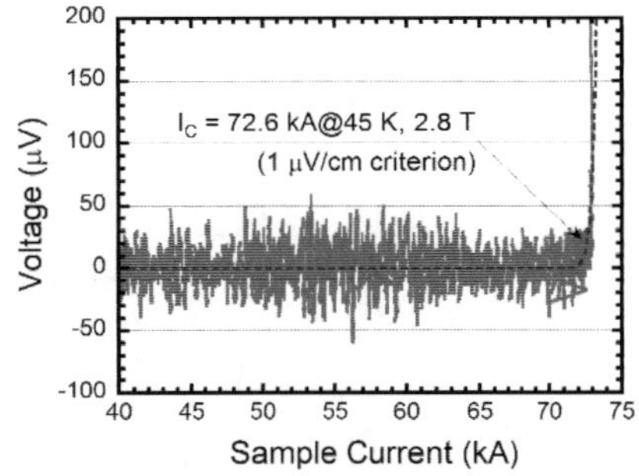

Figure 4. Voltage waveform of the continuous part of the 100-kA conductor sample as a function of the sample current.

3. Numerical analysis of the critical current

The critical current of the sample conductor was examined by the following numerical analysis to evaluate the critical current of the sample. In this analysis, the critical current of the conductor sample was calculated using the critical current density of a single HTS tape dependent on the local magnetic field, its orientation and temperature. In our first analysis [9], we considered only the local self-field perpendicular to the wide face of the tape. The calculated critical currents were in fairly good agreements with the experimentally observed values. After that, we performed the analysis considering the local magnetic field, its orientation and the bias magnetic field [14]. As a result, the critical current was calculated at various temperature and magnetic field.

3.1 Calculation model

In this experiment, it is considered that a non-uniform current distribution in the conductor sample is formed due to imbalance of the impedance among HTS tapes. The sample current, however, reached the critical current without having a premature quench. We may then consider that the inhomogeneous currents among tapes are stably redistributed when the whole conductor current approaches the critical current, owing to the high cryogenic stability of the HTS tapes. Figure 5 shows the calculation region of this analysis. The calculated cross-section was determined as a cross-section at the highest external magnetic field. The bundle of GdBCO tapes is regarded as a one-body conductor, which is divided into 405 elements. The origin of the coordinate is given at the center of the sample. At first each element is given an initial current, and the self-field of the sample at each element is calculated by the Biot-Savart's law taking into account the whole sample geometry (the calculation point of the field is at the center of each element). Then, the critical current of each element is evaluated by the magnetic field intensity (= self-field + bias field), its orientation and the sample temperature using the critical current characteristics of a single tape given by the model proposed by F. Grilli et al. [15]. If the current of any element exceeds the critical current of that element, the current is reduced to the critical current and the residual current is equally distributed to other elements as long as each of these current is lower than the critical current. This process is iterated, by developing non-uniform current distribution, until the current of every element reaches the critical current. Then, the critical current of the conductor sample is defined as the summation of the currents in all the elements.

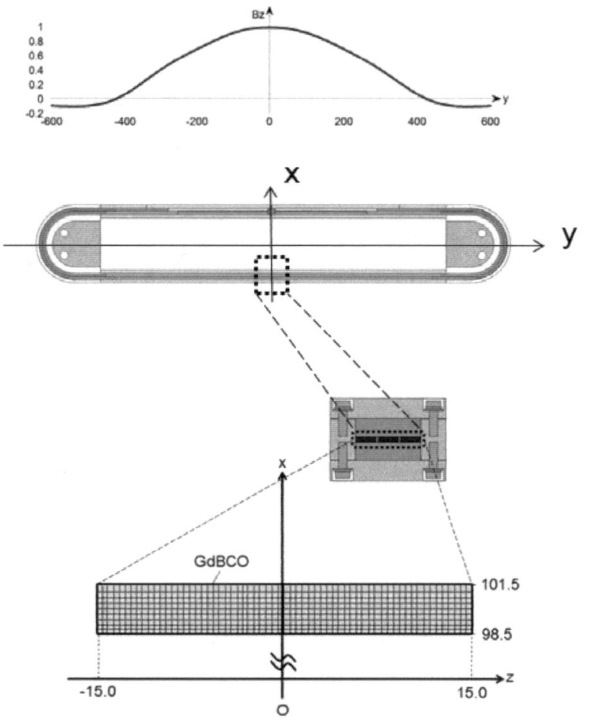

Figure 5. Schematic illustration of the model for the critical current analysis. The units of the *x*- and *z*-axis are in millimeter. The normalized external magnetic field distribution is also shown in this figure.

3.2 Critical current characteristics of HTS

In the previous study [14], the critical current characteristics were evaluated by percolation model [16]. The necessary parameters of the percolation model referenced from the literature were parameters obtained another GdBCO tape instead of the actual GdBCO tape used in this experiment. In this case, the critical current characteristics of the GdBCO tape were evaluated by the model proposed by F. Grilli et al [15] for precise analysis. The model is an useful method to describe the critical current density and the magnetic field characteristics of HTS tapes with few scaling parameters. Parameters are obtained from the actual GdBCO tape provided by the tape supplier. The characteristics are formulated by the following equations:

$$J_C(B_\parallel, B_\perp) = \frac{J_{C0}}{\left(1 + \sqrt{(kB_\parallel)^2 + {B_\perp}^2}/B_C\right)^b} \tag{2}$$

where J_{C0}, B_C, b and k are numerical parameters.

3.3 Validations of experimental results by numerical analysis for 30 kA-class HTS conductor sample

The validation of the above mentioned model was performed for the experimental results for a 30 kA class HTS conductor sample having 20 GdBCO tapes (2-rows and 10-layers). The details of the 30-kA sample were described in [9]. Figure 6 shows the critical currents of the 30-kA sample obtained by the experiment and numerical analysis as a function of the bias magnetic field at each temperature. Symbols and curves in the figure show the results of the experiment and the analysis, respectively. The experimental and the analytical results are in fairly good agreement at the lower field region, whereas there is still a discrepancy between the results at the higher field region. This may be due to the difference of the critical current characteristics of the GdBCO tape used in the experiment and the analysis.

3.4 Calculation results for the 100 kA-class HTS conductor sample

The critical current analysis for the 100-kA sample was performed. Figure 7 shows the critical currents of the 100-kA sample obtained by the experiment and numerical analysis as a function of the bias magnetic field. In this case, the experimental results at the temperature reading of 45 K are located between the calculated results of 30 K and 40 K. As stated above, the tape temperature is considered to be lower than the temperature indicated by the sensors attached on the stainless-steel jacket. So, a numerical analysis to evaluate the actual temperature of tapes was performed. Figure 8 shows schematic illustrations of a calculated region in the sample. The finite element method was used in this analysis using ANSYS^{TM}. The boundary conditions are also shown in figure 8. Physical properties were quoted by CryoComp, which provides the physical property data of various materials. A physical property of a HTS tape was determined by the mixture rule of copper and stainless-steel. The calculated results are shown in figure 9. A red zone in the figure corresponds to the position of the heater. The average temperature in the tape from the center of the straight part within 100 mm is about 38 K. This seems to support the result of the critical current analysis of the 100-kA conductor sample. Thus, it is confirmed that the 100-kA sample did not show a premature quench.

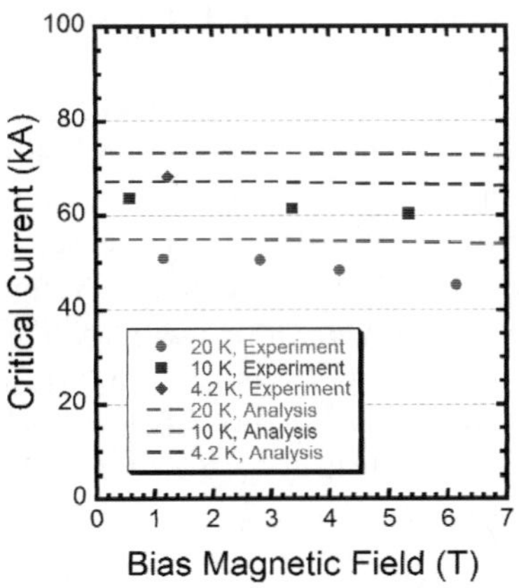

Figure 6. Critical currents of the 30-kA-class conductor sample obtained by the experiment and numerical analysis as a function of the bias magnetic field at each temperature. Symbols and curves show the results of the experiment and analysis, respectively.

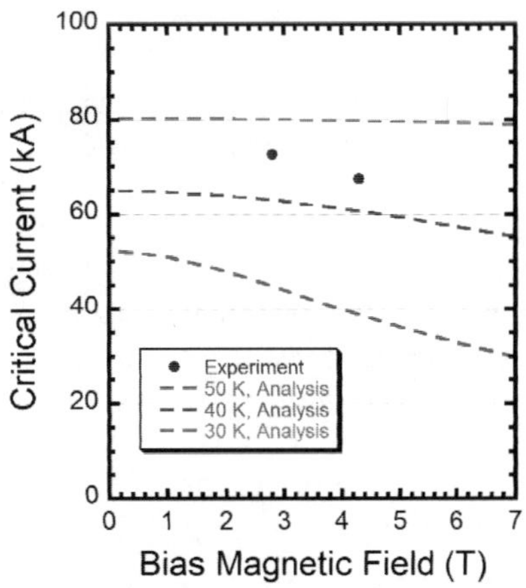

Figure 7. Critical currents of the 100-kA-class conductor sample obtained by the experiment and numerical analysis as a function of the bias magnetic field. Symbols and curves show the results of the experiment and analysis, respectively.

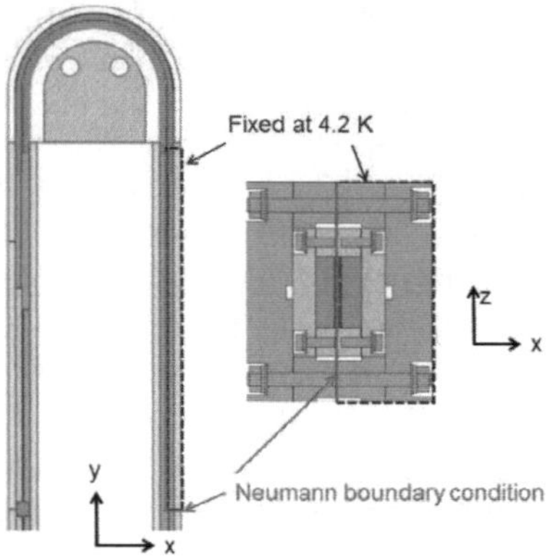

Figure 8. Schematic illustration of a calculated region of the temperature analysis. The boundary conditions were also shown in the figure.

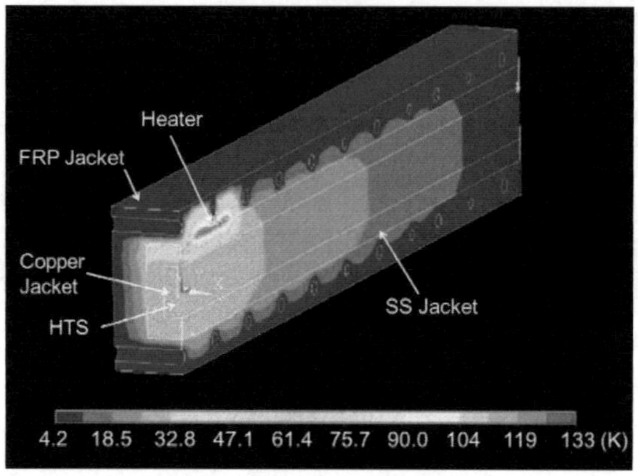

Figure 9. Temperature distribution in the conductor analyzed by ANSYS.

4. Summary

A 100 kA-class HTS STARS conductor sample as a prototype conductor for the helical fusion reactor FFHR-d1 was fabricated by simply stacking GdBCO tapes. The current was induced in the short-circuit sample by changing the bias magnetic field. The critical current of 67.4 kA and 72.6 kA were measured at the bias magnetic field of 4.3 T and 2.8 T, respectively, at the temperature of ~38 K.

A numerical analysis of the critical current of the sample was performed by solving self-consistently the spatial distribution of the magnetic field and current density in the sample. The calculated values for the 30-kA sample agrees well with the experimentally obtained values at the lower field region, whereas there is still some discrepancy at the higher field region. This may be caused by the difference of the critical current characteristics of the tape used in the experiments and the analysis. As a result of the analysis for the 100-kA sample, it is confirmed that the sample did not show a premature quench. Thus, we consider that the STARS conductor is applicable for large-current DC-operating conductors such as for the helical coils of FFHR-d1.

Acknowledgment

The authors would like to thank J. Miyazawa, T. Goto, T. Obana, S. Imagawa and K. Takahata for their valuable comments. Special thanks are given to H. Noguchi and Y. Seino.

References

[1] Sagara A, Tamura H, Tanaka T, Yanagi N, Miyazawa J, Goto T, Sakamoto R, Yagi J, Watanabe T, Takayama S and The FFHR design group 2014 *Fusion Eng. Des.* **89** 2114
[2] Imagawa S, Takahata K, Tamura H, Yanagi N, Mito T, Obana T and Sagara A 2009 *Nucl. Fus.* **49** 075017
[3] Takahata K, Mito T, Tamura H, Imagawa S and Sagara A 2007 *Fus. Eng. and Des.* **82** 1487
[4] Yanagi N, Mito T, Champailler R, Bansal G, Tamura H and Sagara A 2011 *Fus. Sci. Technol.* **60** 648-652
[5] Hashizume H, Ito S, Yagi K, Kitajima S 2002 *Fus. Eng. and Des.* **63** 449-454
[6] Bansal G., Yanagi N, Hemmi T, Takahata K, Mito T and Sagara A 2008 *Plas. and Fus. Res.* **3** S1049
[7] Terazaki Y, Yanagi N, Tomida S, Noguchi H, Natsume K, Mito T, Ito S, Hashizume H and Sagara A 2012 *Plas. and Fus. Res.* **7** 2405027

[8] Yanagi N, Terazaki Y, Ito S, Kawai K, Seino Y, Ohinata T, Tanno Y, Natsume K, Hamaguchi S, Noguchi H, Tamura H, Mito T, Hashizume H and Sagara A 2014 *IEEE Trans. Appl. Supercond.* **24** 4202805

[9] Terazaki Y, Yanagi N, Ito S, Kawai K, Seino Y, Ohinata T, Tanno Y, Natsume K, Hamaguchi S, Noguchi H, Tamura H, Mito T, Hashizume H. and Sagara A 2014 *IEEE Trans. Appl. Supercond.* **24** 4801305

[10] Ito S, Kawai K, Seino Y, Ohinata T, Tanno Y, Yanagi N, Terazaki Y, Natsume K, Hamaguchi S, Noguchi H, Tamura H, Mito T, Sagara A and Hashizume H 2014 *IEEE Trans. Appl. Supercond.* **24** 4602305

[11] Ito S and Hashizume H 2012 *IEEE Trans. Appl. Supercond.* **22** 6400104

[12] Kawai K, Ito S, Seino Y, Yanagi N, Tamura H, Sagara A and Hashizume H 2013 *IEEE Trans. Appl. Supercond.* **23** 4801704

[13] Ito S, Seino Y, Yanagi N, Terazaki Y, Sagara A and Hashizume H 2014 *Plasma and Fusion Res.* **9** 3405086

[14] Terazaki Y, Yanagi N, Ito S, Seino Y, Hamaguchi S, Tamura H, Mito T, Hashizume H and Sagara A 2015 *IEEE Trans. Appl. Supercond.* **25** 4602905

[15] Grilli F, Sirois F, Zermeno R M V, Vojenciak M 2014 *IEEE Trans. Appl. Supercond.* **24** 8000508

[16] Kiss T, Inoue M, Egashira S, Kuga T, Ishimaru M, Takeo M, Matsushita T, Iijima Y, Kakimoto K, Saitoh T, Awaji S, Watanabe K and Shiohara Y 2003 *IEEE Trans. Appl. Supercond.* **13** 2670

IOP Conf. Series: Journal of Physics: Conf. Series **871** (2017) 012100 doi:10.1088/1742-6596/871/1/012100

Detection of Local Temperature Change on HTS Cables via Time-Frequency Domain Reflectometry

Su Sik Bang[1], Geon Seok Lee[1], Gu-Young Kwon[1], Yeong Ho Lee[1], Gyeong Hwan Ji[1], Songho Sohn[2], Kijun Park[2] and Yong-June Shin[1]

[1] School of Electrical and Electronic Engineering, Yonsei University, Seoul, 03722, Korea
[2] Korea Electric Power Corporation Research Insitute, Daejeon, 34056, Korea

E-mail: *yongjune@yonsei.ac.kr*

Abstract. High temperature superconducting (HTS) cables are drawing attention as transmission and distribution cables in future grid, and related researches on HTS cables have been conducted actively. As HTS cables have come to the demonstration stage, failures of cooling systems inducing quench phenomenon of the HTS cables have become significant. Several diagnosis of the HTS cables have been developed but there are still some limitations of the experimental setup. In this paper, a non-destructive diagnostic technique for the detection of the local temperature change point is proposed. Also, a simulation model of HTS cables with a local temperature change point is suggested to verify the proposed diagnosis. The performance of the diagnosis is checked by comparative analysis between the proposed simulation results and experiment results of a real-world HTS cable. It is expected that the suggested simulation model and diagnosis will contribute to the commercialization of HTS cables in the power grid.

1. Introduction

High temperature superconducting (HTS) cables, which are expected as transmission and distribution cables in future power grid, are capable of high density power transmission with low loss and small dimension compared to conventional cables. Therefore, research of HTS cables and verification test are actively under-way to utilize the HTS cables to the power grid [1-3]. Especially, the reliability of cooling systems are significant to HTS systems because a failure of the system leads to quench phenomenon which can seriously impact on stability of the power grid. In order to prevent extreme accidents caused by the quench, diagnosis and protection techniques for HTS cables have been developed [4, 5]; however, it is difficult to reflect various types of defects in verification test of the techniques.

Modeling and simulations of HTS cables have been conducted in several ways to research in power grid [6-8]. Yet, the conventional simulations are not suitable for electrical signals with the high frequency range. In order to research the diagnostic technologies by simulation, the model of HTS cables should consider the frequency characteristic of the cables [9]. In addition, it is necessary to reflect thermal characteristic of HTS cables in the simulation model.

In this paper, a novel simulation model of a local temperature change on HTS cables is proposed to simulate various types of quench situation. The simulation model reflects thermal and frequency characteristics of HTS cables which influence accuracy of the simulation. Also, a non-destructive diagnostics applying time-frequency domain reflectometry (TFDR) [10] is proposed to detect a local temperature change point. TFDR simulation using the proposed model is explained in Section 3. In

Section 4, TFDR is applied to an experimental setup with a real-world HTS cable and the results are analyzed. Finally, conclusion is presented in Section 5.

2. Time-frequency domain reflectometry

An incident signal of TFDR is the Gaussian enveloped chirp signal which is designed on both time and frequency domain while incident signals of time domain reflectomery (TDR) and frequency domain reflectometry (FDR) are designed only on time or frequency domain. The Gaussian enveloped chirp signal is represented as:

$$s(t) = \left(\frac{\alpha}{\pi}\right)^{1/4} e^{-\frac{\alpha(t-t_0)^2}{2}+j\frac{\beta(t-t_0)^2}{2}+j\omega_0(t-t_0)} \tag{1}$$

where α is a coefficient determining the time duration, β determines the sweep rate of frequency, t_0 is center time and ω_0 is center frequency [10]. The propagation characteristic of a signal depends on structures of a cable and frequency range of the signal. Therefore, an optimal incident signal of TFDR can be designed by customizing the parameters of the Gaussian envelope chirp signal which consider characteristics of the target cable. The incident signal propagating on the cable is reflected at impedance discontinuities which have potential for defects. Thus, it is possible to detect the defects by measuring and analyzing the reflected signal.

In order to analyze a time signal s(t) on time-frequency domain, TFDR considers the Wigner-Ville distribution (WVD) represented by:

$$W(t,\omega) = \frac{1}{2\pi}\int_{-\infty}^{\infty} s^*\left(t-\frac{1}{2}\tau\right) s\left(t+\frac{1}{2}\tau\right) e^{-j\tau\omega} d\tau \tag{2}$$

Denote the WVD of the incident signal and the reflected signal as $W_s(t,\omega)$ and $W_r(t,\omega)$ respectively. Then a time-frequency cross-correlation function $C_{sr}(t)$ is evaluated as follows:

$$C_{sr}(t) = \frac{\iint_{T_s} W_s(t',\omega) W_r(t'-t,\omega)\, d\omega dt'}{\iint_{T_s} W_s(t',\omega)\, d\omega dt' \iint_{T_s} W_r(t'-t,\omega)\, d\omega dt'} \tag{3}$$

where T_s is the time duration of the incident signal. Values of the time-frequency cross-correlation function are bounded between 0 and 1 because the function is normalized by each energy of WVDs of the incident and reflected signals [10]. The results of the function represent the similarity between the incident signal and the reflected signal. Therefore, TFDR detects the faults through the evaluation of the time-frequency cross-correlation function results.

3. Simulation

In this paper, a simulation model of a HTS cable which considers the dielectric loss tangent and high frequency range is used to apply reflectometries on HTS cables suffering a local temperature change. The electrical impedances of HTS cables under ambient and superconductive condition are different. The parameters of the simulation model, permittivity, permeability, and dielectric loss tangent are obtained by analysis of scattering parameter (S-parameter) under ambient and superconductive conditions [9]. Figure 1 shows the reflectometry simulation model for a local temperature change on a

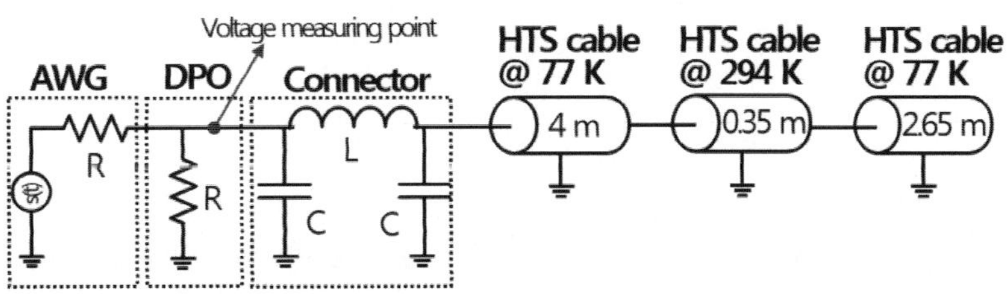

Figure 1. Reflectometry simulation model for a local temperature change on a HTS cable

Table 1. Parameters in reflectometry simulation model

Parameter		Value	Unit
Inductance L in connector		43.6	nH
Capacitance C in connector		6.4	pF
Resistance R in AWG and DPO		50	ohm
Radius of inner conductor of HTS cable		10	mm
Inner radius of outer conductor of HTS cable		18	mm
Outer radius of outer conductor of HTS cable		19	mm
Conductivity of conductor of HTS cable		6.3×10^7	S/m
Relative permeability of dielectric of HTS cable		0.69	-
Relative permittivity of dielectric of HTS cable	at 294 K	2.7	-
	at 77 K	2.664	-
Dielectric loss tangent of HTS cable	at 294 K	0.07	-
	at 77 K	0.0424	-

HTS cable. The HTS cable with 7 m of the total length is under superconductive condition without a local region from 4 m to 4.35 m. The length of the local region is determined by 5% of the total length of the HTS cable. In order to conduct accurate simulations of a local temperature change, the parameters of HTS cables as internal temperature are required. However, it is hard to obtain the accurate parameters on internal temperatures excluding room temperature and liquid nitrogen temperature because of limits in control and measurement for internal temperature without damage to the cable. Therefore, temperature of 294 K, which is room temperature, as a local temperature change is set in the simulation. In addition, the simulation model reflects the influences of the arbitrary waveform generator (AWG), digital phosphor oscilloscope (DPO), and connector to obtain the simulation results similar to actual experimental results. Each of the parameters referred by [9] in Figure 1 is shown in the Table 1.

In order to compare the performance of TFDR for detection of the local temperature change point, additionally TDR is conducted in this paper. Step signal and Gaussian enveloped chirp signal with center frequency of 90 MHz, bandwidth of 160 MHz and time duration of 30 ns are used as the incident signals of TDR and TFDR respectively. The plus and minus of the incident signal are applied to conductor and shield layer of the HTS cables respectively. The incident signal of TFDR is designed by considering

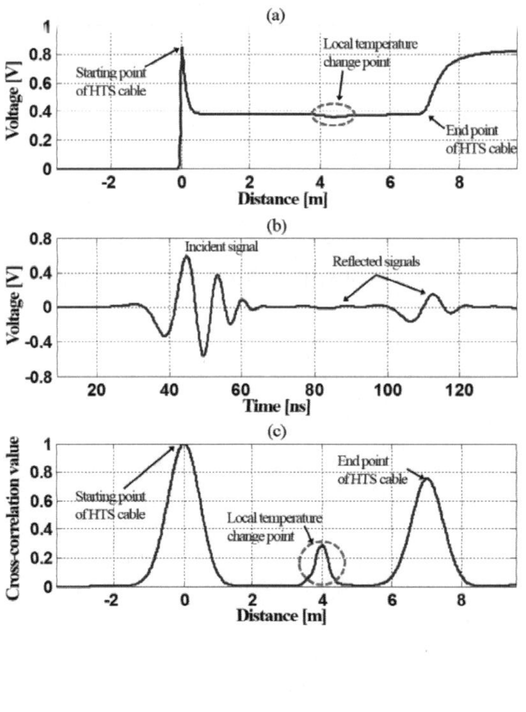

Figure 2. The simulation results; (a) TDR, (b) TFDR measured signal, (c) The values of time-frequency cross-correlation of TFDR

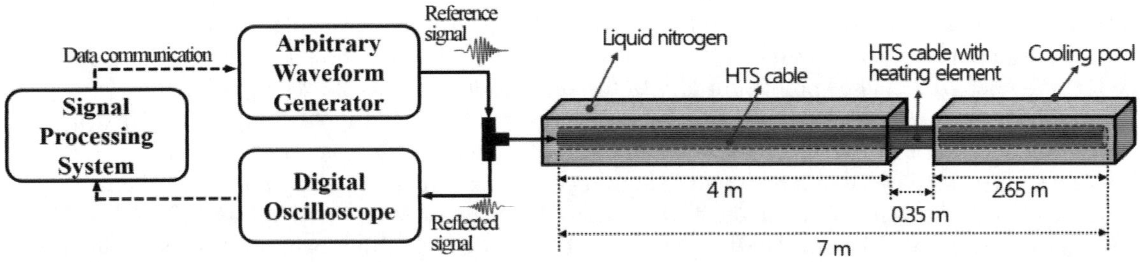

Figure 3. Experimental setup for reflectometries applying to real-world HTS cables with a local temperature change point.

structures of the HTS cable and the uncertainty principle [5]. The results of the TDR and TFDR simulations are represented in Figure 2. Figure 2(a) and (b) are the measured signals at DPO in Figure 1 using the step signal and the Gaussian envelope chirp signal respectively as incident signal. Also, the result of time-frequency cross-correlation calculated by (2) and (3) is shown in Figure 2(c). Both TDR and TFDR clearly detect the end point of HTS cable because of a large impedance discontinuity. However, although the local temperature change point can be classified finely in TDR result as represented in Figure 2(a), exact location of the local point is clearly distinguished only by the result of the time-frequency cross-correlation as shown in Figure 2(c). The distance value of peak included in red dotted line in Figure 2(c) perfectly matches the position of the local temperature change, 4 m without error, in the simulation model. In TDR result, it will be difficult to detect the local temperature change point if a noise is added in the measured TDR data. The noise analysis in TDR and TFDR will be discussed in the next section.

4. Applying to real-world HTS cable

In order to verify simulation results of TDR and TFDR for detection of a local temperature change on HTS cables, TDR and TFDR apply to a real-world HTS cable as shown in Figure 3. A 7 m single-phase 22.9 kV/50 MVA HTS cable (1G) is used. The HTS cable is cooled with liquid nitrogen in cooling pools without a local region from 4 m to 4.35 m to conduct the experiment under same condition of the simulation in Section 3. The heating element is also installed on the HTS cable sheath of the local region separated with the cooling pools to maintain the temperature of the local region.

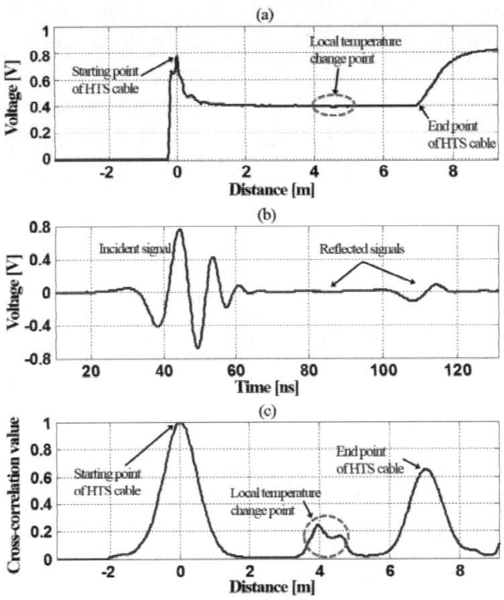

Figure 4. The experiment results; (a) TDR, (b) TFDR measured signal, (c) The values of time-frequency cross-correlation of TFDR

The incident signals used in the simulation is applied to the setup of real-world HTS cable. The experiment results are represented in Figure 4. The end point of the HTS cable is both detected by TDR and TFDR, but only the time-frequency cross-correlation of TFDR as shown in Figure 4(c) can display the local temperature change point similarly to the simulation result as shown in Figure 2.

The reason of the peak at staring point of HTS cable as shown in Figure 2(a) and Figure 4(a) is the connection accessories between the devices and the HTS cable. Reflectometry accessories and devices generate noise signals in the experiment. Therefore, the peak shape in the TDR experiment result as shown in Figure 4(a) is noisy unlike the TDR simulation result as represented in Figure 2(a). In addition, in measured results of TDR and TFDR as shown in Figure 4(a) and (b), it is hard to distinguish the local temperature change point from the noise. On the other hand, as shown in Figure 4(c), the result of time-frequency cross-correlation calculated by (2) and (3) using the data of Figure 4(b) clearly detects the local temperature change point. The distance value of peak included in red dotted line in Figure 4(c) is 3.934 m which corresponds to 1.65% of error rate of detection accuracy.

The time-frequency cross-correlation value of the end point of HTS cable in the experiment is smaller than that in the simulation because of the noise and attenuation from the connection accessories. Also, the shape of the graph at the local temperature change point in Figure 4(c) is distorted. It is possible to produce two peaks at the local temperature change because the reflected signal generated by the temperature change is distorted by the noise. However, this paper verifies that TFDR can detect the local temperature change point where TDR cannot distinguish from the noise in the real-world HTS cable.

5. Conclusion

In this paper, a non-destructive diagnostic technology, TFDR, is introduced for detection of local temperature change points on HTS cables. A simulation model of HTS cable with a local temperature change is proposed and the simulation considers the characteristics of electrical wave propagation and the effects of devices, AWG, DPO and connector. TFDR for detection of local temperature change points on HTS cables is examined by the simulation results. In addition, the performance of TFDR is verified using experimental setup of a real-world HTS cable which is similar to the simulation. The resolution and compatibility of TFDR for distinction of the local temperature change region is confirmed by comparative analysis with TDR. In this paper, the room temperature is considered as a local temperature change. If a lower the temperature change occurs, a lower reflected signal is generated. In theory, a low magnitude of reflected signal has not an effect on the performance of the proposed technique because the time-frequency cross-correlation function is normalized by each energy of WVDs of the incident and reflected signals as represented in equation (3). However, in order to apply for a low temperature changes on real-world HTS cable systems, other techniques such as noise filtering and compensation filter will be required to measure the low magnitude of reflected signal because of severe noise and distortion of reflected signal in the field. By utilizing the proposed simulation model, numerous kinds of failure as well as local temperature change on HTS cables, which is hard to emulate in real-world test-bed, are able to be simulated. It is expected that the analysis of the simulation results will improve the proposed diagnostics and the proposed technique will contribute to prevent accidents by quench in HTS cables in future power grid.

Acknowledgments

This work was supported by the National Research Foundation of Korea (NRF) grant funded by the Ministry of Science, ICT & Future Planning, #NRF-2017R1A2A1A05001022. Also, this research was supported by Korea Electric Power Corporation Research Institute.

References

[1] Sohn S H, Hwang S D, Lim J H, Yim S W, Hyun O B, Kim H R, Yatsuka K, Isojima S, Watanabe M, Ryoo H S, Yang H S and Kim D L 2007 *Physica C* **463-5** 1146-49

[2] Sohn S H, Lim J H, Yang B M, Lee S K, Jang H M, Kim Y H, Yang H S, Kim D L, Kim H R, Yim S W, Won Y J and Hwang S D 2010 *Physica C* **470** 1567-71

[3] Lim J H, Sohn S H, Yang H S, Hwang S D, Kim D L, Ryoo H S and Choi H O 2010 *Physica C* **470** 1597-600
[4] Lee G S, Kwon G -Y, Chang S J, Lee C -K, Bang S S, Lee Y H, Park J B and Shin Y -J 2015 *Proc. Int. Conf. on Insulated Power Cables (Versilles, France)* C 7-4
[5] Lee G S, Kwon G -Y, Bang S S, Lee Y H, Chang S J, Sohn S H, Park K and Shin Y -J 2016 *IEEE Trans. Appl. Supercond.* **26** 5401005
[6] Kim J G, Kim A -R, Kim D, Park M, Yu I -K, Cho J, Sim K -D, Kim S, Lee J K and Won Y -J 2010 *IEEE Trans. Appl. Supercond.* **20** 1284-87
[7] Del-Rosario-Calaf G, Lloberas-Valls J, Sumper A, Granados X and Villafafila-Robles R 2013 *IEEE Trans. Appl. Supercond.* **23** 5401204
[8] Li J, Zhao Z, Shu B, Han X, Ma X, Bian B, Liang Z, Li J and Jiang W 2014 *IEEE Trans. Appl. Supercond.* **24** 4803205
[9] Bang S S, Lee G S, Kwon G Y, Lee Y H, Chang S J, Lee C -K, Sohn S H, Park K and Shin Y -J 2016 *Physica C* **530** 142-6
[10] Shin Y -J, Powers E -J, Choe T -S, Hong C -Y, Song E -S, Yook J -G and Park J B 2005 *IEEE Trans. Instrum. Meas.* **54** 2493-500

IOP Conf. Series: Journal of Physics: Conf. Series **871** (2017) 012101 doi:10.1088/1742-6596/871/1/012101

Basic study for a large AC current supply with a single phase air-core Bi2223 high temperature superconducting transformer

N Nanato, N Kishi, Y Tanaka, M Kondo

Graduate School of Natural Science and Technology, Okayama University,
3-1-1, Tsushima Naka, Kita-ku, Okayama, 700-8530, Japan

nanato@okayama-u.ac.jp

Abstract. The authors have been developing a compact power supply with a single-phase Bi2223 high temperature superconducting (HTS) transformer. The conventional transformer has an iron-core for enhancing magnetic coupling between its primary coil and secondary one. However, the iron-core has great majority of size and weight of the transformer and therefore it is desirable to be removed for a smaller and lighter transformer. In this paper, the authors propose an air-core HTS transformer for a more compact power supply than the conventional one. As experimental results, it is shown that appropriate design of the air-core transformer has a possibility to decrease the weight and volume of the large AC current supply.

1. Introduction

It is very important to grasp current conduction characteristics of high temperature superconductors for developing high temperature superconducting (HTS) apparatus such as HTS electric cables, motors and SMES. HTS tapes used in the apparatus have large critical currents and therefore a large current supply is needed to grasp the current conduction characteristics. However, commercial power supplies especially for large AC current are generally very large and heavy and are difficult to handle in ordinary laboratories. The authors have been developing a compact power supply with a single-phase Bi2223 HTS transformer [1]. Some HTS transformers have been developed for large power supply such as a distribution power grid [2] and railway rolling stock [3]. The transformer for the compact power supply is different from these transformers in that it is used for very low voltage and large current. From viewpoint of the very low voltage, in this paper, the authors study a possibility of an air-core transformer which has no iron-core resulting in a more compact supply. The air-core transformer needs larger excitation current than that of a general iron-core transformer, however the current is originally small due to the very low voltage. Also no copper loss is consumed in the transformer unlike a general transformer with copper windings. Therefore the transformer has a possibility to make the supply small and light without increasing loss. The authors show comparison of current carrying properties of an air-core transformer and a conventional iron-core one. As experimental results, it is shown that appropriate design of the air-core transformer has a possibility to decrease the weight and volume of the large AC current supply.

2. A compact power supply with a single-phase Bi2223 HTS transformer

Figure 1 shows an overview of a proposed compact power supply with a single-phase Bi2223 HTS transformer. Number of turns of its primary coil is more than those of its secondary coil. A power supply connected to the primary coil of the transformer supplies small current i_1 to the primary coil and then the secondary coil outputs large current i_2 to a load Z which is a superconducting wire. The primary power supply is a variable voltage and variable frequency (VVVF) one and the transformer can output a large current with variable frequency.

Figure 2 shows an overview of a conventional HTS transformer with an iron-core. The primary and secondary coils are Bi2223 HTS coils with cylindrical and solenoidal shapes. They are wound with DI-BSCCO Type H fabricated by Sumitomo Electric Industries, Ltd. Specifications of the wire is shown in Table 1. Both coils are co-axially arranged and have an iron core. Only the coils were cooled in liquid nitrogen and the iron core was always in room temperature air. Specifications of the transformer are shown in Table 2 [1]

Figure 3 shows an overview of the iron-core. Its shape was decided so that space for the cooling container was secured. The iron-core was made by layering oriented steel sheets. The steel sheet has thickness of 0.3 mm, density of 7,650 kg/m^3, iron loss less than 1.1 W/kg at 1.7 T and 50 Hz, and saturation magnetic flux density of over 1.88 T. Weight of the iron-core is 10 kg which is great part of that of the transformer 12.6 kg.

Figure 4 shows an overview of a proposed air-core HTS transformer and its specifications are shown in Table 2. Its structure is almost same as that of the iron-core transformer but it has no iron-core and its cooling container has no cylindrical bore. Weight and volume of the air-core transformer are respectively 2.6 kg and 4.2×10^{-3} m^3, which are only about 20 % and 36 % of those of the iron-core transformer, respectively.

Table 1. Specifications of the DI-BSCCO Type H wire

Width	4.3 mm
Thickness	0.23 mm
Critical current (at 77 K, Self-field)	143 A

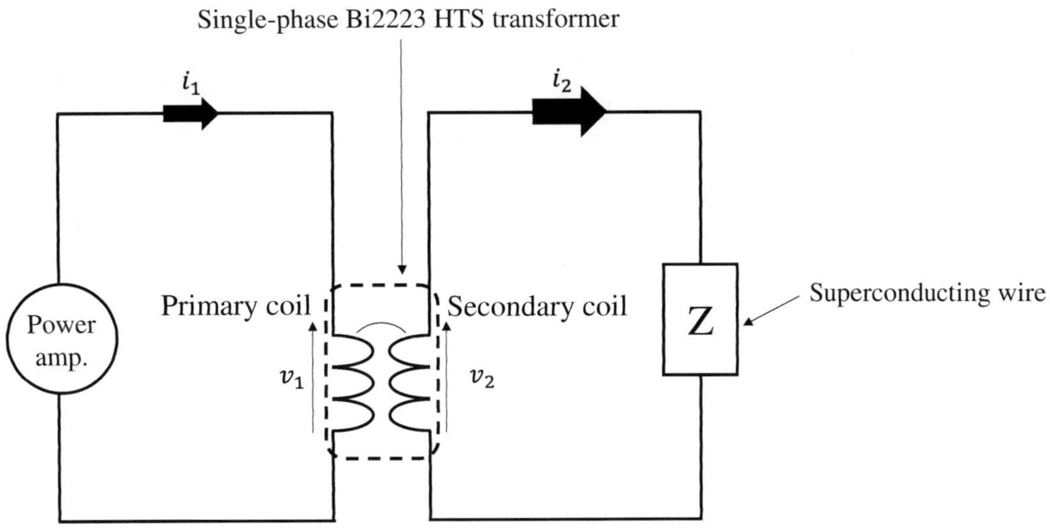

Figure 1. An overview of a compact power supply with a single-phase Bi2223 HTS transformer.

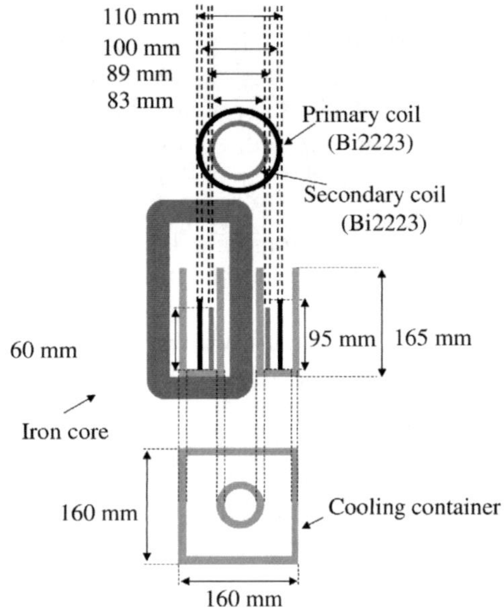

Volume : 11.8×10^{-3} m^3
Weight : 12.6 kg

Figure 2. An overview of an HTS transformer with an iron-core.

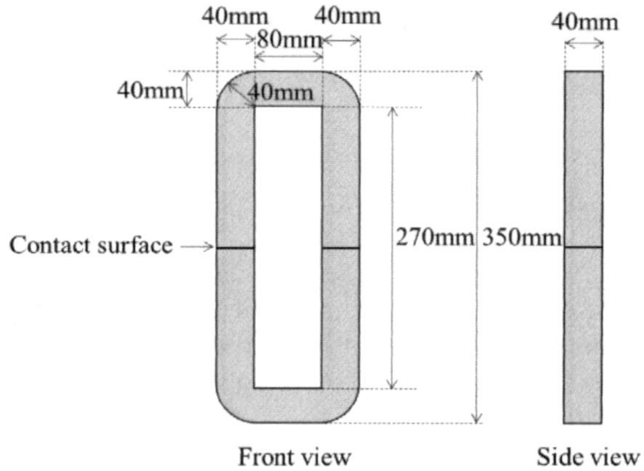

Weight: 10.0 kg

Figure 3. An overview of the iron-core.

Table 2. Specifications of a single-phase Bi2223 HTS transformer.

	Primary coil	Secondary coil
Number of turns	120	5
Inductance		
Iron-core	448 mH	760 μH
Air-core	1208 μH	2.67 μH
I_C	143 A	760 A

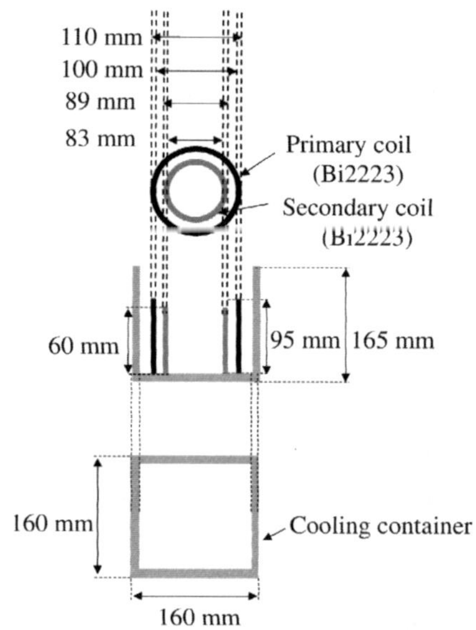

Volume : 4.2×10^{-3} m^3
Weight : 2.6 kg

Figure 4. An overview of an air-core HTS transformer.

3. Current carrying tests

The authors carried out experiments to measure current ratio i_2/i_1 and losses of the air-core transformer compared with the iron-core one.

A test circuit is shown in Figure 5. A VVVF power amplifier supplied current to the transformer. A resistor (= Z, 13 μΩ) was connected to the secondary coil. The transformer was cooled at 77 K in liquid nitrogen.

Losses of the transformers were measured as follows.

$$P_1 = \frac{1}{T}\int_0^T v_1 \times i_1 dt \tag{1}$$

$$P_2 = \frac{1}{T}\int_0^T v_2 \times i_2 dt \tag{2}$$

$$P_T = P_1 - P_2 \tag{3}$$

P_1 is summation of loss in a transformer and that in a load resistor, P_2 is loss in the load and P_T is loss in the transformer. The losses were measured for secondary currents of 10 A to 200 A, 60 Hz, and 60 Hz to 1 kHz, 200 A. As shown in Figure 5, the currents were measured by non-contact type current sensors (Hall sensors), and data of the currents and voltages were acquired by a digital oscilloscope through isolation amplifiers.

Experimental results for measuring current ratio are shown in Figures 6 and 7. Current ratio i_2/i_1 was 24 for the iron-core transformer shown in Figure 6 and 12.5 for the air-core transformer shown in Figure 7. The current ratio of the air-core transformer was about half of that of the iron-core one because magnetic coupling of the air-core transformer was reduced due to no iron-core. Therefore the air-core transformer needed about twice primary current for same secondary current compared with the iron-core one.

Experimental results for measuring losses are shown in Figures 8 and 9. Figure 8 shows losses of the air-core transformer and iron-core one as functions of the secondary current and Figure 9 shows losses of those transformers as functions of frequency. The air-core transformer needed about twice primary current as mentioned above, however, the loss of the air-core transformer was slightly different from that of the iron-core one. The iron-core transformer has iron loss in its iron-core and AC loss in its superconducting winding and the air-core transformer has only AC loss. Moreover, leakage magnetic flux of the air-core transformer is larger than that of the iron-core one. Therefore it is supposed that AC loss in the air-core transformer was larger than that in the iron-core one and then total loss in the iron-core transformer was almost same as that in the air-core one. Hereafter the authors will analyze these losses in detail.

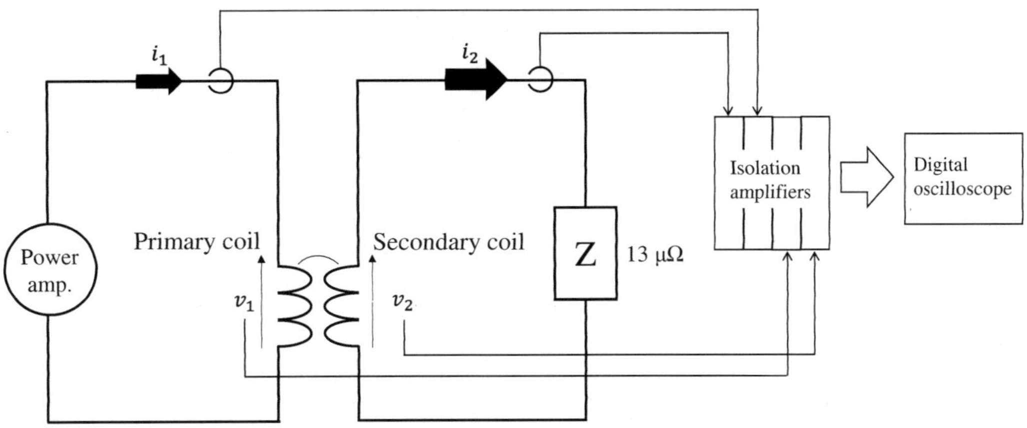

Figure 5. A test circuit.

IOP Conf. Series: Journal of Physics: Conf. Series **871** (2017) 012101 doi:10.1088/1742-6596/871/1/012101

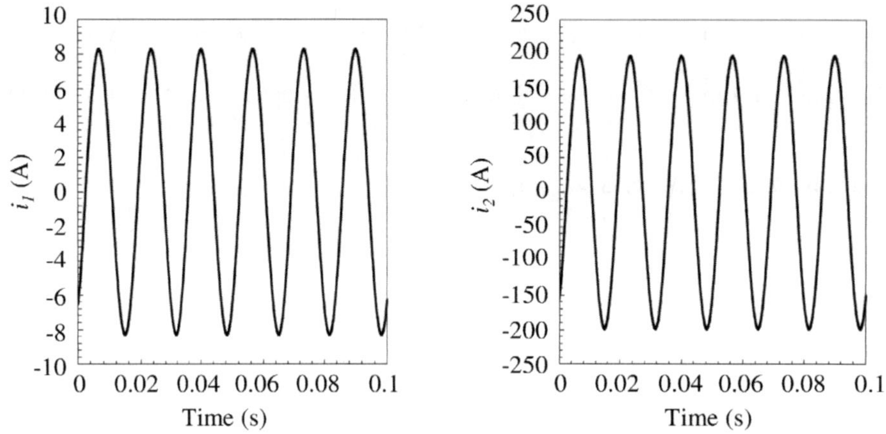

(a) i_1 (b) i_2

Figure 6. Experimental results for measuring current ratio of the iron-core transformer.

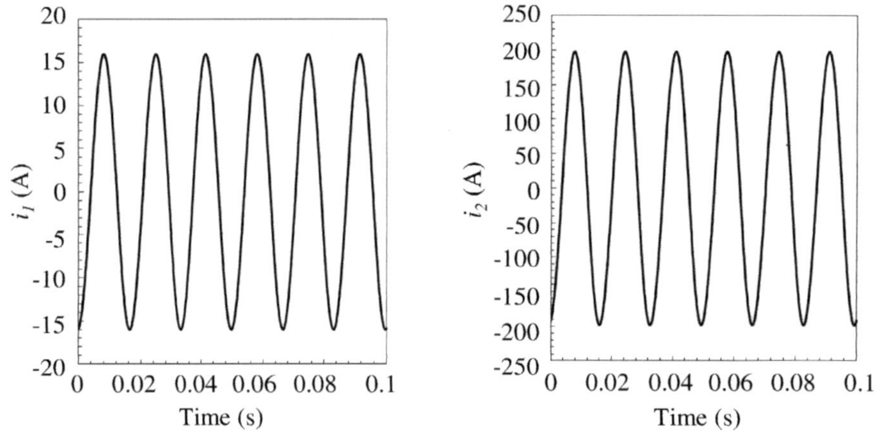

(a) i_1 (b) i_2

Figure 7. Experimental results for measuring current ratio of the air-core transformer.

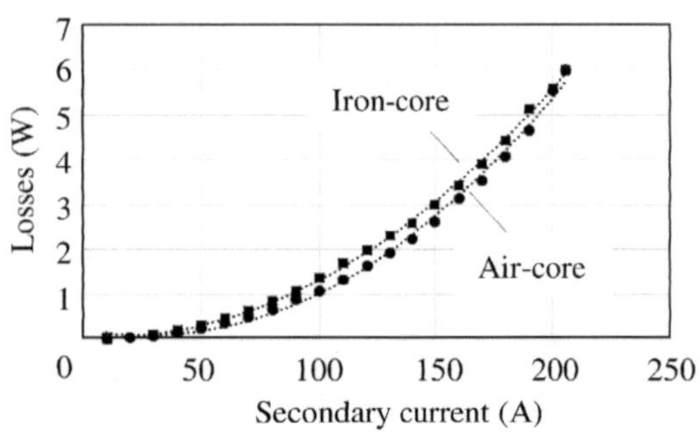

Figure 8. Losses of the air-core transformer and iron-core one as functions of the secondary current.

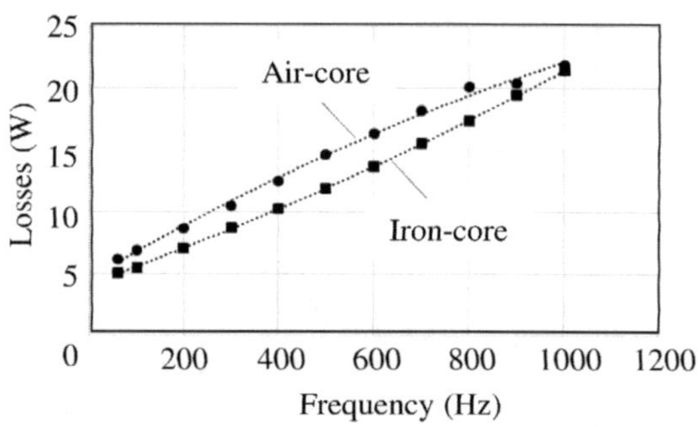

Figure 9. Losses of the air-core transformer and iron-core one as functions of frequency.

This study shows that appropriate design of the air-core transformer has a possibility to decrease the weight and volume of the large AC current supply because the loss of the transformer with and without the iron core in this study was equivalent. Therefore the authors will study the appropriate design hereafter.

4. Conclusion

The authors have been developing a compact power supply with a single-phase Bi2223 HTS transformer for outputting a large AC current. In this paper, the authors measured current carrying characteristics of an air-core HTS transformer. Compared with an iron-core transformer, the air-core transformer with same primary and secondary coils as those of the iron-core one had almost same loss and 36 % volume and 20 % weight. That shows the air-core transformer is supposed to be useful for a highly efficient and compact supply. Hereafter the authors will analyze losses in the air-core transformer in detail and study on the appropriate design of the air-core transformer for a more compact supply.

Acknowledgements

This work was supported by JSPS KAKENHI Grant Number 15K18020.

References

[1] Nanato N, Kobayashi Y 2014 Quench Detection and Protection for High Temperature Superconducting Transformers by Using the Active Power Method *Physics Procedia* **58** 264

[2] Iwakuma M et. al. 2015 Development of a 3φ-66/6.9 kV-2 MVA REBCO Superconducting Transformer *IEEE Trans. Appl. Supercond.* **25** Art. no. 5500206

[3] Kamijo H et. al. 2007 Tests of Superconducting Traction Transformer for Railway Rolling Stock *IEEE Trans. Appl. Supercond.* **17** 1927

IOP Conf. Series: Journal of Physics: Conf. Series **871** (2017) 012102 doi:10.1088/1742-6596/871/1/012102

Volume reduction of cesium contaminated soil by magnetic separation − Pretreatment of organic matters −

Hiroki Horie[1], Kazuki Yukumatsu[1], Fumihito Mishima[2] Yoko Akiyama[1], Shigehiro Nishijima[1], Tomio Sekiyama[3], Seiichiro Mitsui[3] and Mitsugu Kato[3]

[1] Graduate School of Engineering Osaka university, 2-1 Yamadaoka, Suita, Osaka, Japan

[2] Fukui University of Technology 3-6-1 Gakuen, Fukui, Fukui, Japan

[3] Japan Atomic Energy Agency (JAEA) 2-10 Fukasaku, Miharumachi, Tamuragun, Fusima, Japan

E-mail address horie@qb.see.eng.osaka-u.ac.jp

Abstract. By the accident of Fukushima Daiichi Nuclear Power Plant, a large amount of soil was contaminated by radioactive cesium. We developed a new volume reduction method of contaminated soil combining classification and magnetic separation. In magnetic separation, 2:1 type clay minerals, which adsorb cesium strongly and show paramagnetism, are removed from soil suspension of silt and clay, and then the contaminated soil can be separated into two groups that is high and low dose soil. However, there is an issue that the clay aggregates induced by organic matters prevent 2:1 type clay minerals from selective separation magnetically. The purpose of this study is to disperse aggregates by means of the alkaline K_2CO_3 solution treatment for selective separation of 2:1 type clay minerals. Firstly, particle size distribution was measured and the dispersion by K_2CO_3 treatment was investigated. Moreover, the radioactivity of passed soils after magnetic separation was measured to investigate the effect of dispersion treatment before magnetic separation. The result showed the possibility of more selective separation for 2:1 type clay minerals by treatment of organic matters.

1. Introduction

By the accident of Fukushima Daiichi Nuclear Power Plant, many kind of radioactive substances were released and discharged around the plant, which caused increase in air dose rate. Cs^{137} is one of the radioactive substances that has a long half-life (approximately 30 years) and released amount was large. In Fukushima prefecture, the surface of the soil was removed by the decontamination works to reduce air dose rate. According to Ministry of the Environment, the total quantity of contaminated soils discharged through the work is expected to be 16-22 million m^3[1]. Therefore, volume reduction of removed contaminated soil is needed.

We examined the volume reduction system combining wet classification and magnetic separation with superconducting magnet. The flow diagram of the decontamination process is shown in figure 1. Firstly, in the process of classification, soil is separated depending on their particle size. The soil radioactivity is reduced by removing silt and clay minerals from the soil. Secondly, in the process of magnetic separation with superconducting magnet, silt and clay minerals are separated into low dose soil mainly consists of 1:1 type clay minerals and high dose soil mainly consists of 2:1 type clay minerals. However, there is an issue that clay aggregates by organic matters could prevent 2:1 type clay minerals from selective separation magnetically [2]. The purpose of this study is to propose a method of disaggregation of 2:1 type clay minerals and other soil components. 2:1 type clay minerals show paramagnetism. Therefore, a superconducting magnet as high magnetic field source is required for this method.

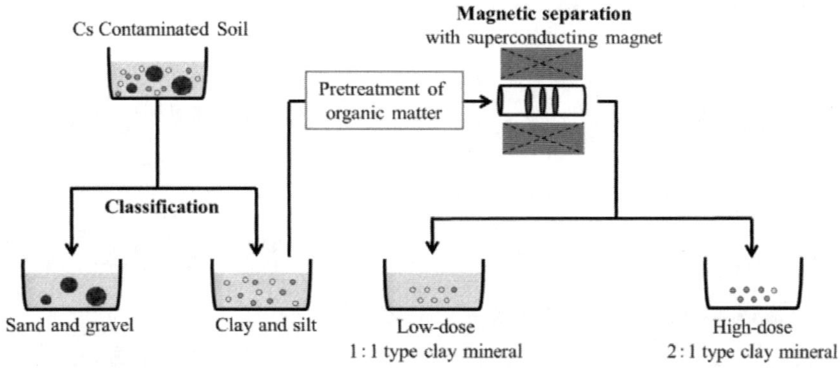

Figure 1. Flow diagram of soil decontamination by magnetic separation.

2. Theory

2.1. Properties of clay minerals

Clay minerals are divided into two groups; 1:1 and 2:1 type clay. They are constructed by the combination of the layers of Al octahedral and Si tetrahedral structure. Two kinds of minerals are much different in Cs ions adsorption ability [3].

1:1 type clay mineral mainly adsorbs Cs ions by variable charge on the surface and its Cs adsorption ability is low. On the other hand, 2:1 type clay mineral adsorbs Cs ions by negative permanent charge and its Cs adsorption ability is high. In addition, 2:1 type clay mineral has cavities called "frayed edge site" which can fix Cs ions strongly in the interlayer space [4].

Moreover, these clay minerals are different in magnetic susceptibility. The magnetic susceptibilities of kaolinite and vermiculite, which are 1:1 and 2:1 type clay minerals respectively are shown in table 1 [5]. Kaolinite and vermiculite show diamagnetic and paramagnetic properties, respectively.

Thus, selective separation of 2:1 type clay mineral can effectively reduce the dose of silt and clay fraction. The separation of these clay minerals can not be realized without solenoidal superconducting magnet for two reasons. One is that the superconducting magnet can generate high magnetic field and magnetic field gradient to capture paramagnetic 2:1 type clay mineral. The other is that the superconducting magnet has wide magnetic field area to separate a large amount of contaminated soil. The magnet makes it possible to reduce the radioactivity of soil passed through the magnet by removing 2:1 type clay minerals.

Table 1. Magnetic susceptibility of clay minerals. [5]

	Volume Susceptibility (-)
Kaolinite (1:1 type clay mineral)	-6.0×10^{-5}
Vermiculite (2:1 type clay mineral)	7.0×10^{-4}

2.2. Reaction between organic matters and clay minerals

Minerals and organic matters such as humic substance are usually aggregated in the soil. Reaction between humic substance and clay minerals is shown in figure 2 and 3. Humic substance in the soil has a large number of hydroxyl groups and carboxyl groups. These functional groups are negatively charged by dissociation in water. Therefore, humic substance is combined with metal ions such as Al^{3+}

to make insoluble substance in neutral solution as figure 2. On the other hand, aggregates of humic substance and clay minerals is disaggregated in alkaline solution and Al^{3+} combined with organic matters changes into $(Al(OH)_4)^-$. In this way, humic substance is dissolved into the solution as shown in figure 3[6]. For that reason, it can be presumed that dispersion of soils is improved by treatment of soil with alkaline solution. In this study, K_2CO_3, whose solution shows alkaline, was used as dispersing agent for the soils. K_2CO_3 is used as a food additive and fertilizer that have low environmental burden.

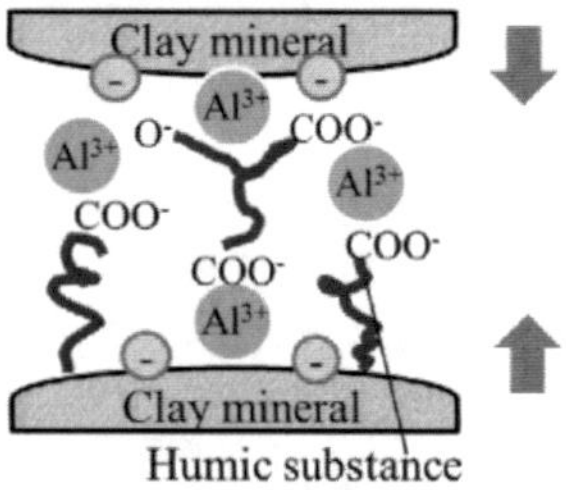

Figure 2. Reaction between organic matters and clay mineral in neutral solution.

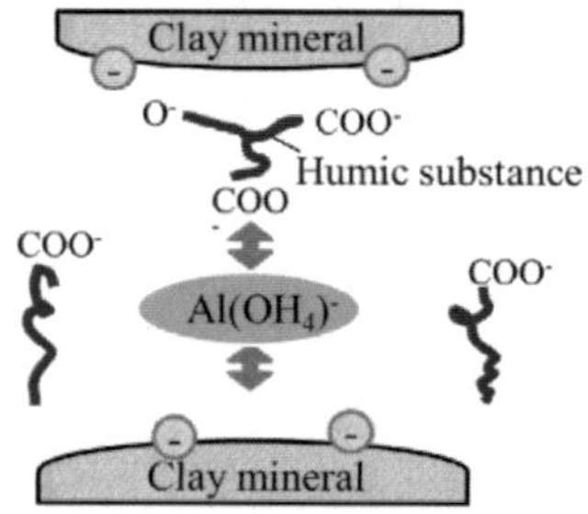

Figure 3. Reaction between organic matters and clay mineral in alkaline solution.

2.3. High gradient magnetic separation (HGMS)

HGMS is the method that enables to separate and capture ferromagnetic or paramagnetic particles on the magnetic filters made of ferromagnetic wires by making sharp magnetic gradient around them. Magnetic force and drag force acting on the particles in the fluid in magnetic separation method is shown in equations (1) and (2).

$$\mathbf{F_M} = \frac{4}{3}\pi r_p^3 (\mathbf{m} \cdot \nabla)\mathbf{B} \quad (1)$$

$$\mathbf{F_D} = 6\pi \eta r_p (\mathbf{v_f} - \mathbf{v_p}) \quad (2)$$

Here, $\mathbf{F_M}$ [N] and $\mathbf{F_D}$ [N] indicates magnetic and drag force, $\mathbf{m}$ [A m^{-1}], $\mathbf{B}$ [T], η [Pa·s], r_p [m], $\mathbf{v_f}$[m s^{-1}] and $\mathbf{v_p}$[m s^{-1}] respectively indicates the magnetization of the particle, the magnetic field strength, the viscosity of the solvent, the radius of the particle, the velocity of the solvent and the particle. However, aggregates of paramagnetic and diamagnetic particles might prevent them from selective separation as shown in figure 4(a). In order to capture only paramagnetic 2:1 clay particles, it is necessary to disaggregation as shown in figure 4(b).

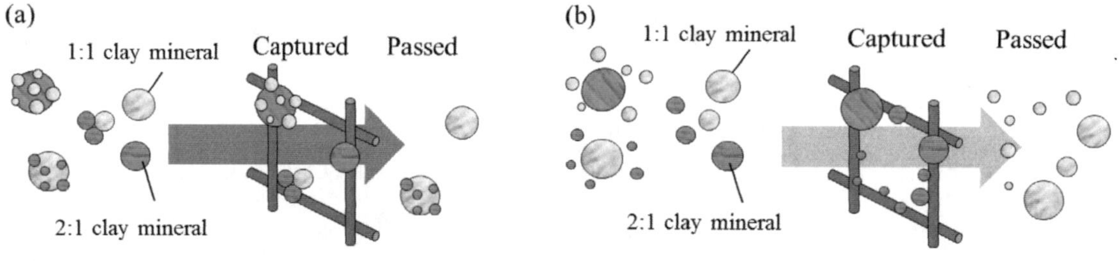

Figure 4. Schematic illustration of effect of disaggregation of clay mineral.

3. Treatment of organic matters

3.1. Materials and method

In order to investigate the dispersion effect of soil by organic matters treatment, non-contaminated soil was used. The soil was rich in organic matters and was collected from Tochigi prefecture. The particles smaller than 75 μm were separated from the soil with a grading sieve. 1 g of the classified soil (under 75 μm) was added into 100 mL of pure water and 0.1 mol L^{-1} K_2CO_3, respectively. The soil suspensions were shaken at 120 rpm for 1 hour. The particle size distribution for each suspension was measured by particle size analyser (LA-920, HORIBA). As a reference experiment, H_2O_2 treatment used in the sequential extraction of soil, that can resolve the organic matters absolutely, was also conducted. The classified soil was added into 50 mL of 30 wt% H_2O_2 (pH was adjusted to 2 with HNO_3), keeping the temperature stable with a water bath for 2 hours at 85 ℃.

3.2. Results and discussions

Figure 5 shows the particle size distribution before and after K_2CO_3 and H_2O_2 treatments. The particle distribution in water had a peak at around 30 μm. In case of organic matters treatment, the height of the particles around 30 μm decreased and smaller particles increased. The result indicates that particle aggregates was dispersed by the treatment. In addition, the distribution in H_2O_2 was almost as same as that in K_2CO_3. The result showed that K_2CO_3 has same dispersion ability as H_2O_2 and that the soil particles can be effectively dispersed by K_2CO_3.

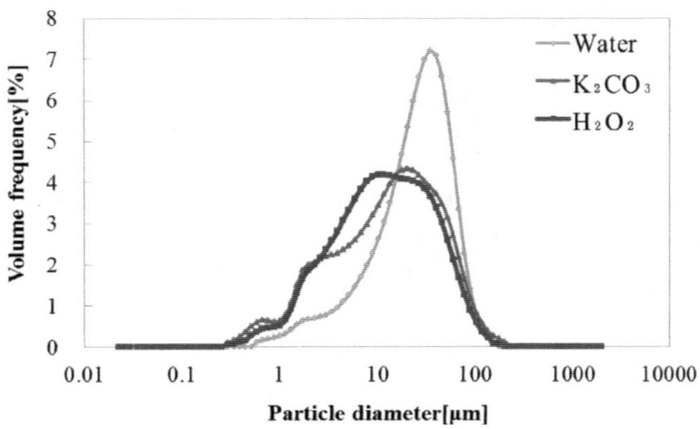

Figure 5. Particle size distribution of soils by treatment of organic matters.

4. Magnetic separation of Cs contaminated soil

4.1. Materials and method

It became clear by the above experiment that clay aggregates by organic matters can be dispersed by K_2CO_3 solution treatment. In this section, the effect of the treatment on the magnetic separation efficiency was examined. The soil was collected in agricultural field in Fukushima prefecture. Here, the particles smaller than 75 μm (silt and clay) were separated from contaminated soil with a grading sieve. After the classification, the suspension was adjusted at solid-liquid ratio of 1:100. In order to disperse aggregates, K_2CO_3 was added into the suspension whose concentration was adjusted to be 0.1 mol L^{-1}. Then, the suspension was shaken for 1 hour at 120 rpm and was subjected to the magnetic separation. The conditions of magnetic separation is shown in table 2. For reference, water suspension without treatment of organic matters was also prepared and the magnetic separation was performed in the same condition. Passed and captured soil after magnetic separation were respectvely divided into 4 size fractions (0-2 μm, 2-10 μm, 10-30 μm and 30-75 μm) with grading sieves. Mass and radioactivity for each fraction were measured by NaI scintillation detector (EMF211 EMF Japan Co., Ltd.).

Table 2. Experimental condition of magnetic separation.

Filter material	SUS430
Wire diameter of filter	0.34 mm
Applied magnetic field	6 T
Number of mesh	20 mesh
Number of filter	100

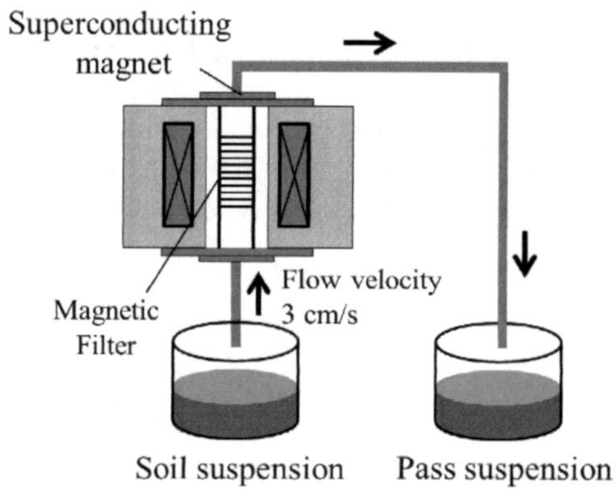

Figure 6. Schematic setup of experimental system of magnetic separation.

4.2. Results and discussions

Figure 7 shows change in radioactivity of each fractions of the passed suspension. Without the treatment of organic matters, the radioactivity in the fraction of 30-75 μm and 10-30 μm decreased by 44 % and 34 % compared with that before magnetic separation. After K_2CO_3 treatment, radioactivity decreased by 63 % and 40 % for each fractions. It indicates that the selective separation of 2:1 type clay minerals was enhanced by K_2CO_3 treatment. On the other hand, in the fraction of 2-10 μm and 0-2 μm, the radioactivity was not decreased by the magnetic separation both in water and in K_2CO_3 solution. It indicates that 2:1 type clay minerals having smaller size of 10 μm could not be captured in the magnetic separation. Therefore, it is necessary to improve the magnetic separation efficiency to capture smaller particles than 10 μm.

Moreover, figure 8 shows change of mass ratio of the passed suspension. The mass ratio in the fraction of 0-2 μm increased by particle dispersion treatment. It indicates that it is more important to capture 0-10 μm particles for volume reduction of the soil after particle dispersion treatment. Thus, the treatment of organic matters and the improving magnetic separation efficiency could make it possible to reduce radioactivity.

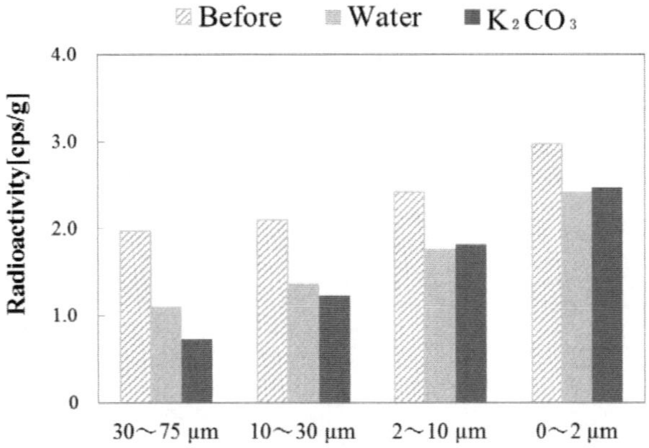

Figure 7. Radioactivity of passed soil.

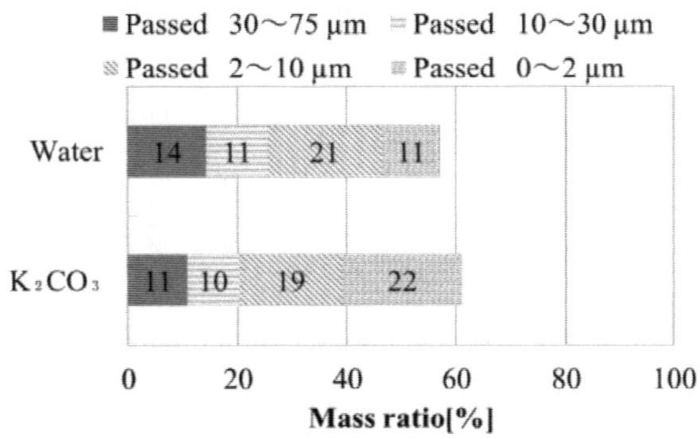

Figure 8. Mass ratio of passed soil.

5. Conclusion

In this study, we aimed to disperse clay aggregates with organic matters for selective separation of 1:1 and 2:1 type clay minerals magnetically. The particle size distribution of the particles of silt and clay in K_2CO_3 indicated that aggregates can be dispersed effectively by treatment of organic matters. In the fraction larger than 10 μm, 1:1 and 2:1 type clay mineral could be separated selectively by magnetic separation after treatment of organic matters. On the other hand, in the fraction smaller than 10 μm, 2:1 type clay might be passed through the magnetic filters even after K_2CO_3 treatment. It is concluded that the pretreatment of organic matters in combination with improvement of magnetic separation will be achieved effective volume reduction of contaminated soil. In the practical application, the solenoidal superconducting magnet which has larger bore diameter could enable to increase the procssing amount of soil.

References

[1] Ministry of the Environment, Japan, 2016. "Progress on Off-site Cleanup and Interim Storage in Japan,"
[2] H Mukai, T Hatta, H Kitazawa, H Yamada, T Yaita, T Kogure 2014 *Environmental. Science and Technology.* **48** 13053–59
[3] A Nomura, S Ogasawara, O Sanob, T Ito, J Yanai 2014 *Science of the Total Environment.* **468–9** 523–9
[4] N Yamaguchi *et al.* 2012 *Bulletin of National Institute for Agro-Environmental Sciences* **31** 75-129

[5] K Sekiya, H Kuwahara, Y Yoshida, S Igarashi, N Nomura, F Mishima, Y Akiyama, S Nishijia 2014 *IEEE Transactions on Applied Superconductivity* **24** 3700205
[6] M Kimura *et al*. 1994 Soil biochemistry (Japan: Asakura Publishing Co., Ltd.) p 77

Separation of Flame and Nonflame-retardant Plastics Utilizing Magneto-Archimedes Method

Kohei Misawa[1], Takayuki Kobayashi[1], Tatsuya Mori[1], Fumihito Mishima[2], Yoko Akiyama[1] and Shigehiro Nishijima[1]

[1] Division of Sustainable Energy and Environmental Engineering, Graduate school of engineering, Osaka University, 2-1 Yamadaoka, Suita, Osaka 565-0871, Japan

[2] Department of Applied Nuclear Technology, Fukui University of technology, 3-6-1 Gakuen, Fukui City, Fukui 910-8505, Japan

Email: misawa@qb.see.eng.osaka-u.ac.jp

Abstract. In physical recycling process, the quality of recycled plastics becomes usually poor in case various kinds of plastic materials are mixed. In order to solve the problem, we tried to separate flame and nonflame-retardant plastics used for toner cartridges as one example of mixed plastics by using magneto-Archimedes method. By using this method, we can control levitation and settlement of the particles in the medium by controlling the density and magnetic susceptibility of the medium and the magnetic field. In this study, we introduced the separation system of plastics by the combination of wet type specific gravity separation and magneto-Archimedes separation. In addition, we examined continuous and massive separation by introducing the system which can separate the plastics continuously in the flowing fluid.

1. Introduction

1.1. Background

Plastics are used for various purposes in our daily life and industries. Due to the mass production and consumption of plastics, we face the exhaustion of natural resources and the environmental problems.

In order to solve the problems, physical recycling has been paid particular attention for sustainable use of natural resources and CO_2 reduction. However, in physical recycling process, the quality of recycled plastics becomes poor when the various kinds of plastic materials are mixed together.

To solve the problem, we focused on magneto-Archimedes method where the difference of densities and magnetic susceptibilities of mixed plastics are utilized. If we can separate the mixed plastics accurately and massively by using the method, physical recycling which is superior to chemical recycling in the sense of environmental load, can be promoted furthermore. In this study, we conducted the separation experiments using magneto-Archimedes method as a fundamental experiment. Based on the results, we examined the possibility of the continuous separation system by using a superconducting solenoidal magnet.

1.2. Separation objects

The materials suitable for the physical recycling are those satisfy the following conditions, that is, 1)industrial waste having large amount and 2)including small amount of contaminant or foreign

IOP Conf. Series: Journal of Physics: Conf. Series **871** (2017) 012103 doi:10.1088/1742-6596/871/1/012103

substances. Thus, we separated flame-retardant plastics (black) and nonflame-retardant plastics (black, translucent, blue, orange, gray, white) used for toner cartridges as one example of mixed plastics. Flame-retardant plastics are hardly combustible because they contain flame retarders such as aromatic brominate compounds. Once flame and nonflame-retardant plastics are mixed, the flame-retardant effect is reduced. In this study, we separated flame-retardant plastics (black) selectively from the mixture of flame and nonflame-retardant plastics by using magneto-Archimedes method.

1.3. Purpose

We tried to separate flame and nonflame-retardant plastics accurately and massively. The target purity and processing speed were set to 90 % and 200 kg/h, respectively. By employing 10 magnetic separation apparatuses developed in this study, the processing speed of 2000 kg/h will be realized, which corresponds to the annual amount of waste plastics of toner cartridges in Japan[1].

First, we examined separation process based on the density and magnetic susceptibility of each plastic. Table 1 shows the density and magnetic susceptibility of each plastic. The density of the flame-retardant plastic is larger than those of some nonflame-retardant plastics, whereas is smaller than those of residual nonflame-retardant plastics. Therefore, in order to separate flame-retardant plastics only by wet type specific gravity separation, two-stage wet type specific gravity separation by using two different mediums is required. In addition, in case the difference in densities is small, separation accuracy becomes lower. Thus, we conducted magnet-Archimedes method which can separate each object accurately, focusing on the difference in magnetic susceptibilities among the plastics.

Table 1. Densities and magnetic susceptibilities of plastics used for toner cartridges.

Color		Type	Density (gcm^{-3})	Magnetic susceptibility (-)
Flame-retardant black	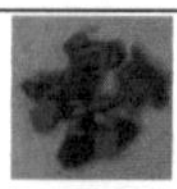	HIPS (High Impact Polystyrene)	1.085	1.06×10^{-5}
Nonflame-retardant translucent	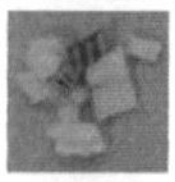	POM (PolyOxyMethylene)	1.412	-3.85×10^{-6}
Nonflame-retardant black		HIPS	1.250	2.90×10^{-6}
Nonflame-retardant blue		ABS (Acrylonitrile Butadiene Styrene Resin)	1.050	-5.39×10^{-6}
Nonflame-retardant orange		HIPS	1.044	-5.39×10^{-6}
Nonflame-retardant gray		HIPS	1.042	-6.41×10^{-6}
Nonflame-retardant white		HIPS	1.040	-4.36×10^{-6}

2. Magneto-Archimedes method[2]

Archimedes' principle is that a particle in a fluid is subjected to buoyancy which equals to the weight of the fluid replaced by the particle. This means that it is possible to increase buoyancy by increasing the weight of the fluid apparently. When we use a paramagnetic medium as the fluid, the weight of the fluid becomes apparently heavier due to the magnetic force. In this way, we can increase the buoyancy acting on the particle in the fluid.

In addition to buoyancy, gravity and magnetic force act on the particle. Due to the resultant force, the particle levitates depending on its density and magnetic susceptibility. Magneto-Archimedes separation can be performed by utilizing the difference in levitating position of particles. Figure 1 shows the conceptual illustration of the magnetic Archimedes effect (one-dimensional representation). Figure 1(a) shows the forces acting on a paramagnetic particle in a paramagnetic medium. The z-axis is oriented vertically upward and the magnetic field is applied in the positive direction of z-axis.

Buoyancy F_b acts on the paramagnetic particle in the positive direction of z-axis, whereas gravity F_g and magnetic force F_m acts in the negative direction. Figure 1(b) shows the forces acting on a diamagnetic particle. Buoyancy F_b and magnetic force F_m act on the diamagnetic particle in the positive direction of z-axis and gravity F_g acts in the negative direction.

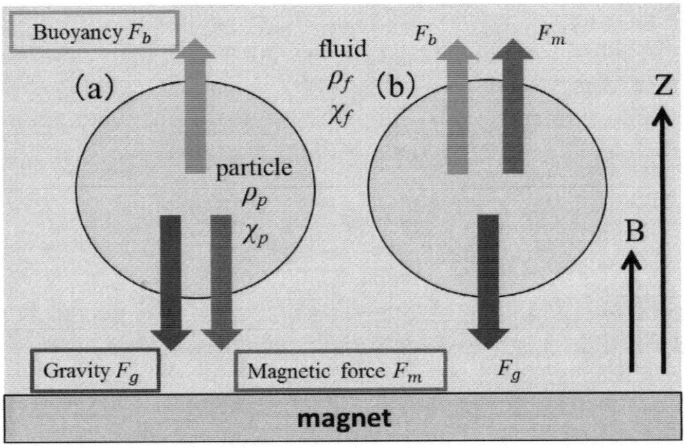

Figure 1. Conceptual illustration of magneto-Archimedes effect (one-dimensional representation). (a) paramagnetic particle (b) diamagnetic particle ρ_p : Density of particle, ρ_f : Density of medium χ_p : Magnetic susceptibility of particle, χ_f : Magnetic susceptibility of medium

Firstly, we examined the resultant force of the paramagnetic particle ($\chi_p > 0$) in the paramagnetic medium. The resultant force F_z, which is sum of gravity F_g, magnetic force F_m and buoyancy F_b acting on the paramagnetic particle is expressed as (1).

$$F_z = F_b - F_m - F_g \tag{1}$$

Both gravity F_g and magnetic force F_m acting on the paramagnetic particle are in the negative direction of z-axis as shown in figure 1(a), and hence these signs are negative. Regarding the term of gravity, when we define g as gravitational acceleration, gravity per unit volume acting on the particle is in the negative direction of z-axis and expressed as (2).

$$-F_g = -\rho_p g \tag{2}$$

Regarding the term of magnetic force F_m, magnetic force per unit volume acting on the particle is denoted one-dimensionally as (3), where B is magnetic flux density and μ_0 is space permeability.

$$F_m = \frac{\chi_p}{\mu_0} B \frac{dB}{dz} \tag{3}$$

Magnetic force field (the product of magnetic flux density and magnetic field gradient) always negative, because $\frac{dB}{dz} < 0$ is held. The magnetic force per unit volume acting on the paramagnetic particle is expressed as (4).

$$-F_m = \frac{\chi_p}{\mu_0} B \frac{dB}{dz} \tag{4}$$

IOP Conf. Series: Journal of Physics: Conf. Series **871** (2017) 012103 doi:10.1088/1742-6596/871/1/012103

Buoyancy F_b is the inverse resultant force of gravity and magnetic force acting on the displaced medium by the paramagnetic particle. Consequently, buoyancy F_b per unit volume acting on the paramagnetic particle is expressed as (5)

$$F_b = \rho_f g - \frac{\chi_f}{\mu_0} B \frac{dB}{dz} \tag{5}$$

Therefore, the resultant force F_z per unit volume acting on the paramagnetic particle in the paramagnetic medium is expressed as (6), which is derived from (1)(2)(4)(5).

$$F_z = \left(\rho_f - \rho_p\right) g + \frac{\chi_p - \chi_f}{\mu_0} B \frac{dB}{dz} \tag{6}$$

Next, we examined the resultant force of the diamagnetic particle ($\chi_p < 0$) in the paramagnetic medium. The resultant force F_z is expressed as (7).

$$F_z = F_b + F_m - F_g \tag{7}$$

In addition, magnetic force is expressed as (3). Therefore, (6) is provided by (2)(3)(5)(7). After all, the resultant force F_z per unit volume acting on the paramagnetic or diamagnetic particle in the paramagnetic medium is expressed in the same formula as shown in (6). When the resultant force $F_z > 0$, the particle levitates, whereas the resultant force $F_z < 0$, the particle settles. Just when the resultant force $F_z = 0$, the levitating position of particle is defined. When the resultant force $F_z = 0$, magnetic force field $B\frac{dB}{dz}$ is expressed as (8), derived from (6)

$$B\frac{dB}{dz} = \frac{\rho_p - \rho_f}{\chi_p - \chi_f} \mu_0 g \tag{8}$$

The density and magnetic susceptibility of materials are fixed and hence the levitating heights of particles are uniquely defined. In magneto-Archimedes method, the objects are separated depending on the difference in levitating heights.

3. Examination of separation conditions

3.1. Examination of separation process

Although we can separate each object accurately by using magneto-Archimedes method, there is a problem of processing speed. To solve the problem, we introduced wet type specific gravity separation, which can separate each object massively before the magneto-Archimedes separation. In addition, in order to realize continuous separation, we used a common medium in wet type specific gravity separation and magneto-Archimedes separation. In this study, we used manganese chloride aqueous solution (hereinafter referred as $MnCl_2aq.$), because it shows paramagnetism and soluble in water. The separation process is shown in figure 2. First, we settled flame and nonflame-retardant plastics whose densities are larger than the medium. Next, we separated flame-retardant plastics selectively from the mixture of the settled plastics by utilizing magneto-Archimedes levitation. In this way, we can control levitation and settlement of the plastics by controlling the concentration of the medium and the strength of magnetic field.

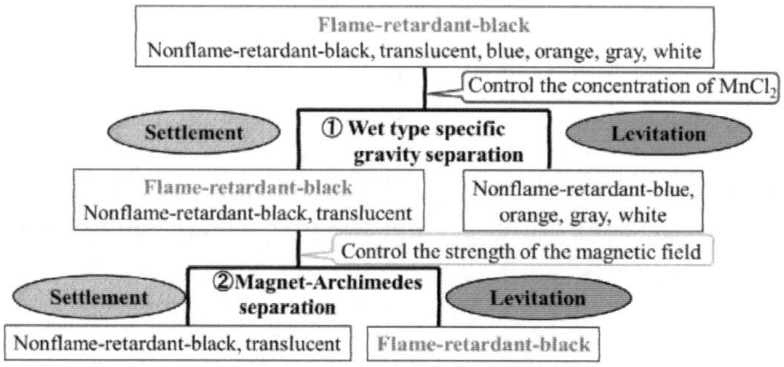

Figure 2. Separation process of separating Flame and non-flame-retardant plastics.

3.2. Batch type separation using a Halbach array magnet

We conducted batch type separation experiment according to the separation process of figure 2. In wet type specific gravity separation process, the result of calculation showed that flame-retardant (black), nonflame-retardant (black and translucent) settled in 7 wt.% $MnCl_2$aq..

Next, we examined the concentration of $MnCl_2$aq. suitable for the magneto-Archimedes separation. Here, we used a Halbach array magnet as fundamental experiment before introducing a superconducting solenoidal magnet. The appearance of the Halbach array magnet and its specifications are shown in figure 3 and table 2, respectively. Figure 4 shows the levitating height of each plastic when the maximum magnetic flux density at the bottom of the separation vessel was 980 mT and the concentration of $MnCl_2$aq. was 6.7 wt.%. The levitating heights were obtained by the graph showing the vertical distribution of the magnetic force field of the Halbach array magnet and calculation of the magnetic force field at the levitating height. Figure 4 indicates that the difference in heights between flame-retardant (black) and nonflame-retardant (black) was 21.8 mm.

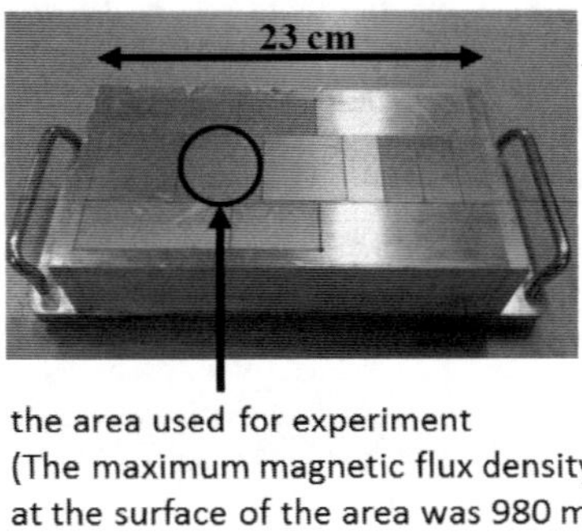

Figure 3. Halbach array magnet.

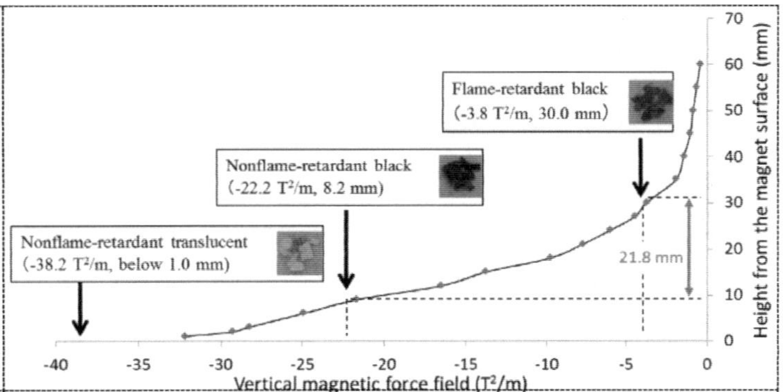

Figure 4. Levitating height above the Halbach array magnet (observed values).

Table 2. Specifications of the Halbach array magnet.

Drawing number	Width (m)	Depth (m)	Maximum magnetic field (T)	Maximum magnetic field product (T^2m^{-1})
12E136-A (NEOMAX ENGINEERING Co.,Ltd.)	0.23	0.13	1.4	-140

Figure 5 (a) and (b) show the results of wet type specific gravity separation and magneto-Archimedes separation conducted under the conditions mentioned above, respectively. Magnetic field is applied vertically upward by placing the separation vessel on the Halbach array magnets after wet type specific gravity separation. The results showed that flame-retardant plastics (black), nonflame-retardant plastics (black, translucent) were settled by wet type specific gravity separation, and then only flame-retardant plastics (black) were levitated among the three settled plastics by magneto-Archimedes separation. The results of this experiment showed the possibility of separation for any kinds of plastics.

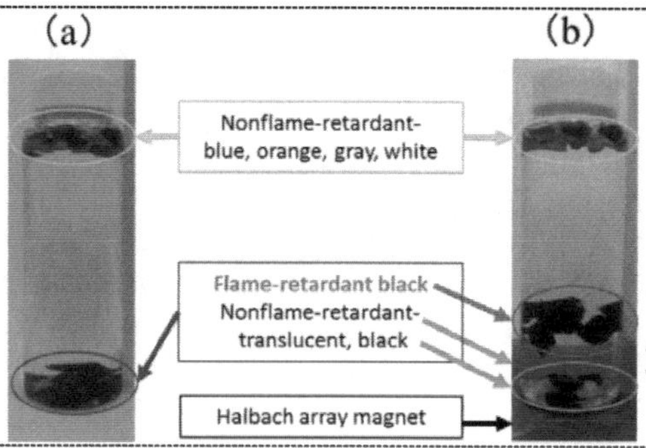

Figure 5. Bach type separation using the Halbach array magnet.
(a)wet type specific gravity separation (b)magneto-Archimedes separation

3.3. Batch type separation using a superconducting solenoidal magnet

In order to introduce the continuous separation apparatus, it is important to improve separation speed and accuracy. We used a superconducting solenoidal magnet that has a wide range of magnetic field and can increase the magnetic force field gradient. The appearance of the superconducting solenoidal magnet and its specifications are shown in figure 6 and table 3, respectively.

Figure 6. Cryogen free superconducting solenoidal magnet.

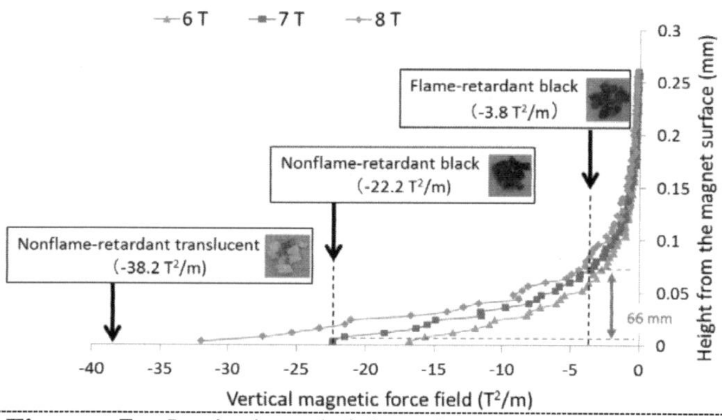

Figure 7. Levitating height above superconducting solenoidal magnet (observed values).

Table 3. Specifications of the superconducting solenoidal magnet.

Model number	Bore diameter (m)	Length of bore (m)	Maximum magnetic field (T)	Maximum magnetic field product (T^2m^{-1})
JMTD-10T100E3	0.10	0.46	10	-350

In the same way as the separation experiment using the Halbach array magnets, we examined the concentration of $MnCl_2$aq. and the strength of magnetic field. Figure 7 shows the levitating heights of each plastic when the concentration of the $MnCl_2$aq. was 6.7 wt.%, and the maximum magnetic field was set to 6 T, 7 T and 8 T at the center of the magnet. When the maximum magnetic field was 7 T at the center of the magnet, it was found that the difference in heights between flame-retardant plastics (black) and nonflame-retardant plastics (black) was 66 mm, which showed the maximum.

Figure 8 shows the results of magneto-Archimedes separation under the conditions mentioned above. When the maximum magnetic field was 7 T at the center of the magnet, and the difference in heights of flame-retardanct plastics (black) and nonflame-retardant plastics (black) was 40 mm. It was twice as large as that with the Halbach array magnet. The results showed the possibility of accurate separation by superconducting solenoidal magnets.

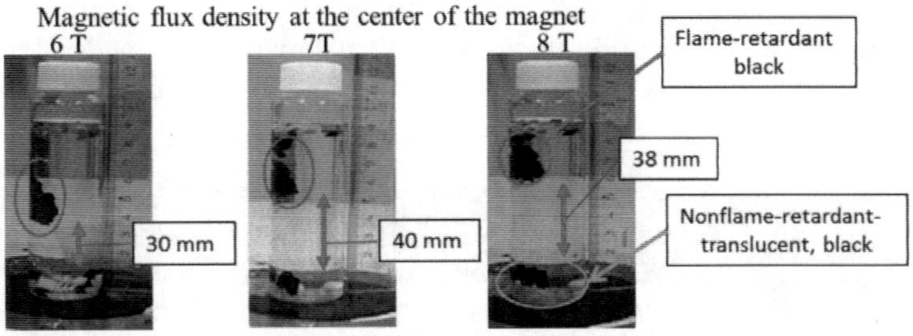

Figure 8. Bach type separation using the superconducting solenoidal magnet. (These samples were put outside the magnet)

3.4. General plastics separation method

Based on the results of batch type separation, we designed the process of flowchart for separating general plastics as shown in figure 9. In this flowchart, when the density of separation object is larger than 1 gcm^{-3}, we use water as the solvent of $MnCl_2$. When the density of separation object is less than 1 gcm^{-3}, we use ethanol as the solvent of $MnCl_2$. By selecting the solvent in this way, we can always settle separation objects. Furthermore, we can levitate the object selectively from the mixture of plastics by using magneto-Archimedes method.

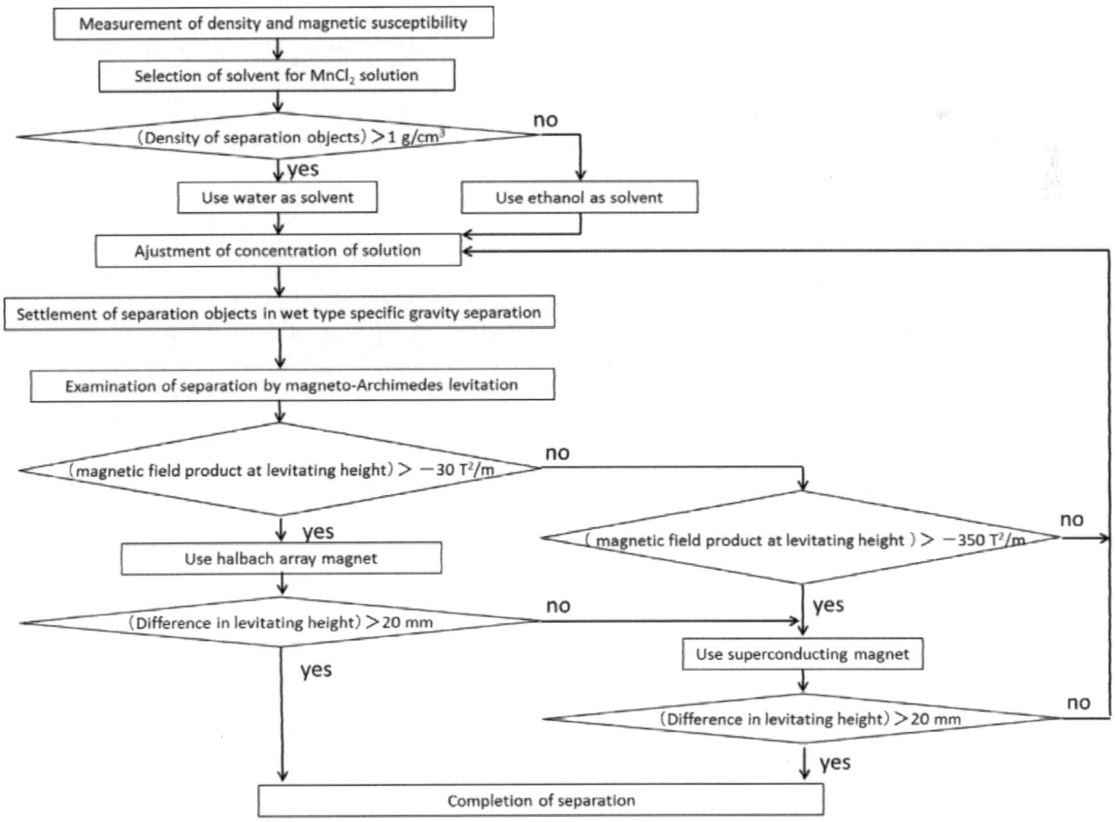

Figure 9. Flowchart of separation for any kinds of plastics.

4. Examination of separation conditions

Finally, we examined massive separation by using continuous separation apparatus and the Halbach array magnet as shown in figure 10. In this system, plastics in the fluid flowed into the wet type specific gravity separation area and floated plastics were trapped with the filter.

Then, the settled plastics passed through the filter flowed into the magneto-Archimedes separation area. Here, they were separated by the partition depending on magneto-Archimedes levitation heights. In order to improve the separation speed and accuracy, we are trying to make a larger continuous separation apparatus using the superconducting solenoidal magnet.

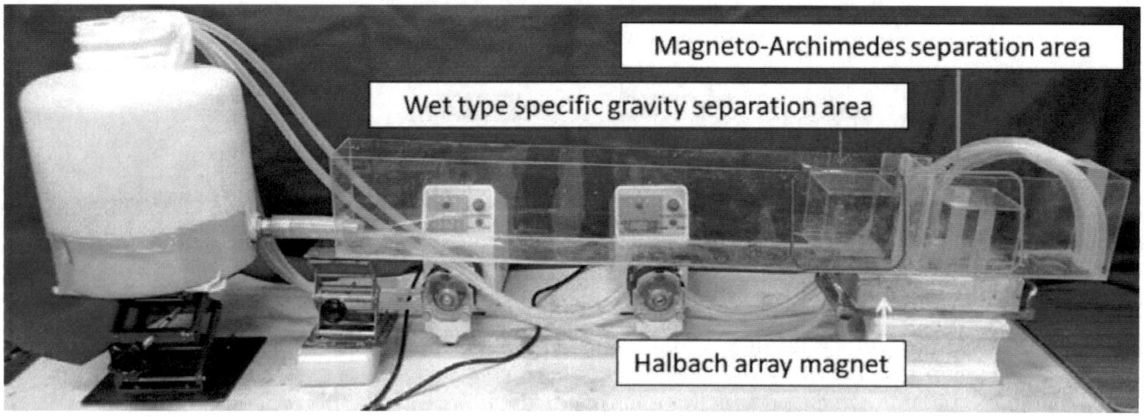

Figure 10. Continuous separation apparatus using the Halbach array magnet.

5. Conclusion

By utilizing the difference in density and magnetic susceptibility of plastics, we conducted wet type specific gravity separation and magneto-Archimedes separation using common medium in both separation processes. The result showed the possibilities to separate any kinds of plastics with high accuracy. In addition, the result of separation experiment by using continuous separation apparatus indicated the prospect of massive separation. For the next stage, we are trying to design a larger continuous separation apparatus using superconducting magnet for further improvement of the separation speed and accuracy.

Reference

[1] Association of Japan Cartridge Re-manufactures (AJCR) 2016 *Fact Book* vol 7 p 2
[2] Ikezoe Y et.al. 2002 *Energy Conversion and Management* **43** 417-425

IOP Conf. Series: Journal of Physics: Conf. Series **871** (2017) 012104 doi:10.1088/1742-6596/871/1/012104

Fundamental study on the magnetic field control method using multiple HTS coils for Magnetic Drug Delivery System

R Hirano, S B Kim, T Nakagawa, Y Tomisaka and H Ueda

Graduate School of Natural Science and Technology, Okayama University, 3-1-1, Tsushimanaka, Kita-ku, Okayama, 700-8530, Japan

kim@ec.okayama-u.ac.jp

Abstract. The magnetic drug delivery system (MDDS) is a key technology to reduce the side effects in the medical applications, and the magnetic force control is very important issue in MDDS. In this application, the strength of magnetic field and gradient required to MDDS devices are 54 mT and 5.5 T/m, respectively. We proposed the new magnetic force control system that consists of the multiple racetrack HTS magnets. We can control the magnetic field gradient along the longitudinal direction by the arrangement of the multiple racetrack HTS magnets and operating current of each magnet. When the racetrack HTS magnets were used, the critical current was reduced by the self-magnetic field. Therefore, the shape design of HTS magnet to reduce the magnet field into the surface of HTS tapes was required. Therefore, the electromagnetic analysis based on finite element method (FEM) was carried out to design and optimize the shape of multiple racetrack HTS magnet. We were able to suppress the reduction of critical current by placing the magnetic substance at upper and lower side of the HTS magnets. It was confirmed that obtained maximum values of magnetic field strength and field gradient were 33 mT and 0.18 T/m, respectively.

1. Introduction

Recently, in the field of regenerative medicine, it is known that magnetic field applied to the human body can control the transplanted cells in the body. The techniques for magnetic field concentration are very important for controlling magnetic nanoparticles, novel electromagnets and trapped HTS bulk magnets were fabricated to apply the cell culture and the magnetic drug delivery system (MDDS) [1-3]. In these applications, we need not only the strength of the magnetic field but also the magnetic field gradient in order to enlarge the medicinal effects. The large magnetic moment is required with increasing the size of the drug particles. However, when the drug particle diameter is 400 nm or less, the magnetic moment is not obtained enough. In that case, we need to increase of the magnetic gradient. In general, the strength of magnetic field and gradient required to MDDS devices are 54 mT and 5.5 T/m, respectively [4]. A neodymium magnet has the highest remnant magnetization in the permanent magnet commercially available. However, the magnetic field generated by neodymium magnet is not sufficiently high for the MDDS. Furthermore, in the electromagnet using a metal wire, a large size coil is required for a high magnetic field generation. On the other hand, it is contemplated that the HTS bulk magnet is used as the source of high magnetic field. However, the HTS bulk magnetic is difficult to control of magnetic field for the MDDS. In this paper, we proposed a new field control method to generate a high magnetic field and field gradient along the longitudinal direction using the multiple racetrack HTS magnets. The multiple racetrack HTS magnets were proposed and designed in order to increase the operating current and generate the high field gradient along the longitudinal direction. We were able to amplify the critical current and high field gradient. In the new MDDS using the multiple racetrack HTS magnets, it is possible to control magnetic field by changing the operating current value.

2. Multiple racetrack HTS coils for MDDS

2.1. Analytical model

The multiple racetrack HTS magnets were proposed and designed in order to generate the high field gradient along the longitudinal direction by controlling the current of HTS magnet. The system consists of the 5 multiple racetrack HTS magnets using GdBCO tape as shown in figure 1, hereinafter referred as to Model (A). The inner curvature diameter of curve section in racetrack coil was 30 mm and the number of turns was 20. Five HTS magnets are individually operated. The critical current is reduced by the self-magnetic field entering HTS magnets. Especially, HTS tape is significantly influenced by the magnetic field in the perpendicular direction to it. Therefore, we studied the way to reduce the magnetic field entering magnets by placing the magnetic substance at upper and lower side of the coil as shown in Figure 1, hereinafter referred as to Model (B). There were air gaps with 1.0 mm between the coils and magnetic substance. Iron (35H210) was used as a magnetic substance. In the Model (C), the magnetic substance located in the center of magnet system (Model (B)), was removed so that the magnetic field reaches to high position from magnet surface.

We performed an electromagnetic analysis based on three-dimensional finite element method (3D FEM) to confirm the critical current and to find the suitable configuration for magnetic control system with high field strength and high field gradient in axial and longitudinal directions.

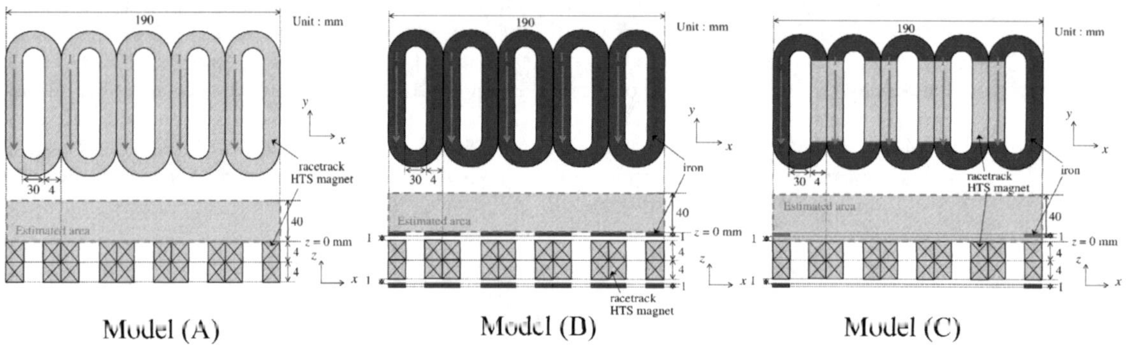

Figure 1. Schematic drawing of top view and cross-sectional view of five racetrack HTS magnets analytical model. Model (A), racetrack HTS magnets. Model (B), racetrack HTS magnets +whole iron. Model (C), racetrack HTS magnets + partial iron.

2.2. Analytical result

Figure 2 shows the calculated magnetic field profiles along the longitudinal center line at 0, 10, 20, 30 mm above the racetrack magnet at a transport current of 25 A. The magnetic field is obviously increased in case of the Model (B) and (C).

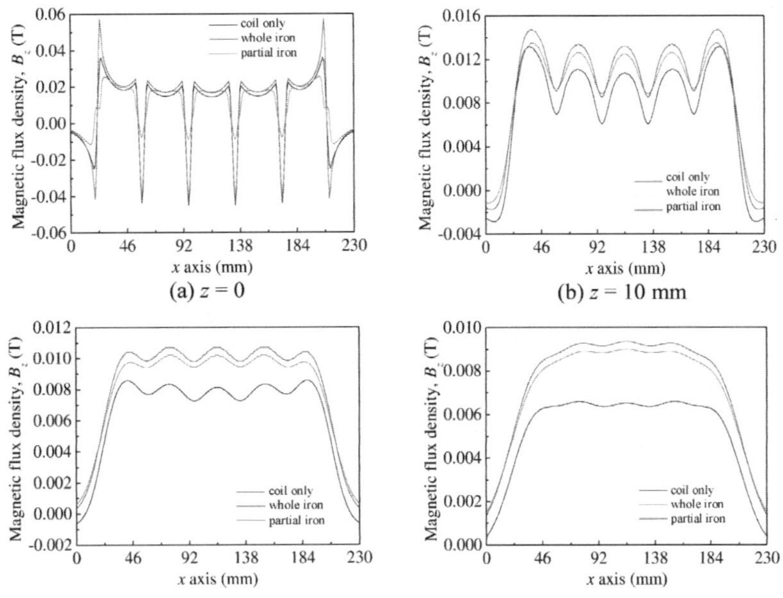

Figure 2. Calculated the magnetic field profiles along the longitudinal direction at (a) $z = 0$ (b) $z = 10$ mm (c) $z = 20$ mm (d) $z = 30$ mm at 25 A transported in case of Model (A), (B), (C).

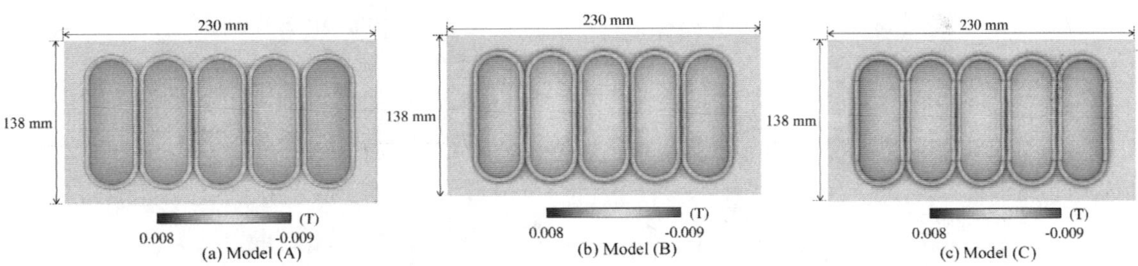

Figure 3. Magnetic field contour maps at top surface of magnet at a transport current of 25 A in case of Model (A), (B), (C).

Figure 3 shows the contour map of magnetic field in the *x*-*y* plane at $z = 0$ mm. Figure 4 shows the magnetic field dependence of critical current and the load line for the magnetic field applied perpendicular or parallel to the wide face of the GdBCO tape. The magnetic field in the perpendicular direction to the HTS tape was reduced by placing iron. From figure 4 (b), the critical current is 88 A, 103 A and 102 A in the case of Model (A), (B) and (C), respectively. The critical current for Model (B) and (C) were almost the same, but improved compared with the Model (A). Therefore, we would like to discuss the magnet design based on Model (C) with possible to generate more effective magnetic field.

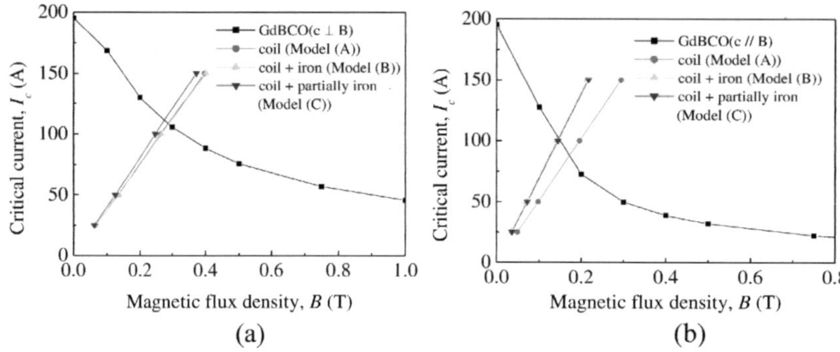

Figure 4. The magnetic field dependence of critical current of GdBCO tape and the load line for the magnetic field applied (a) parallel and (b) perpendicular to the face of the HTS tape in case of Model (A), (B), (C).

3. Improvement of magnet design to obtain the high critical current

To increase the critical current, we placed iron at inner side of HTS racetrack magnets as shown in figure 5 (a). In addition, we inserted iron cores into the racetrack HTS magnets as shown in figure 5 (b). Figure 6 shows the magnetic field dependence of the critical current and load line. The critical current of the racetrack HTS magnets was increased. Table 1 shows the critical current of each model.

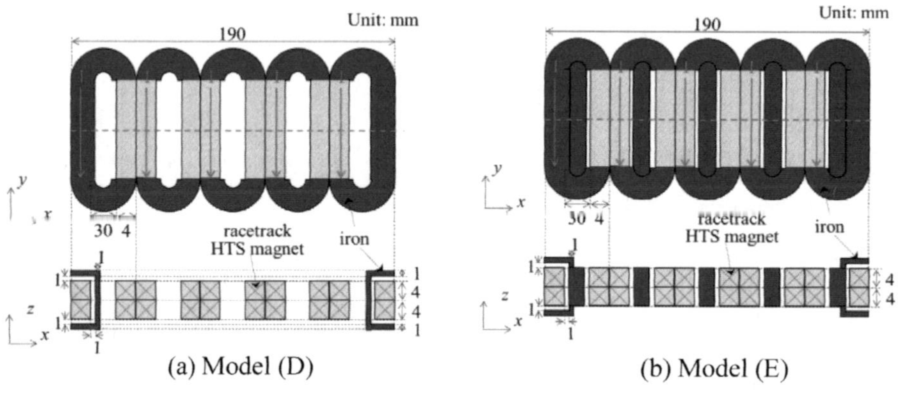

Figure 5. Schematic drawing of top view and side view of five racetrack HTS magnets. Model (D), with inner side iron. Model (E), with inner and iron cores model.

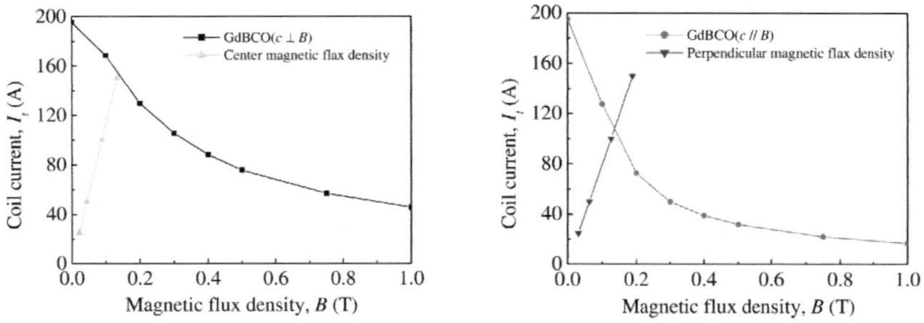

(a) Model (D)

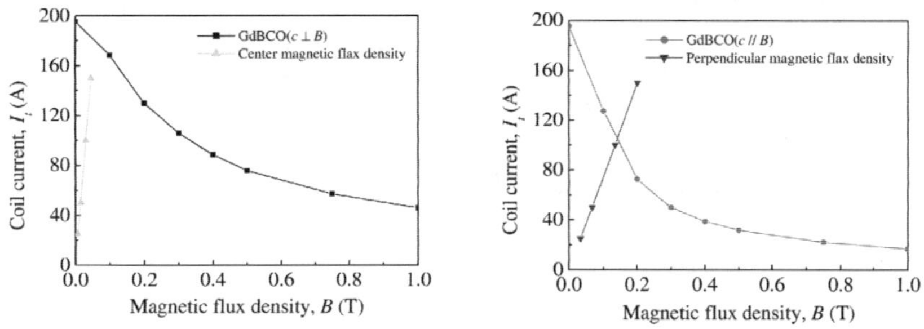

(b) Model (E)

Figure 6. The magnetic field dependence of critical current of GdBCO tape and the load line in case of Model (D), (E).

Table 1. Critical current of each model

model	racetrack HTS magnets(**Model (A)**)	With inner side and partial iron (**Model (D)**)	With inner side and partial iron + iron cores **(Model (E))**
I_c (A)	88	108	105

4. magnetic field generating for MDDS

We evaluated the magnetic field and magnetic field gradient in control of operating current of coils in Model (A), (D), and (E). The operating current of each racetrack coil in the magnet system was individually controlled then, for generation of the magnetic field gradient, operating current was set to be linearly large value in order from the left-end coil under the condition the maximum value of operating current was 90% of critical current. Figure 9 shows the calculated magnetic field profiles of the racetrack HTS magnets at z = 10, 20 and 30 mm. Table 2 shows maximum magnetic field density and magnetic gradient at z = 30 mm. The gradient was maximum in Model (E). Therefore, in this study, Model (E) was the most effective for MDDS.

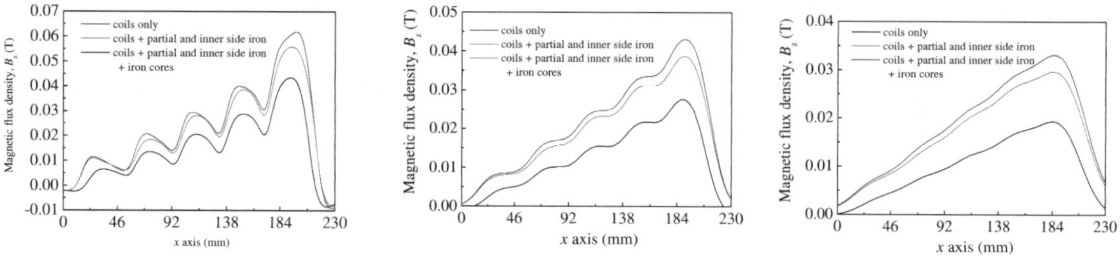

(a) z = 10 mm (b) z = 20 mm (c) z = 30 mm

Figure 7. The calculated magnetic field profiles at each height.

Table 2. Magnetic flux density and magnetic field gradient at z = 30 mm of each model.

Analysis model	Magnetic flux density (mT)	Magnetic field gradient (T/m)
HTS racetrack Magnets(**Model (A)**)	19	0.103
With inner side and partial iron(**Model (B)**)	30	0.158
Coils + inner side and partial iron + iron cores (**Model(E)**)	33	0.179

5. Conclusion

In this study, we proposed the multiple racetrack HTS magnets for MDDS, it was possible to control magnetic field by changing the current value. We investigated and developed the magnetic field source for the MDDS by the multiple racetrack HTS magnets using the electromagnetic field analysis. The results in this study were summarized as below,

- The critical current of the racetrack HTS magnets was increased 17A and the magnetic field gradient was improved by placing the iron plate and core compared with only HTS racetrack coils.
- The magnetic field of 33 mT and field gradient of 0.179 T/m were obtained above 30 mm from the magnets surface , but they were still not enough to apply for the MDDS devices.

References

[1] Kobayashi T, Ochi M, Yanada S, Ishikawa M, Adachi N, Deie M and Arihiro K 2008 *Arthroscopy* A novel cell delivery system using magnetically labeled mesenchymal stem cells and an external magnetic device for clinical cartilage repair **24** 69-76

[2] Nishijima S, Mishima F, Tabata Y, Iseki H, Muragaki Y, Sasaki A and Saho N 2009 *IEEE Trans. Appl. Supercond.* Research and development of magnetic drug delivery system using bulk high temperature superconducting magnet **19** 2257-2260

[3] Kiyoshi T, Choi S, Matsumoto S, Asano T and Uglietti D 2009 *IEEE Trans. Appl. Supercond.* Magnetic Flux Concentrator Using Gd-Ba-Cu-O Bulk Superconductors **19** 2174-2177

[4] Jho J, Mishima F, Takeda S, Yamamoto M, Muragaki Y, Iseki H, Saho N, Kubota J, Sasaki A, Nishijima S and Tabata Y 2007 *Drug Delivery System* Magnetic nanoparticles-based DDS therapeutic system of next generation for deep body site **22** 558-568

Institute of Physics
Dirac House, Temple Back
Bristol BS1 6BE UK

ISSN: 1742-6588
ISBN 978-1-5108-4558-9

CURRAN ASSOCIATES INC.
proceedings.com